8.	RNA: Transcription and Processing	replication time, p. 242 replisome, p. 242 replication origins, p. 244	RNA polymerase II, p. 265 introns, p. 267	Volkin–Astrachan pulse chase experiment, p. 257 stages of transcription, p. 261 promoter sequences, p. 261 sigma factors, p. 262 termination, p. 263 gene density, p. 264		gene density, p. 264	
9.	Proteins and Their Synthesis	Yanofsky and colinearity, p. 278 Crick codon length experiment, p. 279 elucidation of genetic code, p. 281	abundance of RNA transcripts, p. 275				
10.	Regulation of Gene Transcription	Jacob–Monod *lac* operon experiments, p. 307 tryptophan pathway, p. 316	position-effect variegation, p. 326 lack of methylation, p. 324 components of gene regulation, p. 327 SWI-SNF mutants, p. 327 chromatin remodeling, p. 329 **Model Organism box, p. 328**	enhancers, p. 320 misregulation by gene fusion, p. 321 position effect variegation, p. 326 epigenetic silencing, p. 332		parental imprinting, p. 324	
11.	Gene Isolation and Manipulation	restriction enzymes, p. 344 use in cloning, p. 346 cloning with bacteriophage vectors, p. 348 cloning with Ti plasmids, p. 369	genetic engineering with yeast vectors, p. 366	transgenesis in, p. 371	genome size and cloning, p. 351 transgenesis in, p. 370	transgenesis in, p. 372 targeted gene knockouts, p. 372 gene therapy, p. 375	**Organism:** *Aspergillus* identifying alkaptonuria gene, p. 362
12.	Genomics		filling sequence gaps, p. 398 finding genes through comparative genomics, p. 411	method of sequencing, p. 395 repeats in, p. 396 codon bias, p. 408	vectors for mapping, p. 396 method of sequencing, p. 397	human–mouse comparative genomics, p. 412	
13.	The Dynamic Genome: Transposable Elements	IS elements, pp. 429–433	Ty elements, p. 434 lack of DNA transposons, p. 438 transposon targeting, p. 442	*copia*-like elements, p. 435 *P* elements, p. 436 hybrid dysgenesis, p. 436 use in tagging/transgenesis, p. 439 transposon targeting, p. 443 transposition and telomeres, p. 444		mouse mammary tumor virus, p. 434 retrotransposon activity in, p. 442	**Organism:** *Zea maize* *Ac/Dc* system, pp. 425, 445 **Model Organism box, p. 428**
14.	Mutation, Repair, and Recombination	SOS repair, p. 459 Luria–Delbrück	translesion DNA synthesis, p. 459	homologous recombination, p. 474			

(Continued)

(Continued)

		E. coli	S. cerevisiae	N. crassa	D. melanogaster	C. elegans	M. musculus	Other Featured Organisms
14.	(Continued)	fluctuation test, p. 461; hot spots, p. 465; deletion hot spots, p. 466; methyltransferase, p. 468; nucleotide excision repair, p. 470	transcription-coupled repair, p. 470; homologous recombination, p. 474					
15.	Large-Scale Chromosomal Changes		descendant of an autopolyploid, p. 487		experimental polyploids, p. 490; nondisjunction, p. 491; aneuploidy, p. 494; dosage compensation, p. 495; inversions, p. 500; balancer chromosomes, p. 500; position-effect variegation, p. 502; pseudodominance and deletion mapping, p. 504	balancer chromosomes, p. 500		
16.	Dissection of Gene Function	mutant selection, p. 524; lacZ genetic dissection system, p. 528	dissection of cell cycle, p. 527	filtration enrichment, p. 525; dissection of morphogenesis, p. 527; RIP, p. 532	Muller's X-ray screen, p. 523; mutagens, p. 523; light response mutants, p. 526; enhancer traps, p. 530; RNAi, p. 533	RNAi, p. 534		**Organism:** *Aspergillus* dissection of nuclear division, p. 528; **Organism:** Zebra fish dissection of development, p. 529
17.	Genetic Regulation of Cell Number: Normal and Cancer Cells		cell cycle, p. 551; **Model Organism box, p. 551**			apoptosis in cell lineages, p. 554; **Model Organism box, p. 554**		
18.	The Genetic Basis of Development				general steps in development, p. 576; determination of germ line, p. 581; establishment of body axes, pp. 583–590; establishing cell fates, pp. 591–597; genome compared to human, p. 600; transcriptional profiling with microarray, p. 605; **Model Organism box, p. 584**	determination of germ lines, p. 581; vulva development, p. 598	homeobox genes: HOX, p. 601	**Organism:** *Arabidopsis thaliana* homeobox genes in flower development, p. 603; **Model Organism box, p. 604**
19.	Population Genetics				restriction site variation, p. 618; synonymous base-pair polymorphisms, p. 620			
20.	Quantitative Genetics				norms of reactions, p. 651; testing for heritability, p. 656; distinguishing genetic mutations among similar phenotypes, p. 664			
21.	Evolutionary Genetics	relationship of genetics to functional change, p. 694	genetic divergence, p. 694		selection to change in wing size, p. 685; canalized characters, p. 688; species distinctions, p. 689; genetics of species isolation, p. 690; gene origin from transposable elements, p. 694; genetic evidence of common ancestry, p. 697		human–mouse comparative genomics, p. 700	
	A Brief Guide to Model Organisms	p. 708	p. 710	p. 712	p. 718	p. 716	p. 720	*Arabidopsis thaliana*, p. 714

Introduction to Genetic Analysis

About the Authors

Anthony Griffiths is Professor of Botany at the University of British Columbia, where he teaches Introductory Genetics to a class of 550 students. The challenges of teaching that course have led to a lasting interest in how students learn genetics. Griffiths's research focuses on the developmental genetics of fungi, using the model fungus *Neurospora crassa*. He also loves to dabble in the population genetics of local plants. Griffiths was President of the Genetics Society of Canada from 1987 to 1989, receiving its Award of Excellence in 1997. For the past ten years, he has been Secretary-General of the International Genetics Federation.

Susan Wessler is Distinguished Research Professor of Plant Biology and Genetics at the University of Georgia, where she has been since 1983. She teaches courses in introductory biology and plant genetics to both graduate students and undergraduates. She is coauthor of *The Mutants of Maize* (Cold Spring Harbor Laboratory Press) and of more than 100 research articles. Her scientific interest focuses on the subject of plant transposable elements and the structure and evolution of plant genomes. She was elected to membership in the National Academy of Sciences in 1998.

Richard Lewontin is the Alexander Agassiz Research Professor at Harvard University. He has taught genetics, statistics, and evolution at North Carolina State University, the University of Rochester, and the University of Chicago. His chief area of research is population and evolutionary genetics; he introduced molecular methods into population genetics in 1966. Since then, he has concentrated on the study of genetic variation in proteins and DNA within species. Lewontin has been President of the Society for the Study of Evolution, the American Society of Naturalists, and the Society for Molecular Biology and Evolution, and, for some years, he was coeditor of *The American Naturalist*.

William Gelbart is Professor of Molecular and Cellular Biology at Harvard University, a position that he has held since 1976. He teaches courses in genetics and genomics to both undergraduate and graduate students. Gelbart's research focuses on the structure of genes and genomes, bioinformatics, and developmental genetics. He is Principal Investigator of FlyBase, the genetic and genomic database of *Drosophila*. Gelbart's laboratory discovered many of the key components of the decapentaplegic (*dpp*)/BMP morphogen signaling pathway.

INTRODUCTION TO GENETIC ANALYSIS

Eighth Edition

Anthony J. F. Griffiths
University of British Columbia

Susan R. Wessler
University of Georgia

Richard C. Lewontin
Harvard University

William M. Gelbart
Harvard University

David T. Suzuki
University of British Columbia

Jeffrey H. Miller
University of California, Los Angeles

■▪ W. H. Freeman and Company • New York

Acquisitions Editor: Jason Noe
Developmental Editor: Susan Moran
Project Editor: Mary Louise Byrd
Cover and Text Designer: Diana Blume
Illustration Coordinator: Bill Page
Illustrations: Fine Line Illustrations, Network Graphics
Photo Editors: Meg Kuhta, Bianca Moscatelli
Production Coordinator: Julia DeRosa
New Media and Supplements Editors: Jeff Ciprioni, Elaine Palucki
Composition: Progressive Information Technologies
Manufacturing: RR Donnelley & Sons Company
Marketing Manager: Sarah Martin

COVER IMAGE: The human set of 46 chromosomes during nuclear division, "painted" with molecular probes in such a way that each pair of chromosomes appears in a different color. The image is a suitable icon for the subject matter of genetic analysis: the hereditary material, its function, and its transmission through time. [Image by Linda Barenboim Stapleton and Thomas Ried, Genetics branch, Center for Cancer Research, NIH, Bethesda, Maryland. Filter set by Chroma (Germany), imaging system SKYview (Applied Spectral Imaging, California) on a Leica epifluorescent microscope.]

Library of Congress Control Number: 2004102340

© 2005, 2000, 1996, 1993, 1989, 1986, 1981, 1976 by
W. H. Freeman and Company

ISBN 0-7167-4939-4
EAN 9780716749394

Printed in the United States of America

First printing
W. H. Freeman and Company
41 Madison Avenue
New York, NY 10010
Houndmills, Basingstoke RG21 6XS, England

www.whfreeman.com

Contents in Brief

Preface xi

1 Genetics and the Organism 1

Part I: TRANSMISSION GENETIC ANALYSIS

2 Patterns of Inheritance 27
3 The Chromosomal Basis of Inheritance 73
4 Eukaryote Chromosome Mapping by Recombination 115
5 The Genetics of Bacteria and Their Viruses 151

Part II: THE RELATIONSHIP OF DNA AND PHENOTYPE

6 From Gene to Phenotype 185
7 DNA: Structure and Replication 227
8 RNA: Transcription and Processing 255
9 Proteins and Their Synthesis 273
10 Regulation of Gene Transcription 301

Part III: GENOME STRUCTURE AND ENGINEERING

11 Gene Isolation and Manipulation 341
12 Genomics 389
13 The Dynamic Genome: Transposable Elements 423

Part IV: THE NATURE OF HERITABLE CHANGE

14 Mutation, Repair, and Recombination 451
15 Large-Scale Chromosomal Changes 481

Part V: FROM GENES TO PROCESSES

16 Dissection of Gene Function 521
17 Genetic Regulation of Cell Number: Normal and Cancer Cells 545
18 The Genetic Basis of Development 575

Part VI: THE IMPACT OF GENETIC VARIATION

19 Population Genetics 611
20 Quantitative Genetics 643
21 Evolutionary Genetics 679

A Brief Guide to Model Organisms 707

Appendix A: Genetic Nomenclature 723

Appendix B: Bioinformatics Resources for Genetics and Genomics 724

Glossary 727

Answers to Selected Problems 749

Index 763

Contents

Preface xi

1 Genetics and the Organism 1
1.1 Genes as determinants of the inherent properties of species 3
1.2 Genetic variation 8
1.3 Methodologies used in genetics 11
1.4 Model organisms 13
1.5 Genes, the environment, and the organism 16

Part I: TRANSMISSION GENETIC ANALYSIS

2 Patterns of Inheritance 27
2.1 Autosomal inheritance 29
2.2 Sex chromosomes and sex-linked inheritance 48
2.3 Cytoplasmic inheritance 55

3 The Chromosomal Basis of Inheritance 73
3.1 Historical development of the chromosome theory 75
3.2 The nature of chromosomes 80
3.3 Mitosis and meiosis 90
3.4 Chromosome behavior and inheritance patterns
     in eukaryotes 96
3.5 Organelle chromosomes 103

**4 Eukaryote Chromosome Mapping
    by Recombination** 115
4.1 The discovery of the inheritance patterns of linked genes 117
4.2 Recombination 121
4.3 Linkage maps 124
4.4 Using the chi-square test in linkage analysis 131
4.5 Using Lod scores to assess linkage in human pedigrees 133
4.6 Accounting for unseen multiple crossovers 134

5 The Genetics of Bacteria and Their Viruses 151
5.1 Working with microorganisms 153
5.2 Bacterial conjugation 155
5.3 Bacterial transformation 166
5.4 Bacteriophage genetics 166
5.5 Transduction 170
5.6 Physical maps versus linkage maps 174

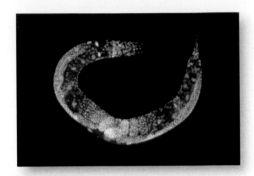

Figure 1-15d

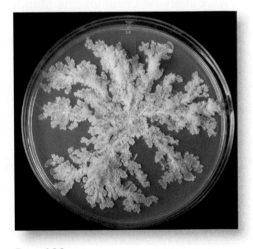

Page 100

Part II: THE RELATIONSHIP OF DNA AND PHENOTYPE

6   From Gene to Phenotype 185
6.1   Genes and gene products 187
6.2   Interactions between the alleles of one gene 192
6.3   Interacting genes and proteins 197
6.4   Applications of chi-square (χ^2) test to gene interaction ratios 209

7   DNA: Structure and Replication 227
7.1   DNA: the genetic material 229
7.2   The DNA structure 231
7.3   Semiconservative replication 237
7.4   Overview of DNA replication 240
7.5   The replisome: a remarkable replication machine 242
7.6   Assembling the replisome: replication initiation 244
7.7   Telomeres and telomerase: replication termination 247

8   RNA: Transcription and Processing 255
8.1   RNA 257
8.2   Transcription 259
8.3   Transcription in eukaryotes 264

9   Proteins and Their Synthesis 273
9.1   Protein structure 275
9.2   Collinearity of gene and protein 277
9.3   The genetic code 278
9.4   tRNA: the adapter 282
9.5   Ribosomes 285
9.6   Posttranslational events 291

10   Regulation of Gene Transcription 301
10.1 Prokaryotic gene regulation 303
10.2 Discovery of the *lac* system of negative control 307
10.3 Catabolite repression of the *lac* operon: positive control 312
10.4 Dual positive and negative control: the arabinose operon 315
10.5 Metabolic pathways 316
10.6 Transcriptional regulation in eukaryotes 316
10.7 Chromatin's role in eukaryotic gene regulation 322

Part III: GENOME STRUCTURE AND ENGINEERING

11   Gene Isolation and Manipulation 341
11.1 Generating recombinant molecules 343
11.2 DNA amplification in vitro: the polymerase chain reaction 360

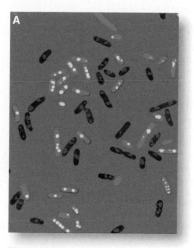

Figure 5-9a

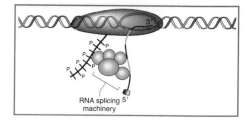

Figure 8-13b

Figure 10-30

11.3 Zeroing in on the gene for alkaptonuria: another case study 362

11.4 Detecting human disease alleles: molecular genetic diagnostics 364

11.5 Genetic engineering 366

12 Genomics 389

12.1 The nature of genomics 391

12.2 The sequence map of a genome 392

12.3 Creating genomic sequence maps 394

12.4 Using genomic sequence to find a specific gene 398

12.5 Bioinformatics: meaning from genomic sequence 406

12.6 Take-home lessons from the genomes 410

12.7 Functional genomics 413

13 The Dynamic Genome: Transposable Elements 423

13.1 Discovery of transposable elements in maize 425

13.2 Transposable elements in prokaryotes 429

13.3 Transposable elements in eukaryotes 434

13.4 The dynamic genome: more transposable elements than ever imagined 440

13.5 Host regulation of transposable elements 445

Part IV: THE NATURE OF HERITABLE CHANGE

14 Mutation, Repair, and Recombination 451

14.1 Point mutations 453

14.2 Spontaneous mutation 461

14.3 Biological repair mechanisms 468

14.4 The mechanism of meiotic crossing-over 473

15 Large-Scale Chromosomal Changes 481

15.1 Changes in chromosome number 483

15.2 Changes in chromosome structure 496

15.3 Overall incidence of human chromosome mutations 506

Part V: FROM GENES TO PROCESSES

16 Dissection of Gene Function 521

16.1 Forward genetics 523

16.2 Reverse genetics 530

16.3 Analysis of recovered mutations 535

16.4 Broader applications of functional dissection 537

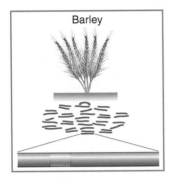

Figure 13-25

Figure 16-19

Figure 16-24

17   Genetic Regulation of Cell Number: Normal and Cancer Cells 545

17.1   The balance between cell loss and proliferation 547

17.2   The cell-proliferation machinery of the cell cycle 548

17.3   The machinery of programmed cell death 552

17.4   Extracellular signals 553

17.5   Cancer: the genetics of aberrant cell number regulation 558

17.6   Applying genomic approaches to cancer research, diagnosis, and therapies   566

18   The Genetic Basis of Development 575

18.1   The logic of building the body plan 577

18.2   Binary fate decisions: the germ line versus the soma 578

18.3   Forming complex pattern: the logic of the decision-making process 582

18.4   Forming complex pattern: establishing positional information 583

18.5   Forming complex pattern: utilizing positional information to establish cell fates 591

18.6   Refining the pattern 597

18.7   The many parallels in vertebrate and insect pattern formation 600

18.8   The genetics of sex determination in humans 602

18.9   Do the lessons of animal development apply to plants? 603

18.10 Genomic approaches to understanding pattern formation 605

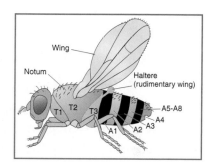

Figure 18-24a

Part VI: THE IMPACT OF GENETIC VARIATION

19   Population Genetics 611

19.1   Variation and its modulation 612

19.2   Effect of sexual reproduction on variation 621

19.3   Sources of variation 626

19.4   Selection 629

19.5   Balanced polymorphism 634

19.6   Random events 636

20   Quantitative Genetics 643

20.1   Genes and quantitative traits 645

20.2   Some basic statistical notions 647

20.3   Genotypes and phenotypic distribution 649

20.4   Norm of reaction and phenotypic distribution 651

20.5   Determining norms of reaction 652

20.6   The heritability of a quantitative character 654

20.7   Quantifying heritability 656

20.8   Locating genes 664

Statistical appendix 667

Figure 20-1

21 Evolutionary Genetics 679

21.1 A synthesis of forces: variation and divergence
of populations 682

21.2 Multiple adaptive peaks 685

21.3 Heritability of variation 687

21.4 Observed variation within and between populations 688

21.5 The process of speciation 689

21.6 Origin of new genes 691

21.7 Rate of molecular evolution 695

21.8 Genetic evidence of common ancestry in evolution 696

21.9 Comparative genomics and proteomics 698

A Brief Guide to Model Organisms 707

Appendix A: Genetic Nomenclature 723

**Appendix B: Bioinformatics Resources
for Genetics and Genomics** 724

Glossary 727

Answers to Selected Problems 749

Index 763

Preface

In the past quarter century, *Introduction to Genetic Analysis* has evolved to reflect the changing face of genetics. In this edition, the molecular core of the book—eight chapters—has been extensively reworked to bring modern genetic thinking to the forefront. Toward this end, we are delighted to welcome Susan R. Wessler to the author team. Susan is a plant molecular geneticist specializing in mobile elements. She brings not only a botanical perspective but also expertise on a fascinating class of genetic elements greatly affecting many areas of genetics, including mutation, genomics, and evolutionary genetics. Her responses to the following questions help to explain her contributions and highlight some new directions taken by this edition.

In your research, you focus on transposable elements—the "jumping genes." Why this specific area of research?

Because transposable elements are fun, they are mysterious, and they can make up more than half the genome. Now that many genomes are being and have been sequenced, we're finding much there that wasn't predicted, including the discovery that transposable elements are often the major component—by far. It's been said that the genome is more than just a blueprint, that it also contains our genetic history. In this regard, transposable elements are a significant part of how we got to be humans.

What appealed to you about working on a textbook, especially this one?

I love explaining what I do to a new audience, and I've always enjoyed writing. I'd been hoping to write more for students when this opportunity came along. Even though *Introduction to Genetic Analysis* was already an authoritative, respected textbook, I felt that I had something to contribute, some perspectives on where the field is heading and how to relate to students. Some people think that the more we learn, the more complicated things become, and the harder it is to teach. I think just the opposite: the more we know, the better the pieces fit, and the easier it becomes to tell the story. There are some amazing things going on, some really advanced stuff that can be communicated to students if you draw them in a step at a time and tie things together.

What specific contributions have you made to the new edition?

In regard to updating, I wanted to call attention to, among other things, the growing focus on RNA and the shift from prokaryotes to eukaryotes that we've seen in the field. In regard to communicating, I saw opportunities to tell stories that tie together material previously dispersed into different sections, such as our new single chapter on repair, mutation, and recombination. We also have running themes throughout the book, another way of helping students see how everything fits. One theme is the role of biological machines in coordinating fundamental cellular processes such as replication, transcription, and translation. Another is the use of model organisms, which we always look at in context. If we just catalog the organisms, there's no reason for students to care about them. But if we introduce them when they matter—such as showing the role of maize in understanding how transposable elements work or what yeast has taught us about eukaryotic gene regulation—it makes a more lasting impression.

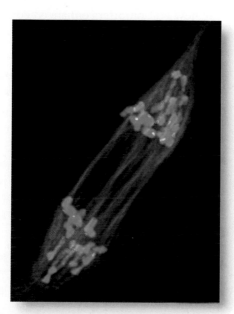

p. 428 Maize chromosomes (green) and spindle fibers (blue)

Molecular Chapters Extensively Revised

In recent decades, we have witnessed "the molecular revolution" and with it the rise of molecular genetics and recombinant DNA technology, culminating in the relatively new discipline of genomics. The present edition incorporates significant new developments in these areas. The newly developed theme of "molecular machines" introduces students to genetic systems such as translation, transcription, and regulation as they are understood in the laboratory today—as integrated, unified processes. Some examples of how the new edition has enhanced and updated molecular coverage are

- Entirely rewritten discussions of the mechanisms of DNA replication, transcription, and translation (Chapters 7 through 9)
- The first molecular-level glimpses of the ribosome and its transactions (Chapter 9)
- Entirely new discussion of eukaryotic gene regulation, with a focus on the role of chromatin and epigenetic mechanisms (Chapter 10)
- Completely reconceived and rewritten genomics chapter, presenting the current strategies for genome sequencing and bioinformatic analyses (Chapter 12)
- New discussions of how the host controls transposable elements and how transposable elements influence genome evolution in the freshly reconceived chapter on "The Dynamic Genome: Transposable Elements" (Chapter 13).
- New chapter titled "Dissection of Gene Function," examining the latest experimental techniques such as RNAi (Chapter 16)

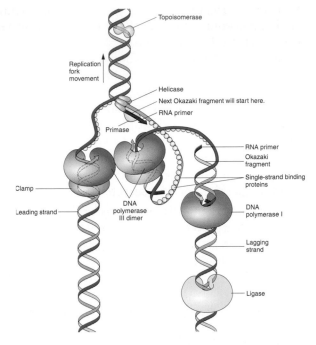

Figure 7-18 **The replisome**

New Coverage of Model Organisms

Building on the book's traditionally integrated coverage of model organisms, the new edition offers enhanced coverage of model systems in formats that are practical and flexible for both students and instructors.

- Chapter 1 includes a new section on model organisms.
- Model Organism boxes at key locations in the text provide additional information about the organism in nature and its use experimentally.
- A Brief Guide to Model Organisms, at the back of the book, provides quick access to essential, practical information about the uses of specific model organisms in research studies.
- An Index to Model Organisms, on the endpapers in the front of the book, provides chapter-by-chapter page references to discussions of specific organisms in the text, enabling instructors and students to easily assemble comparative information across organisms.

New Ways to Engage Students

Our aim is to tell a story that is engaging, coherent, and meaningful to students. To ensure an accessible text, we have reexamined every explanation, looking for ways to improve clarity. In addition, we have made the following improvements.

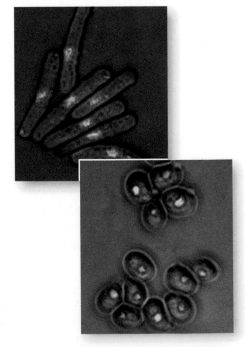

Figure 16-9 **Yeast mutants**

Added Theme of Molecular Machines

The easily accessible analogy of molecular machines imposes order on complex multistep processes, helping students understand how the components of genetic systems are integrated.

Streamlined Presentation

By reconceiving the relations among its parts, and with a careful eye toward eliminating outdated material, we have reduced the number of chapters in this edition from 26 to 21, making it approximately 10 percent briefer than the preceding edition.

Realignment of Topics

We have made some large translocations of material between and within chapters to bring related topics into better alignment. These new arrangements allow the reader to see the connections within one general area and to better grasp the big picture.

New and Revised Illustration Program, developed by Lynne Hunter, University of Pennsylvania

More than 100 new illustrations and more than 200 illustrations that have been redrawn offer students better accuracy, clarity, and simplicity. A bolder, more vivid color palette ensures the clarity of illustrations projected in lecture presentations.

Additional Pedagogical Features

The authors introduce several new features designed to make the chapters more accessible to students, such as

Chapter-Opening Key Questions

Chapter Overview Art (see Figure 16-1 at right)

Chapter Overview Sections

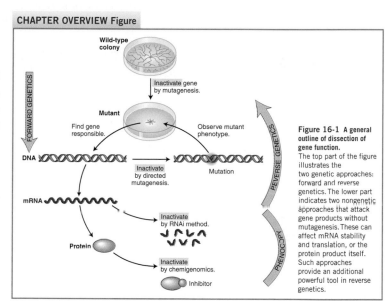

CHAPTER OVERVIEW Figure

Figure 16-1 A general outline of dissection of gene function. The top part of the figure illustrates the two genetic approaches: forward and reverse genetics. The lower part indicates two nongenetic approaches that attack gene products without mutagenesis. These can affect mRNA stability and translation, or the protein product itself. Such approaches provide an additional powerful tool in reverse genetics.

Superior Problem Solving

No matter how clear the exposition, deep understanding requires the student to personally engage with the material. Hence our continuing efforts to encourage student problem solving. Building on its highly praised focus on genetic analysis, the new edition provides students even greater opportunities to practice problem-solving skills—both in the text and on-line through the following features:

Versatile Problem Sets

Problems span the full range of degrees of difficulty. In this edition, problems are categorized according to level of difficulty—basic or challenging.

More Basic Problems

End-of-chapter problem sets include 20 percent more basic problems, giving students more opportunities to put fundamental concepts into practice.

Interactive Genetics Tutorials

These exercises, available on the CD-ROM that accompanies the solutions "MegaManual," help students develop critical thinking skills as they work through interactive experiments that focus on the most important model systems in genetic research.

Exploring Genomes Interactive Bioinformatics Tutorials

These tutorials guide students through live searches and analyses on the NCBI database. Activities introduce students to the basics of BLAST, Entrez, Pub-Med, OMIM, and more.

Media and Supplements

FOR INSTRUCTORS

Access

The Password-Protected Instructor's Resource Web site at
www.whfreeman.com/iga8e includes all of the electronic resources listed here. Contact your W. H. Freeman sales representative to learn how to log on as an instructor.

The Instructor's Resource CD-ROM (ISBN 0-7167-8661-3) contains all text images in PowerPoint and JPEG formats, with and without labels, and all 45 FLASH animations.

Electronic Images

• **All Images from the Text**
More than 500 illustrations in multiple downloadable formats, including PowerPoint, on the Instructor's Resource CD-ROM and Web site. Images are provided in two versions: one with enlarged labels that project more clearly for lecture-hall presentation; the other with the labels removed for easy customization in PowerPoint.

• **PowerPoint Lecture Outlines**

• **45 Step-Through and Continuous Play FLASH Animations**

Overhead Transparency Set (ISBN 0-7167-8664-8)

The full-color overhead transparency set contains 150 key illustrations from the text with enlarged labels that project more clearly for lecture-hall presentation.

Assessment and Teaching

NEW! Assessment Bank

This new resource brings together a wide selection of genetics problems for use in testing, homework assignments, or in-class activities. Searchable by topic and level and provided in Microsoft Word format on the Instructor's Resource Web site, the assessment bank offers a new level of flexibility, as well as problems for every purpose.

Understanding Genetics: Strategies for Teachers and Learners in Universities and High Schools, Anthony Griffiths and Jolie Mayer-Smith, University of British Columbia (ISBN 0-7167-5216-6)
This collection of articles focuses on student learning problems and describes methods for helping students improve their ability to process and integrate new information. The authors provide specific solutions to problems faced in a genetics classroom.

Student Solutions Manual
Available on the Instructor's Web site as easy-to-print Word files.

FOR STUDENTS

Student Web Site at www.whfreeman.com/iga8e

- **On-Line Sample Tests.** Students can test their understanding by answering on-line questions. All questions were chosen and often written by the textbook and assessment bank author teams to ensure coverage of the core concepts in each chapter. Questions are page-referenced to the text for easy review of the material.

- **Interactive Unpacking the Problem** by Craig Berezowsky, University of British Columbia. One problem (indicated in the text by an icon) from the Problems set in each chapter is available on the Web site in interactive form. As with the text versions, each Web-based Unpacking the Problem poses a series of questions to take students through the thought processes needed to reach a solution. The on-line versions offer immediate feedback to students as they work through the problems, as well as convenient tracking and grading functions.

- **Step-Through and Continuous Play FLASH Animations.** There are 45 animations created by Tony Griffiths with BioStudio Visual Communications. Each illustration in the text that has a related animation is tabbed for ready reference. See the endpapers at the back of the book for descriptions of each animation and the location in the text to which it is connected.

- **Exploring Genomes: Web-Based Bioinformatics Tutorials.** These ten interactive tutorials, written by Paul G. Young of Queens University, teach students the basics of how to use the vast genomic resources of the National Center for Biotechnology Information (NCBI) on its Web site. The tutorials guide students through exercises, including live searches, that illustrate the potential applications of these bioinformatic resources in modern genetic analyses.

Solutions MegaManual (ISBN 0-7167-8665-6) includes

- **Solutions Manual,** by William Fixsen, Harvard University. The *Solutions Manual* contains complete worked-out solutions to all the problems in the textbook, including the Unpacking the Problem exercises. Each solution has been thoroughly reviewed for accuracy and clarity.

- **Interactive Genetics CD-ROM,** by Lianna Johnson and John Merriam, University of California, Los Angeles. With this program, students can walk through solutions to a number of genetics problems step by step, getting specific feedback at every part. The program addresses common mistakes and misunderstandings and offers useful problem-solving strategies.

- **Exploring Genomes: Web-Based Bioinformatics Tutorials,** by Paul G. Young, Queen's University. Used in conjunction with the on-line tutorials found at www.whfreeman.com/young, Exploring Genomes guides students through live searches and analyses on the commonly used National Center for Biotechnology Information (NCBI) database. Includes the following tutorials:

    1. Introduction to Genomic Databases
    2. Learning to Use Entrez
    3. Learning to Use BLAST
    4. Using BLAST to Compare Nucleic Acid Sequences
    5. Learning to Use PubMed
    6. Online Mendelian Inheritance in Man (OMIM) and Huntington Disease
    7. Finding Conserved Domains
    8. Determining Protein Structure
    9. Exploring the Cancer Genome Anatomy Project
    10. Measuring Phylogenetic Distance

Acknowledgments

We extend our thanks and gratitude to our colleagues who reviewed this edition and whose insights and advice were most helpful:

James B. Anderson, *University of Toronto*
Spencer A. Benson, *University of Maryland*
Edward Berger, *Dartmouth College*
Susan E. Bergeson, *University of Texas at Austin*
Daniel R. Bergey, *Black Hills State University*
John L. Bowman, *University of California, Davis*
Mary C. Colavito, *Santa Monica College*
Doreen Cupo, *George Mason University*
Tamara L. Davis, *Bryn Mawr College*
Alyce DeMarais, *University of Puget Sound*
Richard E. Duhrkopf, *Baylor University*
Robert G. Fowler, *San Jose State University*
Andreas Fritz, *Emory University*
T. J. Gill, *University of Houston*
Michael A. Goldman, *San Francisco State University*
Gary Grothman, *St. Mary's College*
Nancy A. Guild, *University of Colorado*
Stanley Hattman, *University of Rochester*
David C. Hinkle, *University of Rochester*
Stanton Hoegerman, *College of William and Mary*
Nancy M. Hollingsworth, *State University of New York at Stony Brook*
Richard B. Imberski, *University of Maryland*
John B. Jenkins, *Swarthmore College*
Maurice Kernan, *State University of New York at Stony Brook*
Sidney R. Kushner, *University of Georgia*
John Locke, *University of Alberta*
Maggie Lopes, *Queen's University*

Kim S. McKim, *Rutgers University*
Philip M. Meneely, *Haverford College*
Beth A. Montelone, *Kansas State University*
Bryan Ness, *Pacific Union College*
Todd Nickle, *Mount Royal College*
Robin E. Owen, *Mount Royal College*
Inés Pinto, *University of Arkansas*
James Price, *Utah Valley State College*
Mitch Price, *Penn State University*
Todd Rainey, *Bluffton College*
Dennis T. Ray, *University of Arizona*
Mark Rose, *Princeton University*
Charles Rozek, *Case Western University*
Inder Saxena, *University of Texas, Austin*
Malcolm D. Schug, *University of North Carolina, Greensboro*
Trudi Schupbach, *Princeton University*
Rodney J. Scott, *Wheaton College*
Mark Seeger, *The Ohio State University*
Nava Segev, *University of Illinois, Chicago*
David Sheppard, *University of Delaware*
Richard N. Sherwin, *University of Pittsburgh*
Carol Hopkins Sibley, *University of Washington*
Laurie G. Smith, *University of California, San Diego*
Justin Thackeray, *Clark University*
James N. Thompson, Jr., *University of Oklahoma*
Harold Vaessin, *The Ohio State University*
Esther Verheyen, *Simon Fraser University*
Rick Ward, *Michigan State University*
David B. Wing, *Shepherd College*
Paul Wong, *University of Alberta*
Andrew J. Wood, *Southern Illinois University*
S. M. Wood, *Queen's University*

The authors would like to express their deep gratitude for the work of the people at W. H. Freeman and Company. Special thanks must be given to developmental editor Susan Moran, whose imagination, dedication, and hard work have substantially improved the readability and accessibility of the book. Thanks also to acquisitions editor Jason Noe for steering the project through the publishing rapids so ably and with such good humor. The clarity of the illustrations has benefited a great deal from the detailed revisions suggested by Lynne Hunter at the University of Pennsylvania. Manuscript editors William O'Neal and Patricia Zimmerman substantially improved the flow of the text and winkled out a large number of inconsistencies. Bill Fixsen, "the answers man," made a great contribution to the pedagogical usefulness of the problem sets in the book through his well-crafted selected solutions and answer manual. In the final stages of production, Mary Louise Byrd was in large part responsible for getting the book out on time; her dedication and sharp eye for error contributed greatly to the quality of the final book. We also thank Julia DeRosa, production manager; Diana Blume, designer; Bill Page, illustration coordinator; and Meg Kuhta and Bianca Moscatelli, photo editors.

Introduction to Genetic Analysis

1

GENETICS AND THE ORGANISM

Genetic variation in the color of corn kernels. Each kernel represents a separate individual with a distinct genetic makeup. The photograph symbolizes the history of humanity's interest in heredity. Humans were breeding corn thousands of years before the advent of the modern discipline of genetics. Extending this heritage, corn today is one of the main research organisms in classical and molecular genetics. [William Sheridan, University of North Dakota; photograph by Travis Amos.]

KEY QUESTIONS

- What is the hereditary material?

- What is the chemical and physical structure of DNA?

- How is DNA copied in the formation of new cells and in the gametes that will give rise to the offspring of an individual?

- What are the functional units of DNA that carry information about development and physiology?

- What molecules are the main determinants of the basic structural and physiological properties of an organism?

- What are the steps in translating the information in DNA into protein?

- What determines the differences between species in their physiology and structure?

- What are the causes of variation between individuals within species?

- What is the basis of variation in populations?

OUTLINE

1.1 Genes as determinants of the inherent properties of species

1.2 Genetic variation

1.3 Methodologies used in genetics

1.4 Model organisms

1.5 Genes, the environment, and the organism

1

CHAPTER OVERVIEW

Why study genetics? There are two basic reasons. First, genetics occupies a pivotal position in the entire subject of biology. Therefore, for any serious student of plant, animal, or microbial life, an understanding of genetics is essential. Second, genetics, like no other scientific discipline, is central to numerous aspects of human affairs. It touches our humanity in many different ways. Indeed, genetic issues seem to surface daily in our lives, and no thinking person can afford to be ignorant of its discoveries. In this chapter, we take an overview of the science of genetics, showing how it has come to occupy its crucial position. In addition, we provide a perspective from which to view the subsequent chapters.

First, we need to define what *genetics* is. Some define it as the "study of heredity," but hereditary phenomena were of interest to humans long before biology or genetics existed as the scientific disciplines that we know today. Ancient peoples were improving plant crops and domesticated animals by selecting desirable individuals for breeding. They also must have puzzled about the inheritance of individuality in humans and asked such questions as "Why do children resemble their parents?" and "How can various diseases run in families?" But these people could not be called "geneticists." Genetics as a set of principles and analytical procedures did not begin until the 1860s, when an Augustinian monk named Gregor Mendel (Figure 1-1) performed a set of experiments that pointed to the existence of biological elements that we now call *genes*. The word *genetics* comes from the word "gene," and genes are the focus of the subject. Whether geneticists study at the molecular, cellular, organismal, family, population, or evolutionary level, genes are always central in their studies. Simply stated, genetics is the study of genes.

What is a **gene?** A gene is a section of a threadlike double-helical molecule called **deoxyribonucleic acid,** abbreviated **DNA.** The discovery of genes and the understanding of their molecular structure and function have been sources of profound insight into two of the biggest mysteries of biology:

1. What makes a species what it is? We know that cats always have kittens and people always have babies. This commonsense observation naturally leads to questions about the determination of the properties of a species. The determination must be hereditary because, for example, the ability to have kittens is inherited by every generation of cats.

2. What causes variation within a species? We can distinguish one another as well as our own pet cat from other cats. Such differences within a species require explanation. Some of these distinguishing features are clearly familial; for example, animals of a certain unique color often have offspring with the same color, and in human families, certain features, such as the shape of the nose, definitely "run in the family." Hence we might suspect that a hereditary component explains at least some of the variation within a species.

The answer to the first question is that genes dictate the inherent properties of a species. The products of most genes are specific **proteins.** Proteins are the main macromolecules of an organism. When you look at an organism, what you see is either a protein or something that has been made by a protein. The amino acid sequence of a protein is encoded in a gene. The timing and rate of production of proteins and other cellular components are a function both of the genes within the cells and of the environment in which the organism is developing and functioning.

The answer to the second question is that any one gene can exist in several forms that differ from one another, generally in small ways. These forms of a gene are called **alleles.** Allelic variation causes hereditary variation within a species. At the protein level, allelic variation becomes protein variation.

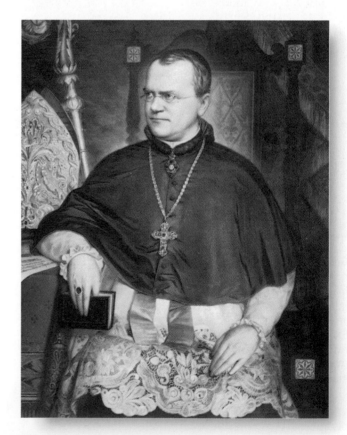

Figure 1-1 Gregor Mendel. [Moravian Museum, Brno.]

The next sections of this chapter show how genes influence the inherent properties of a species and how allelic variation contributes to variation within a species. These sections are an overview; most of the details will be presented in later chapters.

1.1 Genes as determinants of the inherent properties of species

What is the nature of genes, and how do they perform their biological roles? Three fundamental properties are required of genes and the DNA of which they are composed.

1. *Replication.* Hereditary molecules must be capable of being copied at two key stages of the life cycle (Figure 1-2). The first stage is the production of the cell type that will ensure the continuation of a species from one generation to the next. In plants and animals, these cells are the gametes: egg and sperm. The other stage is when the first cell of a new organism undergoes multiple rounds of division to produce a multicellular organism. In plants and animals, this is the stage at which the fertilized egg, the **zygote,** divides repeatedly to produce the complex organismal appearance that we recognize.

2. *Generation of form.* The working structures that make up an organism can be thought of as form or substance, and DNA has the essential "information" needed to create form.

3. *Mutation.* A gene that has changed from one allelic form into another has undergone mutation—an event that happens rarely but regularly. Mutation is not only a basis for variation within a species, but also, over the long term, the raw material for evolution.

We will examine replication and the generation of form in this section and mutation in the next.

DNA and its replication

An organism's basic complement of DNA is called its **genome.** The somatic cells of most plants and animals contain two copies of their genome (Figure 1-3); these organisms are **diploid.** The cells of most fungi, algae, and bacteria contain just one copy of the genome; these organisms are **haploid.** The genome itself is made up of one or more extremely long molecules of DNA that are organized into chromosomes. Genes are simply the regions of chromosomal DNA that are involved in the

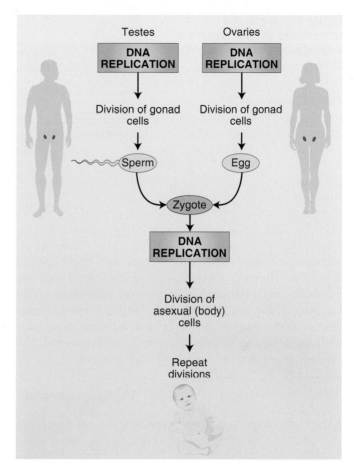

Figure 1-2 DNA replication is the basis of the perpetuation of life through time.

cell's production of proteins. Each chromosome in the genome carries a different array of genes. In diploid cells, each chromosome and its component genes are present twice. For example, human somatic cells contain two sets of 23 chromosomes, for a total of 46 chromosomes. Two chromosomes with the same gene array are said to be **homologous.** When a cell divides, all its chromosomes (its one or two copies of the genome) are replicated and then separated, so that each daughter cell receives the full complement of chromosomes.

To understand replication, we need to understand the basic nature of DNA. DNA is a linear, double-helical structure that looks rather like a molecular spiral staircase. The double helix is composed of two intertwined chains made up of building blocks called **nucleotides.** Each nucleotide consists of a phosphate group, a deoxyribose sugar molecule, and one of four different nitrogenous bases: adenine, guanine, cytosine, or thymine. Each of the four nucleotides is usually designated by the first letter of the base it contains: A, G, C, or T. Each nucleotide chain is held together by bonds between the sugar and phosphate portions of the

Figure 1-3 Successive enlargements bringing the genetic material of an organism into sharper focus.

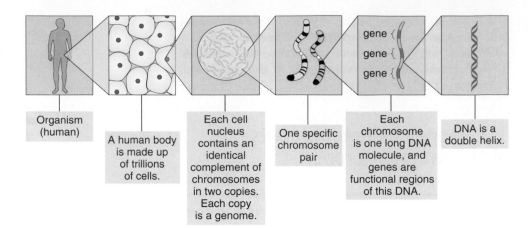

Organism (human)

A human body is made up of trillions of cells.

Each cell nucleus contains an identical complement of chromosomes in two copies. Each copy is a genome.

One specific chromosome pair

Each chromosome is one long DNA molecule, and genes are functional regions of this DNA.

DNA is a double helix.

consecutive nucleotides, which form the "backbone" of the chain. The two intertwined chains are held together by weak bonds between bases on opposite chains (Figure 1-4). There is a "lock-and-key" fit between the bases on the opposite strands, such that adenine pairs only with thymine and guanine pairs only with cytosine. The

bases that form base pairs are said to be **complementary.** Hence a short segment of DNA drawn with arbitrary nucleotide sequence might be

$$\cdots\cdot\text{CAGT}\cdots\cdot$$
$$\cdots\cdot\text{GTCA}\cdots\cdot$$

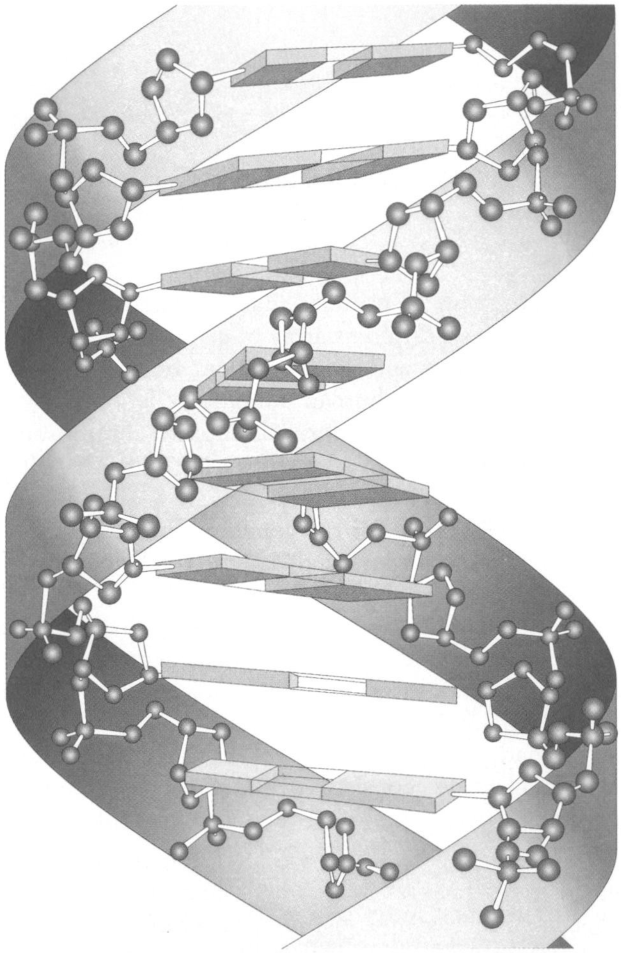

> **MESSAGE** DNA is composed of two nucleotide chains held together by complementary pairing of A with T and G with C.

For replication of DNA to take place, the two strands of the double helix must come apart, rather like the opening of a zipper. The two exposed nucleotide chains then act as alignment guides, or templates, for the deposition of free nucleotides, which are then joined together by the enzyme DNA polymerase to form a new strand. The crucial point illustrated in Figure 1-5 is that because of base complementarity, the two daughter DNA molecules are identical with each other and with the original molecule.

> **MESSAGE** DNA is replicated by the unwinding of the two strands of the double helix and the building up of a new complementary strand on each of the separated strands of the original double helix.

Generation of form

If DNA represents information, what constitutes form at the cellular level? The simple answer is "protein" because the great majority of structures in a cell are protein or have been made by protein. In this section, we trace the steps through which information becomes form.

The biological role of most genes is to carry information specifying the chemical composition of proteins or the regulatory signals that will govern their production by the cell. This information is encoded by the sequence of nucleotides. A typical gene contains the information for one specific protein. The collection of

Figure 1-4 Ribbon representation of the DNA double helix.
Blue = sugar-phosphate backbone; brown = paired bases.

Figure 1-5 DNA replication in process. Blue = nucleotides of the original double helix; gold = new nucleotides being polymerized to form daughter chains. S = sugar; P = phosphate group.

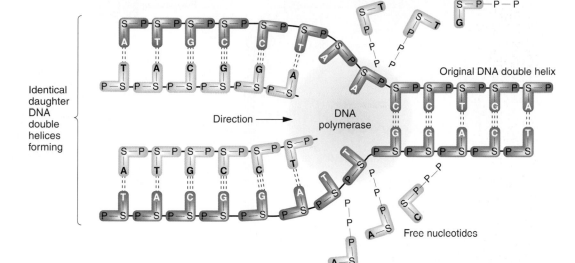

proteins an organism can synthesize, as well as the timing and amount of production of each protein, is an extremely important determinant of the structure and physiology of organisms. A protein generally has one of two basic functions, depending on the gene. First, the protein may be a structural component, contributing to the physical properties of cells or organisms. Examples of **structural proteins** are microtubule, muscle, and hair proteins. Second, the protein may be an active agent in cellular processes—such as an active-transport protein or an **enzyme** that catalyzes one of the chemical reactions of the cell.

The primary structure of a protein is a linear chain of amino acids, called a **polypeptide.** The sequence of amino acids in the primary chain is specified by the sequence of nucleotides in the gene. The completed primary chain is coiled and folded—and in some cases, associated with other chains or small molecules—to form a functional protein. A given amino acid sequence may fold in a large number of stable ways. The final folded state of a protein depends both on the sequence of amino acids specified by its gene and on the physiology of the cell during folding.

> **MESSAGE** The sequence of nucleotides in a gene specifies the sequence of amino acids that is put together by the cell to produce a polypeptide. This polypeptide then folds under the influence of its amino acid sequence and other molecular conditions in the cell to form a protein.

TRANSCRIPTION The first step taken by the cell to make a protein is to copy, or **transcribe,** the nucleotide sequence in one strand of the gene into a complementary single-stranded molecule called **ribonucleic acid (RNA).** Like DNA, RNA is composed of nucleotides, but these nucleotides contain the sugar ribose instead of deoxyribose. Furthermore, in place of thymine, RNA

contains uracil (U), which like thymine, pairs with adenine. Hence the RNA bases are A, G, C, and U. The transcription process, which occurs in the cell nucleus, is very similar to the process for replication of DNA because the DNA strand serves as the template for making the RNA copy, which is called a **transcript.** The RNA transcript, which in many species undergoes some structural modifications, becomes a "working copy" of the information in the gene, a kind of "message" molecule called **messenger RNA (mRNA).** The mRNA then enters the cytoplasm, where it is used by the cellular machinery to direct the manufacture of a protein. Figure 1-6 summarizes the process of transcription.

> **MESSAGE** During transcription, one of the DNA strands of a gene acts as a template for the synthesis of a complementary RNA molecule.

TRANSLATION The process of producing a chain of amino acids based on the sequence of nucleotides in the mRNA is called **translation.** The nucleotide sequence of an mRNA molecule is "read" from one end of the mRNA to the other, in groups of three successive bases. These groups of three are called **codons.**

AUU	CCG	UAC	GUA	AAU	UUG
codon	codon	codon	codon	codon	codon

Because there are four different nucleotides, there are $4 \times 4 \times 4 = 64$ different codons possible, each one coding for an amino acid or a signal to terminate translation. Because only 20 kinds of amino acids are used in the polypeptides that make up proteins, more than one codon may correspond to the same amino acid. For instance, AUU, AUC, and AUA all encode isoleucine, while UUU and UUC code for phenylalanine, and UAG is a translation termination ("stop") codon.

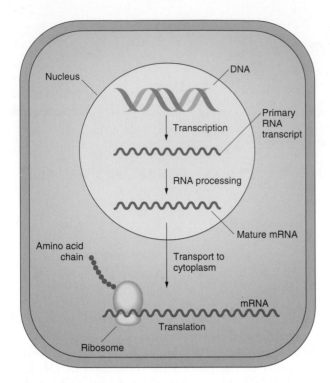

Figure 1-6 Transcription and translation in a eukaryotic cell. The mRNA is processed in the nucleus, then transported to the cytoplasm for translation into a polypeptide chain.

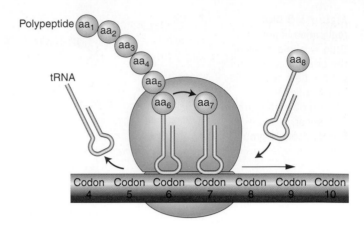

Figure 1-7 Translation. An amino acid (aa) is added to a growing polypeptide chain in the translation of mRNA.

Protein synthesis takes place on cytoplasmic organelles called **ribosomes.** A ribosome attaches to one end of an mRNA molecule and moves along the mRNA, catalyzing the assembly of the string of amino acids that will constitute the primary polypeptide chain of the protein. Each kind of amino acid is brought to the assembly process by a small RNA molecule called **transfer RNA (tRNA),** which is complementary to the mRNA codon that is being read by the ribosome at that point in the assembly.

Trains of ribosomes pass along an mRNA molecule, each member of a train making the same type of polypeptide. At the end of the mRNA, a termination codon causes the ribosome to detach and recycle to another mRNA. The process of translation is shown in Figure 1-7.

> **MESSAGE** The information in genes is used by the cell in two steps of information transfer: DNA is transcribed into mRNA, which is then translated into the amino acid sequence of a polypeptide. The flow of information from DNA to RNA to protein is a central focus of modern biology.

GENE REGULATION Let's take a closer look at the structure of a gene, which determines the final form of the RNA "working copy" as well as the timing of transcription in a particular tissue. Figure 1-8 shows the general structure of a gene. At one end, there is a regulatory region to which various proteins involved in the regulation of the gene's transcription bind, causing the gene to be transcribed at the right time and in the right amount. A region at the other end of the gene signals the end point of the gene's transcription. Between these two end regions lies the DNA sequence that will be transcribed to specify the amino acid sequence of a polypeptide.

Gene structure is more complex in eukaryotes than in prokaryotes. **Eukaryotes,** which include all the multicellular plants and animals, are those organisms whose cells have a membrane-bound nucleus. **Prokaryotes** are organisms with a simpler cellular structure lacking a nucleus, such as bacteria. In the genes of many eukaryotes, the protein-encoding sequence is interrupted by one or more stretches of DNA called **introns.** The origin and functions of introns are still unclear. They are excised from the primary transcript during the formation of mRNA. The segments of coding sequence between the introns are called **exons.**

Some protein-encoding genes are transcribed more or less constantly; these are the "housekeeping" genes that are always needed for basic reactions. Other genes may be rendered unreadable or readable to suit the functions of the organism at particular times and under particular external conditions. The signal that masks or

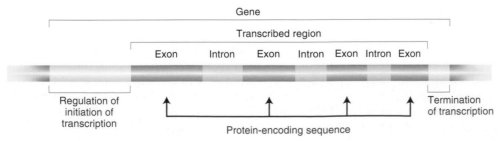

Figure 1-8 Generalized structure of a eukaryotic gene. This example has three introns and four exons.

unmasks a gene may come from outside the cell, for example, from a steroid hormone or a nutrient. Alternatively, the signal may come from within the cell as the result of the reading of other genes. In either case, special regulatory sequences in the DNA are directly affected by the signal, and they in turn affect the transcription of the protein-encoding gene. The regulatory substances that serve as signals bind to the regulatory region of the target gene to control the synthesis of transcripts.

Figure 1-9 illustrates the essentials of gene action in a generalized eukaryotic cell. Outside the nucleus of the cell is a complex array of membranous structures, including

Figure 1-9 Simplified view of gene action in a eukaryotic cell. The basic flow of genetic information is from DNA to RNA to protein. Four types of genes are shown. Gene 1 responds to external regulatory signals and makes a protein for export; gene 2 responds to internal signals and makes a protein for use in the cytoplasm; gene 3 makes a protein to be transported into an organelle; gene 4 is part of the organelle DNA and makes a protein for use inside its own organelle. Most eukaryotic genes contain introns, regions (generally noncoding) that are cut out in the preparation of functional messenger RNA. Note that many organelle genes have introns and that an RNA-synthesizing enzyme is needed for organelle mRNA synthesis. These details have been omitted from the diagram of the organelle for clarity. (Introns will be explained in detail in subsequent chapters.)

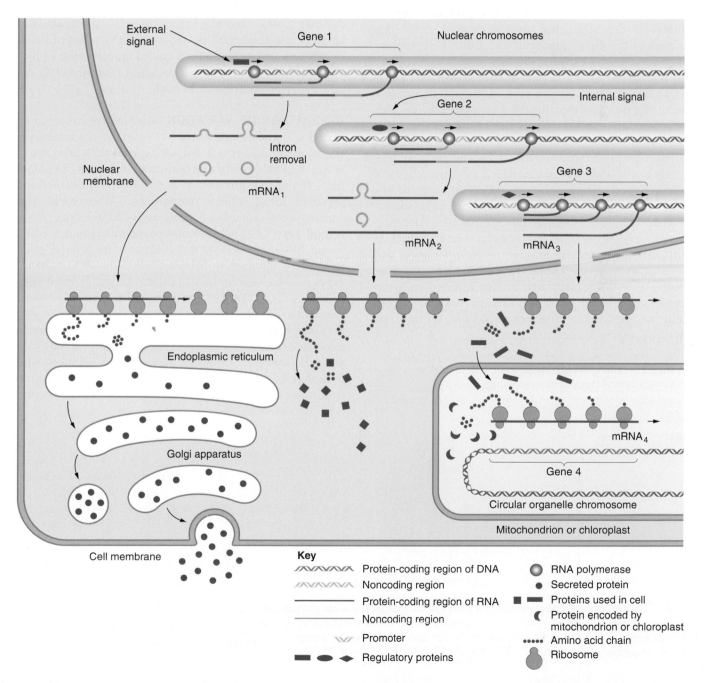

the endoplasmic reticulum and Golgi apparatus, and organelles such as mitochondria and chloroplasts. The nucleus contains most of the DNA, but note that mitochondria and chloroplasts also contain small chromosomes.

Each gene encodes a separate protein, each with specific functions either within the cell (for example, the purple-rectangle proteins in Figure 1-9) or for export to other parts of the organism (the purple-circle proteins). The synthesis of proteins for export (secretory proteins) takes place on ribosomes that are located on the surface of the rough endoplasmic reticulum, a system of large, flattened membrane vesicles. The completed amino acid chains are passed into the lumen of the endoplasmic reticulum, where they fold up spontaneously to take on their three-dimensional structure. The proteins may be modified at this stage, but they eventually enter the chambers of the Golgi apparatus and from there, the secretory vessels, which eventually fuse with the cell membrane and release their contents to the outside.

Proteins destined to function in the cytoplasm and most of the proteins that function in mitochondria and chloroplasts are synthesized in the cytoplasm on ribosomes not bound to membranes. For example, proteins that function as enzymes in the glycolysis pathway follow this route. The proteins destined for organelles are specially tagged to target their insertion into specific organelles. In addition, mitochondria and chloroplasts have their own small circular DNA molecules. The synthesis of proteins encoded by genes on mitochondrial or chloroplast DNA takes place on ribosomes inside the organelles themselves. Therefore the proteins in mitochondria and chloroplasts are of two different origins: either encoded in the nucleus and imported into the organelle or encoded in the organelle and synthesized within the organelle compartment.

> **MESSAGE** The flow of information from DNA to RNA to protein is a central focus of modern biology.

1.2 Genetic variation

If all members of a species have the same set of genes, how can there be genetic variation? As indicated earlier, the answer is that genes come in different forms called *alleles*. In a population, for any given gene there can be from one to many different alleles; however, because most organisms carry only one or two chromosome sets per cell, any individual organism can carry only one or two alleles per gene. The alleles of one gene will always be found in the same position along the chromosome. Allelic variation is the basis for hereditary variation.

Types of variation

Because a great deal of genetics concerns the analysis of variants, it is important to understand the types of variation found in populations. A useful classification is into *discontinuous* and *continuous* variation (Figure 1-10). Allelic variation contributes to both.

DISCONTINUOUS VARIATION Most of the research in genetics in the past century has been on discontinuous variation because it is a simpler type of variation, and it is easier to analyze. In **discontinuous variation,** a character is found in a population in two or more distinct and separate forms called **phenotypes.** "Blue eyes" and "brown eyes" are phenotypes, as is "blood type A" or "blood type O." Such alternative phenotypes are often found to be encoded by the alleles of one gene. A good example is albinism in humans, which concerns phenotypes of the character of skin pigmentation. In most people, the cells of the skin can make a dark-brown or black pigment called *melanin*, the substance that gives our skin its color ranging from tan color in people of European ancestry to brown or black in those of tropical and subtropical ancestry. Although always rare, albinos, who completely lack pigment in their skin and hair, are found in all races (Figure 1-11). The difference between pig-

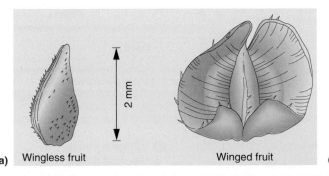

(a) Wingless fruit   Winged fruit   **(b)**

Figure 1-10 Examples of discontinuous and continuous variation in natural populations.
(a) Fruits of the sea blush, *Plectritis congesta*, have one of two distinct forms. Any one plant has either all winged or all wingless fruits. (b) Variation in height, branch number, and flower number in the herb *Achillea*. [Part b, Carnegie Institution of Washington.]

Figure 1-11 An albino. The phenotype is caused by two doses of a recessive allele, *a/a*. The dominant allele *A* determines one step in the chemical synthesis of the dark pigment melanin in the cells of skin, hair, and eye retinas. In *a/a* individuals, this step is nonfunctional, and the synthesis of melanin is blocked. [Copyright Yves Gellie/Icone.]

mented and unpigmented skin is caused by different alleles of a gene that encodes an enzyme involved in melanin synthesis.

The alleles of a gene are conventionally designated by letters. The allele that codes for the normal form of the enzyme involved in making melanin is called *A*, and the allele that codes for an inactive form of that enzyme (resulting in albinism) is designated *a*, to show that they are related. The allelic constitution of an organism is its **genotype,** which is the hereditary underpinning of the phenotype. Because humans have two sets of chromosomes in each cell, genotypes can be either *A/A*, *A/a*, or *a/a* (the slash shows that the two alleles are a pair). The phenotype of *A/A* is pigmented, that of *a/a* is albino, and that of *A/a* is pigmented. The *ability* to make pigment is expressed over *inability* (*A* is said to be dominant, as we shall see in Chapter 2).

Although allelic differences cause phenotypic differences such as pigmented and albino coloration, this does not mean that only one gene affects skin color. It is known that there are several, although the identity and number of these genes are currently unknown. However, the *difference* between pigmented, of whatever shade, and albinism is caused by the *difference* in the alleles of one gene—the gene that determines the ability to make melanin; the allelic composition of other genes is irrelevant.

In some cases of discontinuous variation, there is a predictable one-to-one relation between genotype and phenotype under most conditions. In other words, the two phenotypes (and their underlying genotypes) can almost always be distinguished. In the albinism example, the *A* allele always allows some pigment formation, whereas the *a* allele always results in albinism when present in two copies. For this reason, discontinuous variation has been successfully used by geneticists to identify the underlying alleles and their role in cellular functions.

Geneticists distinguish two categories of discontinuous variation. In a natural population, the existence of two or more common discontinuous variants is called **polymorphism** (Greek; many forms). The various forms are called **morphs.** It is often found that different morphs are determined by different alleles of a single gene. Why do populations show genetic polymorphism? Special types of natural selection can explain a few cases, but, in other cases, the morphs seem to be selectively neutral.

Rare, exceptional discontinuous variants are called **mutants,** whereas the more common "normal" phenotype is called the **wild type.** Figure 1-12 shows an example of a mutant phenotype. Again, in many cases, the wild-type and mutant phenotypes are determined by different alleles of one gene. Both mutants and polymorphisms originally arise from rare changes in DNA (mutations), but somehow the mutant alleles of a polymorphism become common. These rare changes in DNA may be nucleotide-pair substitutions or small deletions or duplications. Such mutations change the amino acid composition of the protein. In the case of albinism, for example, the DNA of a gene that encodes an enzyme involved in melanin synthesis is changed, such that a crucial amino acid is replaced by another amino acid or lost, yielding a nonfunctioning enzyme. Mutants (such as those that produce albinism) can occur spontaneously in nature, or they can be produced by treatment with mutagenic chemicals or radiation. Geneticists regularly induce mutations artificially to

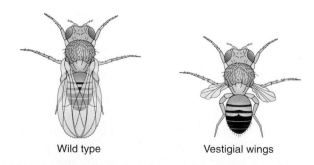

Wild type   Vestigial wings

Figure 1-12 Wild type and mutant *Drosophila*. A *Drosophila* mutant with abnormal wings and a normal fly (wild type) for comparison. In both cases, the two phenotypes are caused by the alleles of one gene.

carry out genetic analysis because mutations that affect some specific biological function under study identify the various genes that interact in that function.

> **MESSAGE** In many cases, an allelic difference at a single gene may result in discrete phenotypic forms that make it easy to study the gene and its associated biological function.

CONTINUOUS VARIATION A character showing **continuous variation** has an unbroken range of phenotypes in a population (see Figure 1-10b). Measurable characters such as height, weight, and skin or hair color are good examples of such variation. Intermediate phenotypes are generally more common than extreme phenotypes. In some cases, all the variation is environmental and has no genetic basis, as in the case of the different languages spoken by different human groups. In other cases, such as that of the various shades of human eye color, the differences are caused by allelic variation in one or many genes. For most continuously variable characters, both genetic and environmental variation contribute to differences in phenotype. In continuous variation, there is no one-to-one correspondence of genotype and phenotype. For this reason, little is known about the types of genes underlying continuous variation, and only recently have techniques become available for identifying and characterizing them.

Continuous variation is encountered more commonly than discontinuous variation in everyday life. We can all identify examples of continuous variation, such as variation in size or shape, in plant or animal populations that we have observed—many examples exist in human populations. One area of genetics in which continuous variation is important is in plant and animal breeding. Many of the characters that are under selection in breeding programs, such as seed weight or milk production, arise from many gene differences interacting with environmental variation, and the phenotypes show continuous variation in populations. We shall return to the specialized techniques for analyzing continuous variation in Chapter 20, but for the greater part of the book, we shall be dealing with the genes underlying discontinuous variation.

Molecular basis of allelic variation

Consider the difference between the pigmented and the albino phenotypes in humans. The dark pigment melanin has a complex structure that is the end product of a biochemical synthetic pathway. Each step in the pathway is a conversion of one molecule into another, with the progressive formation of melanin in a step-by-step manner. Each step is catalyzed by a separate enzyme protein encoded by a specific gene. Most cases of albinism result from changes in one of these enzymes—tyrosinase. The enzyme tyrosinase catalyzes the last step of the pathway, the conversion of tyrosine into melanin.

To perform this task, tyrosinase binds to its substrate, a molecule of tyrosine, and facilitates the molecular changes necessary to produce the pigment melanin. There is a specific "lock-and-key" fit between tyrosine and the active site of the enzyme. The **active site** is a pocket formed by several crucial amino acids in the polypeptide. If the DNA of the tyrosinase-encoding gene changes in such a way that one of these crucial amino acids is replaced by another amino acid or is lost, then there are several possible consequences. First, the enzyme might still be able to perform its functions but in a less efficient manner. Such a change may have only a small effect at the phenotypic level, so small as to be difficult to observe, but it might lead to a reduction in the amount of melanin formed and, consequently, a lighter skin coloration. Note that the protein is still present more or less intact, but its ability to convert tyrosine into melanin has been compromised. Second, the enzyme might be incapable of any function, in which case the mutational event in the DNA of the gene would have produced an albinism allele, referred to earlier as an *a* allele. Hence a person of genotype *a/a* is an albino. The genotype *A/a* is interesting. It results in normal pigmentation because transcription of one copy of the wild-type allele (*A*) can provide enough tyrosinase for synthesis of normal amounts of melanin. Genes are termed *haplosufficient* if roughly normal function is obtained when there is only a single copy of the normal gene. Wild-type alleles commonly appear to be haplosufficient, in part because small reductions in function are not vital to the organism. Alleles that fail to code for a functional protein are called **null** ("nothing") **alleles** and are generally not expressed in combination with functional alleles (in individuals of genotype *A/a*). The molecular basis of albinism is represented in Figure 1-13. Third, more rarely, the altered protein may perform its function more efficiently and thus be the basis for future evolution by natural selection.

The mutational site in the DNA can be of a number of types. The simplest and most common type is **nucleotide-pair substitution,** which can lead to amino acid substitution or to premature stop codons. Small **deletions** and **duplications** also are common. Even a single base deletion or insertion produces widespread damage at the protein level; because mRNA is read from one end "in frame" in groups of three, a loss or gain of one nucleotide pair shifts the reading frame, and all the amino acids translationally downstream will be incorrect. Such mutations are called **frameshift mutations.**

At the protein level, mutation changes the amino acid composition of the protein. The most important outcomes are change in protein shape and size. Such

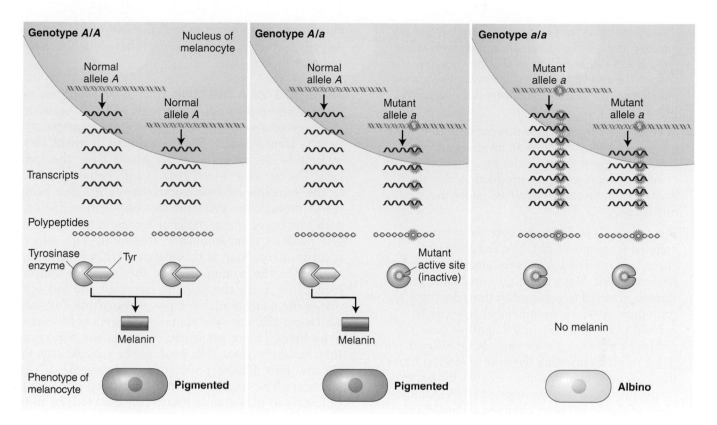

Figure 1-13 Molecular basis of albinism. *Left:* Melanocytes (pigment-producing cells) containing two copies of the normal tyrosinase allele (*A*) produce the tyrosinase enzyme, which converts the amino acid tyrosine into the pigment melanin *Center:* Melanocytes containing one copy of the normal allele make enough tyrosinase to allow production of melanin and the pigmented phenotype. *Right:* Melanocytes containing two copies of the mutant null allele (*a*) are unable to produce any of the enzyme.

change in shape or size can result in an absence of biological function (which would be the basis of a null allele) or reduced function. More rarely, mutation can lead to new function of the protein product.

> **MESSAGE** New alleles formed by mutation can result in no function, less function, more function, or new function at the protein level.

1.3 Methodologies used in genetics

An overview

The study of genes has proved to be a powerful approach to understanding biological systems. Because genes affect virtually every aspect of the structure and function of an organism, being able to identify and determine the role of genes and the proteins that they specify is an important step in charting the various processes that underlie a particular character under investigation. It is interesting that geneticists study not only hereditary mechanisms, but *all* biological mechanisms. Many different methodologies are used to study genes and gene activities, and these methodologies can be summarized briefly as follows:

1. Isolation of mutations affecting the biological process under study. Each mutant gene reveals a genetic component of the process, and together the mutant genes show the range of proteins that interact in that process.

2. Analysis of progeny of controlled matings ("crosses") between mutants and wild-type individuals or other discontinuous variants. This type of analysis identifies genes and their alleles, their chromosomal locations, and their inheritance patterns. These methods will be introduced in Chapter 2.

3. Genetic analysis of the cell's biochemical processes. Life is basically a complex set of chemical reactions, so studying the ways in which genes are relevant to these reactions is an important way of dissecting this complex chemistry. Mutant alleles underlying defective function (see method 1) are invaluable in

this type of analysis. The basic approach is to find out how the cellular chemistry is disturbed in the mutant individual and, from this information, deduce the role of the gene. The deductions from many genes are assembled to reveal the larger picture.

4. Microscopic analysis. Chromosome structure and movement have long been an integral part of genetics, but new technologies have provided ways of labeling genes and gene products so that their locations can be easily visualized under the microscope.

5. Direct analysis of DNA. Because the genetic material is composed of DNA, the ultimate characterization of a gene is the analysis of the DNA sequence itself. Many techniques, including gene cloning, are used to accomplish this. Cloning is a procedure by which an individual gene can be isolated and amplified (copied multiple times) to produce a pure sample for analysis. One way of doing this is by inserting the gene of interest into a small bacterial chromosome and allowing bacteria to do the job of copying the inserted DNA. After the clone of a gene has been obtained, its nucleotide sequence can be determined, and hence important information about its structure and function can be obtained.

Entire genomes of many organisms have been sequenced by extensions of the above techniques, thereby giving rise to a new discipline within genetics called **genomics,** the study of the structure, function, and evolution of whole genomes. Part of genomics is **bioinformatics,** the mathematical analysis of the information content of genomes.

Detecting specific molecules of DNA, RNA, and protein

Whether studying genes individually or as genomes, geneticists often need to detect the presence of a specific molecule each of DNA, RNA, or protein, the main macromolecules of genetics. These techniques will be described fully in Chapter 11, but we need a brief overview of them that can be used in earlier chapters.

How can specific molecules be identified among the thousands of types in the cell? The most extensively used method for detecting specific macromolecules in a mixture is **probing.** This method makes use of the specificity of intermolecular binding, which we have already encountered several times. A mixture of macromolecules is exposed to a molecule—the probe—that will bind only with the sought-after macromolecule. The probe is labeled in some way, either by a radioactive atom or by a fluorescent compound, so that the site of binding can easily be detected. Let's look at probes for DNA, RNA, and protein.

PROBING FOR A SPECIFIC DNA A cloned gene can act as a probe for finding segments of DNA that have the same or a very similar sequence. For example, if a gene G from a fungus has been cloned, it might be of interest to determine whether plants have the same gene. The use of a cloned gene as a probe takes us back to the principle of base complementarity. The probe works through the principle that, in solution, the random motion of probe molecules enables them to find and bind to complementary sequences. The experiment must be done with separated DNA strands, because then the bonding sites of the bases are unoccupied. DNA from the plant is extracted and cut with one of the many available types of **restriction enzymes,** which cut DNA at specific target sequences of four or more bases. The target sequences are at the same positions in all the plant cells used, so the enzyme cuts the genome into defined populations of segments of specific sizes. The fragments can be separated into groups of fragments of the same length (fractionated) by using electrophoresis.

Electrophoresis fractionates a population of nucleic acid fragments on the basis of size. The cut mixture is placed in a small well in a gelatinous slab (a gel), and the gel is placed in a powerful electrical field. The electricity causes the molecules to move through the gel at speeds inversely proportional to their size. After fractionation, the separated fragments are blotted onto a piece of porous membrane, where they maintain the same relative positions. This procedure is called a **Southern blot.** After having been heated to separate the DNA strands and hold the DNA in position, the membrane is placed in a solution of the probe. The single-stranded probe will find and bind to its complementary DNA sequence. For example,

TAGGTATCG   Probe
ACTAATCCATAGCTTA   Genomic fragment

On the blot, this binding concentrates the label in one spot, as shown in the left panel of Figure 1-14.

PROBING FOR A SPECIFIC RNA It is often necessary to determine whether a gene is being transcribed in some particular tissue. For this purpose, a modification of the Southern analysis is useful. Total mRNA is extracted from the tissue, fractionated electrophoretically, and blotted onto a membrane (this procedure is called a **Northern blot**). The cloned gene is used as a probe, and its label will highlight the mRNA in question if it is present (middle panel of Figure 1-14).

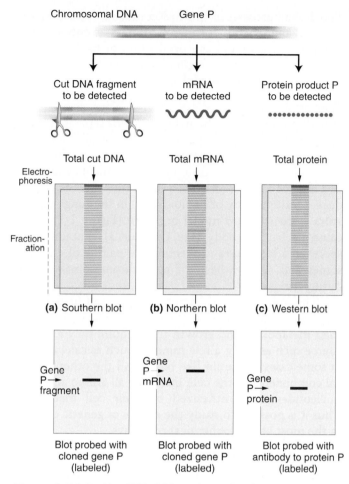

Figure 1-14 Probing DNA, RNA, and protein mixtures.

PROBING FOR A SPECIFIC PROTEIN Probing for proteins is generally performed with antibodies because an antibody has a specific lock-and-key fit with its protein target, or antigen. The protein mixture is separated into bands of distinct proteins by electrophoresis and blotted onto a membrane (this procedure is a **Western blot**). The position of a specific protein on the membrane is revealed by bathing the membrane in a solution of antibody, obtained from a rabbit or other host into which the protein has been injected. The position of the protein is revealed by the position of the label that the antibody carries (right-hand panel of Figure 1-14).

1.4 Model organisms

The science of genetics discussed in this book is meant to provide an understanding of features of inheritance and development that are characteristic of organisms in general. Some of these features, especially at the molecular level, are true of all known living forms. For others there is some variation between large groups of organisms, for example, between bacteria and all multicellular

species. Even for the features that vary, however, that variation is always between major groups of living forms, so that we do not have to investigate the basic phenomena of genetics over and over again for every species. In fact, all the phenomena of genetics have been investigated by experiments on a small number of species, **model organisms,** whose genetic mechanisms are common either to all species or to a large group of related organisms.

Lessons from the first model organisms

The use of model organisms goes back to the work of Gregor Mendel, who used crosses between horticultural varieties of the garden pea, *Pisum sativum*, to establish the basic rules of inheritance. Mendel's use of these varieties of the garden pea is instructive for our understanding of both the strengths and weaknesses of studying model organisms. Mendel studied the inheritance of three character differences: tall versus short plant height, purple versus white flowers, and round versus wrinkled seeds. These are all inherited as simple, single-gene differences. Hybrids between the varieties with contrasting characters were always identical with one of the two parents, while the hybrids produced some offspring showing one of the original parental types and some showing the other parental type in repeatable ratios. So, a cross between a purple variety and a white variety produced purple hybrids, while a cross between hybrids produced purple and white progeny in a ratio of 3:1. Moreover, if two varieties differed in two of the traits, one trait difference, say, purple versus white, was independent in its inheritance of the other trait, say, tall versus short. As a result of his observations, Mendel proposed three "laws" of inheritance:

1. The law of segregation: alternative trait "factors" that came together in the offspring separate again when the offspring produce gametes.

2. The law of dominance: hybrids between two alternative forms of a trait resemble one of the parental types.

3. The law of independent assortment: differences for one trait are inherited independently of differences for another trait.

These laws were the foundation for genetics and, in particular, established that the mechanism of inheritance was based on discrete particles in the gametes that come together in an offspring and then separate again when the offspring produces gametes, rather than by the mixing of a continuous fluid. But Mendel could not have inferred this mechanism had he studied height variation in most plant varieties, where such variation is continuous, because it depends on many gene differences. Moreover, the law of dominance does not hold true for many trait differences in many species. Indeed, had Mendel studied flower color in the sweet pea, *Lathyrus odoratus*, he would have observed pink-flowered offspring from the cross between a red and a white variety, and would not have observed the existence of dominance. Finally, many traits, even in the garden pea, do not show independent inheritance, but are linked together on chromosomes.

The need for a variety of model organisms

While the use of a particular model organism can reveal quite general features of inheritance and development, we cannot know how general such features are unless experiments are carried out on a variety of inherited traits in a variety of model organisms with very different patterns of reproduction and development.

Model organisms have been chosen partly for their different basic biological properties, and partly for small size of individuals, short generation time, and the ease with which they can be grown and mated under simple controlled conditions. For the study of vertebrate genetics, mice are to be preferred to elephants.

The need to study a wide range of biological and genetic traits has led to an array of model organisms from each of the basic biological groups (Figure 1-15).

VIRUSES These are simple nonliving particles that lack all metabolic machinery. They infect a host cell and divert its biosynthetic apparatus to the production of more virus, including the replication of viral genes. The viruses infecting bacteria, called *bacteriophage*, are the standard model (Figure 1-15a). The chief use of viruses has been to study the physical and chemical structure of DNA and the fundamental mechanics of DNA replication and mutation.

PROKARYOTES These single-celled living organisms have no nuclear membrane and lack intracellular compartments. While there is a special form of mating and genetic exchange between prokaryotic cells, they are essentially haploid throughout their lifetimes. The gut bacterium *Escherichia coli* is the common model. So convinced were some *E. coli* geneticists of the general applicability of their model organism that one university department's postage meter printed a stylized *E. coli* cell rearranged to look like an elephant.

EUKARYOTES All other cellular life is made up of one or more cells with a nuclear membrane and cellular compartments.

Yeasts Yeasts are single-celled fungi that usually reproduce by division of haploid cells to form colonies, but may also reproduce sexually by the fusion of two cells.

The diploid product of this fusion may reproduce by cell division and colony formation, but is eventually followed by meiosis and the production of haploid spores that give rise to new haploid colonies. *Saccharomyces cerevisiae* is the usual model species.

Filamentous fungi In these fungi nuclear division and growth produces long, branching threads separated irregularly into "cells" by membranes and cell walls, but a single such cellular compartment may contain more than one haploid nucleus. A fusion of two filaments will result in a diploid nucleus that then undergoes meiosis to produce a fruiting body of haploid cells. In the fungi, *Neurospora* is the standard model organism (Figure 1-15b) because its fruiting body (see Chapter 3) contains eight spores in a linear array, reflecting the pairing of chromosomes and the synthesis of new chromosomal strands during meiosis.

The importance of bacteria, yeasts, and filamentous fungi for genetics lies in their basic biochemistry. For their metabolism and growth they require only a carbon source such as sugar, a few minerals such as calcium, and in some cases a vitamin like biotin. All the other chemical components of the cell, including all amino acids and nucleotides, are synthesized by their cell machinery. Thus it is possible to study the effects of genetic changes in the most basic biochemical pathways.

Multicellular organisms For the genetic study of the differentiation of cells, tissues, and organs, as well as the development of body form, it is necessary to use more complex organisms. These organisms must be easy to culture under controlled conditions, have life cycles short enough to allow breeding experiments over many generations, and be small enough to make the production of large numbers of individuals practical. The main model organisms that fill these requirements are

- *Arabidopsis thaliana*, a small flowering plant that can be cultured in large numbers in the greenhouse or laboratory (Figure 1-15c). It has a small genome contained in only five chromosomes. It is an ideal model for studying the development of higher plants and the comparison of animal and plant development and genome structure.

- *Drosophila melanogaster*, a fruit fly with only four chromosomes. In the larval stage these chromosomes have a well-marked pattern of banding that makes it possible to observe physical changes such as deletions and duplications, which can then be correlated with genetic changes in morphology and biochemistry. The development of *Drosophila* produces body segments in an anterior-posterior order that is an example of the basic body plan common to invertebrates and vertebrates.

Figure 1-15 Some model organisms. (a) Bacteriophage λ attached to an infected *E. coli* cell; progeny phage particles are maturing inside the cell. (b) *Neurospora* growing on a burnt tree after a forest fire. (c) *Arabidopsis*. (d) *Caenorhabditis elegans*. [Part a, Lee D. Simon/Science Source/Photo Researchers; part b, courtesy of David Jacobson; part c, Wally Eberhart/Visuals Unlimited; part d, AFP/CORBIS.]

- *Caenorhabditis elegans*, a tiny roundworm with a total of only a few thousand adult cells. These form a nervous system; a digestive tract with a mouth, pharynx, and anus; and a reproductive system that can produce both eggs and sperm (Figure 1-15d).

- *Mus musculus*, the house mouse, the model organism for vertebrates. It has been studied to compare the genetic basis of vertebrate and invertebrate development as well as to explore the genetics of antigen-antibody systems, of maternal-fetal interactions in utero, and in understanding the genetics of cancer.

The genomes of all the model organisms discussed above have been sequenced. Despite the great differences in biology there are many similarities in their genomes. Figure 1-16 is a comparison of the genomes of eukaryotes, prokaryotes, and viruses.

At the end of the book we summarize and compare the inferences made about genetics from the use of the various models.

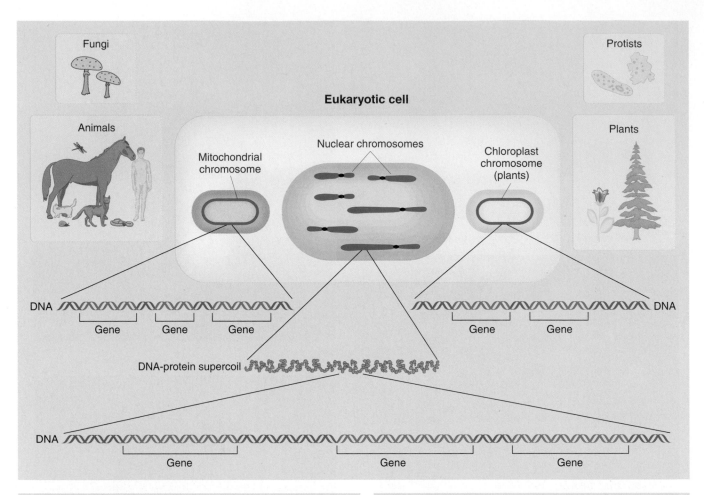

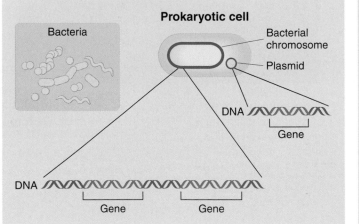

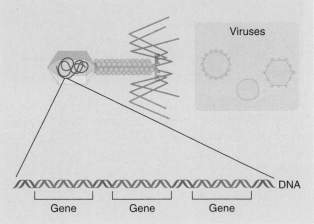

Figure 1-16 Structural comparison of the genome components of eukaryotes, prokaryotes, and viruses.

1.5 Genes, the environment, and the organism

Genes cannot dictate the structure of an organism by themselves. The other crucial component in the formula is the environment. The environment influences gene action in many ways, about which we shall learn in subsequent chapters. Most concretely perhaps, the environment pro-

vides the raw materials for the synthetic processes controlled by genes. An acorn becomes an oak tree, by using in the process only water, oxygen, carbon dioxide, some inorganic materials from the soil, and light energy.

Model I: genetic determination

It is clear that virtually all the differences between species are determined by the differences in their genomes. There is no environment in which a lion will give birth to a lamb. An acorn develops into an oak, whereas the spore of a moss develops into a moss, even if both are growing side by side in the same forest. The two plants that result from these developmental processes resemble their parents and differ from each other, even though they have access to the same narrow range of materials from the environment.

Even within species, some variation is entirely a consequence of genetic differences that cannot be modified by any change in what we normally think of as environment. The children of African slaves brought to North America had dark skins, unchanged by the relocation of their parents from tropical Africa to temperate Maryland. The possibility of much of experimental genetics depends on the fact that many of the phenotypic differences between mutant and wild-type individuals resulting from allelic differences are insensitive to environmental conditions. The determinative power of genes is often demonstrated by differences in which one allele is normal and the other abnormal. The human inherited disease sickle-cell anemia is a good example. The underlying cause of the disease is a variant of hemoglobin, the oxygen-transporting protein molecule found in red blood cells. Normal people have a type of hemoglobin called *hemoglobin-A*, the information for which is encoded in a gene. A single nucleotide change in the DNA of this gene, leading to a change in a single amino acid in the polypeptide, results in the production of a slightly changed hemoglobin, called *hemoglobin-S*. In people possessing only hemoglobin-S, the ultimate effect of this small change in DNA is severe ill health—sickle-cell anemia—and often death.

Such observations, if generalized, lead to a model, shown in Figure 1-17, of how genes and the environment interact. In this view, the genes act as a set of instructions for turning more or less undifferentiated environmental materials into a specific organism, much as blueprints specify what form of house is to be built from basic materials. The same bricks, mortar, wood, and nails can be made into an A-frame or a flat-roofed house, according to different plans. Such a model implies that the

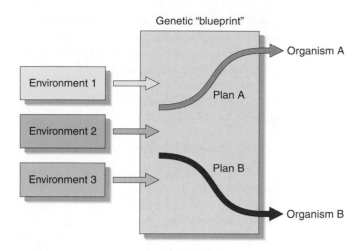

Figure 1-17 A model of determination that emphasizes the role of genes.

genes are really the dominant elements in phenotypic determination; the environment simply supplies the undifferentiated raw materials.

Model II: environmental determination

Consider two monozygotic ("identical") twins, the products of a single fertilized egg that divided and produced two complete babies with identical genes. Suppose that the twins are born in England but are separated at birth and taken to different countries. If one is reared in China by Chinese-speaking foster parents, she will speak Chinese, whereas her sister reared in Budapest will speak Hungarian. Each will absorb the cultural values and customs of her environment. Although the twins begin life with identical genetic properties, the different cultural environments in which they live will produce differences between them (and differences from their parents). Obviously, the differences in this case are due to the environment, and genetic effects are of no importance in determining the differences.

This example suggests the model of Figure 1-18, which is the converse of that shown in Figure 1-17. In the model in Figure 1-18, the genes impinge on the system, giving certain general signals for development, but

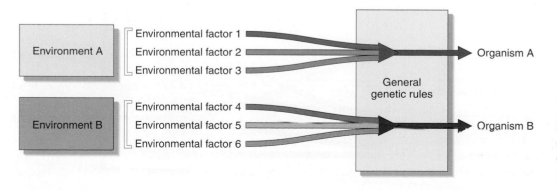

Figure 1-18 A model of determination that emphasizes the role of the environment.

the environment determines the actual course of development. Imagine a set of specifications for a house that simply calls for a "floor that will support 300 pounds per square foot" or "walls with an insulation factor of 15 inches"; the actual appearance and other characteristics of the structure would be determined by the available building materials.

Model III: genotype-environment interaction

In general, we deal with organisms that differ in both genes and environment. If we wish to predict how a living organism will develop, we need to know both the genetic constitution that it inherits from its parents and the historical sequence of environments to which it has been exposed. Every organism has a developmental history from conception to death. What an organism will become in the next moment depends critically both on its present state and on the environment that it encounters during that moment. It makes a difference to an organism not only what environments it encounters, but also in what sequence it encounters them. A fruit fly *(Drosophila melanogaster)*, for example, develops normally at 25°C. If the temperature is briefly raised to 37°C early in its pupal stage of development, the adult fly will be missing part of the normal vein pattern on its wings. However, if this "temperature shock" is administered just 24 hours later, the vein pattern develops normally. A general model in which genes and the environment jointly determine (by some rules of development) the actual characteristics of an organism is depicted in Figure 1-19.

> **MESSAGE** As an organism transforms developmentally from one stage to another, its genes interact with its environment at each moment of its life history. The interaction of genes and environment determines what organisms are.

The use of genotype and phenotype

In light of the preceding discussion we can now better understand the use of the terms *genotype* and *phenotype*.

A typical organism resembles its parents more than it resembles unrelated individuals. Thus, we often speak as if

the individual characteristics themselves are inherited: "He gets his brains from his mother" or "She inherited diabetes from her father." Yet the preceding section shows that such statements are inaccurate. "His brains" and "her diabetes" develop through long sequences of events in the life histories of the affected people, and both genes and environment play roles in those sequences. In the biological sense, individuals inherit only the molecular structures of the fertilized eggs from which they develop. Individuals inherit their genes, not the end products of their individual developmental histories.

To prevent such confusion between genes (which are inherited) and developmental outcomes (which are not), geneticists make the fundamental distinction between the genotype and the phenotype of an organism. Organisms have the same genotype in common if they have the same set of genes. Organisms have the same phenotype if they look or function alike.

Strictly speaking, the genotype describes the complete set of genes inherited by an individual, and the phenotype describes all aspects of the individual's morphology, physiology, behavior, and ecological relations. In this sense, no two individuals ever belong to the same phenotype, because there is always some difference (however slight) between them in morphology or physiology. Additionally, except for individuals produced from another organism by asexual reproduction, any two organisms differ at least a little in genotype. In practice, we use the terms *genotype* and *phenotype* in a more restricted sense. We deal with some partial phenotypic description (say, eye color) and with some subset of the genotype (say, the genes that affect eye pigmentation).

> **MESSAGE** When we use the terms *phenotype* and *genotype*, we generally mean "partial phenotype" and "partial genotype," and we specify one or a few traits and genes that are the subsets of interest.

Note one very important difference between genotype and phenotype: the genotype is essentially a fixed character of an individual organism; the genotype remains constant throughout life and is essentially un-

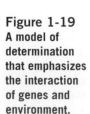

**Figure 1-19
A model of
determination
that emphasizes
the interaction
of genes and
environment.**

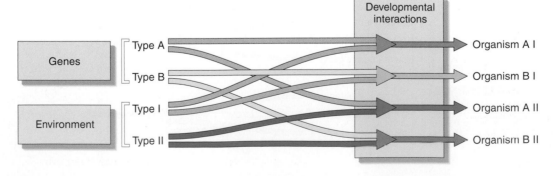

changed by environmental effects. Most phenotypes change continually throughout the life of an organism as its genes interact with a sequence of environments. Fixity of genotype does not imply fixity of phenotype.

Norm of reaction

How can we quantify the relation between the genotype, the environment, and the phenotype? For a particular genotype, we could prepare a table showing the phenotype that would result from the development of that genotype in each possible environment. Such a set of environment-phenotype relations for a given genotype is called the **norm of reaction** of the genotype. In practice, we can make such a tabulation only for a partial genotype, a partial phenotype, and some particular aspects of the environment. For example, we might specify the eye sizes that fruit flies would have after developing at various constant temperatures; we could do this for several different eye-size genotypes to get the norms of reaction of the species.

Figure 1-20 represents just such norms of reaction for three eye-size genotypes in the fruit fly *Drosophila melanogaster.* The graph is a convenient summary of more extensive tabulated data. The size of the fly eye is measured by counting its individual facets, or cells. The vertical axis of the graph shows the number of facets (on a logarithmic scale); the horizontal axis shows the constant temperature at which the flies develop.

Three norms of reaction are shown on the graph. When flies of the wild-type genotype that is characteristic of flies in natural populations are raised at higher temperatures, they develop eyes that are somewhat smaller than those of wild-type flies raised at cooler temperatures. The graph shows that wild-type phenotypes range from more than 700 to 1000 facets—the wild-type norm of reaction. A fly that has the *ultrabar* genotype has smaller eyes than those of wild-type flies regardless of temperature during development. Temperatures have a stronger effect on development of *ultrabar* genotypes than on wild-type genotypes, as we see by noticing that the *ultrabar* norm of reaction slopes more steeply than the wild-type norm of reaction. Any fly of the *infrabar* genotype also has smaller eyes than those of any wild-type fly, but temperatures have the opposite effect on flies of this genotype; *infrabar* flies raised at higher temperatures tend to have larger eyes than those

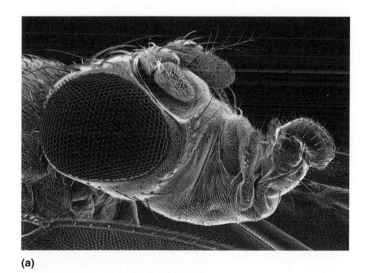

(a)

Figure 1-20 Norms of reaction to temperature for three different eye-size genotypes in *Drosophila.*
(a) Close-up showing how the normal eye comprises hundreds of units called *facets*. The number of facets determines eye size. (b) Relative eye sizes of wild-type, *infrabar*, and *ultrabar* flies raised at the higher end of the temperature range.
(c) Norm-of-reaction curves for the three genotypes. [Part a, Don Rio and Sima Misra, University of California, Berkeley.]

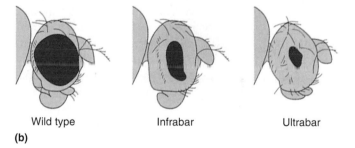

Wild type   Infrabar   Ultrabar

(b)

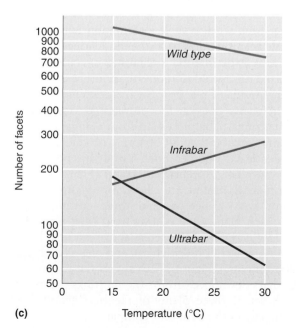

(c)

of flies raised at lower temperatures. These norms of reaction indicate that the relation between genotype and phenotype is complex rather than simple.

> **MESSAGE** A single genotype may produce different phenotypes, depending on the environment in which organisms develop. The same phenotype may be produced by different genotypes, depending on the environment.

If we know that a fruit fly has the wild-type genotype, this information alone does not tell us whether its eye has 800 or 1000 facets. On the other hand, the knowledge that a fruit fly's eye has 170 facets does not tell us whether its genotype is *ultrabar* or *infrabar*. We cannot even make a general statement about the effect of temperature on eye size in *Drosophila*, because the effect is opposite in two different genotypes. We see from Figure 1-20 that some genotypes do differ unambiguously in phenotype, no matter what the environment: any wild-type fly has larger eyes than any *ultrabar* or *infrabar* fly. But other genotypes overlap in phenotypic expression: the eyes of an *ultrabar* fly may be larger or smaller than those of an *infrabar* fly, depending on the temperatures at which the individuals developed.

To obtain a norm of reaction such as the norms of reaction in Figure 1-20, we must allow different individuals of identical genotype to develop in many different environments. To carry out such an experiment, we must be able to obtain or produce many fertilized eggs with identical genotypes. For example, to test a human genotype in 10 environments, we would have to obtain genetically identical sibs and raise each individual in a different milieu. However, that is possible neither biologically nor ethically. At the present time, we do not know the norm of reaction of any human genotype for any character in any set of environments. Nor is it clear how we can ever acquire such information without the unacceptable manipulation of human individuals.

For a few experimental organisms, special genetic methods make it possible to replicate genotypes and thus to determine norms of reaction. Such studies are particularly easy in plants that can be propagated vegetatively (that is, by cuttings). The pieces cut from a single plant all have the same genotype, so all offspring produced in this way have identical genotypes. Such a study has been done on the yarrow plant, *Achillea millefolium* (Figure 1-21a). The experimental results are shown in Figure 1-21b. Many plants were collected, and three cuttings were taken from each plant. One cutting from each plant was planted at low elevation (30 meters above sea level), one at medium elevation (1400 meters), and one at high elevation (3050 meters). Figure 1-21b shows the mature individuals that developed from the cuttings of seven plants; each set of three plants of identical genotype is aligned vertically in the figure for comparison.

(a)

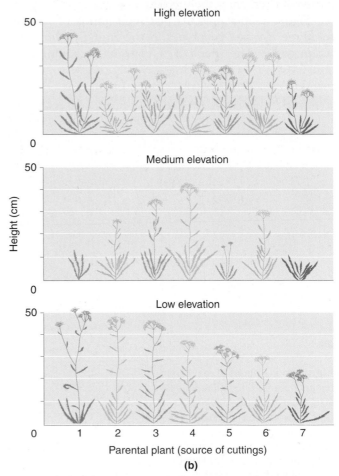

(b)

Figure 1-21 A comparison of genotype versus phenotype in yarrow. (a) *Achillea millefolium.* (b) Norms of reaction to elevation for seven different *Achillea* plants (seven different genotypes). A cutting from each plant was grown at low, medium, and high elevations. [Part a, Harper Horticultural Slide Library; part b, Carnegie Institution of Washington.]

First, we note an average effect of environment: in general, the plants grew poorly at the medium elevation. This is not true for every genotype, however; the cutting of plant 4 grew best at the medium elevation. Second, we note that no genotype is unconditionally superior in growth to all others. Plant 1 showed the best growth at low and high elevations but showed the poorest growth at the medium elevation. Plant 6 showed the second-worst growth at low elevation and the second-best at high elevation. Once again, we see the complex relation between genotype and phenotype. Figure 1-22 graphs the norms of reaction derived from the results shown in Figure 1-21b. Each genotype has a different norm of reaction, and the norms cross one another; so we cannot identify either a "best" genotype or a "best" environment for *Achillea* growth.

We have seen two different patterns of reaction norms. The difference between the wild-type and the other eye-size genotypes in *Drosophila* is such that the corresponding phenotypes show a consistent difference, regardless of the environment. Any fly of wild-type genotype has larger eyes than any fly of the other geno-types; so we could (imprecisely) speak of "large eye" and "small eye" genotypes. In this case, the differences in phenotype between genotypes are much greater than the variation within a genotype caused by development in different environments. But the variation for a single *Achillea* genotype in different environments is so great that the norms of reaction cross one another and form no consistent pattern. In this case, it makes no sense to identify a genotype with a particular phenotype except in regard to response to particular environments.

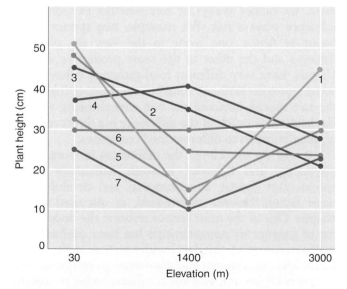

Figure 1-22 Graphic representation of the complete set of results of the type shown in Figure 1-21. Each line represents the norm of reaction of one plant.

Developmental noise

Thus far, we have assumed that a phenotype is uniquely determined by the interaction of a specific genotype and a specific environment. But a closer look shows some further unexplained variation. According to Figure 1-20, a *Drosophila* of wild-type genotype raised at 16°C has 1000 facets in each eye. In fact, this is only an average value; one fly raised at 16°C may have 980 facets, and another may have 1020. Perhaps these variations are due to slight fluctuations in the local environment or slight differences in genotypes. However, a typical count may show that a wild-type fly has, say, 1017 facets in the left eye and 982 in the right eye. In another wild-type fly raised in the same experimental conditions, the left eye has slightly fewer facets than the right eye. Yet the left and right eyes of the same fly are genetically identical. Furthermore, under typical experimental conditions, the fly develops as a larva (a few millimeters long) burrowing in homogeneous artificial food in a laboratory bottle and then completes its development as a pupa (also a few millimeters long) glued vertically to the inside of the glass high above the food surface. Surely the environment does not differ significantly from one side of the fly to the other. If the two eyes experience the same sequence of environments and are identical genetically, then why is there any phenotypic difference between the left and right eyes?

Differences in shape and size are partly dependent on the process of cell division that turns the zygote into a multicellular organism. Cell division, in turn, is sensitive to molecular events within the cell, and these events may have a relatively large random component. For example, the vitamin biotin is essential for *Drosophila* growth, but its average concentration is only one molecule per cell. Obviously, the rate of any process that depends on the presence of this molecule will fluctuate as the concentration varies. Fewer eye facets may develop if the availability of biotin by chance fluctuates downward within the relatively short developmental period during which the eye is being formed. Thus, we would expect random variation in such phenotypic characters as the number of eye cells, the number of hairs, the exact shape of small features, and the connections of neurons in a very complex central nervous system—even when the genotype and the environment are precisely fixed. Random events in development lead to variation in phenotype called **developmental noise**.

MESSAGE In some characteristics, such as eye cells in *Drosophila*, developmental noise is a major source of the observed variations in phenotype.

Adding developmental noise to our model of phenotypic development, we obtain something like Figure 1-23. With a given genotype and environment, there is a range

Figure 1-23 A model of phenotypic determination that shows how genes, environment, and developmental noise interact to produce a phenotype.

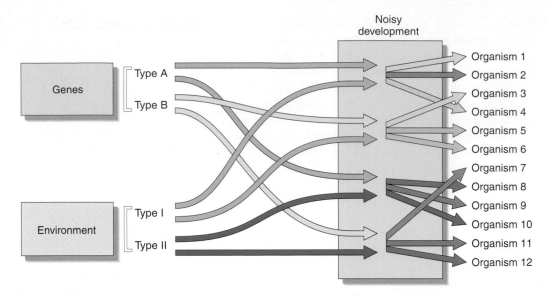

of possible outcomes for each developmental step. The developmental process does contain feedback systems that tend to hold the deviations within certain bounds so that the range of deviation does not increase indefinitely through the many steps of development. However, this feedback is not perfect. For any given genotype developing in any given sequence of environments, there remains some uncertainty regarding the exact phenotype that will result.

Three levels of development

Chapter 18 of this book is concerned with the way in which genes mediate development, but nowhere in that chapter do we consider the role of the environment or the influence of developmental noise. How can we, at the beginning of the book, emphasize the joint role of genes, environment, and noise in influencing phenotype, yet in our later consideration of development ignore the environment? The answer is that modern developmental genetics is concerned with very basic processes of differentiation that are common to all individual members of a species and, indeed, are common to animals as different as fruit flies and mammals. How does the front end of an animal become differentiated from the back end, the ventral from the dorsal side? How does the body become segmented, and why do limbs form on some segments and not on others? Why do eyes form on the head and not in the middle of the abdomen? Why do the antennae, wings, and legs of a fly look so different even though they are derived in evolution from appendages that looked alike in the earliest ancestors of insects? At this level of development, which is constant across individuals and species, normal environmental variation plays no role, and we can speak correctly of genes "determining" the phenotype. Precisely because the effects of genes can be isolated at this level of development and precisely because the processes seem to be general across a wide variety of organisms, they are easier to study than are characteristics for which environmental

variation is important, and developmental genetics has concentrated on understanding them.

At a second level of development, there are variations on the basic developmental themes that are different between species but are constant within species, and these too could be understood by concentrating on genes, although at the moment they are not part of the study of developmental genetics. So, although both lions and lambs have four legs, one at each corner, lions always give birth to lions and lambs to lambs, and we have no difficulty in distinguishing between them in any environment. Again, we are entitled to say that genes "determine" the difference between the two species, although we must be more cautious here. Two species may differ in some characteristic because they live in quite different environments, and until we can raise them in the same environment, we cannot always be sure whether environmental influence plays a role. For example, two species of baboons in Africa, one living in the very dry plains of Ethiopia and the other in the more productive areas of Uganda, have very different food-gathering behavior and social structure. Without actually transplanting colonies of the two species between the two environments, we cannot know how much of the difference is a direct response of these primates to different food conditions.

It is at the third level, the differences in morphology, physiology, and behavior between individuals within species, that genetic, environmental, and developmental noise factors become intertwined, as discussed in this chapter. One of the most serious errors in the understanding of genetics by nongeneticists has been confusion between variation at this level and variation at the higher levels. The experiments and discoveries to be discussed in Chapters 18 are not, and are not meant to be, models for the causation of individual variation. They apply directly only to those characteristics, deliberately chosen, that are general features of development and for which environmental variation appears to be irrelevant.

KEY QUESTIONS REVISITED

- **What is the hereditary material?**

The molecule DNA.

- **What is the chemical and physical structure of DNA?**

It is a double helix of two chains of nucleotides, intertwined in opposite orientation. At the center of the helix, nucleotides with G pair with C, and those with A pair with T.

- **How is DNA copied in the formation of new cells and in the gametes that will give rise to the offspring of an individual?**

The strands separate, and each is used as a template for building a new strand.

- **What are the functional units of DNA that carry information about development and physiology?**

The functional units are genes, regional units bearing the appropriate signal sequences for being transcribed into RNA.

- **What molecules are the main determinants of the basic structural and physiological properties of an organism?**

Many thousands of different types of proteins.

- **What are the steps in translating the information in DNA into protein?**

For most genes, first, transcription of the DNA sequence into mRNA, then translation of the RNA sequence into the amino acid sequence of protein.

- **What determines the differences between species in their physiology and structure?**

For the most part, differences between species are the consequences of species differences in DNA, reflected in species differences in the structure of proteins and the timing and cell localization of their production.

- **What are the causes of variation between individuals within species?**

Variation between individuals within species is the result of genetic differences, environmental differences, and the interaction between genes and environment, as well as developmental noise.

- **What is the basis of variation in populations?**

Variation among individuals is a consequence of both genetic and environmental variation and of random events during cell division during development.

SUMMARY

Genetics is the study of genes at all levels from molecules to populations. As a modern discipline, it began in the 1860s with the work of Gregor Mendel, who first formulated the idea that genes exist. We now know that a gene is a functional region of the long DNA molecule that constitutes the fundamental structure of a chromosome. DNA is composed of four nucleotides, each containing deoxyribose sugar, phosphate, and one of four bases: adenine (A), thymine (T), guanine (G), and cytosine (C). DNA is two nucleotide chains, oriented in opposite directions (antiparallel) and held together by bonding A with T and G with C. In replication, the two chains separate, and their exposed bases are used as templates for the synthesis of two identical daughter DNA molecules.

Most genes encode the structure of a protein (proteins are the main determinants of the properties of an organism). To make protein, DNA is first transcribed by the enzyme RNA polymerase into a single-stranded working copy called *messenger RNA (mRNA)*. The nucleotide sequence in the mRNA is translated into an amino acid sequence that constitutes the primary structure of a protein. Amino acid chains are synthesized on ribosomes. Each amino acid is brought to the ribosome by a tRNA molecule that docks by the binding of its triplet (called the *anticodon*) to a triplet codon in mRNA.

The same gene may have alternative forms called *alleles*. Individuals may be classified by genotype (their al-

lelic constitution) or by phenotype (observable characteristics of appearance or physiology). Both genotypes and phenotypes show variation within a population. Variation is of two types: discontinuous, showing two or more distinct phenotypes, and continuous, showing phenotypes with a wide range of quantitative values. Discontinuous variants are often determined by alleles of one gene. For example, people with normal skin pigmentation have the functional allele coding for the enzyme tyrosinase, which converts tyrosine into the dark pigment melanin, whereas albinos have a mutated form of the gene that codes for a protein that can no longer make the conversion.

The relation of genotype to phenotype across an environmental range is called the *norm of reaction*. In the laboratory, geneticists study discontinuous variants under conditions where there is a one-to-one correspondence between genotype and phenotype. However, in natural populations, where environment and genetic background vary, there is generally a more complex relation, and genotypes can produce overlapping ranges of phenotypes. As a result, discontinuous variants have been the starting point for most experiments in genetic analysis.

The main tools of genetics are breeding analysis of variants, biochemistry, microscopy, and direct analysis of DNA using cloned DNA. Cloned DNA can provide useful probes for detecting the presence of related DNA and RNA.

KEY TERMS

active site (p. 10)

alleles (p. 2)

bioinformatics (p. 12)

codon (p. 5)

complementary (p. 4)

continuous variation (p. 10)

deletion (p. 10)

deoxyribonucleic acid (DNA) (p. 2)

developmental noise (p. 21)

diploid (p. 3)

discontinuous variation (p. 8)

duplication (p. 10)

enzyme (p. 5)

eukaryotes (p. 6)

exons (p. 6)

frameshift mutation (p. 10)

gene (p. 2)

genome (p. 3)

genomics (p. 12)

genotype (p. 9)

haploid (p. 3)

homologous (p. 3)

introns (p. 6)

messenger RNA (mRNA) (p. 5)

model organisms (p. 13)

morphs (p. 9)

mutants (p. 9)

norm of reaction (p. 19)

Northern blot (p. 12)

nucleotides (p. 3)

nucleotide-pair substitution (p. 10)

null allele (p. 10)

phenotypes (p. 8)

polymorphism (p. 9)

polypeptide (p. 5)

probing (p. 12)

prokaryotes (p. 6)

proteins (p. 2)

restriction enzymes (p. 12)

ribonucleic acid (RNA) (p. 5)

ribosome (p. 6)

Southern blot (p. 12)

structural protein (p. 5)

transcribe (p. 5)

transcript (p. 5)

transfer RNA (tRNA) (p. 6)

translation (p. 5)

Western blot (p. 13)

wild type (p. 9)

zygote (p. 3)

PROBLEMS

BASIC PROBLEMS

1. Define *genetics*. Do you think the ancient Egyptian racehorse breeders were geneticists? How might their approaches have differed from those of modern geneticists?

2. How does DNA dictate the general properties of a species?

3. What are the features of DNA that suit it for its role as a hereditary molecule? Can you think of alternative types of hereditary molecules that might be found in extraterrestrial life-forms?

 4. How many different DNA molecules 10 nucleotide pairs long are possible?

5. If thymine makes up 15 percent of the bases in a certain DNA sample, what percentage of the bases must be cytosine?

6. If the G + C content of a DNA sample is 48 percent, what are the proportions of the four different nucleotides?

7. Each cell of the human body contains 46 chromosomes.

   a. How many DNA molecules does this statement represent?

   b. How many different types of DNA molecules does it represent?

8. A certain segment of DNA has the following nucleotide sequence in one strand:

   ATTGGTGCATTACTTCAGGCTCT

   What must the sequence in the other strand be?

9. In a single strand of DNA, is it ever possible for the number of adenines to be greater than the number of thymines?

10. In normal double-helical DNA, is it true that

    a. A plus C will always equal G plus T?

    b. A plus G will always equal C plus T?

11. Suppose that the following DNA molecule replicates to produce two daughter molecules. Draw these daughter molecules by using black for previously polymerized nucleotides and red for newly polymerized nucleotides.

    TTGGCACGTCGTAAT
    AACCGTGCAGCATTA

12. In the DNA molecule in Problem 11, assume that the bottom strand is the template strand and draw the RNA transcript.

13. Draw Northern and Western blots of the three genotypes in Figure 1-13. (Assume the probe used in the Northern blot to be a clone of the tyrosinase gene.)

14. What is a gene? What are some of the problems with your definition?

15. The gene for the human protein albumin spans a chromosomal region 25,000 nucleotide pairs (25 kilobases, or kb) long from the beginning of the protein-coding sequence to the end of the protein-coding sequence, but the messenger RNA for this protein is only 2.1 kb long. What do you think accounts for this huge difference?

16. DNA is extracted from cells of *Neurospora*, a fungus that has one set of seven chromosomes; pea, a plant that has two sets of seven chromosomes; and housefly, an animal that has two sets of six chromosomes. If powerful electrophoresis is used to separate the DNA on a gel, how many bands will each of these three species produce?

17. Devise a formula that relates size of RNA to gene size, number of introns, and average size of introns.

18. If a codon in mRNA is UUA, draw the tRNA anticodon that would bind to this codon.

19. Two mutations arise in separate cultures of a normally red fungus (which has only one set of chromosomes). The mutations are found to be in different genes. Mutation 1 gives an orange color, and mutation 2 gives a yellow color. Biochemists working on the synthesis of the red pigment in this species have already described the following pathway:

colorless precursor $\xrightarrow[\text{enzyme A}]{}$ yellow pigment $\xrightarrow[\text{enzyme B}]{}$

orange pigment $\xrightarrow[\text{enzyme C}]{}$ red pigment

a. Which enzyme is defective in mutant 1?

b. Which enzyme is defective in mutant 2?

c. What would be the color of a double mutant (1 and 2)?

CHALLENGING PROBLEMS

20. In sweet peas, the purple color of the petals is controlled by two genes, *B* and *D*. The pathway is

colorless precursor $\xrightarrow[]{\text{gene B}}$ red anthocyanin pigment

colorless precursor $\xrightarrow[\text{gene D}]{}$ blue anthocyanin pigment

⎫⎬⎭ purple

a. What color petals would you expect in a plant that carries two copies of a null mutation for *B*?

b. What color petals would you expect in a plant that carries two copies of a null mutation for *D*?

c. What color petals would you expect in a plant that is a double mutant; that is, it carries two copies of a null mutation for both *B* and *D*?

UNPACKING PROBLEM 20

In many chapters, one or more of the problems will ask "unpacking" questions. These questions are designed to assist in working the problem by showing the often quite extensive content that is inherent in the problem. The same approach can be applied to other questions by the solver. If you have trouble with Problem 20, try answering the following unpacking questions. If necessary, look up material that you cannot remember. Then try to solve the problem again by using the information emerging from the unpacking.

1. What are sweet peas, and how do they differ from edible peas?

2. What is a pathway in the sense used here?

3. How many pathways are evident in this system?

4. Are the pathways independent?

5. Define the term *pigment*.

6. What does colorless mean in this problem? Think of an example of any solute that is colorless.

7. What would a petal that contained only colorless substances look like?

8. Is color in sweet peas anything like mixing paint?

9. What is a mutation?

10. What is a null mutation?

11. What might be the cause of a null mutation at the DNA level?

12. What does "two copies" mean? (How many copies of genes do sweet peas normally have?)

13. What is the relevance of proteins to this problem?

14. Does it matter whether genes *B* and *D* are on the same chromosome?

15. Draw a representation of the wild-type allele of *B* and a null mutant at the DNA level.

16. Repeat for gene *D*.

17. Repeat for the double mutant.

18. How would you explain the genetic determination of petal color in sweet peas to a gardener with no scientific training?

21. Twelve null alleles of an intronless *Neurospora* gene are examined, and all the mutant sites are found to cluster in a region occupying the central third of the gene. What might be the explanation for this finding?

22. An albino mouse mutant is obtained whose pigment lacks melanin, normally made by an enzyme T. Indeed, the tissue of the mutant lacks all detectable activity for enzyme T. However, a Western blot clearly shows that a protein with immunological properties identical with those of enzyme T is present in the cells of the mutant. How is this possible?

23. In Norway in 1934, a mother with two mentally retarded children consulted the physician Asbjørn
Følling. In the course of the interview, Følling
learned that the urine of the children had a curious
odor. He later tested their urine with ferric chloride
and found that, whereas normal urine gives a
brownish color, the children's urine stained green.
He deduced that the chemical responsible must be
phenylpyruvic acid. Because of chemical similarity
to phenylalanine, it seemed likely that this substance
had been formed from phenylalanine in the blood,
but there was no assay for phenylalanine. However,
a certain bacterium could convert phenylalanine
into phenylpyruvic acid; therefore, the level of
phenylalanine could be measured by using the ferric
chloride test. The children were indeed found to
have high levels of phenylalanine in their blood, and
the phenylalanine was probably the source of the
phenylpyruvic acid. This disease, which came to be
known as phenylketonuria (PKU), was shown to be
inherited and caused by a recessive allele.

It became clear that phenylalanine was the
culprit and that this chemical accumulated in PKU
patients and was converted into high levels of
phenylpyruvic acid, which then interfered with the
normal development of nervous tissue. This finding
led to the formulation of a special diet low in
phenylalanine, which could be fed to newborn babies diagnosed with PKU and which allowed normal
development to continue without retardation. Indeed, it was found that after the child's nervous system had developed, the patient could be taken off
the special diet. However, tragically, many PKU
women who had developed normally with the special diet were found to have babies who were born
mentally retarded, and the special diet had no effect
on these children.

a. Why do you think the babies of the PKU mothers were born retarded?

b. Why did the special diet have no effect on them?

c. Explain the reason for the difference in the results between the PKU babies and the babies of
PKU mothers.

d. Propose a treatment that might allow PKU mothers to have unaffected children.

e. Write a short essay on PKU, integrating concepts
at the genetic, diagnostic, enzymatic, physiological,
pedigree, and population levels.

24. Normally the thyroid growth hormone thyroxine is
made in the body by an enzyme as follows:

$$\text{tyrosine} \xrightarrow{\text{enzyme}} \text{thyroxine}$$

If the enzyme is deficient, the symptoms are called
genetic goiterous cretinism (GGC), a rare syndrome
consisting of slow growth, enlarged thyroid (called a
goiter), and mental retardation.

a. If the normal allele is haplosufficient, would you
expect GGC to be inherited as a dominant or a recessive phenotype? Explain.

b. Speculate on the nature of the GGC-causing allele, comparing its molecular sequence with that of
the normal allele. Show why it results in an inactive
enzyme.

c. How might the symptoms of GGC be alleviated?

d. At birth, infants with GGC are perfectly normal
and develop symptoms only later. Why do you think
this is so?

25. Compare and contrast the processes by which information becomes form in an organism and in house
building.

26. Try to think of exceptions to the statement made in
this chapter that "When you look at an organism,
what you see is either a protein or something that
has been made by a protein."

27. What is the relevance of norms of reaction to
phenotypic variation within a species?

28. What are the types and the significance of phenotypic variation within a species?

29. Is the formula "genotype + environment = phenotype" accurate?

2

PATTERNS
OF INHERITANCE

Gregor Mendel's monastery. A statue of Mendel is visible in the background. Today, this part of Mendel's monastery is a museum, and the curators have planted red and white begonias in an array that graphically represents the type of inheritance patterns Mendel obtained with peas. [Anthony Griffiths.]

KEY QUESTIONS

- How is it possible to tell if a phenotypic variant has a genetic basis?

- Are phenotypic variants inherited in consistent patterns through the generations?

- At the gene level, what is the explanation for the patterns by which phenotypic variants are inherited?

- Is the pattern of inheritance influenced by the location of the relevant gene or genes in the genome?

- Is the pattern for one phenotype independent of that for phenotypes of other characters?

OUTLINE

2.1 Autosomal inheritance

2.2 Sex chromosomes and sex-linked inheritance

2.3 Cytoplasmic inheritance

CHAPTER OVERVIEW

Most of genetics is based on hereditary variation. In Chapter 1 we saw that within a species individuals may vary in their characteristics. Even though the members of a species share most of the attributes of that species, there are often differences that allow us to tell individuals apart. Differences can involve any property, such as eye color, body size, shape, disease, or behavior. The differing individuals are called **variants.** Genetic analysis commonly begins with at least two such variants. In the prototypic genetic analysis carried out by Gregor Mendel, described in this chapter, the two variants were purple- and white-flowered pea plants.

In any given case of variation, a crucial step is to decide whether or not the observed variation is hereditary. A corollary issue is the genetic basis for the inheritance. As part of the latter question we need to know the number of genes involved, and how they are passed on through the generations.

Answering these questions requires an analysis of *inheritance pattern.* The two different variants are mated, and their descendants are followed through several generations. What patterns do we look for? The first issue is whether or not the variant characteristics show up in future generations. If they do, an important issue is the proportions in which they occur.

There are a limited number of standard inheritance patterns, and from these one or more genes can be identified that form the basis of the inherited variation. Subsequent studies might zero in on the molecular nature of the genes involved.

In this chapter we explore the inheritance patterns encountered in the analysis of *discontinuous* variants, those that can be grouped into two or more distinct forms, or phenotypes. Simple examples might be red versus blue petal color in some species of plant, or black versus brown coat colors in mice. In many cases such variation is inherited and can be explained by differences in a single gene. These single gene differences

CHAPTER OVERVIEW Figure

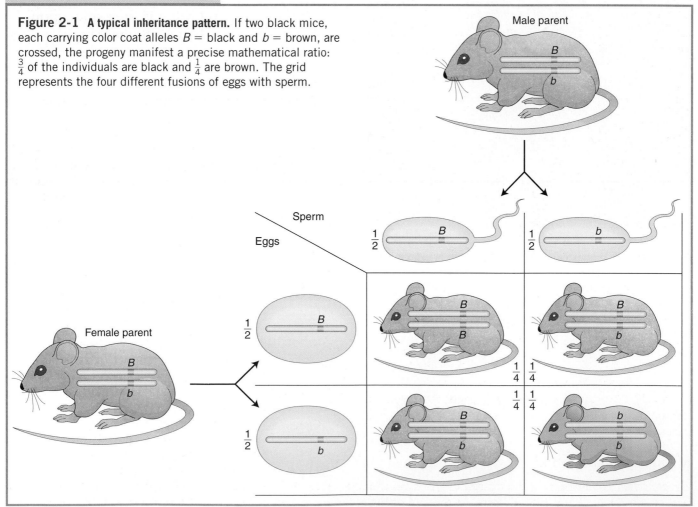

Figure 2-1 A typical inheritance pattern. If two black mice, each carrying color coat alleles B = black and b = brown, are crossed, the progeny manifest a precise mathematical ratio: $\frac{3}{4}$ of the individuals are black and $\frac{1}{4}$ are brown. The grid represents the four different fusions of eggs with sperm.

form the basis of a great deal of genetics described in this text. Inheritance patterns fall into three general categories, based on the different possible locations of that single gene:

- *Autosomal inheritance*, based on variation of single genes on regular chromosomes (autosomes)
- *Sex-linked inheritance*, based on variation of single genes on the sex-determining chromosomes
- *Cytoplasmic inheritance*, based on variation of single genes on organelle chromosomes

(Patterns of inheritance of *continuous* variation are more complex and are dealt with in Chapter 20.)

The chapter summary diagram (Figure 2-1) shows a typical pattern of inheritance that we shall encounter in this chapter. It is in fact an example of autosomal inheritance, which is the type most commonly encountered, because autosomes carry most of the genes. The figure shows an experimental mating between two black mice of a specific genetic type. The main grid of the figure shows a commonly encountered inheritance pattern manifested in the progeny of this mating: a 3/4 to 1/4 ratio of animals with black and brown coats. The gene symbols accompanying each mouse show the genetic mechanism that accounts for the inheritance pattern. The parental mice are postulated to each carry two different forms (alleles) of an autosomal gene that affects coat color, written symbolically *B* and *b*, and the separation of these into the gametes and their combination in the progeny accounts perfectly for the observed inheritance pattern.

2.1 Autosomal inheritance

We begin with the pattern of inheritance observed for genes located on the **autosomal chromosomes,** which are all the chromosomes in the cell nucleus except the sex chromosomes.

The pattern we now call *autosomal inheritance* was discovered in the nineteenth century by the monk Gregor Mendel. His work has great significance because not only did he discover this pattern, but by analyzing it he was able to deduce the very existence of genes and their alternative forms. Furthermore, his approach to gene discovery is still used to this day. For these reasons Mendel is known as the founder of genetics. We will use Mendel's historical experiments to illustrate autosomal inheritance.

Mendel's experimental system

Gregor Mendel was born in the district of Moravia, then part of the Austro-Hungarian Empire. At the end of high

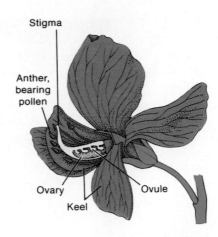

Figure 2-2 Reproductive parts of the pea flower. A pea flower with the keel cut and opened to expose the reproductive parts. The ovary is shown in a cutaway view. [After J. B. Hill, H. W. Popp, and A. R. Grove, Jr., *Botany.* Copyright 1967 by McGraw-Hill.]

school, he entered the Augustinian monastery of St. Thomas in the city of Brünn, now Brno of the Czech Republic. His monastery was dedicated to teaching science and to scientific research, so Mendel was sent to university in Vienna to obtain his teaching credentials. However, he failed his examinations and returned to the monastery at Brünn. There he embarked on the research program of plant hybridization that was posthumously to earn him the title of founder of the science of genetics.

Mendel's studies constitute an outstanding example of good scientific technique. He chose research material well suited to the study of the problem at hand, designed his experiments carefully, collected large amounts of data, and used mathematical analysis to show that the results were consistent with his explanatory hypothesis. He then tested the predictions of the hypothesis in a new round of experimentation.

Mendel studied the garden pea *(Pisum sativum),* which he chose as his object of study for two main reasons. First, peas were available from seed merchants in a wide array of distinct variant shapes and colors that could be easily identified and analyzed. Second, peas can either self-pollinate **(self)** or cross-pollinate **(cross).** A plant is said to self when it reproduces through the fertilization of its eggs by sperm from its own pollen. In all flowering plants, the male parts of a flower (anthers) release pollen containing the sperm, and the sperm fertilize eggs released from ovules in the ovary, the female part of a flower. Garden peas are amenable to selfing because anthers and ovaries of the flower are enclosed by two petals fused to form a compartment called a *keel* (Figure 2-2), and if left alone the pollen simply falls on the stigma of its own flower. Alternatively, the gardener

Figure 2-3 One of the techniques of artificial cross-pollination, demonstrated with the *Mimulus guttatus*, the yellow monkey flower. To transfer pollen, the experimenter touches anthers from the male parent to the stigma of an emasculated flower, which acts as the female parent. [Anthony Griffiths.]

or experimenter can cross any two pea plants at will. The anthers from one plant are removed before they have shed their pollen, preventing selfing. Pollen from the other plant is then transferred to the receptive stigma with a paintbrush or on anthers themselves (Figure 2-3). Thus, the experimenter can choose either to self or to cross the pea plants.

Other practical reasons for Mendel's choice of peas were that they are inexpensive and easy to obtain, take up little space, have a short generation time, and produce many offspring. Such considerations often enter into the choice of organism in genetic research.

Crosses of plants differing in one character

Mendel chose seven different *characters* to study. The word *character* in this regard means a specific property of an organism; geneticists use this term as a synonym for characteristic.

For each of the characters that he chose, Mendel obtained lines of plants, which he grew for two years to make sure that they were pure. A **pure line** is a population that breeds true for (shows no variation in) the particular character being studied; that is, all offspring produced by selfing or crossing within the population are identical in this character. By making sure that his lines bred true, Mendel had made a clever beginning: he had established a fixed baseline for his future studies so that any changes observed subsequent to deliberate manipulation in his research would be scientifically meaningful. In effect, he had set up a control experiment.

Two of the pea lines studied by Mendel bred true for the character of flower color. One line bred true for purple flowers; the other, for white flowers. Any plant in the purple-flowered line—when selfed or when crossed with others from the same line—produced seeds that all grew into plants with purple flowers. When these plants in turn were selfed or crossed within the line, their progeny also had purple flowers, and so forth. The white-flowered line similarly produced only white flowers through all generations. Mendel obtained seven pairs of pure lines for seven characters, with each pair differing in only one character (Figure 2-4).

Figure 2-4 The seven character differences studied by Mendel. [After S. Singer and H. Hilgard, *The Biology of People.* Copyright 1978 by W. H. Freeman and Company.]

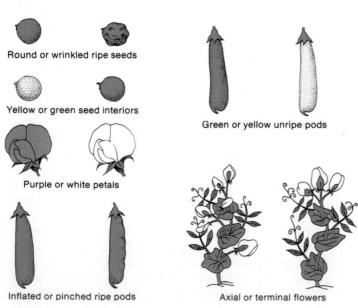

Round or wrinkled ripe seeds

Yellow or green seed interiors

Purple or white petals

Inflated or pinched ripe pods

Green or yellow unripe pods

Axial or terminal flowers

Long or short stems

Each pair of Mendel's plant lines can be said to show a **character difference**—a contrasting difference between two lines of organisms (or between two organisms) in one particular character. Such contrasting differences for a particular character are the starting point for any genetic analysis. The differing lines (or individuals) represent the different forms that the character may take: the differing forms themselves can be called *character forms, variants*, or **phenotypes**. The term *phenotype* (derived from Greek) literally means the "form that is shown"; it is the term used by geneticists today. Even though such words as *gene* and *phenotype* were not coined or used by Mendel, we shall use them in describing Mendel's results and hypotheses.

We turn now to Mendel's analysis of the lines breeding true for flower color. In one of his early experiments, Mendel pollinated a purple-flowered plant with pollen from a white-flowered plant. We call the plants from the pure lines the **parental generation (P)**. All the plants resulting from this cross had purple flowers (Figure 2-5). This progeny generation is called the **first filial generation (F_1)**. (The subsequent generations, produced by selfing, are symbolized F_2, F_3, and so forth.)

Mendel made **reciprocal crosses.** In most plants, any cross can be made in two ways, depending on which

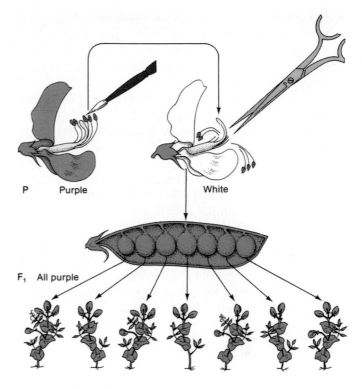

Figure 2-6 Mendel's cross of white-flowered ♀ × purple-flowered ♂ showed that all progeny of a reciprocal cross were also purple-flowered.

phenotype is used as male (♂) or female (♀). For example, the following two crosses

$$\text{phenotype A ♂} \times \text{phenotype B ♀}$$
$$\text{phenotype B ♂} \times \text{phenotype A ♀}$$

are reciprocal crosses. Mendel made a reciprocal cross in which he pollinated a white flower with pollen from a purple-flowered plant. This reciprocal cross produced the same result (all purple flowers) in the F_1 as the original cross had (Figure 2-6). He concluded that it makes no difference which way the cross is made. If one pure-breeding parent is purple-flowered and the other is white-flowered, all plants in the F_1 have purple flowers. The purple flower color in the F_1 generation is identical with that in the purple-flowered parental plants.

Next, Mendel selfed the F_1 plants, allowing the pollen of each flower to fall on its own stigma. He obtained 929 pea seeds from this selfing (the F_2 individuals) and planted them. Interestingly, some of the resulting plants were white-flowered; the white phenotype, which was not observed in the F_1, had reappeared in the F_2. He inferred that somehow "whiteness" must have been present but unexpressed in the F_1. He coined the terms **dominant** and **recessive** to describe the phenomenon. In the present discussion, the purple phenotype was dominant, whereas white was recessive. (We shall

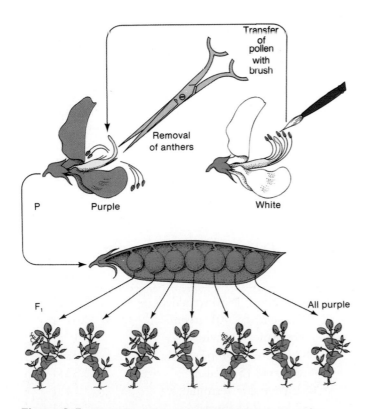

Figure 2-5 Mendel's cross of purple-flowered ♀ × white-flowered ♂ yielded all purple-flowered progeny.

see later that it is the genetic determinants of the phenotypes that are either dominant or recessive.) The operational definition of dominance is provided by the phenotype of an F_1 established by intercrossing two contrasting pure lines. The parental phenotype that is expressed in such F_1 individuals is by definition the dominant phenotype.

Mendel then did something that, more than anything else, marks the birth of modern genetics: he *counted* the numbers of F_2 plants with each phenotype. Quantification had never been applied to studies on inheritance before Mendel's work. Indeed, others had obtained remarkably similar results in breeding studies but had failed to count the numbers in each class. In the F_2, Mendel counted 705 purple-flowered plants and 224 white-flowered plants. He noted that the ratio of $705:224$ is almost exactly a $3:1$ ratio.

Mendel repeated the crossing procedures for the six other pairs of pea character differences. He found the same $3:1$ ratio in the F_2 generation for each pair (Table 2-1). By this time, he was undoubtedly beginning to attribute great significance to the $3:1$ ratio and to think up an explanation for it. In all cases, one parental phenotype disappeared in the F_1 and reappeared in one-fourth of the F_2.

Mendel went on to thoroughly test the class of F_2 individuals showing the dominant phenotype. He found (apparently unexpectedly) that there were in fact two genetically distinct subclasses. In this case, he was working with the two phenotypes of seed color. In peas, the color of the seed is determined by the genetic constitution of the seed itself, not by the maternal parent as in some plant species. This autonomy is convenient because the investigator can treat each pea as an individual and can observe its phenotype directly without having to grow a plant from it, as must be done for flower color. It also means that because peas are small, much larger numbers can be examined, and studies can be extended into subsequent generations. The seed color phenotypes that Mendel used were yellow and green. Mendel crossed a pure yellow line with a pure green line and observed that the F_1 peas that appeared were all yellow. Symbolically,

$$P \quad \text{yellow} \times \text{green}$$
$$\downarrow$$
$$F_1 \quad \text{all yellow}$$

Therefore, by definition, yellow is the dominant phenotype and green is recessive.

Mendel grew F_1 plants from these F_1 peas and then selfed the plants. The peas that developed on the F_2 plants constituted the F_2 generation. He observed that, in the pods of the F_1 plants, three-fourths of the F_2 peas were yellow and one-fourth were green:

$$F_1 \quad \text{all yellow (self)}$$
$$F_2 \quad \begin{cases} \frac{3}{4} \text{ yellow} \\ \frac{1}{4} \text{ green} \end{cases}$$

The ratio in the F_2 is simply the $3:1$ phenotypic ratio that we encountered earlier. To further test the members of the F_2, Mendel selfed many individual plants. He took a sample consisting of 519 yellow F_2 peas and grew plants from them. These yellow-pea F_2 plants were selfed individually, and the colors of peas that developed on them were noted.

Mendel found that 166 of the plants bore only yellow peas, and each of the remaining 353 plants bore a mixture of yellow and green peas, again in a $3:1$ ratio. Plants from green F_2 peas were also grown and selfed and were found to bear only green peas. In summary, all the F_2 green peas were evidently pure-breeding, like the green parental line, but, of the F_2 yellow peas, two-thirds were like the F_1 yellow peas (producing yellow and green seeds in a $3:1$ ratio) and one-third were like the pure-breeding yellow parent. Thus the study of the individual selfings revealed that underlying the $3:1$

Table 2-1 Results of All Mendel's Crosses in Which Parents Differed in One Character

Parental phenotype	F_1	F_2	F_2 ratio
1. Round × wrinkled seeds	All round	5474 round; 1850 wrinkled	2.96:1
2. Yellow × green seeds	All yellow	6022 yellow; 2001 green	3.01:1
3. Purple × white petals	All purple	705 purple; 224 white	3.15:1
4. Inflate × pinched pods	All inflated	882 inflated; 299 pinched	2.95:1
5. Green × yellow pods	All green	428 green; 152 yellow	2.82:1
6. Axial × terminal flowers	All axial	651 axial; 207 terminal	3.14:1
7. Long × short stems	All long	787 long; 277 short	2.84:1

phenotypic ratio in the F_2 generation was a more fundamental $1:2:1$ ratio:

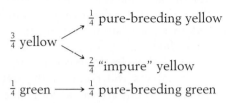

$\frac{3}{4}$ yellow
  - $\frac{1}{4}$ pure-breeding yellow
  - $\frac{2}{4}$ "impure" yellow

$\frac{1}{4}$ green $\longrightarrow \frac{1}{4}$ pure-breeding green

Further studies showed that such $1:2:1$ ratios underlie all the $3:1$ F_2 phenotypic ratios that Mendel had observed. Thus, his challenge was to explain not the $3:1$ ratio, but the $1:2:1$ ratio. Mendel's explanation was a classic example of a creative model or hypothesis derived from observation and suitable for testing by further experimentation. His model contained the following concepts:

1. *The existence of genes.* Contrasting phenotypic differences and their inheritance patterns were attributed to certain hereditary particles. We now call these particles *genes.*

2. *Genes are in pairs.* The gene may have different forms, each corresponding to an alternative phenotype of a character. The different forms of one type of gene are called **alleles.** In adult pea plants, each type of gene is present twice in each cell, constituting a **gene pair.** In different plants, the gene pair can be of the same alleles or of different alleles of that gene. Mendel's reasoning here was probably that the F_1 plants must have had one allele that was responsible for the dominant phenotype and another allele that was responsible for the recessive phenotype, which showed up only in later generations.

3. *Halving of gene pairs in gametes.* Each gamete carries only one member of each gene pair. In order to be expressed in later generations, obviously the alleles must find their way into the gametes—the eggs and sperm. However, to prevent the number of genes from doubling every time gametes fused, he had to propose that during gamete formation the gene pair halved.

4. *Equal segregation.* The members of the gene pairs segregate (separate) equally into the gametes. The key word "equal" means that 50% of the gametes will carry one member of a gene pair, and 50% will carry the other.

5. *Random fertilization.* The union of one gamete from each parent to form the first cell **(zygote)** of a new progeny individual is random—that is, gametes combine without regard to which allele is carried.

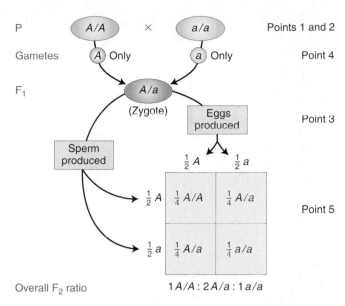

Figure 2-7 Mendel's explanation of the 1:2:1 ratio. Mendel's model of the hereditary determinants of a character difference in the P, F_1, and F_2 generations explains the $1:2:1$ ratio. The sections of the figure that illustrate each of Mendel's five points are indicated on the right.

These points can be illustrated symbolically for a general case by using A to represent the allele that determines the dominant phenotype and a to represent the allele for the recessive phenotype (as Mendel did). The use of A and a is similar to the way in which a mathematician uses symbols to represent abstract entities of various kinds. In Figure 2-7, these symbols are used to illustrate how Mendel's model explains the $1:2:1$ ratio. As mentioned in Chapter 1, the members of a gene pair are separated by a slash (/). This slash is used to show us that they are indeed a pair.

The whole model made logical sense of the data. However, many beautiful models have been knocked down under test. Mendel's next job was to test his model. A key part of the model is the nature of the F_1 individuals, which are proposed to carry two different alleles, and these segregate equally into gametes. Therefore Mendel took an F_1 plant (that grew from a yellow seed) and crossed it with a plant grown from a green seed. The principle of equal segregation means that a $1:1$ ratio of yellow to green seeds could be predicted in the next generation. If we let Y stand for the allele that determines the dominant phenotype (yellow seeds) and y stand for the allele that determines the recessive phenotype (green seeds), we can diagram Mendel's predictions, as shown in Figure 2-8. In this experiment, Mendel obtained 58 yellow (Y/y) and 52 green (y/y), a very close approximation to the predicted $1:1$ ratio attributed to the equal segregation of Y and y in the F_1 individual.

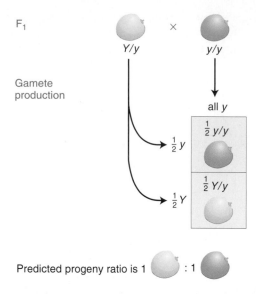

Figure 2-8 The cross of a presumed heterozygote *Y/y* to a presumed homozygous recessive *y/y* produces progeny in the ratio 1:1.

The concept of **equal segregation** has been given formal recognition as **Mendel's first law:** *The two members of a gene pair segregate from each other into the gametes; so half the gametes carry one member of the pair and the other half of the gametes carry the other member of the pair.*

Now we need to introduce some more terms. The individuals represented as *A/a* are called **heterozygotes** or, sometimes, **hybrids,** whereas the individuals in pure lines are called **homozygotes.** In such words, *hetero-* means "different" and *homo-* means "identical." Thus, an *A/A* plant is said to be **homozygous dominant;** an *a/a* plant is homozygous for the recessive allele, or **homozygous recessive.** As stated in Chapter 1, the designated genetic constitution of the character or characters under study is called the **genotype.** Thus, *Y/Y* and *Y/y*, for example, are different genotypes even though the seeds of both types are of the same phenotype (that is, yellow). In such a situation, the phenotype is viewed simply as the outward manifestation of the underlying genotype. Note that, underlying the 3:1 phenotypic ratio in the F_2, there is a 1:2:1 genotypic ratio of $Y/Y : Y/y : y/y$.

As we noted before, the expressions *dominant* and *recessive* are properties of the phenotype. The dominant phenotype is established in analysis by the appearance of the F_1. However, a phenotype (which is merely a description) cannot really exert dominance. Mendel showed that the dominance of one phenotype over another is in fact due to the dominance of one member of a gene pair over the other.

Let's pause to let the significance of this work sink in. What Mendel did was to develop an analytic scheme that could be used to identify genes affecting any bio-

logical character or function. Let's take petal color as an example. Starting with two different phenotypes (purple and white) of one character (petal color), Mendel was able to show that the difference was caused by one gene pair. Modern geneticists would say that Mendel's analysis had identified a gene for petal color. What does this mean? It means that, in these organisms, there is a gene that greatly affects the color of the petals. This gene can exist in different forms: a dominant form of the gene (represented by C) causes purple petals, and a recessive form of the gene (represented by *c*) causes white petals. The forms C and *c* are alleles (alternative forms) of that gene for petal color. Use of the same letter designation shows that the alleles are forms of one gene. We can express this idea in another way by saying that there is a gene, called phonetically a "see" gene, with alleles C and *c*. Any individual pea plant will always have two "see" genes, forming a gene pair, and the actual members of the gene pair can be C/C, C/*c*, or *c*/*c*. Notice that, although the members of a gene pair can produce different effects, they both affect the same character.

> **MESSAGE** The existence of genes was originally inferred by observing precise mathematical ratios in the descendants of two parental individuals that show contrasting phenotypes. These standard inheritance patterns are still used to deduce the existence of specific genes affecting specific characters.

The cellular and molecular basis of Mendelian genetics

Having examined the way that Mendel identified the existence of genes in peas, we can now translate his notion of the gene into a modern context. To Mendel the gene was an invented entity needed to explain a pattern of inheritance. However, today the gene is very much a reality, as a result of a great volume of research carried out for the very purpose of deducing its nature. We will examine such research throughout this book, but for the present let us summarize the modern view of the gene.

Mendel proposed that genes come in different forms we now call *alleles*. What is the molecular nature of alleles? When alleles such as *A* and *a* are examined at the DNA level by using modern technology, they are generally found to be identical in most of their sequences and differ only at one or a few nucleotides of the thousands of nucleotides that make up the gene. Therefore, we see that the alleles are truly different versions of the same basic gene. Looked at another way, *gene* is the generic term and allele is specific. (The pea-color gene has two alleles coding for yellow and green.) The following diagram represents the DNA of two alleles of one gene; the letter "x" represents a difference in the nucleotide sequence:

In studying any kind of variation such as allelic variation, it is often helpful to have a standard to act as a fixed reference point. What could be used as the "standard" allele of a gene? In genetics today the "wild type" is the allele used as the standard; it is the form of any particular gene that is found in the wild, in other words in natural populations. In the case of garden peas, the wild type is not familiar to most of us. We must turn to the long and interesting history of the garden pea to determine its wild type. Archaeological research has shown that peas were one of the first plant species brought into cultivation, in the Near and Middle East, possibly as early as 8000 B.C. The original wild pea was probably *Pisum elatium*, which is quite different from the *Pisum sativum* studied by Mendel. In this wild precursor, humans observed naturally arising variants and selected those with higher yield, better taste, and so on. The net result of all these selections over the generations was *Pisum sativum*. Hence the genes of *Pisum elatium* are probably the closest we can come to defining wild-type alleles of garden peas. For example, wild peas have purple petals, and thus the allele that gives purple petals is the wild-type allele.

Let's consider the cellular basis for the gene affecting pea petal color. How does the presence of an allele cause its phenotype to appear? For example, how does the presence of the wild-type allele for pea color cause petals to be purple? The purple color of wild peas is caused by a pigment called *anthocyanin*, which is a chemical made in petal cells as the end product of a series of consecutive chemical conversions, rather like a chemical assembly line. Typically each of the conversions is controlled by a specific enzyme (a biological catalyst), and the structure of each of these enzymes (essentially its amino acid sequence) is dictated by the nucleotide sequence of a specific gene. If the nucleotide sequence of *any* of the genes in the pathway changes as a result of a rare chemical "accident", a new allele is created. Such changes are called *mutations*: they can occur anywhere along the nucleotide sequence of a gene. One common outcome of a mutation is that one or more amino acids in the relevant enzyme may be changed. If the change is at a crucial site, the result is a loss of enzyme function. If such a mutant allele is homozygous, one point in the chemical assembly pathway becomes blocked, and no purple pigment is produced. Lack of pigment means that no wavelength of light is absorbed by the petal, and it will reflect sunlight and look white. If the plant is het-

erozygous, the one functional copy generally provides enough enzyme function to allow the synthesis of enough pigment to make the petals purple. Hence the allele for purple is dominant, as Mendel found (see Table 2-1). If we represent a wild-type allele as *A*, and the corresponding recessive mutant allele as *a*, we can summarize the molecular picture as follows:

$$A/A \longrightarrow \text{active enzyme} \longrightarrow \text{purple pigment}$$
$$A/a \longrightarrow \text{active enzyme} \longrightarrow \text{purple pigment}$$
$$a/a \longrightarrow \text{no active enzyme} \longrightarrow \text{white}$$

Note that inactivation of *any* one of the genes in the anthocyanin pathway could lead to the white phenotype. Hence the symbol *A* could represent any one of those genes. Note also that there are many ways a gene can be inactivated by mutation. For one thing, the mutational damage could occur at many different sites. We could represent the situation as follows, where dark blue indicates normal wild-type DNA sequence and red with the letter X represents altered sequence resulting in nonfunctional enzyme:

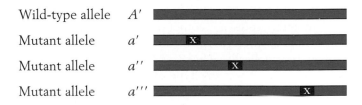

When geneticists use the symbol *A* to represent a wild-type allele, it means one specific sequence of DNA. But when they use the symbol *a* to represent a recessive allele, it is a shorthand that can represent any one of the possible types of damage that can lead to nonfunctional recessive alleles.

We have seen above that the concept of alleles is explained at the molecular level as variant DNA sequences. Can we also explain Mendel's first law? To do so, we must look at the behavior of the chromosomes on which the alleles are found. In diploid organisms, there are two copies of each chromosome, each containing one of the two alleles. How is Mendel's first law, the equal segregation of alleles at gamete formation, accomplished at the cell level? In a diploid organism such as peas, all the cells of the organism contain two chromosome sets. Gametes, however, are haploid, containing one chromosome set. Gametes are produced by specialized cell divisions in the diploid cells in the germinal tissue (ovaries and anthers). The nucleus also divides during these specialized cell divisions, in a process called **meiosis.** Highly programmed chromosomal movements in meiosis carry each allele of a chromosome pair to a separate gamete. In meiosis in a heterozygote *A/a*, the chromosome carrying *A* is pulled in the opposite direction from the chromosome carrying a; so half the resulting gametes carry A and the other half carry a. The situation can be summarized in a simplified form as follows (meiosis will be revisited in detail in Chapter 3):

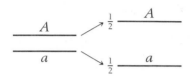

The force that moves the chromosomes to the cell poles is generated by the nuclear spindle, a set of microtubules, made of the protein tubulin, that attach to the centromeres of chromosomes and pull them to the poles of the cell. The orchestration of these molecular interactions is complex, yet constitutes the basis of the laws of hereditary transmission in eukaryotes.

Crosses of plants differing in two characters

Mendel's experiments described so far stemmed from two pure-breeding parental lines that differed in one character. As we have seen, such lines produce F_1 progeny that are heterozygous for one gene (genotype A/a). Such heterozygotes are sometimes called **monohybrids**. The selfing or intercross of identical heterozygous F_1 individuals (symbolically $A/a \times A/a$) is called a **monohybrid cross,** and it was this type of cross that provided the interesting 3:1 progeny ratios that suggested the principle of equal segregation. Mendel went on to analyze the descendants of pure lines that differed in *two* characters. Here we need a general symbolism to represent genotypes including two genes. If two genes are on different chromosomes, the gene pairs are separated by a semicolon—for example, $A/a \; ; \; B/b$. If they are on the same chromosome, the alleles on one chromosome are written adjacently with no punctuation and are separated from those on the other chromosome by a slash—for example, $A \, B/a \, b$ or $A \, b/a \, B$. An accepted symbolism does not exist for situations in which it is not known whether the genes are on the same chromosome or on different chromosomes. For this situation, we will separate the genes with a dot—for example, $A/a \cdot B/b$. A double heterozygote such as $A/a \cdot B/b$ is also known as a **dihybrid.** From studying **dihybrid crosses** ($A/a \cdot B/b \times A/a \cdot B/b$), Mendel came up with another important principle of heredity.

The two specific characters that he began working with were seed shape and seed color. We have already followed the monohybrid cross for seed color ($Y/y \times Y/y$), which gave a progeny ratio of 3 yellow:1 green. The seed-shape phenotypes (Figure 2-9) were round (determined by allele R) and wrinkled (determined by allele r). The monohybrid cross $R/r \times R/r$ gave a progeny ratio of 3 round:1 wrinkled

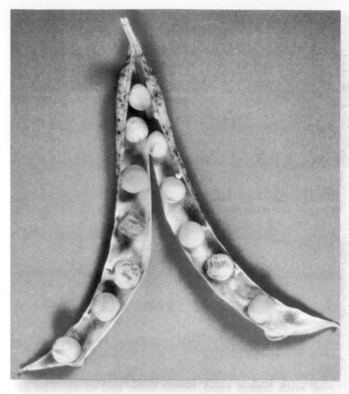

Figure 2-9 Round (R/R or R/r) and wrinkled (r/r) peas in a pod of a selfed heterozygous plant (R/r). The phenotypic ratio in this pod happens to be precisely the 3:1 ratio expected on average in the progeny of this selfing. (Molecular studies have shown that the wrinkled allele used by Mendel is produced by insertion into the gene of a segment of mobile DNA of the type to be discussed in Chapter 13.) [Madan K. Bhattacharyya.]

as expected (see Table 2-1). To perform a dihybrid cross, Mendel started with two parental pure lines. One line had yellow, wrinkled seeds. Because Mendel had no concept of the chromosomal location of genes, we must use the dot representation to write this genotype as $Y/Y \cdot r/r$. The other line had green, round seeds, with genotype $y/y \cdot R/R$. The cross between these two lines produced dihybrid F_1 seeds of genotype $Y/y \cdot R/r$, which he discovered were yellow and round. This result showed that the dominance of Y over y and R over r was unaffected by the condition of the other gene pair in the $Y/y \cdot R/r$ dihybrid. Next Mendel made the dihybrid cross by selfing the dihybrid F_1 to obtain the F_2 generation.

The F_2 seeds were of four different types in the following proportions:

$\frac{9}{16}$ round, yellow

$\frac{3}{16}$ round, green

$\frac{3}{16}$ wrinkled, yellow

$\frac{1}{16}$ wrinkled, green

as shown in Figure 2-10. This rather unexpected 9:3:3:1 ratio seems a lot more complex than the simple 3:1 ratios of the monohybrid crosses. What could be the explanation? Before attempting to explain the ratio, Mendel made dihybrid crosses that included several other combinations of characters and found that *all* of the dihybrid F_1 individuals produced 9:3:3:1 progeny ratios similar to that obtained for seed shape and color. The 9:3:3:1 ratio was another consistent inheritance pattern that required the development of a new idea to explain it.

Mendel added up the numbers of individuals in certain F_2 phenotypic classes (the numbers are shown in Figure 2-10) to determine if the monohybrid 3:1 F_2 ratios were still present. He noted that, in regard to seed shape, there were 423 round seeds (315 + 108) and 133 wrinkled seeds (101 + 32). This result is close to a 3:1 ratio. Next, in regard to seed color, there were 416 yellow seeds (315 + 101) and 140 green (108 + 32), also very close to a 3:1 ratio. The presence of these two 3:1 ratios hidden in the 9:3:3:1 ratio was undoubtedly a source of the insight that Mendel needed to explain the 9:3:3:1 ratio, because he realized that it was nothing more than two different 3:1 ratios combined at random. One way of visualizing the random combination of these two ratios is with a branch diagram, as follows:

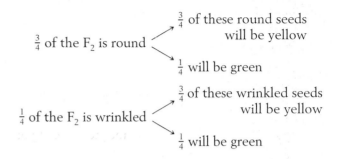

The combined proportions are calculated by multiplying along the branches in the diagram. For example, $\frac{3}{4}$ of $\frac{3}{4}$ is calculated as $\frac{3}{4} \times \frac{3}{4}$, which equals $\frac{9}{16}$. These multiplications give us the following four proportions:

$$\frac{3}{4} \times \frac{3}{4} = \frac{9}{16} \text{ round, yellow}$$

$$\frac{3}{4} \times \frac{1}{4} = \frac{3}{16} \text{ round, green}$$

$$\frac{1}{4} \times \frac{3}{4} = \frac{3}{16} \text{ wrinkled, yellow}$$

$$\frac{1}{4} \times \frac{1}{4} = \frac{1}{16} \text{ wrinkled, green}$$

These proportions constitute the 9:3:3:1 ratio that we are trying to explain. However, is this not merely number juggling? What could the combination of the two 3:1 ratios mean biologically? The way that Mendel phrased his explanation does in fact amount to a biological mechanism. In what is now known as **Mendel's second law,** he concluded that *different gene pairs assort independently in gamete formation.* The consequence is, that for two heterozygous gene pairs *A/a* and *B/b,* the *b* allele is just as likely to end up in a gamete with an *a* allele as with an *A* allele, and likewise for the *B* allele. With hindsight about the chromosomal location of genes, we now know that this "law" is true only in some cases. Most cases of independence are observed for genes on different chromosomes. Genes on the same chromosome generally do not assort independently, because they are held together on the chromosome. Hence the modern version of Mendel's second law is stated as follows: *gene pairs on different chromosome pairs assort independently at meiosis.*

We have explained the 9:3:3:1 phenotypic ratio as two combined 3:1 phenotypic ratios. But the second law pertains to packing alleles into gametes. Can the 9:3:3:1 ratio be explained on the basis of gametic genotypes? Let us consider the gametes produced by the F_1 dihybrid *R/r ; Y/y* (the semicolon shows that we are now assuming the genes to be on different chromosomes). Again, we will use the branch diagram to get us started because it illustrates independence visually. Combining Mendel's laws of

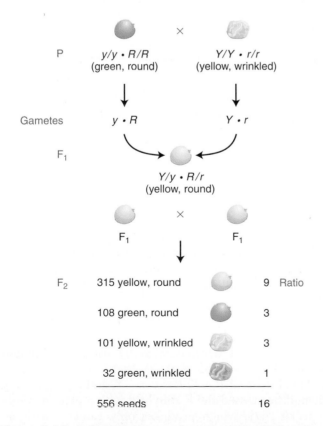

Figure 2-10 A dihybrid cross produces F_2 progeny in the ratio 9:3:3:1.

equal segregation and independent assortment, we can predict that

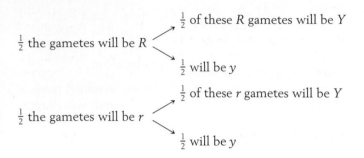

$\frac{1}{2}$ the gametes will be R

- $\frac{1}{2}$ of these R gametes will be Y
- $\frac{1}{2}$ will be y

$\frac{1}{2}$ the gametes will be r

- $\frac{1}{2}$ of these r gametes will be Y
- $\frac{1}{2}$ will be y

Multiplication along the branches gives us the gamete proportions:

$$\frac{1}{4} R ; Y$$
$$\frac{1}{4} R ; y$$
$$\frac{1}{4} r ; Y$$
$$\frac{1}{4} r ; y$$

These proportions are a direct result of the application of the two Mendelian laws. However, we still have not arrived at the $9:3:3:1$ ratio. The next step is to recognize that both the male and the female gametes will show the same proportions just given, because Mendel did not specify different rules for male and female gamete formation. The four female gametic types will be fertilized randomly by the four male gametic types to obtain the F_2. The best way of showing this graphically is to use a 4×4 grid called a *Punnett square,* which is depicted in Figure 2-11. Grids are useful in genetics because their proportions can be drawn according to genetic proportions or ratios being considered; thereby a visual representation of the data is obtained. In the Punnett square in Figure 2-11, for example, we see that the areas of the 16 boxes representing the various gametic fusions are each one-sixteenth of the total area of the grid, simply because the rows and columns were drawn to correspond to the gametic proportions of each. As the Punnett square shows, the F_2 contains a variety of genotypes, but there are only four phenotypes and their proportions are in the $9:3:3:1$ ratio. So we see that when we work at the biological level of gamete formation, Mendel's laws explain not only the F_2 phenotypes, but also the genotypes underlying them.

Mendel was a thorough scientist; he went on to test his principle of independent assortment in a number of ways. The most direct way zeroed in on the $1:1:1:1$ gametic ratio hypothesized to be produced by the F_1 dihybrid $R/r ; Y/y$, because this ratio sprang from his principle of independent assortment and was the biological basis of the $9:3:3:1$ ratio in the F_2,

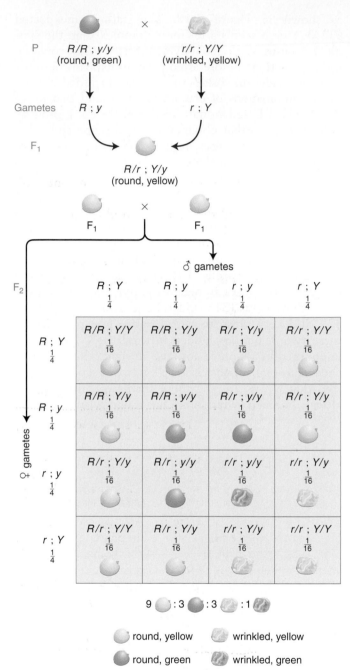

Figure 2-11 Using a Punnett square to predict the result of a dihybrid cross. A Punnett shows the predicted genotypic and phenotypic constitution of the F_2 generation from a dihybrid cross.

as we have just demonstrated by using the Punnett square. He reasoned that, if there were in fact a $1:1:1:1$ ratio of $R ; Y, R ; y, r ; Y$, and $r ; y$ gametes, then, if he crossed the F_1 dihybrid with a plant of genotype $r/r ; y/y$, which produces only gametes with recessive alleles (genotype $r ; y$), the progeny proportions of this cross should be a direct manifestation of the gametic proportions produced by the dihybrid; in other words,

$$\frac{1}{4}\, R/r \,;\, Y/y$$

$$\frac{1}{4}\, R/r \,;\, y/y$$

$$\frac{1}{4}\, r/r \,;\, Y/y$$

$$\frac{1}{4}\, r/r \,;\, y/y$$

These proportions were the result that he obtained, perfectly consistent with his expectations. He obtained similar results for all the other dihybrid crosses that he made, and these and other types of tests all showed that he had in fact devised a robust model to explain the inheritance patterns observed in his various pea crosses.

The type of cross just considered, of an individual of unknown genotype with a fully recessive homozygote, is now called a **testcross**. The recessive individual is called a **tester**. Because the tester contributes only recessive alleles, the gametes of the unknown individual can be deduced from progeny phenotypes.

In the early 1900s, Mendel's principles were tested in a wide spectrum of eukaryotic organisms. The results of these tests showed that Mendelian principles were generally applicable. Mendelian ratios (such as $3:1$, $1:1$, $9:3:3:1$, and $1:1:1:1$) were extensively reported, suggesting that equal segregation and independent assortment are fundamental hereditary processes found throughout nature. Mendel's laws are not merely laws about peas, but laws about the genetics of eukaryotic organisms in general. The experimental approach used by Mendel can be extensively applied in plants. However, in some plants and in most animals, the technique of selfing is impossible. This problem can be circumvented by crossing identical genotypes. For example, animals from differing pure lines are mated to produce an F_1 generation, as above. Instead of selfing, an F_1 animal can be mated to its F_1 siblings (brothers or sisters) to produce an F_2. The F_1 individuals are identical for the genes in question, so the F_1 cross is equivalent to a selfing.

Using Mendelian ratios

Mendelian ratios pop up in many aspects of genetic analysis. In this section we shall examine two important procedures revolving around Mendelian ratios.

PREDICTING PROGENY OF CROSSES USING MENDELIAN RATIOS

An important part of genetics is concerned with predicting the types of progeny that emerge from a cross and calculating their expected frequency—in other words, their probability. We have already examined two methods for doing so—Punnett squares and branch diagrams. Punnett squares can be used to show

hereditary patterns based on one gene pair, two gene pairs (as in Figure 2-11), or more. Such grids are a good graphic device for representing progeny, but drawing them is time-consuming. Even the 16-compartment Punnett square in Figure 2-11 takes a long time to write out, but, for a trihybrid, there are 2^3, or 8, different gamete types, and the Punnett square has 64 compartments. The branch diagram (below) is easier and is adaptable for phenotypic, genotypic, or gametic proportions, as illustrated for the dihybrid $A/a \,;\, B/b$. (The dash "—" in the genotypes means that the allele can be present in either form, that is, dominant or recessive.)

Note that the "tree" of branches for genotypes is quite unwieldy even in this case, which uses two gene pairs, because there are $3^2 = 9$ genotypes. For three gene pairs, there are 3^3, or 27, possible genotypes.

The application of simple statistical rules is a third method for calculating the probabilities (expected frequencies) of specific phenotypes or genotypes coming from a cross. The two probability rules needed are the **product rule** and the **sum rule,** which we will consider in that order.

> **MESSAGE** The product rule states that the probability of independent events occurring together is the product of the probabilities of the individual events.

The possible outcomes of rolling dice follow the product rule because the outcome on each separate die is independent of the others. As an example, let us consider two dice and calculate the probability of rolling a pair of 4's. The probability of a 4 on one die is $\frac{1}{6}$ because the die has six sides and only one side carries the 4. This probability is written as follows:

$$p \text{ (of a 4)} = \tfrac{1}{6}$$

Therefore, with the use of the product rule, the probability of a 4 appearing on both dice is $\tfrac{1}{6} \times \tfrac{1}{6} = \tfrac{1}{36}$, which is written

$$p \text{ (of two 4's)} = \tfrac{1}{6} \times \tfrac{1}{6} = \tfrac{1}{36}$$

Now for the sum rule:

> **MESSAGE** The sum rule states that the probability of either of two mutually exclusive events occurring is the sum of their individual probabilities.

In the product rule, the focus is on outcomes A **and** B. In the sum rule, the focus is on the outcome A **or** B. Dice can also be used to illustrate the sum rule. We have already calculated that the probability of two 4's is $\tfrac{1}{36}$, and with the use of the same type of calculation, it is clear that the probability of two 5's will be the same, or $\tfrac{1}{36}$. Now we can calculate the probability of either two 4's *or* two 5's. Because these outcomes are mutually exclusive, the sum rule can be used to tell us that the answer is $\tfrac{1}{36} + \tfrac{1}{36}$, which is $\tfrac{1}{18}$. This probability can be written as follows:

$$p \text{ (two 4's or two 5's)} = \tfrac{1}{36} + \tfrac{1}{36} = \tfrac{1}{18}$$

Now we can consider a genetic example. Assume that we have two plants of genotypes A/a ; b/b ; C/c ; D/d ; E/e and A/a ; B/b ; C/c ; d/d ; E/e. From a cross between these plants, we want to recover a progeny plant of genotype a/a ; b/b ; c/c ; d/d ; e/e (perhaps for the purpose of acting as the tester strain in a testcross). To estimate how many progeny plants need to be grown to stand a reasonable chance of obtaining the desired genotype, we need to calculate the proportion of the progeny that is expected to be of that genotype. If we assume that all the gene pairs assort independently, then we can do this calculation easily by using the product rule. The five different gene pairs are considered individually, as if five separate crosses, and then the appropriate probabilities are multiplied together to arrive at the answer:

From $A/a \times A/a$, one-fourth of the progeny will be a/a (see Mendel's crosses).
From $b/b \times B/b$, half the progeny will be b/b.
From $C/c \times C/c$, one-fourth of the progeny will be c/c.
From $D/d \times d/d$, half the progeny will be d/d.
From $E/e \times E/e$, one-fourth of the progeny will be e/e.

Therefore, the overall probability (or expected frequency) of progeny of genotype a/a ; b/b ; c/c ; d/d ; e/e will be $\tfrac{1}{4} \times \tfrac{1}{2} \times \tfrac{1}{4} \times \tfrac{1}{2} \times \tfrac{1}{4} = \tfrac{1}{256}$. So we learn that we need to examine 200 to 300 hundred progeny to stand a chance of obtaining at least one of the desired genotype. This probability calculation can be extended to predict phenotypic frequencies or gametic frequencies. Indeed, there are thousands of other uses for this method in genetic analysis, and we will encounter many in later chapters.

USING THE CHI-SQUARE (χ^2) TEST ON MENDELIAN RATIOS In genetics generally, a researcher is often confronted with results that are close to but not identical with an expected ratio. But how close is enough? A statistical test is needed to check such ratios against expectations, and the **chi-square test** fulfills this role.

In which experimental situations is the χ^2 test generally applicable? Research results often involve items in several distinct classes or categories; red, blue, male, female, lobed, unlobed, and so on. Furthermore it is often necessary to compare the observed numbers of items in the different categories with numbers that are predicted on the basis of some hypothesis. This is the general situation in which the χ^2 test is useful: comparing observed results with those predicted by a hypothesis. In a simple genetic example, suppose you have bred a plant that you hypothesize on the basis of previous analysis to be a heterozygote, A/a. To test this hypothesis you could make a cross to a tester of genotype a/a and count the numbers of $A/-$ and a/a phenotypes in the progeny. Then you will need to assess whether the numbers you obtain constitute the expected $1:1$ ratio. If there is a close match, then the hypothesis is deemed consistent with the result, while if there is a poor match, the hypothesis is rejected. As part of this process a judgment has to be made about whether the observed numbers are close *enough* to those expected. Very close matches and blatant mismatches generally present no problem, but inevitably there are gray areas in which the match is not obvious.

The χ^2 test is simply a way of quantifying the various deviations expected by chance if a hypothesis is true. Take the above simple hypothesis predicting a $1:1$ ratio, for example. Even if the hypothesis is true, we would not always expect an exact $1:1$ ratio. We can model this idea with a barrel full of equal numbers of red and white marbles. If we blindly remove samples of 100 marbles, on the basis of chance we would expect samples to show small deviations such as 52 red:48 white quite commonly, and larger deviations such as 60 red:40 white less commonly. Even 100 red marbles is a possible outcome, at a very low probability of $(1/2)^{100}$. However, if all levels of deviation are expected with different probabilities even if the hypothesis is true, how can we ever reject a hypothesis? It has become a general scientific convention that if there is a probability of less than 5 percent of observing a deviation from expectations at least this large, the hypothesis will be rejected as false. The hypothesis might still be true, but we have to make a decision somewhere and 5 percent is the conventional decision line. The implication is that although results this far from expectations are anticipated 5 percent of the time even when the hypothesis is true, we will mistakenly reject the hypothesis in only 5 percent of cases and we are willing to take this chance of error.

Let's look at some real data. We will test our above hypothesis that a plant is a heterozygote. We will let A stand for red petals and a stand for white. Scientists test a hypothesis by making predictions based on the hypothesis. In the present situation, one possibility is to predict the results of a testcross. Assume we testcross the presumed heterozygote. Based on the hypothesis, Mendel's first law of equal segregation predicts that we should have 50 percent A/a and 50 percent a/a. Assume that in reality we obtain 120 progeny, and find that 55 are red and 65 are white. These numbers differ from the precise expectations, which would have been 60 red and 60 white. The result seems a bit far off the expected ratio; this raises uncertainty, so we need to use the chi-square test. It is calculated using the following formula:

$$\chi^2 = \Sigma\, (O - E)^2/E \quad \text{for all classes}$$

in which E = expected number in a class, O = observed number in a class, and Σ means "sum of."

The calculation is most simply performed using a table:

Class	O	E	$(O - E)^2$	$(O - E)^2/E$
Red	55	60	25	25/60 = 0.42
White	65	60	25	25/60 = 0.42
				Total = χ^2 = 0.84

Now we must look up this χ^2 value in a table (Table 2-2), which will give us the probability value we want. The lines in the table represent different values of *degrees of freedom (df)*. The number of degrees of freedom is the number of independent variables in the data. In the present context this is simply the number of phenotypic classes minus 1. In this case df = 2 − 1 = 1. So we look only at the 1 df line. We see that our χ^2 value of 0.84 lies somewhere between the columns marked 0.5 and 0.1, in other words between 50 percent and 10 percent. This probability value is much greater than the cutoff value of 5 percent, so we would accept the observed results as being compatible with the hypothesis.

Some important notes on the application of this test:

1. What does the probability value actually mean? It is the probability of observing a deviation from the expected results *at least this large* on the basis of chance, if the hypothesis is correct. (Not *exactly* this deviation.)

Table 2-2 Critical Values of the χ^2 Distribution

df					P					df
	0.995	0.975	0.9	0.5	0.1	0.05	0.025	0.01	0.005	
1	.000	.000	0.016	0.455	2.706	3.841	5.024	6.635	7.879	1
2	0.010	0.051	0.211	1.386	4.605	5.991	7.378	9.210	10.597	2
3	0.072	0.216	0.584	2.366	6.251	7.815	9.348	11.345	12.838	3
4	0.207	0.484	1.064	3.357	7.779	9.488	11.143	13.277	14.860	4
5	0.412	0.831	1.610	4.351	9.236	11.070	12.832	15.086	16.750	5
6	0.676	1.237	2.204	5.348	10.645	12.592	14.449	16.812	18.548	6
7	0.989	1.690	2.833	6.346	12.017	14.067	16.013	18.475	20.278	7
8	1.344	2.180	3.490	7.344	13.362	15.507	17.535	20.090	21.955	8
9	1.735	2.700	4.168	8.343	14.684	16.919	19.023	21.666	23.589	9
10	2.156	3.247	4.865	9.342	15.987	18.307	20.483	23.209	25.188	10
11	2.603	3.816	5.578	10.341	17.275	19.675	21.920	24.725	26.757	11
12	3.074	4.404	6.304	11.340	18.549	21.026	23.337	26.217	28.300	12
13	3.565	5.009	7.042	12.340	19.812	22.362	24.736	27.688	29.819	13
14	4.075	5.629	7.790	13.339	21.064	23.685	26.119	29.141	31.319	14
15	4.601	6.262	8.547	14.339	22.307	24.996	27.488	30.578	32.801	15

2. Now that the above results have "passed" the chi-square test because $p > 0.05$, it does not mean that the hypothesis is true, merely that the results are compatible with that hypothesis. However, if we had obtained a p value of < 0.05, we would have been forced to reject the hypothesis. Science is all about falsifiable hypotheses, not "truth."

3. We must be careful about the wording of the hypothesis, because often there are tacit assumptions buried within it. The present hypothesis is a case in point; if we were to carefully state it, we would have to say it is that the "individual under test is a heterozygote A/a and the A/a and a/a progeny are of equal viability." We will investigate allele effects on viability in Chapter 6, but for the time being we must keep them in mind as a possible complication because differences in survival would affect the sizes of the various classes. The problem is that if we reject a hypothesis that has hidden components, we do not know which of the components we are rejecting.

4. The outcome of the chi-square test is heavily dependent on sample sizes (numbers in the classes). Hence the test must use *actual numbers*, not proportions or percentages. Also, the larger the samples, the more reliable is the test.

Any of the familiar Mendelian ratios discussed in this chapter can be tested using the chi-square test—for example, 3:1 (1 df), 1:2:1 (2 df), 9:3:3:1 (3 df), and 1:1:1:1 (3 df). We will return to more applications of the chi-square test in Chapters 4 and 6.

Autosomal inheritance in humans

Human matings, like those of experimental organisms, show many examples of the inheritance patterns described above. Because controlled experimental crosses cannot be made with humans, geneticists must resort to scrutinizing records in the hope that informative matings have been made by chance. Such a scrutiny of records of matings is called **pedigree analysis**. A member of a family who first comes to the attention of a geneticist is called the **propositus**. Usually the phenotype of the propositus is exceptional in some way (for example, the propositus might suffer from some type of disorder). The investigator then traces the history of the phenotype in the propositus back through the history of the family and draws a family tree, or pedigree, by using the standard symbols given in Figure 2-12.

Many variant phenotypes of humans are determined by the alleles of single autosomal genes, in the same manner we encountered in peas. Human pedigrees often show inheritance patterns of this simple Mendelian type. However, the patterns in the pedigree have to be interpreted differently, depending on whether one of the

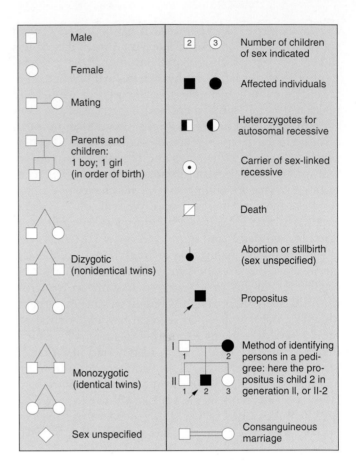

Figure 2-12 Symbols used in human pedigree analysis. [After W. F. Bodmer and L. L. Cavalli-Sforza, *Genetics, Evolution, and Man.* Copyright 1976 by W. H. Freeman and Company.]

contrasting phenotypes is a rare disorder or whether both phenotypes of a pair are common morphs of a polymorphism. Most pedigrees are drawn up for medical reasons and hence inherently concern medical disorders that are almost by definition rare. Let's look first at rare recessive disorders caused by recessive alleles of single autosomal genes.

PEDIGREE ANALYSIS OF AUTOSOMAL RECESSIVE DISORDERS The affected phenotype of an autosomal recessive disorder is determined by a recessive allele, and hence the corresponding unaffected phenotype must be determined by the corresponding dominant allele. For example, the human disease phenylketonuria (PKU) is inherited in a simple Mendelian manner as a recessive phenotype, with PKU determined by the allele p and the normal condition by P. Therefore, sufferers from this disease are of genotype p/p, and people who do not have the disease are either P/P or P/p. What patterns in a pedigree would reveal such an inheritance? The two key points are that (1) generally the disease appears in the progeny of unaffected parents and (2) the affected progeny include both males and females. When we know that both male and female progeny are affected, we can assume that we are most likely dealing with simple Mendelian inheritance of a gene on an autosome, rather than a gene on a sex chromosome. The following typical pedigree illustrates the key point that affected children are born to unaffected parents:

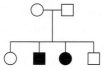

From this pattern, we can deduce simple Mendelian inheritance of the recessive allele responsible for the exceptional phenotype (indicated in black). Furthermore, we can deduce that the parents are both heterozygotes, say A/a; both must have an a allele because each contributed an a allele to each affected child, and both must have an A allele because they are phenotypically normal. We can identify the genotypes of the children (in the order shown) as $A/-$, a/a, a/a, and $A/-$. Hence, the pedigree can be rewritten as follows:

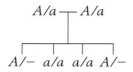

[Once you have read the section on sex-linked inheritance, you will realize that this pedigree does not support the hypothesis of X-linked recessive inheritance, because, under that hypothesis, an affected daughter must have a heterozygous mother (possible) and a hemizygous father, which is clearly impossible, because he would have expressed the phenotype of the disorder.]

Notice an interesting feature of pedigree analysis: even though Mendelian rules are at work, Mendelian ratios are rarely observed in families because the sample size is too small. In the preceding example, we observe a 1:1 phenotypic ratio in the progeny of a monohybrid cross. If the couple were to have, say, 20 children, the ratio would be something like 15 unaffected children and 5 with PKU (a 3:1 ratio), but in a sample of 4 children, any ratio is possible, and all ratios are commonly found.

The pedigrees of autosomal recessive disorders tend to look rather bare, with few black symbols. A recessive condition shows up in groups of affected siblings, and the people in earlier and later generations tend not to be affected. To understand why this is so, it is important to have some understanding of the genetic structure of populations underlying such rare conditions. By definition, if the condition is rare, most people do not carry the abnormal allele. Furthermore, most of those people who do carry the abnormal allele are heterozygous for it rather than homozygous. The basic reason that heterozy-

gotes are much more common than recessive homozygotes is that to be a recessive homozygote, both parents must have had the a allele, but to be a heterozygote, only one parent must carry it.

The formation of an affected person usually depends on the chance union of unrelated heterozygotes. However, inbreeding (mating between relatives) increases the chance that two heterozygotes will mate. An example of a marriage between cousins is shown in Figure 2-13. Individuals III-5 and III-6 are first cousins and produce two homozygotes for the rare allele. You can see from Figure 2-13 that an ancestor who is a heterozygote may produce many descendants who also are heterozygotes. Hence two cousins can carry the *same* rare recessive allele inherited from a common ancestor. For two *unrelated* persons to be heterozygous, they would have to inherit the rare allele from *both* their families. Thus matings between relatives generally run a higher risk of producing recessive disorders than do matings between nonrelatives. For this reason, first-cousin marriages contribute a large proportion of the sufferers of recessive diseases in the population.

What are some examples of human recessive disorders? PKU has already served as an example of pedigree analysis, but what kind of phenotype is it? PKU is a disease caused by abnormal processing of the amino acid phenylalanine, a component of all proteins in the food that we eat. Phenylalanine is normally converted into

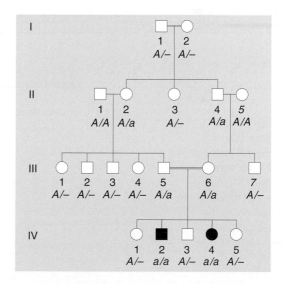

Figure 2-13 Pedigree of a rare recessive phenotype determined by a recessive allele *a*. Gene symbols are normally not included in pedigree charts, but genotypes are inserted here for reference. Individuals II-1 and II-5 marry into the family; they are assumed to be normal because the heritable condition under scrutiny is rare. Note also that it is not possible to be certain of the genotype in some persons with normal phenotype; such persons are indicated by $A/-$.

the amino acid tyrosine by the enzyme phenylalanine hydroxylase:

$$\text{phenylalanine} \xrightarrow{\text{phenylalanine hydroxylase}} \text{tyrosine}$$

However, if a mutation in the gene encoding this enzyme alters the amino acid sequence in the vicinity of the enzyme's active site, the enzyme cannot bind phenylalanine (its substrate) or convert it to tyrosine. Therefore phenylalanine builds up in the body and is converted instead into phenylpyruvic acid. This compound interferes with the development of the nervous system, leading to mental retardation.

$$\text{phenylalanine} \xrightarrow[]{\text{phenylananine hydroxylase}} \!\!\!\!\times\!\!\!\! \longrightarrow \text{tyrosine}$$
$$\qquad\quad \longrightarrow \text{phenylpyruvic acid}$$

Babies are now routinely tested for this processing deficiency at birth. If the deficiency is detected, phenylalanine can be withheld by use of a special diet and the development of the disease can be arrested.

Cystic fibrosis is another disease inherited according to Mendelian rules as a recessive phenotype. Cystic fibrosis is a disease whose most important symptom is the secretion of large amounts of mucus into the lungs, resulting in death from a combination of effects but usually precipitated by infection of the respiratory tract. The mucus can be dislodged by mechanical chest thumpers, and pulmonary infection can be prevented by antibiotics; thus, with treatment, cystic fibrosis patients can live to adulthood. The allele that causes cystic fibrosis was isolated in 1989, and the sequence of its DNA was determined. This line of research eventually revealed that the disorder is caused by a defective protein that transports chloride ions across the cell membrane. The resultant alteration of the salt balance changes the constitution of the lung mucus. This new understanding of gene function in affected and unaffected persons has given hope for more effective treatment.

Albinism, which served as a model of how differing alleles determine contrasting phenotypes in Chapter 1, also is inherited in the standard autosomal recessive manner.

Figure 2-14 shows how a mutation in one allele leads to a simple pattern of autosomal recessive inheritance in a pedigree. In this example, the recessive allele *a* is caused by a base-pair change that introduces a stop codon into the middle of the gene, resulting in a truncated protein. The mutation, by chance, also introduces a new target site for a restriction enzyme. Hence, a probe for the gene detects two fragments in the case of *a* and only one in *A*. (Other types of mutations would produce different effects at the level detected by Southern, Northern, and Western analyses.)

In all the examples considered so far, the disorder is caused by an allele that codes for a defective protein. In

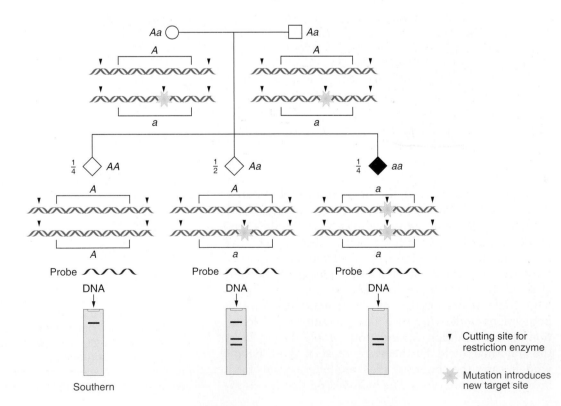

Figure 2-14 The inheritance of an autosomal recessive at the molecular level.
A Southern blot using a probe that binds to the region spanning the mutation causing albinism detects one DNA fragment in homozygous normal individuals (*A/A*) and two fragments in albino individuals (*a/a*). The three fragments detected in heterozygous individuals (*A/a*) are due to the presence of the normal and mutant alleles.

heterozygotes, the single functional allele provides enough active protein for the cell's needs. This situation is called *haplosufficiency*. Thus the amount of protein is insufficient only if the mutant allele is present in two copies, producing the recessive trait.

> **MESSAGE** In human pedigrees, an autosomal recessive disorder is revealed by the appearance of the disorder in the male and female progeny of unaffected persons.

PEDIGREE ANALYSIS OF AUTOSOMAL DOMINANT DISORDERS What pedigree patterns are expected from autosomal dominant disorders? Here the normal allele is recessive, and the abnormal allele is dominant. It may seem paradoxical that a rare disorder can be dominant, but remember that dominance and recessiveness are simply properties of how alleles act and are not defined in terms of how common they are in the population. A good example of a rare dominant phenotype with Mendelian inheritance is pseudoachondroplasia, a type of dwarfism (Figure 2-15). In regard to this gene, people with normal stature are genotypically d/d, and the dwarf phenotype in principle could be D/d or D/D. However, it is believed that the two "doses" of the D allele in the D/D genotype produce such a severe effect that this genotype is lethal. If this is true, all dwarf individuals are heterozygotes.

In pedigree analysis, the main clues for identifying an autosomal dominant disorder with Mendelian inheritance are that the phenotype tends to appear in every generation of the pedigree and that affected fathers and mothers transmit the phenotype to both sons and daughters. Again, the equal representation of both sexes among the affected offspring rules out inheritance via the sex chromosomes. The phenotype appears in every generation because generally the abnormal allele carried by a person must have come from a parent in the preceding generation. (Abnormal alleles can also arise de novo by the process of mutation. This event is relatively rare but must be kept in mind as a possibility.) A typical pedigree for a dominant disorder is shown in Figure 2-16. Once again, notice that Mendelian ratios are not necessarily observed in families. As with recessive disorders, persons bearing one copy of the rare A allele (A/a) are much more common than those bearing two copies (A/A), so most affected people are heterozygotes, and virtually all matings that produce progeny with dominant disorders are $A/a \times a/a$. Therefore, when the progeny of such matings are totaled, a $1:1$ ratio is expected of unaffected (a/a) to affected (A/a) persons.

Huntington disease is another example of a disease inherited as a dominant phenotype determined by an allele of a single gene. The phenotype is one of neural degeneration, leading to convulsions and premature death.

Figure 2-15 The human pseudoachondroplasia phenotype, illustrated by a family of five sisters and two brothers. The phenotype is determined by a dominant allele, which we can call D, that interferes with bone growth during development. This photograph was taken upon the arrival of the family in Israel after the end of the Second World War.
[UPI/Bettmann News Photos.]

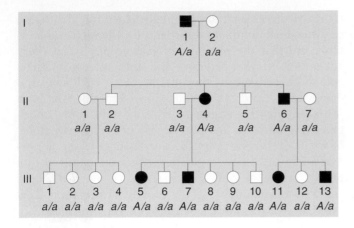

Figure 2-16 Pedigree of a dominant phenotype determined by a dominant allele A. In this pedigree, all the genotypes have been deduced.

However, it is a late-onset disease, the symptoms generally not appearing until after the person has begun to have children (Figure 2-17). Each child of a carrier of the abnormal allele stands a 50 percent chance of inheriting the allele and the associated disease. This tragic pattern has inspired a great effort to find ways of identifying people who carry the abnormal allele before they experience the onset of the disease. The application of molecular techniques has resulted in useful screening procedures.

Some other rare dominant conditions are polydactyly (extra digits), shown in Figure 2-18, and piebald spotting, shown in Figure 2-19.

> **MESSAGE** Pedigrees of Mendelian autosomal dominant disorders show affected males and females in each generation; they also show that affected men and women transmit the condition to equal proportions of their sons and daughters.

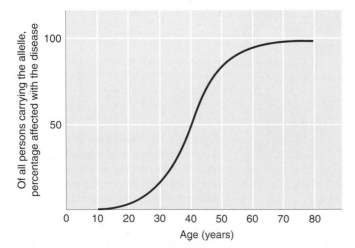

Figure 2-17 The age of onset of Huntington disease. The graph shows that people carrying the allele generally do not express the disease until after childbearing age.

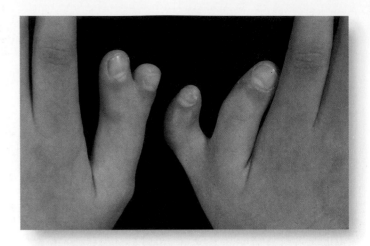

(a)

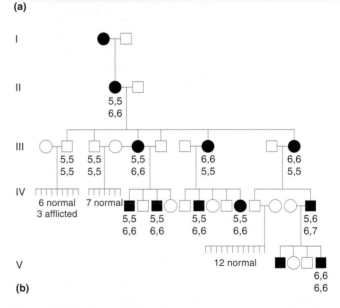

(b)

Figure 2-18 A rare dominant phenotype of the human hand. (a) Polydactyly, a dominant phenotype characterized by extra fingers, toes, or both, determined by an allele P. The numbers in the accompanying pedigree (b) give the number of fingers in the upper lines and the number of toes in the lower. (Note the variation in expression of the P allele.) [Part a, photograph © Biophoto Associates/Science Source.]

PEDIGREE ANALYSIS OF AUTOSOMAL POLYMOR-PHISMS Recall from Chapter 1 that a polymorphism is the coexistence of two or more common phenotypes of a character in a population. The alternative phenotypes of polymorphisms are often inherited as alleles of a single autosomal gene in the standard Mendelian manner. In humans, there are many examples; consider, for example, the dimorphisms brown versus blue eyes, dark versus blonde hair, chin dimples versus none, widow's peak versus none, and attached versus free earlobes.

The interpretation of pedigrees for polymorphisms is somewhat different from that of rare disorders because, by definition, the morphs are common. Let's look

Figure 2-19 Piebald spotting, a rare dominant human phenotype.
Although the phenotype is encountered sporadically in all races, the patterns show up best in those with dark skin. (a) The photographs show front and back views of affected persons IV-1, IV-3, III-5, III-8, and III-9 from (b) the family pedigree. Notice the variation between family members in expression of the

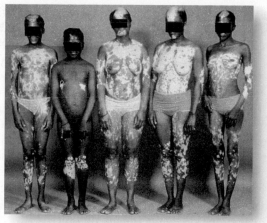
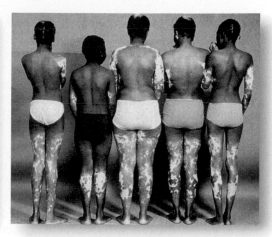
(a)

piebald gene. It is believed that the patterns are caused by the dominant allele interfering with the migration of melanocytes (melanin-producing cells) from the dorsal to the ventral surface in the course of development. The white forehead blaze is particularly characteristic and is often accompanied by a white forelock in the hair.

Piebaldism is not a form of albinism; the cells in the light patches have the genetic potential to make melanin, but because they are not melanocytes, they are not developmentally programmed to do so. In true albinism, the cells lack the potential to make melanin. (Piebaldism is caused by mutations in c-*kit*, a type of gene called a *protooncogene*, to be discussed in Chapter 17.) [Parts a and b from I. Winship, K. Young, R. Martell, R. Ramesar, D. Curtis, and P. Beighton, "Piebaldism: An Autonomous Autosomal Dominant Entity," *Clinical Genetics* 39, 1991, 330.]

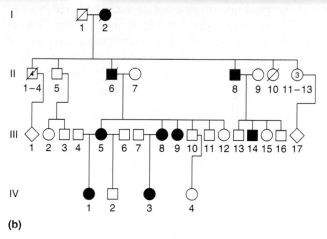
(b)

at a pedigree for an interesting human dimorphism. Most human populations are dimorphic for the ability to taste the chemical phenylthiocarbamide (PTC); that is, people can either detect it as a foul, bitter taste, or—to the great surprise and disbelief of tasters—cannot taste it at all. From the pedigree in Figure 2-20, we can see that two tasters sometimes produce nontaster children, which makes it clear that the allele that confers the ability to taste is dominant and that the allele for nontasting is recessive. Notice that almost all people who marry into this family carry the recessive allele either in heterozygous or in homozygous condition. Such a pedigree thus differs from those of rare recessive disorders, for which it is conventional to assume that all who marry into a family are homozygous normal. Because both PTC alleles are common, it is not surprising that all but one of the family members in this pedigree married persons with at least one copy of the recessive allele.

Polymorphism is an interesting genetic phenomenon. Population geneticists have been surprised at how much polymorphism there is in natural populations of plants and animals generally. Furthermore, even though the genetics of polymorphisms is straightforward, there are very few polymorphisms for which there is satisfac-

tory explanation for the coexistence of the morphs. But polymorphism is rampant at every level of genetic analysis, even at the DNA level; indeed, polymorphisms observed at the DNA level have been invaluable as land-

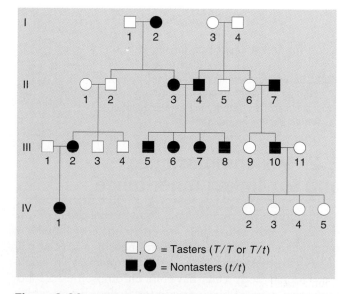

Figure 2-20 Pedigree for the ability to taste the chemical PTC.

marks to help geneticists find their way around the chromosomes of complex organisms. The population and evolutionary genetics of polymorphisms is considered in Chapters 19 and 21.

One useful type of molecular chromosomal landmark, or marker, is a restriction fragment length polymorphism (RFLP). In Chapter 1, we learned that restriction enzymes are bacterial enzymes that cut DNA at specific base sequences in the genome. The target sequences have no biological significance in organisms other than bacteria—they occur purely by chance. Although the target sites generally occur quite consistently at specific locations, sometimes, on any one chromosome, a specific target site is missing or there is an extra site. If the presence or absence of such a restriction site flanks the sequence hybridized by a probe, then a Southern hybridization will reveal a length polymorphism, or RFLP. Consider this simple example in which one chromosome of one parent contains an extra site not found in the other chromosomes of that type in that cross:

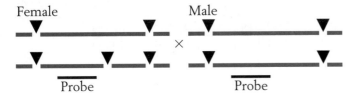

The Southern hybridizations will show two bands in the female and only one in the male. The "heterozygous" fragments will be inherited in exactly the same way as a gene. The preceding cross could be written as follows:

$$\text{long/short} \times \text{long/long}$$

and the progeny will be

$$\tfrac{1}{2} \text{ long/short}$$
$$\tfrac{1}{2} \text{ long/long}$$

according to the law of equal segregation.

> **MESSAGE** Populations of plants and animals (including humans) are highly polymorphic. Contrasting morphs are often determined by alleles inherited in a simple Mendelian manner.

2.2 Sex chromosomes and sex-linked inheritance

Most animals and many plants show sexual dimorphism; in other words, individuals are either male or female. In most of these cases, sex is determined by a special pair of **sex chromosomes.** Let's look at humans as an example. Human body cells have 46 chromosomes: 22 homologous pairs of autosomes plus 2 sex chromosomes.

In females, there is a pair of identical sex chromosomes called the **X chromosomes.** In males, there is a nonidentical pair, consisting of one X and one Y. The **Y chromosome** is considerably shorter than the X. At meiosis in females, the two X chromosomes pair and segregate like autosomes so that each egg receives one X chromosome. Hence with regard to sex chromosomes the gametes are of only one type, and the female is said to be the **homogametic sex.** At meiosis in males, the X and the Y chromosomes pair over a short region, which ensures that the X and Y separate to opposite ends of the meiotic cell, creating two types of sperm, half with an X and the other half with a Y. Therefore the male is called the **heterogametic sex.**

The inheritance patterns of genes on the sex chromosomes are different from those of autosomal genes. These sex chromosome inheritance patterns were first investigated in the fruit fly *Drosophila melanogaster.* This insect has been one of the most important research organisms in genetics; its short, simple life cycle contributes to its usefulness in this regard. Fruit flies have three pairs of autosomes plus a pair of sex chromosomes again referred to as X and Y. As in mammals, *Drosophila* females have the constitution XX and males are XY. However, the mechanism of sex determination in *Drosophila* differs from that in mammals. In *Drosophila,* the *number of X chromosomes* determines sex: two X's result in a female and one X results in a male. In mammals, the *presence of the Y* determines maleness and the absence of a Y determines femaleness. This difference is demonstrated by the sexes of the abnormal chromosome types XXY and XO, as shown in Table 2-3. We must postpone a full discussion of this topic until Chapter 15, but for the present discussion it is important to note that despite this somewhat different basis for sex determination, the inheritance of genes on the sex chromosomes shows remarkably similar patterns in *Drosophila* and mammals.

Vascular plants show a variety of sexual arrangements. **Dioecious** species are those showing animal-like sexual dimorphism, with female plants bearing flowers containing only ovaries and male plants bearing flowers containing only anthers (Figure 2-21). Some, but not all, dioecious plants have a nonidentical pair of chromosomes associated with (and almost certainly determining) the

Table 2-3 Chromosomal Determination of Sex in *Drosophila* and Humans

Species	Sex chromosomes			
	XX	XY	XXY	XO
Drosophila	♀	♂	♀	♂
Human	♀	♂	♂	♀

Note: O indicates absence of a chromosome.

(a)

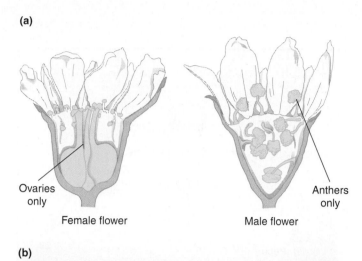

Ovaries
only

Female flower

Anthers
only

Male flower

(b)

♀ plants   ♂ plants

Figure 2-21 Two dioecious plant species. (a) *Osmaronia dioica*; (b) *Aruncus dioicus*. [Part a, Leslie Bohm; part b, Anthony Griffiths.]

sex of the plant. Of the species with nonidentical sex chromosomes, a large proportion have an XY system. For example, the dioecious plant *Melandrium album* has 22 chromosomes per cell: 20 autosomes plus 2 sex chromosomes, with XX females and XY males. Other dioecious plants have no visibly different pair of chromosomes; they may still have sex chromosomes, but not visibly distinguishable types.

Sex-linked patterns of inheritance

Cytogeneticists divide the X and Y chromosomes into homologous and differential regions. Again, let's use humans as an example (Figure 2-22). The *homologous* regions contain DNA sequences that are substantially similar on both sex chromosomes. The *differential* regions contain genes that have no counterparts on the other sex chromosome. Hence in males these genes in the differential regions are thus said to be **hemizygous** ("half zy-

gous"). The X chromosome contains many hundreds of genes; most of these genes are not involved in sexual function, and most have no counterparts on the Y. The Y chromosome contains only a few dozen genes. Some of these genes have counterparts on the X, but some do not. Most of the latter type are involved in male sexual function. One of these genes, SRY, determines maleness itself. Several other genes are specific for sperm production in males.

Genes in the differential region of the X show an inheritance pattern called **X linkage;** those in the differential region of the Y show **Y linkage.** In general, genes on the differential regions are said to show **sex linkage.** A gene that is sex-linked shows patterns of inheritance related to sex. This pattern contrasts with the inheritance patterns of genes on the autosomes, which are not connected to sex. In autosomal inheritance, male and female progeny show inherited phenotypes in exactly the same proportions, as typified by Mendel's results (for example, both sexes of an F_2 might show a 3:1 ratio). In contrast, crosses performed to track the inheritance of genes on the sex chromosomes often produce male and female progeny that show different phenotypic ratios. In fact, for studies of genes of unknown chromosomal location, this pattern is a diagnostic of location on the sex chromosomes.

The human X and Y chromosomes have two homologous regions, one at each end (see Figure 2-22). In being homologous these regions are autosomal-like, so they are called *pseudoautosomal regions 1* and *2*. One or both of these regions pair during meiosis, and undergo crossing over (see Chapter 3 for a discussion of crossing over). In this way, the X and the Y can act as a pair and segregate into equal numbers of sperm.

X-linked inheritance

For our first example of X linkage we examine eye color in *Drosophila*. The wild-type eye color of *Drosophila* is dull red, but pure lines with white eyes are available

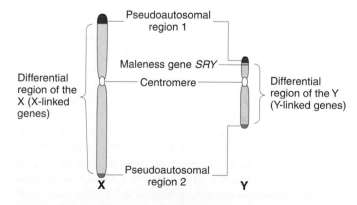

Figure 2-22 Differential and pairing regions of human sex chromosomes. The regions were located by observing where the chromosomes paired up in meiosis and where they did not.

MODEL ORGANISM   *Drosophila*

Drosophila melanogaster was one of the first model organisms to be used in genetics. It is readily available from ripe fruit, has a short life cycle, and is simple to culture and cross. Sex is determined by X and Y sex chromosomes (XX = female, XY = male), and males and females are easily distinguished. Mutant phenotypes regularly arise in lab populations, and their frequency can be increased by treatment with radiation or chemicals. It is a diploid organism, with four pairs of homologous chromosomes ($2n = 8$). In salivary glands and certain other tissues, multiple rounds of DNA replication without chromosomal division result in "giant chromosomes," each with a unique banding pattern that provides geneticists with landmarks for the study of chromosome mapping and rearrangement. There are many species and races of *Drosophila*, which have been important raw material for the study of evolution.

"Time flies like an arrow; fruit flies like a banana." (Groucho Marx)

Drosophila melanogaster, the common fruit fly. [SPL/Photo Researchers, Inc.]

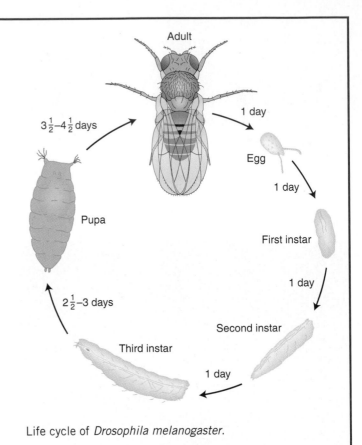

Life cycle of *Drosophila melanogaster*.

(Figure 2-23). This phenotypic difference is determined by two alleles of a gene located on the differential region of the X chromosome. In *Drosophila* and many other organisms, the convention is to name a gene after the first mutant allele found, and then designate the wild-type allele with the mutant symbol plus a superscript cross. Hence the mutant allele in the present case is w for white eyes (the lowercase shows it is recessive), and the corresponding wild-type allele is w^+. When white-eyed males are crossed with red-eyed females, all the F_1 progeny have red eyes, showing that the allele for white is recessive. Crossing these red-eyed F_1 males and females produces a 3:1 F_2 ratio of red-eyed to white-eyed flies, but all the white-eyed flies are males. This inheritance pattern is explained by the inheritance of a gene on the differential region of the X chromosome, with a dominant wild-type allele for redness, and a recessive allele for whiteness. In other words, this is a case of X linkage. The genotypes are shown in Figure 2-24. The reciprocal cross gives a different result. A reciprocal cross between white-eyed females and red-eyed males gives an F_1 in which all the females are red-eyed, but all the males

are white-eyed. In this case, every female inherited the dominant w^+ allele from the father's X chromosome, whereas every male inherited a single X chromosome from the mother bearing the recessive w allele. The F_2

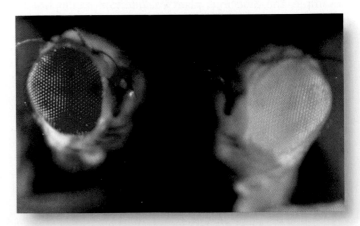

Figure 2-23 Red-eyed and white-eyed *Drosophila*. [Carolina Biological Supply.]

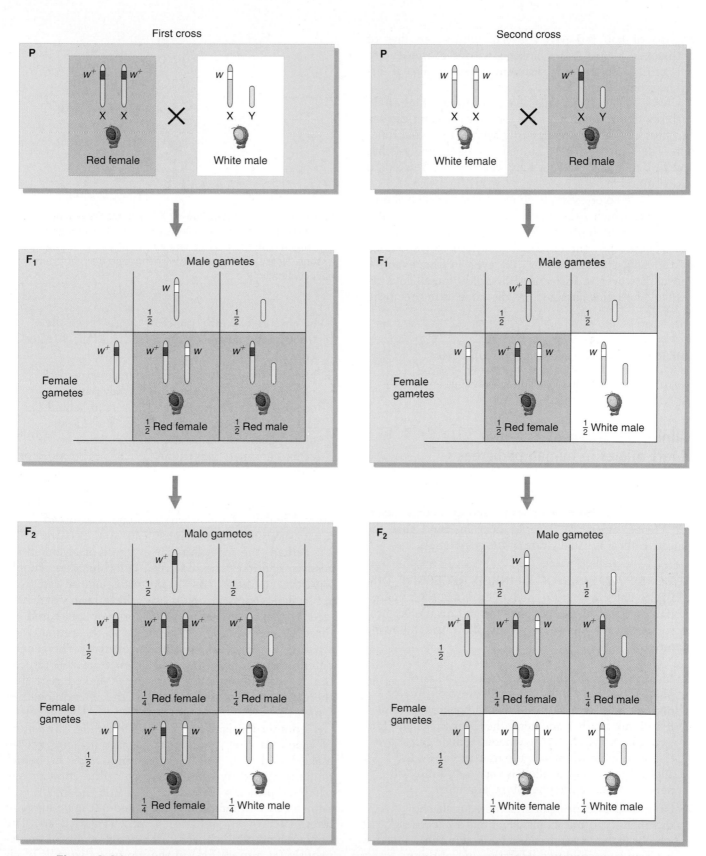

Figure 2-24 Reciprocal crosses between red-eyed (red) and white-eyed (white) *Drosophila* give different results. The alleles are X-linked, and the inheritance of the X chromosome explains the phenotypic ratios observed, which are different from those of autosomal genes. (In *Drosophila* and many other experimental systems, a superscript plus sign is used to designate the normal, or wild-type, allele. Here w^+ = red and w = white.)

consists of half red-eyed and half white-eyed flies of both sexes. Hence in sex linkage, we see examples not only of different ratios in different sexes, but also of differences between reciprocal crosses.

Note that in *Drosophila*, eye color has nothing to do with sex determination, so we see that genes on the sex chromosomes are not necessarily related to sexual function. The same is true in humans: pedigree analysis has revealed many X-linked genes, yet few are related to sexual function.

The eye color example concerned a recessive abnormal allele, which must have originally arisen by mutation. Dominant mutant alleles of genes on the X also arise. These show the inheritance pattern corresponding to the wild-type allele for red eyes in the example above. However, in such cases the wild-type allele is recessive. The ratios obtained are the same as in the above example.

> **MESSAGE** Sex-linked inheritance regularly shows different phenotypic ratios in the two sexes of progeny, as well as different ratios in reciprocal crosses.

X-linked inheritance of rare alleles in human pedigrees

As in the analysis of autosomal genes, human pedigrees of genes on the X chromosome are generally drawn up to follow the inheritance of some kind of medical disorder. Hence, again we have to bear in mind that the causative allele is usually rare in the population.

PEDIGREE ANALYSIS OF X-LINKED RECESSIVE DISORDERS Let's look at the pedigrees of disorders caused by rare recessive alleles of genes located on the X chromosome. Such pedigrees typically show the following features:

1. Many more males than females show the rare phenotype under study. This is because of the product law: a female will show the phenotype only if both her mother *and* her father bear the allele (for example, $X^A X^a \times X^a Y$), whereas a male can show the phenotype when *only* the mother carries the allele. If the recessive allele is very rare, almost all persons showing the phenotype are male.

2. None of the offspring of an affected male show the phenotype, but all his daughters are "carriers," who bear the recessive allele masked in the heterozygous condition. Half the sons of these carrier daughters show the phenotype (Figure 2-25). (Note that in the case of phenotypes caused by *common* recessive X-linked alleles, this pattern might

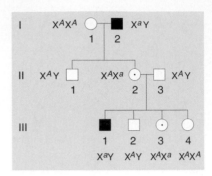

Figure 2-25 Pedigree showing X-linked recessive alleles expressed in males. These alleles are carried unexpressed by daughters in the next generation, to be expressed again in sons. Note that III-3 and III-4 cannot be distinguished phenotypically.

be obscured by inheritance of the recessive allele from a heterozygous mother as well as the affected father.)

3. None of the sons of an affected male show the phenotype under study, nor will they pass the condition to their offspring. The reason behind this lack of male-to-male transmission is that a son obtains his Y chromosome from his father, so he cannot normally inherit the father's X chromosome, too.

In the pedigree analysis of rare X-linked recessives, a normal female of unknown genotype is assumed to be homozygous unless there is evidence to the contrary.

Perhaps the most familiar example of X-linked recessive inheritance is red-green colorblindness. People with this condition are unable to distinguish red from green. The genes for color vision have been characterized at the molecular level. Color vision is based on three different kinds of cone cells in the retina, each sensitive to red, green, or blue wavelengths. The genetic determinants for the red and green cone cells are on the X chromosome. As with any X-linked recessive, there are many more males with the phenotype than females.

Another familiar example is *hemophilia*, the failure of blood to clot. Many proteins act in sequence to make blood clot. The most common type of hemophilia is caused by the absence or malfunction of one of these proteins, called *factor VIII*. The most well known cases of hemophilia are found in the pedigree of interrelated royal families in Europe (Figure 2-26). The original hemophilia allele in the pedigree arose spontaneously (as a mutation) in the reproductive cells of either Queen Victoria's parents or Queen Victoria herself. The son of the last czar of Russia, Alexis, inherited the allele ultimately from Queen Victoria, who was the

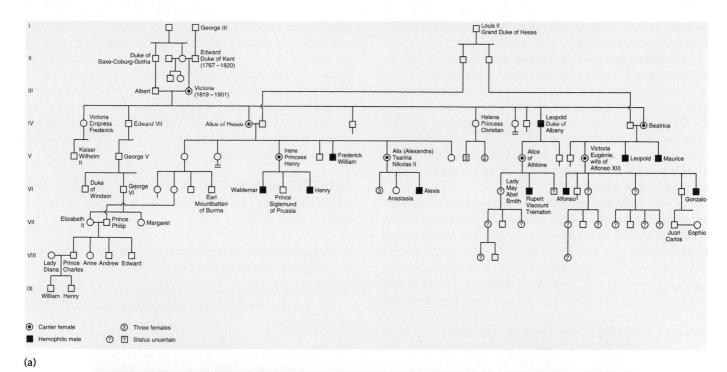

(a)

(b)

Figure 2-26 The inheritance of the X-linked recessive condition hemophilia in the royal families of Europe. A recessive allele causing hemophilia (failure of blood clotting) arose in the reproductive cells of Queen Victoria, or one of her parents, through mutation. This hemophilia allele spread into other royal families by intermarriage. (a) This partial pedigree shows affected males and carrier females (heterozygotes). Most spouses marrying into the families have been omitted from the pedigree for simplicity. Can you deduce the likelihood of the present British royal family's harboring the recessive allele? (b) A painting showing Queen Victoria surrounded by her numerous descendants. [Part a, after C. Stern, *Principles of Human Genetics*, 3d ed. Copyright 1973 by W. H. Freeman and Company; part b, Royal Collection, St. James's Palace. Copyright Her Majesty Queen Elizabeth II.]

grandmother of his mother Alexandra. Nowadays, hemophilia can be treated medically, but it was formerly a potentially fatal condition. It is interesting to note that in the Jewish Talmud there are rules about exemp-

tions to male circumcision that show clearly that the mode of transmission of the disease through unaffected carrier females was well understood in ancient times. For example, one exemption was for the sons of women whose sisters' sons had bled profusely when they were circumcised.

Duchenne muscular dystrophy is a fatal X-linked recessive disease. The phenotype is a wasting and atrophy of muscles. Generally the onset is before the age of 6, with confinement to a wheelchair by 12, and death by 20. The gene for Duchenne muscular dystrophy has now been isolated and shown to encode the muscle protein dystrophin. This discovery holds out hope for a better understanding of the physiology of this condition and, ultimately, a therapy.

A rare X-linked recessive phenotype that is interesting from the point of view of sexual differentiation is a condition called *testicular feminization syndrome*, which has a frequency of about 1 in 65,000 male births. People afflicted with this syndrome are chromosomally males, having 44 autosomes plus an X and a Y, but they develop as females (Figure 2-27). They have female external genitalia, a blind vagina, and no uterus. Testes may be present either in the labia or in the abdomen. Although

many such persons marry, they are sterile. The condition is not reversed by treatment with the male hormone androgen, so it is sometimes called *androgen insensitivity syndrome*. The reason for the insensitivity is that a mutation in the androgen receptor gene causes the receptor to malfunction, so the male hormone can have no effect on the target organs that contribute to maleness. In humans, femaleness results when the male-determining system is not functional.

PEDIGREE ANALYSIS OF X-LINKED DOMINANT DISORDERS These disorders have the following characteristics (Figure 2-28):

1. Affected males pass the condition to all their daughters but to none of their sons

2. Affected heterozygous females married to unaffected males pass the condition to half their sons and daughters.

There are few examples of X-linked dominant phenotypes in humans. One example is *hypophosphatemia*, a type of vitamin D–resistant rickets.

Y-linked inheritance

Only males inherit genes on the differential region of the human Y chromosome, with fathers transmitting the genes to their sons. The gene that plays a primary role in maleness is the **SRY gene,** sometimes called the *testis-determining factor*. The SRY gene has been located and mapped on the differential region of the Y chromosome (see Chapter 18). Hence maleness itself is Y-linked, and shows the expected pattern of exclusively male-to-male transmission. Some cases of male sterility have been shown to be caused by deletions of Y chro-

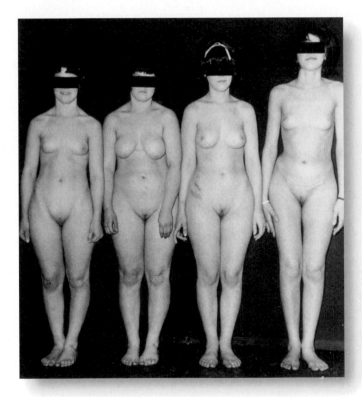

Figure 2-27 Four siblings with testicular feminization syndrome (congenital insensitivity to androgens). All four subjects in this photograph have 44 autosomes plus an X and a Y chromosome, but they have inherited the recessive X-linked allele conferring insensitivity to androgens (male hormones). One of their sisters (not shown), who was genetically XX, was a carrier and bore a child who also showed testicular feminization syndrome. [Leonard Pinsky, McGill University.]

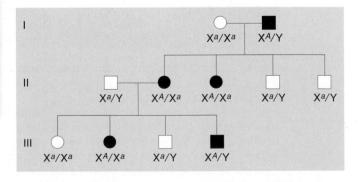

Figure 2-28 Pedigree of an X-linked dominant condition. All the daughters of a male expressing an X-linked dominant phenotype will show the phenotype. Females heterozygous for an X-linked dominant allele will pass the condition on to half their sons and daughters.

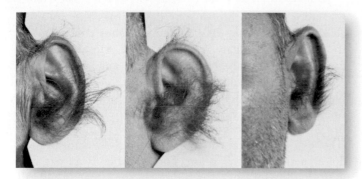

Figure 2-29 Hairy ear rims. This phenotype has been proposed to be caused by an allele of a Y-linked gene. [From C. Stern, W. R. Centerwall, and S. S. Sarkar, *The American Journal of Human Genetics* 16, 1964, 467. By permission of Grune & Stratton, Inc.]

mosome regions containing sperm-promoting genes. Male sterility is not heritable, but interestingly the fathers of these men have normal Y chromosomes, showing that the deletions are new.

There have been no convincing cases of nonsexual phenotypic variants associated with the Y. Hairy ear rims (Figure 2-29) has been proposed as a possibility. The phenotype is extremely rare among the populations of most countries but more common among the populations of India. In some (but not all) families hairy ear rims are transmitted exclusively from father to son.

> **MESSAGE** Inheritance patterns with an unequal representation of phenotypes in males and females can locate the genes concerned to one of the sex chromosomes.

2.3 Cytoplasmic inheritance

Mitochondria and chloroplasts are specialized organelles located in the cytoplasm. They contain small circular chromosomes that carry a defined subset of the total cell genome. Mitochondrial genes are concerned with the mitochondrion's task of energy production, whereas chloroplast genes are needed for the chloroplast to carry out its function of photosynthesis. However, neither organelle is genetically independent, because each relies to some extent on nuclear genes for function. Why some of the necessary genes are in the organelles themselves while others are in the nucleus is still something of a mystery, which we will not address here.

Another peculiarity of organelle genes is the large number of copies present in the cell. Each organelle is present in many copies per cell, and furthermore each organelle contains many copies of the chromosome. Hence each cell can contain hundreds or thousands of organelle chromosomes. For the time being we shall

assume that all copies within a cell are identical, but we will have to relax this assumption later.

Organelle genes show their own special mode of inheritance called uniparental inheritance; that is, progeny inherit organelle genes exclusively from one parent. In most cases, that parent is the mother: **maternal inheritance.** Why only the mother? The answer lies in the fact that the organelle chromosomes are located in the cytoplasm rather than the nucleus and the fact that male and female gametes do not contribute cytoplasm equally to the zygote. In the case of nuclear genes, we have seen that both parents do contribute equally to the zygote. However, the egg contributes the bulk of the cytoplasm and the sperm essentially none. Therefore, because organelles reside in the cytoplasm, the female parent contributes the organelles along with the cytoplasm and essentially none of the organelle DNA in the zygote is from the male parent.

Are there organelle mutations we can use to observe patterns of inheritance? Some phenotypic variants are caused by a mutant allele of an organelle gene. Once again we will assume temporarily that the mutant allele is present in all copies, a situation that is often found. In a cross, the variant phenotype will be transmitted to progeny if the variant used is the female parent, but not if it is the male parent. Hence generally cytoplasmic inheritance shows the following pattern:

Mutant ♀ × wild-type ♂ ⟶ progeny all mutant
Wild-type ♀ × mutant ♂ ⟶ progeny all wild type

Maternal inheritance can be clearly demonstrated in certain mutants of haploid fungi. (Although we have not specifically covered haploid cycles yet, we do not need to worry for this example because it does not concern the nuclear genome.) For example in the fungus *Neurospora*, a mutant called *poky* has a slow-growth phenotype. Neurospora can be crossed in such a way that one parent acts as the maternal parent, contributing the cytoplasm. The results of reciprocal crosses suggest that the mutant gene or genes reside in the mitochondria (fungi have no chloroplasts):

Poky ♀ × wild-type ♂ ⟶ progeny all poky
Wild-type ♀ × poky ♂ ⟶ progeny all wild type

The poky mutation is now known to be in mitochondrial DNA.

> **MESSAGE** Variant phenotypes caused by mutations in cytoplasmic organelle DNA are generally inherited maternally.

Since a mutation such as poky must have arisen originally in one mitochondrial "chromosome," how can it come to occupy all the mitochondrial chromosomes in a poky mutant? The process, which is quite common among organelle mutations, is not well understood. In some cases it appears to be a series of random chances. In other cases, the mutant chromosome seems to possess some competitive advantage in replication.

In some cases, cells contain mixtures of mutant and normal organelles. These cells are called *cytohets* or *heteroplasmons*. In these mixtures, a type of **cytoplasmic segregation** can be detected, in which the two types apportion themselves into different daughter cells. The process most likely stems from chance partitioning during cell division. Plants provide a good example. Many cases of white leaves are caused by mutations in chloroplast genes that control the production and deposition of the green pigment chlorophyll. Since chlorophyll is necessary for the plant to live, this type of mutation is lethal, and white-leaved plants cannot be obtained for experimental crosses. However, some plants are variegated, bearing both green and white patches, and these plants are viable. Thus variegated plants provide a way of demonstrating cytoplasmic segregation.

Figure 2-30 shows a commonly observed variegated leaf and branch phenotype that demonstrates inheritance of a mutant allele of chloroplast gene. The mutant allele causes chloroplasts to be white; in turn the color of the chloroplasts determines the color of cells and hence the color of the branches composed of those cells. Variegated branches are mosaics of all-green and all-white cells. Flowers can develop on green, white, or variegated branches, and the chloroplast genes of the flower's cells will be those of the branch on which it grows. Hence, when crossed (Figure 2-31), it is the maternal gamete within the flower (the egg cell) that determines the leaf and branch color of the progeny plant. For example, if an egg cell is from a flower on a green branch, the progeny will all be green, regardless of the origin of the pollen. A white branch will have white chloroplasts, and the resulting progeny plants will be white. (Because of lethality, white descendants would not live beyond the seedling stage.)

The variegated zygotes (bottom of figure) demonstrate cytoplasmic segregation. These variegated progeny come from eggs that are cytoplasmic mixtures of two chloroplast types. Interestingly, when such a zygote divides, the white and green chloroplasts often segregate; that is, they sort themselves into separate cells, yielding the distinct green and white sectors that cause the variegation in the branches. Here then is a direct demonstration of cytoplasmic segregation.

> **MESSAGE** Organelle populations that contain mixtures of two genetically distinct chromosomes often show segregation of the two types into the daughter cells at cell division. This is called *cytoplasmic segregation.*

Are there cytoplasmic mutations in humans? Some human pedigrees show transmission of a specific rare phenotype only through females and never through males. This pattern strongly suggests cytoplasmic inheritance, and would point to the phenotype being caused by a mutation in mitochondrial DNA. The disease MERRF (myoclonic epilepsy and ragged red fiber) is such a phenotype, resulting from a single base change in mitochondrial DNA. It is a muscle disease, but the symptoms also include eye and hearing disorders. Another example is Kearns-Sayre syndrome, a constellation of symptoms affecting eyes, heart, muscles and brain that is caused by loss of part of the mitochondrial DNA. In some of these cases, the cells of a sufferer contain mixtures of normal and mutant chromosomes, and the proportions of each passed on to progeny can vary as a result of cytoplasmic segregation. The proportions in one individual can also vary in different tissues or over time. It has been proposed that the accumulation of certain types of mitochondrial mutations over time is one possible cause of aging.

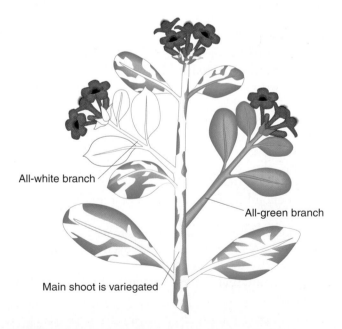

All-white branch

All-green branch

Main shoot is variegated

Figure 2-30 Leaf variegation in *Mirabilis jalapa*, the four-o'clock plant. Flowers can form on any branch (variegated, green, or white), and these flowers can be used in crosses.

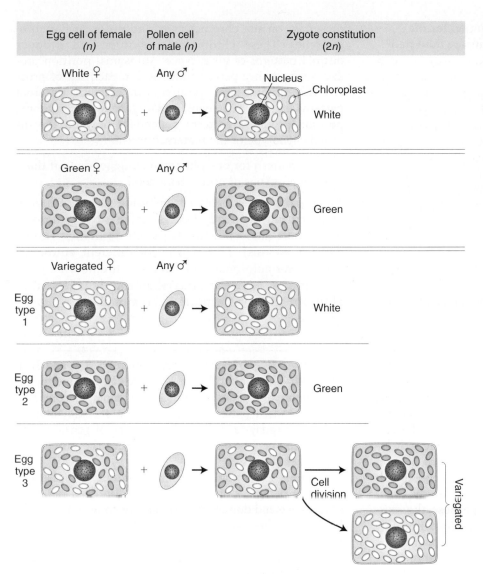

Egg cell of female (n)	Pollen cell of male (n)	Zygote constitution (2n)

Figure 2-31 A model explaining the results of the *Mirabilis jalapa* crosses in regard to autonomous chloroplast inheritance. The large, dark spheres are nuclei. The smaller bodies are chloroplasts, either green or white. Each egg cell is assumed to contain many chloroplasts, and each pollen cell is assumed to contain no chloroplasts. The first two crosses exhibit strict maternal inheritance. If, however, the maternal branch is variegated, three types of zygotes can result, depending on whether the egg cell contains only white, only green, or both green and white chloroplasts. In the last case, the resulting zygote can produce both green and white tissue, so a variegated plant results.

KEY QUESTIONS REVISITED

- **How is it possible to tell if a phenotypic variant has a genetic basis?**

In this chapter we have been dealing with discontinuous variants, those that form distinct and nonoverlapping subdivisions of a character, such as the division of the character eye color into the phenotypes red eyes and white eyes. The key to answering the question is that such hereditary variant phenotypes (in this case red and white) are transmitted with more or less constant expression (for example, the same amount of redness) down through the generations. This in itself strongly suggests a hereditary basis. If the phenotypes are observed in consistent proportions (see next question) the hereditary hypothesis is strengthened.

In reality we must be careful with such inference. Some variants that are *non*hereditary might *appear* to show some degree of inheritance. For example a plant might show a variant leaf color such as yellow because it has been infected with a virus. Some of its progeny might also show the same symptoms because the virus is still present in the greenhouse. However, such situations would not show the specific proportions that are shown by hereditary transmission.

- **Are phenotypic variants inherited in consistent patterns through the generations?**

We have seen that there are indeed consistent patterns. Not only do the hereditary phenotypes keep appearing, but they appear in consistent ratios such as $3:1$, $1:1$, $1:2:1$, $9:3:3:1$, and so on. These we have identified as standard patterns of inheritance; they can be observed in all organisms.

• **At the gene level, what is the explanation for the patterns by which phenotypic variants are inherited?**

We have seen that many cases of discontinuous variation are caused by variant alleles of a single gene: one allele causes one phenotype, and another allele of that same gene causes a different phenotype. The inheritance of such a pair of variants is governed by rules laid down by Mendel: genes come in pairs; the pairs segregate equally during gamete formation; gametes therefore carry one member of each pair; and zygotes are formed by random fusion of male and female gametes. These simple rules govern the production of the standard ratios. Which ratio is observed depends on which genotypes are mated.

• **Is the pattern of inheritance influenced by the location of the relevant gene or genes in the genome?**

We have seen that this indeed is the case. Three broad genomic locations can be delineated: on autosomal chromosomes, on sex chromosomes, and on organelle chromosomes. These chromosomal locations all produce distinct patterns of inheritance. Autosomal position produces inheritance patterns identical in each sex of progeny, sex chromosome position can produce different patterns in the two sexes, and organelle chromosome position produces patterns dependent only on the organelle genotype of the maternal parent.

• **Is the pattern for one phenotype independent of that for phenotypes of other characters?**

Allele pairs on different chromosome pairs are indeed inherited independently because different chromosome pairs assort independently at gamete formation (essentially at meiosis). This is true for different autosomal pairs, or for autosomal pairs and sex chromosome pairs. Furthermore, organelle gene inheritance is independent of that of genes on the nuclear chromosomes.

SUMMARY

Within a species, variant phenotypes are common. Discrete, discontinuous variants for a specific character are often found to be caused by the alleles of a single gene, with one allele designating one phenotype and another allele the other phenotype. Such discontinuous hereditary phenotypes are passed down through the generations in standard patterns of inheritance. In this connection "pattern" means precise, specific ratios of individuals with each phenotype. Indeed it was these patterns of inheritance that led Gregor Mendel to propose the existence of genes. Mendel's hypothesis contained not only the notion that genes account for discrete phenotypic difference, but also a mechanism of inheritance of these discrete differences. The essence of Mendel's thesis was that genes are in pairs; these segregate equally into the gametes, which come to contain one of each pair (Mendel's first law); and gene pairs on different chromosome pairs assort independently (a modern statement of Mendel's second law).

For any one gene, alleles may be classified as dominant or recessive, corresponding to the dominant and recessive phenotypes. The term *dominant* is given to the allele expressed in any given heterozygote. For any pair of dominant and recessive alleles, there are three genotypes: homozygous dominant (A/A), heterozygous (A/a), and homozygous recessive (a/a).

Inheritance patterns are influenced by the type of chromosome on which the gene is located. Most genes reside on the autosomal chromosomes. Here the transmission of genes follows Mendel's ideas exactly. A commonly encountered example of autosomal inheritance is the so-called monohybrid cross, $A/a \times A/a$, giving progeny that are 1/4 A/A, 1/2 A/a, and 1/4 a/a. This is a result of the equal segregation of A and a in each parent. A dihybrid cross A/a ; $B/b \times A/a$; B/b gives progeny that are 9/16 $A/-$; $B/-$, 3/16 $A/-$; b/b, 3/16 a/a ; $B/-$, and 1/16 a/a ; b/b. This dihybrid ratio is a result of two independent monohybrid ratios, according to Mendel's second law. These standard patterns of inheritance are still used by geneticists today to deduce the presence of genes, and to predict the genotypes and phenotypes of the progeny of experimental crosses.

Simple autosomal inheritance patterns are also observed in pedigrees of human genetic disorders, and cases of recessive and dominant disorders are common. Because human families are small, precise Mendelian ratios are rarely observed in any one mating.

A small proportion of genes reside on the X chromosome (they are said to be *X-linked*), and they show an inheritance pattern that is often different in the two sexes of the progeny. These differences exist essentially because a male has a single X chromosome, derived exclusively from his mother, whereas the XX pair of a female is obtained from both her father and mother. Pedigrees of some recessive and dominant human disorders show X-linked inheritance.

Any of the ratios observed in autosomal or X-linked inheritance can be tested against expectations using the chi-square test. This test tells us the probability of obtaining results this far off expectations purely on the basis of chance. P values < 5 percent lead to rejection of the hypothesis that generated the expected ratio.

An even smaller proportion of genes is found on the mitochondrial chromosome (animals and plants) or on the chloroplast chromosome (plants only). These chromosomes are passed on only through the cytoplasm of the egg. Thus variants caused by mutations on the genes of these organelles are inherited only through the mother, an inheritance pattern quite different from the genes of the nuclear chromosomes.

KEY TERMS

allele (p. 33)

autosomal chromosomes (p. 29)

character difference (p. 31)

chi-square test (p. 40)

cross (p. 29)

cytoplasmic segregation (p. 56)

dihybrid (p. 36)

dihybrid cross (p. 36)

dioecious (p. 48)

dominant (p. 31)

equal segregation (p. 34)

first filial generation (F_1) (p. 31)

gene pair (p. 33)

genotype (p. 34)

hemizygous (p. 49)

heterogametic sex (p. 48)

heterozygote (p. 34)

homogametic sex (p. 48)

homozygote (p. 34)

homozygous dominant (p. 34)

homozygous recessive (p. 34)

hybrid (p. 34)

maternal inheritance (p. 55)

meiosis (p. 35)

Mendel's first law (p. 34)

Mendel's second law (p. 37)

monohybrid (p. 36)

monohybrid cross (p. 36)

parental generation (P) (p. 31)

pedigree analysis (p. 42)

phenotype (p. 31)

product rule (p. 39)

propositus (p. 42)

pure line (p. 30)

recessive (p. 31)

reciprocal cross (p. 31)

self (p. 29)

sex chromosome (p. 48)

sex linkage (p. 49)

SRY gene (p. 54)

sum rule (p. 39)

testcross (p. 39)

tester (p. 39)

variant (p. 28)

X chromosome (p. 48)

X linkage (p. 49)

Y chromosome (p. 48)

Y linkage (p. 49)

zygote (p. 33)

SOLVED PROBLEMS

This section in each chapter contains a few solved problems that show how to approach the problem sets that follow. The purpose of the problem sets is to challenge your understanding of the genetic principles learned in the chapter. The best way to demonstrate an understanding of a subject is to be able to use that knowledge in a real or simulated situation. Be forewarned that there is no machinelike way of solving these problems. The three main resources at your disposal are the genetic principles just learned, common sense, and trial and error.

Here is some general advice before beginning. First, it is absolutely essential to read and understand all of the question. Find out exactly what facts are provided, what assumptions have to be made, what clues are given in the question, and what inferences can be made from the available information. Second, be methodical. Staring at the question rarely helps. Restate the information in the question in your own way, preferably using a diagrammatic representation or flowchart to help you think out the problem. Good luck.

1. Crosses were made between two pure lines of rabbits that we can call A and B. A male from line A was mated with a female from line B, and the F_1 rabbits were subsequently intercrossed to produce an F_2. It was discovered that $\frac{3}{4}$ of the F_2 animals had white subcutaneous fat and had $\frac{1}{4}$ yellow subcutaneous fat. Later, the F_1 was examined and was found to have white fat. Several years later, an attempt was made to repeat the experiment by using the same male from line A and the same female from line B. This time, the F_1 and all the F_2 (22 animals) had white fat. The only difference between the original experiment and the repeat that seemed relevant was that, in the original, all the animals were fed fresh vegetables, whereas in the repeat they were fed commercial rabbit chow. Provide an explanation for the difference and a test of your idea.

Solution

The first time that the experiment was done, the breeders would have been perfectly justified in proposing that a pair of alleles determine white versus yellow body fat because the data clearly resemble Mendel's results in peas. White must be dominant, so we can represent the white allele as W and the yellow allele as w. The results can then be expressed as follows:

$$P \quad W/W \times w/w$$

$$F_1 \quad W/w$$

$$F_2 \quad \tfrac{1}{4} W/W$$

$$\tfrac{1}{2} W/w$$

$$\tfrac{1}{4} w/w$$

No doubt, if the parental rabbits had been sacrificed, it would have been predicted that one (we cannot tell which) would have white fat and the other yellow. Luckily, this was not done, and the same animals were bred again, leading to a very interesting, different result. Often in science, an unexpected observation can lead to a novel principle, and rather than moving on to something else, it is useful to try to explain the inconsistency. So why did the 3:1 ratio disappear? Here are some possible explanations.

First, perhaps the genotypes of the parental animals had changed. This type of spontaneous change affecting the whole animal, or at least its gonads, is very unlikely, because even common experience tells us that organisms tend to be stable to their type.

Second, in the repeat, the sample of 22 F_2 animals did not contain any yellow simply by chance ("bad luck"). This again seems unlikely, because the sample was quite large, but it is a definite possibility.

A third explanation draws on the principle covered in Chapter 1 that genes do not act in a vacuum; they depend on the environment for their effects. Hence, the useful catchphrase "Genotype + environment = phenotype" arises. A corollary of this catchphrase is that genes can act differently in different environments; so

Genotype 1 + environment 1 = phenotype 1

and

Genotype 1 + environment 2 = phenotype 2

In the present question, the different diets constituted different environments, so a possible explanation of the results is that the recessive allele w produces yellow fat only when the diet contains fresh vegetables. This explanation is testable. One way to test it is to repeat the experiment again and use vegetables as food, but the parents might be dead by this time. A more convincing way is to interbreed several of the white-fatted F_2 rabbits from the second experiment. According to the original interpretation, about $\frac{3}{4}$ would bear at least one recessive w allele for yellow fat, and if their progeny are raised on vegetables, yellow should appear in Mendelian proportions. For example, if we choose two rabbits, W/w and w/w, the progeny would be $\frac{1}{2}$ white and $\frac{1}{2}$ yellow.

If this outcome did not happen and no yellow progeny appeared in any of the F_2 matings, one would be forced back to explanations 1 or 2. Explanation 2 can be tested by using larger numbers, and if this explanation doesn't work, we are left with number 1, which is difficult to test directly.

As you might have guessed, in reality the diet was the culprit. The specific details illustrate environmental effects beautifully. Fresh vegetables contain yellow substances called xanthophylls, and the dominant allele W gives rabbits the ability to break down these substances to a colorless ("white") form. However, w/w animals lack this ability, and the xanthophylls are deposited in the fat, making it yellow. When no xanthophylls have been ingested, both $W/-$ and w/w animals end up with white fat.

2. Consider three yellow, round peas, labeled A, B, and C. Each was grown into a plant and crossed to a plant grown from a green, wrinkled pea. Exactly 100 peas issuing from each cross were sorted into phenotypic classes as follows:

A:	51 yellow, round
	49 green, round
B:	100 yellow, round
C:	24 yellow, round
	26 yellow, wrinkled
	25 green, round
	25 green, wrinkled

What were the genotypes of A, B, and C? (Use gene symbols of your own choosing; be sure to define each one.)

Solution

Notice that each of the crosses is

Yellow, round $\times$ green, wrinkled

$\downarrow$

Progeny

Because A, B, and C were all crossed to the same plant, all the differences between the three progeny populations must be attributable to differences in the underlying genotypes of A, B, and C.

You might remember a lot about these analyses from the chapter, which is fine, but let's see how much we can deduce from the data. What about dominance? The key cross for deducing dominance is B. Here, the inheritance pattern is

Yellow, round $\times$ green, wrinkled

$\downarrow$

All yellow, round

So yellow and round must be dominant phenotypes because dominance is literally defined in terms of the phenotype of a hybrid. Now we know that the green, wrinkled parent used in each cross must be fully recessive; we have a very convenient situation because it means that each cross is a testcross, which is generally the most informative type of cross.

Turning to the progeny of A, we see a 1:1 ratio for yellow to green. This ratio is a demonstration of Mendel's first law (equal segregation) and shows that, for the character of color, the cross must have been heterozygote × homozygous recessive. Letting Y = yellow and y = green, we have

$$Y/y \; \times \; y/y$$
$$\downarrow$$
$$\tfrac{1}{2} \, Y/y \text{ (yellow)}$$
$$\tfrac{1}{2} \, y/y \text{ (green)}$$

For the character of shape, because all the progeny are round, the cross must have been homozygous dominant × homozygous recessive. Letting R = round and r = wrinkled, we have

$$R/R \; \times \; r/r$$
$$\downarrow$$
$$R/r \text{ (round)}$$

Combining the two characters, we have

$$Y/y \; ; \; R/R \; \times \; y/y \; ; \; r/r$$
$$\downarrow$$
$$\tfrac{1}{2} \, Y/y \; ; \; R/r$$
$$\tfrac{1}{2} \, y/y \; ; \; R/r$$

Now, cross B becomes crystal clear and must have been

$$Y/Y \; ; \; R/R \; \times \; y/y \; ; \; r/r$$
$$\downarrow$$
$$Y/y \; ; \; R/r$$

because any heterozygosity in pea B would have given rise to several progeny phenotypes, not just one.

What about C? Here, we see a ratio of 50 yellow : 50 green (1 : 1) and a ratio of 49 round : 51 wrinkled (also 1 : 1). So both genes in pea C must have been heterozygous, and cross C was

$$Y/y \; ; \; R/r \; \times \; y/y \; ; \; r/r$$

which is a good demonstration of Mendel's second law (independent behavior of different genes).

How would a geneticist have analyzed these crosses? Basically, the same way that we just did but with fewer intervening steps. Possibly something like this: "yellow and round dominant; single-gene segregation in A; B homozygous dominant; independent two-gene segregation in C."

3. Phenylketonuria (PKU) is a human hereditary disease resulting from the inability of the body to process the chemical phenylalanine, which is contained in the protein that we eat. PKU is manifested in early infancy and, if it remains untreated, generally leads to mental retardation. PKU is caused by a recessive allele with simple Mendelian inheritance.

A couple intends to have children but consults a genetic counselor because the man has a sister with PKU and the woman has a brother with PKU. There are no other known cases in their families. They ask the genetic counselor to determine the probability that their first child will have PKU. What is this probability?

Solution

What can we deduce? If we let the allele causing the PKU phenotype be p and the respective normal allele be P, then the sister and brother of the man and woman, respectively, must have been p/p. To produce these affected persons, all four grandparents must have been heterozygous normal. The pedigree can be summarized as follows:

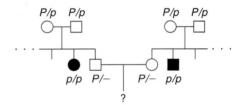

When these inferences have been made, the problem is reduced to an application of the product rule. The only way in which the man and woman can have a PKU child is if both of them are heterozygotes (it is obvious that they themselves do not have the disease). Both the grandparental matings are simple Mendelian monohybrid crosses expected to produce progeny in the following proportions:

$$\left.\begin{array}{l} \tfrac{1}{4}\,P/P \\ \tfrac{1}{2}\,P/p \end{array}\right\} \quad \text{Normal } (\tfrac{3}{4})$$
$$\tfrac{1}{4}\,p/p \qquad \text{PKU}(\tfrac{1}{4})$$

We know that the man and the woman are normal, so the probability of their being a heterozygote is $\tfrac{2}{3}$, because within the $P/-$ class, $\tfrac{2}{3}$ are P/p and $\tfrac{1}{3}$, are P/P.

The probability of *both* the man and the woman being heterozygotes is $\tfrac{2}{3} \times \tfrac{2}{3} = \tfrac{4}{9}$. If they are both heterozygous, then one-quarter of their children would have PKU, so the probability that their first child will have PKU is $\tfrac{1}{4}$ and the probability of their being heterozygous *and* of their first child having PKU is $\tfrac{4}{9} \times \tfrac{1}{4} = \tfrac{4}{36} = \tfrac{1}{9}$, which is the answer.

4. A rare human disease afflicted a family as shown in the accompanying pedigree.

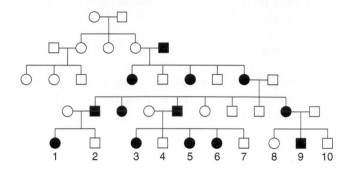

a. Deduce the most likely mode of inheritance.

b. What would be the outcomes of the cousin marriages 1 × 9, 1 × 4, 2 × 3, and 2 × 8?

Solution

a. The most likely mode of inheritance is X-linked dominant. We assume that the disease phenotype is dominant because, after it has been introduced into the pedigree by the male in generation II, it appears in every generation. We assume that the phenotype is X-linked because fathers do not transmit it to their sons. If it were autosomal dominant, father-to-son transmission would be common.

In theory, autosomal recessive could work, but it is improbable. In particular, note the marriages between affected members of the family and unaffected outsiders. If the condition were autosomal recessive, the only way in which these marriages could have affected offspring is if each person marrying into the family were a heterozygote; then the matings would be a/a (affected) × A/a (unaffected). However, we are told that the disease is rare; in such a case, it is highly unlikely that heterozygotes would be so common. X-linked recessive inheritance is impossible, because a mating of an affected woman with a normal man could not produce affected daughters. So we can let A represent the disease-causing allele and a represent the normal allele.

b. 1 × 9: Number 1 must be heterozygous A/a because she must have obtained a from her normal mother. Number 9 must be A/Y. Hence, the cross is $A/a\ ♀ × A/Y\ ♂$.

Female gametes	Male gametes	Progeny
$\frac{1}{2}A$	$\frac{1}{2}A \longrightarrow$	$\frac{1}{4}A/A\ ♀$
	$\frac{1}{2}Y \longrightarrow$	$\frac{1}{4}A/Y\ ♂$
$\frac{1}{2}a$	$\frac{1}{2}A \longrightarrow$	$\frac{1}{4}A/a\ ♀$
	$\frac{1}{2}Y \longrightarrow$	$\frac{1}{4}a/Y\ ♂$

1 × 4: Must be $A/a\ ♀ × a/Y\ ♂$.

Female gametes	Male gametes	Progeny
$\frac{1}{2}A$	$\frac{1}{2}a \longrightarrow$	$\frac{1}{4}A/a\ ♀$
	$\frac{1}{2}Y \longrightarrow$	$\frac{1}{4}A/Y\ ♂$
$\frac{1}{2}a$	$\frac{1}{2}a \longrightarrow$	$\frac{1}{4}a/a\ ♀$
	$\frac{1}{2}Y \longrightarrow$	$\frac{1}{4}a/Y\ ♂$

2 × 3: Must be $a/Y\ ♂ × A/a\ ♀$ (same as 1 × 4).
2 × 8: Must be $a/Y\ ♂ × a/a\ ♀$ (all progeny normal).

PROBLEMS

Basic Problems

1. What are Mendel's laws?

2. If you had a fruit fly *(Drosophila melanogaster)* that was of phenotype A, what test would you make to determine if it was A/A or A/a?

3. Two black guinea pigs were mated and over several years produced 29 black and 9 white offspring. Explain these results, giving the genotypes of parents and progeny.

4. Look at the Punnett square in Figure 2-11.

a. How many genotypes are there in the 16 squares of the grid?

b. What is the genotypic ratio underlying the 9:3:3:1 phenotypic ratio?

c. Can you devise a simple formula for the calculation of the number of progeny genotypes in dihybrid, trihybrid, and so forth, crosses? Repeat for phenotypes.

d. Mendel predicted that within all but one of the phenotypic classes in the Punnett square there should be several different genotypes. In particular, he performed many crosses to identify the underlying genotypes of the round, yellow phenotype. Show two different ways that could be used to identify the various genotypes underlying the round, yellow phenotype. (Remember, all the round, yellow peas look identical.)

5. You have three dice: one red (R), one green (G), and one blue (B). When all three dice are rolled at

the same time, calculate the probability of the following outcomes:

a. 6(R) 6(G) 6(B)

b. 6(R) 5(G) 6(B)

c. 6(R) 5(G) 4(B)

d. No sixes at all

e. 2 sixes and 1 five on any dice

f. 3 sixes or 3 fives

g. The same number on all dice

h. A different number on all dice

6. In the accompanying pedigree, the black symbols represent individuals with a very rare blood disease.

If you had no other information to go on, would you think it most likely that the disease was dominant or recessive? Give your reasons.

7. a. The ability to taste the chemical phenylthiocarbamide is an autosomal dominant phenotype, and the inability to taste it is recessive. If a taster woman with a nontaster father marries a taster man who in a previous marriage had a nontaster daughter, what is the probability that their first child will be

(1) A nontaster girl

(2) A taster girl

(3) A taster boy

b. What is the probability that their first two children will be tasters of either sex?

8. John and Martha are contemplating having children, but John's brother has galactosemia (an autosomal recessive disease) and Martha's great-grandmother also had galactosemia. Martha has a sister who has three children, none of whom have galactosemia. What is the probability that John and Martha's first child will have galactosemia?

 UNPACKING PROBLEM 8

In some chapters, we expand a specific problem with a list of exercises that help mentally process the principles and other knowledge surrounding the subject area of the problem. You can make up similar exercises yourself for other problems. Before attempting a solution to Problem 8, consider some questions such as the following, which are meant only as examples.

1. Can the problem be restated as a pedigree? If so, write one.

2. Can parts of the problem be restated by using Punnett squares?

3. Can parts of the problem be restated by using branch diagrams?

4. In the pedigree, identify a mating that illustrates Mendel's first law.

5. Define all the scientific terms in the problem, and look up any other terms that you are uncertain about.

6. What assumptions need to be made in answering this problem?

7. Which unmentioned family members must be considered? Why?

8. What statistical rules might be relevant, and in what situations can they be applied? Do such situations exist in this problem?

9. What are two generalities about autosomal recessive diseases in human populations?

10. What is the relevance of the rareness of the phenotype under study in pedigree analysis generally, and what can be inferred in this problem?

11. In this family, whose genotypes are certain and whose are uncertain?

12. In what way is John's side of the pedigree different from Martha's side? How does this difference affect your calculations?

13. Is there any irrelevant information in the problem as stated?

14. In what way is solving this kind of problem similar to or different from solving problems that you have already successfully solved?

15. Can you make up a short story based on the human dilemma in this problem?

Now try to solve the problem. If you are unable to do so, try to identify the obstacle and write a sentence or two describing your difficulty. Then go back to the expansion questions and see if any of them relate to your difficulty.

9. Holstein cattle are normally black and white. A superb black and white bull, Charlie, was purchased by a farmer for $100,000. All the progeny sired by Charlie were normal in appearance. However, certain pairs of his progeny, when interbred, produced red and white progeny at a frequency of about 25 percent. Charlie was soon removed from the stud lists of the Holstein breeders. Use symbols to explain precisely why.

10. Suppose that a husband and wife are both heterozygous for a recessive allele for albinism. If they have

dizygotic (two-egg) twins, what is the probability that both the twins will have the same phenotype for pigmentation?

11. The plant blue-eyed Mary grows on Vancouver Island and on the lower mainland of British Columbia. The populations are dimorphic for purple blotches on the leaves—some plants have blotches and others don't. Near Nanaimo, one plant in nature had blotched leaves. This plant, which had not yet flowered, was dug up and taken to a laboratory, where it was allowed to self. Seeds were collected and grown into progeny. One randomly selected (but typical) leaf from each of the progeny is shown in the accompanying illustration.

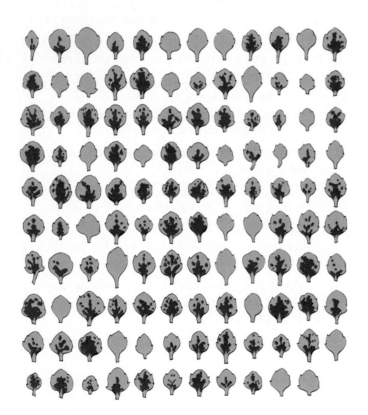

a. Formulate a concise genetic hypothesis to explain these results. Explain all symbols and show all genotypic classes (and the genotype of the original plant).

b. How would you test your hypothesis? Be specific.

12. Can it ever be proved that an animal is *not* a carrier of a recessive allele (that is, not a heterozygote for a given gene)? Explain.

13. In nature, the plant *Plectritis congesta* is dimorphic for fruit shape; that is, individual plants bear either wingless or winged fruits, as shown in the illustration. Plants were collected from nature before

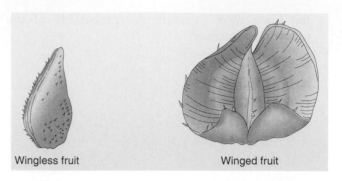

Wingless fruit   Winged fruit

flowering and were crossed or selfed with the following results:

| | Number of progeny | |
Pollination	Winged	Wingless
Winged (selfed)	91	1*
Winged (selfed)	90	30
Wingless (selfed)	4*	80
Winged × wingless	161	0
Winged × wingless	29	31
Winged × wingless	46	0
Winged × winged	44	0
Winged × winged	24	0

*Phenotype probably has a nongenetic explanation.

Interpret these results, and derive the mode of inheritance of these fruit-shaped phenotypes. Use symbols. What do you think is the nongenetic explanation for the phenotypes marked by asterisks in the table?

14. The accompanying pedigree is for a rare, but relatively mild, hereditary disorder of the skin.

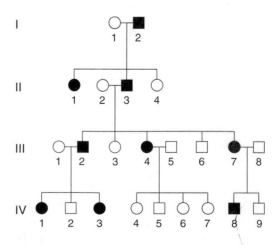

a. Is the disorder inherited as a recessive or a dominant phenotype? State reasons for your answer.

b. Give genotypes for as many individuals in the

pedigree as possible. (Invent your own defined allele symbols.)

c. Consider the four unaffected children of parents III-4 and III-5. In all four-child progenies from parents of these genotypes, what proportion is expected to contain all unaffected children?

15. Four human pedigrees are shown in the accompanying illustration. The black symbols represent an abnormal phenotype inherited in a simple Mendelian manner.

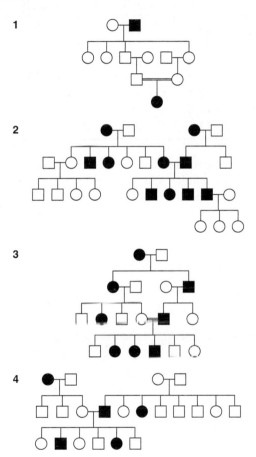

a. For each pedigree, state whether the abnormal condition is dominant or recessive. Try to state the logic behind your answer.

b. For each pedigree, describe the genotypes of as many persons as possible.

16. Tay-Sachs disease ("infantile amaurotic idiocy") is a rare human disease in which toxic substances accumulate in nerve cells. The recessive allele responsible for the disease is inherited in a simple Mendelian manner. For unknown reasons, the allele is more common in populations of Ashkenazi Jews of eastern Europe. A woman is planning to marry her first cousin, but the couple discovers that their shared grandfather's sister died in infancy of Tay-Sachs disease.

a. Draw the relevant parts of the pedigree, and show all the genotypes as completely as possible.

b. What is the probability that the cousins' first child will have Tay-Sachs disease, assuming that all people who marry into the family are homozygous normal?

17. The accompanying pedigree was obtained for a rare kidney disease.

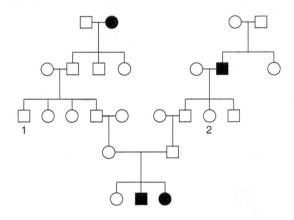

a. Deduce the inheritance of this condition, stating your reasons.

b. If individuals 1 and 2 marry, what is the probability that their first child will have the kidney disease?

18. The accompanying pedigree is for Huntington disease (HD), a late-onset disorder of the nervous system. The slashes indicate deceased family members.

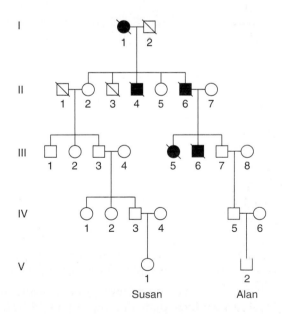

a. Is this pedigree compatible with the mode of inheritance for HD mentioned in the chapter?

b. Consider two newborn children in the two arms of the pedigree, Susan in the left arm and Alan in the right arm. Study the graph in Figure 2-17 and come up with an opinion on the likelihood that they will develop HD. Assume for the sake of the discussion that parents have children at age 25.

19. Consider the accompanying pedigree of a rare autosomal recessive disease, PKU.

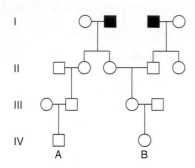

a. List the genotypes of as many of the family members as possible.

b. If individuals A and B marry, what is the probability that their first child will have PKU?

c. If their first child is normal, what is the probability that their second child will have PKU?

d. If their first child has the disease, what is the probability that their second child will be unaffected?

(Assume that all people marrying into the pedigree lack the abnormal allele.)

20. A man lacks earlobes, whereas his wife does have earlobes. Their first child, a boy, lacks earlobes.

a. If it is assumed that the phenotypic difference is due to two alleles of a single gene, is it possible that the gene is X-linked?

b. Is it possible to decide if the lack of earlobes is dominant or recessive?

21. A rare, recessive allele inherited in a Mendelian manner causes the disease cystic fibrosis. A phenotypically normal man whose father had cystic fibrosis marries a phenotypically normal woman from outside the family, and the couple consider having a child.

a. Draw the pedigree as far as described.

b. If the frequency in the population of heterozygotes for cystic fibrosis is 1 in 50, what is the chance that the couple's first child will have cystic fibrosis?

c. If the first child does have cystic fibrosis, what is the probability that the second child will be normal?

22. The allele *c* causes albinism in mice (C causes mice to be black). If the cross is made *C/c* × *c/c*, and there are 10 progeny, what is the probability of their all being black?

23. In dogs, dark coat color is dominant over albino and short hair is dominant over long hair. Assume that these effects are caused by two independently assorting genes, and write the genotypes of the parents in each of the crosses shown here, in which D and A stand for the dark and albino phenotypes, respectively, and S and L stand for the short-hair and long-hair phenotypes.

	Number of progeny			
Parental phenotypes	D, S	D, L	A, S	A, L
a. D, S × D, S	89	31	29	11
b. D, S × D, L	18	19	0	0
c. D, S × A, S	20	0	21	0
d. A, S × A, S	0	0	28	9
e. D, L × D, L	0	32	0	10
f. D, S × D, S	46	16	0	0
g. D, S × D, L	30	31	9	11

Use the symbols C and *c* for the dark and albino coat-color alleles and the symbols *S* and *s* for the short-hair and long-hair alleles, respectively. Assume homozygosity unless there is evidence otherwise.

(Problem 25 reprinted by permission of Macmillan Publishing Co., Inc., from *Genetics* by M. Strickberger. Copyright 1968 by Monroe W. Strickberger.)

24. In tomatoes, two alleles of one gene determine the character difference of purple (P) versus green (G) stems, and two alleles of a separate, independent gene determine the character difference of "cut" (C) versus "potato" (Po) leaves. The results for five matings of tomato-plant phenotypes are as follows:

	Parental	*Number of progeny*			
Mating	phenotypes	P, C	P, Po	G, C	G, Po
1	P, C × G, C	321	101	310	107
2	P, C × P, Po	219	207	64	71
3	P, C × G, C	722	231	0	0
4	P, C × G, Po	404	0	387	0
5	P, Po × G, C	70	91	86	77

a. Determine which alleles are dominant.

b. What are the most probable genotypes for the parents in each cross?

(Problem 24 from A. M. Srb, R. D. Owen, and R. S. Edgar, *General Genetics*, 2d ed. Copyright 1965 by W. H. Freeman and Company.)

25. The recessive allele *s* causes *Drosophila* to have small wings and the *s¹* allele causes normal wings. This gene is known to be X-linked. If a small-winged male is crossed with a homozygous wild-type female, what ratio of normal to small-winged flies can be expected in each sex in the F_1? If F_1 flies are intercrossed, what F_2 progeny ratios are expected? What progeny ratios are predicted if F_1 females are backcrossed with their father?

26. An X-linked dominant allele causes hypophosphatemia in humans. A man with hypophosphatemia marries a normal woman. What proportion of their sons will have hypophosphatemia?

27. Duchenne muscular dystrophy is sex-linked and usually affects only males. Victims of the disease become progressively weaker, starting early in life.

   a. What is the probability that a woman whose brother has Duchenne's disease will have an affected child?

   b. If your mother's brother (your uncle) had Duchenne's disease, what is the probability that you have received the allele?

   c. If your father's brother had the disease, what is the probability that you have received the allele?

28. The accompanying pedigree is concerned with an inherited dental abnormality, amelogenesis imperfecta.

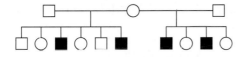

   a. What mode of inheritance *best* accounts for the transmission of this trait?

   b. Write the genotypes of all family members according to your hypothesis.

29. A sex-linked recessive allele *c* produces a red-green colorblindness in humans. A normal woman whose father was colorblind marries a colorblind man.

   a. What genotypes are possible for the mother of the colorblind man?

   b. What are the chances that the first child from this marriage will be a colorblind boy?

   c. Of the girls produced by these parents, what proportion can be expected to be colorblind?

   d. Of all the children (sex unspecified) of these parents, what proportion can be expected to have normal color vision?

30. Male house cats are either black or orange; females are black, orange, or calico.

   a. If these coat-color phenotypes are governed by a sex-linked gene, how can these observations be explained?

   b. Using appropriate symbols, determine the pheno-

types expected in the progeny of a cross between an orange female and a black male.

   c. Repeat part b for the reciprocal of the cross described there.

   d. Half the females produced by a certain kind of mating are calico, and half are black; half the males are orange, and half are black. What colors are the parental males and females in this kind of mating?

   e. Another kind of mating produces progeny in the following proportions: $\frac{1}{4}$ orange males, $\frac{1}{4}$ orange females, $\frac{1}{4}$ black males, and $\frac{1}{4}$ calico females. What colors are the parental males and females in this kind of mating?

31. The accompanying pedigree concerns a certain rare disease that is incapacitating but not fatal.

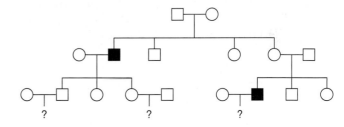

   a. Determine the most likely mode of inheritance of this disease.

   b. Write the genotype of each family member according to your proposed mode of inheritance.

   c. If you were this family's doctor, how would you advise the three couples in the third generation about the likelihood of having an affected child?

32. A mutant allele in mice causes a bent tail. Six pairs of mice were crossed. Their phenotypes and those of their progeny are given in the following table. N is normal phenotype; B is bent phenotype. Deduce the mode of inheritance of this phenotype.

	Parents		*Progeny*	
Cross	♀	♂	♀	♂
1	N	B	All B	All N
2	B	N	$\frac{1}{2}$ B, $\frac{1}{2}$ N	$\frac{1}{2}$ B, $\frac{1}{2}$ N
3	B	N	All B	All B
4	N	N	All N	All N
5	B	B	All B	All B
6	B	B	All B	$\frac{1}{2}$ B, $\frac{1}{2}$ N

   a. Is it recessive or dominant?

   b. Is it autosomal or sex-linked?

   c. What are the genotypes of all parents and progeny?

33. The normal eye color of *Drosophila* is red, but strains in which all flies have brown eyes are available. Similarly, wings are normally long, but there are strains with short wings. A female from a pure line with brown eyes and short wings is crossed with a male from a normal pure line. The F_1 consists of normal females and short-winged males. An F_2 is then produced by intercrossing the F_1. *Both* sexes of F_2 flies show phenotypes as follows:

$\frac{3}{8}$ red eyes, long wings

$\frac{3}{8}$ red eyes, short wings

$\frac{1}{8}$ brown eyes, long wings

$\frac{1}{8}$ brown eyes, short wings

Deduce the inheritance of these phenotypes, using clearly defined genetic symbols of your own invention. State the genotypes of all three generations and the genotypic proportions of the F_1 and F_2.

  UNPACKING PROBLEM 33

Before attempting a solution to this problem, try answering the following questions:

1. What does the word "normal" mean in this problem?

2. The words "line" and "strain" are used in this problem. What do they mean and are they interchangeable?

3. Draw a simple sketch of the two parental flies showing their eyes, wings, and sexual differences.

4. How many different characters are there in this problem?

5. How many phenotypes are there in this problem, and which phenotypes go with which characters?

6. What is the full phenotype of the F_1 females called "normal"?

7. What is the full phenotype of the F_1 males called "short-winged"?

8. List the F_2 phenotypic ratios for each character that you came up with in question 4.

9. What do the F_2 phenotypic ratios tell you?

10. What major inheritance pattern distinguishes sex-linked inheritance from autosomal inheritance?

11. Do the F_2 data show such a distinguishing criterion?

12. Do the F_1 data show such a distinguishing criterion?

13. What can you learn about dominance in the F_1? the F_2?

14. What rules about wild-type symbolism can you use in deciding which allelic symbols to invent for these crosses?

15. What does "deduce the inheritance of these phenotypes" mean?

Now try to solve the problem. If you are unable to do so, make a list of questions about the things that you do not understand. Inspect the key concepts at the beginning of the chapter and ask yourself which are relevant to your questions. If this doesn't work, inspect the messages of this chapter and ask yourself which might be relevant to your questions.

34. In a natural population of annual plants, a single plant is found that is sickly looking and has yellowish leaves. The plant is dug up and brought back to the laboratory. Photosynthesis rates are found to be very low. Pollen from a normal dark green-leaved plant is used to fertilize emasculated flowers of the yellowish plant. A hundred seeds result. Of these, only 60 germinate. The resulting plants are all sickly yellow in appearance.

a. Propose a genetic explanation for the inheritance pattern.

b. Suggest a simple test for your model.

c. Account for the reduced photosynthesis, sickliness, and yellowish appearance.

35. What is the basis for the green-white color variegation in the leaves of *Mirabilis*? If the following cross is made:

Variegated ♀ × green ♂

what progeny types can be predicted? What about the reciprocal cross?

36. In *Neurospora*, the mutant *stp* exhibits erratic stop-start growth. The mutant site is known to be in the mitochondrial DNA. If an *stp* strain is used as the female parent in a cross to a normal strain acting as the male, what type of progeny can be expected? What about the progeny from the reciprocal cross?

37. Two corn plants are studied. One is resistant (R and the other is susceptible (S) to a certain pathogenic fungus. The following crosses are made, with the results shown:

S ♀ × R ♂ ⟶ all progeny S
R ♀ × S ♂ ⟶ all progeny R

What can you conclude about the location of the genetic determinants of R and S?

38. A plant geneticist has two pure lines, one with purple petals and one with blue. She hypothesizes that the phenotypic difference is due to two alleles of one gene. To test this idea, she aims to look for a $3:1$ ratio in the F_2. She crosses the lines and finds that all

the F_1 progeny are purple. The F_1 plants are selfed and 400 F_2 plants obtained. Of these, 320 are purple and 80 are blue. Use the chi-square test to determine if these results fit her hypothesis.

39. From a presumed testcross $A/a \times a/a$, in which A = red and a = white, use the chi-square test to find out which of the following possible results would fit the expectations:

    a. 120 red, 100 white
    b. 5000 red, 5400 white
    c. 500 red, 540 white
    d. 50 red, 54 white

40. A presumed dihybrid in *Drosophila* B/b ; F/f is testcrossed to b/b ; f/f. (B = black body; b = brown body; F = forked bristles; f = unforked bristles.) The results were

Black, forked	230
Black, unforked	210
Brown, forked	240
Brown, unforked	250

    Use the chi-square test to determine if these results fit the results expected from testcrossing the hypothesized dihybrid.

41. Are the progeny numbers below consistent with the results expected from selfing a plant presumed to be a dihybrid of two independently assorting genes, H/h ; R/r? (H = hairy leaves; h = smooth leaves; R = round ovary; r = elongated ovary.)

Hair, round	178
Hairy, elongated	62
Smooth, round	56
Smooth, elongated	24

CHALLENGING PROBLEMS

42. You have three jars containing marbles, as follows:

jar 1	600 red	and	400 white
jar 2	900 blue	and	100 white
jar 3	10 green	and	990 white

    a. If you blindly select one marble from each jar, calculate the probability of obtaining

    (1) a red, a blue, and a green
    (2) three whites
    (3) a red, a green, and a white
    (4) a red and two whites
    (5) a color and two whites
    (6) at least one white

    b. In a certain plant, R = red and r = white. You self a red R/r heterozygote with the express purpose of obtaining a white plant for an experiment. What minimum number of seeds do you have to grow to be at least 95 percent certain of obtaining at least one white individual? (**Hint:** consider your answer to part a(6).)

    c. When a woman is injected with an egg fertilized in vitro, the probability of its implanting successfully is 20 percent. If a woman is injected with five eggs simultaneously, what is the probability that she will become pregnant?

    (Part c from Margaret Holm.)

43. A man's grandfather has galactosemia. This is a rare autosomal recessive disease caused by inability to process galactose, leading to muscle, nerve, and kidney malfunction. The man married a woman whose sister had galactosemia. The woman is now pregnant with their first child.

    a. Draw the pedigree as described.

    b. What is the probability that this child will have galactosemia?

    c. If the first child does have galactosemia, what is the probability a second child will have it.

44. A curious polymorphism in human populations has to do with the ability to curl up the sides of the tongue to make a trough ("tongue rolling"). Some people can do this trick, and others simply cannot. Hence it is an example of a dimorphism. Its significance is a complete mystery. In one family, a boy was unable to roll his tongue but, to his great chagrin, his sister could. Furthermore, both his parents were rollers, and so were both grandfathers, one paternal uncle, and one paternal aunt. One paternal aunt, one paternal uncle, and one maternal uncle could not roll their tongues.

    a. Draw the pedigree for this family, defining your symbols clearly, and deduce the genotypes of as many individual members as possible.

    b. The pedigree that you drew is typical of the inheritance of tongue rolling and led geneticists to come up with the inheritance mechanism that no doubt you came up with. However, in a study of 33 pairs of identical twins, both members of 18 pairs could roll, neither member of 8 pairs could roll, and one of the twins in 7 pairs could roll but the other could not. Because identical twins are derived from the splitting of one fertilized egg into two embryos, the members of a pair must be genetically identical. How can the existence of the seven discordant pairs be reconciled with your genetic explanation of the pedigree?

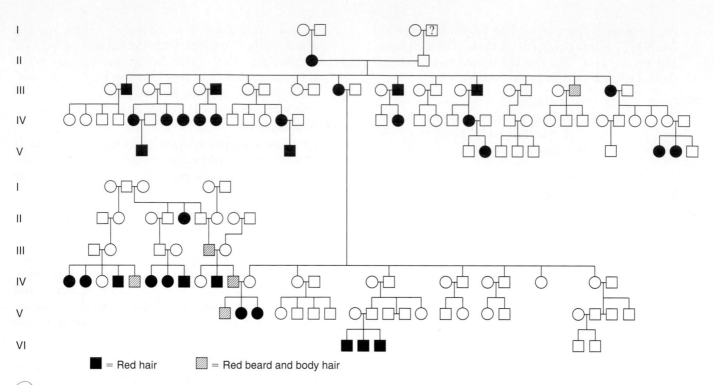

■ = Red hair ▨ = Red beard and body hair

45. Red hair runs in families, and the accompanying illustration shows a large pedigree for red hair. (Pedigree from W. R. Singleton and B. Ellis, *Journal of Heredity* 55, 1964, 261.)

a. Does the inheritance pattern in this pedigree suggest that red hair could be caused by a dominant or a recessive allele of a gene that is inherited in a simple Mendelian manner?

b. Do you think that the red-hair allele is common or rare in the population as a whole?

46. When many families were tested for the ability to taste the chemical PTC, the matings were grouped into three types and the progeny totaled, with the results shown below:

		Children	
Parents	Number of families	Tasters	Non-tasters
Taster × taster	425	929	130
Taster × nontaster	289	483	278
Nontaster × nontaster	86	5	218

Assuming that PTC tasting is dominant (*P*) and nontasting is recessive (*p*), how can the progeny ratios in each of the three types of mating be accounted for?

47. In tomatoes, red fruit is dominant to yellow, two-loculed fruit is dominant to many-loculed fruit, and tall vine is dominant to dwarf. A breeder has two pure lines: red, two-loculed, dwarf and yellow, many- loculed, tall. From these two lines, he wants to produce a new pure line for trade that is yellow, two-loculed, and tall. How exactly should he go

about doing this? Show not only which crosses to make, but also how many progeny should be sampled in each case.

48. We have dealt mainly with only two genes, but the same principles hold for more than two genes. Consider the following cross:

$$A/a \; ; \; B/b \; ; \; C/c \; ; \; D/d \; ; \; E/e \; \times$$
$$a/a \; ; \; B/b \; ; \; c/c \; ; \; D/d \; ; \; e/e$$

a. What proportion of progeny will *phenotypically* resemble (1) the first parent, (2) the second parent, (3) either parent, and (4) neither parent?

b. What proportion of progeny will be *genotypically* the same as (1) the first parent, (2) the second parent, (3) either parent, and (4) neither parent?

Assume independent assortment.

49. The accompanying pedigree shows the pattern of transmission of two rare human phenotypes: cataract and pituitary dwarfism. Family members with cataract are shown with a solid *left* half of the symbol; those with pituitary dwarfism are indicated by a solid *right* half.

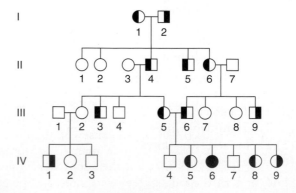

a. What is the most likely mode of inheritance of each of these phenotypes? Explain.

b. List the genotypes of all members in generation III as far as possible.

c. If a hypothetical mating took place between IV-1 and IV-5, what is the probability of the first child's being a dwarf with cataracts? A phenotypically normal child?

(Problem 49 after J. Kuspira and R. Bhambhani, *Compendium of Problems in Genetics.* Copyright 1994 by Wm. C. Brown.)

50. A corn geneticist has three pure lines of genotypes *a/a* ; *B/B* ; *C/C*, *A/A* ; *b/b* ; *C/C*, and *A/A* ; *B/B* ; *c/c*. All the phenotypes determined by *a*, *b*, and *c* will increase the market value of the corn, so naturally he wants to combine them all in one pure line of genotype *a/a* ; *b/b* ; *c/c*.

a. Outline an effective crossing program that can be used to obtain the *a/a* ; *b/b* ; *c/c* pure line.

b. At each stage, state exactly which phenotypes will be selected and give their expected frequencies.

c. Is there more than one way to obtain the desired genotype? Which is the best way?

(Assume independent assortment of the three gene pairs. **Note:** Corn will self- or cross-pollinate easily.)

51. A condition known as icthyosis hystrix gravior appeared in a boy in the early eighteenth century. His skin became very thick and formed loose spines that were sloughed off at intervals. When he grew up, this "porcupine man" married and had six sons, all of whom had this condition, and several daughters, all of whom were normal. For four generations, this condition was passed from father to son. From this evidence, what can you postulate about the location of the gene?

52. The wild-type (W) *Abraxas* moth has large spots on its wings, but the lacticolor (L) form of this species has very small spots. Crosses were made between strains differing in this character, with the following results:

| Cross | Parents | | Progeny | |
	♀	♂	F₁	F₂
1	L	W	♀ W	♀ $\frac{1}{2}$L, $\frac{1}{2}$W
			♂ W	♂ W
2	W	L	♀ L	♀ $\frac{1}{2}$W, $\frac{1}{2}$L
			♂ W	♂ $\frac{1}{2}$W, $\frac{1}{2}$L

Provide a clear genetic explanation of the results in these two crosses, showing the genotypes of all individuals.

53. The pedigree below shows the inheritance of a rare human disease. Is the pattern best explained as being caused by an X-linked recessive allele or by an autosomal dominant allele with expression limited to males?

(Pedigree modified from J. F. Crow, *Genetics Notes*, 6th ed. Copyright 1967 by Burgess Publishing Co., Minneapolis.)

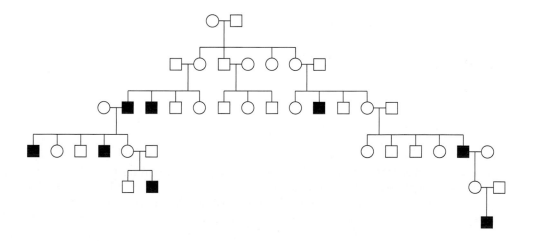

54. In humans, color vision depends on genes encoding three pigments. The *R* (red pigment) and *G* (green pigment) genes are on the X chromosome, whereas the *B* (blue pigment) gene is autosomal. A mutation in any one of these genes can cause colorblindness. Suppose that a colorblind man married a woman with normal color vision. All their sons were colorblind, and all their daughters were normal. Specify the genotypes of both parents and all possible children, explaining your reasoning. (A pedigree drawing will probably be helpful.)

(Problem by Rosemary Redfield.)

55. A certain type of deafness in humans is inherited as an X-linked recessive. A man who suffers from this type of deafness marries a normal woman, and they are ex-

pecting a child. They find out that they are distantly related. Part of the family tree is shown here.

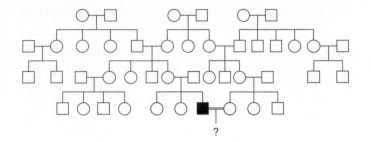

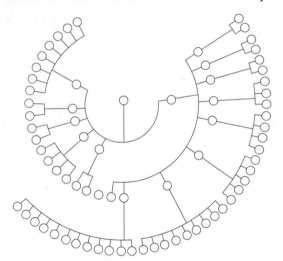

?

How would you advise the parents about the probability of their child's being a deaf boy, a deaf girl, a normal boy, or a normal girl? Be sure to state any assumptions that you make.

56. The accompanying pedigree shows a very unusual inheritance pattern that actually did exist. All progeny are shown, but the fathers in each mating have been omitted to draw attention to the remarkable pattern.

a. State concisely exactly what is unusual about this pedigree.

b. Can the pattern be explained by
(i) Chance (if so, what is the probability)?
(ii) Cytoplasmic factors?
(iii) Mendelian inheritance?
Explain.

57. Consider the accompanying pedigree for a rare human muscle disease.

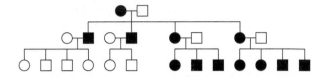

a. What is the unusual feature that distinguishes this pedigree from those studied earlier in this chapter?

b. Where in the cell do you think the mutant DNA resides that is responsible for this phenotype?

INTERACTIVE GENETICS   MegaManual CD-ROM Tutorial

Mendelian Analysis

For additional practice with Mendelian inheritance problems, refer to the Interactive Genetics CD-ROM included with the Solutions MegaManual. The Mendelian Analysis activity includes 12 interactive problems, which focus on using inheritance, pedigrees, and probability data to solve problems.

EXPLORING GENOMES   A Web-Based Bioinformatics Tutorial

OMIM and Huntington Disease

The Online Mendelian Inheritance in Man (OMIM) program collects data on the genetics of human genes. In the Genomics tutorial at www.whfreeman.com/iga, you will learn how to search OMIM for data on gene locus and inheritance pattern for conditions such as Huntington disease.

3

THE CHROMOSOMAL BASIS OF INHERITANCE

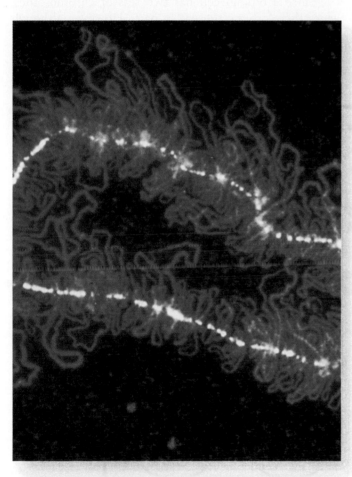

Lampbrush chromosomes. The chromosomes of some animals take on this lampbrush appearance during meiotic diplotene in females. The lampbrush structure is thought to be reminiscent of the underlying organization of all chromosomes: a central scaffold (here stained brightly) and projecting lateral loops (stained red) formed by a folded continuous strand of DNA with associated histone proteins. [M. Roth and J. Gall.]

KEY QUESTIONS

- How do we know that genes are parts of chromosomes?

- How are genes arranged on chromosomes?

- Does a chromosome contain material other than genes?

- How is chromosome number maintained constant through the generations?

- What is the chromosomal basis for Mendel's law of equal segregation?

- What is the chromosomal basis for Mendel's law of independent assortment?

- How does all the DNA fit into a tiny nucleus?

OUTLINE

3.1 Historical development of the chromosome theory

3.2 The nature of chromosomes

3.3 Mitosis and meiosis

3.4 Chromosome behavior and inheritance patterns in eukaryotes

3.5 Organelle chromosomes

CHAPTER OVERVIEW

The elegance of Mendel's analysis is that to assign alleles to phenotypic differences and to predict the outcomes of crosses it is not necessary to know what genes are, nor how they control phenotypes, nor how the laws of segregation and independent assortment are accomplished inside the cell. We simply represent genes as abstract, hypothetical factors using symbols, and move these around in crosses without any concern for their molecular structures or their locations in a cell. Nevertheless, our interest naturally turns to the location of genes in cells and the mechanisms by which segregation and independent assortment are achieved at the cellular level.

We shall see in this chapter that the key components to answering our questions about the cellular basis of heredity are the chromosomes. In eukaryotic cells, most of the chromosomes are worm-shaped structures found in the nuclei. In most of the larger organisms we are familiar with, there are two complete homologous sets of chromosomes in each nucleus, a condition known as *diploidy*. Genes are functional regions in the long, continuous coiled piece of DNA that constitutes a chromosome. There are other DNA segments between genes, and this intergenic material differs in its extent and nature in different species. In higher organisms much of this intergenic DNA is repetitive and of unknown function.

Since a diploid cell has two sets of chromosomes, genes therefore are present in pairs. However, although

CHAPTER OVERVIEW Figure

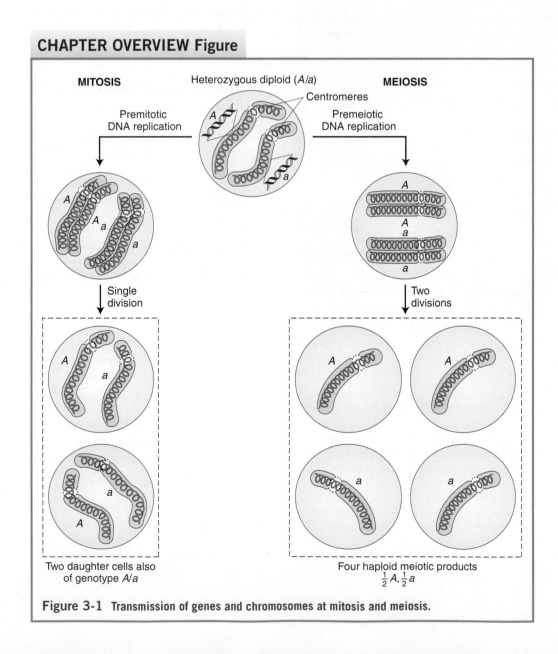

Figure 3-1 Transmission of genes and chromosomes at mitosis and meiosis.

in diploid organisms genes and chromosomes come in sets of two, the chromosomes are not physically paired in the cells of the body. When body cells divide, each chromosome also divides in an accompanying nuclear division called *mitosis*. Homologous chromosomes physically pair only during the nuclear division that occurs during gamete formation—the two nuclear divisions called *meiosis*.

At mitosis and meiosis duplicated chromosomes are partitioned into the daughter cells by molecular ropes called *spindle fibers*. These fibers attach to specialized regions of the chromosomes called *centromeres*.

When a heterozygote (*A/a*) undergoes mitosis, the *A* and the *a* chromosomes replicate, and the copies are pulled into opposite daughter cells, which then both have a genotype identical with the original (*A/a*). This process is summarized in Figure 3-1.

Meiosis is two consecutive nuclear divisions, starting with a pair of replicated chromosomes. When a cell undergoes meiosis, the pulling apart of the replicated homologous chromosomes by the spindle fibers leads to the creation of four haploid cells. When a heterozygote *A/a* undergoes meiosis, half the haploid cells are *A* and half *a* (see Figure 3-1). This separation of alleles into different haploid cells is the basis for Mendel's first law of equal segregation.

Since the spindle fiber action for different chromosome pairs is totally independent, in a dihybrid *A/a* ; *B/b* the pulling apart processes are independent for each gene pair, resulting in independent assortment (Mendel's second law).

Mitochondrial and chloroplast chromosomes are mostly circular. They are much smaller than nuclear chromosomes and do not have the condensed, wormlike appearance of nuclear chromosomes. They contain small sets of genes, also arranged in a continuous array, but with little space between them. Genes on these chromosomes do not obey Mendelian laws, but it is their unique location and abundance that generate their special inheritance patterns.

3.1 Historical development of the chromosome theory

The theory and practice of genetics took a major step forward in the early part of the twentieth century with the development of the notion that the genes, as identified by Mendel, are parts of specific cellular structures, the chromosomes. This simple concept has become known as the **chromosome theory of heredity.** Although simple, the idea has had enormous implications, providing a means of correlating the results of breeding experiments such as Mendel's with the behavior of structures that can be seen under the microscope. This fusion between genetics and cytology is still an essential part of genetic analysis today and has important applications in

medical genetics, agricultural genetics, and evolutionary genetics. First, we shall consider the history of the idea.

Evidence from cytology

How did the chromosome theory take shape? Evidence accumulated from a variety of sources. One of the first lines of evidence came from observations of how chromosomes behave during the division of a cell's nucleus.

Mendel's results lay unnoticed in the scientific literature until 1900, when they were repeated independently by other researchers. In the late 1800s, biologists were keenly interested in heredity, even though they were unaware of Mendel's work. One key issue of that era was the location of the hereditary material in the cell. An obvious place to look was in the gametes, because they are the only connecting link between generations. Egg and sperm were believed to contribute equally to the genetic endowment of offspring, even though they differ greatly in size. Because an egg has a great volume of cytoplasm and a sperm has very little, the cytoplasm of gametes seemed an unlikely seat of the hereditary structures. The nuclei of egg and sperm, however, were known to be approximately equal in size, so the nuclei were considered good candidates for harboring hereditary material.

What was known about the contents of cell nuclei? It became clear that their most prominent components were the chromosomes. Between cell divisions, the contents of nuclei appear densely packed and difficult to resolve into shapes. During cell division, however, the nuclear chromosomes appear as worm-shaped structures easily visible under the microscope. Chromosomes proved to possess unique properties that set them apart from all other cellular structures. A property that especially intrigued biologists was that the number of chromosomes is constant from cell to cell within an organism, from organism to organism within any one species, and from generation to generation within that species. The question therefore arose: How is the chromosome number maintained? The question was first answered by observing under the microscope the orderly behavior of chromosomes during mitosis, the nuclear division that accompanies simple cell division. These studies showed the way by which chromosome number is maintained from cell to cell. Similarities between chromosome behavior and that of genes gave rise to the idea that chromosomes are the structures that contain the genes.

Credit for the chromosome theory of heredity—the concept that genes are parts of chromosomes—is usually given to Walter Sutton (an American who at the time was a graduate student) and Theodor Boveri (a German biologist). Working this time with meiosis, in 1902 these investigators recognized independently that the behavior of Mendel's hypothetical particles during the production of gametes in peas precisely parallels the behavior of chromosomes at meiosis. Consider: genes are in pairs (so are chromosomes); the alleles of a gene segregate equally into gametes (so do the members of a pair of homologous chromosomes); different genes act independently (so do different chromosome pairs). Both investigators reached the same conclusion, which was that the parallel behavior of genes and chromosomes strongly suggests that genes are located on chromosomes.

It is worth considering some of the objections raised to the Sutton-Boveri theory. For example, at the time, chromosomes could not be detected in interphase (the stage between cell divisions). Boveri had to make some very detailed studies of chromosome position before and after interphase before he could argue persuasively that chromosomes retain their physical integrity through interphase, even though they are cytologically invisible at that time. It was also pointed out that, in some organisms, several different pairs of chromosomes look alike, making it impossible to say from visual observation that they are not all pairing randomly, whereas Mendel's laws absolutely require the orderly pairing and segregation of alleles. However, in species in which chromosomes do differ in size and shape, it was verified that chromosomes come in pairs and that these two homologous chromosomes physically pair and segregate in meiosis.

In 1913, Elinor Carothers found an unusual chromosomal situation in a certain species of grasshopper—a situation that permitted a direct test of whether different chromosome pairs do indeed segregate independently. Studying grasshopper testes, she made use of a highly unusual situation to make deductions about the usual, an approach that has become standard in genetic analysis. She found a grasshopper in which one chromosome "pair" had nonidentical members. Such a pair is called a *heteromorphic* pair; presumably the chromosomes show only partial homology. In addition, the same grasshopper had another chromosome, unrelated to the heteromorphic pair, that had no pairing partner at all. Carothers was able to use these unusual chromosomes as visible cytological markers of the behavior of chromosomes during meiosis. By observing many meioses, she could count the number of times that a given member of the heteromorphic pair migrated to the same pole as the chromosome with no pairing partner (Figure 3-2). She observed the two patterns of chromosome behavior with equal frequency. Although these unusual chromosomes are not typical, the results do suggest that nonhomologous chromosomes assort independently.

Other investigators argued that, because all chromosomes appear as stringy structures, qualitative differences between them are of no significance. It was suggested that perhaps all chromosomes were just more or less made of the same stuff. It is worth introducing a

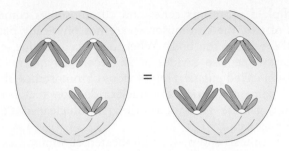

Figure 3-2 Random assortment demonstrated at meiosis. Two equally frequent patterns by which a heteromorphic pair and an unpaired chromosome move into gametes, as observed by Carothers.

study out of historical sequence that effectively counters this objection. In 1922, Alfred Blakeslee performed a study on the chromosomes of jimsonweed *(Datura stramonium)*, which has 12 chromosome pairs. He obtained 12 different strains, each of which had the normal 12 chromosome pairs plus an extra representative of one pair. Blakeslee showed that each strain was phenotypically distinct from the others (Figure 3-3). This result would not be expected if there were no genetic differences between the chromosomes.

All these results pointed indirectly to the chromosomes as the location of genes. The Sutton-Boveri theory was attractive, but there was as yet no direct evidence that genes are located on chromosomes. The argument was based simply on correlation. Further observations, however, did provide the desired evidence, and they began with the discovery of sex linkage.

Evidence from sex linkage

Most of the early crosses analyzed gave the same results in reciprocal crosses, as shown by Mendel. The first exception to this pattern was discovered in 1906 by L. Doncaster and G. H. Raynor. They were studying wing color in the magpie moth *(Abraxas)* by using two different lines, one with light wings and the other with dark wings. They made the following reciprocal crosses:

Light-winged females × dark-winged males

↓

All the progeny have dark wings

(showing that the allele for light wings is recessive)

Dark-winged female × light-winged male

↓

All the female progeny have light wings
All the male progeny have dark wings

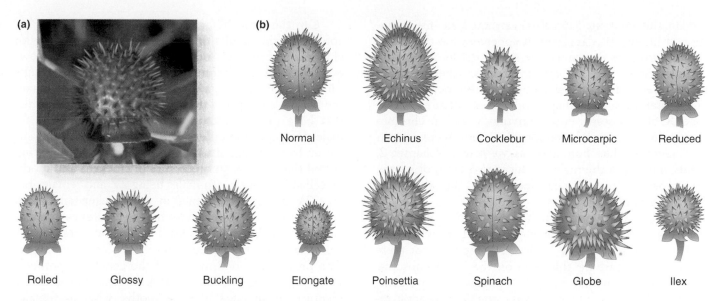

(a) **(b)**

Normal Echinus Cocklebur Microcarpic Reduced

Rolled Glossy Buckling Elongate Poinsettia Spinach Globe Ilex

Figure 3-3 Fruits from *Datura* plants. (a) *Datura* fruit. (b) Each *Datura* plant has one different extra chromosome. Their characteristic appearances suggest that each chromosome is different. [From E. W. Sinnott, L. C. Dunn, and T. Dobzhansky, *Principles of Genetics*, 5th ed. McGraw-Hill Book Company, New York.]

Hence, this pair of reciprocal crosses does not give similar results, and the wing phenotypes in the second cross are associated with the sex of the moths. Note that the female progeny of this second cross are phenotypically similar to their fathers, as the males are to their mothers. Later William Bateson found that in chickens the inheritance pattern of a feather phenotype called *barred* was exactly the same as that of dark wing color in *Abraxas*.

The explanation for such results came from the laboratory of Thomas Hunt Morgan, who in 1909 began studying inheritance in a fruit fly *(Drosophila melanogaster)*. The choice of *Drosophila* as a research organism was a very fortunate one for geneticists—and especially for Morgan, whose work earned him a Nobel Prize in 1934.

The normal eye color of *Drosophila* is dull red. Early in his studies, Morgan discovered a male with completely white eyes. He found that reciprocal crosses gave different results and that phenotypic ratios differed by sex of progeny, as discussed in Chapter 2 (see Figure 2-24). This result was similar to the outcomes in the ex-

amples of moths and chickens, except that the two sexes were reversed.

Before turning to Morgan's explanation of the *Drosophila* results, we should look at some of the cytological information that he was able to use in his interpretations, because no new ideas are born in a vacuum.

In 1905, Nettie Stevens found that males and females of the beetle *Tenebrio* have the same number of chromosomes, but one of the chromosome pairs in males is heteromorphic. One member of the heteromorphic pair appears identical with the members of a pair in the female; Stevens called this the X chromosome. The other member of the heteromorphic pair is never found in females; Stevens called this the Y chromosome (Figure 3-4). She found a similar situation in *Drosophila melanogaster*, which has four pairs of chromosomes, with one of the pairs being heteromorphic in males (Figure 3-5).

X

Y

Figure 3-4 Segregating chromosomes of a male *Tenebrio* beetle. Segregation of the heteromorphic chromosome pair (X and Y) in meiosis in a *Tenebrio* male. The X and Y chromosomes are being pulled to opposite poles during anaphase I. [From A. M. Srb, R. D. Owen, and R. S. Edgar, *General Genetics*, 2d ed. Copyright 1965 by W. H. Freeman and Company.]

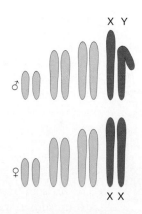

X Y

♂

♀

X X

Figure 3-5 Male and female chromosomes in *Drosophila*.

In this context, Morgan constructed an interpretation of his genetic data. First, it appeared that the X and Y chromosomes determine the sex of the fly. *Drosophila* females have four chromosome pairs, whereas males have three matching pairs plus a heteromorphic pair. Thus, meiosis in the female produces eggs that each bear one X chromosome. Although the X and Y chromosomes in males are heteromorphic, they seem to pair and segregate like homologs, as we saw in Chapter 2. Thus, meiosis in the male produces two types of sperm, one type bearing an X chromosome and the other bearing a Y chromosome.

Morgan then turned to the problem of eye color. He postulated that the alleles for red or white eye color are present on the X chromosome but that there is no counterpart for this gene on the Y chromosome. Thus, females would have two alleles for this gene, whereas males would have only one. The genetic results were completely consistent with the known meiotic behavior of the X and Y chromosomes. This experiment strongly supports the notion that genes are located on chromosomes. However, again it is only a correlation; it does not provide direct support for the Sutton-Boveri theory.

Can the same XX and XY chromosome theory be applied to the results of the earlier crosses made with chickens and moths? You will find that it cannot. However, Richard Goldschmidt realized that these results can be explained with a similar hypothesis, if one makes the simple assumption that in these cases it is the *males* that have pairs of identical chromosomes, and the females that have the different pair. To distinguish this situation from the XY situation in *Drosophila*, Morgan suggested that the sex chromosomes in chickens and moths be called W and Z, with males being ZZ and females being ZW. Thus, if the genes in the chicken and moth crosses are on the Z chromosome, the crosses can be diagrammed as shown in Figure 3-6. The interpretation is consistent with the genetic data. In this case, cytological data provided a confirmation of the genetic hypothesis. In 1914, J. Seiler verified that both chromosomes are identical in all pairs in male moths, whereas females have one different pair.

> **MESSAGE** The special inheritance pattern of some genes makes it likely that they are borne on the sex chromosomes, which show a parallel pattern of inheritance.

(a) MOTHS

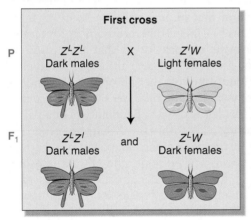
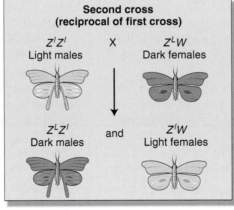

Figure 3-6 WZ inheritance pattern. The inheritance pattern of genes on the sex chromosomes of two species having the WZ mechanism of sex determination.

(b) CHICKENS

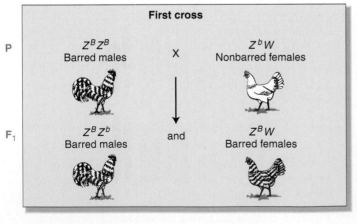
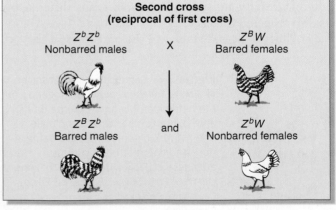

Evidence from abnormal chromosome segregation

The critical analysis that firmly established the chromosome theory of heredity came from one of Morgan's students, Calvin Bridges. He was able to predict that, if genes are located on chromosomes, certain unusual and unexpected genetic results should be explained by the occurrence of abnormal chromosome arrangements. He subsequently observed the abnormal chromosomes under the microscope, exactly as he had predicted.

Bridges's work began with a fruit fly cross that we have considered before:

White-eyed females × red-eyed males

We can write the parental genotypes by using the conventional symbolism as $X^w X^w$ (white-eyed) ♀ × $X^{w+}Y$ (red-eyed) ♂. We have learned that the expected progeny are $X^{w+}X^w$ (red-eyed females) and X^wY (white-eyed males). However, when Bridges made the cross on a large scale, he observed a few exceptions among the large number of progeny. About 1 of every 2000 F_1 progeny was a white-eyed female or a red-eyed male. Collectively, these individuals were called *primary exceptional progeny*. All the primary exceptional males proved to be sterile. How did Bridges explain these exceptional progeny?

Like all *Drosophila* females, the exceptional females should have two X chromosomes. The exceptional females must get both of these chromosomes from their mothers because only the mothers' chromosomes can

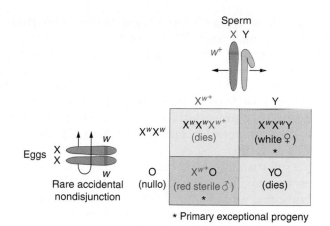

Figure 3-7 Proposed explanation of primary exceptional progeny. Through nondisjunction of the X chromosomes, the maternal parent generates gametes with either two X chromosomes or no X chromosome (O, or nullo). Red, red-eyed; white, white-eyed *Drosophila.*

provide the two *w* alleles that give a fly white eyes. Similarly, exceptional males must get their X chromosomes from their fathers because these chromosomes carry w^+. Bridges hypothesized rare mishaps in the course of meiosis in the female whereby the paired X chromosomes failed to separate in either the first or the second division. This failure would result in meiotic nuclei containing either two X chromosomes or no X at all. Such a failure to separate is called **nondisjunction.** Fertilization of an egg having this type of nucleus by sperm from a wild-type male produces four zygotic classes-XXX, XXY, XO, and YO (Figure 3-7).

BOX 3-1 An Aside on Genetic Symbols

In *Drosophila*, a special symbolism was introduced to distinguish variant alleles from a designated "normal" allele. This system is now used by many geneticists and is especially useful in genetic dissection. For a given *Drosophila* character, the allele that is found most frequently in natural populations (or, alternatively, the allele that is found in standard laboratory stocks) is designated as the standard, or **wild type.** All other alleles are then mutant alleles. The symbol for a gene comes from the first mutant allele found. In Morgan's *Drosophila* experiment, this allele was for white eyes, symbolized by *w.* The wild-type counterpart allele is conventionally represented by adding a superscript plus sign; so the normal red-eye-determining allele is written w^+.

In a polymorphism, several alleles might be common in nature and all might be regarded as wild type. In this case, the alleles can be distinguished by using superscripts. For example, populations of *Drosophila* have two common forms of the enzyme alcohol dehydrogenase. These two forms move at different speeds on an electrophoretic gel. The alleles coding for these forms are designated Adh^F (fast) and Adh^S (slow).

The wild-type allele can be dominant or recessive to a mutant allele. For the two alleles w^+ and w, the use of the lowercase letter indicates that the wild-type allele is dominant over the one for white eyes (that is, w is recessive to w^+). As another example, consider the character wing shape. The wild-type phenotype of a fly's wing is straight and flat, and a mutant allele causes the wing to be curled. Because the latter allele is dominant over the wild-type allele, it is written Cy (short for Curly), whereas the wild-type allele is written Cy^+. Here note that the capital letter indicates that Cy is dominant over Cy^+. (Also note from these examples that the symbol for a single gene may consist of more than one letter.)

To follow Bridges's logic, recall that in *Drosophila* XXY is a female and XO is a male. Bridges assumed that XXX and YO zygotes die before development is complete, so the two types of viable exceptional progeny are expected to be $X^w X^w Y$ (white-eyed female) and X^{w+} O (red-eyed sterile male). What about the sterility of the primary exceptional males? This sterility makes sense if we assume that a male must have a Y chromosome to be fertile.

To summarize, Bridges explained the primary exceptional progeny by postulating rare abnormal meioses that gave rise to viable XXY females and XO males. To test this model, he examined microscopically the chromosomes of the primary exceptional progeny, and indeed they were of the type that he had predicted, XXY and XO. Therefore, by assuming that the eye-color gene was located on the chromosomes, Bridges was able to accurately predict several unusual chromosome rearrangements as well as a previously unknown genetic process, nondisjunction.

3.2 The nature of chromosomes

What is the view of the chromosome today, a century after Sutton and Boveri's original speculations that chromosomes must contain the genes? A recent photograph of a set of chromosomes obtained using modern technology is shown in Figure 3-8. In this photograph each pair of homologous chromosomes has been stained a different color using a special procedure. In the upper part of the figure a nondividing nucleus is shown. Notice that the chromosomes appear to be tightly packed but occupy distinct domains within the nucleus. The lower portion of the figure shows the three pairs of chromosomes during cell division.

What is the substructure of such chromosomes? We will start by relating chromosomes to DNA.

There is one DNA molecule per chromosome

If eukaryotic cells are broken and the contents of their nuclei are examined under the electron microscope, each chromosome appears as a mass of spaghetti-like fibers with a diameter of about 30 nm. An example is shown in the electron micrograph in Figure 3-9. In the 1960s, Ernest DuPraw studied such chromosomes carefully and found that there are no ends protruding from the fibrillar mass. This finding suggests that each chromosome is one, long, fine fiber folded up in some way. If the fiber somehow corresponds to a DNA molecule, then we arrive at the idea that each chromosome is one densely folded DNA molecule.

In 1973, Ruth Kavenoff and Bruno Zimm performed experiments that showed this was most likely the case. They studied *Drosophila* DNA by using a viscoelastic recoil technique to measure the size of DNA

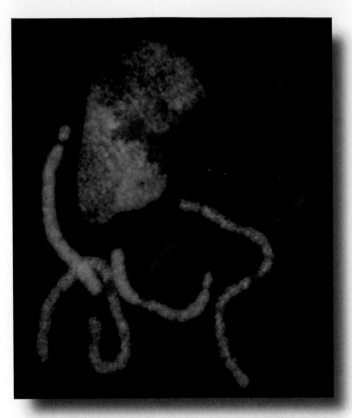

Figure 3-8 Nuclear genome in cells of a female Indian muntjac, a type of small deer ($2n = 6$). The six visible chromosomes are from a cell caught in the process of nuclear division. The three pairs of chromosomes have been stained with chromosome-specific DNA probes, each tagged with a different fluorescent dye ("chromosome paint"). A nucleus derived from another cell is at the stage between divisions. [Photo provided by Fengtang Yang and Malcolm Ferguson-Smith of Cambridge University. Appeared as the cover of *Chromosome Research*, vol. 6, no. 3, April 1998.]

molecules in solution. Put simply, the procedure is analogous to stretching out a coiled spring and measuring how long it takes to return to its fully coiled state. DNA is stretched by spinning a paddle in a DNA solution and then allowed to recoil into its relaxed state. The recoil time is known to be proportional to the size of the largest DNA molecules present. In their study Kavenoff and Zimm obtained a value of 41×10^9 daltons for the largest DNA molecule in the wild-type genome. Then they studied two chromosomal rearrangements of *Drosophila* that produced chromosomes larger than normal and showed that the increase of viscoelasticity was proportional to increased chromosome size. Therefore, it looked as if the chromosome was indeed one strand of DNA, continuous from one end, through the centromere, to the other end. Kavenoff and Zimm were also able to piece together electron micrographs of DNA molecules about 1.5 cm long, each presumably corresponding to a *Drosophila* chromosome (Figure 3-10).

ecule, we might expect the number of bands to be greater than the number of chromosomes. Such separations cannot be made for organisms with large chromosomes (such as humans and *Drosophila*) because the DNA molecules are too large to move through the gel. However, now that whole genomes have been sequenced, all doubt has been removed because the number of two-ended linear units in the sequence equals the chromosome number. Indeed, all the evidence now supports the general principle that a chromosome contains one DNA molecule.

> **MESSAGE** Each eukaryotic chromosome contains a single, long, folded DNA molecule.

The arrangement of genes on chromosomes

The genes are the functional regions along the DNA molecule that constitutes a chromosome—those regions that are transcribed to produce RNA. The location of any individual gene can be shown by using a cloned and labeled copy of the gene as a probe (Figure 3-11). But chromosomes must contain many genes—that had been clear ever since the time of Morgan, when it was shown that many combinations of genes did not segregate

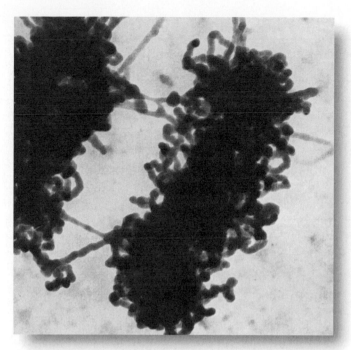

Figure 3-9 Electron micrograph of metaphase chromosomes from a honeybee. The chromosomes appear to be composed of one continuous fiber 30 nm wide. [From E. J. DuPraw, *Cell and Molecular Biology*. Copyright 1968 by Academic Press.]

Eventually geneticists demonstrated directly that certain chromosomes contain single DNA molecules. One strategy was to use pulsed field gel electrophoresis, a specialized electrophoretic technique for separating very long DNA molecules by size. The extracted DNA of an organism with relatively small chromosomes, such as the fungus *Neurospora*, is subjected to electrophoresis for long periods of time in this apparatus. The number of bands that appear on the gel is equal to the number of chromosomes (seven, for *Neurospora*). If each chromosome contained more than one DNA mol-

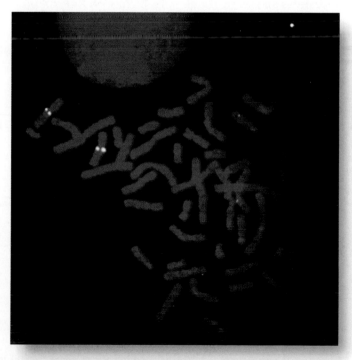

Figure 3-11 Chromosomes probed in situ with a fluorescent probe to locate a gene. The probe is specific for a gene present in a single copy in each chromosome set—in this case, a muscle protein. Only one locus shows a fluorescent spot, corresponding to the probe bound to the muscle protein gene. [From Peter Lichter et al., *Science* 247, 1990, 64.]

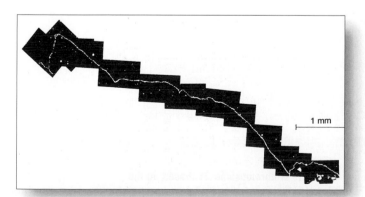

Figure 3-10 Composite electron micrograph of a single DNA molecule constituting one *Drosophila* chromosome. The overall length is 1.5 cm. [From R. Kavenoff, L. C. Klotz, and B. H. Zimm, *Cold Spring Harbor Symp. Quant. Biol.*, 38, 1974, 4.]

Table 3-1 The Relationship Between Gene Size and mRNA Size

Species	Average exon number	Average gene length (kb)	Average mRNA length (kb)
Hemophilus influenzae	1	1.0	1.0
Methanococcus jannaschii	1	1.0	1.0
S. cerevisiae	1	1.6	1.6
Filamentous fungi	2	1.5	1.5
Caenorhabditis elegans	5	5.0	3.0
D. melanogaster	4	11.3	2.7
Chicken	9	13.9	2.4
Mammals	7	16.6	2.2

Source: Based on B. Lewin, *Genes 5*, Table 2-2. Oxford University Press. 1994.

independently. These cases showed inheritance patterns that suggested that some combinations of genes were inherited together, in other words as though they were on the same chromosome (see Chapter 4). Also, the thousands of characteristics of an organism seemed to require thousands of genes. So, well before the advent of DNA sequencing techniques, it was clear that a chromosome represents large numbers of genes in a specific linear array, and that this array is different for different chromosomes. Molecular techniques available before the advent of large-scale sequencing also supported this general principle.

Genomic sequencing did however show the details that were previously lacking. In particular, sequencing identified features of the DNA between genes, not only the sizes of intergenic segments, but also the presence there of repeating segments. Sequencing also showed that genes vary enormously in size, both between and within species. Most of this variation is caused by differences in the size and number of introns that interrupt the coding sequences of a gene (its exons). Some examples of these gene sizes are shown in Table 3-1. The distances between genes are also highly variable in length both between and within species. Most of this variation is caused by repetitive DNA elements. There are many types of such elements, and we will learn more about them in Chapters 12 and 13. Two specific regions of a human chromosome are shown in Figure 3-12 as an example of gene arrangement in our species. Some different species are compared in Figure 3-13.

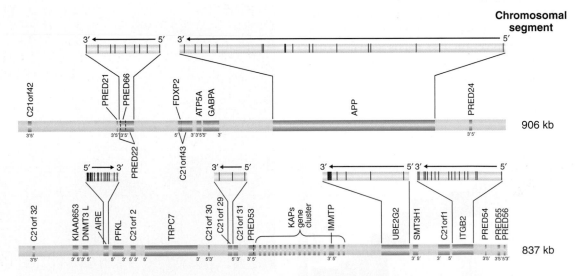

Figure 3-12 Transcribed regions of genes (green) in two segments of chromosome 21, based on the complete sequence for this chromosome. (Two genes, FDXP2 and IMMTP, have been colored orange to distinguish them from neighboring genes.) Some genes are expanded to show exons (black bars) and introns (light green). Vertical labels are gene names (some of known, some of unknown, function). The 5′ and 3′ labels show the direction of transcription of the genes. [Modified from M. Hattori et al., *Nature* 405, 2000, 311–319.]

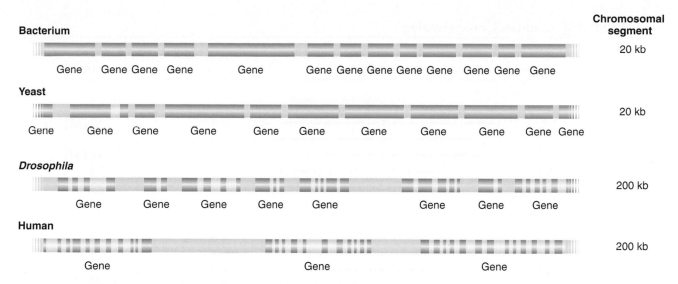

Figure 3-13 Differences in gene topography in four species. Light green = introns; dark green = exons; white = regions between the coding sequences (including regulatory region plus "spacer" DNA). Note the different scales of the top two and bottom two illustrations.

Visible chromosomal landmarks

Chromosomes themselves are highly varied in size and shape. Cytogeneticists are able to identify specific chromosomes under the microscope by studying these distinctive features, which act as chromosomal "landmarks." In this section, we shall consider such features, which allow cytogeneticists to distinguish one chromosome set from another, as well as one chromosome from another.

CHROMOSOME NUMBER Different species have highly characteristic numbers of chromosomes. Chromosome number is the product of two other numbers, the haploid number and the number of sets. The haploid number, represented as **n,** is the number of chromosomes in the basic genomic set. In most fungi and algae, the cells of the visible structures have only one chromosome set and therefore are called **haploid.** In most familiar animals and plants, the cells of the body have two sets of chromosome; such cells are called **diploid** and represented as 2n. The range of the haploid number is immense, from two in some flowering plants to many hundreds in certain ferns. Examples are shown in Table 3-2.

CHROMOSOME SIZE The chromosomes of a single genome may differ considerably in size. In the human genome, for example, there is about a three- to four-fold range in size from chromosome 1 (the biggest) to chromosome 21 (the smallest), as shown in Table 3-3.

Table 3-2 Numbers of Pairs of Chromosomes in Different Species of Plants and Animals

Common name	Scientific name	Number of chromosome pairs	Common name	Scientific name	Number of chromosome pairs
Mosquito	*Culex pipiens*	3	Wheat	*Triticum aestivum*	21
Housefly	*Musca domestica*	6	Human	*Homo sapiens*	23
Garden onion	*Allium cepa*	8	Potato	*Solanum tuberosum*	24
Toad	*Bufo americanus*	11	Cattle	*Bos taurus*	30
Rice	*Oryza sativa*	12	Donkey	*Equus asinus*	31
Frog	*Rana pipiens*	13	Horse	*Equus caballus*	32
Alligator	*Alligator mississipiensis*	16	Dog	*Canis familiaris*	39
Cat	*Felis domesticus*	19	Chicken	*Gallus domesticus*	39
House mouse	*Mus musculus*	20	Carp	*Cyprinus carpio*	52
Rhesus monkey	*Macaca mulatta*	21			

Table 3-3   Human Chromosomes

Group	Number	Diagrammatic representation	Relative length*	Centromeric index†
Large chromosomes				
A	1		8.4	48 (M)
	2		8.0	39
	3		6.8	47 (M)
B	4		6.3	29
	5		6.1	29
Medium chromosomes				
C	6		5.9	39
	7		5.4	39
	8		4.9	34
	9		4.8	35
	10		4.6	34
	11		4.6	40
	12		4.7	30
D	13		3.7	17 (A)
	14		3.6	19 (A)
	15		3.5	20 (A)
Small chromosomes				
E	16		3.4	41
	17		3.3	34
	18		2.9	31
F	19		2.7	47 (M)
	20		2.6	45 (M)
G	21		1.9	31
	22		2.0	30
Sex chromosomes				
	X		5.1 (group C)	40
	Y		2.2 (group G)	27 (A)

* Percentage of the total combined length of a haploid set of 22 autosomes.

† Percentage of a chromosome's length spanned by its short arm. The four most metacentric chromosomes are indicated by an (M); the four most acrocentric by an (A).

HETEROCHROMATIN The general material that collectively composes a chromosome is called **chromatin** by cytogeneticists. When chromosomes are treated with chemicals that react with DNA, such as Feulgen stain, distinct regions with different staining characteristics become visible. Densely staining regions are called **heterochromatin**; poorly staining regions are said to be **euchromatin.** The distinction is the result of the degree of compactness, or coiling, of the DNA in the chromosome. The position of much of the heterochromatin on the chromosome is constant and is, in this sense, a hereditary feature. Look ahead to Figure 4-14 (page 130) for good examples of heterochromatin in tomato.

We now know that most of the active genes are located in euchromatin. Euchromatin stains less densely because it is packed less tightly, and the general idea is that the looser packing makes genes more accessible for transcription and hence gene activity. The question of how euchromatin and heterochromatin are maintained in more or less constant position is under current investigation.

MESSAGE Euchromatin contains most of the active genes. Heterochromatin is more condensed and densely staining.

CENTROMERES The centromere is the region of the chromosome to which spindle fibers attach. The chromosome usually appears to be constricted at the centromere region. The position of this constriction defines the ratio between the lengths of the two chromosome arms. This ratio is a useful characteristic for distinguishing chromosomes (see Table 3-3). Centromere positions are categorized as **telocentric** (at one end), **acrocentric** (off center), or **metacentric** (in the middle).

When genomic DNA is spun for a long time in a cesium chloride density gradient in an ultracentrifuge, the DNA settles into one prominent visible band. However, satellite bands are often visible, distinct from the main DNA band. Such **satellite DNA** consists of multiple tandem repeats of short nucleotide sequences, stretching to as much as hundreds of kilobases in length. Probes may be prepared from such simple-sequence DNA and allowed to bind to partially denatured chromosomes. The great bulk of the satellite DNA is found to reside in the heterochromatic regions flanking the centromeres. There can be either one or several basic repeating units, but usually they are less than 10 bases long. For example, in *Drosophila melanogaster*, the sequence AATAACATAG is found in tandem arrays around all centromeres. Similarly, in the guinea pig, the shorter sequence CCCTAA is arrayed flanking the centromeres. In situ labeling of a mouse satellite DNA is shown in Figure 3-14.

Because the centromeric repeats are a nonrepresentative sample of the genomic DNA, the G + C content can be significantly different from the rest of the DNA. For this reason, the DNA forms a separate satellite band in an ultracentrifuge density gradient. There is no demonstrable function for centromeric repetitive DNA, nor is there any understanding of its relation to heterochromatin or to genes in the heterochromatin. Some organisms have staggering amounts of this DNA; for example, as much as 50 percent of kangaroo DNA can be centromeric satellite DNA.

NUCLEOLAR ORGANIZERS Nucleoli are organelles within the nucleus that contain ribosomal RNA, an important component of ribosomes. Different organisms are differently endowed with nucleoli, which range in number from one to many per chromosome set. The diploid cells of many species have two nucleoli. The nucleoli reside next to slight constrictions of the chromosomes, called **nucleolar organizers** (**NO;** Figure 3-15), which have highly specific positions in the chromosome set. Nucleolar organizers contain the genes that code for ribosomal RNA. The NO does not stain with normal chromatin stains. The NOs on the *Drosophila* X and Y chromosomes contain 250 and 150 tandem copies of rRNA genes, respectively. One human NO has about 250 copies. Such redundancy is one way of ensuring a large amount of rRNA per cell.

(a)

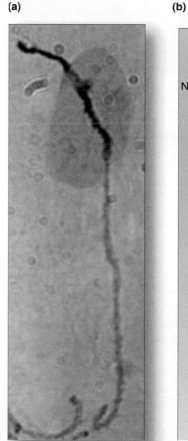

(b)

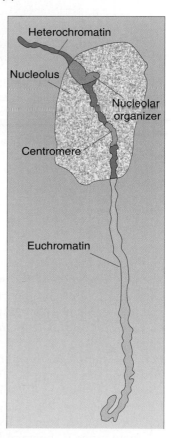

Figure 3-15 Chromosome 2 of tomato, showing the nucleolus and the nucleolar organizer. (a) Photograph; (b) interpretation. [Photo Peter Moens; from P. Moens and L. Butler, "The Genetic Location of the Centromere of Chromosome 2 in the Tomato," *Can. J. Genet. Cytol.* 5, 1963, 364–370.]

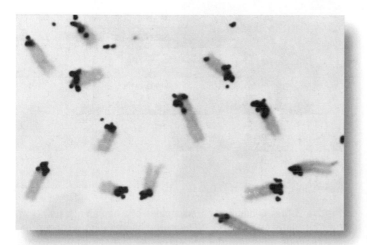

Figure 3-14 Location of satellite DNA in mouse chromosomes. In situ hybridization of mouse chromosomes localizes satellite DNA (black dots) to centromeres. Note that all mouse chromosomes have their centromeres at one end. [From M. L. Pardue and J. G. Gall, *Science* 168, 1970, 1356.]

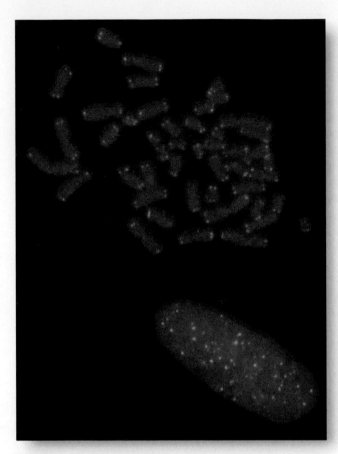

Figure 3-16 Visualization of telomeres. The DNA strands of the chromosomes are slightly separated and treated with a short segment of single-stranded DNA that specifically binds to the telomeres by complementary base pairing. The short segment has been coupled to a substance that can fluoresce yellow under the microscope. The chromosomes have formed sister chromatids. An unbroken nucleus is shown at the bottom. [Robert Moyzis.]

TELOMERES Telomeres are the ends of chromosomes. Generally there is no visible structure that represents the telomere, but at the DNA level it can be distinguished by the presence of distinct nucleotide sequences. The ends of chromosomes represent a special challenge to the chromosomal replication mechanism, and this problem is overcome by the presence at the tips of chromosomes of tandem arrays of simple DNA sequences that do not encode an RNA or a protein product. For example, in the ciliate *Tetrahymena* there is repetition of the sequence TTGGGG, and in humans the repeated sequence is TTAGGG. Chapter 7 will explain how the telomeric repeats solve the problem of replicating the ends of linear DNA molecules. Binding of a telomere probe is shown in Figure 3-16.

BANDING PATTERNS Special chromosome-staining procedures have revealed sets of intricate bands (transverse stripes) in many different organisms. The positions and sizes of the **chromosome bands** are constant and specific to the individual chromosome. One of the basic

chromosomal banding patterns is that produced by Giemsa reagent, a DNA stain applied after mild proteolytic digestion of the chromosomes. This reagent produces patterns of light-staining (G-light) regions and dark-staining (G-dark) regions. An example of G bands in human chromosomes is shown in Figure 3-17. In the

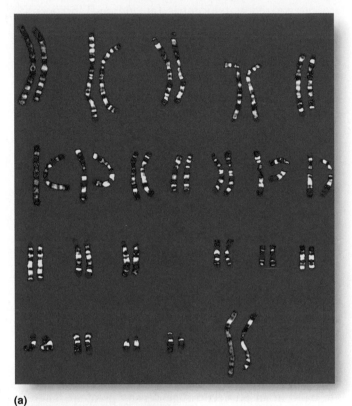

(a)

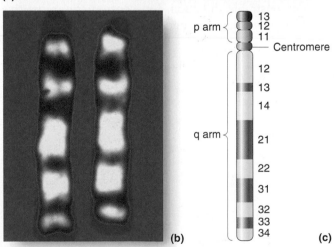

(b)   **(c)**

Figure 3-17 G-banding chromosomes of a human female. (a) Complete set (44A XX). The chromosomes are arranged in homologous pairs in order of decreasing size, starting with the largest autosome (chromosome 1) and ending with the relatively large X. (b) Enlargement of chromosome pair 13. (c) Labeling for G bands of chromosome 13. Note the convention of naming the short and long arms p and q, respectively. [Parts a and b from L. Willatt, East Anglian Regional Genetics Service/Science Photo Library/Photo Researchers.]

complete set of 23 human chromosomes, there are approximately 850 G-dark bands visible during a stage of mitosis called *metaphase*, which is just before the chromosome pairs are pulled apart. These bands have provided a useful way of subdividing the various regions of chromosomes, and each band has been assigned a specific number.

The difference between dark- and light-staining regions was formerly believed to be caused by differences in the relative proportions of bases: the G-light bands were thought to be relatively GC-rich, and the G-dark bands AT-rich. However, it is now thought that the differences are too small to account for banding patterns. The crucial factor again appears to be chromatin packing density: the G-dark regions are packed more densely, with tighter coils. Thus there is a higher density of DNA to take up the stain.

In addition, G bands have been correlated to various other properties. For example, nucleotide-labeling studies showed that G-light bands are early replicating. Furthermore, if polysomal (poly*ribo*somal) mRNA (representing genes being actively transcribed) is used to label chromosomes in situ, then most label binds to the G-light regions, suggesting that these regions contain most of the active genes. From such an analysis, it was presumed that the density of active genes is higher in the G-light bands.

Our view of chromosome banding is based largely on how chromosomes stain when they are in mitotic metaphase. Nevertheless, the regions revealed by metaphase banding must still be in the same relative position in interphase.

A rather specialized kind of banding, which has been used extensively by cytogeneticists for many years, is characteristic of the so-called **polytene chromosomes** in certain organs of the dipteran insects (the two-winged flies). Polytene chromosomes develop in secretory tissues, such as the Malpighian tubules, rectum, gut, footpads, and salivary glands of the dipterans. The chromosomes involved replicate their DNA many times without actually separating. As the number of replicas in a chromosome increases, the chromosome elongates and thickens. This bundle of replicas becomes the polytene chromosome. We can look at *Drosophila* as an example. This insect has a 2*n* number of 8, but the special organs contain only four polytene chromosomes (Figure 3-18). There are four and not eight because in the replication process, the homologs unexpectedly become tightly paired. Furthermore, all four polytene chromosomes become joined at a structure called the **chromocenter,** which is a coalescence of the heterochromatic areas around the

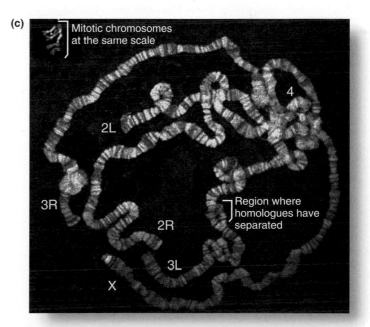

Figure 3-18 The chromosomes of *Drosophila*. Polytene chromosomes form a chromocenter in a *Drosophila* salivary gland. (a) The basic chromosome set as seen in dividing cells, with arms represented by different colors. (b) In salivary glands, heterochromatin coalesces to form the chromocenter. (c) Photograph of polytene chromosomes. [Courtesy of Brian Harmon and John Sedat, University of California, San Francisco.]

centromeres of all four chromosome pairs. The chromocenter of *Drosophila* salivary gland chromosomes is shown in Figure 3-18b, in which the letters L and R stand for arbitrarily assigned left and right arms.

Along the length of a polytene chromosome are transverse stripes called **bands.** Polytene bands are much more numerous than G bands, numbering in the hundreds on each chromosome (see Figure 3-18c). The bands differ in width and morphology, so the banding pattern of each chromosome is unique and characteristic of that chromosome. Recent molecular studies have shown that, in any chromosomal region of *Drosophila*, there are more genes than there are polytene bands, so there is not a one-to-one correspondence of bands and genes as was once believed.

Another tool that is useful for distinguishing chromosomes is to label them with specific tags that are joined to different colored fluorescent dyes, a procedure called **FISH (fluorescent in situ hybridization).** This procedure was used to obtain the image in Figures 3-8 and 3-11. By using all the available chromosomal landmarks together, cytogeneticists can distinguish each of the chromosomes in many species. As an example, Figure 3-19 is a map of the chromosomal landmarks of the genome of corn. Notice how the landmarks enable each of the 10 chromosomes to be distinguished under the microscope.

> **MESSAGE** Features such as size, arm ratio, heterochromatin, number and position of thickenings, number and location of nucleolar organizers, and banding pattern identify the individual chromosomes within the set that characterizes a species.

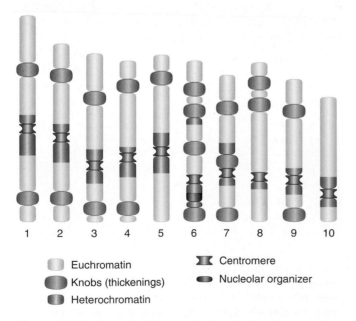

1   2   3   4   5   6   7   8   9   10

Euchromatin   Centromere
Knobs (thickenings)   Nucleolar organizer
Heterochromatin

Figure 3-19 The landmarks that distinguish the chromosomes of corn.

Three-dimensional structure of chromosomes

How much DNA is there in a chromosome set? The single chromosome of the bacterium *Escherichia coli* is about 1.3 mm of DNA. In stark contrast, a human cell contains about 2 m of DNA (1 m per chromosome set). The human body consists of approximately 10^{13} cells and therefore contains a total of about 2×10^{13} m of DNA. Some idea of the extreme length of this DNA can be obtained by comparing it with the distance from the earth to the sun, which is 1.5×10^{11} m. You can see that the DNA in your body could stretch to the sun and back about 50 times. This peculiar fact makes the point that the DNA of eukaryotes is efficiently packed. In fact, the 2 m of DNA in a human cell is packed into 46 chromosomes, all in a nucleus 0.006 mm in diameter. In this section, we have to translate what we have learned about the structure and function of eukaryotic genes into the "real world" of the nucleus. We must come to grips with the fact that the inside of a nucleus must be very much like the inside of a densely wound ball of wool.

What are the mechanisms that pack DNA into chromosomes? How is the very long DNA thread converted into the worm-shaped structure that is a chromosome? Chromatin, the material that makes up chromosomes, is composed of a mixture of DNA and protein. If chromatin is extracted and treated with differing concentrations of salt, different degrees of compaction, or condensation, are observed under the electron microscope. With low salt concentrations, a structure about 10 nm in diameter that resembles a bead necklace is seen. The string between the beads of the necklace can be digested away with the enzyme DNase, so the string can be inferred to be DNA. The beads on the necklace are called **nucleosomes,** and they consist of special chromosomal proteins called **histones** and DNA. Histone structure is remarkably conserved across the gamut of eukaryotic organisms, and nucleosomes are always found to contain an octamer composed of two units each of histones H2A, H2B, H3, and H4. The DNA is wrapped twice around the octamer, as shown in Figure 3-20. When salt concentrations are higher, the nucleosome bead necklace gradually assumes a coiled form called a **solenoid** (see Figure 3-20b). This solenoid produced in vitro is 30 nm in diameter and probably corresponds to the in vivo spaghetti-like structures that we first encountered in Figure 3-9. The solenoid is thought to be stabilized by another histone, H1, that runs down the center of the structure, as Figure 3-20b shows.

We see, then, that to achieve its first level of packaging, DNA winds onto histones, which act somewhat like spools. Further coiling results in the solenoid conformation. However, it takes at least one more level of packaging to convert the solenoids into the three-dimensional structure that we call the *chromosome.* Whereas the diameter of the solenoids is 30 nm, the diameter of the

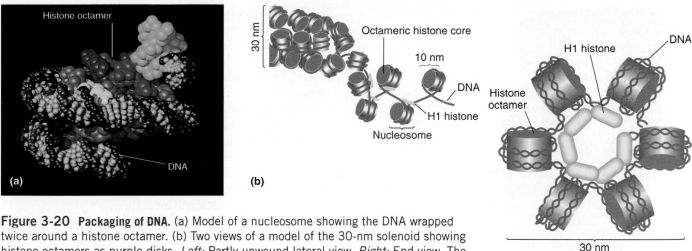

Figure 3-20 **Packaging of DNA.** (a) Model of a nucleosome showing the DNA wrapped twice around a histone octamer. (b) Two views of a model of the 30-nm solenoid showing histone octamers as purple disks. *Left:* Partly unwound lateral view. *Right:* End view. The additional histone H1 is shown running down the center of the coil, probably acting as a stabilizer. With increasing salt concentrations, the nucleosomes close up to form a solenoid with six nucleosomes per turn. [Part a, Alan Wolffe and Van Moudrianakis; part b from H. Lodish, D. Baltimore, A. Berk, S. L. Zipursky, P. Matsudaira, and J. Darnell, *Molecular Cell Biology,* 3d ed. Copyright 1995 by Scientific American Books.]

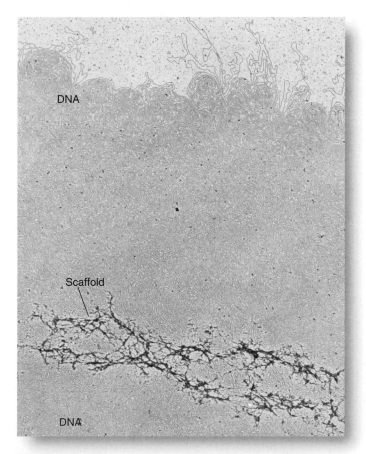

Figure 3-21 **Electron micrograph of a chromosome from a dividing human cell.** Note the central core, or scaffold, from which the DNA strands extend outward. No free ends are visible at the outer edge *(top).* [From W. R. Baumbach and K. W. Adolph, *Cold Spring Harbor Symp. Quant. Biol.,* 1977.]

coils in the next level of condensation is the same as the diameter of the chromosome during cell division, often about 700 nm. What produces these supercoils? One clue comes from observing mitotic metaphase chromosomes from which the histone proteins have been removed chem- ically. After such treatment, the chromosomes have a densely staining central core of nonhistone protein called a **scaffold,** as shown in Figure 3-21 and in the electron micrograph on the first page of this chapter. Projecting laterally from this protein scaffold are loops of DNA. At high magnifications, it is clear from electron micrographs that each DNA loop begins and ends at the scaffold. The central scaffold in metaphase chromosomes is largely composed of the enzyme topoisomerase II. This enzyme has the ability to pass a strand of DNA through another cut strand. Presumably, this central scaffold manipulates the vast skein of DNA during replication, preventing many possible problems that could hinder the unwinding of DNA strands at this crucial stage.

Now let us return to the question of how the supercoiling of the chromosome is produced. The best evidence suggests that the solenoids arrange in loops emanating from the central scaffold matrix, which itself is in the form of a spiral. We see the general idea in Figure 3-22. How do the loops attach to the scaffold? There appear to be special regions along the DNA called **scaffold attachment regions (SARs).** The evidence for these regions is as follows. When histoneless chromatin is treated with restriction enzymes, the DNA loops are cut off the scaffold, but special regions of DNA remain attached to it. These regions have been shown to have protein bound to them. When the protein is digested away, the remaining

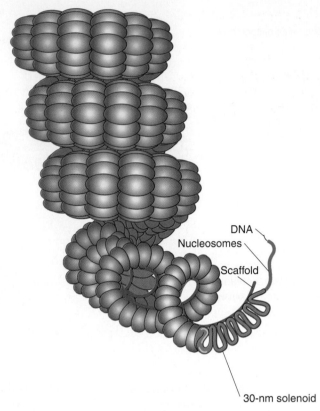

Figure 3-22 Model of a supercoiled chromosome during cell division. The loops are so densely packed that only their tips are visible. At the free ends, the solenoids are shown uncoiled to give an approximation of relative scale.

DNA regions can be analyzed and have been shown in *Drosophila* to contain sequences that are known to be specific for the binding of the enzyme topoisomerase. This finding makes it likely that these regions are the SARs that glue the loops onto the scaffold. The SARs are only in nontranscribed regions of the DNA.

Recent studies on newt mitotic chromosomes have challenged the existence of a central scaffold, at least in this organism. A stretched chromosome showed elasticity until it was sprayed with DNA-digesting enzyme. This suggested that it is the DNA itself that accounts for the structural integrity of the chromosome, and not a scaffold. The authors proposed that there are DNA cross-linking proteins but these are arranged throughout the chromosome and not as a central scaffold.

> **MESSAGE** In the progressive levels of chromosome packing
> 1. DNA winds onto nucleosome spools.
> 2. The nucleosome chain coils into a solenoid.
> 3. The solenoid forms loops, and the loops attach to a central scaffold.
> 4. The scaffold plus loops arrange themselves into a giant supercoil.

3.3 Mitosis and meiosis

When cells divide, the chromosomes must also make copies of themselves (replicate) to maintain the appropriate chromosome number in the descendant cells. In eukaryotes, the chromosomes replicate in two main types of nuclear divisions, called *mitosis* and *meiosis*. Even though these two types of divisions are quite different and have different functions, some of the molecular features are held in common. Three common molecular processes to focus on are DNA *replication*, *adhesion* of replicated chromosomes, and orderly *movement* of chromosomes into descendant cells.

Mitosis is the nuclear division associated with the asexual division of cells. In multicellular organisms mitosis takes place during the division of somatic cells, the cells of the body. In single-celled eukaryotes such as yeasts, mitosis occurs in the cell divisions that cause population growth. Since asexual cell division is aimed at straightforward reproduction of cell type, it is necessary for the set of chromosomes to be maintained con-

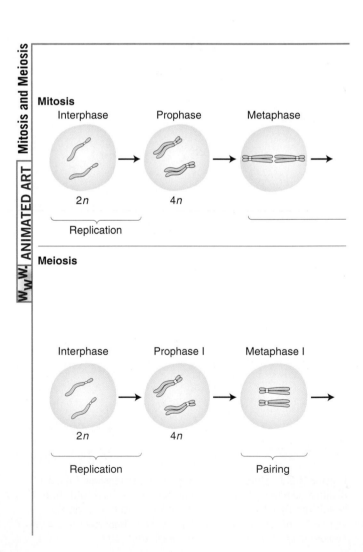

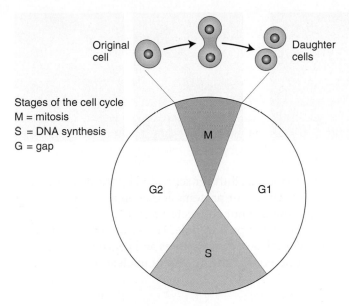

Stages of the cell cycle
M = mitosis
S = DNA synthesis
G = gap

Figure 3-23 Stages of the cell cycle.

stant down through the cell generations, and this is what mitosis achieves.

The stages of the cell division cycle (Figure 3-23) are similar in most organisms. The two basic parts of the cycle are **interphase** (comprising gap 1, synthesis, and gap 2) and mitosis. An event essential for the propagation of genotype takes place in the **S phase** (synthesis phase) because it is here that the actual replication of the DNA of each chromosome occurs. As a result of DNA replication, each chromosome becomes two side-by-side units called **sister chromatids.** The sister chromatids stay attached through the action of specific adherence proteins.

Follow the stages of mitosis using the simplified version shown in Figure 3-24.

1. **Prophase:** The pairs of sister chromatids, which cannot be seen during interphase, become visible. The chromosomes contract into a shorter, thicker shape that is more easily moved around.

2. **Metaphase:** The sister chromatid pairs come to lie in the equatorial plane of the cell.

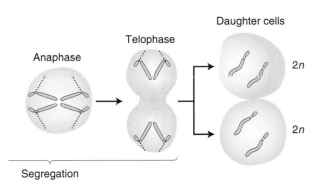

Figure 3-24 Simplified representation of mitosis and meiosis in diploid cells (2n, diploid; n, haploid). (Detailed versions are shown in Figures 3-28 and 3-29.)

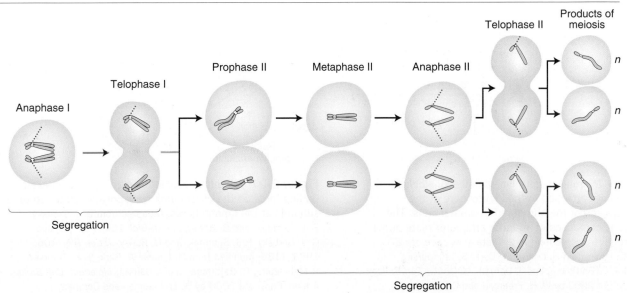

Figure 3-25 Fluorescent label of the nuclear spindle (green) and chromosomes (blue) in mitosis: (a) before the chromatids are pulled apart; (b) during the pulling apart. [From J. C. Waters, R. W. Cole, and C. L. Rieder, *J. Cell Biol.* 122, 1993, 361; courtesy of C. L. Rieder.]

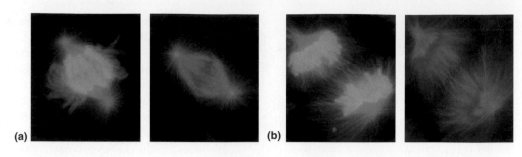
(a)   (b)

3. **Anaphase:** The sister chromatids are pulled to opposite ends of the cell by microtubules that attach to the centromeres. The microtubules are part of the **nuclear spindle,** a set of parallel fibers running from one pole of the cell to the other.

Nuclear spindle fibers provide the motive force that pulls apart the chromosomes or chromatids in mitosis and meiosis (Figure 3-25). In nuclear division, spindle fibers form parallel to the cell axis, connected to one of the cell poles. These spindle fibers are polymers of a protein called *tubulin*. Each centromere acts as a site to which a multiprotein complex called the **kinetochore** binds (Figure 3-26). The kinetochore acts as the site for attachment to spindle fiber microtubules. From one to many microtubules from one pole attach to one kinetochore, and a similar number from the opposite pole attach to the kinetochore on the homologous chromatid. Although the microtubules of the spindle appear like ropes, their action is not ropelike. Instead the tubulin depolymerizes at the kinetochores, shortening the microtubule and thereby pulling the sister chromatids apart (Figure 3-27). Later, the chromatids are further separated by the action of molecular motor proteins acting on another set of microtubules not connected to the kinetochore but running from pole to pole. The spindle apparatus and the complex of kinetochores and centromeres are what determine the fidelity of nuclear division.

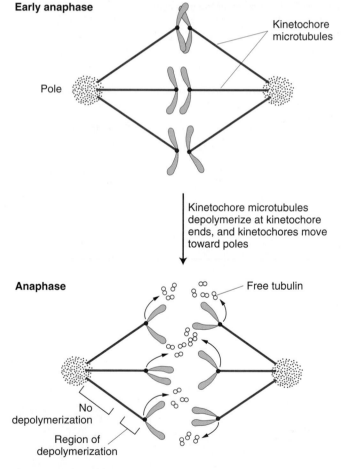

Figure 3-27 Microtubule action. The microtubules exert pulling force on the chromatids by depolymerizing into tubulin subunits at the kinetochores. [Adapted from G. J. Gorbsky, P. J. Sammak, and G. Borisy, *J. Cell Biol.* 104, 1987, 9; and G. J. Gorbsky, P. J. Sammak, and G. Borisy, *J. Cell Biol.* 106, 1988, 1185; modified from H. Lodish, A. Berk, S. L. Zipursky, P. Matsudaira, D. Baltimore, and J. Darnell, *Molecular Cell Biology,* 4th ed. Copyright 2000 by W. H. Freeman and Company.]

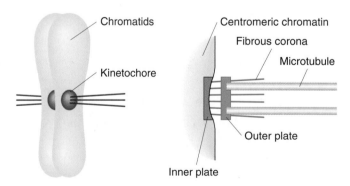

Figure 3-26 Microtubule attachment to the kinetochore. Microtubules are attached to the kinetochore at the centromere region of the chromatid in animal cells. The kinetochore is composed of an inner and outer plate and a fibrous corona. [Adapted from A. G. Pluta et al., *Science* 270, 1995, 1592; taken from H. Lodish, A. Berk, S. L. Zipursky, P. Matsudaira, D. Baltimore, and J. Darnell, *Molecular Cell Biology,* 4th ed. Copyright 2000 by W. H. Freeman and Company.]

4. **Telophase:** Chromatids have arrived at the poles and the pulling-apart process is complete. A nuclear membrane reforms around each nucleus, and the cell divides into two **daughter cells.** Each daughter cell inherits one of each pair of sister chromatids, which now become chromosomes in their own right.

Thus overall, the main events of mitosis are *replication* and sister chromatid *adhesion,* followed by *segregation* of the sister chromatids into each daughter cell. In a diploid cell, for any chromosomal type the number of copies goes from $2 \rightarrow 4 \rightarrow 2$. A full description of mitosis in a plant is given for reference in Figure 3-28.

Even though early investigators did not know about DNA or that it is replicated during interphase, it was still evident from observing mitosis under the microscope that mitosis is the way in which the chromosome number is maintained during cell division. But the sex cycle still presented a puzzle. Two gametes join in the fertilization event. The early investigators knew that in this process two nuclei fuse but that the chromosome number of the fusion product nevertheless is the stan-dard for that species. What prevents the doubling of the chromosome number at each generation? This puzzle was resolved by the prediction of a special kind of nuclear division that *halved* the chromosome number. This special division, which was eventually discovered in the gamete-producing tissues of plants and animals, is called *meiosis.* A simplified representation of meiosis is shown in the lower panel of Figure 3-24.

Meiosis is the general name given to *two* successive nuclear divisions called *meiosis I* and *meiosis II.* Meiosis takes place in special diploid cells called **meiocytes.** Because of the two successive divisions, each meiocyte cell gives rise to four cells, 1 cell $\rightarrow$ 2 cells $\rightarrow$ 4 cells. The four cells are called **products of meiosis.** In animals and plants, the products of meiosis become the haploid **gametes.** In humans and other animals, meiosis takes place in the gonads, and the products of meiosis are the gametes—sperm (more properly, spermatozoa) and eggs (ova). In flowering plants, meiosis takes place in the anthers and ovaries, and the products of meiosis are **meiospores,** which eventually give rise to gametes.

Before meiosis, an S phase duplicates each chromosome's DNA to form sister chromatids, just as in mitosis.

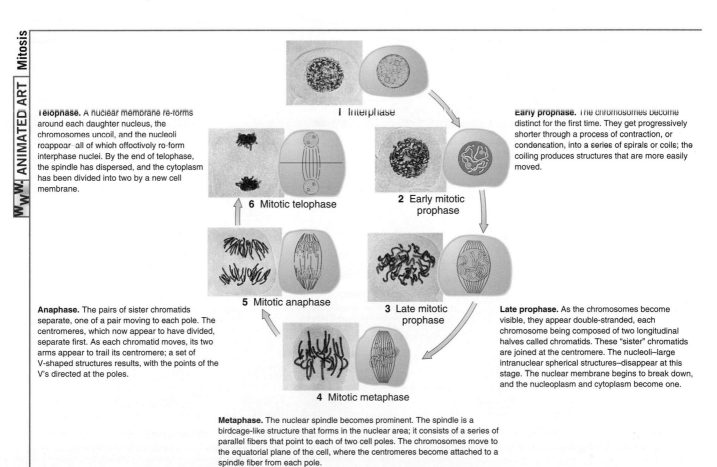

Telophase. A nuclear membrane re-forms around each daughter nucleus, the chromosomes uncoil, and the nucleoli reappear—all of which effectively re-form interphase nuclei. By the end of telophase, the spindle has dispersed, and the cytoplasm has been divided into two by a new cell membrane.

6 Mitotic telophase

1 Interphase

Early prophase. The chromosomes become distinct for the first time. They get progressively shorter through a process of contraction, or condensation, into a series of spirals or coils; the coiling produces structures that are more easily moved.

2 Early mitotic prophase

Anaphase. The pairs of sister chromatids separate, one of a pair moving to each pole. The centromeres, which now appear to have divided, separate first. As each chromatid moves, its two arms appear to trail its centromere; a set of V-shaped structures results, with the points of the V's directed at the poles.

5 Mitotic anaphase

3 Late mitotic prophase

Late prophase. As the chromosomes become visible, they appear double-stranded, each chromosome being composed of two longitudinal halves called chromatids. These "sister" chromatids are joined at the centromere. The nucleoli–large intranuclear spherical structures–disappear at this stage. The nuclear membrane begins to break down, and the nucleoplasm and cytoplasm become one.

4 Mitotic metaphase

Metaphase. The nuclear spindle becomes prominent. The spindle is a birdcage-like structure that forms in the nuclear area; it consists of a series of parallel fibers that point to each of two cell poles. The chromosomes move to the equatorial plane of the cell, where the centromeres become attached to a spindle fiber from each pole.

Figure 3-28 Mitosis. The photographs show nuclei of root-tip cells of *Lilium regale.* [After J. McLeish and B. Snoad, *Looking at Chromosomes.* Copyright 1958, St. Martin's, Macmillan.]

ANIMATED ART Mitosis

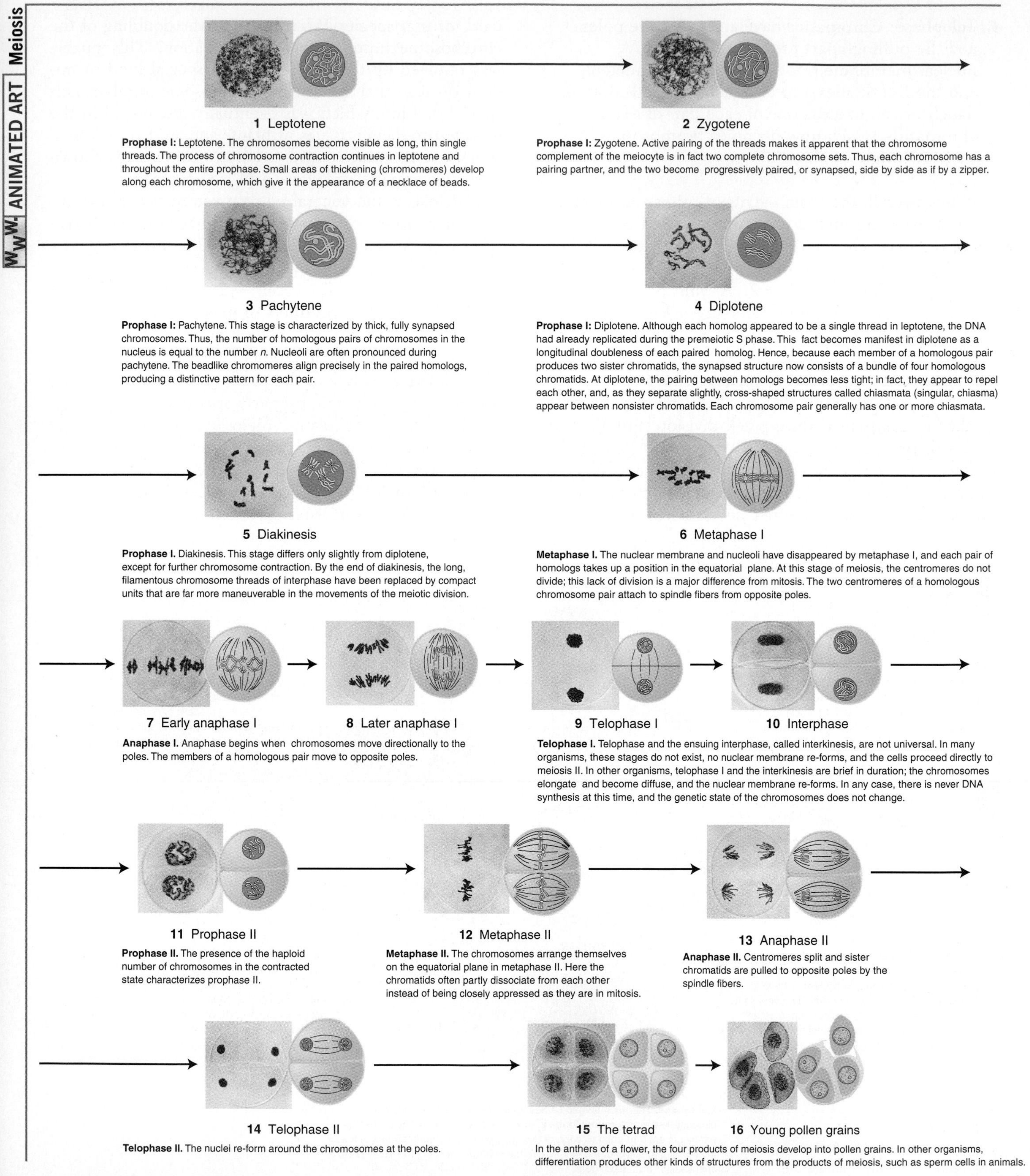

1 Leptotene

Prophase I: Leptotene. The chromosomes become visible as long, thin single threads. The process of chromosome contraction continues in leptotene and throughout the entire prophase. Small areas of thickening (chromomeres) develop along each chromosome, which give it the appearance of a necklace of beads.

2 Zygotene

Prophase I: Zygotene. Active pairing of the threads makes it apparent that the chromosome complement of the meiocyte is in fact two complete chromosome sets. Thus, each chromosome has a pairing partner, and the two become progressively paired, or synapsed, side by side as if by a zipper.

3 Pachytene

Prophase I: Pachytene. This stage is characterized by thick, fully synapsed chromosomes. Thus, the number of homologous pairs of chromosomes in the nucleus is equal to the number *n*. Nucleoli are often pronounced during pachytene. The beadlike chromomeres align precisely in the paired homologs, producing a distinctive pattern for each pair.

4 Diplotene

Prophase I: Diplotene. Although each homolog appeared to be a single thread in leptotene, the DNA had already replicated during the premeiotic S phase. This fact becomes manifest in diplotene as a longitudinal doubleness of each paired homolog. Hence, because each member of a homologous pair produces two sister chromatids, the synapsed structure now consists of a bundle of four homologous chromatids. At diplotene, the pairing between homologs becomes less tight; in fact, they appear to repel each other, and, as they separate slightly, cross-shaped structures called chiasmata (singular, chiasma) appear between nonsister chromatids. Each chromosome pair generally has one or more chiasmata.

5 Diakinesis

Prophase I. Diakinesis. This stage differs only slightly from diplotene, except for further chromosome contraction. By the end of diakinesis, the long, filamentous chromosome threads of interphase have been replaced by compact units that are far more maneuverable in the movements of the meiotic division.

6 Metaphase I

Metaphase I. The nuclear membrane and nucleoli have disappeared by metaphase I, and each pair of homologs takes up a position in the equatorial plane. At this stage of meiosis, the centromeres do not divide; this lack of division is a major difference from mitosis. The two centromeres of a homologous chromosome pair attach to spindle fibers from opposite poles.

7 Early anaphase I   **8 Later anaphase I**

Anaphase I. Anaphase begins when chromosomes move directionally to the poles. The members of a homologous pair move to opposite poles.

9 Telophase I   **10 Interphase**

Telophase I. Telophase and the ensuing interphase, called interkinesis, are not universal. In many organisms, these stages do not exist, no nuclear membrane re-forms, and the cells proceed directly to meiosis II. In other organisms, telophase I and the interkinesis are brief in duration; the chromosomes elongate and become diffuse, and the nuclear membrane re-forms. In any case, there is never DNA synthesis at this time, and the genetic state of the chromosomes does not change.

11 Prophase II

Prophase II. The presence of the haploid number of chromosomes in the contracted state characterizes prophase II.

12 Metaphase II

Metaphase II. The chromosomes arrange themselves on the equatorial plane in metaphase II. Here the chromatids often partly dissociate from each other instead of being closely appressed as they are in mitosis.

13 Anaphase II

Anaphase II. Centromeres split and sister chromatids are pulled to opposite poles by the spindle fibers.

14 Telophase II

Telophase II. The nuclei re-form around the chromosomes at the poles.

15 The tetrad   **16 Young pollen grains**

In the anthers of a flower, the four products of meiosis develop into pollen grains. In other organisms, differentiation produces other kinds of structures from the products of meiosis, such as sperm cells in animals.

Figure 3-29 Meiosis and pollen formation. The photographs are of *Lilium regale*. Note: For simplicity, multiple chiasmata are drawn between only two chromatids; in reality, all four chromatids can take part. [After J. McLeish and B. Snoad, *Looking at Chromosomes.* Copyright 1958, St. Martin's, Macmillan.]

Meiosis then proceeds through the following stages (Figures 3-24 and 3-29):

1. **Prophase I:** As in mitosis, the sister chromatids become visible, closely adhered side by side. However, in contrast with mitosis, the sister chromatids (although fully replicated at the DNA level) show an apparently undivided centromere. The sister chromatid pairs at this stage are called **dyads,** from the Greek word for "two."

2. **Metaphase I:** The homologous dyads now pair to form structures called **bivalents.** Thus, any one bivalent contains a total of four chromatids, sometimes referred to as a **tetrad** (Greek; four). This stage represents the most obvious difference from mitosis.

   Pairing of the dyads to form a bivalent is accomplished by molecular assemblages called **synaptonemal complexes** along the middle of the tetrads (Figure 3-30). Although the existence of synaptonemal complexes has been known for some time, the precise working of these structures is still a topic of research.

   Nonsister chromatids of the tetrad engage in a breakage-and-reunion process called **crossing over,** to be discussed in detail in Chapter 4. The crossover is visible as two chromatids crossing each other to form a structure called a **chiasma** (plural, chiasmata). For those species that have been carefully studied, a minimum of one crossover per tetrad is known to be

a necessary prelude to subsequent orderly chromosome separation.

3. **Anaphase I:** Each of the two pairs of sister chromatids (dyads) is pulled to a different pole.

4. **Telophase I:** A nucleus forms at each pole.

5. **Prophase II:** The dyads reappear.

6. **Metaphase II:** The dyads move to the equatorial plane.

7. **Anaphase II:** Each of the sister chromatids of a dyad is pulled into a different daughter nucleus as the cells divide for a second time.

We see therefore that the fundamental events of meiosis are DNA *replication* and sister chromatid *adhesion,* followed by homologous *pairing, segregation,* and then another *segregation.* Hence, within a single cell, the number of copies of a chromosome of the same type goes from $2 \rightarrow 4 \rightarrow 2 \rightarrow 1$, and each *product of meiosis* must therefore contain one chromosome of each type, half the number of the original meiocyte.

MESSAGE In *mitosis,* each chromosome replicates to form sister chromatids, which segregate into the daughter cells. In *meiosis,* each chromosome replicates to form sister chromatids. Homologous chromosomes physically pair and segregate at the first division. Sister chromatids segregate at the second division.

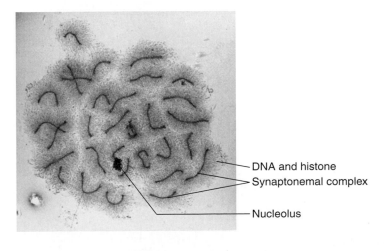

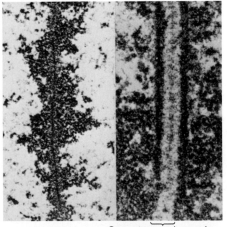

Synaptonemal complex

(a)

(b)

Figure 3-30 Synaptonemal complexes. (a) In *Hyalophora cecropia,* a silk moth, the normal male chromosome number is 62, giving 31 synaptonemal complexes. In the individual shown here, one chromosome *(center)* is represented three times; such a chromosome is termed *trivalent.* The DNA is arranged in regular loops around the synaptonemal complex. (b) Regular synaptonemal complex in *Lilium tyrinum.* Note *(right)* the two lateral elements of the synaptonemal complex and *(left)* an unpaired chromosome, showing a central core corresponding to one of the lateral elements. [Courtesy of Peter Moens.]

3.4 Chromosome behavior and inheritance patterns in eukaryotes

Equipped with our knowledge of the general structure and behavior of chromosomes, we can now interpret the inheritance patterns of the previous chapter more clearly.

The basic life cycles

Any general model for patterns of inheritance must take into account an organism's life cycle. Eukaryotes have three basic types of life cycles, as follows:

Diploids: organisms that are in the diploid state for most of their life cycle (Figure 3-31); that is, for most of their life cycle they consist of cells having two sets of homologous chromosomes. Animals are examples of organisms with this type of life cycle; in most species diploid cells arise from a fertilized egg. Meiosis takes place in special diploid meiocytes set aside for this purpose in the gonads (testes and ovaries) and results in haploid gametes. These are the sperm and eggs that unite when the egg is fertilized, producing the zygote. The zygote subsequently goes through repeated mitotic division to produce the multicellular state.

Haploids: organisms that are in the haploid state for most of their life cycle (Figure 3-32). Common examples are molds and yeasts, both fungi. An organism arises as a haploid spore, which then through mitotic divisions produces a branching network of end-to-end haploid cells (as in molds) or a population of identical cells (as in yeasts). How can meiosis take place in a haploid organism? After

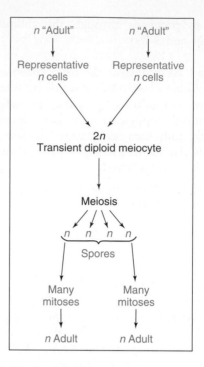

Figure 3-32 The haploid life cycle.

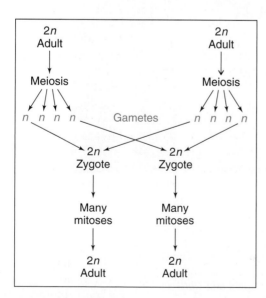

Figure 3-31 The diploid life cycle.

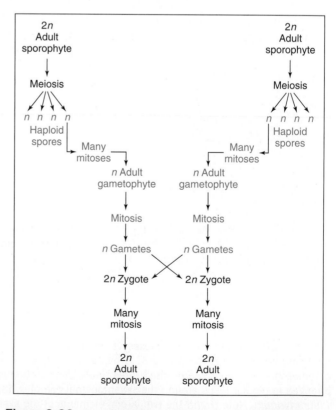

Figure 3-33 The alternation of diploid and haploid stages in the life cycle of plants.

all, meiosis requires the pairing of two homologous chromosome sets. The answer is that two haploid cells of two parental strains fuse to form a temporary diploid meiocyte. Meiosis takes place in the meiocyte, forming haploid spores.

Organisms with alternating haploid-diploid generations: organisms that are haploid for part of the life cycle and diploid for part of the life cycle. Both haploid and diploid parts grow by mitosis, but meiosis occurs only in the diploid stage. Plants show such alternation of haploid and diploid generations: the organism during the haploid stage of the cycle is called the *gametophyte* and during the diploid part the *sporophyte*, as shown in Figure 3-33. Plants such

as ferns and mosses have separate free-living haploid and diploid stages. Flowering plants are predominantly diploid but do have a small haploid gametophytic stage parasitic on the diploid within the flower (see Figure 3-34, which shows the corn life cycle). However, for most genetic purposes plants can be treated simply as showing a diploid cycle.

Fates of specific genotypes after mitosis and meiosis

With these cycles in mind, let us return to the fate of specific genotypes when cells divide. The first question is how constancy of genotype is maintained during asexual division. We have seen from the above discussions that mitosis can take place in diploid or haploid cells. Mitosis in diploid and haploid cells of specific genotypes is shown in the two leftmost columns in Figure 3-35. The diagrams show clearly that the two daughter cells have the same genotype as the progenitor in each case. The success of mitosis in maintaining genotype depends on the fidelity of the replication process that underlies the production of sister chromatids during the S phase. Figure 3-36 illustrates how faithful replication of the chromosomes produces identical DNA molecules.

What is the fate of a specific genotype of a meiocyte that undergoes meiosis? The inheritance patterns originally observed by Mendel for the pea plant and extended into many other plants and animals were based on the production of male and female gametes in meiosis and their subsequent union. Next we look at how the specific events of meiosis produce the inheritance patterns—the Mendelian ratios—predicted by Mendel's laws.

The mechanism that leads to Mendel's first law (the law of equal segregation) is the orderly segregation of a pair of homologous dyads during meiosis. This is illustrated in the right-hand column of Figure 3-35. Looked at another way, this complex cellular choreography is simply the partitioning of the four DNA copies that constitute the tetrad. If we start with a diploid meiocyte of genotype *A/a*, then replication of chromosomes results in two dyads of type *A/A* and *a/a*, which is equivalent to a four-chromatid tetrad that we can represent *A/A/a/a*. The outcome of the two cell divisions of meiosis is simply to place one of these chromatids into each product of meiosis; hence the number of *A* products must equal the number of *a* products, and so there is a 1:1 ratio of *A* and *a*.

Is there any direct demonstration of this mechanism acting at the level of an *individual* meiocyte? Recall that Mendel illustrated equal segregation by crossing *A/a* × *a/a* and observing a 1:1 ratio in the progeny. However, this demonstration is based on the behavior of

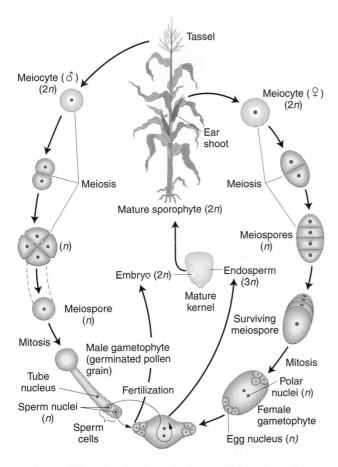

Figure 3-34 Alternation of generations of corn. The male gametophyte arises from a meiocyte in the tassel. The female gametophyte arises from a meiocyte in the ear shoot. One sperm cell from the male gametophyte fuses with an egg nucleus of the female gametophyte, and the diploid zygote thus formed develops into the embryo. The other sperm cell fuses with the two polar nuclei in the center of the female gametophyte, forming a triploid (3*n*) cell that generates the endosperm tissue surrounding the embryo. The endosperm provides nutrition to the embryo during seed germination. Which parts of the diagram represent the haploid stage? Which parts represent the diploid stage?

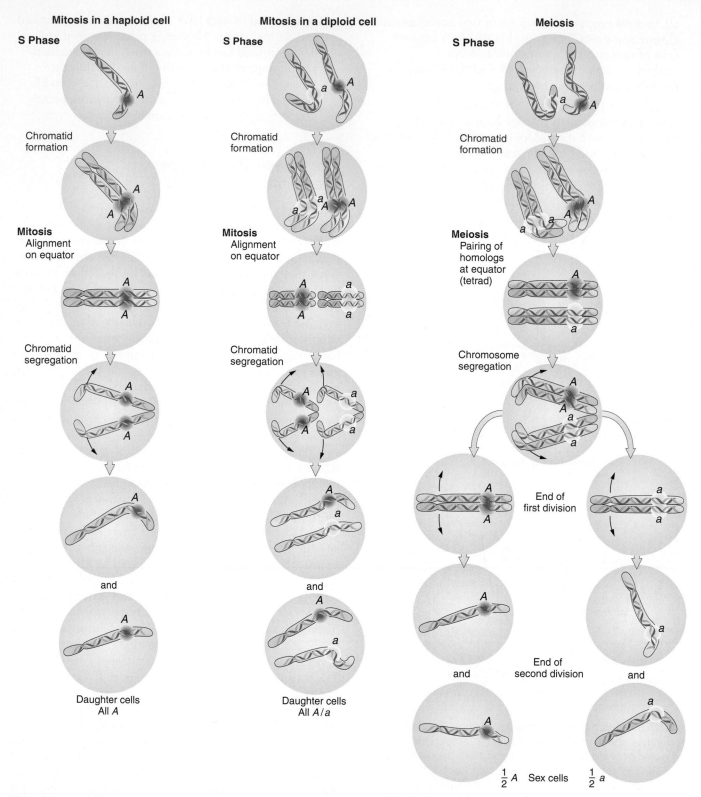

Figure 3-35 DNA and gene transmission during mitosis and meiosis in eukaryotes. S phase and the main stages of mitosis and meiosis are shown. Mitotic divisions *(first two panels)* conserve the genotype of the original cell. In the third panel, the two successive meiotic divisions that occur during the sexual stage of the life cycle have the net effect of halving the number of chromosomes. The alleles *A* and *a* of one gene are used to show how genotypes are transmitted during cell division.

DNA replication to form chromatids

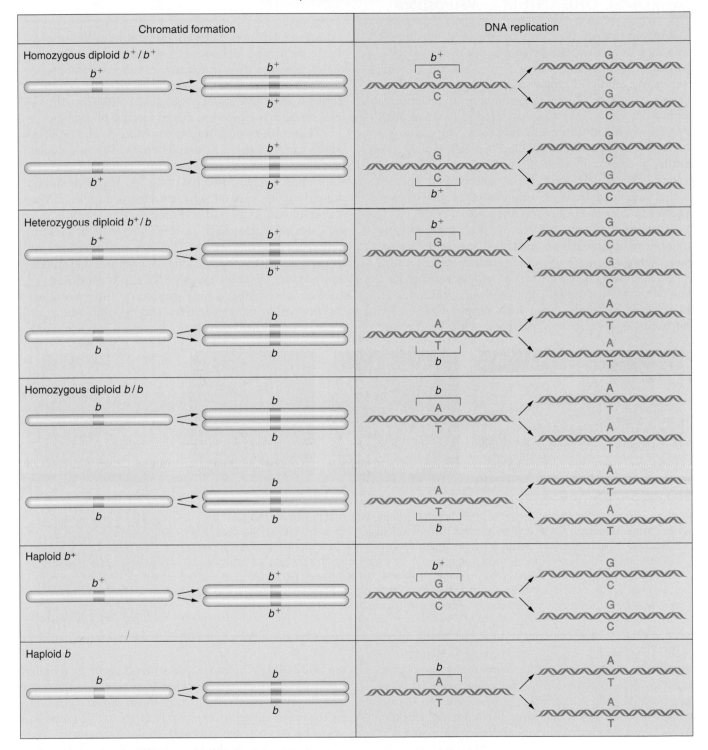

Figure 3-36 Chromatid formation and the underlying DNA replication. *Left:* Each chromosome divides longitudinally into two chromatids; *(right)* at the molecular level, the single DNA molecule of each chromosome replicates, producing two DNA molecules, one for each chromatid. Also shown are various combinations of a gene with normal allele b^+ and mutant form b, caused by a change of a single base pair from GC to AT. Notice that at the DNA level the two chromatids produced when a chromosome replicates are always identical with each other and with the original chromosome.

MODEL ORGANISM   *Neurospora*

Neurospora crassa was one of the first eukaryotic microbes to be adopted by geneticists as a model organism. It is a haploid fungus ($n = 7$) found growing on dead vegetation in many parts of the world. When an asexual spore (haploid) germinates, it produces a tubular structure that extends rapidly by tip growth and throws off multiple side branches. The result is a mass of branched threads (called *hyphae*), which constitute a colony. Hyphae have no cross walls, so a colony is essentially one cell containing many haploid nuclei. A colony buds off millions of asexual spores, which can disperse and repeat the asexual cycle.

Asexual colonies are easily and inexpensively maintained in the lab on a defined medium of inorganic salts plus an energy source such as sugar. (An inert gel such as agar is added to provide a firm surface.) The fact that *Neurospora* can chemically synthesize all its essential molecules from such simple medium led biochemical geneticists (beginning with Beadle and

Tatum, see Chapter 6) to choose it for studies of synthetic pathways. Geneticists worked out the steps in these pathways by introducing mutations and observing their effects. The haploid state of *Neurospora* is ideal for such mutational analysis because mutant alleles are always expressed directly in the phenotype.

There are two mating types MAT-A and MAT-a, which can be viewed as simple "sexes." When colonies of different mating type come into contact, their cell walls and nuclei fuse, resulting in many transient diploid nuclei, each of which undergoes meiosis. The four haploid products of one meiosis stay together in a sac called an *ascus*. Each of these products of meiosis undergoes a further mitotic division, resulting in eight ascospores within each ascus. Ascospores germinate and produce colonies exactly like those produced by asexual spores. Hence such *ascomycete* fungi are ideal for the study of segregation and recombination of genes in individual meioses.

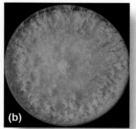
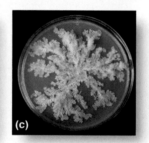

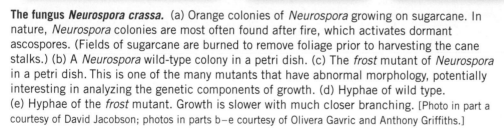

The fungus *Neurospora crassa*. (a) Orange colonies of *Neurospora* growing on sugarcane. In nature, *Neurospora* colonies are most often found after fire, which activates dormant ascospores. (Fields of sugarcane are burned to remove foliage prior to harvesting the cane stalks.) (b) A *Neurospora* wild-type colony in a petri dish. (c) The *frost* mutant of *Neurospora* in a petri dish. This is one of the many mutants that have abnormal morphology, potentially interesting in analyzing the genetic components of growth. (d) Hyphae of wild type. (e) Hyphae of the *frost* mutant. Growth is slower with much closer branching. [Photo in part a courtesy of David Jacobson; photos in parts b–e courtesy of Olivera Gavric and Anthony Griffiths.]

a *population* of gametes arising from many meiocytes. The *most likely* explanation is that equal segregation is taking place in each individual *A/a* meiocyte, but it cannot be observed *directly* in plants (or animals). Luckily, there is a way to visualize equal segregation directly. Haploid fungi called *ascomycetes* are unique in that for any given meiocyte the spores that are the products of meiosis are held together in a membranous sac called an **ascus.** Thus for these organisms it is possible to see the products of a single meiosis. In the pink bread mold *Neurospora*, the nuclear spindles of meiosis I and II do not overlap within the cigar-shaped ascus, so the four products of a single meiocyte lie in a straight row (Fig-

ure 3-37a). Furthermore for some reason not understood there is a *postmeiotic mitosis*, which also shows no spindle overlap, resulting in a linear ascus containing eight spores called *ascospores*. A cross in *Neurospora* is made by mixing two parental haploid strains of opposite mating type, as shown in Figure 3-38. Mating type is a simple form of sex, determined by two alleles of one gene, called MAT-A and MAT-a.

Let's make a cross between the normal pink wild type and an albino strain caused by a single mutation in a pigment gene. We will assume they are of opposite mating type. Hence the cross between the haploid parents can be represented very simply as follows, where

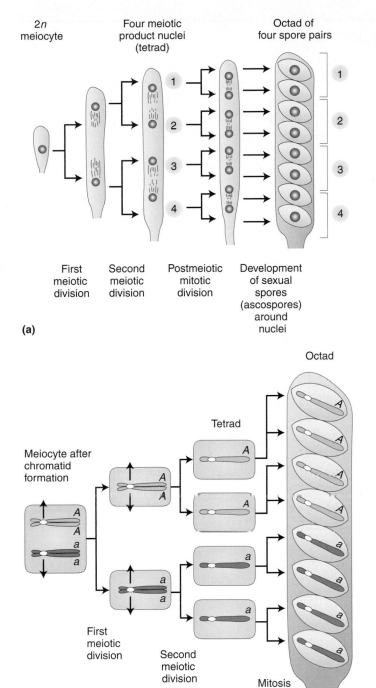

First meiotic division | Second meiotic division | Postmeiotic mitotic division | Development of sexual spores (ascospores) around nuclei

(a)

(b)

al^+ represents the wild-type allele coding for pigment production, and al the albino-causing allele:

$$\begin{array}{ccc} \text{Pink wild type} & & \text{albino} \\ al^+ & \times & al \end{array}$$

In the haploid sexual cycle, we have seen that haploid parental cells of each type fuse to form transient diploid meiocytes, each one of which produces an eight-spore ascus. Every ascus from the above cross will have four pink (al^+) and four white (al) ascospores, directly demonstrating the Mendelian law of equal segregation at the level of a single meiocyte (Figure 3-37b).

Figure 3-37 *Neurospora* is an ideal model system for studying allelic segregation at meiosis. (a) The four products of meiosis (tetrad) undergo mitosis to produce an octad. The products are contained within an ascus. (b) An *A/a* meiocyte undergoes meiosis, then mitosis, resulting in equal numbers of *A* and *a* products, demonstrating the principle of equal segregation.

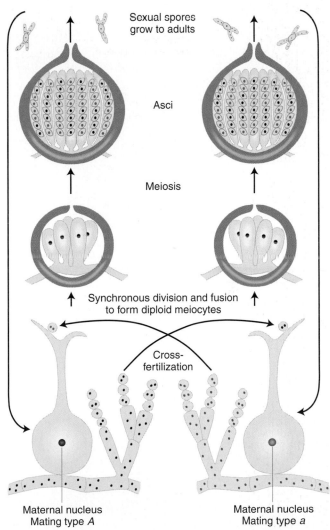

Figure 3-38 The life cycle of *Neurospora crassa*, the orange bread mold. Self-fertilization is not possible in this species: there are two mating types, determined by the alleles *A* and *a* of one gene. A cross will succeed only if it is *A* × *a*. An asexual spore from the opposite mating type fuses with a receptive hair, and a nucleus travels down the hair to pair with a nucleus in the knot of cells. The *A* and *a* pair then undergo synchronous mitoses, finally fusing to form diploid meiocytes.

WWW ANIMATED ART Meiotic recombination between unlinked genes by independent assortment

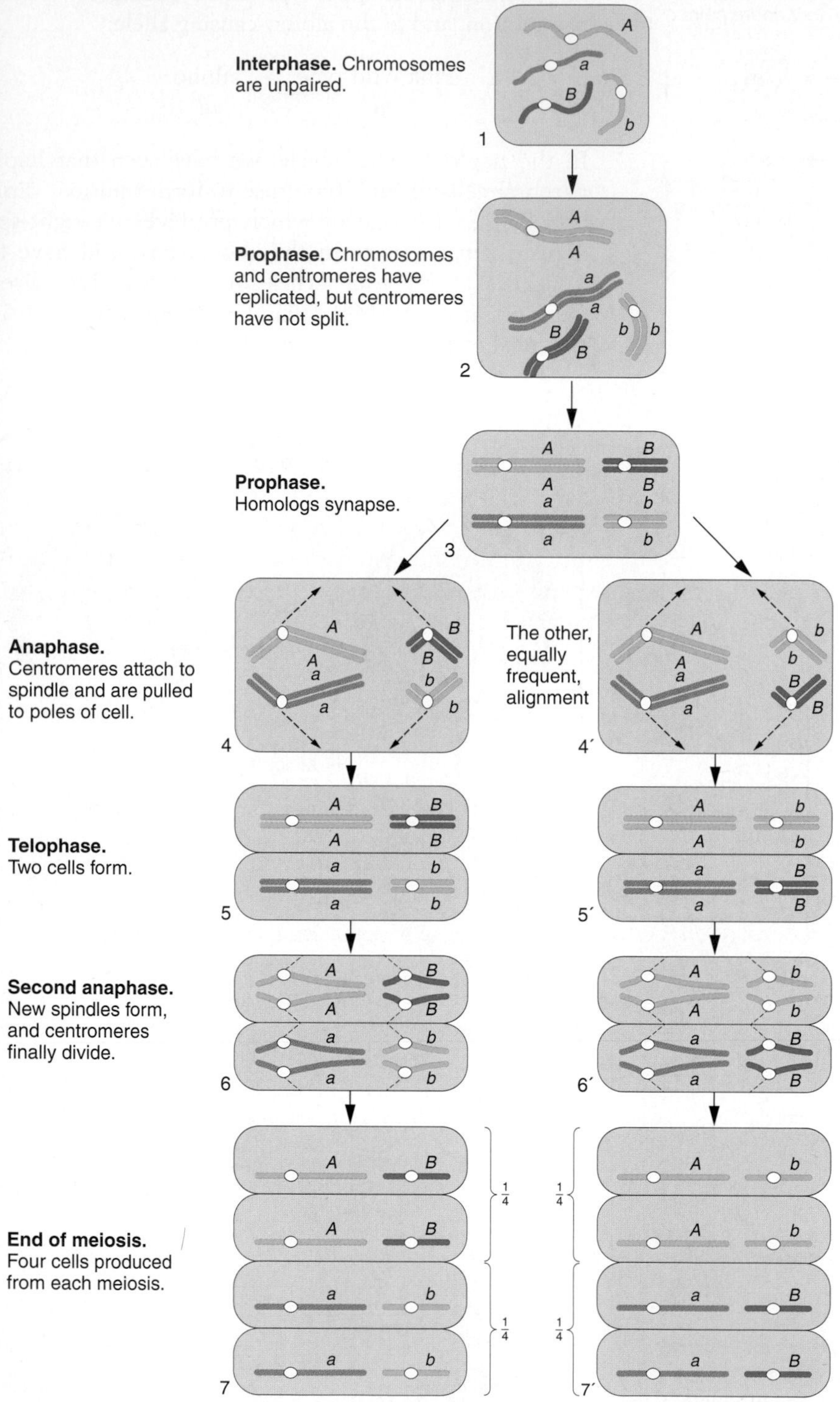

Interphase. Chromosomes are unpaired.
1
Prophase. Chromosomes and centromeres have replicated, but centromeres have not split.
2
Prophase. Homologs synapse.
3
Anaphase. Centromeres attach to spindle and are pulled to poles of cell.
4
The other, equally frequent, alignment
4′
Telophase. Two cells form.
5
5′
Second anaphase. New spindles form, and centromeres finally divide.
6
6′
End of meiosis. Four cells produced from each meiosis.
7
7′
1/4
1/4
1/4
1/4
A
a
B
b

Figure 3-39 Meiosis in a diploid cell of genotype *A/a* ; *B/b*. The diagram shows how the segregation and assortment of different chromosome pairs give rise to the 1 : 1 : 1 : 1 Mendelian gametic ratio.

Now we leave *Neurospora* to consider the law of independent assortment. The situation is diagrammed in Figure 3-39. The figure illustrates how the separate behavior of two different chromosome pairs gives rise to the 1 : 1 : 1 : 1 Mendelian ratios of gametic types diagnostic of independent assortment. The genotype of the meiocytes is *A/a* ; *B/b*, and the two allelic pairs, *A/a* and *B/b*, are shown on two different chromosome pairs. The hypothetical cell has four chromosomes: a pair of homologous long chromosomes and a pair of homologous short ones. Parts 4 and 4′ of Figure 3-39 show the key step in understanding the effects of Mendel's laws: they illustrate both equal segregation and independent assortment. There are two different allelic segregation patterns, one shown in 4 and one in 4′, that result from two equally frequent spindle attachments to the centromeres in the first anaphase. Meiosis then produces four cells of the genotypes shown from each of these segregation patterns. Segregation patterns 4 and 4′ are equally common, and therefore the meiotic product cells of genotypes A ; B, a ; b, A ; b, and a ; B are produced in equal frequencies. In other words, the frequency of each of the four genotypes is $\frac{1}{4}$. This gametic distribution is that postulated by Mendel for a dihybrid, and it is the one that we insert along one edge of the Punnett square. The random fusion of these gametes results in the $9:3:3:1$ F_2 phenotypic ratio.

> **MESSAGE** Mendelian laws apply to meiosis in any organism and may be generally stated as follows:
>
> 1. At meiosis, the alleles of a gene segregate equally into the haploid products of meiosis.
> 2. At meiosis, the alleles of one gene segregate independently of the alleles of genes on other chromosome pairs.

3.5 Organelle chromosomes

Organelle genomes are neither haploid nor diploid. Consider chloroplasts, for example. Any green cell of a plant has many chloroplasts, and each chloroplast contains many identical circular DNA molecules, the so-called chloroplast chromosomes. Hence the number of chloro-

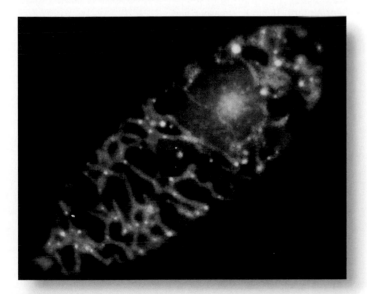

Figure 3-40 Fluorescent staining of a cell of *Euglena gracilis.* With the dyes used, the nucleus appears red because of the fluorescence of large amounts of nuclear DNA. The mitochondria fluoresce green, and within mitochondria the concentrations of mtDNA (nucleoids) fluoresce yellow. [From Y. Huyashi and K. Veda, *Journal of Cell Science* 93, 1989, 565.]

plast chromosomes per cell can number in the hundreds or thousands, and the number can even vary somewhat from cell to cell. The DNA is packaged into suborganellar structures called *nucleoids*, which become visible if stained with a DNA-binding dye (Figure 3-40). The DNA is folded within the nucleoid, but does not have the type of histone-associated coiling shown by nuclear chromosomes. The same arrangement is true for mitochondria.

Many organelle chromosomes have now been sequenced. Some examples of relative gene size and spacing in mitochondrial DNA (**mtDNA**) and chloroplast DNA (**cpDNA**) are shown in Figure 3-41. Organelle genes are very closely spaced, and in some organisms genes can contain introns. Note that most genes are concerned with the chemical reactions going on within the organelle itself; photosynthesis in chloroplasts, and oxidative phosphorylation in mitochondria. However, organelle chromosomes are not self-sufficient: many proteins that act within the organelle are encoded by nuclear genes (see Figure 1-9).

Cytoplasmic organelle chromosomes are embraced by the chromosome theory of heredity, but we must be careful not to associate their inheritance patterns with mitosis or meiosis, which are both nuclear processes. A zygote inherits its organelles from the cytoplasm of the egg, and thus organelle inheritance is generally maternal, as described in Chapter 2. Hence, where there is only one copy of a nuclear chromosome per gamete, there are many copies of an organellar chromosome. In a wild-type cross it is a *population* of identical organelle chromosomes that is transmitted to the offspring through the egg. Phenotypic variants caused by organelle gene mutations are also inherited maternally. Hence the "ratio" produced is $1:0$ or $0:1$ depending on which parent is female. But note that although this progeny ratio can be observed in the same individuals that result from meiosis, it has nothing to do with meiosis. This transmission pattern is shown for the *Neurospora* mutant *poky* in Figure 3-42.

Given that a cell is a population of organelle molecules, how is it ever possible to obtain a "pure" mutant cell, containing only mutant chromosomes? Most likely, pure mutants are created in asexual cells as follows: The variants arise by mutation of a single gene in a single chromosome. Then in some cases the mutation-bearing chromosome may by chance rise in frequency in the population within the cell. This process is called *random genetic drift*. Cells with mixtures of organelle genotypes have been termed **cytohets**. A cell that is a cytohet may have, say, 60 percent A chromosomes and 40 percent a chromosomes. When this cell divides, sometimes all the A chromosomes go into one daughter, and all the a chromosomes into the other (again, by chance). More often this partitioning requires several subsequent

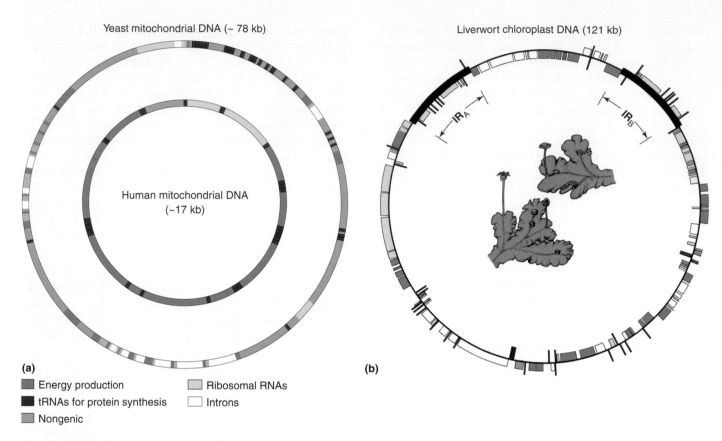

Figure 3-41 DNA maps for mitochondria and chloroplasts. Many of the organelle genes encode proteins that carry out the energy-producing functions of these organelles (green), whereas others (red and orange) function in protein synthesis. (a) Maps of yeast and human mtDNAs. (Note that the human map is not drawn to the same scale as the yeast map.) (b) The 121-kb chloroplast genome of the liverwort *Marchantia polymorpha.* Genes shown inside the map are transcribed clockwise, and those outside are transcribed counterclockwise. IR$_A$ and IR$_B$ indicate inverted repeats. The drawing in the center of the map depicts a male *(upper)* and a female *(lower) Marchantia* plant. [From K. Umesono and H. Ozeki, *Trends in Genetics* 3, 1987.]

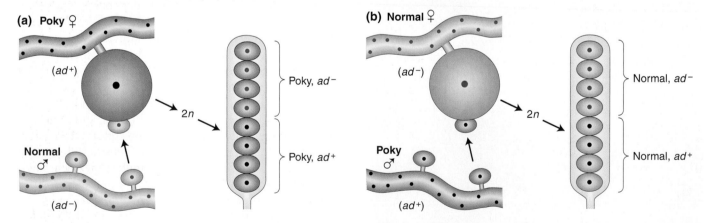

Figure 3-42 Explanation of the different results from reciprocal crosses of poky and normal *Neurospora*. The parent contributing most of the cytoplasm of the progeny cells is called female. Brown shading represents cytoplasm with mitochondria containing the poky mutation, and green shading cytoplasm with normal mitochondria. Note that in (a) the progeny will all be poky, whereas in (b) the progeny are all normal. Hence both crosses show maternal inheritance. The nuclear gene with the alleles ad^+ (black) and ad^- (red) is used to illustrate the segregation of the nuclear genes in the 1 : 1 Mendelian ratio expected for this haploid organism.

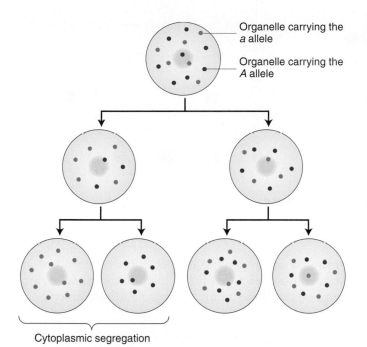

generations of cell division to be complete (Figure 3-43). Hence as a result of these chance events both alleles are expressed in different daughter cells, and this separation will continue through the descendants of these cells. This type of genetic segregation is called **cytoplasmic segregation.** Note that it is not a mitotic process; it takes place in dividing asexual cells, but it is unrelated to mitosis. In chloroplasts, cytoplasmic segregation is a common mechanism for producing variegated (green/white) plants, as we saw in Chapter 2. In the *poky* mutant of the fungus *Neurospora*, also introduced in Chapter 2, the original mutation in one mtDNA molecule must have accumulated and undergone cytoplasmic segregation to produce the strain expressing the poky symptoms.

Figure 3-44 shows some of the mutations in human mitochondrial genes that can lead to disease when, by random drift and cytoplasmic segregation, they rise in frequency to such an extent that cell function is impaired. The inheritance of a human mitochondrial disease is

Figure 3-43 Cytoplasmic segregation.
By chance, genetically distinct organelles may segregate into separate cells over successive cell divisions. Red and blue dots represent populations of genetically distinguishable organelles, such as mitochondria with and without a mutation.

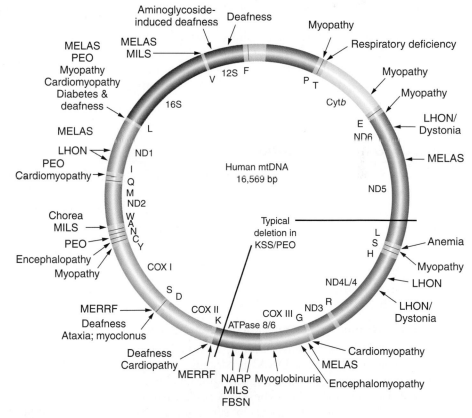

Figure 3-44 **Map of human mitochondrial DNA (mtDNA) showing loci of mutations leading to cytopathies.**
The tRNA genes are represented by single-letter amino acid abbreviations; ND = NADH dehydrogenase; COX = cytochrome oxidase; and 12S and 16S refer to ribosomal RNAs. [After S. DiMauro et al., "Mitochondria in Neuromuscular Disorders," *Biochim. Biophys. Acta* 1366, 1998, 199–210.]

Diseases:

MERRF	Myoclonic epilepsy and ragged red fiber disease
LHON	Leber hereditary optic neuropathy
NARP	Neurogenic muscle weakness, ataxia, and retinitis pigmentosum
MELAS	Mitochondrial encephalomyopathy, lactic acidosis, and strokelike symptoms
MMC	Maternally inherited myopathy and cardiomyopathy
PEO	Progressive external opthalmoplegia
KSS	Kearns-Sayre syndrome
MILS	Maternally inherited Leigh syndrome

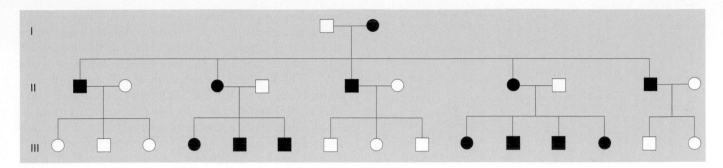

Figure 3-45 Pedigree showing maternal inheritance of a human mitochondrial disease.

shown in Figure 3-45. Note that the condition is always passed on through mothers and never fathers. Occasionally a mother will produce an unaffected child (not shown), probably reflecting cytoplasmic segregation in the gamete-forming tissue. Cytoplasmic segregation of accumulated deleterious mutations in certain tissues such as muscle and brain has been proposed as a mechanism of aging.

In certain special systems, cytohets that are "dihybrid" have been obtained (say, *A B* in one organelle chromosome and *a b* in another). In such cases, rare crossover-like processes can occur, but this must be con-

sidered a minor genetic phenomenon. Independent assortment is a term that is not relevant to organelle chromosomes, as there is only one organelle chromosome.

> **MESSAGE** Alleles on organelle chromosomes
> 1. In sexual crosses are inherited from one parent only (generally the maternal parent) and hence show no segregation ratios of the type nuclear genes do
> 2. Asexual cells can show cytoplasmic segregation
> 3. Asexual cells can occasionally show processes analogous to crossing-over

ANSWERS TO KEY QUESTIONS

• **How do we know that genes are parts of chromosomes?**

Originally this was shown by the exact match between the patterns of gene inheritance and chromosome behavior during meiosis. DNA sequencing has confirmed that genes are parts of chromosomes.

• **How are genes arranged on chromosomes?**

In a linear array from one end of the single DNA molecule to the other.

• **Does a chromosome contain material other than genes?**

There are two parts to the answer. First, a chromosome contains protein in addition to DNA, in particular histone protein, which promotes coiling and packing of DNA into the nucleus. Furthermore, within the DNA itself, there is nongenic DNA between the genes. In some organisms genes are close together with little intervening DNA. In others, there is substantial distance between genes, mostly consisting of various types of repetitive DNA.

• **How is chromosome number maintained constant through the generations?**

In asexual division, DNA replicates before cell division, and a copy passes to each daughter cell at mitosis. In the

production of gametes in a diploid organism the DNA replicates before division of the diploid meiocyte, then two successive divisions result in haploid meiotic products. Fusion of haploid products restores diploidy.

• **What is the chromosomal basis for Mendel's law of equal segregation?**

When a heterozygote *A/a* undergoes meiosis, the pairing of *A* and *a* homologs followed by their pulling apart during meiosis ensures that $\frac{1}{2}$ the products will be *A* and $\frac{1}{2}$ *a*.

• **What is the chromosomal basis for Mendel's law of independent assortment?**

In a dihybrid *A/a* ; *B/b*, spindle fiber attachment and pulling action are separate and independent for both chromosome pairs. Hence *A* (for example) can go to the same pole as *B* in some meioses, or the same pole as *b* in others.

• **How does all the DNA fit into a tiny nucleus?**

By coiling and folding. DNA is coiled and supercoiled, and the supercoils are arranged as loops upon a central scaffold.

SUMMARY

When Mendel's principles were found to be widespread in application, scientists set out to discover what structures within cells correspond to Mendel's hypothetical units of heredity, which we now call genes. Recognizing that the behavior of chromosomes during meiosis parallels the behavior of genes, Sutton and Boveri suggested that genes were located in or on the chromosomes. Morgan's evidence from sex-linked inheritance strengthened the hypothesis. Proof that genes are on chromosomes came from Bridges's use of aberrant chromosome behavior to explain inheritance anomalies.

Chromosomes are distinguished by a number of topological features such as centromere and nucleolar position, size, and banding pattern. The stuff that chromosomes are made of is chromatin, composed of DNA and protein. Each chromosome is one DNA molecule wrapped around octamers of histone proteins. Between cell divisions, the chromosomes are in a relatively extended state, although still associated with histones. At cell division, the chromosomes condense by a tightening of the coiling. This condensed state permits easy manipulation by the nuclear spindle fibers during cell division.

The DNA in eukaryotic genomes consists partly of transcribed sequences (genes) and partly of repetitive DNA of unknown function. There are many classes of repetitive DNA, with a large range of unit sizes. One type of highly repetitive DNA consists of short-sequence repeats located around the centromeres.

Mitosis is the nuclear division that results in two daughter nuclei whose genetic material is identical with that of the original nucleus. Mitosis can take place in diploid or haploid cells during asexual cell division.

Meiosis is the nuclear division by which a diploid meiocyte divides twice to produce four meiotic products, each of which is haploid (has only one set of chromosomes).

We now know which chromosome behaviors produce Mendelian ratios. Mendel's first law (equal segregation) results from the separation of a pair of homologous chromosomes into opposite cells at the first division of meiosis. Mendel's second law (independent assortment) results from independent behavior of separate pairs of homologous chromosomes.

Because Mendelian laws are based on meiosis, Mendelian inheritance characterizes any organism with a meiotic stage in its life cycle, including diploid organisms, haploid organisms, and organisms with alternating haploid and diploid generations.

Genes on organelle chromosomes show their own inheritance patterns, which are different from those of nuclear genes. Maternal inheritance and cytoplasmic segregation of heterogenic organelle mixtures are the two main types observed.

KEY TERMS

acrocentric (p. 85)

anaphase (p. 92)

ascus (p. 100)

bands (p. 88)

bivalents (p. 95)

chiasma (p. 95)

chromatin (p. 84)

chromocenter (p. 87)

chromosome bands (p. 86)

chromosome theory
 of heredity (p. 75)

cpDNA (p. 103)

crossing-over (p. 95)

cytohet (p. 103)

cytoplasmic segregation (p. 105)

daughter cell (p. 93)

diploid (p. 83)

dyads (p. 95)

euchromatin (p. 84)

FISH (fluorescent in situ
 hybridization) (p. 88)

gamete (p. 93)

haploid (p. 83)

heterochromatin (p. 84)

histones (p. 88)

interphase (p. 91)

kinetochore (p. 92)

meiocytes (p. 93)

meiosis (p. 93)

meiospores (p. 93)

metacentric (p. 85)

metaphase (p. 91)

mitosis (p. 90)

mtDNA (p. 103)

n (p. 83)

nondisjunction (p. 79)

nuclear spindle (p. 92)

nucleolar organizers (NO) (p. 85)

nuclcoli (p. 85)

nucleosome (p. 88)

polytene chromosomes (p. 87)

products of meiosis (p. 93)

prophase (p. 91)

S phase (p. 91)

satellite DNA (p. 85)

scaffold (p. 89)

scaffold attachment regions
 (SARs) (p. 89)

sister chromatids (p. 91)

solenoid (p. 88)

synaptonemal complexes (p. 95)

telocentric (p. 85)

telomeres (p. 86)

telophase (p. 93)

tetrad (p. 95)

wild type (p. 79)

SOLVED PROBLEMS

1. Two *Drosophila* flies that had normal (transparent, long) wings were mated. In the progeny, two new phenotypes appeared: dusky wings (having a semi-opaque appearance) and clipped wings (with squared ends). The progeny were as follows:

Females	179	transparent, long
	58	transparent, clipped
Males	92	transparent, long
	89	dusky, long
	28	transparent, clipped
	31	dusky, clipped

a. Provide a chromosomal explanation for these results, showing chromosomal genotypes of parents and of all progeny classes under your model.

b. Design a test for your model.

Solution

a. The first step is to state any interesting features of the data. The first striking feature is the appearance of two new phenotypes. We encountered the phenomenon in Chapter 2, where it was explained as recessive alleles masked by their dominant counterparts. So first we might suppose that one or both parental flies have recessive alleles of two different genes. This inference is strengthened by the observation that some progeny express only one of the new phenotypes. If the new phenotypes always appeared together, we might suppose that the same recessive allele determines both.

However, the other striking aspect of the data, which we cannot explain by using the Mendelian principles from Chapter 2, is the obvious difference between the sexes; although there are approximately equal numbers of males and females, the males fall into four phenotypic classes, but the females constitute only two. This fact should immediately suggest some kind of sex-linked inheritance. When we study the data, we see that the long and clipped phenotypes are segregating in both males and females, but only males have the dusky phenotype. This observation suggests that the inheritance of wing transparency differs from the inheritance of wing shape. First, long and clipped are found in a 3 : 1 ratio in both males and females. This ratio can be explained if the parents are both heterozygous for an autosomal gene; we can represent them as L/l, where L stands for long and l stands for clipped.

Having done this partial analysis, we see that it is only the inheritance of wing transparency that is associated with sex. The most obvious possibility is that the alleles for transparent (D) and dusky (d) are on the X chromosome, because we have seen in Chapter 2 that gene location on this chromosome gives inheritance patterns correlated with sex. If this suggestion is true, then

the parental female must be the one sheltering the d allele, because, if the male had the d, he would have been dusky, whereas we were told that he had transparent wings. Therefore, the female parent would be D/d and the male D. Let's see if this suggestion works: if it is true, all female progeny would inherit the D allele from their father, so all would be transparent winged. This was observed. Half the sons would be D (transparent) and half d (dusky), which was also observed.

So, overall, we can represent the female parent as D/d ; L/l and the male parent as D ; L/l. Then the progeny would be

Females

$$\frac{1}{2}\,D/D \begin{cases} \frac{3}{4}\,L/- \longrightarrow \frac{3}{8}\,D/D;\,L/- \\ \frac{1}{4}\,l/l \longrightarrow \frac{1}{8}\,D/D\,;\,l/l \end{cases}$$
$$\frac{1}{2}\,D/d \begin{cases} \frac{3}{4}\,L/- \longrightarrow \frac{3}{8}\,D/d\,;\,L/- \\ \frac{1}{4}\,l/l \longrightarrow \frac{1}{8}\,D/d\,;\,l/l \end{cases}$$

$\frac{3}{4}$ transparent, long

$\frac{1}{4}$ transparent, clipped

Males

$$\frac{1}{2}\,D \begin{cases} \frac{3}{4}\,L/- \longrightarrow \frac{3}{8}\,D\,;\,L/- \quad \text{transparent, long} \\ \frac{1}{4}\,l/l \longrightarrow \frac{1}{8}\,D\,;\,l/l \quad \text{transparent, clipped} \end{cases}$$
$$\frac{1}{2}\,d \begin{cases} \frac{3}{4}\,L/- \longrightarrow \frac{3}{8}\,d\,;\,L/- \quad \text{dusky, long} \\ \frac{1}{4}\,l/l \longrightarrow \frac{1}{8}\,d\,;\,l/l \quad \text{dusky, clipped} \end{cases}$$

b. Generally, a good way to test such a model is to make a cross and predict the outcome. But which cross? We have to predict some kind of ratio in the progeny, so it is important to make a cross from which a unique phenotypic ratio can be expected. Notice that using one of the female progeny as a parent would not serve our needs: we cannot say from observing the phenotype of any one of these females what her genotype is. A female with transparent wings could be D/D or D/d, and one with long wings could be L/L or L/l. It would be good to cross the parental female of the original cross with a dusky, clipped son, because the full genotypes of both are specified under the model that we have created. According to our model, this cross is:

$$D/d \; ; \; L/l \; \times \; d \; ; \; l/l$$

From this cross, we predict

Females

$$\frac{1}{2}\,D/d \begin{cases} \frac{1}{2}\,L/l \longrightarrow \frac{1}{4}\,D/d\,;\,L/l \\ \frac{1}{2}\,l/l \longrightarrow \frac{1}{4}\,D/d\,;\,l/l \end{cases}$$
$$\frac{1}{2}\,d/d \begin{cases} \frac{1}{2}\,L/l \longrightarrow \frac{1}{4}\,d/d\,;\,L/l \\ \frac{1}{2}\,l/l \longrightarrow \frac{1}{4}\,d/d\,;\,l/l \end{cases}$$

Males

$$\frac{1}{2}D \begin{cases} \frac{1}{2}L/l \longrightarrow \frac{1}{4}D;L/l \\ \frac{1}{2}l/l \longrightarrow \frac{1}{4}D;l/l \end{cases}$$

$$\frac{1}{2}d \begin{cases} \frac{1}{2}L/l \longrightarrow \frac{1}{4}d;L/l \\ \frac{1}{2}l/l \longrightarrow \frac{1}{4}d;l/l \end{cases}$$

2. Two corn plants are studied; one is A/a and the other is a/a. These two plants are intercrossed in two ways: using A/a as female and a/a as male, and using a/a as female and A/a as male. Recall from Figure 3-34 that the endosperm is $3n$ and is formed by the union of a sperm cell with the two polar nuclei of the female gametophyte.

a. What endosperm genotypes does each cross produce? In what proportions?

b. In an experiment to study the effects of "doses" of alleles, you wish to establish endosperms with genotypes $a/a/a$, $A/a/a$, $A/A/a$, and $A/A/A$ (carrying 0, 1, 2, and 3 doses of A, respectively). What crosses would you make to obtain these endosperm genotypes?

Solution

a. In such a question, we have to think about meiosis and mitosis at the same time. The meiospores are produced by meiosis; the nuclei of the male and female gametophytes in higher plants are produced by the mitotic division of the meiospore nucleus. We also need to study the corn life cycle to know what nuclei fuse to form the endosperm.

First cross: A/a ♀ × a/a ♂

Here, the female meiosis will result in spores of which half will be A and half will be a. Therefore, similar proportions of haploid female gametophytes will be produced. Their nuclei will be either all A or all a, because

mitosis reproduces genetically identical genotypes. Likewise, all nuclei in every male gametophyte will be a. In the corn life cycle, the endosperm is formed from two female nuclei plus one male nucleus, so two endosperm types will be formed as follows:

♀ spore	♀ polar nuclei	♂ sperm	$3n$ endosperm
$\frac{1}{2}A$	A and A	a	$\frac{1}{2}A/A/a$
$\frac{1}{2}a$	a and a	a	$\frac{1}{2}a/a/a$

Second cross: a/a ♀ × A/a ♂

♀ spore	♀ polar nuclei	♂ sperm	$3n$ endosperm
All a	All a and a	$\frac{1}{2}A$	$\frac{1}{2}A/a/a$
		$\frac{1}{2}a$	$\frac{1}{2}a/a/a$

The phenotypic ratio of endosperm characters would still be Mendelian, even though the underlying endosperm genotypes are slightly different. (However, none of these problems arise in embryo characters, because embryos are diploid.)

b. This kind of experiment has been very useful in studying plant genetics and molecular biology. In answering the question, all we need to realize is that the two polar nuclei contributing to the endosperm are genetically identical. To obtain endosperms, all of which will be $a/a/a$, any a/a × a/a cross will work. To obtain endosperms, all of which will be $A/a/a$, the cross must be a/a ♀ × A/A ♂. To obtain embryos, all of which will be $A/A/a$, the cross must be A/A ♀ × a/a ♂. For $A/A/A$, any A/A × A/A cross will work. These endosperm genotypes can be obtained in other crosses but only in combination with other endosperm genotypes.

PROBLEMS

BASIC PROBLEMS

1. Name the key function of mitosis.

2. Name two key functions of meiosis.

3. Can you design a different nuclear division system that would achieve the same outcome as meiosis?

4. In a possible future scenario, male fertility drops to zero, but, luckily, scientists develop a way for women to produce babies by virgin birth. Meiocytes are converted directly (without undergoing meiosis) into zygotes, which implant in the usual way. What would be the short- and long-term effects in such a society?

5. In what ways does the second division of meiosis differ from mitosis?

6. a. In a cell, two homologous dyads are seen, one at each of the poles. What stage of nuclear division is this?

b. In a meiocyte where $2n = 14$, how many bivalents will be visible?

7. In the fungus *Neurospora*, the mutant allele *lys-5* causes the ascospores bearing that allele to be white, whereas the wild-type allele *lys-5⁺* results in black ascospores. (Ascospores are the spores that constitute the octad.) Draw a linear octad from each of the following crosses:

a. *lys-5* × *lys-5⁺*

b. *lys-5* × *lys-5*

c. *lys-5⁺* × *lys-5⁺*

8. Make up mnemonics for remembering the five stages of prophase I of meiosis and the four stages of mitosis.

9. In *Neurospora* 100 meiocytes develop in the usual way. How many ascospores will result?

10. Normal mitosis takes place in a diploid cell of genotype *A/a ; B/b*. Which of the following genotypes might represent possible daughter cells?

   a. *A ; B*   e. *A/A ; B/B*

   b. *a ; b*   f. *A/a ; B/b*

   c. *A ; b*   g. *a/a ; b/b*

   d. *a ; B*

11. In an attempt to simplify meiosis for the benefit of students, mad scientists develop a way of preventing premeiotic S phase and making do with just having one division, including pairing, crossing-over, and segregation. Would this system work, and would the products of such a system differ from those of the present system?

12. In a diploid organism of $2n = 10$, assume that you can label all the centromeres derived from its female parent and all the centromeres derived from its male parent. When this organism produces gametes, how many male- and female-labeled centromere combinations are possible in the gametes?

13. In corn, the DNA in several nuclei is measured on the basis of its light absorption. The measurements were

$$0.7, 1.4, 2.1, 2.8, \text{ and } 4.2$$

Which cells could have been used for these measurements?

14. Draw a haploid mitosis of the genotype *a⁺ ; b*.

15. Peas are diploid and $2n = 14$. *Neurospora* is haploid and $n = 7$. If it were possible to fractionate genomic DNA from both by using pulsed field electrophoresis, how many distinct bands would be visible in each species?

16. The broad bean (*Vicia faba*) is diploid and $2n = 18$. Each haploid chromosome set contains approximately 4 m of DNA. The average size of each

chromosome during the metaphase of mitosis is 13 μm. What is the average packing ratio of DNA at metaphase? (Packing ratio = length of chromosome/length of DNA molecule therein.) How is this packing achieved?

17. Boveri said, "The nucleus doesn't divide; it is divided." What was he getting at?

18. Galton, a geneticist of the pre-Mendelian era, devised the principle that half of our genetic makeup is derived from each parent, one-quarter from each grandparent, one-eighth from each great grandparent, and so forth. Was he right? Explain.

19. If children obtain half their genes from one parent and half from the other parent, why aren't siblings identical?

20. In corn, the allele *s* causes sugary endosperm, whereas *S* causes starchy. What endosperm genotypes result from the following crosses?

   s/s female × *S/S* male

   S/S female × *s/s* male

   S/s female × *S/s* male

21. In moss, the genes *A* and *B* are expressed only in the gametophyte. A sporophyte of genotype *A/a ; B/b* is allowed to produce gametophytes.

   a. What proportion of the gametophytes will be *A ; B*?

   b. If fertilization is random, what proportion of sporophytes in the next generation will be *A/a ; B/b*?

22. When a cell of genotype *A/a ; B/b ; C/c* having all the genes on separate chromosome pairs divides mitotically, what are the genotypes of the daughter cells?

23. In the haploid yeast *Saccharomyces cerevisiae* the two mating types are known as MATa and MATα. You cross a purple (*ad⁻*) strain of mating type *a* and a white (*ad⁺*) strain of mating type *α*. If *ad⁻* and *ad⁺* are alleles of one gene, and *a* and *α* are alleles of an independently inherited gene on a separate chromosome pair, what progeny do you expect to obtain? In what proportions?

24. State where cells divide mitotically and where they divide meiotically in a fern, a moss, a flowering plant, a pine tree, a mushroom, a frog, a butterfly, and a snail.

25. Human cells normally have 46 chromosomes. For each of the following stages, state the number of nuclear DNA molecules present in a human cell:

a. Metaphase of mitosis

b. Metaphase I of meiosis

c. Telophase of mitosis

d. Telophase I of meiosis

e. Telophase II of meiosis

26. Four of the following events arc part of both meiosis and mitosis, but one is only meiotic. Which one? (1) Chromatid formation, (2) spindle formation, (3) chromosome condensation, (4) chromosome movement to poles, (5) synapsis.

27. Suppose you discover two interesting *rare* cytological abnormalities in the karyotype of a human male. (A karyotype is the total visible chromosome complement.) There is an extra piece (or satellite) on *one* of the chromosomes of pair 4, and there is an abnormal pattern of staining on one of the chromosomes of pair 7. Assuming that all the gametes of this male are equally viable, what proportion of his children will have the same karyotype as his?

28. Suppose that meiosis occurs in the transient diploid stage of the cycle of a haploid organism of chromosome number *n*. What is the probability that an individual haploid cell resulting from the meiotic division will have a complete parental set of centromeres (that is, a set all from one parent or all from the other parent)?

29. Pretend that the year is 1868. You are a skilled young lens maker working in Vienna. With your superior new lenses, you have just built a microscope that has better resolution than any others available. In your testing of this microscope, you have been observing the cells in the testes of grasshoppers and have been fascinated by the behavior of strange elongated structures that you have seen within the dividing cells. One day, in the library, you read a recent journal paper by G. Mendel on hypothetical "factors" that he claims explain the results of certain crosses in peas. In a flash of revelation, you are struck by the parallels between your grasshopper studies and Mendel's, and you resolve to write him a letter. What do you write?

(Based on an idea by Ernest Kroeker.)

30. A dark female moth is crossed to a dark male. The male progeny are all dark, but half the females are light and the rest are dark. Propose an explanation for this pattern of inheritance.

31. Reciprocal crosses and selfs were performed between the two moss species *Funaria mediterranea* and *F. hygrometrica*. The sporophytes and the leaves of the gametophytes are shown in the accompanying diagram.

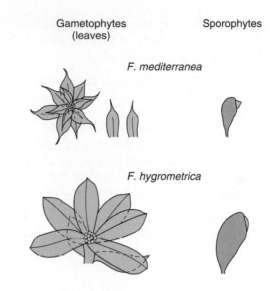

Gametophytes (leaves)   Sporophytes

F. mediterranea

F. hygrometrica

The crosses are written with the female parent first.

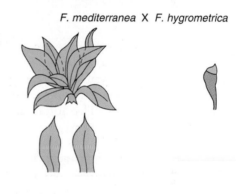

F. mediterranea X *F. hygrometrica*

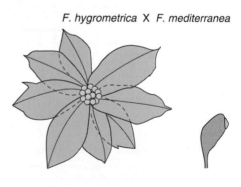

F. hygrometrica X *F. mediterranea*

a. Describe the results presented, summarizing the main findings.

b. Propose an explanation of the results.

c. Show how you would test your explanation; be sure to show how it could be distinguished from other explanations.

32. Assume that diploid plant A has a cytoplasm genetically different from that of plant B. To study nuclear–cytoplasmic relations, you wish to obtain a

plant with the cytoplasm of plant A and the nuclear genome predominantly of plant B. How would you go about producing such a plant?

33. You are studying a plant with tissue comprising both green and white sectors. You wish to decide whether this phenomenon is due (1) to a chloroplast mutation of the type considered in this chapter or (2) to a dominant nuclear mutation that inhibits chlorophyll production and is present only in certain tissue layers of the plant as a mosaic. Outline the experimental approach that you would use to resolve this problem.

34. Early in the development of a plant, a mutation in cpDNA removes a specific *Bg*III restriction site (*B*) as follows:

Normal cpDNA
$$- - +\overset{B}{\vert}\quad\quad\overset{B}{\vert}\overset{B}{\vert}\quad\quad\quad\quad\overset{B}{\vert} + -$$

Mutant cpDNA
$$- - +\overset{B}{\vert}\quad\quad\quad\overset{B}{\vert}\quad\quad\quad\quad\overset{B}{\vert} + -$$

$$\vdash\!\!-\!\!\underset{P}{-\!\!-}\!\!\dashv$$

In this species cpDNA is inherited maternally. Seeds from the plant are grown, and the resulting progeny plants are sampled for cpDNA. The cpDNAs are cut with *Bg*III, and Southern blots are hybridized with the probe P shown. The autoradiograms show three patterns of hybridization:

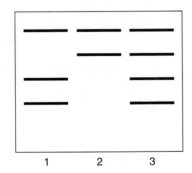

Explain the production of these three seed types.

CHALLENGING PROBLEMS

35. The plant *Haplopappus gracilis* has a 2*n* of 4. A diploid cell culture was established and, at premitotic S phase, a radioactive nucleotide was added and was incorporated into newly synthesized DNA. The cells were then removed from the radioactivity, washed, and allowed to proceed through mitosis. Radioactive chromosomes or chromatids can be detected by placing photographic emulsion on the cells; radioactive chromosomes or chromatids appeared covered with spots of silver from the emulsion. (The chromosomes "take their own photograph.") Draw the chromo-

somes at prophase and telophase of the first and second mitotic divisions after the radioactive treatment. If they are radioactive, show it in your diagram. If there are several possibilities, show them, too.

36. In the species of Problem 35, you can introduce radioactivity by injection into the anthers at the S phase before meiosis. Draw the four products of meiosis with their chromosomes and show which are radioactive.

37. The DNA double helices of chromosomes can be partly unwound in situ by special treatments. What pattern of radioactivity is expected if such a preparation is bathed in a radioactive pulse for

  a. a unique gene?

  b. dispersed repetitive DNA?

  c. ribosomal DNA?

  d. telomeric DNA?

  e. simple-repeat heterochromatic DNA?

38. If genomic DNA is cut with a restriction enzyme and fractionated by size by electrophoresis, what pattern of Southern hybridization is expected for the probes cited in Problem 37?

39. In corn, the allele f' causes floury endosperm, and f'' causes flinty endosperm. In the cross f'/f' ♀ × f''/f'' ♂, all the progeny endosperms are floury, but in the reciprocal cross, all the progeny endosperms are flinty. What is a possible explanation? (Check the corn life cycle.)

40. The plant *Haplopappus gracilis* is diploid and 2*n* = 4. There are one long pair and one short pair of chromosomes. The accompanying diagrams (see page 113) represent anaphases ("pulling apart" stages) of individual cells in meiosis or mitosis in a plant that is genetically a dihybrid (*A/a* ; *B/b*) for genes on different chromosomes. The lines represent chromosomes or chromatids, and the points of the V's represent centromeres. In each case, indicate if the diagram represents a cell in meiosis I, meiosis II, or mitosis. If a diagram shows an impossible situation, say so.

41. The accompanying pedigree shows the recurrence of a rare neurological disease (large black symbols) and spontaneous fetal abortion (small black symbols) in one family. (Slashes mean that the individual is deceased.) Provide an explanation for this pedigree in regard to cytoplasmic segregation of defective mitochondria.

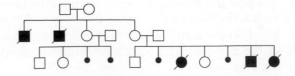

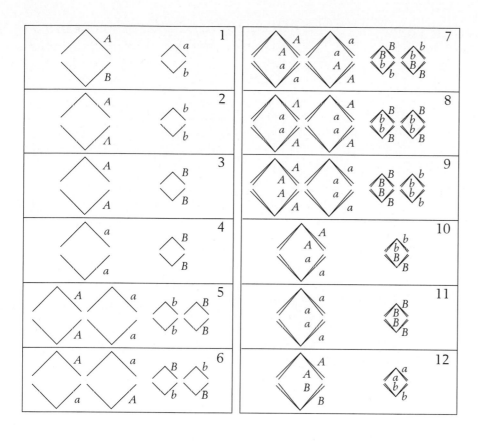

Diagram for Problem 40.

42. One form of male sterility in corn is maternally transmitted. Plants of a male-sterile line crossed with normal pollen give male-sterile plants. In addition, some lines of corn are known to carry a dominant nuclear restorer gene (*Rf*) that restores pollen fertility in male-sterile lines.

a. Research shows that the introduction of restorer genes into male-sterile lines does not alter or affect the maintenance of the cytoplasmic factors for male sterility. What kind of research results would lead to such a conclusion?

b. A male-sterile plant is crossed with pollen from a plant homozygous for gene *Rf*. What is the genotype of the F_1? The phenotype?

c. The F_1 plants from part b are used as females in a testcross with pollen from a normal plant (*rf/rf*). What would be the result of this testcross? Give genotypes and phenotypes, and designate the kind of cytoplasm.

d. The restorer gene already described can be called *Rf-1*. Another dominant restorer, *Rf-2*, has been found. *Rf-1* and *Rf-2* are located on different chromosomes. Either or both of the restorer alleles will give pollen fertility. With the use of a male-sterile plant as a tester, what would be the result of a cross in which the male parent is:

(i) heterozygous at both restorer loci?

(ii) homozygous dominant at one restorer locus and homozygous recessive at the other?

(iii) heterozygous at one restorer locus and homozygous recessive at the other?

(iv) heterozygous at one restorer locus and homozygous dominant at the other?

INTERACTIVE GENETICS MegaManual CD-ROM Tutorial

Chromosomal Inheritance

For a interactive review of mitosis and meiosis, refer to this activity on the Interactive Genetics CD-ROM included with the Solutions MegaManual. The activity also features interactive problems focusing on X-linked inheritance and a six-problem FlyLab that allows you to solve genetics problems by performing your own crosses using *Drosophila*.

EUKARYOTE CHROMOSOME MAPPING BY RECOMBINATION

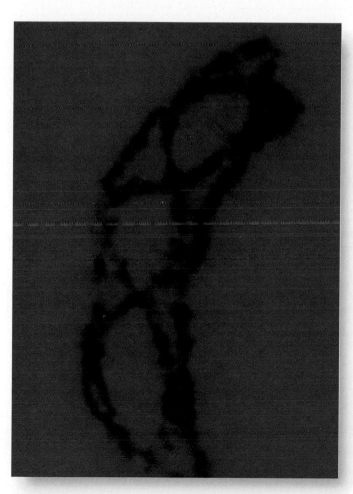

Chiasmata, the visible manifestations of crossing-over, photographed during meiosis in a grasshopper testis.
[John Cabisco/Visuals Unlimited.]

KEY QUESTIONS

- For genes on the same chromosome (known as *linked genes*), can new combinations of alleles be detected in the progeny of a dihybrid?

- If new combinations of alleles arise, by what cellular mechanism does this happen?

- Can the frequency of new combinations of alleles for linked genes be related to their distance apart on the chromosome?

- If we do not know whether two genes are linked, is there a diagnostic test that can be made?

OUTLINE

4.1 The discovery of the inheritance patterns of linked genes

4.2 Recombination

4.3 Linkage maps

4.4 Using the chi-square test in linkage analysis

4.5 Using Lod scores to assess linkage in human pedigrees

4.6 Accounting for unseen multiple crossovers

CHAPTER OVERVIEW

This chapter is about using crossing analysis to map the positions of genes on chromosomes. The map position of a gene is a key piece of information that is required to analyze its function. You might suppose that the sequencing of complete genomes has made crossing analysis redundant because such sequencing reveals genes and their positions on the chromosome. However, most of the genes identified by sequencing are of unknown function, and their impact on the organism is not understood. Hence the gene as identified by phenotypic analysis must be tied to the gene as identified by sequencing. This is where chromosome maps are crucial.

Mendel's experiments showed that allele pairs for genes influencing different pea characters such as size and color assorted independently, giving rise to precise ratios of progeny. To explain independent assortment, we noted that the genes influencing color and texture resided on different chromosomes, and that different chromosome pairs behave independently at meiosis. But what happens when allele pairs of different genes are on

CHAPTER OVERVIEW Figure

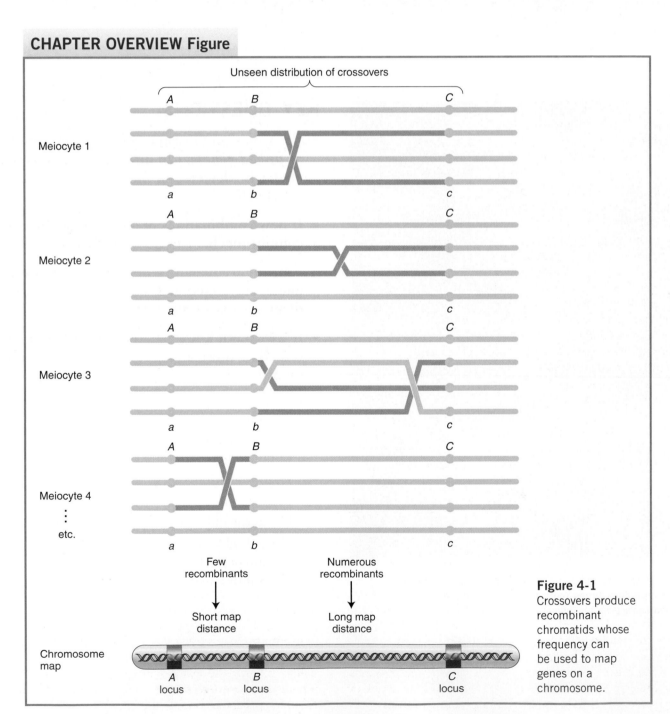

Figure 4-1
Crossovers produce recombinant chromatids whose frequency can be used to map genes on a chromosome.

the *same* chromosome? Do such allele pairs show any regularities in their patterns of inheritance? Are there diagnostic ratios or other patterns that are associated with such genes? The answer to these questions is "Yes, there are simple patterns that are unique to genes that reside on the same chromosome." These patterns are the topic of the present chapter.

Once again, a key element of the analysis is the dihybrid—an individual that is heterozygous for two genes of interest (say, *A/a* and *B/b*). Suppose that the two genes are on the same chromosome. Such dihybrids could be represented as follows:

$$\frac{A \qquad B}{a \qquad b}$$

At the simplest level, this chapter analyzes the inheritance patterns produced in the descendants of such dihybrids.

What would be the result of crossing the above dihybrid? Rather than sorting independently, *A* and *B* will most likely be inherited together, as will *a* and *b* (they are after all attached by the chromosome segment between them). As a consequence, the inheritance pattern that results will be different from those resulting from independent assortment. In some cases, however, the combinations of alleles on the parental chromosomes *can* be broken up, so that in a case like that above, *A* is inherited with *b* and *a* with *B*. The mechanism for this is a precise breakage and union of parental chromatids during meiosis. This cut-and-paste process is called *crossing-over*. It occurs at the stage when the four chromatids produced from a chromosome pair are grouped together as a tetrad. Crossover events occur more or less randomly along the tetrad; in some meiocytes they will be in certain positions and in other meiocytes at different positions.

The general situation is illustrated in Figure 4-1. This diagram is a "snapshot" of four different meiocytes that typify the ways that crossovers can distribute themselves among the four chromatids. (Many other arrangements are possible.) The figure shows in color the chromatids that carry the new combinations of alleles arising from crossovers. Crossovers have a very useful property: the larger the region between two genes, the more likely it is that a crossover will occur in that region. Hence crossovers are more likely to occur between genes that are far apart on the chromosome, and less likely to occur between genes that are close together. Thus the frequency of new combinations tells us whether two genes are far apart or close together and can be used to map the positions of the genes on chromosomes. Such a map is shown at the bottom of Figure 4-1, with the positions of the genes in question labeled "loci." These types of maps are of central importance in genomic analysis.

4.1 The discovery of the inheritance patterns of linked genes

Today the analysis of the inheritance patterns of genes on the same chromosome is routine in genetic research. The way in which early geneticists deduced these inheritance patterns is a useful lead-in, as it introduces most of the key ideas and procedures in the analysis.

Deviations from independent assortment

In the early 1900s, William Bateson and R. C. Punnett were studying inheritance of two genes in the sweet pea. In a standard self of a dihybrid F_1, the F_2 did not show the $9:3:3:1$ ratio predicted by the principle of independent assortment. In fact Bateson and Punnett noted that certain combinations of alleles showed up more often than expected, almost as though they were physically attached in some way. They had no explanation for this discovery.

Later, Thomas Hunt Morgan found a similar deviation from Mendel's second law while studying two autosomal genes in *Drosophila*. Morgan proposed a hypothesis to explain the phenomenon of apparent allele association.

Let's look at Morgan's data. One of the genes affected eye color (*pr*, purple, and *pr*$^+$, red), and the other wing length (*vg*, vestigial, and *vg*$^+$, normal). The wild-type alleles of both genes are dominant. Morgan performed a cross to obtain dihybrids, then followed with a testcross:

P $pr/pr \cdot vg/vg \times pr^+/pr^+ \cdot vg^+/vg^+$

F_1 $pr^+/pr \cdot vg^+/vg$
                                      Dihybrid

Testcross

        $pr^+/pr \cdot vg^+/vg \; ♀ \times pr/pr \cdot vg/vg \; ♂$
        F_1 dihybrid female   Tester male

His use of the testcross is important. As we saw in Chapter 2, because the tester parent contributes gametes carrying only recessive alleles, the phenotypes of the offspring directly reveal the alleles contributed by the gametes of the dihybrid parent. Hence, the analyst can concentrate on meiosis in one parent (the dihybrid) and essentially forget about the other (the tester). This contrasts with the analysis of progeny from an F_1 self, where there are two sets of meioses to consider: one in the male parent and one in the female.

Morgan's testcross results were as follows (listed as the gametic classes from the dihybrid):

$pr^+ \cdot vg^+$	1339
$pr \cdot vg$	1195
$pr^+ \cdot vg$	151
$pr \cdot vg^+$	154
	2839

Obviously, these numbers deviate drastically from the Mendelian prediction of a $1:1:1:1$ ratio, and they clearly indicate an association of certain alleles. The alleles associated with each other are those in the two largest classes, the combinations $pr^+ \cdot vg^+$ and $pr \cdot vg$. These are the very same allele combinations introduced by the original homozygous parental flies. The testcross also reveals another new finding: there is approximately a $1:1$ ratio not only between the two parental combinations ($1339 \approx 1195$), but also between the two nonparental combinations ($151 \approx 154$).

Now let's look at another cross Morgan made using the same alleles but in a different combination. In this cross, each parent is homozygous for the dominant allele of one gene and the recessive allele of the other. Again F_1 females were testcrossed:

$$P \qquad pr^+/pr^+ \cdot vg/vg \times pr/pr \cdot vg^+/vg^+$$

$$\downarrow$$

$$F_1 \qquad\qquad pr^+/pr \cdot vg^+/vg$$

Testcross   Dihybrid

$$pr^+/pr \cdot vg^+/vg\ ♀ \times pr/pr \cdot vg/vg\ ♂$$

F$_1$ dihybrid female   Tester male

The following progeny were obtained from the testcross:

$pr^+ \cdot vg^+$	157
$pr \cdot vg$	146
$pr^+ \cdot vg$	965
$pr \cdot vg^+$	1067
	2335

Again, these results are not even close to a $1:1:1:1$ Mendelian ratio. Now, however, the largest classes are the converse of those in the first analysis. But notice that once again the most frequent classes in the testcross progeny correspond to the allele combinations that were originally contributed to the F_1 by the parental flies.

Morgan suggested that the two genes in his analyses are located *on the same pair of homologous chromosomes*. Thus, in the first analysis, when *pr* and *vg* are introduced from one parent, they are physically located on the same chromosome, whereas *pr*$^+$ and *vg*$^+$ are on the homologous chromosome from the other parent (Figure 4-2). In

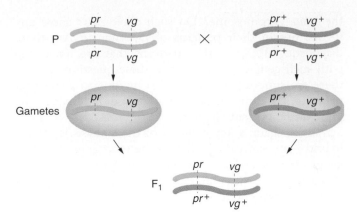

Figure 4-2 Inheritance of linked genes. Simple inheritance of two pairs of alleles located on the same chromosome pair.

the second analysis, one parental chromosome carries *pr* and *vg*$^+$ and the other carries *pr*$^+$ and *vg*.

Results like those just presented are commonly encountered in analysis, and are not exceptions but part of the general fabric of genetics. They indicate that the genes under analysis are on the same chromosome. In modern parlance, genes on the same chromosome are said to be **linked**.

> **MESSAGE** When two genes are close together on the same chromosome pair (i.e., linked), they do not assort independently.

Linkage symbolism and terminology

The work of Morgan showed that linked genes in a dihybrid may be present in one of two basic conformations. In one, the two dominant or wild-type alleles are present on the same homolog (as in Figure 4-2); this arrangement is called a **cis conformation**. In the other, they are on different homologs, in what is called a **trans conformation**. The two conformations are written as follows:

Cis	*A B/a b*
Trans	*A b/a B*

Note the following conventions:

1. Alleles on the same homolog have no punctuation between them.

2. A slash symbolically separates the two homologs.

3. Alleles are always written in the same order on each homolog.

4. As we have seen in earlier chapters, genes known to be on different chromosomes (unlinked genes) are shown separated by a semicolon—for example, *A/a ; C/c*.

5. In this book, genes of unknown linkage are shown separated by a dot, *A/a · D/d*.

New combinations of alleles arise from crossovers

The linkage hypothesis explains why allele combinations from the parental generations remain together—because they are physically attached by the segment of chromosome between them. But how do we explain the appearance of the minority class of nonparental combinations? Morgan suggested that when homologous chromosomes pair in meiosis, the chromosomes occasionally break and exchange parts in a process called **crossing-over.** Figure 4-3 illustrates this physical exchange of chromosome segments. The two new combinations are called **crossover products.**

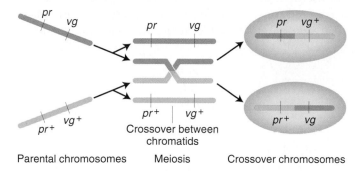

Parental chromosomes   Meiosis   Crossover chromosomes

Figure 4-3 Crossing-over in meiosis. The exchange of parts by crossing-over may produce gametic chromosomes whose allelic combinations differ from the parental combinations.

Is there any cytologically observable process that could account for crossing-over? We saw in Chapter 3 that in meiosis, when duplicated homologous chromosomes pair with each other, a cross-shaped structure called a *chiasma* (pl., chiasmata) often forms between two nonsister chromatids. Chiasmata are shown well in the chapter-opening figure. To Morgan, the appearance of the chiasmata visually corroborated the concepts of crossing-over. (Note that the chiasmata seem to indicate it is *chromatids*, not unduplicated chromosomes, that participate in a crossover. We shall return to this point later.)

> **MESSAGE** Chiasmata are the visible manifestations of crossovers.

Evidence that crossing-over is a breakage-and-rejoining process

The idea that recombinants were produced by some kind of exchange of material between homologous chromosomes was a compelling one. But experimentation was necessary to test this hypothesis. One of the first

steps was to find a case where the exchange of parts between chromosomes would be visible under the microscope. Several investigators approached this problem in the same way, and one of these analyses follows.

In 1931, Harriet Creighton and Barbara McClintock were studying two genes of corn that they knew were both located on chromosome 9. One affected seed color (C, colored; c, colorless), and the other affected endosperm composition (*Wx*, waxy; *wx*, starchy). However, in one plant the chromosome 9 carrying the alleles C and *Wx* was unusual in that it carried a large, densely staining element (called a *knob*) on the C end and a longer piece of chromosome on the *Wx* end; thus, the heterozygote was

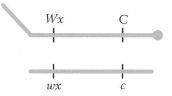

In the progeny of a testcross of this plant they compared the chromosomes carrying new allele combinations with those carrying parental alleles. They found that all the recombinants inherited one or the other of the two following chromosomes, depending on their recombinant makeup:

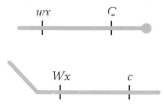

Thus, there was a precise correlation between the *genetic* event of the appearance of new allele combinations and the *chromosomal* event of crossing-over. Consequently, the chiasmata appeared to be the sites of exchange, although the final proof did not come until 1978.

What can we say about the molecular mechanism of chromosome exchange in a crossover event? The short answer is that a crossover results from breakage and reunion of DNA. Two parental chromosomes break at the same position, and then each piece joins up with the neighboring piece from the *other* chromosome. In Chapter 14 we will study models of the molecular processes that allow DNA to break and rejoin in a precise manner such that no genetic material is lost or gained.

> **MESSAGE** A crossover is the breakage of two DNA molecules at the same position and their rejoining in two reciprocal nonparental combinations.

Evidence that crossing-over occurs at the four-chromatid stage

As previously noted, the diagrammatic representation of crossing-over in Figure 4-3 shows a crossover taking place at the four-chromatid stage of meiosis. However, it was *theoretically* possible that crossing-over occurred at the *two*-chromosome stage (before replication). This uncertainty was resolved through the genetic analysis of organisms whose four products of meiosis remain together in groups of four called **tetrads.** These organisms, which we met in Chapter 3, are fungi and unicellular algae. The products of meiosis in a single tetrad can be isolated, which is equivalent to isolating all four chromatids from a single meiocyte. Tetrad analyses of crosses *in which genes are linked* show many tetrads that contain four different allele combinations. For example, from the cross

$$A\,B \ \times\ a\,b$$

some (but not all) tetrads contain four genotypes:

$$A\,B$$
$$A\,b$$
$$a\,B$$
$$a\,b$$

This result can be explained only if crossovers occur at the four-chromatid stage because if crossovers occurred at the two-chromosome stage, there could only ever be a maximum of two different genotypes in an individual tetrad. This reasoning is illustrated in Figure 4-4.

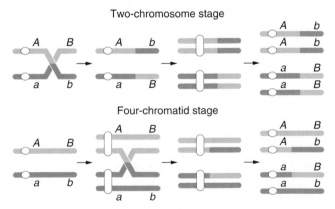

Two-chromosome stage

Four-chromatid stage

Figure 4-4 Crossing-over occurs at the four-chromatid stage. Because more than two different products of a single meiosis can be seen in some tetrads, crossing-over cannot occur at the two-strand stage (prior to DNA replication). The white circle designates the position of the centromere. When sister chromatids are visible, the centromere appears unreplicated.

Multiple crossovers can involve more than two chromatids

Tetrad analysis can also show two other important features of crossing-over. First, in one meiocyte several crossovers can occur along a chromosome pair. Second, in any one meiocyte these multiple crossovers can involve more than two chromatids. To think about this issue, we need to look at the simplest case: double crossovers. To study double crossovers, we need three linked genes. For example, in a cross such as

$$A\,B\,C \ \times\ a\,b\,c$$

many different tetrad types are possible, but some types are useful in the present connection because they can be accounted for only by double crossovers involving more than two chromatids. Consider the following tetrad as an example:

$$A\,B\,c$$
$$A\,b\,C$$
$$a\,B\,C$$
$$a\,b\,c$$

This tetrad must be explained by two crossovers involving *three* chromatids, as shown in Figure 4-5a. Furthermore the following type of tetrad shows that all *four* chromatids can participate in crossing-over in the same meiosis (see Figure 4-5b):

$$A\,B\,c$$
$$A\,b\,c$$
$$a\,B\,C$$
$$a\,b\,C$$

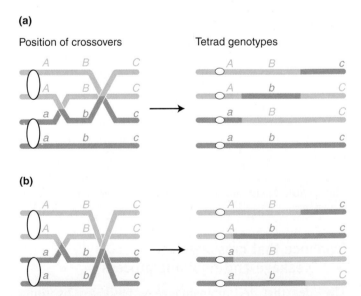

(a)

Position of crossovers   Tetrad genotypes

(b)

Figure 4-5 Double crossovers that involve (a) three chromatids or (b) four chromatids.

Therefore, for any pair of homologous chromosomes, two, three, or four chromatids can take part in crossing-over events in a single meiocyte.

You might be wondering about crossovers between *sister* chromatids. These do occur but are rare. They do not produce new allele combinations so normally are not considered.

4.2 Recombination

The production of new allele combinations is formally called *recombination*. Crossovers are one mechanism for recombination, and so is independent assortment. In this section we define recombination in such a way that we would recognize it in experimental results, and we lay out the way that recombination is analyzed and interpreted.

Recombination is observed in a variety of biological situations, but for the present let's define it in relation to meiosis. **Meiotic recombination** *is defined as any meiotic process that generates a haploid product with new combinations of the alleles carried by the haploid genotypes that united to form the dihybrid meiocyte.* This seemingly complex definition is actually quite simple; it makes the important point that we detect recombination by comparing the *output* genotypes of meiosis with the *input* genotypes (Figure 4-6). The input genotypes are the two haploid genotypes that combine to form the meiocyte,

the diploid cell that undergoes meiosis. For humans, the input is the parental egg and sperm that unite to form a diploid zygote and, hence, all the body cells, including the meiocytes that are set aside within the gonads. The output genotypes are the haploid products of meiosis. In humans, these are the individual's own eggs or sperm. Any meiotic product that has a new combination of the alleles provided by the two input genotypes is by definition a **recombinant.**

> **MESSAGE** At meiosis, recombination generates recombinants, which are haploid meiotic products with new combinations of those alleles borne by the haploid genotypes that united to form the meiocyte.

It is straightforward to detect recombinants in organisms with haploid life cycles such as fungi or algae. The input and output types in haploid life cycles are the genotypes of individuals and may thus be inferred directly from phenotypes. Figure 4-6 can be viewed as summarizing the simple detection of recombinants in organisms with haploid life cycles. Detecting recombinants in organisms with diploid life cycles is trickier. The input and output types in diploid cycles are gametes. Thus we must know the genotypes of both input and output gametes to detect recombinants in an organism with a diploid cycle. We cannot detect the genotypes of input or output gametes directly; hence to know the input gametes it is necessary to use pure-breeding diploid parents because they can produce only one gametic type. To detect recombinant output gametes, we must testcross the diploid individual and observe its progeny (Figure 4-7). If a testcross offspring is shown to have arisen from a recombinant product of meiosis, it too is called a *recombinant*. Notice again that the testcross allows us to concentrate on *one* meiosis and prevent ambiguity. From a self of the F_1 in Figure 4-7, for example, a recombinant $A/A \cdot B/b$ offspring could not be distinguished from $A/A \cdot B/B$ without further crosses.

Recombinants are produced by two different cellular processes: independent assortment and crossing-over. The proportion of recombinants is the key idea here because it is the diagnostic value that will tell us whether or not genes are linked. We shall deal with independent assortment first.

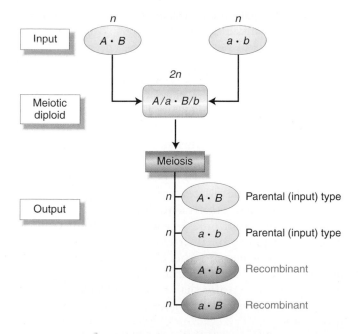

Figure 4-6 Meiotic recombination. Recombinants are those products of meiosis with allele combinations different from those of the haploid cells that formed the meiotic diploid.

Recombination of unlinked genes is by independent assortment

In a dihybrid with genes on separate chromosomes, recombinants are produced by independent assortment, as shown

Figure 4-7 Detection of recombination in diploid organisms. Recombinant products of a diploid meiosis are most readily detected in a cross of a heterozygote and a recessive tester. Note that Figure 4-6 is repeated as part of this diagram.

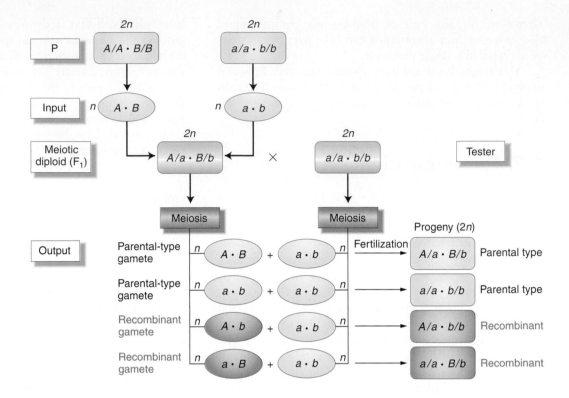

in Figure 4-8. Let's reconstruct a standard Mendelian analysis to illustrate how recombinants are produced:

P	A/A ; B/B × a/a ; b/b
Gametes	A ; B a ; b
F_1	A/a ; B/b
Testcross	A/a ; B/b × a/a ; b/b
	F_1 dihybrid Tester
Progeny	$\frac{1}{4}$ A/a ; B/b
	$\frac{1}{4}$ a/a ; b/b
	$\frac{1}{4}$ A/a ; b/b recombinant
	$\frac{1}{4}$ a/a ; B/b recombinant

The last two genotypes must be recombinant because they were formed from gametes of the dihybrid (output) that differed from the gametes that formed the F_1 (input). Note that the frequency of recombinants from independent assortment must be 50 percent ($\frac{1}{4} + \frac{1}{4}$).

A recombinant frequency of 50 percent in a testcross suggests that the two genes under study assort independently. The simplest and most likely interpretation of independent assortment is that the two genes are on separate chromosome pairs. (However, we must note that genes that are very far apart on the *same* chromosome pair can assort virtually independently and produce the same result; see the end of this chapter.)

Recombination of linked genes is by crossing-over

In Morgan's study, the linked genes could not assort independently, but there were still some recombinants, which as we have seen, must have been produced by crossovers. Fungal tetrad analysis has shown that for any two specific linked genes, crossovers occur between them in some, but not all, meiocytes. The general situation is shown in Figure 4-9. (Multiple crossovers are rarer; we will deal with them later.)

In general, for genes close together on the same chromosome pair, recombinant frequencies are significantly lower than 50 percent (Figure 4-10, page 124). We saw an example of this situation in Morgan's data (see page 117), where the recombinant frequency was $(151 + 154) \div 2839 = 10.7$ percent. This is obviously much less than the 50 percent that we would expect with independent assortment. The recombinant frequency arising from linked genes ranges from 0 to 50 percent, depending on their closeness (see below). The farther apart genes are, the closer they approach 50 percent recombinant frequency. What about recombinant frequencies greater than 50 percent? The answer is that such frequencies are *never* observed, as we shall prove later.

Note in Figure 4-9 that a single crossover generates two reciprocal recombinant products, which explains why the reciprocal recombinant classes are generally approximately equal in frequency.

Meiotic recombination between unlinked genes by independent assortment

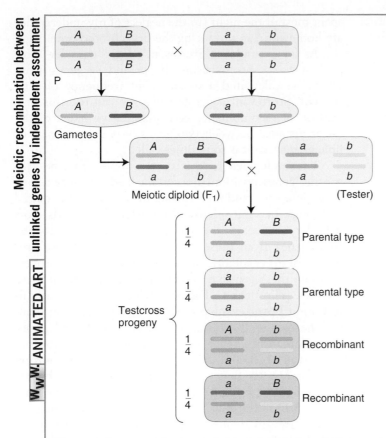

WWW ANIMATED ART

Figure 4-8 Recombination between unlinked genes by independent assortment. This diagram shows two chromosome pairs of a diploid organism with A and a on one pair and B and b on the other. Independent assortment produces a recombinant frequency of 50 percent. Note that we could represent the haploid situation by removing the parental (P) cross and the testcross.

Meiotic recombination between linked genes by crossing-over

	Meiotic chromosomes		Meiotic products		
Meioses with no crossover between the genes	A	B	A	B	Parental
	A	B	A	B	Parental
	a	b	a	b	Parental
	a	b	a	b	Parental
Meioses with a crossover between the genes	A	B	A	B	Parental
	A	B	A	b	Recombinant
	a	b	a	B	Recombinant
	a	b	a	b	Parental

WWW ANIMATED ART

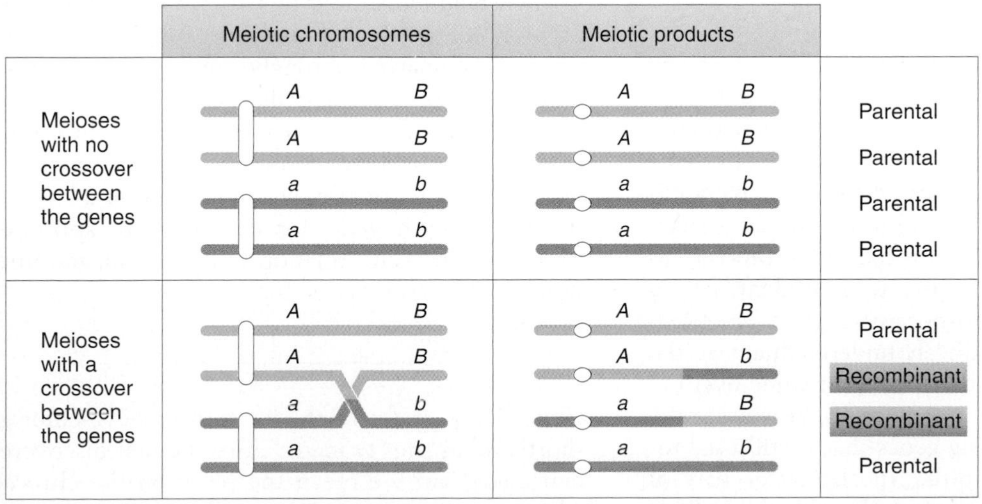

Figure 4-9 Recombination between linked genes by crossing-over. Recombinants arise from meioses in which a crossover occurs between nonsister chromatids in a region under study.

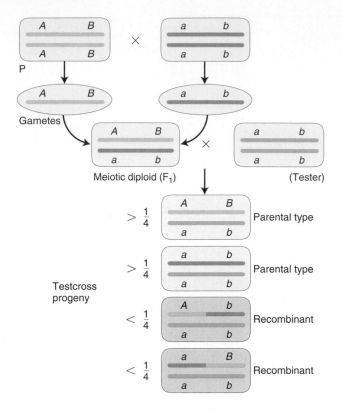

Figure 4-10 Frequencies of recombinants arising from crossing-over. The frequencies of such recombinants are less than 50 percent.

> **MESSAGE** A recombinant frequency significantly less than 50 percent indicates that the genes are linked. A recombinant frequency of 50 percent generally means that the genes are unlinked and reside on separate chromosomes.

4.3 Linkage maps

As Morgan studied more and more linked genes, he saw that the proportion of recombinant progeny varied considerably, depending on which linked genes were being studied, and he thought that such variation in recombinant frequency might somehow indicate the actual distances separating genes on the chromosomes. Morgan assigned the study of this problem to a student, Alfred Sturtevant, who (like Bridges) became a great geneticist. Morgan asked Sturtevant, still an undergraduate at the time, to make some sense of the data on crossing-over between different linked genes. In one night, Sturtevant developed a method for mapping genes that is still used today. In Sturtevant's own words, "In the latter part of 1911, in conversation with Morgan, I suddenly realized that the variations in strength of linkage, already attributed by Morgan to differences in the spatial separation of genes, offered the possibility of determining sequences in the linear dimension of a chromosome. I went home and

spent most of the night (to the neglect of my undergraduate homework) in producing the first chromosome map."

As an example of Sturtevant's logic, consider Morgan's testcross results using the *pr* and *vg* genes, from which he calculated a recombinant frequency of 10.7 percent. Sturtevant suggested that we can use this percentage of recombinants as a quantitative index of the linear distance between two genes on a genetic map, or **linkage map,** as it is sometimes called.

The basic idea here is quite simple. Imagine two specific genes positioned a certain fixed distance apart. Now imagine random crossing-over along the paired homologs. In some meioses, nonsister chromatids cross over by chance in the chromosomal region between these genes; from these meioses, recombinants are produced. In other meiotic divisions, there are no crossovers between these genes; no recombinants result from these meioses. (Flip back to Figure 4-1 for a diagrammatic illustration.) Sturtevant postulated a rough proportionality: the greater the distance between the linked genes, the greater the chance of crossovers in the region between the genes and, hence, the greater the proportion of recombinants that would be produced. Thus, by determining the frequency of recombinants, we can obtain a measure of the map distance between the genes. In fact, Sturtevant defined one **genetic map unit (m.u.)** as that distance between genes for which one product of meiosis in 100 is recombinant. Put another way, a **recombinant frequency (RF)** of 0.01 (1 percent) is defined as 1 m.u. A map unit is sometimes referred to as a **centimorgan (cM)** in honor of Thomas Hunt Morgan.

A direct consequence of the way in which map distance is measured is that, if 5 map units (5 m.u.) separate genes *A* and *B* whereas 3 m.u. separate genes *A* and *C*, then *B* and *C* should be either 8 or 2 m.u. apart (Figure 4-11). Sturtevant found this to be the case. In other words, his analysis strongly suggested that genes are arranged in some linear order.

The place on the map—and on the chromosome—where a gene is located is called the **gene locus** (plural, loci). The locus of the eye-color gene and the locus of the wing-length gene, for example, are approximately 11 m.u. apart. The relation is usually diagrammed this way:

$$\underset{pr}{\vdash}\qquad 11.0 \qquad\underset{vg}{\dashv}$$

Generally we refer to the locus of this eye-color gene in shorthand as the "*pr* locus," after the first discovered mutant allele, but we mean the place on the chromosome where *any* allele of this gene will be found.

Linkage analysis works in both directions. Given a known genetic distance in map units, we can predict frequencies of progeny in different classes. For example, the genetic distance between the *pr* and *vg* loci in *Drosophila*

Figure 4-11 Map distances are additive. A chromosome region containing three linked genes. Calculation of $A-B$ and $A-C$ distances leaves us with the two possibilities shown for the $B-C$ distance.

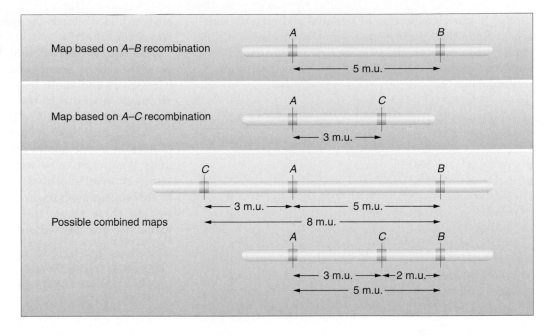

is approximately 11 map units. So, in the progeny from a testcross of a female dihybrid in cis conformation ($pr\ vg/pr^+\ vg^+$) heterozygote, we know that there will be 11 percent recombinants. These recombinants will consist of two reciprocal recombinants of equal frequency: thus, $5\frac{1}{2}$ percent will be $pr\ vg^+/pr\ vg$ and $5\frac{1}{2}$ percent will be $pr^+\ vg/pr\ vg$. Of the progeny from a testcross of a dihybrid in trans conformation (female $pr\ vg^+/pr^+\ vg$), heterozygote, $5\frac{1}{2}$ percent will be $pr\ vg/pr\ vg$ and $5\frac{1}{2}$ percent will be $pr^+\ vg^+/pr\ vg$.

There is a strong implication that the "distance" on a linkage map is a physical distance along a chromosome, and Morgan and Sturtevant certainly intended to imply just that. But we should realize that the linkage map is an entity constructed from a purely genetic analysis. The linkage map could have been derived without even knowing that chromosomes existed. Furthermore, at this point in our discussion, we cannot say whether the "genetic distances" calculated by means of recombinant frequencies in any way represent actual physical distances on chromosomes. However, cytogenetic analysis and genomic sequencing have shown that genetic distances are, in fact, roughly proportional to chromosome distances. Nevertheless, it must be emphasized that the hypothetical structure (the linkage map) was developed with a very real structure (the chromosome) in mind. In other words, the chromosome theory provided the framework for the conceptual development of linkage mapping.

MESSAGE Recombination between linked genes can be used to map their distance apart on the chromosome. The unit of mapping (1 m.u.) is defined as a recombinant frequency of 1 percent.

Three-point testcross

So far, we have looked at linkage in crosses of dihybrids (double heterozygotes) to doubly recessive testers. The next level of complexity is a cross of a trihybrid (triple heterozygote) to a triply recessive tester. This kind of cross, called a **three-point testcross,** is a commonly used format in linkage analysis. The goal is to deduce whether the three genes are linked and, if they are, to deduce their order and the map distances between them.

Let's look at an example, also from *Drosophila*. In our example, the mutant alleles are v (vermilion eyes), cv (crossveinless, or absence of a crossvein on the wing), and ct (cut, or snipped, wing edges). The analysis is carried out by performing the following crosses:

P $v^+/v^+ \cdot cv/cv \cdot ct/ct \times v/v \cdot cv^+/cv^+ \cdot ct^+/ct^+$

Gametes $v^+ \cdot cv \cdot ct$ $v \cdot cv^+ \cdot ct^+$

F$_1$ trihybrid $v/v^+ \cdot cv/cv^+ \cdot ct/ct^+$

Trihybrid females are testcrossed to triple recessive males:

$v/v^+ \cdot cv/cv^+ \cdot ct/ct^+$ ♀   × $v/v \cdot cv/cv \cdot ct/ct$ ♂

F$_1$ trihybrid female   Tester male

From any trihybrid, only $2 \times 2 \times 2 = 8$ gamete genotypes are possible. These are the genotypes seen in the testcross progeny. The chart (on the following page) shows the number of each of the eight gametic genotypes counted out of a sample of 1448 progeny flies. The columns alongside show which genotypes are recombinant (R) for the loci taken two at a time. We must be careful in our classification of parental and recombinant types. Note that the parental input genotypes for the triple heterozygotes are $v^+ \cdot cv \cdot ct$ and $v \cdot cv^+ \cdot ct^+$; any combination other than these two constitutes a recombinant.

Gametes		*Recombinant for loci*		
		v and *cv*	*v* and *ct*	*cv* and *ct*
$v \cdot cv^+ \cdot ct^+$	580			
$v^+ \cdot cv \cdot ct$	592			
$v \cdot cv \cdot ct^+$	45	R		R
$v^+ \cdot cv^+ \cdot ct$	40	R		R
$v \cdot cv \cdot ct$	89	R	R	
$v^+ \cdot cv^+ \cdot ct^+$	94	R	R	
$v \cdot cv^+ \cdot ct$	3		R	R
$v^+ \cdot cv \cdot ct^+$	5		R	R
	1448	268	191	93

Let's analyze the loci two at a time, starting with the *v* and *cv* loci. In other words, in the gamete listings we look at just the first two columns and cover up the other one. Since the parentals are $v \cdot cv^+$ and $v^+ \cdot cv$, we know that the recombinants are by definition $v \cdot cv$ and $v^+ \cdot cv^+$. There are $45 + 40 + 89 + 94 = 268$ of these recombinants. Of a total of 1448 flies, this number gives an RF of 18.5 percent.

For the *v* and *ct* loci, the recombinants are $v \cdot ct$ and $v^+ \cdot ct^+$. There are $89 + 94 + 3 + 5 = 191$ of these recombinants among 1448 flies, so the RF = 13.2 percent.

For *ct* and *cv*, the recombinants are $cv \cdot ct^+$ and $cv^+ \cdot ct$. There are $45 + 40 + 3 + 5 = 93$ of these recombinants among the 1448, so the RF = 6.4 percent.

All the loci are linked, because the RF values are all considerably less than 50 percent. Because the *v* and *cv* loci show the largest RF value, they must be farthest apart; therefore, the *ct* locus must lie between them. A map can be drawn as follows:

```
v                              ct          cv
+------------------------------+-----------+
         13.2                      6.4
<----------------------------->|<--------->
         m.u.                      m.u.
```

The testcross can be rewritten as follows, now that we know the linkage arrangement:

$$v^+ \, ct \, cv / v \, ct^+ \, cv^+ \; \times \; v \, ct \, cv / v \, ct \, cv$$

Note several important points here. First, we have deduced a gene order that is different from that used in our list of the progeny genotypes. Because the point of the exercise was to determine the linkage relation of these genes, the original listing was of necessity arbitrary; the order simply was not known before the data were analyzed. Henceforth the genes must be written in correct order.

Second, we have definitely established that *ct* is between *v* and *cv*, and we have established the distances between *ct* and these loci in map units. But we have arbitrarily placed *v* to the left and *cv* to the right; the map could equally well be inverted.

Third, note that linkage maps merely map the loci in relation to one another, with the use of standard map units. We do not know where the loci are on a chromosome—or even which specific chromosome they are on. In subsequent analyses as more loci are mapped in relation to these three, the complete chromosome map would become "fleshed out."

> **MESSAGE** Three-point (and higher) testcrosses enable geneticists to evaluate linkage between three (or more) genes in one cross.

A final point to note is that the two smaller map distances, 13.2 m.u. and 6.4 m.u., add up to 19.6 m.u., which is greater than 18.5 m.u., the distance calculated for *v* and *cv*. Why is this so? The answer to this question lies in the way in which we have treated the two rarest classes of progeny (totaling 8) with respect to recombination of *v* and *cv*. Now that we have the map, we can see that these two rare classes are in fact double recombinants, arising from two crossovers (Figure 4-12). However, when we calculated the RF value for *v* and *cv* we did not count the $v \, ct \, cv^+$ and $v^+ \, ct^+ \, cv$ genotypes; after all, with regard to *v* and *cv*, they are parental combinations ($v \, cv^+$ and $v^+ \, cv$). In the light of our map, however, we see that this oversight led us to underestimate the distance between the *v* and *cv* loci. Not only should we have counted the two rarest classes, we should have counted each of them *twice* because each represents a double recombinant class. Hence, we can correct the value by adding the numbers $45 + 40 + 89 + 94 + 3 + 3 + 5 + 5 = 284$. Of the total of 1448, this number is exactly 19.6 percent, which is identical with the sum of the two component values.

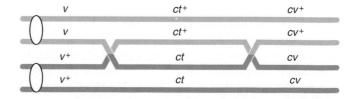

Figure 4-12 Example of a double crossover involving two chromatids. Notice that a double crossover produces double recombinant chromatids that have the parental allele combinations at the outer loci. The position of the centromere cannot be determined from the data. It has been added for completeness.

Deducing gene order by inspection

Now that we have had some experience with the three-point testcross, we can look back at the progeny listing and see that for trihybrids of linked genes it is usually possible to deduce gene order by inspection, without a recombinant frequency analysis. Only three gene orders are possible, each with a different gene in the middle position. It is generally true that the double recombinant classes are the smallest ones. Only one order is compatible with the smallest classes' having been formed by double crossovers, as shown in Figure 4-13; that is, only one order gives double recombinants of genotype $v\ ct\ cv^+$ and $v^+\ ct^+\ cv$.

Interference

Knowing the existence of double crossovers permits us to ask questions about their possible interdependence. We can ask: Are the crossovers in adjacent chromosome regions independent events, or does a crossover in one region affect the likelihood of there being a crossover in an adjacent region? The answer is that generally crossovers inhibit each other in an interaction called **interference**. Double recombinant classes can be used to deduce the extent of this interference.

Interference can be measured in the following way. If the crossovers in the two regions are independent, we can use the product rule (see page 39) to predict the frequency of double recombinants: that frequency would equal the product of the recombinant frequencies in the adjacent regions. In the v-ct-cv recombination data, the v-ct RF value is 0.132 and the ct-cv value is 0.064, so if there is no interference double recombinants might be expected at the frequency $0.132 \times 0.064 = 0.0084$ (0.84 percent). In the sample of 1448 flies, $0.0084 \times 1448 = 12$ double recombinants are expected. But the data show that only 8 were actually observed. If this deficiency of double recombinants were consistently observed, it would show us that the two regions are not independent and suggest that the distribution of crossovers favors singles at the expense of doubles. In other words, there is some kind of interference: a crossover does reduce the probability of a crossover in an adjacent region.

Interference is quantified by first calculating a term called the **coefficient of coincidence (c.o.c.)**, which is the ratio of observed to expected double recombinants. Interference (I) is defined as $1 - $ c.o.c. Hence

$$I = 1 - \frac{\text{observed frequency, or number, of double recombinants}}{\text{expected frequency, or number, of double recombinants}}$$

In our example

$$I = 1 - \tfrac{8}{12} = \tfrac{4}{12} = \tfrac{1}{3},\ \text{or 33 percent}$$

In some regions, there are never any observed double recombinants. In these cases, c.o.c. = 0, so I = 1 and interference is complete. Most of the time, the interference values that are encountered in mapping chromosome loci are between 0 and 1.

You may have wondered why we always use heterozygous females for testcrosses in *Drosophila*. The explanation lies in an unusual feature of *Drosophila* males. When $pr\ vg/pr^+\ vg^+$ *males* are crossed with $pr\ vg/pr\ vg$ females, only $pr\ vg/pr^+\ vg^+$ and $pr\ vg/pr\ vg$ progeny are recovered. This result shows that there is no crossing-over in *Drosophila* males! However, this absence of crossing-over in one sex is limited to certain species; it is not the case for males of all species (or for the heterogametic sex). In other organisms, there is crossing-over in XY males and in WZ females. The reason for the absence of crossing-over in *Drosophila* males is that they have an unusual prophase I, with no synaptonemal complexes. Incidentally, there is a recombination difference between human sexes as well. Women show higher recombinant frequencies for the same loci than do men.

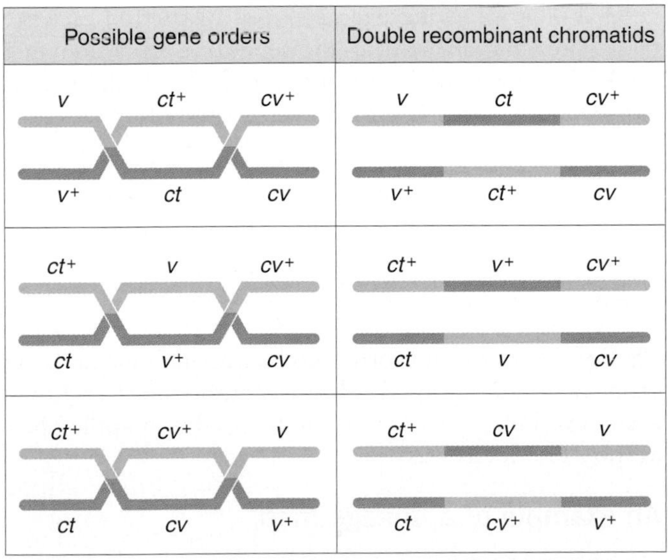

Mapping with molecular markers

Genes themselves are of course important in the analysis of how genes function, but in the analysis of linkage they are often used simply as "markers" for chromosome maps, rather like milestones on a road. For a gene to be useful as a marker, at least two alleles must exist to provide a

Possible gene orders	Double recombinant chromatids
v ct^+ cv^+ v^+ ct cv	v ct cv^+ v^+ ct^+ cv
ct^+ v cv^+ ct v^+ cv	ct^+ v^+ cv^+ ct v cv
ct^+ cv^+ v ct cv v^+	ct^+ cv v ct cv^+ v^+

Figure 4-13 Each gene order gives unique double recombinant genotypes. The three possible gene orders and the resulting products of a double crossover are shown. Only the first possibility is compatible with the data in the text. Note that only the nonsister chromatids involved in the double crossover are shown.

heterozygote for mapping analysis. In the early years of building genetic maps, the markers were genes with variant alleles producing detectably different phenotypes. As organisms became more intensively studied, a larger range of mutant alleles was found and hence larger numbers of genes could be used as markers for mapping studies. However, even when maps appeared to be "full" of loci of known phenotypic effect, measurements showed that those genes were separated by vast amounts of DNA. These gaps could not be mapped by linkage analysis, because no phenotypes had been matched to genes in those regions. Large numbers of additional genetic markers were needed that could be used to fill in the gaps to provide a higher-resolution map. The discovery of various kinds of molecular markers provided a solution.

A **molecular marker** is a site of heterozygosity for some type of DNA change not associated with any measurable phenotypic change. These are called *silent* changes. Such a heterozygous site can be mapped by linkage analysis just as a conventional heterozygous allele pair can be. Because molecular markers can be easily detected and are so numerous in a genome, when mapped, they fill the voids between genes of known phenotype. The two basic types of molecular markers are those based on nucleotide differences and those based on differences in the amount of repetitive DNA.

USE OF NUCLEOTIDE POLYMORPHISMS IN MAPPING Some positions in DNA are occupied by a different nucleotide in different homologous chromosomes. These differences are called *single nucleotide polymorphisms*, or SNPs (pronounced "snips"). Although mostly silent, they provide markers for mapping. There are several ways of detecting these polymorphisms, of which the simplest to visualize is through restriction enzyme analysis. Bacterial restriction enzymes cut DNA at specific target sequences that exist by chance in the DNA of other organisms. Generally, the target sites are found in the same position in the DNA of different individuals in a population; that is, in the DNA of homologous chromosomes. However, quite commonly, a specific site might be negated as a result of a silent mutation of one or more nucleotides. The mutation might be within a gene or in a noncoding area between genes. If an individual is heterozygous for the presence and absence of the restriction site $(+/-)$, that locus can be used in mapping. One way to determine the $+/-$ sites is to apply Southern analysis using a probe derived from DNA of that region. A typical heterozygous marker would be

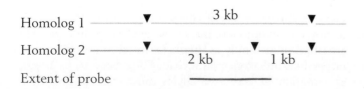

Homolog 1 3 kb

Homolog 2 2 kb   1 kb

Extent of probe

On a Southern hybridization of such an individual, the probe would highlight three fragments, of size 3, 2, and 1kb. These multiple forms of this region constitute a **restriction fragment length polymorphism (RFLP).** You can see that the RFLP in this case is based on the SNP at the central site. The heterozygous RFLP may be linked to a heterozygous gene, as shown below for the gene *D* in cis conformation with the (1–2) morph:

d   3

D   2   1

Crossovers between these sites would produce recombinant products that are detectable as D–3 and d–2–1. In this way, the RFLP locus can be mapped in relation to genes or to other molecular markers, thereby providing another milestone for chromosomal navigation.

USE OF TANDEM REPEATED SEQUENCES IN MAPPING In many organisms some DNA segments are repeated in tandem at precise locations. However the precise number of repeated units in such a tandem array may be variable, and hence they are called **VNTRs** (*variable number tandem repeats*). Individuals may be heterozygous for VNTRs: the site on one homolog may have, say, eight repeating units and the site on the other homolog, say, five. The mechanisms that produce this variation need not concern us at present. The important fact is that heterozygous can be detected, and the heterozygous site can be used as a molecular marker in mapping. A probe that binds to the repetitive DNA is needed. The following example uses restriction enzyme target sites that are outside the repetitive array, allowing the VNTR region to be cut out as a block. The basic unit of the tandem array is shown as an arrow.

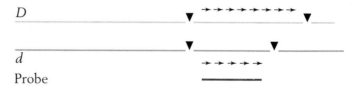

D

d

Probe

This VNTR locus will form two bands, one long and one short, on a Southern hybridization autoradiogram. Once again, this heterozygous site can be used in mapping just as the RFLP locus was.

An example of a linkage map

Chromosome maps are essential to the experimental genetic study of any organism. They are the prelude to any serious piece of genetic manipulation. Why is mapping so important? The types of genes that an organism has and their positions in the chromosome set are fundamental components of genetic analysis. The main reasons for mapping are to understand gene *function* and gene *evolution*, and to facilitate *strain building*.

1. *Gene function.* How can knowing the map position of a gene contribute to understanding its function? Essentially map position provides a way of "zeroing in" on a piece of DNA. If the genome of an organism has not been sequenced, the map position can provide a way of physically isolating the gene (this is called *positional cloning*, described in Chapter 11). Even if the genome has been completely sequenced, the phenotypic impact of most of the genes in the sequence is not understood, and mapping provides a way of correlating the position of an allele of known phenotypic affect with a candidate gene in the genomic sequence. Map position can be important in another way. A gene's location is sometimes important for gene function because the location can affect the gene's expression, a phenomenon generally called a "neighborhood effect." Genes of related function are often clustered next to one another in bacterial chromosomes, generally because they are transcribed as one unit. In eukaryotes, the position of a gene in or near heterochromatin can affect its expression.

2. *Genome evolution.* A knowledge of gene position is useful in evolutionary studies because, from the relative positions of the same genes in related organisms, investigators can deduce the rearrangement of chromosomes in the course of evolution. Special mention must be made of human-directed plant and animal evolution (breeding) for the purpose of obtaining profitable genotypes.

3. *Strain building.* In designing strains of a complex genotype for genetic research, it is helpful to know if the alleles that need to be united are linked or not.

Many organisms have had their chromosomes intensively mapped. The resultant maps represent a vast amount of genetic analysis achieved by collaborative efforts of research groups throughout the world. Figure 4-14 shows an example of a linkage map from the tomato. Tomatoes have been interesting from the perspectives of both basic and applied genetic research, and the tomato genome is one of the best mapped of plants.

The different panels of Figure 4-14 illustrate some of the stages of understanding through which research arrives at a comprehensive map. First, although chromosomes are visible under the microscope, there is initially no way to locate genes on them. However, the chromosomes can be individually identified and numbered on the basis of their inherent landmarks such as staining patterns and centromere positions, as has been

done in Figure 4-14a and b. Next, analysis of recombinant frequencies generates groups of linked genes that must correspond to chromosomes, although at this point the linkage groups cannot necessarily be correlated with specific numbered chromosomes. At some stage, other types of analyses allow the linkage groups to be assigned to specific chromosomes. Today this is mainly accomplished by molecular approaches. For example, a cloned gene known to be in a certain linkage group can be used as a probe against a partially denatured chromosome set (in situ hybridization; see page 403). The probe binds to the chromosome corresponding to that linkage group.

Figure 4-14c shows a tomato map made in 1952, indicating the linkages of the genes known at that time. Each locus is represented by the two alleles used in the original mapping experiments. As more and more loci became known, they were mapped in relation to the loci shown in the figure, so today the map shows hundreds of loci. Some of the chromosome numbers shown in Figure 4-14c are tentative and do not correspond to the modern chromosome numbering system. Notice that genes with related functions (for example, fruit shape) are scattered.

Centromere mapping using linear tetrads

In most eukaryotes, recombination analysis cannot be used to map the loci of the special DNA sequences called centromeres, because they show no heterozygosity that can enable them to be used as markers. However in fungi that produce linear tetrads (see Chapter 3, page 100) centromeres *can* be mapped. We shall use the fungus *Neurospora* as an example. Recall that in fungi such as *Neurospora* (a haploid) the meiotic divisions take place along the long axis of the ascus so that each meiocyte produces a linear array of eight ascospores (an **octad**). These eight represent the four products of meiosis (a tetrad) plus a postmeiotic mitosis.

In its simplest form, centromere mapping considers a gene locus and asks how far this locus is from the centromere. The method is based on the fact that a different pattern of alleles will appear in a tetrad that arises from a meiosis with a crossover between a gene and its centromere. Consider a cross between two individuals, each having a different allele at a locus (say, $a \times A$). Mendel's first law of equal segregation dictates that there will always be four ascospores of genotype a and four of A. If there has been no crossover in the region between a/A and the centromere, in the linear octad there will two adjacent blocks of four ascospores (see Figure 3-37). However if there has been a crossover in that region, in the octad there will be one of four different patterns, each of which shows

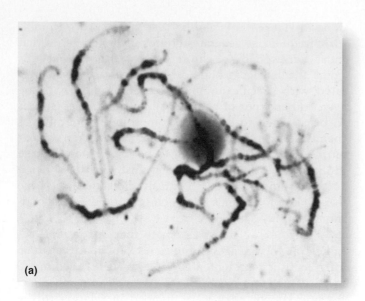

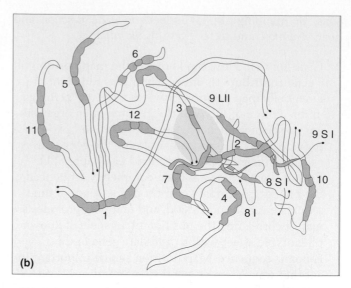

Figure 4-14 Mapping the 12 chromosomes of tomatoes.
(a) Photomicrograph of a meiotic prophase I (pachytene) from anthers, showing the 12 pairs of chromosomes. (b) Illustration of the 12 chromosomes shown in (a). The chromosomes are labeled with the currently used chromosome numbering system. The centromeres are shown in orange, and the flanking, densely staining regions (heterochromatin) in green. (c) 1952 linkage map. Each locus is flanked by drawings of the normal and variant phenotypes. Interlocus map distances are shown in map units. [Parts a and b from C. M. Rick, "The Tomato." Copyright 1978 by Scientific American, Inc. All rights reserved. Part c from L. A. Butler.]

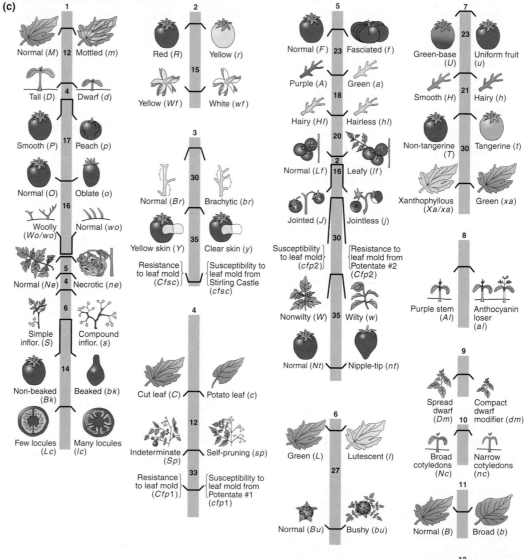

(c)

some blocks of two. Some data from an actual cross of $A \times a$ are shown in the table below.

Octads					
A	*a*	*A*	*a*	*A*	*a*
A	a	A	a	A	a
A	a	A	a	A	a
A	a	a	A	a	A
A	a	a	A	a	A
a	A	A	a	a	A
a	A	A	a	a	A
a	A	a	A	A	a
a	A	a	A	A	a
126	132	9	11	10	12

Total = 300

The first two columns are from meioses with *no* crossover in the region between the *A* locus and the centromere. These are called **first-division segregation patterns (M_I patterns)** because the two different alleles segregate into the two daughter nuclei at the first division of meiosis. The other four patterns are all from meiocytes *with* a crossover. These patterns are called **second-division segregation patterns (M_{II})** because, as a result of crossover in the centromere-to-locus region, the *A* and *a* alleles are still together in the nuclei at the end of the first division of meiosis (Figure 4-15). There has been no first division segregation. However, the second meiotic division does move the *A* and *a* alleles into separate nuclei. Figure 4-15 shows how one of these M_{II} patterns is produced. The other patterns are produced similarly; the difference is that the chromatids move in different directions at the second division (Figure 4-16).

You can see that the frequency of octads with an M_{II} pattern should be proportional to the size of the centromere–*a/A* region, and could be used as a measure of the size of that region. In our example the M_{II} frequency is 42/300 = 14 percent. Does this percentage mean that the mating-type locus is 14 map units from the centromere? The answer is no, but this value can be used to calculate the number of map units. The 14 percent value is a percentage of *meioses*, which is not the way that map units are defined. Map units are defined in terms of the percentage of recombinant *chromatids* issuing from meiosis. Because a crossover in any meiosis re-

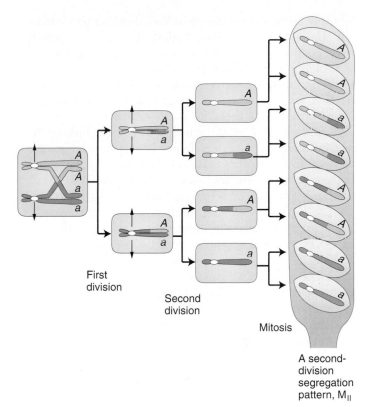

Figure 4-15 Second-division segregation pattern. *A* and *a* segregate into separate nuclei at the second meiotic division when there is a crossover between the centromere and the *A* locus.

sults in only 50 percent recombinant chromatids (4 out of 8; see Figure 4-15), we must divide the 14 percent by 2 to convert the M_{II} frequency (a frequency of *meioses*) into map units (a frequency of recombinant *chromatids*). Hence this region must be 7 map units in length, and this measurement can be introduced into the map of that chromosome.

4.4 Using the chi-square test in linkage analysis

In linkage analysis, the question often arises "Are these two genes linked?" Sometimes the answer is obvious, and sometimes not. But in either situation it is helpful

Figure 4-16 Four second-division segregation patterns in linear asci. During the second meiotic division, the centromeres attach to the spindle at random, producing the four arrangements shown. The four arrangements are equally frequent.

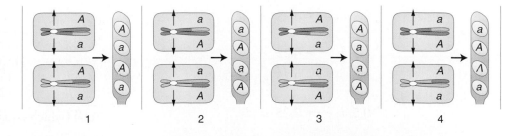

to apply an objective statistical test that can support or not support one's intuitive feeling. The χ^2 test, which we encountered first in Chapter 2, provides a useful way of deciding if two genes are linked. How is the χ^2 test applied to linkage? We have learned earlier in this chapter that we can infer that two genes are linked on the same chromosome if the RF is less than 50 percent. But how much less? It is not possible to test for linkage directly, because a priori we do not have a precise linkage distance to use to formulate our expectations if the hypothesis is true. Are the genes 1 m.u. apart? 10 m.u.? 45 m.u.? The only genetic criterion for linkage that we can use to make a precise prediction is the presence or absence of independent assortment. Consequently it is necessary to test the hypothesis of the *absence* of linkage. If the observed results cause us to reject the hypothesis of *no linkage*, then we can infer linkage. This type of hypothesis, called a **null hypothesis,** is generally useful in χ^2 analysis because it provides a precise experimental prediction that can be tested.

Let's test a specific set of data for linkage using χ^2 analysis. Assume that we have crossed pure-breeding parents of genotypes $A/A \cdot B/B$ and $a/a \cdot b/b$ and obtained a dihybrid $A/a \cdot B/b$, which we testcross to $a/a \cdot b/b$. A total of 500 progeny are classified as follows (written as gametes from the dihybrid):

142	$A \cdot B$	parental
133	$a \cdot b$	parental
113	$A \cdot b$	recombinant
112	$a \cdot B$	recombinant
500		Total

From these data the recombinant frequency is 225/500 = 45 percent. On the face of it this seems like a case of linkage because the RF is less than the 50 percent expected from independent assortment. However, it is possible that the recombinant classes are less than 50 percent merely on the basis of chance. Therefore, we need to perform a χ^2 test to calculate the likelihood of this result based on chance.

As is usual for the χ^2 test, the first step is to calculate the expectations E for each class. As we saw above, the hypothesis that must be tested in this case is that the two loci assort independently (that is, there is *no linkage*). How can we calculate gametic E values? One way might be to make a simple prediction based on Mendel's first and second laws, as follows:

E values

```
          0.5 B ——→ 0.25 A ; B
   0.5 A<
          0.5 b ——→ 0.25 A ; b

          0.5 B ——→ 0.25 a ; B
   0.5 a<
          0.5 b ——→ 0.25 a ; b
```

Hence we might assert that if the allele pairs of the dihybrid are assorting independently, there should be a $1:1:1:1$ ratio of gametic types. Therefore it seems reasonable to use 1/4 of 500, or 125, as the expected proportion of each gametic class. However, note that the $1:1:1:1$ ratio is expected only if all genotypes are equally viable. It is often the case that genotypes are *not* equally viable because individuals that carry certain alleles do not survive to adulthood. Therefore instead of alleles ratios of 0.5:0.5, used above, we might see (for example) ratios of 0.6 A:0.4 a, or 0.45 B:0.55 b. We should use these ratios in our predictions of independence.

Let's arrange the observed genotypic classes in a grid to reveal the allele proportions more clearly.

OBSERVED VALUES		*Segregation of A and a*		
		A	a	Total
Segregation of	B	142	112	254
B and b	b	113	133	246
	Total	255	245	500

We see that the allele proportions are 255/500 for A, 245/500 for a, 254/500 for B, and 246/500 for b. Now we calculate the values expected under independent assortment simply by multiplying these allelic proportions. For example, to find the expected number of A B genotypes in the sample if the two ratios are combined randomly, we simply multiply the following terms:

$$\text{Expected }(E)\text{ value for }A\,B$$
$$= (255/500) \times (254/500) \times 500 = 129.54$$

Using this approach the entire grid of E values can be completed, as follows:

EXPECTED VALUES		*Segregation of A and a*		
		A	a	Total
Segregation of	B	129.54	124.46	254
B and b	b	125.46	120.56	246
	Total	255	245	500

The value of χ^2 is calculated as follows:

Genotype	O	E	$(O-E)^2/E$
$A\,B$	142	129.54	1.19
$a\,b$	133	120.56	1.29
$A\,b$	113	125.46	1.24
$a\,B$	112	124.46	1.25

Total (which equals the χ^2 value) = 4.97

The obtained value of χ^2 (4.97) is used to find a corresponding probability value p, using the χ^2 table (see Table 2-2). Generally in a statistical test the number of degrees of freedom is the number of nondependent values. Working through the following "thought experiment" will show what this means in the present application. In 2×2 grids of data (of the sort we used above), since the column and row totals are given from the experimental results, specifying any one value within the grid automatically dictates the other three values. Hence there is only one nondependent value and, therefore, only one degree of freedom. A rule of thumb useful for larger grids is that the number of degrees of freedom is equal to the number of classes represented in the rows minus one, times the number of classes represented in the columns minus one. Applying that rule in the present example gives

$$df = (2 - 1) \times (2 - 1) = 1$$

Therefore, using Table 2-2, we look along the row corresponding to one degree of freedom until we locate our χ^2 value of 4.97. Not all values of χ^2 are shown in the table, but 4.97 is close to the value 5.021. Hence the corresponding probability value is very close to 0.025, or 2.5 percent. This p value is the probability value we seek, that of obtaining a deviation from expectations this large or larger. Since this probability is less than 5 percent, the hypothesis of independent assortment must be rejected. Thus, having rejected the hypothesis of no linkage, we are left with the inference that indeed the loci are probably linked.

4.5 Using Lod scores to assess linkage in human pedigrees

Humans have thousands of autosomally inherited phenotypes, and it might seem that it should be relatively straightforward to map the loci of the genes causing these phenotypes by using the techniques developed in this chapter. However, progress in mapping these loci was initially slow for several reasons. First, it is not possible to make controlled crosses in humans, and geneticists had to try to calculate recombinant frequencies from the occasional dihybrids that were produced by chance in human matings. Crosses that were the equivalent of testcrosses were extremely rare. Second, human matings generally produce only small numbers of progeny, making it difficult to obtain enough data to calculate statistically reliable map distances. Third, the human genome is immense, which means that on average the distances between the known genes are large.

To obtain reliable RF values, large sample sizes are necessary. However, even where the number of progeny of any individual mating is small, a more reliable estimate can be made by combining the results of many identical matings. The standard way of doing this is to calculate **Lod scores.** (*Lod* stands for "log of odds.") The method

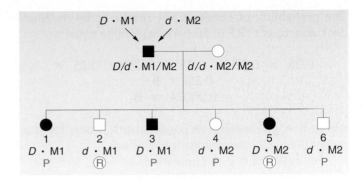

Figure 4-17 Pedigree amounting to a dihybrid testcross.
D/d are alleles of a disease gene; M1 and M2 are molecular "alleles," such as two forms of an RFLP. P, parental (nonrecombinant); R, recombinant.

simply calculates two different probabilities for obtaining a set of results in a family. The first probability is calculated assuming independent assortment and the second assuming a specific degree of linkage. Then the ratio (odds) of the two probabilities is calculated, and the logarithm of this number is calculated, which is the Lod value. Because logarithms are exponents, this approach has the useful feature that we can add Lod scores from different matings for which the same markers are used, hence providing a cumulative set of data either supporting or not supporting some particular linkage value. Let's look at a simple example of how the calculation works.

Assume that we have a family that amounts to a dihybrid testcross. Also assume that, for the dihybrid individual, we can deduce the input gametes and hence assess that individual's gametes for recombination. The dihybrid is heterozygous for a dominant disease allele (*D/d*) and for a molecular marker (M1/M2). Assume that it is a man and that the gametes that united to form him were $D \cdot M1$ and $d \cdot M2$. His wife is $d/d \cdot M2/M2$. The pedigree in Figure 4-17 shows their six children, categorized as parental or recombinant with respect to the father's contributing gamete. Of the six, there are two recombinants, which would give an RF of 33 percent. However, it is possible that the genes are assorting independently and the children constitute a nonrandom sample. Let's calculate the probability of this outcome under several hypotheses. The following display shows the expected proportions of parental (P) and recombinant (R) genotypes under three RF values and under independent assortment:

	RF			
	0.5	0.4	0.3	0.2
P	0.25	0.3	0.35	0.4
P	0.25	0.3	0.35	0.4
R	0.25	0.2	0.15	0.1
R	0.25	0.2	0.15	0.1

The probability of obtaining the results under independent assortment (RF of 50 percent) will be equal to

$$0.25 \times 0.25 \times 0.25 \times 0.25 \times 0.25 \times$$
$$0.25 \times B$$
$$= 0.00024 \times B$$

where B = the number of possible birth orders for four parental and two recombinant individuals.

For an RF of 0.2, the probability is

$$0.4 \times 0.1 \times 0.4 \times 0.4 \times 0.1 \times 0.4 \times B$$
$$= 0.00026 \times B$$

The ratio of the two is 0.00026/0.00024 = 1.08 (note that the B's cancel out). Hence, on the basis of these data, the hypothesis of an RF of 0.2 is 1.08 times as likely as the hypothesis of independent assortment. Finally, we take the logarithm of the ratio to obtain the Lod value. Some other ratios and their Lod values are shown in the following table:

	RF			
	0.5	0.4	0.3	0.2
Probability	0.00024	0.00032	0.00034	0.00026
Ratio	1.0	1.33	1.41	1.08
Lod	0	0.12	0.15	0.03

These numbers confirm our original suspicions that the RF is about 30 to 40 percent because these hypotheses generate the largest Lod scores. However, these data alone do not provide convincing support for any model of linkage. Conventionally, a Lod score of at least 3, obtained by adding the scores from many matings, is considered convincing support for a specific RF value. Note that a Lod score of 3 represents an RF value that is 1000 times (that is, 10^3 times) as likely as the hypothesis of no linkage.

4.6 Accounting for unseen multiple crossovers

In the discussion about the three-point testcross, we saw that some parental (nonrecombinant) chromatids resulted from *double* crossovers. These crossovers would not be counted in the recombinant frequency, skewing the results. This possibility leads to the worrisome notion that *all* map distances based on recombinant frequency might be underestimates of physical chromosomal distances because multiple crossovers might have occurred, some of whose products would not be recombinant. Several mathematical approaches have been designed to get around the multiple crossover problem. We will look at two methods. First we examine a method originally worked out by J. B. S. Haldane in the early years of genetics.

A mapping function

The approach worked out by Haldane was to devise a **mapping function,** a formula that relates RF value to "real" physical distance. An accurate measure of physical distance is the *mean number of crossovers* that occur in that segment per meiosis. Let's call this *m*. Hence our goal is to find a function that relates RF to *m*.

To find this formula, we must first think about outcomes of the various crossover possibilities. In any chromosomal region we might expect meioses with zero, one, two, three, four, or more multiple crossovers. Surprisingly, the only class that is really crucial is the zero class. To see why, consider the following. It is a curious but nonintuitive fact that *any number* of crossovers produces a frequency of 50 percent recombinants *within those meioses*. Figure 4-18 proves this for single and double crossovers as examples, but it is true for any number of crossovers. Hence the true determinant of RF is the relative sizes of the classes with no crossovers, versus the classes with any nonzero number of crossovers.

Now the task is to calculate the size of the zero class. The occurrence of crossovers in a specific chromosomal region is well described by a statistical distribution called the **Poisson distribution.** The Poisson distribution describes the distribution of "successes" in samples when the average probability of successes is low. One illustrative example is dipping a child's net into a pond of fish: most dips will produce no fish, a smaller proportion will produce one fish, an even smaller proportion two, and so on. One can translate this analogy directly into a chromosomal region, which in different meioses will have 0, 1, 2, etc., crossover "successes." The Poisson formula (below) will tell us the proportion of the classes with different numbers of crossovers.

$$f_i = (e^{-m} m^i)/i!$$

The terms in the formula have the following meanings:

e = the base of natural logarithms (approximately 2.7)

m = the mean number of successes in a defined sample size

i = the actual number of successes in a sample of that size

f_i = the frequency of samples with *i* successes in them

! = the factorial symbol (e.g., 5! = 5 × 4 × 3 × 2 × 1)

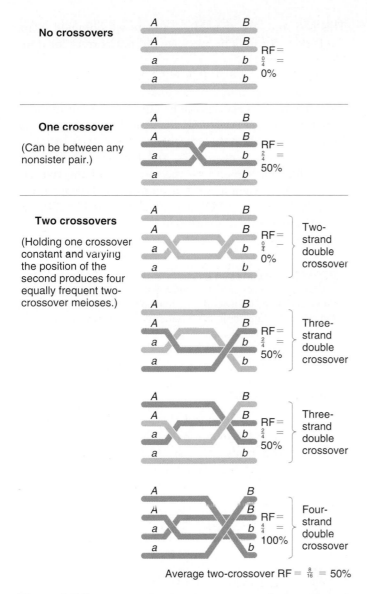

Figure 4-18 Demonstration that the average RF is 50 percent for meioses in which the number of crossovers is not zero. Recombinant chromatids are brown. Two-strand double crossovers produce all parental types, so all the chromatids are orange. Note that all crossovers are between nonsister chromatids. Try the triple crossover class for yourself.

The Poisson distribution tells us that the $i = 0$ class (the key one) is

$$e^{-m} \frac{m^0}{0!}$$

and since m^0 and $0!$ both equal 1, the formula reduces to e^{-m}.

Now we can write a function that relates RF to m. The frequency of the class with any nonzero number of crossovers will be $1 - e^{-m}$, and in these meioses 1/2 of the products will be recombinant, so

$$RF = \tfrac{1}{2}(1 - e^{-m})$$

and this formula is the mapping function we have been seeking.

Let's look at an example of how it works. Assume that in one testcross we obtain an RF value of 27.5 percent (0.275). Plugging this into the function allows us to solve for m:

$$0.275 = \tfrac{1}{2}(1 - e^{-m})$$

so

$$e^{-m} = 1 - (2 \times 0.275) = 0.45$$

Using a calculator we can deduce that $m = 0.8$. That is, on average there are 0.8 crossovers per meiosis in that chromosomal region.

The final step is to convert this measure of physical map distance to "corrected" map units. The following thought experiment reveals how:

*"In very small genetic regions, RF is expected to be an accurate measure of physical distance because there aren't any multiple crossovers. In fact meioses will show either no crossovers or one crossover. The frequency of crossovers (m) will then be translatable into a 'correct' recombinant fraction of m/2 because the recombinants will be 1/2 of the chromatids arising from the single-crossover class. This de*fines a general relationship *between m and a corrected re*combinant fraction, so for any size *of region a 'corrected' re*combinant fraction can be thought of as m/2."*

Hence in the numerical example above, the m value of 0.8 can be converted into a corrected recombinant fraction of $0.8/2 = 0.4$ (40 percent), or 40 map units. We see that indeed this value is substantially larger than the 27.5 map units that we would have deduced from the observed RF.

Note that the mapping function explains why the maximum RF value for linked genes is 50 percent. As m gets very large, e^{-m} tends to zero and the RF tends to 1/2, or 50 percent.

The Perkins formula

Another way of compensating for multiple crossovers can be applied in fungi, using tetrad analysis. In tetrad analysis of "dihybrids" generally, only three types of tetrads are possible, based on the presence of parental and recombinant genotypes in the products. In a cross $A\,B \times a\,b$ they are

Parental ditype (PD)	Tetratype (T)	Nonparental ditype (NPD)
$A \cdot B$	$A \cdot B$	$A \cdot b$
$A \cdot B$	$A \cdot b$	$A \cdot b$
$a \cdot b$	$a \cdot B$	$a \cdot B$
$a \cdot b$	$a \cdot b$	$a \cdot B$

The recombinant genotypes are shown in red. If the genes are linked, a simple approach to mapping their distance apart might be to use the following formula:

$$\text{map distance} = RF = 100(NPD + 1/2\ T)$$

because this formula gives the percentage of all recombinants. However, in the 1960s David Perkins developed a formula that compensates for the effects of double crossovers, which are the most common multiple crossovers. Perkins' formula thus provides a more accurate estimate of map distance:

$$\text{corrected map distance} = 50(T + 6\ NPD)$$

We will not go through the derivation of this formula other than to say that it is based on the totals of the PD, T, and NPD classes expected from meioses with 0, 1, and 2 crossovers (it assumes higher numbers are vanishingly rare). Let's look at an example of its use. We assume that in our hypothetical cross of $A\ B \times a\ b$, the observed frequencies of the tetrad classes are 0.56 PD, 0.41 T, and 0.03 NPD. Using the Perkins formula, the corrected map distance between the a and b loci is

$$50[0.41 + (6 \times 0.03)] = 50(0.59) = 29.5\ \text{m.u.}$$

Let us compare this value with the uncorrected value obtained directly from the RF.

Using the same data:

$$\begin{aligned} \text{map distance} &= 100(1/2\ T + NPD) \\ &= 100(0.205 + 0.03) \\ &= 23.5\ \text{m.u.} \end{aligned}$$

This is 6 m.u. less than the estimate we obtained using the Perkins formula because we did not correct for double crossovers.

As an aside, what PD, NPD and T values are expected when dealing with unlinked genes? The sizes of the PD and NPD classes will be equal as a result of independent assortment. The T class can be produced only from a crossover between either of the two loci and their respective centromeres, and therefore the size of the T class will depend on the total size of the two regions lying between locus and centromere. However, the formula (T + 1/2 NPD) should always yield 0.50, reflecting independent assortment.

> **MESSAGE** The inherent tendency of multiple crossover to lead to an underestimate of map distance can be circumvented by the use of map functions (in any organism), and by the Perkins formula (in tetrad-producing organisms such as fungi).

Summary of ratios

The figure below shows all the main phenotypic ratios encountered so far in the book, for monohybrids, dihybrids, and trihybrids. You can read the ratios from the relative widths of the colored boxes in a row. Note that in cases of linkage, the sizes of the classes depend on map distances.

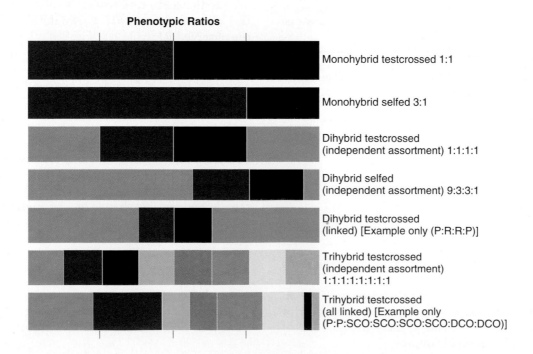

Phenotypic Ratios

Monohybrid testcrossed 1:1

Monohybrid selfed 3:1

Dihybrid testcrossed (independent assortment) 1:1:1:1

Dihybrid selfed (independent assortment) 9:3:3:1

Dihybrid testcrossed (linked) [Example only (P:R:R:P)]

Trihybrid testcrossed (independent assortment) 1:1:1:1:1:1:1:1

Trihybrid testcrossed (all linked) [Example only (P:P:SCO:SCO:SCO:SCO:DCO:DCO)]

KEY QUESTIONS REVISITED

- **For genes on the same chromosome (known as _linked genes_), can new combinations of alleles be detected in the progeny of a dihybrid?**

Yes, new allele combinations (recombinants) regularly arise from such dihybrids. These are routinely detected in testcrosses to homozygous recessive testers. Their frequency is variable and depends on which genes are being studied. For any two genes a consistent value is obtained.

- **If new combinations of alleles arise, by what cellular mechanism does this happen?**

Crossovers are the cellular mechanism responsible. Crossovers occur more or less randomly along the chromosome at the four-chromatid stage of meiosis and can result in recombinants. Any one region may experience a crossover in one meiocyte and none in another meiocyte. Any pair of nonsister chromatids can take part in a crossover. However, some double crossovers result in nonrecombinants.

- **Can the frequency of new combinations of alleles for linked genes be related to their distance apart on the chromosome?**

Yes, in general if we let the frequency of new allele combinations (recombinants) be a measure of distance on the chromosome, we can plot a consistent map of the relative positions of gene loci on the chromosome. The map distances are more or less additive, especially over shorter regions.

- **If we do not know whether two genes are linked, is there a diagnostic test that can be made?**

Yes, the diagnostic test is whether or not the recombinant frequency is 50 percent, which indicates independent assortment (most often, nonlinkage), or significantly less than 50 percent, which indicates linkage. The chi-square test is used for decision making in borderline situations.

SUMMARY

In a dihybrid testcross in _Drosophila_, Thomas Hunt Morgan found a deviation from Mendel's law of independent assortment. He postulated that the two genes were located on the same pair of homologous chromosomes. This relation is called _linkage_.

Linkage explains why the parental gene combinations stay together but not how the recombinant (nonparental) combinations arise. Morgan postulated that in meiosis there may be a physical exchange of chromosome parts by a process now called _crossing-over_. Thus, there are two types of meiotic recombination. Recombination by Mendelian independent assortment results in a recombinant frequency of 50 percent. Crossing-over results in a recombinant frequency generally less than 50 percent.

As Morgan studied more linked genes, he discovered many different values for recombinant frequency (RF) and wondered if these corresponded to the actual distances between genes on a chromosome. Alfred Sturtevant, a student of Morgan's, developed a method of determining the distance between genes on a linkage map, based on the RF. The easiest way to measure RF is with a testcross of a dihybrid or trihybrid. RF values cal-

culated as percentages can be used as map units to construct a chromosomal map showing the loci of the genes analyzed. Silent DNA variation is now used as a source of markers for chromosome mapping. In ascomycete fungi, centromeres can also be located on the map by measuring second-division segregation frequencies.

Although the basic test for linkage is deviation from independent assortment, in a testcross such a deviation may not be obvious and a statistical test is needed. The χ^2 test, which tells how often observations deviate from expectations purely by chance, is particularly useful in determining whether loci are linked.

Crossing-over is the result of physical breakage and reunion of chromosome parts and occurs at the four-chromatid stage of meiosis.

Sample sizes in human pedigree analysis are too small to permit mapping, but cumulative data, expressed as Lod scores, can demonstrate linkage.

Some multiple crossovers result in nonrecombinant chromatids, leading to an underestimate of map distance based on RF. The mapping function corrects for this tendency. The Perkins formula has the same use in tetrad analysis.

KEY TERMS

centimorgan (cM) (p. 124)

cis conformation (p. 118)

coefficient of coincidence (c.o.c.) (p. 127)

crossing-over (p. 119)

crossover products (p. 119)

first-division segregation patterns (M_I patterns) (p. 131)

gene locus (p. 124)

genetic map unit (m.u.) (p. 124)

interference (p. 127)

linkage map (p. 124)

linked (p. 118)
Lod scores (p. 133)
mapping function (p. 134)
meiotic recombination (p. 121)
molecular marker (p. 128)
null hypothesis (p. 132)

octad (p. 129)
Poisson distribution (p. 134)
recombinant (p. 121)
recombinant frequency (RF) (p. 124)
restriction fragment length
  polymorphism (RFLP) (p. 128)

second-division segregation patterns
  (M_{II}) (p. 131)
tetrads (p. 120)
three-point testcross (p. 125)
trans conformation (p. 118)
VNTRs (p. 128)

SOLVED PROBLEMS

1. A human pedigree shows people affected with the rare nail-patella syndrome (misshapen nails and kneecaps) and gives the ABO blood group genotype of each individual. Both loci concerned are autosomal. Study the accompanying pedigree.

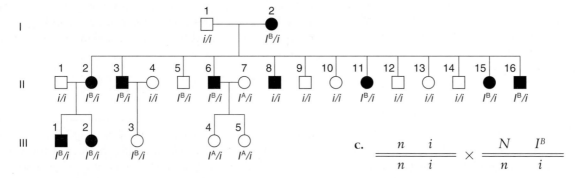

a. Is the nail-patella syndrome a dominant or recessive phenotype? Give reasons to support your answer.

b. Is there evidence of linkage between the nail-patella gene and the gene for ABO blood type, as judged from this pedigree? Why or why not?

c. If there is evidence of linkage, then draw the alleles on the relevant homologs of the grandparents. If there is no evidence of linkage, draw the alleles on two homologous pairs.

d. According to your model, which descendants are recombinants?

e. What is the best estimate of RF?

f. If man III-1 marries a normal woman of blood type O, what is the probability that their first child will be blood type B with nail-patella syndrome?

Solution

a. Nail-patella syndrome is most likely dominant. We are told that it is a rare abnormality, so it is unlikely that the unaffected people marrying into the family carry a presumptive recessive allele for nail-patella syndrome. Let N be the causative allele. Then all people with the syndrome are heterozygotes N/n because all (probably including the grandmother, too) result from a mating to an n/n normal person. Notice that the syndrome appears in all three generations—another indication of dominant inheritance.

b. There is evidence of linkage. Notice that most of the affected people—those that carry the N allele—also carry the I^B allele; most likely, these alleles are linked on the same chromosome.

c.
$$\frac{n \qquad i}{n \qquad i} \times \frac{N \qquad I^B}{n \qquad i}$$

(The grandmother must carry both recessive alleles to produce offspring of genotype i/i and n/n.)

d. Notice that the grandparental mating is equivalent to a testcross, so the recombinants in generation II are

$$\text{II-5}: n\,I^B/n\,i \quad \text{and} \quad \text{II-8}: N\,i/n\,i$$

whereas all others are nonrecombinants, being either $N\,I^B/n\,i$ or $n\,i/n\,i$.

e. Notice that the grandparental cross and the first two crosses in generation II are identical and are all testcrosses. Three of the total 16 progeny are recombinant (II-5, II-8, and III-3). This gives a recombinant frequency of RF = $\frac{3}{16}$ = 18.8 percent. (We cannot include the cross of II-6 × II-7, because the progeny cannot be designated as recombinant or not.)

f. (III-1 ♂)
$$\frac{N \qquad I^B}{n \qquad i} \times \frac{n \qquad i}{n \qquad i} \quad \begin{array}{l}\text{(normal}\\\text{type O ♀)}\end{array}$$

$$\downarrow$$

Gametes

$$81.2\% \begin{cases} N\,I^B & 40.6\% \quad \longleftarrow \text{ nail-patella,} \\ n\,i & 40.6\% \qquad \qquad \text{blood type B} \end{cases}$$

$$18.8\% \begin{cases} N\,i & 9.4\% \\ n\,I^B & 9.4\% \end{cases}$$

The two parental classes are always equal, and so are the two recombinant classes. Hence, the probability that the first child will have nail-patella syndrome and blood type B is 40.6 percent.

2. The allele b gives *Drosophila* flies a black body, and b^+ gives brown, the wild-type phenotype. The allele wx of a separate gene gives waxy wings, and wx^+ gives nonwaxy, the wild-type phenotype. The allele cn of a third gene gives cinnabar eyes, and cn^+ gives red, the wild-type phenotype. A female heterozygous for these three genes is test-crossed, and 1000 progeny are classified as follows: 5 wild type; 6 black, waxy, cinnabar; 69 waxy, cinnabar; 67 black; 382 cinnabar; 379 black, waxy; 48 waxy; and 44 black, cinnabar. Note that a progeny group may be specified by listing only the mutant phenotypes.

   a. Explain these numbers.

   b. Draw the alleles in their proper positions on the chromosomes of the triple heterozygote.

   c. If it is appropriate according to your explanation, calculate interference.

Solution

a. One general piece of advice is to be methodical. Here it is a good idea to write out the genotypes that may be inferred from the phenotypes. The cross is a testcross of type

$$b^+/b \cdot wx^+/wx \cdot cn^+/cn \times b/b \cdot wx/wx \cdot cn/cn$$

Notice that there are distinct pairs of progeny classes in regard to frequency. Already, we can guess that the two largest classes represent parental chromosomes, that the two classes of about 68 represent single crossovers in one region, that the two classes of about 45 represent single crossovers in the other region, and that the two classes of about 5 represent double crossovers. We can write out the progeny as classes derived from the female's gametes, grouped as follows:

$b^+ \cdot wx^+ \cdot cn$		382
$b \cdot wx \cdot cn^+$		379
$b^+ \cdot wx \cdot cn$		69
$b \cdot wx^+ \cdot cn^+$		67
$b^+ \cdot wx \cdot cn^+$		48
$b \cdot wx^+ \cdot cn$		44
$b \cdot wx \cdot cn$		6
$b^+ \cdot wx^+ \cdot cn^+$		5
		1000

Writing the classes out this way confirms that the pairs of classes are in fact reciprocal genotypes arising from zero, one, or two crossovers.

At first, because we do not know the parents of the triple heterozygous female, it looks as if we cannot apply the definition of recombination in which gametic genotypes are compared with the two input genotypes that form an individual. But on reflection, the only parental types that make sense in regard to the data presented are $b^+/b^+ \cdot wx^+/wx^+ \cdot cn/cn$ and $b/b \cdot wx/wx \cdot cn^+/cn^+$ because these are still the most common classes.

Now we can calculate the recombinant frequencies. For $b-wx$,

$$\text{RF} = \frac{69 + 67 + 48 + 44}{1000} = 22.8\%$$

for $b-cn$,

$$\text{RF} = \frac{48 + 44 + 6 + 5}{1000} = 10.3\%$$

and for $wx-cn$,

$$\text{RF} = \frac{69 + 67 + 6 + 5}{1000} = 14.7\%$$

The map is therefore

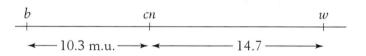

b. The parental chromosomes in the triple heterozygote were

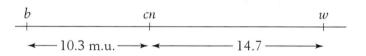

c. The expected number of double recombinants is $0.103 \times 0.147 \times 1000 = 15.141$. The observed number is $6 + 5 = 11$, so interference can be calculated as

$$I = 1 - 11/15.141 = 1 - 0.726 = 0.274$$
$$= 27.4\%$$

3. A cross is made between a haploid strain of *Neurospora* of genotype $nic^+ \, ad$ and another haploid strain of genotype $nic \, ad^+$. From this cross, a total of

1	2	3	4	5	6	7
$nic^+ \cdot ad$	$nic^+ \cdot ad^+$	$nic^+ \cdot ad^+$	$nic^+ \cdot ad$	$nic^+ \cdot ad$	$nic^+ \cdot ad^+$	$nic^+ \cdot ad^+$
$nic^+ \cdot ad$	$nic^+ \cdot ad^+$	$nic^+ \cdot ad^+$	$nic^+ \cdot ad$	$nic^+ \cdot ad$	$nic^+ \cdot ad^+$	$nic^+ \cdot ad^+$
$nic^+ \cdot ad$	$nic^+ \cdot ad^+$	$nic^+ \cdot ad$	$nic \cdot ad$	$nic \cdot ad^+$	$nic \cdot ad$	$nic \cdot ad$
$nic^+ \cdot ad$	$nic^+ \cdot ad^+$	$nic^+ \cdot ad$	$nic \cdot ad$	$nic \cdot ad^+$	$nic \cdot ad$	$nic \cdot ad$
$nic \cdot ad^+$	$nic \cdot ad$	$nic \cdot ad^+$	$nic^+ \cdot ad^+$	$nic^+ \cdot ad$	$nic^+ \cdot ad^+$	$nic^+ \cdot ad$
$nic \cdot ad^+$	$nic \cdot ad$	$nic \cdot ad^+$	$nic^+ \cdot ad^+$	$nic^+ \cdot ad$	$nic^+ \cdot ad^+$	$nic^+ \cdot ad$
$nic \cdot ad^+$	$nic \cdot ad$	$nic \cdot ad$	$nic \cdot ad^+$	$nic \cdot ad^+$	$nic \cdot ad$	$nic \cdot ad^+$
$nic \cdot ad^+$	$nic \cdot ad$	$nic \cdot ad$	$nic \cdot ad^+$	$nic \cdot ad^+$	$nic \cdot ad$	$nic \cdot ad^+$
808	1	90	5	90	1	5

1000 linear asci are isolated and categorized as in the above table. Map the *ad* and *nic* loci in relation to centromeres and to each other.

Solution

What principles can we draw on to solve this problem? It is a good idea to begin by doing something straightforward, which is to calculate the two locus-to-centromere distances. We do not know if the *ad* and the *nic* loci are linked, but we do not need to know. The frequencies of the M_{II} patterns for each locus give the distance from locus to centromere. (We can worry about whether it is the same centromere later.)

Remember that an M_{II} pattern is any pattern that is not two blocks of four. Let's start with the distance between the *nic* locus and the centromere. All we have to do is add the ascus types 4, 5, 6, and 7, because they are all M_{II} patterns for the *nic* locus. The total is $5 + 90 + 1 + 5 = 101$ out of 1000, or 10.1 percent. In this chapter, we have seen that to convert this percentage into map units, we must divide by 2, which gives 5.05 m.u.

We do the same thing for the *ad* locus. Here the total of the M_{II} patterns is given by types 3, 5, 6, and 7 and is $90 + 90 + 1 + 5 = 186$ out of 1000, or 18.6 percent, which is 9.3 m.u.

ad

9.30 m.u.

Now we have to put the two together and decide between the following alternatives, all of which are compatible with the preceding locus-to-centromere distances:

a.

nic   ad

5.05 m.u.   9.30 m.u.

b.

nic   ad

5.05 m.u.   9.30 m.u.

c.

nic   ad

5.05 m.u.

9.30 m.u.

Here a combination of common sense and simple analysis tells us which alternative is correct. First, an inspection of the asci reveals that the most common single type is the one labeled 1, which contains more than 80 percent of all the asci. This type contains only $nic^+ \cdot ad$ and $nic \cdot ad^+$ genotypes, and they are *parental* genotypes. So we know that recombination is quite low and the loci are certainly linked. This rules out alternative a.

Now consider alternative c; if this alternative were correct, a crossover between the centromere and the *nic* locus would generate not only an M_{II} pattern for that locus, but also an M_{II} pattern for the *ad* locus, because it is farther from the centromere than *nic*. The ascus pattern produced by alternative c should be

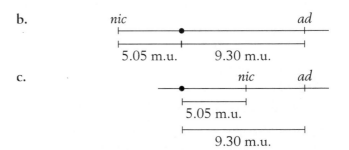

nic^+	ad
nic^+	ad
nic	ad^+
nic	ad^+
nic^+	ad
nic^+	ad
nic	ad^+
nic	ad^+

Remember that the *nic* locus shows M_{II} patterns in asci types 4, 5, 6, and 7 (a total of 101 asci); of them, type 5 is the very one that we are talking about and contains 90 asci. Therefore, alternative c appears to be correct because ascus type 5 comprises about 90 percent of the M_{II} asci for the *nic* locus. This relation would not hold if alternative b were correct, because crossovers on either side of the centromere would generate the M_{II} patterns for the *nic* and the *ad* loci independently.

Is the map distance from *nic* to *ad* simply 9.30 − 5.05 = 4.25 m.u.? Close, but not quite. The best way of calculating map distances between loci is always by measuring the recombinant frequency (RF). We could go through the asci and count all the recombinant ascospores, but it is simpler to use the formula RF = $\frac{1}{2}$ T + NPD. The T asci are classes 3, 4, and 7, and the NPD asci are classes 2 and 6. Hence, RF + [$\frac{1}{2}$(100) + 2] / 1000 = 5.2 percent, or 5.2 m.u., and a better map is

nic   ad
●————┼————————┼————
    5.05 m.u. ┆ 5.2 m.u.
├————————————┼————————┤
        10.25 m.u.

The reason for the underestimate of the *ad*-to-centromere distance calculated from the M_{II} frequency is the occurrence of double crossovers, which can produce an M_i pattern for *ad*, as in ascus type 4:

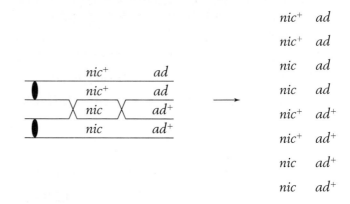

nic^+	ad
nic^+	ad
nic	ad
nic	ad
nic^+	ad^+
nic^+	ad^+
nic	ad^+
nic	ad^+

PROBLEMS

BASIC PROBLEMS

1. A plant of genotype

$$\frac{A \quad B}{a \quad b}$$

is testcrossed to

$$\frac{a \quad b}{a \quad b}$$

If the two loci are 10 m.u. apart, what proportion of progeny will be *A B/a b*?

2. The *A* locus and the *D* locus are so tightly linked that no recombination is ever observed between them. If *A d/A d* is crossed to *a D/a D*, and the F_1 is intercrossed, what phenotypes will be seen in the F_2 and in what proportions?

3. The *R* and *S* loci are 35 m.u. apart. If a plant of genotype

$$\frac{R \quad S}{r \quad s}$$

is selfed, what progeny phenotypes will be seen and in what proportions?

4. The cross *E/E · F/F × e/e · f/f* is made, and the F_1 is then backcrossed to the recessive parent. The progeny genotypes are inferred from the phenotypes. The progeny genotypes, written as the gametic contributions of the heterozygous parent, are in the following proportions:

$E \cdot F$	$\frac{2}{6}$
$E \cdot f$	$\frac{1}{6}$
$e \cdot F$	$\frac{1}{6}$
$e \cdot f$	$\frac{2}{6}$

Explain these results.

5. A strain of *Neurospora* with the genotype *H · I* is crossed with a strain with the genotype *h · i*. Half the progeny are *H · I*, and half are *h · i*. Explain how this is possible.

6. A female animal with genotype *A/a · B/b* is crossed with a double-recessive male (*a/a · b/b*). Their progeny include 442 *A/a · B/b*, 458 *a/a · b/b*, 46 *A/a · b/b*, and 54 *a/a · B/b*. Explain these results.

7. If *A/A · B/B* is crossed to *a/a · b/b*, and the F_1 is testcrossed, what percent of the testcross progeny will be *a/a · b/b* if the two genes are (a) unlinked; (b) completely linked (no crossing-over at all); (c) 10 map units apart; (d) 24 map units apart?

8. In a haploid organism, the C and *D* loci are 8 m.u. apart. From a cross C *d* × *c D*, give the proportion of each of the following progeny classes: (a) C *D*; (b) *c d*; (c) C *d*; (d) all recombinants.

9. A fruit fly of genotype *B R/b r* is testcrossed to *b r/b r*. In 84 percent of the meioses, there are no chiasmata between the linked genes; in 16 percent of the meioses, there is one chiasma between the genes. What proportion of the progeny will be *B r/br*?

10. A three-point testcross was made in corn. The results and a recombination analysis are shown in the following display, which is typical of three-point testcrosses (p = purple leaves, $+$ = green; v = virus-resistant seedlings, $+$ = sensitive; b = brown midriff to seed, $+$ =plain). Study the display and answer parts a–c.

P $+/+ \cdot +/+ \cdot +/+ \times p/p \cdot v/v \cdot b/b$

Gametes $+ \cdot + \cdot +$ $p \cdot v \cdot b$

F_1 $+/p \cdot +/v \cdot +/b \times p/p \cdot v/v \cdot b/b$ (tester)

Class	Progeny phenotypes	F_1 gametes	Numbers	Recombinant for $p–b$	Recombinant for $p–v$	Recombinant for $v–b$
1	gre sen pla	$+ \cdot + \cdot +$	3,210			
2	pur res bro	$p \cdot v \cdot b$	3,222			
3	gre res pla	$+ \cdot v \cdot +$	1,024		R	R
4	pur sen bro	$p \cdot + \cdot b$	1,044		R	R
5	pur res pla	$p \cdot v \cdot +$	690	R		R
6	gre sen bro	$+ \cdot + \cdot b$	678	R		R
7	gre res bro	$+ \cdot v \cdot b$	72	R	R	
8	pur sen pla	$p \cdot + \cdot +$	60	R	R	
		Total	10,000	1,500	2,200	3,436

a. Determine which genes are linked.

b. Draw a map that shows distances in map units.

c. Calculate interference, if appropriate.

UNPACKING THE PROBLEM

1. Sketch cartoon drawings of the parent P, F_1, and tester corn plants, and use arrows to show exactly how you would perform this experiment. Show where seeds are obtained.

2. Why do all the $+$'s look the same, even for different genes? Why does this not cause confusion?

3. How can a phenotype be purple and brown (for example) at the same time?

4. Is it significant that the genes are written in the order p-v-b in the problem?

5. What is a tester and why is it used in this analysis?

6. What does the column marked "Progeny phenotypes" represent? In class 1, for example, state exactly what "gre sen pla" means.

7. What does the line marked "Gametes" represent, and how is this different from the column marked "F_1 gametes"? In what way is comparison of these two types of gametes relevant to recombination?

8. Which meiosis is the main focus of study? Label it on your drawing.

9. Why are the gametes from the tester not shown?

10. Why are there only eight phenotypic classes? Are there any classes missing?

11. What classes (and in what proportions) would be expected if all the genes are on separate chromosomes?

12. To what do the four pairs of class sizes (very big, two intermediates, very small) correspond?

13. What can you tell about gene order simply by inspecting the phenotypic classes and their frequencies?

14. What will be the expected phenotypic class distribution if only two genes are linked?

15. What does the word "point" refer to in a three-point testcross? Does this word usage imply linkage? What would a four-point testcross be like?

16. What is the definition of *recombinant*, and how is it applied here?

17. What do the "Recombinant for" columns mean?

18. Why are there only three "Recombinant for" columns?

19. What do the R's mean, and how are they determined?

20. What do the column totals signify? How are they used?

21. What is the diagnostic test for linkage?

22. What is a map unit? Is it the same as a centimorgan?

23. In a three-point testcross such as this one, why aren't the F_1 and the tester considered to be parental in calculating recombination? (They *are* parents in one sense.)

24. What is the formula for interference? How are the "expected" frequencies calculated in the coefficient of coincidence formula?

25. Why does part *c* of the problem say "if appropriate"?

26. How much work is it to obtain such a large progeny size in corn? Which of the three genes would take the most work to score? Approximately how many progeny are represented by one corn cob?

11. You have a *Drosophila* line that is homozygous for autosomal recessive alleles *a*, *b*, and *c*, linked in that order. You cross females of this line with males homozygous for the corresponding wild-type alleles. You then cross the F_1 heterozygous males with their heterozygous sisters. You obtain the following F_2 phenotypes (where letters denote recessive phenotypes and pluses denote wild-type phenotypes): 1364 + + +, 365 *a b c*, 87 *a b* +, 84 + + *c*, 47 *a* + +, 44 + *b c*, 5 *a* + *c*, and 4 + *b* +.

   a. What is the recombinant frequency between *a* and *b*? Between *b* and *c*? (Remember, there is no crossing over in *Drosophila* males.)

   b. What is the coefficient of coincidence?

12. R. A. Emerson crossed two different pure-breeding lines of corn and obtained a phenotypically wild-type F_1 that was heterozygous for three alleles that determine recessive phenotypes: *an* determines anther; *br*, brachytic; and *f*, fine. He testcrossed the F_1 to a tester that was homozygous recessive for the three genes and obtained these progeny phenotypes: 355 anther; 339 brachytic, fine; 88 completely wild type; 55 anther, brachytic, fine; 21 fine; 17 anther, brachytic; 2 brachytic; 2 anther, fine.

   a. What were the genotypes of the parental lines?

   b. Draw a linkage map for the three genes (include map distances).

   c. Calculate the interference value.

13. Chromosome 3 of corn carries three loci (*b* for plant-color booster, *v* for virescent, and *lg* for liguleless). A testcross of triple recessives with F_1 plants heterozygous for the three genes yields progeny having the following genotypes: 305 + *v lg*, 275 *b* + +, 128 *b* + *lg*, 112 + *v* +, 74 + + *lg*, 66 *b v* +, 22 + + +, and 18 *b v lg*. Give the gene sequence on the chromosome, the map distances between genes, and the coefficient of coincidence.

14. Groodies are useful (but fictional) haploid organisms that are pure genetic tools. A wild-type groody has a fat body, a long tail, and flagella. Mutant lines are known that have thin bodies, or are tailless, or do not have flagella. Groodies can mate with each other (although they are so shy that we do not know how) and produce recombinants. A wild-type groody

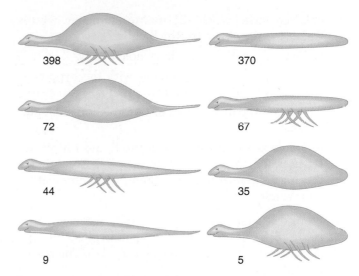

398   370

72   67

44   35

9   5

mates with a thin-bodied groody lacking both tail and flagella. The 1000 baby groodies produced are classified as shown in the accompanying illustration. Assign genotypes, and map the three genes.

(Problem 14 from Burton S. Guttman.)

15. In *Drosophila*, the allele dp^+ determines long wings and *dp* determines short ("dumpy") wings. At a separate locus, e^+ determines gray body and *e* determines ebony body. Both loci are autosomal. The following crosses were made, starting with pure-breeding parents:

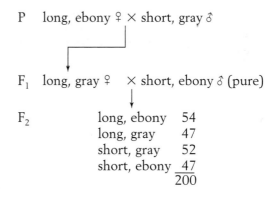

P   long, ebony ♀ × short, gray ♂

F_1 long, gray ♀ × short, ebony ♂ (pure)

F_2

long, ebony	54
long, gray	47
short, gray	52
short, ebony	47
	$\overline{200}$

Use the χ^2 test to determine if these loci are linked. In doing so, indicate (**a**) the hypothesis, (**b**) calculation of χ^2, (**c**) *p* value, (**d**) what the *p* value means, (**e**) your conclusion, (**f**) the inferred chromosomal constitutions of parents, F_1, tester, and progeny.

16. The mother of a family with 10 children has blood type Rh^+. She also has a very rare condition (elliptocytosis, phenotype E) that causes red blood cells to be oval rather than round in shape but that produces no adverse clinical effects. The father is Rh^- (lacks the Rh^+ antigen) and has normal red cells (phenotype e). The children are 1 Rh^+ c, 4 Rh^+ E, and 5 Rh^- e. Information is available on the mother's parents, who are Rh^+ E and Rh^- e. One of the 10

children (who is Rh⁺ E) marries someone who is Rh⁺e, and they have an Rh⁺ E child.

a. Draw the pedigree of this whole family.

b. Is the pedigree in agreement with the hypothesis that the Rh^+ allele is dominant and Rh^- is recessive?

c. What is the mechanism of transmission of elliptocytosis?

d. Could the genes governing the E and Rh phenotypes be on the same chromosome? If so, estimate the map distance between them, and comment on your result.

17. From several crosses of the general type $A/A \cdot B/B \times a/a \cdot b/b$ the F_1 individuals of type $A/a \cdot B/b$ were testcrossed to $a/a \cdot b/b$. The results are as follows:

	Testcross progeny			
Testcross of F_1 from cross	$A/a \cdot B/b$	$a/a \cdot b/b$	$A/a \cdot b/b$	$a/a \cdot B/b$
1	310	315	287	288
2	36	38	23	23
3	360	380	230	230
4	74	72	50	44

For each set of progeny, use the χ^2 test to decide if there is evidence of linkage.

18. In the two pedigrees diagrammed here, a vertical bar in a symbol stands for steroid sulfatase deficiency, and a horizontal bar stands for ornithine transcarbamylase deficiency.

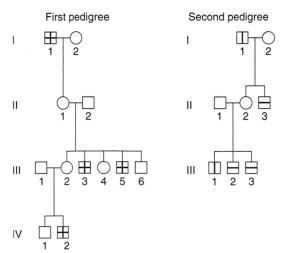

a. Is there any evidence in these pedigrees that the genes determining the deficiencies are linked?

b. If the genes are linked, is there any evidence in the pedigree of crossing-over between them?

c. Draw genotypes of these individuals as far as possible.

19. In the accompanying pedigree, the vertical lines stand for protan colorblindness, and the horizontal lines stand for deutan colorblindness. These are separate conditions causing different misperceptions of colors; each is determined by a separate gene.

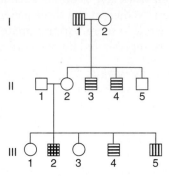

a. Does the pedigree show any evidence that the genes are linked?

b. If there is linkage, does the pedigree show any evidence of crossing-over? Explain both your answers with the aid of the diagram.

c. Can you calculate a value for the recombination between these genes? Is this recombination by independent assortment or crossing-over?

20. In corn, a triple heterozygote was obtained carrying the mutant alleles s (shrunken), w (white aleurone), and y (waxy endosperm), all paired with their normal wild-type alleles. This triple heterozygote was testcrossed, and the progeny contained 116 shrunken, white; 4 fully wild-type; 2538 shrunken; 601 shrunken, waxy; 626 white; 2708 white, waxy; 2 shrunken, white, waxy; and 113 waxy.

a. Determine if any of these three loci are linked and, if so, show map distances.

b. Show the allele arrangement on the chromosomes of the triple heterozygote used in the testcross.

c. Calculate interference, if appropriate.

21. **a.** A mouse cross $A/a \cdot B/b \times a/a \cdot b/b$ is made, and in the progeny there are

$$25\%\ A/a \cdot B/b,\ \ 25\%\ a/a \cdot b/b,$$
$$25\%\ A/a \cdot b/b,\ \ 25\%\ a/a \cdot B/b$$

Explain these proportions with the aid of simplified meiosis diagrams.

b. A mouse cross $C/c \cdot D/d \times c/c \cdot d/d$ is made, and in the progeny there are

$$45\%\ C/c \cdot d/d,\ \ 45\%\ c/c \cdot D/d,$$
$$5\%\ c/c \cdot d/d,\ \ 5\%\ C/c \cdot D/d$$

Explain these proportions with the aid of simplified meiosis diagrams.

22. In the tiny model plant *Arabidopsis*, the recessive allele *hyg* confers seed resistance to the drug hygromycin, and *her*, a recessive allele of a different gene, confers seed resistance to herbicide. A plant that was homozygous *hyg/hyg · her/her* was crossed to wild type, and the F₁ was selfed. Seeds resulting from the F₁ self were placed on petri dishes containing hygromycin and herbicide.

    a. If the two genes are unlinked, what percentage of seeds are expected to grow?

    b. In fact, 13 percent of the seeds grew. Does this percentage support the hypothesis of no linkage? Explain. If not, calculate the number of map units between the loci.

    c. Under your hypothesis, if the F₁ is testcrossed, what proportion of seeds will grow on the medium containing hygromycin and herbicide?

23. In the pedigree in Figure 4-17, calculate the Lod score for a recombinant frequency of 34 percent.

24. In a diploid organism of genotype *A/a ; B/b ; D/d*, the allele pairs are all on different chromosome pairs. The accompanying diagrams purport to show anaphases ("pulling apart" stages) in individual cells. A line represents a chromosome or a chromatid, and the dot indicates the position of the centromere. State whether each drawing represents mitosis, meiosis I, or meiosis II, or is impossible for this particular genotype.

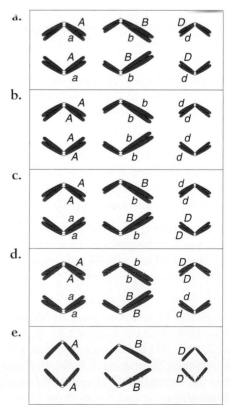

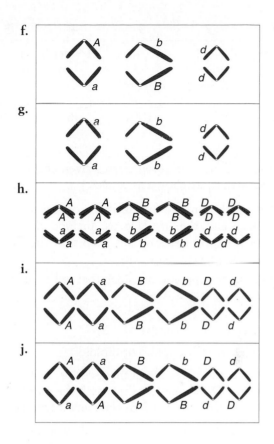

25. The *Neurospora* cross *al-2⁺ × al-2* is made. A linear tetrad analysis reveals that the second-division segregation frequency is 8 percent.

    a. Draw two examples of second-division segregation patterns in this cross.

    b. What can be calculated using the 8 percent value?

26. From the fungal cross *arg-6 · al-2 × arg-6⁺ · al-2⁺*, what will the spore genotypes be in unordered tetrads that are

    (a) parental ditypes?

    (b) tetratypes?

    (c) nonparental ditypes?

27. For a certain chromosomal region, the mean number of crossovers at meiosis is calculated to be two per meiosis. In that region, what proportion of meioses are predicted to have

    (a) no crossovers?

    (b) one crossover?

    (c) two crossovers?

28. A *Neurospora* cross was made between one strain that carried the mating-type allele *A* and the mutant allele *arg-1* and another strain that carried the mating-type allele *a* and the wild-type allele for *arg-1* (+). Four hundred linear octads were isolated,

1	2	3	4	5	6	7
$A \cdot arg$	$A \cdot +$	$A \cdot arg$	$A \cdot arg$	$A \cdot arg$	$A \cdot +$	$A \cdot +$
$A \cdot arg$	$A \cdot +$	$A \cdot +$	$a \cdot arg$	$a \cdot +$	$a \cdot arg$	$a \cdot arg$
$a \cdot +$	$a \cdot arg$	$a \cdot arg$	$A \cdot +$	$A \cdot arg$	$A \cdot +$	$A \cdot arg$
$a \cdot +$	$a \cdot arg$	$a \cdot +$	$a \cdot +$	$a \cdot +$	$a \cdot arg$	$a \cdot +$
127	125	100	36	2	4	6

and they fell into the seven classes given above. (For simplicity, they are shown as tetrads.)

a. Deduce the linkage arrangement of the mating-type locus and the *arg-1* locus. Include the centromere or centromeres on any map that you draw. Label *all* intervals in map units.

b. Diagram the meiotic divisions that led to class 6. Label clearly.

 UNPACKING THE PROBLEM

1. Are fungi generally haploid or diploid?

2. How many ascospores are in the ascus of *Neurospora*? Does your answer match the number presented in this problem? Explain any discrepancy.

3. What is mating type in fungi? How do you think it is determined experimentally?

4. Do the symbols *A* and *a* have anything to do with dominance and recessiveness?

5. What does the symbol *arg-1* mean? How would you test for this genotype?

6. How does the *arg-1* symbol relate to the symbol *+*?

7. What does the expression *wild type* mean?

8. What does the word *mutant* mean?

9. Does the biological function of the alleles shown have anything to do with the solution of this problem?

10. What does the expression *linear octad analysis* mean?

11. In general, what more can be learned from linear tetrad analysis that cannot be learned from unordered tetrad analysis?

12. How is a cross made in a fungus such as *Neurospora*? Explain how to isolate asci and individual ascospores. How does the term *tetrad* relate to the terms *ascus* and *octad*?

13. Where does meiosis take place in the *Neurospora* life cycle? (Show it on a diagram of the life cycle.)

14. What does Problem 28 have to do with meiosis?

15. Can you write out the genotypes of the two parental strains?

16. Why are only four genotypes shown in each class?

17. Why are there only seven classes? How many ways have you learned for classifying tetrads generally? Which of these classifications can be applied to both linear and unordered tetrads? Can you apply these classifications to the tetrads in this problem? (Classify each class in as many ways as possible.) Can you think of more possibilities in this cross? If so, why are they not shown?

18. Do you think there are several different spore orders within each class? Why would these different spore orders not change the class?

19. Why is the following class not listed?

$$
\begin{array}{l}
a \cdot + \\
a \cdot + \\
A \cdot arg \\
A \cdot arg
\end{array}
$$

20. What does the expression *linkage arrangement* mean?

21. What is a genetic *interval*?

22. Why does the problem state "centromere or centromeres" and not just "centromere"? What is the general method for mapping centromeres in tetrad analysis?

23. What is the total frequency of $A \cdot +$ ascospores? (Did you calculate this frequency by using a formula or by inspection? Is this a recombinant genotype? If so, is it the only recombinant genotype?)

24. The first two classes are the most common and are approximately equal in frequency. What does this information tell you? What is their content of parental and recombinant genotypes?

29. A geneticist studies 11 different pairs of *Neurospora* loci by making crosses of the type $a \cdot b \times a^+ \cdot b^+$ and then analyzing 100 linear asci from each cross. For the convenience of making a table, the geneticist organizes the data as if all 11 pairs of genes had the same designation—*a* and *b*—as shown here:

NUMBER OF ASCI OF TYPE

Cross	$a \cdot b$ $a \cdot b$ $a^+ \cdot b^+$ $a^+ \cdot b^+$	$a \cdot b^+$ $a \cdot b^+$ $a^+ \cdot b$ $a^+ \cdot b$	$a \cdot b$ $a \cdot b^+$ $a^+ \cdot b^+$ $a^+ \cdot b$	$a \cdot b$ $a^+ \cdot b$ $a^+ \cdot b^+$ $a \cdot b^+$	$a \cdot b$ $a^+ \cdot b^+$ $a^+ \cdot b^+$ $a \cdot b$	$a \cdot b^+$ $a^+ \cdot b$ $a^+ \cdot b$ $a \cdot b^+$	$a \cdot b^+$ $a^+ \cdot b$ $a^+ \cdot b^+$ $a \cdot b$
1	34	34	32	0	0	0	0
2	84	1	15	0	0	0	0
3	55	3	40	0	2	0	0
4	71	1	18	1	8	0	1
5	9	6	24	22	8	10	20
6	31	0	1	3	61	0	4
7	95	0	3	2	0	0	0
8	6	7	20	22	12	11	22
9	69	0	10	18	0	1	2
10	16	14	2	60	1	2	5
11	51	49	0	0	0	0	0

For each cross, map the loci in relation to each other and to centromeres.

30. Three different crosses in *Neurospora* are analyzed on the basis of unordered tetrads. Each cross combines a different pair of linked genes. The results are shown in the following table:

Cross	Parents	Parental ditypes (%)	Tetra-types (%)	Non-parental ditypes (%)
1	$a \cdot b^+ \times a^+ \cdot b$	51	45	4
2	$c \cdot d^+ \times c^+ \cdot d$	64	34	2
3	$e \cdot f^+ \times e^+ \cdot f$	45	50	5

For each cross, calculate:

a. The frequency of recombinants (RF).

b. The uncorrected map distance, based on RF.

c. The corrected map distance, based on tetrad frequencies.

CHALLENGING PROBLEMS

31. An individual heterozygous for four genes, $A/a \cdot B/b \cdot C/c \cdot D/d$, is testcrossed to $a/a \cdot b/b \cdot c/c \cdot d/d$, and 1000 progeny are classified by the gametic contribution of the heterozygous parent as follows:

$a \cdot B \cdot C \cdot D$	42
$A \cdot b \cdot c \cdot d$	43
$A \cdot B \cdot C \cdot d$	140
$a \cdot b \cdot c \cdot D$	145
$a \cdot B \cdot c \cdot D$	6
$A \cdot b \cdot C \cdot d$	9
$A \cdot B \cdot c \cdot d$	305
$a \cdot b \cdot C \cdot D$	310

a. Which genes are linked?

b. If two pure-breeding lines had been crossed to produce the heterozygous individual, what would their genotypes have been?

c. Draw a linkage map of the linked genes, showing the order and the distances in map units.

d. Calculate an interference value, if appropriate.

32. There is an autosomal allele N in humans that causes abnormalities in nails and patellae (kneecaps) called the *nail-patella syndrome*. Consider marriages in which one partner has the nail-patella syndrome and blood type A and the other partner has normal nails and patellae and blood type O. These marriages produce some children who have both the nail-patella syndrome and blood type A. Assume that unrelated children from this phenotypic group mature, intermarry, and have children. Four phenotypes are observed in the following percentages in this second generation:

nail-patella syndrome, blood type A	66%
normal nail-patella, blood type O	16%
normal nail-patella, blood type A	9%
nail-patella syndrome, blood type O	9%

Fully analyze these data, explaining the relative frequencies of the four phenotypes.

33. Assume that three pairs of alleles are found in *Drosophila*: x^+ and x, y^+ and y, and z^+ and z. As shown by the symbols, each non-wild-type allele is recessive to its wild-type allele. A cross between females heterozygous at these three loci and wild-type males yields progeny having the following genotypes: 1010 $x^+ \cdot y^+ \cdot z^+$ females, 430 $x \cdot y^+ \cdot z$ males,

441 $x^+ \cdot y \cdot z^+$ males, 39 $x \cdot y \cdot z$ males, 32 $x^+ \cdot$ $y^+ \cdot z$ males, 30 $x^+ \cdot y^+ \cdot z^+$ males, 27 $x \cdot y \cdot z^+$ males, 1 $x^+ \cdot y \cdot z$ male, and 0 $x \cdot y^+ \cdot z^+$ males.

a. On what chromosome of *Drosophila* are the genes carried?

b. Draw the relevant chromosomes in the heterozygous female parent, showing the arrangement of the alleles.

c. Calculate the map distances between the genes and the coefficient of coincidence.

34. From the five sets of data given in the following table, determine the order of genes by inspection—that is, without calculating recombination values. Recessive phenotypes are symbolized by lowercase letters and dominant phenotypes by pluses.

Phenotypes observed in 3-point testcross	Data sets				
	1	2	3	4	5
+ + +	317	1	30	40	305
+ + c	58	4	6	232	0
+ b +	10	31	339	84	28
+ b c	2	77	137	201	107
a + +	0	77	142	194	124
a + c	21	31	291	77	30
a b +	72	4	3	235	1
a b c	203	1	34	46	265

35. From the phenotype data given in the following table for two three-point testcrosses for (1) *a*, *b*, and *c* and (2) *b*, *c*, and *d*, determine the sequence of the four genes *a*, *b*, *c*, and *d*, and the three map distances between them. Recessive phenotypes are symbolized by lowercase letters and dominant phenotypes by pluses.

1		2	
+ + +	669	b c d	8
a b +	139	b + +	441
a + +	3	b + d	90
+ + c	121	+ c d	376
+ b c	2	+ + +	14
a + c	2280	+ + d	153
a b c	653	+ c +	65
+ b +	2215	b c +	141

36. The father of Mr. Spock, first officer of the starship *Enterprise*, came from planet Vulcan; Spock's mother came from Earth. A Vulcan has pointed ears (determined by allele *P*), adrenals absent (determined by *A*), and a right-sided heart (determined by *R*). All these alleles are dominant to normal Earth alleles.

The three loci are autosomal, and they are linked as shown in this linkage map:

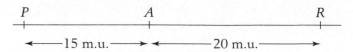

If Mr. Spock marries an Earth woman and there is no (genetic) interference, what proportion of their children will have

a. Vulcan phenotypes for all three characters?

b. Earth phenotypes for all three characters?

c. Vulcan ears and heart but Earth adrenals?

d. Vulcan ears but Earth heart and adrenals?

(Problem 36 from D. Harrison, *Problems in Genetics*. Addison-Wesley, 1970.)

37. In a certain diploid plant, the three loci *A*, *B*, and *C* are linked as follows:

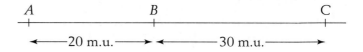

One plant is available to you (call it the parental plant). It has the constitution *A b c/a B C*.

a. Assuming no interference, if the plant is selfed, what proportion of the progeny will be of the genotype *a b c/a b c*?

b. Again assuming no interference, if the parental plant is crossed with the *a b c/a b c* plant, what genotypic classes will be found in the progeny? What will be their frequencies if there are 1000 progeny?

c. Repeat part b, this time assuming 20 percent interference between the regions.

38. The pedigree at the top of the next page shows a family with two rare abnormal phenotypes: blue sclerotic (a brittle bone defect), represented by a black-bordered symbol, and hemophilia, represented by a black center in a symbol. Individuals represented by completely black symbols have both disorders. The numbers in some symbols are the numbers of those types.

a. What pattern of inheritance is shown by each condition in this pedigree?

b. Provide the genotypes of as many family members as possible.

c. Is there evidence of linkage?

d. Is there evidence of independent assortment?

e. Can any of the members be judged as recombinants (that is, formed from at least one recombinant gamete)?

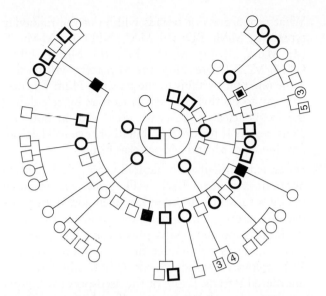

39. The human genes for colorblindness and for hemophilia are both on the X chromosome, and they show a recombinant frequency of about 10 percent. Linkage of a pathological gene to a relatively harmless one can be used for genetic prognosis. Shown below is part of a more extensive pedigree. Blackened symbols indicate that the subjects had hemophilia, and crosses indicate colorblindness. What information could be given to women III-4 and III-5 about the likelihood of their having sons with hemophilia?

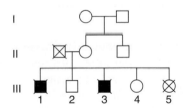

(Problem 39 adapted from J. F. Crow, *Genetics Notes: An Introduction to Genetics*. Burgess, 1983.)

40. A geneticist mapping the genes *A, B, C, D,* and *E* makes two 3-point testcrosses. The first cross of pure lines is

$A/A \cdot B/B \cdot C/C \cdot D/D \cdot E/E$
$\qquad \times \ a/a \cdot b/b \cdot C/C \cdot d/d \cdot E/E$

The geneticist crosses the F_1 with a recessive tester and classifies the progeny by the gametic contribution of the F_1:

A	B	C	D	E	316
a	b	C	d	E	314
A	B	C	d	E	31
a	b	C	D	E	39
A	b	C	d	E	130
a	B	C	D	E	140
A	b	C	D	E	17
a	B	C	d	E	13
					1000

The second cross of pure lines is

$A/A \cdot B/B \cdot C/C \cdot D/D \cdot E/E$
$\qquad \times \ a/a \cdot B/B \cdot c/c \cdot D/D \cdot e/e$

The geneticist crosses the F_1 from this cross with a recessive tester and obtains:

A	B	C	D	E	243
a	B	c	D	e	237
A	B	c	D	e	62
a	B	C	D	E	58
A	B	C	D	e	155
a	B	c	D	E	165
a	B	C	D	e	46
A	B	c	D	E	34
					1000

The geneticist also knows that genes *D* and *E* assort independently.

a. Draw a map of these genes, showing distances in map units wherever possible.

b. Is there any evidence of interference?

41. In the plant *Arabidopsis*, the loci for pod length (*L*, long; *l*, short) and fruit hairs (*H*, hairy; *h*, smooth) are linked 16 map units apart on the same chromosome. The following crosses were made:

$$(\text{i}) \ L\,H/L\,H \ \times \ l\,h/l\,h \longrightarrow F_1$$
$$(\text{ii}) \ L\,h/L\,h \ \times \ l\,H/l\,H \longrightarrow F_1$$

If the F_1's from (i) and (ii) are crossed,

a. What proportion of the progeny are expected to be $l\,h/l\,h$?

b. What proportion of the progeny are expected to be $L\,h/l\,h$?

42. In corn *(Zea mays)*, the genetic map of part of chromosome 4 is as follows, where *w, s,* and *e* represent recessive mutant alleles affecting the color and shape of the pollen:

If the following cross is made

$$+ + +/+ + + \ \times \ w\,s\,e/w\,s\,e$$

and the F_1 is testcrossed to $w\,s\,e/w\,s\,e$, and if it is assumed that there is no interference on this region of the chromosome, what proportion of progeny will be of genotypes?

a. + + +

b. w s e

c. + s e

d. w + +

e. + + e

f. w s +

g. w + e

h. + s +

43. Every Friday night, genetics student Jean Allele, exhausted by her studies, goes to the student union's bowling lane to relax. But even there, she is haunted by her genetic studies. The rather modest bowling lane has only four bowling balls: two red and two blue. They are bowled at the pins and are then collected and returned down the chute in random order, coming to rest at the end stop. Over the evening, Jean notices familiar patterns of the four balls as they come to rest at the stop. Compulsively, she counts the different patterns. What patterns did she see, what were their frequencies, and what is the relevance of this matter to genetics?

44. In a tetrad analysis, the linkage arrangement of the p and q loci is as follows:

Assume that,

- in region i, there is no crossover in 88 percent of meioses and a single crossover in 12 percent of meioses;

- in region ii, there is no crossover in 80 percent of meioses and a single crossover in 20 percent of meioses;

- there is no interference (in other words, the situation in one region does not affect what is going on in the other region).

What proportions of tetrads will be of the following types? (a) $M_I M_I$, PD; (b) $M_I M_I$, NPD; (c) $M_I M_{II}$, T; (d) $M_{II} M_I$, T; (e) $M_{II} M_{II}$, PD; (f) $M_{II} M_{II}$, NPD; (g) $M_{II} M_{II}$, T. (**Note:** Here the M pattern written first is the one that pertains to the p locus.) **Hint:** The easiest way to do this problem is to start by calculating the frequencies of asci with crossovers in both regions, region 1, region 2, and neither region. Then determine what M_I and M_{II} patterns result.

45. For an experiment with haploid yeast, you have two different cultures. Each will grow on minimal medium to which arginine has been added, but neither will grow on minimal medium alone. (Minimal medium is inorganic salts plus sugar.) Using appropriate methods, you induce the two cultures to mate. The diploid cells then divide meiotically and form unordered tetrads. Some of the ascospores will grow on minimal medium. You classify a large number of these tetrads for the phenotypes ARG⁻ (arginine-requiring) and ARG⁺ (arginine-independent) and record the following data:

Segregation of ARG⁻ : ARG⁺	Frequency (%)
4:0	40
3:1	20
2:2	40

a. Using symbols of your own choosing, assign genotypes to the two parental cultures. For each of the three kinds of segregation, assign genotypes to the segregants.

b. If there is more than one locus governing arginine requirement, are these loci linked?

INTERACTIVE GENETICS   MegaManual CD-ROM Tutorial

Linkage Analysis

For additional practice in solving mapping and linkage problems, refer to the Linkage Analysis activity on the Interactive Genetics CD-ROM included with the Solutions MegaManual. Six interactive problems covering test-crosses and map distances are provided to increase your understanding.

Molecular Markers

This activity on the Interactive Genetics CD-ROM includes an interactive tutorial on how markers are used in genetic research. Six exercises explore how markers are used by scientists in fields such as forensic science and medicine.

5

THE GENETICS OF BACTERIA AND THEIR VIRUSES

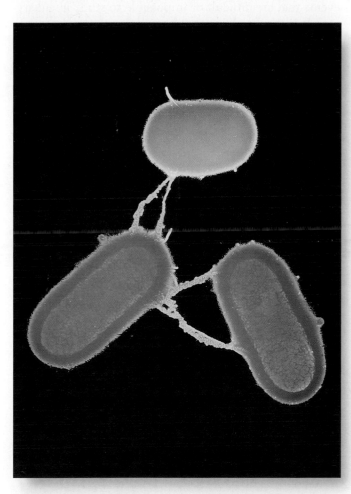

Sexual union of bacteria. Cells of *Escherichia coli* that have become attached by pili prior to DNA transfer between donor and recipient cell types. [Dr. L. Caro/Science Photo Library/Photo Researchers.]

KEY QUESTIONS

- Do bacterial cells ever pair up for any type of sexual cycle?

- Do bacterial genomes ever show recombination?

- If so, in what ways do genomes become associated to permit recombination?

- Does bacterial recombination resemble eukaryote recombination?

- Do the genomes of bacterial viruses ever show recombination?

- Do bacterial and viral genomes interact physically in any way?

- Can bacterial and viral chromosomes be mapped using recombination?

OUTLINE

5.1 Working with microorganisms

5.2 Bacterial conjugation

5.3 Bacterial transformation

5.4 Bacteriophage genetics

5.5 Transduction

5.6 Physical maps versus linkage maps

CHAPTER OVERVIEW

A large part of the history of genetics and current molecular genetics is concerned with bacteria and their viruses. Although bacteria have genes composed of DNA arranged in a long series on a "chromosome," their genetic material is not organized in the same way as that of eukaryotes. They belong to a class of organisms known as **prokaryotes,** which includes the blue-green algae, now classified as *cyanobacteria,* and the bacteria. One of the key defining features of prokaryotes is that they do not have membrane-bound nuclei.

Viruses are also very different from the organisms we have been studying so far. Viruses share some of the properties of organisms; for example, their genetic material is DNA or RNA, constituting a short "chromosome." However, most biologists regard viruses as nonliving because they cannot grow or multiply alone. To reproduce, they must parasitize living cells and use the molecular machinery of these cells. The viruses that parasitize bacteria are called **bacteriophages,** or simply **phages.**

When scientists began studying bacteria and phages, they were naturally curious about their hereditary systems. Clearly they must have hereditary systems because they show a constant appearance and function from one generation to the next (they are true to type). But how

do these hereditary systems work? Bacteria, like unicellular eukaryotic organisms, reproduce asexually by cell growth and division; one cell becoming two. This is quite easy to demonstrate experimentally. However is there ever a union of different types for the purpose of sexual reproduction? Furthermore, how do the much smaller phages reproduce—do they ever unite for a sexlike cycle? These questions are the subject of this chapter.

We shall see that there are a variety of hereditary processes in bacteria and phages. These processes are interesting because of the basic biology of these forms, but studies of their genetics are also providing insights into genetic processes at work in *all* organisms. For a geneticist, the attraction of these forms is that because they are so small, they can be cultured in very large numbers. This makes it possible to detect and study very rare events that are difficult or impossible to study in eukaryotes. It is worth adding that bacterial and phage genetics are the foundation of genetic engineering for the genomes of all organisms because these simpler forms are used as convenient vectors to carry the DNA of higher organisms.

What hereditary processes are observed in prokaryotes? Compared with eukaryotes, bacteria and viruses have simple chromosomes. Generally there is only one chromosome, present in only one copy. Because the cells

CHAPTER OVERVIEW Figure

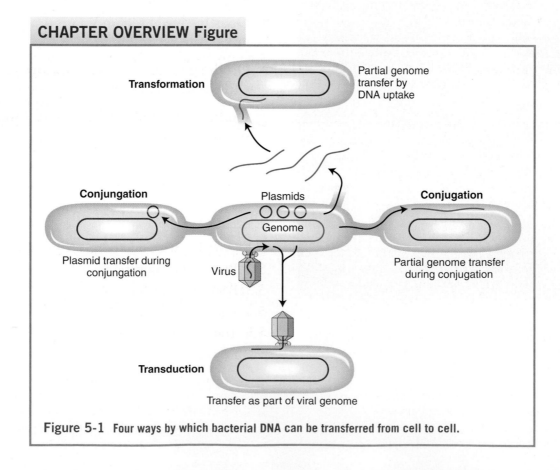

Figure 5-1 **Four ways by which bacterial DNA can be transferred from cell to cell.**

and their chromosomes are so small, possible sexlike fusion events are difficult to observe, even with a microscope. Therefore the general approach has been a genetic one based on the detection of recombinants. The logic is that if different genomes ever do get together, they should occasionally produce recombinants. Conversely, if recombinants are detected, with marker A from one parent and B from another, then there must have been some type of "sexual" union. Hence, even though bacteria and phages do not undergo meiosis, the approach to the genetic analysis of these forms is surprisingly similar to that for eukaryotes.

The opportunity for genetic recombination in bacteria can arise in several different ways, but in all cases two DNA molecules are brought together. The possibilities are outlined in Figure 5-1. The first process to be examined here is *conjugation*: one bacterial cell transfers DNA in one direction to another cell by direct cell-to-cell contact. The transferred DNA may be part of or all the bacterial genome, or it may be an extragenomic DNA element called a *plasmid*. A genomic fragment may recombine with the recipient's chromosome after entry.

A bacterial cell can also acquire a piece of DNA from the environment and incorporate this DNA into its own chromosome; this procedure is called *transformation*. In addition, certain phages can pick up a piece of DNA from one bacterial cell and inject it into another, where it can be incorporated into the chromosome, in a process known as *transduction*.

Phages themselves can undergo recombination when two different genotypes both infect the same bacterial cell (**phage recombination**).

5.1 Working with microorganisms

Bacteria are fast-dividing and take up little space, so they are very convenient to use as genetic model organisms. They can be cultured in a liquid medium, or on a solid surface such as an agar gel, so long as basic nutrients are supplied. Each bacterial cell divides from $1 \rightarrow 2 \rightarrow 4 \rightarrow 8 \rightarrow 16$, and so, on until the nutrients are exhausted or until toxic waste products accumulate to levels that halt the population growth. A small amount of a liquid culture can be pipetted onto a petri plate containing solid agar medium and spread evenly on the surface with a sterile spreader, in a process called **plating** (Figure 5-2). The cells divide, but because they cannot travel far on the surface of the gel, all the cells remain together in a clump. When this mass reaches more than 10^7 cells, it becomes visible to the naked eye as a **colony**. Each distinct colony on the plate will be derived from a single original cell. Members of a colony that share a single genetic ancestor are known as a **cell clone**.

Bacterial mutants are also convenient. Nutritional mutants are a good example. Wild-type bacteria are

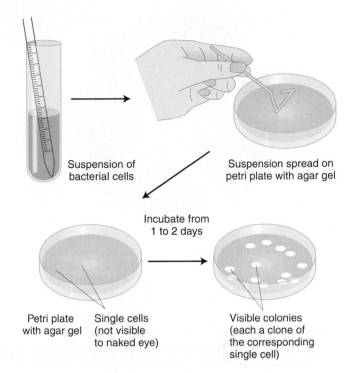

Figure 5-2 Methods of growing bacteria in the laboratory. Bacteria can be grown in liquid media containing nutrients. A small number of bacteria from liquid suspension can also be spread on agar medium containing nutrients. Each cell will give rise to a colony. All cells in a colony have the same genotype and phenotype.

Suspension of bacterial cells

Suspension spread on petri plate with agar gel

Incubate from 1 to 2 days

Petri plate with agar gel

Single cells (not visible to naked eye)

Visible colonies (each a clone of the corresponding single cell)

prototrophic. This means they can grow and divide on **minimal medium**—a substrate containing only inorganic salts, a carbon source for energy, and water. From a prototrophic culture, **auxotrophic** mutants can be obtained: these are cells that will not grow unless the medium contains one or more specific cellular building blocks such as adenine, threonine, or biotin. Another type of useful mutant differs from wild type in the ability to use a specific energy source; for example, the wild type *can* use lactose, whereas a mutant may *not* be able to (Figure 5-3). In another mutant category, whereas wild types are susceptible to an inhibitor, such as the antibiotic streptomycin, **resistant mutants** can divide and form colonies in the presence of the inhibitor. All these types of mutants allow the geneticist to distinguish different individual strains, thereby providing **genetic markers** (marker alleles) to keep track of genomes and cells in experiments. Table 5-1 summarizes some mutant bacterial phenotypes and their genetic symbols.

The following sections document the discovery of the various processes by which bacterial genomes recombine. The historical methods are interesting in themselves but also serve to introduce the diverse processes of recombination, as well as analytical techniques that are still applicable today.

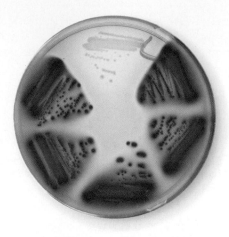

Figure 5-3 Bacterial colonies on staining medium. The colonies stained red contain wild-type bacteria able to use lactose as an energy source (*lac⁺*). The unstained cells are mutants unable to use lactose (*lac⁻*). [Jeffrey H. Miller.]

Table 5-1 Some Genotypic Symbols Used in Bacterial Genetics

Symbol	Character or phenotype associated with symbol
bio^-	Requires biotin added as a supplement to minimal medium
arg^-	Requires arginine added as a supplement to minimal medium
met^-	Requires methionine added as a supplement to minimal medium
lac^-	Cannot utilize lactose as a carbon source
gal^-	Cannot utilize galactose as a carbon source
str^r	Resistant to the antibiotic streptomycin
str^s	Sensitive to the antibiotic streptomycin

Note: Minimal medium is the basic synthetic medium for bacterial growth without nutrient supplements.

 MODEL ORGANISM *Escherichia coli*

The seventeenth-century microscopist Antony van Leeuwenhoek was probably the first to see bacterial cells, and to appreciate their small size: "there are more living in the scum on the teeth in a man's mouth than there are men in the whole kingdom." However bacteriology did not begin in earnest until the nineteenth century. In the 1940s Joshua Lederberg and Edward Tatum made the discovery that launched bacteriology into the burgeoning field of genetics: they discovered that in a certain bacterium there was a type of sexual cycle including a crossing-over-like process. The organism they chose for this experiment has become the model not only for prokaryote genetics, but in a sense for all of genetics. It was *Escherichia coli*, a bacterium named after its discov-

erer, the nineteenth-century German bacteriologist Theodore Escherich.

The choice of *E. coli* was fortunate, as it has proved to have many features suitable for genetic research, not the least of which is that it is easily obtained, since it lives in the gut of humans and other animals. In the gut it is a benign symbiont, but occasionally it causes urinary tract infections and diarrhea.

E. coli has a single circular chromosome 4.6 megabases in length. Of its 4000 intron-free genes, about 35 percent are of unknown function. The sexual cycle is made possible by the action of an extragenomic plasmid called *F*, which determines a type of "maleness." Other plasmids carry genes whose functions equip the cell for life in specific environments, such as drug-resistance genes. These plasmids have been adapted as gene vectors, forming the basis of the gene transfers that are at the center of modern genetic engineering.

E. coli is unicellular and grows by simple cell division. Because of its small size (~1μ in length) it can be grown in large numbers and subjected to intensive selection and screening for rare genetic events. *E. coli* research represents the beginning of "black box" reasoning in genetics: through the selection and analysis of mutants, the workings of the genetic machinery could be deduced even though it was too small to be seen. Phenotypes such as colony size, drug resistance, carbon source utilization, and colored dye production took the place of the visible phenotypes of eukaryotic genetics.

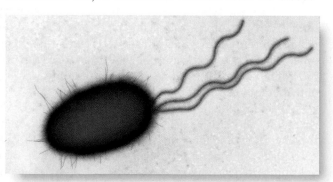

An *E. coli* cell. An electron micrograph of *E. coli* showing long flagella, used for locomotion, and fimbriae, proteinaceous hairs that are important in anchoring the cells to animal tissues. (Sex pili are not shown in this photo.) [Dr. Dennis Kunkel/Visuals Unlimited.]

5.2 Bacterial conjugation

The earliest studies in bacterial genetics revealed the unexpected process of cell conjugation.

Discovery of conjugation

Do bacteria possess any processes similar to sexual reproduction and recombination? The question was answered by the elegantly simple experimental work of Joshua Lederberg and Edward Tatum, who in 1946 discovered a sexlike process in bacteria. They were studying two strains of *Escherichia coli* with different sets of auxotrophic mutations. Strain A would grow only if

the medium were supplemented with methionine and biotin; strain B would grow only if it were supplemented with threonine, leucine, and thiamine. Thus, we can designate the strains as

strain A: *met⁻ bio⁻ thr⁺ leu⁺ thi⁺*
strain B: *met⁺ bio⁺ thr⁻ leu⁻ thi⁻*

Figure 5-4a displays in simplified form the design of their experiment. Strains A and B were mixed together, incubated for a while, and then plated on minimal medium, on which neither auxotroph could grow. A small minority of the cells (1 in 10^7) was found to grow as prototrophs and hence must have been wild type,

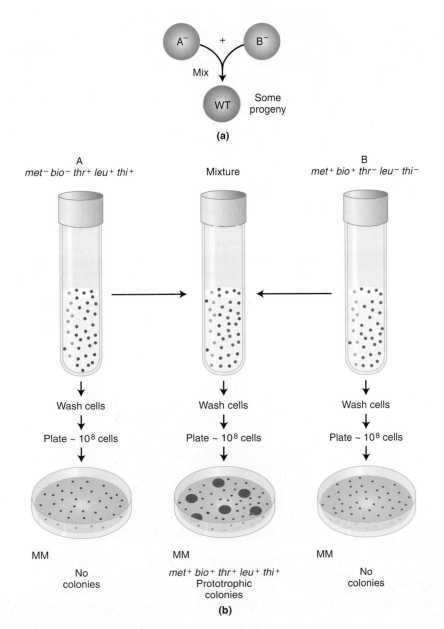

Figure 5-4 Lederberg and Tatum's demonstration of genetic recombination between bacterial cells. (a) The basic concept: two auxotrophic cultures (A⁻ and B⁻) are mixed, yielding prototrophic wild types (WT). (b) Cells of type A or type B cannot grow on an unsupplemented (minimal) medium (MM), because A and B each carry mutations that cause the inability to synthesize constituents needed for cell growth. When A and B are mixed for a few hours and then plated, however, a few colonies appear on the agar plate. These colonies derive from single cells in which an exchange of genetic material has occurred; they are therefore capable of synthesizing all the required constituents of metabolism.

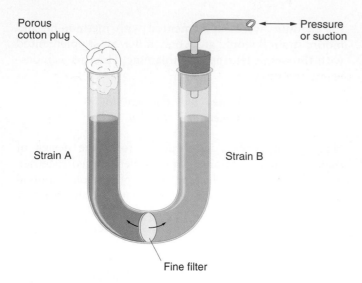

Figure 5-5 Physical contact between bacterial cells is required for genetic recombination. Auxotrophic bacterial strains A and B are grown on either side of a U-shaped tube. Liquid may be passed between the arms by applying pressure or suction, but the bacterial cells cannot pass through the filter. After incubation and plating, no recombinant colonies grow on minimal medium.

having regained the ability to grow without added nutrients. Some of the dishes were plated only with strain A bacteria and some only with strain B bacteria to act as controls, but from these no prototrophs arose. Figure 5-4b illustrates the experiment in more detail. These results suggested that some form of recombination of genes had taken place between the genomes of the two strains to produce the prototrophs.

It could be argued that the cells of the two strains do not really exchange genes but instead leak substances that the other cells can absorb and use for growing. This possibility of "cross feeding" was ruled out by Bernard Davis in the following way. He constructed a U-tube in which the two arms were separated by a fine filter. The pores of the filter were too small to allow bacteria to pass through but large enough to allow easy passage of any dissolved substances (Figure 5-5). Strain A was put in one arm, strain B in the other. After the strains had been incubated for a while, Davis tested the contents of each arm to see if there were any prototrophic cells, but none were found. In other words, *physical contact* between the two strains was needed for wild-type cells to form. It looked as though some kind of genome union had taken place, and genuine recombinants produced. The physical union of bacterial cells can be confirmed under an electron microscope, and is now called **conjugation.**

Discovery of the fertility factor (F)

In 1953, William Hayes discovered that in the above types of "crosses" the conjugating parents acted *unequally* (later we shall see ways to demonstrate this). It seemed that one parent (and *only* that parent) transferred some of or all its genome into another cell. Hence one cell acts as **donor,** and the other cell as a **recipient.** This is quite different from eukaryotic crosses in which parents contribute nuclear genomes equally.

> **MESSAGE** The transfer of genetic material in *E. coli* conjugation is not reciprocal. One cell, the donor, transfers part of its genome to the other cell, which acts as the recipient.

By accident, Hayes discovered a variant of his original donor strain that would not produce recombinants on crossing with the recipient strain. Apparently, the donor-type strain had lost the ability to transfer genetic material and had changed into a recipient-type strain. In working with this "sterile" donor variant, Hayes found that it could regain the ability to act as a donor by association with other donor strains. Indeed the donor ability was transmitted rapidly and effectively between strains during conjugation. A kind of "infectious transfer" of some factor seemed to be taking place. He suggested that donor ability is itself a hereditary state, imposed by a **fertility factor (F).** Strains that carry F can donate, and are designated **F⁺.** Strains that lack F cannot donate and are recipients, designated **F⁻.**

We now know much more about F. It is an example of a small, nonessential circular DNA molecule called a **plasmid** that can replicate in the cytoplasm independent of the host chromosome. Figures 5-6 and 5-7 show how bacteria can transfer plasmids such as F.

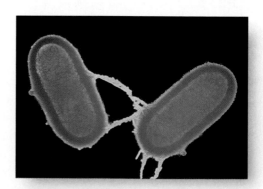

Figure 5-6 Bacteria can transfer plasmids (circles of DNA) through conjugation. A donor cell extends one or more projections—pili—that attach to a recipient cell and pull the two bacteria together. [Oliver Meckes/MPI-Tübingen, Photo Researchers.]

Figure 5-7 Conjugation.
(a) During conjugation, the pilus pulls two bacteria together.
(b) Next, a bridge (essentially a pore) forms between the two cells. A single-stranded copy of plasmid DNA is produced in the donor cell and then passes into the recipient bacterium, where the single strand, serving as a template, is converted to the double-stranded helix.

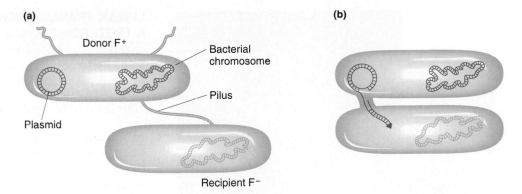

The F plasmid directs the synthesis of pili, projections that initiate contact with a recipient (Figure 5-6) and draw it closer. The F DNA in the donor cell makes a single-stranded copy of itself in a peculiar mechanism called **rolling circle replication.** The circular plasmid "rolls," and as it turns, it reels out the single-stranded copy like fishing line. This copy passes through a pore into the recipient cell, where the other strand is synthesized, forming a double helix. Hence a copy of F remains in the donor and another appears in the recipient, as shown in Figure 5-7. Note in the figure that the *E. coli* genome is depicted as a single circular chromosome. (We will examine the evidence for this later.) Most bacterial genomes are circular, a feature quite different from eukaryotic nuclear chromosomes. We shall see that this feature leads to many idiosyncrasies of bacterial genetics.

Hfr strains

An important breakthrough came when Luca Cavalli-Sforza discovered a derivative of an F+ strain with two unusual properties:

1. On crossing with F− strains this new strain produced 1000 times as many recombinants as a normal F+ strain. Cavalli-Sforza designated this derivative an **Hfr** strain to symbolize its ability to promote a *high frequency* of *recombination.*

2. In Hfr × F− crosses, virtually none of the F− parents were converted into F+ or into Hfr. This result is in contrast with F+ × F− crosses, in which, as we have seen, infectious transfer of F results in a large proportion of the F− parents being converted into F+.

It became apparent that an Hfr strain results from the integration of the F factor into the chromosome, as pictured in Figure 5-8. We can now explain the first unusual property of Hfr strains. During conjugation the F factor inserted in the chromosome efficiently drives

part or all of that chromosome into the F− cell. The chromosomal fragment can then engage in recombination with the recipient chromosome. The rare recombinants observed by Lederberg and Tatum in F+ × F− crosses were due to the spontaneous, but rare, formation of Hfr cells in the F+ culture. Cavalli-Sforza isolated examples of these rare cells from F+ cultures, and found that indeed they now acted as true Hfr's.

Does an Hfr cell die after donating its chromosomal material to an F− cell? The answer is no. Just like the F plasmid, during conjugation the Hfr chromosome replicates and transfers a single strand to the F− cell. The single-stranded nature of the transferred DNA can be demonstrated visually using special strains and antibodies, as shown in Figure 5-9. The replication of the chromosome ensures a complete chromosome for the donor cell after mating. The transferred strand is converted into a double helix in the recipient cell, and donor genes may become incorporated in the recipient's chromosome through crossovers, creating a recombinant cell (Figure 5-10). If there is no recombination, the transferred fragments of DNA are simply lost in the course of cell division.

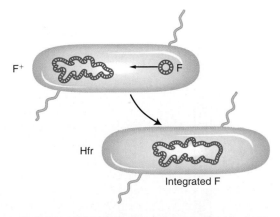

Figure 5-8 Formation of an Hfr. Occasionally, the independent F factor combines with the *E. coli* chromosome, creating an Hfr strain.

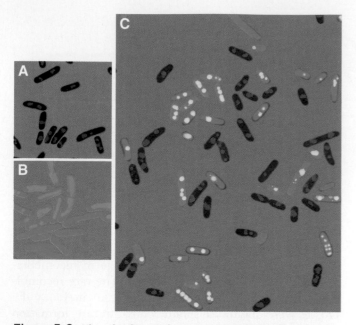

Figure 5-9 Visualization of single-stranded DNA transfer in conjugating *E. coli* cells, using special fluorescent antibodies. Parental Hfr strains (A) are black with red DNA. The red is from binding of an antibody to a protein normally attached to DNA. The recipient F⁻ cells (B) are green due to the presence of the GFP protein; and because they are mutant for a certain gene they do not bind the special protein that binds to antibody. When single-stranded DNA enters the recipient, it promotes atypical binding of the special protein, which fluoresces yellow in this background. Part C shows Hfrs (unchanged) and exconjugants with yellow transferred DNA. A few unmated F⁻ cells are visible. [From Masamichi Kohiyama, Sota Hiraga, Ivan Matic, and Miroslav Radman, "Bacterial Sex: Playing Voyeurs 50 Years Later," *Science* 8 August 2003, p. 803, Figure 1.]

LINEAR TRANSMISSION OF THE HFR GENES FROM A FIXED POINT A clearer view of the behavior of Hfr strains was obtained in 1957, when Elie Wollman and François Jacob investigated the pattern of transmission of Hfr genes to F⁻ cells during a cross. They crossed

$$\text{Hfr } azi^r\ ton^r\ lac^+\ gal^+\ str^s \times \text{F}^-\ azi^s\ ton^s\ lac^-\ gal^-\ str^r$$

At specific times after mixing, they removed samples, which were each put in a kitchen blender for a few seconds to separate the mating cell pairs. This procedure is called **interrupted mating.** The sample was then plated onto a medium containing streptomycin to kill the Hfr donor cells, which bore the sensitivity allele *str*ˢ. The surviving *str*ʳ cells then were tested for the presence of alleles from the donor genome. Any *str*ʳ cell bearing a donor allele must have taken part in conjugation; such cells are called **exconjugants.** Figure 5-11a shows a plot of the results, showing a time course of entry of each donor allele *azi*ʳ, *ton*ʳ, *lac*⁺, and *gal*⁺. Figure 5-11b portrays the transfer of Hfr alleles.

The key elements in these results are

1. Each donor allele first appears in the F⁻ recipients at a specific time after mating began.
2. The donor alleles appear in a specific sequence.
3. Later donor alleles are present in fewer recipient cells.

Putting all these observations together, Wollman and Jacob deduced that in the conjugating Hfr, single-stranded DNA transfer begins from a fixed point on the donor chromosome, termed the **origin (O),** and continues in a linear fashion. The point O is now known to be the site at which the F plasmid is inserted. The farther an gene is from O, the later it is transferred to the F⁻. The transfer process will

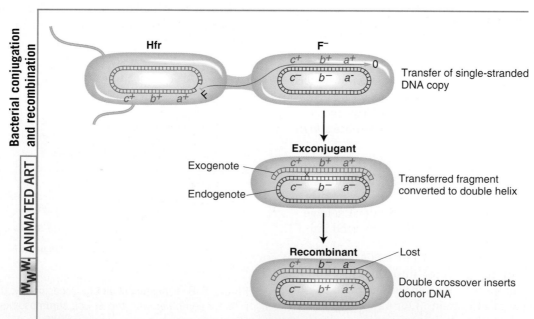

Figure 5-10 Bacterial conjugation and recombination. Transfer of single-stranded fragment of donor chromosome and recombination with recipient chromosome.

(a)

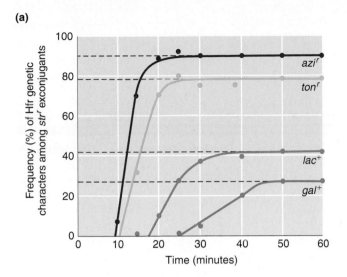

(b)

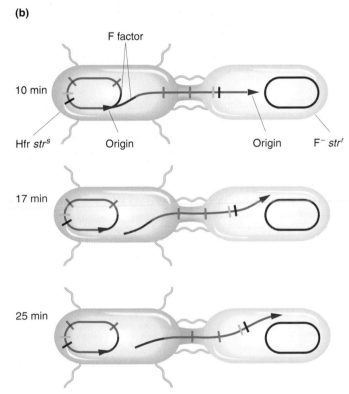

Figure 5-11 Interrupted-mating conjugation experiments. F⁻ streptomycin-resistant cells with mutations in *azi*, *ton*, *lac*, and *gal* are incubated for varying times with Hfr cells that are sensitive to streptomycin and carry wild-type alleles for these genes. (a) A plot of the frequency of donor alleles in exconjugants as a function of time after mating. (b) A schematic view of the transfer of markers (shown in different colors) over time. [Part a after E. L. Wollman, F. Jacob, and W. Hayes, *Cold Spring Harbor Symp. Quant. Biol.* 21, 1956, 141.]

generally stop before the farthermost genes are transferred, resulting in their being included in fewer exconjugants.

How can we explain the second unusual property of Hfr crosses, that F⁻ exconjugants are rarely converted into Hfr or F⁺? When Wollman and Jacob allowed Hfr × F⁻ crosses to continue for as long as 2 hours before disruption, they found that in fact a few of the exconjugants were converted into Hfr. In other words, the part of F that confers donor ability was eventually transmitted but at a very low frequency. The rareness of Hfr exconjugants suggested that the inserted F was transmitted as the *last* element of the linear chromosome. We can summarize this with the following map, in which the arrow indicates the process of transfer, beginning with O:

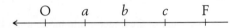

Thus almost none of the F⁻ recipients are converted because the fertility factor is the last element transmitted, and usually the transmission process will have stopped before getting that far.

> **MESSAGE** The Hfr chromosome, originally circular, unwinds and is transferred to the F⁻ cell in a linear fashion, with the F factor entering last.

INFERRING INTEGRATION SITES OF F AND CHROMOSOME CIRCULARITY Wollman and Jacob went on to shed more light on how and where the F plasmid integrates to form an Hfr, and in doing so deduced the circularity of the chromosome. They performed interrupted-mating experiments using different, separately derived Hfr strains. Significantly, the order of transmission of the alleles differed from strain to strain, as in the following examples:

Hfr strain

H	O *thr pro lac pur gal his gly thi* F
1	O *thr thi gly his gal pur lac pro* F
2	O *pro thr thi gly his gal pur lac* F
3	O *pur lac pro thr thi gly his gal* F
AB 312	O *thi thr pro lac pur gal his gly* F

Each line can be considered a map showing the order of alleles on the chromosome. At first glance, there seems to be a random shuffling of genes. However, when the identical alleles of the different Hfr maps are lined up, the similarity in sequence becomes clear.

H	F *thi gly his gal pur lac pro thr* O
(written backwards)	
1	O *thr thi gly his gal pur lac pro* F
2	O *pro thr thi gly his gal pur lac* F
3	O *pur lac pro thr thi gly his gal* F
AB 312	F *gly his gal pur lac pro thr thi* O
(written backwards)	

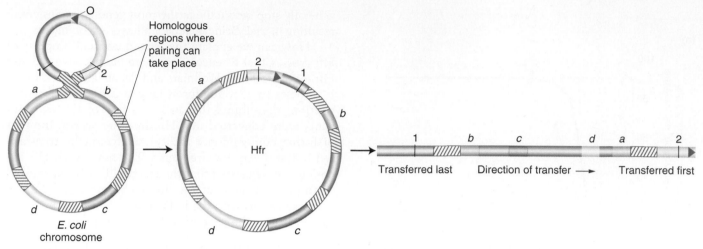

Figure 5-12 Insertion of the F factor into the *E. coli* chromosome by crossing-over.
Hypothetical markers 1 and 2 are shown on F to depict the direction of insertion. The origin
(O) is the mobilization point where insertion into the *E. coli* chromosome occurs; the
pairing region is homologous with a region on the *E. coli* chromosome; *a–d* are
representative genes in the *E. coli* chromosome. Fertility genes on F are responsible for the
F⁺ phenotype. Pairing regions (hatched) are identical in plasmid and chromosome. They are
derived from mobile elements called *insertion sequences* (see Chapter 13). In this example,
the Hfr cell created by the insertion of F would transfer its genes in the order *a, d, c, b.*

The relationship of the sequences to one another is
explained if each map is the segment of a circle. This was
the first indication that bacterial chromosomes are circular.
Furthermore, Allan Campbell proposed a startling hypothesis that accounted for the different Hfr maps. He proposed

that if F is a ring, then insertion might be by simple crossover between F and the bacterial chromosome (Figure
5-12). That being the case, any of the linear Hfr chromosomes could be generated simply by insertion of F into the
ring in the appropriate place and orientation (Figure 5-13).

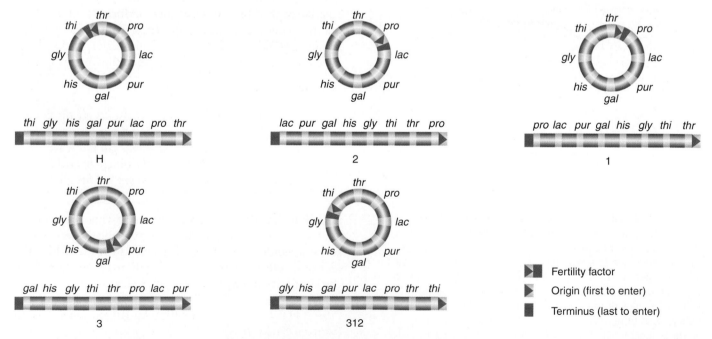

Figure 5-13 Order of gene transfer. The five *E. coli* Hfr strains shown each have different
F factor insertion points and orientations. All strains share the same order of genes on
the *E. coli* chromosome. The orientation of the F factor determines which gene enters the
recipient cell first. The gene closest to the terminus enters last.

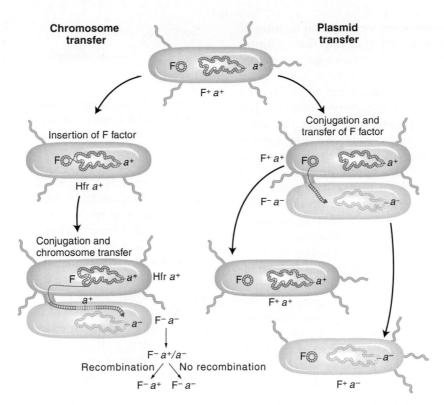

Figure 5-14 Conjugation summary. Summary of the various events that take place in the conjugational cycle of *E. coli*.

Several hypotheses—later supported—followed from Campbell's proposal.

1. One end of the integrated F factor would be the **origin,** where transfer of the Hfr chromosome begins. The **terminus** would be at the other end of F.

2. The orientation in which F is inserted would determine the order of entry of donor alleles. If the circle contains genes *A, B, C,* and *D,* then insertion between *A* and *D* would give the order *ABCD* or *DCBA,* depending on orientation. Check the different orientations of the insertions in Figure 5-12.

How is it possible for F to integrate at different sites? If F DNA had a region homologous to any of several regions on the bacterial chromosome, any one of these could act as a pairing region at which pairing could be followed by a crossover. These regions of homology are now known to be mainly segments of transposable elements called *insertion sequences.* For a full explanation of these, see Chapter 13.

The fertility factor thus exists in two states:

1. The plasmid state: as a free cytoplasmic element F is easily transferred to F⁻ recipients.

2. The integrated state: as a contiguous part of a circular chromosome F is transmitted only very late in conjugation.

The *E. coli* conjugation cycle is summarized in Figure 5-14.

Mapping of bacterial chromosomes

BROAD-SCALE CHROMOSOME MAPPING USING TIME OF ENTRY Wollman and Jacob realized that it would be easy to construct linkage maps from the interrupted-mating results, using as a measure of "distance" the times at which the donor alleles first appear after mating. The units of map distance in this case are minutes. Thus, if b^+ begins to enter the F⁻ cell 10 minutes after a^+ begins to enter, then a^+ and b^+ are 10 units apart. Like eukaryotic maps based on crossovers, these linkage maps were originally purely genetic constructions. At the time they were originally devised, there was no way of testing their physical basis.

FINE-SCALE CHROMOSOME MAPPING BY RECOMBINANT FREQUENCY For an exconjugant to acquire donor genes as a permanent feature of its genome, the donor fragment must recombine with the recipient chromosome. However, note that time-of-entry mapping is not based on recombinant frequency. Indeed the units are minutes, not RF. Nevertheless it is possible to use recombinant frequency for a more fine scale type of mapping in bacteria, and this is the method to which we now turn.

First we need to understand some special features of the recombination event in bacteria. Note that recombination does not take place between two whole genomes, as it does in eukaryotes. In contrast, it takes place between one *complete* genome, from the F⁻, called the **endogenote,** and an *incomplete* one, derived from the Hfr donor and called the **exogenote.** The cell at this stage has two copies of one segment of DNA—one copy is the exogenote and one copy is part of the endogenote. Thus at this stage the cell is a *partial* diploid, called a **merozygote.** Bacterial genetics is merozygote genetics. A single crossover in a merozygote would break the ring and thus not produce viable recombinants, as shown in Figure 5-15. To keep the circle intact, there must be an even number of crossovers. An even number of crossovers produces a circular, intact chromosome and a fragment. Although such recombination events are represented in a shorthand way as double crossovers, the actual molecular mechanism is somewhat different, more like an invasion of the endogenote by an internal section of the exogenote. The other product of the "double crossover," the fragment, is generally lost in subsequent cell growth. Hence, only one of the reciprocal products of recombination survives. Therefore, another unique feature of bacterial recombination is that we must forget about reciprocal exchange products in most cases.

> **MESSAGE** Recombination during conjugation results from a double crossover–like event, which gives rise to reciprocal recombinants of which only one survives.

With this understanding we can examine recombination mapping. Suppose that we want to calculate map distances separating three close loci: *met, arg,* and *leu.* Assume that an interrupted-mating experiment has shown that the order is *met, arg, leu,* with *met* transferred first and *leu* last. Now we want to examine the recombination of these genes, but we can study this only in "trihybrids," exconjugants that have received all three donor markers. In other words, we want to set up the merozygote diagrammed here:

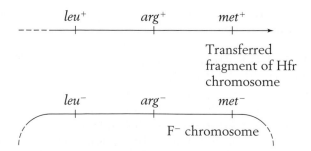

To do this, we must first select stable exconjugants bearing the *last* donor allele, which in this case is *leu*⁺. Why? Because, if we select for the last marker, then we know

Figure 5-15 Crossover between exogenote and endogenote in a merozygote. A single crossover would lead to a linear, partly diploid chromosome.

that such cells at some stage must also have contained the earlier markers too—namely, *arg*⁺ and *met*⁺.

The goal now is to count the frequencies of crossovers at different locations. The *leu*⁺ recombinants that we select may or may not have incorporated the other donor markers, depending on where the double crossover occurred. Hence the procedure is to first select *leu*⁺ exconjugants and then isolate and test a large sample of these to see which of the other markers were integrated. Let's look at an example. In the cross Hfr *met*⁺ *arg*⁺ *leu*⁺ *str*ˢ × F⁻ *met*⁻ *arg*⁻ *leu*⁻ *str*ʳ, we would select *leu*⁺ recombinants and then examine them for the *arg*⁺ and *met*⁺ alleles, called the **unselected markers.** Figure 5-16 depicts the types of double-crossover events expected. One crossover must be on the left side of the *leu* marker and the second must be on the right side. Let's assume the *leu*⁺ exconjugants are of the following types and frequencies:

leu⁺ *arg*⁻ *met*⁻	4%
leu⁺ *arg*⁺ *met*⁻	9%
leu⁺ *arg*⁺ *met*⁺	87%

The double crossovers needed to produce these genotypes are shown in Figure 5-16. The first two classes are the key because they require a crossover between *leu* and *arg* in the first case, and between *arg* and *met* in the second. Hence the relative frequencies of these classes reflect the sizes of these two regions. We would conclude that the *leu–arg* region is 4 map units, and *arg–met* is 9 map units.

In a cross such as the one just described, one class of potential recombinants, of genotype *leu*⁺ *arg*⁻ *met*⁺, requires four crossovers instead of two (see the bottom of Figure 5-16). These recombinants are rarely recovered because their frequency is very low compared with the other classes of recombinants.

F plasmids that carry genomic fragments

The F factor in Hfr strains is generally quite stable in its inserted position. However, occasionally an F factor exits from the chromosome cleanly by a reversal of the recombination process that inserted it in the first place. The two homologous pairing regions on either side repair, and a crossover occurs to liberate the F plasmid.

Bacterial conjugation and mapping by recombination

W.W.W. ANIMATED ART

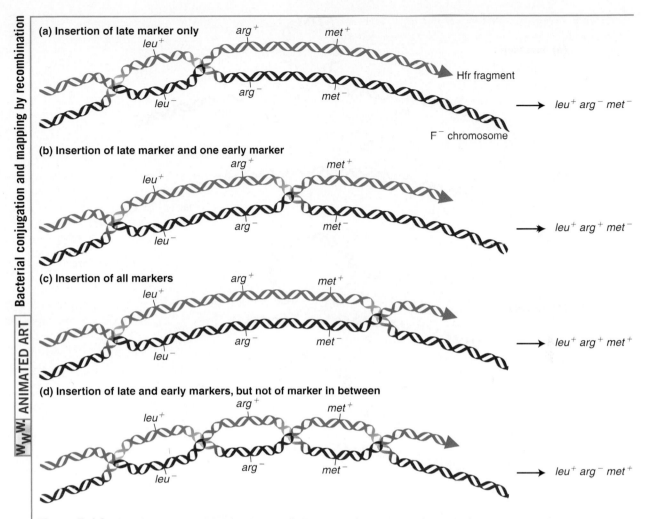

(a) Insertion of late marker only

leu⁺   arg⁺   met⁺

Hfr fragment

arg⁻   met⁻

leu⁻

F⁻ chromosome

→ leu⁺ arg⁻ met⁻

(b) Insertion of late marker and one early marker

arg⁺   met⁺

leu⁺

arg⁻   met⁻

leu⁻

→ leu⁺ arg⁺ met⁻

(c) Insertion of all markers

arg⁺   met⁺

leu⁺

arg⁻   met⁻

leu⁻

→ leu⁺ arg⁺ met⁺

(d) Insertion of late and early markers, but not of marker in between

arg⁺   met⁺

leu⁺

arg⁻   met⁻

leu⁻

→ leu⁺ arg⁻ met⁺

Figure 5-16 Mapping by recombination in *E. coli*. After a cross, selection is made for the *leu⁺* marker, which is donated late. The early markers (*arg⁺* and *met⁺*) may or may not be inserted, depending on the site where recombination between the Hfr fragment and the F⁻ chromosome takes place. The frequencies of events diagrammed in parts a and b are used to obtain the relative sizes of the *leu–arg* and *arg–met* regions. Note that in each case only the DNA inserted into the F⁻ chromosome survives; the other fragment is lost.

However, sometimes the exit is not clean, and the plasmid carries with it a part of the bacterial chromosome. An F plasmid carrying bacterial genomic DNA is called an **F′ (F prime) plasmid.**

The first evidence of this process came from experiments in 1959 by Edward Adelberg and François Jacob. One of their key observations was of an Hfr in which the F factor was integrated near the *lac⁺* locus. Starting with this Hfr *lac⁺* strain, Jacob and Adelberg found an F⁺ derivative that in crosses transferred *lac⁺* to F⁻ *lac⁻* recipients at a very high frequency. (These transferrants could be detected by plating on medium lacking lactose.) Furthermore, these F⁺ *lac⁺* exconjugants occasionally gave rise to F⁻ *lac⁻* daughter cells, at a frequency of 1 × 10⁻³. Thus, the genotype of these recipients appeared to be F⁺ *lac⁺* / F⁻ *lac⁻*. In other words the *lac⁺*

exconjugants seemed to carry an F plasmid with a piece of the donor chromosome incorporated. The origin of this F′ plasmid is shown in Figure 5-17. Note that the faulty excision occurs because there is another homologous region nearby that pairs with the original. The F′ in our example is also called F′ *lac* because the piece of host chromosome that it picked up has the *lac* gene on it. F′ factors have been found carrying many different chromosomal genes and have been named accordingly. For example, F′ factors carrying *gal* or *trp* are called F′ *gal* and F′ *trp*, respectively. Because F *lac⁺*/*lac⁻* cells are Lac⁺ in phenotype, we know that *lac⁺* is dominant over *lac⁻*.

Partial diploids made using F′ strains are useful for some aspects of routine bacterial genetics, such as the study of dominance or of allele interaction. Some F′ strains can carry very large parts (up to one-quarter)

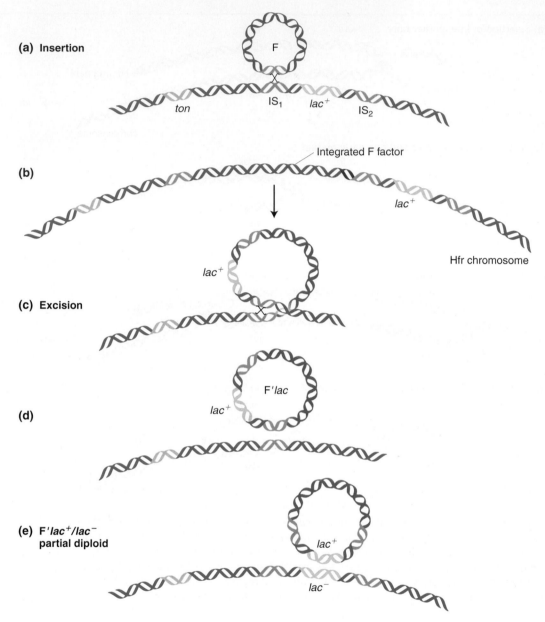

Figure 5-17 Origin of an F′ factor. (a) F is inserted in an Hfr strain at a repetitive element labeled IS$_1$ between the *ton* and *lac*$^+$ alleles. (b) The inserted F factor. (c) Abnormal "outlooping" by crossing-over with a different element, IS$_2$, to include the *lac* locus. (d) The resulting F′ *lac*$^+$ particle. (e) F′ *lac*$^+$ / *lac*$^-$ partial diploid produced by the transfer of the F′ *lac*$^+$ particle to an F$^-$ *lac*$^-$ recipient. [From G. S. Stent and R. Calendar, *Molecular Genetics*, 2d ed. Copyright 1978 by W. H. Freeman and Company.]

of the bacterial chromosome. These plasmids have been useful as vectors for carrying DNA in sequencing large genomes (see Chapters 11 and 12).

> **MESSAGE** The DNA of an F′ plasmid is part F factor and part bacterial genome. F′s transfer rapidly like F, and can be used to establish partial diploids for studies of bacterial dominance and gene interaction.

R plasmids

An alarming property of pathogenic bacteria first came to light through studies in Japanese hospitals in the 1950s. Bacterial dysentery is caused by bacteria of the genus *Shigella*. This bacterium initially was sensitive to a wide array of antibiotics that were used to control the disease. In the Japanese hospitals, however, *Shigella* isolated from patients with dysentery proved to be simultaneously resis-

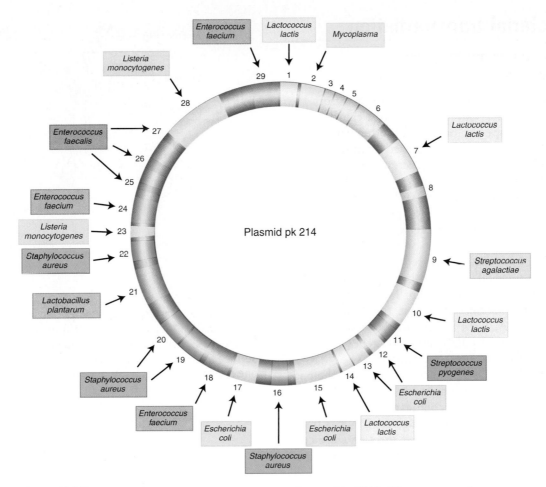

Figure 5-18 Origins of genes of the *Lactococcus lactis* plasmid pK214. The genes are from many different bacteria. [Data from Table 1 in V. Perreten, F. Schwarz, L. Cresta, M. Boeglin, G. Dasen, and M. Teuber, *Nature* 389, 1997, 801–802.]

tant to many of these drugs, including penicillin, tetracycline, sulfanilamide, streptomycin, and chloramphenicol. This resistance to multiple drugs was inherited as a single genetic package, and it could be transmitted in an infectious manner—not only to other sensitive *Shigella* strains, but also to other related species of bacteria. This talent, which resembles the mobility of the *E. coli* F plasmid, is an extraordinarily useful one for the pathogenic bacterium because resistance can rapidly spread through a population. However, its implications for medical science are dire because the bacterial disease suddenly becomes resistant to treatment by a large range of drugs.

From the point of view of the geneticist, however, the mechanism has proved interesting, and useful in genetic engineering. The vectors carrying these multiple resistances proved to be another group of plasmids called **R plasmids.** They are transferred rapidly on cell conjugation, much like the F plasmid in *E. coli.*

In fact, the R plasmids in *Shigella* proved to be just the first of many similar ones to be discovered. All exist in the plasmid state in the cytoplasm. These elements have

Table 5-2 Genetic Determinants Borne by Plasmids

Characteristic	Plasmid examples
Fertility	F, R1, Col
Bacteriocin production	Col E1
Heavy-metal resistance	R6
Enterotoxin production	Ent
Metabolism of camphor	Cam
Tumorigenicity in plants	T1 (in *Agrobacterium tumefaciens*)

been found to carry many different kinds of genes in bacteria. Table 5-2 shows some of the characteristics that can be borne by plasmids. Figure 5-18 shows an example of a well-traveled plasmid isolated from the dairy industry.

R plasmids become important in designing strains for use in genetic engineering because the plasmids are easily shuttled between cells, and the R genes can be used to keep track of them.

5.3 Bacterial transformation

Transformation as another type of bacterial gene transfer

Some bacteria can take up fragments of DNA from the external medium. The source of the DNA can be other cells of the same species or cells of other species. In some cases the DNA has been released from dead cells; in other cases the DNA has been secreted from live bacterial cells. The DNA taken up integrates into the recipient's chromosome. If this DNA is of a different genotype from the recipient, the genotype of the recipient can become permanently changed, a process aptly termed **transformation.**

Transformation was discovered in the bacterium *Streptococcus pneumoniae* in 1928 by Frederick Griffith. In 1944, Oswald T. Avery, Colin M. MacLeod, and Maclyn McCarty demonstrated that the "transforming principle" was DNA. Both results are milestones in the elucidation of the molecular nature of genes. We consider this work in more detail in Chapter 7.

The transforming DNA is incorporated into the bacterial chromosome by a process analogous to the production of recombinant exconjugants in Hfr × F⁻ crosses. Note, however, that in *conjugation* DNA is transferred from one living cell to another through close contact, whereas in *transformation* isolated pieces of external DNA are taken up by a cell through the cell wall and plasma membrane. Figure 5-19 shows one way in which this process can occur.

Transformation has been a handy tool in several areas of bacterial research because the genotype of a strain can be deliberately changed in a very specific way by transforming with an appropriate DNA fragment. For example, transformation is used widely in genetic engineering. More recently it has been found that even eukaryotic cells can be transformed, using quite similar procedures, and this technique has been invaluable for modifying eukaryotic cells.

Chromosome mapping using transformation

Transformation can be used to provide information on bacterial gene linkage. When DNA (the bacterial chromosome) is extracted for transformation experiments, some breakage into smaller pieces is inevitable. If two donor genes are located close together on the chromosome, there is a good chance that sometimes they will be carried on the same piece of transforming DNA. Hence both will be taken up, causing a **double transformation.** Conversely, if genes are widely separated on the chromosome, they will be carried on separate transforming segments. Any double transformants will most likely arise from separate independent transformations. Hence in the case of widely separated genes,

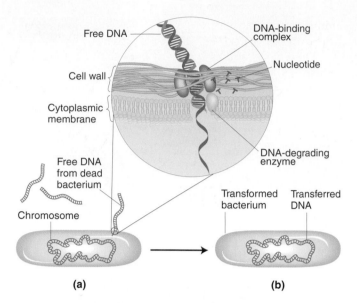

Figure 5-19 Transformation. Bacterium undergoing transformation (a) picks up free DNA released from a dead bacterial cell. As DNA-binding complexes on the bacterial surface take up the DNA (inset), enzymes break down one strand into nucleotides; a derivative of the other strand may integrate into the bacterium's chromosome (b). [After R. V. Miller, "Bacterial Gene Swapping in Nature." Copyright 1998 by Scientific American, Inc. All rights reserved.]

the frequency of double transformants will equal the product of the single-transformation frequencies. Therefore it should be possible to test for close linkage by testing for a departure from the product rule. In other words, if genes are linked, then the proportion of double transformants will be greater than the product of single transformants.

Unfortunately, the situation is made more complex by several factors—the most important is that not all cells in a population of bacteria are competent to be transformed. Nevertheless you can sharpen your skills in transformation analysis in one of the problems at the end of the chapter, which assumes that 100 percent of the recipient cells are competent.

> **MESSAGE** Bacteria can take up DNA fragments from the medium, which once inside the cell can integrate into the chromosome.

5.4 Bacteriophage genetics

The word *bacteriophage*, which is a name for bacterial viruses, means "eater of bacteria." Pioneering work on the genetics of bacteriophages in the second half of the twentieth century formed the foundation of more recent research on tumor-causing viruses and other kinds of animal and plant viruses. In this way bacterial viruses have provided an important model system.

These viruses, which parasitize and kill bacteria, can be used in two different types of genetic analysis. First, two distinct phage genotypes can be crossed to measure recombination and hence map the viral genome. Second, bacteriophages can be used as a way of bringing bacterial genes together for linkage and other genetic studies. We will study this in the next section. In addition, as we shall see in Chapter 11, phages are used in DNA technology as carriers, or vectors, of foreign DNA inserts from any organism. Before we can understand phage genetics, we must first examine the infection cycle of phages.

Infection of bacteria by phages

Most bacteria are susceptible to attack by bacteriophages. A phage consists of a nucleic acid "chromosome" (DNA or RNA) surrounded by a coat of protein molecules. Phage types are identified not by species names but by symbols, for example, phage T4, phage λ, and so forth. Figures 5-20 and 5-21 show the structure of phage T4.

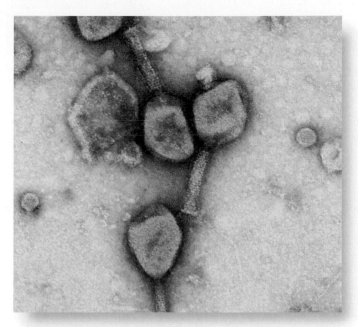

Figure 5-21 Bacteriophage T4. Enlargement of the *E. coli* phage T4 showing details of structure: note head, tail, and tail fibers. [Photograph from Jack D. Griffith.]

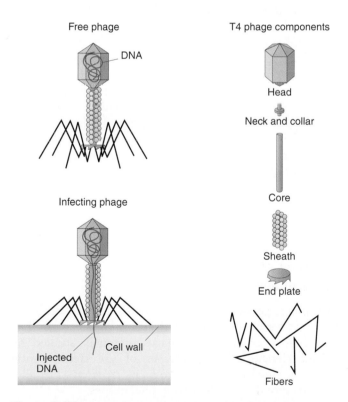

Figure 5-20 An infecting phage injects DNA through its core structure into the cell. The left half of the figure shows bacteriophage T4 as a free phage, then in the process of infecting an *E. coli* cell. The major structural components of T4 are shown on the right. [After R. S. Edgar and R. H. Epstein, "The Genetics of a Bacterial Virus." Copyright 1965 by Scientific American, Inc. All rights reserved.]

During infection, a phage attaches to a bacterium and injects its genetic material into the bacterial cytoplasm (Figure 5-22). The phage genetic information then takes over the machinery of the bacterial cell by turning off the synthesis of bacterial components and redirecting the bacterial synthetic machinery to make more phage components. Newly made phage heads are individually stuffed with replicates of the phage chromosome. Ultimately, many phage descendants are made, and these are released when the bacterial cell wall breaks open. This breaking-open process is called **lysis.**

How can we study inheritance in phages when they are so small that they are visible only under the electron

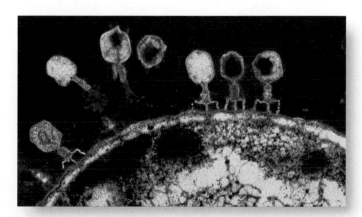

Figure 5-22 Micrograph of bacteriophages attaching to a bacterium and injecting their DNA. [Dr. L. Caro/Science Photo Library, Photo Researchers.]

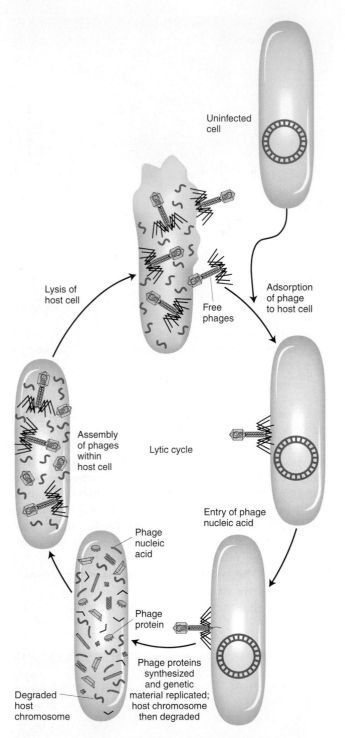

Figure 5-23 A generalized bacteriophage lytic cycle. [After J. Darnell, H. Lodish, and D. Baltimore, *Molecular Cell Biology*. Copyright 1986 by W. H. Freeman and Company.]

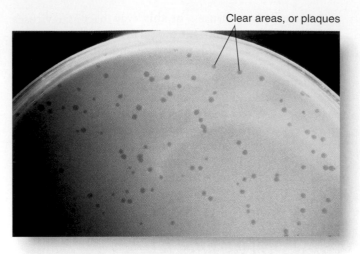

Figure 5-24 Phage plaques. Through repeated infection and production of progeny phage, a single phage produces a clear area, or plaque, on the opaque lawn of bacterial cells. [Barbara Morris, Novagen.]

microscope? In this case, we cannot produce a visible colony by plating, but we can produce a visible manifestation of a phage by taking advantage of several phage characters.

Let's look at the consequences of a phage's infecting a single bacterial cell. Figure 5-23 shows the sequence of events in the infectious cycle that leads to the release of progeny phages from the lysed cell.

After lysis, the progeny phages infect neighboring bacteria. Repetition of this cycle through progressive rounds of infection results in an exponential increase in the number of lysed cells. Within 15 hours after one single phage particle infects a single bacterial cell, the effects are visible to the naked eye as a clear area, or **plaque,** in the opaque lawn of bacteria covering the surface of a plate of solid medium (Figure 5-24). Such plaques can be large or small, fuzzy or sharp, and so forth, depending on the phage genotype. Thus, *plaque morphology* is a phage character that can be analyzed at the genetic level. Another phage phenotype that we can analyze genetically is *host range*, because phages may differ in the spectra of bacterial strains that they can infect and lyse. For example, one specific strain of bacteria might be immune to phage 1 but susceptible to phage 2.

Mapping phage chromosomes using phage crosses

Two phage genotypes can be crossed in much the same way that we cross organisms. A phage cross can be illustrated by a cross of T2 phages originally studied by Alfred Hershey. The genotypes of the two parental strains in Hershey's cross were $h^- r^+ \times h^+ r^-$. The alleles correspond to the following phenotypes:

h^-: can infect two different *E. coli* strains (which we can call strains 1 and 2)

h^+: can infect only strain 1

r^-: rapidly lyses cells, thereby producing large plaques

r^+: slowly lyses cells, producing small plaques

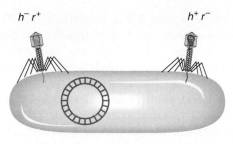

$h^- r^+$ $h^+ r^-$

E. coli strain 1

Figure 5-25 A double infection of *E. coli* by two phages.

To make the cross, *E. coli* strain 1 is infected with both parental T2 phage genotypes. This kind of infection is called a **mixed infection** or a **double infection** (Figure 5-25). After an appropriate incubation period, the phage lysate (the progeny phages) is then analyzed by spreading it onto a bacterial lawn composed of a mixture of *E. coli* strains 1 and 2. Four plaque types are then distinguishable (Figure 5-26). Large plaques indicate rapid lysis (r^-) and small plaques slow lysis (r^+). Phage plaques with the allele h^- will infect both hosts, forming a clear plaque, whereas h^+ results in a cloudy plaque

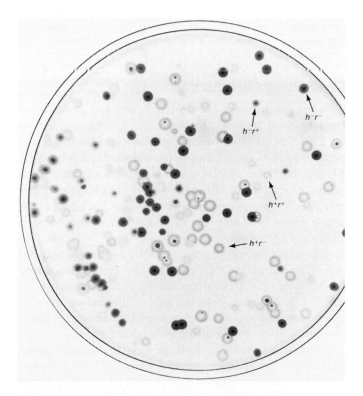

$h^- r^+$

$h^- r^-$

$h^+ r^+$

$h^+ r^-$

Figure 5-26 Plaque phenotypes produced by progeny of the cross $h^- r^+ \times h^+ r^-$. Four plaque phenotypes can be differentiated, representing two parental types and two recombinants. [From G. S. Stent, *Molecular Biology of Bacterial Viruses.* Copyright 1963 by W. H. Freeman and Company.]

because one host is not infected. Thus the four genotypes can easily be classified as parental ($h^- r^+$ and $h^+ r^-$) and recombinant ($h^+ r^+$ and $h^- r^-$), and a recombinant frequency can be calculated as follows:

$$RF = \frac{(h^+ \ r^+) + (h^- \ r^-)}{\text{total plaques}}$$

If we assume that the recombining phage chromosomes are linear, then single crossovers produce viable reciprocal products. However, phage crosses are subject to some analytical complications. First, several rounds of exchange can occur within the host: a recombinant produced shortly after infection may undergo further recombination in the same cell or in later infection cycles. Second, recombination can occur between genetically similar phages as well as between different types. Thus, if we let P_1 and P_2 refer to general parental genotypes, crosses of $P_1 \times P_1$ and $P_2 \times P_2$ occur in addition to $P_1 \times P_2$. For both these reasons, recombinants from phage crosses are a consequence of a *population* of events rather than defined, single-step exchange events. Nevertheless, *all other things being equal,* the RF calculation does represent a valid index of map distance in phages.

Because astronomically large numbers of phages can be used in phage recombination analyses, it is possible to detect very rare crossover events. In the 1950s, Seymour Benzer made use of such rare crossover events to map the mutant sites *within* the rII gene of phage T4, a gene that controls lysis. For different rII mutant alleles arising spontaneously, the mutant site is usually at different positions within the gene. Therefore when two different rII mutants are crossed, a few rare crossovers may occur between the mutant sites, producing wild-type recombinants, as shown here:

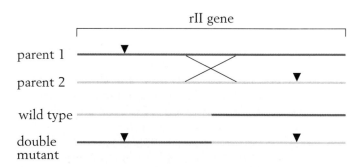

rII gene

parent 1

parent 2

wild type

double mutant

Because such crossover events are more likely the larger the distance between the mutant sites, the frequency of rII$^+$ recombinants is a measure of that distance within the gene. (The reciprocal product is a double mutant and indistinguishable from the parentals.)

In passing, it is noteworthy that Benzer used a clever approach to detect the very rare rII$^+$ recombinants. He made use of the fact that rII mutants will not infect a strain of *E. coli* called K. Therefore he made the rII × rII cross on another strain and then plated the phage lysate on a lawn of strain K. Only rII⁺ recombinants will form plaques on this lawn. This way of finding a rare genetic event (in this case a recombinant) is a **selective system:** *only* the desired rare event can produce a certain visible outcome. Contrast this with **screens,** systems in which large numbers of individuals are visually scanned to seek the rare "needle in the haystack."

This same approach can be used to map mutant sites within genes for any organism from which large numbers of cells can be obtained, and for which it is possible to easily distinguish wild-type and mutant phenotypes. However, this sort of intragenic mapping has been largely superseded by the advent of cheap chemical methods for DNA sequencing, which identify the positions of mutant sites directly.

> **MESSAGE** Recombination between phage chromosomes can be studied by bringing the parental chromosomes together in one host cell through mixed infection. Progeny phages can be examined for parental versus recombinant genotypes.

5.5 Transduction

Some phages are able to pick up bacterial genes and carry them from one bacterial cell to another: a process known as **transduction.** Thus, transduction joins the battery of modes of transfer of genomic material between bacteria—along with Hfr chromosome transfer, F′ plasmid transfer, and transformation.

Discovery of transduction

In 1951, Joshua Lederberg and Norton Zinder were testing for recombination in the bacterium *Salmonella typhimurium* by using the techniques that had been successful with *E. coli*. The researchers used two different strains: one was *phe⁻ trp⁻ tyr⁻*, and the other was *met⁻ his⁻*. We won't worry about the nature of these alleles except to note that they are all auxotrophic. When either strain was plated on a minimal medium, no wild-type cells were observed. However, after the two strains were mixed, wild-type prototrophs appeared at a frequency of about 1 in 10⁵. Thus far, the situation seems similar to that for recombination in *E. coli*.

However, in this case, the researchers also recovered recombinants from a U-tube experiment, in which conjugation was prevented by a filter separating the two arms. By varying the size of the pores in the filter, they found that the agent responsible for gene transfer was the same size as a known phage of *Salmonella*, called phage P22. Furthermore, the filterable agent and P22 were identical in sensitivity to antiserum and immunity

to hydrolytic enzymes. Thus, Lederberg and Zinder had discovered a new type of gene transfer, mediated by a virus. They were the first to call this process *transduction*. As a rarity in the lytic cycle, virus particles sometimes pick up bacterial genes and transfer them when they infect another host. Transduction has subsequently been demonstrated in many bacteria.

To understand the process of transduction we need to distinguish two types of phage cycle. **Virulent phages** are those that immediately lyse and kill the host. **Temperate phages** can remain within the host cell for a period without killing it. Their DNA either integrates into the host chromosome to replicate with it or replicates like a plasmid, separately in the cytoplasm. A phage integrated into the bacterial genome is called a **prophage.** A bacterium harboring a quiescent phage is called **lysogenic.** Occasionally a lysogenic bacterium lyses spontaneously. A resident temperate phage confers resistance to infection by other phages of that type.

Only temperate phages can transduce. There are two kinds of transduction: generalized and specialized. *Generalized* transducing phages can carry *any* part of the bacterial chromosome, whereas *specialized* transducing phages carry only certain *specific* parts.

> **MESSAGE** Virulent phages cannot become prophages; they are always lytic. Temperate phages can exist within the bacterial cell as prophages, allowing their hosts to survive as lysogenic bacteria; they are also capable of occasional bacterial lysis.

Generalized transduction

By what mechanisms can a phage carry out **generalized transduction?** In 1965, K. Ikeda and J. Tomizawa threw light on this question in some experiments on the *E. coli* phage P1. They found that when a donor cell is lysed by P1, the bacterial chromosome is broken up into small pieces. Occasionally, the newly forming phage particles mistakenly incorporate a piece of the bacterial DNA into a phage head in place of phage DNA. This event is the origin of the transducing phage.

A phage carrying bacterial DNA can infect another cell. That bacterial DNA can then be incorporated into the recipient cell's chromosome by recombination (Figure 5-27). Because genes on any of the cut-up parts of the host genome can be transduced, this type of transduction is by necessity of the generalized type.

Phages P1 and P22 both belong to a phage group that shows generalized transduction. P22 DNA inserts into the host chromosome, whereas P1 DNA remains free, like a large plasmid. However, both transduce by faulty head stuffing.

Generalized transduction can be used to obtain bacterial linkage information when genes are close enough that the phage can pick them up and transduce them in

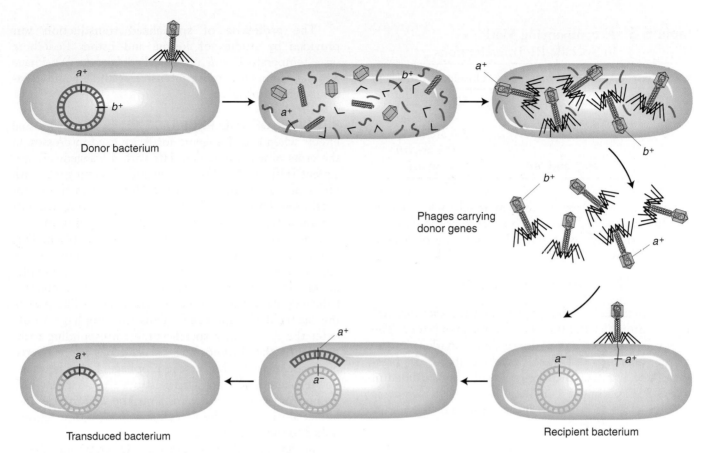

Figure 5-27 The mechanism of generalized transduction. In reality, only a very small minority of phage progeny (1 in 10,000) carry donor genes.

a single piece of DNA. For example, suppose that we wanted to find the linkage between *met* and *arg* in *E. coli*. We could grow phage P1 on a donor *met*⁺ *arg*⁺ strain, and then allow P1 phages from lysis of this strain to infect a *met*⁻ *arg*⁻ strain. First, one donor allele is selected, say, *met*⁺. Then, the percentage of *met*⁺ colonies that is also *arg*⁺ is measured. Strains transduced to both *met*⁺ and *arg*⁺ are called **cotransductants.** The *greater* the cotransduction frequency, the *closer* two genetic markers must be (the opposite of most mapping correlations). Linkage values are usually expressed as cotransduction frequencies (Figure 5-28).

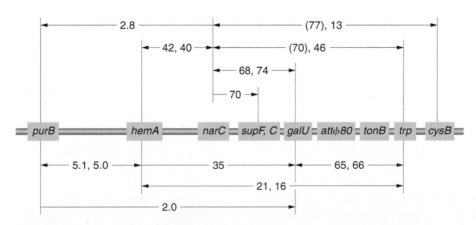

Figure 5-28 Genetic map of the *purB*-to-*cysB* region of *E. coli* determined by P1 cotransduction. The numbers given are the averages in percent for cotransduction frequencies obtained in several experiments. The values in parentheses are considered unreliable. [After J. R. Guest, *Molecular and General Genetics* 105, 1969, 285.]

Table 5-3 Accompanying Markers in Specific P1 Transductions

Experiment	Selected marker	Unselected markers
1	*leu*+	50% are *azi*r; 2% are *thr*+
2	*thr*+	3% are *leu*+; 0% are *ari*r
3	*leu*+ and *thr*+	0% are *azi*r

Using an extension of this approach, we can estimate the *size* of the piece of host chromosome that a phage can pick up, as in the following type of experiment, which uses P1 phage:

donor *leu*+ *thr*+ *azi*r ⟶ recipient *leu*− *thr*− *azi*s

In this experiment, P1 phage grown on the *leu*+ *thr*+ *azi*r donor strain infect the *leu*− *thr*− *azi*s recipient strain. The strategy is to select one or more donor markers in the recipient and then test these transductants for the presence of the unselected markers. Results are outlined in Table 5-3. Experiment 1 in Table 5-3 tells us that *leu* is relatively close to *azi* and distant from *thr*, leaving us with two possibilities:

```
    thr         leu   azi
  --+-----------+-----+--
```

or

```
    thr         azi   leu
  --+-----------+-----+--
```

Experiment 2 tells us that *leu* is closer to *thr* than *azi* is, so the map must be

```
    thr         leu   azi
  --+-----------+-----+--
```

By selecting for *thr*+ and *leu*+ together in the transducing phages in experiment 3, we see that the transduced piece of genetic material never includes the *azi* locus because the phage head cannot carry a fragment of DNA that big. P1 can only cotransduce genes less than approximately 1.5 minutes apart on the *E. coli* chromosome map.

Specialized transduction

We have seen that phage P22, a generalized transducer, picks up fragments of broken host DNA at random. How are other phages, which act as specialized transducers, able to carry only certain host genes to recipient cells? The short answer is that specialized transducers insert into the bacterial chromosome at one position only. When they exit, a faulty outlooping occurs (similar to the type that produces F′ plasmids). Hence they can pick up and transduce only genes that are close by.

The prototype of **specialized transduction** was provided by studies of Joshua and Esther Lederberg on a temperate *E. coli* phage called *lambda* (λ). Phage λ has become the most intensively studied and best-characterized phage.

BEHAVIOR OF THE PROPHAGE Phage λ has unusual effects when cells lysogenic for it are used in crosses. In the cross of a nonlysogenic Hfr with a lysogenic F− recipient [Hfr × F− (λ)], lysogenic F− exconjugants with Hfr genes are readily recovered. However, in the reciprocal cross Hfr(λ) × F−, the *early* genes from the Hfr chromosome are recovered among the exconjugants, but recombinants for *late* markers are not recovered. Furthermore, lysogenic exconjugants are almost never recovered from this reciprocal cross. What is the explanation? The observations make sense if the λ prophage is behaving like a bacterial gene locus (that is, like part of the bacterial chromosome). Thus the prophage would enter the F− cell at a specific time corresponding to its position in the chromosome. Earlier genes are recovered because they enter before the prophage. Later genes are not recovered, because lysis destroys the recipient cell. In interrupted-mating experiments, the λ prophage does in fact always enter the F− cell at a specific time, closely linked to the *gal* locus.

In an Hfr(λ) × F− cross, the entry of the λ prophage into the cell immediately triggers the prophage into a lytic cycle; this process is called **zygotic induction** (Figure 5-29). However, in the cross of *two* lysogenic cells Hfr(λ) × F− (λ), there is no zygotic induction. The presence of any prophage prevents another infecting virus from causing lysis. The prophage

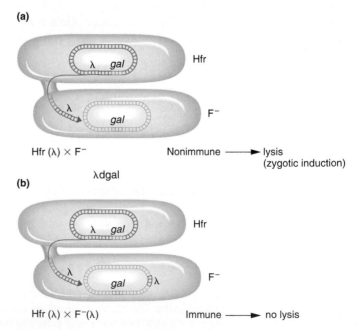

(a)

Hfr (λ) × F−   Nonimmune ⟶ lysis (zygotic induction)

λdgal

(b)

Hfr (λ) × F−(λ)   Immune ⟶ no lysis

Figure 5-29 Zygotic induction.

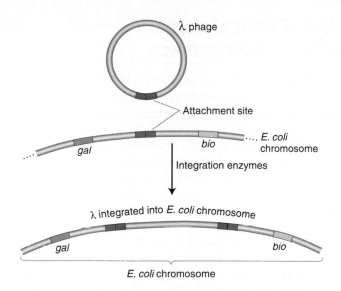

Figure 5-30 **Model for the integration of phage λ into the** **_E. coli_ chromosome.** Reciprocal recombination takes place between a specific attachment site on the circular λ DNA and a specific region on the bacterial chromosome between the _gal_ and _bio_ genes.

produces a cytoplasmic factor that represses the multiplication of the virus. (The phage-directed cytoplasmic repressor nicely explains the immunity of the lysogenic bacteria, because a phage would immediately encounter a repressor and be inactivated.)

λ INSERTION The interrupted-mating experiments described above showed that the λ prophage is part of the lysogenic bacterium's chromosome. How is the λ prophage inserted into the bacterial genome? In 1962, Allan Campbell proposed that it inserts by a single crossover between a circular λ chromosome and the circular _E. coli_ chromosome, as shown in Figure 5-30. The crossover point would be between a specific site in λ, the λ **attachment site,** and an attachment site in the bacterial chromosome located between the genes _gal_ and _bio_, because λ integrates at that position in the _E. coli_ chromosome. The crossing-over is mediated by a phage-encoded recombination system.

One attraction of Campbell's proposal is that from it follow predictions that geneticists can test. For example, integration of the prophage into the _E. coli_ chromosome should increase the genetic distance between flanking bacterial markers, as can be seen in Figure 5-30 for _gal_ and _bio_. In fact, studies show that lysogeny _does_ increase time-of-entry or recombination distances between the bacterial genes. This unique location of λ accounts for its specialized transduction.

Mechanism of specialized transduction

As a prophage, λ always inserts between the _gal_ region and the _bio_ region of the host chromosome (see Figure 5-31),

and in transduction experiments, as expected, λ can transduce only the _gal_ and _bio_ genes.

How does λ carry away neighboring genes? The explanation lies again in an imperfect reversal of the Campbell insertion mechanism, as for generalized transduction. The recombination event between specific regions of λ and the bacterial chromosome is catalyzed by a specialized enzyme system. The λ attachment site and the enzyme that uses this site as a substrate dictate that λ integrates only at that point in the chromosome (Figure 5-31a). Furthermore, during lysis the λ prophage normally excises at precisely the correct point to produce a normal circular λ chromosome as seen in Figure 5-31b(i). Very rarely, excision is abnormal owing to faulty outlooping and can result in phage

(a) **Production of lysogen**

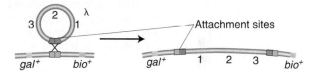

(b) **Production of initial lysate**

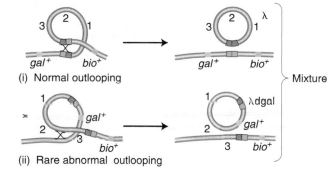

(c) **Transduction by initial lysate**

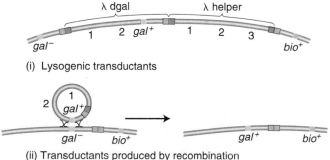

Figure 5-31 **Specialized transduction mechanism in phage λ.** (a) A crossover at the specialized attachment site produces a lysogenic bacterium. (b) Lysogenic bacterium can produce a (i) normal λ or, rarely, (ii) λdgal, a transducing particle containing the _gal_ gene. (c) _gal+_ transductants can be produced by either (i) coincorporation of λdgal and λ (acting as a helper) or, rarely, (ii) crossovers flanking the _gal_ gene. The blue double boxes are the bacterial attachment site, the purple double boxes are the λ attachment site, and the pairs of blue and purple boxes are hybrid integration sites, derived partly from _E. coli_ and partly from λ.

particles that now carry a nearby gene and leave behind some phage genes [see Figure 5-31b(ii)]. The resulting phage genome is defective because of the genes left behind, but it has also gained a bacterial gene *gal* or *bio*. These phages are referred to as λdgal (λ-defective *gal*) or λdbio. The abnormal DNA carrying nearby genes can be packaged into phage heads and can infect other bacteria. In the presence of a second, normal phage particle in a double infection, the λdgal can integrate into the chromosome at the λ attachment site (Figure 5-31c). In this manner, the *gal* genes in this case are transduced into the second host.

> **MESSAGE** Transduction occurs when newly forming phages acquire host genes and transfer them to other bacterial cells. *Generalized transduction* can transfer any host gene. It occurs when phage packaging accidentally incorporates bacterial DNA instead of phage DNA. *Specialized transduction* is due to faulty outlooping of the prophage from the bacterial chromosome, so the new phage includes both phage and bacterial genes. The transducing phage can transfer only specific host genes.

5.6 Physical maps versus linkage maps

Some very detailed chromosomal maps for bacteria have been obtained by combining the mapping techniques of interrupted mating, recombination mapping, transformation, and transduction. Today, new genetic markers are typically mapped first into a segment of about 10 to 15 map minutes by using interrupted mating. This method allows the selection of markers to be used for mapping by P1 cotransduction or by recombination.

By 1963, the *E. coli* map (Figure 5-32) already detailed the positions of approximately 100 genes. After 27 years of further refinement, the 1990 map depicted the positions of more than 1400 genes. Figure 5-33 shows a 5-minute section of the 1990 map (which is adjusted to a scale of 100 minutes). The complexity of these maps illustrates the power and sophistication of genetic analysis. How well do these maps correspond to physical reality? In 1997, the DNA sequence of the entire *E. coli*

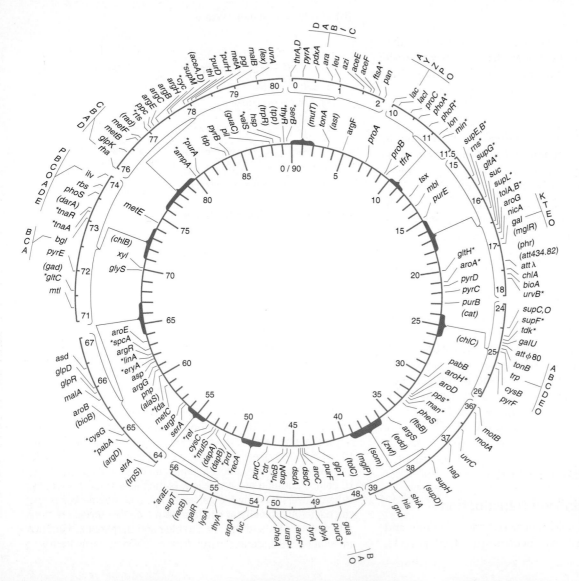

Figure 5-32 The 1963 genetic map of *E. coli*. Units are minutes, based on interrupted-mating experiments, timed from an arbitrarily located origin. Asterisks refer to map positions that are not as precise as the other positions. [From G. S. Stent, *Molecular Biology of Bacterial Viruses.* Copyright 1963 by W. H. Freeman and Company.]

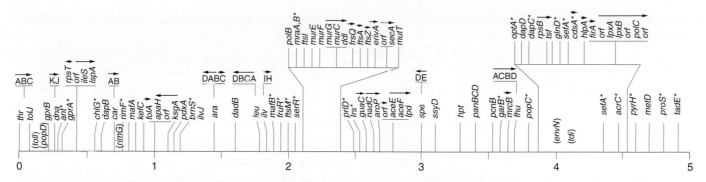

Figure 5-33 Linear scale drawing of a 5-minute section of the 100-minute 1990 *E. coli* linkage map. The parentheses and asterisks indicate markers for which the exact location was unknown at the time of publication. Arrows above genes and groups of genes indicate the direction of transcription. [From B. J. Bachmann, "Linkage Map of *Escherichia coli* K-12, Edition 8," *Microbiological Reviews* 54, 1990, 130–197.]

genome was completed, allowing us to compare the exact position of genes on the genetic map with the corresponding position of the respective coding sequence on the linear DNA sequence. Figure 5-34 makes this comparison for a segment of both maps. Clearly, the genetic map is a close match to the physical map.

As an aside in closing, it is interesting that many of the historical experiments revealing circularity of bacter-

ial and plasmid genomes coincided with the publication and popularization of J. R. R. Tolkien's *The Lord of the Rings*. Consequently a review of bacterial genetics at that time led off with the following quotation from the trilogy:

*"One Ring to rule them all, One Ring to find them,
One Ring to bring them all and in the darkness bind them."*

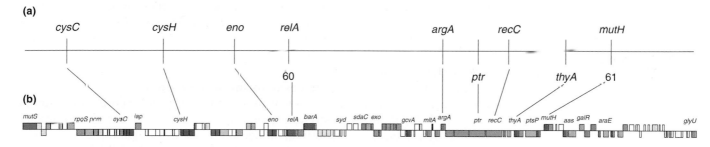

Figure 5-34 Correlation of the genetic and physical maps. (a) Markers on the 1990 genetic map in the region near 60 and 61 minutes. (b) The exact positions of every gene, based on the complete sequence of the *E. coli* genome. (Not every gene is named in this figure, for simplicity.) The elongated boxes are genes and putative genes. Each color represents a different type of function. For example, red denotes regulatory functions, and dark blue denotes functions in DNA replication, recombination, and repair. The correspondence of the order of genes on both maps is indicated.

KEY QUESTIONS REVISITED

• **Do bacterial cells ever pair up for any type of sexual cycle?**

Yes, there is a mating-like process called *conjugation*, governed by the F plasmid. All matings are between one cell with F and one cell without F.

• **Do bacterial genomes ever show recombination?**

Yes, recombinants can be detected. They are relatively rare but can be demonstrated using selection procedures:

for example, selection on plates for prototrophic recombinants between two auxotrophic parents.

• **If so, in what ways do genomes become associated to permit recombination?**

Conjugation, transformation, and transduction are all ways of bringing together genes from different bacterial parents.

• **Does bacterial recombination resemble eukaryote recombination?**

Yes, in the sense that *A B* can be produced from *A b* and *a B*. No, in the sense that recombination always takes place in a partial diploid (merozygote) as opposed to a true diploid. Hence the recombination is formally a double-crossover event, needed to maintain the circular bacterial chromosome intact.

• **Do the genomes of bacterial viruses ever show recombination?**

Yes, if a bacterial host is infected simultaneously by two phage types, recombinant phage are found in the lysate.

• **Do bacterial and viral genomes interact physically in any way?**

Yes, temperate phages can pick up bacterial genome fragments during lysis and transfer them to other bacteria when they reinfect. Some phages pick up only one specific region; others can pick up any segment that can be stuffed into a phage head.

• **Can bacterial and viral chromosomes be mapped using recombination?**

Yes, all bacterial merozygotes can be used for recombinant frequency-based mapping. There are also some unusual methods, for example, mapping by time of entry during conjugation, and frequency of cotransduction by phage. Virus chromosomes can be mapped, too.

SUMMARY

Advances in microbial genetics within the past 50 years have provided the foundation for recent advances in molecular biology (discussed in subsequent chapters). Early in this period, gene transfer and recombination were found to take place between different strains of bacteria. In bacteria, however, genetic material is passed in only one direction—from a donor cell (F$^+$ or Hfr) to a recipient cell (F$^-$). Donor ability is determined by the presence in the cell of a fertility factor (F), a type of plasmid.

On occasion, the F factor present in the free state in F$^+$ cells can integrate into the *E. coli* chromosome and form an Hfr cell. When this occurs, gene transfer and subsequent recombination take place. Furthermore, because the F factor can insert at different places on the host chromosome, investigators were able to show that the *E. coli* chromosome is a single circle, or ring. Interruption of the transfer at different times has provided geneticists with a new method for constructing a linkage map of the single chromosome of *E. coli* and other similar bacteria.

Genetic traits can also be transferred from one bacterial cell to another in the form of pieces of DNA taken into the cell from the extracellular environment. This process of *transformation* in bacterial cells was the first demonstration that DNA is the genetic material. For transformation to occur, DNA must be taken into a recipient cell, and recombination between a recipient chromosome and the incorporated DNA must then take place.

Bacteria can be infected by bacteriophages. In one method of infection, the phage chromosome may enter the bacterial cell and, using the bacterial metabolic machinery, produce progeny phage that burst the host bacterium. The new phages can then infect other cells. If two phages of different genotypes infect the same host, recombination between their chromosomes can take place in this lytic process.

In another mode of infection, lysogeny, the injected phage lies dormant in the bacterial cell. In many cases, this dormant phage (the prophage) incorporates into the host chromosome and replicates with it. Either spontaneously or under appropriate stimulation, the prophage can arise from its latency and can lyse the bacterial host cell.

Phages can carry bacterial genes from a donor to a recipient. In generalized transduction, random host DNA is incorporated alone into the phage head during lysis. In specialized transduction, faulty excision of the prophage from a unique chromosomal locus results in the inclusion of specific host genes as well as phage DNA in the phage head.

KEY TERMS

attachment site (p. 173)

auxotrophic (p. 153)

bacteriophages (p. 152)

clone (p. 153)

colony (p. 153)

conjugation (p. 156)

cotransductants (p. 171)

donor (p. 156)

double infection (p. 169)

double transformation (p. 166)

endogenote (p. 162)

exconjugants (p. 158)

exogenote (p. 162)

F′ plasmid (p. 163)

fertility factor (F) (p. 156)

generalized transduction (p. 170)

genetic markers (p. 153)

Hfr (p. 157)

interrupted mating (p. 158)

lysis (p. 167)

lysogenic bacterium (p. 170)

mcrozygote (p. 162)

minimal medium (p. 153)

mixed infection (p. 169)

origin (O) (p. 158)

phage (p. 152)

phage recombination (p. 153)

plaque (p. 168)

plasmid (p. 156)

plating (p. 153)

prokaryotes (p. 152)

prophage (p. 170)

prototrophic (p. 153)

R plasmids (p. 165)

recipient (p. 156)

resistant mutants (p. 153)

rolling circle replication (p. 157)

screen (p. 170)

selective system (p. 170)

specialized transduction (p. 172)

temperate phage (p. 170)

terminus (p. 161)

transduction (p. 170)

transformation (p. 166)

unselected markers (p. 162)

virulent phage (p. 170)

viruses (p. 152)

zygotic induction (p. 172)

SOLVED PROBLEMS

1. Suppose that a cell were unable to carry out generalized recombination (*rec*⁻). How would this cell behave as a recipient in generalized and in specialized transduction? First compare each type of transduction and then determine the effect of the *rec*⁻ mutation on the inheritance of genes by each process.

Solution

Generalized transduction entails the incorporation of chromosomal fragments into phage heads, which then infect recipient strains. Fragments of the chromosome are incorporated randomly into phage heads, so any marker on the bacterial host chromosome can be transduced to another strain by generalized transduction. In contrast, specialized transduction entails the integration of the phage at a specific point on the chromosome and the rare incorporation of chromosomal markers near the integration site into the phage genome. Therefore, only those markers that are near the specific integration site of the phage on the host chromosome can be transduced.

Markers are inherited by different routes in generalized and specialized transduction. A generalized transducing phage injects a fragment of the donor chromosome into the recipient. This fragment must be incorporated into the recipient's chromosome by recombination, with the use of the recipient recombination system. Therefore, a *rec*⁻ recipient will not be able to incorporate fragments of DNA and cannot inherit markers by generalized transduction. On the other hand, the major route for the inheritance of markers by specialized transduction is by integration of the specialized transducing particle into the host chromosome at the specific phage integration site. This integration, which sometimes requires an additional wild-type (helper) phage, is mediated by a phage-specific enzyme system that is independent of the normal recombination en-

zymes. Therefore, a *rec*⁻ recipient can still inherit genetic markers by specialized transduction.

2. In *E. coli*, four Hfr strains donate the following genetic markers, shown in the order donated:

Strain 1:	Q	W	D	M	T
Strain 2:	A	X	P	T	M
Strain 3:	B	N	C	A	X
Strain 4:	B	Q	W	D	M

All these Hfr strains are derived from the same F⁺ strain. What is the order of these markers on the circular chromosome of the original F⁺?

Solution

A two-step approach works well: (1) determine the underlying principle and (2) draw a diagram. Here the principle is clearly that each Hfr strain donates genetic markers from a fixed point on the circular chromosome and that the earliest markers are donated with the highest frequency. Because not all markers are donated by each Hfr, only the early markers must be donated for each Hfr. Each strain allows us to draw the following circles:

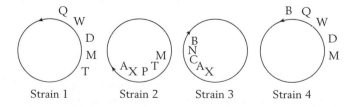

From this information, we can consolidate each circle into one circular linkage map of the order Q, W, D, M, T, P, X, A, C, N, B, Q.

3. In an Hfr × F⁻ cross, *leu⁺* enters as the first marker, but the order of the other markers is unknown. If the Hfr is wild-type and the F⁻ is auxotrophic for each marker in question, what is the order of the markers in a cross where *leu⁺* recombinants are selected if 27 percent are *ile⁺*, 13 percent are *mal⁺*, 82 percent are *thr⁺*, and 1 percent are *trp⁺*?

Solution

Recall that spontaneous breakage creates a natural gradient of transfer, which makes it less and less likely for a recipient to receive later and later markers. Because we have selected for the earliest marker in this cross, the frequency of recombinants is a function of the order of entry for each marker. Therefore, we can immediately determine the order of the genetic markers simply by looking at the percentage of recombinants for any marker among the *leu⁺* recombinants. Because the inheritance of *thr⁺* is the highest, this must be the first marker to enter after *leu*. The complete order is *leu, thr, ile, mal, trp*.

4. A cross is made between an Hfr that is *met⁺ thi⁺ pur⁺* and an F⁻ that is *met⁻ thi⁻ pur⁻*. Interrupted-mating studies show that *met⁺* enters the recipient last, so *met⁺* recombinants are selected on a medium containing supplements that satisfy only the *pur* and *thi* requirements. These recombinants are tested for the presence of the *thi⁺* and *pur⁺* alleles. The following numbers of individuals are found for each genotype:

met⁺ thi⁺ pur⁺	280
met⁺ thi⁺ pur⁻	0
met⁺ thi⁻ pur⁺	6
met⁺ thi⁻ pur⁻	52

a. Why was methionine (Met) left out of the selection medium?

b. What is the gene order?

c. What are the map distances in recombination units?

Solution

a. Methionine was left out of the medium to allow selection for *met⁺* recombinants, because *met⁺* is the last marker to enter the recipient. The selection for *met⁺* ensures that all the loci that we are considering in the cross will have already entered each recombinant that we analyze.

b. Here it is helpful to diagram the possible gene orders. Because we know that *met* enters the recipient last, there are only two possible gene orders if the first marker enters on the right: *met, thi, pur* or *met, pur, thi*. How can we distinguish between these two orders? Fortunately, one of the four possible classes of recombinants requires two additional crossovers. Each possible order predicts a different class that arises by four crossovers

rather than two. For instance, if the order were *met, thi, pur*, then *met⁺ thi⁻ pur⁺* recombinants would be very rare. On the other hand, if the order were *met, pur, thi*, then the four-crossover class would be *met⁺ pur⁻ thi⁺*. From the information given in the table, it is clear that the *met⁺ pur⁻ thi⁺* class is the four-crossover class and therefore that the gene order *met, pur, thi* is correct.

c. Refer to the following diagram:

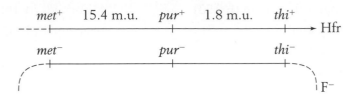

To compute the distance between *met* and *pur*, we compute the percentage of *met⁺ pur⁻ thi⁻*, which is 52/388 = 15.4 m.u. The distance between *pur* and *thi* is, similarly, 6/338 = 1.8 m.u.

5. Compare the mechanism of transfer and inheritance of the *lac⁺* genes in crosses with Hfr, F⁺, and F′-*lac⁺* strains. How would an F⁻ cell that cannot undergo normal homologous recombination (*rec⁻*) behave in crosses with each of these three strains? Would the cell be able to inherit the *lac⁺* gene?

Solution

Each of these three strains donates genes by conjugation. In the Hfr and F⁺ strains, the *lac⁺* genes on the host chromosome are donated. In the Hfr strain, the F factor is integrated into the chromosome in every cell, so efficient donation of chromosomal markers can occur, particularly if the marker is near the integration site of F and is donated early. The F⁺ cell population contains a small percentage of Hfr cells, in which F is integrated into the chromosome. These cells are responsible for the gene transfer displayed by cultures of F⁺ cells. In the Hfr- and F⁺-mediated gene transfer, inheritance requires the incorporation of a transferred fragment by recombination (recall that two crossovers are needed) into the F⁻ chromosome. Therefore, an F⁻ strain that cannot undergo recombination cannot inherit donor chromosomal markers even though they are transferred by Hfr strains or Hfr cells in F⁺ strains. The fragment cannot be incorporated into the chromosome by recombination. Because these fragments do not possess the ability to replicate within the F⁻ cell, they are rapidly diluted out during cell division.

Unlike Hfr cells, F′ cells transfer genes carried on the F′ factor, a process that does not require chromosome transfer. In this case, the *lac⁺* genes are linked to the F′ and are transferred with the F′ at a high efficiency. In the F⁻ cell, no recombination is required, because the F′ *lac⁺* strain can replicate and be maintained in the dividing F⁻ cell population. Therefore, the *lac⁺* genes are inherited even in a *rec⁻* strain.

PROBLEMS

BASIC PROBLEMS

1. Describe the state of the F factor in an Hfr, F$^+$, and F$^-$ strain.

2. How does a culture of F$^+$ cells transfer markers from the host chromosome to a recipient?

3. With respect to gene transfer and integration of the transferred gene into the recipient genome compare the following:

    a. Hfr crosses by conjugation and generalized transduction.

    b. F$'$ derivatives such as F$'$ *lac* and specialized transduction.

4. Why is generalized transduction able to transfer any gene, but specialized transduction is restricted to only a small set?

5. A microbial geneticist isolates a new mutation in *E. coli* and wishes to map its chromosomal location. She uses interrupted-mating experiments with Hfr strains and generalized-transduction experiments with phage P1. Explain why each technique, by itself, is insufficient for accurate mapping.

6. In *E. coli*, four Hfr strains donate the following markers, shown in the order donated:

Strain 1:	M	Z	X	W	C
Strain 2:	L	A	N	C	W
Strain 3:	A	L	B	R	U
Strain 4:	Z	M	U	R	B

    All these Hfr strains are derived from the same F$^+$ strain. What is the order of these markers on the circular chromosome of the original F$^+$?

7. You are given two strains of *E. coli*. The Hfr strain is *arg$^+$ ala$^+$ glu$^+$ pro$^+$ leu$^+$ T^s*; the F$^-$ strain is *arg$^-$ ala$^-$ glu$^-$ pro$^-$ leu$^-$ T^r*. The markers are all nutritional except *T*, which determines sensitivity or resistance to phage T1. The order of entry is as given, with *arg$^+$* entering the recipient first and *T^s* last. You find that the F$^-$ strain dies when exposed to penicillin (*pen^s*), but the Hfr strain does not (*pen^r*). How would you locate the locus for *pen* on the bacterial chromosome with respect to *arg*, *ala*, *glu*, *pro*, and *leu*? Formulate your answer in logical, well-explained steps and draw explicit diagrams where possible.

www.

8. A cross is made between two *E. coli* strains: Hfr *arg$^+$ bio$^+$ leu$^+$* × F$^-$ *arg$^-$ bio$^-$ leu$^-$*. Interrupted-mating studies show that *arg$^+$* enters the recipient last, so *arg$^+$* recombinants are selected on a medium containing *bio* and *leu* only. These recombinants are tested for the presence of *bio$^+$* and *leu$^+$*. The following numbers of individuals are found for each genotype:

arg$^+$ bio$^+$ leu$^+$	320
arg$^+$ bio$^+$ leu$^-$	8
arg$^+$ bio$^-$ leu$^+$	0
arg$^+$ bio$^-$ leu$^-$	48

  a. What is the gene order?

  b. What are the map distances in recombination percentages?

9. Linkage maps in an Hfr bacterial strain are calculated in units of minutes (the number of minutes between genes indicates the length of time it takes for the second gene to follow the first in conjugation). In making such maps, microbial geneticists assume that the bacterial chromosome is transferred from Hfr to F$^-$ at a constant rate. Thus, two genes separated by 10 minutes near the origin end are assumed to be the same physical distance apart as two genes separated by 10 minutes near the F-attachment end. Suggest a critical experiment to test the validity of this assumption.

10. A particular Hfr strain normally transmits the *pro$^+$* marker as the last one in conjugation. In a cross of this strain with an F$^-$ strain, some *pro$^+$* recombinants are recovered early in the mating process. When these *pro$^+$* cells are mixed with F$^-$ cells, the majority of the F$^-$ cells are converted into *pro$^+$* cells that also carry the F factor. Explain these results.

11. F$'$ strains in *E. coli* are derived from Hfr strains. In some cases, these F$'$ strains show a high rate of integration back into the bacterial chromosome of a second strain. Furthermore, the site of integration is often the same site that the sex factor occupied in the original Hfr strain (before production of the F$'$ strains). Explain these results.

12. You have two *E. coli* strains, F$^-$ *str^r ala$^-$* and Hfr *str^s ala$^+$*, in which the F factor is inserted close to *ala$^+$*. Devise a screening test to detect strains carrying F$'$ *ala$^+$*.

13. Five Hfr strains A through E are derived from a single F$^+$ strain of *E. coli*. The following chart shows the entry times of the first five markers into an F$^-$ strain when each is used in an interrupted-conjugation experiment:

A		B		C		D		E	
mal$^+$	(1)	*ade$^+$*	(13)	*pro$^+$*	(3)	*pro$^+$*	(10)	*his$^+$*	(7)
str^s	(11)	*his$^+$*	(28)	*met$^+$*	(29)	*gal$^+$*	(16)	*gal$^+$*	(17)
ser$^+$	(16)	*gal$^+$*	(38)	*xyl$^+$*	(32)	*his$^+$*	(26)	*pro$^+$*	(23)
ade$^+$	(36)	*pro$^+$*	(44)	*mal$^+$*	(37)	*ade$^+$*	(41)	*met$^+$*	(49)
his$^+$	(51)	*met$^+$*	(70)	*str^s*	(47)	*ser$^+$*	(61)	*xyl$^+$*	(52)

a. Draw a map of the F$^+$ strain, indicating the positions of all genes and their distances apart in minutes.

b. Show the insertion point and orientation of the F plasmid in each Hfr strain.

c. In the use of each of these Hfr strains, state which gene you would select to obtain the highest proportion of Hfr exconjugants.

14. *Streptococcus pneumoniae* cells of genotype *str*^{s} *mtl*$^-$ are transformed by donor DNA of genotype *str*^{r} *mtl*$^+$ and (in a separate experiment) by a mixture of two DNAs with genotypes *str*^{r} *mtl*$^-$ and *str*^{s} *mtl*$^+$. The accompanying table shows the results.

	Percentage of cells transformed into		
Transforming DNA	*str*^{r} *mtl*$^-$	*str*^{s} *mtl*$^+$	*str*^{r} *mtl*$^+$
str^r *mtl*$^+$	4.3	0.40	0.17
str^r *mtl*$^-$ + *str*^{s} *mtl*$^+$	2.8	0.85	0.0066

a. What does the first line of the table tell you? Why?

b. What does the second line of the table tell you? Why?

15. Recall that in Chapter 4 we considered the possibility that a crossover event may affect the likelihood of another crossover. In the bacteriophage T4, gene *a* is 1.0 m.u. from gene *b*, which is 0.2 m.u. from gene *c*. The gene order is *a, b, c*. In a recombination experiment, you recover five double crossovers between *a* and *c* from 100,000 progeny viruses. Is it correct to conclude that interference is negative? Explain your answer.

16. You have infected *E. coli* cells with two strains of T4 virus. One strain is minute (*m*), rapid-lysis (*r*), and turbid (*tu*); the other is wild-type for all three markers. The lytic products of this infection are plated and classified. The resulting 10,342 plaques were distributed among eight genotypes, as follows:

m r tu	3467	*m + +*	520
+ + +	3729	*+ r tu*	474
m r +	853	*+ r +*	172
m + tu	162	*+ + tu*	965

a. Determine the linkage distances between *m* and *r*, between *r* and *tu*, and between *m* and *tu*.

b. What linkage order would you suggest for the three genes?

c. What is the coefficient of coincidence (see Chapter 4) in this cross, and what does it signify?

(Problem 16 is reprinted with the permission of Macmillan Publishing Co., Inc., from Monroe W. Strickberger, *Genetics*. Copyright 1968 by Monroe W. Strickberger.)

17. With the use of P22 as a generalized transducing phage grown on a *pur*$^+$ *pro*$^+$ *his*$^+$ bacterial donor, a recipient strain of genotype *pur*$^-$ *pro*$^-$ *his*$^-$ is infected and incubated. Afterward, transductants for *pur*$^+$, *pro*$^+$, and *his*$^+$ are selected individually in experiments I, II, and III, respectively.

a. What media are used for these selection experiments?

b. The transductants are examined for the presence of unselected donor markers, with the following results:

I		II		III	
pro$^-$ *his*$^-$	87%	*pur*$^-$ *his*$^-$	43%	*pur*$^-$ *pro*$^-$	21%
pro$^+$ *his*$^-$	0%	*pur*$^+$ *his*$^-$	0%	*pur*$^+$ *pro*$^-$	15%
pro$^-$ *his*$^+$	10%	*pur*$^-$ *his*$^+$	55%	*pur*$^-$ *pro*$^+$	60%
pro$^+$ *his*$^+$	3%	*pur*$^+$ *his*$^+$	2%	*pur*$^+$ *pro*$^+$	4%

What is the order of the bacterial genes?

c. Which two genes are closest together?

d. On the basis of the order that you proposed in part c, explain the relative proportions of genotypes observed in experiment II.

(Problem 17 is from D. Freifelder, *Molecular Biology and Biochemistry*. Copyright 1978 by W. H. Freeman and Company, New York.)

18. Although most λ-mediated *gal*$^+$ transductants are inducible lysogens, a small percentage of these transductants in fact are not lysogens (that is, they contain no integrated λ). Control experiments show that these transductants are not produced by mutation. What is the likely origin of these types?

19. An *ade*$^+$ *arg*$^+$ *cys*$^+$ *his*$^+$ *leu*$^+$ *pro*$^+$ bacterial strain is known to be lysogenic for a newly discovered phage, but the site of the prophage is not known. The bacterial map is

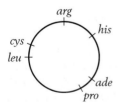

The lysogenic strain is used as a source of the phage, and the phages are added to a bacterial strain of genotype *ade*$^-$ *arg*$^-$ *cys*$^-$ *his*$^-$ *leu*$^-$ *pro*$^-$. After a short incubation, samples of these bacteria are plated on six different media, with the supplementations indicated in the table below. The table also shows whether colonies were observed on the various media.

Medium	Nutrient supplementation in medium						Presence of colonies
	Ade	Arg	Cys	His	Leu	Pro	
1	−	+	+	+	+	+	N
2	+	−	+	+	+	+	N
3	+	+	−	+	+	+	C
4	+	+	+	−	+	+	N
5	+	+	+	+	−	+	C
6	+	+	+	+	+	−	N

(In this table, a plus sign indicates the presence of a nutrient supplement, a minus sign indicates supplement not present, N indicates no colonies, and C indicates colonies present.)

a. What genetic process is at work here?

b. What is the approximate locus of the prophage?

20. In a generalized transduction system using P1 phage, the donor is *pur*⁺ *nad*⁺ *pdx*⁻ and the recipient is *pur*⁻ *nad*⁻ *pdx*⁺. The donor allele *pur*⁺ is initially selected after transduction, and 50 *pur*⁺ transductants are then scored for the other alleles present. Here are the results:

Genotype	Number of colonies
nad⁺ *pdx*⁺	3
nad⁺ *pdx*⁻	10
nad⁻ *pdx*⁺	24
nad⁻ *pdx*⁻	13
	50

a. What is the cotransduction frequency for *pur* and *nad*?

b. What is the cotransduction frequency for *pur* and *pdx*?

c. Which of the unselected loci is closest to *pur*?

d. Are *nad* and *pdx* on the same side or on opposite sides of *pur*? Explain. (Draw the exchanges needed to produce the various transformant classes under either order to see which requires the minimum number to produce the results obtained.)

21. In a generalized transduction experiment, phages are collected from an *E. coli* donor strain of genotype *cys*⁺ *leu*⁺ *thr*⁺ and used to transduce a recipient of genotype *cys*⁻ *leu*⁻ *thr*⁻. Initially, the treated recipient population is plated on a minimal medium supplemented with leucine and threonine. Many colonies are obtained.

a. What are the possible genotypes of these colonies?

b. These colonies are then replica-plated onto three different media: (1) minimal plus threonine only, (2) minimal plus leucine only, and (3) minimal.

What genotypes could, in theory, grow on these three media?

c. It is observed that 56 percent of the original colonies grow on medium 1, 5 percent grow on medium 2, and no colonies grow on medium 3. What are the actual genotypes of the colonies on media 1, 2, and 3?

d. Draw a map showing the order of the three genes and which of the two outer genes is closer to the middle gene.

22. Deduce the genotypes of the following four *E. coli* strains:

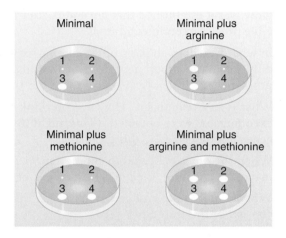

23. In an interrupted-conjugation experiment in *E. coli*, it is established that the *pro* gene enters after the *thi* gene. A *pro*⁺ *thi*⁺ Hfr is crossed with a *pro*⁻ F⁻ strain, and exconjugants are plated on medium containing thiamine but no proline. A total of 360 colonies are observed, and they are isolated and cultured on fully supplemented medium. These cultures are then tested for their ability to grow on medium containing no proline or thiamine (minimal medium), and it is found that 320 of the cultures can grow but the remainder cannot.

a. Deduce the genotypes of the two types of cultures.

b. Draw the crossover events required to produce these genotypes.

c. Calculate the distance between the *pro* and *thi* genes in recombination units.

UNPACKING PROBLEM 23

1. What type of organism is *E. coli*?

2. What does a culture of *E. coli* look like?

3. On what sort of substrates does *E. coli* generally grow in its natural habitat?

4. What are the minimal requirements for *E. coli* cells to divide?

5. Define the terms *prototroph* and *auxotroph*.

6. Which cultures in this experiment are prototrophic and which are auxotrophic?

7. Given some strains of unknown genotype regarding thiamine and proline, how would you test their genotypes? Give precise experimental details, including equipment.

8. What kinds of chemicals are proline and thiamine? Does this matter in this experiment?

9. Draw a schematic diagram showing the full set of manipulations performed in the experiment.

10. Why do you think the experiment was done?

11. How was it established that *pro* enters after *thi*? Give precise experimental steps.

12. In what way does an interrupted-mating experiment differ from the experiment described in this problem?

13. What is an exconjugant? How do you think that exconjugants were obtained? (This might involve genes not described in this problem.)

14. When the *pro* gene is said to enter after *thi*, does it mean the *pro* allele, the *pro⁺* allele, either, or both?

15. What is "fully supplemented medium" in the context of this question?

16. Some exconjugants did not grow on minimal medium. On what medium would they grow?

17. State the types of crossovers that are involved in Hfr × F⁻ recombination. How do these crossovers differ from crossovers in eukaryotes?

18. What is a recombination unit in the context of the present analysis? How does it differ from the map units used in eukaryote genetics?

24. A generalized transduction experiment uses a *metE⁺ pyrD⁺* strain as donor and *metE⁻ pyrD⁻* as recipient. *MetE⁺* transductants are selected and then tested for the *pyrD⁺* allele. The following numbers were obtained:

$$metE^+ \ pyrD^- \quad 857$$
$$metE^+ \ pyrD^+ \quad 1$$

Do these results suggest that these loci are closely linked? What other explanations are there for the lone "double"?

CHALLENGING PROBLEMS

25. Four *E. coli* strains of genotype $a^+ b^-$ are labeled 1, 2, 3, and 4. Four strains of genotype $a^- b^+$ are labeled 5, 6, 7, and 8. The two genotypes are mixed in all possible combinations and (after incubation) are plated to determine the frequency of $a^+ b^+$ recombinants. The following results are obtained, where M = many recombinants, L = low numbers of recombinants, and 0 = no recombinants:

	1	2	3	4
5	0	M	M	0
6	0	M	M	0
7	L	0	0	M
8	0	L	L	0

On the basis of these results, assign a sex type (either Hfr, F⁺, or F⁻) to each strain.

26. An Hfr strain of genotype $a^+ b^+ c^+ d^- str^s$ is mated with a female strain of genotype $a^- b^- c^- d^+ str^r$. At various times, the culture is shaken vigorously to separate mating pairs. The cells are then plated on agar of the following three types, where nutrient A allows the growth of a^- cells; nutrient B, of b^- cells; nutrient C, of c^- cells; and nutrient D, of d^- cells (a plus indicates the presence of, and a minus the absence of, streptomycin or a nutrient):

Agar type	Str	A	B	C	D
1	+	+	+	−	+
2	+	−	+	+	+
3	+	+	−	+	+

a. What donor genes are being selected on each type of agar?

b. The table below shows the number of colonies on each type of agar for samples taken at various times after the strains are mixed. Use this information to determine the order of genes *a*, *b*, and *c*.

Time of sampling (minutes)	Number of colonies on agar of type		
	1	2	3
0	0	0	0
5	0	0	0
7.5	100	0	0
10	200	0	0
12.5	300	0	75
15	400	0	150
17.5	400	50	225
20	400	100	250
25	400	100	250

c. From each of the 25-minute plates, 100 colonies are picked and transferred to a dish containing agar with all the nutrients except D. The numbers of colonies that grow on this medium are 89 for the sample from agar type 1, 51 for the sample from agar type 2, and 8 for the sample from agar type 3. Using these data, fit gene *d* into the sequence of *a*, *b*, and *c*.

d. At what sampling time would you expect colonies to first appear on agar containing C and streptomycin but no A or B?

(Problem 26 is from D. Freifelder, *Molecular Biology and Biochemistry*. Copyright 1978 by W. H. Freeman and Company.)

27. In the cross Hfr aro^+ arg^+ ery^r str^s × F⁻ aro^- arg^- ery^s str^r, the markers are transferred in the order given (with aro^+ entering first), but the first three genes are very close together. Exconjugants are plated on a medium containing Str (streptomycin, to kill Hfr cells), Ery (erythromycin), Arg (arginine), and Aro (aromatic amino acids). The following results are obtained for 300 colonies isolated from these plates and tested for growth on various media: on Ery only, 263 strains grow; on Ery + Arg, 264 strains grow; on Ery + Aro, 290 strains grow; on Ery + Arg + Aro, 300 strains grow.

a. Draw up a list of genotypes, and indicate the number of individuals in each genotype.

b. Calculate the recombination frequencies.

c. Calculate the ratio of the size of the *arg*-to-*aro* region to the size of the *ery*-to-*arg* region.

28. A transformation experiment is performed with a donor strain that is resistant to four drugs: A, B, C, and D. The recipient is sensitive to all four drugs. The treated recipient cell population is divided up and plated on media containing various combinations of the drugs. The table below shows the results.

Drugs added	Number of colonies	Drugs added	Number of colonies
None	10,000	BC	51
A	1156	BD	49
B	1148	CD	786
C	1161	ABC	30
D	1139	ABD	42
AB	46	ACD	630
AC	640	BCD	36
AD	942	ABCD	30

a. One of the genes is obviously quite distant from the other three, which appear to be tightly (closely) linked. Which is the distant gene?

b. What is the probable order of the three tightly linked genes?

(Problem 28 is from Franklin Stahl, *The Mechanics of Inheritance*, 2d ed. Copyright 1969, Prentice Hall, Englewood Cliffs, NJ. Reprinted by permission.)

29. You have two strains of λ that can lysogenize *E. coli*; the following figure shows their linkage maps:

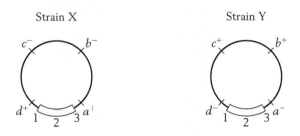

The segment shown at the bottom of the chromosome, designated 1–2–3, is the region responsible for pairing and crossing-over with the *E. coli* chromosome. (Keep the markers on all your drawings.)

a. Diagram the way in which λ strain X is inserted into the *E. coli* chromosome (so that the *E. coli* is lysogenized).

b. It is possible to superinfect the bacteria that are lysogenic for strain X by using strain Y. A certain percentage of these superinfected bacteria become "doubly" lysogenic (that is, lysogenic for both strains). Diagram how this will occur. (Don't worry about how double lysogens are detected.)

c. Diagram how the two λ prophages can pair.

d. It is possible to recover crossover products between the two prophages. Diagram a crossover event and the consequences.

30. You have three strains of *E. coli*. Strain A is F' cys^+ $trp1/cys^+$ $trp1$ (that is, both the F' and the chromosome carry cys^+ and $trp1$, an allele for tryptophan requirement). Strain B is F⁻ cys^- $trp2$ Z (this strain requires cysteine for growth and carries $trp2$, another allele causing a tryptophan requirement; strain B is lysogenic for the generalized transducing phage Z). Strain C is F⁻ cys^+ $trp1$ (it is an F⁻ derivative of strain A that has lost the F'). How would you determine whether $trp1$ and $trp2$ are alleles of the same locus? (Describe the crosses and the results expected.)

31. A generalized transducing phage is used to transduce an a^- b^- c^- d^- e^- recipient strain of *E. coli* with an a^+ b^+ c^+ d^+ $e^!$ donor. The recipient culture is plated

on various media with the results shown in the table below. (Note that a^- indicates a requirement for A as a nutrient, and so forth.) What can you conclude about the linkage and order of the genes?

Compounds added to minimal medium	Presence (+) or absence (−) of colonies
C D E	−
B D E	−
B C E	+
B C D	+
A D E	−
A C E	−
A C D	−
A B E	−
A B D	+
A B C	−

32. In 1965, Jon Beckwith and Ethan Signer devised a method of obtaining specialized transducing phages carrying the *lac* region. They knew that the integration site, designated *att80*, for the temperate phage $\phi80$ (a relative of phage λ) was located near a gene termed *tonB* that confers resistance to the virulent phage T1:

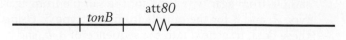

They used an F' *lac⁺* plasmid that could not replicate at high temperatures in a strain carrying a deletion of the *lac* genes. By forcing the cell to remain *lac⁺* at high temperatures, the researchers could select strains in which the plasmid had integrated into the chromosome, thereby allowing the F' *lac* to be maintained at high temperatures. By combining this selection with a simultaneous selection for resistance to T1 phage infection, they found that the only survivors were cells in which the F' *lac* had integrated into the *tonB* locus, as shown in the figure below.

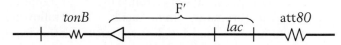

This placed the *lac* region near the integration site for phage $\phi80$. Describe the subsequent steps that the researchers must have followed to isolate the specialized transducing particles of phage $\phi80$ that carried the *lac* region.

INTERACTIVE GENETICS   MegaManual CD-ROM Tutorial

Bacterial Genetics

For additional coverage of the topics in this chapter, refer to the Interactive Genetics CD-ROM included with the Solutions MegaManual. The Bacterial Genetics activity includes animated tutorials on conjugation and transduction, along with interactive problems to help you hone your skills.

6

FROM GENE TO PHENOTYPE

The colors of peppers are determined by the interaction of several genes. An allele *Y* promotes early elimination of chlorophyll (a green pigment), whereas *y* does not. *R* determines red and *r* yellow carotenoid pigments. Alleles *c1* and *c2* of two different genes down-regulate the amounts of carotenoids, causing the lighter shades. Orange is down-regulated red. Brown is green plus red. Pale yellow is down-regulated yellow. [Anthony Griffiths.]

KEY QUESTIONS

- How do individual genes exert their effect on an organism's makeup?

- In the cell, do genes act directly or through some sort of gene product?

- What is the nature of gene products?

- What do gene products do?

- Is it correct to say that an allele of a gene determines a specific phenotype?

- In what way or ways do genes interact at the cellular level?

- How is it possible to dissect complex gene interaction using a mutational approach?

OUTLINE

6.1 Genes and gene products

6.2 Interactions between the alleles of one gene

6.3 Interacting genes and proteins

6.4 Applications of chi-square (χ^2) test to gene interaction ratios

CHAPTER OVERVIEW

Much of the early success of genetics can be attributed to the correlation of phenotypes and alleles, as when Mendel equated *Y* with yellow peas and *y* with green. However, from this logic there arises a natural tendency to view alleles as somehow *determining* phenotypes. Although this is a useful mental shorthand, we must now examine the relationship between genes and phenotypes more carefully. The fact is that there is no way a gene can do anything alone. (Imagine a gene—a single segment of DNA—alone in a test tube.) For a gene to have any influence on a phenotype it must act in concert with many other genes and with the external and internal environment. So an allele like *Y* cannot produce yellow color without the participation of many other genes and environmental inputs. In this chapter we examine the ways in which these interactions take place.

Even though such interactions represent a higher level of complexity, there are standard approaches that can be used to help elucidate the type of interaction oc-

curring in any one case. The main ones used in genetics are as follows:

1. *Genetic analysis* is the focus of this chapter. The genes interacting in a specific phenotype are identified by going on a hunt for all the different kinds of mutants that affect that phenotype.

2. *Functional genomics* (Chapter 12) provides powerful ways of defining the set of genes that participate in any defined system. For example, the genes that collaborate in some specific process can be deduced from finding the set of RNA transcripts present when that process is going on.

3. *Proteomics* (also Chapter 12) assays protein interaction directly. The essence of the technique is to use one protein as "bait" and find out which other cellular proteins attach to it, suggesting the components of a multiprotein cellular "machine."

How does the genetic analysis approach work? The mutants collected in the mutant hunt identify a set of genes that represent the individual components of the

CHAPTER OVERVIEW Figure

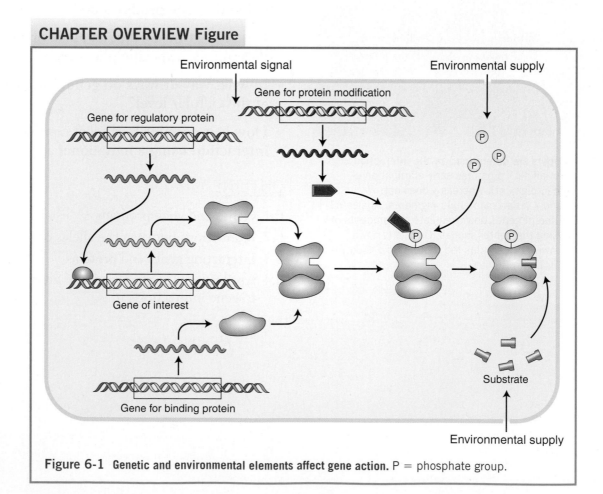

Figure 6-1 Genetic and environmental elements affect gene action. P = phosphate group.

biological system that underlies the specific phenotype under scrutiny. The ways in which they interact can often be deduced by crossing different mutants to create double mutants. The phenotype of a double mutant, and the phenotypic ratios produced when it is crossed, suggests certain types of known interactions. In the cell, gene interaction is manifested by physical interaction between proteins or between proteins and DNA or RNA.

The physical ways in which genes interact with one another and with the environment are summarized by the model in Figure 6-1. Some of the interactions shown in the figure are as follows:

- Transcription of one gene may be turned on or off by other genes called *regulatory genes*. The regulatory proteins they encode generally bind to a region in front of the regulated gene.

- Proteins encoded by one gene may bind to proteins from other genes to form an active complex that performs some function. These complexes, which can be much larger than that shown in the figure, have become known as *molecular machines* because they have several interacting functional parts just like a machine.

- Proteins encoded by one gene may modify the proteins encoded by a second gene in order to activate or deactivate protein function. For example, proteins may be modified through the addition of phosphate groups.

- The environment engages with the system in several ways. In the case of an enzyme, its activity may depend on the availability of a substrate supplied by the environment. Signals from the environment can also set in motion a chain of consecutive gene-controlled steps that follow one another like a cascade of falling dominoes. The chain of events initiated by an environmental signal is called *signal transduction*.

Formal names have been given to certain commonly encountered types of interactions between mutations of different genes: Figure 6-1 can be used to illustrate some of those covered in this chapter. If mutation of one gene prevents expression of alleles of another, the former is said to be *epistatic*. An example would be a mutation of the regulatory gene because, if its protein is defective, any allele of a gene that it regulates could not be transcribed. Sometimes mutation in one gene can restore wild-type expression to a mutation in another gene; in this case the restoring mutation is said to be a *suppressor*. An example is seen in proteins that bind to each other: a mutation causing shape change in one protein might

lead to malfunction because it now cannot bind to its partner protein. However, a mutation in the gene for the partner protein might cause a shape change that would now permit binding to the abnormal protein of the first gene, hence restoring an active complex.

6.1 Genes and gene products

The first clues about how genes function came from studies of humans. Early in the twentieth century, Archibald Garrod, an English physician (Figure 6-2), noted that several recessive human diseases show metabolic defects—harmful alterations in basic body chemistry. This observation led to the notion that such genetic diseases are "inborn errors of metabolism." For example, phenylketonuria (PKU), which is caused by an autosomal recessive allele, results from an inability to convert phenylalanine into tyrosine. Consequently, phenylalanine accumulates and is spontaneously converted into a toxic compound, phenylpyruvic acid. In another example, introduced in Chapter 1, the inability to convert tyrosine into the pigment melanin produces an albino. Garrod's observations focused attention on metabolic control by genes.

One-gene–one-enzyme hypothesis

A landmark study by George Beadle and Edward Tatum in the 1940s clarified the role of genes. They later received a Nobel prize for their study, which marks the beginning of molecular biology. Beadle and Tatum did their work on the haploid fungus *Neurospora*, which we have met in our discussions of octad analysis. They first irradiated *Neurospora* to produce mutations and then

Figure 6-2 Archibald Garrod.

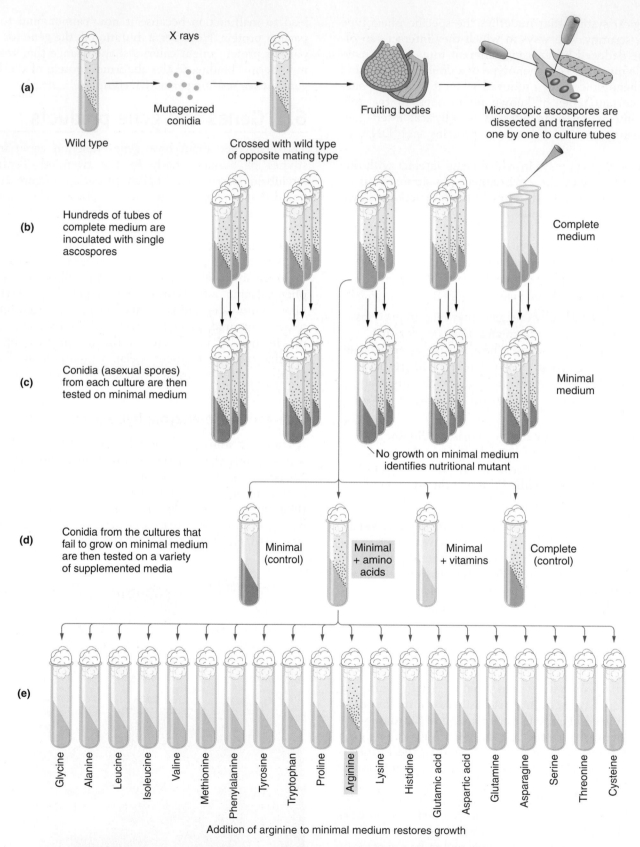

Figure 6-3 Experimental approach used by Beadle and Tatum for generating large numbers of mutants in *Neurospora*. Shown is the isolation of an *arg* mutant. [After Peter J. Russell, *Genetics*, 2d ed. Scott, Foresman.]

tested cultures from ascospores for interesting mutant phenotypes. They found numerous auxotrophic mutants, of the type we have learned about in bacteria. Beadle and Tatum deduced that each mutation that generated the auxotrophic requirement was inherited as a single-gene mutation because each gave a 1:1 ratio when crossed with a wild type. Letting *aux* represent an auxotrophic mutation,

$$+ \times aux$$
$$\downarrow$$

$$\text{progeny} \qquad \tfrac{1}{2} +$$

$$\tfrac{1}{2} \ aux$$

Figure 6-3 depicts the experimental procedure that Beadle and Tatum used. One particular group of mutant strains specifically required the amino acid arginine to grow, and these arginine auxotrophs formed the basis of their subsequent analysis. They found that the arginine-requiring auxotrophic mutations mapped to three different loci on separate chromosomes. Let's call the genes at the three loci the *arg-1*, *arg-2*, and *arg-3* genes. A key breakthrough was their discovery that the auxotrophs for each of the three loci differed in their response to the structurally related compounds ornithine and citrulline (Figure 6-4). The *arg-1* mutants grew when supplied with any one of the chemicals, ornithine, citrulline, or arginine. The *arg-2* mutants grew when given arginine or citrulline but not ornithine. The *arg-3* mutants grew only when arginine was supplied. We can see this more easily by looking at Table 6-1.

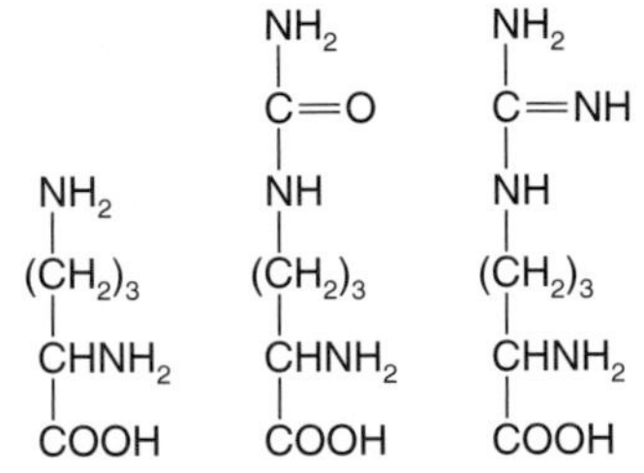

Figure 6-4 Chemical structures of arginine and the related compounds citrulline and ornithine.

It was already known that cellular enzymes interconvert related compounds such as these. On the basis of the properties of the *arg* mutants, Beadle and Tatum and their colleagues proposed a biochemical pathway for such conversions in *Neurospora*:

$$\text{precursor} \xrightarrow{\text{enzyme X}} \text{ornithine} \xrightarrow{\text{enzyme Y}}$$

$$\text{citrulline} \xrightarrow{\text{enzyme Z}} \text{arginine}$$

Table 6-1 Growth of *arg* Mutants in Response to Supplements

| Mutant | Supplement | | |
	Ornithine	Citrulline	Arginine
arg-1	+	+	+
arg-2	−	+	+
arg-3	−	−	+

Note: A plus sign means growth; a minus sign means no growth.

This pathway nicely explains the three classes of mutants shown in Table 6-1. Under the model, the *arg-1* mutants have a defective enzyme X, so they are unable to convert the precursor into ornithine as the first step in producing arginine. However, they have normal enzymes Y and Z, and so the *arg-1* mutants are able to produce arginine if supplied with either ornithine or citrulline. Similarly, the *arg-2* mutants lack enzyme Y, and the *arg-3* mutants lack enzyme Z. Thus, a mutation at a particular gene is assumed to interfere with the production of a single enzyme. The defective enzyme creates a block in some biosynthetic pathway. The block can be circumvented by supplying to the cells any compound that normally comes after the block in the pathway.

We can now diagram a more complete biochemical model:

$$\begin{array}{ccc} arg\text{-}1^+ & & arg\text{-}2^+ \\ \downarrow & & \downarrow \\ \text{precursor} \xrightarrow{\text{enzyme X}} \text{ornithine} & \xrightarrow{\text{enzyme Y}} \end{array}$$

$$\begin{array}{c} arg\text{-}3^+ \\ \downarrow \\ \text{citrulline} \xrightarrow{\text{enzyme Z}} \text{arginine} \end{array}$$

This brilliant model, which has become known as the *one-gene–one-enzyme hypothesis*, was the source of the first exciting insight into the functions of genes: genes somehow were responsible for the function of enzymes, and each gene apparently controlled one specific enzyme. Other researchers obtained similar results for other biosynthetic pathways, and the hypothesis soon achieved general acceptance. It was also found that all proteins, whether or not they are enzymes, are encoded by genes, so the phrase was refined to become **one-gene–one-protein,** or more accurately **one-gene–one-polypeptide.** (Recall that a polypeptide is the simplest type of protein, a single chain of amino acids.) It soon became clear that a gene encodes the *physical structure* of a protein, which in turn dictates its function. Beadle and Tatum's hypothesis became one of the great unifying concepts in biology, because it provided a bridge that brought together the two major research areas of genetics and biochemistry.

We must add that although the great majority of genes encode proteins, it is now known that some encode RNAs that have special functions. All genes are transcribed to make RNA. Protein-coding genes are transcribed to messenger RNA (mRNA), which is then translated into protein. However, for a minority of genes their RNA is never translated to protein because the RNA itself has a unique function. We shall call these **functional RNAs.** Some examples are transfer RNAs, ribosomal RNAs, and small cytoplasmic RNAs—more of these in later chapters.

> **MESSAGE** The majority of genes exert their influence on biological properties at a purely chemical level, by coding for the structures of cellular proteins. Each gene of this type encodes one polypeptide, the simplest protein (a single chain of amino acids). A few genes encode functional RNAs that are destined never to become proteins.

Mutant genes and proteins

If genes code for proteins, how does a mutant allele of a gene affect the protein product? Let us explore the topic using phenylketonuria (PKU), the human disease that we first encountered in Chapter 2. Recall that PKU

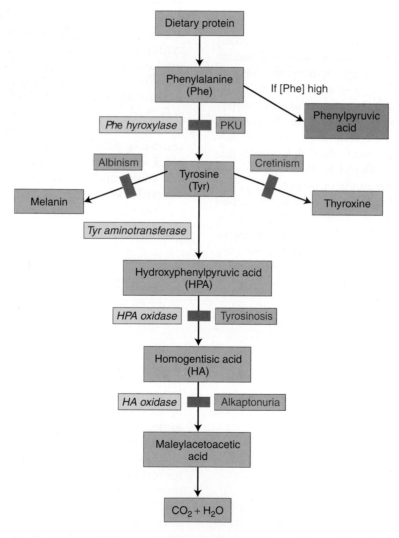

Figure 6-5 A section of the phenylalanine metabolic pathway, including diseases associated with enzyme blockages. The disease PKU is produced when the enzyme phenylalanine hydroxylase malfunctions. Accumulation of phenylalanine results in an increase in phenylpyruvic acid, which interferes with the development of the nervous system. [After I. M. Lerner and W. J. Libby, *Heredity, Evolution, and Society,* 2d ed. Copyright 1976 by W. H. Freeman and Company.]

is an autosomal recessive disease caused by a defective allele of the gene coding for the liver enzyme phenylalanine hydroxylase (PAH). In the absence of normal PAH the phenylalanine entering the body in food is not broken down and hence accumulates. Under such conditions, phenylalanine is converted into phenylpyruvic acid, which is transported to the brain via the bloodstream and there impedes normal development, leading to mental retardation. The section of the metabolic pathway responsible for PKU symptoms is shown in Figure 6-5. (This figure also shows other inborn errors of metabolism caused by blocks at other steps in the reaction sequence.)

The PAH enzyme is a single polypeptide. Recent sequencing of the mutant alleles from many patients has revealed a plethora of mutations at different sites along the gene, summarized in Figure 6-6. What these alleles have in common is that each encodes a defective PAH. They all inactivate some essential part of the protein encoded by the gene. The positions in an enzyme at which function can be adversely affected in general are shown in Figure 6-7. These are likely to fall in the sequences coding for the protein's active site, the region of the protein that catalyzes the chemical reaction breaking down phenylalanine. Mutations may fall anywhere in the gene, but those away from the active site are more likely to be

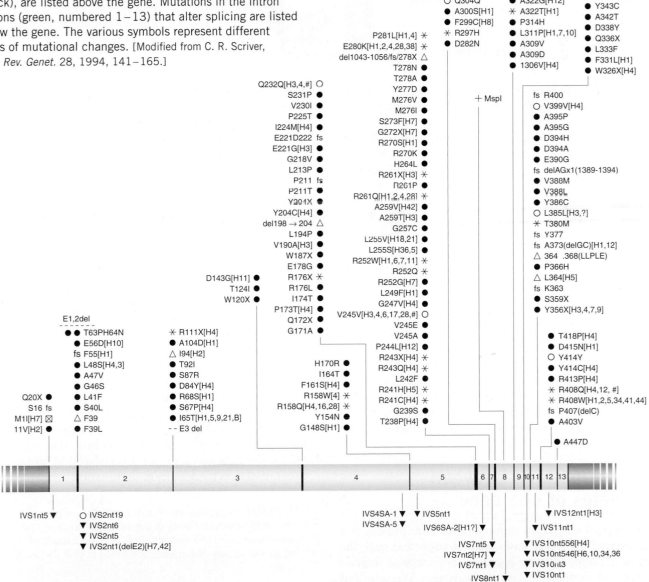

Figure 6-6 Structure of the human phenylalanine hydroxylase gene and a summary of mutations resulting in enzyme malfunction. Mutations in the exons, or protein-coding regions (black), are listed above the gene. Mutations in the intron regions (green, numbered 1–13) that alter splicing are listed below the gene. The various symbols represent different types of mutational changes. [Modified from C. R. Scriver, *Ann. Rev. Genet.* 28, 1994, 141–165.]

Figure 6-7
Representative positions of mutant sites and their functional consequences.

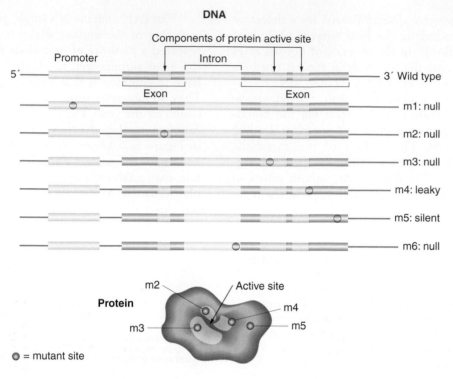

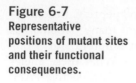

silent mutations that do not result in a defective protein, and therefore produce the wild-type phenotype.

Many of the mutant alleles are of a type generally called **nulls**: the proteins they encode completely lack PAH function. Either there is no protein product or one exists but is nonfunctional. Other mutant alleles produce proteins showing some low residual function: these are called **leaky** mutations. Hence it is a very powerful insight that the reason people express disease symptoms (the disease phenotype) is that they lack some key function that fits into a crucial slot in the overall chemistry of the cell. The same is true for mutations generally in any organism—altering the structure of a gene alters the function of its product, generally producing a decreased or zero function. The absence of a properly functioning protein disturbs cell chemistry, ultimately yielding a mutant phenotype.

The PKU case also makes the point that **multiple alleles** are possible at one locus. However, generally they can be grouped into two main categories: the normally functioning wild type, represented P or p^+, and all the defective recessive mutations, null and leaky, represented as p. The complete set of known alleles of one gene is called an **allelic series**.

6.2 Interactions between the alleles of one gene

What happens when two different alleles are present in a heterozygote? In many cases, one is expressed and the other isn't. Formally, these responses are a type of interaction we call dominance and recessiveness. We saw that

PKU and many other single-gene diseases are recessive. Some single-gene diseases such as achondroplasia are dominant. What is the general explanation for dominance and recessiveness in terms of gene products?

PKU is a good general model for recessive mutations. The reason that a defective PAH allele is recessive is that one "dose" of the wild-type allele P is sufficient to produce wild-type phenotype. The PAH gene is said to be **haplo-sufficient**. Hence both P/P (two doses) and P/p (one dose) have enough PAH activity to result in the normal cellular chemistry. Of course p/p individuals have zero doses of PAH activity. Figure 6-8 illustrates this general notion.

How can we explain dominant mutations? There are several molecular mechanisms for dominance, but a common one is that the wild-type allele of a gene is **haploinsufficient.** This means that one wild-type dose is *not* enough to achieve normal levels of function. Assume that 16 units of a gene's product are needed for normal chemistry, and that each wild-type allele can make 10 units. Two wild-type alleles will produce 20 units of product, well over the minimum. A null mutation in combination with a single wild-type allele would produce $10 + 0 = 10$ units, well below the minimum. Hence the heterozygote (wild type/null) is mutant, and the mutation is by definition dominant.

MESSAGE Recessiveness of a mutant allele is generally a result of haplosufficiency of the wild-type allele of that gene. Dominance of a mutant allele is often a result of haplo*in*sufficiency of the wild-type allele of that particular gene.

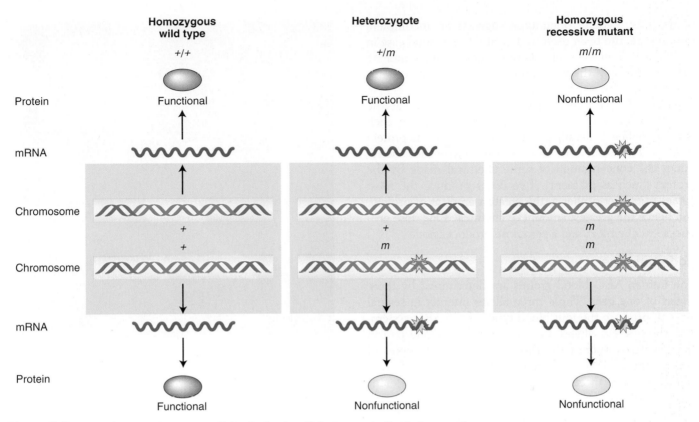

Figure 6-8 Recessiveness of a mutant allele of a haplosufficient gene. In the heterozygote, even though the mutated copy of the gene produces nonfunctional protein, the wild-type copy generates enough functional protein to produce the wild-type phenotype.

The type of dominance we have just described is called **full**, or **complete, dominance**. In cases of full dominance, the homozygous dominant cannot be distinguished from the heterozygote; that is, $A/A = A/a$. However, there are variations on this theme, as shown in the next sections.

Incomplete dominance

Four-o'clocks are plants native to tropical America. Their name comes from the fact that their flowers open in the late afternoon. When a pure-breeding wild-type four-o' clock line with red petals is crossed with a pure line with white petals, the F_1 has pink petals. If an F_2 is produced by selfing the F_1, the result is

$\frac{1}{4}$ of the plants have red petals

$\frac{1}{2}$ of the plants have pink petals

$\frac{1}{4}$ of the plants have white petals

Figure 6-9 shows these phenotypes. From this $1:2:1$ ratio in the F_2, we can deduce that the inheritance pattern is based on two alleles of a single gene. However, the heterozygotes (the F_1 and half the F_2) are intermediate in phenotype. By inventing allele symbols, we can list the genotypes of the four-o'clocks in this experiment as c^+/c^+ (red), c/c (white), and c^+/c (pink). The occurrence

Molecular allele interactions

WWW. ANIMATED ART

Figure 6-9 Red, pink, and white phenotypes of four o'clock plants. The pink heterozygote demonstrates incomplete dominance. [R. Calentine/Visuals Unlimited.]

of the intermediate phenotype suggests an **incomplete dominance,** the term used to describe the general case in which the phenotype of a heterozygote is intermediate between those of the two homozygotes, on some quantitative scale of measurement.

How do we explain incomplete dominance at the molecular level? In cases of incomplete dominance, each wild-type allele generally produces a set dose of its protein product. The number of doses of a wild-type allele determines the concentration of some chemical made by the protein (such as pigment). Two doses produce the most copies of transcript, hence the greatest amount of protein, and hence the greatest amount of chemical. One dose produces less chemical, and a zero dose produces none.

Codominance

The human ABO blood groups are determined by three alleles of one gene. These three alleles interact in several ways to produce the four blood types of the ABO system. The three major alleles are i, I^A, and I^B, but one person can have only two of the three alleles or two copies of one of them. The combinations result in six different genotypes: the three homozygotes and three different types of heterozygotes.

Genotype	Blood type
I^A/I^A, I^A/i	A
I^B/I^B, I^B/i	B
I^A/I^B	AB
i/i	O

In this allelic series, the alleles determine the presence and form of an antigen, a cell-surface molecule that can be recognized by the immune system. The alleles I^A and I^B determine two different forms of this antigen, which is deposited on the surface of the red blood cells. However, the allele i results in no antigenic protein of this type (it is a null allele). In the genotypes I^A/i and I^B/i, the alleles I^A and I^B are fully dominant to i. However, in the genotype I^A/I^B, each of the alleles produces its own form of antigen, so they are said to be **codominant.** Formally, codominance is defined as the expression in a heterozygote of *both* the phenotypes normally shown by the two alleles.

The human disease sickle-cell anemia is a source of another interesting insight into dominance. The gene concerned affects the molecule hemoglobin, which transports oxygen and is the major constituent of red blood cells. The three genotypes have different phenotypes, as follows:

Hb^A/Hb^A: Normal; red blood cells never sickle

Hb^S/Hb^S: Severe, often fatal anemia; abnormal hemoglobin causes red blood cells to have sickle shape

HB^A/Hb^S: No anemia; red blood cells sickle only under low oxygen concentrations

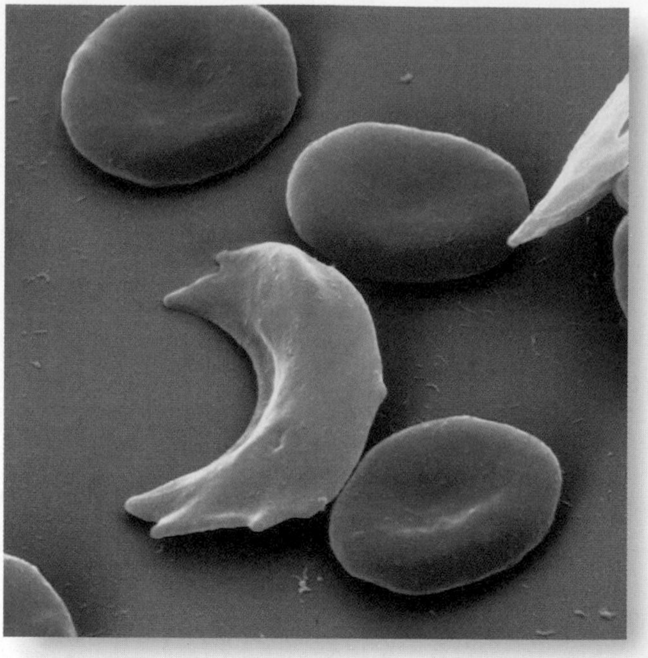

Figure 6-10 An electron micrograph of a sickle-shaped red blood cell. Other, more rounded cells appear almost normal. [Meckes/Ottawa/Photo Researchers.]

Figure 6-10 shows an electron micrograph of blood cells including some sickled cells. In regard to the presence or absence of anemia, the Hb^A allele is dominant. A single Hb^A produces enough functioning hemoglobin to prevent anemia. In regard to blood-cell shape, however, there is incomplete dominance, as shown by the fact that many of the cells have a slight sickle shape. Finally, in regard to hemoglobin itself, there is codominance. The alleles Hb^A and Hb^S code for two different forms of hemoglobin differing by a single amino acid, and both these forms are synthesized in the heterozygote. The A and S forms of hemoglobin can be separated by electrophoresis, because it happens that they have different charges (Figure 6-11). We see that homozygous normal people have one type of hemoglobin (A) and anemics have another (type S), which moves more slowly in the electric field. The heterozygotes have both types, A and S. In other words, there is codominance at the molecular level. The fascinating population genetics of the Hb^A and Hb^S alleles will be considered in Chapter 19.

Sickle-cell anemia illustrates that the terms *dominance, incomplete dominance,* and *codominance* are somewhat arbitrary. The type of dominance inferred depends on the phenotypic level at which the assay is made—organismal, cellular, or molecular. Indeed the same caution can be applied to many of the categories that scientists use to classify structures and processes; these categories are devised by humans for convenience of analysis.

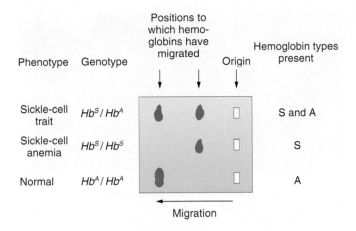

Figure 6-11 Electrophoresis of normal and mutant hemoglobins. Shown are hemoglobin from a heterozygote (with "sickle-cell trait"), a person with sickle-cell anemia, and a normal person. The smudges show the positions to which the hemoglobins migrate on the starch gel.

> **MESSAGE** The type of dominance is determined by the molecular functions of the alleles of a gene and by the investigative level of analysis.

The leaves of clover plants show several variations on the dominance theme. Clover is the common name for plants of the genus *Trifolium*. There are many species. Some are native to North America, whereas others grow there as introduced weeds. Much genetic research has been done with white clover, which shows considerable variation among individuals in the curious V, or chevron, pattern on the leaves. The different chevron forms (and the absence of chevrons) are determined by a series of alleles of one gene, as seen in Figure 6-12. The figure shows the many different types of interactions that are possible, even for one allele.

Recessive lethal alleles

Many mutant alleles are capable of causing the death of an organism; such alleles are called **lethal alleles.** The human disease alleles provide examples. A gene whose mutations may be lethal is clearly an essential gene. The ability to determine whether a gene is essential is an important aid to research on experimental organisms, especially when working on a gene of unknown function. However, maintaining stocks bearing lethal alleles for laboratory use is a challenge. In diploids, recessive lethal alleles can be maintained as heterozygotes. In haploids, heat-sensitive lethal alleles are useful. These are members of a general class of **temperature-sensitive (ts) mutations.** Their phenotype is wild type at the **permissive temperature** (often room temperature), but mutant at the **restrictive temperature.** Temperature-sensitive alleles

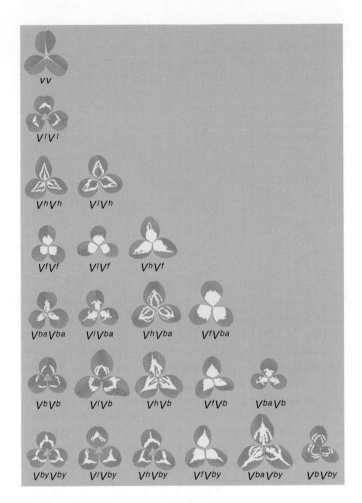

Figure 6-12 Multiple alleles determine the chevron pattern on the leaves of white clover. The genotype of each plant is shown below it. [After photograph by W. Ellis Davies.]

are thought to be caused by mutations that make the protein prone to twist or bend its shape to an inactive conformation at the restrictive temperature. Research stocks can be maintained easily under permissive conditions, and the mutant phenotype assayed by a switch to the restrictive conditions.

A good example of a recessive lethal allele is an allele of a coat color gene in mice (see Model Organism box on page 196). Normal wild-type mice have coats with a rather dark overall pigmentation. A mutation called *yellow* (a lighter coat color) shows a curious inheritance pattern. If any yellow mouse is mated to a homozygous wild-type mouse, a 1 : 1 ratio of yellow to wild-type mice is always observed in the progeny. This result suggests that a yellow mouse is always heterozygous for the yellow allele and that the yellow allele is dominant to wild type. However, if any two yellow mice are crossed with each other, the result is always as follows:

$$\text{yellow} \times \text{yellow} \longrightarrow \tfrac{2}{3} \text{ yellow}, \tfrac{1}{3} \text{ wild type}$$

Figure 6-13 shows a typical litter from a cross between yellow mice.

How can the 2 : 1 ratio be explained? The results make sense if it is assumed that the yellow allele is lethal when homozygous. It is known that the yellow allele is of a coat-color gene called A. Let's call it A^Y. Hence the results of crossing two yellow mice are

$$A^Y/A \ \times \ A^Y/A$$

Progeny $\frac{1}{4} A^Y/A^Y$ lethal

$\frac{1}{2} A^Y/A$ yellow

$\frac{1}{4} A/A$ wild type

Figure 6-13 A litter from a cross between two mice heterozygous for the yellow coat-color allele. The allele is lethal in a double dose. Not all progeny are visible. [Anthony Griffiths.]

MODEL ORGANISM   Mouse

The laboratory mouse is descended from the house mouse *Mus musculus*. The pure lines used today as standards are derived from mice bred in past centuries by mouse "fanciers." Among model organisms it is the one whose genome most closely resembles the human genome. Its diploid chromosome number is 40 (compared to 46 in humans), and the genome is slightly smaller than that of humans (3000 Mb) and approximately the same number of genes (current estimates about 30,000). Furthermore, it seems that all mouse genes have a counterpart in humans. A large proportion of genes are arranged in blocks in exactly the same positions as those of humans.

Research on the Mendelian genetics of mice began early in the twentieth century. One of the most important early contributions was the elucidation of the genes that control coat color and pattern. Genetic control of the mouse coat has provided a model for all mammals, including cats, dogs, horses and cattle. Also a great deal of work was done on mutations induced by radiation and chemicals. Mouse genetics has been of great significance in medicine. A large proportion of human genetic diseases have a mouse counterpart useful for experimental study (these are called "mouse models"). The mouse has played a particularly important role in the development of our current understanding of the genes underlying cancer.

The genome of mice can be modified by the insertion of specific fragments of DNA into a fertilized egg or into somatic cells. The mice in the photo at the right have received a jellyfish gene for green fluorescent protein (GFP) that makes them glow green. Gene knockouts and replacements are also possible.

A major limitation of mouse genetics is its cost. Whereas working with a million individuals of *E. coli* or *S. cerevisiae* is a trivial matter, a million mice requires a factory-sized building. Furthermore, although mice do breed rapidly (compared with humans), they cannot compete with microorganisms for speedy life cycle. Hence, the large-scale selections and screens necessary to detect rare genetic events are not possible.

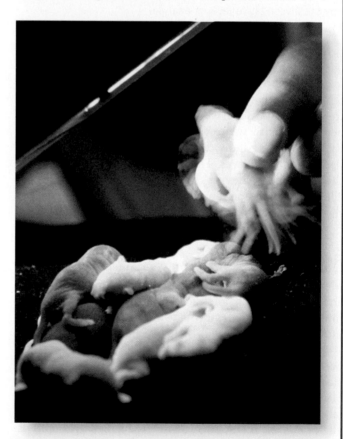

Green-glowing genetically modified mice embryos. The jellyfish gene for green fluorescent protein has been inserted into the chromosomes of the glowing mice. Normal mice are darker in the photo. [Kyodo News.]

Figure 6-14 Manx cat. A dominant allele causing taillessness is lethal in the homozygous state. The phenotype of two eye colors is unrelated to taillessness. [Gerard Lacz/NHPA.]

Geneticists commonly encounter situations in which expected phenotypic ratios are consistently skewed in one direction because one allele of a gene reduces viability. For example, in the cross $A/a \times a/a$, we predict a progeny ratio of 50 percent A/a and 50 percent a/a, but we might consistently observe a ratio such as 55 percent : 45 percent or 60 percent : 40 percent. In such a case, the recessive allele is said to be *sublethal* because the lethality is expressed in only some of and not all the homozygous individuals. Thus, lethality may range from 0 to 100 percent, depending on the gene itself, the rest of the genome, and the environment.

MESSAGE A gene can have several different states or forms—called *multiple alleles*. The alleles are said to constitute an allelic series, and the members of a series can show various degrees of dominance to one another.

The expected monohybrid ratio of 1 : 2 : 1 would be found among the zygotes, but it is altered to a 2 : 1 ratio in the progeny actually seen at birth because zygotes with a lethal A^Y/A^Y genotype do not survive to be counted. This hypothesis is supported by the removal of uteri from pregnant females of the yellow $\times$ yellow cross; one-fourth of the embryos are found to be dead.

The A^Y allele produces effects on two characters: coat color and survival. It is entirely possible, however, that both effects of the A^Y allele result from the same basic cause, which promotes yellowness of coat in a single dose and death in a double dose.

The tailless Manx phenotype in cats (Figure 6-14) also is produced by an allele that is lethal in the homozygous state. A single dose of the Manx allele, M^L, severely interferes with normal spinal development, resulting in the absence of a tail in the M^L/M heterozygote. But in the M^L/M^L homozygote, the double dose of the gene produces such an extreme abnormality in spinal development that the embryo does not survive.

Whether an allele is lethal or not often depends on the environment in which the organism develops. Whereas certain alleles are lethal in virtually any environment, others are viable in one environment but lethal in another. Human hereditary diseases provide some examples. Cystic fibrosis and sickle-cell anemia are diseases that would be lethal without treatment. Furthermore, many of the alleles favored and selected by animal and plant breeders would almost certainly be eliminated in nature as a result of competition with the members of the natural population. The dwarf mutant varieties of grain, which are very high-yielding, provide good examples; only careful nurturing by farmers has maintained such alleles for our benefit.

6.3 Interacting genes and proteins

One of the first indications that individual genes do not act alone was the observation in the early 1900s that a single gene mutation may have multiple effects. Even though the impact of a mutation is seen mainly in the character under investigation, many mutations produce effects in other characters. (We just saw an example in the yellow coat allele of mice. The allele affected both coast color and chances of survival.) Such multiple effects are called **pleiotropic effects.** Some pleiotropic effects can be subtle, but others quite strong. The interpretation is that pleiotropy is based on the complexity of gene interactions in the cell.

Let's reexamine PKU in light of the complexity of genes interacting to produce a particular phenotype. The simple model of PKU as a single-gene disease was extremely useful medically. Notably, it provided a successful treatment and therapy for the disease—simply restrict the dietary intake of phenylalanine. Many PKU patients have benefited from this treatment. However, there are some interesting complications that draw attention to the underlying complexity of the genetic system involved. For example, some cases of elevated phenylalanine level and its symptoms are associated not with the PAH locus, but with other genes. Also, some people who have PKU and its associated elevated phenylalanine level do not show abnormal cognitive development. These apparent exceptions to the model show that the expression of the symptoms of PKU depends not only on the PAH locus, but also on many other genes, and on the environment.

This complex situation is summarized in simple form in Figure 6-15. The figure shows that there are many steps in the pathway leading from the ingestion of phenylalanine to impaired cognitive development, and any one of them can show variation. First, the amount of phenylalanine in the diet is obviously of key importance. Then the phenylalanine must be transported into the appropriate sites in the liver, the "chemical factory" of the body. In the liver, PAH must act in concert with its cofactor, tetrahydrobiopterin. If excess phenylpyruvic acid is produced, to affect cognitive development it must be transported to the brain in the bloodstream, and then pass through the blood-brain barrier. Inside the brain, developmental processes must be susceptible to the detrimental action of phenylpyruvic acid. Each of these multiple steps is a possible site at which genetic or environmental variation may be found. Hence what seems to be a simple "monogenic" disease is actually de-

pendent on a complex set of processes. The example well illustrates the idea that individual genes do not "determine" phenotype. We also see how exceptions to the simple PKU model can be explained. For example, we see how mutations in genes other than PAH can cause elevated levels of phenylalanine—a gene needed for tetrahydrobiopterin synthesis is one such gene.

How experimental genetics dissects complexity

There is a standard genetic methodology for identifying the interacting genes that contribute to a particular biological property. Briefly, the approach is as follows:

Step 1. Treat cells with mutation-causing agents (mutagens), such as ultraviolet radiation. This treatment produces a large set of mutants with an abnormal expression of the property under study.

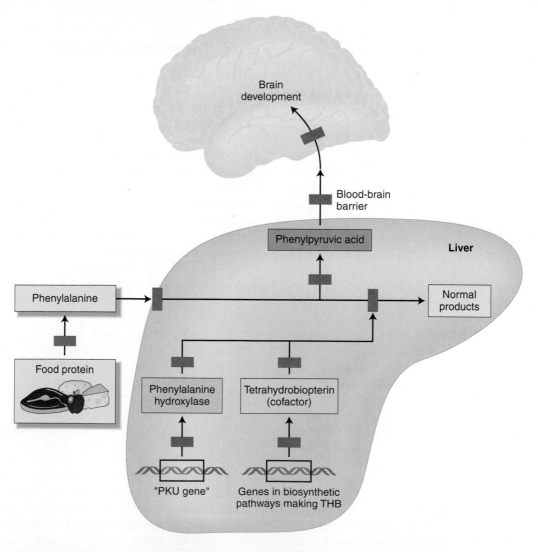

Figure 6-15 The determination of the disease PKU involves a complex series of steps.
The red rectangles indicate those steps where variation or blockage is possible.

Step 2. Test these mutants to determine how many gene loci are involved, and which mutations are alleles of the same gene.

Step 3. Combine the mutations pairwise by means of crosses to form **double mutants** to see if they interact. Remember, gene interaction implies that gene products interact in the cell. We will see that classes of progeny appear in specific ratios, as with the dihybrid crosses in Chapter 2. These ratios provide a clue to the type of gene interaction.

Hence there are three steps: the mutant "hunt"; tests for allelism; and tests for gene interaction. The techniques of mutant induction and selection are dealt with in Chapter 16. For the present, let's assume we have assembled a set of single-gene mutations affecting some biological property of interest. We now need to determine *how many* genes are represented in that set (step 2), and this is done using the **complementation test.**

Complementation

Let's illustrate the complementation test with an example from harebell plants (genus *Campanula*). The wild-type flower color of this plant is blue. Let's assume that by applying mutagenic radiation we have induced three white-petaled mutants and that they are available as homozygous pure-breeding strains. They all look the same, so we do not know a priori whether they are genetically identical or not. We can call the mutant strains $, £, and ¥, to avoid any symbolism using letters, which might imply dominance. When crossed with wild type, each mutant gives the same results in the F_1 and F_2, as follows:

white $ × blue ⟶ F_1, all blue ⟶ F_2, $\frac{3}{4}$ blue, $\frac{1}{4}$ white

white £ × blue ⟶ F_1, all blue ⟶ F_2, $\frac{3}{4}$ blue, $\frac{1}{4}$ white

white ¥ × blue ⟶ F_1, all blue ⟶ F_2, $\frac{3}{4}$ blue, $\frac{1}{4}$ white

A harebell plant (*Campanula* species). [Gregory G. Dimijian/Photo Researchers.]

In each case, the results show that the mutant condition is determined by the recessive allele of a single gene. However are they three alleles of *one* gene, or of two genes, or of three genes? The question can be answered by asking if the mutants *complement* one another. Complementation is defined as the production of a wild-type phenotype when two recessive mutant alleles are brought together in the same cell.

> **MESSAGE** Complementation is the production of a wild-type phenotype when two haploid genomes bearing different recessive mutations are united in the same cell.

In a diploid organism, the complementation test is performed by intercrossing two individuals that are homozygous for different recessive mutants. The next step is to observe whether the progeny have the wild-type phenotype. If recessive mutations are alleles of the same gene, they will *not* produce wild-type progeny, because the progeny are essentially homozygotes. Such alleles can be thought of generally as a' and a'', using primes to distinguish between two different mutant alleles of a gene whose wild-type allele is a^+. These alleles could have different mutant sites within the same gene, but they would both be nonfunctional. The heterozygote a'/a'' would be

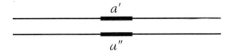

However, two recessive mutations in *different* genes would have wild-type function provided by the respective wild-type alleles. Here we can name the genes $a1$ and $a2$, after their mutant alleles. We can represent the heterozygotes as follows, depending on whether the genes are on the same or different chromosomes:

Different chromosomes

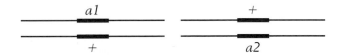

Same chromosome (shown in trans)

Let us return to the harebell example and intercross the mutants to test for complementation. Assume that the results of intercrossing mutants $, £, and ¥ are as follows:

white $ × white £ ⟶ F_1, all white

white $ × white ¥ ⟶ F_1, all blue

white £ × white ¥ ⟶ F_1, all blue

Figure 6-16 The molecular basis of genetic complementation. Three phenotypically identical white mutants—$, £, and ¥—are intercrossed. Mutations in the same gene (such as $ and £) cannot complement, because the F_1 has one gene with two mutant alleles. The pathway is blocked and the flowers are white. When the mutations are in different genes (such as £ and ¥), complementation of the wild-type alleles of each gene occurs in the F_1 heterozygote. Pigment is synthesized and the flowers are blue. (What would you predict to be the result of crossing $ and ¥?)

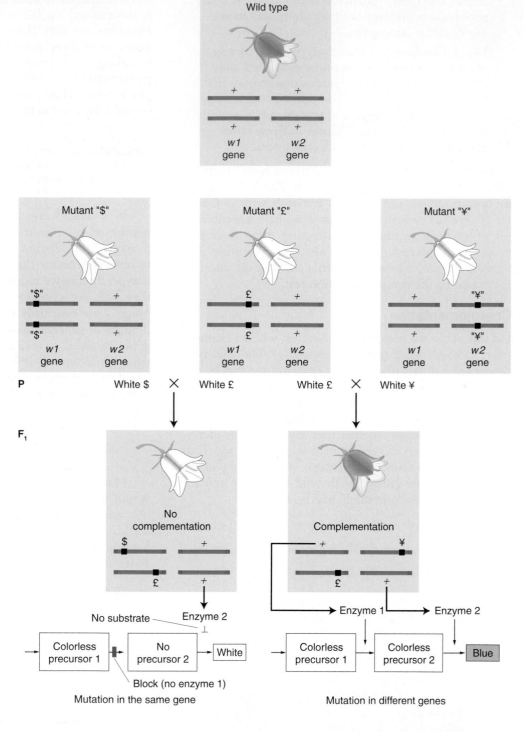

From this set of results, we can conclude that mutants $ and £ must be caused by alleles of one gene (say, *w1*) because they do not complement; but ¥ must be caused by a mutant allele of another gene (*w2*) because complementation is seen.

How does complementation work at the molecular level? The normal blue color of the harebell flower is caused by a blue pigment called *anthocyanin*. Pigments are chemicals that absorb certain parts of the visible

spectrum; in the case of the harebell, the anthocyanin absorbs all wavelengths except blue, which is reflected into the eye of the observer. However, this anthocyanin is made from chemical precursors that are not pigments; that is, they do not absorb light of any specific wavelength and simply reflect back the white light of the sun to the observer, giving a white appearance. The blue pigment is the end product of a series of biochemical conversions of nonpigments. Each step is catalyzed by a specific enzyme encoded by a specific gene. We can explain the results with a pathway as follows:

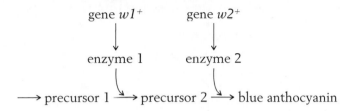

A homozygous mutation in either of the genes will lead to the accumulation of a precursor that will simply make the plant white. Now the mutant designations could be written as follows:

$$
\begin{array}{ll}
\$ & w1_\$/w1_\$ \cdot w2^+/w2^+ \\
£ & w1_£/w1_£ \cdot w2^+/w2^+ \\
¥ & w1^+/w1^+ \cdot w2_¥/w2_¥
\end{array}
$$

However, in practice, the subscript symbols would be dropped and the genotypes written as follows:

$$
\begin{array}{ll}
\$ & w1/w1 \cdot w2^+/w2^+ \\
£ & w1/w1 \cdot w2^+/w2^+ \\
¥ & w1^+/w1^+ \cdot w2/w2
\end{array}
$$

Hence an F_1 from $\$ \times £$ will be

$$
w1/w1 \cdot w2^+/w2^+
$$

These F_1 individuals will thus have two defective alleles for $w1$ and will therefore be blocked at step 1. Even though enzyme 2 is fully functional, it has no substrate to act on, so no blue pigment will be produced and the phenotype will be white.

The F_1's from the other crosses, however, will have the wild-type alleles for both the enzymes needed to take the intermediates to the final blue product. Their genotypes will be

$$
w1^+/w1 \cdot w2^+/w2
$$

Hence we see that complementation is actually a result of the cooperative interaction of the *wild-type* alleles of the two genes. Figure 6-16 is a summary diagram of the interaction of the complementing and noncomplementing white mutants.

In a haploid organism, the complementation test cannot be performed by intercrossing. In fungi, there is an alternative way to bring mutant alleles together to test complementation: one makes a **heterokaryon** (Figure 6-17). Fungal cells fuse readily. When two different strains fuse, the haploid nuclei from the different strains occupy one cell, which is called a *heterokaryon* (Greek; "different kernels"). The nuclei in a heterokaryon generally do not fuse. In one sense, this condition is a "mimic" diploid.

Assume that in different strains there are mutations in two different genes conferring the same mutant phenotype—for example, an arginine requirement. We can call these genes *arg-1* and *arg-2*. The genotypes of the two strains can be represented as *arg-1* · *arg-2*⁺ and *arg-1*⁺ · *arg-2*. These two strains can be fused to form a heterokaryon with the two nuclei in a common cytoplasm:

$$
\text{Nucleus 1 is } arg\text{-}1 \cdot arg\text{-}2^+
$$
$$
\text{Nucleus 2 is } arg\text{-}1^+ \cdot arg\text{-}2
$$

Because gene products are made in a common cytoplasm, the two wild-type alleles can exert their dominant effect and cooperate to produce a heterokaryon of wild-type phenotype. In other words, the two mutations complement, just as they would in a diploid. If the mutations had been alleles of the same gene, there would have been no complementation.

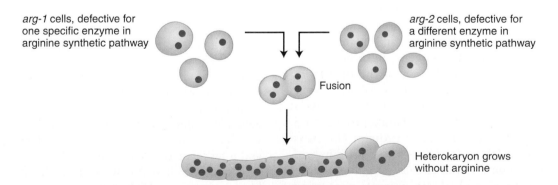

Figure 6-17 Formation of a heterokaryon of *Neurospora* mimics a diploid state. When vegetative cells fuse, haploid nuclei share the same cytoplasm in a heterokaryon. In this example, haploid nuclei with mutations in different genes in the arginine synthetic pathway complement to produce a *Neurospora* that no longer requires arginine.

MESSAGE When two independently derived recessive mutant alleles producing similar recessive phenotypes fail to complement, the alleles must be of the same gene.

A modified F_2 ratio can be useful in supporting a hypothesis of gene complementation. As an example, let's consider what F_2 ratio will result from crossing the dihybrid F_1 harebell plants. The F_2 shows both blue and white plants in a ratio of 9:7. How can these results be explained? The 9:7 ratio is clearly a modification of the dihybrid 9:3:3:1 ratio with the 3:3:1 combined to make 7. The cross of the two white lines and subsequent generations can be represented as follows:

$$w1/w1 \; ; \; w2^+/w2^+ \text{ (white)} \; \times \; w1^+/w1^+ \; ; \; w2/w2 \text{ (white)}$$

$$\downarrow$$

$$F_1 \quad w1^+/w1 \; ; \; w2^+/w2 \text{ (blue)}$$
$$w1^+/w1 \; ; \; w2^+/w2 \; \times \; w1^+/w1 \; ; \; w2^+/w2$$

$$\downarrow$$

$$F_2 \quad \begin{array}{ll} 9 \; w1^+/- \; ; \; w2^+/- \text{ (blue)} & 9 \\ 3 \; w1^+/- \; ; \; w2/w2 \text{ (white)} \\ 3 \; w1/w1 \; ; \; w2^+/- \text{ (white)} & \left.\right\} \; 7 \\ 1 \; w1/w1 \; ; \; w2/w2 \text{ (white)} \end{array}$$

The results show that a plant will have white petals if it is homozygous for the recessive mutant allele of *either* gene or *both* genes. To have the blue phenotype, a plant must have at least one of the dominant allele of both genes because both are needed to complement each other and complete the sequential steps in the pathway. Thus, three of the genotypic classes will produce the same phenotype, so overall only two phenotypes result.

The example of complementation in harebells involved different steps in a biochemical pathway. Similar results can come from gene regulation. A regulatory gene often functions by producing a protein that binds to a regulatory site upstream of the target gene, facilitating the transcription of the gene by RNA polymerase (Figure 6-18). In the absence of the regulatory protein, the target gene would be transcribed at very low levels, inadequate for cellular needs. Let's cross a pure line r/r defective for the regulatory protein to a pure line a/a defective for the target protein. The cross is $r/r \; ; \; a^+/a^+ \; \times \; r^+/r^+ \; ; \; a/a$. The $r^+/r \; ; \; a^+/a$ dihybrid will show complementation between the mutant genotypes because both r^+ and a^+ are present, permitting normal transcription of the wild-type allele. When selfed the F_1 dihybrid will also result in a 9:7 phenotypic ratio in the F_2:

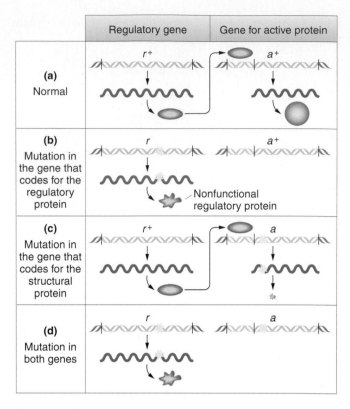

Regulatory gene	Gene for active protein
(a) Normal	r^+	a^+
(b) Mutation in the gene that codes for the regulatory protein	r — Nonfunctional regulatory protein	a^+
(c) Mutation in the gene that codes for the structural protein	r^+	a
(d) Mutation in both genes	r	a

Figure 6-18 **Interaction between a regulating gene and its target.** The r^+ gene codes for a regulatory protein, and the a^+ gene codes for a structural protein. Both must be normal for a functional ("active") structural protein to be synthesized.

Proportion	Genotype	Functional a^+ protein	Ratio
$\frac{9}{16}$	$r^+/- \; ; \; a^+/-$	Yes	9
$\frac{3}{16}$	$r^+/- \; ; \; a/a$	No	
$\frac{3}{16}$	$r/r \; ; \; a^+/-$	No	7
$\frac{1}{16}$	$r/r \; ; \; a/a$	No	

An example of interacting genes from different pathways is the inheritance of skin coloration in corn snakes. The snake's natural color is a repeating black-and-orange camouflage pattern, as shown in Figure 6-19a. The phenotype is produced by two separate pigments, both of which are under genetic control. One gene determines the orange pigment, and the alleles that we shall consider are o^+ (presence of orange pigment) and o (absence of orange pigment). Another gene determines the black pigment, with alleles b^+ (presence of black pigment) and b (absence of black pigment). These two genes are unlinked. The natural pattern is produced by the genotype $o^+/- \; ; \; b^+/-$. A snake that is $o/o \; ; \; b^+/-$ is black because it lacks orange pigment (Figure 6-19b),

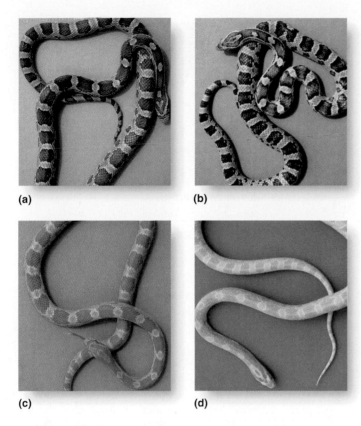

(a)   **(b)**

(c)   **(d)**

Figure 6-19 Skin pigmentation patterns in the corn snake. Combinations of orange and black pigments determine the four phenotypes shown. (a) A wild-type black and orange camouflaged snake synthesizes both black and orange pigments. (b) A black snake does not synthesize orange pigment. (c) An orange snake does not synthesize black pigment. (d) An albino snake synthesizes neither black nor orange pigments. [Anthony Griffiths.]

and a snake that is $o^+/-$; b/b is orange because it lacks the black pigment (Figure 6-19c). The double homozygous recessive o/o ; b/b is albino (Figure 6-19d). Notice, however, that the faint pink color of the albino is from yet another pigment, the hemoglobin of the blood that is visible through this snake's skin when the other pigments are absent. The albino snake also clearly shows that there is another element to the skin-pigmentation pattern in addition to pigment: the repeating motif in and around which pigment is deposited.

If a homozygous orange and a homozygous black snake are crossed, the F_1 is wild type (camouflaged), demonstrating complementation:

female o^+/o^+ ; b/b × male o/o ; b^+/b^+
(orange) | (black)

F_1 o^+/o ; b^+/b
(camouflaged)

Here, however, an F_2 shows a standard $9:3:3:1$ ratio:

female o^+/o ; b^+/b × male o^+/o ; b^+/b
(camouflaged) | (camouflaged)

F_2 9 $o^+/-$; $b^+/-$ (camouflaged)
            3 $o^+/-$; b/b (orange)
            3 o/o ; $b^+/-$ (black)
            1 o/o ; b/b (albino)

The $9:3:3:1$ ratio is produced because of the independence of the interacting genes at the level of cellular action:

$$\text{precursor} \xrightarrow{b^+} \text{black pigment} \left.\vphantom{\begin{array}{c}a\\a\end{array}}\right\} \text{camouflaged}$$
$$\text{precursor} \xrightarrow{o^+} \text{orange pigment}$$

Epistasis

In trying to find evidence of gene interaction, one approach is to look for cases of a type of gene interaction called **epistasis.** This word means "stand upon," referring to the ability of a mutation at one locus to override a mutation at another in a double mutant. The overriding mutation is called *epistatic,* whereas the overridden one is *hypostatic.* Epistasis results from genes being in the same cellular pathway. In the case of a simple biochemical pathway, the epistatic mutation is of a gene that is farther "upstream" (earlier in the pathway) than that of the hypostatic. The mutant phenotype of the upstream gene takes precedence, no matter what is going on later in the pathway.

Epistasis is difficult to screen for in any way other than combining candidate mutations pairwise to form double mutants. How is the double mutant identified? In fungi, tetrad analysis is useful. For example an ascus containing half its products as wild type must contain double mutants. Consider the cross

$$a \cdot b^+ \times a^+ \cdot b$$

A tetrad showing chance cosegregation of a and b (a nonparental ditype ascus) would show the following phenotypes:

wild type	$a^+ \cdot b^+$
wild type	$a^+ \cdot b^+$
double mutant	$a \cdot b$
double mutant	$a \cdot b$

Hence the double mutant must be the non-wild-type genotype.

In diploids, the double mutant is more difficult to identify, but ratios can help us identify epistasis. Let's look at an example using petal pigment synthesis in the plant blue-eyed Mary *(Collinsia parviflora)*. From the blue wild type, we'll start with two pure mutant lines, one with white (w/w) and one with magenta petals (m/m). The w and m genes are not linked. The F_1 and F_2 are as follows:

$$w/w \; ; \; m^+/m^+ \text{ (white) } \times \; w^+/w^+ \; ; \; m/m \text{ (magenta)}$$
$$F_1 \qquad\qquad w^+/w \; ; \; m^+/m \qquad\qquad \text{(blue)}$$

$$w^+/w \; ; \; m^+/m \times w^+/w \; ; \; m^+/m$$

$$
\begin{array}{lll}
F_2 & 9 \; w^+/- \; ; \; m^+/- \text{ (blue)} & 9 \\
 & 3 \; w^+/- \; ; \; m/m \text{ (magenta)} & 3 \\
 & 3 \; w/w \; ; \; m^+/- \text{ (white)} & \\
 & 1 \; w/w \; ; \; m/m \text{ (white)} & \} \; 4
\end{array}
$$

Complementation results in a wild-type F_1. However, in the F_2, a $9:3:4$ phenotypic ratio is produced. This ratio is the mark of epistasis: the ratio tells us that the double mutant must be white, so white must be epistatic to magenta. To find the double mutant for subsequent study, white F_2 individuals would have to be testcrossed.

At the cellular level, we can account for the epistasis by the following type of pathway (see also Figure 6-20).

$$\text{colorless} \xrightarrow{\text{gene } w^+} \text{magenta} \xrightarrow{\text{gene } m^+} \text{blue}$$

> **MESSAGE** Epistasis is inferred when a mutant allele of one gene masks expression of alleles of another gene and expresses its own phenotype instead.

Another case of recessive epistasis is the yellow coat color of some Labrador retriever dogs. Two alleles, B and b, stand for black and brown coats, respectively. The two alleles produce black and brown melanin. The allele e of another gene is epistatic on these alleles, giving a yellow coat (Figure 6-21). Therefore the genotypes $B/-$; e/e and b/b ; e/e both produce a yellow phenotype, whereas $B/-$; $E/-$ and b/b ; $E/-$ are black and brown, respectively. This case of epistasis is *not* caused by an upstream block in a pathway leading to dark pigment. Yellow dogs can make black or brown pigment, as can be seen in their noses and lips. The action of the allele e is to prevent deposition of the pigment in hairs. In this case, the epistatic gene is *developmentally downstream*; it represents a kind of developmental target that has to be of E genotype before pigment can be deposited.

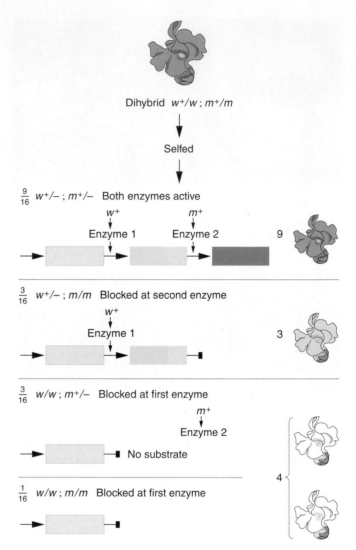

Figure 6-20 A molecular mechanism for recessive epistasis. Wild-type alleles of two genes (w^+ and m^+) encode enzymes catalyzing successive steps in the synthesis of a blue petal pigment. Homozygous m/m plants produce magenta flowers and homozygous w/w plants produce white flowers. The double mutant w/w ; m/m also produces white flowers, indicating that white is epistatic to magenta.

> **MESSAGE** Epistasis points to interaction of genes in some biochemical or developmental sequence.

Suppressors

A type of gene interaction that can be detected more easily is suppression. A suppressor is a mutant allele of one gene that reverses the effect of a mutation of another gene, resulting in a wild-type or near wild-type phenotype. For example, assume that an allele a^+ produces the normal phenotype, whereas a recessive mutant allele a results in abnormality. A recessive mutant

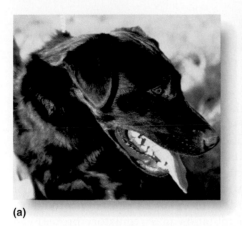

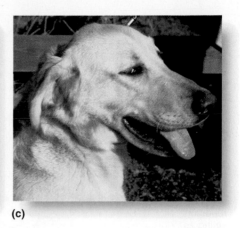

(a)   **(b)**   **(c)**

Figure 6-21 Coat-color inheritance in Labrador retrievers. Two alleles *B* and *b* of a pigment gene determine (a) black and (b) brown, respectively. At a separate gene, *E* allows color deposition in the coat, and *e/e* prevents deposition, resulting in (c) the gold phenotype. This is a case of recessive epistasis. [Anthony Griffiths.]

allele *s* at another gene suppresses the effect of *a*, so that the genotype *a/a · s/s* will have wild-type (*a*⁺-like) phenotype. Suppressor alleles sometimes have no effect in the absence of the other mutation; in such a case, the phenotype of $a^+/a^+ · s/s$ would be wild type. In other cases, the suppressor allele produces its own abnormal phenotype.

Again, a genetic interaction (suppression) implies that gene products interact somehow, as well, so suppressors are useful in revealing such interactions. Screening for suppressors is quite straightforward. Start with a mutant in some process of interest, expose this mutant to mutation-causing agents such as high-energy radiation, and screen the descendants for wild types. In haploids such as fungi, screening is accomplished by simply plating mutagenized cells and looking for colonies with wild-type phenotypes. Many wild types arising in this way are merely reversals of the original mutational event; these are called **revertants**. However, many will be double mutants, in which one of the mutations is a suppressor. Revertant and suppressed states can be distinguished by appropriate crossing. For example in yeast, the two results would be distinguished as follows:

Revertant a^+ × standard wild type a^+
Progeny   all a^+

Suppressed mutant $a · s$ × standard wild type $a^+ · s^+$
Progeny $a^+ · s^+$ wild type
           $a^+ · s$ wild type
           $a^+ · s^+$ original mutant
           $a · s$ wild type (suppressed)

The appearance of the original mutant phenotype identifies the parent as a suppressed mutant.

In diploids, suppressors produce specific F_2 ratios, which are useful in confirming suppression. Let's look at a real-life example from *Drosophila*. The recessive allele *pd* will result in purple eye color when unsuppressed. A recessive allele *su* has no detectable phenotype itself, but suppresses the unlinked recessive allele *pd*. Hence *pd/pd ; su/su* is wild type in appearance and has red eyes. The following analysis illustrates the inheritance pattern. A homozygous purple-eyed fly is crossed to a homozygous red-eyed stock carrying the suppressor.

$pd/pd ; su^+/su^+$ (purple) × $pd^+/pd^+ ; su/su$ (red)

F_1 all $pd^+/pd ; su^+/su$ (red)
$pd^+/pd ; su^+/su$ (red) × $pd^+/pd ; su^+/su$ (red)

F_2 9 $pd^+/-$; $su^+/-$ red
         3 $pd^+/-$; su/su red   } 13
         1 pd/pd ; su/su red
         3 pd/pd ; $su^+/-$ purple   3

The overall ratio in the F_2 is 13 red:3 purple. This ratio is characteristic of a recessive suppressor acting on a recessive mutation. Both recessive and dominant suppressors are found, and they can act on recessive or dominant mutations. These possibilities result in a variety of different phenotypic ratios.

Suppression is sometimes confused with epistasis. However, the key difference is that a suppressor cancels the expression of a mutant allele and restores the corresponding wild-type phenotype. The modified ratio is an indicator of this type of interaction. Furthermore, often only two phenotypes segregate (as in the preceding examples), not three, as in epistasis.

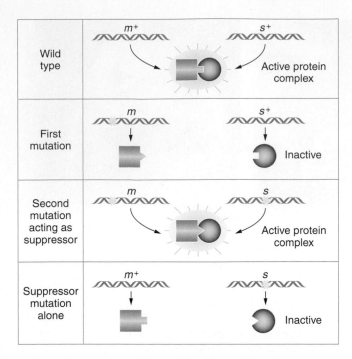

Figure 6-22 A molecular mechanism for suppression.

How do suppressors work at the molecular level? There are many possible mechanisms. A particularly useful type of suppression is based on the physical binding of gene products in the cell, for example protein-protein binding, as in the case of protein machines. Two proteins normally fit together to provide some type of cellular function. When a mutation causes a shape change in one protein, it no longer fits together with the other, and hence the function is lost (Figure 6-22). However, a suppressor mutation that causes a compensatory shape change in the second protein can restore fit and hence normal function. In this way, the interacting proteins of cellular machines can be pieced together.

Alternatively, in situations in which a mutation causes a block in a metabolic pathway, the suppressor finds some way of bypassing the block—for example, by rerouting into the blocked pathway intermediates similar to those beyond the block. In the following example, the suppressor provides an intermediate B to circumvent the block.

No suppressor

$$A \longrightarrow \!\!\!\!\times\!\!\!\!X \longrightarrow product$$

With suppressor

$$A \longrightarrow \!\!\!\!\times B \longrightarrow product$$
$$B$$

MESSAGE Suppressors cancel the expression of a mutant allele of another gene, resulting in normal wild-type phenotype.

Modifiers

As the name suggests, a modifier mutation at a second locus changes the expression of a mutated gene at a first. Regulatory genes provide a simple illustration. As we saw in an earlier example, regulatory proteins bind to the sequence of the DNA upstream to the start site for transcription. These proteins regulate the level of transcription. In the discussion of complementation we considered a null mutation of a regulatory gene that almost completely prevented transcription. However, some regulatory mutations change the level of transcription of the target gene so that more or less protein is produced. In other words, a mutation in a regulatory protein can "downregulate" or "upregulate" the transcribed gene. Let's look at an example using a downregulating regulatory mutation b, affecting a gene A in a fungus such as yeast. We make a cross of a leaky mutation a, to the regulatory mutation b:

leaky mutant $a \cdot b^+$ $\times$ inefficient regulator $a^+ \cdot b$

Progeny	Phenotype
$a^+ \cdot b^+$	wild type
$a^+ \cdot b$	defective (low transcription)
$a \cdot b^+$	defective (defective protein A)
$a \cdot b$	extremely defective (low transcription of defective protein)

Hence the action of the modifier is seen in the appearance of two grades of mutant phenotypes *within* the *a* progeny.

Synthetic lethals

In some cases, when two viable single mutants are intercrossed, the resulting double mutants are lethal. These **synthetic lethals** can be considered a special category of gene interaction. They can point to specific types of interactions of gene products. For instance, genome analysis has revealed that evolution has produced many duplicate systems within the cell. One advantage of these duplicates might be to provide "backups." If there are null mutations in genes in both duplicate systems, then a faulty system will have no backup, and the individual will lack essential function and die. In another instance, a leaky mutation in one step of a pathway may cause the pathway to slow down, but leave enough function for life. However, if double mutants combine, each with a leaky mutation in a different step, the whole pathway grinds to a halt. One version of the latter interaction is two mutations in a protein machine, as shown in Figure 6-23.

Building a protein machine is partly a matter of constituent proteins finding each other by random molecu-

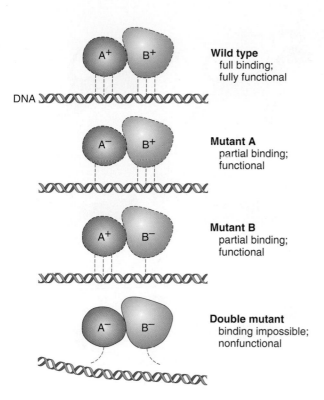

Figure 6-23 One genetic mechanism for the origin of synthetic lethality. Two interacting proteins perform some essential function on some substrate such as DNA, but must first bind to it. Reduced binding of either protein allows some functions to remain, but reduced binding of both is lethal.

lar motion and binding through complementary shape. However, some steps in assembly require energy and enzymes. Such an example is shown in Figure 6-24. Any of these interacting components, whether machine components or associated enzymes, can be dissected by analysis of synthetic lethals.

Penetrance and expressivity

In the preceding examples, the dependence of one gene on another is deduced from clear genetic ratios. In such cases, we can use the phenotype to distinguish mutant and wild-type genotypes with 100 percent certainty. In these cases, we say that the mutation is 100 percent penetrant. However, many mutations show *incomplete* penetrance: not every individual with the genotype expresses the corresponding phenotype. Thus **penetrance** is defined as the percentage of *individuals* with a given allele who exhibit the phenotype associated with that allele.

Why would an organism have a particular genotype and yet not express the corresponding phenotype? There are several possible reasons:

1. The influence of the environment. We saw in Chapter 1 that individuals with the same genotype

may show a range of phenotypes depending on the environment. It is possible that the range of phenotypes for mutant and wild-type individuals will overlap: the phenotype of a mutant individual raised in one set of circumstances may match the phenotype of a wild-type individual raised in a separate set of circumstances. Should this happen, it becomes impossible to distinguish mutant from wild type.

2. The influence of other genes. Modifiers, epistatic genes, or suppressors in the rest of the genome may act to prevent the expression of the typical phenotype.

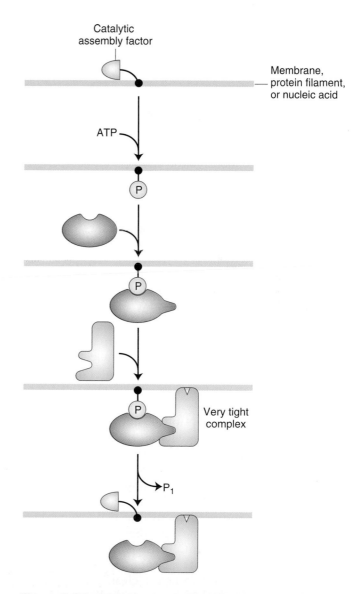

Figure 6-24 Building a protein machine. Phosphorylation activates an assembly factor, enabling protein machines to be assembled in situ on a membrane, filament, or nucleic acid. [B. Alberts, "The Cell as a Collection of Protein Machines," *Cell*, 92, 1998, 291–294.]

Phenotypic expression
(each oval represents an individual)

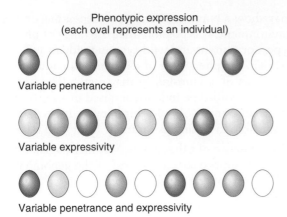

Variable penetrance

Variable expressivity

Variable penetrance and expressivity

Figure 6-25 Pigment intensity as an example of penetrance and expressivity. Assume that all the individuals shown have the same pigment allele (*P*) and possess the same potential to produce pigment. Effects from the rest of the genome and the environment may suppress or modify pigment production in any one individual. The color reflects the level of expression.

3. The subtlety of the mutant phenotype. The subtle effects brought about by the absence of a gene function may be difficult to measure in a laboratory situation.

Another measure for describing the range of phenotypic expression is called **expressivity**. Expressivity measures the degree to which a given allele is expressed at the phenotypic level; that is, expressivity measures the intensity of the phenotype. For example, "brown" animals (genotype *b/b*) from different stocks might show very different intensities of brown pigment from light to dark. Different degrees of expression in different individuals may be due to variation in the allelic constitution of the rest of the genome or to environmental factors. Figure 6-25 illustrates the distinction between penetrance and expressivity. Like penetrance, expressivity is integral to the concept of the norm of reaction. An example of variable expressivity in dogs is found in Figure 6-26.

The phenomena of incomplete penetrance and variable expressivity can make any kind of genetic analysis substantially more difficult, including human pedigree analysis and predictions in genetic counseling. For example, it is often the case that a disease-causing allele is not fully penetrant. Thus someone could have the allele but not show any signs of the disease. If that is the case, it is difficult to give a clean genetic bill of health to any individual in a disease pedigree (for example, individual *R* in Figure 6-27). On the other hand, pedigree analysis can sometimes identify individuals who do not express but almost certainly do have a disease genotype (for example, individual *Q* in Figure 6-27).

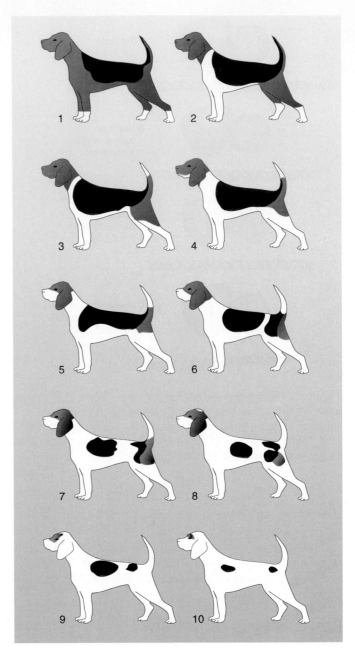

Figure 6-26 Variable expressivity shown by 10 grades of piebald spotting in beagles. Each of these dogs has S^P, the allele responsible for piebald spots in dogs. [After Clarence C. Little, *The Inheritance of Coat Color in Dogs.* Cornell University Press, 1957; and Giorgio Schreiber, *Journal of Heredity* 9, 1930, 403.]

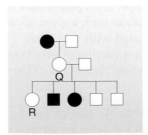

Figure 6-27 Pedigree for a dominant allele that is not fully penetrant. Individual Q does not display the phenotype but passed the dominant allele to at least two progeny. Since the allele is not fully penetrant, the other progeny (for example, R) may or may not have inherited the dominant allele.

Similarly, variable expressivity can complicate counseling because individuals with low expressivity might be misdiagnosed.

> **MESSAGE** The terms *penetrance* and *expressivity* quantify the modification of gene expression by varying environment and genetic background; they measure, respectively, the percentage of cases in which the gene is expressed and the level of expression.

6.4 Applications of chi-square (χ^2) test to gene interaction ratios

Often, the observed ratios of some specific gene interaction don't conform precisely to those expected. The geneticist uses the χ^2 test to decide if the ratios are close enough to the expected results to indicate the presence of the suspected interaction. Let's look at an example. We cross two pure lines of plants, one with yellow petals and one with red. The F_1 are all orange. When the F_1 is selfed to give an F_2, we find the following result:

orange	182
yellow	61
red	77
Total	320

What hypothesis can we invent to explain the results? There are at least two possibilities:

Hypothesis 1: Incomplete dominance

$$G^1/G^1 \text{ (yellow)} \times G^2/G^2 \text{ (red)}$$

$$\downarrow$$

F_1 $\qquad\qquad G^1/G^2$ (orange)

		Expected numbers
F_2	$\frac{1}{4}$ G^1/G^1 (yellow)	80
	$\frac{1}{2}$ G^1/G^2 (orange)	160
	$\frac{1}{4}$ G^2/G^2 (red)	80

Hypothesis 2: Recessive epistasis of r (red) on Y (orange) and y (yellow)

$$y/y \; ; \; R/R \text{ (yellow)} \times Y/Y \; ; \; r/r \text{ (red)}$$

$$\downarrow$$

F_1 $\qquad\qquad Y/y \; ; \; R/r$ (orange)

		Expected numbers
F_2	$\frac{9}{16}$ $Y/-$; $R/-$ (yellow)	180
	$\frac{3}{16}$ y/y ; $R/-$ (orange)	60
	$\frac{3}{16}$ $Y/-$; r/r (red)	
	$\frac{1}{16}$ y/y ; r/r (red)	80

Recall that the general formula for calculating χ^2 is

$$\chi^2 = \Sigma(O - E)^2/E \text{ for all classes}$$

For hypothesis 1, the calculation is as follows:

	O	E	$(O - E)^2$	$(O - E)^2/E$
orange	182	160	484	3.0
yellow	61	80	361	4.5
red	77	80	9	0.1
				$\chi^2 = 7.6$

To convert the χ^2 value into a probability, we use Table 2-2, page 41, which shows χ^2 values for different degrees of freedom (df). In this case, there are two degrees of freedom. Looking along the 2-df line, we find that the χ^2 value places the probability at less than .025, or 2.5 percent. This means that, if the hypothesis is true, then deviations from expectations this large or larger are expected approximately 2.5 percent of the time. As mentioned earlier, by convention the 5 percent level is used as the cutoff line. When values of less than 5 percent are obtained, the hypothesis is rejected as being too unlikely. Hence the incomplete dominance hypothesis must be rejected.

For hypothesis 2, the calculation is set up as follows:

	O	E	$(O - E)^2$	$(O - E)^2/E$
orange	182	180	4	0.02
yellow	61	60	1	0.02
red	77	80	9	0.11
				$\chi^2 = 0.15$

The probability value (for 2 df) this time is greater than .9, or 90 percent. Hence a deviation this large or larger is expected approximately 90 percent of the time—in other words, very frequently. Formally, because 90 percent is greater than 5 percent, we conclude that the results uphold the hypothesis of recessive epistasis.

KEY QUESTIONS REVISITED

- **How do individual genes exert their effect on an organism's makeup?**

Each gene is part of a set of genes needed to produce a certain property during development. These genes interact with the environment, which provides signals, nutrients, and various other necessary conditions.

- **In the cell, do genes act directly or through some sort of gene product?**

Through their gene products.

- **What is the nature of gene products?**

For most genes the product is a polypeptide (a single-chain protein), but some genes have functional RNA as their final product, and this RNA is never translated into protein.

- **What do gene products do?**

They control cellular chemistry. The best example is enzymes, which catalyze reactions that would otherwise occur far too slowly.

- **Is it correct to say that an allele of a gene determines a specific phenotype?**

No, a gene is but a single part of the set of genes necessary to produce a particular phenotype. Variants of a single gene can produce phenotypic variants, but even here they rely on all the other genes and on the environment.

- **In what way or ways do genes interact at the cellular level?**

Some important ways are by producing separate components for the same pathway, by regulating one another, and by producing components of multimolecular assemblies ("machines").

- **How is it possible to dissect complex gene interaction using a mutational approach?**

Mutants affecting one specific character of interest are assembled, grouped into genes using the complementation test, and then paired to produce double mutation, which might reveal interactive effects. Suppressors can be screened directly.

SUMMARY

Genes act through their products, in most cases proteins, but in some cases functional (and untranslated) RNA. Mutations in genes may alter the function of these products, producing a change in phenotype. Mutations are changes in the DNA sequence of a gene. These can be of a variety of types and occupy many different positions, resulting in multiple alleles. Recessive mutations are often a result of haplosufficiency of the wild-type allele, whereas dominant mutations are often the result of the haploinsufficiency of the wild type. Some homozygous mutations cause severe effects or even death (lethal mutations).

Although it is possible by genetic analysis to isolate a single gene whose alleles dictate two alternative phenotypes for one character, this gene does not control that character by itself; the gene must interact with many other genes in the genome and the environment. Genetic dissection of complexity begins by amassing mutants affecting a character of interest. The complementation test decides if two separate recessive mutations are of one gene or of two different genes. The mutant genotypes are brought together in an F_1 individual, and if the phenotype is mutant, then no complementation has occurred and the two alleles must be of the same gene. If complementation is observed, the alleles must be of different genes.

The interaction of different genes can be detected by testing double mutants, because allele interaction implies interaction of gene products at the functional level. Some key types of interaction are epistasis, suppression, and synthetic lethality. Epistasis is the replacement of a mutant phenotype produced by one mutation with a mutant phenotype produced by mutation of another gene. The observation of epistasis suggests a common pathway. A suppressor is a mutation of one gene that can restore wild-type phenotype to a mutation at another gene. Suppressors often reveal physically interacting proteins or nucleic acids. Some combinations of viable mutants are lethal, a result known as *synthetic lethality*. Synthetic lethals can reveal a variety of interactions, depending on the nature of the mutations.

The different types of gene interactions produce F_2 dihybrid ratios that are modifications of the standard $9:3:3:1$. For example, recessive epistasis results in a $9:3:4$ ratio.

An observed modified phenotypic ratio can be assessed against that expected from a specific hypothesis of gene interaction by using the chi-square test.

KEY TERMS

allelic series (p. 192)

codominant (p. 194)

complementation test (p. 199)

double mutants (p. 199)

epistasis (p. 203)

expressivity (p. 208)

full, or complete, dominance (p. 193)

functional RNAs (p. 190)

haploinsufficient (p. 192)

haplosufficient (p. 192)

heterokaryon (p. 201)

incomplete dominance (p. 194)

leaky (p. 192)

lethal allele (p. 195)

multiple alleles (p. 192)

nulls (p. 192)

one-gene–one-polypeptide hypothesis (p. 189)

one-gene–one protein hypothesis (p. 189)

penetrance (p. 207)

permissive temperature (p. 195)

pleiotropic effects (p. 197)

restrictive temperature (p. 195)

revertants (p. 205)

synthetic lethal (p. 206)

temperature-sensitive (ts) mutations (p. 195)

SOLVED PROBLEMS

1. Most pedigrees show polydactyly (see Figure 2-18) to be inherited as a rare autosomal dominant, but the pedigrees of some families do not fully conform to the patterns expected for such inheritance. Such a pedigree is shown below. (The unshaded diamonds stand for the specified number of unaffected persons of unknown sex.)

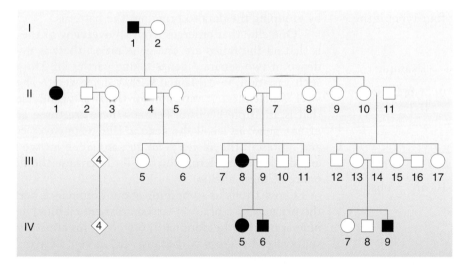

a. What irregularity does this pedigree show?

b. What genetic phenomenon does this pedigree illustrate?

c. Suggest a specific gene interaction mechanism that could produce such a pedigree, showing genotypes of pertinent family members.

Solution

a. The normal expectation for an autosomal dominant is for each affected individual to have an affected parent, but this expectation is not seen in this pedigree, which constitutes the irregularity. What are some possible explanations?

Could some cases of polydactyly be caused by a different gene, one that is an X-linked dominant? This suggestion is not useful, because we still have to explain the absence of the condition in persons II-6 and II-10. Furthermore, postulating recessive inheritance, whether autosomal or sex-linked, requires many people in the pedigree to be heterozygotes, which is inappropriate because polydactyly is a rare condition.

b. Thus we are left with the conclusion that polydactyly must sometimes be incompletely penetrant. We have learned in this chapter that some individuals who have the genotype for a particular phenotype do not express it. In this pedigree, II-6 and II-10 seem to belong in this category; they must carry the polydactyly gene inherited from I-1 because they transmit it to their progeny.

c. We have seen in the chapter that environmental suppression of gene expression can cause incomplete penetrance, as can suppression by another gene. To give the requested genetic explanation, we must come up with a genetic hypothesis. What do we need to explain? The key is that I-1 passes the gene on to two types of progeny, represented by II-1, who expresses the gene, and by II-6 and II-10, who do not. (From the pedigree, we cannot tell whether the other children of I-1 have the gene.) Is genetic suppression at work? I-1 does not have a suppressor allele, because he expresses polydactyly. So the only person from whom a suppressor could come is I-2. Furthermore, I-2 must be heterozygous for the suppressor gene because at least one of her children does express polydactyly. We have thus formulated the hypothesis that the mating in generation I must have been

$$(I\text{-}1)\ P/p \cdot s/s \times (I\text{-}2)\ p/p \cdot S/s$$

where S is the suppressor and P is the allele responsible for polydactyly. From this hypothesis, we predict that the progeny will comprise the following four types if the genes assort:

Genotype	Phenotype	Example
$P/p \cdot S/s$	normal (suppressed)	II-6, II-10
$P/p \cdot s/s$	polydactylous	II-1
$p/p \cdot S/s$	normal	
$p/p \cdot s/s$	normal	

If S is rare, the matings of II-6 and II-10 are probably giving

Progeny genotype	Example
$P/p \cdot S/s$	III-13
$P/p \cdot s/s$	III-8
$p/p \cdot S/s$	
$p/p \cdot s/s$	

We cannot rule out the possibilities that II-2 and II-4 have the genotype $P/p \cdot S/s$ and that by chance none of their descendants are affected.

2. Beetles of a certain species may have green, blue, or turquoise wing covers. Virgin beetles were selected from a polymorphic laboratory population and mated to determine the inheritance of wing-cover color. The crosses and results were as given in the following table:

Cross	Parents	Progeny
1	blue × green	all blue
2	blue × blue	$\frac{3}{4}$ blue : $\frac{1}{4}$ turquoise
3	green × green	$\frac{3}{4}$ green : $\frac{1}{4}$ turquoise
4	blue × turquoise	$\frac{1}{2}$ blue : $\frac{1}{2}$ turquoise
5	blue × blue	$\frac{3}{4}$ blue : $\frac{1}{4}$ green
6	blue × green	$\frac{1}{2}$ blue : $\frac{1}{2}$ green
7	blue × green	$\frac{1}{2}$ blue : $\frac{1}{4}$ green $\frac{1}{4}$ turquoise
8	turquoise × turquoise	all turquoise

a. Deduce the genetic basis of wing-cover color in this species.

b. Write the genotypes of all parents and progeny as completely as possible.

Solution

a. These data seem complex at first, but the inheritance pattern becomes clear if we consider the crosses one at a time. A general principle of solving such problems, as we have seen, is to begin by looking over all the crosses and by grouping the data to bring out the patterns.

One clue that emerges from an overview of the data is that all the ratios are one-gene ratios: there is no evidence of two separate genes taking part at all. How can such variation be explained with a single gene? The answer is that there is variation for the single gene itself—that is, multiple allelism. Perhaps there are three alleles of one gene; let's call the gene w (for wing-cover color) and represent the alleles as w^g, w^b, and w^t. Now we have an additional problem, which is to determine the dominance of these alleles.

Cross 1 tells us something about dominance because the progeny of a blue × green cross are all blue; hence, blue appears to be dominant to green. This conclusion is supported by cross 5, because the green determinant must have been present in the parental stock to appear in the progeny. Cross 3 informs us about the turquoise determinants, which must have been present, although unexpressed, in the parental stock because there are turquoise wing covers in the progeny. So green must be dominant to turquoise. Hence, we have formed a model in which the dominance is $w^b > w^g > w^t$. Indeed, the inferred position of the w^t allele at the bottom of the dominance series is supported by the results of cross 7, where turquoise shows up in the progeny of a blue × green cross.

b. Now it is just a matter of deducing the specific genotypes. Notice that the question states that the parents were taken from a polymorphic population; this means that they could be either homozygous or heterozygous.

A parent with blue wing covers, for example, might be homozygous (w^b/w^b) or heterozygous (w^b/w^g or w^b/w^t). Here, a little trial and error and common sense are called for, but by this stage the question has essentially been answered, and all that remains is to "cross the t's and dot the i's." The following genotypes explain the results. A dash indicates that the genotype may be *either* homozygous or heterozygous in having a second allele farther down the allelic series.

Cross	Parents	Progeny
1	$w^b/w^b \times w^g/-$	w^b/w^g or $w^b/-$
2	$w^b/w^t \times w^b/w^t$	$\frac{3}{4} w^b/- : \frac{1}{4} w^t/w^t$
3	$w^g/w^t \times w^g/w^t$	$\frac{3}{4} w^g/- : \frac{1}{4} w^t/w^t$
4	$w^b/w^t \times w^t/w^t$	$\frac{1}{2} w^b/w^t : \frac{1}{2} w^t/w^t$
5	$w^b/w^g \times w^b/w^g$	$\frac{3}{4} w^b/- : \frac{1}{4} w^g/w^g$
6	$w^b/w^g \times w^g/w^g$	$\frac{1}{2} w^b/w^g : \frac{1}{2} w^g/w^g$
7	$w^b/w^t \times w^g/w^t$	$\frac{1}{2} w^b/- : \frac{1}{4} w^g/w^t : \frac{1}{4} w^t/w^t$
8	$w^t/w^t \times w^t/w^t$	all w^t/w^t

3. The leaves of pineapples can be classified into three types: spiny (S), spiny tip (ST), and piping (non-spiny; P). In crosses between pure strains followed by intercrosses of the F_1, the following results appeared:

		Phenotypes	
Cross	Parental	F_1	F_2
1	ST × S	ST	99 ST : 34 S
2	P × ST	P	120 P : 39 ST
3	P × S	P	95 P : 25 ST : 8 S

a. Assign gene symbols. Explain these results in regard to the genotypes produced and their ratios.

b. Using the model from part a, give the phenotypic ratios that you would expect if you crossed (1) the F_1 progeny from piping × spiny with the spiny parental stock and (2) the F_1 progeny of pipin × spiny with the F_1 progeny of spiny × spiny tip.

Solution

a. First, let's look at the F_2 ratios. We have clear 3:1 ratios in crosses 1 and 2, indicating single-gene segregations. Cross 3, however, shows a ratio that is almost certainly a 12:3:1 ratio. How do we know this? Well, there are simply not that many complex ratios in genetics, and trial and error brings us to the 12:3:1 quite quickly. In the 128 progeny total, the numbers of 96:24:8 are expected, but the actual numbers fit these expectations remarkably well.

One of the principles of this chapter is that modified Mendelian ratios reveal gene interactions. Cross 3

gives F_2 numbers appropriate for a modified dihybrid Mendelian ratio, so it looks as if we are dealing with a two-gene interaction. This seems the most promising place to start; we can go back to crosses 1 and 2 and try to fit them in later.

Any dihybrid ratio is based on the phenotypic proportions 9:3:3:1. Our observed modification groups them as follows:

$$\left.\begin{array}{l} 9 \; A/- \; ; \; B/- \\ 3 \; A/- \; ; \; b/b \end{array}\right\} \; 12 \text{ piping}$$
$$3 \; a/a \; ; \; B/- \qquad 3 \text{ spiny tip}$$
$$1 \; a/a \; ; \; b/b \qquad 1 \text{ spiny}$$

So without worrying about the name of the type of gene interaction (we are not asked to supply this anyway), we can already define our three pineapple-leaf phenotypes in relation to the proposed allelic pairs A/a and B/b:

$$\text{piping} = A/- \; (B/b \text{ irrelevant})$$
$$\text{spiny tip} = a/a \; ; \; B/-$$
$$\text{spiny} = a/a \; ; \; b/b$$

What about the parents of cross 3? The spiny parent must be $a/a \; ; \; b/b$, and because the B gene is needed to produce F_2 spiny-tip individuals, the piping parent must be $A/A \; ; \; B/B$. (Note that we are *told* that all parents are pure, or homozygous.) The F_1 must therefore be $A/a \; ; \; B/b$.

Without further thought, we can write out cross 1 as follows:

$$a/a \; ; \; B/B \times a/a \; ; \; b/b \longrightarrow$$

$$a/a \; ; \; B/b \begin{array}{l} \nearrow \frac{3}{4} a/a \; ; \; B/- \\ \searrow \frac{1}{4} a/a \; ; \; b/b \end{array}$$

Cross 2 can be partly written out without further thought by using our arbitrary gene symbols:

$$A/A \; ; \; -/- \times a/a \; ; \; B/B \longrightarrow$$

$$A/a \; ; \; B/- \begin{array}{l} \nearrow \frac{3}{4} A/- \; ; \; -/- \\ \searrow \frac{1}{4} a/a \; ; \; B/- \end{array}$$

We know that the F_2 of cross 2 shows single-gene segregation, and it seems certain now that the A/a allelic pair has a role. But the B allele is needed to produce the spiny tip phenotype, so all individuals must be homozygous B/B:

$$A/A \; ; \; B/B \times a/a \; ; \; B/B \longrightarrow$$

$$A/a \; ; \; B/B \begin{array}{l} \nearrow \frac{3}{4} A/- \; ; \; B/B \\ \searrow \frac{1}{4} a/a \; ; \; B/B \end{array}$$

Notice that the two single-gene segregations in crosses 1 and 2 do not show that the genes are *not* interacting. What is shown is that the two-gene interaction is not *revealed* by these crosses—only by cross 3, in which the F_1 is heterozygous for both genes.

b. Now it is simply a matter of using Mendel's laws to predict cross outcomes:

(1) A/a ; B/b × a/a ; $b/b \longrightarrow$

(independent assortment in a standard testcross)

$\frac{1}{4} A/a$; B/b ⎫
$\frac{1}{4} A/a$; b/b ⎬ piping

$\frac{1}{4} a/a$; B/b spiny tip

$\frac{1}{4} a/a$; b/b spiny

(2) A/a ; B/b × a/a ; $B/b \longrightarrow$

$\frac{1}{2} A/a$ ⟨ $\frac{3}{4} B/- \longrightarrow \frac{3}{8}$ ⎫
                  $\frac{1}{4} b/b \longrightarrow \frac{1}{8}$ ⎬ $\frac{1}{2}$ piping

$\frac{1}{2} a/a$ ⟨ $\frac{3}{4} B/- \longrightarrow \frac{3}{8}$ spiny tip
                  $\frac{1}{4} b/b \longrightarrow \frac{1}{8}$ spiny

PROBLEMS

BASIC PROBLEMS

1. In humans, the disease galactosemia causes mental retardation at an early age because lactose in milk cannot be broken down, and this failure affects brain function. How would you provide a secondary cure for galactosemia? Would you expect this phenotype to be dominant or recessive?

2. In humans, PKU (phenylketonuria) is a disease caused by an enzyme inefficiency at step A in the following simplified reaction sequence, and AKU (alkaptonuria) is due to an enzyme inefficiency in one of the steps summarized as step B here:

$$\text{phenylalanine} \xrightarrow{A} \text{tyrosine} \xrightarrow{B} CO_2 + H_2O$$

A person with PKU marries a person with AKU. What phenotypes do you expect for their children? All normal, all having PKU only, all having AKU only, all having both PKU and AKU, or some having AKU and some having PKU?

3. In *Drosophila*, the autosomal recessive *bw* causes a dark-brown eye, and the unlinked autosomal recessive *st* causes a bright scarlet eye. A homozygote for both genes has a white eye. Thus, we have the following correspondences between genotypes and phenotypes:

st^+/st^+ ; bw^+/bw^+ = red eye (wild type)
st^+/st^+ ; bw/bw = brown eye
st/st ; bw^+/bw^+ = scarlet eye
st/st ; bw/bw = white eye

Construct a hypothetical biosynthetic pathway showing how the gene products interact and why the different mutant combinations have different phenotypes.

4. Several mutants are isolated, all of which require compound G for growth. The compounds (A to E) in the biosynthetic pathway to G are known but their order in the pathway is not known. Each compound is tested for its ability to support the growth of each mutant (1 to 5). In the following table, a plus sign indicates growth and a minus sign indicates no growth:

| | Compound tested | | | | | |
	A	B	C	D	E	G
Mutant 1	−	−	−	+	−	+
2	−	+	−	+	−	+
3	−	−	−	−	−	+
4	−	+	+	+	−	+
5	+	+	+	+	−	+

a. What is the order of compounds A to E in the pathway?

b. At which point in the pathway is each mutant blocked?

c. Would a heterokaryon composed of double mutants 1,3 and 2,4 grow on a minimal medium? 1,3 and 3,4? 1,2 and 2,4 and 1,4?

5. In a certain plant, the flower petals are normally purple. Two recessive mutations arise in separate plants and are found to be on different chromosomes. Mutation 1 (m_1) gives blue petals when homozygous (m_1/m_1). Mutation 2 (m_2) gives red petals when homozygous (m_2/m_2). Biochemists working on the synthesis of flower pigments in this species have already described the following pathway:

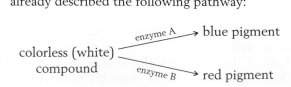

a. Which mutant would you expect to be deficient in enzyme A activity?

b. A plant has the genotype M_1/m_1 ; M_2/m_2. What would you expect its phenotype to be?

c. If the plant in part b is selfed, what colors of progeny would you expect, and in what proportions?

d. Why are these mutants recessive?

6. In sweet peas, the synthesis of purple anthocyanin pigment in the petals is controlled by two genes, B and D. The pathway is

$$\text{white intermediate} \xrightarrow[\text{enzyme}]{\text{gene } B} \text{blue intermediate} \xrightarrow[\text{enzyme}]{\text{gene } D} \text{anthocyanin (purple)}$$

a. What color petals would you expect in a pure-breeding plant unable to catalyze the first reaction?

b. What color petals would you expect in a pure-breeding plant unable to catalyze the second reaction?

c. If the plants in parts a and b are crossed, what color petals will the F_1 plants have?

d. What ratio of purple : blue : white plants would you expect in the F_2?

7. If a man of blood group AB marries a woman of blood group A whose father was of blood group O, to what different blood groups can this man and woman expect their children to belong?

8. Erminette fowls have mostly light-colored feathers with an occasional black one, giving a flecked appearance. A cross of two erminettes produced a total of 48 progeny, consisting of 22 erminettes, 14 blacks, and 12 pure whites. What genetic basis of the erminette pattern is suggested? How would you test your hypotheses?

9. Radishes may be long, round, or oval, and they may be red, white, or purple. You cross a long, white variety with a round, red one and obtain an oval, purple F_1. The F_2 shows nine phenotypic classes as follows: 9 long, red; 15 long, purple; 19 oval, red; 32 oval, purple; 8 long, white; 16 round, purple; 8 round, white; 16 oval, white; and 9 round, red.

a. Provide a genetic explanation of these results. Be sure to define the genotypes and show the constitution of parents, F_1, and F_2.

b. Predict the genotypic and phenotypic proportions in the progeny of a cross between a long, purple radish and an oval, purple one.

10. In the multiple allele series that determines coat color in rabbits, c^+ codes for agouti, c^{ch} for chinchilla (a beige coat color), and c^h for Himalayan. Dominance is in the order $c^+ > c^{ch} > c^h$. In a cross of $c^+/c^{ch} \times c^{ch}/c^h$, what proportion of progeny will be chinchilla?

11. Black, sepia, cream, and albino are all coat colors of guinea pigs. Individual animals (not necessarily from pure lines) showing these colors were intercrossed; the results are tabulated as follows, where the abbreviations A (albino), B (black), C (cream), and S (sepia) represent the phenotypes:

Cross	Parental phenotypes	Phenotypes of progeny			
		B	S	C	A
1	B × B	22	0	0	7
2	B × A	10	9	0	0
3	C × C	0	0	34	11
4	S × C	0	24	11	12
5	B × A	13	0	12	0
6	B × C	19	20	0	0
7	B × S	18	20	0	0
8	B × S	14	8	6	0
9	S × S	0	26	9	0
10	C × A	0	0	15	17

a. Deduce the inheritance of these coat colors and use gene symbols of your own choosing. Show all parent and progeny genotypes.

b. If the black animals in crosses 7 and 8 are crossed, what progeny proportions can you predict by using your model?

12. In a maternity ward, four babies become accidentally mixed up. The ABO types of the four babies are known to be O, A, B, and AB. The ABO types of the four sets of parents are determined. Indicate which baby belongs to each set of parents: **(a)** AB × O, **(b)** A × O, **(c)** A × AB, **(d)** O × O.

13. Consider two blood polymorphisms that humans have in addition to the ABO system. Two alleles L^M and L^N determine the M, N, and MN blood groups. The dominant allele R of a different gene causes a person to have the Rh$^+$ (rhesus positive) phenotype, whereas the homozygote for r is Rh$^-$ (rhesus negative). Two men took a paternity dispute to court, each claiming three children to be his own. The blood groups of the men, the children, and their mother were as follows:

Person	Blood group		
husband	O	M	Rh$^+$
wife's lover	AB	MN	Rh$^-$
wife	A	N	Rh$^+$
child 1	O	MN	Rh$^+$
child 2	A	N	Rh$^+$
child 3	A	MN	Rh$^-$

From this evidence, can the paternity of the children be established?

14. On a fox ranch in Wisconsin, a mutation arose that gave a "platinum" coat color. The platinum color proved very popular with buyers of fox coats, but the breeders could not develop a pure-breeding platinum strain. Every time two platinums were crossed, some normal foxes appeared in the progeny. For example, the repeated matings of the same pair of platinums produced 82 platinum and 38 normal progeny. All other such matings gave similar progeny ratios. State a concise genetic hypothesis that accounts for these results.

15. For a period of several years, Hans Nachtsheim investigated an inherited anomaly of the white blood cells of rabbits. This anomaly, termed the *Pelger anomaly*, is the arrest of the segmentation of the nuclei of certain white cells. This anomaly does not appear to seriously inconvenience the rabbits.

    a. When rabbits showing the typical Pelger anomaly were mated with rabbits from a true-breeding normal stock, Nachtsheim counted 217 offspring showing the Pelger anomaly and 237 normal progeny. What appears to be the genetic basis of the Pelger anomaly?

    b. When rabbits with the Pelger anomaly were mated to each other, Nachtsheim found 223 normal progeny, 439 showing the Pelger anomaly, and 39 extremely abnormal progeny. These very abnormal progeny not only had defective white blood cells, but also showed severe deformities of the skeletal system; almost all of them died soon after birth. In genetic terms, what do you suppose these extremely defective rabbits represented? Why do you suppose there were only 39 of them?

    c. What additional experimental evidence might you collect to support or disprove your answers to part b?

    d. In Berlin, about 1 human in 1000 shows a Pelger anomaly of white blood cells very similar to that described in rabbits. The anomaly is inherited as a simple dominant, but the homozygous type has not been observed in humans. Can you suggest why, if you are permitted an analogy with the condition in rabbits?

    e. Again by analogy with rabbits, what phenotypes and genotypes might be expected among the children of a man and woman who both show the Pelger anomaly?

    (Problem 15 from A. M. Srb, R. D. Owen, and R. S. Edgar, *General Genetics*, 2d ed. W. H. Freeman and Company, 1965.)

16. Two normal-looking fruit flies were crossed, and in the progeny there were 202 females and 98 males.

    a. What is unusual about this result?

    b. Provide a genetic explanation for this anomaly.

    c. Provide a test of your hypothesis.

17. You have been given a virgin *Drosophila* female. You notice that the bristles on her thorax are much shorter than normal. You mate her with a normal male (with long bristles) and obtain the following F_1 progeny: $\frac{1}{3}$ short-bristled females, $\frac{1}{3}$ long-bristled females, and $\frac{1}{3}$ long-bristled males. A cross of the F_1 long-bristled females with their brothers gives only long-bristled F_2. A cross of short-bristled females with their brothers gives $\frac{1}{3}$ short-bristled females, $\frac{1}{3}$ long-bristled females, and $\frac{1}{3}$ long-bristled males. Provide a genetic hypothesis to account for all these results, showing genotypes in every cross.

18. A dominant allele H reduces the number of body bristles that *Drosophila* flies have, giving rise to a "hairless" phenotype. In the homozygous condition, H is lethal. An independently assorting dominant allele S has no effect on bristle number except in the presence of H, in which case a single dose of S suppresses the hairless phenotype, thus restoring the hairy phenotype. However, S also is lethal in the homozygous (S/S) condition.

    a. What ratio of hairy to hairless flies would you find in the live progeny of a cross between two hairy flies both carrying H in the suppressed condition?

    b. When the hairless progeny are backcrossed with a parental hairy fly, what phenotypic ratio would you expect to find among their live progeny?

19. After irradiating wild-type cells of *Neurospora* (a haploid fungus), a geneticist finds two leucine-requiring auxotrophic mutants. He combines the two mutants in a heterokaryon and discovers that the heterokaryon is prototrophic.

    a. Were the mutations in the two auxotrophs in the *same* gene in the pathway for synthesizing leucine, or in two *different* genes in that pathway? Explain.

    b. Write the genotype of the two strains according to your model.

    c. What progeny and in what proportions would you predict from crossing the two auxotrophic mutants? (Assume independent assortment.)

20. A yeast geneticist irradiates haploid cells of a strain that is an adenine-requiring auxotrophic mutant, caused by mutation of the gene *ade1*. Millions of the irradiated cells are plated on minimal medium, and a small number of cells divide and produce prototrophic colonies. These colonies are crossed indi-

vidually to a wild-type strain. Two types of results are obtained:

prototroph $\times$ wild type $\cdot$ progeny all prototrophic
prototroph $\times$ wild type $\cdot$ progeny 75% prototrophic, 25% adenine-requiring auxotrophs

a. Explain the difference between these two types of results.

b. Write the genotypes of the prototrophs in each case.

c. What progeny phenotypes and ratios do you predict from crossing a prototroph of type 2 by the original *ade1* auxotroph?

21. It is known that in roses the synthesis of red pigment is by two steps in a pathway, as follows:

$$\text{colorless intermediate} \xrightarrow{\text{gene } P}$$

$$\text{magenta intermediate} \xrightarrow{\text{gene } Q} \text{red pigment}$$

a. What would the phenotype be of a plant homozygous for a null mutation of gene *P*?

b. What would the phenotype be of a plant homozygous for a null mutation of gene *Q*?

c. What would the phenotype be of a plant homozygous for null mutations of genes *P* and *Q*?

d. Write the genotypes of the three strains in **a**, **b**, and **c**.

e. What F_2 ratio is expected from crossing plants from **a** and **b**? (Assume independent assortment.)

22. Because snapdragons *(Antirrhinum)* possess the pigment anthocyanin, they have reddish-purple petals. Two pure anthocyanin-less lines of *Antirrhinum* were developed, one in California and one in Holland. They looked identical in having no red pigment at all, manifested as white (albino) flowers. However, when petals from the two lines were ground up together in buffer in the same test tube, the solution, which appeared colorless at first, gradually turned red.

a. What control experiments should an investigator conduct before proceeding with further analysis?

b. What could account for the production of the red color in the test tube?

c. According to your explanation for part b, what would be the genotypes of the two lines?

d. If the two white lines were crossed, what would you predict the phenotypes of the F_1 and F_2 to be?

23. The frizzle fowl is much admired by poultry fanciers. It gets its name from the unusual way that its feathers curl up, giving the impression that it has been (in the memorable words of animal geneticist F. B. Hutt) "pulled backwards through a knothole." Unfortunately, frizzle fowls do not breed true; when two frizzles are intercrossed, they always produce 50 percent frizzles, 25 percent normal, and 25 percent with peculiar woolly feathers that soon fall out, leaving the birds naked.

a. Give a genetic explanation for these results, showing genotypes of all phenotypes, and provide a statement of how your explanation works.

b. If you wanted to mass-produce frizzle fowls for sale, which types would be best to use as a breeding pair?

24. The petals of the plant *Collinsia parviflora* are normally blue, giving the species its common name, blue-eyed Mary. Two pure-breeding lines were obtained from color variants found in nature; the first line had pink petals, and the second line had white petals. The following crosses were made between pure lines, with the results shown:

Parents	F_1	F_2
blue $\times$ white	blue	101 blue, 33 white
blue $\times$ pink	blue	192 blue, 63 pink
pink $\times$ white	blue	272 blue, 121 white, 89 pink

a. Explain these results genetically. Define the allele symbols that you use, and show the genetic constitution of parents, F_1, and F_2.

b. A cross between a certain blue F_2 plant and a certain white F_2 plant gave progeny of which $\frac{3}{8}$ were blue, $\frac{1}{8}$ were pink, and $\frac{1}{2}$ were white. What must the genotypes of these two F_2 plants have been?

 UNPACKING PROBLEM 24

1. What is the character being studied?

2. What is the wild-type phenotype?

3. What is a variant?

4. What are the variants in this problem?

5. What does "in nature" mean?

6. In what way would the variants have been found in nature? (Describe the scene.)

7. At which stages in the experiments would seeds be used?

8. Would the way of writing a cross "blue $\times$ white" (for example) mean the same as "white $\times$ blue"? Would you expect similar results? Why or why not?

9. In what way do the first two rows in the table differ from the third row?

10. Which phenotypes are dominant?

11. What is complementation?

12. Where does the blueness come from in the progeny of the pink × white cross?

13. What genetic phenomenon does the production of a blue F_1 from pink and white parents represent?

14. List any ratios that you can see.

15. Are there any monohybrid ratios?

16. Are there any dihybrid ratios?

17. What does observing monohybrid and dihybrid ratios tell you?

18. List four modified Mendelian ratios that you can think of.

19. Are there any modified Mendelian ratios in the problem?

20. What do modified Mendelian ratios indicate generally?

21. What does the specific modified ratio or ratios in this problem indicate?

22. Draw chromosomes representing the meioses in the parents in the cross blue × white and meiosis in the F_1.

23. Repeat for the cross blue × pink.

25. A woman who owned a purebred albino poodle (an autosomal recessive phenotype) wanted white puppies, so she took the dog to a breeder, who said he would mate the female with an albino stud male, also from a pure stock. When six puppies were born, they were all black, so the woman sued the breeder, claiming that he replaced the stud male with a black dog, giving her six unwanted puppies. You are called in as an expert witness, and the defense asks you if it is possible to produce black offspring from two pure-breeding recessive albino parents. What testimony do you give?

26. A snapdragon plant that bred true for white petals was crossed to a plant that bred true for purple petals, and all the F_1 had white petals. The F_1 was selfed. Among the F_2, three phenotypes were observed in the following numbers:

white	240
solid-purple	61
spotted-purple	19
Total	320

a. Propose an explanation for these results, showing genotypes of all generations (make up and explain your symbols).

b. A white F_2 plant was crossed to a solid-purple F_2 plant, and the progeny were

white	50%
solid-purple	25%
spotted-purple	25%

What were the genotypes of the F_2 plants crossed?

27. Most flour beetles are black, but several color variants are known. Crosses of pure-breeding parents produced the following results in the F_1 generation, and intercrossing the F_1 from each cross gave the ratios shown for the F_2 generation. The phenotypes are abbreviated Bl, black; Br, brown; Y, yellow; and W, white.

Cross	Parents	F_1	F_2
1	Br × Y	Br	3 Br : 1 Y
2	Bl × Br	Bl	3 Bl : 1 Br
3	Bl × Y	Bl	3 Bl : 1 Y
4	W × Y	Bl	9 Bl : 3 Y : 4 W
5	W × Br	Bl	9 Bl : 3 Br : 4 W
6	Bl × W	Bl	9 Bl : 3 Y : 4 W

a. From these results deduce and explain the inheritance of these colors.

b. Write the genotypes of each of the parents, the F_1, and the F_2 in all crosses.

28. Two albinos marry and have four normal children. How is this possible?

29. Consider production of flower color in the Japanese morning glory (Pharbitis nil). Dominant alleles of either of two separate genes ($A/-$ · b/b or a/a · $B/-$) produce purple petals. $A/-$ · $B/-$ produces blue petals, and a/a · b/b produces scarlet petals. Deduce the genotypes of parents and progeny in the following crosses:

Cross	Parents	Progeny
1	blue × scarlet	$\frac{1}{4}$ blue : $\frac{1}{2}$ purple : $\frac{1}{4}$ scarlet
2	purple × purple	$\frac{1}{4}$ blue : $\frac{1}{2}$ purple : $\frac{1}{4}$ scarlet
3	blue × blue	$\frac{3}{4}$ blue : $\frac{1}{4}$ purple
4	blue × purple	$\frac{3}{8}$ blue : $\frac{4}{8}$ purple : $\frac{1}{8}$ scarlet
5	purple × scarlet	$\frac{1}{2}$ purple : $\frac{1}{2}$ scarlet

30. Corn breeders obtained pure lines whose kernels turn sun red, pink, scarlet, or orange when exposed to sunlight (normal kernels remain yellow in sunlight). Some crosses between these lines produced the following results. The phenotypes are abbreviated O, orange; P, pink; Sc, scarlet; and SR, sun red.

Cross	Phenotypes		
	Parents	F₁	F₂
1	SR × P	all SR	66 SR : 20 P
2	O × SR	all SR	998 SR : 314 O
3	O × P	all O	1300 O : 429 P
4	O × Sc	all Y	182 Y : 80 O : 58 Sc

Analyze the results of each cross, and provide a unifying hypothesis to account for *all* the results. (Explain all symbols that you use.)

31. Many kinds of wild animals have the agouti coloring pattern, in which each hair has a yellow band around it.

a. Black mice and other black animals do not have the yellow band; each of their hairs is all black. This absence of wild agouti pattern is called *nonagouti*. When mice of a true-breeding agouti line are crossed with nonagoutis, the F₁ is all agouti and the F₂ has a 3:1 ratio of agoutis to nonagoutis. Diagram this cross, letting *A* represent the allele responsible for the agouti phenotype and *a*, nonagouti. Show the phenotypes and genotypes of the parents, their gametes, the F₁, their gametes, and the F₂.

b. Another inherited color deviation in mice substitutes brown for the black color in the wild-type hair. Such brown-agouti mice are called *cinnamons*. When wild-type mice are crossed with cinnamons, the F₁ is all wild type and the F₂ has a 3:1 ratio of wild type to cinnamon. Diagram this cross as in part a, letting *B* stand for the wild-type black allele and *b* stand for the cinnamon brown allele.

c. When mice of a true-breeding cinnamon line are crossed with mice of a true-breeding nonagouti (black) line, the F₁ is all wild type. Use a genetic diagram to explain this result.

d. In the F₂ of the cross in part c, a fourth color called *chocolate* appears in addition to the parental cinnamon and nonagouti and the wild type of the F₁. Chocolate mice have a solid, rich-brown color. What is the genetic constitution of the chocolates?

e. Assuming that the *A/a* and *B/b* allelic pairs assort independently of each other, what do you expect to be the relative frequencies of the four color types in the F₂ described in part d? Diagram the cross of parts c and d, showing phenotypes and genotypes (including gametes).

f. What phenotypes would be observed in what proportions in the progeny of a backcross of F₁ mice from part c to the cinnamon parental stock? To the nonagouti (black) parental stock? Diagram these backcrosses.

g. Diagram a testcross for the F₁ of part c. What colors would result and in what proportions?

h. Albino (pink-eyed white) mice are homozygous for the recessive member of an allelic pair *C/c* which assorts independently of the *A/a* and *B/b* pairs. Suppose that you have four different highly inbred (and therefore presumably homozygous) albino lines. You cross each of these lines with a true-breeding wild-type line, and you raise a large F₂ progeny from each cross. What genotypes for the albino lines can you deduce from the following F₂ phenotypes?

(Problem 31 adapted from A. M. Srb, R. D. Owen, and R. S. Edgar, *General Genetics*, 2d ed. W. H. Freeman and Company, 1965.)

 32. An allele A that is not lethal when homozgous causes rats to have yellow coats. The allele *R* of a separate gene that assorts independently produces a black coat. Together, *A* and *R* produce a grayish coat, whereas *a* and *r* produce a white coat. A gray male is crossed with a yellow female, and the F₁ is $\frac{3}{8}$ yellow, $\frac{3}{8}$ gray, $\frac{1}{8}$ black, and $\frac{1}{8}$ white. Determine the genotypes of the parents.

33. The genotype *r/r* ; *p/p* gives fowl a single comb, *R/–* ; *P/–* gives a walnut comb, *r/r* ; *P/–* gives a pea comb, and *R/–* ; *p/p* gives a rose comb (see the illustrations).

Single　　　Walnut　　　Pea　　　Rose

a. What comb types will appear in the F₁ and in the F₂ in what proportions if single-combed birds are crossed with birds of a true-breeding walnut strain?

b. What are the genotypes of the parents in a walnut × rose mating from which the progeny are $\frac{3}{8}$ rose, $\frac{3}{8}$ walnut, $\frac{1}{8}$ pea, and $\frac{1}{8}$ single?

c. What are the genotypes of the parents in a walnut × rose mating from which all the progeny are walnut?

d. How many genotypes produce a walnut phenotype? Write them out.

34. The production of eye-color pigment in *Drosophila* requires the dominant allele *A*. The dominant allele *P* of a second independent gene turns the pigment to purple, but its recessive allele leaves it red. A fly producing no pigment has white eyes. Two pure lines were crossed with the following results:

P   red-eyed female × white-eyed male

F₁   purple-eyed females
              red-eyed males
              F₁ × F₁

F₂   both males and females: $\frac{3}{8}$ purple-eyed
                              $\frac{3}{8}$ red-eyed
                              $\frac{2}{8}$ white-eyed

Explain this mode of inheritance and show the genotypes of the parents, the F₁, and the F₂.

35. When true-breeding brown dogs are mated with certain true-breeding white dogs, all the F₁ pups are white. The F₂ progeny from some F₁ × F₁ crosses were 118 white, 32 black, and 10 brown pups. What is the genetic basis for these results?

36. Wild-type strains of the haploid fungus *Neurospora* can make their own tryptophan. An abnormal allele *td* renders the fungus incapable of making its own tryptophan. An individual of genotype *td* grows only when its medium supplies tryptophan. The allele *su* assorts independently of *td*; its only known effect is to suppress the *td* phenotype. Therefore, strains carrying both *td* and *su* do not require tryptophan for growth.

a. If a *td ; su* strain is crossed with a genotypically wild-type strain, what genotypes are expected in the progeny, and in what proportions?

b. What will be the ratio of tryptophan-dependent to tryptophan-independent progeny in the cross of part a?

37. Mice of the genotypes *A/A ; B/B ; C/C ; D/D ; S/S* and *a/a ; b/b ; c/c ; d/d ; s/s* are crossed. The progeny are intercrossed. What phenotypes will be produced in the F₂, and in what proportions? (The allele symbols stand for the following: *A* = agouti, *a* = solid (nonagouti); *B* = black pigment, *b* = brown; *C* = pigmented, *c* = albino; *D* = nondilution, *d* = dilution (milky color); *S* = unspotted, *s* = pigmented spots on white background.)

38. Consider the genotypes of two lines of chickens: the pure-line mottled Honduran is *i/i ; D/D ; M/M ; W/W*, and the pure-line leghorn is *I/I ; d/d ; m/m ; w/w*, where

    I = white feathers, *i* = colored feathers
    D = duplex comb, *d* = simplex comb
    M = bearded, *m* = beardless
    W = white skin, *w* = yellow skin

These four genes assort independently. Starting with these two pure lines, what is the fastest and most convenient way of generating a pure line of birds that has colored feathers, has a simplex comb, is beardless, and has yellow skin? Make sure that you show

a. The breeding pedigree.

b. The genotype of each animal represented.

c. How many eggs to hatch in each cross, and why this number.

d. Why your scheme is the fastest and the most convenient.

39. The following pedigree is for a dominant phenotype governed by an autosomal gene. What does this pedigree suggest about the phenotype, and what can you deduce about the genotype of individual A?

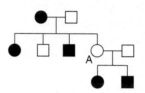

40. Petal coloration in foxgloves is determined by three genes. *M* codes for an enzyme that synthesizes anthocyanin, the purple pigment seen in these petals; *m/m* produces no pigment, resulting in the phenotype albino with yellowish spots. *D* is an enhancer of anthocyanin, resulting in a darker pigment; *d/d* does not enhance. At the third locus, *w/w* allows pigment deposition in petals, but *W* prevents pigment deposition except in the spots and, so, results in the white, spotted phenotype. Consider the following two crosses:

Cross	Parents		Progeny
1	dark-purple	× white with yellowish spots	$\frac{1}{2}$ dark-purple: $\frac{1}{2}$ light-purple
2	white with yellowish spots	× light-purple	$\frac{1}{2}$ white with purple spots: $\frac{1}{4}$ dark-purple: $\frac{1}{4}$ light-purple

In each case, give the genotypes of parents and progeny with respect to the three genes.

41. In one species of *Drosophila*, the wings are normally round in shape, but you have obtained two pure lines, one of which has oval wings and the other sickle-shaped wings. Crosses between pure lines reveal the following results:

Parents		F_1	
Female	Male	Female	Male
sickle	round	sickle	sickle
round	sickle	sickle	round
sickle	oval	oval	sickle

a. Provide a genetic explanation of these results, defining all allele symbols.

b. If the F_1 oval females from cross 3 are crossed to the F_1 round males from cross 2, what phenotypic proportions are expected in each sex of progeny?

42. Mice normally have one yellow band on their hairs, but variants with two or three bands are known. A female mouse with one band was crossed to a male who had three bands. (Neither animal was from a pure line.) The progeny were

Females	$\frac{1}{2}$	one band
	$\frac{1}{2}$	three bands
Males	$\frac{1}{2}$	one band
	$\frac{1}{2}$	two bands

a. Provide a clear explanation of the inheritance of these phenotypes.

b. Under your model, what would be the outcome of a cross between a three-banded daughter and a one-banded son?

43. In minks, wild types have an almost black coat. Breeders have developed many pure lines of color variants for the mink coat industry. Two such pure lines are platinum (blue-gray) and aleutian (steel-gray). These lines were used in crosses, with the following results:

Cross	Parents	F_1	F_2
1	wild × platinum	wild	18 wild, 5 platinum
2	wild × aleutian	wild	27 wild, 10 aleutian
3	platinum × aleutian	wild	133 wild
			41 platinum
			46 aleutian
			17 sapphire (new)

a. Devise a genetic explanation of these three crosses. Show genotypes for parents, F_1, and F_2 in the three crosses, and make sure that you show the alleles of each gene that you hypothesize in every individual.

b. Predict the F_1 and F_2 phenotypic ratios from crossing sapphire with platinum and aleutian pure lines.

44. In *Drosophila*, an autosomal gene determines the shape of the hair, with B giving straight and b bent hairs. On another autosome, there is a gene of which a dominant allele I inhibits hair formation so that the fly is hairless (i has no known phenotypic effect).

a. If a straight-haired fly from a pure line is crossed with a fly from a pure-breeding hairless line known to be an inhibited bent genotype, what will the genotypes and phenotypes of the F_1 and the F_2 be?

b. What cross would give the ratio 4 hairless:3 straight:1 bent?

45. The following pedigree concerns eye phenotypes in *Tribolium* beetles. The solid symbols represent black eyes, the open symbols represent brown eyes, and the cross symbols (X) represent the "eyeless" phenotype, in which eyes are totally absent.

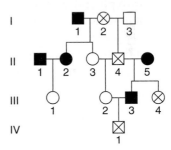

a. From these data, deduce the mode of inheritance of these three phenotypes.

b. Using defined gene symbols, show the genotype of individual II-3.

46. A plant believed to be heterozygous for a pair of alleles B/b (where B encodes yellow and b encodes bronze) was selfed, and in the progeny there were 280 yellow and 120 bronze individuals. Do these results support the hypothesis that the plant is B/b?

47. A plant thought to be heterozygous for two independently assorting genes (P/p ; Q/q) was selfed, and the progeny were

88	$P/-$; $Q/-$
32	$P/-$; q/q
25	p/p ; $Q/-$
14	p/p ; q/q

Do these results support the hypothesis that the original plant was P/p ; Q/q?

48. A plant of phenotype 1 was selfed, and in the progeny there were 100 individuals of phenotype 1 and 60 of an alternative phenotype 2. Are these numbers compatible with expected ratios of 9:7, 13:3, and 3:1? Formulate a genetic hypothesis based on your calculations.

49. Four homozygous recessive mutant lines of *Drosophila melanogaster* (labeled 1 through 4) showed abnormal leg coordination, which made their walking highly erratic. These lines were intercrossed; the phenotypes of the F_1 flies are shown in the following grid, in which "+" represents wild-type walking and "−" represents abnormal walking:

	1	2	3	4
1	−	+	+	+
2	+	−	−	+
3	+	−	−	+
4	+	+	+	−

a. What type of test does this analysis represent?

b. How many different genes were mutated in creating these four lines?

c. Invent wild-type and mutant symbols and write out full genotypes for all four lines and for the F_1's.

d. Do these data tell us which genes are linked? If not, how could linkage be tested?

e. Do these data tell us the total number of genes involved in leg coordination in this animal?

50. Three independently isolated tryptophan-requiring mutants of haploid yeast are called *trpB*, *trpD*, and *trpE*. Cell suspensions of each are streaked on a plate of nutritional medium supplemented with just enough tryptophan to permit weak growth for a *trp* strain. The streaks are arranged in a triangular pattern so that they do not touch one another. Luxuriant growth is noted at both ends of the *trpE* streak and at one end of the *trpD* streak (see the accompanying figure).

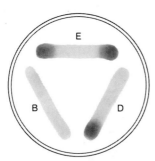

a. Do you think complementation is involved?

b. Briefly explain the pattern of luxuriant growth.

c. In what order in the tryptophan-synthesizing pathway are the enzymatic steps that are defective in *trpB*, *trpD*, and *trpE*?

d. Why was it necessary to add a small amount of tryptophan to the medium in order to demonstrate such a growth pattern?

CHALLENGING PROBLEMS

51. A pure-breeding strain of squash that produced disk-shaped fruits (see the accompanying illustration) was crossed with a pure-breeding strain having long fruits. The F_1 had disk fruits, but the F_2 showed a new phenotype, sphere, and was composed of the following proportions:

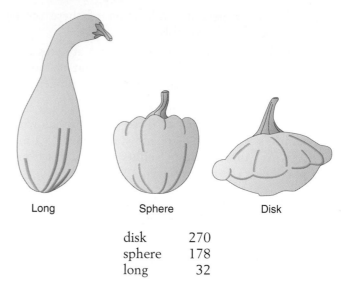

Long	Sphere	Disk

disk	270
sphere	178
long	32

Propose an explanation for these results, and show the genotypes of the P, F_1, and F_2 generations.

(Illustration from P. J. Russell, *Genetics*, 3d ed. Harper-Collins, 1992.)

52. Marfan's syndrome is a disorder of the fibrous connective tissue, characterized by many symptoms, including long, thin digits; eye defects; heart disease; and long limbs. (Flo Hyman, the American volleyball star, suffered from Marfan's syndrome. She died soon after a match from a ruptured aorta.)

a. Use the accompanying pedigree (see the top of page 223) to propose a mode of inheritance for Marfan's syndrome.

b. What genetic phenomenon is shown by this pedigree?

c. Speculate on a reason for such a phenomenon.

(Illustration from J. V. Neel and W. J. Schull, *Human Heredity.* University of Chicago Press, 1954.)

53. In corn, three dominant alleles, called *A*, *C*, and *R*, must be present to produce colored seeds. Genotype *A/−* ; *C/−* ; *R/−* is colored; all others are colorless. A colored plant is crossed with three tester plants of known genotype. With tester *a/a* ; *c/c* ; *R/R*, the colored plant produces 50 percent colored seeds; with *a/a* ; *C/C* ; *r/r*, it produces 25 percent colored; and with *A/A* ; *c/c* ; *r/r*, it produces 50 percent colored. What is the genotype of the colored plant?

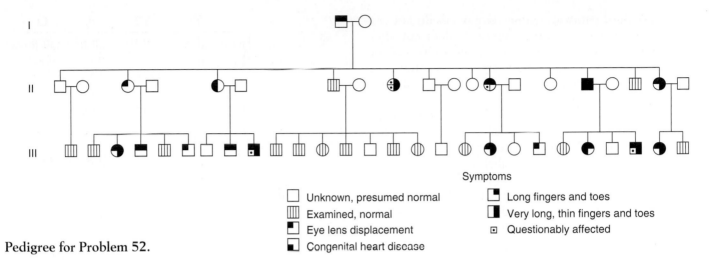

Symptoms	
☐ Unknown, presumed normal	▉ Long fingers and toes
▦ Examined, normal	▉ Very long, thin fingers and toes
◨ Eye lens displacement	▫ Questionably affected
◨ Congenital heart disease	

Pedigree for Problem 52.

54. The production of pigment in the outer layer of seeds of corn requires each of the three independently assorting genes *A*, *C*, and *R* to be represented by at least one dominant allele, as specified in Problem 53. The dominant allele *Pr* of a fourth independently assorting gene is required to convert the biochemical precursor into a purple pigment, and its recessive allele *pr* makes the pigment red. Plants that do not produce pigment have yellow seeds. Consider a cross of a strain of genotype *A/A* ; *C/C* ; *R/R* ; *pr/pr* with a strain of genotype *a/a* ; *c/c* ; *r/r* ; *Pr/Pr*.

a. What are the phenotypes of the parents?

b. What will be the phenotype of the F_1?

c. What phenotypes, and in what proportions, will appear in the progeny of a selfed F_1?

d. What progeny proportions do you predict from the testcross of an F_1?

55. The allele *B* gives mice a black coat, and *b* gives a brown one. The genotype *e/e* of another, independently assorting gene prevents expression of *B* and *b*, making the coat color beige, whereas *E/–* permits expression of *B* and *b*. Both genes are autosomal. In the following pedigree, black symbols indicate a black coat, pink symbols indicate brown, and white symbols indicate beige.

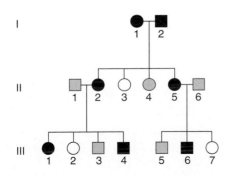

a. What is the name given to the type of gene interaction in this example?

b. What are the genotypes of the individuals in the pedigree? (If there are alternative possibilities, state them.)

56. A researcher crosses two white-flowered lines of *Antirrhinum* plants as follows and obtains the following results:

$$\text{pure line 1} \times \text{pure line 2}$$
$$\downarrow$$
$$F_1 \quad \text{all white}$$
$$F_1 \times F_1$$
$$\downarrow$$
$$F_2 \quad \begin{array}{l}\text{131 white}\\\text{29 red}\end{array}$$

a. Deduce the inheritance of these phenotypes, using clearly defined gene symbols. Give the genotypes of the parents, F_1, and F_2.

b. Predict the outcome of crosses of the F_1 to each parental line.

57. Assume that two pigments, red and blue, mix to give the normal purple color of petunia petals. Separate biochemical pathways synthesize the two pigments, as shown in the top two rows of the accompanying diagram. "White" refers to compounds that are not pigments. (Total lack of pigment results in a white petal.) Red pigment forms from a yellow intermediate that normally is at a concentration too low to color petals.

pathway I $\cdots\longrightarrow$ white$_1$ $\xrightarrow{E}$ blue

pathway II $\cdots\longrightarrow$ white$_2$ $\xrightarrow{A}$ yellow $\xrightarrow{B}$ red

$\uparrow$
C

pathway III $\cdots\longrightarrow$ white$_3$ $\xrightarrow{D}$ white$_4$

A third pathway, whose compounds do not contribute pigment to petals, normally does not affect the blue and red pathways, but if one of its intermediates (white$_3$) should build up in concentration, it can be converted into the yellow intermediate of the red pathway.

In the diagram, A to E represent enzymes; their corresponding genes, all of which are unlinked, may be symbolized by the same letters.

Assume that wild-type alleles are dominant and code for enzyme function and that recessive alleles result in lack of enzyme function. Deduce which combinations of true-breeding parental genotypes could be crossed to produce F$_2$ progenies in the following ratios:

a. 9 purple : 3 green : 4 blue

b. 9 purple : 3 red : 3 blue : 1 white

c. 13 purple : 3 blue

d. 9 purple : 3 red : 3 green : 1 yellow

(**Note:** Blue mixed with yellow makes green; assume that no mutations are lethal.)

58. The flowers of nasturtiums *(Tropaeolum majus)* may be single (S), double (D), or superdouble (Sd). Superdoubles are female sterile; they originated from a double-flowered variety. Crosses between varieties gave the progenies as listed in the following table, where *pure* means "pure-breeding."

Cross	Parents	Progeny
1	pure S × pure D	All S
2	cross 1 F$_1$ × cross 1 F$_1$	78 S : 27 D
3	pure D × Sd	112 Sd : 108 D
4	pure S × Sd	8 Sd : 7 S
5	pure D × cross 4 Sd progeny	18 Sd : 19 S
6	pure D × cross 4 S progeny	14 D : 16 S

Using your own genetic symbols, propose an explanation for these results, showing

a. All the genotypes in each of the six rows.

b. The proposed origin of the superdouble.

59. In a certain species of fly, the normal eye color is red (R). Four abnormal phenotypes for eye color were found: two were yellow (Y1 and Y2), one was brown (B), and one was orange (O). A pure line was established for each phenotype, and all possible combinations of the pure lines were crossed. Flies of each F$_1$ were intercrossed to produce an F$_2$. The F$_1$'s and F$_2$'s are shown within the following square; the pure lines are given in the margins.

		Y1	Y2	B	O
Y1	F$_1$	all Y	all R	all R	all R
	F$_2$	all Y	9 R	9 R	9 R
			7 Y	4 Y	4 O
				3 B	3 Y
Y2	F$_1$		all Y	all R	all R
	F$_2$		all Y	9 R	9 R
				4 Y	4 Y
				3 B	3 O
B	F$_1$			all B	all R
	F$_2$			all B	9 R
					4 O
					3 B
O	F$_1$				all O
	F$_2$				all O

a. Define your own symbols and show genotypes of all four pure lines.

b. Show how the F$_1$ phenotypes and the F$_2$ ratios are produced.

c. Show a biochemical pathway that explains the genetic results, indicating which gene controls which enzyme.

60. In common wheat, *Triticum aestivum*, kernel color is determined by multiply duplicated genes, each with an *R* and an *r* allele. Any number of *R* alleles will give red, and the complete lack of *R* alleles will give the white phenotype. In one cross between a red pure line and a white pure line, the F$_2$ was $\frac{63}{64}$ red and $\frac{1}{64}$ white.

a. How many *R* genes are segregating in this system?

b. Show genotypes of the parents, the F$_1$, and the F$_2$.

c. Different F$_2$ plants are backcrossed to the white parent. Give examples of genotypes that would give the following progeny ratios in such backcrosses: (1) 1 red : 1 white, (2) 3 red : 1 white, (3) 7 red : 1 white.

d. What is the formula that generally relates the number of segregating genes to the proportion of red individuals in the F$_2$ in such systems?

61. The accompanying pedigree shows the inheritance of deaf-mutism.

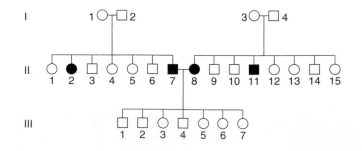

a. Provide an explanation for the inheritance of this rare condition in the two families in generations I and II, showing genotypes of as many individuals as possible, using symbols of your own choosing.

b. Provide an explanation for the production of only normal individuals in generation III, making sure that your explanation is compatible with the answer to part a.

62. The accompanying pedigree is for blue sclera (bluish thin outer wall to the eye) and brittle bones.

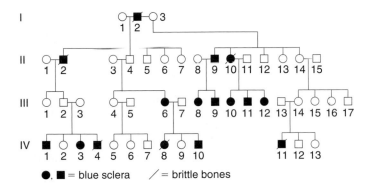

●, ■ = blue sclera   / = brittle bones

a. Are these two abnormalities caused by the same gene or separate genes? State your reasons clearly.

b. Is the gene (or genes) autosomal or sex-linked?

c. Does the pedigree show any evidence of incomplete penetrance or expressivity? If so, make the best calculations that you can of these measures.

63. Workers of the honeybee line known as *Brown* (nothing to do with color) show what is called "hygienic behavior"; that is, they uncap hive compartments containing dead pupae and then remove the dead pupae. This behavior prevents the spread of infectious bacteria through the colony. Workers of the *Van Scoy* line, however, do not perform these actions, and therefore this line is said to be "nonhygienic." When a queen from the *Brown* line was mated with *Van Scoy* drones, the F_1 were all nonhygienic. When drones from this F_1 inseminated a queen from the *Brown* line, the progeny behaviors were as follows:

$\frac{1}{4}$ hygienic

$\frac{1}{4}$ uncapping but no removing of pupae

$\frac{1}{2}$ nonhygienic

However, when the nonhygienic individuals were examined further, it was found that if the compart-

ment of dead pupae was uncapped by the beekeeper, about half the individuals removed the dead pupae, but the other half did not.

a. Propose a genetic hypothesis to explain these behavioral patterns.

b. Discuss the data in relation to epistasis, dominance, and environmental interaction.

(**Note:** Workers are sterile, and all bees from one line carry the same alleles.)

64. The normal color of snapdragons is red. Some pure lines showing variations of flower color have been found. When these pure lines were crossed, they gave the following results:

Cross	Parents	F_1	F_2
1	orange × yellow	orange	3 orange : 1 yellow
2	red × orange	red	3 red : 1 orange
3	red × yellow	red	3 red : 1 yellow
4	red × white	red	3 red : 1 white
5	yellow × white	red	9 red : 3 yellow : 4 white
6	orange × white	red	9 red : 3 orange : 4 white
7	red × white	red	9 red : 3 yellow : 4 white

a. Explain the inheritance of these colors.

b. Write the genotypes of the parents, the F_1, and the F_2.

65. Consider the following F_1 individuals in different species and the F_2 ratios produced by selfing:

F_1	Phenotypic ratio in the F_2		
1 cream	$\frac{12}{16}$ cream	$\frac{3}{16}$ black	$\frac{1}{16}$ gray
2 orange	$\frac{9}{16}$ orange	$\frac{7}{16}$ yellow	
3 black	$\frac{13}{16}$ black	$\frac{3}{16}$ white	
4 solid red	$\frac{9}{16}$ solid red	$\frac{3}{16}$ mottled red	$\frac{4}{16}$ small red dots

If each F_1 were testcrossed, what phenotypic ratios would result in the progeny of the testcross?

66. To understand the genetic basis of locomotion in the diploid nematode *Caenorhabditis elegans*, recessive mutations were obtained, all making the worm "wiggle" ineffectually instead of moving with its usual smooth gliding motion. These mutations presumably affect the nervous or muscle systems. Twelve homozygous mutants were intercrossed, and the F_1 hybrids were examined to see

if they wiggled. The results were as follows, where a plus sign means that the F_1 hybrid was wild type (gliding) and "w" means that the hybrid wiggled.

	1	2	3	4	5	6	7	8	9	10	11	12
1	w	+	+	+	w	+	+	+	+	+	+	+
2		w	+	+	+	w	+	w	+	w	+	+
3			w	w	+	+	+	+	+	+	+	+
4				w	+	+	+	+	+	+	+	+
5					w	+	+	+	+	+	+	+
6						w	+	w	+	w	+	+
7							w	+	+	+	w	w
8								w	+	w	+	+
9									w	+	+	+
10										w	+	+
11											w	w
12												w

a. Explain what this experiment was designed to test.

b. Use this reasoning to assign genotypes to all 12 mutants.

c. Explain why the F_1 hybrids between mutants 1 and 2 had a different phenotype from that of the hybrids between mutants 1 and 5.

67. A geneticist working on a haploid fungus makes a cross between two slow-growing mutants call *mossy* and *spider* (referring to abnormal appearance of the colonies). Tetrads from the cross are of three types (A, B, C), but two of them contain spores that do not germinate.

Spore	A	B	C
1	wild type	wild type	spider
2	wild type	spider	spider
3	no germination	mossy	mossy
4	no germination	no germination	mossy

Devise a model to explain these genetic results, and propose a molecular basis for your model.

INTERACTIVE GENETICS   MegaManual CD-ROM Tutorial

Biochemical Genetics

For additional coverage of the topics in this chapter, refer to the Interactive Genetics CD-ROM included with the Solutions MegaManual. The Biochemical Genetics activity contains animated tutorials on the model organism *Neurospora*, and how it has been used to determine the relationships between genes and enzymes. Five interactive problems are also provided for extra practice.

DNA: STRUCTURE AND REPLICATION

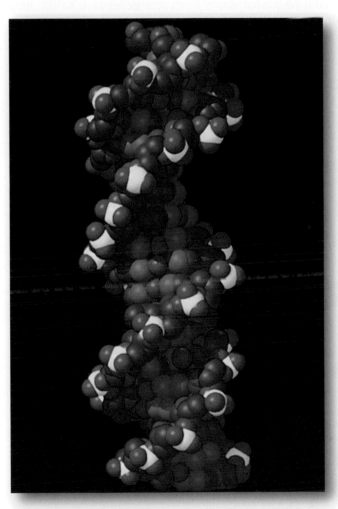

Computer model of DNA. [J. Newdol, Computer Graphics Laboratory, University of California, San Francisco. Copyright by Regents, University of California.]

KEY QUESTIONS

- Before the discovery of the double helix, what was the experimental evidence that DNA is the genetic material?

- What data were used to deduce the double-helix model of DNA?

- How does the double-helical structure suggest a mechanism for DNA replication?

- Why are the proteins that replicate DNA called a biological machine?

- How can DNA replication be both rapid and accurate?

- What special mechanism replicates chromosome ends?

OUTLINE

7.1 DNA: the genetic material

7.2 The DNA structure

7.3 Semiconservative replication

7.4 Overview of DNA replication

7.5 The replisome: a remarkable replication machine

7.6 Assembling the replisome: replication initiation

7.7 Telomeres and telomerase: replication termination

CHAPTER OVERVIEW

In this chapter we focus on DNA, its structure, and the production of DNA copies in a process called replication. James Watson (an American microbial geneticist) and Francis Crick (an English physicist) solved the structure of DNA in 1953. Their model of the structure of DNA was revolutionary. It proposed a definition for the gene in chemical terms and, in doing so, paved the way for an understanding of gene action and heredity at the molecular level.

The story begins in the first half of the twentieth century when the results of several experiments led scientists to conclude that DNA, not some other biological molecule (such as carbohydrates or fats), is the genetic material. DNA is a simple molecule made up of only four different building blocks (the four nucleotides). It was thus necessary to understand how this very simple molecule could be the blueprint for the incredible diversity of organisms on earth.

The model of the double helix proposed by Watson and Crick was built on the results of scientists before them. They relied on earlier discoveries of the chemical composition of DNA and the ratios of its bases. In addition, X-ray diffraction pictures of DNA revealed to the trained eye that DNA is a helix of precise dimensions. Watson and Crick concluded that DNA is a double helix composed of two strands of linked nucleotides that wind around each other.

The proposed structure of the hereditary material immediately suggested how it could serve as a blueprint and how this blueprint could be passed down through the generations. First, the information for making an organism was encoded in the sequence of the nucleotide bases composing the two DNA strands of the helix. Second, because of the rules of base complementarity discovered by Watson and Crick, the sequence of one strand dictated the sequence of the other strand. In this way, the genetic information in the DNA sequence could be passed down from one generation to the next by having each of the separated strands of DNA serve as a template for producing new copies of the molecule, as shown in Figure 7-1.

Precisely how DNA is replicated is still an active area of research 50 years after the discovery of the double helix. Our current understanding of the mechanism of replication gives a central role to a protein machine, called the replisome. This complex of associated proteins coordinates the numerous reactions that are necessary for the rapid and accurate replication of DNA.

CHAPTER OVERVIEW Figure

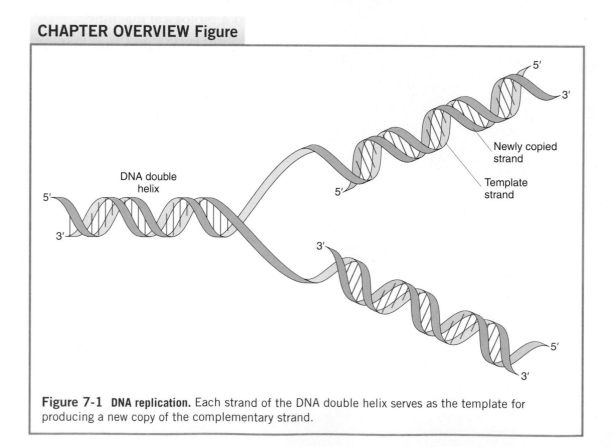

Figure 7-1 DNA replication. Each strand of the DNA double helix serves as the template for producing a new copy of the complementary strand.

7.1 DNA: the genetic material

Before we see how Watson and Crick solved the structure of DNA, let's review what was known about genes and DNA at the time that they began their historic collaboration:

1. Genes—the hereditary "factors" described by Mendel—were known to be associated with specific character traits, but their physical nature was not understood. Similarly, mutations were known to alter gene function, but precisely what a mutation is also was not understood.

2. The one-gene–one-protein theory (described in Chapter 6) postulated that genes control the structure of proteins.

3. Genes were known to be carried on chromosomes.

4. The chromosomes were found to consist of DNA and protein.

5. The results of a series of experiments beginning in the 1920s revealed that DNA was the genetic material. These experiments, described next, showed that bacterial cells that express one phenotype can be transformed into cells that express a different phenotype and that the transforming agent is DNA.

Discovery of transformation

Frederick Griffith made a puzzling observation in the course of experiments on the bacterium *Streptococcus pneumoniae* performed in 1928. This bacterium, which causes pneumonia in humans, is normally lethal in mice. However, some strains of this bacterial species have evolved to be less virulent (less able to cause disease or death). Griffith's experiments are summarized in Figure 7-2. In these experiments, Griffith used two strains that are distinguishable by the appearance of their colonies when grown in laboratory cultures. One strain was a normal virulent type deadly to most laboratory animals. The cells of this strain are enclosed in a polysaccharide capsule, giving colonies a smooth appearance; hence, this strain is identified as S. Griffith's other strain was a mutant nonvirulent type that grows in mice but is not lethal. In this strain, the polysaccharide coat is absent, giving colonies a rough appearance; this strain is called R.

Griffith killed some virulent cells by boiling them. He then injected the heat-killed cells into mice. The mice

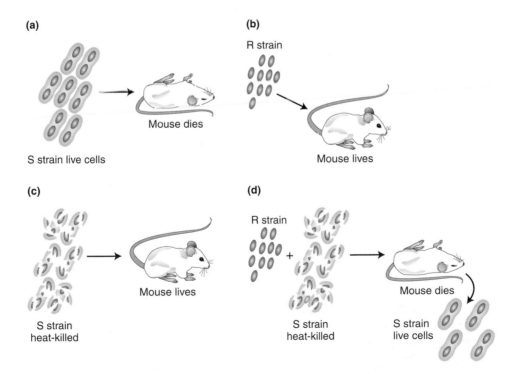

Figure 7-2 The presence of heat-killed S cells transforms live R cells into live S cells.
(a) Mouse dies after injection with the virulent S strain. (b) Mouse survives after injection with the R strain. (c) Mouse survives after injection with heat-killed S strain. (d) Mouse dies after injection with a mixture of heat-killed S strain and live R strain. The heat-killed S strain somehow transforms the R strain into virulence. [Adapted from G. S. Stent and R. Calendar, *Molecular Genetics*, 2d ed. Copyright 1978 by W. H. Freeman and Company. After R. Sager and F. J. Ryan, *Cell Heredity*. Wiley, 1961.]

survived, showing that the carcasses of the cells do not cause death. However, mice injected with a mixture of heat-killed virulent cells and live nonvirulent cells did die. Furthermore, live cells could be recovered from the dead mice; these cells gave smooth colonies and were virulent on subsequent injection. Somehow, the cell debris of the boiled S cells had converted the live R cells into live S cells. The process, already discussed in Chapter 5, is called *transformation*.

The next step was to determine which chemical component of the dead donor cells had caused this transformation. This substance had changed the genotype of the recipient strain and therefore might be a candidate for the hereditary material. This problem was solved by experiments conducted in 1944 by Oswald Avery and two colleagues, C. M. MacLeod and M. McCarty (Figure 7-3). Their approach to the problem was to chemically destroy all the major categories of chemicals in the extract of dead cells one at a time and find out if the extract had lost the ability to transform. The virulent cells had a smooth polysaccharide coat, whereas the nonvirulent cells did not; hence, polysaccharides were an obvious candidate for the transforming agent. However, when polysaccharides were destroyed, the mixture could still transform. Proteins, fats, and ribonucleic acids (RNA) were all similarly shown not to be the transforming agent.

The mixture lost its transforming ability only when the donor mixture was treated with the enzyme deoxyribonuclease (DNase), which breaks up DNA. These results strongly implicate DNA as the genetic material. It is now known that fragments of the transforming DNA that confer virulence enter the bacterial chromosome and replace their counterparts that confer nonvirulence.

> **MESSAGE** The demonstration that DNA is the transforming principle was the first demonstration that genes (the hereditary material) are composed of DNA.

Hershey–Chase experiment

The experiments conducted by Avery and his colleagues were definitive, but many scientists were very reluctant to accept DNA (rather than proteins) as the genetic material. Additional evidence was provided in 1952 by Alfred Hershey and Martha Chase. Their experiment made use of phage T2, a virus that infects bacteria. They reasoned that infecting phage must inject into the bacterium the specific information that dictates the reproduction of new viral particles. If they could find out what material the phage was injecting into the bacterial host, they would have determined the genetic material of phage.

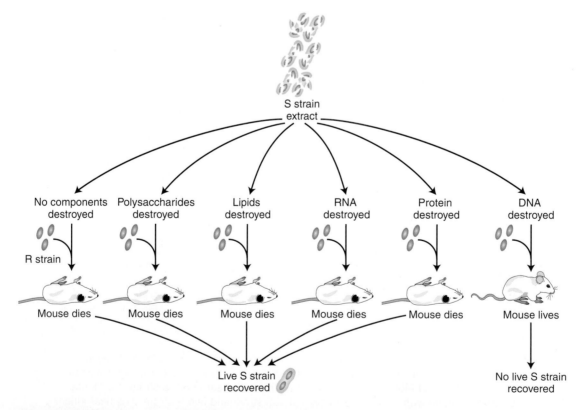

Figure 7-3 DNA is the agent transforming the R strain into virulence. If the DNA in an extract of heat-killed S strain cells is destroyed, then mice injected with a mixture of the heat-killed cells and the live nonvirulent strain R are no longer killed.

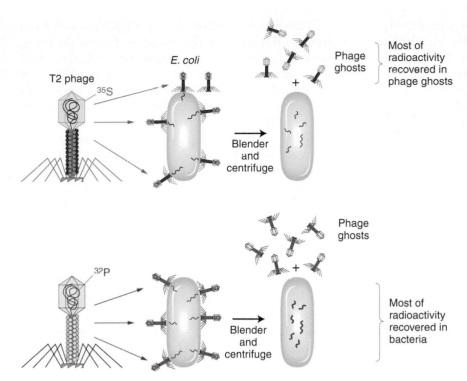

Figure 7-4 The Hershey–Chase experiment demonstrated that the genetic material of phage is DNA, not protein. The experiment uses two sets of T2 bacteriophage. In one set, the protein coat is labeled with radioactive sulfur (^{35}S), not found in DNA. In the other set, the DNA is labeled with radioactive phosphorus (^{32}P), not found in protein. Only the ^{32}P is injected into the *E. coli*, indicating that DNA is the agent necessary for the production of new phage.

The phage is relatively simple in molecular constitution. Most of its structure is protein, with DNA contained inside the protein sheath of its "head." Hershey and Chase decided to label the DNA and protein by using radioisotopes so that they could track the two materials during infection. Phosphorus is not found in proteins but is an integral part of DNA; conversely, sulfur is present in proteins but never in DNA. Hershey and Chase incorporated the radioisotope of phosphorus (^{32}P) into phage DNA and that of sulfur (^{35}S) into the proteins of a separate phage culture. As shown in Figure 7-4, they then infected two *E. coli* cultures with many virus particles per cell: one *E. coli* culture received phage labeled with ^{32}P and the other received phage labeled with ^{35}S. After allowing sufficient time for infection to take place, they sheared the empty phage carcasses (called *ghosts*) off the bacterial cells by agitation in a kitchen blender. They separated the bacterial cells from the phage ghosts in a centrifuge and then measured the radioactivity in the two fractions. When the ^{32}P-labeled phage were used to infect *E. coli*, most of the radioactivity ended up inside the bacterial cells, indicating that the phage DNA entered the cells. When the ^{35}S-labeled phage were used, most of the radioactive material ended up in the phage ghosts, indicating that the phage protein never entered the bacterial cell. The conclusion is inescapable: DNA is the hereditary material. The phage proteins are mere structural packaging that is discarded after delivering the viral DNA to the bacterial cell.

7.2 The DNA structure

Even before the structure of DNA was elucidated, genetic studies indicated that the hereditary material had to have three key properties:

1. Because essentially every cell in the body of an organism has the same genetic makeup, it is crucial that the genetic material be faithfully replicated at every cell division. Thus the structural features of DNA *must allow faithful replication*. These structural features will be considered later in this chapter.

2. Because it must encode the constellation of proteins expressed by an organism, the genetic material *must have informational content*. How the information coded in DNA is deciphered to produce proteins will be the subject of Chapters 8 and 9.

3. Because hereditary changes, called mutations, provide the raw material for evolutionary selection, the genetic material *must be able to change* on rare occasion. Nevertheless, the structure of DNA must be relatively stable so that organisms can rely on its encoded information. We will consider the mechanisms of mutation in Chapter 14.

DNA structure before Watson and Crick

Consider the discovery of the double-helical structure of DNA by Watson and Crick as the solution to a complicated three-dimensional puzzle. Incredibly, Watson and Crick were able to put this puzzle together without doing a single experiment. Rather, they used a process called "model building" in which they assembled the results of earlier experiments (the puzzle pieces) to form the three-dimensional puzzle (the double-helix model). To understand how they did so, we first need to know what pieces of the puzzle were available to Watson and Crick in 1953.

THE BUILDING BLOCKS OF DNA The first piece of the puzzle was knowledge of the basic building blocks of DNA. As a chemical, DNA is quite simple. It contains three types of chemical components: (1) **phosphate,** (2) a sugar called **deoxyribose,** and (3) four nitrogenous **bases**—adenine, guanine, cytosine, and thymine. The carbon atoms in the bases are assigned numbers for ease of reference. The carbon atoms in the sugar group also are assigned numbers, in this case followed by a prime (1′, 2′, and so forth). The sugar in DNA is called "deoxyribose" because it has only a hydrogen atom (H) at

the 2′ carbon atom, unlike ribose (of RNA), which has a hydroxyl (OH) group at that position. Two of the bases, adenine and guanine, have a double-ring structure characteristic of a type of chemical called a **purine.** The other two bases, cytosine and thymine, have a single-ring structure of a type called a **pyrimidine.** The chemical components of DNA are arranged into groups called **nucleotides,** each composed of a phosphate group, a deoxyribose sugar molecule, and any one of the four bases (Figure 7-5). It is convenient to refer to each nucleotide by the first letter of the name of its base: A, G, C, or T. The nucleotide with the adenine base is called deoxyadenosine 5′-monophosphate, where the 5′ refers to the position of the carbon atom in the sugar ring to which the single (mono) phosphate group is attached.

CHARGAFF'S RULES OF BASE COMPOSITION The second piece of the puzzle used by Watson and Crick came from work done several years earlier by Erwin Chargaff. Studying a large selection of DNAs from different organisms (Table 7-1), Chargaff established cer-

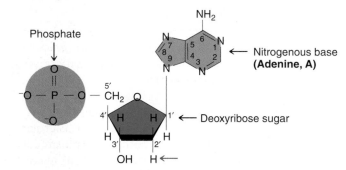

Figure 7-5 Chemical structures of the four nucleotides found in DNA. These nucleotides, two with purine bases and two with pyrimidine bases, are the fundamental building blocks of DNA. The sugar is called *deoxyribose* because it is a variation of a common sugar, ribose, that has one more oxygen atom (position indicated by the red arrow).

Table 7-1 Molar Properties of Bases* in DNAs from Various Sources

Organism	Tissue	Adenine	Thymine	Guanine	Cytosine	$\dfrac{A + T}{G + C}$
Escherichia coli (K12)	—	26.0	23.9	24.9	25.2	1.00
Diplococcus pneumoniae	—	29.8	31.6	20.5	18.0	1.59
Mycobacterium tuberculosis	—	15.1	14.6	34.9	35.4	0.42
Yeast	—	31.3	32.9	18.7	17.1	1.79
Paracentrotus lividus (sea urchin)	Sperm	32.8	32.1	17.7	18.4	1.85
Herring	Sperm	27.8	27.5	22.2	22.6	1.23
Rat	Bone marrow	28.6	28.4	21.4	21.5	1.33
Human	Thymus	30.9	29.4	19.9	19.8	1.52
Human	Liver	30.3	30.3	19.5	19.9	1.53
Human	Sperm	30.7	31.2	19.3	18.8	1.62

*Defined as moles of nitrogenous constituents per 100 g-atoms phosphate in hydrolysate.
Source: E. Chargaff and J. Davidson, eds., *The Nucleic Acids.* Academic Press, 1995.

tain empirical rules about the amounts of each type of nucleotide found in DNA:

1. The total amount of pyrimidine nucleotides (T + C) always equals the total amount of purine nucleotides (A + G).

2. The amount of T always equals the amount of A, and the amount of C always equals the amount of G. But the amount of A + T is not necessarily equal to the amount of G + C, as can be seen in the last column of Table 7-1. This ratio varies among different organisms but is virtually the same in different tissues of the same organism.

X-RAY DIFFRACTION ANALYSIS OF DNA The third and most important piece of the puzzle came from X-ray diffraction data on DNA structure that were collected by Rosalind Franklin when she was in the laboratory of Maurice Wilkins (Figure 7-6). In such experiments, X rays are fired at DNA fibers, and the scatter of the rays from the fibers is observed by catching the rays on photographic film, on which the X rays produce spots. The angle of scatter represented by each spot on the film gives information about the position of an atom or certain groups of atoms in the DNA molecule. This procedure is not simple to carry out (or to explain), and the interpretation of the spot patterns is very difficult. The available data suggested that DNA is long and skinny and that it has two similar parts that are parallel to each other and run along the length of the molecule. The X-ray data showed the molecule to be helical (spiral-like). Other regularities were present in the spot patterns, but no one had yet thought of a three-dimensional structure that could account for just those spot patterns.

The double helix

A 1953 paper by Watson and Crick in the journal *Nature* began with two sentences that ushered in a new age of biology: "We wish to suggest a structure for the salt of deoxyribose nucleic acid (D.N.A.). This structure has novel features which are of considerable biological interest." The structure of DNA had been a subject of great debate since the experiments of Avery and co-workers in 1944. As we have seen, the general composition of DNA was known, but how the parts fit together was not known. The structure had to fulfill the main requirements for a hereditary molecule: the ability to store information, the ability to be replicated, and the ability to mutate.

The three-dimensional structure derived by Watson and Crick is composed of two side-by-side chains ("strands") of nucleotides twisted into the shape of a **double helix** (Figure 7-7). The two nucleotide strands are held together by weak association between the bases of each strand, forming a structure like a spiral staircase (Figure 7-8a). The backbone of each strand is formed of alternating phosphate and deoxyribose sugar units that are connected by phosphodiester linkages. (Figure 7-8b). We can use these linkages to describe how a nucleotide chain is organized. As already mentioned, the carbon atoms of the sugar groups are numbered 1' through 5'. A phosphodiester linkage connects the 5' carbon atom of one deoxyribose to the 3' carbon atom of the adjacent deoxyribose. Thus, each sugar-phosphate backbone is said to have a 5'-to-3' polarity, or direction, and understanding this polarity is essential in understanding

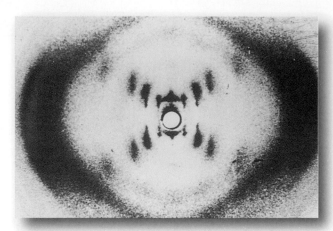

Figure 7-6 Rosalind Franklin *(left)* and her X-ray diffraction pattern of DNA *(right)*. [*(Left)* courtesy of National Portrait Gallery, London. *(Right)* Rosalind Franklin/Science Source/ Photo Researchers.]

Figure 7-7 James Watson and Francis Crick with their DNA model. [Camera Press.]

how DNA fulfills its roles. In the double-stranded DNA molecule, the two backbones are in opposite, or **antiparallel,** orientation (see Figure 7-8b).

Each base is attached to the 1′ carbon atom of a deoxyribose sugar in the backbone of each strand and faces inward toward a base on the other strand. Hydrogen bonds between pairs of bases hold the two strands of the DNA molecule together. The hydrogen bonds are indicated by dashed lines in Figure 7-8b.

Two nucleotide strands paired in an antiparallel manner automatically assume a double-helical confor-

mation (Figure 7-9), mainly through interaction of the base pairs. The base pairs, which are flat planar structures, stack on top of one another at the center of the double helix (Figure 7-9a). Stacking adds to the stability of the DNA molecule by excluding water molecules from the spaces between the base pairs. The most stable form that results from base stacking is a double helix with two distinct sizes of grooves running in a spiral: the **major groove** and the **minor groove,** which can be seen in both the ribbon and the space-filling (Figure 7-9b) models. A single strand of nucleotides has no helical structure; the helical shape of DNA depends entirely on the pairing and stacking of the bases in the antiparallel strands. DNA is a right-handed helix; in other words, it has the same structure as that of a screw that would be screwed into place by using a clockwise turning motion.

The double helix accounted nicely for the X-ray data and successfully accounted for Chargaff's data. Studying models that they made of the structure, Watson and Crick realized that the observed radius of the double helix (known from the X-ray data) would be explained if a purine base always pairs (by hydrogen bonding) with a pyrimidine base (Figure 7-10, page 236). Such pairing would account for the $(A + G) = (T + C)$ regularity observed by Chargaff, but it would predict four possible pairings: T···A, T···G, C···A, and C···G. Chargaff's data, however, indicate that T pairs only with A and C pairs only with G. Watson and Crick concluded that each base pair consists of one purine base and one pyrimidine base, paired according to the following rule: G pairs with C, and A pairs with T.

Note that the G–C pair has three hydrogen bonds, whereas the A–T pair has only two. We would predict

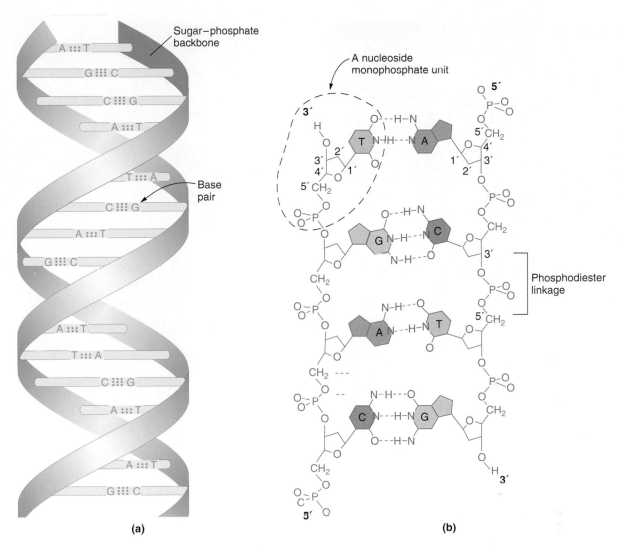

Figure 7-8 The structure of DNA. (a) A simplified model showing the helical structure of DNA. The sticks represent base pairs, and the ribbons represent the sugar-phosphate backbones of the two antiparallel chains. (b) An accurate chemical diagram of the DNA double helix, unrolled to show the sugar-phosphate backbones (blue) and base-pair rungs (red). The backbones run in opposite directions; the 5′ and 3′ ends are named for the orientation of the 5′ and 3′ carbon atoms of the sugar rings. Each base pair has one purine base, adenine (A) or guanine (G), and one pyrimidine base, thymine (T) or cytosine (C), connected by hydrogen bonds *(dotted lines)*. [From R. E. Dickerson, "The DNA Helix and How It Is Read." Copyright 1983 by Scientific American, Inc. All rights reserved.]

that DNA containing many G–C pairs would be more stable than DNA containing many A–T pairs. In fact, this prediction is confirmed. Heat causes the two strands of DNA double helix to separate (a process called DNA melting or DNA denaturation); it can be shown that DNAs with higher G + C content require higher temperatures to melt because of the greater attraction of the G–C pairing.

> **MESSAGE** DNA is a double helix composed of two nucleotide chains held together by complementary pairing of A with T and G with C.

Watson and Crick's discovery of the structure of DNA is considered by some to be the most important biological discovery of the twentieth century. The reason is that the double helix, in addition to being consistent with earlier data about DNA structure, fulfilled the three requirements for a hereditary substance.

1. The double-helical structure suggested how the genetic material might determine the structure of proteins. Perhaps the *sequence* of nucleotide pairs in DNA dictates the sequence of amino acids in the protein specified by that gene. In other words, some sort of **genetic code** may write information in DNA

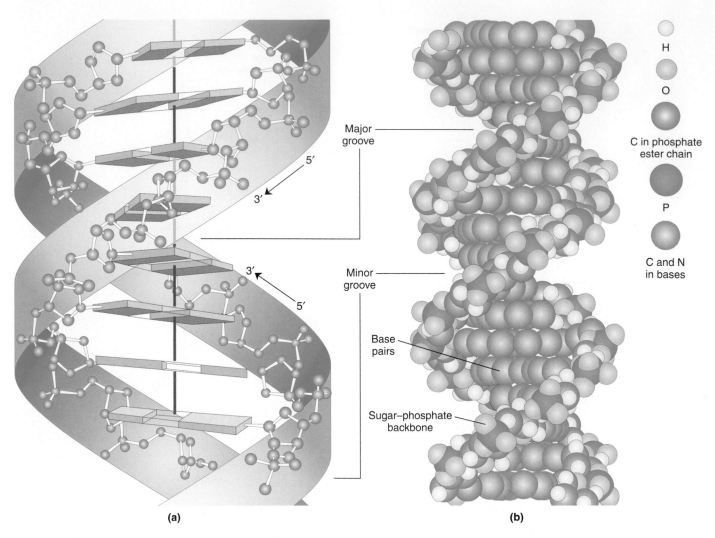

Major
groove

5′

3′

3′

Minor
groove

5′

Base
pairs

Sugar–phosphate
backbone

H

O

C in phosphate
ester chain

P

C and N
in bases

(a)   **(b)**

Figure 7-9 Two representations of the DNA double helix. [Part b from C. Yanofsky, "Gene Structure
and Protein Structure." Copyright 1967 by Scientific American, Inc. All rights reserved].

as a sequence of nucleotides and then translate it into
a different language of amino acid sequences
in protein. Just how it is done is the subject of
Chapter 9.

2. If the base sequence of DNA specifies the amino acid
   sequence, then mutation is possible by the
   substitution of one type of base for another at one or
   more positions. Mutations will be discussed in
   Chapter 14.

3. As Watson and Crick cryptically stated in the
   concluding words of their 1953 *Nature* paper that
   reported the double-helical structure of DNA: "It
   has not escaped our notice that the specific pairing
   we have postulated immediately suggests a
   possible copying mechanism for the genetic
   material." To geneticists at the time, the meaning of
   this statement was clear, as we see in the next
   section.

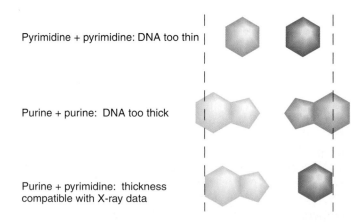

Pyrimidine + pyrimidine: DNA too thin

Purine + purine: DNA too thick

Purine + pyrimidine: thickness
compatible with X-ray data

Figure 7-10 Base pairing in DNA. The pairing of purines with
pyrimidines accounts exactly for the diameter of the DNA
double helix determined from X-ray data. That diameter is
indicated by the vertical dashed lines. [From R. E. Dickerson,
"The DNA Helix and How It Is Read." Copyright 1983 by Scientific
American, Inc. All rights reserved.]

7.3 Semiconservative replication

The copying mechanism to which Watson and Crick referred is called semiconservative and is diagrammed in Figure 7-11. The sugar-phosphate backbones are represented by thick ribbons, and the sequence of base pairs is random. Let's imagine that the double helix is like a zipper that unzips, starting at one end (at the bottom in Figure 7-11). We can see that, if this zipper analogy is valid, the unwinding of the two strands will expose single bases on each strand. Each exposed base has the potential to pair with free nucleotides in solution. Because the DNA structure imposes strict pairing requirements, each exposed base will pair only with its **complementary base,** A with T and G with C. Thus, each of the two single strands will act as a **template,** or mold, to direct the assembly of complementary bases to reform a double helix identical with the original. The newly added nucleotides are assumed to come from a pool of free nucleotides that must be present in the cell.

If this model is correct, then each daughter molecule should contain one parental nucleotide chain and one newly synthesized nucleotide chain. However, a little thought shows that there are at least three different ways in which a parental DNA molecule might be related to the daughter molecules. These hypothetical modes of replication are called semiconservative (the Watson–Crick model), conservative, and dispersive (Figure 7-12). In **semiconservative replication,** the double helix of each daughter DNA molecule contains one strand from the original DNA molecule and one newly synthesized strand. However, in **conservative replication,** the parent DNA molecule is conserved, and a single daughter double helix is produced consisting of two newly synthesized strands. In **dispersive replication,** daughter molecules consist of strands *each* containing segments of *both* parental DNA and newly synthesized DNA.

Meselson–Stahl experiment

The first problem in understanding DNA replication was to figure out whether the mechanism of replication was semiconservative, conservative, or dispersive.

In 1958, two young scientists, Matthew Meselson and Franklin Stahl, set out to discover which of these possibilities correctly described DNA replication. Their idea was to allow parental DNA molecules containing nucleotides of one density to replicate in medium containing nucleotides of different density. If DNA replicated semiconservatively, the daughter molecules should be half old and half new and therefore of intermediate density. To carry out their experiment, they grew *E. coli* cells in a medium containing the heavy isotope of

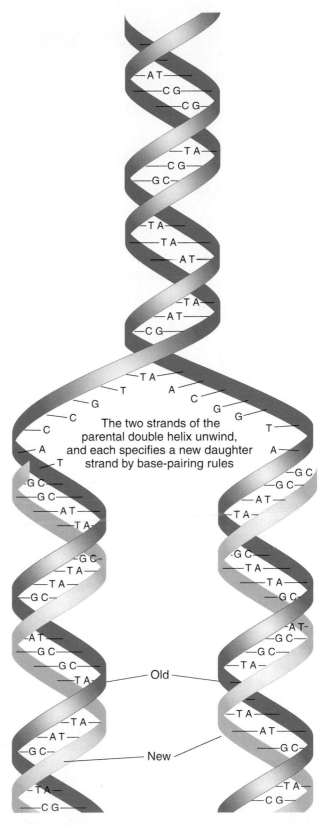

Figure 7-11 Semiconservative DNA replication. The model of DNA replication proposed by Watson and Crick is based on the hydrogen-bonded specificity of the base pairs. Parental strands, shown in blue, serve as templates for polymerization. The newly polymerized strands, shown in gold, have base sequences that are complementary to their respective templates.

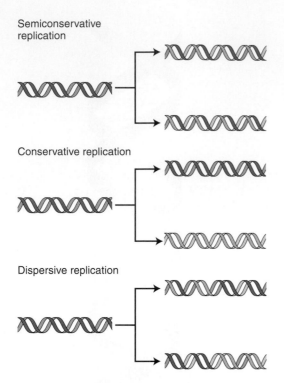

Semiconservative
replication

Conservative replication

Dispersive replication

Figure 7-12 Three alternative patterns for DNA replication.
The Watson–Crick model would produce the first (semi-conservative) pattern. Gold lines represent the newly synthesized strands.

nitrogen (^{15}N) rather than the normal light (^{14}N) form. This isotope was inserted into the nitrogen bases, which then were incorporated into newly synthesized DNA strands. After many cell divisions in ^{15}N, the DNA of the cells were well labeled with the heavy isotope. The cells were then removed from the ^{15}N medium and put into a ^{14}N medium; after one and two cell divisions, samples were taken and the DNA was isolated from each sample.

Meselson and Stahl were able to distinguish DNA of different densities because the molecules can be separated from each other by a procedure called *cesium chloride gradient centrifugation*. If cesium chloride is spun in a centrifuge at tremendously high speeds (50,000 rpm) for many hours, the cesium and chloride ions tend to be pushed by centrifugal force toward the bottom of the tube. Ultimately, a gradient of ions is establshed in the tube, with the highest ion concentration, or density, at the bottom. DNA centrifuged with the cesium chloride forms a band at a position identical with its density in the gradient. DNA of different densities will form bands at different places. Cells initially grown in the heavy isotope ^{15}N showed DNA of high density. This DNA is shown in red at the left-hand side of Figure 7-13a. After growing these cells in the light isotope ^{14}N for one generation, the researchers found that the DNA was of intermediate density, shown half red (^{15}N) and half green (^{14}N) in the central part. After two generations, both

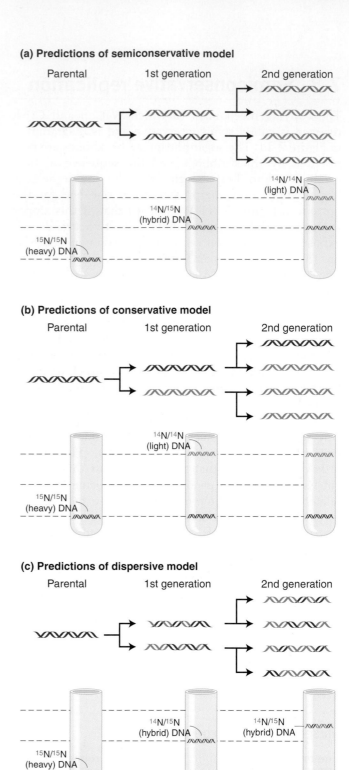

(a) Predictions of semiconservative model

Parental 1st generation 2nd generation

^{14}N/^{14}N (light) DNA
^{14}N/^{15}N (hybrid) DNA
^{15}N/^{15}N (heavy) DNA

(b) Predictions of conservative model

Parental 1st generation 2nd generation

^{14}N/^{14}N (light) DNA
^{15}N/^{15}N (heavy) DNA

(c) Predictions of dispersive model

Parental 1st generation 2nd generation

^{14}N/^{15}N (hybrid) DNA
^{14}N/^{15}N (hybrid) DNA
^{15}N/^{15}N (heavy) DNA

Figure 7-13 The Meselson–Stahl experiment demonstrates that DNA is copied by semiconservative replication. DNA centrifuged in a cesium chloride (CsCl) gradient will form bands according to its density. (a) When the cells grown in ^{15}N are transferred to a ^{14}N medium, the first generation produces a single intermediate DNA band and the second generation produces two bands: one intermediate and one light. This result matches the predictions of the semiconservative model of DNA replication. (b and c) The results predicted for conservative and dispersive replication, shown here, were *not* found.

intermediate- and low-density DNA was observed (right-hand side of Figure 7-13a), precisely as predicted by the Watson–Crick model.

> **MESSAGE** DNA is replicated by the unwinding of the two strands of the double helix and the building up of a new complementary strand on each of the separated strands of the original double helix.

The replication fork

Another prediction of the Watson–Crick model of DNA replication is that a replication zipper, or *fork*, will be found in the DNA molecule during replication. This fork is the location at which the double helix is unwound to produce the two single strands that serve as templates for copying. In 1963, John Cairns tested this prediction by allowing replicating DNA in bacterial cells to incorporate tritiated thymidine ([³H]thymidine)— the thymine nucleotide labeled with a radioactive hydrogen isotope called tritium. Theoretically, each newly synthesized daughter molecule should then contain one radioactive ("hot") strand (with ³H) and another nonradioactive ("cold") strand (with ²H). After varying intervals and varying numbers of replication cycles in a "hot" medium, Cairns carefully lysed the bacteria and allowed the cell contents to settle onto a piece of filter paper, which was put on a microscope slide. Finally, Cairns covered the filter with photographic emulsion and exposed it in the dark for 2 months. This procedure, called autoradiography, allowed Cairns to develop a picture of the location of ³H in the cell material. As ³H decays, it emits a beta particle (an energetic electron). The photographic emulsion detects a chemical reaction that takes place wherever a beta particle strikes the emulsion. The emulsion can then be developed like a photographic print so that the emission track of the beta particle appears as a black spot or grain.

After one replication cycle in [³H]thymidine, a ring of dots appeared in the autoradiograph. Cairns interpreted this ring as a newly formed radioactive strand in a circular daughter DNA molecule, as shown in Figure 7-14a. It is thus apparent that the bacterial chromosome is circular—a fact that also emerged from genetic analysis described earlier (Chapter 5). In the second replication cycle, the forks predicted by the model were indeed seen. Furthermore, the density of grains in the three segments was such that the interpretation shown in Figure 7-14b could be made: the thick curve of dots cutting through the interior of the circle of DNA would be the newly synthesized daughter strand, this time consisting of *two* radioactive strands. Cairns saw all sizes of these moon-shaped, autoradiographic patterns, corresponding to the progressive movement of the replication forks,

(a) Chromosome after one round of replication

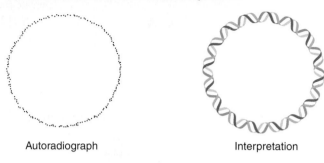

Autoradiograph   Interpretation

(b) Chromosome during second round of replication

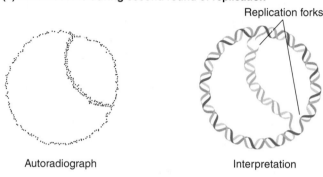

Autoradiograph   Interpretation

Figure 7-14 A replicating bacterial chromosome has two replication forks. (a) *Left:* Autoradiograph of a bacterial chromosome after one replication in tritiated thymidine. According to the semiconservative model of replication, one of the two strands should be radioactive. *Right:* Interpretation of the autoradiograph. The gold helix represents the tritiated strand. (b) *Left:* Autoradiograph of a bacterial chromosome in the second round of replication in tritiated thymidine. In this theta (θ) structure, the newly replicated double helix that crosses the circle could consist of two radioactive strands (if the parental strand were the radioactive one). *Right:* The double thickness of the radioactive tracing on the autoradiogram appears to confirm the interpretation shown here.

around the ring. Structures of the sort shown in Figure 7-14b are called **theta (θ) structures.**

DNA polymerases

A problem confronted by scientists was to understand just how the bases are brought to the double-helix template. Although scientists suspected that enzymes played a role, that possibility was not proved until 1959, when Arthur Kornberg isolated DNA polymerase. This enzyme adds deoxyribnucleotides to the 3′ end of a growing nucleotide chain, using for its template a single strand of DNA that has been exposed by localized unwinding of the double helix (Figure 7-15). The substrates for DNA polymerase are the triphosphate forms of the deoxyribonucleotides, dATP, dGTP, dCTP, and dTTP.

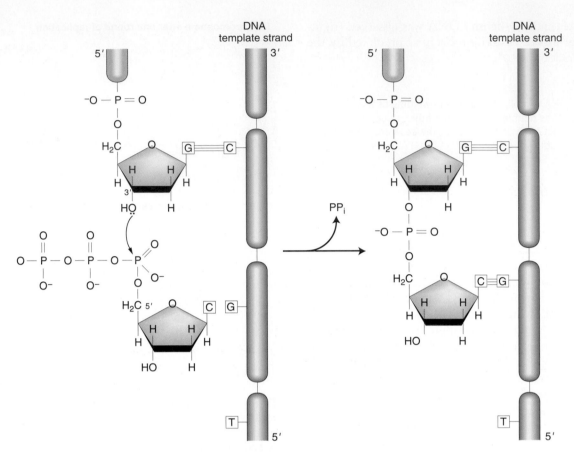

Figure 7-15 Chain-elongation reaction catalyzed by DNA polymerase. Energy for the reaction comes from breaking the high-energy phosphate bond of the triphosphate substrate.

We now know that there are three DNA polymerases in *E. coli*. The first enzyme that Kornberg purified is now called DNA polymerase I, or pol I. Although pol I has a role in DNA replication (see next section), DNA pol III carries out the majority of DNA synthesis.

7.4 Overview of DNA replication

As DNA pol III moves forward, the double helix is continuously unwinding ahead of the enzyme to expose further lengths of single DNA strands that will act as templates (Figure 17-16). DNA pol III acts at the **replication fork,** the zone where the double helix is unwinding. However, because DNA polymerase always adds nucleotides at the 3′ *growing tip*, only one of the two antiparallel strands can serve as a template for replication in the direction of the replication fork. For this strand, synthesis can take place in a smooth continuous manner in the direction of the fork; the new strand synthesized on this template is called the **leading strand.**

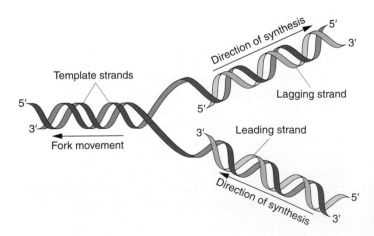

Figure 7-16 DNA replication at the growing fork. The replication fork moves in DNA synthesis as the double helix continuously unwinds. Synthesis of the leading strand can proceed smoothly without interruption in the direction of movement of the replication fork, but synthesis of the lagging strand must proceed in the opposite direction, away from the replication fork.

1. Primase synthesizes short RNA oligonucleotides (primer) copied from DNA.

2. DNA polymerase III elongates RNA primers with new DNA.

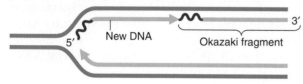

3. DNA polymerase I removes RNA at 5′ end of neighboring fragment and fills gap.

4. DNA ligase connects adjacent fragments.

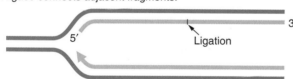

Figure 7-17 Steps in the synthesis of the lagging strand. DNA synthesis proceeds by continuous synthesis on the leading strand and discontinuous synthesis on the lagging strand.

Synthesis on the other template also takes place at 3′ growing tips, but this synthesis is in the "wrong" direction, because, for this strand, the 5′-to-3′ direction of synthesis is away from the replication fork (see Figure 7-16). As we will see, the nature of the replication machinery requires that synthesis of both strands take place in the region of the replication fork. Therefore, synthesis moving away from the growing fork cannot go on for long. It must be in short segments: polymerase synthesizes a segment, then moves back to the segment's 5′ end, where the growing fork has exposed new template, and begins the process again. These short (1000–2000 nucleotides) stretches of newly synthesized DNA are called **Okazaki fragments.**

Another problem in DNA replication arises because DNA polymerase can extend a chain but cannot start a chain. Therefore, synthesis of both the leading strand and each Okazaki fragment must be initiated by a **primer,** or short chain of nucleotides, that binds with the template strand to form a segment of duplex DNA. The primer in DNA replication can be seen in Figure 7-17. The primers

are synthesized by a set of proteins called a **primosome,** of which a central component is an enzyme called **primase,** a type of RNA polymerase. Primase synthesizes a short (~8–12 nucleotides) stretch of RNA complementary to a specific region of the chromosome. On the leading strand, only one initial primer is needed because, after the initial priming, the growing DNA strand serves as the primer for continuous addition. However, on the lagging strand, every Okazaki fragment needs its own primer. The RNA chain composing the primer is then extended as a DNA chain by DNA pol III.

A different DNA polymerase, pol I, removes the RNA primers and fills in the resulting gaps with DNA. As mentioned earlier, pol I is the enzyme originally purified by Kornberg. Another enzyme, **DNA ligase,** joins the 3′ end of the gap-filling DNA to the 5′ end of the downstream Okazaki fragment. The new strand thus formed is called the **lagging strand.** DNA ligase joins broken pieces of DNA by catalyzing the formation of a phosphodiester bond between the 5′-phosphate end of one fragment and the adjacent 3′-OH group of another fragment.

It is the only enzyme that can seal DNA chains.

MESSAGE DNA replication takes place at the replication fork, where the double helix is unwinding and the two strands are separating. DNA replication proceeds continuously in the direction of the unwinding replication fork on the leading strand. DNA is synthesized in short segments, in the direction away from the replication fork, on the lagging strand. DNA polymerase requires a primer, or short chain of nucleotides, to be already in place to begin synthesis.

One of the hallmarks of DNA replication is its accuracy, also called fidelity: overall, there is less than one error per 10^{10} nucleotides inserted. Part of the reason for the accuracy of DNA replication is that both DNA pol I and DNA pol III possess 3′-to-5′ exonuclease activity, which serves a "proofreading" function by excising mismatched bases that were inserted erroneously. Strains lacking a functional 3′-to-5′ exonuclease have a higher rate of mutation. In addition, because primase lacks a proofreading function, the RNA primer is more likely than DNA to contain errors. To maintain the high fidelity of replication, it is essential that the RNA primers at the ends of Okazaki fragments be removed and replaced with DNA by DNA pol I. The subject of DNA repair will be covered in detail in Chapter 14.

7.5 The replisome: a remarkable replication machine

The second hallmark of DNA replication is speed. The time needed for *E. coli* to replicate its chromosome can be as short as 40 minutes. Therefore, its genome of about 5 million base pairs must be copied at a rate of about *2000 nucleotides per second.* From the experiment of Cairns, we know that *E. coli* uses only two replication forks to copy its entire genome. Thus, each fork must be able to move at a rate of as many as *1000 nucleotides per second.* What is remarkable about the entire process of

DNA replication is that it does not sacrifice speed for accuracy. How can it maintain both speed and accuracy, given the complexity of the reactions that need to be carried out at the replication fork? The answer is that DNA polymerase is actually part of a large "nucleoprotein" complex that coordinates the activities at the replication fork. This complex, called the **replisome,** is an example of a "molecular machine". You will encounter other examples in later chapters. The discovery that most of the major functions of cells—replication, transcription, and translation, for example—are carried out by large multi-subunit complexes has changed the way that we think about the cell. To begin to understand why, let's look at the replisome more closely.

Some of the interacting components of the replisome in *E. coli* are shown in Figure 7-18. At the replication fork, the catalytic core of DNA pol III is actually part of a much larger complex, called the **pol III holoenzyme,** which consists of two catalytic cores and

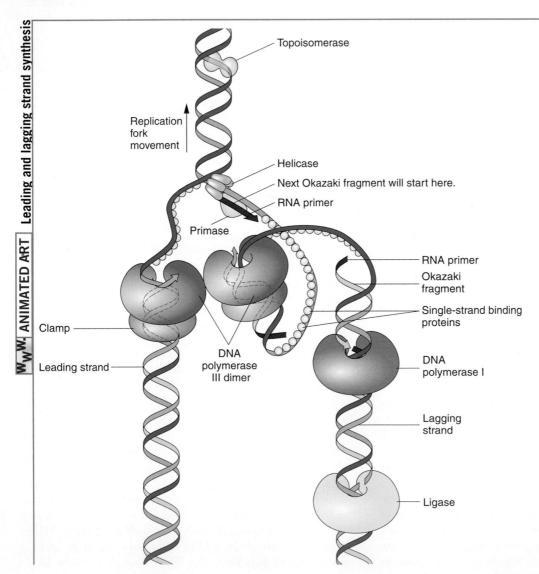

Leading and lagging strand synthesis

ANIMATED ART www

Topoisomerase

Replication fork movement

Helicase

Next Okazaki fragment will start here.

RNA primer

Primase

Clamp

Leading strand

DNA polymerase III dimer

RNA primer

Okazaki fragment

Single-strand binding proteins

DNA polymerase I

Lagging strand

Ligase

Figure 7-18 Details of the replisome and accessory proteins at the replication fork. Topoisomerase and helicase unwind and open the double helix in preparation for DNA replication. Once unwound, single-strand binding proteins prevent the double helix from re-forming. The figure is a representation of the so-called trombone model (named for its resemblance to a trombone owing to the looping of the lagging strand) showing how the two catalytic cores of the replisome are envisioned to interact to coordinate the numerous events of leading- and lagging-strand replication. [Adapted from Geoffrey Cooper, *The Cell.* Sinauer Associates, 2000.]

many **accessory proteins.** One of the catalytic cores handles the synthesis of the leading strand while the other handles lagging-strand synthesis. Some of the accessory proteins (not visible in Figure 7-18) form a connection that bridges the two catalytic cores, thus coordinating the synthesis of the leading and lagging strands. The lagging strand is shown looping around so that the replisome can coordinate the synthesis of both strands and move in the direction of the replication fork. Also shown is an important accessory protein called the **sliding clamp,** which encircles the DNA like a donut. Its association with the clamp protein keeps pol III attached to the DNA molecule. Thus, pol III is transformed from an enzyme that can add only 10 nucleotides before falling off the template (termed a **distributive enzyme**) to an enzyme that stays at the moving fork and adds tens of thousands of nucleotides (a **processive enzyme**). In sum, through the action of accessory proteins, synthesis of both the leading and the lagging strands is rapid and highly coordinated. Note also that primase, the enzyme that synthesizes the RNA primer, is not touching the clamp protein. Therefore, primase will act as a distributive enzyme—it adds only a few ribonucleotides before dissociating from the template. This mode of action makes sense because the primer need be only long enough to form a suitable duplex starting point for DNA pol III.

Unwinding the double helix

When the double helix was proposed in 1953, a major objection was that the replication of such a structure would require the unwinding of the double helix at the replication fork and the breaking of the hydrogen bonds that hold the strands together. How could DNA be unwound so rapidly and, even if it could, wouldn't that overwind the DNA behind the fork and make it hopelessly tangled? We now know that the replisome contains two classes of proteins that open the helix and prevent overwinding: they are **helicases** and **topoisomerases,** respectively. Helicases are enzymes that disrupt the hydrogen bonds that hold the two strands of the double helix together. Like the clamp protein, the helicase fits like a donut around the DNA; from this position, it rapidly unzips the double helix ahead of DNA synthesis. The unwound DNA is stabilized by **single-strand binding (SSB) proteins,** which bind to single-stranded DNA and prevent the duplex from re-forming.

Circular DNA can be twisted and coiled, much like the extra coils that can be introduced into a rubber band. Untwisting of the replication fork by helicases causes extra twisting at other regions, and coils called supercoils form to release the strain of the extra twisting. Both the twists and the supercoils must be removed

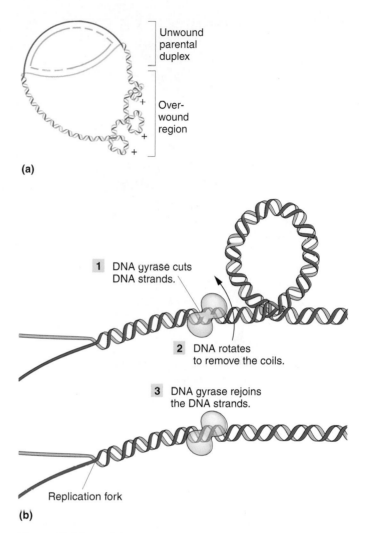

(a)

(b)

1 DNA gyrase cuts DNA strands.

2 DNA rotates to remove the coils.

3 DNA gyrase rejoins the DNA strands.

Replication fork

Figure 7-19 Activity of DNA gyrase, a topoisomerase, during replication. (a) Extra-twisted (positively supercoiled) regions accumulate ahead of the fork as the parental strands separate for replication. (b) A topoisomerase such as DNA gyrase removes these regions, by cutting the DNA strands, allowing them to rotate, and then rejoining the strands. [Part a from A. Kornberg and T. A. Baker, *DNA Replication*, 2d ed. Copyright 1992 by W. H. Freeman and Company.]

to allow replication to continue. This supercoiling can be created or relaxed by enzymes termed topoisomerases, an example of which is DNA gyrase (Figure 7-19). Topoisomerases relax supercoiled DNA by breaking either a single DNA strand or both strands, which allows DNA to rotate into a relaxed molecule. Topoisomerase finishes by rejoining the strands of the now relaxed DNA molecule.

MESSAGE A molecular machine called the replisome carries out DNA synthesis. It includes two DNA polymerase units to handle synthesis on each strand and coordinates the activity of accessory proteins required for priming, unwinding the double helix, and stabilizing the single strands.

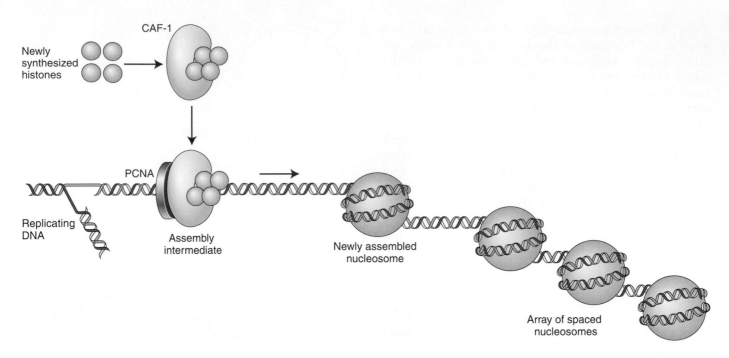

Figure 7-20 Assembling nucleosomes during DNA replication. The protein CAF-1 brings histones to the replication fork, where they are assembled to form nucleosomes. For simplicity, nucleosome assembly is shown on only one of the replicated DNA molecules.

The eukaryotic replisome

DNA replication in both prokaryotes and eukaryotes uses a semiconservative mechanism and employs leading- and lagging-strand synthesis. For this reason, it should not come as a surprise that the components of the replisome in prokaryotes and eukaryotes are very similar. However, as organisms increase in complexity, the number of replisome components also increases. There are now known to be 13 components of the *E. coli* replisome and at least 27 in the replisomes of yeast and mammals. One reason for the added complexity of the eukaryotic replisome is the higher complexity of the eukaryotic template. Recall that, unlike the bacterial chromosome, eukaryotic chromosomes exist in the nucleus as chromatin. As described in Chapter 3, the basic unit of chromatin is the **nucleosome,** which consists of DNA wrapped around histone proteins. Thus, the replisome has to not only copy the parental strands, but also disassemble the nucleosomes in the parental strands and reassemble them in the daughter molecules. This is done by randomly distributing the old histones (from the existing nucleosomes) to daughter molecules and delivering new histones in association with a protein called **chromatin assembly factor 1 (CAF-1)** to the replisome. CAF-1 binds to histones and targets them to the replication fork, where they can be assembled together with newly synthesized DNA.

CAF-1 and its cargo of histones arrive at the replication fork by binding to the eukaryotic version of the clamp protein, called **proliferating cell nuclear antigen (PCNA)** (Figure 7-20).

> **MESSAGE** The eukaryotic replisome performs all the functions of the prokaryotic replisome, and in addition it must disassemble and reassemble the protein–DNA complexes called nucleosomes.

7.6 Assembling the replisome: replication initiation

Assembly of the replisome in both prokaryotes and eukaryotes is an orderly process that begins at precise sites on the chromosome (called **origins**) and takes place only at certain times in the life of the cell.

Prokaryotic origins of replication

E. coli replication begins from a fixed origin (called *oriC*) and then proceeds in both directions (with moving forks at both ends, as previously shown in Figure 7-14) until the forks merge. Figure 7-21a shows the process. The first step in the assembly of the replisome is the binding of a

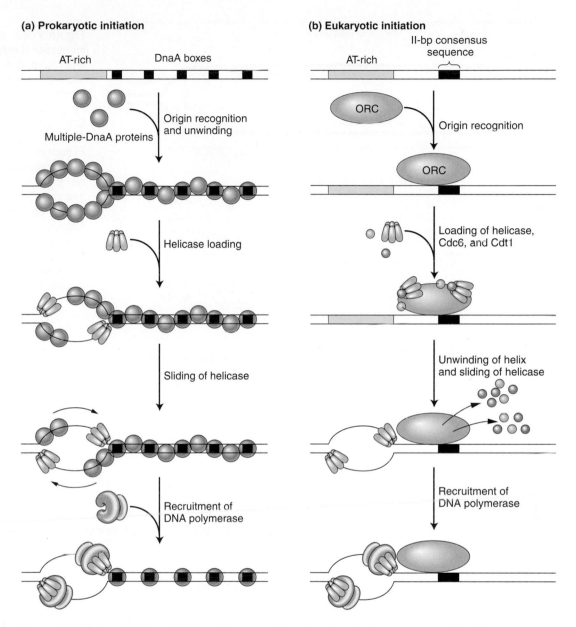

Figure 7-21 The initiation of DNA synthesis at origins of replication in prokaryotes (a) and in yeast, a simple eukaryote (b). In both cases, proteins bind to the origin (*oriC* and *ORC*), where they separate the two strands of the double helix and recruit replisome components to the two replication forks. In eukaryotes, replication is linked to the cell cycle through the availability of two proteins: Cdc6 and Cdt1.

protein called **DnaA** to a specific 13-bp sequence (called a "DnaA box") that is repeated five times in *oriC*. In response to the binding of DnaA, the origin is unwound at a cluster of A and T nucleotides. Recall that AT base pairs are held together only with two hydrogen bonds, whereas GC base pairs are held together by three. Thus, it is easier to separate **(melt)** the double helix at stretches of DNA that are enriched in A and T bases.

After unwinding begins, additional DnaA proteins bind to the newly unwound single-stranded regions.

With DnaA coating the origin, two helicases (the **DnaB** protein) now bind and slide in a 5′-to-3′ direction to begin unzipping the helix at the replication fork. Primase and DNA pol III holoenzyme are now recruited to the replication fork by protein–protein interactions and DNA synthesis begins. You may be wondering why DnaA is not present in Figure 7-18 (the replisome machine). The answer is that, although it is necessary for the assembly of the replisome, it is not part of the replication machinery. Rather, its job is to bring the

DNA replication: replication of a chromosome

WWW ANIMATED ART

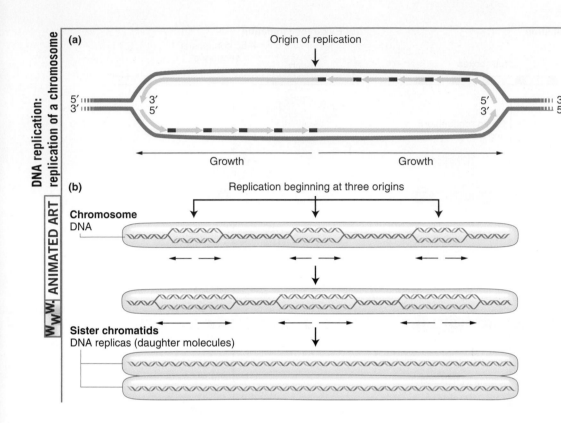

Figure 7-22 The bidirectional nature of DNA replication. Black arrows show the direction of growth of daughter DNA molecules. (a) Starting at the origin, DNA polymerases move outward in both directions. Long orange arrows represent leading strands and short joined orange arrows represent lagging strands. (b) How replication proceeds at the chromosome level. Three origins of replication are shown in this example.

replisome to the correct place in the circular chromosome for the initiation of replication.

Eukaryotic origins of replication

Bacteria such as *E. coli* usually complete a replication–division cycle in from 20 to 40 minutes but, in eukaryotes, the cycle can vary from 1.4 hours in yeast to 24 hours in cultured animal cells and may last from 100 to 200 hours in some cells. Eukaryotes have to solve the problem of coordinating the replication of more than one chromosome, as well as the problem of replicating the complex structure of the chromosome itself.

The origins of the simple eukaryote, yeast, are very much like *oriC* in *E. coli*. They have AT-rich regions that melt when an initiator protein binds to adjacent binding sites. Origins of replication are not well characterized in higher organisms, but they are known to be much longer, possibly as long as thousands or tens of thousands of nucleotides. Unlike prokaryotic chromosomes, each chromosome has many replication origins in order to replicate the much larger eukaryotic genomes quickly. Approximately 400 replication origins are dispersed throughout the 16 chromosomes of yeast, and there are estimated to be thousands of growing forks in the 23 chromosomes of humans. Thus, in eukaryotes, replication proceeds in both directions from multiple points of

origin (Figure 7-22). The double helices that are being produced at each origin of replication elongate and eventually join one another. When replication of the two strands is complete, two identical **daughter molecules** of DNA result.

> **MESSAGE** Where and when replication takes place is carefully controlled by the ordered assembly of the replisome at a precise site called the origin. Replication proceeds in both directions from a single origin on the circular prokaryotic chromosome. Replication proceeds in both directions from hundreds or thousands of origins on each of the longer, linear eukaryotic chromosomes.

DNA replication and the eukaryotic cell cycle

We know from Chapter 3 that DNA synthesis takes place in only one stage of the eukaryotic cell cycle, the S (synthesis) phase (Figure 7-23). How is the onset of DNA synthesis limited to this single stage? In yeast, the best-characterized eukaryotic system, the method of control is to link replisome assembly to the cell cycle. Figure 7-21b shows the process. In yeast, three proteins are required to begin assembly of the replisome. The **origin recognition complex (ORC)** first binds to sequences in yeast origins, much as DnaA protein does

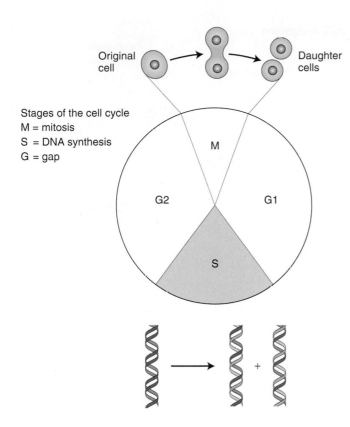

Figure 7-23 Stages of the cell cycle.

in *E. coli*. Although first discovered in yeast, a similar complex has been found in all eukaryotes studied. The presence of ORC at the origin serves to recruit two other proteins, Cdc6 and Cdt1. Both proteins plus ORC then recruit the replicative helicase, called the MCM complex, and the other components of the replisome. Binding of helicase is said to **"license"** the origin. Origins must be licensed first before they are able to support the assembly of the replisome and begin DNA synthesis.

Replication is linked to the cell cycle through the availability of Cdc6 and Cdt1. In yeast, these proteins

are synthesized during late mitosis and gap1 and are destroyed by proteolysis after synthesis has begun. In this way, the replisome can be assembled only before the S phase. Once replication begins, new replisomes cannot form at the origins, because Cdc6 and Cdt1 are degraded during the S phase and are no longer available. That is, origins cannot be licensed once the S phase is under way.

7.7 Telomeres and telomerase: replication termination

Replication of the linear DNA molecule in a eukaryotic chromosome proceeds in both directions from numerous replication origins, as shown in Figure 7-22. This process replicates most of the chromosomal DNA, but there is an inherent problem in replicating the two ends of linear DNA molecules, the regions called **telomeres.** Continuous synthesis on the leading strand can proceed right up to the very tip of the template. However, lagging-strand synthesis requires primers ahead of the process; so, when the last primer is removed, a single-stranded tip remains in one of the daughter DNA molecules (Figure 7-24). If the daughter chromosome with this DNA molecule replicated again, the strand missing sequences at the end would become a shortened double-stranded molecule after replication. At each subsequent replication cycle, the telomere would continue to shorten, until eventually essential coding information would be lost.

Cells have devised a specialized system to prevent this loss. They add multiple copies of a simple noncoding sequence to the DNA at the chromosome tips to prevent shortening. For example, in the single-celled ciliate *Tetrahymena*, copies of the sequence TTGGGG are added to the 3′ end of each chromosome. In humans, copies of the sequence TTAGGG are added. This lengthening of the DNA molecule creates noncoding DNA that can be "sacrificed" to the replication process.

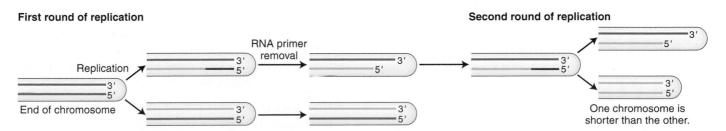

First round of replication

Replication

End of chromosome

RNA primer removal

Second round of replication

One chromosome is shorter than the other.

Figure 7-24 The replication problem at chromosome ends. Once the primer for the last section of the lagging strand is removed, there is no way to polymerize that segment, and a shortened chromosome would result when the chromosome containing the incomplete strand replicated.

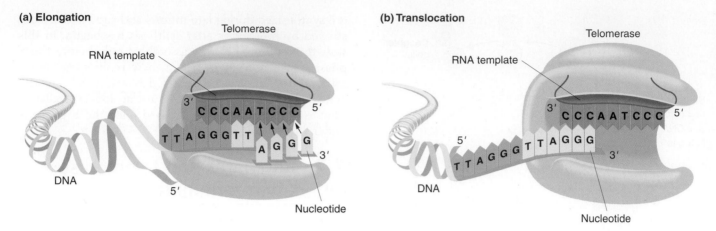

(a) Elongation

(b) Translocation

Figure 7-25 Telemere lengthening. (a) Telomerase carries a short RNA molecule that acts as a template for the addition of the complementary DNA sequence, which is added one nucleotide at a time to the 3' end of the double helix. In humans, the DNA sequence added (often in many repeats) is TTAGGG. (b) To add another repeat, the telomerase translocates to the end of the repeat that it just added.

An enzyme called **telomerase** handles the addition of this noncoding sequence to the 3' end. The telomerase protein carries a small RNA molecule, part of which acts as a template for the polymerization of the telomeric repeat unit. In humans, the RNA sequence 3'-AATCCC-5' acts as the template for the 5'-TTAGGG-3' repeat unit by a mechanism shown in Figure 7-25. Figure 3-16 demonstrates the positions of the telomeric DNA through a special chromosome-labeling technique.

KEY QUESTIONS REVISITED

• **Before the discovery of the double helix, what was the experimental evidence that DNA is the genetic material?**

Avery and co-workers demonstrated that DNA is the chemical component of dead cells capable of transforming living bacteria from nonvirulent to virulent. Hershey and Chase extended this finding to bacterial viruses (phage) by showing that bacterial infection follows the injection of phage DNA, not protein, into bacterial cells.

• **What data were used to deduce the double-helix model of DNA?**

First, the components of DNA and how they formed single chains were known. Second, it was known from Chargaff's rules that (1) the total amount of pyrimidine nucleotides (T + C) always equals the total amount of purine nucleotides (A + G) and (2) the amount of T always equals the amount of A and, similarly, the amount of G equals the amount of C, but the amount of A + T does not always equal the amount of G + C. Finally, the X-ray diffraction patterns of Rosalind Franklin showed that DNA was organized as a double helix and provided the dimensions of that helix. Knowledge of Chargaff's rules and the dimensions of the double helix were used by Watson and Crick to determine that A always pairs with T and G always pairs with C.

• **How does the double-helical structure suggest a mechanism for DNA replication?**

The two halves of the double helix must separate and, because of the specificity of base pairing, they both act as templates for the polymerization of new strands, thus forming two identical daughter double helices.

• **Why are the proteins that replicate DNA called a biological machine?**

Many of the proteins taking part in DNA replication interact through protein–protein contacts and are assembled only where (at the origins) and when needed to coordinate a complex biological process.

• **How can DNA replication be both rapid and accurate?**

Replication is rapid because DNA pol III remains attached to its template substrate (processive) and is accurate because of the proofreading activity of DNA pol III and because the multiple steps of leading- and lagging-strand synthesis, including the removal of misincorporated nucleotides, are coordinated by the replisome.

• **What special mechanism replicates chromosome ends?**

Chromosome ends (telomeres) are extended by a special enzyme (telomerase) that is composed of both protein and RNA.

SUMMARY

Experimental work on the molecular nature of hereditary material has demonstrated conclusively that DNA (not protein, lipids, or some other substance) is indeed the genetic material. Using data supplied by others, Watson and Crick deduced a double-helical model with two DNA strands, wound around each other, running in antiparallel fashion. Specificity of binding the two strands together is based on the fit of adenine (A) to thymine (T) and guanine (G) to cytosine (C). The former pair is held by two hydrogen bonds; the latter, by three.

The Watson–Crick model shows how DNA can be replicated in an orderly fashion—a prime requirement for genetic material. Replication is accomplished semiconservatively in both prokaryotes and eukaryotes. One double helix is replicated into two identical helices, each with identical linear orders of nucleotides; each of the two new double helices is composed of one old and one newly polymerized strand of DNA.

The DNA double helix is unwound at a replication fork, and the two single strands thus produced serve as templates for the polymerization of free nucleotides. Nucleotides are polymerized by the enzyme DNA polymerase, which adds new nucleotides only to the 3′ end of a growing DNA chain. Because addition is only at 3′ ends, polymerization on one template is continuous, producing the leading strand and, on the other, it is discontinuous in short stretches (Okazaki fragments), producing the lagging strand. Synthesis of the leading strand and of every Okazaki fragment is primed by a short RNA primer (synthesized by primase) that provides a 3′ end for deoxyribonucleotide addition.

The multiple events that have to occur accurately and rapidly at the replication fork are processed by the replisome, a biological machine that includes two DNA polymerase units that act on the leading and lagging strands. In this way, the additional time required to synthesize and join the Okazaki fragments into a continuous strand can be temporally coordinated with the less complicated synthesis of the leading strand. Where and when replication takes place is carefully controlled by the ordered assembly of the replisome at certain sites on the chromosome called origins. Eukaryotic genomes may have tens of thousands of origins at which the assembly of replisomes can only take place at a specific time in the cell cycle. The ends of linear chromosomes (telomeres) present a problem for the replication system because there is always a short stretch on one strand that cannot be primed. Adding a number of short, repetitive sequences to maintain length is achieved by the enzyme telomerase, which carries a short RNA that acts as the template for the synthesis of the telomeric repeats.

KEY TERMS

accessory protein (p. 243)

antiparallel (p. 234)

base (p. 232)

chromatin assembly factor 1 (CAF-1) (p. 244)

complementary base (p. 237)

conservative replication (p. 237)

daughter molecule (p. 246)

deoxyribose (p. 232)

dispersive replication (p. 237)

distributive enzyme (p. 243)

double helix (p. 233)

DnaA (p. 245)

DnaB (p. 245)

DNA ligase (p. 241)

genetic code (p. 235)

helicase (p. 243)

lagging strand (p. 241)

leading strand (p. 240)

license (p. 247)

major groove (p. 234)

melt (p. 245)

minor groove (p. 234)

nucleosome (p. 244)

nucleotide (p. 232)

Okazaki fragment (p. 241)

origin (p. 244)

*ori*C (p. 244)

origin recognition complex
  (ORC) (p. 246)

phosphate (p. 232)

pol III holoenzyme (p. 242)

primase (p. 241)

primosome (p. 241)

primer (p. 241)

processive enzyme (p. 243)

proliferating cell nuclear antigen
  (PCNA) (p. 244)

purine (p. 232)

pyrimidine (p. 232)

replication fork (p. 240)

replisome (p. 242)

semiconservative
  replication (p. 237)

single-strand binding (SSB)
  protein (p. 243)

sliding clamp (p. 243)

telomerase (p. 248)

telomere (p. 247)

template (p.237)

theta θ structure (p. 239)

topoisomerase (p. 243)

SOLVED PROBLEMS

1. Mitosis and meiosis were presented in Chapter 3. Considering what we have covered in this chapter concerning DNA replication, draw a graph showing DNA content against time in a cell that undergoes mitosis and then meiosis. Assume a diploid cell.

Solution

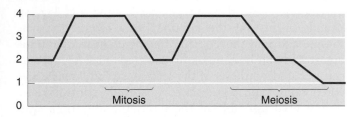

2. If the GC content of a DNA molecule is 56 percent, what are the percentages of the four bases (A, T, G, and C) in this molecule?

Solution

If the GC content is 56 percent, then, because G = C, the content of G is 28 percent and the content of C is 28 percent. The content of AT is 100 − 56 = 44 percent. Because A = T, the content of A is 22 percent and the content of T is 22 percent.

3. Describe the expected pattern of bands in a CsCl gradient for *conservative* replication in the Meselson–Stahl experiment. Draw a diagram.

Solution

Refer to Figure 7-13 for an additional explanation. In conservative replication, if bacteria are grown in the presence of ^{15}N and then shifted to ^{14}N, one DNA molecule will be all ^{15}N after the first generation and the other molecule will be all ^{14}N, resulting in one heavy band and one light band in the gradient. After the second generation, the ^{15}N DNA will yield one molecule with all ^{15}N and one molecule with all ^{14}N, whereas the ^{14}N DNA will yield only ^{14}N DNA. Thus, only all ^{14}N or all ^{15}N DNA is generated, again yielding a light band and a heavy band:

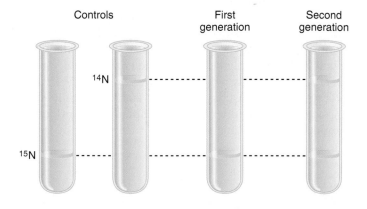

PROBLEMS

Basic Problems

1. Describe the types of chemical bonds in the DNA double helix.

2. Explain what is meant by the terms *conservative* and *semiconservative replication*.

3. What is meant by a *primer*, and why are primers necessary for DNA replication?

4. What are helicases and topoisomerases?

5. Why is DNA synthesis continuous on one strand and discontinuous on the opposite strand?

6. If the four deoxynucleotides showed nonspecific base pairing (A to C, A to G, T to G, etc.), would the unique information contained in a gene be maintained through round after round of replication? Explain.

7. If the helicases were missing during replication, what would happen to the replication process?

8. Both strands of a DNA molecule are replicated simultaneously in a continuous fashion on one strand and a discontinuous one on the other. Why can't one strand be replicated in its entirety (from end to end) before replication of the other is initiated?

9. What would happen if during replication the topoisomerases were unable to reattach the DNA fragments of each strand after unwinding (relaxing) the DNA molecule?

10. What would happen if DNA synthesis were discontinuous on both strands?

    a. The DNA fragments from the two new strands could become mixed, producing possible mutations.

    b. DNA synthesis would not occur because the appropriate enzymes to carry out discontinuous replication on both strands would not be present.

    c. DNA synthesis might take longer, but otherwise there would be no noticeable difference.

    d. DNA synthesis would not occur, as the entire length of the chromosome would have to be unwound before both strands could be replicated in a discontinuous fashion.

11. Which of the following is *not* a key property of hereditary material?

    a. It must be capable of being copied accurately.

    b. It must encode the information necessary to form proteins and complex structures.

    c. It must occasionally mutate.

    d. It must be able to adapt itself to each of the body's tissues.

12. It is essential that RNA primers at the ends of Okazaki fragments be removed and replaced by DNA because otherwise

    a. the RNA might not be accurately read during transcription, thus interfering with protein synthesis.

    b. the RNA would be more likely to contain errors because primase lacks a proofreading function.

    c. the stretches of RNA would destabilize and begin to break up into ribonucleotides, thus creating gaps in the sequence.

    d. the RNA primers would be likely to hydrogen bond to each other, forming complex structures that might interfere with the proper formation of the DNA helix.

13. Polymerases usually add only about 10 nucleotides to a DNA strand before dissociating. However, during replication, DNA pol III can add tens of thousands of nucleotides at a moving fork. How is this accomplished?

14. At each origin of replication there are two bidirectional replication forks. What would happen if a mutant arose having only one function fork per replication bubble? (See diagram below.)

| Normal | Mutant |

    a. No change at all in replication.

    b. Replication would occur only on one half of the chromosome.

    c. Replication would be complete only on the leading strand.

    d. Replication would take twice as long.

15. In a diploid cell in which $2n = 14$, how many telomeres are there in each of the following phases of the cell cycle: (a) G_1; (b) G_2; (c) mitotic prophase; (d) mitotic telophase?

16. If thymine makes up 15 percent of the bases in a specific DNA molecule, what percentage of the bases is cytosine?

17. If the GC content of a DNA molecule is 48 percent, what are the percentages of the four bases (A, T, G, and C) in this molecule?

18. Assume that a certain bacterial chromosome has one origin of replication. Under some conditions of rapid cell division, replication could start from the origin before the preceding replication cycle is complete. How many replication forks would be present under these conditions?

19. A molecule of composition

$$5'\text{-AAAAAAAAAAA-}3'$$
$$3'\text{-TTTTTTTTTTTTT-}5'$$

is replicated in a solution of adenine nucleoside triphosphate with all its phosphorus atoms in the form of the radioactive isotope ^{32}P. Will both daughter molecules be radioactive? Explain. Then repeat the question for the molecule

5'-ATATATATATATAT-3'
3'-TATATATATATATA-5'

20. Would the Meselson and Stahl experiment have worked if diploid eukaryotic cells had been used instead?

21. Consider the following segment of DNA, which is part of a much longer molecule constituting a chromosome:

5'.....ATTCGTACGATCGACTGACTGACAGTC.....3'
3'.....TAAGCATGCTAGCTGACTGACTGTCAG.....5'

If the DNA polymerase starts replicating this segment from the right,

a. which will be the template for the leading strand?

b. Draw the molecule when the DNA polymerase is halfway along this segment.

c. Draw the two complete daughter molecules.

d. Is your diagram in part b compatible with bidirectional replication from a single origin, the usual mode of replication?

22. The DNA polymerases are positioned over the following DNA segment (which is part of a much larger molecule) and moving from right to left. If we assume that an Okazaki fragment is made from this segment, what will be its sequence? Label its 5' and 3' ends.

5'.....CCTTAAGACTAACTACTTACTGGGATC.....3'
3'.....GGAATTCTGATTGATGAATGACCCTAG.....5'

23. *E. coli* chromosomes in which every nitrogen atom is labeled (that is, every nitrogen atom is the heavy isotope ^{15}N instead of the normal isotope ^{14}N) are allowed to replicate in an environment in which all the nitrogen is ^{14}N. Using a solid line to represent a heavy polynucleotide chain and a dashed line for a light chain, sketch each of the following descriptions:

a. The heavy parental chromosome and the products of the first replication after transfer to a ^{14}N medium, assuming that the chromosome is one DNA double helix and that replication is semiconservative.

b. Repeat part a, but now assume that replication is conservative.

c. Repeat part a, but assume that the chromosome is in fact two side-by-side double helices, each of which replicates semiconservatively.

d. Repeat part c, but assume that each side-by-side double helix replicates conservatively and that the overall *chromosome* replication is semiconservative.

e. If the daughter chromosomes from the first division in ^{14}N are spun in a cesium chloride (CsCl) density gradient and a single band is obtained, which of possibilities in parts a through d can be ruled out? Reconsider the Meselson and Stahl experiment: What does it *prove*?

24. A student in Griffith's lab found three cell samples marked "A," "B," and "C." Not knowing what each sample contained, the student decided to try injecting these samples into some of her mice, both singly and in combinations, to see if she could determine what each sample contained. She observed the responses of the infected mice after an incubation period and recovered blood samples from each test group to look for the possible presence of infecting cells. She recorded her observations in the table below. Assuming that each sample contained only one thing in a pure form, what do you think was in samples "A," "B," and "C"?

Sample injected	Response of mice	Type of cells recovered from mice
A	dead	live S cells
B	none	none
C	none	live R cells
A + B	dead	live S cells
A + C	dead	live R and live S cells
B + C	dead	live S cells
A + B + C	dead	live S cells

25. If in the experiment in Problem 24, the cell coat proteins were the transforming factors, what results would you expect? Fill in the table below. (Remember that the samples are the same as in Problem 24.)

Sample injected	Response of mice	Type of cells recovered from mice
A		
B		
C		
A + B		
A + C		
B + C		
A + B + C		

CHALLENGING PROBLEMS

26. If a mutation that inactivated telomease occurred in a cell (telomerase activity in the cell = zero) what do you expect the outcome to be?

27. On the planet of Rama, the DNA is of six nucleotide types: A, B, C, D, E, and F. A and B are called *marzines*, C and D are *orsines*, and E and F are *pirines*. The following rules are valid in all Raman DNAs:

$$\text{Total marzines} = \text{total orsines} = \text{total pirines}$$
$$A = C = E$$
$$B = D = F$$

a. Prepare a model for the structure of Raman DNA.

b. On Rama, mitosis produces three daughter cells. Bearing this fact in mind, propose a replication pattern for your DNA model.

c. Consider the process of meiosis on Rama. What comments or conclusions can you suggest?

28. If you extract the DNA of the coliphage φX174, you will find that its composition is 25 percent A, 33 percent T, 24 percent G, and 18 percent C. Does this composition make sense in regard to Chargaff's rules? How would you interpret this result? How might such a phage replicate its DNA?

RNA: TRANSCRIPTION AND PROCESSING

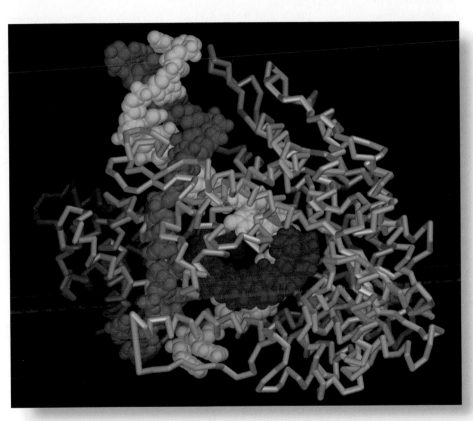

RNA polymerase in action. A very small RNA polymerase (blue), made by the bacteriophage T7, transcribes DNA into a strand of RNA (red). The enzyme separates the DNA double helix (yellow, orange), exposing the template strand to be copied into RNA. [David S. Goodsell, Scripps Research Institute.]

KEY QUESTIONS

- How does the structure of RNA differ from that of DNA?

- What are the different classes of RNA in the cell?

- How is RNA polymerase correctly positioned to start transcription in prokaryotes?

- How is eukaryotic RNA synthesized by RNA polymerase II modified before leaving the nucleus?

- Why is the discovery of self-splicing introns considered by some to be as important as the discovery of the DNA double helix?

OUTLINE

8.1 RNA

8.2 Transcription

8.3 Transcription in eukaryotes

CHAPTER OVERVIEW

In this chapter, we see the first steps in the transfer of information from genes to gene products. Within the DNA sequence of any organism's genome is encrypted information specifying each of the gene products that the organism can make. Not only do these DNA sequences code for the structure of these products, but they also contain information specifying when, where, and how much of the product is made. However, this information is static, embedded in the sequence of the DNA. To utilize the information, an intermediate molecule that is a copy of a discrete gene must be synthesized with the use of the DNA sequence as a guide. This molecule is RNA and the process of its synthesis from DNA is called *transcription*.

Figure 8-1 shows an overview of the main ideas of this chapter applied to both prokaryotic and eukaryotic systems. We can think of gene action as a process of copying and deciphering the information encrypted in the gene. The transfer of information from gene to gene product takes place in several steps. The first step, which is the focus of this chapter, is to copy *(transcribe)* the information into a strand of RNA with the use of DNA as an alignment guide, or template. In prokaryotes, the information in RNA is almost immediately converted into an amino acid chain (polypeptide) by a process called *translation*. This second step is the focus of Chapter 9. In eukaryotes, transcription and translation are spatially separated—transcription takes place in the nucleus and translation in the cytoplasm. However, before RNAs are ready to be transported into the cytoplasm for translation, they undergo extensive processing, including the removal of introns and the addition of a special 5' cap and a 3' tail of adenine nucleotides. A fully processed RNA is called *messenger* RNA (mRNA). Like DNA replication, transcription is carried out by a molecular machine, which coordinates the synthesis and processing of mRNA. For a minority of genes, the RNA is the final product and, in these cases, the RNA is never translated into protein.

DNA and RNA function is based on two principles:

1. Complementarity of bases is responsible for determining the sequence of a new DNA strand in replication and of the RNA transcript in transcription. Through the matching of complementary bases, DNA is replicated and the information encoded in the DNA passes into RNA (and ultimately protein).

2. Certain proteins recognize particular base sequences in DNA. These nucleic acid–binding proteins bind to these sequences and act on them.

CHAPTER OVERVIEW Figure

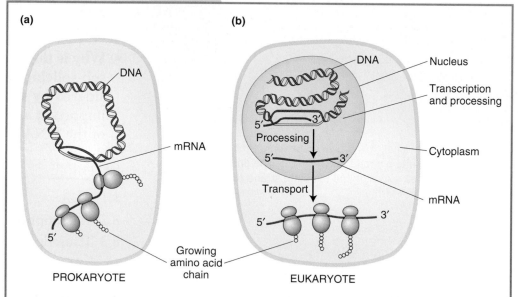

Figure 8-1 Transcription and translation. These two processes take place in the same cellular compartment in prokaryotes but in different compartments in eukaryotes. Moreover, unlike prokaryotic RNA transcripts, eukaryotic transcripts undergo extensive processing before they can be translated into proteins. [After J. Darnell, H. Lodish, and D. Baltimore, *Molecular Cell Biology*, 2d ed. Copyright 1990 by Scientific American Books, Inc. All rights reserved.]

We shall see these two principles at work throughout the detailed discussions of transcriptions and translation that follow in this chapter and in chapters to come.

> **MESSAGE** The transactions of DNA and RNA take place through the complementarity of base sequences and the binding of various proteins to specific sites on the DNA or RNA.

8.1 RNA

Early investigators had good reason for thinking that information is not transferred directly from DNA to protein. In a eukaryotic cell, DNA is found in the nucleus, whereas protein is synthesized in the cytoplasm. An intermediate is needed.

Early experiments suggest an RNA intermediate

In 1957, Elliot Volkin and Lawrence Astrachan made a significant observation. They found that one of the most striking molecular changes when *E. coli* is infected with the phage T2 is a rapid burst of RNA synthesis. Furthermore, this phage-induced RNA "turns over" rapidly; that is, its lifetime is brief. Its rapid appearance and disappearance suggested that RNA might play some role in the expression of the T2 genome necessary to make more virus particles.

Volkin and Astrachan demonstrated the rapid turnover of RNA by using a protocol called a **pulse-chase experiment**. To conduct a pulse-chase experiment, the infected bacteria are first fed (pulsed with) radioactive uracil (a molecule needed for the synthesis of RNA but not DNA). Any RNA synthesized in the bacteria from then on is "labeled" with the readily detectable radioactive uracil. After a short period of incubation, the radioactive uracil is washed away and replaced (chased) by uracil that is not radioactive. This procedure "chases" the label out of the RNA because, as the RNA breaks down, only the unlabeled precursors are available to synthesize new RNA molecules. The RNA recovered shortly after the pulse is labeled, but that recovered somewhat longer after the chase is unlabeled, indicating that the RNA has a very short lifetime.

A similar experiment can be done with eukaryotic cells. Cells are first pulsed with radioactive uracil and, after a short period of time, they are transferred to medium with unlabeled uracil. In samples taken after the pulse, most of the label is in the nucleus. In samples taken after the chase, the labeled RNA is found in the cytoplasm (Figure 8-2). Apparently, in eukaryotes, the RNA is synthesized in the nucleus and then moves into the cytoplasm, where proteins are synthesized. Thus, RNA is a good candidate for an information-transfer intermediary between DNA and protein.

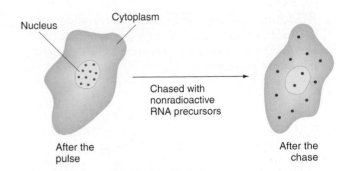

Figure 8-2 RNA synthesized in the nucleus moves to the cytoplasm. Cells are grown briefly in radioactive uracil to label newly synthesized RNA (pulse). Cells are washed to remove the radioactive uracil and are then grown in excess nonradioactive uracil (chase). The red dots indicate the location of the RNA containing radioactive uracil over time.

Properties of RNA

Let's consider the general features of RNA. Although both RNA and DNA are nucleic acids, RNA differs from DNA in several important ways:

1. RNA is usually a single-stranded nucleotide chain, not a double helix like DNA. A consequence is that RNA is more flexible and can form a much greater variety of complex three-dimensional molecular shapes than can double-stranded DNA. An RNA strand can bend in such a way that some of its own bases pair with each other. Such *intramolecular* base pairing is an important determinate of RNA shape.

2. RNA has **ribose** sugar in its nucleotides, rather than the deoxyribose found in DNA. As the names suggest, the two sugars differ in the presence or absence of just one oxygen atom. The RNA sugar groups contain an oxygen–hydrogen pair bound to the 2′ carbon, whereas only a hydrogen atom is bound to the 2′ carbon in DNA sugar groups.

Like an individual DNA strand, a strand of RNA is formed of a sugar–phosphate backbone, with a base covalently linked at the 1′ position on each ribose. The sugar–phosphate linkages are made at the 5′ and 3′ positions of the sugar, just as in DNA; so an RNA chain will have a 5′ end and a 3′ end.

3. RNA nucleotides (called ribonucleotides) contain the bases adenine, guanine, and cytosine, but the pyrimidine base **uracil** (abbreviated **U**) is present instead of thymine.

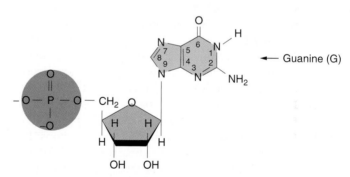

Uracil

Uracil forms hydrogen bonds with adenine just as thymine does. Figure 8-3 shows the four ribonucleotides found in RNA.

4. RNA—like protein, but unlike DNA—can catalyze important biological reactions. The RNA molecules that function like protein enzymes are called **ribozymes.**

Classes of RNA

RNAs can be grouped into two general classes. One class of RNAs is an intermediary in the process of decoding genes into polypeptide chains. We will refer to these "informational" RNAs as messenger RNAs because they pass information, like a messenger, from DNA to protein. In the remaining minority of genes, the RNA itself is the final functional product. We will refer to these RNAs as "functional RNAs."

MESSENGER RNA The steps through which a gene influences phenotype are called *gene expression*. For the vast majority of genes, the RNA transcript is only an intermediate necessary for the synthesis of a protein, which is the ultimate functional product that influences phenotype.

FUNCTIONAL RNA As more is learned about the intimate details of cell biology, it has become apparent that functional RNAs fall into a variety of classes that play diverse roles. Again, it is important to emphasize that functional RNAs are active as RNA; they are never

Purine ribonucleotides

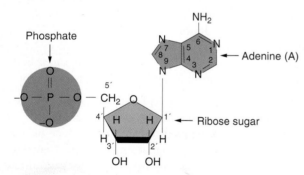

Adenosine 5′-monophosphate (AMP)

Guanosine 5′-monophosphate (GMP)

Pyrimidine ribonucleotides

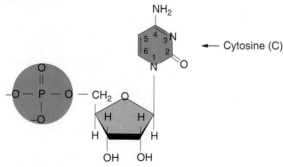

Cytidine 5′-monophosphate (CMP)

Uridine 5′-monophosphate (UMP)

Figure 8-3 The four ribonucleotides found in RNA.

translated into polypeptides. Each class of functional RNA is encoded by a small number of genes (a few tens to a few hundred at most). However, though the genes that encode them are relatively few, some functional RNAs account for a very large percentage of the RNA in the cell because they are both stable and transcribed in many copies.

The main classes of functional RNAs contribute to various steps in the informational processing of DNA into protein. Two such classes of functional RNAs are found in prokaryotes and eukaryotes: transfer RNAs and ribosomal RNAs.

- **Transfer RNA (tRNA)** molecules are responsible for bringing the correct amino acid to the mRNA in the process of translation.

- **Ribosomal RNAs (rRNAs)** are the major components of ribosomes, which are large macromolecular machines that guide the assembly of the amino acid chain by the mRNA and tRNAs.

Another class of functional RNAs participates in the processing of RNA and is specific to eukaryotes.

- **Small nuclear RNAs (snRNAs)** are part of a system that further processes RNA transcripts in eukaryotic cells. Some snRNAs guide the modification of rRNAs. Others unite with several protein subunits to form the ribonucleoprotein processing complex (called the *spliceosome*) that removes introns from eukaryotic mRNAs.

> **MESSAGE** There are two types of RNAs, those coding for proteins (the majority) and those that are functional as RNA.

8.2 Transcription

The first step of information transfer from gene to protein is to produce an RNA strand whose base sequence matches the base sequence of a DNA segment, sometimes followed by modification of that RNA to prepare it for its specific cellular roles. Hence RNA is produced by a process that copies the nucleotide sequence of DNA. Because this process is reminiscent of transcribing (copying) written words, the synthesis of RNA is called **transcription**. The DNA is said to be transcribed into RNA, and the RNA is called a **transcript.**

Overview: DNA as transcription template

How is the information encrypted in the DNA molecule transferred to the RNA transcript? Transcription relies on the complementary pairing of bases. Consider

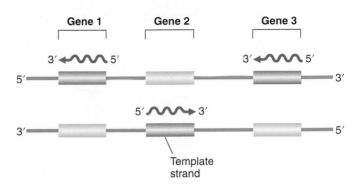

Figure 8-4 DNA strands used as templates for transcription. The direction of transcription is always the same for any gene and starts from the 3′ end of the template and the 5′ end of the RNA transcript. Hence genes transcribed in different directions use opposite strands of the DNA as templates.

the transcription of a chromosomal segment that constitutes a gene. First, the two strands of the DNA double helix separate locally, and one of the separated strands acts as a **template** for RNA synthesis. In the chromosome overall, both DNA strands are used as templates; *but, in any one gene, only one strand is used,* and, in that gene, it is always the same strand (Figure 8-4). Next, ribonucleotides that have been chemically synthesized elsewhere in the cell form stable pairs with their complementary bases in the template. The ribonucleotide A pairs with T in the DNA, G with C, C with G, and U with A. Each ribonucleotide is positioned opposite its complementary base by the enzyme **RNA polymerase,** which attaches to the DNA and moves along it, linking the aligned ribonucleotides together to make an ever-growing RNA molecule, as shown in Figure 8-5a. Hence, we already see the two principles of base complementarity and nucleic acid–protein binding in action (in this case, the binding of RNA polymerase).

We have seen that RNA has a 5′ end and a 3′ end. During synthesis, RNA growth is always in the 5′-to-3′ direction; in other words, nucleotides are always added at a 3′ growing tip, as shown in Figure 8-5b. Because complementary nucleic acid strands are oppositely oriented, the fact that RNA is synthesized 5′ to 3′ means that the template strand must be oriented 3′ to 5′.

As an RNA polymerase molecule moves along the gene, it unwinds the DNA double helix ahead of it and rewinds the DNA that has already been transcribed. As the RNA molecule progressively lengthens, the 5′ end of the RNA is displaced from the template as the transcription bubble closes behind the polymerase. "Trains" of RNA polymerases, each synthesizing an RNA molecule, move along the gene. The multiple RNA strands can be viewed coming off a single DNA molecule under the

Transcription

WWW ANIMATED ART

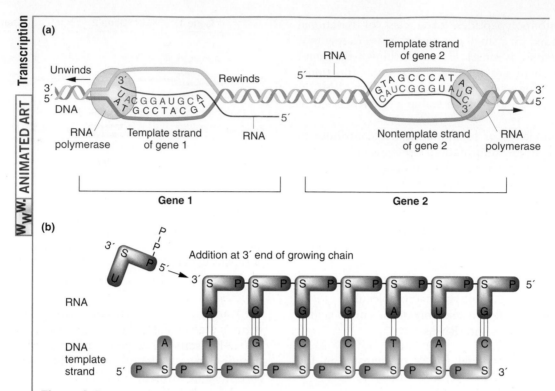

Figure 8-5 Overview of transcription. (a) Transcription of two genes in opposite directions. Genes 1 and 2 from Figure 8-4 are shown. Gene 1 is transcribed from the bottom strand. The RNA polymerase migrates to the left, reading the template strand in a 3′-to-5′ direction and synthesizing RNA in a 5′-to-3′ direction. Gene 2 is transcribed in the opposite direction, to the right, because the top strand is the template. As transcription proceeds, the 5′ end of the RNA is displaced from the template as the transcription bubble closes behind the polymerase. (b) As gene 1 is transcribed, the phosphate group on the 5′ end of the entering ribonucleotide (U) attaches to the 3′ end of the growing RNA chain.

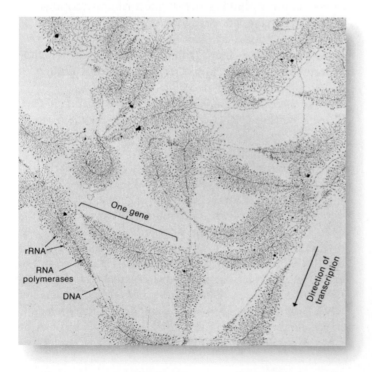

Figure 8-6 Transcription of ribosomal RNA genes repeated in tandem in the nucleus of the amphibian *Triturus viridiscens*. Along each gene, many RNA polymerases are transcribing in one direction. The growing RNA transcripts appear as threads extending outward from the DNA backbone. The shorter transcripts are close to the start of transcription; the longer ones are near the end of the gene. The "Christmas tree" appearance is the result. [Photograph from O. L. Miller, Jr., and Barbara A. Hamkalo.]

electron microscope, providing a way to visualize the progressive enlargement of RNA strands (Figure 8-6).

We have also seen that the bases in transcript and template are complementary. Consequently, the nucleotide sequence in the RNA must be the same as that in the nontemplate strand of the DNA, except that the T's are replaced by U's, as shown in Figure 8-7. When DNA base sequences are cited in scientific literature, by convention it is the sequence of the nontemplate strand that is given, because this sequence is the same as that

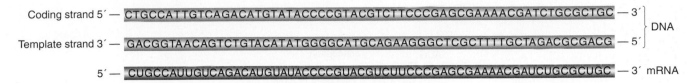

Coding strand 5′ — CTGCCATTGTCAGACATGTATACCCCGTACGTCTTCCCGAGCGAAAACGATCTGCGCTGC — 3′ } DNA

Template strand 3′ — GACGGTAACAGTCTGTACATATGGGGCATGCAGAAGGGCTCGCTTTTGCTAGACGCGACG — 5′

5′ — CUGCCAUUGUCAGACAUGUAUACCCCGUACGUCUUCCCGAGCGAAAACGAUCUGCGCUGC — 3′ mRNA

Figure 8-7 Comparison of DNA and mRNA sequences of a transcribed region of DNA. The mRNA sequence is complementary to the DNA template strand from which it is transcribed, and therefore matches the sequence of the coding strand (except that the RNA has U where the DNA has T). This sequence is from the gene for the enzyme β-galactosidase.

found in the RNA. For this reason, the nontemplate strand of the DNA is referred to as the **coding strand.** This distinction is extremely important to keep in mind during discussions of transcription.

> **MESSAGE** Transcription is asymmetrical: only one strand of the DNA of a gene is used as a template for transcription. This strand is in 3′-to-5′ orientation, and RNA is synthesized in the 5′-to-3′ direction.

Stages of transcription

The protein-coding sequence in a gene is a relatively small segment of DNA embedded in a much longer DNA molecule (the chromosome). How is the appropriate segment transcribed into a single-stranded RNA molecule of correct length and nucleotide sequence? Because the DNA of a chromosome is a continuous unit, the transcriptional machinery must be directed to the start of a gene to begin transcribing at the right place, continue transcribing the length of the gene, and finally stop transcribing at the other end. These three distinct stages of transcription are called **initiation, elongation, and termination.** Although the overall process of transcription is remarkably similar in prokaryotes and eukaryotes, there are important differences. For this reason, we will follow the three stages first in prokaryotes (by using the gut bacterium *E. coli* as an example) and then in eukaryotes.

INITIATION IN PROKARYOTES How does RNA polymerase find the correct starting point for transcription? In prokaryotes, RNA polymerase usually binds to a specific DNA sequence called a **promoter,** located close to the start of the transcribed region. A promoter is an important part of the regulatory region of a gene. Remember that, because the synthesis of an RNA transcript begins at its 5′ end and continues in the 5′-to-3′ direction, the convention is to draw and refer to the orientation of the gene in the 5′-to-3′ direction, too. Generally, the 5′ end is drawn at the left and the 3′ at the right. With this view, because the promoter must be near the end of the gene where transcription begins, it is said to be at the 5′

end of the gene; thus the promoter region is also called the 5′ regulatory region (Figure 8-8a).

Figure 8-8b shows the promoter sequences of seven different genes in the *E. coli* genome. Because the same RNA polymerase binds to the promoter sequences of these different genes, it is not surprising that there are similarities among the promoters. In particular, two regions of great similarity appear in virtually every case. These regions have been termed the −35 (minus 35) and −10 *regions* because they are located 35 base pairs and 10 base pairs, respectively, ahead of (commonly referred to as **upstream** of) the first transcribed base. They are shown in yellow in Figure 8-8b. As you can see, the −35 and −10 regions from different genes do not have to be identical to perform a similar function. Nonetheless, it is possible to arrive at a sequence of nucleotides that is in agreement with most sequences, called a **consensus sequence.** The *E. coli* promoter consensus sequence is shown at the bottom of Figure 8-8b. An RNA polymerase holoenzyme (see next paragraph) binds to the DNA at this point, then unwinds the DNA double helix, and begins the synthesis of an RNA molecule. The first transcribed base is always at the same location, designated the *initiation site*, numbered +1. Note in Figure 8-8a that transcription starts *before* the protein-coding segment of the gene (usually at the sequence ATG, which is where, as you will see in Chapter 9, translation usually starts). Hence a transcript has what is called a **5′ untranslated region (5′ UTR).**

The bacterial RNA polymerase that scans the DNA for a promoter sequence is called the **RNA polymerase holoenzyme** (Figure 8-9). This multisubunit complex is composed of the four subunits of the **core enzyme** (two subunits of α, one of β, and one of β′) plus a subunit called **sigma factor (σ).** The σ subunit binds to the −10 and −35 regions, thus positioning the holoenzyme to initiate transcription correctly at the start site (Figure 8-9a). The σ subunit also has a role in separating (melting) the DNA strands around the −10 region so that the core enzyme can bind tightly to the DNA in preparation for RNA synthesis. Once the core enzyme is bound, transcription begins and the σ subunit dissociates from the rest of the complex (Figure 8-9b).

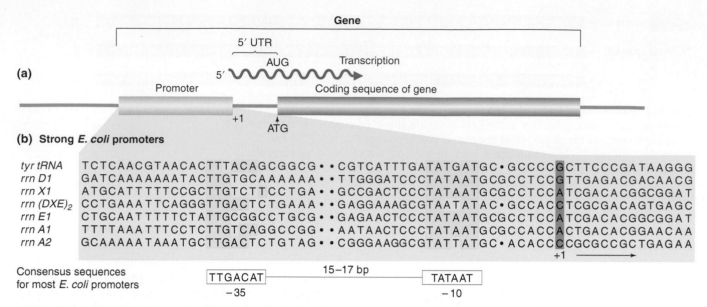

Figure 8-8 Promoter sequence. (a) The promoter lies "upstream" (toward 5′ end) of the initiation point and coding sequences. (b) Promoters have regions of similar sequences, as indicated by the yellow shading in seven different promoter sequences in *E. coli.* Spaces (dots) are inserted in the sequences to optimize the alignment of the common sequences. Numbers refer to the number of bases before (−) or after (+) the RNA synthesis initiation point. The consensus sequence for most *E. coli* promoters is at the bottom. [After H. Lodish, D. Baltimore, A. Berk, S. L. Zipursky, P. Matsudaira, and J. Darnell, *Molecular Cell Biology,* 3d ed. Copyright 1995 by Scientific American Books, Inc. All rights reserved. See W. R. McClure, *Annual Review of Biochemistry* 54, 1985, 171, Consensus Sequences.]

E. coli, like most other bacteria, has several different σ factors. One, called σ^{70} because its mass in kilodaltons is 70, is the primary σ subunit used to initiate the transcription of the vast majority of *E. coli* genes. Other σ factors recognize different promoter sequences. Thus, by associating with different σ factors, the same core enzyme can recognize different promoter sequences and transcribe different sets of genes.

ELONGATION As the RNA polymerase moves along the DNA, it unwinds the DNA ahead of it and rewinds the DNA that has already been transcribed. In this way, it maintains a region of single-stranded DNA, called a **transcription bubble,** within which the template strand is exposed. In the bubble, polymerase monitors the binding of a free ribonucleoside triphosphate to the next exposed base on the DNA template and, if there is a complementary match, adds it to the chain. The energy for the addition of a nucleotide is derived from splitting the high-energy triphosphate and releasing inorganic diphosphate, according to the following general formula:

$$\text{NTP} + (\text{NMP})_n \xrightarrow[\substack{\text{Mg}^{2+} \\ \text{RNA polymerase}}]{\text{DNA}} (\text{NMP})_{n+1} + PP_i$$

Figure 8-10a gives a physical picture of elongation. Inside the bubble, the last 10 nucleotides added to the RNA chain form an RNA:DNA hybrid by complementary base pairing with the template strand.

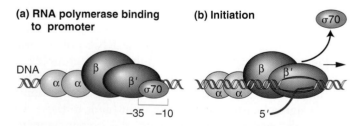

Figure 8-9 Transcription initiation in prokaryotes and the subunit composition of prokaryotic RNA polymerase. (a) Binding of the σ subunit to the −10 and −35 regions positions the other subunits for correct initiation. (b) Shortly after RNA synthesis begins, the σ subunit dissociates from the other subunits, which continue transcription. [After B. M.Turner, *Chromatin and Gene Regulation.* Copyright 2001 by Blackwell Science Ltd.]

(a) Elongation

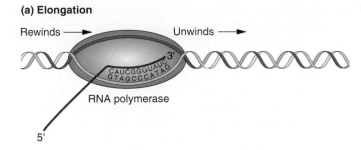

RNA polymerase

(b) Termination: intrinsic mechanism

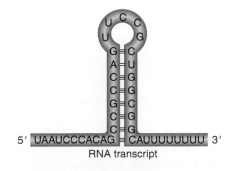

RNA being released

Hairpin loop

Figure 8-10 Elongation and termination of transcription.
The four subunits of RNA polymerase are shown as a single ellipse-like shape surrounding the transcription bubble.
(a) *Elongation:* Synthesis of an RNA strand complementary to the single-strand region of the DNA template strand is in the 5′-to-3′ direction. DNA that is unwound ahead of RNA polymerase is rewound after it has been transcribed.
(b) *Termination:* The intrinsic mechanism shown here is one of two ways used to end RNA synthesis and release the completed RNA transcript and RNA polymerase from the DNA. In this case, the formation of a hairpin loop sets off their release. For both the intrinsic and the rho-mediated mechanism, termination first requires the synthesis of certain RNA sequences.

TERMINATION The transcription of an individual gene is terminated beyond the protein-coding segment of the gene, creating a **3′ untranslated region (3′ UTR)** at the end of the transcript. Elongation continues until RNA polymerase recognizes special nucleotide sequences that act as a signal for chain termination. The encounter with the signal nucleotides initiates the release of the nascent RNA and the enzyme from the template (Figure 8-10b). The two major mechanisms for termination in *E. coli* (and other bacteria) are called **intrinsic** and **rho dependent.**

In the first mechanism, the termination is direct. The terminator sequences contain about 40 base pairs, ending in a GC-rich stretch that is followed by a run of six or more A's. Because G and C in the template will give C and G, respectively, in the transcript, the RNA in this region also is GC rich. These C and G bases are able to form complementary hydrogen bonds with each other, resulting in a **hairpin loop** (Figure 8-11). Recall that the G–C base pair is more stable than the A–T pair because it is hydrogen bonded at three sites, whereas

the A–T (or A–U) pair is held together by only two hydrogen bonds. Hairpin loops that are largely G–C pairs are more stable than loops that are largely A–U pairs. The loop is followed by a run of about eight U's that correspond to the A residues on the DNA template.

Normally, in the course of transcription elongation, RNA polymerase will pause if the short DNA-RNA hybrid in the transcription bubble is weak and will backtrack to stabilize the hybrid. Like hairpins, the strength of the hybrid is determined by the relative number of G–C compared with A–U (or A–T in RNA-DNA hybrids) base pairs. In the intrinsic mechanism, the polymerase is believed to pause after synthesizing the U's (which form a weak DNA-RNA hybrid). However, the backtracking polymerase encounters the hairpin loop which prevents it from finding a stable hybrid. This roadblock sets off the release of RNA from the polymerase and the release of the polymerase from the DNA template.

The second type of termination mechanism requires the help of the rho factor for RNA polymerase to recognize the termination signals. RNAs with rho-dependent termination signals do not have the string of U residues at the end of the RNA and usually do not have hairpin loops. Instead, they have a sequence of about 40 to 60 nucleotides that is rich in C residues and poor in G residues and includes an upstream part called the *rut* (rho *u*tilization) site. Rho is a hexamer consisting of six identical subunits that bind a nascent RNA chain at the *rut* site. These sites are located just upstream from (recall that upstream means 5′ of) sequences at which the RNA polymerase tends to pause. After binding, rho facilitates the release of the RNA from RNA polymerase. Thus, rho-dependent termination entails the binding of rho to *rut*, the pausing of polymerase, and rho-mediated dissociation of the RNA from the RNA polymerase.

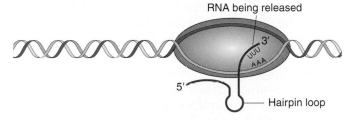

Figure 8-11 The structure of a termination site for RNA polymerase in bacteria. The hairpin structure forms by complementary base pairing within a GC-rich RNA strand. Most of the RNA base pairing is between G and C, but there is one single A–U pair.

8.3 Transcription in eukaryotes

We saw in Chapter 7 that the replication of DNA in eukaryotes, although more complicated, is very similar to the replication of DNA in prokaryotes. In some ways, this is also true for transcription because eukaryotes retain many of the events associated with initiation, elongation, and termination in prokaryotes. DNA replication is more complex in eukaryotes in large part because there is a lot more DNA to copy. Transcription is more complicated in eukaryotes for three primary reasons.

1. The larger eukaryotic genomes have many more genes to be recognized and transcribed. Whereas bacteria usually have a few thousand genes, eukaryotes have tens of thousands of genes. Furthermore, there is much more noncoding DNA in eukaryotes. Noncoding DNA originates by a variety of mechanisms that will be discussed in Chapter 13. So, even though eukaryotes have more genes than prokaryotes do, their genes are, on average, farther apart. For example, whereas the gene density (average number of genes per length of DNA) in *E. coli* is 900 genes per million base pairs, that number drops to 110 genes per million base pairs for the fruit fly *Drosophila*, and it is only 9 genes per million base pairs for humans. This density makes transcription, specifically the initiation step, a much more complicated process. In the genomes of multicellular eukaryotes, finding the start of a gene can be like finding a needle in a haystack.

    As you will see, eukaryotes deal with this situation in several ways. First, they have divided the job of transcription among three different polymerases.

    a. RNA polymerase I transcribes rRNA genes (excluding 5S rRNA).

    b. RNA polymerase II transcribes all protein-coding genes, for which the ultimate transcript is mRNA, and transcribes some snRNAs.

    c. RNA polymerase III transcribes the small functional RNA genes (such as the genes for tRNA, some snRNAs, and 5S rRNA).

    In this section, we will focus our attention on RNA polymerase II.

    Second, they require the assembly of many proteins at a promoter before RNA polymerase II can begin to synthesize RNA. Some of these proteins, called **general trancription factors (GTFs)**, bind before RNA polymerase II binds, whereas others bind afterward. The role of the GTFs and their interaction with RNA polymerase II will be described in the section on transcription initiation in eukaryotes.

2. A significant difference between eukaryotes and prokaryotes is the presence of a nucleus in eukaryotes. RNA is synthesized in the nucleus where the DNA is located and must be modified in several ways before it can be exported out of the nucleus into the cytoplasm for translation. These modifications are collectively referred to as **RNA processing.** As you will see, the 5′ half of the RNA undergoes processing while the 3′ half is still being synthesized. One of the reasons that RNA polymerase II is a more complicated multisubunit enzyme (it is considered to be another molecular machine) than prokaryotic RNA polymerase is that it must synthesize RNA while simultaneously coordinating a diverse array of processing events. To distinguish the RNA before and after processing, newly synthesized RNA is called the **primary transcript** or **pre-mRNA** and the terms **mRNA** or **mature RNA** are reserved for the fully processed transcript that can be exported out of the nucleus. The coordination of RNA processing and synthesis by RNA polymerase II will be discussed in the section on transcription elongation in eukaryotes.

3. Finally, the template for transcription, genomic DNA, is organized into chromatin in eukaryotes (see Chapter 3), whereas it is virtually "naked" in prokaryotes. As you will learn in Chapter 10, certain chromatin structures can block access of RNA polymerase to the DNA template. This feature of chromatin has evolved into a very sophisticated mechanism for regulating eukaryotic gene expression. However, a discussion of the influence of chromatin on the ability of RNA polymerase II to initiate transcription will be put aside for now as we focus on the events that take place *after* RNA polymerase II gains access to the DNA template.

Transcription initiation in eukaryotes

As stated earlier, transcription starts in prokaryotes when the σ subunit of the RNA polymerase holoenzyme recognizes the -10 and -35 regions in the promoter of a gene. After transcription begins, the σ subunit dissociates and the core polymerase continues to synthesize RNA within a transcription bubble that moves along the DNA. Similarly, in eukaryotes, the core of RNA polymerase II also cannot recognize promoter sequences on its own. However, unlike the case for bacteria, where σ factor is an integral part of the polymerase holoenzyme, GTFs are required in eukaryotes to bind to regions in the promoter *before* the binding of the core enzyme.

The initiation of transcription in eukaryotes has some features that are reminiscent of the initiation of replication at origins of replication. Recall from Chapter 7 that proteins that are not part of the replisome initiate the assembly of the replication machine. DnaA in *E. coli* and ORC in yeast, for example, first recognize and bind to origin DNA sequences. These proteins serve to attract replication proteins, including DNA polymerase III, through protein–protein interactions. Similarly, GTFs, which do not take part in RNA synthesis, recognize and bind to sequences in the promoter or to one another and serve to attract the RNA polymerase II core and position it at the correct site to start transcription. The GTFs are designated TFIIA, TFIIB, and so forth (for *transcription factor* of RNA polymerase *II*).

The GTFs and the RNA polymerase II core constitute the **preinitiation complex (PIC)**. This complex is quite large: it contains six GTFs, *each* of which is a multiprotein complex, plus the RNA polymerase II core, which is made up a dozen or more protein subunits. The sequence of amino acids of some of the RNA polymerase II core subunits is conserved from yeast to humans. This conservation can be dramatically demonstrated by replacing some yeast RNA polymerase II subunits with their human counterparts to form a functional *chimeric* RNA polymerase II complex (named after a fire-breathing creature from Greek mythology that had a lions head, a goat's body, and a serpent's tail).

Like prokaryotic promoters, eukaryotic promoters are located on the 5' side (upstream) of the transcription start site. In an alignment of eukaryotic promoter regions, it can be seen that the sequence TATA is often located about 30 base pairs (−30 bp) from the transcription start site (Figure 8-12). This sequence, called the **TATA box,** is the site of the first event in transcription: the binding of the **TATA binding protein (TBP)**. TBP is part of the TFIID complex, which is one of the six GTFs. When bound to the TATA box, TBP attracts other GTFs and the RNA polymerase II core to the promoter, thus forming the preinitiation complex. After transcription has been initiated, RNA polymerase II dissociates from most of the GTFs to elongate the primary RNA transcript. Some of the GTFs remain at the promoter to attract the next RNA polymerase core. In this way, multiple RNA polymerase II enzymes can be synthesizing transcripts from a single gene at one time.

How is the RNA polymerase II core able to separate from the GTFs and start transcription? Although the details of this process are still being worked out, what is known is that a protein tail of the β subunit of RNA polymerase II, called the **carboxyl tail domain (CTD),** participates in initiation and in several other critical phases of RNA synthesis and processing. The CTD is strategically located near the site at which nascent RNA will emerge from the polymerase. The initiation phase ends and the elongation phase begins after the CTD is phosphorylated by one of the GTFs. This phosphorylation is thought to somehow weaken the connection of RNA polymerase II to the other proteins of the preinitiation complex and permit elongation.

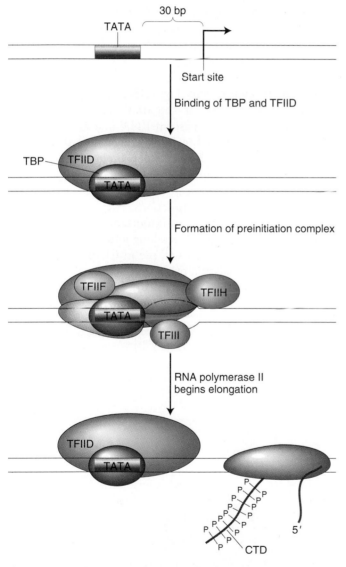

Figure 8-12 Transcription initiation in eukaryotes. Formation of the preinitiation complex usually begins with the binding of the TATA binding protein (TBP), which then recruits the other general transcription factors and RNA polymerase II to the transcription start site. Transcription begins after phosphorylation of the carboxyl tail domain (CTD) of RNA polymerase II. [After "RNA Polymerase II Holoenzyme and Transcription Factors," *Encyclopedia of Life Sciences.* Copyright 2001, Macmillan Publishing Group Ltd./Nature Publishing Group.]

MESSAGE Eukaryotic promoters are first recognized by general transcription factors whose function is to attract the core RNA polymerase II so that it is positioned to begin RNA synthesis at the transcription start site.

Elongation, termination, and pre-mRNA processing in eukaryotes

Elongation takes place inside the transcription bubble essentially as described for the synthesis of prokaryotic RNA. However, nascent RNA has very different fates in prokaryotes and eukaryotes. In prokaryotes, translation begins at the 5′ end of the nascent RNA while the 3′ half is still being synthesized. The process of translation will be described in greater detail in Chapter 9. In contrast, the RNA of eukaryotes must undergo further processing before it can be translated. This processing includes (1) the addition of a cap at the 5′ end, (2) the addition of a 3′ tail of adenine nucleotides (polyadenylation), and (3) splicing to eliminate introns.

Like DNA replication, the synthesis and processing of pre-mRNA to mRNA requires that many steps be performed rapidly and accurately. At first most of the processing of eukaryotic pre-mRNA was thought to take place after RNA synthesis was complete. Processing after RNA synthesis is complete is said to be **posttranscriptional**. However, experimental evidence now indicates that processing actually takes place during RNA synthesis; it is **cotranscriptional**. Therefore, the partly synthesized (nascent) RNA is undergoing processing reactions as it emerges from the RNA polymerase II complex.

The CTD of eukaryotic RNA polymerase II plays a central role in coordinating all processing events. The CTD is composed of many repeats of a sequence of seven amino acids. These repeats serve as binding sites for some of the enzymes and other proteins that are re-quired for RNA capping, splicing, and cleavage followed by polyadenylation. The CTD is located near the site where nascent RNA emerges from the polymerase so that it is in an ideal place to orchestrate the binding and release of proteins needed to process the nascent RNA transcript while RNA synthesis continues. In the various phases of transcription, the amino acids of the CTD are reversibly modified—usually through the addition and removal of phosphate groups (called phosphorylation and dephosphorylation, respectively). The phosphorylation state of the CTD determines which processing proteins can bind. In this way, the CTD determines what task is to be performed on the RNA as it emerges from the polymerase. The processing events and the role of CTD in executing them are shown in Figure 8-13 and considered next.

Processing 5′ and 3′ ends Figure 8-13a depicts the processing of the 5′ end of the transcript of a protein-coding gene. When the nascent RNA first emerges from RNA polymerase II, a special structure, called a **cap**, is added to the 5′ end by several proteins that interact with the CTD. The cap consists of a 7-methylguanosine residue linked to the transcript by three phosphate groups. The cap has two functions. First, it protects the RNA from degradation—an important step considering that a eukaryotic mRNA has a long journey before being translated. Second, as you will see in Chapter 9, the cap is required for translation of the mRNA.

RNA elongation continues until the conserved sequence, AAUAAA or AUUAAA, near the 3′ end is rec-

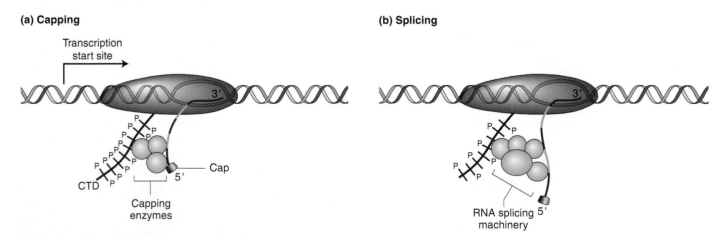

(a) Capping   **(b) Splicing**

Figure 8-13 Cotranscriptional processing of RNA. Cotranscriptional processing is coordinated by the carboxyl tail domain (CTD) of the β subunit of RNA polymerase II. Reversible phosphorylation of the amino acids of the CTD (indicated by the P's) creates binding sites for the different processing enzymes and factors required for (a) capping and (b) splicing. [After R. I. Drapkin and D. F. Reinberg, "RNA Synthesis," *Encyclopedia of Life Sciences.* Copyright 2002, Macmillan Publishing Group Ltd./Nature Publishing Group.]

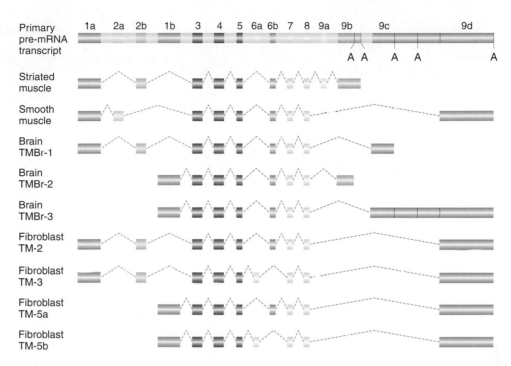

Figure 8-14 Complex patterns of eukaryotic mRNA splicing. The pre-mRNA transcript of the α-tropomyosin gene is alternatively spliced in different cell types. The light green boxes represent introns; the other colors represent exons. Polyadenylation signals are indicated by an A. Dashed lines in the mature mRNAs indicate regions that have been removed by splicing. TM, tropomyosin. [After J. P. Lees et al., *Molecular and Cellular Biology* 10, 1990, 1729–1742.]

ognized by an enzyme that cuts off the end of the RNA approximately 20 bases farther down. To this cut end, a stretch of 150 to 200 adenine nucleotides called a **poly(A) tail** is added. Hence the AAUAAA sequence of protein-coding genes is called a *polyadenylation signal.*

RNA splicing, the removal of introns The vast majority of eukaryotic genes contain **introns,** segments of unknown function that do not code for polypeptides. Introns are present not only in protein-coding genes but also in some rRNA and even tRNA genes. Introns are removed from the primary transcript while RNA is still being synthesized and after the cap has been added, but before the transcript is transported into the cytoplasm. The removal of introns and the joining of exons is called **splicing,** because it is reminiscent of the way in which videotape or movie film can be cut and rejoined to edit out a specific segment. Splicing brings together the coding regions, the **exons,** so that the mRNA now contains a coding sequence that is completely colinear with the protein that it encodes.

The number and size of introns varies from gene to gene and from species to species. For example, only about 235 of the 6000 genes in yeast have introns, whereas typical genes in mammals, including humans, have several. The average size of a mammalian intron is about 2000 nucleotides; thus, a larger percentage of the DNA in mammals encodes introns, not exons. An extreme example is the human Duchenne muscular distrophy gene. This gene has 79 exons and 78 introns spread across 2.5 million base pairs. When spliced together, its 79 exons produce an mRNA of 14,000 nucleotides, which means that introns account for the vast majority of the 2.5 million base pairs.

Alternative splicing Alternative pathways of splicing can produce different mRNAs and, subsequently, different proteins from the same primary transcript. The altered forms of the same protein that are generated by alternative spicing are usually used in different cell types or at different stages of development. Figure 8-14 shows the myriad combinations produced by different splicings of the primary RNA transcript of the α-tropomyosin gene. These alternative splicings will ultimately generate a set of related proteins that function optimally in each cell type.

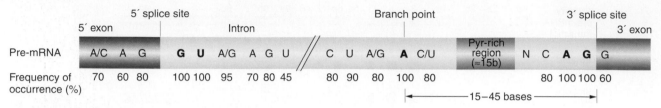

	5' splice site					Branch point			3' splice site		

	5' exon		Intron									3' exon

Pre-mRNA: A/C A G | G U A/G A G U // C U A/G A C/U | Pyr-rich region (≈15b) | N C A G G

Frequency of occurrence (%): 70 60 80 | 100 100 95 70 80 45 | 80 90 80 100 80 | 80 100 100 60

15–45 bases

Figure 8-15 Conserved sequences related to intron splicing. The numbers below the nucleotides indicate the percentage of similarity among organisms. Of particular importance are the G and U residues at the 5' end, the A and G residues at the 3' end, and the A residue labeled "branch point" (see Figure 8-16 for a view of the branch structure). N represents any base.

Mechanism of exon splicing Figure 8-15 shows the exon–intron junctions of pre-mRNAs. These junctions are the sites at which the splicing reactions take place. At these junctions, certain specific nucleotides are identical across genes and across species; they have been conserved because they participate in the splicing reactions. Each intron will be cut at each end, and these ends almost always have GU at the 5' end and AG at the 3' end (the **GU–AG rule**). Another invariant site is an A residue between 15 and 45 nucleotides upstream from

Pre-mRNA
GU   A AG
Exon 1 ⌐————————————————————⌐ Exon 2
                    Intron

Spliceosome attached to pre-mRNA

Spliceosome

SnRNA

Spliced exons

Lariat

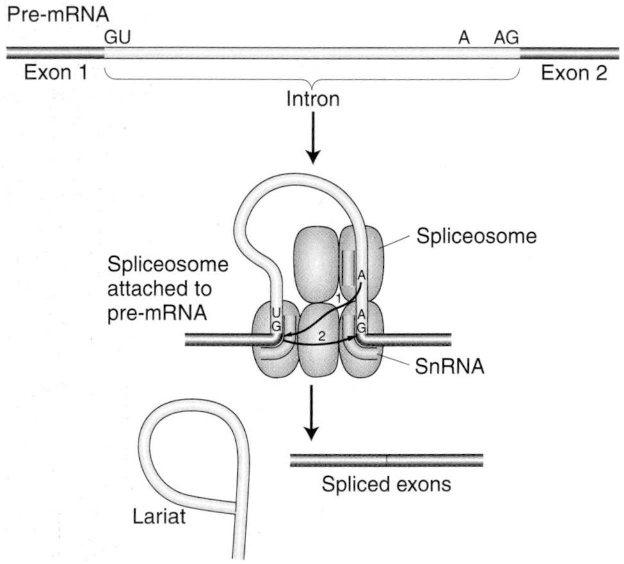

Figure 8-16 The structure and function of a spliceosome. The spliceosome is composed of several snRNPs that attach sequentially to the RNA, taking up positions roughly as shown. Alignment of the snRNPs results from hydrogen bonding of their snRNA molecules to the complementary sequences of the intron. In this way, the reactants are properly aligned and the splicing reactions (1 and 2) can take place. The P-shaped loop, or lariat structure, formed by the excised intron is joined through the central adenine nucleotide.

the 3' splice site. This residue takes part in an intermediate reaction that is necessary for intron excision. Other, less well conserved nucleotides are found flanking the highly conserved ones.

These conserved nucleotides in the transcript are recognized by small nuclear ribonucleoprotein particles (snRNPs), which are complexes of protein and small nuclear RNA. A functional splicing unit is composed of a team of snRNPs called a **spliceosome.** Components of the spliceosome interact with the CTD and attach to intron and exon sequences, as shown in Figures 8-12 and 8-16. The snRNAs in the spliceosome help to align the splice sites by forming hydrogen bonds to the conserved intron sequences. Then the spliceosome catalyzes the removal of the intron through two consecutive splicing steps, labeled 1 and 2 in Figure 8-16. Splice 1 attaches one end of the intron to the conserved internal adenine, forming a structure having the shape of a cowboy's lariat. Splice 2 releases the lariat and joins the two adjacent exons. Figure 8-17 portrays the chemistry behind intron excision. Chemically, steps 1 and 2 are transesterification reactions between the conserved nucleotides.

Self-splicing introns and the RNA world

One exceptional case of RNA splicing led to a discovery considered by some to be as important as that of the double-helical structure of DNA. In 1981, Tom Cech and coworkers reported that, in a test tube, the primary transcript of an rRNA from the ciliate protozoan *Tetrahymena* can excise a 413-nucleotide intron *from itself* without the addition of any protein. Subsequently, other introns have been shown to have this property and they have come to be known as **self-splicing introns.** Cech's finding is considered a landmark discovery in biology because it marked the first time that a biological molecule, other than protein, was shown to catalyze a reaction. This discovery and similar ones have provided solid evidence for a theory called the **RNA world,** which

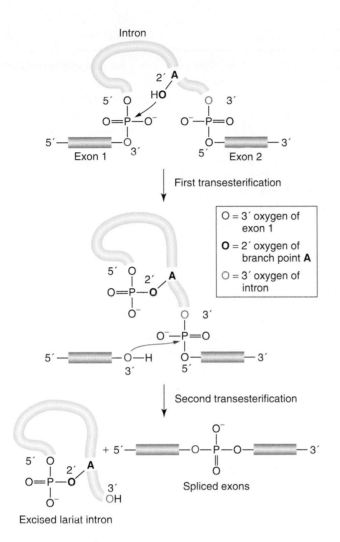

Intron

Exon 1   Exon 2

First transesterification

O = 3′ oxygen of exon 1

O = 2′ oxygen of branch point **A**

O = 3′ oxygen of intron

Second transesterification

+ 5′ — Spliced exons — 3′

Excised lariat intron

Figure 8-17 The reactions in the splicing of an intron from surrounding exons. Two transesterification reactions take place, first to join the GU end to the internal branch point (reaction 1 in Figure 8-16) and second to join the two exons together (reaction 2 in Figure 8-16). [After H. Lodish, A. Berk, S. L. Zipurski, P. Matsudaira, D. Baltimore, and J. Darnell, *Molecular Cell Biology*, 4th ed. Copyright 2000 by W. H. Freeman and Company.]

holds that RNA must have been the genetic material in the first cells because only RNA is known to both encode genetic information and catalyze biological reactions.

The discovery of self-splicing introns has led to a reexamination of the role of the snRNAs in the spliceosome. The question currently under investigation is whether intron excision is catalyzed by the RNA or the protein component of the splicesome. As you will see in Chapter 9, the RNAs (the rRNAs) in the ribosome, not the ribosomal proteins, are now thought to have the central role in most of the important events of protein synthesis.

MESSAGE Eukaryotic pre-mRNA is extensively processed by 5′ capping, 3′ polyadenylation, and the removal of introns and splicing of exons before it can be transported as mRNA to the cytoplasm for translation into protein. These events are cotranscriptional and are coordinated by part of the RNA polymerase II complex.

KEY QUESTIONS REVISITED

• **How does the structure of RNA differ from that of DNA?**

Whereas DNA is a double helix, RNA usually exists in the cell as a single chain. In addition, as the name indicates, the sugar component of RNA is ribose, whereas that of DNA is deoxyribose. Finally, where DNA has a thymine base, RNA contains a uracil base.

• **What are the different classes of RNA in the cell?**

RNAs either code for protein (they are mRNAs) or are themselves functional. Functional RNAs (for example, rRNA, tRNA, and snRNA) are active as RNA; they are never translated into polypeptides.

• **How is RNA polymerase correctly positioned to start transcription in prokaryotes?**

The σ subunit of the RNA polymerase holoenzyme recognizes the −10 and −35 regions of the promoter. After

transcription initiation, the σ subunit dissociates from the core polymerase, which continues RNA synthesis.

• **How is eukaryotic RNA synthesized by RNA polymerase II modified before leaving the nucleus?**

A cap is added to the 5′ end, introns are removed by the spliceosome, the 3′ end is cleaved by an endonuclease, and a poly(A) tail is added to the 3′ end.

• **Why is the discovery of self-splicing introns considered by some to be as important as the discovery of the DNA double helix?**

The discovery that the primary transcript of *Tetrahymena* rRNA excised its own intron was the first demonstration that RNA could function as a biological catalyst. Many more RNA catalysts, called ribozymes, have been identified, providing support for the view that RNA was the genetic material in the first cells.

SUMMARY

We know that information is not transferred directly from DNA to protein, because, in a eukaryotic cell, DNA is in the nucleus, whereas protein is synthesized in the cytoplasm. Information transfer from DNA to protein requires an intermediate. That intermediate is RNA.

Although DNA and RNA are nucleic acids, RNA differs from DNA in that (1) it is usually single stranded rather than a double helix, (2) it has the sugar ribose rather than deoxyribose in its nucleotides, (3) it has the pyrimidine base uracil rather than thymine, and (4) it can serve as a biological catalyst.

The similarity of RNA to DNA suggests that the flow of information from DNA to RNA relies on the complementarity of bases, which is also the key to DNA replication. RNA is copied, or transcribed, from a template DNA strand into either a functional RNA (such as tRNA or rRNA), which is never translated into polypeptides, or a messenger RNA, from which proteins are synthesized.

In prokaryotes, all classes of RNA are transcribed by a single RNA polymerase. This multisubunit enzyme initiates transcription by binding to the DNA at promoters that contain specific sequences at -35 and -10 bases before the transcription start site at $+1$. Once bound, RNA polymerase locally unwinds the DNA and begins incorporating ribonucleotides that are complementary to the template DNA strand. The chain grows in the 5′-to-3′ direction until one of two mechanisms, intrinsic or rho dependent, leads to the dissociation of the polymerase and the RNA from the DNA template. As we will see in Chapter 9, in the absence of a nucleus, prokaryotic RNAs that code for proteins are translated while they are being transcribed.

In eukaryotes, there are three different RNA polymerases; only RNA polymerase II transcribes mRNAs. Overall, the phases of initiation, elongation, and termination of RNA synthesis in eukaryotes resembles those in prokaryotes. However, there are important differences. RNA polymerase II does not bind directly to promoter DNA, but rather to general transcription factors, one of which recognizes the TATA sequence in most eukaryotic promoters. RNA polymerase II is a much larger molecule than its prokaryotic counterpart. It contains numerous subunits that function not only to elongate the primary RNA transcript, but also to coordinate the extensive processing events that are necessary to produce the mature mRNA. These processing events are 5′ capping, intron removal and exon splicing by spliceosomes, and 3′ cleavage followed by polyadenylation. Part of the RNA polymerase II core, the carboxyl tail domain, is positioned ideally to interact with the nascent RNA as it emerges from polymerase. In this way, RNA polymerase II coordinates the numerous events of RNA synthesis and processing.

The discovery of self-splicing introns demonstrated that RNA can function as a catalyst, much like proteins. Since the discovery of these ribozymes, the scientific community has begun to pay much more attention to RNA. What was once thought to be a lowly messenger is now recognized as a versatile and dynamic participant in many cellular processes. You will learn more about the diverse roles of RNA in subsequent chapters.

KEY TERMS

cap (p. 266)

carboxyl tail domain (CTD) (p. 265)

coding strand (p. 261)

consensus sequence (p. 261)

core enzyme (p. 261)

cotranscriptional processing (p. 266)

elongation (p. 261)

exon (p. 267)

general transcription factor (GTF) (p. 264)

GU–AG rule (p. 268)

hairpin loop (p. 263)

initiation (p. 261)

intrinsic mechanism (p. 263)

intron (p.267)

mature RNA (mRNA) (p. 264)

poly(A) tail (p. 267)

posttranscriptional processing (p. 266)

preinitiation complex (PIC) (p. 265)

primary transcript (pre-mRNA) (p. 264)

promoter (p. 261)

pulse-chase experiment (p. 257)

rho-dependent mechanism (p. 263)

ribose (p. 257)

ribosomal RNA (rRNA) (p. 259)

ribozyme (p. 258)

RNA polymerase (p. 259)

RNA polymerase holoenzyme (p. 261)

RNA processing (p. 264)

RNA world (p. 268)

self-splicing intron (p. 268)

sigma factor (σ) (p. 261)

small nuclear RNA (snRNA) (p. 259)

spliceosome (p. 268)

splicing (p. 262)

TATA binding protein (TBP) (p. 265)

TATA box (p. 265)

template (p. 259)

termination (p. 261)

transcript (p. 259)

transcription (p. 259)

transcription bubble (p. 262)

transfer RNA (tRNA) (p. 259)

3′ untranslated region (3′ UTR)
  (p. 263)

5′ untranslated region (5′ UTR)
  (p. 261)

upstream (p. 261)

uracil (U) (p. 257)

PROBLEMS

BASIC PROBLEMS

1. The two strands of λ phage differ from each other in their GC content. Owing to this property, they can be separated in an alkaline cesium chloride gradient (the alkalinity denatures the double helix). When RNA synthesized by λ phage is isolated from infected cells, it is found to form DNA-RNA hybrids with both strands of λ DNA. What does this finding tell you? Formulate some testable predictions.

2. In both prokaryotes and eukaryotes, describe what else is happening to the RNA while RNA polymerase is synthesizing a transcript from the DNA template.

3. List three examples of proteins that act on nucleic acids.

4. What is the primary function of the sigma factor? Is there a protein in eukaryotes analogous to the sigma factor?

5. You have identified a mutation in yeast, a unicellular eukaryote, that prevents the capping of the 5′ end of the RNA transcript. However, much to your surprise, all the enzymes required for capping are normal. Instead, you determine that the mutation is in one of the subunits of RNA polymerase II. What subunit is mutant and how does this mutation result in failure to add a cap to yeast RNA?

6. Why is RNA produced only from the template DNA strand and not from both strands?

7. A linear plasmid contains only two genes, which are transcribed in opposite directions, each one from the end, toward the center of the plasmid. Draw diagrams that show

    a. The plasmid DNA, showing 5′ and 3′ ends of the nucleotide strands

    b. The template strand for each gene

    c. The positions of the transcription initiation site

    d. The transcripts, showing 5′ and 3′ ends

8. Are there similarities between the DNA replication bubbles and the transcription bubbles found in eukaryotes? Explain.

9. Which of the following are true about eukaryotic mRNA?

    a. The sigma factor is essential for correct initiation of transcription.

    b. Processing of the nascent mRNA may begin before its transcription is complete.

    c. Processing occurs in the cytoplasm.

    d. Termination occurs via a hairpin loop or use of rho factor.

    e. Many RNAs can be transcribed simultaneously from one DNA template.

10. A researcher was mutating prokaryotic cells by inserting segments of DNA. In this way she made the following mutation:

Original	TTGACAT 15 to 17 bp TATAAT
Mutant	TATAAT 15 to 17 bp TTGACAT

    a. What does this sequence represent?

    b. What do you predict will be the effect of such a mutation? Explain.

11. You will learn more about genetic engineering in Chapter 11, but for now, put on your genetic engineer cap and try this problem: E. coli is widely used in laboratories to produce proteins from other organisms.

    a. You have isolated a yeast gene that encodes a metabolic enzyme and want to produce this enzyme in E. coli. You suspect that the yeast promoter will not work in E. coli. Why?

    b. After replacing the yeast promoter with an E. coli promoter, you are pleased to detect RNA from the yeast gene but are confused because the RNA is almost twice the length of the mRNA from this gene isolated from yeast. Provide an explanation for why this might have occurred.

12. Draw a prokaryotic gene and its RNA product. Be sure to include the promoter, transcription start site, transcription termination site, untranslated regions, and labeled 5′ and 3′ ends.

13. Draw a eukaryotic gene that has two introns and its pre-mRNA and mRNA products. Be sure to include all the features of the prokaryotic gene in your answer to Problem 12, plus the processing events required to produce the mRNA.

14. A certain *Drosophila* protein-coding gene has one intron. If a large sample of null alleles of this gene is examined, will any of the mutant sites be expected

    a. In the exons?

    b. In the intron?

    c. In the promoter?

    d. In the intron-exon boundary?

CHALLENGING PROBLEMS

15. The following data represent the base compositions of double-stranded DNA from two different bacterial species and their RNA products obtained in experiments conducted in vitro:

Species	(A+T)/(G+C)	(A+U)/(G+C)	(A+G)/(U+C)
Bacillus subtilis	1.36	1.30	1.02
E. coli	1.00	0.98	0.80

a. From these data, can you determine whether the RNA of these species is copied from a single strand or from both strands of the DNA? How? Drawing a diagram will make it easier to solve this problem.

b. Explain how you can tell whether the RNA itself is single-stranded or double-stranded.

(Problem 15 is reprinted with permission of Macmillan Publishing Co., Inc., from M. Strickberger, *Genetics*. Copyright 1968, Monroe W. Strickberger.)

16. A human gene was initially identified as having three exons and two introns. The exons are 456, 224, and 524 bp, while the introns are 2.3 kb and 4.6 kb.

a. Draw this gene, showing the promoter, introns, exons, and transcription start and stop sites.

b. Surprisingly, it is found that this gene encodes not one but two mRNAs that have only 224 nucleotides in common. The original mRNA is 1204 nucleotides, while the new mRNA is 2524 nucleotides. Use your drawing to show how it is possible for this one region of DNA to encode these two transcripts.

INTERACTIVE GENETICS   **MegaManual CD-ROM Tutorial**

Molecular Biology: Gene Expression

For additional coverage of the material in this chapter, refer to the Molecular Biology activity on the Interactive Genetics CD-ROM included with the Solutions MegaManual. In the Gene Expression unit, you will find animated tutorials on transcription and splicing, which allow you to visualize the molecular machinery involved in the process. Problems 1 and 2 in this section will help you apply your knowledge of these processes to typical experiments done by geneticists.

9

PROTEINS AND THEIR SYNTHESIS

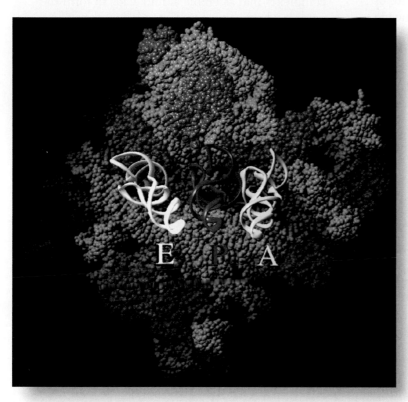

Ribosome structure. This image shows at atomic resolution a surface of the ribosome from the bacteria *Haloarcula marismortui*, deduced from X-ray crystallography. The portion of the ribosome consisting of RNA is shown in blue; that consisting of protein is shown in purple. The white, red, and yellow structures in the center are tRNAs at the E, P, and A binding sites, their acceptor stems disappearing into a cleft in the ribosome. [From P. Nissen, J. Hansen, N. Ban, P. B. Moore, and T. A. Steitz, "The Structural Basis of Ribosome Activity in Peptide Bond Synthesis," *Science* 289, 2000, 920–930, Figure 10A, at p. 926.]

KEY QUESTIONS

- How are the sequences of a gene and its protein related?

- Why is it said that the genetic code is nonoverlapping and degenerate?

- How is the correct amino acid paired with each mRNA codon?

- Why is the attachment of an amino acid to the correct tRNA considered to be such an important step in protein synthesis?

- What is the evidence that the ribosomal RNA, not the ribosomal proteins, carries out the key steps in translation?

- How does the initiation of translation differ in prokaryotes and eukaryotes?

- What is posttranslational processing, and why is it important for protein function?

OUTLINE

9.1 Protein structure

9.2 Colinearity of gene and protein

9.3 The genetic code

9.4 tRNA: the adapter

9.5 Ribosomes

9.6 Posttranslational events

CHAPTER OVERVIEW

We saw in Chapters 7 and 8 how DNA is copied from generation to generation and how RNA is synthesized from specific regions of DNA. We can think of these processes as two stages of information transfer: *replication* (the synthesis of DNA) and *transcription* (the synthesis of an RNA copy of a part of the DNA). In this chapter, we will learn about the final stage of information transfer: *translation* (the synthesis of a polypeptide directed by the RNA sequence).

As we learned in Chapter 8, RNA transcribed from genes is classified as either messenger RNA (mRNA) or functional RNA. In this chapter, we will see the fate of both RNA classes. The vast majority of genes encode mRNAs whose function is to serve as an intermediate in the synthesis of the ultimate gene product, protein. In contrast, recall that *functional* RNAs are active as RNAs; they are never translated into proteins. The main classes of functional RNAs are important actors in protein synthesis. They include transfer RNAs and ribosomal RNAs.

- **Transfer RNA (tRNA)** molecules are the adapters that translate the three-nucleotide codon in the mRNA into the corresponding amino acid, which is brought to the ribosome in the process of translation. The tRNAs are general components of the translation machinery; a tRNA molecule can bring an amino acid to the ribosome for the purpose of translating *any* mRNA.

- **Ribosomal RNAs (rRNAs)** are the major components of **ribosomes,** which are large macromolecular complexes that assemble amino acids to form the protein whose sequence is encoded in a specific mRNA. Ribosomes are composed of several types of rRNA and scores of different proteins. Like tRNA, ribosomes are general in function in the sense that they can be used to translate the mRNAs of *any* protein-coding gene.

CHAPTER OVERVIEW Figure

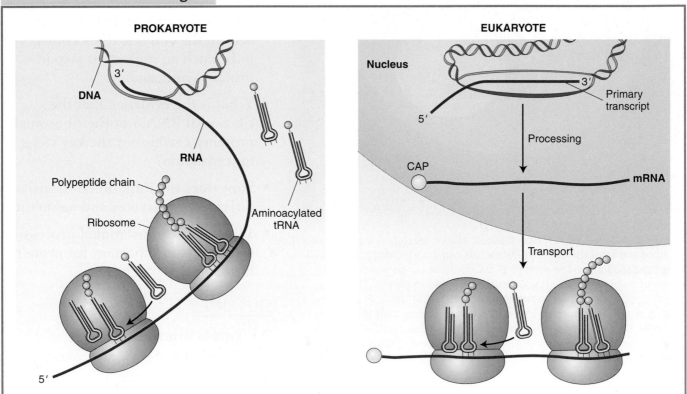

Figure 9-1 The transfer of genetic information from DNA to RNA to protein in prokaryotes and eukaryotes. Transcription and translation occur in separate compartments in the eukaryotic cell but in the same compartment in the prokaryotic cell. Otherwise the process of translation is similar in all organisms: An RNA called tRNA brings each amino acid to the ribosome, a molecular machine that joins amino acids into a chain according to the information provided by mRNA.

Although most genes encode mRNAs, functional RNAs make up, by far, the largest fraction of total cellular RNA. In a typical actively dividing eukaryotic cell, rRNA and tRNA account for almost 95 percent of the total RNA, whereas mRNA accounts for only about 5 percent. Two factors explain the abundance of rRNAs and tRNAs. First, they are much more stable than mRNAs, and so these molecules remain intact much longer. Second, the transcription of rRNA and tRNA genes constitutes more than half of the total nuclear transcription in active eukaryotic cells and almost 80 percent of transcription in yeast cells.

The components of the translational machinery and the process of translation are very similar in prokaryotes and eukaryotes. Nevertheless, there are several differences, which are highlighted in Figure 9-1. These differences are largely due to differences in where transcription and translation take place in the cell: the two processes take place in the same compartment in prokaryotes, whereas they are physically separated in eukaryotes. After extensive processing, eukaryotic mRNAs are exported from the nucleus for translation on ribosomes that reside in the cytoplasm. In contrast, transcription and translation are coupled in prokaryotes: translation of an RNA begins at its 5′ end while the rest of the mRNA is still being transcribed.

9.1 Protein structure

When a primary transcript has been fully processed into a mature mRNA molecule, translation into protein can take place. Before considering how proteins are made, we need to understand protein structure.

Proteins are the main determinants of biological form and function. These molecules heavily influence the shape, color, size, behavior, and physiology of organisms. Because genes function by encoding proteins, understanding the nature of proteins is essential to understanding gene action.

A protein is a polymer composed of monomers called **amino acids.** In other words, a protein is a chain of amino acids. Because amino acids were once called *peptides*, the chain is sometimes referred to as a **polypeptide.** Amino acids all have the general formula

$$H_2N-\overset{\overset{\displaystyle H}{|}}{\underset{\underset{\displaystyle R}{|}}{C}}-COOH$$

All amino acids have a side chain, or R (reactive) group. There are 20 amino acids known to exist in proteins, each having a different R group that gives the amino acid its unique properties. The side chain can be anything from a hydrogen atom (as in the amino acid

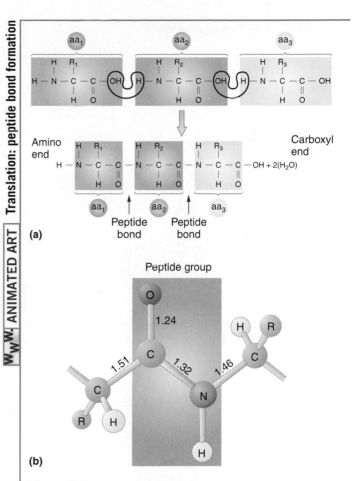

ANIMATED ART Translation: peptide bond formation

(a)

(b)

Figure 9-2 The peptide bond. (a) A polypeptide is formed by the removal of water between amino acids to form peptide bonds. Each aa indicates an amino acid. R_1, R_2, and R_3 represent R groups (side chains) that differentiate the amino acids. (b) The peptide group is a rigid planar unit with the R groups projecting out from the C–N backbone. Standard bond distances (in angstroms) are shown. [Part b from L. Stryer, *Biochemistry*, 4th ed. Copyright 1995 by Lubert Stryer.]

glycine) to a complex ring (as in the amino acid tryptophan). In proteins, the amino acids are linked together by covalent bonds called peptide bonds. A peptide bond is formed by the linkage of the **amino end** (NH_2) of one amino acid with the **carboxyl end** (COOH) of another amino acid. One water molecule is removed during the reaction (Figure 9-2). Because of the way in which the peptide bond forms, a polypeptide chain always has an amino end (NH_2) and a carboxyl end (COOH), as shown in Figure 9-2a.

Proteins have a complex structure that has four levels of organization, illustrated in Figure 9-3. The linear sequence of the amino acids in a polypeptide chain constitutes the **primary structure** of the protein. The **secondary structure** of a protein is the specific shape taken by the polypeptide chain by folding. This shape arises from the bonding forces between amino acids that are close

(a) Primary structure

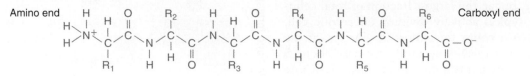

(b) Secondary structure

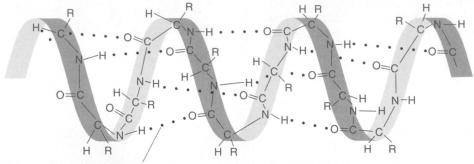

Hydrogen bonds between amino acids
at different locations in polypeptide chain

α helix

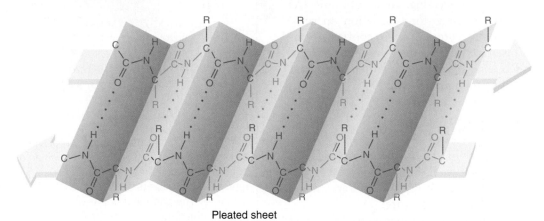

Pleated sheet

(c) Tertiary structure

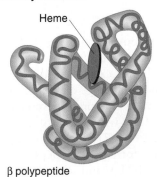

Heme

β polypeptide

(d) Quaternary structure

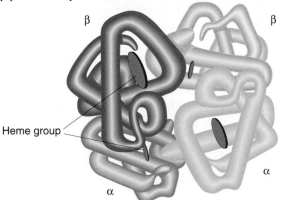

β

β

Heme group

α

α

Figure 9-3 Levels of protein structure. (a) Primary structure. (b) Secondary structure. The polypeptide can form a helical structure (an α helix) or a zigzag structure (a pleated sheet). The pleated sheet has two polypeptide segments arranged in opposite polarity, as indicated by the arrows. (c) Tertiary structure. The heme group is a nonprotein ring structure with an iron atom at its center. (d) Quaternary structure illustrated by hemoglobin, which is composed of four polypeptide subunits: two α subunits and two β subunits.

together in the linear sequence. These forces include several types of weak bonds, notably hydrogen bonds, electrostatic forces, and van der Waals forces. The most common secondary structures are the α helix and the pleated sheet. Different proteins show either one or the other or sometimes both within their structures. **Tertiary structure** is produced by the folding of the secondary structure. Some proteins have **quaternary structure:** such a protein is composed of two or more separate folded polypeptides, also called **subunits,** joined together by weak bonds. The quaternary association can be between different types of polypeptides (resulting in a heterodimer) or between identical polypeptides (making a homodimer). Hemoglobin is an example of a heterotetramer, composed of two copies each of two different polypeptides, shown in green and purple in Figure 9-3.

Many proteins are compact structures; they are called **globular proteins.** Enzymes and antibodies are among the best-known globular proteins. Proteins with linear shape, called **fibrous proteins,** are important components of such structures as hair and muscle.

Shape is all-important to a protein because its specific shape enables it to do its specific job in the cell. A protein's shape is determined by its primary amino acid sequence and by conditions in the cell that promote the folding and bonding necessary to form higher-level structures. The folding of proteins into their correct conformation will be discussed at the end of this chapter. The amino acid sequence also determines which R groups are present at specific positions and thus available to bind with other specific cellular components.

The active sites of enzymes are good illustrations of the precise interactions of R groups. Each enzyme has a pocket called the **active site** into which its substrate or substrates can fit (Figure 9-4). Within the active site, the R groups of certain amino acids are strategically positioned to interact with a substrate and catalyze a variety of chemical reactions.

At present, the rules by which primary structure is converted into higher-order structure are imperfectly understood. However, from knowledge of the primary amino sequence of a protein, the functions of specific regions can be predicted. For example, some characteristic protein sequences are the contact points with membrane phospholipids that position a protein in a membrane. Other characteristic sequences act to bind the protein to DNA. Amino acid sequences that are associated with particular functions are called **domains.** A protein may contain one or more separate domains.

9.2 Colinearity of gene and protein

The one-gene–one-enzyme hypothesis of Beadle and Tatum (see Chapter 6) was the source of the first exciting insight into the functions of genes: genes were somehow responsible for the function of enzymes, and each gene apparently controlled one specific enzyme. This hypothesis became one of the great unifying concepts in biology, because it provided a bridge that brought together the concepts and research techniques of genetics and biochemistry. When the structure of DNA was deduced in 1953, it seemed likely that there must be a linear correspondence between the nucleotide sequence in DNA and the amino acid sequence in protein (such as an enzyme). However, not until 1963 was an experimental demonstration of this colinearity obtained. In that year, Charles Yanofsky of Stanford University led one of two research groups that demonstrated the linear correspondence between the nucleotides in DNA and the amino acids in protein.

Yanofsky probed the relation between altered genes and altered proteins by studying the enzyme tryptophan synthetase and its gene. Tryptophan synthetase is a heterotetramer composed of two α and two β subunits. It catalyzes the conversion of indole glycerol phosphate into tryptophan.

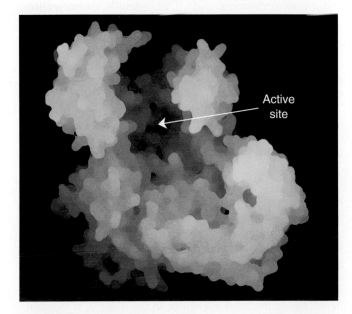

Figure 9-4 Structural model of an enzyme. Substrate binds to the enzyme in the active site. [From H. Lodish, D. Baltimore, A. Berk, S. L. Zipursky, P. Matsudaira, and J. Darnell, *Molecular Cell Biology*, 3d ed. Copyright 1995 by Scientific American Books, Inc.]

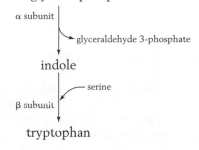

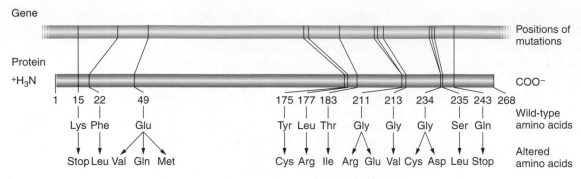

Figure 9-5 Colinearity of *trpA* mutations and amino acid changes. Although the order of mutation on the gene map and the amino acid positions are the same, the relative positions differ because the gene map is derived from recombination frequencies, which are not uniform along the length of the gene. [After C. Yanofsky, "Gene Structure and Protein Structure." Copyright 1967 by Scientific American. All rights reserved.]

Yanofsky induced 16 mutant alleles of the *trpA* gene of *E. coli*, whose wild-type allele was known to code for the α subunit of the enzyme. All 16 mutant alleles produced inactive forms of the enzyme. Yanofsky showed that all the mutant alleles were substantially similar in structure to the wild-type alleles but differed at 16 different mutant sites. He went on to map the positions of the mutant sites. In 1963, DNA sequencing had not been invented, but the mutant sites could be mapped by the method of recombination analysis, which was described in Chapter 4.

Through biochemical analysis of the TrpA protein, Yanofsky showed that each of the 16 mutations resulted in an amino acid substitution at a different position in the protein. Most exciting, he showed that the mutational sites in the gene map of the *trpA* gene appeared in the same order as the corresponding altered amino acids in the TrpA polypeptide chain (Figure 9-5). In addition, the distance between mutant sites, as measured by recombination frequency, was shown to correlate with the distance between the altered amino acids in the corresponding mutant proteins. Thus, Yanofsky demonstrated **colinearity**—the correspondence between the linear sequence of the gene and that of the polypeptide. Subsequently, these results were shown to apply to mutations in other proteins generally.

> **MESSAGE** The linear sequence of nucleotides in a gene determines the linear sequence of amino acids in a protein.

9.3 The genetic code

If genes are segments of DNA and if a strand of DNA is just a string of nucleotides, then the sequence of nucleotides must somehow dictate the sequence of amino acids in proteins. How does the DNA sequence dictate the protein sequence? The analogy to a code springs to mind at once. Simple logic tells us that, if the nucleotides are the "letters" in a code, then a combination of letters can form "words" representing different amino acids. First, we must ask how the code is read. Is it overlapping or nonoverlapping? Then, we must ask how many letters in the mRNA make up a word, or **codon,** and which codon or codons represent each amino acid. The cracking of the genetic code is the story told in this section.

Overlapping versus nonoverlapping codes

Figure 9-6 shows the difference between an overlapping and a nonoverlapping code. The example shows a three-

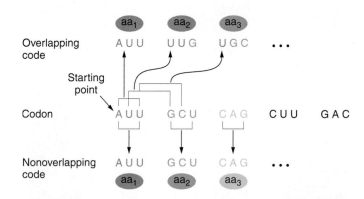

Figure 9-6 Comparison of an overlapping and a nonoverlapping genetic code. The example uses a codon with three nucleotides in the RNA (a triplet code). In an overlapping code, single nucleotides occupy positions in multiple codons. In this figure, the third nucleotide in the RNA, U, is found in three codons. In a nonoverlapping code, a protein is translated by reading nucleotides sequentially in sets of three. A nucleotide is found in only one codon. In this example, the third U in the RNA is only in the first codon.

letter, or **triplet,** code. For a nonoverlapping code, consecutive amino acids are specified by consecutive code words (codons), as shown at the bottom of Figure 9-6. For an overlapping code, consecutive amino acids are specified by codons that have some consecutive bases in common; for example, the last two bases of one codon may also be the first two bases of the next codon. Overlapping codons are shown in the upper part of Figure 9-6. Thus, for the sequence AUUGCUCAG in a nonoverlapping code, the three triplets AUU, GCU, and CAG encode the first three amino acids, respectively. However, in an overlapping code, the triplets AUU, UUG, and UGC encode the first three amino acids if the overlap is two bases, as shown in Figure 9-6.

By 1961, it was already clear that the genetic code was nonoverlapping. Analyses of mutationally altered proteins showed that only a single amino acid changes at one time in one region of the protein. This result is predicted by a nonoverlapping code. As you can see in Figure 9-6, an overlapping code predicts that a single base change will alter as many as three amino acids at adjacent positions in the protein.

Number of letters in the codon

If an mRNA molecule is read from one end to the other, only one of four different bases, A, U, G, or C, can be found at each position. Thus, if the words were one letter long, only four words would be possible. This vocabulary cannot be the genetic code, because we must have a word for each of the 20 amino acids commonly found in cellular proteins. If the words were two letters long, then $4 \times 4 = 16$ words would be possible; for example, AU, CU, or CC. This vocabulary is still not large enough.

If the words are three letters long, then $4 \times 4 \times 4 = 64$ words are possible; for example, AUU, GCG, or UGC. This vocabulary provides more than enough words to describe the amino acids. We can conclude that the code word must consist of at least three nucleotide pairs. However, if all words are "triplets," then the possible words are in considerable excess of the 20 needed to name the common amino acids. We will come back to these excess codons later in the chapter.

Use of suppressors to demonstrate a triplet code

Convincing proof that a codon is, in fact, three letters long (and no more than three) came from beautiful genetic experiments first reported in 1961 by Francis Crick, Sidney Brenner, and their co-workers. These experiments used mutants in the *rII* locus of T4 phage. The use of *rII* mutations in recombination analysis was discussed in Chapter 5. Phage T4 is usually able to grow on two different *E. coli* strains, called B and K. However, mutations in the *rII* gene change the host range of the

phage: mutant phage can still grow on an *E. coli* B host, but they cannot grow on an *E. coli* K host. Mutations causing this rII phenotype were induced by using a chemical called proflavin, which was thought to act by the addition or deletion of single nucleotide pairs in DNA. (This assumption is based on experimental evidence not presented here.) The following examples illustrate the action of proflavin on double-stranded DNA.

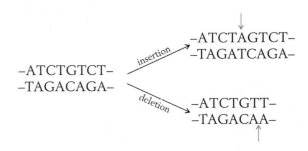

Starting with one particular proflavin-induced mutation called FCO, Crick and his colleagues found "reversions" (reversals of the mutation) that were able to grow on *E. coli* strain K. Genetic analysis of these plaques revealed that the "revertants" were not identical with true wild types. Thus the back mutation was not an exact reversal of the original forward mutation. In fact, the reversion was found to be due to the presence of a *second mutation* at a different site from that of FCO, although in the same gene. This second mutation "suppressed" mutant expression of the original FCO. Recall from Chapter 6 that a suppressor mutation counteracts or suppresses the effects of another mutation.

Researchers were able to tell that there were two mutations in one gene by using the mapping approach that Benzer had pioneered with the rII system (see Chapter 5). Crosses showed that the suppressor mutation could be separated from the original forward mutation by recombination, thus demonstrating that the two mutations were in separate locations. But the recombination frequency was very low, which proved that the suppressor itself also was a mutation of the *rII* gene (Figure 9-7).

How can we explain these results? If we assume that the gene is read from one end only, then the original addition or deletion induced by proflavin could be mutant because it interrupts a normal reading mechanism that establishes the group of bases to be read as words. For example, if each three bases on the resulting mRNA make a word, then the "reading frame" might be established by taking the first three bases from the end as the first word, the next three as the second word, and so forth. In that case, a proflavin-induced addition or deletion of a single pair on the DNA would shift the reading frame on the mRNA from that corresponding point on, causing all following words to be misread. Such a frameshift mutation could reduce most of the

Figure 9-7 Demonstration of a suppressor mutation in the same gene as the original mutation. The suppressor of an initial *rII* mutation (FCO) is shown to be a second *rII* mutation after separation by crossing-over. The original mutant, FCO, was induced by proflavin. Later, when the FCO strain was treated with proflavin again, a revertant was found, which on first appearance seemed to be wild type. However, a second mutation within the *rII* region was found to have been induced, and the double mutant *rII*$_x$*rII*$_y$ was shown not to be quite identical with the original wild type.

genetic message to gibberish. However, the proper reading frame could be restored by a compensatory insertion or deletion somewhere else, leaving only a short stretch of gibberish between the two. Consider the following example in which three-letter English words are used to represent the codons:

	THE	FAT	CAT	ATE	THE	BIG	RAT
Delete C:	THE	FAT	ATA	TET	HEB	IGR	AT
Insert A:	THE	FAT	ATA	ATE	THE	BIG	RAT

The insertion suppresses the effect of the deletion by restoring most of the sense of the sentence. By itself, however, the insertion also disrupts the sentence:

THE FAT CAT AAT ETH EBI GRA T

If we assume that the FCO mutant is caused by an addition, then the second (suppressor) mutation would have to be a deletion because, as we have seen, only a deletion would restore the reading frame of the resulting message (a second insertion would not correct the frame). In the following diagrams, we use a hypothetical nucleotide chain to represent RNA for simplicity. We also assume that the code words are three letters long and are read in one direction (left to right in our diagrams).

1. Wild-type message

CAU   CAU   CAU   CAU   CAU

2. *rII*$_a$ message: words after the addition are changed (x) by frameshift mutation (words marked ✓ are unaffected)

Addition—

CAU   ACA   UCA   UCA   UCA   U
 ✓   x   x   x   x

3. *rII*$_a$*rII*$_b$ message: few words wrong, but reading frame restored for later words

Deletion—

CAU   ACA   UCU   CAU   CAU
 ✓   x   x   ✓   ✓

The few wrong words in the suppressed genotype could account for the fact that the "revertants" (suppressed phenotypes) that Crick and his associates recovered did not look exactly like the true wild types in phenotype.

We have assumed here that the original frameshift mutation was an addition, but the explanation works just as well if we assume that the original FCO mutation is a deletion and the suppressor is an addition. If the FCO mutation is defined as plus, then suppressor mutations are automatically minus. The results of experiments have confirmed that a plus cannot suppress a plus and a minus cannot suppress a minus. In other words, two mutations of the same sign never act as suppressors of each other.

Very interestingly, combinations of *three* pluses or *three* minuses have been shown to act together to restore a wild-type phenotype. This observation provided the first experimental confirmation that a word in the genetic code consists of three successive nucleotides, or a triplet. The reason is that three additions or three deletions within a gene automatically restore the reading frame in the mRNA if the words are triplets. For example,

Deletions

CAU   CAU   CAU   CAU   CAU   CAU   CAU
CAU   ACA   UAU   CAU   CAU   CAU
 ✓   x   x   ✓   ✓   ✓

Proof that the genetic deductions about proflavin were correct also came from an analysis of proflavin-induced mutations. But in this case the mutations were in a gene with a protein product that could be analyzed. George Streisinger worked with the gene that controls the enzyme lysozyme, which has a known amino acid sequence. He induced a mutation in the gene with proflavin and selected for revertants, which were shown genetically to be double mutants (with mutations of opposite sign). When the protein of the double mutant was analyzed, a stretch of different amino acids lay between two wild-type ends, just as predicted:

Wild type:
        –Thr–Lys–Ser–Pro–Ser–Leu–Asn–Ala–

Suppressed mutant:
        –Thr–Lys–Val–His–His–Leu–Met–Ala–

Degeneracy of the genetic code

As already discussed, with four letters to choose from at each position, a three-letter codon could make $4 \times 4 \times 4 = 64$ words. With only 20 words needed for the 20 common amino acids, what are the other words used for, if anything? Crick's work suggested that the genetic code is **degenerate,** meaning that each of the 64 triplets must have some meaning within the code. For this to be true, some of the amino acids must be specified by at least two or more different triplets.

The reasoning goes like this. If only 20 triplets are used, then the other 44 are nonsense in that they do not code for any amino acid. In that case, most frameshift mutations can be expected to produce nonsense words, which presumably stop the protein-building process. If this were the case, then the suppression of frameshift mutations would rarely, if ever, work. However, if all triplets specified some amino acid, then the changed words would simply result in the insertion of incorrect amino acids into the protein. Thus, Crick reasoned that many or all amino acids must have several different names in the base-pair code; this hypothesis was later confirmed biochemically.

MESSAGE The discussion up to this point demonstrates that

1. The genetic code is nonoverlapping.
2. Three bases encode an amino acid. These triplets are termed codons.
3. The code is read from a fixed starting point and continues to the end of the coding sequence. We know this because a single frameshift mutation anywhere in the coding sequence alters the codon alignment for the rest of the sequence.
4. The code is degenerate in that some amino acids are specified by more than one codon.

Cracking the code

The deciphering of the genetic code—determining the amino acid specified by each triplet—was one of the most exciting genetic breakthroughs of the past 50 years. Once the necessary experimental techniques became available, the genetic code was broken in a rush.

One breakthrough was the discovery of how to make synthetic mRNA. If the nucleotides of RNA are mixed with a special enzyme (polynucleotide phosphorylase), a single-stranded RNA is formed in the reaction. Unlike transcription, no DNA template is needed for this synthesis, and so the nucleotides are incorporated at random. The ability to synthesize mRNA offered the exciting prospect of creating specific mRNA sequences and then seeing which amino acids they would specify. The first synthetic messenger obtained was made by reacting only uracil nucleotides with the RNA-synthesizing enzyme, producing–UUUU–[poly(U)]. In 1961, Marshall Nirenberg and Heinrich Matthaei mixed poly(U) with the protein-synthesizing machinery of *E. coli* in vitro and *observed the formation of a protein.* The main excitement centered on the question of the amino acid sequence of this protein. It proved to be polyphenylalanine—a string of phenylalanine molecules attached to form a polypeptide. Thus, the triplet UUU must code for phenylalanine:

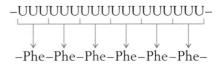

Next, mRNAs containing two types of nucleotides in repeating groups were synthesized. For instance, synthetic mRNA having the sequence $(AGA)_n$, which is a long sequence of AGAAGAAGAAGAAGA, was used to stimulate polypeptide synthesis in vitro (in a test tube that also contained a cell extract with all the components necessary for translation). From the sequence of the resulting polypeptides and the possible triplets that could reside in other synthetic RNAs, many code words could be verified. (This kind of experiment is detailed in Problem 30 at the end of this chapter. In solving it, you can put yourself in the place of H. Gobind Khorana, who received a Nobel Prize for directing the experiments.)

Additional experimental approaches led to the assignment of each amino acid to one or more codons. Recall that the code was proposed to be degenerate, meaning that some amino acids had more than one codon assignment. This degeneracy can be seen clearly in Figure 9-8, which shows the codons and the amino acids that they specify. Virtually all organisms on the planet use this same genetic code. (There are just a few exceptions in which a small number of the codons have different meanings—for example, in mitochondrial genomes.)

Figure 9-8 The genetic code.

	Second letter				
First letter	U	C	A	G	Third letter
U	UUU Phe, UUC Phe, UUA Leu, UUG Leu	UCU, UCC, UCA, UCG Ser	UAU Tyr, UAC Tyr, UAA Stop, UAG Stop	UGU Cys, UGC Cys, UGA Stop, UGG Trp	U C A G
C	CUU, CUC, CUA, CUG Leu	CCU, CCC, CCA, CCG Pro	CAU His, CAC His, CAA Gln, CAG Gln	CGU, CGC, CGA, CGG Arg	U C A G
A	AUU, AUC, AUA Ile, AUG Met	ACU, ACC, ACA, ACG Thr	AAU Asn, AAC Asn, AAA Lys, AAG Lys	AGU Ser, AGC Ser, AGA Arg, AGG Arg	U C A G
G	GUU, GUC, GUA, GUG Val	GCU, GCC, GCA, GCG Ala	GAU Asp, GAC Asp, GAA Glu, GAG Glu	GGU, GGC, GGA, GGG Gly	U C A G

Stop codons

You may have noticed in Figure 9-8 that some codons do not specify an amino acid at all. These codons are stop, or termination, codons. They can be regarded as being similar to periods or commas punctuating the message encoded in the DNA.

One of the first indications of the existence of stop codons came in 1965 from Brenner's work with the T4

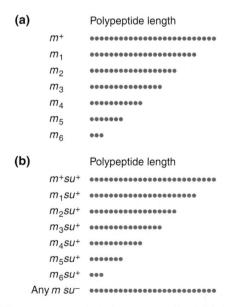

Figure 9-9 Introduction and suppression of amber mutants. (a) Polypeptide chain lengths of phage T4 head protein in wild type (m^+) and various amber mutants (m_1–m_6). The amber mutants have a single nucleotide change that introduces a stop codon. (b) An amber suppressor (su) leads to the phenotypic development of the wild-type chain.

phage. Brenner analyzed certain mutations (m_1–m_6) in a single gene that controls the head protein of the phage. He found that the head protein of each mutant was a shorter polypeptide chain than that of the wild type (Figure 9-9a).

Brenner examined the ends of the shortened proteins and compared them with wild-type protein. For each mutant, he recorded the next amino acid that *would* have been inserted to continue the wild-type chain. The amino acids for the six mutations were glutamine, lysine, glutamic acid, tyrosine, tryptophan, and serine. There is no immediately obvious pattern to these results, but Brenner brilliantly deduced that certain codons for each of these amino acids are similar. Specifically, each of these codons can mutate to the codon UAG by a single change in a DNA nucleotide pair. He therefore postulated that UAG is a stop (termination) codon—a signal to the translation mechanism that the protein is now complete.

UAG was the first stop codon deciphered; it is called the amber codon. Mutants that are defective owing to the presence of an abnormal amber codon are called *amber mutants*. Similarly, UGA is called the opal codon and UAA, the ochre codon. Mutants that are defective because they contain abnormal opal or ochre codons are called opal and ochre mutants, respectively. Stop codons are often called nonsense codons because they designate no amino acid.

Brenner's phage mutants had a second interesting feature in common in addition to a shorter head protein: the presence of a suppressor mutation (su^-) in the host chromosome would cause the phage to develop a head protein of normal (wild-type) chain length despite the presence of the *m* mutation (Figure 9-9b). We shall consider stop codons and their suppressors further after we have dealt with the process of protein synthesis.

9.4 tRNA: the adapter

After it became known that the sequence of amino acids of a protein was determined by the triplet codons of the mRNA, scientists began to wonder how this determination was accomplished. An early model, quickly dismissed as naive and unlikely, proposed that the mRNA codons could fold up and form 20 distinct cavities that directly bind specific amino acids in the correct order. Instead, in 1958 Crick recognized that

> It is therefore a natural hypothesis that the amino acid is carried to the template by an adapter molecule, and that the adapter is the part which actually fits on to the RNA. In its simplest form [this hypothesis] would require twenty adapters, one for each amino acid.

He speculated that the adapter "might contain nucleotides. This would enable them to join on the RNA template by the same 'pairing' of bases as is found in DNA." Furthermore, "a separate enzyme would be required to join each adapter to its own amino acid."

We now know that Crick's "adapter hypothesis" is largely correct. Amino acids are in fact attached to an adapter (recall that adapters constitute a special class of stable RNAs called *transfer RNAs*). Each amino acid becomes attached to a specific tRNA, which then brings that amino acid to the ribosome, the molecular complex that will attach the amino acid to a growing polypeptide.

Codon translation by tRNA

The structure of tRNA holds the secret of the specificity between an mRNA codon and the amino acid that it designates. The single-stranded tRNA molecule has a cloverleaf shape consisting of four double-helical stems and three single-stranded loops (Figure 9-10a). The middle loop of each tRNA is called the anticodon loop because it carries a nucleotide triplet called an **anticodon.**

This sequence is complementary to the codon for the amino acid carried by the tRNA. The anticodon in tRNA and the codon in the mRNA bind by specific RNA-to-RNA base pairing. (Again, we see the principle of nucleic acid complementarity at work, this time in the binding of two different RNAs.) Because codons in mRNA are read in the $5' \rightarrow 3'$ direction, anticodons are oriented and written in the $3' \rightarrow 5'$ direction, as Figure 9-10a shows.

Amino acids are attached to tRNAs by enzymes called **aminoacyl-tRNA synthetases.** The tRNA with an attached amino acid is said to be **charged.** Each amino acid has a specific synthetase that links it only to those tRNAs that recognize the codons for that particular amino acid. An amino acid is attached at the free 3' end of its tRNA, the amino acid alanine in the case shown in Figure 9-10a.

The "flattened" cloverleaf shown in Figure 9-10a is not the normal conformation of tRNA molecules; a tRNA normally exists as an L-shaped folded cloverleaf, as shown in Figure 9-10b. The three-dimensional structure

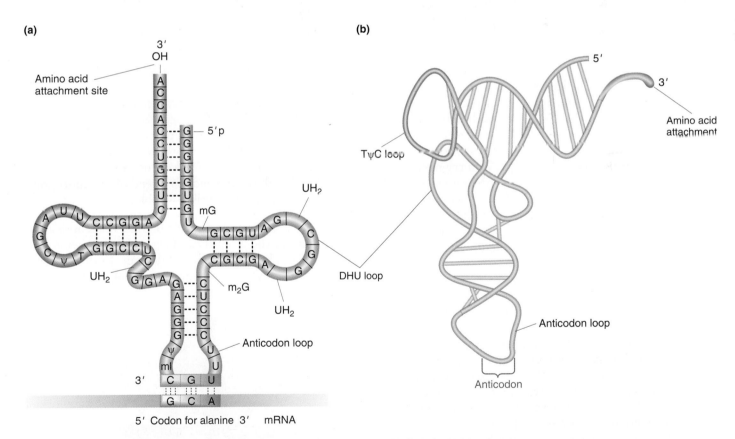

Figure 9-10 The structure of transfer RNA. (a) The structure of yeast alanine tRNA, showing the anticodon of the tRNA binding to its complementary codon in mRNA. (b) Diagram of the actual three-dimensional structure of yeast phenylalanine tRNA. The abbreviations ψ, mG, m$_2$G, mI, and UH$_2$ refer to the modified bases pseudouridine, methylguanosine, dimethylguanosine, methylinosine, and dihydrouridine, respectively. [Part a after S. Arnott, "The Structure of Transfer RNA," *Progress in Biophysics and Molecular Biology* 22, 1971, 186; part b after L. Stryer, *Biochemistry*, 4th ed. Copyright 1995 by Lubert Stryer. Part b is based on a drawing by Sung-Hou Kim.]

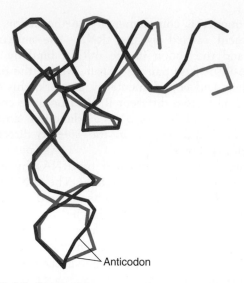

Anticodon

Figure 9-11 Two superimposed tRNAs. When folded into their correct three-dimensional structures, the yeast tRNA for glutamine (blue) almost completely overlaps the yeast tRNA for phenylalanine (red) except for the anticodon loop and aminoacyl end. [After M. A. Rould, J. J. Perona, D. Soll, and T. A. Steitz, "Structure of *E. coli* Glutaminyl–tRNA Synthetase Complexed with tRNA(Gln) and ATP at 2.8 A Resolution," *Science* 246, 1989, 1135–1142.]

of tRNA was determined with the use of X-ray crystallography. In the years since it was used to deduce the double-helical structure of DNA, this technique has been refined so that it can now be used to determine the structure of very complex macromolecules such as the ribosome. Although tRNAs differ in their primary nucleotide sequence, all tRNAs fold into virtually the same L-shaped conformation except for differences in their anticodon and acceptor loops. This similarity of structure can be easily seen in Figure 9-11, which shows two different tRNAs superimposed. Conservation of structure tells us that shape is important for tRNA function. As we will see later in the chapter, this shape is critical to the interaction of tRNA with the ribosome during protein synthesis.

What would happen if the wrong amino acid were covalently attached to a tRNA? The results of a very convincing experiment answered this question. The experiment used cysteinyl-tRNA (tRNA^{Cys}), the tRNA specific for cysteine. This tRNA was "charged" with cysteine, meaning that cysteine was attached to the tRNA. The charged tRNA was treated with nickel hydride, which converted the cysteine (while still bound to tRNA^{Cys}) into another amino acid, alanine, without affecting the tRNA:

$$\text{cysteine}-\text{tRNA}^{Cys} \xrightarrow{\text{nickel hydride}} \text{ananine}-\text{tRNA}^{Cys}$$

Protein synthesized with this hybrid species had alanine wherever we would expect cysteine. Thus, the experi-

ment demonstrated that the amino acids are "illiterate"; they are inserted at the proper position because the tRNA "adapters" recognize the mRNA codons and insert their attached amino acids appropriately. Thus the attachment of the correct amino acid to its cognate tRNA is a critical step in ensuring that a protein is synthesized correctly. If the wrong amino acid is attached, there is no way to prevent it from being incorporated into a growing protein chain. As already mentioned, the attachment of an amino acid to tRNA is carried out by an aminoacyl-tRNA synthetase, which has two binding sites: one for the amino acid and another for the tRNA. There are 20 of these remarkable enzymes in the cell, one for each of the 20 amino acids.

Degeneracy revisited

We saw in Figure 9-8 that the number of codons for a single amino acid varies, ranging from one codon (UGG for tryptophan) to as many as six (UCC, UCU, UCA, UCG, AGC, or AGU for serine). Why the genetic code shows this variation is not exactly clear, but two facts account for it:

1. Most amino acids can be brought to the ribosome by several alternative tRNA types. Each type has a different anticodon that base-pairs with a different codon in the mRNA.

2. Certain charged tRNA species can bring their specific amino acids to any one of several codons. These tRNAs recognize and bind to several alternative codons, not just the one with a complementary sequence, through a loose kind of base pairing at the 3′ end of the codon and the 5′ end of the anticodon. This loose pairing is called **wobble.**

Wobble is a situation in which the third nucleotide of an anticodon (at the 5′ end) can form two alignments (Figure 9-12). This third nucleotide can form hydrogen bonds not only with its normal complementary nu-

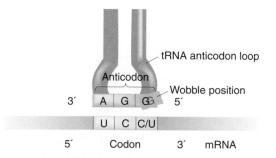

Figure 9-12 Wobble. In the third site (5′ end) of the anticodon, G can take either of two wobble positions, thus being able to pair with either U or C. This ability means that a single tRNA species carrying an amino acid (in this case, serine) can recognize two codons—UCU and UCC—in the mRNA.

Table 9-1	Codon–Anticodon Pairings Allowed by the Wobble Rules	
5′ end of anticodon		3′ end of codon
G		C or U
C		G only
A		U only
U		A or G
I		U, C, or A

cleotide in the third position of the codon but also with a different nucleotide in that position. There are "wobble rules" that dictate which nucleotides can and cannot form hydrogen bonds with alternative nucleotides through wobble (Table 9-1). In the table, the letter *I* stands for inosine, one of the rare bases found in tRNA, often in the anticodon.

Table 9-2	Different tRNAs That Can Service Codons for Serine	
tRNA	Anticodon	Codon
tRNA$^{Ser}_1$	ACG + wobble	UCC UCU
tRNA$^{Ser}_2$	AGU + wobble	UCA UCG
tRNA$^{Ser}_3$	UCG + wobble	AGC AGU

Table 9-2 lists all the codons for serine and shows how three different tRNAs (tRNA$^{Ser}_1$, tRNA$^{Ser}_2$, and tRNA$^{Ser}_3$) can pair with these codons. Some organisms possess an additional tRNA species (which we could represent as tRNA$^{Ser}_4$) that has an anticodon identical with one of the three anticodons shown in Table 9-2 but differs in its nucleotide sequence elsewhere in the molecule. These four tRNAs are called **isoaccepting tRNAs** because they accept the same amino acid, but are transcribed from different tRNA genes.

> **MESSAGE** The degree of degeneracy for a given amino acid is determined by the number of codons for that amino acid that have only one tRNA each plus the number of codons for amino acids that share a tRNA through wobble.

9.5 Ribosomes

Protein synthesis takes place when tRNA and mRNA molecules associate with *ribosomes*. The task of the tRNAs and the ribosome is to translate the sequence of nu-cleotide codons in mRNA into the sequence of amino acids in protein. The term *biological machine* was used in preceding chapters to characterize multisubunit complexes that perform cellular functions. The replisome, for example, is a biological machine that can replicate DNA with precision and speed. The site of protein synthesis, the ribosome, is much larger and more complex than the machines described thus far. Its complexity is due to the fact that it has to perform several jobs with precision and speed. For this reason, it is better to think of the ribosome as a factory containing many machines that act in concert. Let's see how this factory is organized to perform its numerous functions.

In all organisms, the ribosome consists of one small and one large subunit; each made up of RNA (called ribosomal RNA or rRNA) and protein. Each subunit is composed of several rRNA types and as many as 50 proteins. Ribosome subunits were originally characterized by their rate of sedimentation when spun in an ultracentrifuge, and so their names are derived from their sedimentation coefficients in Svedberg (S) units, which is an indication of molecular size. In prokaryotes, the small and large subunits are called 30S and 50S, respectively, and they associate to form a 70S particle (Figure 9-13 top). The eukaryotic counterparts are called 40S and 60S, with 80S for the complete ribosome (Figure 9-13 bottom). Although eukaryotic ribosomes are bigger due to their larger and more numerous components, the components and the steps in protein synthesis are similar overall. The similarities clearly indicate that translation is an ancient process that originated in the common ancestor of eukaryotes and prokaryotes.

When ribosomes were first studied, it came as a surprise that almost two-thirds of their mass is RNA and only one-third is protein. For decades, rRNAs, which fold up by intramolecular base pairing into stable secondary structures (Figure 9-14), were assumed to function as the scaffold or framework necessary for the correct assembly of the ribosomal proteins. According to this model, the ribosomal proteins were solely responsible for carrying out the important steps in protein synthesis. This view changed with the discovery in the 1980s of catalytic RNAs (see Chapter 8). As you will see, scientists now believe that the rRNAs, assisted by the ribosomal proteins, carry out most of the important steps in protein synthesis.

Ribosome features

The ribosome brings together the other important players in protein synthesis—tRNA and mRNA molecules—to translate the nucleotide sequence of an mRNA into the amino acid of a protein. The tRNA and mRNA molecules are positioned in the ribosome so that the codon of the mRNA can interact with the anticodon of the tRNA. The key sites of interaction are

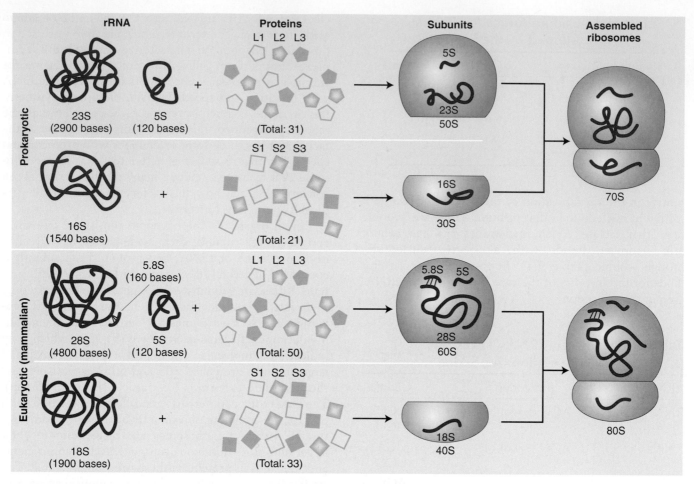

Figure 9-13 A ribosome contains a large and a small subunit. Each subunit contains both rRNA of varying lengths and a set of proteins. There are two principal rRNA molecules in all ribosomes (shown in the column on the left). Prokaryotic ribosomes also contain one 120-base-long rRNA that sediments at 5S, whereas eukaryotic ribosomes have two small rRNAs: a 5S RNA molecule similar to the prokaryotic 5S, and a 5.8S molecule 160 bases long. The proteins of the large subunit are named L1, L2, and so forth, and those of the small subunit S1, S2, and so forth. [After H. Lodish, D. Baltimore, A. Berk, S. L. Zipursky, P. Matsudaira, and J. Darnell, *Molecular Cell Biology*, 3d ed. Copyright 1995 by Scientific American Books, Inc.]

Figure 9-14 The folded structure of the prokaryotic 16S ribosomal RNA of the small ribosomal subunit.

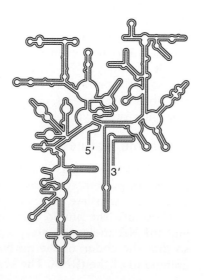

illustrated in Figure 9-15. The binding site for mRNA is completely within the small subunit. There are three binding sites for tRNA molecules. Each bound tRNA bridges the 30S and 50S subunits, with its anticodon end in the former and its aminoacyl end (carrying the amino acid) in the latter. The **A site** (for aminoacyl) binds an incoming aminoacyl tRNA whose anticodon matches the codon in the A site of the 30S subunit. As we move in the 5′ direction on the mRNA, the next codon interacts with the anticodon of the tRNA in the **P site** (for peptidyl) of the 30S subunit. The tRNA in the P site contains the growing peptide chain, part of which fits into a tunnel-like structure in the 50S subunit. The **E site** (for exit) contains a deacylated tRNA (it no longer carries an amino acid) that is ready to be released from the ribosome. Whether codon–anticodon interactions also take place between the mRNA and the tRNA in the E site is not clear.

(a) Computer model

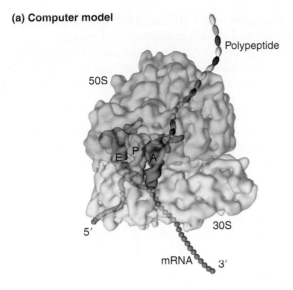

(b) Schematic model

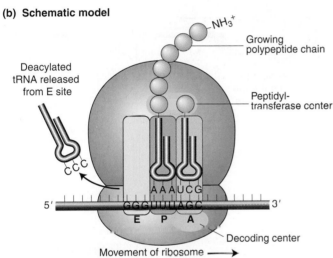

Figure 9-15 Key sites of interaction in a ribosome during the elongation phase of translation. (a) A computer model of the three-dimensional structure of the ribosome including mRNA, tRNAs, and the nascent polypeptide chain as it emerges from the large ribosomal subunit. (b) A schematic model of the ribosome during translation elongation. See text for details.
[Part a from J. Frank, *Bioessays* 23, 2001, 725–732, Figure 2.]

Two additional regions in the ribosome are critical for protein synthesis. The **decoding center** in the 30S subunit ensures that only tRNAs carrying anticodons that match the codon (called *cognate* tRNAs) will be accepted into the A site. Cognate tRNAs associate with the **peptidyl transferase center** in the 50S subunit where peptide-bond formation is catalyzed. Recently, the structure of ribosomes in association with tRNAs was determined at atomic level with the use of several techniques, including X-ray crystallography. The results of these elegant studies clearly show that both centers are composed entirely of regions of rRNA; that is, the important contacts in these centers are tRNA–rRNA contacts. Peptide-bond formation is even thought to be catalyzed

by an active site in the ribosomal RNA and only assisted by ribosomal proteins.

The process of translation can be divided into three phases: initiation, elongation, and termination. Aside from the ribosome, mRNA, and tRNAs, additional proteins are required for the successful completion of each phase. Because certain steps in initiation differ significantly in prokaryotes and eukaryotes, initiation is described separately for the two groups. The elongation and termination phases are described largely as they take place in bacteria, which have been the focus of many recent studies of translation.

Translation initiation

The main task of initiation is to place the first aminoacyl-tRNA in the P site of the ribosome and, in this way, establish the correct reading frame of the mRNA. In most prokaryotes and all eukaryotes, the first amino acid in any newly synthesized polypeptide is methionine, specified by the codon AUG. It is inserted not by tRNA^Met but by a special tRNA called an **initiator,** symbolized tRNA^Met_i. In bacteria, a formyl group is added to the methionine while the amino acid is attached to the initiator, forming *N*-formylmethionine. (The formyl group on *N*-formylmethionine is removed later.)

How does the translation machinery know where to begin? In other words, how is the initiation AUG codon selected from among the many AUG codons in an mRNA molecule? Recall that, in both prokaryotes and eukaryotes, mRNA has a 5′ untranslated region consisting of the sequence between the transcriptional start site and the translational start site. In prokaryotes, initiation codons are preceded by special sequences called **Shine–Dalgarno sequences** that pair with the 3′ end of an rRNA, called the 16S rRNA, in the 30S ribosomal subunit. This pairing correctly positions the initiator codon in the P site where the initiator tRNA will bind (Figure 9-16). The mRNA can pair only with a 30S

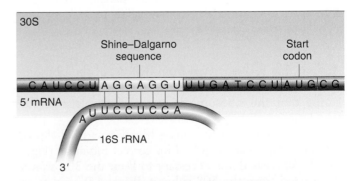

Figure 9-16 Shine–Dalgarno sequence. In bacteria, base complementarity between the 3′ end of the 16S rRNA of the small ribosomal subunit and the Shine–Dalgarno sequence of the mRNA positions the ribosome to correctly initiate translation at the downstream AUG codon.

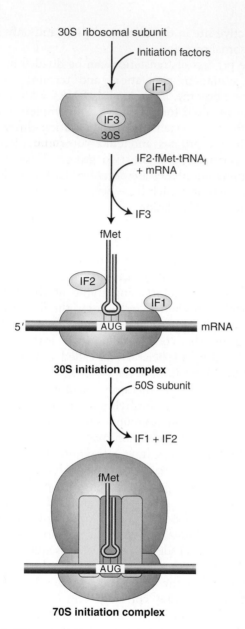

Figure 9-17 Translation initiation in prokaryotes. Initiation factors assist the assembly of the ribosome at the translation start site and then dissociate before translation. [After J. Berg, J. Tymoczko, and L. Stryer, *Biochemistry*, 5th ed. Copyright 2002 by W. H. Freeman and Company.]

subunit that is dissociated from the rest of the ribosome. Note again that rRNA performs the key function in ensuring that the ribosome is at the right place to start translation.

In bacteria three proteins—IF1, IF2, and IF3 (for **initiation factor**)—are required for correct initiation (Figure 9-17). Whereas IF3 is necessary to keep the 30S subunit dissociated from the 50S subunit, IF1 and IF2 act to ensure that only the initiator tRNA enters the P site. The 30S subunit, mRNA, and initiator tRNA constitute the initiation complex. The complete 70S ribosome is formed

by the association of the 50S large subunit with the initiation complex and release of the initiation factors.

Because there is no nuclear compartment in a prokaryote to separate transcription and translation, the prokaryotic initiation complex is able to form at a Shine–Dalgarno sequence near the 5′ end of an RNA that is still being transcribed. In contrast, transcription and translation take place in separate compartments of the eukaryotic cell. As discussed in Chapter 8, eukaryotic mRNAs are transcribed and processed in the nucleus

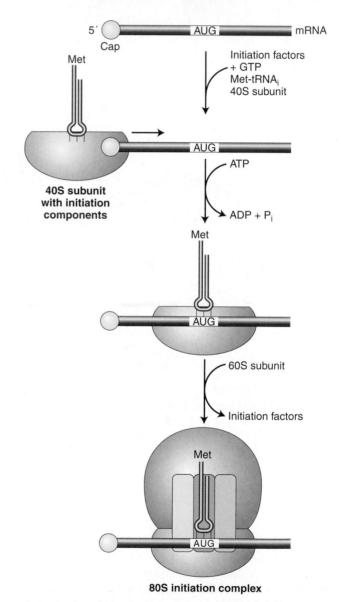

Figure 9-18 Translation initiation in eukaryotes. The initiation complex forms at the 5′ end of the mRNA and then scans in the 3′ direction in search of a start codon. Recognition of the start codon triggers assembly of the complete ribosome and the dissociation of initiation factors (not shown). The hydrolysis of ATP provides an input of energy to drive the scanning process. [After J. Berg, J. Tymoczko, and L. Stryer, *Biochemistry*, 5th ed. Copyright 2002 by W. H. Freeman and Company.]

before export to the cytoplasm for translation. On arrival in the cytoplasm, the mRNA is usually covered with proteins, and regions may be double helical due to intramolecular base pairing. These regions of secondary structure need to be removed to expose the AUG initiator codon. This removal is accomplished by initiation factors (called eIF4A, B, and G) that associate with the cap structure (found at the 5′ end of virtually all eukaryotic mRNAs) and with the 40S subunit and initiator tRNA and then move in the 5′-to-3′ direction to unwind the base-paired regions (Figure 9-18). At the same time, the exposed sequence is "scanned" for an AUG codon where translation can begin when the initiation complex is joined by the 60S subunit to form the 80S ribosome. As in prokaryotes, the eukaryotic initiation factors dissociate from the ribosome before the translation elongation phase.

Elongation

It is during the process of elongation that the ribosome most resembles a factory. The mRNA acts as a blueprint specifying the delivery of cognate tRNAs, each carrying as cargo an amino acid. Each amino acid is added to the growing polypeptide chain while the deacylated tRNA is recycled by the addition of another amino acid. Figure 9-19 details the steps in elongation and the involvement of two protein factors called **elongation factor Tu (EF-Tu)** and **elongation factor G (EF-G)**.

Earlier in this chapter we saw that an aminoacyl-tRNA is formed by the covalent attachment of an amino acid to the 3′ end of a tRNA that contains the correct anticodon. Before aminoacyl-tRNAs can be used in protein synthesis, they associate with the protein factor EF-Tu to form a **ternary complex** (composed of tRNA, amino acid, and EF-Tu). The elongation cycle commences with an initiator tRNA (and its attached methionine) in the P site and with the A site ready to accept a ternary complex (see Figure 9-19). Which of the 20 different ternary complexes to accept is determined by codon–anticodon recognition in the decoding center of the small subunit (see Figure 9-15b). When the correct match has been made, the ribosome changes shape, the EF-Tu leaves the ternary complex, and the two aminoacyl ends are juxtaposed in the peptidyltransferase center of the large subunit (see Figure 9-15b). There, a peptide bond is formed with the transfer of the methionine in the P site to the amino acid in the A site. Through the action of the second protein factor, EF-G,

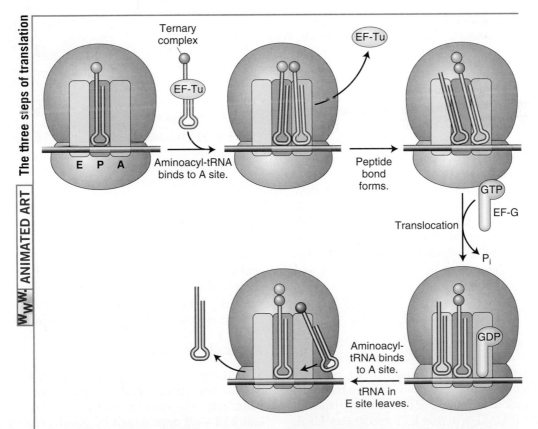

The three steps of translation

WWW. ANIMATED ART

E P A

Ternary complex

EF-Tu

Aminoacyl-tRNA binds to A site.

EF-Tu

Peptide bond forms.

GTP

EF-G

Translocation

P_i

GDP

Aminoacyl-tRNA binds to A site.

tRNA in E site leaves.

Figure 9-19 Steps in elongation. A ternary complex consisting of an aminoacyl-tRNA attached to an EF-Tu factor binds to the A site. When its amino acid has joined the growing polypeptide chain, an EF-G factor binds to the A site while nudging the tRNAs and their mRNA codons into the E and P sites. See text for details.

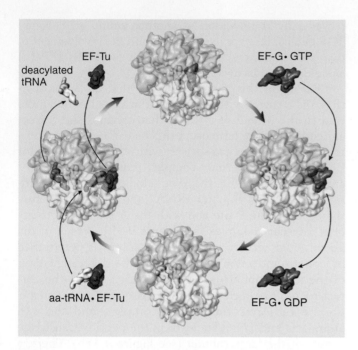

Figure 9-20 Steps in the elongation cycle, showing the three-dimensional structures of the interacting components. Structures were deduced with the use of high-resolution microscopy and X-ray crystallography. Note the similarity in shape and orientation between the aminoacyl-tRNA with its attached EF-Tu binding the A site and the EF-G factor binding the same site. [From J. Frank, "Conformational Proteomics of Macromolecular Architecture." World Scientific Publishing Co., 2003.]

which appears to fit into the A site, the tRNAs in the A and P sites are shifted to the P and E sites, respectively, and the mRNA moves through the ribosome so that the next codon is positioned in the A site (Figure 9-20). When EF-G leaves the ribosome, the A site is open to accept the next ternary complex.

In subsequent cycles, the A site is filled with a new ternary complex as the deacylated tRNA leaves the E site. As elongation progresses, the number of amino acids on the peptidyl-tRNA (at the P site) increases and the amino-terminal end of the growing polypeptide eventually emerges from the tunnel in the 50S subunit and protrudes from the ribosome.

Termination

The cycle continues until the codon in the A site is one of the three stop codons: UGA, UAA, or UAG. Recall that no tRNAs recognize these codons. Instead, proteins called **release factors** (RF1, RF2, and RF3 in bacteria) recognize stop codons (Figure 9-21). In bacteria, RF1 recognizes UAA or UAG, whereas RF2 recognizes UAA or UGA; both are assisted by RF3. The interaction between release factors 1 and 2 and the A site differs from that of the ternary complex in two important ways. First, the stop codons are recognized by tripeptides in

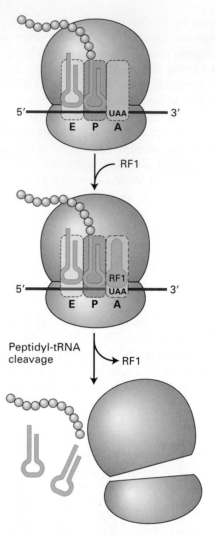

Figure 9-21 Termination of translation. Translation is terminated when release factors recognize stop codons in the A site of the ribosome. [From H. Lodish et al., *Molecular Cell Biology*, 5th ed. Copyright 2004 by W. H. Freeman and Company.]

the RF proteins, not by an anticodon. Second, release factors fit into the A site of the 30S subunit but do not participate in peptide-bond formation. Instead, a water molecule gets into the peptidyltransferase center and leads to the release of the polypeptide from the tRNA in the P site. The ribosomal subunits separate, and the 30S subunit is now ready to form a new initiation complex.

Molecular mimicry

Something remarkable was seen when the three-dimensional structure of the ternary complex was compared with the structures of protein factors EF-G, RF1, and RF2—all determined by X-ray crystallography. The shape of EF-G is amazingly similar to the ternary complex (aminoacylated tRNA with EF-Tu attached), but the shape of the release factors resembles that of deacylated tRNAs (Figure 9-22). The ability of a molecule to

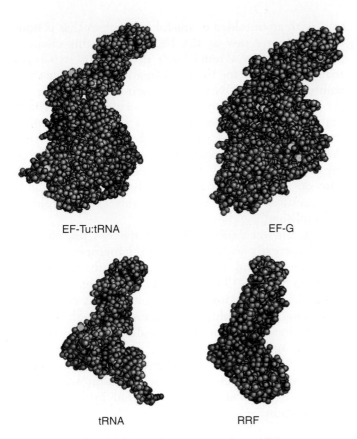

EF-Tu:tRNA

EF-G

tRNA

RRF

Figure 9-22 Molecular mimicry. These space-filling representations of three-dimensional structures illustrate that EF-G mimics a ternary complex formed of EF-Tu, tRNA, and an amino acid (top) and that the release factor RRF mimics an aminoacylated tRNA (bottom). The amino acid bound to the tRNA and ternary complex in this image is phenylalanine. [C. Al Karadaghi, U. Kristensen, and A. Liljas, "A Decade of Progress in Understanding the Structural Basis of Protein Synthesis," *Progress in Biophysics and Molecular Biology* 73, 2000, 167–193, at p. 184.]

assume the structure of another molecule is called *molecular mimicry.* The structural similarities help to explain how these proteins do their jobs during translation. That is, the fact that EF-G looks like the ternary complex means that it interacts with both the 30S and the 50S subunit by replacing the ternary complex in the A site in elongation. Similarly, like the tRNA end of the ternary complex, the release factors can fit into the decoding center but, without the end that binds to the A site in the 50S subunit, protein synthesis will terminate.

> **MESSAGE** Translation is carried out by ribosomes moving along mRNA in the 5′ → 3′ direction. A set of tRNA molecules brings amino acids to the ribosome, their anticodons binding to mRNA codons exposed on the ribosome. An incoming amino acid becomes bonded to the amino end of the growing polypeptide chain in the ribosome.

Nonsense suppressor mutations

It is interesting to consider the suppressors of the nonsense mutations defined by Brenner and co-workers. Recall that mutations in phage called amber mutants replaced wild-type codons with stop codons but that suppressor mutations in the host chromosome counteracted the effects of the amber mutations. We can now say more specifically where the suppressor mutations were located and how they worked. Many of these suppressors are mutations in genes coding for tRNAs. These mutations are known to alter the anticodon loops of specific tRNAs in such a way that a tRNA becomes able to recognize a nonsense codon in mRNA. Thus, an amino acid is inserted in response to the nonsense codon, and translation continues past that triplet. In Figure 9-23, the amber mutation replaces a wild-type codon with the chain-terminating nonsense codon UAG. By itself, the UAG would cause the protein to be prematurely cut off at the corresponding position. The suppressor mutation in this case produces a tRNA^{Tyr} with an anticodon that recognizes the mutant UAG stop codon. The suppressed mutant thus contains tyrosine at that position in the protein.

Could the tRNA produced by a suppressor mutation also bind to normal termination signals at the ends of proteins? Would the presence of a suppressor mutation thus prevent normal termination? Many of the natural termination signals consist of two chain-termination signals in a row. Because of competition with release factors, the probability of suppression at two codons in a row is small. Consequently, very few protein copies carry many extraneous amino acids resulting from translation beyond the natural stop codon.

9.6 Posttranslational events

When released from the ribosome, newly synthesized proteins are usually unable to function. This may come as a surprise to those who believe that the protein sequences encoded in DNA and transcribed into mRNAs are all that is needed to explain how organisms work. As you will see in this section and in subsequent chapters of this book, DNA sequence tells only part of the story. In this case, all newly synthesized proteins need to fold up correctly and the amino acids of some proteins need to be chemically modified. Because protein folding and modification take place after protein synthesis, they are called posttranslational events.

Protein folding inside the cell

The most important posttranslational event is the folding of the **nascent** (newly synthesized) protein into its correct three-dimensional shape. A protein that is folded correctly is said to be in its **native** conformation (in con-

(a) Amber mutation introduces UAG stop codon. Translation stops.

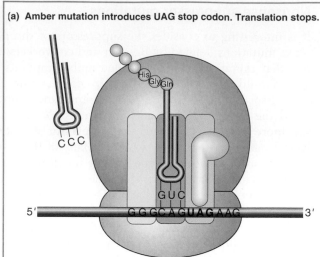

(b) Tyrosine tRNA anticodon is changed to AUC.

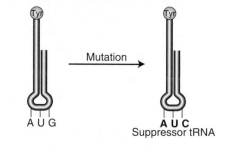

(c) Tyrosine tRNA reads UAG codon. Translation continues.

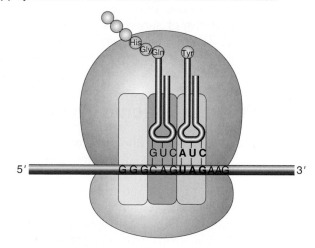

Figure 9-23 Action of a suppressor. (a) Termination of translation. Here the translation apparatus cannot go past a nonsense codon (UAG in this case), because no tRNA can recognize the UAG triplet. Therefore, protein synthesis ends, with the subsequent release of the polypeptide fragment. The release factors are not shown here. (b) The molecular consequences of a mutation that alters the anticodon of a tyrosine tRNA. This tRNA can now read the UAG codon. (c) The suppression of the UAG codon by the altered tRNA, which now permits chain elongation. [After D. Watson, J. Tooze, and D. T. Kurtz, *Recombinant DNA: A Short Course.* Copyright 1983 by W. H. Freeman and Company.]

trast with an unfolded or misfolded protein that is **non-native**). As we saw at the beginning of this chapter, proteins exist in a remarkable diversity of structures that are essential for their enzymatic activity, for their ability to bind to DNA, or for their structural roles in the cell. Although it has been known since the 1950s that the amino acid sequence of a protein determines its three-dimensional structure, it is also known that the aqueous environment inside the cell does not favor the correct folding of most proteins. Given that proteins do in fact fold correctly in the cell, a long-standing question has been, How is this correct folding accomplished?

The answer seems to be that nascent proteins are folded correctly with the help of **chaperones**—a class of proteins found in all organisms from bacteria to plants to humans. One family of chaperones, called the GroE chaperonins, form large multisubunit complexes called **chaperonin folding machines.** Although the precise mechanism is not as yet understood, newly synthesized, unfolded proteins are believed to enter a chamber in the folding machine that provides an electrically neutral microenvironment within which the nascent protein can successfully fold into its native conformation.

Posttranslational modification of amino acid side chains

As already stated, proteins are polymers of 20 amino acids. However, biochemical analysis of many proteins reveals that a variety of molecules can be covalently attached to R groups. For example, enzymes called *kinases* attach phosphate groups to the hydroxyl groups of the amino acids serine and threonine, whereas enzymes called *phosphatases* remove these phosphate groups. The addition of phosphate groups to particular amino acids and their removal are used in the cell as a reversible switch to control protein activity.

Protein targeting

In eukaryotes, all proteins are synthesized on ribosomes that are located in the cytoplasm. However, some of these proteins end up in the nucleus, in the mitochondria, anchored in the membrane, or secreted from the cell. How do these proteins "know" where they are supposed to go? The answer to this seemingly complex problem is actually quite simple: the proteins are synthesized with short sequences that act as bar codes to target the protein to the correct place or cellular compartment. For example, membrane proteins or proteins that are secreted from the cell are synthesized with a short leader peptide, called a **signal sequence,** at the amino-terminal end. This stretch of 15 to 25 amino acids is recognized by membrane proteins that transport the new protein through the cell membrane; in this

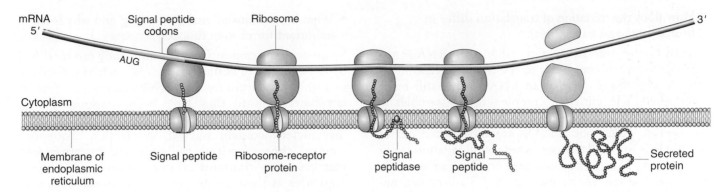

Figure 9-24 Signal sequences. Proteins destined to be secreted from the cell have an amino-terminal sequence that is rich in hydrophobic residues. This signal sequence binds to the membrane and draws the remainder of the protein through the lipid bilayer. The signal sequence is cleaved from the protein in this process by an enzyme called *signal peptidase*.

process, the signal sequence is cleaved by a peptidase (Figure 9-24). A similar phenomenon exists for certain bacterial proteins that are secreted.

In contrast, proteins (such as RNA and DNA polymerases and transcription factors discussed in Chapters 7 and 8) destined for the nucleus contain amino acid sequences embedded in the interior of the proteins that are necessary for transport from the cytoplasm into the nucleus. These **nuclear localization sequences (NLSs)** are recognized by proteins that make up the nuclear pores—sites in the membrane through which large molecules are

able to pass into and out of the nucleus. A protein not normally found in the nucleus will be directed to the nucleus if a nuclear localization sequence is attached to it.

Why are signal sequences cleaved off during targeting, whereas nuclear localization sequences, located in a protein's interior, remain after the protein moves into the nucleus? One explanation might be that, during the nuclear disintegration that accompanies mitosis, proteins localized to the nucleus may find themselves in the cytoplasm. Because such a protein contains a NLS, it can relocate to the nucleus of a daughter cell that results from mitosis.

KEY QUESTIONS REVISITED

- **How are the sequences of a gene and its protein related?**

The order of codons in RNA dictates the order of amino acids in protein. This concept was first demonstrated by showing a linear correspondence between the positions of base changes in mutant *trpA* genes and amino acid changes in the corresponding mutant TrpA proteins.

- **Why is it said that the genetic code is nonoverlapping and degenerate?**

The genetic code is nonoverlapping because consecutive amino acids are specified by consecutive codons. It is degenerate because some amino acids have more than one corresponding tRNA.

- **How is the correct amino acid paired with each codon?**

The tRNA molecule acts as an "adapter" that pairs each codon with the correct amino acid. Think of an adapter plug that makes a monitor compatible with a computer. The cord from the monitor may not fit directly into the computer but will fit if it is attached to the correct adapter, which translates the shape of the monitor cord into a shape that fits into the computer. In a similar way,

one end of the tRNA molecule binds to an amino acid; the anticodon-containing end binds (by base complementarity) to a codon in mRNA. In this way, the tRNA translates the codon of the mRNA into the amino acid of the protein.

- **Why is the attachment of an amino acid to the correct tRNA considered to be such an important step in proteins synthesis?**

The results of experiments have shown that the anticodon of a tRNA charged with an incorrect amino acid still interacts with the complementary codon in mRNA, thus causing the wrong amino acid to be incorporated into the growing polypeptide chain. No proofreading or editing mechanism exists in the ribosome to remove these incorrect amino acids.

- **What is the evidence that the ribosomal RNA, not the ribosomal proteins, carries out the key steps in translation?**

The crystal structure of ribosomes clearly shows that the decoding center in the 30S subunit and the peptidyl-transferase center in the 50S subunit are composed entirely of rRNA.

• **How does the initiation of translation differ in prokaryotes and eukaryotes?**

In prokaryotes, translation starts when 16S rRNA in the 30S ribosomal subunit binds to the Shine–Delgarno sequence near the 5′ end of an RNA that is still being transcribed. The initiation complex then assembles at that site, which is upstream of an initiator AUG codon. In contrast, eukaryotic mRNAs are transported out of the nucleus to the cytoplasm, where an elongation factor binds to the 5′ cap. The initiation complex can assemble and scan the mRNA by moving in a 3′ direction, unwinding the RNA in search of an AUG codon that is in the proper sequence context for translation initiation.

• **What is posttranslational processing, and why is it important for protein function?**

Some proteins cannot function until they are modified by the addition of certain molecules, such as phosphate or carbohydrate groups. In addition, some proteins do not function until they have been transported to a particular cellular compartment. Proteins destined for the nucleus contain nuclear localization sequences, whereas those destined for secretion, insertion into membranes, or transport into organelles contain signal sequences at their amino ends. These signal sequences are cleaved during targeting and are not part of the mature protein.

SUMMARY

This chapter has dealt with the translation of information encoded in the nucleotide sequence of mRNAs into the amino acid sequence of protein. More than any other macromolecule, proteins are who and what we are. They are the enzymes responsible for cell metabolism including DNA and RNA synthesis and are the regulatory factors required for the expression of the genetic program. The versatility of proteins as biological molecules is manifested in the diversity of shapes that they can assume. Furthermore, even after they are synthesized, they can be modified in a variety of ways by the addition of molecules that can alter their function.

Given the central role of proteins in life, it is not surprising that both the genetic code and the machinery of translating this code into protein has been highly conserved from bacteria to humans. The major components of translation are three classes of RNA: tRNA, mRNA, and rRNA. The accuracy of translation depends on the enzymatic linkage of an amino acid with its cognate tRNA, generating a charged tRNA molecule. As adapters, tRNAs are the key molecules in translation. In contrast, the huge ribosome is the factory where mRNA, charged tRNAs, and other protein factors come together for protein synthesis.

The key decision in translation is where to initiate translation. In prokaryotes, the initiation complex assembles on mRNA at the Shine–Dalgarno sequence, just upstream of the AUG start codon. The initiation complex in eukaryotes is assembled at the 5′ cap structure of the mRNA and moves in a 3′ direction until the start codon is recognized. The longest phase of translation is the elongation cycle; in this phase, the ribosome moves along the mRNA, revealing the next codon that will interact with its cognate charged tRNA so that the charged tRNA's amino acid can be added to the growing polypeptide chain. This cycle continues until a stop codon is encountered and release factors facilitate translation termination.

In the past few years, new imaging techniques have been employed to reveal ribosomal interactions at the atomic level. With these new "eyes," we can now see that the ribosome is an incredibly dynamic machine that changes shape in response to the contacts made with tRNAs and with protein factors that mimic tRNA molecules. Furthermore, atomic resolution has revealed that the ribosomal RNAs, not the ribosomal proteins, are intimately associated with the functional centers of the ribosome.

KEY TERMS

A site (p. 286)	amino end (p. 275)	chaperonin folding machine (p. 292)
active site (p. 277)	anticodon (p. 283)	charged tRNA (p. 283)
amino acid (p. 275)	carboxyl end (p. 275)	codon (p. 278)
aminoacyl-tRNA synthetase (p. 283)	chaperone (p. 292)	colinearity (p. 278)

decoding center (p. 287)

degenerate code (p. 281)

domain (p. 277)

E site (p. 286)

elongation factor G (EF-G) (p. 289)

elongation factor Tu (EF-Tu)
  (p. 289)

fibrous protein (p. 277)

globular protein (p. 277)

initiation factor (p. 288)

initiator (p. 287)

isoaccepting tRNA (p. 285)

nascent (p. 291)

native (p. 291)

nonnative (p. 292)

nuclear localization sequence (NLS)
  (p. 293)

P site (p. 286)

peptidyl transferase center (p. 287)

polypeptide (p. 275)

primary structure (p. 275)

quaternary structure (p. 277)

release factor (RF) (p. 290)

ribosomal RNA (rRNA) (p. 274)

ribosome (p. 274)

secondary structure (p. 275)

Shine–Dalgarno sequence
  (p. 287)

signal sequence (p. 292)

subunit (p. 277)

ternary complex (p. 289)

tertiary structure (p. 277)

transfer RNA (tRNA) (p. 274)

triplet (p. 279)

wobble (p. 284)

SOLVED PROBLEMS

1. In this chapter we considered different aspects of protein structure. Refer to Figure 9-8 and explain the effects individually on protein activity of nonsense, missense, and frameshift mutations in a gene encoding a protein.

Solution

Nonsense mutations result in termination of protein synthesis. Only a fragment of the protein is completed. As can be seen from Figure 9-8, fragments of proteins will not be able to adopt the correct configuration to form the active site. Therefore, the protein fragment will be inactive. Missense mutations, on the other hand, result in the exchange of one amino acid for another. If the amino acid is important for the correct folding of the protein or is part of the active site or participates in subunit interaction, then changing that amino acid will often lead to an inactive protein. On the other hand, if the amino acid is on the outside of the protein and does not take part in any of these activities or functions, then many such substitutions are acceptable and will not affect activity. Frameshift mutations change the reading frame and result in the incorporation of amino acids that are different from those encoded in the wild-type reading frame. Unless the frameshift mutation is at the end of the protein, the resulting altered protein will not be able to fold into the correct configuration and will not have activity.

2. Using Figure 9-10, show the consequences on subsequent translation of the addition of an adenine base to the beginning of the following coding sequence:

Ⓐ
↓

–CGA–UCG–GAA–CCA–CGU–GAU–AAG–CAU–
– Arg   Ser – Glu – Pro – Arg – Asp – Lys – His –

Solution

With the addition of A at the beginning of the coding sequence, the reading frame shifts, and a different set of amino acids is specified by the sequence, as shown here (note that a set of nonsense codons is encountered, which results in chain termination):

–ACG–AUC–GGA–ACC–ACG–UGA–UAA–GCA–
– Thr – Ile – Gly – Thr – Thr – stop – stop –

3. A single nucleotide addition followed by a single nucleotide deletion approximately 20 bp apart in DNA causes a change in the protein sequence from

   –His–Thr–Glu–Asp–Trp–Leu–His–Gln–Asp–

   to

   –His–Asp–Arg–Gly–Leu–Ala–Thr–Ser–Asp–

   Which nucleotide has been added and which nucleotide has been deleted? What are the original and the new mRNA sequences? (**Hint:** Consult Figure 9-8.)

Solution

We can draw the mRNA sequence for the original protein sequence (with the inherent ambiguities at this stage):

$$-His-Thr-Glu-Asp-Trp-Leu-His-Gln-Asp$$

$$-CA\genfrac{}{}{0pt}{}{U}{C}-AC\genfrac{}{}{0pt}{}{U}{C}-GA\genfrac{}{}{0pt}{}{A}{G}-GA\genfrac{}{}{0pt}{}{U}{C}-UGG-CUC-CA\genfrac{}{}{0pt}{}{U}{C}-CA\genfrac{}{}{0pt}{}{A}{G}-GA\genfrac{}{}{0pt}{}{U}{C}$$

Because the protein-sequence change given to us at the beginning of the problem begins after the first amino acid (His) owing to a single nucleotide addition, we can deduce that a Thr codon must change to an Asp codon. This change must result from the addition of a G directly before the Thr codon (indicated by a box), which shifts the reading frame, as shown here:

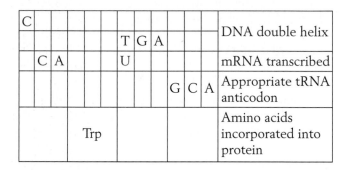

– His – Asp – Arg – Gly – Leu – Ala – Thr – Ser – Asp –

Additionally, because a deletion of a nucleotide must restore the final Asp codon to the correct reading frame, an A or G must have been deleted from the end of the original next-to-last codon, as shown by the arrow. The original protein sequence permits us to draw the mRNA with a number of ambiguities. However, the protein sequence resulting from the frame-shift allows us to determine which nucleotide was in the original mRNA at most of these points of ambiguity. The nucleotide that must have appeared in the original sequence is circled. In only a few cases does the ambiguity remain.

PROBLEMS

BASIC PROBLEMS

1. a. Use the codon dictionary in Figure 9-8 to complete the following table. Assume that reading is from left to right and that the columns represent transcriptional and translational alignments.

C				T	G	A			DNA double helix
	C	A			U				mRNA transcribed
						G	C	A	Appropriate tRNA anticodon
		Trp							Amino acids incorporated into protein

b. Label the 5′ and 3′ ends of DNA and RNA, as well as the amino and carboxyl ends of protein.

2. Consider the following segment of DNA:

5′ GCTTCCCAA 3′

3′ CGAAGGGTT 5′

Assume that the top strand is the template strand used by RNA polymerase.

a. Draw the RNA transcribed.

b. Label its 5′ and 3′ ends.

c. Draw the corresponding amino acid chain.

d. Label its amino and carboxyl ends.

Repeat, assuming the bottom strand to be the template strand.

3. A mutational event inserts an extra nucleotide pair into DNA. Which of the following outcomes do you expect? (1) No protein at all; (2) a protein in which one amino acid is changed; (3) a protein in which three amino acids are changed; (4) a protein in which two amino acids are changed; (5) a protein in which most amino acids after the site of the insertion are changed.

4. Before the true nature of the genetic coding process was fully understood, it was proposed that the message might be read in overlapping triplets. For example, the sequence GCAUC might be read as GCA CAU AUC:

G C A U C

Devise an experimental test of this idea.

5. In protein-synthesizing systems in vitro, the addition of a specific human mRNA to the *E. coli* translational apparatus (ribosomes, tRNA, and so forth) stimulates the synthesis of a protein very much like that specified by the mRNA. What does this result show?

6. Which anticodon would you predict for a tRNA species carrying isoleucine? Is there more than one possible answer? If so, state any alternative answers.

7. a. In how many cases in the genetic code would you *fail* to know the amino acid specified by a codon if you knew only the first two nucleotides of the codon?

b. In how many cases would you fail to know the first two nucleotides of the codon if you knew which amino acid is specified by it?

8. Deduce what the six wild-type codons may have been in the mutants that led Brenner to infer the nature of the amber codon UAG.

9. If a polyribonucleotide contains equal amounts of randomly positioned adenine and uracil bases, what proportion of its triplets will code for **(a)** phenylalanine, **(b)** isoleucine, **(c)** leucine, **(d)** tyrosine?

10. You have synthesized three different messenger RNAs with bases incorporated in random sequence in the following ratios: **(a)** 1U:5C, **(b)** 1A:1C:4U, **(c)** 1A:1C:1G:1U. In a protein-synthesizing system in vitro, indicate the identities and proportions of amino acids that will be incorporated into proteins when each of these mRNAs is tested. (Refer to Figure 9-8.)

11. In the fungus *Neurospora* some mutants were obtained that lacked activity for a certain enzyme. The mutations were found, by mapping, to be in either of two unlinked genes. Provide a possible explanation in terms of quaternary protein structure.

12. A mutant is found that lacks all detectable function for one specific enzyme. If you had a labeled antibody that detects this protein in a Western blot (see Chapter 1), would you expect there to be any protein detectable by the antibody in the mutant? Explain.

13. In a Western blot (see Chapter 1), the enzyme tryptophan synthetase usually shows two bands of different mobility on the gel. Some mutants with no enzyme activity showed exactly the same bands as the wild type. Other mutants with no activity showed just the slow band; still others, just the fast band.

    (a) Explain the different types of mutants at the level of protein structure.

    (b) Why do you think there were no mutants that showed no bands?

14. In the Crick–Brenner experiments described in this chapter, three "plus" or three "minus" changes restored the normal reading frame and the deduction was that the code was read in groups of three. Is this really proved by the experiments? Could a codon have been composed of six bases, for example?

15. A mutant has no activity for the enzyme isocitrate lyase. Does this result prove that the mutation is in the gene coding for isocitrate lyase?

16. A certain nonsense suppressor corrects a nongrowing mutant to a state that is near, but not exactly, wild type (it has abnormal growth). Suggest a possible reason why the reversion is not a full correction.

17. In bacterial genes, as soon as any partial mRNA transcript is produced by the RNA polymerase system the ribosome jumps on it and starts translating. Draw a diagram of this process, labeling 5′ and 3′ ends of mRNA, the COOH and NH$_2$ ends of the protein, the RNA polymerase, and at least one ribosome. (Why couldn't this system work in eukaryotes?)

18. In a haploid, a nonsense suppressor *su1* acts on mutation 1 but not on mutation 2 or 3 of gene *P*. An unlinked nonsense suppressor *su2* works on *P* mutation 2 but not on 1 or 3. Explain this pattern of suppression in terms of the nature of the mutations and the suppressors.

19. In vitro translation systems have been developed in which specific RNA molecules can be added to a test tube containing a bacterial cell extract that includes all the components needed for translation (ribosomes, tRNAs, amino acids). If a radioactively labeled amino acid is included, any protein translated from that RNA can be detected and displayed on a gel. If a eukaryotic mRNA is added to the test tube, would radioactive protein be produced? Explain.

20. In a comparable eukaryotic translation system (containing a cell extract from a eukaryotic cell), would a protein be produced by a bacterial RNA? If not, why not?

21. Would a chimeric translation system containing the large ribosomal subunit from *E. coli* and the small ribosomal subunit from yeast (a unicellular eukaryote) be able to function in protein synthesis?

22. Mutations that change a single amino acid in the active site of an enzyme can result in the synthesis of wild-type amounts of an inactive enzyme. Can you think of other regions in a protein where a single amino-acid change might have the same result?

23. What evidence supports the view that ribosomal RNAs are a more important component of the ribosome than the ribosomal proteins?

CHALLENGING PROBLEMS

24. A single nucleotide addition and a single nucleotide deletion approximately 15 sites apart in the DNA cause a protein change in sequence from

    Lys – Ser – Pro – Ser – Leu – Asn – Ala – Ala – Lys

    to

    Lys – Val – His – His – Leu – Met – Ala – Ala – Lys

    a. What are the old and new mRNA nucleotide sequences? (Use the codon dictionary in Figure 9-8.)

    b. Which nucleotide has been added and which has been deleted?

    (Problem 24 is from W. D. Stansfield, *Theory and Problems of Genetics.* McGraw-Hill, 1969.)

25. You are studying an *E. coli* gene that specifies a protein. A part of its sequence is

    – Ala – Pro – Trp – Ser – Glu – Lys – Cys – His –

    You recover a series of mutants for this gene that show no enzymatic activity. By isolating the mutant enzyme products, you find the following sequences:

Mutant 1:
 –Ala–Pro–Trp–Arg–Glu–Lys–Cys–His–

Mutant 2:
 –Ala–Pro–

Mutant 3:
 –Ala–Pro–Gly–Val–Lys–Asn–Cys–His–

Mutant 4:
 –Ala–Pro–Trp–Phe–Phe–Thr–Cys–His–

What is the molecular basis for each mutation? What is the DNA sequence that specifies this part of the protein?

26. Suppressors of frameshift mutations are now known. Propose a mechanism for their action.

27. Consider the gene that specifies the structure of hemoglobin. Arrange the following events in the most likely sequence in which they would take place.

   a. Anemia is observed.

   b. The shape of the oxygen-binding site is altered.

   c. An incorrect codon is transcribed into hemoglobin mRNA.

   d. The ovum (female gamete) receives a high radiation dose.

   e. An incorrect codon is generated in the DNA of the hemoglobin gene.

   f. A mother (an X-ray technician) accidentally steps in front of an operating X-ray generator.

   g. A child dies.

   h. The oxygen-transport capacity of the body is severely impaired.

   i. The tRNA anticodon that lines up is one of a type that brings an unsuitable amino acid.

   j. Nucleotide-pair substitution occurs in the DNA of the gene for hemoglobin.

28. An induced cell mutant is isolated from a hamster tissue culture because of its resistance to α-amanitin (a poison derived from a fungus). Electrophoresis shows that the mutant has an altered RNA polymerase; *just one* electrophoretic band is in a position different from that of the wild-type polymerase. The cells are presumed to be diploid. What do the results of this experiment tell you about ways in which to detect recessive mutants in such cells?

29. A double-stranded DNA molecule with the sequence shown here produces, in vivo, a polypeptide that is five amino acids long.

TAC ATG ATC ATT TCA CGG AAT TTC TAG CAT GTA
ATG TAC TAG TAA AGT GCC TTA AAG ATC GTA CAT

   a. Which strand of DNA is transcribed and in which direction?

   b. Label the 5′ and the 3′ ends of each strand.

   c. If an inversion occurs between the second and third triplets from the left and right ends, respectively, and the same strand of DNA is transcribed, how long will the resultant polypeptide be?

   d. Assume that the original molecule is intact and that the bottom strand is transcribed from left to right. Give the base sequence, and label the 5′ and 3′ ends of the anticodon that inserts the *fourth* amino acid into the nascent polypeptide. What is this amino acid?

30. One of the techniques used to decipher the genetic code was to synthesize polypeptides in vitro, with the use of synthetic mRNA with various repeating base sequences—for example, $(AGA)_n$, which can be written out as AGAAGAAGAAGAAGA. . . . Sometimes the synthesized polypeptide contained just one amino acid (a homopolymer), and sometimes it contained more than one (a heteropolymer), depending on the repeating sequence used. Furthermore, sometimes different polypeptides were made from the same synthetic mRNA, suggesting that the initiation of protein synthesis in the system in vitro does not always start on the end nucleotide of the messenger. For example, from $(AGA)_n$, three polypeptides may have been made: aa_1 homopolymer (abbreviated aa_1-aa_1), aa_2 homopolymer (aa_2-aa_2), and aa_3 homopolymer (aa_3-aa_3). These polypeptides probably correspond to the following readings derived by starting at different places in the sequence:

   AGA AGA AGA AGA . . .

   GAA GAA GAA GAA . . .

   AAG AAG AAG AAG . . .

The following table shows the actual results obtained from the experiment done by Khorana.

Synthetic mRNA	Polypeptide(s) synthesized
$(UG)_n$	(Ser-Leu)
$(UG)_n$	(Val-Cys)
$(AC)_n$	(Thr-His)
$(AG)_n$	(Arg-Glu)
$(UUC)_n$	(Ser-Ser) and (Leu-Leu) and (Phe-Phe)
$(UUG)_n$	(Leu-Leu) and (Val-Val) and (Cys-Cys)
$(AAG)_n$	(Arg-Arg) and (Lys-Lys) and (Glu-Glu)
$(CAA)_n$	(Thr-Thr) and (Asn-Asn) and (Gln-Gln)
$(UAC)_n$	(Thr-Thr) and (Leu-Leu) and (Tyr-Tyr)
$(AUC)_n$	(Ile-Ile) and (Ser-Ser) and (His-His)
$(GUA)_n$	(Ser-Ser) and (Val-Val)
$(GAU)_n$	(Asp-Asp) and (Met-Met)
$(UAUC)_n$	(Tyr-Leu-Ser-Ile)
$(UUAC)_n$	(Leu-Leu-Thr-Tyr)
$(GAUA)_n$	None
$(GUAA)_n$	None

[**Note:** The order in which the polypeptides or amino acids are listed in the table is not significant except for (UAUC)$_n$ and (UUAC)$_n$.]

a. Why do (GUA)$_n$ and (GAU)$_n$ each encode only two homopolypeptides?

b. Why do (GAUA)$_n$ and (GUAA)$_n$ fail to stimulate synthesis?

c. Assign an amino acid to each triplet in the following list. Bear in mind that there are often several codons for a single amino acid and that the first two letters in a codon are usually the important ones (but that the third letter is occasionally significant). Also remember that some very different looking codons sometimes encode the same amino acid. Try to carry out this task without consulting Figure 9-8.

AUG	GAU	UUG	AAC
GUG	UUC	UUA	CAA
GUU	CUC	AUC	AGA
GUA	CUU	UAU	GAG
UGU	CUA	UAC	GAA
CAC	UCU	ACU	UAG
ACA	AGU	AAG	UGA

To solve this problem requires both logic and trial and error. Don't be disheartened: Khorana received a Nobel Prize for doing it. Good luck!

(Problem 30 is from J. Kuspira and G. W. Walker, *Genetics: Questions and Problems.* McGraw-Hill, 1973.)

INTERACTIVE GENETICS   MegaManual CD-ROM Tutorial

Molecular Biology: Gene Expression

The Molecular Biology activity on the Interactive Genetics CD-ROM included with the Solutions MegaManual provides an animated tutorial on translation as a part of the Gene Expression unit. The visualization of the translation machinery in action is an excellent aid to comprehension.

EXPLORING GENOMES   A Web-Based Bioinformatics Tutorial

Finding Conserved Domains

Conserved protein sequences are a reflection of conservation of amino acid residues necessary for structure, regulation, or catalytic function. Often, groups of residues can be identified as being a pattern or signature of a particular type of enzyme or regulatory domain. In the Genomics tutorial at www.whfreeman.com/iga, you will learn how to find the conserved domains in a complex protein.

Determining Protein Structure

Protein function depends on a three-dimensional structure, which is, in turn, dependent on the primary sequence of the protein. Protein structure is determined experimentally by X-ray crystallography or NMR. In the absence of direct experimental information, powerful programs are used to try to fit primary amino acid sequence data to a three-dimensional model. We'll try one in the Genomics tutorial at www.whfreeman.com/iga.

10

REGULATION OF GENE TRANSCRIPTION

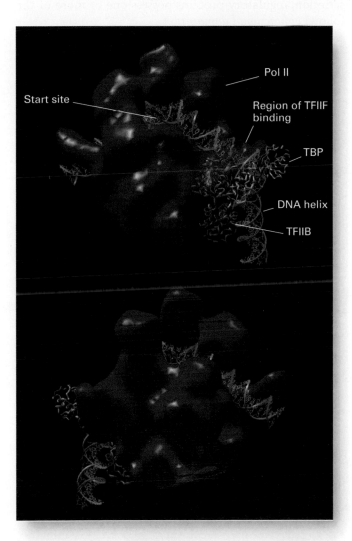

Pol II

Start site

Region of TFIIF binding

TBP

DNA helix

TFIIB

A three-dimensional model of promoter DNA complexed with eukaryotic RNA polymerase (Pol II) and other transcription initiation proteins (TBP, TFIIB). [From H. Lodish, D. Baltimore, A. Berk, S. L. Zipursky, P. Matsudaira, and J. Darnell, *Molecular Cell Biology*, 4th ed. Copyright 2000 by Scientific American Books, p. 383. Adapted from T.-K. Kim et al., *Proceedings of the National Academy of Sciences U.S.A.* 94, 1997, 12268.]

KEY QUESTIONS

- In what ways do the expression levels of genes vary (that is, the output of their RNA and protein products)?

- What factors in a cell or its environment trigger changes in a gene's level of expression?

- What are the molecular mechanisms of gene regulation in prokaryotes and eukaryotes?

- How does a eukaryote achieve thousands of patterns of gene interaction with a limited number of regulatory proteins?

- What determines the degree of chromatin condensation?

- What are two ways in which to change chromatin structure?

OUTLINE

10.1 Prokaryotic gene regulation

10.2 Discovery of the *lac* system of negative control

10.3 Catabolic repression of the *lac* operon: positive control

10.4 Dual positive and negative control: the arabinose operon

10.5 Metabolic pathways

10.6 Transcriptional regulation in eukaryotes

10.7 Chromatin's role in eukaryotic gene regulation

CHAPTER OVERVIEW

The biological properties of each cell are largely determined by the active proteins expressed in it. This constellation of expressed proteins determines much of the cell's architecture, its enzymatic activities, its interactions with its environment, and many other physiological properties. However, at any given time in a cell's life history, only a fraction of the RNAs and proteins encoded in its genome are expressed. At different times, the profile of expressed gene products can differ dramatically, both in regard to which proteins are expressed and at what levels. How are these specific profiles created?

The regulation of the synthesis of a gene's transcript and its protein product is often termed **gene regulation.** As one might expect, if the final product is a protein (as is true for most genes), regulation could be achieved by adjusting the transcription of DNA into RNA or the translation of RNA into protein. In fact, gene regulation takes place at many other levels, including the stability

of the mRNA and posttranslational modifications of proteins. However, most regulation is thought to act at the level of gene transcription; so, in this chapter, the primary focus is on the regulation of transcription. The basic mechanism at work is that molecular signals from outside or inside the cell lead to the binding of regulatory proteins to specific DNA sites adjacent to the protein-coding region, and the binding of these these proteins modulates the rate of transcription. These proteins may directly or indirectly assist RNA polymerase in binding to its transcription initiation site—the *promoter*—or they may repress transcription by preventing the binding of RNA polymerase. To modulate transcription, regulatory proteins possess one or more of the following *functional domains:*

1. A domain that recognizes the correct regulatory element (that is, the protein's DNA docking site)

2. A domain that interacts with one or more proteins of the basal transcription apparatus (RNA

CHAPTER OVERVIEW Figure

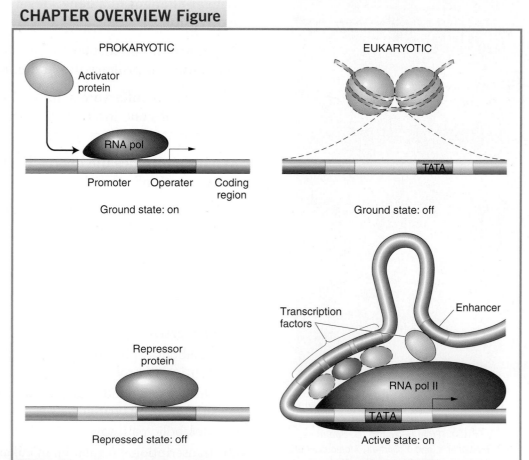

Figure 10-1 Overview of transcriptional regulation in prokaryotes and eukaryotes. In prokaryotes, RNA polymerase can usually begin transcription unless a repressor protein blocks it, whereas in eukaryotes, the packaging of DNA with nucleosomes prevents transcription unless other regulatory proteins are present to nudge nucleosomes aside from key binding sites.

polymerase or an RNA-polymerase-associated protein)

3. A domain that interacts with proteins bound to nearby docking sites, such that they can act cooperatively to regulate transcription

4. A domain that influences chromatin condensation either directly or indirectly

5. A domain that acts as a sensor of physiological conditions within the cell

Regulation has evolved to permit a cell to cope with variations in a particular type of circumstance such as the availability of nutrients, invading infectious agents, changes in temperature or other stresses, and changes in the developmental state of the cell. Thus, some regulatory molecules—typically proteins—must have a sensor domain that interacts with appropriate factors in the cellular environment so that its regulatory role can be responsive to the cell's needs at a given moment.

Although prokaryotes and eukaryotes have many of the mechanisms of gene regulation in common, there are some fundamental differences in the underlying logic. Both the similarities and the differences are summarized in Figure 10-1. Both use regulatory proteins that bind near the protein-coding region to modulate the level of transcription. However, because eukaryote genomes are bigger and their range of properties is larger, inevitably their regulation is more complex, requiring more types of regulatory proteins and more types of interactions with the adjacent regulatory regions. The most important difference is that eukaryotic DNA is packaged into *nucleosomes*, forming *chromatin*, whereas prokaryotic DNA lacks nucleosomes. In eukaryotes, chromatin structure is dynamic and is an essential ingredient in gene regulation.

As shown in Figure 10-1, the ground state of a prokaryotic gene is "on." Thus, RNA polymerase can usually bind to a promoter when no other regulatory proteins are around to bind to the DNA. In prokaryotes, transcription initiation is prevented or reduced if the binding of RNA polymerase is blocked, usually through the binding of a repressor regulatory protein. Activator regulatory proteins increase the binding of RNA polymerase to promoters where a little help is needed. In contrast, the ground state in eukaryotes is "off." Therefore, the basal transcription apparatus (including RNA polymerase II and associated general transcription factors) cannot bind to the promoter in the absence of other regulatory proteins. In many cases, the binding of the transcription apparatus is not possible due to the position of nucleosomes (DNA plus histones) over key binding sites in the promoter. The nucleosome, as you recall, is the basic unit of chromatin; thus, chromatin structure usually has to be

changed to activate eukaryotic transcription. Chromatin structure is heritable just like the DNA sequence. Inheritance of chromatin structure is a form of epigenetic inheritance.

We begin with examples from the simpler prokaryotes. Findings from studies of prokaryotic gene regulation beginning in the 1950s provided the basic model for gene regulation that is still at the center of our thinking today.

10.1 Prokaryotic gene regulation

Despite the simplicity of their form, bacteria have a fundamental need to regulate the expression of their genes. One of the main reasons is that they are nutritional opportunists. Consider how bacteria obtain the many important compounds, such as sugars, amino acids, and nucleotides, needed for metabolism. Bacteria swim in a sea of potential nutrients. They can either acquire these compounds from the environment or synthesize them by enzymatic pathways. Synthesizing the necessary enzymes for these pathways expends energy and cellular resources; so, given the choice, bacteria will take compounds from the environment instead. To be economical, they will synthesize the enzymes necessary to produce these compounds only when there is no other option—in other words, when these compounds are unavailable in their local environment.

Bacteria have evolved regulatory systems that couple the expression of gene products to sensor systems that detect the relevant compound in a bacterium's local environment. The regulation of enzymes taking part in sugar metabolism provides an example. Sugar molecules can be oxidized to provide energy or they can be used as building blocks for a great range of organic compounds. However, there are many different types of sugar that bacteria could use, including lactose, glucose, galactose, and xylose. First, a different import protein is required to allow each of these sugars to enter the cell. Further, a different set of enzymes is required to process each of the sugars. If a cell were to simultaneously synthesize all the enzymes that it might possibly need, the cell would expend much more energy and materials to produce the enzymes than it could ever derive from breaking down prospective carbon sources. The cell has devised mechanisms to shut down (repress) the transcription of all genes encoding enzymes that are not needed at a given time and to turn on (activate) those genes encoding enzymes that are needed. For example, if lactose is in the environment, the cell will shut down transcription of the genes encoding enzymes needed for the import and metabolism of glucose, galactose, xylose, and other sugars. Conversely, the cell will initiate the transcription of the genes encoding enzymes needed for the import and metabolism of lactose. In sum, cells need mechanisms that fulfill two criteria:

1. They must be able to recognize environmental conditions in which they should activate or repress transcription of the relevant genes.

2. They must be able to toggle on or off the transcription of each specific gene or group of genes.

Let's preview the current model for prokaryotic transcriptional regulation and then use a well-understood example—the regulation of the genes in the metabolism of the sugar lactose—to examine it in detail.

The basics of prokaryotic transcriptional regulation

Two types of DNA–protein interactions are required for regulated transcription. Both take place near the site at which gene transcription begins.

One of these DNA–protein interactions determines where transcription begins. The DNA that participates in this interaction is a DNA segment called the **promoter,** and the protein that binds to this site is RNA polymerase. When RNA polymerase binds to the promoter DNA, transcription can initiate a few bases away from the promoter site. Every gene must have a promoter or it cannot be transcribed.

The other type of DNA–protein interaction decides whether promoter-driven transcription takes place. DNA segments near the promoter serve as binding sites for regulatory proteins called **activators** and **repressors.** In bacteria, most of these sites are termed **operators.** For some genes, an activator protein must bind to its target DNA site as a necessary prerequisite for transcription to begin. Such instances are sometimes referred to as *positive regulation* because the *presence* of the bound protein is required for transcription (Figure 10-2). For other genes, a repressor protein must be prevented from binding to its target site as a necessary prerequisite for transcription to

begin. Such cases are sometimes termed *negative regulation* because the *absence* of the bound repressor allows transcription to begin. How do activators and repressors regulate transcription? Often, a DNA-bound activator protein physically helps tether RNA polymerase to its nearby promoter so that polymerase may begin transcribing. A DNA-bound repressor protein typically acts either by physically interfering with the binding of RNA polymerase to its promoter (blocking transcription initiation) or by impeding the movement of RNA polymerase along the DNA chain (blocking transcription).

> **MESSAGE** Genes must contain two kinds of binding sites to permit regulated transcription. First, binding sites for RNA polymerase must be present. Second, binding sites for activator or repressor proteins can be present in the vicinity of the promoter.

Both activator and repressor proteins must be able to recognize when environmental conditions are appropriate for their actions and act accordingly. Thus, for activator or repressor proteins to do their job, each must be able to exist in two states: one that can bind its DNA targets and another that cannot. The binding state must be appropriate to the set of physiological conditions present in the cell and its environment. For many regulatory proteins, DNA binding is effected through the interaction of two different sites in the three-dimensional structure of the protein. One site is the **DNA-binding domain.** The other site, the **allosteric site,** acts as a toggle switch that sets the DNA-binding domain in one of two modes: functional or nonfunctional. The allosteric site interacts with small molecules called *allosteric effectors.* In lactose metabolism, the sugar lactose is an allosteric effector. An **allosteric effector** binds to the allosteric site of the regulatory protein in such a way that it changes the structure of the DNA-binding domain. Some activator or repressor proteins must bind to their allosteric effectors before they can bind DNA.

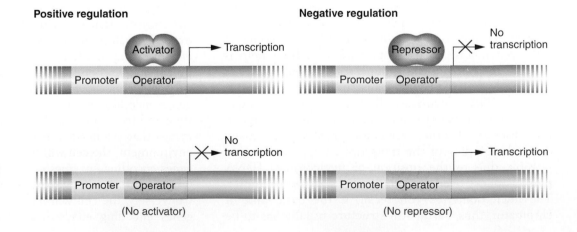

Figure 10-2 The binding of regulatory proteins can either activate or block transcription.

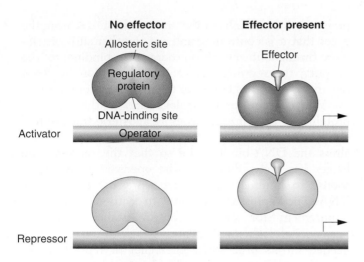

Figure 10-3 The influence of allosteric effectors on the DNA-binding activities of activators and repressors.

Others can bind DNA only in the absence of their allosteric effectors. Two of these situations are shown in Figure 10–3.

> **MESSAGE** Allosteric effectors control the ability of activator or repressor proteins to bind to their DNA target sites.

A first look at the *lac* regulatory circuit

The pioneering work of François Jacob and Jacques Monod in the 1950s showed how lactose metabolism is genetically regulated. Let's examine the system under two conditions: the presence and the absence of lactose. Figure 10-4 is a simplified view of the components of this system. The cast of characters for *lac* regulation includes protein-coding genes and sites on the DNA that are targets for DNA-binding proteins.

THE *lac* STRUCTURAL GENES The metabolism of lactose requires two enzymes: a permease to transport lactose into the cell and β-galactosidase to cleave the

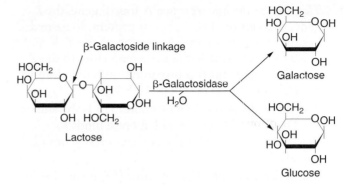

Figure 10-5 The metabolism of lactose. The enzyme β-galactosidase catalyzes a reaction in which water is added to the β-galactoside linkage to break lactose into separate molecules of galactose and glucose.

lactose molecule to yield glucose and galactose (Figure 10-5). The structures of the β-galactosidase and permease proteins are encoded by two adjacent sequences, *Z* and *Y*, respectively. A third contiguous sequence encodes an additional enzyme, termed *transacetylase*, which is not required for lactose metabolism. We will call *Z*, *Y*, and *A structural genes*—in other words, segments coding for protein structure—while reserving judgment on this categorization until later. We will focus mainly on *Z* and *Y*. All three genes are transcribed into a single messenger RNA molecule. Regulation of the production of this mRNA coordinates the synthesis of all three enzymes. That is, either all or none are synthesized.

> **MESSAGE** If the genes encoding proteins comprise a single transcription unit, the expression of all these genes will be coordinately regulated.

REGULATORY COMPONENTS OF THE *lac* SYSTEM Key regulatory components of the lactose metabolic system include a transcription regulatory protein and two docking sites—one site for the regulatory protein and another site for RNA polymerase.

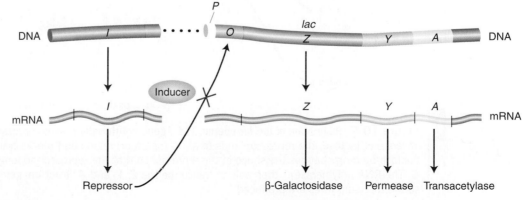

Figure 10-4 A simplified *lac* operon model. Coordinate expression of the *Z*, *Y*, and *A* genes is under the negative control of the product of the *I* gene, the repressor. When the inducer binds the repressor, the operon is fully expressed.

1. *The gene for the Lac repressor.* A fourth gene, the *I* gene, encodes the Lac repressor protein, so named because it can block the expression of the *Z*, *Y*, and *A* genes. The *I* gene happens to map close to the *Z*, *Y*, and *A* genes, but this proximity does not seem to be important to its function.

2. *The lac promoter site.* The promoter (*P*) is the site on the DNA to which RNA polymerase binds to initiate transcription of the *lac* structural genes (*Z*, *Y*, and *A*).

3. *The lac operator site.* The operator (*O*) is the site on the DNA to which the Lac repressor binds. It is located between the promoter and the *Z* gene near the point at which transcription of the multigenic mRNA begins.

THE INDUCTION OF THE *lac* SYSTEM The *P*, *O*, *Z*, *Y*, and *A* segments (shown in Figure 10-6) together constitute an **operon**, defined as a segment of DNA that encodes a multigenic mRNA as well as an adjacent common promoter and regulatory region. The *lacI* gene, encoding the Lac repressor, is *not* considered part of the *lac* operon itself, but the interaction between the Lac repressor and the *lac* operator site is crucial to proper regulation of the *lac* operon. The Lac repressor has a *DNA-binding* site that can recognize the operator DNA sequence and an *allosteric site* that binds lactose or analogs of lactose that are useful experimentally. The re-

pressor will bind only to the site on the DNA near the genes that it is controlling and not to other sites distributed throughout the chromosome. By binding to the operator, the repressor prevents transcription by RNA polymerase that has bound to the adjacent promoter site.

When lactose or its analogs bind to the repressor protein, the protein undergoes an **allosteric transition,** a change in shape. This slight alteration in shape in turn alters the DNA-binding site so that the repressor no longer has high affinity for the operator. Thus, in response to binding lactose, the repressor falls off the DNA. The repressor's response to lactose satisfies one requirement for such a control system—that the presence of lactose stimulates the synthesis of genes needed for its processing. The relief of repression for systems such as *lac* is termed **induction;** lactose and its analogs that allosterically inactivate the repressor and lead to the expression of the *lac* genes are termed **inducers.**

Let's summarize. In the absence of an inducer (lactose or an analog), the Lac repressor binds to the *lac* operator site and prevents transcription of the *lac* operon by blocking the movement of RNA polymerase. In this sense, the Lac repressor acts as a roadblock on the DNA. Consequently, all the structural genes of the *lac* operon (the *Z*, *Y*, and *A* genes) are repressed, and there is no β-galactosidase, β-galactoside permease, or transacetylase in the cell. In contrast, when an inducer is present, it

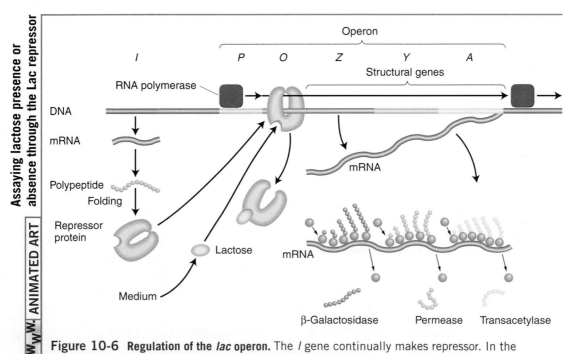

Figure 10-6 Regulation of the *lac* operon. The *I* gene continually makes repressor. In the absence of lactose, the repressor binds to the *O* (operator) region and blocks transcription. Lactose binding changes the shape of the repressor so that the repressor no longer binds to *O*. The RNA polymerase is then able to transcribe the *Z*, *Y*, and *A* structural genes, and so the three enzymes are produced.

binds to the allosteric site of each Lac repressor subunit, thereby inactivating the site that binds to the operator. The Lac repressor falls off the DNA, allowing the transcription of the structural genes of the *lac* operon to begin. The enzymes β-galactosidase, β-galactoside permease, and transacetylase now appear in the cell in a coordinated fashion.

10.2 Discovery of the *lac* system of negative control

To study gene regulation, ideally we need three things: a biochemical assay that lets us measure the amount of mRNA or expressed protein or both, reliable conditions in which differences in the levels of expression occur in a wild-type genotype, and mutations that perturb the levels of expression. In other words, we need a way of describing wild-type gene regulation and we need mutations that can disrupt the wild-type regulatory process. With these elements in hand, we can analyze the expression in mutant genotypes, treating the mutations singly and in combination, to unravel any kind of gene-regulation event. The classical application of this approach was used by Jacob and Monod, who performed the definitive studies of bacterial gene regulation.

Jacob and Monod used the lactose metabolism system of *E. coli* (see Figure 10-4) to genetically dissect the process of enzyme induction—that is, the appearance of a specific enzyme only in the presence of its substrates. This phenomenon had been observed in bacteria for many years, but how could a cell possibly "know" precisely which enzymes to synthesize? How could a particular substrate induce the appearance of a specific enzyme?

In the *lac* system, the presence of the inducer lactose causes cells to produce more than 1000 times as much of the enzyme β-galactosidase as they produced when grown in the absence of lactose. What role did the inducer play in the induction phenomenon? One idea was that the inducer was simply activating a precursor form of β-galactosidase that had accumulated in the cell. However, when Jacob and Monod followed the fate of radioactively labeled amino acids added to growing cells either before or after the addition of an inducer, they found that induction resulted in the synthesis of new enzyme molecules, as indicated by the presence of the radioactive amino acids in the enzymes. These new molecules could be detected as early as 3 minutes after the addition of an inducer. Additionally, withdrawal of the inducer brought about an abrupt halt in the synthesis of the new enzyme. Therefore, it became clear that the cell has a rapid and effective mechanism for turning gene expression on and off in response to environmental signals.

Genes controlled together

When Jacob and Monod induced β-galactosidase, they found that they also induced the enzyme permease, which is required to transport lactose into the cell. The analysis of mutants indicated that each enzyme was encoded by a different gene. The enzyme transacetylase (with a dispensable and as yet unknown function) was also induced together with β-galactosidase and permease and was later shown to be encoded by a separate gene. Therefore, Jacob and Monod could identify three **coordinately controlled genes**. Recombination mapping showed the *Z*, *Y*, and *A* genes were very closely linked on the chromosome.

Genetic evidence for the operator and repressor

Now we come to heart of Jacob and Monod's work: How did they deduce the mechanisms of gene regulation in the *lac* system? Their approach was a classic genetic approach: to examine the physiological consequences of mutations. As we shall see, the properties of mutations in the structural genes and the regulatory elements of the *lac* operon are quite different, providing important clues for Jacob and Monod.

Natural inducers, such as lactose, are not optimal for these experiments, because they are hydrolyzed by β-galactosidase; the inducer concentration decreases during the experiment, and so the measurements of enzyme induction become quite complicated. Instead, for such experiments, Jacob and Monod used synthetic inducers, such as isopropyl-β-D-thiogalactoside (IPTG; Figure 10-7). IPTG is not hydrolyzed by β-galactosidase.

Jacob and Monod found that several different classes of mutations can alter the expression of the structural genes of the *lac* operon. They were interested in the interactions between the new alleles, such as an assessment of dominance. But to perform such tests, one needs diploids, and bacteria are haploid. However, by using F′ factors (see Chapter 5) carrying the *lac* region of the genome, Jacob and Monod were able to produce bacteria that are partially diploid and heterozygous for the desired *lac* mutations. These partial diploids allowed Jacob and Monod to distinguish mutations in

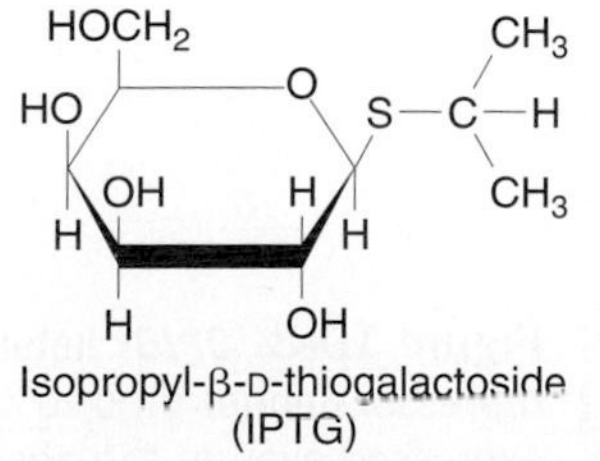

Isopropyl-β-D-thiogalactoside (IPTG)

Figure 10-7 Structure of IPTG, an inducer of the *lac* operon.

TABLE 10-1 Synthesis of β-Galactosidase and Permease in Haploid and Heterozygous Diploid Operator Mutants

Strain	Genotype	β-Galactosidase (Z)		Permease (Y)		Conclusion
		Noninduced	Induced	Noninduced	Induced	
1	$O^+Z^+Y^+$	−	+	−	+	Wild type is inducible
2	$O^+Z^+Y^+/F'O^+Z^-Y^+$	−	+	−	+	Z^+ is dominant to Z^-
3	$O^CZ^+Y^+$	+	+	+	+	O^C is constitutive
4	$O^+Z^-Y^+/F'O^CZ^+Y^-$	+	+	−	+	Operator is cis-acting

Note: Bacteria were grown in glycerol (no glucose present) with and without the inducer IPTG. The presence or absence of enzyme is indicated by + or −, respectively. All strains are I^+.

the regulatory DNA site (the *lac* operator) from mutations in the regulatory protein (the *I* gene encoding the Lac repressor).

We begin by examining mutations that inactivate the structural genes for β-galactosidase and permease (designated Z^- and Y^-, respectively). The first thing that we learn is that Z^- and Y^- are recessive to their respective wild-type alleles (Z^+ and Y^+). For example, strain 2 in Table 10-1 can be induced to synthesize β-galactosidase (like the wild-type haploid strain 1 in this table), even though it is heterozygous for mutant and wild-type Z alleles. This demonstrates that the Z^+ allele is dominant to its Z^- counterpart.

Jacob and Monod first identified two classes of regulatory mutations, called O^C and I^-, which were called **constitutive,** meaning that they caused the *lac* operon structural genes to be expressed regardless of whether inducer was present. Jacob and Monod identified the existence of the operator on the basis of their analysis of the O^C mutations. These mutations make the operator

incapable of binding to repressor and hence the operon is always "on" (Table 10-1, strain 3). Strangely, the constitutive effects of O^C mutations were restricted solely to those *lac* structural genes *on the same chromosome.* For this reason, the operator mutant was said to be **cis-acting,** as demonstrated by the phenotype of strain 4 in Table 10-1. Here, because the wild-type permease (Y^+) gene is cis to the wild-type operator, permease activity is induced only when lactose or an analog is present. In contrast, the wild-type β-galactosidase (Z^+) gene is cis to the O^C mutant operator; hence, β-galactosidase is expressed constitutively. This unusual property of cis action suggested that the operator acts simply as a protein-binding site and makes *no* gene product. The operator-binding site is thus a segment of DNA that influences only the expression of the structural genes linked to it (Figure 10-8).

Jacob and Monod did comparable genetic tests with the I^- mutations (Table 10-2). A comparison of the inducible wild-type I^+ (strain 1) with I^- strains

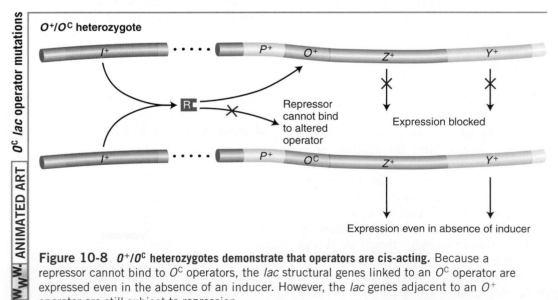

Figure 10-8 O^+/O^C **heterozygotes demonstrate that operators are cis-acting.** Because a repressor cannot bind to O^C operators, the *lac* structural genes linked to an O^C operator are expressed even in the absence of an inducer. However, the *lac* genes adjacent to an O^+ operator are still subject to repression.

TABLE 10-2 Synthesis of β-Galactosidase and Permease in Haploid and Heterozygous Diploid Strains Carrying l^+ and l^-

| Strain | Genotype | β-Galactosidase (Z) | | Permease (Y) | | Conclusion |
		Noninduced	Induced	Noninduced	Induced	
1	$I^+Z^+Y^+$	−	+	−	+	I^+ is inducible
2	$I^-Z^+Y^+$	+	+	+	+	I^- is constitutive
3	$I^+Z^-Y^+/F'I^-Z^+Y^+$	−	+	−	+	I^+ is dominant to I^-
4	$I^-Z^-Y^+/F'I^+Z^+Y^-$	−	+	−	+	I^+ is trans-acting

Note: Bacteria were grown in glycerol (no glucose present) and induced with IPTG. The presence of the maximal level of the enzyme is indicated by a plus sign; the absence or very low level of an enzyme is indicated by a minus sign. (All strains are O^+.)

shows that I^- mutations are constitutive (strain 2). Strain 3 demonstrates that the inducible phenotype of I^+ is dominant to the constitutive phenotype of I^-. This finding showed Jacob and Monod that the amount of wild-type protein encoded by one copy of the gene is sufficient to regulate both copies of the operator in a diploid cell. Most significantly, strain 4 showed them that the I^+ gene product is **trans-acting,** meaning that the gene product can regulate *all* structural *lac* operon genes, whether in cis or in trans (residing on different DNA molecules). Unlike the operator, the action of the I gene is that of a standard protein-coding gene. The protein product of the I gene is able to diffuse and act on both operators in the partial diploid (Figure 10-9).

> **MESSAGE** Operator mutations reveal that such a site is cis-acting; that is, it regulates the expression of an adjacent transcription unit on the same DNA molecule. In contrast, mutations in the gene encoding a repressor protein reveal that this protein is trans-acting; that is, it can act on any copy of the target DNA site in the cell.

Genetic evidence for allostery

Finally, Jacob and Monod were able to demonstrate allostery through the analysis of another class of repressor mutations. Recall that the Lac repressor inhibits transcription of the *lac* operon in the absence of an inducer but permits transcription when the inducer is present. This regulation is accomplished through a second site on the repressor protein, the allosteric site, which binds to the inducer. When bound to the inducer, the repressor undergoes a change in overall structure such that its DNA-binding site can no longer function.

Jacob and Monod isolated another class of repressor mutation, called superrepressor (I^S) mutations. I^S mutations cause repression even in the presence of an inducer (compare strain 2 in Table 10-3 with the inducible wild-type strain 1). Unlike the case for I^-, I^S mutations are dominant to I^+ (see Table 10-3, strain 3). This key observation led Jacob and Monod to speculate that I^S mutations alter the allosteric site so that it can no longer bind to an inducer. As a consequence, I^S-encoded repressor protein continually binds to the operator—preventing transcription of the *lac* operon even when the

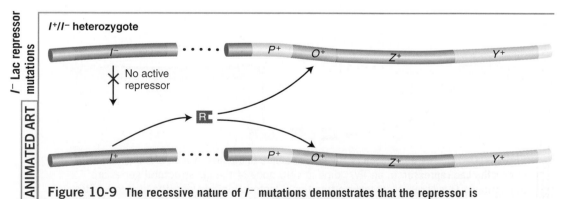

Figure 10-9 The recessive nature of I^- mutations demonstrates that the repressor is trans-acting. Although no active repressor is synthesized from the I^- gene, the wild-type (I^+) gene provides a functional repressor that binds to both operators in a diploid cell and blocks *lac* operon expression (in the absence of an inducer).

TABLE 10-3 Synthesis of β-Galactosidase and Permease by the Wild Type and by Strains Carrying Different Alleles of the *l* Gene

Strain	Genotype	β-Galactosidase (Z)		Permease (Y)		Conclusion
		Noninduced	Induced	Noninduced	Induced	
1	$I^+Z^+Y^-$	−	+	−	+	I^+ is inducible
2	$I^SZ^+Y^+$	−	−	−	−	I^S is always repressed
3	$I^SZ^+Y^+/F'l^+Z^+Y^+$	−	−	−	−	I^S is dominant to I^+

Note: Bacteria were grown in glycerol (no glucose present) with and without the inducer IPTG. Presence of the indicated enzyme is represented by +; absence or low levels, by −.

inducer is present in the cell. On this basis, we can see why I^S is dominant to I^+. Mutant I^S protein will bind to both operators in the cell, even in the presence of an inducer and regardless of the fact that I^+-encoded protein may be present in the same cell (Figure 10-10).

Genetic analysis of the *lac* promoter

Mutational analysis also demonstrated that an element essential for *lac* transcription is located between *I* and *O*. This element, termed the *promoter* (*P*), serves as the initiation site for transcription. There are two binding regions for RNA polymerase in a typical prokaryotic promoter, shown in Figure 10-11 as the two highly conserved regions at −35 and −10. Promoter mutations are cis acting in that they affect the transcription of all adjacent structural genes in the operon. This cis dominance is because promoters, like operators, are sites on the DNA molecule that are bound by proteins and themselves produce no protein product.

Molecular characterization of the Lac repressor and the *lac* operator

Walter Gilbert and Benno Müller-Hill provided a decisive demonstration of the *lac* system in 1966 by monitoring the binding of the radioactively labeled inducer IPTG to purified repressor protein. They first showed that the repressor consists of four identical subunits, and hence contains four IPTG-binding sites. (A more detailed description of the repressor is given later in the chapter.) Second, they showed that, in the test tube, repressor protein binds to DNA containing the operator and comes off the DNA in the presence of IPTG.

Gilbert and his co-workers showed that the repressor can protect specific bases in the operator from chemical reagents. They took operon DNA to which repressor was bound and treated it with the enzyme DNase, which breaks up DNA.

They were able to recover short DNA strands that had been shielded from the enzyme activity by the re-

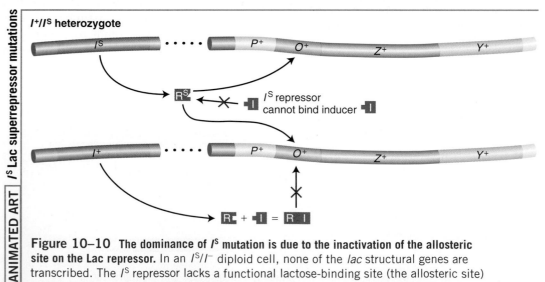

Figure 10–10 The dominance of I^S mutation is due to the inactivation of the allosteric site on the Lac repressor. In an I^S/I^- diploid cell, none of the *lac* structural genes are transcribed. The I^S repressor lacks a functional lactose-binding site (the allosteric site) and thus is not inactivated by an inducer. Therefore, even in the presence of an inducer, the I^S repressor binds irreversibly to all operators in a cell, thereby blocking transcription of the *lac* operon.

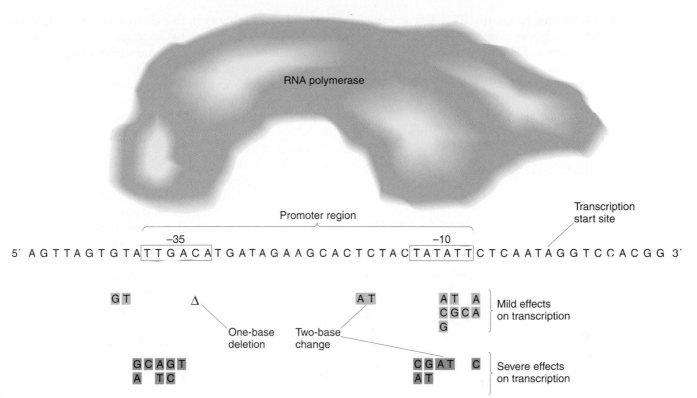

Figure 10-11 Specific DNA sequences are important for efficient transcription of *E. coli* genes by RNA polymerase. The boxed sequences are highly conserved in all *E. coli* promoters, an indication of their role as contact sites on the DNA for RNA polymerase binding. Mutations in these regions have mild (gold) and severe (brown) effects on transcription. The mutations may be changes of single nucleotides or pairs of nucleotides or a deletion (D) may occur. [From J. D. Watson, M. Gilman, J. Witkowski, and M. Zoller, *Recombinant DNA*, 2d ed. Copyright 1992 by James D. Watson, Michael Gilman, Jan Witkowski, and Mark Zoller.]

pressor molecule and that presumably constituted the operator sequence. The base sequence of each strand was determined, and each operator mutation was shown to be a change in the sequence (Figure 10-12). These results showed that the operator locus is a specific sequence of 17 to 25 nucleotides situated just before the structural *Z* gene. They also showed the incredible specificity of repressor–operator recognition, which can be disrupted by a single base substitution. When the sequence of bases in the *lac* mRNA (transcribed from the *lac* operon) was determined, the first 21 bases on the 5′ initiation end proved to be complementary to the operator sequence that Gilbert had determined, showing that the operator sequence is transcribed.

The results of these experiments provided crucial confirmation of the mechanism of repressor action formulated by Jacob and Monod.

Polar mutations

Some of the mutations that mapped to the *Z* and *Y* genes were found to be polar—that is, affecting genes "downstream" in the operon. For example, polar *Z* mu-

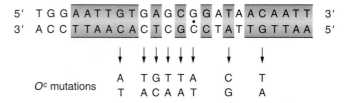

Figure 10-12 The DNA base sequence of the lactose operator and the base changes associated with eight O^c mutations. Regions of twofold rotational symmetry are indicated by color and by a dot at their axis of symmetry. [From W. Gilbert, A. Maxam, and A. Mirzabekov, in N.O. Kjeldgaard and O. Malløe, eds., *Control of Ribosome Synthesis.* Academic Press, 1976. Used by permission of Munksgaard International Publishers, Ltd., Copenhagen.]

tations resulted in null function not only for *Z* but also for *Y* and *A*. Polar mutations in *Y* affected *A* also but not *Z*. These polar mutations were the genetic observations that suggested to Jacob and Monod that the three genes were transcribed from one end as a unit. The polar mutations resulted from stop codons that cause the ribosomes to fall off the transcript. This left a naked stretch of mRNA which was degraded, thereby inactivating downstream genes. (The normal stop and slant codons that cause ribosomes to exit and enter the mRNA between the structural genes do not involve this degradation.)

10.3 Catabolite repression of the *lac* operon: positive control

The existing *lac* system is one that, through a long evolutionary process, has been selected to operate in an optimal fashion for the energy efficiency of the bacterial cell. Presumably to maximize energy efficiency, two environmental conditions have to be satisfied for the lactose metabolic enzymes to be expressed.

One condition is that lactose must be present in the environment. This condition makes sense, because it would be inefficient for the cell to produce the lactose metabolic enzymes if there is no substrate to metabolize. We have already seen that the cell's recognition that lactose is present is accomplished by a repressor protein.

The other condition is that glucose cannot be present in the cell's environment. Because the cell can capture more energy from the breakdown of glucose than it can from the breakdown of other sugars, it is more efficient for the cell to metabolize glucose rather than lactose. Thus, mechanisms have evolved that prevent the cell from synthesizing the enzymes for lactose metabolism when lactose and glucose are present together. The repression of the transcription of lactose-metabolizing genes in the presence of glucose is an example of **catabolite repression**. The transcription of proteins necessary for the metabolism of many different sugars is similarly repressed in the presence of glucose. We shall see that catabolite repression works through an *activator protein*.

The basics of catabolite repression of the *lac* operon: choosing the best sugar to metabolize

If both lactose and glucose are present, the synthesis of β-galactosidase is not induced until all the glucose has been used up. Thus, the cell conserves its energy by metabolizing any existing glucose before going through the energy-expensive process of creating new machinery to metabolize lactose.

The results of studies indicate that a breakdown product, or *catabolite*, of glucose prevents activation of the *lac* operon by lactose—this is the catabolite repression just referred to. The identity of this catabolite is as

yet unknown. However, the glucose catabolite is known to modulate the level of an important cellular constituent—**cyclic adenosine monophosphate (cAMP)**. When glucose is present in high concentrations, the cell's cAMP concentration is low. As the glucose concentration decreases, the cell's concentration of cAMP increases correspondingly. A high concentration of cAMP is necessary for activation of the *lac* operon. Mutants that cannot convert ATP into cAMP cannot be induced to produce β-galactosidase, because the concentration of cAMP is not great enough to activate the *lac* operon. In addition, there are other mutants that do make cAMP but cannot activate the Lac enzymes, because they lack yet another protein, called **CAP (catabolite activator protein)**, made by the *crp* gene. CAP binds to a specific site on the *lac* operon. The DNA-bound CAP is then able to interact physically with RNA polymerase and increases that enzyme's affinity for the *lac* promoter. By itself, CAP cannot bind to the CAP site of the *lac* operon. However, by binding to its allosteric effector, cAMP, CAP is able to bind to the CAP site and activate RNA polymerase. In this way, the catabolite repression system contributes to the selective activation of the *lac* operon (Figure 10-13).

(a) Glucose levels regulate cAMP levels

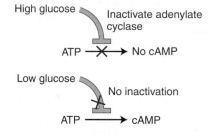

(b) cAMP–CAP complex activates transcription

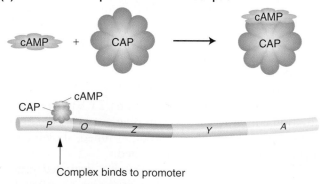

Figure 10-13 Catabolite control of the *lac* operon. (a) Only under conditions of low glucose is adenylate cyclase active and cAMP (cyclic adenosine monophosphate) formed. (b) When cAMP is present, it forms a complex with CAP (catabolite activator protein) that activates transcription by binding to a region within the *lac* promoter.

> **MESSAGE** The *lac* operon has an added level of control so that the operon stays inactive in the presence of glucose even if lactose also is present. An allosteric effector, cAMP, binds to the activator CAP to permit the induction of the *lac* operon. However, high concentrations of glucose catabolites produce low concentrations of cAMP, thus failing to produce cAMP–CAP and thereby failing to activate the *lac* operon.

The structures of target DNA sites

The DNA sequences to which the CAP–cAMP complex binds are now known. These sequences (Figure 10-14) are very different from the sequences to which the Lac repressor binds, although both bind to the 5′ end of the operon. These differences underlie the specificity of DNA binding by these very different regulatory proteins. One property that these sequences do have in common and that is common to many other DNA-binding sites is rotational twofold symmetry. In other words, if we rotate the DNA sequence 180° within the plane of the page, the sequence of the highlighted bases of the binding sites will be identical. The highlighted bases are thought to constitute the important contact sites for protein–DNA interactions. This rotational symmetry corresponds to symmetries within the DNA-binding proteins, many of which are composed of two or four identical subunits. We shall consider the structures of some DNA-binding proteins later in the chapter.

How does the binding of the cAMP–CAP complex to the operon further the binding of RNA polymerase to the *lac* promoter? In Figure 10-15, the DNA is shown as being bent when CAP binds. This bending of DNA may aid the binding of RNA polymerase to the promoter. There is also evidence that CAP makes direct contact

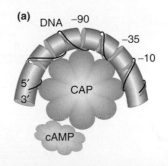

(b)

Figure 10-15 Binding of CAP to DNA. When CAP binds the promoter, it creates a bend greater than 90° in the DNA. Part b is derived from the structural analysis of the CAP–DNA complex. [Part a redrawn from B. Gartenberg and D. M. Crothers, *Nature* 333, 1988, 824. (See H. N. Lie-Johnson et al., *Cell* 47, 1986, 995.) After H. Lodish, D. Baltimore, A. Berk, S. L. Zipursky, P. Matsudaira, and J. Darnell, *Molecular Cell Biology*, 3d ed. Copyright 1995 by Scientific American Books. Part b from L. Schultz and T. A. Steitz.]

with RNA polymerase that is important for the CAP activation effect. The base sequence shows that CAP and DNA polymerase bind directly adjacent to each other on the *lac* promoter (Figure 10-16).

> **MESSAGE** Generalizing from the *lac* operon story, we can envision the chromosome as heavily decorated by regulatory proteins binding to the operator sites that they control. The exact pattern of decorations will depend on which genes are turned on or off and whether activators or repressors regulate particular operons.

A summary of the *lac* operon

We can now fit the CAP–cAMP- and RNA-polymerase-binding sites into the detailed model of the *lac* operon, as shown in Figure 10-17. The presence of glucose prevents lactose metabolism because a glucose breakdown product inhibits maintenance of the high cAMP levels necessary for formation of the CAP–cAMP complex, which in turn is required for the RNA polymerase to attach at the *lac* promoter site. Even when there is a shortage of glucose

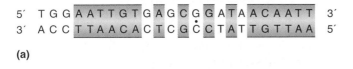

```
5′  T G G A A T T G T G A G C G G A T A A C A A T T  3′
3′  A C C T T A A C A C T C G C C T A T T G T T A A  5′
```

(a)

```
5′  G T G A G T T A G C T C A C  3′
3′  C A C T C A A T C G A G T G  5′
```

(b)

Figure 10-14 The DNA base sequences of (a) the lac operator, to which the Lac repressor binds, and (b) the CAP-binding site, to which the CAP-cAMP complex binds. Sequences exhibiting twofold rotational symmetry are indicated by the colored boxes and by a dot at the center point of symmetry. [Part a from W. Gilbert, A. Maxam, and A. Mirzabekov, in N. O. Kjeldgaard and O. Malløe, eds., *Control of Ribosome Synthesis*. Academic Press, 1976. Used by permission of Munksgaard International Publishers, Ltd., Copenhagen.]

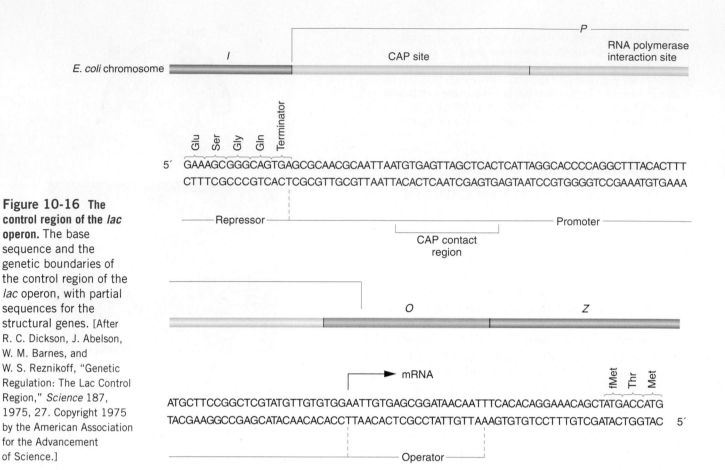

Figure 10-16 The control region of the *lac* operon. The base sequence and the genetic boundaries of the control region of the *lac* operon, with partial sequences for the structural genes. [After R. C. Dickson, J. Abelson, W. M. Barnes, and W. S. Reznikoff, "Genetic Regulation: The Lac Control Region," *Science* 187, 1975, 27. Copyright 1975 by the American Association for the Advancement of Science.]

(a) Glucose present (cAMP low); no lactose; no *lac* mRNA

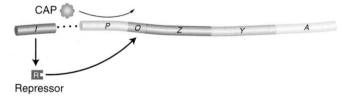

(b) Glucose present (cAMP low); lactose present

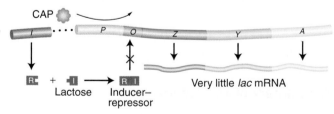

(c) No glucose present (cAMP high); lactose present

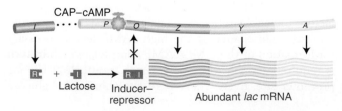

catabolites and CAP–cAMP forms, the mechanism for lactose metabolism will be implemented only if lactose is present. This level of control is accomplished because lactose must bind to the repressor protein to remove it from the operator site and permit transcription of the *lac* operon. Thus, the cell conserves its energy and resources by producing the lactose-metabolizing enzymes only when they are both needed and useful.

Inducer–repressor control of the *lac* operon is an example of repression, or **negative control,** in which expression is normally blocked. In contrast, the CAP–cAMP system is an example of activation, or **posi-**

Figure 10-17 Negative and positive control of the *lac* operon by the Lac repressor and catabolite activator protein (CAP), respectively. Large amounts of mRNA are produced only when lactose is present to inactivate the repressor, and low glucose levels promote the formation of the CAP-cAMP complex, which positively regulates transcription. [Redrawn from B. Gartenberg and D. M. Crothers, *Nature* 333, 1988, 824. (See H. N. Lie-Johnson et al., *Cell* 47, 1986, 995.) After H. Lodish, D. Baltimore, A. Berk, S. L. Zipursky, P. Matsudaira, and J. Darnell, *Molecular Cell Biology*, 3d ed. Copyright 1995 by Scientific American Books.]

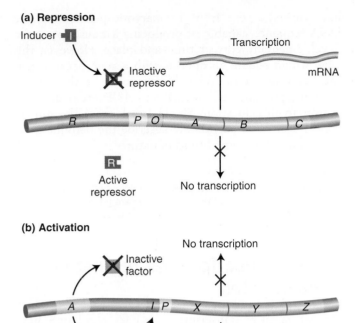

(a) Repression

(b) Activation

Figure 10-18 Comparison of repression and activation. (a) In repression, an active repressor (encoded by the *R* gene in this example) blocks expression of the *A, B, C* operon by binding to an operator site (*O*). (b) In activation, a functional activator is required for gene expression. A nonfunctional activator results in no expression of genes *X, Y, Z*. Small molecules can convert a nonfunctional activator into a functional one that then binds to the control region of the operon, termed *I* in this case. The positions of both *O* and *I* with respect to the promoter *P* in the two examples are arbitrarily drawn.

tive control, because it acts as a signal activating expression—in this case, the activating signal is the interaction of the CAP–cAMP complex with the CAP site. Figure 10-18 outlines these two basic types of control systems.

> **MESSAGE** The *lac* operon is a cluster of structural genes that specify enzymes taking part in lactose metabolism. These genes are controlled by the coordinated actions of cis-acting promoter and operator regions. The activity of these regions is, in turn, determined by repressor and activator molecules specified by separate regulator genes.

10.4 Dual positive and negative control: the arabinose operon

Like the *lac* system, prokaryotic control of transcription is often not purely positive or purely negative; rather, it seems to mix and match different aspects of positive and

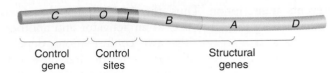

Figure 10-19 Map of the *ara* region. The *B, A,* and *D* genes together with the *I* and *O* sites constitute the *ara* operon. *I* is *araI*.

negative regulation in different ways. The regulation of the arabinose operon provides an example in which a single DNA-binding protein may act as *either* a repressor *or* an activator (Figure 10-19).

The structural genes (*araB, araA,* and *araD*) encode the metabolic enzymes that break down the sugar arabinose. The three genes are transcribed in a unit as a single mRNA. Transcription is activated at *araI*, the **initiator** region, which contains both an operator site and a promoter. The *araC* gene, which maps nearby, encodes an activator protein. When bound to arabinose, this protein activates transcription of the *ara* operon, perhaps by helping RNA polymerase bind to the promoter. In addition, the same CAP–cAMP catabolite repression system that regulates *lac* operon expression also regulates expression of the *ara* operon.

In the presence of arabinose, both the CAP–cAMP complex and the AraC–arabinose complex must bind to the operator region of *araI* in order for RNA polymerase to bind to the promoter and transcribe the *ara* operon (Figure 10-20a). In the absence of arabinose, the AraC protein assumes a different conformation and represses the *ara* operon by binding both to *araI* and to a second operator region, *araO*, thereby forming a loop (Figure 10-20b) that

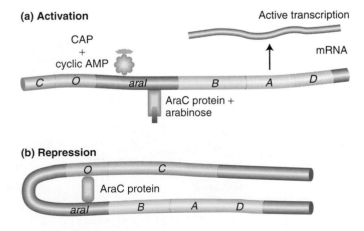

(a) Activation

(b) Repression

Figure 10-20 Dual control of the *ara* operon. (a) In the presence of arabinose, the AraC protein binds to the *araI* region. The CAP–cAMP complex binds to a site adjacent to *araI*. This binding stimulates the transcription of the *araB, araA,* and *araD* genes. (b) In the absence of arabinose, the AraC protein binds to both the *araI* and the *araO* regions, forming a DNA loop. This binding prevents transcription of the *ara* operon.

prevents transcription. Thus, the AraC protein has two conformations, one that acts as an activator and another that acts as a repressor. The two conformations, dependent on whether the allosteric effector arabinose has bound to the protein, differ in their abilities to bind a specific target site in the *ara*O region of the operon.

> **MESSAGE** Operon transcription can be regulated by both activation and repression. Operons regulating the metabolism of similar compounds, such as sugars, can be regulated in quite different ways.

10.5 Metabolic pathways

Coordinate control of genes in bacteria and bacteriophage is widespread. In the preceding section, we looked at examples illustrating the regulation of pathways for the breakdown and processing of specific sugars. In fact, most coordinated gene function in prokaryotes acts through operon mechanisms. In many pathways that synthesize essential molecules from simple inorganic building blocks, the genes that encode the enzymes are organized into operons, complete with multigenic mRNAs. Furthermore, in cases where the sequence of catalytic activity is known, there is a remarkable congruence between the sequence of operon genes on the chromosome and the sequence in which their products act in the metabolic pathway. This congruence is strikingly illustrated by the organization of the tryptophan operon in *E. coli* (Figure 10-21). Overall, the themes in prokaryotic gene regulation are similar, although there are specific variations in each system.

It is worth briefly musing about how the existence of operons affects our definition of a gene. So far, we have viewed a gene from a eukaryotic perspective as a DNA sequence capable of producing a transcript at the developmentally correct time and place. However, this view would equate an operon with a gene because it is the transcribed unit. What, then, would be the regions encoding proteins such as *Z*, *Y*, and *A*? There really is no answer to this question. Single words from the human lexicon can never adequately describe the baroque array of variants on a theme found in nature.

> **MESSAGE** In prokaryotes, genes that encode enzymes that are in the same metabolic pathways are generally organized into operons.

10.6 Transcriptional regulation in eukaryotes

On the surface, transcription and its regulation have many features that are common to both prokaryotes and eukaryotes. Like prokaryotic genes, most eukaryotic genes are controlled at the level of transcription, and some mechanisms of transcriptional regulation are very similar to those found in bacteria. For example, eukaryotes also rely on trans-acting regulatory proteins that bind to cis-acting regulatory target sequences on the DNA molecule. As in some prokaryotic genes that you have learned about, these regulatory proteins determine the level of transcription from a gene by controlling the binding of RNA polymerase to the gene's promoter.

However, the regulation of eukaryotic genes is a very active area of research, and scientists continue to discover new mechanisms. The unique features of eukaryotic gene regulation revealed thus far are the focus

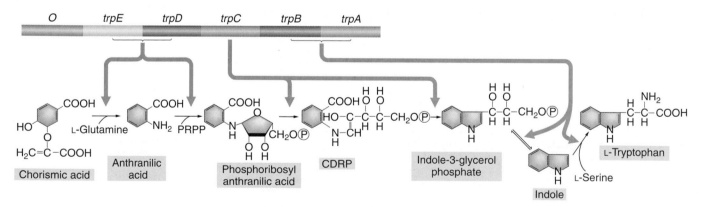

Figure 10-21 The chromosomal order of genes in the *trp* operon of *E. coli* and the sequence of reactions catalyzed by the enzyme products of the *trp* structural genes. The products of genes *trpD and trpE* form a complex that catalyzes specific steps, as do the products of genes *trpB* and *trpA*. Tryptophan synthetase is a tetrameric enzyme formed by the products of *trpB* and *trpA*. It catalyzes a two-step process leading to the formation of tryptophan. [PRPP = phosphoribosylpyrophosphate; CDRP = 1-(*o*-carboxyphenylamino)-1-deoxyribulose 5-phosphate.] [After S. Tanemura and R. H. Bauerle, *Genetics* 95, 1980, 545.]

of the rest of this chapter. Some differences have already been noted in Chapter 8:

1. In prokaryotes, all genes are transcribed into RNA by the same RNA polymerase, whereas three RNA polymerases function in eukaryotes. RNA polymerase II, which transcribes mRNAs, was the focus of Chapter 8 and will be the only polymerase discussed in this chapter.

2. RNA transcripts are extensively processed during transcription in eukaryotes; the 5′ and 3′ ends are modified and introns are spliced out.

3. RNA polymerase II is much larger and more complex than its prokaryotic counterpart. One reason for the added complexity is that RNA polymerase II must synthesize RNA *and* coordinate the special processing events unique to eukaryotes. In this chapter, you will learn that the complexity of RNA polymerase II is also a very important feature of eukaryotic gene regulation.

Eukaryotic gene regulation: an overview

Eukaryotes typically have tens of thousands of genes, one or two orders of magnitude more than the average prokaryote. In addition, patterns of eukaryotic gene expression can be extraordinarily complex. That is, there may be great variation in when a gene is on (transcribed) or off (not transcribed) and how much transcript needs to be made. For example, one gene may be transcribed only during early development and another only in the presence of a viral infection. Finally, the vast majority of the genes in a eukaryotic cell are off at any one time. On the basis of these considerations alone, eukaryotic gene regulation must be able to:

1. turn off the expression of most genes in the genome.
2. generate thousands of patterns of gene expression with a limited number of regulatory proteins.

As you will see later in the chapter, complex and ingenious mechanisms have evolved to ensure that most of the genes in a eukaryotic cell are not transcribed (called **silenced**). Before considering transcriptional silencing, we turn our attention to the second question: How are eukaryotic genes able to exhibit an enormous number and diversity of expression patterns? The key to generating so many patterns is twofold. First, when bound to the promoter by itself, the large RNA polymerase II complex can catalyze transcription only at a very low *(basal)* level. For this reason, the RNA polymerase II complex is also called the **basal transcription apparatus.** In contrast, **activated transcription,** at higher rates, requires the binding of regulatory proteins called transcription factors or activators to cis-acting elements in the DNA around the gene. The second key component for generating many diverse patterns of gene expression is **modularity** and **cooperativity**: complex patterns require many binding sites for different regulatory proteins to interact with each other and with the basal transcription apparatus. This component of gene expression diversity is called **combinatorial interaction** because different combinations of transcription factors can display unique interactions that result in distinct patterns of gene expression. To accommodate all these binding sites for all these transcription factors, the regulatory regions of many eukaryotic genes are often longer than the gene itself. Let's first look at these regulatory regions in detail.

Cis-acting regulatory elements

For RNA polymerase II to transcribe DNA into RNA at maximum speed, multiple cis-acting regulatory elements must act cooperatively. We can distinguish three classes of elements on the basis of their relative locations: (1) near the transcription-initiation site is the *promoter,* the region that binds RNA polymerase II (the binding of RNA polymerase II to the promoter was discussed in Chapter 8); (2) near the promoter are *promoter-proximal* cis-acting sequences that bind to proteins that in turn assist the binding of RNA polymerase II to its promoter; (3) additional cis-acting sequence elements can act at considerable distance, and these elements, which are independent of distance, are termed *enhancers* and *silencers.* Often, an enhancer or silencer element will act in only one or a few cell types in a multicellular eukaryote. The promoters, promoter-proximal elements, and distance-independent elements are all targets for binding by different trans-acting DNA-binding proteins.

THE PROMOTER AND PROMOTER-PROXIMAL ELEMENTS Figure 10-22 is a schematic view of the promoter and promoter-proximal sequence elements. As already mentioned, the binding of RNA polymerase II to this promoter does not produce efficient transcription by itself. Transcription is enhanced somewhat when transcription factors bind to the **promoter-proximal elements** (also called **upstream promoter elements** or **UPEs**) that are found within 100–200 bp of the transcription initiation site. One of

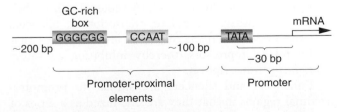

Figure 10-22 The region upstream of the transcription start site in higher eukaryotes.

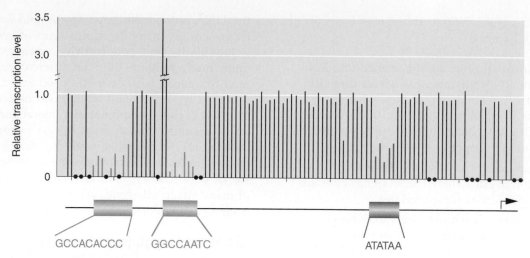

Figure 10-23 Consequences of point mutations in the promoter for the β-globin gene.
Point mutations throughout the promoter region were analyzed for their effects on
transcription rates. The height of each line represents the transcription level relative to a
wild-type promoter (1.0). Only the base substitutions that lie within the three promoter
elements change the level of transcription. Positions with black dots were not tested.
[From T. Maniatis, S. Goodbourn, and J. A. Fischer, *Science* 236, 1987, 1237.]

these UPEs is the CCAAT box, and often another is a
GC-rich segment farther upstream. The transcription
factors that bind to the promoter-proximal elements are
constitutively expressed, in all cells at all times, so that
they can activate transcription in all cell types. Mutations in these sites can have a dramatic effect on transcription, demonstrating how important they are. An example of the consequences of mutating these sequence
elements on transcription rates is shown in Figure 10-23.

DISTANCE-INDEPENDENT CIS-ACTING ELEMENTS
All of the promoter elements discussed thus far are close
to the transcription start site and, as such, are similar to
prokaryotic regulatory elements. However, unlike that in
prokaryotes, binding to proximal regulatory elements is
usually not sufficient to induce high levels of transcription in eukaryotes. Eukaryotic genes are typically expressed at high levels in only a subset of tissues or in response to a signal such as a hormone or a pathogen. In
eukaryotes, we distinguish between two classes of cis-
acting elements that can exert their effects at considerable distance from the promoter. **Enhancers** are DNA
sequences that can greatly *increase* transcription rates
from promoters on the same DNA molecule; thus, they
activate, or positively regulate, transcription. Silencers
have the opposite effect. **Silencers** are sequences that
are bound by repressors, thereby inhibiting activators
and *reducing* transcription.

Enhancers and silencers are similar to promoter-
proximal regions in that they are organized as a series of
sequences that are bound by regulatory proteins. However, they are distinguished from promoter-proximal ele-

ments by their ability to act at a distance, sometimes
50 kb or more, and by their ability to operate either upstream or downstream from the promoter that they control. Enhancer elements are intricately structured and
are themselves composed of multiple binding sites for
many trans-acting regulatory proteins. Interactions between regulatory proteins or between regulatory proteins and the RNA polymerase II complex or both determine the rate of transcription.

How can these distant enhancer and silencer elements regulate transcription? One model for such action
at a distance includes some type of DNA looping. Figure
10-24a details a DNA-looping model for activation of
the initiation complex. In this model, a DNA loop
brings together activator proteins bound to promoter-
proximal elements with activator proteins bound to distant enhancers so that they can interact and stabilize the
RNA polymerase II initiation complex bound to the
TATA box and surrounding DNA. The bending or looping of DNA requires a special class of DNA binding proteins, which have come to be known as *architectural* proteins (Figure 10-24b). A second model for long-distance
activation and silencing includes changes in chromatin
structure and will be described later in the chapter.

Transcription factors and DNA-binding domains

Transcription factors, whether bound to enhancers or to
promoter-proximal elements, need to perform at least
two functions: (1) DNA binding and (2) transcription
activation or repression. For this reason, they are usually

(a)

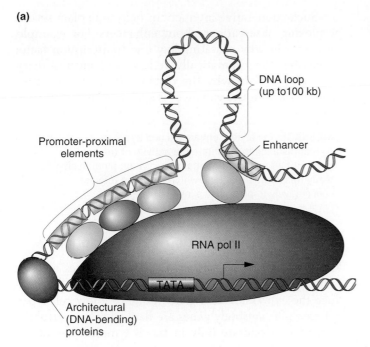

Promoter-proximal elements

DNA loop (up to100 kb)

Enhancer

RNA pol II

TATA

Architectural (DNA-bending) proteins

(b)

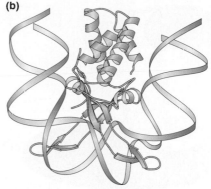

Figure 10-24 Regulation of RNA polymerase II by promoter-proximal elements. (a) DNA forms a loop to bring together an enhancer and promoter-proximal elements. Their interaction with RNA polymerase II initiates transcription. (b) Interaction of the DNA-bending protein with DNA. [Part a after B. Turner, *Chromatin and Gene Regulation.* Blackwell Science, 2001.]

composed of at least two domains, one that binds DNA and another that influences transcription by binding to another bound protein. For example, a transcription factor might have the DNA-binding domain at its amino end and the activation domain at its carboxyl end.

A few characteristic DNA-binding domains have been given descriptive names such as helix-turn-helix, zinc finger, helix-loop-helix, and leucine zipper. The helix-turn-helix domain is the best studied and is found in both prokaryotic and eukaryotic regulatory proteins. As the name indicates, it consists of at least two α helices. Most DNA-binding domains are positively charged so that they are attracted to the negatively charged DNA phosphate backbone. When they get close to the DNA, hydrogen bonding between the bases and amino acids fine-tunes the interaction. Binding domains usually fit into the major groove of the double helix. Figure 10-25 shows how the helix-turn-helix domain interacts with DNA.

DNA-binding domains are conserved from yeast to plants to humans. Recall from the discussion of molec-

ular mimicry that proteins are extremely versatile molecules that can assume a wide range of shapes and charged surfaces. In this case, a few protein domains that were able to fit snugly into the DNA double helix evolved in early life forms and have been used over and over again for a variety of functions. Thus, because of the modular construction of transcripton factors, the same DNA-binding domain in two organisms may be part of transcription factors with very different regulatory roles.

> **MESSAGE** The structures of DNA-binding proteins enables them to contact specific DNA sequences through polypeptide domains that fit into the major groove of the DNA double helix.

Cooperative interactions: the meaning of all those binding sites

The development of a complex organism requires that transcripton levels be regulated over a wide range. Think of a regulation mechanism as more like a rheostat than an on-or-off switch. In eukaryotes, transcription levels are made finely adjustable by the clustering of cis-acting elements into enhancers. Having several copies of the same or different transcription factors binding to adjacent sites (sites that are the correct distance apart) leads to an amplified, or superadditive, effect on activating transcription. When an effect is greater than additive, it is said to be **synergistic.**

The presence of multiple bound sites can catalyze the formation of an **enhanceosome,** a large protein complex that acts synergistically to activate transcription. In

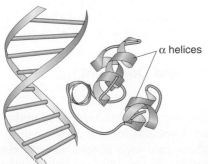

α helices

Figure 10-25 Interaction of helix-turn-helix binding domain with DNA.

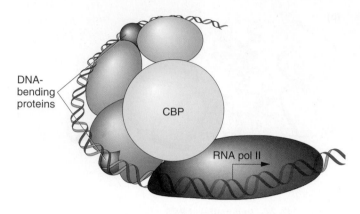

DNA-bending proteins

CBP

RNA pol II

Figure 10-26 A model for enhanceosome action.
The β-interferon enhanceosome. In this case, the transcription factors recruit a coactivator (CBP), which binds both to the transcription factors and to RNA polymerase II, initiating transcription. [After A. J. Courey, *Current Biology* 7, 2001, R250–R253, Figure 1.]

Figure 10-26 you can see how architectural proteins bend the DNA to promote cooperative interactions between the other DNA-binding proteins. In this mode of enhanceosome action, transcription is activated to very high levels only when all the proteins are present and touching each other in just the right way. To better understand what an enhanceosome is and how it acts synergistically, let's look at a specific example.

CASE STUDY: THE β-INTERFERON ENHANCEOSOME

The human β-interferon gene, which codes for the antiviral protein interferon, is one of the best-characterized genes in eukaryotes. It is normally switched off but is activated to very high levels of transcription on viral infection. The key to the activation of this gene is the assembly of transcription factors into an enhanceosome about 100 bp upstream of the TATA box and transcription start site. The regulatory proteins of the β-interferon enhanceosome all bind to the same face of the DNA double helix. Binding to the other side of the helix are several architectural proteins that bend the DNA and allow the different regulatory proteins to touch one another and form an activated complex. When all of the regulatory proteins are bound and interacting correctly, they form a "landing pad," a high-affinity binding site for a coactivator (CBP) which then activates RNA polymerase to high levels of transcription (*see* Figure 10-26). The **coactivator** is a special class of regulatory complex that serves as a bridge to bring together regulatory proteins and RNA polymerase.

Such cooperative interactions help to explain several perplexing observations about enhancers. For example, they explain why mutating any one transcription factor or binding site dramatically reduces enhancer activity. They also explain why the distance between cis-acting elements within the enhancer is such a critical feature.

> **MESSAGE** Eukaryotic enhancers and silencers can act at a great distance to modulate the activity of the basal transcription apparatus. Enhancers contain binding sites for many transcription factors, which bind and interact cooperatively to produce a synergistic response.

Tissue-specific regulation of transcription

Many enhancer elements in higher eukaryotes activate transcription in a tissue-specific manner; that is, they induce the expression of a gene in one or a few cell types. For example, antibody genes are flanked by powerful enhancers that operate only in the B lymphocytes of the immune system. An enhancer can act in a tissue-specific manner if the activator that binds to it is present in only some types of cells.

The expression of some genes can be controlled by simple sets of enhancers. For example, in *Drosophila*, vitellogenins are large egg-yolk proteins made in the female adult's ovary and fat body (an organ that is essentially the fly's liver) and transported into the developing oocyte. Two distinct enhancers regulate the vitellogenin gene. They are located within a few hundred base pairs of the promoter, one driving expression in the ovaries and the other in the fat body.

The array of enhancers for a gene can be quite complex, controlling similarly complex patterns of gene expression. For example, the *dpp* (decapentaplegic) gene in *Drosophila* encodes a protein that mediates signals between cells. The gene contains numerous enhancers, perhaps numbering in the tens or hundreds, dispersed along a 50-kb interval of DNA. Some of these enhancers are located 5′ (upstream) of the transcription initiation site of *dpp*, others are downstream of the promoter, some are in introns, and still others are 3′ of the gene's polyadenylation site. Each of these enhancers regulates the expression of *dpp* in a different site in the developing animal. Some of the better-characterized *dpp* enhancers are shown in Figure 10-27.

Regulatory Elements and Dominant Mutations

The properties of regulatory elements help us to understand certain classes of dominant mutations. We can divide dominant mutations into two general classes. For some dominant mutations, inactivation of one of the

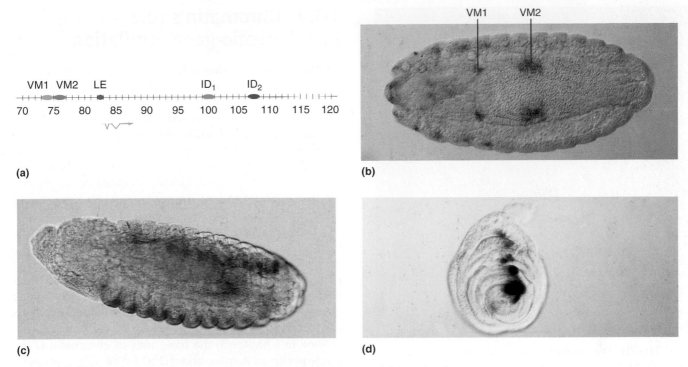

Figure 10-27 The complex tissue-specific regulation of the *dpp* gene. (a) A molecular map of the *dpp* gene over a 50-kb range. The basic transcription unit of the gene is shown below the map coordinate line. The abbreviations above the line mark the sites of tissue-specific enhancers. In parts b through d, expression from the different enhancers was examined by linking enhancers with the coding region of a gene that produces a blue stain when expressed (a reporter gene). (b) VM1 and VM2 enhancer activity detected in two parts of the embryonic visceral mesoderm, the precursor of the gut musculature. (c) LE enhancer activity detected in the lateral ectoderm of an embryo. (d) ID is one of many enhancer elements driving imaginal disk expression of *dpp*. (An imaginal disk is a flat circle of cells in the larva that gives rise to one of the adult appendages.) [Part b courtesy of D. Hursh, part c courtesy of R. W. Padgett, and part d courtesy of R. Blackman and M. Sanicola.]

two copies of a gene reduces the gene product below some critical threshold for producing a normal phenotype; we can think of these mutations as **loss-of-function dominant mutations,** resulting from haploinsufficiency of that gene. In other cases, the dominant phenotype is due to some new property of the mutant gene, not to a reduction in its normal activity; this class comprises **gain-of-function dominant mutations.**

One way that gain-of-function dominant mutations arise is through the fusion of the regulatory elements of one gene to the protein-coding or structural RNA-coding sequences of another. Such fusions can occur at the breakpoints of chromosomal rearrangements such as inversions, translocations, duplications, or deletions (see Chapter 15). For example, a chromosomal rearrangement may juxtapose enhancers of one gene and a transcription unit of another gene. In such cases, the enhancers of the gene at one breakpoint can now regulate the transcription of a gene near the other breakpoint. Often, the result is the inappropriate expression

of the mRNA encoded by the transcription unit in question.

An example of such misregulation through gene fusion is the *Tab (Transabdominal)* mutation in *Drosophila*. *Tab* causes part of the thorax of the adult fly to develop instead as tissue normally characteristic of the sixth abdominal segment (A6; Figure 10-28). *Tab* is associated with a chromosomal inversion: the chromosome breaks in two places, and the segment between the breaks is flipped 180° and reinserted between the breaks. One breakpoint of the inversion falls within an enhancer region of a different gene, the *sr (striped)* gene. The enhancers of the *sr* gene induce gene expression in certain parts of the thorax of the fly. The other breakpoint is near the transcription unit of the *Abd-B (Abdominal-B)* gene. The *Abd-B* gene encodes a transcription factor that is normally expressed only in posterior regions of the animal, and this Abd-B transcription factor is responsible for conferring an abdominal phenotype on any tissues that express it. (We

Figure 10-28 The *Tab* mutation. A wild-type male is on the left. The fly on the right is a *Tab/+* heterozygous mutant male. In the mutant fly, part of the thorax (the black tissue) is changed into tissue normally found in the dorsal part of one of the posterior abdominal segments. [From S. Celniker and E. B. Lewis, *Genes and Development* 1, 1987, 111.]

shall have more to say about genes such as *Abd-B* in the treatment of homeotic genes in Chapter 18.) In the *Tab* inversion, the *sr* enhancer elements controlling thoracic expression are juxtaposed to the *Abd-B* transcription unit, causing the *Abd-B* gene to be activated in exactly those parts of the thorax where *sr* would ordinarily be expressed (see Figure 10-28). Because of the function of the Abd-B transcription factor, its activation in these thoracic cells changes their fate to that of posterior abdomen.

Gene fusions are an extremely important source of genetic variation. Through chromosomal rearrangements, novel patterns of gene expression can be generated. In fact, we can imagine that such fusions might play an important role in the shifts of gene-expression pattern that occur in the divergence and evolution of species (discussed in Chapter 21). In addition to affecting development (discussed in Chapter 18), such mutations can play a pivotal role in the formation and progression of many cancers (discussed in Chapter 17).

> **MESSAGE** The fusion of tissue-specific enhancers to genes not normally under their control can produce dominant gain-of-function mutant phenotypes.

10.7 Chromatin's role in eukaryotic gene regulation

At this point you might be thinking that the mechanisms of eukaryotic gene regulation sound pretty much like those that you read about earlier in the part of the chapter describing prokaryotic regulation. Less than a decade ago, many scientists also thought that eukaryotic regulation was simply a complicated version of what had been discovered in prokaryotes, except that there were more regulatory proteins binding to more cis-acting elements to generate a larger number of diverse patterns of gene expression.

In the past decade, this view has changed dramatically as scientists began to consider the effect of the organization of genomic DNA in eukaryotes. The DNA of prokaryotes is essentially "naked," making it readily accessible to RNA polymerase. In contrast, eukaryotic chromosomes are organized into chromatin, which is composed of DNA and protein (mostly histones). As described in Chapter 3, the basic unit of chromatin is the nucleosome containing about 150 bp of DNA wrapped twice around a histone octamer (Figure 10-29a). The histone octamer is composed of two subunits each of the four histones: histone 2A, 2B, 3, and 4. Nucleosomes can associate into higher-order structures that further condense the DNA. As discussed in Chapter 3, chromatin is not uniform over all the chromosomes; highly condensed chromatin is called **heterochromatin,** and less condensed chromatin is called **euchromatin** (Figure 10-29b). Chromatin condensation also changes in the course of the cell cycle. The chromatin of cells entering mitosis becomes highly condensed as the chromosomes align in preparation for cell division. After cell division, regions forming heterochromatin remain condensed especially around the centromeres and telomeres (called **constitutive heterochromatin**), whereas the regions forming euchromatin becomes less condensed.

Geneticists first suspected a limited role for the influence of chromatin structure on gene regulation early in the history of genetics. At that time, it was noticed that heterochromatic DNA contained few genes, whereas euchromatin was rich in genes. But what is heterochromatin if not genes? Most of the eukaryotic genome is composed of repetitive sequences that do not make protein or structural RNA—sometimes called junk DNA. Thus, the densely packed nucleosomes of heterochromatin were said to form a "closed" structure that was inaccessible to regulatory proteins and inhospitable to gene activity. In contrast, euchromatin, with its more widely spaced nucleosomes, was proposed to assume an "open" structure that permitted transcription. The existence of open and closed regions of chromatin was also suggested as a reason for the 100- to 1000-fold

(a)

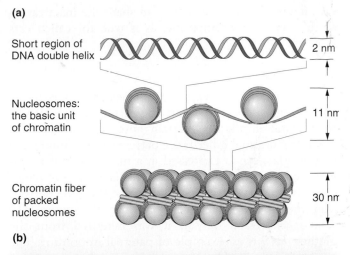

Short region of DNA double helix — 2 nm

Nucleosomes: the basic unit of chromatin — 11 nm

Chromatin fiber of packed nucleosomes — 30 nm

(b)

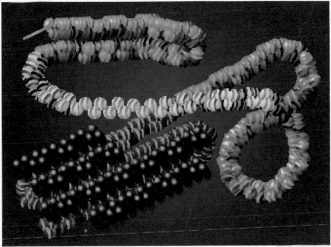

Figure 10-29 **The structure of chromatin.** (a) The nucleosome in decondensed and condensed chromatin. (b) Chromatin structure varies along the length of a chromosome. The least condensed chromatin is shown in yellow, regions of intermediate condensation in orange and blue, and heterochromatin coated with special proteins (purple) in red. [Part b from Peter J. Horn and Craig L. Peterson, "Chromatin Higher Order Folding: Wrapping Up Transcription," *Science* 297, September 13, 2002, p. 1827, Figure 3. Copyright 2002, AAAS.]

difference in recombination frequencies in euchromatin compared with heterochromatin. Euchromatin, with its more open conformation, was hypothesized to be more accessible to proteins needed for DNA recombination. Note that, in these cases, chromatin has a passive role: regions of the genome are either open or closed, and transcription and recombination are largely restricted to regions of open chromatin. However, observations of three phenomena began to change this view. These phenomena demonstrated that euchromatic chromatin could be altered and, more importantly, that active genes in these regions could become inactive. The three

phenomena are X inactivation, imprinting, and position-effect variegation.

> **MESSAGE** The chromatin of eukaryotes is not uniform. Highly condensed heterochromatic regions have fewer genes and lower recombination frequencies than do the less condensed euchromatic regions.

X inactivation in female mammals

In Chapter 15, you will learn about the effects of gene copy number on the phenotype of an organism. For now, it is sufficient to know that the number of transcripts produced by a gene is usually proportional to the number of copies of that gene in a cell. Mammals, for example, are diploid and have two copies of each gene located on their autosomes. However, as discussed in Chapter 2, the number of the X and Y sex chromosomes differs between the sexes, with female mammals having two X chromosomes to only one in males. The mammalian X chromosome is thought to contain about 1000 genes. Females have twice as many copies of these X-linked genes and would normally express twice as much transcript from these genes as males. (Not having a Y chromosome is not a problem for females, because the very few genes on this chromosome are only required for the development of males.) This dosage imbalance is corrected by a process called **dosage compensation,** which makes the amount of most gene products from the two copies of the X chromosome in females equivalent to the single dose of the X chromosome in males. This equivalency is accomplished by random inactivation of one of the two X chromosomes in each cell at an early stage in development. This inactive state is then propagated to all progeny cells. (In the germ line, the second X becomes reactivated in oogenesis). The inactivated chromosome, called a **Barr body,** can be seen in the nucleus as a darkly staining, highly condensed, heterochromatic structure.

Two aspects of X-chromosome inactivation are relevant to a discussion of chromatin and the regulation of gene expression. First, most of the genes on the inactivated X chromosome are turned off (they are said to be *silenced*). In organisms where this phenomenon is better characterized, one can see that alteration in the chromatin structure of this chromosome has silenced previously active genes. Second, genes on the inactivated chromosome remain inactive in all descendants of these cells. Such a heritable alteration, in which the DNA sequence itself is unchanged, is called **epigenetic inheritance.**

A beautiful example of X inactivation is the pattern of fur color on the calico cat. Because the same X chromosome is turned off in all descendants of a cell, large

(a)

(b)

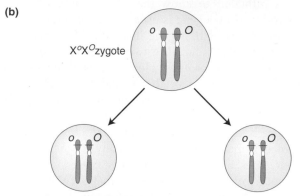

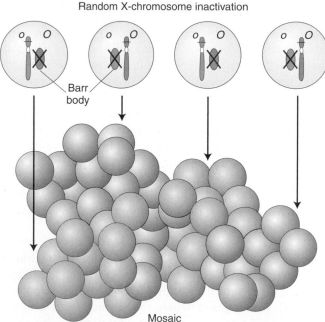

continuous sectors of a tissue can share the inactivation of the same X chromosome, as is true for calico cats (Figure 10-30).

Imprinting

Another example of epigenetic inheritance is **parental imprinting**, discovered about 15 years ago in mammals. In parental imprinting, certain autosomal genes have seemingly unusual inheritance patterns. For example, the mouse *igf2* gene is expressed in a mouse only if it was inherited from the mouse's father. This is an example of maternal imprinting because a copy of the gene derived from the mother is inactive. Conversely, the mouse *H19* gene is expressed only if it was inherited from the mother; *H19* is an example of paternal imprinting. The consequence of parental imprinting is that imprinted genes are expressed as if there were only one copy of the gene present in the cell (they are hemizygous) even though there are two. Furthermore, no changes are observed in the DNA sequences of imprinted genes. Rather, the only change that is seen is the presence of extra methyl (—CH₃) groups on certain bases of the DNA, usually on the cytosine base of a CG dinucleotide. These methyl groups are enzymatically added through the action of special methyltransferases.

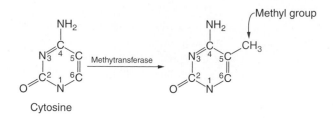

The level of methylation generally correlates with the transcriptional state of a gene: active genes are less methylated than inactive genes. We have seen that bacterial chromosomes are methylated, but this example illustrates that occasional bases of the DNA of higher organisms are methylated as well (exceptions include yeast and *Drosophila*).

Note that parental imprinting can greatly affect pedigree analysis. Because the inherited allele from one parent is inactive, a mutation in the allele inherited from the other parent will appear to be dominant, whereas in fact the allele is expressed because only one of the two homologs is active for this gene.

The results of recent studies suggest that there may be a hundred or so such parentally imprinted autosomal genes in the mammalian genome. Mutational analysis has shown that imprinted genes are important in mammalian development, and that several take part in regulating embryo growth and differentiation. As you will see in Chapter 15, in many insect species such as bees, wasps, and ants, males are able to develop by *parthenogenesis*

Figure 10-30 X inactivation produces the coat pattern of a calico cat. (a) A calico cat. (b) Calico cats are females heterozygous for the alleles *O* (which causes fur to be orange) and *o* (which causes it to be black). Inactivation of the *O*-bearing X chromosome produces a black patch expressing *o*, and inactivation of the *o*-bearing X chromosome produces an orange patch expressing *O*. The white areas are caused by a separate genetic determinant present in calicos. [Part a by Anthony Griffiths.]

(the development of a specialized type of unfertilized egg into an embryo without the need for fertilization). Because imprinted genes are required for development, you might imagine that parthenogenesis might be a problem for mammals. In fact, parthenogenesis does not occur in mammals, largely because of their having imprinted genes.

Some human genetic diseases, such as Prader-Willi syndrome (PWS), are due to parentally imprinted genes. PWS patients have short stature, mild mental retardation, and poor muscle tone, and they are compulsive eaters. In PWS patients, both copies of the gene *SNRPN*, located on human chromosome 15, are inactive. In most cases, the maternal copy is inactive owing to imprinting, whereas the paternal copy is inactive owing to a chance mutation (Figure 10-31).

Many steps are required for imprinting (Figure 10-32). Soon after fertilization, mammals set aside cells that will become their germ cells. Imprints are removed or erased before the germ cells form. Without their distinguishing "mark" (for example, methylated DNA bases), these genes are now said to be "epigenetically equivalent." As these primordial germ cells become fully formed gametes, imprinted genes receive the sex-specific mark that will determine whether the gene will be active or silent after fertilization.

In 1996, the world was shocked by the successful cloning of Dolly the sheep. Dolly's cloning was a surprise because the cloning of animals from somatic cells was thought to be impossible. However, Dolly devel-

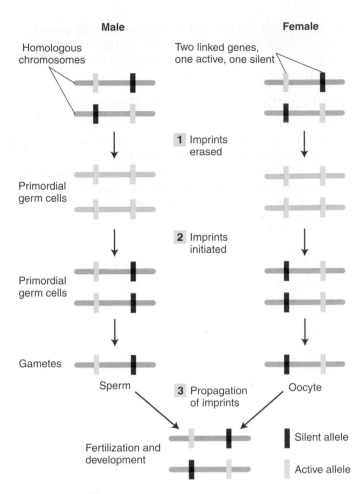

Figure 10-32 Steps required for imprinting The figure shows how two genes can be differentially imprinted in males and females.

oped from adult somatic nuclei that had been implanted into enucleated eggs (nuclei removed). More recently, cows, pigs, mice, and other mammals have been cloned as well (Figure 10-33). Although the transplanted adult nucleus is thought to undergo extensive reprogramming

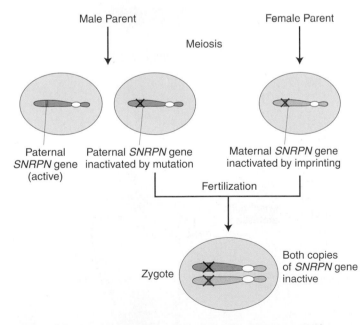

Figure 10-31 The genetic origin of Prader-Willi syndrome. The syndrome can be caused by mutation of the paternally derived *SNRPN* allele and imprinting of the maternally derived *SNRPN* allele.

Figure 10-33 Cloned calves. These five 7 month-old calves on a farm in Iowa were cloned from a single individual. [Advanced Cell Technologies/AP/Wide World Photos.]

of its genome, one of these reprogramming events is probably not the erasure of the marks in imprinted genes, because it would almost certainly have prevented normal embryo development.

Position-effect variegation in *Drosophila*

An interesting genetic phenomenon first discovered in *Drosophila* revealed that epigenetic signals are able to spread from one chromosomal region to adjacent DNA. In these experiments, flies were irradiated with X rays to induce mutations in their germ cells. The progeny of the irradiated flies were screened for unusual phenotypes. A mutation in the *white* gene, near the tip of the X chromosome, will result in progeny with white eyes instead of the wild-type red color. Some of the progeny had very unusual eyes with patches of white and red color. Cytological examination revealed a chromosomal rearrangement in the mutant flies: present in the X chromosome was an inversion of a piece of the chromosome carrying the *white* gene (Figure 10-34). Inversions and other chromosomal rearrangements will be discussed in Chapter 15. In this rearrangement, the *white* gene, which is normally located in a euchromatic region of the X chromosome, now finds itself near the heterochromatic centromere. In some cells, the heterochromatin "spreads" to the neighboring euchromatin and silences the *white* gene. Patches of white tissue

in the eye are derived from the descendants of a single cell in which the *white* gene has been **epigenetically silenced** and remains silenced through future cell divisions. In contrast, the red patches arise from cells in which heterochromatin has not spread to the *white* gene, and so this gene remains active in all its descendants.

Findings from subsequent studies in *Drosophila* and yeast demonstrated that many active genes are silenced in this mosaic fashion when they are relocated to neighborhoods (near centromeres or telomeres) that are heterochromatic. Thus, the ability of heterochromatin to spread into euchromatin and silence genes in the euchromatin is a feature common to many organisms. This phenomenon has been called **position-effect variegation,** or **PEV.** Position-effect variegation provides powerful evidence that chromatin structure is able to regulate the expression of genes—in this case, determining whether genes with identical DNA sequence will be active or silenced.

> **MESSAGE** A heritable alteration, in which the DNA sequence itself is unchanged, is called epigenetic inheritance.

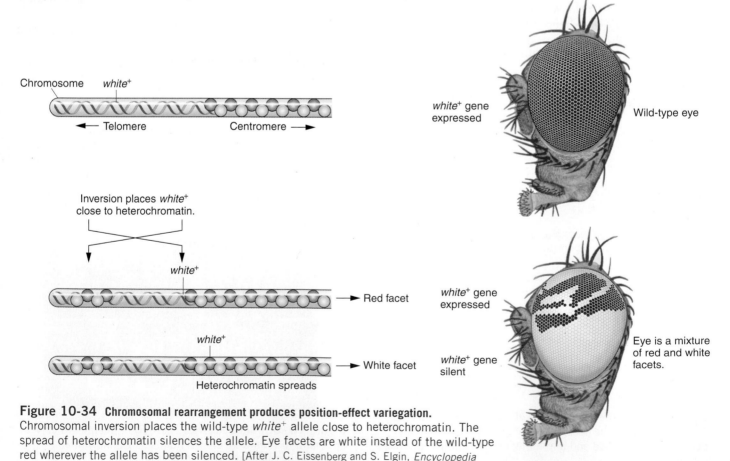

Figure 10-34 Chromosomal rearrangement produces position-effect variegation.
Chromosomal inversion places the wild-type *white*⁺ allele close to heterochromatin. The spread of heterochromatin silences the allele. Eye facets are white instead of the wild-type red wherever the allele has been silenced. [After J. C. Eissenberg and S. Elgin, *Encyclopedia of Life Sciences.* Nature Publishing Group, 2001, p. 3, Figure 1.]

Chromatin remodeling

Imprinting, X inactivation, and position-effect variegation demonstrate that gene expression can be reduced or silenced without changing the DNA sequence of the gene. Furthermore, the existence of these phenomena imply that nucleosome organization is dynamic—that it can respond to changes in cellular metabolism or developmental programs by condensing and decondensing and, in this way, contribute to turning genes on and off.

Our understanding of chromatin structure and its effect on gene expression is continuously evolving as new findings in the laboratory lead to modifications of existing concepts and definitions. For example, although there is a clear difference between euchromatin and constitutive heterochromatin (the extremely condensed genomic regions near the centromeres and telomeres), the precise chromatin state necessary to silence genes is not at all clear. When a gene is silenced, does euchromatin become heterochromatin? Probably not. In actuality there are probably many intermediate states of chromatin condensation between euchromatin and constitutive heterochromatin. The term *facultative heterochromatin* has been used to describe euchromatic regions that can assume a more condensed chromatin structure. Here we will use the term **silent chromatin** to describe these dynamic regions of euchromatin in the condensed state that leads to gene silencing.

One can imagine several ways to alter chromatin structure. For example, one mechanism might be to simply move the nucleosomes along the DNA. In the 1980s, biochemical techniques were developed that allowed researchers to determine the position of nucleosomes in and around genes. In these studies, chromatin was isolated from tissue where a gene was on and compared with chromatin from tissue where the same gene was off. The result for most genes analyzed was that nucleosome positions changed, especially in their regulatory regions. Thus, nucleosomes are dynamic: their positions can shift in the life of an organism. Transcription might be repressed when the TATA box and flanking sequences are wound up in a nucleosome and unable to bind the RNA polymerase II complex. Activation of transcription would require the blocking nucleosome to be removed or moved elsewhere. This change in nucleosome position came to be known as **chromatin remodeling.** Once chromatin remodeling was known to be an integral part of eukaryotic gene expression, the race was on to determine the underlying mechanism(s) and the regulatory proteins responsible.

GENETIC SCREENS IDENTIFY REMODELING PROTEINS Genetic analysis of the model organism yeast has been essential in dissecting the components of eukaryotic gene regulation. As a unicellular eukaryotic organism with a very short generation time, yeast is ideal for mutant isolation and analysis (see the Model Organism box). In this case, two genetic screens for mutants in seemingly unrelated processes led to the same gene. In one screen, mutations were introduced into yeast-cell genes subsequent to their exposure to chemicals that damage DNA. These mutagenized yeast cells were then screened for cells that could not grow well on sucrose (sugar *nonfermenting mutants, snf*). In another screen, mutagenized yeast cells were screened for mutants that were defective in switching their mating type (*switch mutants, swi*) (see the Model Organism box). Many mutants were recovered from both screens, but one mutant gene was found to cause both phenotypes. That is, mutants with a mutation at the so-called *swi-snf* locus (called switch-sniff) could neither utilize sucrose effectively nor switch mating type.

What was the connection between the ability to utilize a sugar and the ability to switch mating types? The SWI-SNF protein was purified and shown to be part of a large, multisubunit complex that could reposition nucleosomes in a test tube assay if ATP was provided as an energy source. As we will see later in the chapter, in some situations, the multisubunit SWI-SNF complex activates transcription by moving nucleosomes that are covering the TATA sequence and, in this way, facilitating the binding of RNA polymerase II (Figure 10-35). Apparently, in

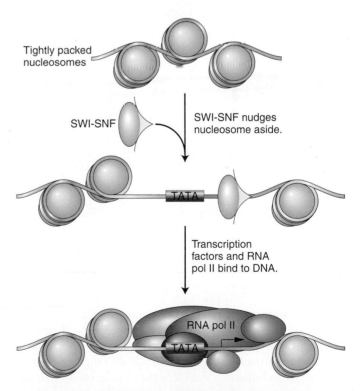

Figure 10-35 Chromatin remodeling. The movement of nucleosomes in response to SWI-SNF activity is shown (see Figure 10–38 for details on how SWI-SNF is recruited to a particular DNA region). In this example, chromatin remodeling has exposed the TATA box, thereby facilitating the binding of the RNA polymerase II complex.

MODEL ORGANISM   Yeast

Saccharomyces cerevisiae, or budding yeast, has emerged in recent years as the premier eukaryotic genetic system. Humans have grown yeast for centuries because it is an essential component of beer, bread, and wine. Yeast has many features that make it an ideal model organism. As a unicellular eukaryote, it can be grown on agar plates and, with a life cycle of just 90 minutes, large quantities can be cultured in liquid media. It has a very compact genome with only about 12 megabase pairs of DNA (compared with 2500 megabase pairs for humans) containing 6000 genes that are distributed on 16 chromosomes. It was the first eukaryote to have its genome sequenced.

The yeast life cycle makes it very versatile for laboratory studies. Cells can be grown as either diploid or haploid. In both cases, the mother cell produces a bud containing an identical daughter cell. Diploid cells either continue to grow by budding or undergo meiosis, which produces four haploid spores held together in an *ascus* (also called a *tetrad*). Haploid spores of opposite mating type (**a** or **α**) will go through fertilization

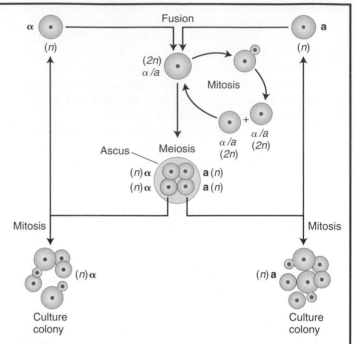

The life cycle of baker's yeast. The nuclear alleles *a* and *α* determine mating type

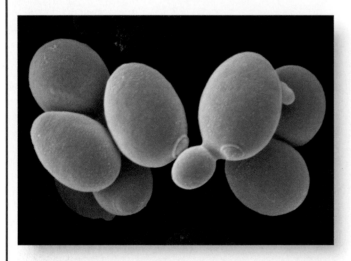

Electron micrograph of budding yeast cells. [SciMAT/Photo Researchers.]

and form a diploid. Spores of the same mating type will continue growth by budding.

Yeast has been called the *E. coli* of eukaryotes because of the ease of mutant analysis. To isolate mutants, haploid cells are mutagenized (with X rays, for example) and screened on plates for mutant phenotypes. This procedure is usually done by first plating cells on rich media where all cells grow and copying, or *replica plating*, the colonies from this master plate onto replica plates containing selective media or special growth conditions. For example, temperature-sensitive mutants will grow on the master plate at the permissive temperature but not on a replica plate at the restrictive temperature. Comparison of the colonies on the master and replica plates will reveal the temperature-sensitive mutants.

a SWI-SNF mutant, some of the genes required for sugar utilization and for switching mating types could not be transcriptionally activated. The fact that SWI-SNF-like proteins have been isolated from a wide range of taxa—from yeast to human—indicates that this protein is an integral component of eukaryotic gene regulation.

Histones and chromatin remodeling

Protein complexes such as SWI-SNF that remodel chromatin by repositioning nucleosomes turned out to be one part of the machinery necessary to change the organization of chromatin. Recall that chromatin is made up of as much protein as DNA. Because the identical DNA sequence can be packaged into different forms of chromatin, it was reasoned that alterations in chromatin proteins in some way determine whether a region of DNA has its nucleosomes widely spaced or tightly packaged. Although chromatin is made up of many different kinds of proteins, the results of genetic experiments led scientists to suspect that the histone proteins were the key to changes in chromatin structure.

HISTONE DEPLETION IN YEAST ALTERS GENE EXPRESSION In addition to its other attributes as a model genetic organism, yeast has only two copies each of the genes encoding the core histones, making them suitable for mutational analysis. In contrast, higher eukaryotes can have hundreds of copies of histone genes.

In the late 1980s, experiments were carried out to determine if a severe reduction in histone 4 (H4) led to changes in chromatin structure and gene expression. The experimental design called for the use of yeast strains carrying mutations in both H4 genes. Furthermore, a copy of an H4 gene had been "engineered" in a test tube so that the experimenter could control the level of gene expression. Normally histone genes are constitutively transcribed; that is, they are on all the time. To control the level of H4 protein, the scientists constructed a chimeric gene by combining in a test tube the H4 coding sequence from the H4 gene with the promoter taken from another gene. They then put this artificial gene (called a gene construct) into the yeast cells carrying mutations in both H4 genes. The chimeric gene was inserted into a small extra DNA molecule (a plasmid) so that it could be maintained extrachromosomally in the nucleus. Such recombinant DNA techniques will be described in more detail in Chapter 11.

Figure 10-36 summarizes the results of the experiment. The gene construct contains the H4 coding sequence connected (*fused*) to the promoter taken from another gene that could be turned on by growing yeast in the sugar galactose and almost completely turned off by growing yeast in glucose without galactose. Thus, strains grown in galactose had normal levels of H4 be-cause the chimeric gene was transcribed at high levels, whereas strains grown in glucose had severely reduced H4 levels because the chimeric gene was transcribed at extremely low levels. Analysis of these strains showed that strains with lower histone 4 levels also displayed altered chromatin structure—in this case, less tightly packed nucleosomes. This outcome is expected because the histone octamer cannot assemble without H4, one of its major components. Most important, the experimenters found that gene expression was altered in these strains in a very interesting way. The expression of constitutively expressed genes was the same regardless of whether the strains were grown in glucose or galactose. However, inducible genes (genes that are normally repressed unless an inducer molecule is present in the medium) were constitutively expressed when H4 levels were low. From this result, it was concluded that nucleosome density had no influence on the expression of genes that were always active. In contrast, chromatin structure—specifically, nucleosome density—is important for the repression of inducible gene expression. The results also suggested that histones could influence chromatin structure.

> **MESSAGE** Mutational analysis of yeast histones and chromatin remodeling proteins has shown that nucleosomes are an integral part of eukaryotic gene regulation.

A HISTONE CODE Let's look at the nucleosome more closely to see if any part of this structure could carry the information necessary to determine the extent of chromatin condensation.

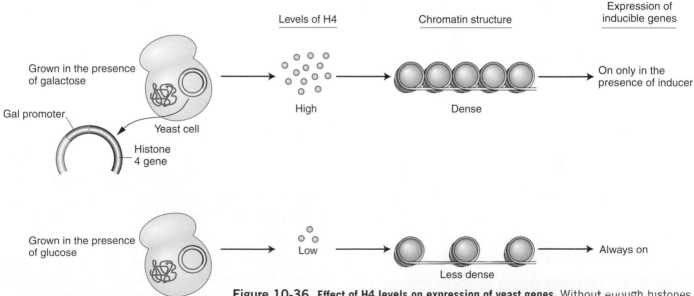

Figure 10-36 Effect of H4 levels on expression of yeast genes. Without enough histones present to synthesize nucleosomes, genes normally repressed in the presence of glucose will remain switched on.

As you have already seen, most nucleosomes are composed of an octamer made up of two copies each of the four core histones. Histones are known to be the most conserved proteins in nature; that is, histones are almost identical in all eukaryotic organisms from yeast to plants to animals. This conservation contributed to the view that histones could not take part in anything more complicated than the packaging of DNA to fit in the nucleus. However, recall that DNA with its four bases also was considered "too dumb" a molecule to carry the blueprint for all organisms on earth. As we saw in Chapter 9, four bases is sufficient to specify all 20 amino acids.

Figure 10-37 shows a model of nucleosome structure that represents contributions from many studies. Of note is that the histone proteins are organized into the core octamer with their N-terminal ends protruding from the nucleosome. These protruding ends are called **histone tails.** Since the early 1960s, it has been known that specific lysine residues in the histone tails can be covalently modified by the attachment of acetyl and methyl groups. These reactions take place after the histone protein has been translated and even after the histone has been incorporated into a nucleosome.

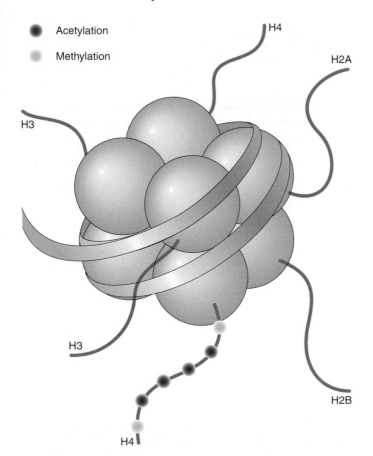

Figure 10-37 Nucleosome structure showing the protruding histone tails. The sites of posttranslational modifications such as acetylation and methylation are shown for one histone tail. In fact, all the tails contain such sites.

The acetylation reaction is the best-characterized histone modification:

Note that the reaction is reversible, which means that acetyl groups can be added and removed from the same histone residue. With 44 histone lysine residues available to accept acetyl groups, the presence or absence of these groups can carry a tremendous amount of information. For this reason, the covalent modification of histone tails is said to be a **histone code.** Scientists coined the term histone code because the covalent modification of histone tails is reminiscent of the genetic code. For the histone code, information is stored in the patterns of histone modification rather than in the sequence of nucleotides. However, before scientists broke the genetic code, they already knew that it determined the protein sequence and that tRNA molecules functioned as the adapters. In contrast, it was not clear how (or if) modified residues in histone tails could influence chromatin structure and control gene expression.

HISTONE ACETYLATION AND GENE EXPRESSION Evidence had been accumulating for years that the histones associated with the nucleosomes of active genes were rich in acetyl groups (said to be **hyperacetylated**), whereas inactive genes were underacetylated **(hypoacetylated).** For example, the inactive mammalian X chromosome is hypoacetylated.

The enzyme responsible for adding acetyl groups, histone acetyltransferase (HAT), proved very difficult to isolate. When it was finally isolated and its protein sequence deduced, it was found to be very similar to a transcriptional activator called GCN5 that is required for activation of a subset of yeast genes. Thus, it was concluded that GCN5 bound to the DNA in the regulatory regions of some genes and activated transcription by acetylating nearby histones. Similarly, the enzyme that removes acetyl groups, histone deacetyltransferase (HDAC), was found to be related to the yeast transcriptional repressor, Rpd3. In this case, the binding of Rpd3 to a gene's regulatory regions was proposed to repress transcription through the deacetylation of nearby histones.

Recall that transcription factors contain two or more domains. As discussed earlier in this chapter, one domain is for DNA binding and the other performs the biological activity of activating or repressing transcription. The discovery that transcription activators and repressors contain domains related to HAT and HDAC, respectively, revealed that some transcription factors function by altering the histone code. But how do alterations in the histone code lead to changes in chromatin structure? Proteins, such as transcription factors, that bind DNA are now thought to recognize not only a specific DNA sequence but also a specific histone code. Thus, acetylated histones may serve to recruit remodeling complexes that decondense chromatin while deacetylated histones recruit other protein complexes that further condense the nucleosomes and produce silent chromatin.

Studies into the meaning of the histone code are still in their infancy. As such, it is still not clear how patterns of histone modifications may result in chromatin alterations. Furthermore, unlike the genetic code, it is unlikely that the histone code will be universal (the same for all organisms).

> **MESSAGE** The histone code refers to the pattern of posttranslational modifications of amino acids in the histone tails. Some transcription factors have a domain that modifies nearby histone tails to effect a change in local chromatin structure.

The β-interferon gene revisited

The enhanceosome of the β-interferon gene was introduced earlier in the chapter (see Figure 10-26). For simplicity, the β-interferon promoter was shown without the nucleosomes in order to focus on the cooperative and synergistic interactions among the transcription factors. It is now known that the enhanceosome is nucleosome-free but is surrounded by two nucleosomes, called nuc I and nuc II in Figure 10-38. One of them, nuc II, is strategically positioned over the TATA box and transcription start site. As discussed before, the transcription factors interact to form a high-affinity binding site for the coactivator, CBP. However, GCN5 binding is now known to actually precede CBP binding. As stated earlier, GCN5 encodes histone acetylatase activity, which acetylates histones on nuc II. This acetylation is followed by recruitment of the coactivator CBP, the RNA pol II holoenzyme, and the SWI-SNF chromatin remodeling complex. SWI-SNF is now positioned to nudge the nucleosome 37 bp off the TATA box, making the TATA box accessible to the TATA-binding protein and allowing transcription to be initiated.

The inheritance of chromatin states

Epigenetic inheritance can now be defined operationally as the inheritance of domains of chromatin from one cell generation to the next. What this inheritance means is that, during DNA replication, both the DNA sequence and the chromatin structure are faithfully passed on to the next cell generation. However, unlike the sequence of DNA, chromatin structure can change in the course of the cell cycle when, for example, transcription factors modify the histone code, causing local changes in nucleosome position.

As mentioned in Chapter 7, the replisome not only copies the parental strands but also disassembles the nucleosomes in the parental strands and reassembles them in both the parental and the daughter strands. This is thought to be accomplished by the random distribution of the old histones from existing nucleosomes to daughter strands

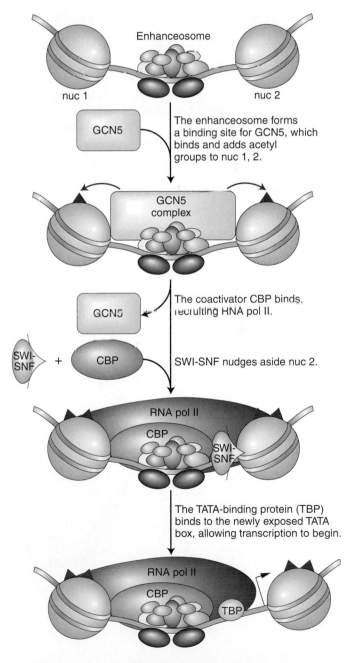

Figure 10-38 Action of the enhanceosome on nucleosomes. The β-interferon enhanceosome acts to move nucleosomes by recruiting the SWI-SNF complex.

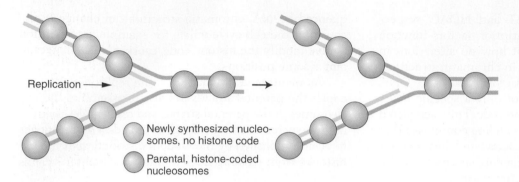

Figure 10-39 The inheritance of chromatin states. During replication, old nucleosomes with their histone codes (purple) are distributed randomly to the daughter strands, where they direct the coding of adjacent newly assembled histones (pink).

and the delivery of new histones to the replisome. In this way, the old histones with their modified tails and the new histones with unmodified tails are assembled into nucleosomes that become associated with both daughter strands. The code carried by the old histones most likely guides the modification of the new histones and the reconstitution of the local chromatin structure that existed prior to DNA synthesis and mitosis (Figure 10-39).

Position-effect variegation revisited

Recall that a variegated eye-color phenotype in *Drosophila* was shown to be due to chromosomal rearrangements that moved the *white* gene next to the heterochromatic DNA (see Figure 10-34). Evidently, heterochromatin can spread to adjacent euchromatic regions and silence *white* gene expression. To find out what proteins might be implicated in the spreading of

heterochromatin, geneticists isolated mutations at a second site that either suppressed or enhanced the variegated pattern (Figure 10-40). Suppressors of variegation [called *Su(var)*] are genes that, when mutated, reduce the spread of heterochromatin, meaning that the wild-type product of these genes is required for spreading. Among more than 50 *Drosophila* gene products identified by these screens was heterochromatin protein-1

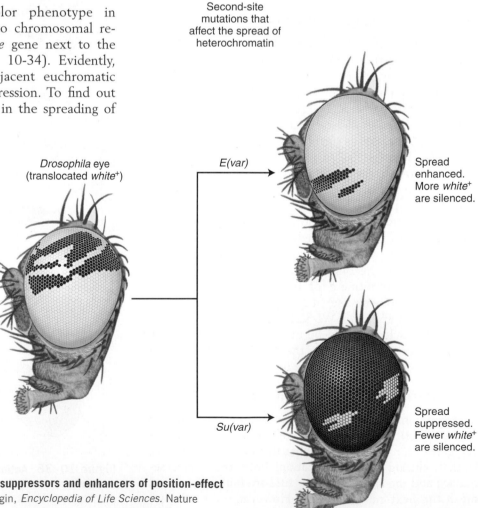

Figure 10-40 Phenotypes produced by suppressors and enhancers of position-effect variegation. [After J. C. Eissenberg and S. Elgin, *Encyclopedia of Life Sciences.* Nature Publishing Group, 2001, p. 3, Figure 1.]

(HP-1), which had previously been found associated with the heterochromatic telomeres and centromeres. Thus it makes sense that a mutation in the HP-1 gene will show up as a *Su(var)* allele because the protein is required in some way to produce or maintain heterochromatin. Proteins similar to HP-1 have been isolated in diverse taxa, suggesting the conservation of an important eukaryotic function.

KEY QUESTIONS REVISITED

- **In what ways do the expression levels of genes vary (that is, in the output of their RNA and protein products)?**

Transcription levels for individual genes vary in time and position within a developing organism. Regulation mechanisms ensure that the specific gene products needed at each stage are produced in a timely manner.

- **What factors in a cell or its environment trigger changes in a gene's level of expression?**

Such factors are numerous. Bacteria need to be able to respond to the presence or absence of specific nutrients in the medium. In eukaryotes, agents such as hormones signal between cells to activate or deactivate specific sets of genes.

- **What are the molecular mechanisms of gene regulation in prokaryotes and eukaryotes?**

Bacteria use regulatory proteins that bind near the promoter to modulate the binding of RNA polymerase. Some of these proteins increase, whereas others decrease, the polymerase's rate of activity.

Like prokaryotes, eukaryotes use regulatory proteins, but the number and complexity of the interactions of these proteins with the promoter region are much greater. Eukaryotes also use epigenetic processes to modulate transcription. Chromatin condensation is one such effect that is influenced in many ways, including the modification of DNA by methylation and the acetylation of histone tails.

- **How does a eukaryote achieve thousands of patterns of gene interaction with a limited number of regulatory proteins?**

Multiple factors interact cooperatively in different combinations to effect different biological responses. The activity of a complex is greater than the additive effects of each component, and so the effects are synergistic.

- **How does epigenetic inheritance differ from genetic inheritance?**

Epigenetic inheritance is the inheritance of chromatin structure, whereas genetic inheritance is the inheritance of DNA sequence.

- **What are two ways in which to change chromatin structure?**

By changing the histone code and through the activity of chromatin remodeling complexes such as SWI-SNF. These mechanisms are probably not independent. That is, changes in the histone code may be necessary to either directly or indirectly bind the remodeling complexes that actually do the "heavy lifting" of moving nucleosomes.

SUMMARY

Gene regulation is often mediated by proteins that react to environmental signals by raising or lowering the transcription rates of specific genes. The logic of this regulation is straightforward. In order for regulation to operate appropriately, the regulatory proteins would have to have built-in sensors that continually monitor cellular conditions. The activities of these proteins would then depend on the right set of environmental conditions.

In prokaryotes, the control of several structural genes may be coordinated by clustering the genes together into operons on the chromosome so that they are transcribed into multigenic mRNAs. Coordinated control simplifies the task for bacteria because one cluster of regulatory sites per operon is sufficient to regulate the expression of all the operon's genes.

In negative regulatory control, a repressor protein blocks transcription by binding to DNA at the operator site. Negative regulatory control is exemplified by the *lac* system. Negative regulation is one very straightforward way for the *lac* system to shut down genes in the absence of appropriate sugars in the environment. In positive regulatory control, protein factors are required to activate transcription. Some prokaryotic gene control, such as that for catabolite repression, and many eukaryotic gene regulatory events operate through positive gene control.

Many regulatory proteins are members of families of proteins that have very similar DNA-binding motifs or some other structural feature. Other parts of the proteins, such as their protein–protein interaction domains, tend to be less similar.

Many aspects of eukaryotic gene regulation resemble the regulation of prokaryotic operons. Both operate largely at the level of transcription and both rely on trans-acting proteins that bind to cis-acting regulatory target sequences on the DNA molecule. These regulatory proteins determine the level of transcription from a gene by controlling the binding of RNA polymerase to the gene's promoter. As complex organisms, multicellular eukaryotes must generate thousands of patterns of gene expression with a limited number of regulatory proteins (transcription factors). They do so through combinatorial interactions among transcription factors. Enhanceosomes are complexes of regulatory proteins that interact in a cooperative and synergistic fashion to promote high levels of transcription through the recruitment of RNA polymerase II to the transcription start site.

The vast majority of the tens of thousands of genes in a typical eukaryotic genome are turned off at any one time. Genes are maintained in a transcriptionally inactive state through the condensation of nucleosomes, which serves to compact the chromatin and prevent the binding of RNA polymerase II. The extent of chromatin condensation is instructed by the histone code, the pattern of posttranslational modifications of the histone tails. The histone code can be altered by transcription factors, which bind to regulatory regions and enzymatically modify adjacent nucleosomes. Chromatin remodeling is accomplished by large multisubunit protein complexes that use the energy of ATP hydrolysis to move or relocate nucleosomes.

DNA replication faithfully copies both the DNA sequence and the chromatin structure from parent to daughter cells. Newly formed cells inherit both genetic information, inherent in the nucleotide sequence of DNA, and epigenetic information, which is thought to be written at least in part in the histone code.

KEY TERMS

activated transcription (p. 317)

activator (p. 304)

allosteric effector (p. 304)

allosteric site (p. 304)

allosteric transition (p. 306)

Barr body (p. 323)

basal transcription apparatus (p. 317)

CAP (catabolite activator protein) (p. 312)

catabolite repression (p. 312)

chromatin remodeling (p. 327)

cis-acting (p. 308)

coactivator (p. 320)

combinatorial interaction (p. 317)

constitutive (p. 308)

constitutive heterochromatin (p. 322)

cooperativity (p. 317)

coordinately controlled genes (p. 307)

cyclic adenosine monophosphate (cAMP) (p. 312)

DNA-binding domain (p. 304)

dosage compensation (p. 323)

enhanceosome (p. 319)

enhancer (p. 318)

epigenetic inheritance (p. 323)

epigenetically silenced (p. 334)

euchromatin (p. 322)

gain-of-function dominant mutation (p. 321)

gene regulation (p. 302)

heterochromatin (p. 322)

histone code (p. 330)

histone tail (p. 330)

hyperacetylated (p. 330)

hypoacetylated (p. 330)

inducer (p. 306)

induction (p. 306)

initiator (p. 315)

loss-of-function dominant mutation (p. 321)

modularity (p. 317)

negative control (p. 314)

operator (p. 304)

operon (p. 306)

parental imprinting (p. 324)

position-effect variegation (PEV) (p. 326)

positive control (p. 314)

promoter (p. 304)

promoter-proximal element (p. 317)

repressor (p. 304)

silenced gene (p. 317)

silencer (p. 318)

silent chromatin (p. 327)

synergistic effect (p. 319)

trans-acting (p. 309)

upstream promoter element (UPE) (p. 317)

SOLVED PROBLEMS

This set of four solved problems, which are similar to Problem 4 in the Basic Problems at the end of this chapter, is designed to test understanding of the operon model. Here, we are given several diploids and are asked to determine whether Z and Y gene products are made in the presence or absence of an inducer. Use a table similar to the one in Problem 4 as a basis for your answers, except that the column headings will be as follows:

	Z gene		Y gene	
Genotype	No inducer	Inducer	No inducer	Inducer

1. $\dfrac{I^-P^-O^CZ^+Y^+}{I^+P^+O^+Z^-Y^-}$

Solution

One way to approach these problems is first to consider each chromosome separately and then to construct a diagram. The following illustration diagrams this diploid:

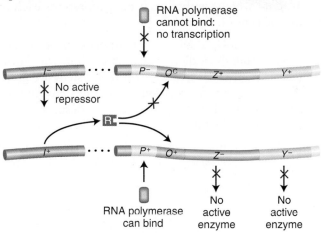

The first chromosome is P^-, and so transcription is blocked and no Lac enzyme can be synthesized from it. The second chromosome (P^+) can be transcribed, and thus transcription is repressible (O^+). However, the structural genes linked to the good promoter are defective; thus, no active Z product or Y product can be generated. The symbols to add to your table are "$-, -, -, -$."

2. $\dfrac{I^+P^-O^+Z^+Y^+}{I^-P^+O^+Z^+Y^-}$

Solution

The first chromosome is P^-, and so no enzyme can be synthesized from it. The second chromosome is O^+, and so transcription is repressed by the repressor supplied from the first chromosome, which can act in trans through the cytoplasm. However, only the Z gene from this chromosome is intact. Therefore, in the absence of an inducer, no enzyme is made; in the presence of an inducer, only the Z gene product, β-galactosidase, is generated. The symbols to add to the table are "$-, +, -, -$."

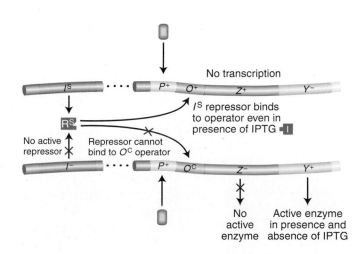

3. $\dfrac{I^+P^+O^CZ^-Y^+}{I^+P^-O^+Z^+Y^-}$

Solution

Because the second chromosome is P^-, we need consider only the first chromosome. This chromosome is O^C, and so enzyme is made in the absence of an inducer, although, because of the Z^- mutation, only active permease is generated. The entries in the table should be "$-, -, +, +$."

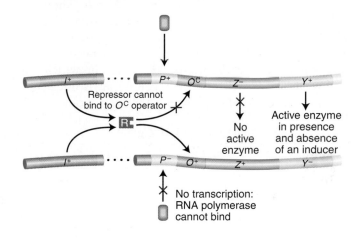

4. $\dfrac{I^SP^+O^+Z^+Y^-}{I^-P^+O^CZ^-Y^+}$

Solution

In the presence of an I^S repressor, all wild-type operators are shut off, both with and without an inducer. Therefore, the first chromosome is unable to produce any enzyme. However, the second chromosome has an altered (O^C) operator and can produce enzyme in both the absence and the presence of an inducer. Only the Y gene is wild type on the O^C chromosome, and so only permease is produced constitutively. The entries in the table should be "$-, -, +, +$."

PROBLEMS

BASIC PROBLEMS

1. Explain why I^- mutations in the *lac* system are normally recessive to I^+ mutations and why I^+ mutations are recessive to I^S mutations.

2. What do we mean when we say that O^C mutations in the *lac* system are cis-acting?

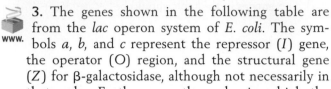

3. The genes shown in the following table are from the *lac* operon system of *E. coli*. The symbols *a*, *b*, and *c* represent the repressor (*I*) gene, the operator (*O*) region, and the structural gene (*Z*) for β-galactosidase, although not necessarily in that order. Furthermore, the order in which the symbols are written in the genotypes is not necessarily the actual sequence in the *lac* operon.

Activity (+) or inactivity (−) of Z gene

Genotype	Inducer absent	Inducer present
$a^- b^+ c^+$	+	+
$a^+ b^+ c^-$	+	+
$a^+ b^- c^-$	−	−
$a^+ b^- c^+/a^- b^+ c^-$	+	+
$a^+ b^+ c^+/a^- b^- c^-$	−	+
$a^+ b^+ c^-/a^- b^- c^+$	−	+
$a^- b^+ c^+/a^+ b^- c^-$	+	+

a. State which symbol (*a*, *b*, or *c*) represents each of the *lac* genes I, O, and Z.

b. In the table, a superscript minus sign on a gene symbol merely indicates a mutant, but you know that some mutant behaviors in this system are given special mutant designations. Use the conventional gene symbols for the *lac* operon to designate each genotype in the table.

(Problem 3 is from J. Kuspira and G. W. Walker, *Genetics: Questions and Problems*. Copyright 1973 by McGraw-Hill.)

4. The map of the *lac* operon is

POZY

The promoter (*P*) region is the start site of transcription through the binding of the RNA polymerase molecule before actual mRNA production. Mutationally altered promoters (P^-) apparently cannot bind the RNA polymerase molecule. Certain predictions can be made about the effect of P^- mutations. Use your predictions and your knowledge of the lactose system to complete the following table. Insert a "+" where an enzyme is produced and a "−" where no enzyme is produced. The first one has been done as an example.

	β-Galactosidase		Permease	
Genotype	No lactose	Lactose	No lactose	Lactose
$I^+ P^+ O^+ Z^+ Y^+/I^+ P^+ O^+ Z^+ Y^+$	−	+	−	+
a. $I^- P^+ O^C Z^+ Y^-/I^+ P^+ O^+ Z^- Y^+$				
b. $I^+ P^- O^C Z^- Y^+/I^- P^+ O^C Z^+ Y^-$				
c. $I^S P^+ O^+ Z^+ Y^-/I^+ P^+ O^+ Z^- Y^+$				
d. $I^S P^+ O^+ Z^+ Y^+/I^- P^+ O^+ Z^+ Y^+$				
e. $I^- P^+ O^C Z^+ Y^-/I^- P^+ O^+ Z^- Y^+$				
f. $I^- P^- O^+ Z^+ Y^+/I^- P^+ O^C Z^+ Y^-$				
g. $I^+ P^+ O^+ Z^- Y^+/I^- P^+ O^+ Z^+ Y^-$				

5. Explain the fundamental differences between negative control and positive control.

6. Mutants that are *lacY$^-$* retain the capacity to synthesize β-galactosidase. However, even though the *lacI* gene is still intact, β-galactosidase can no longer be induced by adding lactose to the medium. Explain.

7. What analogies can you draw between transcriptional trans-acting factors that activate gene expression in eukaryotes and the corresponding factors in prokaryotes? Give an example.

8. Compare the arrangement of cis-acting sites in the control regions of eukaryotes and prokaryotes.

    a. $R^+ O^+ A^+$

    b. $R^- O^+ A^+/R^+ O^+ A^-$

    c. $R^+ O^- A^+/R^+ O^+ A^-$

9. What is meant by the term *epigenetic inheritance*? What are two examples of such inheritance?

BASIC PROBLEMS

10. What is an enhanceosome? Why could a mutation in any one of the enhanceosome proteins severely reduce the transcription rate?

11. Why are mutations in imprinted genes usually dominant?

12. What features distinguish an epigenetically silenced gene from a gene that is not expressed due to an alteration in its DNA sequence?

13. What mechanism is thought to be responsible for the inheritance of epigenetic information?

14. What is the fundamental difference in how pro-karyotic and eukaryotic genes are regulated?

15. Which of the following describes a synergistic interaction?

    a. A transcription factor that activates transcription

    b. Three transcription factors that interact to recruit a coactivator to an enhancer

    c. Binding of the TFIID protein to the TATA box

    d. The acetylation of histone tails by a transcription factor with an acetylase domain

    e. None of the above

16. Why is it said that transcriptional regulation in eukaryotes is characterized by modularity?

CHALLENGING PROBLEMS

17. The transcription of a gene called *YFG* (Your Favorite Gene) is activated when three transcription factors (TFA, TFB, TFC) interact to recruit the coactivator CRX. TFA, TFB, TFC, and CRX and their respective binding sites constitute an enhanceosome located 10 kb from the transcription start site. Draw a diagram showing how you think the enhanceosome functions to recruit RNA polymerase to the promoter of YFG.

18. From Problem 17, a single mutation in one of the transcription factors results in a drastic reduction in YFG transcription. Diagram what this mutant interaction might look like.

19. From Problem 18, diagram the effect of a mutation in the binding site for one of the transcription factors.

20. Yeast strains with reduced levels of the protein histone 4 cannot repress the transcription of certain inducible genes (see Figure 10-36). What do you think would happen to yeast gene expression if an experimenter increased (rather than decreased) the amount of the protein histone 4 in a yeast cell?

21. An interesting mutation in *lacI* results in repressors with 100-fold increased binding to both operator and nonoperator DNA. These repressors display a "reverse" induction curve, allowing β-galactosidase synthesis in the absence of an inducer (IPTG) but partly repressing β-galactosidase expression in the presence of IPTG. How can you explain this? (Note that, when IPTG binds a repressor, it does not completely destroy operator affinity, but rather it reduces affinity 1000-fold. Additionally, as cells divide and new operators are generated by the synthesis of daughter strands, the repressor must find the new operators by searching along the DNA, rapidly binding to nonoperator sequences and dissociating from them.)

22. You are studying a mouse gene that is expressed in the kidneys of male mice. You have already cloned this gene. Now you wish to identify the segments of DNA that control the tissue-specific and sex-specific expression of this gene. Describe an experimental approach that would allow you to do so.

23. In *Neurospora*, all mutants affecting the enzymes carbamyl phosphate synthetase and aspartate trans-carbamylase map at the *pyr*-3 locus. If you induce *pyr*-3 mutations by ICR-170 (a chemical mutagen), you find that either both enzyme functions are lacking or only the transcarbamylase function is lacking; in no case is the synthetase activity lacking when the transcarbamylase activity is present. (ICR-170 is assumed to induce frameshifts.) Interpret these results in regard to a possible operon.

24. Certain *lacI* mutations eliminate operator binding by the Lac repressor but do not affect the aggregation of subunits to make a tetramer, the active form of the repressor. These mutations are partially dominant to wild type. Can you explain the partially dominant I^- phenotype of the I^-/I^+ heterodiploids?

25. You are examining the regulation of the lactose operon in the bacterium *Escherichia coli*. You isolate seven new independent mutant strains that lack the products of all three structural genes. You suspect that some of these mutations are *lacI*^{S} mutations and that other mutations are alterations that prevent the binding of RNA polymerase to the promoter region. Using whatever haploid and partial diploid

genotypes that you think are necessary, describe a set of genotypes that will permit you to distinguish between the *lacI* and *lacP* classes of uninducible mutations.

26. You are studying the properties of a new kind of regulatory mutation of the lactose operon. This mutation, called *S*, leads to the complete repression of the *lacZ*, *lacY*, and *lacA* genes, regardless of whether inducer (lactose) is present. The results of studies of this mutation in partial diploids demonstrate that this mutation is completely dominant to wild type. When you treat bacteria of the *S* mutant strain with a mutagen and select for mutant bacteria that can express the enzymes encoded by *lacZ*, *lacY*, and *lacA* genes in the presence of lactose, some of the mutations map to the *lac* operator region and others to the *lac* repressor gene. On the basis of your knowledge of the lactose operon, provide a molecular genetic explanation for all these properties of the *S* mutation. Include an explanation of the constitutive nature of the "reverse mutations."

27. The *trp* operon in *E. coli* encodes enzymes essential for the biosynthesis of tryptophan. The general mechanism for controlling the *trp* operon is similar to that observed with the *lac* operon: when the repressor binds to the operator, transcription is prevented; when the repressor does not bind the operator, transcription proceeds. The regulation of the *trp* operon differs from the regulation of the *lac* operon in the following way: the enzymes encoded by the *trp* operon are not synthesized when tryptophan is present but rather when it is absent. In the *trp* operon, the repressor has two binding sites: one for DNA and the other for the effector molecule, tryptophan. The *trp* repressor must first bind to a molecule of tryptophan before it can bind effectively to the *trp* operator.

   a. Draw a map of the tryptophan operon, indicating the promoter *(P)*, operator *(O)*, and the first structural gene of the tryptophan operon *(trpA)*. In your drawing, indicate where on the DNA the repressor protein binds when it is bound to tryptophan.

   b. The *trpR* gene encodes the repressor; *trpO* is the operator; *trpA* encodes the enzyme tryptophan synthetase. A *trpR²* repressor cannot bind tryptophan; a *trpO²* operator cannot be bound by the repressor; and the enzyme encoded by a *trpA²* mutant gene is completely inactive. Do you expect to find active tryptophan synthetase in each of the following mutant strains when the cells are grown in the presence of tryptophan? In its absence?

   i. $R^+ O^+ A^+$ (wild type)
   ii. $R^- O^+ A^+/R^+ O^+ A^-$
   iii. $R^+ O^- A^+/R^+ O^+ A^-$

28. The activity of the enzyme β-galactosidase produced by wild-type cells grown in media supplemented with different carbon sources is measured. In relative units, the following is found:

Glucose	Lactose	Lactose + glucose
0	100	1

Predict the relative levels of β-galactosidase activity in cells grown under similar conditions when the cells are *lacI⁻*, *lacIˢ*, *lacO*, and *crp⁻*.

29. The following diagram represents the structure of a gene in *Drosophila melanogaster*; blue segments are exons, and yellow segments are introns.

   a. Which segments of the gene will be represented in the initial RNA transcript?

   b. Which segments of the gene will be removed by RNA splicing?

   c. Which segments would most likely bind proteins that interact with RNA polymerase?

30. You wish to find the cis-regulatory DNA elements responsible for the transcriptional responses of two genes, *c-fos* and *globin*. Transcription of the *c-fos* gene is activated in response to fibroblast growth factor (FGF), but it is inhibited by cortisol (Cort). On the other hand, transcription of the *globin* gene is not affected by either FGF or cortisol, but it is stimulated by the hormone erythropoietin (EP). To find the cis-regulatory DNA elements responsible for these transcriptional responses, you use the following clones of the *c-fos* and *globin* genes, as well as two "hybrid" combinations (fusion genes), as shown in the diagram on the next page. A is the intact *c-fos* gene, D is the intact *globin* gene, and B and C are the *c-fos-globin* gene fusions. The c-fos and globin exons (E) and introns (I) are numbered. For example, E3(f) is the third exon of the *c-fos* gene and I2(g) is the second intron of the *globin* gene. (These labels are provided to help you make your answer clear.) The transcription start sites (black arrows) and polyadenylation sites (red arrows) are indicated.

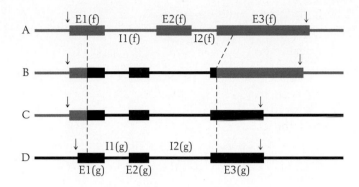

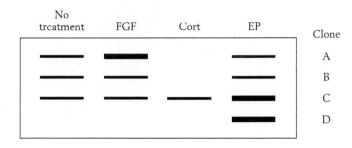

transcript made from a particular clone. (The failure of a band to appear indicates that the level of transcript is undetectable.)

You introduce all four of these clones simultaneously into tissue-culture cells and then stimulate individual aliquots of these cells with one of the three factors. Gel analysis of the RNA isolated from the cells gives the following results. The level of transcripts produced from the introduced genes in response to various treatments are shown; the intensity of these bands is proportional to the amount of

a. Where is the DNA element that permits activation by FGF?

b. Where is the DNA element that permits repression by Cort?

c. Where is the DNA element that permits induction by EP? Explain your answer.

11

GENE ISOLATION AND MANIPULATION

Injection of foreign DNA into an animal cell. The microneedle used for injection is shown at right and a cell-holding pipette at left. [Copyright M. Baret/Rapho/Photo Researchers, Inc.]

KEY QUESTIONS

- How is a gene isolated and amplified by cloning?

- How are specific DNAs or RNAs identified in mixtures?

- How is DNA amplified without cloning?

- How is amplified DNA used in genetics?

- How are DNA technologies applied to medicine?

OUTLINE

11.1 Generating recombinant DNA molecules

11.2 DNA amplification in vitro: the polymerase chain reaction

11.3 Zeroing in on the gene for alkaptonuria: another case study

11.4 Detecting human disease alleles: molecular genetic diagnostics

11.5 Genetic engineering

CHAPTER OVERVIEW

Genes are the central focus of genetics, and so clearly it is desirable to be able to isolate a gene of interest (or any DNA region) from the genome and amplify it to obtain a working amount to study. **DNA technology** is a term that describes the collective techniques for obtaining, amplifying, and manipulating specific DNA fragments. Since the mid-1970s, the development of DNA technology has revolutionized the study of biology, opening many areas of research to molecular investigation. **Genetic engineering,** the application of DNA technology to specific biological, medical, or agricultural problems, is now a well-established branch of technology. **Genomics** is the ultimate extension of the technology to the global analysis of the nucleic acids present in a nucleus, a cell, an organism, or a group of related species (Chapter 12).

How can working samples of individual DNA segments be isolated? That task initially might seem like finding a needle in a haystack. A crucial insight was that researchers could create the large samples of DNA that they needed by tricking the DNA replication machinery to replicate the DNA segment in question. Such replication could be done either within live bacterial cells (in vivo) or in a test tube (in vitro).

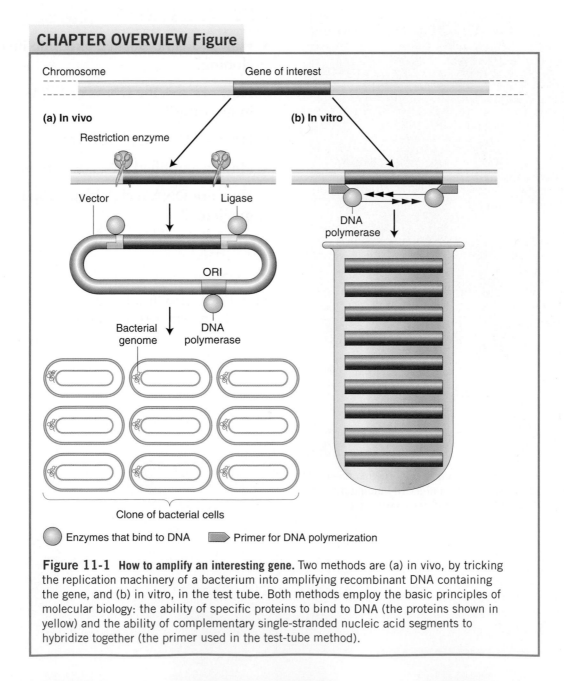

CHAPTER OVERVIEW Figure

Figure 11-1 How to amplify an interesting gene. Two methods are (a) in vivo, by tricking the replication machinery of a bacterium into amplifying recombinant DNA containing the gene, and (b) in vitro, in the test tube. Both methods employ the basic principles of molecular biology: the ability of specific proteins to bind to DNA (the proteins shown in yellow) and the ability of complementary single-stranded nucleic acid segments to hybridize together (the primer used in the test-tube method).

In the in vivo approach (Figure 11-1a), the investigator begins with a sample of DNA molecules containing the gene of interest. This sample is called the **donor DNA** and most often it is an entire genome. Fragments of the donor DNA are inserted into nonessential "accessory" chromosomes (such as plasmids or modified bacterial viruses). These accessory chromosomes will "carry" and amplify the gene of interest and are hence called **vectors**. First, the donor DNA molecules are cut up, by using enzymes called restriction endonucleases as molecular "scissors." These enzymes are a class of DNA-binding proteins that bind to the DNA and cut the sugar–phosphate backbone of each of the two strands of the double helix at a specific sequence. They cut long chromosome-sized DNA molecules into hundreds or thousands of fragments of more manageable size. Next, each fragment is fused with a cut vector chromosome to form **recombinant DNA** molecules. Union with the vector DNA typically depends on short terminal single strands produced by the restriction enzymes. They bond to complementary sequences at the ends of the vector DNA. (The ends act like Velcro to join the different DNA molecules together to produce the recombinant DNA.) The recombinant DNAs are inserted into bacterial cells, and generally only one recombinant molecule is taken up by each cell. Because the accessory chromosome is normally amplified by replication, the recombinant molecule is similarly amplified during the growth and division of the bacterial cell in which the chromosome resides. This process results in a *clone* of identical cells, each containing the recombinant DNA molecule, and so this technique of amplification is called **DNA cloning**. The next stage is finding the rare clone containing the DNA of interest.

In the in vitro approach (Figure 11-1b), a specific gene of interest is amplified chemically by replication machinery extracted from special bacteria. The system "finds" the gene of interest by the complementary binding of specific short primers to the ends of that sequence. These primers then guide the replication process, which cycles exponentially, resulting in a large sample of copies of the gene of interest.

We will see repeatedly that DNA technology depends on two basic foundations of molecular biology research:

- The ability of specific proteins to recognize and bind to specific base sequences, within the DNA double helix (examples are shown in yellow in Figure 11-1).

- The ability of complementary single-stranded DNA or RNA sequences to spontaneously unite to form double-stranded molecules. Examples are the binding of the sticky ends and the binding of the primers.

The remainder of the chapter will explore examples of uses to which we put amplified DNA. These uses range from routine gene isolation for basic biological research to gene-based therapy of human disease.

11.1 Generating recombinant DNA molecules

To illustrate how recombinant DNA is made, let's consider the cloning of the gene for human insulin, a protein hormone used in the treatment of diabetes. Diabetes is a disease in which blood sugar levels are abnormally high either because the body does not produce enough insulin (type I diabetes) or because cells are unable to respond to insulin (type II diabetes). In mild forms of type I, diabetes can be treated by dietary restrictions but, for many patients, daily insulin treatments are necessary. Until about 20 years ago, cows were the major source of insulin protein. The protein was harvested from the pancreases of animals slaughtered in meat-packing plants and purified at large scale to eliminate the majority of proteins and other contaminants in the pancreas extracts. Then, in 1982, the first recombinant human insulin came on the drug market. Human insulin could be made purer, at lower cost, and on an industrial scale because it was produced in bacteria by recombinant DNA techniques. The recombinant insulin is a higher proportion of the proteins in the bacterial cell; hence the protein purification is much easier. We shall follow the general steps necessary for making any recombinant DNA and apply them to insulin.

Type of donor DNA

The choice of DNA to be used as the donor might seem to be obvious, but there are actually three possibilities.

- *Genomic DNA.* This DNA is obtained directly from the chromosomes of the organism under study. It is the most straightforward source of DNA. It needs to be cut up before cloning is possible.

- *cDNA.* **Complementary DNA (cDNA)** is a double-stranded DNA version of an mRNA molecule. In higher eukaryotes, an mRNA is a more useful predictor of a polypeptide sequence than is a genomic sequence, because the introns have been spliced out. Researchers prefer to use cDNA rather than mRNA itself because RNAs are inherently less stable than DNA and techniques for routinely amplifying and purifying individual RNA molecules do not exist. The cDNA is made from mRNA with the use of a special enzyme called *reverse transcriptase*, originally isolated from retroviruses. Using an mRNA molecule as a template, reverse transcriptase synthesizes a single-stranded DNA molecule that can then be used as a template for

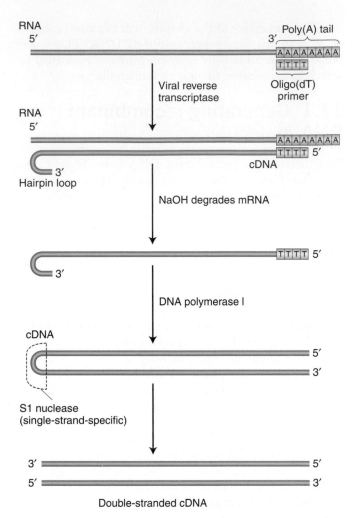

Figure 11-2 The synthesis of double-stranded cDNA from mRNA.
A short oligo(dT) chain is hybridized to the poly(A) tail of an mRNA strand. The oligo(dT) segment serves as a primer for the action of viral reverse transcriptase, an enzyme that uses the mRNA as a template for the synthesis of a complementary DNA strand. The resulting cDNA ends in a hairpin loop. When the mRNA strand has been degraded by treatment with NaOH, the hairpin loop becomes a primer for DNA polymerase I, which completes the paired DNA strand. The loop is then cleaved by S1 nuclease (which acts only on the single-stranded loop) to produce a double-stranded cDNA molecule. [From J. D. Watson, J. Tooze, and D. T. Kurtz, *Recombinant DNA: A Short Course.* Copyright 1983 by W. H. Freeman and Company.]

double-stranded DNA synthesis (Figure 11-2). cDNA does not need to be cut in order to be cloned.

- *Chemically synthesized DNA.* Sometimes, a researcher needs to include in a recombinant DNA molecule a specific sequence that for some reason cannot be isolated from available natural genomic DNA or cDNAs. If the DNA sequence is known (often from a complete genome sequence), then the gene can be synthesized chemically by using automated techniques.

To create bacteria that express human insulin, cDNA was the choice because bacteria do not have the ability to splice out introns present in natural genomic DNA.

Cutting genomic DNA

Most cutting is done using bacterial **restriction enzymes.** These enzymes cut at specific DNA target sequences, called *restriction sites,* and this property is one of the key features that make restriction enzymes suitable for DNA manipulation. Purely by chance, any DNA molecule, be it derived from virus, fly, or human, contains restriction-enzyme target sites. Thus a restriction enzyme will cut the DNA into a set of **restriction fragments** determined by the locations of the restriction sites.

Another key property of some restriction enzymes is that they make "sticky ends." Let's look at an example. The restriction enzyme *Eco*RI (from *E. coli*) recognizes the following sequence of six nucleotide pairs in the DNA of any organism:

5′-GAATTC-3′

3′-CTTAAG-5′

This type of segment is called a DNA **palindrome,** which means that both strands have the same nucleotide

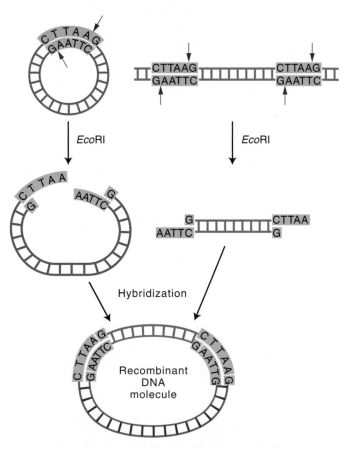

Figure 11-3 Formation of a recombinant DNA molecule.
The restriction enzyme *Eco*RI cuts a circular DNA molecule bearing one target sequence, resulting in a linear molecule with single-stranded sticky ends. Because of complementarity, other linear molecules with *Eco*RI-cut sticky ends can hybridize with the linearized circular DNA, forming a recombinant DNA molecule.

sequence but in antiparallel orientation. Different restriction enzymes cut at different palindromic sequences. Sometimes the cuts are in the same position on each of the two antiparallel strands. However, the most useful restriction enzymes make cuts that are offset, or staggered. For example, the enzyme *Eco*RI makes cuts only between the G and the A nucleotides on each strand of the palindrome:

$$\downarrow$$
$$5'\text{-}G\overset{\downarrow}{A}ATTC\text{-}3'$$
$$3'\text{-}CTTAAG\text{-}5'$$
$$\uparrow$$

These staggered cuts leave a pair of identical sticky ends, each a single strand five bases long. The ends are called

sticky because, being single-stranded, they can base-pair (that is, stick) to a complementary sequence. Single-strand pairing of this type is sometimes called **hybridization**. Figure 11-3 (top left) illustrates the restriction enzyme *Eco*RI making a single cut in a circular DNA molecule such as a plasmid; the cut opens up the circle, and the resulting linear molecule has two sticky ends. It can now hybridize with a fragment of a different DNA molecule having the same complementary sticky ends.

Dozens of restriction enzymes with different sequence specificities are now known, some of which are listed in Figure 11-4. Some enzymes, such as *Eco*RI or *Pst*I, make staggered cuts, whereas others, such as *Sma*I, make flush cuts and leave blunt ends. Even flush cuts, which lack sticky ends, can be used for making

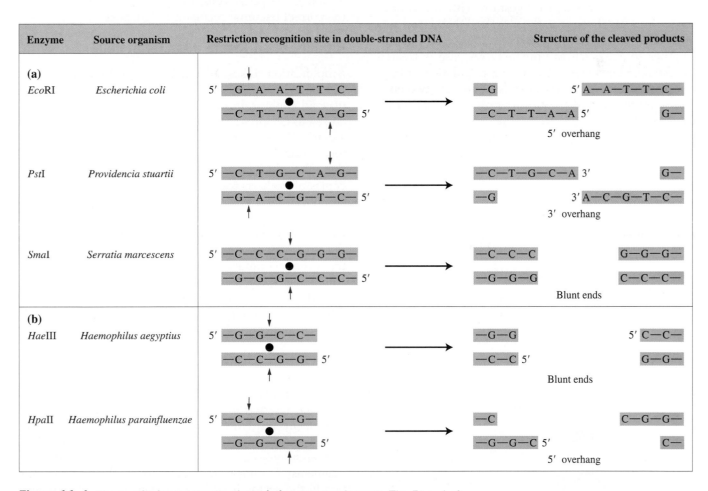

Figure 11-4 The specificity and results of restriction enzyme cleavage. The 5' end of each DNA strand and the site of cleavage (small red arrows) are indicated. The large dot indicates the site of rotational symmetry of each recognition site. Note that the recognition sites differ for different enzymes. In addition, the positions of the cut sites may differ for different enzymes, producing single-stranded overhangs (sticky ends) at the 5' or 3' end of each double-stranded DNA molecule or producing blunt ends if the cut sites are not offset. (a) Three hexanucleotide (six-cutter) recognition sites and the restriction enzymes that cleave them. Note that one site produces a 5' overhang, another a 3' overhang, and the third a blunt end. (b) Examples of enzymes that have tetranucleotide (four-cutter) recognition sites.

recombinant DNA. Special enzymes can join blunt ends together. Other enzymes can make short sticky ends from blunt ends.

> **MESSAGE** Restriction enzymes cut DNA into fragments of manageable size, and many of them generate single-stranded sticky ends suitable for making recombinant DNA.

Attaching donor and vector DNA

Most commonly, both donor and vector DNA are digested by a restriction enzyme that produces complementary sticky ends and are then mixed in a test tube to allow the sticky ends of vector and donor DNA to bind to each other and form recombinant molecules. Figure 11-5a shows a bacterial plasmid DNA that carries a single *Eco*RI restriction site; so digestion with the restriction enzyme *Eco*RI converts the circular DNA into a single linear molecule with sticky ends. Donor DNA from any other source, such as human DNA, also is treated with the *Eco*RI enzyme to produce a population of fragments carrying the same sticky ends. When the two populations are mixed under the proper physiological conditions, DNA fragments from the two sources can hybridize, because double helices form between their sticky ends (Figure 11-5b). There are many opened-up

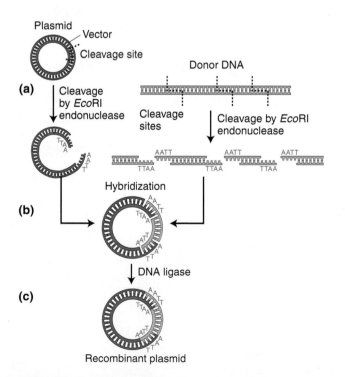

(a)
Plasmid
Vector
Cleavage site

Donor DNA

Cleavage by *Eco*RI endonuclease

Cleavage sites

Cleavage by *Eco*RI endonuclease

(b)
Hybridization

DNA ligase

(c)
Recombinant plasmid

Figure 11-5 Method for generating a recombinant DNA plasmid containing genes derived from donor DNA. [After S. N. Cohen, "The Manipulation of Genes." Copyright 1975 by Scientific American, Inc. All rights reserved.]

plasmid molecules in the solution, as well as many different *Eco*RI fragments of donor DNA. Therefore a diverse array of plasmids recombined with different donor fragments will be produced. At this stage, the hybridized molecules do not have covalently joined sugar–phosphate backbones. However, the backbones can be sealed by the addition of the enzyme **DNA ligase,** which creates phosphodiester linkages at the junctions (Figure 11-5c).

cDNA can be joined to the vector using ligase alone, or short sticky ends can be added to each end of a plasmid and vector.

Another consideration at this stage is that, if the cloned gene is to be transcribed and translated in the bacterial host, it must be inserted next to bacterial regulatory sequences. Hence, to be able to produce human insulin in bacterial cells, the gene must be adjacent to the correct bacterial regulatory sequences.

Amplification inside a bacterial cell

Amplification takes advantage of prokaryotic genetic processes, including those of bacterial transformation, plasmid replication, and bacteriophage growth, all discussed in Chapter 5. Figure 11-6 illustrates the cloning of a donor DNA segment. A single recombinant vector enters a bacterial cell and is amplified by the replication that takes place in cell division. There are generally many copies of each vector in each bacterial cell. Hence, after amplification, a colony of bacteria will typically contain billions of copies of the single donor DNA insert fused to its accessory chromosome. This set of amplified copies of the single donor DNA fragment within the cloning vector is the recombinant DNA *clone.* The replication of recombinant molecules exploits the normal mechanisms that the bacterial cell uses to replicate chromosomal DNA. One basic requirement is the presence of an origin of DNA replication (as described in Chapter 7).

CHOICE OF CLONING VECTORS Vectors must be small molecules for convenient manipulation. They must be capable of prolific replication in a living cell in order to amplify the inserted donor fragment. They must also have convenient restriction sites at which the DNA to be cloned may be inserted. Ideally, the restriction site should be present only once in the vector because then restriction fragments of donor DNA will insert only at that one location in the vector. It is also important that there be a way to identify and recover the recombinant molecule quickly. Numerous cloning vectors are in current use, suitable for different sizes of DNA insert or for different uses of the clone. Some general classes of cloning vectors follow.

Plasmid vectors As described earlier, bacterial plasmids are small circular DNA molecules that replicate their DNA independent of the bacterial chromosome. The

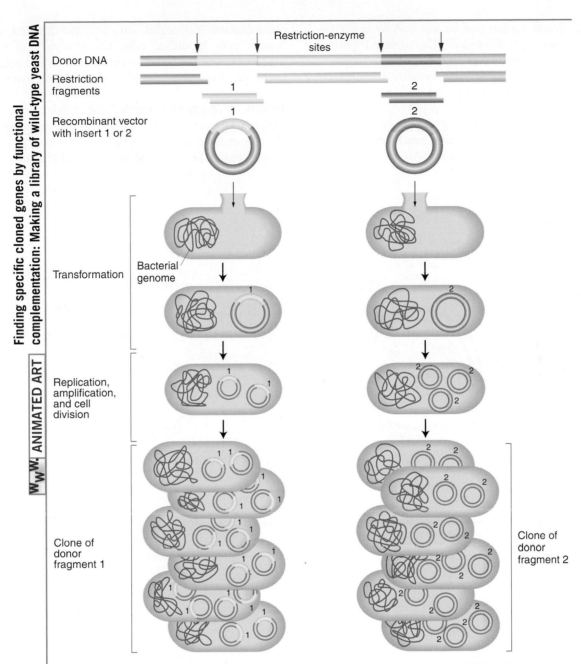

Figure 11-6 **How amplification works.** Restriction-enzyme treatment of donor DNA and vector allows the insertion of single fragments into vectors. A single vector enters a bacterial host, where replication and cell division result in a large number of copies of the donor fragment.

plasmids that are routinely used as vectors are those that carry genes for drug resistance. These drug-resistance genes provide a convenient way to select for cells transformed by plasmids: those cells still alive after exposure to the drug must carry the plasmid vectors containing the DNA insert, as shown at the left in Figure 11-7. Plasmids are also an efficient means of amplifying cloned DNA because there are many copies per cell, as many as several hundred for some plasmids. Examples of some specific plasmid vectors are shown in Figure 11-7.

Bacteriophage vectors Different classes of bacteriophage vectors can carry different sizes of donor DNA insert. A given bacteriophage can harbor a standard amount of DNA as an insert "packaged" inside the phage particle. Bacteriophage λ (lambda) is an effective cloning vector for double-stranded DNA inserts as long as about 15 kb. Lambda phage heads can package DNA molecules no larger than about 50 kb in length (the size of a normal λ chromosome). The central part of the phage genome is not required for replication or packaging of λ DNA molecules in *E. coli* and so can be cut out by using restriction enzymes and discarded. The deleted central part is then replaced by inserts of donor DNA. An insert will be from 10 to 15 kb in length because this size insert brings the total chromosome size back to its normal 50 kb (Figure 11-8).

As Figure 11-8 shows, the recombinant molecules can be directly packaged into phage heads in vitro and then introduced into the bacterium. Alternatively, the recombinant molecules can be transformed directly into *E. coli*. In either case, the presence of a phage plaque on the bacterial lawn automatically signals the presence of recombinant phage bearing an insert.

Vectors for larger DNA inserts The standard plasmid and phage λ vectors just described can accept donor DNA of sizes as large as 25 to 30 kb. However, many experiments require inserts well in excess of this upper limit.

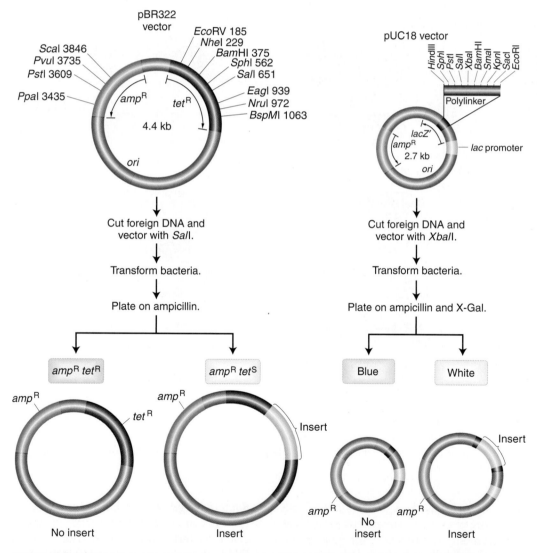

Figure 11-7 Two plasmids designed as vectors for DNA cloning, showing general structure and restriction sites. Insertion into pBR322 is detected by inactivation of one drug-resistance gene (*tet*R), indicated by the *tet*s (sensitive) phenotype. Insertion into pUC18 is detected by inactivation of the β-galactosidase function of *lacZ'*, resulting in an inability to convert the artificial substrate X-Gal into a blue dye. The polylinker has several alternative restriction sites into which donor DNA can be inserted.

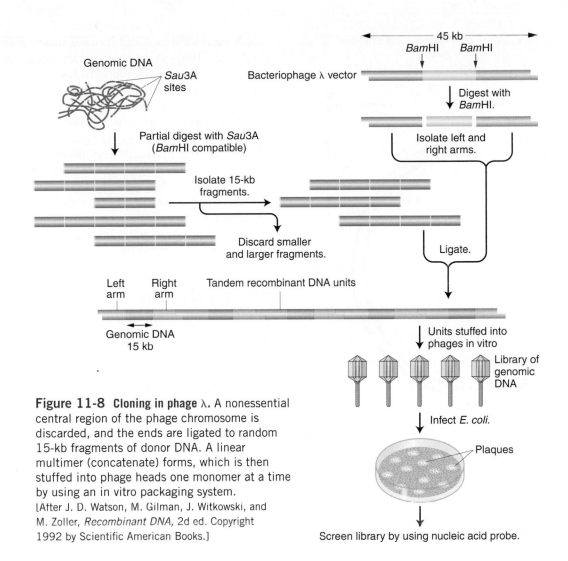

Figure 11-8 Cloning in phage λ. A nonessential central region of the phage chromosome is discarded, and the ends are ligated to random 15-kb fragments of donor DNA. A linear multimer (concatenate) forms, which is then stuffed into phage heads one monomer at a time by using an in vitro packaging system. [After J. D. Watson, M. Gilman, J. Witkowski, and M. Zoller, *Recombinant DNA,* 2d ed. Copyright 1992 by Scientific American Books.]

To meet these needs, the following special vectors have been engineered. In each case, after the DNAs have been delivered into the bacterium, they replicate as large plasmids.

Cosmids are vectors that can carry 35- to 45-kb inserts. They are engineered hybrids of λ phage DNA and bacterial plasmid DNA. Cosmids are inserted into λ phage particles, which act as the "syringes" that introduce these big pieces of recombinant DNA into recipient *E. coli* cells. The plasmid component of the cosmid provides sequences necessary for the cosmid's replication. Once in the cell, these hybrids form circular molecules that replicate extrachromosomally in the same manner as plasmids do. **PAC (P1 artificial chromosome)** vectors deliver DNA by a similar system but can accept inserts ranging from 80 to 100 kb. In this case, the vector is a derivative of bacteriophage P1, a type that naturally has a larger genome than that of λ. **BAC (bacterial artificial chromosome)** vectors, derived from the F plasmid, can carry inserts ranging from 150 to 300 kb (Figure 11-9). The DNA to be cloned is inserted into the plasmid, and this large circular recombinant DNA is introduced into the bacterium by a special type of transformation. BACs are the "workhorse" vectors for the extensive cloning required by large-scale genome-sequencing projects (discussed in Chapter 12). Finally, inserts larger than 300 kb require a eukaryotic vector system called **YACs (yeast artificial chromosomes,** described later in the chapter).

For cloning the gene for human insulin, a plasmid host was selected to carry the relatively short cDNA inserts of approximately 450 bp. This host was a special type of plasmid called a plasmid *expression vector.* Expression vectors contain bacterial promoters that will initiate transcription at high levels when the appropriate allosteric regulator is added to the growth medium. The expression vector induces each plasmid-containing bacterium to produce large amounts of recombinant human insulin.

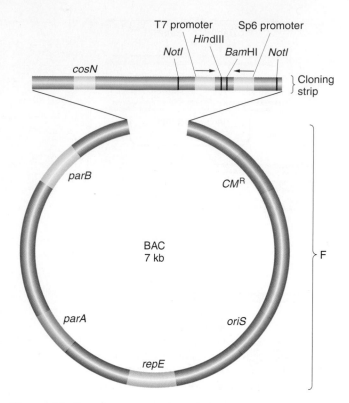

Figure 11-9 **Structure of a bacterial artificial chromosome (BAC), used for cloning large fragments of donor DNA.** CM^R is a selectable marker for chloramphenicol resistance. *oriS*, *repE*, *parA*, and *parB* are F genes for replication and regulation of copy number. *cosN* is the *cos* site from λ phage. *Hind*III and *Bam*HI are cloning sites at which donor DNA is inserted. The two promoters are for transcribing the inserted fragment. The *Not*I sites are used for cutting out the inserted fragment.

which enters the cell and forms a plasmid chromosome (Figure 11-10a). When phages are used, the recombinant molecule is combined with the phage head and tail proteins. These engineered phages are then mixed with the bacteria, and they inject their DNA cargo into the bacterial cells. Whether the result of injection will be the introduction of a new recombinant plasmid (Figure 11-10b) or the production of progeny phages carrying the recombinant DNA molecule (Figure 11-10c) depends on the vector system. If the latter, the resulting free phage particles then infect nearby bacteria. When λ phage is used, through repeated rounds of reinfection, a plaque full of phage particles, each containing a copy of the original recombinant λ chromosome, forms from each initial bacterium that was infected.

Recovery of amplified recombinant molecules

The recombinant DNA packaged into phage particles is easily obtained by collecting phage lysate and isolating the DNA that they contain. For plasmids, the bacteria are chemically or mechanically broken apart. The recombinant DNA plasmid is separated from the much larger main bacterial chromosome by centrifugation, electrophoresis, or other selective techniques.

> **MESSAGE** Gene cloning is carried out through the introduction of single recombinant vectors into recipient bacterial cells, followed by the amplification of these molecules as a result of the natural tendency of these vectors to replicate.

Entry of recombinant molecules into the bacterial cell

Foreign DNA molecules can enter a bacterial cell by two basic paths: transformation and transducing phages (Figure 11-10). In transformation, bacteria are bathed in a solution containing the recombinant DNA molecule,

Making genomic and cDNA libraries

We have seen how to make and amplify individual recombinant DNA molecules. Any one clone represents a small part of the genome of an organism or only one of thousands of mRNA molecules that the organism can synthesize. To ensure that we have cloned the DNA

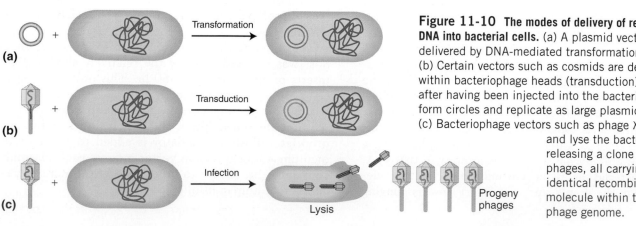

Figure 11-10 **The modes of delivery of recombinant DNA into bacterial cells.** (a) A plasmid vector is delivered by DNA-mediated transformation. (b) Certain vectors such as cosmids are delivered within bacteriophage heads (transduction); however, after having been injected into the bacterium, they form circles and replicate as large plasmids. (c) Bacteriophage vectors such as phage λ infect and lyse the bacterium, releasing a clone of progeny phages, all carrying the identical recombinant DNA molecule within the phage genome.

segment of interest, we have to make large collections of DNA segments that are all-inclusive. For example, we take all the DNA from a genome, break it up into segments of the right size for our cloning vector, and insert each segment into a different copy of the vector, thereby creating a collection of recombinant DNA molecules that, taken together, represent the entire genome. We then transform or transduce these molecules into separate bacterial recipient cells, where they are amplified. The resulting collection of recombinant DNA-bearing bacteria or bacteriophages is called a **genomic library.** If we are using a cloning vector that accepts an average insert size of 10 kb and if the entire genome is 100,000 kb in size (the approximate size of the genome of the nematode *Caenorhabditis elegans*), then 10,000 independent recombinant clones will represent one genome's worth of DNA. To ensure that all sequences of the genome that can be cloned are contained within a collection, genomic libraries typically represent an average segment of the genome at least five times (and so, in our example, there will be 50,000 independent clones in the genomic library). This multifold representation makes it highly unlikely that, by chance, a sequence is not represented at least once in the library.

Similarly, representative collections of cDNA inserts require tens or hundreds of thousands of independent cDNA clones; these collections are **cDNA libraries** and represent only the protein-coding regions of the genome. A comprehensive cDNA library is based on mRNA samples from different tissues, different developmental stages, and organisms grown in different environmental conditions.

Whether we choose to construct a genomic DNA library or a cDNA library depends on the situation. If we are seeking a specific gene that is active in a specific type of tissue in a plant or animal, then it makes sense to construct a cDNA library from a sample of that tissue. For example, suppose we want to identify cDNAs corresponding to insulin mRNAs. The B-islet cells of the pancreas are the most abundant source of insulin, and so mRNAs from pancreas cells are the appropriate source for a cDNA library because these mRNAs should be enriched for the gene in question. A cDNA library represents a subset of the transcribed regions of the genome; so it will inevitably be smaller than a complete genomic library. Although genomic libraries are bigger, they do have the benefit of containing genes in their native form, including introns and untranscribed regulatory sequences. A genomic library is necessary at some stage as a prelude to cloning an entire gene or an entire genome.

> **MESSAGE** The task of isolating a clone of a specific gene begins with making a library of genomic DNA or cDNA—if possible, enriched for sequences containing the gene in question.

Finding a specific clone of interest

The production of a library as heretofore described is sometimes referred to as "shotgun" cloning because the experimenter clones a large sample of fragments and hopes that one of the clones will contain a "hit"—the desired gene. The task then is to find that particular clone, considered next.

FINDING SPECIFIC CLONES BY USING PROBES A library might contain as many as hundreds of thousands of cloned fragments. This huge collection of fragments must be screened to find the recombinant DNA molecule containing the gene of interest to a researcher. Such screening is accomplished by using a specific **probe** that will find and mark only the desired clone. There are two types of probes: those that recognize a specific nucleic acid sequence and those that recognize a specific protein.

Probes for finding DNA Probing for DNA makes use of the power of base complementarity. Two single-stranded nucleic acids with full or partial complementary base sequence will "find" each other in solution by random collision. Once united, the double-stranded hybrid so formed is stable. This provides a powerful approach to finding specific sequences of interest. In the case of DNA, all molecules must be made single stranded by heating. A single-stranded probe, labeled radioactively or chemically, is sent out to find its complementary target sequence in a population of DNAs such as a library. Probes as small as 15 to 20 base pairs will hybridize to specific complementary sequences within much larger cloned DNAs. Thus, probes can be thought of as "bait" for identifying much larger "prey."

The identification of a specific clone in a library is a two-step procedure (Figure 11-11). First, colonies or plaques of the library on a petri dish are transferred to an absorbent membrane (often nitrocellulose) by simply laying the membrane on the surface of the medium. The membrane is peeled off, colonies or plaques clinging to the surface are lysed in situ, and the DNA is denatured. Second, the membrane is bathed with a solution of a single-stranded probe that is specific for the DNA being sought. Generally, the probe is itself a cloned piece of DNA that has a sequence homologous to that of the desired gene. The probe must be labeled with either a radioactive isotope or a fluorescent dye. Thus the position of a positive clone will become clear from the position of the concentrated radioactive or fluorescent label. For radioactive labels, the membrane is placed on a piece of X-ray film, and the decay of the radioisotope produces subatomic particles that "expose" the film, producing a dark spot on the film adjacent to the location of the radioisotope concentration. Such an exposed film is called an **autoradiogram.** If a fluorescent dye is used as a label, the membrane is exposed to the correct

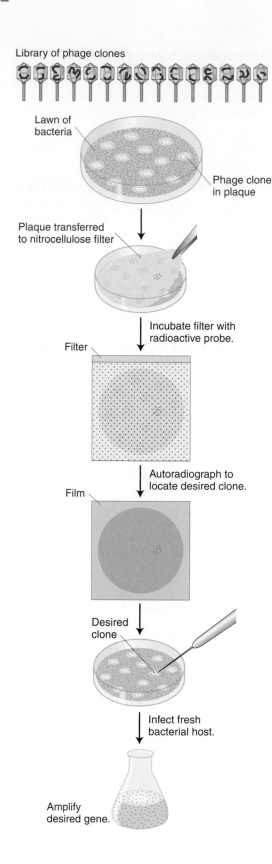

Library of phage clones

Lawn of bacteria

Phage clone in plaque

Plaque transferred to nitrocellulose filter

Filter

Incubate filter with radioactive probe.

Autoradiograph to locate desired clone.

Film

Desired clone

Infect fresh bacterial host.

Amplify desired gene.

Figure 11-11 Using DNA or RNA probes to identify the clone carrying a gene of interest. The clone is identified by probing a genomic library, in this case made by cloning genes in λ bacteriophages, with DNA or RNA known to be related to the desired gene. A radioactive probe hybridizes with any recombinant DNA incorporating a matching DNA sequence, and the position of the clone having the DNA is revealed by autoradiography. Now the desired clone can be selected from the corresponding spot on the petri dish and transferred to a fresh bacterial host so that a pure gene can be manufactured. [After R. A. Weinberg, "A Molecular Basis of Cancer," and P. Leder, "The Genetics of Antibody Diversity." Copyright 1983, 1982 by Scientific American, Inc. All rights reserved.]

wavelength of light to activate the dye's fluorescence, and a photograph is taken of the membrane to record the location of the fluorescing dye.

Where does the DNA to make a probe come from? The DNA can come from one of several sources.

- *cDNA from tissue that expresses a gene of interest at a high level.* For the insulin gene, the pancreas would be the obvious choice.

- *A homologous gene from a related organism.* This method depends on the evolutionary conservation of DNA sequences through time. Even though the probe DNA and the DNA of the desired clone might not be identical, they are often similar enough to promote hybridization. The method is jokingly called "clone by phone" because, if you can phone a colleague who has a clone of your gene of interest from a related organism, then your job of cloning is made relatively easy.

- *The protein product of the gene of interest.* The amino acid sequence of part of the protein is back-translated, by using the table of the genetic code in reverse (from amino acid to codon), to obtain the DNA sequence that encoded it. A synthetic DNA probe that matches that sequence is then designed. Recall, however, that the genetic code is degenerate — that is, most amino acids are encoded by multiple codons. Thus several possible DNA sequences could in theory encode the protein in question, but only one of these DNA sequences is present in the gene that actually encodes the protein. To get around this problem, a short stretch of amino acids with minimal degeneracy is selected. A mixed set of probes is then designed containing all possible DNA sequences that can encode this amino acid sequence. This "cocktail" of oligonucleotides is used as a probe. The correct strand within this cocktail finds the gene of interest. About 20 nucleotides embody enough specificity to hybridize to one unique complementary DNA sequence in the library.

- *Labeled free RNA.* This type of probe is possible only when a nearly pure population of identical molecules of RNA can be isolated, such as rRNA.

Probes for finding proteins If the protein product of a gene is known and isolated in pure form, then this protein can be used to detect the clone of the corresponding gene in a library. The process, described in Figure 11-12, requires two components. First, it requires an expression library, made by using expression vectors. To make the library, cDNA is inserted into the vector in the correct triplet reading frame with a bacterial protein (in this case, β-galactosidase), and cells containing the vec-

tor and its insert produce a "fusion" protein that is partly a translation of the cDNA insert and partly a part of the normal β-galactosidase. Second, the process requires an **antibody** to the specific protein product of the gene of interest. (An antibody is a protein made by an animal's immune system that binds with high affinity to a given molecule.) The antibody is used to screen the expression library for that protein. A membrane is laid over the surface of the medium and removed so that some of the cells of each colony are now attached to the membrane at locations that correspond to their positions on the original petri dish (see Figure 11-12). The imprinted membrane is then dried and bathed in a solution of the antibody, which will bind to the imprint of any colony that contains the fusion protein of interest. Positive clones are revealed by a labeled secondary antibody that binds to the first antibody. By detecting the correct protein, the antibody effectively identifies the clone containing the gene that must have synthesized that protein and therefore contains the desired cDNA.

> **MESSAGE** A cloned gene can be selected from a library by using probes for the gene's DNA sequence or for the gene's protein product.

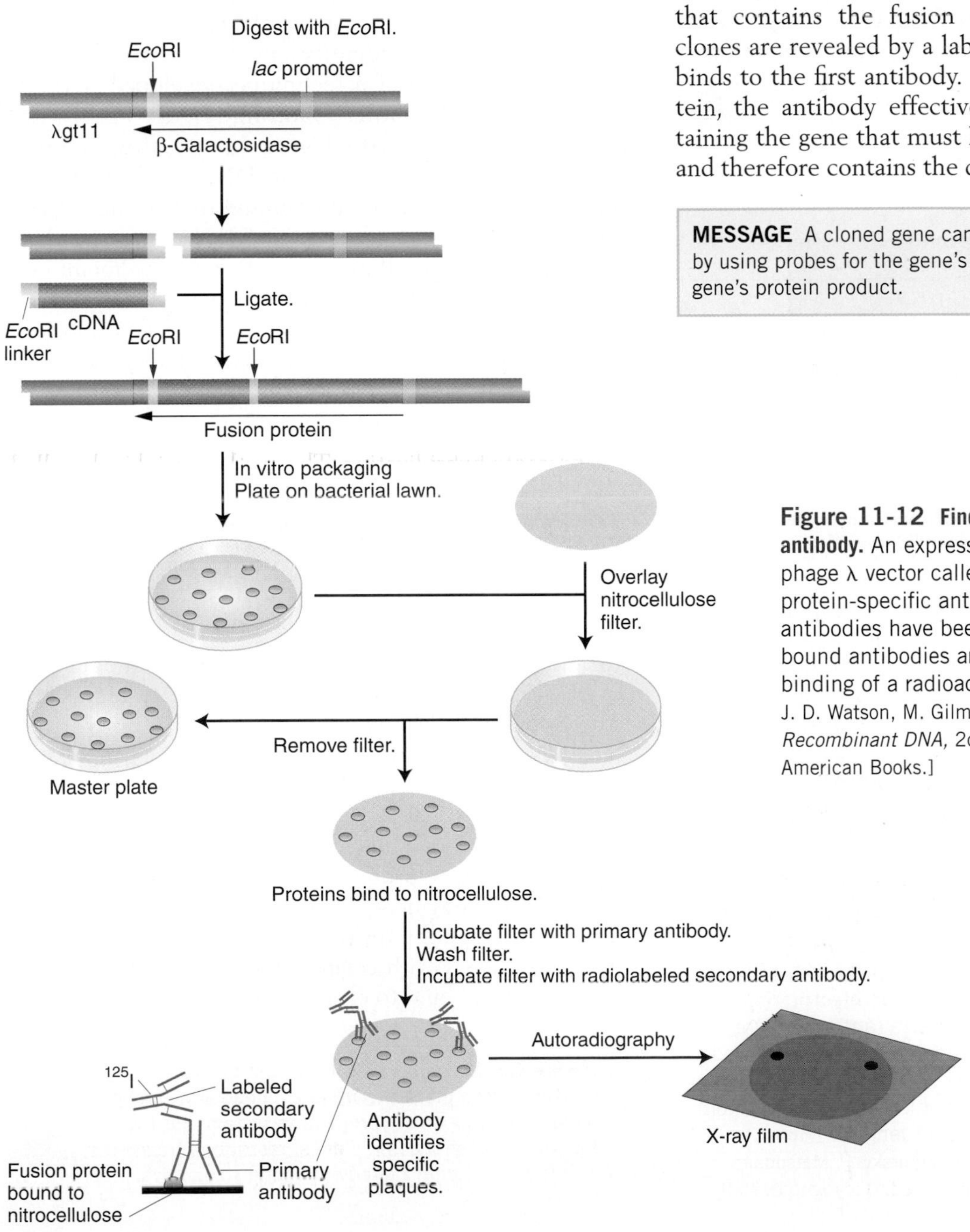

Figure 11-12 Finding the clone of interest by using antibody. An expression library made with special phage λ vector called λgt11 is screened with a protein-specific antibody. After the unbound antibodies have been washed off the filter, the bound antibodies are visualized through the binding of a radioactive secondary antibody. [After J. D. Watson, M. Gilman, J. Witkowski, and M. Zoller, *Recombinant DNA*, 2d ed. Copyright 1992 by Scientific American Books.]

PROBING TO FIND A SPECIFIC NUCLEIC ACID IN A MIXTURE As we shall see later in the chapter, in the course of gene and genome manipulation, it is often necessary to detect and isolate a specific DNA or RNA molecule from among a complex mixture.

The most extensively used method for detecting a molecule within a mixture is blotting. Blotting starts by separating the molecules in the mixture by **gel electrophoresis.** Let's look at DNA first. A mixture of linear DNA molecules is placed into a well cut into an agarose gel, and the well is attached to the cathode of an electric field. Because DNA molecules contain charges, the fragments will migrate through the gel to the anode at speeds inversely dependent on their size (Figure 11-13). Therefore, the fragments in distinct size classes will form distinct bands on the gel. The bands

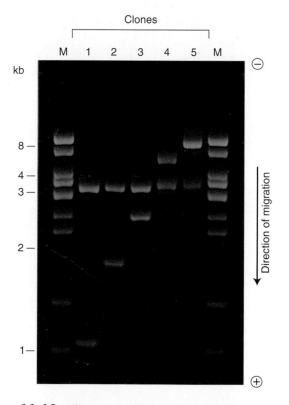

Figure 11-13 Mixtures of different-sized DNA fragments separated electrophoretically on an agarose gel. The samples are five recombinant vectors treated with *Eco*RI. The mixtures are applied to wells at the top of the gel, and fragments move under the influence of an electric field to different positions dependent on size (and, therefore, number of charges). The DNA bands have been visualized by staining with ethidium bromide and photographing under UV light. (*M* represents lanes containing standard fragments acting as markers for estimating DNA length.) [From H. Lodish, D. Baltimore, A. Berk, S. L. Zipursky, P. Matsudaira, and J. Darnell, *Molecular Cell Biology*, 3d ed. Copyright 1995 by Scientific American Books.]

can be visualized by staining the DNA with ethidium bromide, which causes the DNA to fluoresce in ultraviolet light. The absolute size of each restriction fragment in the mixture can be determined by comparing its migration distance with a set of standard fragments of known sizes. If the bands are well separated, an individual band can be cut from the gel, and the DNA sample can be purified from the gel matrix. Therefore DNA electrophoresis can be either diagnostic (showing sizes and relative amounts of the DNA fragments present) or preparative (useful in isolating specific DNA fragments).

Genomic DNA digested by restriction enzymes generally yields so many fragments that electrophoresis produces a continuous smear of DNA and no discrete bands. A probe can identify one fragment in this mixture, with the use of a technique developed by E. M. Southern called **Southern blotting** (Figure 11-14). Like clone identification (see Figure 11-11), this technique entails getting an imprint of DNA molecules on a membrane by using the membrane to blot the gel after electrophoresis is complete. The DNA must be denatured first, which allows it to stick to the membrane. Then the membrane is hybridized with labeled probe. An autoradiogram or a photograph of fluorescent bands will reveal the presence of any bands on the gel that are complementary to the probe. If appropriate, those bands can be cut out of the gel and further processed.

The Southern-blotting technique can be extended to detect a specific *RNA* molecule from a mixture of RNAs fractionated on a gel. This technique is called **Northern blotting** (thanks to some scientist's sense of humor) to contrast it with the Southern-blotting technique used for *DNA* analysis. The electrophoresed RNA is blotted onto a membrane and probed in the same way as DNA is blotted and probed for Southern blotting. One application of Northern analysis is to determine whether a specific gene is transcribed in a certain tissue or under certain environmental conditions.

Hence we see that cloned DNA finds widespread application as a probe, used for detecting a specific clone, DNA fragment, or RNA molecule. In all these cases, note that the technique again exploits the ability of nucleic acids with *complementary* nucleotide sequences to find and bind to each other.

> **MESSAGE** Recombinant DNA techniques that depend on complementarity to a cloned DNA probe include blotting and hybridization systems for the identification of specific clones, restriction fragments, or mRNAs or for measurement of the size of specific DNAs or RNAs.

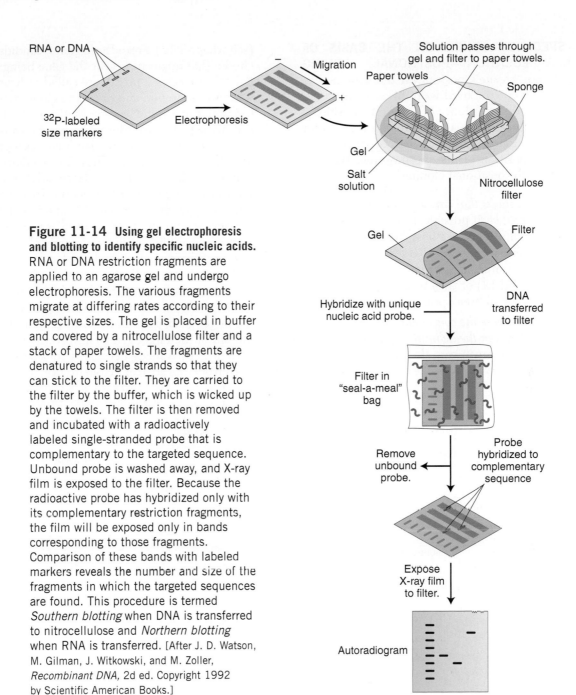

Figure 11-14 Using gel electrophoresis and blotting to identify specific nucleic acids. RNA or DNA restriction fragments are applied to an agarose gel and undergo electrophoresis. The various fragments migrate at differing rates according to their respective sizes. The gel is placed in buffer and covered by a nitrocellulose filter and a stack of paper towels. The fragments are denatured to single strands so that they can stick to the filter. They are carried to the filter by the buffer, which is wicked up by the towels. The filter is then removed and incubated with a radioactively labeled single-stranded probe that is complementary to the targeted sequence. Unbound probe is washed away, and X-ray film is exposed to the filter. Because the radioactive probe has hybridized only with its complementary restriction fragments, the film will be exposed only in bands corresponding to those fragments. Comparison of these bands with labeled markers reveals the number and size of the fragments in which the targeted sequences are found. This procedure is termed *Southern blotting* when DNA is transferred to nitrocellulose and *Northern blotting* when RNA is transferred. [After J. D. Watson, M. Gilman, J. Witkowski, and M. Zoller, *Recombinant DNA*, 2d ed. Copyright 1992 by Scientific American Books.]

FINDING SPECIFIC CLONES BY FUNCTIONAL COMPLEMENTATION In many cases, we don't have a probe for the gene to start with, but we do have a recessive mutation in the gene of interest. If we are able to introduce functional DNA back into the species bearing this allele (see Section 11.5, Genetic engineering), we can detect specific clones in a bacterial or phage library through their ability to restore the function eliminated by the recessive mutation in that organism. This procedure is called **functional complementation** or **mutant rescue.** The general outline of the procedure is as follows:

Make a bacterial or phage library containing wild-type a^+ recombinant donor DNA inserts.

↓

Transform cells of recessive mutant cell-line a^- by using the DNA from individual clones in the library.

↓

Identify clones from the library that produce transformed cells with the dominant a^+ phenotype.

↓

Recover the a^+ gene from the successful bacterial or phage clone.

FINDING SPECIFIC CLONES ON THE BASIS OF GENETIC-MAP LOCATION—POSITIONAL CLONING

Information about a gene's position in the genome can be used to circumvent the hard work of assaying an entire library to find the clone of interest. **Positional cloning** is a term that can be applied to any method for finding a specific clone that makes use of information about the gene's position on its chromosome. Two elements are needed for positional cloning:

- *Some genetic landmarks that can set boundaries on where the gene might be.* If possible, landmarks on either side of the gene of interest are best, because they delimit the possible location of that gene. Landmarks might be RFLPs or other molecular polymorphisms (see Chapters 4 and 12) or they might be well-mapped chromosomal break points (Chapter 15).

- *The ability to investigate the continuous segment of DNA extending between the delimiting genetic landmarks.* In model organisms, the genes in this block of DNA are known from the genome sequence

(see Chapter 12). From these genes, candidates can be chosen that might represent the gene being sought. For other species, a procedure called a **chromosome walk** is used to find and order the clones falling between the genetic landmarks. Figure 11-15 summarizes the procedure. The basic idea is to use the sequence of the nearby landmark as a probe to identify a second set of clones that overlaps the marker clone containing the landmark but extends out from it in one of two directions (toward the target or away from the target). End fragments from the new set of clones can be used as probes for identifying a third set of overlapping clones from the genomic library. In this step-by-step fashion, a set of clones representing the region of the genome extending out from the marker clone can be assayed until one obtains clones that can be shown to include the target gene, perhaps by showing that it rescues a mutant of the target gene. This process is called chromosome walking because it consists of a series of steps from one adjacent clone to the next.

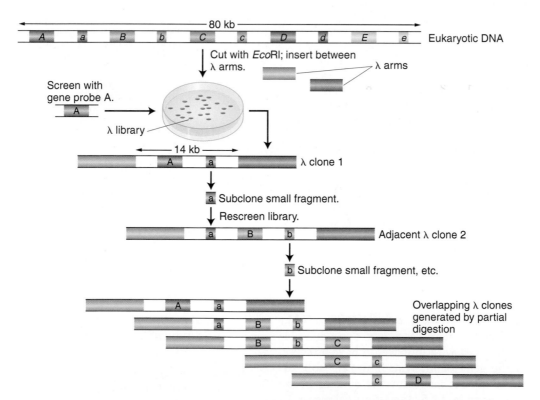

Figure 11-15 Chromosome walking. One recombinant phage obtained from a phage library made by the partial *Eco*RI digestion of a eukaryotic genome can be used to isolate another recombinant phage containing a neighboring segment of eukaryotic DNA. This walk illustrates how to start at molecular landmark A and get to target gene *D*. [After J. D. Watson, J. Tooze, and D. T. Kurtz, *Recombinant DNA: A Short Course.* Copyright 1983 by W. H. Freeman and Company.]

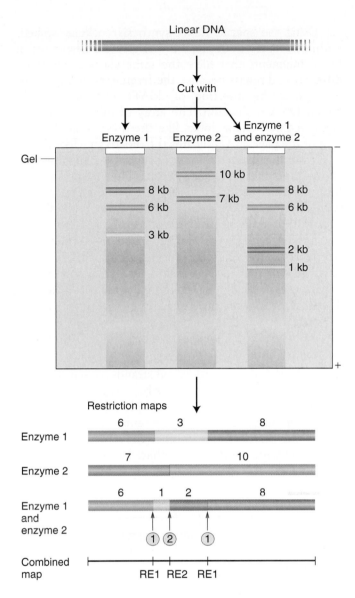

Figure 11-16 Restriction mapping by comparing electrophoretic separations of single and multiple digests. In this simplified example, digestion with enzyme 1 shows that there are two restriction sites for this enzyme but does not reveal whether the 3-kb segment generated by this enzyme is in the middle or on one of the ends of the digested sequence, which is 17 kb long. Combined digestion by both enzyme 1 (RE1) and enzyme 2 (RE2) leaves the 6- and 8-kb segments generated by enzyme 1 intact but cleaves the 3-kb fragment, showing that enzyme 2 cuts at a site within the 3-kb fragment, showing that the 3-kb fragment is in the middle. If the 3-kb segment were at one of the ends of the 17-kb sequence, digestion of the 17-kb sequence by enzyme 2 alone would yield a 1- or 2-kb fragment by cutting at the same site at which this enzyme cut to cleave the 3-kb fragment in the combined digestion by enzymes 1 and 2. Because this result is not the case, of the three restriction fragments produced by enzyme 1, the 3-kb fragment must lie in the middle. That the RE2 site lies closer to the 6-kb section than to the 8-kb section can be inferred from the 7- and 10-kb lengths of the enzyme 2 digestion.

The key to efficient chromosome walking is to know how the array of clones that hybridize to a given probe overlap each other. This is accomplished by comparing the restriction maps of the clones. A **restriction map** is a linear map showing the order and distances of restriction endonuclease cut sites in a segment of DNA. The restriction sites represent small landmarks within the clone. An example of one method used to create a restriction map of a clone is shown in Figure 11-16.

As an aside, it is worth noting that there are many other applications of restriction mapping. In a sense, the restriction map is a partial sequence map of a DNA segment, because every restriction site is one at which a particular short DNA sequence resides (depending on which restriction enzyme cuts at that site). Restriction maps are very important in many aspects of DNA cloning, because the distribution of restriction-endonuclease-cut sites determines where a recombinant DNA engineer can create a clonable DNA fragment with sticky ends.

Determining the base sequence of a DNA segment

After we have cloned our desired gene, the task of trying to understand its function begins. The ultimate language of the genome is composed of strings of the nucleotides A, T, C, and G. Obtaining the complete nucleotide sequence of a segment of DNA is often an important part of understanding the organization of a gene and its regulation, its relation to other genes, or the function of its encoded RNA or protein. Indeed, for the most part, translating the nucleic acid sequence of a cDNA to discover the amino sequence of its encoded polypeptide chain is simpler than directly sequencing the polypeptide itself. In this section, we consider the techniques used to read the nucleotide sequence of DNA.

As with other recombinant DNA technologies, DNA sequencing exploits base-pair complementarity together with an understanding of the basic biochemistry of DNA replication. Several techniques have been developed, but one of them is by far most used. It is called **dideoxy sequencing** or, sometimes, **Sanger sequencing** after its inventor. The term *dideoxy* comes from a special modified nucleotide, called a dideoxynucleotide triphosphate (generically, a ddNTP). This modified nucleotide is key to the Sanger technique because of its ability to block continued DNA synthesis. What is a dideoxynucleotide triphosphate? And how does it block DNA synthesis? A dideoxynucleotide lacks the 3'-hydroxyl group as well as the 2'-hydroxyl group, which is also absent in a deoxynucleotide (Figure 11-17). For DNA synthesis to take place, the DNA polymerase must catalyze a condensation reaction between the 3'-hydroxyl group of the last nucleotide added to the growing chain and the 5'-phosphate group of the next nucleotide to be

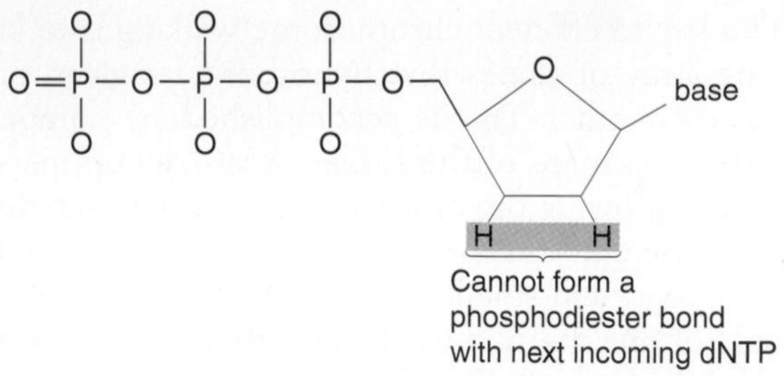

Figure 11-17 The structure of 2′,3′-dideoxynucleotides, which are employed in the Sanger DNA-sequencing method.

added, releasing water and forming a phosphodiester linkage with the 3′-carbon atom of the adjacent sugar. Because a dideoxynucleotide lacks the 3′-hydroxyl group, this reaction cannot take place, and therefore DNA synthesis is blocked at the point of addition.

The logic of dideoxy sequencing is straightforward. Suppose we want to read the sequence of a cloned DNA segment of, say, 5000 base pairs. First, we denature the two strands of this segment. Next, we create a primer for DNA synthesis that will hybridize to exactly one location on the cloned DNA segment and then add a special "cocktail" of DNA polymerase, normal nucleotide triphosphates (dATP, dCTP, dGTP, and dTTP), and a small amount of a special dideoxynucleotide for one of the four bases (for example, dideoxyadenosine triphosphate, abbreviated ddATP). The polymerase will begin to synthesize the complementary DNA strand, starting from the primer, but will stop at any point at which the dideoxynucleotide triphosphate is incorporated into the growing DNA chain in place of the normal nucleotide triphosphate. Suppose the DNA sequence of the DNA segment that we're trying to sequence is:

5′ ACGGGATAGCTAATTGTTTACCGCCGGAGCCA 3′

We would then start DNA synthesis from a complementary primer:

```
5′ ACGGGATAGCTAATTGTTTACCGCCGGAGCCA 3′
3′                          CGGCCTCGGT 5′
              ←────── Direction of DNA synthesis
```

Using the special DNA synthesis cocktail "spiked" with ddATP, for example, we will create a nested set of DNA fragments that have the same starting point but different end points because the fragments stop at whatever point the insertion of ddATP instead of dATP halted DNA replication. The array of different ddATP-arrested DNA chains looks like the diagram at the bottom of the page. (*A indicates the dideoxynucleotide).

We can generate an array of such fragments for each of the four possible dideoxynucleotide triphosphates in four separate cocktails (one spiked with ddATP, one with ddCTP, one with ddGTP, and one with ddTTP). Each will produce a different array of fragments, with no two spiked cocktails producing fragments of the same size. Further, if we add up the results of all four cocktails, we will see that the fragments can be ordered in length, with the lengths increasing by one base at a time. The final steps of the process are:

1. Display the fragments in size order by using by gel electrophoresis.

2. Label the newly synthesized strands so that they can be visualized after they have been separated according to size by gel electrophoresis. Do so by either radioactively or fluorescently labeling the primer (initiation labeling) or the individual dideoxynucleotide triphosphate (termination labeling).

The products of such dideoxy sequencing reactions are shown in Figure 11-18. That result is a ladder of labeled DNA chains increasing in length by one, and so all we need do is read up the gel to read the DNA sequence of the synthesized strand in the 5′-to-3′ direction.

If the tag is a fluorescent dye and a different fluorescent color emitter is used for each of the four ddNTP reactions, then the four reactions can take place in the same test tube and the four sets of nested DNA chains can undergo electrophoresis together. Thus, four times as many sequences can be produced in the same time as can be produced by running the reactions separately. This logic is used in fluorescence detection by auto-

```
5′  ATGGGATAGCTAATTGTTTACCGCCGGAGCCA  3′   Template DNA clone
3′                           CGGCCTCGGT  5′   Primer for synthesis
                        ←──────────             Direction of DNA synthesis
3′              *ATGGCGGCCTCGGT  5′   Dideoxy fragment 1
3′             *AATGGCGGCCTCGGT  5′   Dideoxy fragment 2
3′            *AAATGGCGGCCTCGGT  5′   Dideoxy fragment 3
3′           *ACAAATGGCGGCCTCGGT  5′   Dideoxy fragment 4
3′          *AACAAATGGCGGCCTCGGT  5′   Dideoxy fragment 5
3′         *ATTAACAAATGGCGGCCTCGGT  5′   Dideoxy fragment 6
3′        *ATCGATTAACAAATGGCGGCCTCGGT  5′   Dideoxy fragment 7
3′ *ACCCTATCGATTAACAAATGGCGGCCTCGGT  5′   Dideoxy fragment 8
```

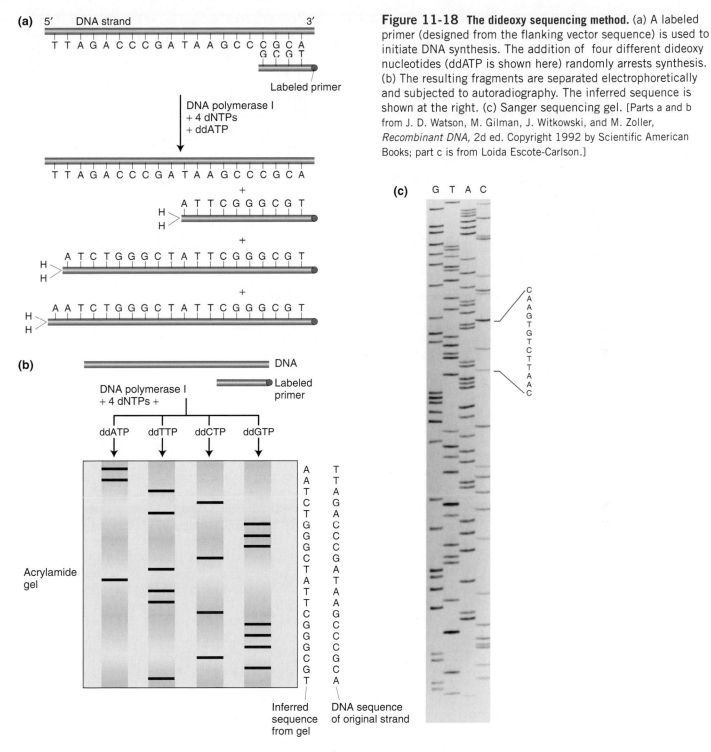

Figure 11-18 The dideoxy sequencing method. (a) A labeled primer (designed from the flanking vector sequence) is used to initiate DNA synthesis. The addition of four different dideoxy nucleotides (ddATP is shown here) randomly arrests synthesis. (b) The resulting fragments are separated electrophoretically and subjected to autoradiography. The inferred sequence is shown at the right. (c) Sanger sequencing gel. [Parts a and b from J. D. Watson, M. Gilman, J. Witkowski, and M. Zoller, *Recombinant DNA*, 2d ed. Copyright 1992 by Scientific American Books; part c is from Loida Escote-Carlson.]

mated DNA-sequencing machines. Thanks to these machines, DNA sequencing can proceed at a massive level, and sequences of whole genomes can be obtained by scaling up the procedures discussed in this section. Figure 11-19 illustrates a readout of automated sequencing. Each colored peak represents a different-size fragment of DNA, ending with a fluorescent base that was detected by the fluorescent scanner of the automated DNA sequencer; the four different colors represent the four bases of DNA. Applications of automated sequencing technology on a genomewide scale is a major focus of Chapter 12.

MESSAGE A cloned DNA segment can be sequenced by characterizing a serial set of truncated synthetic DNA fragments, each terminated at different positions corresponding to the incorporation of a dideoxynucleotide.

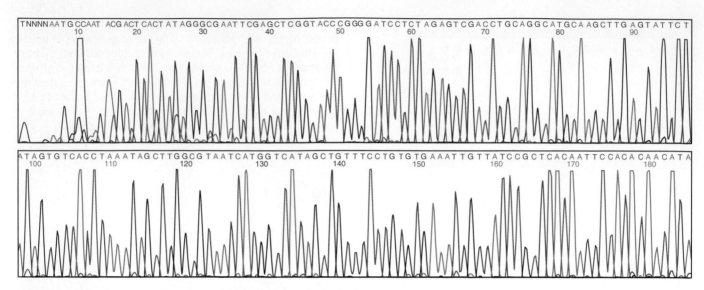

TNNNN AA TG CCAAT ACG ACT CACT AT A GGG CG AAT T CG AGCT CGG T ACC CGG G GA T CC T CT AG AGT C GA CC T GCA GG CA TG CA AG CT T G AGT A TT C T
          10   20   30   40   50   60   70   80   90

A T A GT G TC ACC T A A A T AG CT T GG CG T AAT CAT GG T CA T AG CT G T TT CCT G TG T GA AAT T G T T AT CCG CT CAC AA T T C CAC A CA AC AT A
      100   110   120   130   140   150   160   170   180

Figure 11-19 Printout from an automatic sequencer that uses fluorescent dyes. Each of the four colors represents a different base. N represents a base that cannot be assigned, because peaks are too low. Note that, if this were a gel as in Figure 11-18c, each of these peaks would correspond to one of the dark bands on the gel; in other words, these colored peaks represent a different readout of the same sort of data as are produced on a sequencing gel.

Suppose we have determined the nucleotide sequence of a cloned DNA fragment. How can we tell whether it contains one or more genes? The nucleotide sequence is fed into a computer, which then scans all six reading frames (three in each direction) in the search for possible protein-coding regions that begin with an ATG initiation codon, end with a stop codon, and are long enough that an uninterrupted sequence of its length is unlikely to have arisen by chance. These stretches are called **open reading frames (ORFs)**. They represent sequences that are candidate genes. Figure 11-20 shows such an analysis in which two candidate genes have been identified as ORFs. The use of experimental and computational techniques to search for genes within DNA sequences are discussed in Chapter 12.

11.2 DNA amplification in vitro: the polymerase chain reaction

If we know the sequence of at least some parts of the gene or sequence of interest, we can amplify it in a test tube. The procedure is called the **polymerase chain reaction (PCR)**. The basic strategy of PCR is outlined in Figure 11-21. The process uses multiple copies of a pair of short chemically synthesized primers, from 15 to 20 bases long, each binding to a different end of the gene or region to be amplified. The two primers bind to opposite DNA strands, with their 3′ ends pointing at each other. Polymerases add bases to these primers, and the polymerization process shuttles back and forth between them, forming an exponentially growing number of double-stranded DNA molecules. The details are as follows.

Figure 11-20 Scanning for open reading frames. Any piece of DNA has six possible reading frames, three in each direction. Here the computer has scanned a 9-kb fungal plasmid sequence in looking for ORFs (potential genes). Two large ORFs, 1 and 2, are the most likely candidates as potential genes. The yellow ORFs are too short to be genes.

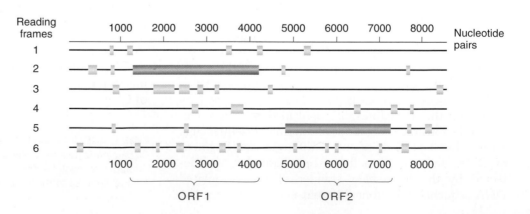

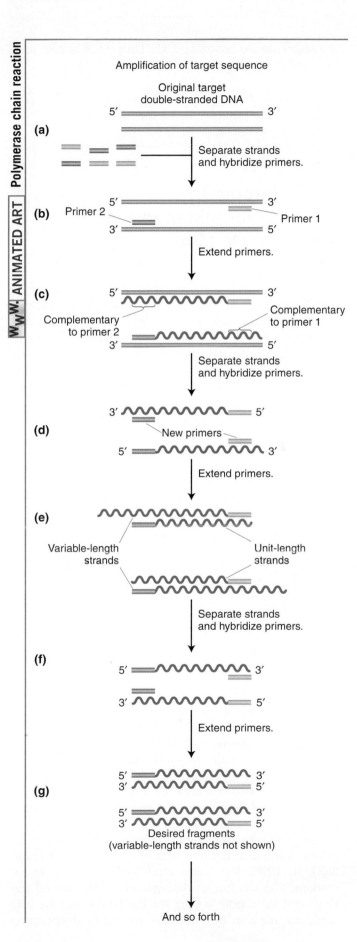

Amplification of target sequence

Figure 11-21 **The polymerase chain reaction.** (a) Double-stranded DNA containing the target sequence. (b) Two chosen or created primers have sequences complementing primer-binding sites at the 3′ ends of the target gene on the two strands. The strands are separated by heating, allowing the two primers to anneal to the primer-binding sites. Together, the primers thus flank the targeted sequence. (c) *Taq* polymerase then synthesizes the first set of complementary strands in the reaction. These first two strands are of varying length, because they do not have a common stop signal. They extend beyond the ends of the target sequence as delineated by the primer-binding sites. (d) The two duplexes are heated again, exposing four binding sites. (For simplicity, only the two new strands are shown.) The two primers again bind to their respective strands at the 3′ ends of the target region. (e) *Taq* polymerase again synthesizes two complementary strands. Although the template strands at this stage are variable in length, the two strands just synthesized from them are precisely the length of the target sequence desired. This precise length is achieved because each new strand begins at the primer-binding site, at one end of the target sequence, and proceeds until it runs out of template, at the other end of the sequence. (f) Each new strand now begins with one primer sequence and ends with the primer-binding sequence for the other primer. Subsequent to strand separation, the primers again anneal and the strands are extended to the length of the target sequence. (The variable-length strands of part c also are producing target-length strands, which, for simplicity, is not shown.) (g) The process can be repeated indefinitely, each time creating two double-stranded DNA molecules identical with the target sequence. [After J. D. Watson, M. Gilman, J. Witkowski, and M. Zoller, *Recombinant DNA*, 2d ed. Copyright 1992 by Scientific American Books.]

We start with a solution containing the DNA source, the primers, the four deoxyribonucleotide triphosphates, and a special DNA polymerase. The DNA is denatured by heat, resulting in single-stranded DNA molecules. Primers hybridize to their complementary sequences in the single-stranded DNA molecules in cooled solutions. A special heat-tolerant DNA polymerase replicates the single-stranded DNA segments extending from a primer. The DNA polymerase *Taq* polymerase, from the bacterium *Thermus aquaticus*, is one such enzyme commonly used. (This bacterium normally grows in thermal vents and so has evolved proteins that are extremely heat resistant. Thus it is able to survive the high temperatures required to denature the DNA duplex, which would denature and inactivate DNA polymerase from most species.) Complementary new strands are synthesized as in normal DNA replication in cells, forming two double-stranded DNA molecules identical with the parental double-stranded molecule. After the replication of the segment between the two primers is completed (one cycle), the two new duplexes are again heat denatured to generate single-stranded templates, and a second cycle of replication is carried out by lowering the temperature in the presence of all the components necessary for the polymerization. Repeated cycles of denaturation, annealing, and synthesis result in an exponential increase in the number of segments replicated. Amplifications by as much as a millionfold can be readily achieved within 1 to 2 hours.

The great advantage of PCR is that fewer procedures are necessary compared with cloning because the location of the primers determines the specificity of the DNA segment that is amplified. If the sequences corresponding to the primers are each present only once in the genome and are sufficiently close together (maximum distance, about 2 kb), the *only* DNA segment that can be amplified is the one between the two primers. This will be true even if this DNA segment is present at very low levels (for example, one part in a million) in a complex mixture of DNA fragments such as might be generated from a preparation of human genomic DNA.

Because PCR is a very sensitive technique, it has many other applications in biology. It can amplify target sequences that are present in extremely low copy numbers in a sample, as long as primers specific to this rare sequence are used. For example, crime investigators can amplify segments of human DNA from the few follicle cells surrounding a single pulled-out hair.

Although PCR's sensitivity and specificity are clear advantages, the technique does have some significant limitations. To design the PCR primers, at least some sequence information must be available for the piece of DNA that is to be amplified; in the absence of such information, PCR amplification cannot be applied. The polymerase amplifies DNA segments reliably only when

the segments are less than 2 kb. Thus, PCR is best used for small fragments of recombinant DNA.

> **MESSAGE** The polymerase chain reaction uses specially designed primers for direct amplification of specific short regions of DNA in a test tube.

11.3 Zeroing in on the gene for alkaptonuria: another case study

Earlier we used the human insulin gene as an example of cloning. A great deal was known about insulin before this cloning exercise, and the main reason for the cloning was to produce insulin as a drug. In most cases of cloning, little is known about the gene before cloning it; indeed, that is the purpose of the cloning exercise. An example of the latter type is the cloning of the human gene defective in alkaptonuria (follow this story in Figure 11-22). The process brings together techniques that have already been discussed: gene cloning in vivo and PCR in vitro.

Alkaptonuria is a human disease with several symptoms, but the most conspicuous is that the urine turns black when exposed to air. In 1898, an English doctor named Archibald Garrod showed that the substance responsible for the black color is homogentisic acid, which is excreted in abnormally large amounts into the urine of alkaptonuria patients. In 1902, early in the post-Mendelian era, Garrod suggested, on the basis of pedigree patterns, that alkaptonuria is inherited as a Mendelian recessive. Soon after, in 1908, he proposed that the disorder was caused by the lack of an enzyme that normally splits the aromatic ring of homogentisic acid to convert it into maleylacetoacetic acid. Because of this enzyme deficiency, he reasoned, homogentisic acid accumulates. Thus alkaptonuria was among the earliest proposed cases of an "inborn error of metabolism," an enzyme deficiency caused by a defective gene. There was a 50-year delay before others were able to show that, in the livers of patients with alkaptonuria, activity for the enzyme that normally splits homogentisic acid, an enzyme called homogentisate 1,2-dioxygenase (HGO), is indeed totally absent. Therefore it seemed likely that the enzyme HGO was normally encoded by the alkaptonuria gene.

In 1992, the alkaptonuria gene was mapped genetically to band 2 of the long arm of chromosome 3 (band 3q2). In 1995, Jose Fernández-Cañón and colleagues, working with the fungus *Aspergillus nidulans*, cloned and characterized a gene coding for the HGO enzyme (the same enzyme that in humans is missing in alkaptonuria

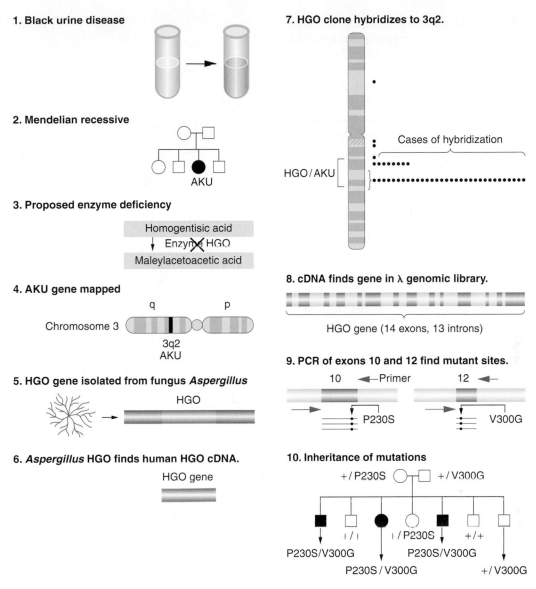

Figure 11-22 The steps in unraveling the biochemical, genetic, and molecular basis of alkaptonuria.

patients). In 1996, they performed a computer search through a large number of sequenced fragments of a human cDNA, looking for a match to the inferred amino acid sequence of the *Aspergillus* gene. They identified a positive clone that contained a human gene coding for 445 amino acids, which showed 52 percent similarity to the *Aspergillus* gene. When this human gene was expressed in an *E. coli* expression vector, its product had HGO activity. The human HGO was then used as a probe for hybridization to chromosomes in which the DNA had been partly denatured (in situ hybridization—see Chapter 12). The probe bound to band 3q2, the known location of the alkaptonuria gene.

After identifying the alkaptonuria gene, researchers turned to the question, What are the mutation or mutations that disable that gene? The cDNA clone was used

to recover the full-length gene from a genomic library. The gene was found to have 14 exons and spanned a total of 60 kb. Investigators then tested a family of seven in which three children suffered from alkaptonuria for mutations in this gene. They amplified all the exons individually by PCR analysis and sequenced the amplified products. One parent was found to be heterozygous for a proline → serine substitution at position 230 in exon 10 (mutation P230S). The other parent was heterozygous for a valine → glycine substitution at position 300 in exon 12 (mutation V300G). All three children with alkaptonuria were of the constitution P230S / V300G, as expected if these positions were the mutant sites inactivating the HGO enzyme. By this means, researchers unambiguously identified that part of the genome that encodes the alkaptonuria/HGO gene.

Here we see how information on sequence, chromosomal position, and evolutionary conservation between species all contributed to the successful identification of the AKU gene clone.

The preceding sections have introduced the fundamental techniques that have revolutionized genetics. The remainder of the chapter will focus on the application of these techniques to human disease diagnosis and to genetic engineering.

11.4 Detecting human disease alleles: molecular genetic diagnostics

A contributing factor in more than 500 human genetic diseases is a recessive mutant allele of a single gene. For families at risk for such diseases, it is important to detect heterozygous prospective parents to permit proper counseling. It is also necessary to be able to detect homozygous progeny early, ideally in the fetal stage, so that doctors can apply drug or dietary therapies early. In the future, there may even be the possibility of gene therapy. Dominant disorders also can require genetic diagnosis. For example, people at risk for the late-onset Huntington disease need to know whether they carry the disease allele before they have children.

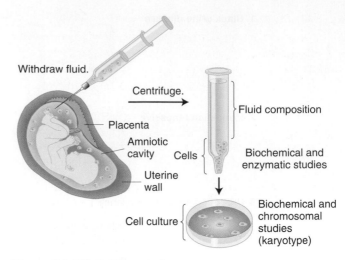

Figure 11-23 Amniocentesis.

Widely used tests are able to detect homozygous defective alleles in fetal cells. The fetal cells can be taken from the amniotic fluid, separated from other components, and cultured to allow the analysis of chromosomes, proteins, enzymatic reactions, and other biochemical properties. This process, **amniocentesis** (Figure 11-23), can identify a number of known disorders; Table 11-1 lists some examples. **Chorionic villus**

Table 11-1 Some Common Genetic Diseases

Inborn errors of metabolism	Approximate incidence among live births
1. Cystic fibrosis (defective chloride channel protein)	1/1600 Caucasians
2. Duchenne muscular dystrophy (defective muscle protein, dystrophin)	1/3000 boys (X linked)
3. Gaucher disease (defective glucocerebrosidase)	1/2500 Ashkenazi Jews; 1/75,000 others
4. Tay-Sachs disease (defective hexosaminidase A)	1/3500 Ashkenazi Jews; 1/35,000 others
5. Essential pentosuria (a benign condition)	1/2000 Ashkenazi Jews; 1/50,000 others
6. Classic hemophilia (defective clotting factor VIII)	1/10,000 boys (X linked)
7. Phenylketonuria (defective phenylalanine hydroxylase)	1/5000 Celtic Irish; 1/15,000 others
8. Cystinuria (defective membrane transporter of cystine)	1/15,000
9. Metachromatic leukodystrophy (defective arylsulfatase A)	1/40,000
10. Galactosemia (defective galactose 1-phosphate uridyl transferase)	1/40,000
11. Sickle-cell anemia (defective β-globin chain)	1/400 U.S. blacks. In some West African populations, the frequency of heterozygotes is 40 percent.
12. Thalassemia (reduced or absent globin chain)	1/400 among some Mediterranean populations

Note: Although a vast majority of more than 500 recognized recessive genetic diseases are extremely rare, in combination they constitute an enormous burden of human suffering. As is consistent with Mendelian mutations, the incidence of some of these diseases is much higher in certain racial groups than in others.

Source: J. D. Watson, M. Gilman, J. Witkowski, and M. Zoller, *Recombinant DNA*, 2d ed. Copyright 1992 Scientific American Books.

sampling (CVS) is a related technique in which a small sample of cells from the placenta is aspirated out with a long syringe. CVS can be performed earlier in the pregnancy than can amniocentesis, which must await the development of a large enough volume of amniotic fluid.

Traditionally, these screening procedures have only identified disorders that can be detected as a chemical defect in the cultured cells. However, with recombinant DNA technology, the DNA can be analyzed directly. In principle, the appropriate fetal gene could be cloned and its sequence compared with that of a cloned normal gene to see if the fetal gene is normal. However, this procedure would be lengthy and impractical, and so shortcuts have been devised. The following sections explain several of the useful techniques that have been developed for this purpose.

Diagnosing mutations on the basis of restriction-site differences

Sometimes a mutation responsible for a specific disease happens to remove a restriction site that is normally present. Conversely, occasionally a mutation associated with a disease alters the normal sequence such that a restriction site is created. In either case, the presence or absence of the restriction site becomes a convenient assay for a disease-causing genotype. For example, sickle-cell anemia is a genetic disease that is commonly caused by a well-characterized mutation in the gene for hemoglobin. Affecting approximately 0.25 percent of African Americans, the disease results from a hemoglobin that has been altered such that valine replaces glutamic acid at amino acid position 6 in the β-globin chain. The GAG-to-GTG change that is responsible for the substitution eliminates a cut site for the restriction enzyme *Mst*II, which cuts the sequence CCTNAGG (in which N represents any of the four bases). The change from CCTGAGG to CCTGTGG can thus be recognized by Southern analysis by using labeled β-globin cDNA as a probe, because the DNA derived from persons with sickle-cell disease lacks one fragment contained in the DNA of normal persons and contains a large (uncleaved) fragment not seen in normal DNA (Figure 11-24).

Diagnosing mutations by probe hybridization

Most disease-causing mutations are not associated with restriction-site changes. For these cases, techniques exist that distinguish mutant and normal alleles by whether a probe hybridizes with the allele. Synthetic oligonucleotide probes can be designed that detect a difference in a single base pair. A good example is the test for α$_1$-antitrypsin deficiency, which greatly increases the probability of developing pulmonary em-

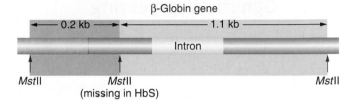

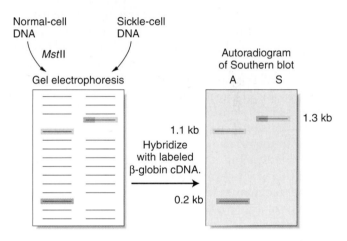

Figure 11-24 Detection of the sickle-cell globin gene by Southern blotting. The base change (A → T) that causes sickle-cell anemia destroys an *Mst*II target site that is present in the normal β-globin gene. This difference can be detected by Southern blotting. [After J. D. Watson, M. Gilman, J. Witkowski, and M. Zoller, *Recombinant DNA*, 2d ed. Copyright 1992 Scientific American Books.]

physema. The condition results from a single base change at a known position. A synthetic oligonucleotide probe is prepared that contains the wild-type sequence. That probe is applied to a Southern-blot analysis to determine whether the DNA contains the wild-type or the mutant sequence. At higher temperatures, a complementary sequence will hybridize, whereas a sequence containing even a single mismatched base will not.

Diagnosing with PCR tests

Because PCR allows an investigator to zero in on any desired sequence, it can be used to amplify and later sequence any potentially defective DNA sequence. In an even simpler approach, primers can be designed that hybridize to the normal allele and therefore prime its amplification but do not hybridize to the mutant allele. This technique can diagnose diseases caused by the presence of a specific mutational site.

> **MESSAGE** Recombinant DNA technology provides many sensitive techniques for testing for defective alleles.

11.5 Genetic engineering

Thanks to recombinant DNA technology, genes can be isolated in a test tube and characterized as specific nucleotide sequences. But even this achievement is not the end of the story. We shall see next that knowledge of a sequence is often the beginning of a fresh round of genetic manipulation. When characterized, a sequence can be manipulated to alter an organism's genotype. The introduction of an altered gene into an organism has become a central aspect of basic genetic research, but it also finds wide commercial application. Two examples of the latter are (1) goats that secrete in their milk antibiotics derived from a fungus and (2) plants kept from freezing by the incorporation of arctic fish "antifreeze" genes into their genomes. The use of recombinant DNA techniques to alter an organism's genotype and phenotype in this way is termed *genetic engineering.*

The techniques of genetic engineering developed originally in bacteria and described in the first part of this chapter needed to be extended to model eukaryotes, which constitute a large proportion of model research

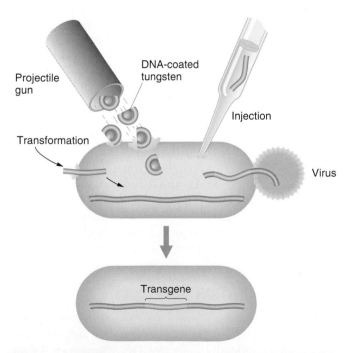

Figure 11-25 Some of the different ways of introducing foreign DNA into a cell.

organisms. Eukaryotic genes are still typically cloned and sequenced in bacterial hosts, but eventually they are introduced into a eukaryote, either the original donor species or a completely different one. The gene transferred is called a **transgene,** and the engineered product is called a **transgenic organism.**

The transgene can be introduced into a eukaryotic cell by a variety of techniques, including transformation, injection, bacterial or viral infection, or bombardment with DNA-coated tungsten or gold particles (Figure 11-25). When the transgene enters a cell, it is able to travel to the nucleus, where to become a stable part of the genome it must insert into a chromosome or (in a few species only) replicate as part of a plasmid. If insertion occurs, it can either replace the resident gene or insert ectopically—that is, at other locations in the genome. Transgenes from other species typically insert ectopically.

> **MESSAGE** Transgenesis can introduce new or modified genetic material into eukaryotic cells.

We now turn to some examples in fungi, plants, and animals and to attempts at human gene therapy.

Genetic engineering in *Saccharomyces cerevisiae*

It is fair to say that *S. cerevisiae* is the most sophisticated eukaryotic genetic model. Most of the techniques used for eukaryotic genetic engineering in general were developed in yeast; so let's consider the general routes for transgenesis in yeast.

INTEGRATIVE PLASMIDS The simplest yeast vectors are yeast integrative plasmids (YIps), derivatives of bacterial plasmids into which the yeast DNA of interest has been inserted (Figure 11-26a). When transformed into yeast cells, these plasmids insert into yeast chromosomes, generally by homologous recombination with the resident gene, either by a single or a double crossover (Figure 11-27). As a result, either the entire plasmid is inserted or the targeted allele is replaced by the allele on the plasmid. The latter is an example of *gene replacement*—in this case, the substitution of an engineered gene for the gene originally in the yeast cell. Gene replacement can be used to delete a gene or substitute a mutant allele for its wild-type counterpart or, conversely, to substitute a wild-type allele for a mutant. Such substitutions can be detected by plating cells on a medium that selects for a marker allele on the plasmid.

The bacterial origin of replication is different from eukaryotic origins, and so bacterial plasmids do not replicate in yeast. Therefore, the only way such vectors can generate a stable modified genotype is if they are integrated into the yeast chromosome.

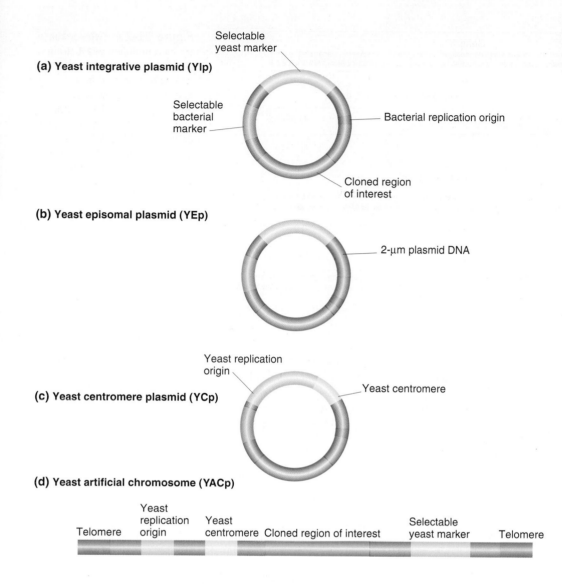

(a) Yeast integrative plasmid (YIp)

Selectable yeast marker

Selectable bacterial marker

Bacterial replication origin

Cloned region of interest

(b) Yeast episomal plasmid (YEp)

2-μm plasmid DNA

(c) Yeast centromere plasmid (YCp)

Yeast replication origin

Yeast centromere

(d) Yeast artificial chromosome (YACp)

Telomere

Yeast replication origin

Yeast centromere

Cloned region of interest

Selectable yeast marker

Telomere

Figure 11-26 Simplified representations of four different kinds of plasmids used in yeast. Each is shown acting as a vector for some genetic region of interest, which has been inserted into the vector. The function of such segments can be studied by transforming a yeast strain of suitable genotype. Selectable markers are needed for the routine detection of the plasmid in bacteria or yeast. Origins of replication are sites needed for the bacterial or yeast replication enzymes to initiate the replication process. (DNA derived from the 2-μm natural yeast plasmid has its own origins of replication.)

AUTONOMOUSLY REPLICATING VECTORS Some yeast strains harbor a circular 6.3-kb natural yeast plasmid that resides in the nucleus and segregates into most daughter cells at meiosis and mitosis. This plasmid, which has a circumference of 2 μm, has become known as the "2 micron" plasmid. If a plasmid containing the transgene also carries the replication origin from the 2-μm plasmid, then that plasmid will be able to replicate as an accessory molecule in the nucleus. This type of plasmid is called a yeast episomal plasmid (YEp) (Figure 11-26b). Although a YEp can replicate autonomously, it occasionally recombines with the homologous chromosomal sequences just like a YIp. Some YEp elements also carry a bacterial replication origin. These elements are very useful "shuttle vectors" because they can be tested in one species and moved immediately to another.

YEAST ARTIFICIAL CHROMOSOMES With any autonomously replicating plasmid, there is the possibility that a daughter cell will not inherit a copy, because the partitioning of plasmid copies to daughter cells depends on where the plasmids are in the cell when the new cell wall is formed. However, if yeast chromosomal centromere and replication origins are added to the plasmid (Figure 11-26c), then the nuclear spindle that ensures the proper segregation of chromosomes will treat the resulting yeast centromere plasmid (YCp) in somewhat the same way as it would treat a chromosome and partition it into daughter cells at cell division. The addition of a centromere is one step toward the creation of an artificial chromosome. A further step is to change a plasmid containing a centromere from circular to linear form and to add the DNA from yeast telomeres to the ends (Figure 11-26d). If this construct contains yeast replication origins (also called autonomous replication sequences), then it constitutes a **yeast artificial chromosome (YAC),** which behaves in many ways like a small yeast chromosome at mitosis and meiosis. For example, when two haploid cells—one bearing a ura^+ YAC and another bearing a ura^- YAC—are brought together to form a diploid, many tetrads will show the clean 2:2 segregations expected if these two elements are behaving as regular chromosomes.

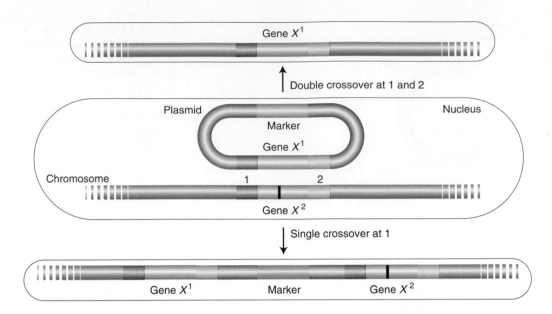

Figure 11-27 Two ways in which a recipient yeast strain bearing a defective gene (X^-) can be transformed by a plasmid bearing an active allele (gene X^+). The mutant site of gene X^- is represented as a vertical black bar. Single crossovers at position 2 also are possible but are not shown.

Recall that YACs were discussed briefly in the context of cloning vectors that carry large inserts. YACs have been extensively used in this regard. Consider, for example, that the size of the region encoding blood-clotting factor VIII in humans is known to span about 190 kb and that the gene for Duchenne muscular dystrophy spans more than 1000 kb. Currently, YACs offer one of the few ways to manipulate such genes intact for genetic engineering.

> **MESSAGE** Yeast vectors can be integrative, can autonomously replicate, or can resemble artificial chromosomes, allowing genes to be isolated, manipulated, and reinserted in molecular genetic analysis.

Genetic engineering in plants

Because of their economic significance in agriculture, many plants have been the subject of genetic analysis aimed at developing improved varieties. Recombinant DNA technology has introduced a new dimension to this effort because the genome modifications made possible by this technology are almost limitless. No longer is genetic diversity achieved solely by selecting variants within a given species. DNA can now be introduced from other species of plants, animals, or even bacteria. In response to new possibilities, a sector of the public has expressed concern that the introduction of **genetically modified organisms (GMOs)** into the food supply may produce unanticipated health problems. The concern about GMOs is one facet of an ongoing public debate about complex public health, safety, ethical, and educational issues raised by the new genetic technologies.

THE TI PLASMID SYSTEM A vector routinely used to produce transgenic plants is the **Ti plasmid,** a natural plasmid derived from a soil bacterium called *Agrobacterium tumefaciens*. This bacterium causes what is known as *crown gall disease*, in which the infected plant produces uncontrolled growths (tumors, or galls), normally at the base (crown) of the stem of the plant. The key to tumor production is a large (200-kb) circular DNA plasmid—*the Ti (tumor-inducing) plasmid*. When the bacterium infects a plant cell, a part of the Ti plasmid—a region called *T-DNA* for transfer DNA—is transferred and inserted, apparently more or less at random, into the genome of the host plant (Figure 11-28). The structure of a Ti plasmid is shown in Figure 11-29. The genes whose products catalyze this T-DNA transfer reside in a region of the Ti plasmid separate from the T-DNA region itself. The T-DNA region encodes several interesting functions that contribute to the bacterium's ability to grow and divide inside the plant cell. These functions include enzymes that contribute to the production of the tumor and other proteins that direct the synthesis of compounds called *opines*, which are important substrates for the bacterium's growth. One important opine is nopaline. Opines are actually synthesized by the infected plant cells, which express the opine-synthesizing genes located in the transferred T-DNA region. The opines are imported into the bacterial cells of the growing tumor and metabolized by enzymes encoded by the bacterium's opine-utilizing genes on the Ti plasmid.

The natural behavior of the Ti plasmid makes it well suited to the role of a vector for plant genetic engineering. If the DNA of interest could be spliced into the T-DNA, then the whole package would be inserted in a stable state into a plant chromosome. This system has indeed been made to work essentially in this way but with some necessary modifications. Let us examine one protocol.

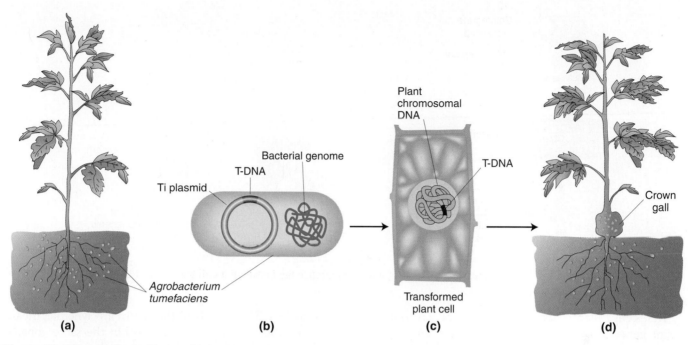

Figure 11-28 Infection by Ti plasmid. In the process of causing crown gall disease, the bacterium *Agrobacterium tumefaciens* inserts a part of its Ti plasmid—a region called T-DNA—into a chromosome of the host plant.

Ti plasmids are too large to be easily manipulated and cannot be readily made smaller, because they contain few unique restriction sites and because much of the plasmid is necessary for either its replication or for the infection and transfer process. Therefore, a properly engineered Ti plasmid is created in steps. The first cloning steps take place in *E. coli*, using an intermediate vector considerably smaller than Ti. The intermediate vector carries the transgene into the T-DNA. This intermediate vector can then be recombined with a "disarmed" Ti plasmid, forming a *cointegrate plasmid* that can be introduced into a plant cell by *Agrobacterium* infection and transformation. An important element on the cointegrate plasmid is a selectable marker that can be used for detecting transformed cells. Kanamycin resistance is one such marker.

As Figure 11-30 shows, bacteria containing the cointegrate plasmid are used to infect cut segments of plant tissue, such as punched-out leaf disks. In infected cells, any genetic material between flanking T-DNA sequences can be inserted into a plant chromosome. If the leaf disks are placed on a medium containing kanamycin, the only plant cells that will undergo cell division are those that have acquired the kan^R gene engineered into the cointegrate plasmid. The transformed cells grow into a clump, or callus, that can be induced to form shoots and roots. These calli are transferred to soil, where they develop into transgenic plants (see Figure 11-30). Typically, only a single copy of the T-DNA region inserts into a given plant genome, where it segregates at meiosis like a regular Mendelian allele (Figure 11-31). The presence of the insert can be verified by screening the transgenic tissue for transgenic genetic markers or the presence of nopaline or by screening purified DNA with a T-DNA probe in a Southern hybridization.

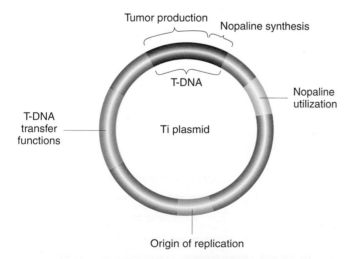

Figure 11-29 Simplified representation of the major regions of the Ti plasmid of *A. tumefaciens*. The T-DNA, when inserted into the chromosomal DNA of the host plant, directs the synthesis of nopaline, which is then utilized by the bacterium for its own purposes. T-DNA also directs the plant cell to divide in an uncontrolled manner, producing a tumor.

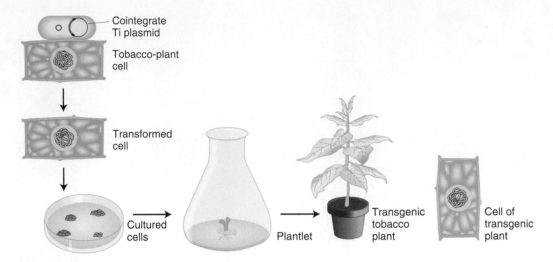

Figure 11-30 The generation of a transgenic plant through the growth of a cell transformed by T-DNA.

Transgenic plants carrying any one of a variety of foreign genes are in current use, including crop plants carrying genes that confer resistance to certain bacterial or fungal pests, and many more are in development. Not only are the qualities of plants themselves being manipulated, but, like microorganisms, plants are also being used as convenient "factories" to produce proteins encoded by foreign genes.

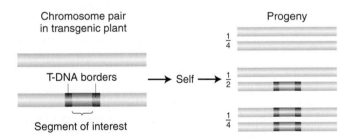

Figure 11-31 Pattern of transmission of T-DNA. The T-DNA region and any DNA inserted into a plant chromosome in a transgenic plant are transmitted in a Mendelian pattern of inheritance.

Genetic engineering in animals

Transgenic technologies are now being employed with many animal model systems. We will focus on the three animal models most heavily used for basic genetic research: the nematode *Caenorhabditis elegans*, the fruit fly *Drosophila melanogaster*, and the mouse *Mus musculus*. Versions of many of the techniques considered so far can also be applied in these animal systems.

TRANSGENESIS IN *C. ELEGANS* The method used to introduce trangenes into *C. elegans* is simple: transgenic DNAs are injected directly into the organism, typically as plasmids, cosmids, or other DNAs cloned in bacteria. The injection strategy is determined by the worm's reproduc-

tive biology. The gonads of the worm are syncitial, meaning that there are many nuclei within the same gonadal cell. One syncitial cell is a large proportion of one arm of the gonad, and the other syncitial cell is the bulk of the other arm (Figure 11-32a). These nuclei do not form individual cells until meiosis, when they begin their transformation into individual eggs or sperm. A solution of DNA is injected into the syncitial region of one of the arms, thereby exposing more than 100 nuclei to the transforming DNA. By chance, a few of these nuclei will incorporate the DNA (remember, the nuclear membrane breaks down in the course of division, and so the cytoplasm into which the DNA is injected becomes continuous with the nucleoplasm). Typically, the transgenic DNA forms multi-copy *extrachromosomal arrays* (Figure 11-32b) that exist as independent units outside the chromosomes. More

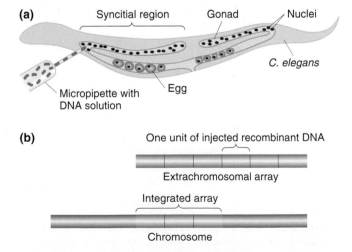

Figure 11-32 Creation of *C. elegans* transgenes. (a) Method of injection. (b) The two main types of transgenic results: extrachromosomal arrays and arrays integrated in ectopic chromosomal locations.

rarely, the transgenes will become integrated into an ectopic position in a chromosome, still as a multicopy array. Unfortunately, sequences may become scrambled within the arrays, complicating the work of the researcher.

TRANSGENESIS IN _D. MELANOGASTER_ Transgenesis in _D. melanogaster_ requires a more complex technique but avoids the difficulties of multicopy arrays. It proceeds by a mechanism that differs from those discussed so far, based on the properties of a **transposable element** called the P element, which acts as the vector. A transposable element is a DNA segment that is capable of moving from one location in the genome to other locations. We will consider transposable elements and how they move in much more detail in Chapter 13.

For our purposes here, all we need to know is that P elements come in two types (Figure 11-33a):

- One type of element, 2912 bp long, encodes a protein called a **transposase** that is necessary for P elements to move to new positions in the genome. This type of element is termed "autonomous" because it can be transposed through the action of its own transposase enzyme.

- The transposase has been deleted from the second type of element, called a nonautonomous element. Still, a nonautonomous element can move to a new genomic location if transposase is supplied by an autonomous element. The one requirement is that the nonautonomous element contain the first 200 bp and final 200 bp of the autonomous element, which includes the sequences that the transposase needs to recognize for transposition. Moreover, any DNA inserted in between the ends of a nonautonomous P element will be transposed as well.

As with _C. elegans_, the DNA is injected into a syncitium—in this case, the early _Drosophila_ embryo (Figure 11-33b). More precisely, the DNA is injected at the site of germ-cell formation, at the posterior pole of the embryo. The adults that grow from the injected embryo will typically not express the transgene but will contain some transgenic germ cells, and these cells will be expressed in the offspring.

What type of vector carries the injected DNA? To produce transgenic _Drosophila_, we must inject _two_ separate bacterial recombinant plasmids. One contains the autonomous P element that supplies the coding sequences for the transposase. This element is the P helper plasmid. The other, the P-element vector, is an engineered nonautonomous element containing the ends of the P element and, inserted between these ends, the piece of cloned DNA that we want to incorporate as a transgene into the fly genome. A DNA solution containing both of these plasmids is injected into the posterior pole of the syncitial embryo. The P transposase

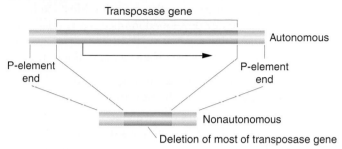

(a) P elements

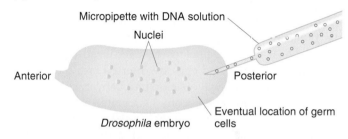

(b) Method of injection

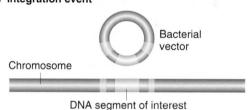

(c) Integration event

Figure 11-33 Creation of _D. melanogaster_ transgenes. (a) The overall structure of autonomous and nonautonomous P transposable elements. (b) Method of injection. (c) The circular P-element vector *(right)* and a typical integration event at an ectopic chromosomal location *(left)*. Note that the bacterial vector sequences do not become integrated into the genome; rather, in integration, exactly one copy of the DNA segment is contained between the P-element ends.

expressed from the injected P helper plasmid catalyzes the insertion of the P-element vector into the fly genome. The nature of the transposase enzymatic reaction guarantees that only a single copy of the element inserts at a given location (Figure 11-33c).

How can we detect the progeny that develop from gametes that successfully receive the cloned DNA? Typically, they are detected because they express a dominant wild-type transgenic allele of a gene for which the recipient strain carries a recessive mutant allele.

Transposable elements are widely used in transgenesis, in plants as well as insects. Perhaps the best-known plant example is the Activator transposable-element system first described in _Zea mays_ (corn), which has been developed into a transgenic cloning vector for use in many plants.

TRANSGENESIS IN *M. MUSCULUS* Mice are the most important models for mammalian genetics. Most exciting, much of the technology developed in mice is potentially applicable to humans. There are two strategies for transgenesis in mice, each of which has its advantages and disadvantages:

- *Ectopic insertions.* Transgenes are inserted randomly in the genome, usually as multicopy arrays.

- *Gene targeting.* The transgene sequence is inserted into a location occupied by a homologous sequence in the genome. That is, the transgene replaces its normal homologous counterpart.

Ectopic insertions To insert transgenes in random locations, the procedure is simply to inject a solution of bacterially cloned DNA into the nucleus of a fertilized egg (Figure 11-34a). Several injected eggs are inserted into the female oviduct, where some will develop into baby mice. At some later stage, the transgene becomes integrated into the chromosomes of random nuclei. On occasion, the transgenic cells form part of the germ line, and in these cases an injected embryo will develop into a mouse adult whose germ cells contain the transgene inserted at some random position in one of the chromosomes (Figure 11-34b). Some of the progeny of these adults will inherit the transgene in all cells. There will be an array of multiple gene copies at each point of insertion, but the location, size, and structure of the arrays will be different for each integration event. The technique does give rise to some problems: (1) the expression pattern of the randomly inserted genes may be abnormal (called a **position effect**) because the local chromosome environment lacks the gene's normal regulatory sequences, and (2) DNA rearrangements can occur inside the multicopy arrays (in essence, mutating the sequences). Nonetheless, this technique is much more efficient and less laborious than gene targeting.

Gene targeting Gene targeting enables us to eliminate or modify the function encoded by a gene. In one application,

a mutant allele can be repaired through **gene replacement** in which a wild-type allele substitutes for a mutant one in its normal chromosomal location. Gene replacement avoids both the position effect and DNA rearrangements associated with ectopic insertion, because a single copy of the gene is inserted in its normal chromosomal environment.

Gene targeting in the mouse is carried out in cultured embryonic stem cells (ES cells). In general, stem cells are undifferentiated cells in a given tissue or organ that divide asymmetrically to produce a progeny stem cell and a cell that will differentiate into a terminal cell type. ES cells are special stem cells that can differentiate to form any cell type in the body—including, most importantly, the germ line.

To illustrate the process of gene targeting, we look at how it achieves one of its typical outcomes—namely, the substitution of an inactive gene for the normal gene. Such a targeted inactivation is called a **gene knockout.** First, a cloned, disrupted gene that is inactive is targeted to replace the functioning gene in a culture of ES cells, producing ES cells containing a gene knockout (Figure 11-35a). DNA constructs containing the defective gene are injected into the nuclei of cultured ES cells. The defective gene inserts far more frequently into nonhomologous (ectopic) sites than into homologous sites (Figure 11-35b), and so

---

Figure 11-35 Producing cells that contain a mutation in one specific gene, known as a targeted mutation or a gene knockout. (a) Copies of a cloned gene are altered in vitro to produce the targeting vector. The gene shown here has been inactivated by the insertion of the neomycin-resistance gene (*neo*^R) into a protein-coding region (exon 2) of the gene and had been inserted into a vector. The *neo*^R gene will serve later as a marker to indicate that the vector DNA took up residence in a chromosome. The vector has also been engineered to carry a second marker at one end: the herpes *tk* gene. These markers are standard, but others could be used instead. When a vector, with its dual markers, is complete, it is introduced into cells isolated from a mouse embryo. (b) When homologous recombination occurs *(left),* the homologous regions on the vector, together with any DNA in between but excluding the marker at the tip, take the place of the original gene. This event is important because the vector sequences serve as a useful tag for detecting the presence of this mutant gene. In many cells, though, the full vector (complete with the extra marker at the tip) inserts ectopically *(middle)* or does not become integrated at all *(bottom).* (c) To isolate cells carrying a targeted mutation, all the cells are put into a medium containing selected drugs—here a neomycin analog (G418) and ganciclovir. G418 is lethal to cells unless they carry a functional *neo*^R gene, and so it eliminates cells in which no integration of vector DNA has taken place (yellow). Meanwhile, ganciclovir kills any cells that harbor the *tk* gene, thereby eliminating cells bearing a randomly integrated vector (red). Consequently, virtually the only cells that survive and proliferate are those harboring the targeted insertion (green). [After M. R. Capecchi, "Targeted Gene Replacement." Copyright 1994 by Scientific American, Inc. All rights reserved.]

(a)

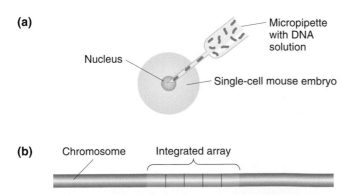

Micropipette with DNA solution

Nucleus

Single-cell mouse embryo

(b)   Chromosome   Integrated array

Figure 11-34 Creation of *M. musculus* transgenes inserted in ectopic chromosomal locations. (a) Method of injection. (b) A typical ectopic integrant, with multiple copies of the recombinant transgene inserted in an array.

(a) Production of ES cells with a gene knockout

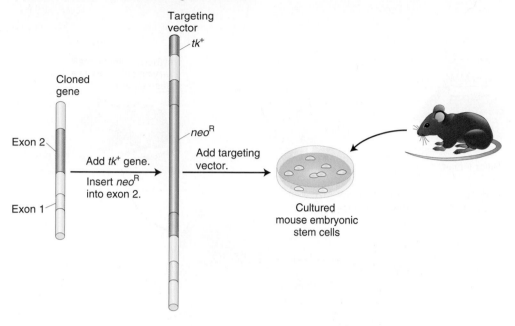

Targeting vector

tk^+

Cloned gene

Exon 2

Exon 1

Add tk^+ gene.

Insert neo^R into exon 2.

neo^R

Add targeting vector.

Cultured mouse embryonic stem cells

(b) Targeted insertion of vector DNA by homologous recombination

Possible outcomes

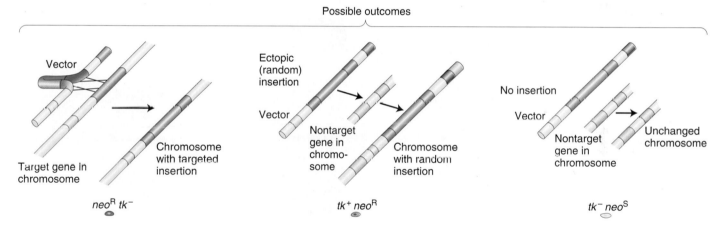

Vector

Target gene in chromosome

Chromosome with targeted insertion

$neo^R\ tk^-$

Ectopic (random) insertion

Vector

Nontarget gene in chromosome

Chromosome with random insertion

$tk^+\ neo^R$

No insertion

Vector

Nontarget gene in chromosome

Unchanged chromosome

$tk^-\ neo^S$

(c) Selection of cells with gene knockout

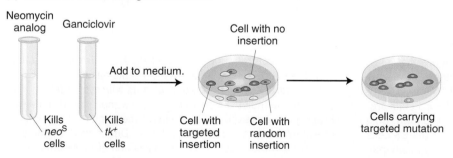

Neomycin analog

Ganciclovir

Add to medium.

Kills neo^S cells

Kills tk^+ cells

Cell with no insertion

Cell with targeted insertion

Cell with random insertion

Cells carrying targeted mutation

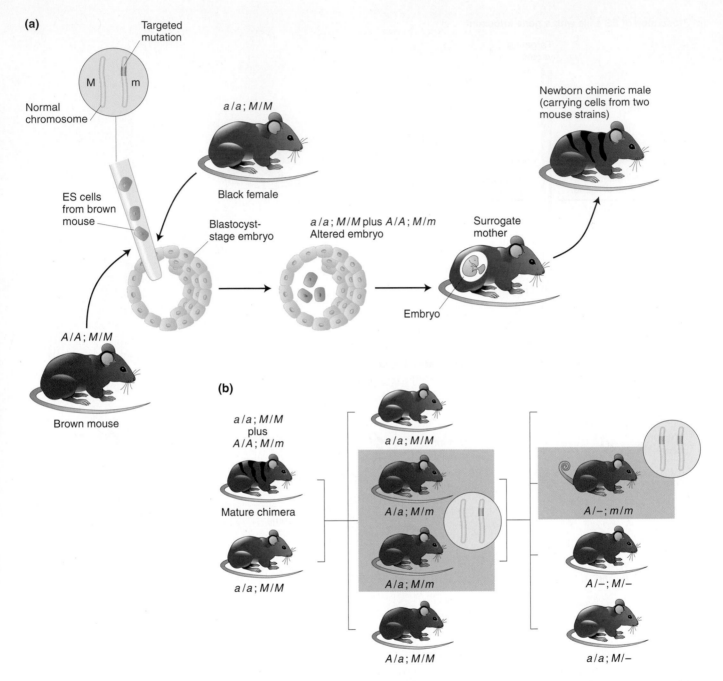

Figure 11-36 Producing a knockout mouse carrying the targeted mutation. (a) Embryonic stem (ES) cells are isolated from an agouti (brown) mouse strain (*A/A*) and altered to carry a targeted mutation (*m*) in one chromosome. The ES cells are then inserted into young embryos, one of which is shown. Coat color of the future newborns is a guide to whether the ES cells have survived in the embryo. Hence, ES cells are typically put into embryos that, in the absence of the ES cells, would acquire a totally black coat. Such embryos are obtained from a black strain that lacks the dominant agouti allele (*a/a*). The embryos containing the ES cells grow to term in surrogate mothers. Agouti shading intermixed with black indicates those newborns in which the ES cells have survived and proliferated. (Such mice are called *chimeras* because they contain cells derived from two different strains of mice.) Solid black coloring, in contrast, indicates that the ES cells have perished, and these mice are excluded. *A* represents agouti, *a* black; *m* is the targeted mutation, and *M* is its wild-type allele. (b) Chimeric males are mated with black (nonagouti) females. Progeny are screened for evidence of the targeted mutation (green in *inset*) in the gene of interest. Direct examination of the genes in the agouti mice reveals which of those animals *(boxed)* inherited the targeted mutation. Males and females carrying the mutation are mated with one another to produce mice whose cells carry the chosen mutation in both copies of the target gene *(inset)* and thus lack a functional gene. Such animals *(boxed)* are identified definitively by direct analyses of their DNA. The knockout in this case results in a curly-tail phenotype. [After M. R. Capecchi, "Targeted Gene Replacement." Copyright 1994 by Scientific American, Inc. All rights reserved.]

the next step is to select the rare cells in which the defective gene has replaced the functioning gene as desired. How is it possible to select ES cells that contain a rare gene replacement? The genetic engineer can include drug-resistant alleles in the DNA construct arranged in such a way that replacements can be distinguished from ectopic insertions. An example is shown in Figure 11-35c.

In the second part of the procedure, the ES cells that contain one copy of the disrupted gene of interest are injected into an early embryo (Figure 11-36a). Adults grown from these embryos are crossed with normal mates. The resulting progeny are chimeric, having some tissue derived from the original lines and some from the transplanted ES lines. Chimeric mice are then mated with their siblings to produce homozygous mice with the knockout in every copy of the gene (Figure 11-36b). Mice containing the targeted transgene in each of their cells are identified by molecular probes for sequences unique to the transgene.

> **MESSAGE** Germ-line transgenic techniques have been developed for all well-studied eukaryotic species. These techniques depend on an understanding of the reproductive biology of the recipient species.

Human gene therapy

A boy is born with a disease that makes his immune system ineffective. Diagnostic testing determines that he has a recessive genetic disorder called SCID (severe combined immunodeficiency disease), more commonly known as *bubble-boy disease*. This disease is caused by a mutation in the gene coding for the blood enzyme adenosine deaminase (ADA). As a result of the loss of this enzyme, the precursor cells that give rise to one of the cell types of the immune system are missing. Because this boy has no ability to fight infection, he has to live in a completely isolated and sterile environment—that is, a bubble in which the air is filtered for sterility (Figure 11-37). No pharmaceutical or other conventional therapy is available to treat this disease. Giving the boy a tissue transplant containing the precursor cells from another person would not work, because such cells would end up creating an immune response against the boy's own tissues (graft versus host disease). In the past two decades, techniques have been developed that offer the possibility of a different kind of transplantation therapy—**gene therapy**—in which, in the present case, a normal ADA gene is "transplanted" into cells of the boy's immune system, thereby permitting their survival and normal function.

The general goal of gene therapy is to attack the genetic basis of disease at its source: to "cure" or correct an abnormal condition caused by a mutant allele by intro-

Figure 11-37 A boy with SCID living in a protective bubble.
[UPI/Bettmann/Corbis.]

ducing a transgenic wild-type allele into the cells. This technique has been successfully applied in many experimental organisms and has the potential in humans to correct some hereditary diseases, particularly those associated with single-gene differences. Although gene therapy has been attempted for several such diseases, thus far there are no clear instances of success. However, the implications of gene therapy are so far-reaching that the approach merits consideration here.

To understand the approach, consider an example showing how gene therapy corrected a growth-hormone deficiency in mice (Figure 11-38). Mice with the recessive mutation *little (lit)* are dwarves because they lack a protein (the growth-hormone-releasing hormone receptor, or GHRHR) that is necessary to induce the pituitary to secrete mouse growth hormone into the circulatory system. The initial step in correcting this deficiency was to inject about 5000 copies of a transgene construct into homozygous *lit/lit* eggs. This construct was a 5-kb linear DNA fragment that contained coding sequences for rat growth hormone *(RGH)* fused to regulatory sequences for the mouse metallothionein gene. These regulatory sequences lead to the expression of any immediately adjacent gene in the presence of heavy metals. The eggs were then implanted into the uteri of surrogate mother mice, and the baby mice were born and raised. About 1 percent of these babies turned out to be transgenic, showing increased size when heavy metals were administered in the course of development. A representative transgenic mouse was then crossed with a homozygous *lit/lit* female. The ensuing pedigree is shown in Figure 11-38a. Here we see that mice from two to three times the weight of their *lit/lit* relatives are produced in subsequent generations (Figure 11-38b). These larger mice are always heterozygous in this pedigree—showing that the rat growth-hormone transgene acts as a dominant allele.

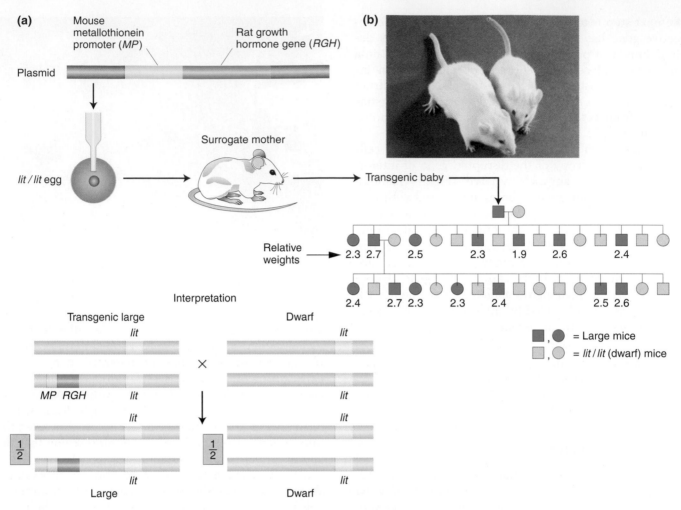

Figure 11-38 Gene therapy in mice. (a) The rat growth-hormone gene (*RGH*), under the control of a mouse promoter region that is responsive to heavy metals, is inserted into a plasmid and used to produce a transgenic mouse. *RGH* compensates for the inherent dwarfism *(lit/lit)* in the mouse. *RGH* is inherited in a Mendelian dominant pattern in the ensuing mouse pedigree. (b) Transgenic mouse. The mice are siblings, but the mouse on the left was derived from an egg transformed by injection with a new gene composed of the mouse metallothionein promoter fused to the rat growth-hormone structural gene. (This mouse weighs 44 g, and its untreated sibling weighs 29 g.) The new gene is passed on to progeny in a Mendelian manner and so is proved to be chromosomally integrated. [From R. L. Brinster.]

Thus, the introduction of the *RGH* transgene achieved a "cure" in the sense that progeny did not show the abnormal phenotype.

This particular example makes some important points about the gene therapy process. The genetic defect occurs in GHRHR, the gene that encodes a regulator of mouse growth-hormone production. However, the gene therapy is not an attempt to correct the original defect in the GHRHR gene. Rather, the gene therapy works by bypassing the need for GHRHR and producing growth hormone by another route, specifically by expressing rat growth hormone under the control of an inducible promoter and in tissues where GHRHR is not needed for growth-hormone release. (You may ask why

rat growth hormone was used instead of mouse growth hormone. The recombinant rat growth-hormone gene produced both mRNA and protein with sequences distinguishable from the mouse versions, and so both molecules could be directly measured.)

Let us now turn to the status of various technical approaches. Two basic types of gene therapy can be applied to humans: germ line and somatic. The goal of **germ-line gene therapy** (Figure 11-39a) is the more ambitious: to introduce transgenic cells into the germ line as well as into the somatic-cell population. Not only would this type of therapy achieve a cure of the person treated, but his or her children also would carry the therapeutic transgene. The cure of the mouse *lit* reces-

(a) Germ-line therapy

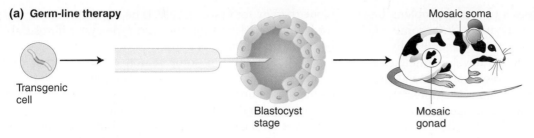

Transgenic cell

Blastocyst stage

Mosaic soma

Mosaic gonad

(b) Somatic therapy

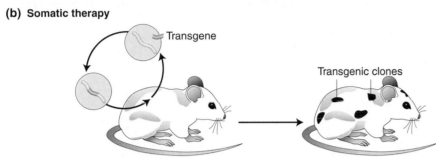

Transgene

Transgenic clones

Figure 11-39 **Types of gene therapy in mammals.**

sive defect is an example of germ-line gene therapy. At present, these technologies depend on ectopic integration or gene replacement occurring by chance, and these events are sufficiently infrequent to make germ-line gene therapy impractical for now.

Somatic gene therapy (Figure 11-39b) attempts to correct a disease phenotype by treating *some* somatic cells in the affected person. No transgenes get into the germ line. At present, it is not possible to render an entire body transgenic, and so the method addresses diseases caused by genes that are expressed predominantly in one tissue. In such cases, it is likely that not all the cells of that tissue need to become transgenic; a portion of cells carrying the transgene can relieve the overall disease symptoms. The method proceeds by removing some cells from a patient with the defective genotype and making these cells transgenic by introducing copies of the cloned wild-type gene. The transgenic cells are then reintroduced into the patient's body, where they provide normal gene function.

Let us return to the boy with severe combined immunodeficiency disease described at the beginning of this section. In his case, the defect is in stem cells of the immune system, which can be isolated from bone marrow. If the defect in adenosine deaminase in these stem cells can be repaired by the introduction of a normal ADA gene, then the progeny of these repaired cells will populate his immune system and cure the SCID condition. Because only a small set of stem cells needs to be repaired to cure the disease, SCID is ideally suited for gene therapy. How has gene therapy been attempted for SCID?

The method uses a specific kind of virus (a retrovirus) containing the normal ADA transgene spliced into its genome, replacing most of the viral genes; this retrovirus is unable to form progeny viruses and is thus aviru-

lent, or "disarmed." The natural cycle of retroviruses includes the integration of the viral genome at some location in one of the host cell's chromosomes. The viral genome will carry the ADA transgene along with it into the chromosome. Blood stem cells are removed from the bone marrow of the person who has SCID, the retroviral vector containing the ADA transgene is added, and the transgenic cells are reintroduced into the blood system. Thus far, no long-term cure has been achieved in any cases, but there have been some encouraging results (Figure 11-40).

Figure 11-40 **Ashanti de Silva, the first person to receive gene therapy.** She was treated for SCID, and her symptoms have been ameliorated. [Courtesy of Van de Silva.]

The retroviral vector poses a potential problem, because the integrating virus may insert into some unknown resident gene and inactivate it. Several individuals with an X-linked form of SCID developed leukemia after gene therapy, possibly as a result of gene inactivation. Another problem is that a retrovirus infects only proliferating cells, such as blood cells, and thus cannot be used to treat the many heritable disorders that affect tissues in which cells rarely or never divide.

Another vector used in human gene therapy is the adenovirus. This virus normally infects respiratory epithelia, injecting its genome into the epithelial cells lining the surface of the lung. The viral genome does not integrate into a chromosome but persists extrachromosomally in the cells, which eliminates the problem of the vector inactivating a resident gene. Another advantage of the adenovirus as a vector is that it attacks nondividing cells, and so most tissues are susceptible in principle. Adenovirus is an appropriate choice of vector for treating cystic fibrosis, a disease of the respiratory epithelium.

Gene therapy for cystic fibrosis is being attempted by introducing viruses bearing the wild-type cystic fibrosis allele through the nose as a spray.

Although there is some reason to believe that the technical hurdles of somatic gene therapy will be overcome, these hurdles are considerable. One hurdle is how to target the transgenic delivery system to the appropriate tissue for a given disease. Another is how to build the transgene to ensure consistently high levels of expression. Still another is how to protect against potentially harmful side effects, such as might be caused by misexpression of the transgenic gene. These are major areas of gene-therapy research.

> **MESSAGE** The technologies of transgenesis are currently being applied to humans with the specific goal of applying gene therapy to the correction of certain heritable disorders. The technical, societal, and ethical challenges of these technologies are considerable and are active areas of research and debate.

KEY QUESTIONS REVISITED

• **How is a gene isolated and amplified by cloning?**

Genomic DNA is cut up with restriction enzymes and spliced into a vector chromosome, which is then replicated in a bacterial cell.

• **How are specific DNAs or RNAs identified in mixtures?**

Most simply by probing with a cloned sequence that will hybridize to the molecule in question (both have to be denatured—that is, single stranded).

• **How is DNA amplified without cloning?**

The polymerase chain reaction is used. Two specific primers that flank the region in question are hybridized

to denatured DNA. Then DNA polymerases shuttle back and forth between the primers, amplifying the flanked sequence exponentially.

• **How is amplified DNA used in genetics?**

The many uses include obtaining the sequence of a region or an entire genome, as a probe, and as a sequence to be inserted as a transgene to modify a recipient genome.

• **How are DNA technologies applied to medicine?**

Two applications are in diagnosing hereditary disease and in gene therapy of hereditary disease.

SUMMARY

The methodologies of recombinant DNA rely on the two fundamental principles of molecular biology: (1) hydrogen bonding of complementary antiparallel nucleotide sequences and (2) interactions between specific proteins and specific nucleotide sequences. Examples of the applications of the principles are numerous. We exploit complementarity to join together DNA fragments with complementary sticky ends; to probe for specific sequences in clones and in Southern and Northern blots; and to prime cDNA synthesis, PCR, and DNA-sequencing reactions. The specificity of interactions between proteins and nucleotide sequences allows restriction endonucleases to cut at specific target-recognition sites and transposases to transpose specific transposons.

Recombinant DNA is made by cutting donor DNA into fragments that are each pasted into an individual vec-

tor DNA. The vector DNA is often a bacterial plasmid or viral DNA. Donor DNA and vector DNA are cut by the same restriction endonuclease at specific sequences. The most useful restriction enzymes for DNA cloning are those that cut at palindromic sequences and that cut the two DNA strands at slightly offset positions, leaving a single-strand terminal overhang at each cut end. Each overhang has a DNA sequence that is characteristic for a given restriction enzyme. Vector and donor DNA are joined in a test tube by complementary binding of the overhangs under conditions that permit complementary single-strand overhangs to hydrogen bond stably. The strands held together by base-pair complementarity of the overhangs are covalently bonded through the action of DNA ligase to form covalent phosphodiester linkages, making an intact phosphate–sugar backbone for each DNA strand.

The vector–donor DNA construct is amplified inside host cells by tricking the basic replication machinery of the cell into replicating the recombinant molecules. Thus the vector must contain all the necessary signals for proper replication and segregation in that host cell. For plasmid-based systems, the vector must include an origin of replication and selectable markers such as drug resistance that can be used to ensure that the plasmid is not lost from the host cell. In bacteriophage, the vector must include all sequences necessary for carrying the bacteriophage (and the hitchhiking foreign DNA) through the lytic growth cycle. The result of amplification is multiple copies of each recombinant DNA construct, called clones.

Often, finding a specific clone requires screening a full genomic library. A genomic library is a set of clones, packaged in the same vector, that together represents all regions of the genome of the organism in question. The number of clones that constitute a genomic library depends on (1) the size of the genome in question and (2) the insert size tolerated by the particular cloning vector system. Similarly, a cDNA library is a representation of the total mRNA set produced by a given tissue or developmental stage in a given organism. A comparison of a genomic region and its cDNA can be a source of insight into the locations of transcription start and stop sites and boundaries between introns and exons.

Labeled single-stranded DNA or RNA probes are important "bait" for fishing out similar or identical sequences from complex mixtures of molecules, either in genomic or cDNA libraries or in Southern and Northern blotting. The general principle of the technique for identifying clones or gel fragments is to create a filter-paper "image" of the colonies or plaques on an agar petri dish culture or of the nucleic acids that have been separated in an electric field passed through a gel matrix. The DNA or RNA is then denatured and mixed with a denatured probe, labeled with a fluorescent dye or a radioactive label. After unbound probe has been washed off, the location of the probe is detected either by observing its fluorescence or, if radioactive, by exposing the sample to X-ray film. The locations of the probe correspond to the locations of the relevant DNA or RNA in the original petri dish or electrophoresis gel.

The polymerase chain reaction (PCR) is a powerful method for the direct amplification of a relatively small sequence of DNA from within a complex mixture of DNA, without the need of a host cell or very much starting material. The key is to have primers that are complementary to flanking regions on each of the two DNA strands. These regions act as sites for polymerization. Multiple rounds of denaturation, priming, and polymerization amplify the sequence of interest exponentially.

Recombinant DNA molecules can be used to assess the risk of a genetic disease. One class of diagnostics uses restriction fragments as markers for the presence of a variant of a gene associated with a hereditary disease. In such cases, the method can be adapted to identify the presence of a mutant allele or its wild-type counterpart or both.

Transgenes are engineered DNA molecules that are introduced and expressed in eukaryotic cells. They can be used to demonstrate an association between a recessive mutation and a specific DNA sequence by rescuing that mutation with a wild-type transgene. They can also be used to engineer a novel mutation or to study the regulatory sequences that constitute part of a gene. Transgenes can be introduced as extrachromosomal molecules or they can be integrated into a chromosome, either in random (ectopic) locations or in place of the homologous gene, depending on the system. Typically, the mechanisms used to introduce a transgene depend on an understanding and exploitation of the reproductive biology of the organism.

Gene therapy is the extension of transgenic technology to the treatment of human diseases. To correct disease conditions, somatic-cell gene therapy attempts to introduce into specific *somatic* tissues a transgene that either replaces the mutant allele or suppresses the mutant phenotype. Germ-line gene therapy attempts to introduce a transgene into the germ line that either corrects the mutant defect or bypasses it.

KEY TERMS

amniocentesis (p. 364)

antibody (p. 353)

autoradiogram (p. 351)

BAC (bacterial artificial chromosome) (p. 349)

cDNA library (p. 351)

chorionic villus sampling (CVS) (p. 364)

chromosome walk (p. 356)

complementary DNA (cDNA) (p. 343)

cosmid (p. 349)

dideoxy sequencing (p. 357)

DNA cloning (p. 343)

DNA ligase (p. 346)

DNA technology (p. 342)

donor DNA (p. 343)

functional complementation (p. 355)

gel electrophoresis (p. 354)

gene knockout (p. 372)

gene replacement (p. 372)

gene therapy (p. 375)

genetic engineering (p. 342)

genetically modified organism (GMO) (p. 368)

genomic library (p. 351)

genomics (p. 342)

germ-line gene therapy (p. 376)

hybridization (p. 345)

mutant rescue (p. 355)

Northern blotting (p. 354)

open reading frame (ORF) (p. 360)

PAC (P1 artificial chromosome)
   (p. 349)

palindrome (p. 344)

polymerase chain reaction
   (PCR) (p. 360)

position effect (p. 372)

positional cloning (p. 356)

probe (p. 351)

recombinant DNA (p. 343)

restriction enzyme (p. 344)

restriction fragment (p. 344)

restriction map (p. 357)

Sanger sequencing (p. 357)

somatic gene therapy (p. 377)

Southern blotting (p. 354)

Ti plasmid (p. 368)

transgene (p. 366)

transgenic organism (p. 366)

transposable element (p. 371)

transposase (p. 371)

vector (p. 343)

yeast artificial chromosome
   (YAC) (p. 349)

SOLVED PROBLEMS

1. In Chapter 9, we studied the structure of tRNA molecules. Suppose that you want to clone a fungal gene that encodes a certain tRNA. You have a sample of the purified tRNA and an *E. coli* plasmid that contains a single *Eco*RI cutting site in a tet^R (tetracycline-resistance) gene, as well as a gene for resistance to ampicillin (amp^R). How can you clone the gene of interest?

Solution

You could use the tRNA itself or a cloned cDNA copy of it to probe for the DNA containing the gene. One method is to digest the genomic DNA with *Eco*RI and then mix it with the plasmid, which you also have cut with *Eco*RI. After transformation of an $amp^S tet^S$ recipient, Amp^R colonies are selected, indicating successful transformation. Of these Amp^R colonies, select the colonies that are Tet^S. These Tet^S colonies will contain vectors with inserts in the tet^R gene, and a great number of them are needed to make the library. Test the library by using the tRNA as the probe. Those clones that hybridize to the probe will contain the gene of interest.

Alternatively, you can subject *Eco*RI-digested genomic DNA to gel electrophoresis and then identify the correct band by probing with the tRNA. This region of the gel can be cut out and used as a source of enriched DNA to clone into the plasmid cut with *Eco*RI. You then probe these clones with the tRNA to confirm that these clones contain the gene of interest.

2. The restriction enzyme *Hin*dIII cuts DNA at the sequence AAGCTT, and the restriction enzyme *Hpa*II cuts DNA at the sequence CCGG. On average, how frequently will each enzyme cut double-stranded DNA? (In other words, what is the average spacing between restriction sites?)

Solution

We need consider only one strand of DNA, because both sequences will be present on the opposite strand at the same site owing to the symmetry of the sequences.

The frequency of the six-base-long *Hin*dIII sequence is $(1/4)^6$, or $1/4096$, because there are four possibilities at each of the six positions. Therefore, the average spacing between *Hin*dIII sites is approximately 4 kb. For *Hpa*II, the frequency of the four-base-long sequence is $(1/4)^4$, or $1/256$. The average spacing between *Hpa*II sites is approximately 0. 25 kb.

3. A yeast plasmid carrying the yeast $leu2^+$ gene is used to transform haploid $leu2^-$ yeast cells. Several leu^+-transformed colonies appear on a medium lacking leucine. Thus, $leu2^+$ DNA presumably has entered the recipient cells, but now we have to decide what has happened to it inside these cells. Crosses of transformants to $leu2^-$ testers reveal that there are three types of transformants, A, B, and C, representing three different fates of the $leu2^+$ gene in the transformation. The results are:

$$\text{type A} \times leu2^- \longrightarrow \tfrac{1}{2}\, leu^-$$
$$\tfrac{1}{2}\, leu^+ \times \text{standard } leu2^+$$
$$\longrightarrow \tfrac{3}{4}\, leu^+$$
$$\tfrac{1}{4}\, leu^-$$

$$\text{type B} \times leu2^- \longrightarrow \tfrac{1}{2}\, leu^-$$
$$\tfrac{1}{2}\, leu^+ \times \text{standard } leu2^+$$
$$\longrightarrow 100\%\ leu^+$$
$$0\%\ leu^-$$

$$\text{type C} \times leu2^- \longrightarrow 100\%\ leu^+$$

What three different fates of the $leu2^+$ DNA do these results suggest? Be sure to explain *all* the results according to your hypotheses. Use diagrams if possible.

Solution

If the yeast plasmid does not integrate, then it replicates independently of the chromosomes. In meiosis, the daughter plasmids would be distributed to the daughter cells, resulting in 100 percent transmission. This percentage was observed in transformant type C.

If one copy of the plasmid is inserted, the resulting offspring from a cross with a *leu2⁻* line would have a ratio of 1 *leu⁺* :1 *leu⁻*. This ratio is seen in type A and type B.

When the resulting *leu⁺* cells are crossed with standard *leu2⁻* lines, the data from type A cells suggest that the inserted gene is segregating independently of the standard *leu2⁺* gene, and so the *leu2⁺* transgene has inserted ectopically into another chromosome.

Type A

$leu2⁻$ $leu2⁺$

When the resulting cells are crossed with a standard wild-type strain,

$leu2⁺$

then the following segregation results:

$$\frac{1}{2}\,leu2⁻ \Bigg\langle \begin{array}{l} \frac{1}{2}\,leu2⁺ \longrightarrow \frac{1}{4}+ \\ \frac{1}{2}\,\text{no allele} \longrightarrow \frac{1}{4}- \end{array}$$

$$\frac{1}{2}\,leu2⁺ \Bigg\langle \begin{array}{l} \frac{1}{2}\,leu2⁺ \longrightarrow \frac{1}{4}+ \\ \frac{1}{2}\,\text{no allele} \longrightarrow \frac{1}{4}+ \end{array}$$

The data from type B cells suggest that the inserted gene has replaced the standard *leu2⁺* allele at its normal locus.

Type B

$leu2⁺$

When crossed with a standard wild type, all the progeny will be *leu⁺*.

PROBLEMS

BASIC PROBLEMS

1. From this chapter, make a list of all the examples of (a) the hybridization of single-stranded DNAs and (b) proteins that bind to and then act on DNA.

2. What is sodium hydroxide used for in making cDNA?

3. Compare and contrast the use of the word *recombinant* as used in the phrases (a) recombinant DNA and (b) recombinant frequency.

4. Why is ligase needed to make recombinant DNA? What would be the immediate consequence in the cloning process if someone forgot to add it?

5. Assume that bases in DNA are positioned randomly (not actually true). If we select a seven-base-pair sequence randomly, what is the probability that it will be a palindrome?

6. In the PCR process, if we assume that each cycle takes 5 minutes, how many fold amplification would be accomplished in 1 hour?

7. The position of the gene for the protein actin in the haploid fungus *Neurospora* is known from the complete genome sequence. If you had a slow-growing mutant that you suspected of being an actin mutant and you wanted to verify that it was one, would you (a) clone the mutant by using convenient restriction sites flanking the actin gene and then sequence it or (b) amplify the mutant sequence by using PCR and then sequence it?

8. Arrange the following cloning vectors in order of the size of the DNA inserts that they can carry: (a) BAC; (b) cosmid; (c) YAC; (d) plasmid pBR322.

9. In what way was a fungus important in the cloning of the human gene for alkaptonurea?

10. A plasmid is cut by a certain restriction enzyme and run on a gel. Bands revealed by ethidium bromide staining are of sizes 3, 5, and 10 kb. Can you *definitely* conclude that the plasmid has a total size of 18 kb?

11. You obtain the DNA sequence of a mutant of a 2-kb gene in which you are interested and it shows base differences at three positions, all in different codons. One is a silent change, but the other two are missense changes (they code for new amino acids). How would you demonstrate that these changes are real mutations and not sequencing errors? (Assume that sequencing is about 99.9 percent accurate.)

12. In the diagnosis of sickle-cell anemia, assume that the β-globin intron is cut out, cloned, and used as a probe in a Southern analysis of *MstII*–cut genomic DNA from a certain fetus. The blot shows two bands, one 1.1 kb and another 1.3 kb in size. What is the genotype of the fetus?

13. In a T-DNA transformation of a plant with a transgene from a fungus (not found in plants), the presumptive transgenic plant does not express the expected phenotype of the transgene. How would you demonstrate that the transgene is in fact present?

14. How would you produce a mouse that is homozygous for a rat growth-hormone transgene?

15. Why was cDNA and not genomic DNA used in the commercial cloning of the human insulin gene?

16. Calculate the average distances (in nucleotide pairs) between the restriction sites in organism X for the

following restriction enzymes, assuming an AT:GC ratio of 50:50.

*Alu*I	5′	AGCT	3′
	3′	TCGA	5′
*Eco*RI	5′	GAATTC	3′
	3′	CTTAAG	5′
*Acy*I	5′	G Pu CG Py C	3′
	3′	C Py GC Pu G	5′

(**Note:** Py = any pyrimidine; Pu = any purine.)

17. A circular bacterial plasmid (pBP1) has a single *Hind*III restriction-enzyme site in the middle of a tetracycline-resistance gene (*tet*R). Fruit fly genomic DNA is digested with *Hind*III, and a library is made in the plasmid vector pBP1. Probing reveals that clone 15 contains a specific *Drosophila* gene of interest. Clone 15 is studied by restriction analysis with *Hind*III and another restriction enzyme, *Eco*RV. The ethidium bromide–stained electrophoretic gel shows bands as in the accompanying diagram (the control is the plasmid pBP1 vector without an insert). The sizes of the bands (in kilobases) are shown alongside. (**Note:** Circular molecules do not give intense bands on this type of gel; so you can assume that all bands represent linear molecules.)

*Hind*III		*Eco*RV		*Hind*III and *Eco*RV	
Control	15	Control	15	Control	15
5 —	5 —	5 —			
			4.5 —		
	2.5 —		3 —	3 —	3 —
				2 —	2 —
					1.5 —
					1 —

a. Draw restriction maps for plasmid pBP1 with and without the insert, showing the sites of the target sequences and the approximate position of the *tet*R gene.

b. If the same *tet*R gene cloned in a completely nonhomologous vector is made radioactive and used as a probe in a Southern blot of this gel, which bands do you expect to appear radioactive on an autoradiogram?

c. If the same gene of interest from a fly closely related to *Drosophila* has been cloned in a nonhomologous vector and this clone is used as a probe for the same gel, what bands do you expect to see on the autoradiogram?

 UNPACKING PROBLEM 17

1. Which plasmid described in this chapter seems closest in type to pBP1?

2. What is the importance of the single *Hind*III restriction site?

3. Why is it important that the single site is in a resistance gene? Would it be useful if not?

4. What is the effect of the insertion of donor DNA into the resistance gene? Is this effect important for this problem?

5. What is a library? What type was used in this experiment and does it matter for this problem?

6. What kind of probing would have shown that clone 15 contains the gene of interest? and is it relevant to the present problem?

7. What is an electrophoretic gel?

8. What function does ethidium bromide serve in this experiment?

9. Does the gel shown represent a Southern blot or a Northern blot or neither?

10. Generically, what types of molecules are visible on the gel?

11. How many fragments are produced if a circular molecule is cut once?

12. How many fragments are produced if a circular molecule is cut twice?

13. Can you write a simple formula relating the number of restriction enzyme sites in a circular molecule to the number of fragments produced?

14. If one enzyme produces *n* fragments and another produces *m* fragments, how many fragments are produced if both enzymes are used?

15. In the diagram, at what positions were the DNA samples loaded into the gel?

16. Why are the smaller-molecular-weight fragments at the bottom of the gel?

17. What is the total molecular weight of the fragments in all the lanes? What patterns do you see?

18. Is it a coincidence that the 3- and 2-kb fragments together equal the 5-kb fragment in size?

19. Is it a coincidence that the 1.5- and 1-kb fragments together equal the 2.5-kb fragment in size?

20. If a fragment produced by one enzyme disappears when the DNA is treated with that same enzyme plus another enzyme, what does the disappearance signify?

21. What determines whether a probe will hybridize to a DNA blot (denatured)?

22. In part c, why is it stressed that a nonhomologous vector is used?

Now attempt to solve the problem.

18. You have a purified DNA molecule, and you wish to map restriction-enzyme sites along its length. After digestion with *Eco*RI, you obtain four fragments: 1, 2, 3, and 4. After digestion of each of these fragments

with *Hind*II, you find that fragment 3 yields two sub-fragments (3_1 and 3_2) and that fragment 2 yields three (2_1, 2_2, and 2_3). After digestion of the entire DNA molecule with *Hind*II, you recover four pieces: A, B, C, and D. When these pieces are treated with *Eco*RI, piece D yields fragments 1 and 3_1, A yields 3_2 and 2_1, and B yields 2_3 and 4. The C piece is identical with 2_2. Draw a restriction map of this DNA.

19. After *Drosophila* DNA has been treated with a restriction enzyme, the fragments are inserted into plasmids and selected as clones in *E. coli*. With the use of this "shotgun" technique, every DNA sequence of *Drosophila* in a library can be recovered.

    a. How would you identify a clone that contains DNA coding for the protein actin, whose amino acid sequence is known?

    b. How would you identify a clone coding for a specific tRNA?

20. You have isolated and cloned a segment of DNA that is known to be a unique sequence in the human genome. It maps near the tip of the X chromosome and is about 10 kb in length. You label the 5′ ends with ^{32}P and cleave the molecule with *Eco*RI. You obtain two fragments: one is 8.5 kb long; the other is 1.5 kb. You split a solution containing the 8.5-kb fragment into two samples, partly digesting one with *Hae*III and the other with *Hind*II. You then separate each sample on an agarose gel. By autoradiography, you obtain the following results:

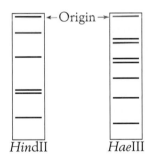

Draw a restriction-enzyme map of the complete 10-kb molecule.

21. A linear fragment of DNA is cleaved with the individual restriction enzymes *Hind*III and *Sma*I and then with a combination of the two enzymes. The fragments obtained are:

*Hind*III	2.5 kb, 5.0 kb
*Sma*I	2.0 kb, 5.5 kb
*Hind*III and *Sma*I	2.5 kb, 3.0 kb, 2.0 kb

    a. Draw the restriction map.

    b. The mixture of fragments produced by the combined enzymes is cleaved with the enzyme *Eco*RI, resulting in the loss of the 3-kb fragment (band stained with ethidium bromide on an agarose gel) and the appearance of a band stained with ethidium bromide representing a 1.5-kb fragment. Mark the *Eco*RI cleavage site on the restriction map.

    (Problem 21 courtesy of Joan McPherson. From A. J. F. Griffiths and J. McPherson, *1001 Principles of Genetics.* W. H. Freeman and Company, 1989.)

22. The gene for β-tubulin has been cloned from *Neurospora* and is available. List a step-by-step procedure for cloning the same gene from the related fungus *Podospora*, using as the cloning vector the pBR *E. coli* plasmid shown here, where *kan* = kanamycin and *tet* = tetracycline:

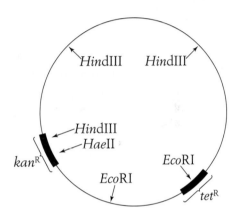

23. In any particular transformed eukaryotic cell (say, of *Neurospora*), how could you tell if the transforming DNA (carried on a circular bacterial vector)

    a. replaced the resident gene of the recipient by double crossing-over or single crossing-over?

    b. was inserted ectopically?

24. In an electrophoretic gel across which is applied a powerful electric alternating pulsed field, the DNA of the haploid fungus *Neurospora crassa* ($n = 7$) moves slowly but eventually forms seven bands, which represent DNA fractions that are of different sizes and hence have moved at different speeds. These bands are presumed to be the seven chromosomes. How would you show which band corresponds to which chromosome?

25. The protein encoded by the alkaptonuria gene is 445 amino acids long, yet the gene spans 60 kb. How is this possible?

26. In yeast, you have sequenced a piece of wild-type DNA and it clearly contains a gene, but you do not know what gene it is. Therefore, to investigate further, you would like to find out its mutant phenotype. How would you use the cloned wild-type gene to do so? Show your experimental steps clearly.

CHALLENGING PROBLEMS

27. A circular bacterial plasmid containing a gene for tetracycline resistance was cut with restriction-enzyme *Bgl*II. Electrophoresis showed one band of 14 kb.

a. What can be deduced from this result?

The plasmid was cut with *Eco*RV and electrophoresis produced two bands, one of 2.5 kb and the other 11.5 kb.

b. What can be deduced from this result?

Digestion with both enzymes together resulted in three bands of 2.5, 5.5, and 6 kb.

c. What can be deduced from this result?

Plasmid DNA cut with *Bgl*II was mixed and ligated with donor DNA fragments, also cut with *Bgl*II, to make recombinant DNA molecules. All recombinant clones proved to be tetracycline sensitive.

d. What can be deduced from this result?

One recombinant clone was cut with *Bgl*II, and fragments of 4 and 14 kb were observed.

e. Explain this result.

The same clone was treated with *Eco*RV and fragments of 2.5, 7, and 8.5 were observed.

f. Explain these results by showing a restriction map of the recombinant DNA.

28. a. A fragment of mouse DNA with *Eco*RI sticky ends carries the gene *M*. This DNA fragment, which is 8 kb long, is inserted into the bacterial plasmid pBR322 at the *Eco*RI site. The recombinant plasmid is cut with three different restriction-enzyme treatments. The patterns of ethidium bromide fragments, after electrophoresis on agarose gels, are shown in this diagram:

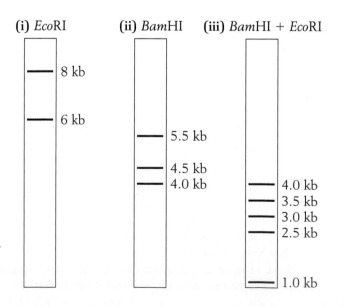

(i) *Eco*RI — 8 kb, 6 kb
(ii) *Bam*HI — 5.5 kb, 4.5 kb, 4.0 kb
(iii) *Bam*HI + *Eco*RI — 4.0 kb, 3.5 kb, 3.0 kb, 2.5 kb, 1.0 kb

A Southern blot is prepared from gel iii. Which fragments will hybridize to a probe (^{32}P) consisting of pBR plasmid DNA?

b. Gene *X* is carried on a plasmid consisting of 5300 nucleotide pairs (5300 bp). Cleavage of the plasmid with the restriction enzyme *Bam*HI gives fragments 1, 2, and 3, as indicated in the following diagram (*B* = *Bam*HI restriction site). Tandem copies of gene *X* are contained within a single *Bam*HI fragment. If gene *X* encodes a protein X of 400 amino acids, indicate the approximate positions and orientations of the gene *X* copies.

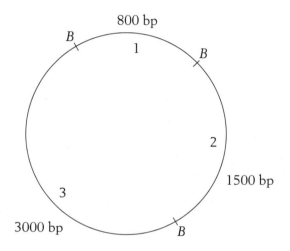

(Problem 28 courtesy of Joan McPherson. From A. J. F. Griffiths and J. McPherson, *1001 Principles of Genetics.* W. H. Freeman and Company, 1989.)

29. Prototrophy is often the phenotype selected to detect transformants. Prototrophic cells are used for donor DNA extraction; then this DNA is cloned and the clones are added to an auxotrophic recipient culture. Successful transformants are identified by plating the recipient culture on minimal medium and looking for colonies. What experimental design would you use to make sure that a colony that you hope is a transformant is not, in fact,

a. a prototrophic cell that has entered the recipient culture as a contaminant?

b. a revertant (mutation back to prototrophy by a second mutation in the originally mutated gene) of the auxotrophic mutation?

30. Two children are investigated for the expression of a gene (*D*) that encodes an important enzyme for muscle development. The results of the studies of the gene and its product follow.

Child 1

Southern blot

Probe: ³²P gene *D*: Autoradiograms
Northern blots
(samples at different
stages of development)

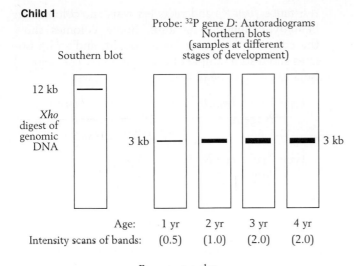

	Age:	1 yr	2 yr	3 yr	4 yr
Intensity scans of bands:		(0.5)	(1.0)	(2.0)	(2.0)

Enzyme samples

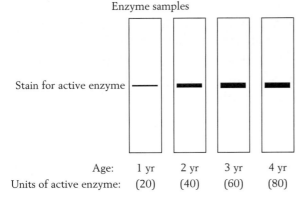

	Age:	1 yr	2 yr	3 yr	4 yr
Units of active enzyme:		(20)	(40)	(60)	(80)

Child 2

Southern blot   Northern blots

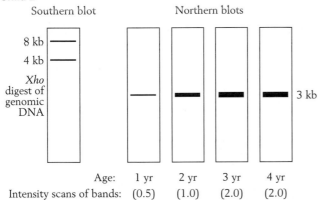

	Age:	1 yr	2 yr	3 yr	4 yr
Intensity scans of bands:		(0.5)	(1.0)	(2.0)	(2.0)

Enzyme samples

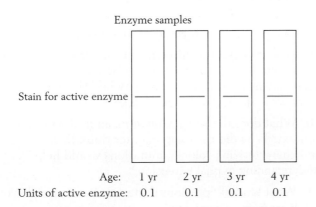

	Age:	1 yr	2 yr	3 yr	4 yr
Units of active enzyme:		0.1	0.1	0.1	0.1

For child 2, the enzyme activity of each stage was very low and could be estimated only at approximately 0.1 unit at ages 1, 2, 3, and 4.

a. For both children, draw graphs representing the developmental expression of the gene. (Fully label both axes.)

b. How can you explain the very low levels of active enzyme for child 2? (Protein degradation is only one possibility.)

c. How might you explain the difference in the Southern blot for child 2 compared with that for child 1?

d. If only one mutant gene has been detected in family studies of the two children, define the individual children as either homozygous or heterozygous for gene *D*.

(Problem 30 courtesy of Joan McPherson. From A. J. F. Griffiths and J. McPherson, *1001 Principles of Genetics.* W. H. Freeman and Company, 1989.)

31. A cloned fragment of DNA was sequenced by using the dideoxy method. A part of the autoradiogram of the sequencing gel is represented here.

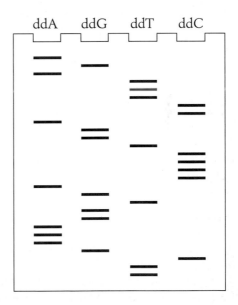

a. Deduce the nucleotide sequence of the DNA nucleotide chain synthesized from the primer. Label the 5′ and 3′ ends.

b. Deduce the nucleotide sequence of the DNA nucleotide chain used as the template strand. Label the 5′ and 3′ ends.

c. Write out the nucleotide sequence of the DNA double helix (label the 5′ and 3′ ends).

d. How many of the six reading frames are "open" as far as you can tell?

32. The cDNA clone for the human gene encoding tyrosinase was radioactively labeled and used in a Southern analysis of *Eco*R1-digested genomic DNA of wild-type mice. Three mouse fragments were found to be radioactive (were bound by the probe). When albino mice were used in this Southern analysis, no genomic fragments bound to the probe. Explain these results in relation to the nature of the wild-type and mutant mouse alleles.

33. Transgenic tobacco plants were obtained in which the vector Ti plasmid was designed to insert the gene of interest plus an adjacent kanamycin-resistance gene. The inheritance of chromosomal insertion was followed by testing progeny for kanamycin resistance. Two plants typified the results obtained generally. When plant 1 was backcrossed with wild-type tobacco, 50 percent of the progeny were kanamycin resistant and 50 percent were sensitive. When plant 2 was backcrossed with the wild type, 75 percent of the progeny were kanamycin resistant, and 25 percent were sensitive. What must have been the difference between the two transgenic plants? What would you predict about the situation regarding the gene of interest?

34. A cystic fibrosis mutation in a certain pedigree is due to a single nucleotide-pair change. This change destroys an *Eco*RI restriction site normally found in this position. How would you use this information in counseling members of this family about their likelihood of being carriers? State the precise experiments needed. Assume that you find that a woman in this family is a carrier, and it transpires that she is married to an unrelated man who also is a heterozygote for cystic fibrosis, but, in his case, it is a different mutation in the same gene. How would you counsel this couple about the risks of a child's having cystic fibrosis?

35. Bacterial glucuronidase converts a colorless substance called X-Gluc into a bright-blue indigo pigment. The gene for glucuronidase also works in plants if given a plant promoter region. How would you use this gene as a reporter gene to find out in which tissues a plant gene that you have just cloned is normally active? (Assume that X-Gluc is easily taken up by the plant tissues.)

36. A *Neurospora* geneticist is interested in the genes that control hyphal extension. He decides to clone a sample of these genes. From previous mutational analyses of *Neurospora*, a common type of mutant is known to have a small-colony ("colonial") phenotype on plates, caused by abnormal hyphal extension. Therefore, he decides to do a tagging experiment by using transforming DNA to produce colonial mutants by insertional mutagenesis. He transforms *Neurospora* cells by using a bacterial plasmid carrying a gene for benomyl

resistance *(ben-R)* and recovers resistant colonies on benomyl-containing medium. Some colonies show the colonial phenotype being sought, and so he isolates and tests a sample of these colonies. The colonial isolates prove to be of two types as follows:

type 1 *col . ben-R* × wild type (*+ . ben-S*)
Progeny   1/2 *col . ben-R*
                       1/2 *+ . ben-S*

type 2 *col . ben-R* × wild type (*+ . ben-S*)
Progeny   1/4 *col . ben-R*
                       1/4 *col . ben-S*
                       1/4 *+ . ben-R*
                       1/4 *+ . ben-S*

a. Explain the difference between these two types of results.

b. Which type should he use to try to clone the genes affecting hyphal extension?

c. How should he proceed with the tagging protocol?

d. If a probe specific for the bacterial plasmid is available, which progeny should be hybridized by this probe?

 UNPACKING PROBLEM 36

1. What is hyphal extension, and why do you think anyone would find it interesting?

2. How does the general approach of this experiment fit in with the general genetic approach of mutational dissection?

3. Is *Neurospora* haploid or diploid? And is this property relevant to the problem?

4. Is it appropriate to transform a fungus (a eukaryote) with a bacterial plasmid? Does it matter?

5. What is transformation? and in what way is it useful in molecular genetics?

6. How are cells prepared for transformation? (Do you think their cell walls need to be removed enzymatically?)

7. What is the fate of transforming DNA if successful trasformation takes place?

8. Draw a successful plasmid entry into the host cell, and draw a representation of a successful stable transformation.

9. Does it make any difference to the protocol to know what benomyl is? What role is the benomyl-resistance gene playing in this experiment? Would the experiment work with another resistance marker?

10. What does the word *colonial* mean in the present context? Why did the experimenter think that finding and characterizing colonial mutations would help in understanding hyphal extension?

11. What kind of "previous mutational analyses" do you think are being referred to?

12. Draw the appearance of a typical petri dish after transformation and selection. Pay attention to the appearance of the colonies.

13. What is tagging? How does it relate to insertional mutagenesis? How does insertion cause mutation?

14. How are crosses made and progeny isolated in *Neurospora*?

15. Is recombination relevant to this question? Can an RF value be calculated? What does it mean?

16. Why are there only two colonial types? Is it possible to predict which type will be more common?

17. What is a probe? How are probes useful in molecular genetics? How would the probing experiment be done?

18. How would you obtain a probe specific to the bacterial plasmid?

37. The plant *Arabidopsis thaliana* was transformed by using the Ti plasmid into which a kanamycin-resistance gene had been inserted in the T-DNA region. Two kanamycin-resistant colonies (A and B) were selected, and plants were regenerated from them. The plants were allowed to self-pollinate, and the results were as follows:

Plant A selfed $\longrightarrow \frac{3}{4}$ progeny resistant to kanamycin

$\frac{1}{4}$ progeny sensitive to kanamycin

Plant B selfed $\longrightarrow \frac{15}{16}$ progeny resistant to kanamycin

$\frac{1}{16}$ progeny sensitive to kanamycin

a. Draw the relevant plant chromosomes in both plants.

b. Explain the two different ratios.

38. Two different circular yeast plasmid vectors (YP1 and YP2) were used to transform *leu*$^-$ cells into *leu*$^+$. The resulting *leu*$^+$ cultures from both experiments were crossed with the same *leu*$^-$ cell of opposite mating type. Typical results were as follows:

YP1 *leu*$^+$ × *leu*$^-$ $\longrightarrow$ all progeny *leu*$^+$ and the DNA of all these progeny showed positive hybridization to a probe specific to the vector YP1

YP2 *leu*$^+$ × *leu*$^-$ $\longrightarrow$ $\frac{1}{2}$ progeny *leu*$^+$ and hybridize to vector probe to YP2
$\frac{1}{2}$ progeny *leu*$^-$ and do not hybridize to YP2 probe

a. Explain the different actions of these two plasmids during transformation.

b. If total DNA is extracted from YP1 and YP2 transformants and digested with an enzyme that cuts once within the vector (but not within the insert), predict the results of electrophoresis and Southern analyses of the DNA; use the specific plasmid as a probe in each case.

39. A linear 9-kb *Neurospora* plasmid, mar1, has the following restriction map:

*Bgl*II	*Xba*I	*Bgl*II

This plasmid was suspected to sometimes integrate into genomic DNA. To test this idea, the central large *Bgl*II fragment was cloned into a pUC plasmid vector and was used as a probe in a Southern analysis of *Xba*I-digested genomic DNA from a mar1-containing strain. Predict what the autoradiogram will look like if

a. the plasmid never integrates.

b. the plasmid occasionally integrates into genomic DNA.

INTERACTIVE GENETICS MegaManual CD-ROM Tutorial

Molecular Biology: Gene Cloning

For additional coverage of the topics in this chapter, refer to the Interactive Genetics CD-ROM included with the Solutions MegaManual. The section on Gene Cloning in the Molecular Biology activity includes animated tutorials on restriction enzymes, sequencing, Southern blotting techniques, and PCR, as well as four interactive problems to sharpen your skills.

EXPLORING GENOMES A Web-Based Bioinformatics Tutorial

Finding Conserved Domains

Conserved protein sequences are a reflection of the conservation of amino acid residues necessary for structure, regulation, or catalytic function. Often, groups of residues can be identified as being a pattern or signature of a particular type of enzyme or regulatory domain. In the Genomics tutorial at www.whfreeman.com/mga, you will learn how to find the conserved domains in a complex protein.

12

GENOMICS

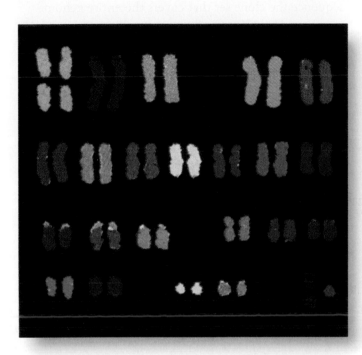

The human nuclear genome viewed as a set of labeled DNA.
The DNA of each chromosome has been labeled with a dye
that emits fluorescence at one specific wavelength (producing
a specific color). [Evelin Schrock and Thomas Ried.]

KEY QUESTIONS

- What is genomics?

- How are sequence maps of genomes
  produced?

- How can genomic sequence maps help find
  a gene of interest?

- What is the nature of the information
  in the genome?

- What new questions can be addressed
  by genome-level analysis?

OUTLINE

12.1 The nature of genomics

12.2 The sequence map of a genome

12.3 Creating genomic sequence maps

12.4 Using genomic sequence
     to find a specific gene

12.5 Bioinformatics: meaning from
     genomic sequence

12.6 Take-home lessons from the genomes

12.7 Functional genomics

CHAPTER OVERVIEW

Science sometimes advances at alarming and unexpected rates. Most older geneticists of today began their careers trying to understand genes by working exclusively with their mutant phenotypes. They only dreamed that in their lifetimes they might see the hypothetical concept of a gene turned into clear reality, as both DNA sequence and function. The complete sequence of whole genomes was not even on their intellectual horizons. Yet today many genomes have been sequenced, with more on the way, and the use of these sequences has become routine in genetic analysis. Indeed the knowledge of entire genomes has revolutionized not

only genetics but most fields of biological research. In this chapter, we examine the development and the operation of this exciting new field, broadly called **genomics**—the study of genomes in their entirety.

A crucial step in genomics is to obtain a full genome sequence, which is done by the automated sequencing of many clones. There are two basic strategies. The first is to randomly sequence large numbers of clones and assemble them on the basis of overlapping sequence. The second is to assemble overlapping clones by aligning various molecular markers such as restriction sites and then sequence the clone set that covers the entire genome.

After the genome sequence has been obtained, the real biology begins. For example, Figure 12-1 shows

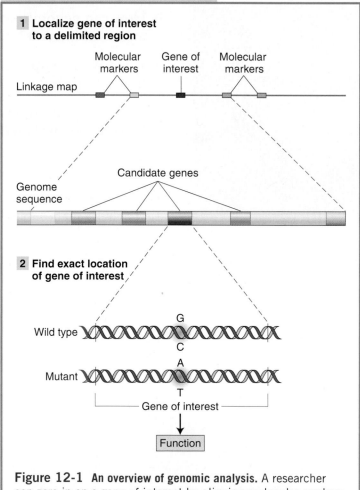

CHAPTER OVERVIEW Figure

1 Localize gene of interest to a delimited region

2 Find exact location of gene of interest

Figure 12-1 An overview of genomic analysis. A researcher can zero in on a gene of interest by aligning molecular markers in a linkage map with molecular markers on the genome sequence map. First, the researcher narrows the location of the gene to a delimited region of the genome *(top)*. Then the researcher uses other techniques to find the exact location of the gene *(bottom)*.

some of the steps in locating a gene of interest in a genome sequence. The genome sequence can be used to find the DNA sequence of a gene with an interesting mutant phenotype. This can be accomplished by matching the genome sequence with other maps such as recombination maps or cytogenetic maps. For example, if the position of a gene on a recombination map is known, then one can zero in on the appropriate gene candidates in that location of the genome sequence, possibly by PCR. This method is faster than standard gene cloning.

The DNA sequence of the genome becomes the taking-off point for a whole new set of analyses aimed at the structure, function, and evolution of the genome and its components.

- *Bioinformatics* analyses the information content of entire genomes. This information includes the numbers and types of genes and gene products, as well as docking sites on DNA and RNA that allow functional product to be produced at the correct time and place.
- *Comparative genomics* considers the genomes of closely and distantly related species for evolutionary insight.
- *Functional genomics* uses various automated procedures to delineate networks of interacting genes active during some developmental process.

12.1 The nature of genomics

After the development of recombinant DNA technology in the 1970s, research laboratories typically undertook the cloning and sequencing of "one gene at a time" (as discussed in Chapter 11), and then only after having had first found out something interesting about that gene from a classic mutational analysis. In the 1980s, some scientists realized that a large team of researchers making a concerted effort could clone and sequence the *entire* genome of a selected organism. Such **genome projects** would then make the clones and the sequence publicly available as resources. One appeal of this resource is that once researchers become interested in a gene of a species whose genome had been sequenced, they need only find out where that gene is located on the map of the genome to be able to zero in on its sequence and potential function. By this means, a gene could be characterized much more rapidly than by cloning and sequencing it from scratch, a project that could take several years to carry out. Indeed, this quicker approach is now a reality for most model organisms. In a similar way, in human genetics, the genome sequence can aid in the pinpointing of disease-causing genes.

In a broader perspective, the genome projects had the appeal of providing some glimmer of the principles on which genomes are built. Obtaining a genome sequence is like having unearthed some ancient tablet in a language that we can't decipher. The human genome, for example, is 24 strings of base pairs, representing the X and Y chromosomes and the 22 autosomes. In total, the human genome contains 3 billion base pairs of DNA. Although we might convince ourselves that we understand a single gene of interest, there are undoubtedly important insights into human biology encoded within the overall patterns that the genes show as a whole. The major challenge of genomics today is genomic literacy: how do we read the storehouse of information enciphered in the sequence of such genomes?

The basic techniques needed for sequencing entire genomes were already available in the 1980s—cloning vectors for making genomic libraries, PCR for amplifying genes, and DNA sequencing machines (see Chapter 11). But the scale that was needed to sequence a complex genome was, as an engineering project, far beyond the capacity of the research community then. Genomics in the late 1980s and the 1990s evolved out of large research centers that could integrate these elemental technologies into an industrial-level production line. These centers developed robotics and automation to carry out the many thousands of cloning steps and millions of sequencing reactions necessary to assemble the sequence of a complex organism. With these centers in place, the late 1990s and 2000s have been the golden age of genome sequencing. At current rates, a genome center takes a day to sequence a bacterial species, a week for a fungus, from 1 to 2 months for an insect, and from 1 to 2 years for a mammal.

Genomics has already had a major effect on how biological research is performed. Just as NASA's lunar landing project produced all sorts of spinoffs in science and engineering, such as miniaturization of electronics and computers, genome projects have produced many scientific and technological spinoffs. Genomics has encouraged researchers to develop ways of experimenting on, and computationally analyzing, the genome as a whole, rather than simply one gene at a time. It has also demonstrated the value of collecting large-scale data sets in advance of their use, with great potential to address specific research problems. Finally, it has changed the sociology of biological research, demonstrating the value of large collaborative research networks as a complement to the small, independent research laboratories (which still flourish). These effects will only increase as more information, technology, and insight emerge. In the final section of this chapter, we will explore some ways that genomics affects basic and applied research in the early years of the twenty-first century. Other applications will emerge in succeeding chapters.

12.2 The sequence map of a genome

When people encounter new territory, one of their first activities is to create a map that they can use as a common reference when discussing that territory. This practice has been true for geographers, oceanographers, and astronomers and it is equally true for geneticists. Geneticists use many kinds of maps to explore the terrain of a genome. In other chapters, we saw examples such as linkage maps based on inheritance patterns of genes with alleles causing phenotypic differences, cytogenetic maps based on the location of microscopically visible features such as rearrangement breakpoints, and restriction maps of restriction-endonuclease-sensitive sites in DNA.

The highest-resolution map is the complete DNA sequence of the genome—that is, the complete sequence of nucleotides A, T, C, and G of each double helix in the genome. Because making a complete sequence map of a genome is such a massive undertaking, of a sort not seen before in biology, new strategies must be used, all based on automation.

Turning sequence reads into a sequence map

You've probably seen a magic act in which the magician cuts up a newspaper page into a great many pieces, mixes it in his hat, says a few magic words and *viola!* an intact newspaper page reappears. Basically, that's how genomic sequence maps are produced. The approach is to (1) break a genome up into thousands to millions of more or less random little pieces, (2) read the sequence of each little piece, (3) overlap the little pieces where their sequences are identical, and (4) continue overlapping ever-larger pieces until you've accounted for all the little pieces (Figure 12-2). At that point, you've assembled a sequence map of a genome.

Why does this process require automation? To understand why, let's consider the human genome, which contains about 3×10^9 bp (or 3 Gigabase pairs = 3 Gbp) of DNA. Suppose we could purify the DNA intact from each of the 24 human chromosomes (X, Y, and the 22 autosomes), separately put each of these 24 DNA samples into a sequencing machine, and read their sequences directly from one telomere to the other. Creating a complete sequence map would be utterly straightforward, like reading a book with 24 chapters—albeit a very, very long book with 3 billion characters. Unfortu-

nately, such a sequencing machine does not exist. Rather, automated fluorescence-based sequencing of the sort discussed in Chapter 11 is the current state of the art in DNA sequencing technology. Individual sequencing reactions (called *sequencing reads*) provide letter strings that are generally about 600 bases long. Such lengths are tiny compared with the DNA of a single chromosome. For example, an individual read is only 0.0002 percent of the longest human chromosome (about 3×10^8 bp of DNA) and only about 0.00002 percent of the entire human genome. Even within a 300,000-bp BAC clone, an individual sequence read is only about 0.2 percent of the length of that clone. Thus, one major challenge facing a genome project is **sequence assembly**—that is, building up all of the individual reads into a **consensus sequence,** a sequence for which there is consensus (or agreement) that it is an authentic representation of the sequence for each of the DNA molecules in that genome. (The consensus is partly between researchers and partly between the different data sets.)

Let's look at these numbers in a somewhat different way to understand the scale of the problem. As with any experimental observation, automated sequencing machines do not always give perfectly accurate sequence reads. The error rate is not constant; it depends on such factors as the dyes that are attached to the sequenced molecules, the purity and homogeneity of the starting DNA sample, and the specific sequence of base pairs in the DNA sample. Thus, to ensure accuracy, genome projects conventionally obtain 10 independent sequence reads for each base pair in a genome. Tenfold coverage ensures that chance errors in the reads do not give a false reconstruction of the consensus sequence. Given an average sequence read of about 600 bases of DNA and a human genome of 3 billion base pairs, 50 million successful independent reads are required to give us our 10-fold average coverage of each base pair. However, not all reads are successful; the failure rate is about 20 percent, and so the real numbers are about 60 million attempted reads to cover the human genome. Thus, the amount of information and material to be tracked is enormous. To try to minimize both human error and the need for people to carry out highly repetitive tasks, genome project laboratories have implemented automation, computer tracking using bar coding, and computer analysis systems wherever practical.

For these reasons, the preparation of clones, DNA isolation, electrophoresis, and sequencing protocols have all been adapted to automation. For example, one of the recent breakthroughs has been the development of production-line sequencing machines that run around the clock without human intervention, producing about 96 reads every 3 hours. A genome center with 200 of these sequencing machines can produce about 150,000 reads in a single day. These figures show how a single sequencing center has the capacity to assemble the

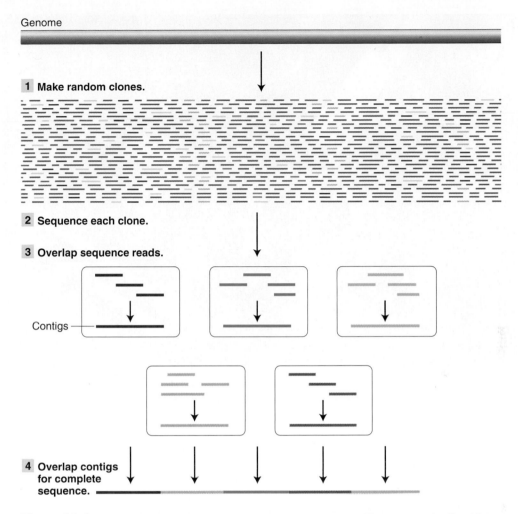

Figure 12-2 The logic of creating a sequence map of the genome. The genome is diced into small pieces that are then cloned and sequenced. The resulting sequence reads are overlapped by matching identical sequences in different clones until a consensus sequence of each DNA double helix in the genome is produced.

sequence of a mammal (3 Gbp) in 1 to 2 years. Figure 12-3 shows a sequencing assembly line.

What are the goals of sequencing the genome? First, we strive to produce a consensus sequence that is a true and accurate representation of the genome, starting with one individual or standard strain from which the DNA was obtained. This sequence will then serve as a reference sequence for the species. We now know that there are many differences in DNA sequence between different individual organisms within a species and even between the maternally and the paternally contributed genomes within a single diploid individual organism. Thus, no one genome sequence truly represents the genome of the entire species. Nonetheless, the genome sequence serves as a standard or reference with which other sequences can be compared, and it can be analyzed to determine the information encoded within the DNA, such as the array of encoded RNAs and polypeptides.

When is a genome sequence complete?

Like written manuscripts, genome sequences can range from *draft* quality (the general outline is there, but there are lots of typos, badly formed sentences, sections that need rearranging, and so forth), to *finished* quality (a very low rate of typos, some missing sections but you've done everything that is currently possible to fill in these sections) to truly *complete* (no typos, every base pair absolutely correct from telomere to telomere). Whether a genome is sequenced to "draft" or "finished" standards is a cost–benefit judgment. It is relatively easy to create a draft but very hard to create a "finished" sequence by current methods. In the following sections, we will consider the general methods for producing draft and finished genome sequence assemblies, as well as some of the features of a genome that challenge genome-sequencing projects.

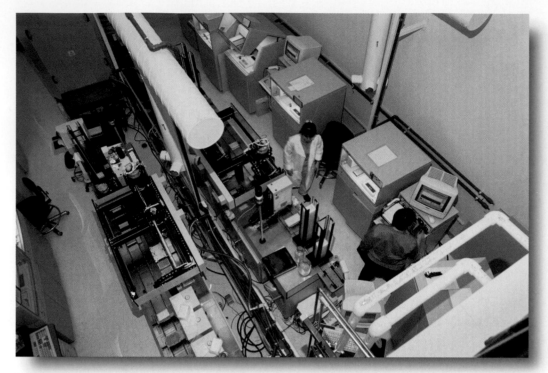

Figure 12-3 Part of the automated production line of a major human genome sequencing center.
All this equipment is used for rapid processing of huge numbers of clones for DNA sequencing.

The problem of repetitive DNA in genome sequencing

A big stumbling block in reassembling a consensus sequence of a eukaryotic genome is the existence of numerous classes of repeated sequences, some arranged in tandem and others dispersed. Why are they a problem for genome sequencing? Not infrequently, a tandem repetitive sequence is in total longer than the length of a maximum sequence read. In that case, there is no way to bridge the gap between adjacent unique sequences. Sometimes, the repeat sequence blocks are relatively short, and their entire length can be included in one sequencing read. Dispersed repetitive elements can cause erroneous alignment of reads from different chromosomes.

> **MESSAGE** The landscape of eukaryotic chromosomes includes a panoply of repetitive DNA segments. These segments are difficult to align as sequence reads.

12.3 Creating genomic sequence maps

All current genome-sequencing strategies are clone based. First, a library of clones is made, and then for each clone the sequence of the insert immediately adjacent to the vector is obtained. However, there are two ways to assemble a consensus sequence of a genome. One is called *whole genome shotgun sequencing* and the other is called *ordered clone sequencing*.

Sequencing a simple genome by using the whole genome shotgun approach

The logic of whole genome shotgun (WGS) sequencing is: sequence first, map later. First, sequence reads are obtained from clones randomly selected from a whole genome library without any information on where these clones map in the genome. Such a library is called a *shotgun library*. Then, these sequence reads are assembled into a consensus sequence covering the whole genome by matching homologous sequences shared by reads from overlapping clones.

Bacterial DNA is essentially *single-copy* DNA, with no repeating sequences. Therefore, any given DNA sequence read from a bacterial genome will come from one unique place in that genome. In addition, a typical bacterial genome is only a few megabase pairs of DNA in size. Owing to these properties, WGS sequencing can be applied to bacterial genomes in a straightforward manner.

How are sequences obtained? Recall from Chapter 11 that a sequencing reaction starts from a primer of known sequence. Because the sequence of a cloned insert is not known (and is the goal of the exercise), primers are based on the sequence of adjacent vector

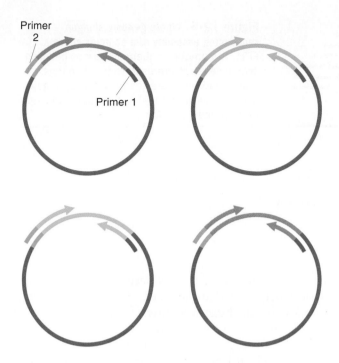

Figure 12-4 The production of terminal insert sequencing reads. The use of two different sequence priming sites, one at each end of the vector, enables the sequencing of as many as 600 base pairs at each end of the genomic insert. If both ends of the same clone are sequenced, the two resulting sequence reads are called *paired-end reads.*

DNA. These primers are used to guide the sequencing reaction into the insert. Hence, short regions at one or both ends of the genomic inserts can be sequenced (Figure 12-4). After sequencing, the output is a large collection of random short sequences, some of them overlapping. The sequences of overlapping reads are assembled into units called **sequence contigs** (sequences that are contiguous, or touching). Each contig covers a large region of the bacterial genome. Occasional gaps are encountered whenever a region of the genome is by chance not found in the shotgun library—some DNA fragments do not grow well in particular cloning vectors. Such gaps are filled in by techniques such as **primer walking**—that is, by using the end of a cloned sequence as a primer to sequence into adjacent uncloned fragments. With the use of the WGS approach, by July 2003, 112 prokaryotic species had been fully sequenced, and more than 200 additional prokaryotic sequencing projects were in progress.

Using the whole genome shotgun approach to create a draft sequence of a complex genome

As already stated, the problem in sequencing complex genomes (apart from greater genome size) is the pres-

ence of repetitive DNA. WGS sequencing is particularly good at producing draft-quality sequences of complex genomes. As an example, we will consider the genome of the fruit fly *Drosophila melanogaster*, which was initially sequenced by this method. The project began with the sequencing of libraries of genomic clones of different sizes (2 kb, 10 kb, 150 kb). Sequence reads from *both* ends of genomic clone inserts were obtained and aligned by a logic identical with that used for prokaryotic WGS sequencing. Through this logic, homologous sequence overlaps were identified and clones placed in order, producing sequence contigs—consensus sequences for these single-copy stretches of the genome. However, unlike the situation in bacteria, where there is only single-copy DNA, the contigs eventually ran into a repetitive DNA segment such as a mobile genetic element that prevented unambiguous assembly of the contigs into a whole genome. The sequence-contigs had an average size of about 150 kb. The challenge then was how to glue the thousands of such sequence-contigs together in their correct order and orientation.

The solution to this problem was to use the knowledge of which pairs of sequence reads came from the opposite ends of the genomic inserts in the same clone—these reads are called **paired-end reads.** The idea was to find paired-end reads that spanned the gaps between two sequence contigs (Figure 12-5). In other words, if one end of an insert was part of one contig and the other end was part of a second contig, then this insert must span the gap between two contigs, and the two contigs were clearly near each other. Indeed, because the size of each clone was known (that is, it came from a library containing genomic inserts of uniform size, either the 2-kb, 100-kb, or 150-kb library), the distance between the end reads was known. Further, aligning the sequences of the two contigs using paired-end reads automatically determines the relative orientation of the two contigs. In this manner, single-copy contigs could be joined together, albeit with gaps where the repetitive elements reside. These gapped collections of

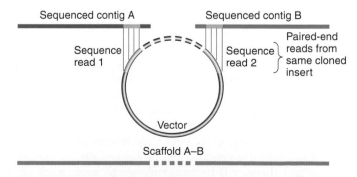

Figure 12-5 The logic of using paired-end reads to join two sequence contigs into a single ordered and oriented scaffold.

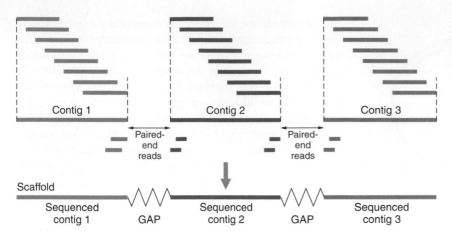

Figure 12-6 Whole genome shotgun sequencing assembly of a complex genome. First, the unique sequence overlaps between sequence reads are used to build contigs. Paired-end reads are then used to span gaps and to order and orient the contigs into larger units called *scaffolds*.

joined-together sequence contigs are called **scaffolds** (sometimes also referred to as **supercontigs**). Because most *Drosophila* repeats are large (3–8 kb) and widely spaced (one repeat approximately every 150 kb), this technique was extremely effective at producing a correctly assembled draft sequence of the single-copy DNA. A summary of the logic of this approach is shown in Figure 12-6).

Using the Ordered Clone Approach to Sequence a Complex Genome

The logic of ordered clone sequencing is the opposite of that for the whole genome shotgun approach: map first, sequence later. Individual cloned inserts from a genomic library are screened to look for similarities in restriction-enzyme recognition sites, indicating that two inserts are contiguous in the genome. This results in a set of ordered and oriented clones that together span the genome. Such an ordered and oriented set of clones covering the whole genome is called the **physical map** of the genome. Here, the word "physical" is used in the sense that the map objects are real objects (DNA segments) that can be isolated and studied in a test tube. After the physical map has been obtained, out of all the clones used to construct the map, a set of minimally overlapping clones is chosen that together covers the whole genome. These clones are then fully sequenced by treating each genomic clone as a minigenome-sequencing project, in which multiple sequencing reads for the clone are put together by using the logic of the whole genome shotgun approach. Finally, the clone sequences are assembled into an overall consensus sequence for the genome according to the known order of these clones on the physical map.

Vectors that can carry very large inserts are the most useful because the genome will be broken up into fewer pieces and there will be fewer clones to keep track of. Cosmids, BACs, and PACs have been the main types of such vectors used (see Chapter 11). However, even with the use of vectors that carry large inserts, creating a physical map is a daunting task. Even so-called small genomes contain huge amounts of DNA. Consider, for example, the 100-Mbp genome of the tiny nematode *Caenorhabditis elegans*. Because an average cosmid insert is about 40 kb, at least 2500 cosmids would be required to embrace this genome, and many more to be sure that all segments of the genome were represented. If you had a *C. elegans* BAC library with an average insert site of 200 kb, your task would be fivefold simplified.

> **MESSAGE** Physical maps are maps of the order, overlap, and orientation of *physically* isolated pieces of the genome—in other words, maps of the distribution of cloned genomic DNA from genomic clone libraries.

Now that we have the bones of the approach, let's look at some details. The procedure begins by amassing a large number of randomly cloned inserts. Then each clone is characterized as follows. Any genomic insert has its own unique sequence, which can be used to generate a "DNA fingerprint." Digestion by multiple restriction enzymes will generate a set of bands whose number and positions are a unique fingerprint of that clone. The pattern of bands generated by each separate clone can be digitized, and the bands from different clones can be aligned by computer to determine if there is any overlap between the inserted DNAs. The order of the bands is not determined in this technique, just the proportion of bands that are shared between two clones. The experimenters then determine the proportion of shared bands that indicate a true overlap. Usually from 25 to 50 percent of the bands must be shared to have a true overlap. Finally, these overlaps are used as the guides to constructing **clone contigs**—an ordered and oriented set of clones (Figure 12-7).

> **MESSAGE** Physical maps can be developed by matching and aligning DNA fingerprints of genomic clones.

(a) DNA fingerprint

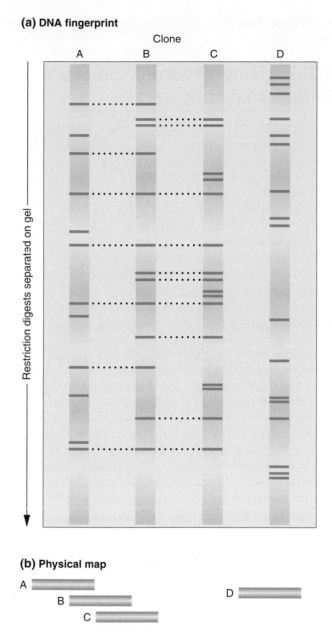

(b) Physical map

Figure 12-7 Creating a physical map by clone fingerprint mapping. (a) Four clones are digested with multiple restriction enzymes, and the resulting complex mixture of restriction fragments is separated on the basis of size by gel electrophoresis. The bands containing the fragments are stained to show their location. The number of identically sized bands for each pair of digests is determined. A and B digests share more than 50 percent of the bands, as do the B and C digests, indicating that they come from overlapping regions of the genome. Several bands are found in all three of A, B, and C, suggesting that some part of the three clones overlap. (b) The physical map derived from the data in part (a). Clone D is from somewhere else in the genome, because it doesn't overlap any of the other three clones.

In the early phases of a genome project, clone contigs representing separate segments of the genome are numerous. But, as more and more clones are characterized, clones are found that overlap two previously separate clone contigs, and these "joining clones" then permit the merger of the two clone contigs into one larger contig. This process of contig merging continues until eventually you have a set of clone contigs that is equal to the number of chromosomes. At this point, if each clone contig extends out to the telomeres of its chromosome, the physical map is complete.

> **MESSAGE** Physical mapping proceeds by assembling clones into overlapping groups called *clone contigs*. As more data accumulate, the clone contigs extend the length of whole chromosomes.

With the physical map complete, we can proceed to sequence the ordered clones. As an example, we will consider the sequencing of the nematode *Caenorhabditis elegans* genome. As shown above, genomic cosmid clones were fingerprinted and arranged into a complete physical map of the *C. elegans* genome. For reasons of economy and efficiency, it was desirable to sequence clones with as little overlap as possible. Thus, as outlined in Figure 12-8, the next step was to select a subset of the cosmid clones in the physical map with clear but minimal overlap. Such a subset of clones is called a **minimum tiling path,** because it contains the minimum number of clones that represents the entirety of the genome. The clones in the minimum tiling path were divided into smaller pieces, and these pieces were inserted into cloning vectors that accept inserts as large as 2 kb, creating sets of "subclones," each set corresponding to one of the cosmid clones. The inserts were then sequenced by using the approach illustrated in Figure 12-4.

After sequencing comes assembly of the sequences. The next step was to assemble the sequence reads of individual subclones into the consensus sequence for their clone. Because the order of the clones was previously determined by the physical mapping project, it was then straightforward to merge the individual consensus sequences with those of their neighbors, eventually producing an overall consensus sequence for the entire genome. This ability to rely on the physical map to order and orient the clone sequences is a major advantage of the ordered clone approach. A second major advantage is its ability to include certain repetitive elements. Although there are many mobile-element families in the *C. elegans* genome, they are dispersed. Thus, it was rare to encounter more than one member of a given mobile element family in the same cosmid. Hence the location of the mobile element within the clone is unambiguous—a major

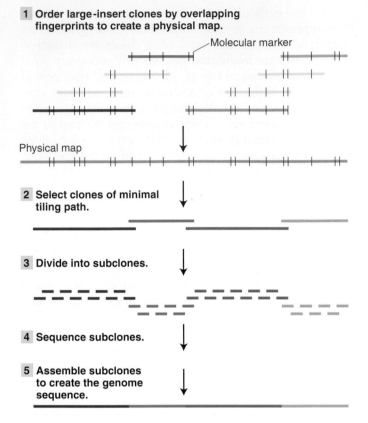

Figure 12-8 Strategy for ordered clone sequencing. By physical mapping, a series of clones that minimally overlap are identified; this series is called a *minimum tiling path*. The clones in the minimum tiling path are divided into subclones, which are sequenced and reassembled.

advantage of creating the consensus sequence clone by clone.

> **MESSAGE** The two basic approaches to genome sequencing are ordered clone sequencing from physical maps by using the minimum tiling path, and whole genome shotgun sequencing.

Filling sequence gaps

In both whole genome shotgun and ordered clone sequencing, some gaps generally remain. Sometimes, the gaps are due to sequences from the genome that are incapable of growth inside of the bacterial cloning host. Special techniques must be used to fill these gaps in the sequence assemblies. If the gaps are short, PCR fragments can be generated from primers based on the ends of the assemblies, and these PCR fragments can be directly sequenced without a cloning step. Successive primer walking can sometimes bridge a gap. If the gaps are longer, attempts can be made to clone these sequences in a different host, such as yeast. If cloning in

a different host fails, then the gaps in the sequence may remain.

12.4 Using genomic sequence to find a specific gene

In genetics, genes are identified as interesting mainly by virtue of the phenotypes that they produce. The molecular nature of such genes of interest can be revealed by isolating that gene and sequencing it (Chapter 11) or by zeroing in on that gene in the genome sequence by using map comparisons. Each of the different types of maps discussed in this text—linkage maps, cytogenetic maps, physical maps, and sequence maps—presents a glimpse of the genome, but from a different perspective. Merging the data from such maps is a powerful way of triangulating on a specific gene.

Merging mutational and sequence maps

Mutations are of two general types: (1) changes within genes and (2) breaks in the structure of the DNA molecules of the chromosomes. Because genomes are large and sequencing a genome is a multimillion-dollar enterprise, finding a gene by sequencing an entire genome to find the altered sequence of a mutation is impractical. Instead, we can use a linkage or cytogenetic map to home in on the region of the genome sequence that contains the gene of interest. However, the resolving power of such maps is low (in humans, 1 percent recombination is equivalent to about 1 Mbp of DNA, and one of the smallest microscopically visible chromosome bands is about the same size). Hence, what is needed is a high density of molecular landmarks surrounding the genes of interest on these low-resolution maps, so that we can narrow the search to as small a region of the physical and genomic sequence maps as possible (Figure 12-9). Largely, the molecular landmarks on linkage maps will be the neutral, polymorphic variants such as RFLPs and VNTRs that were introduced briefly in Chapter 4. In cytogenetic maps, a molecular landmark is any piece of DNA whose position we can locate on the microscopically visible chromosomes through DNA–DNA hybridization techniques. We will now explore both these types of markers.

> **MESSAGE** A specific gene can be found in the genomic sequence by matching linkage and cytological maps with the genome sequence.

Filling in a linkage map with molecular markers

In Chapter 4, we considered the basic approach in creating a linkage map. Such a map can include both

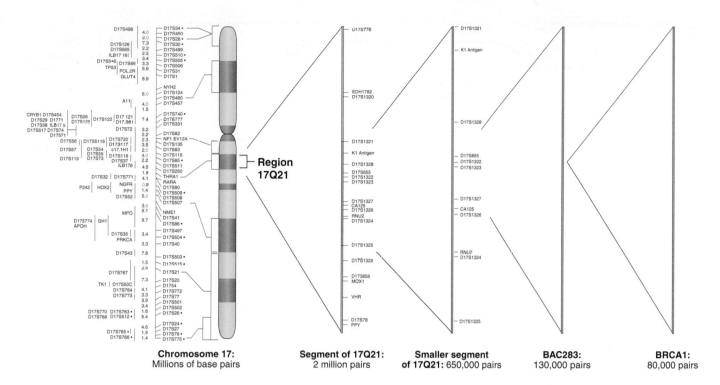

Figure 12-9 Finding a specific gene. A specific gene, the breast cancer gene BRCA1, was found by using the genomic map at increasing levels of resolution. [Copyright 1994 by the New York Times Company. Reprinted by permission.]

molecular markers and classical phenotypic markers. Molecular markers are DNA polymorphisms, variants in the nucleotide sequence of different normal individual members of a population. DNA polymorphisms can be single nucleotide polymorphisms or variants of large DNA segments. These polymorphisms may be present at 0.01 percent to >1 percent of the base pairs in the genome, depending on the species and the particular segment of the genome. Thus, the potential density of molecular markers on a linkage map is much higher than that of classical markers. Let's look at some examples.

RESTRICTION FRAGMENT LENGTH POLYMORPHISMS

Restriction fragment length polymorphisms (RFLPs) represent the presence or absence of restriction-enzyme recognition sites in different individuals. Any two individuals will differ by a great many RFLPs. An RFLP is assayed by using a cloned DNA fragment as a probe. The probe must bind to one site alone, and, in a Southern blot, it will reveal different-sized restriction fragments, thus defining an RFLP "locus."

In model organisms, RFLP mapping is performed on a defined set of strains or individuals that provide standard RFLP loci for mapping that species. In human RFLP mapping, the standard is a defined set of individual members of 61 families with an average of eight children per family. This DNA has been made available

throughout the world. Figure 12-10 shows an example of a human disease allele linked to an RFLP locus. This linkage information provides a way to estimate the chances of a person's having the disease. Because of the close linkage, future generations of children showing the RFLP morph 1 ("allele" 1 of the RFLP locus) can be predicted to have a high chance of inheriting the disease allele *D*. With more similar pedigrees, the linkage of the RFLP and the disease gene can be measured more accurately by using Lod scores (Chapter 4).

SINGLE NUCLEOTIDE POLYMORPHISMS RFLPs are usually based on a type of SNP (single nucleotide polymorphism; Chapter 4). However, there are many more SNPs (pronounced snips) than can be detected with RFLP probes, because most SNPs do not exist in the middle of a known restriction site. SNPs are useful in mapping because they are so numerous throughout the genome. In humans, SNPs are spaced at intervals between 11 and 300 bases. They can be detected in several ways, some of which are adaptable to high throughput. One simple way is simply to sequence one region from several persons and compare the regions for differences. Most SNPs are not within genes and are simply used for marking chromosome regions. A few are within genes and act as specific tags for disease alleleles. At present, there is a massive international drive to find and map humans SNPs.

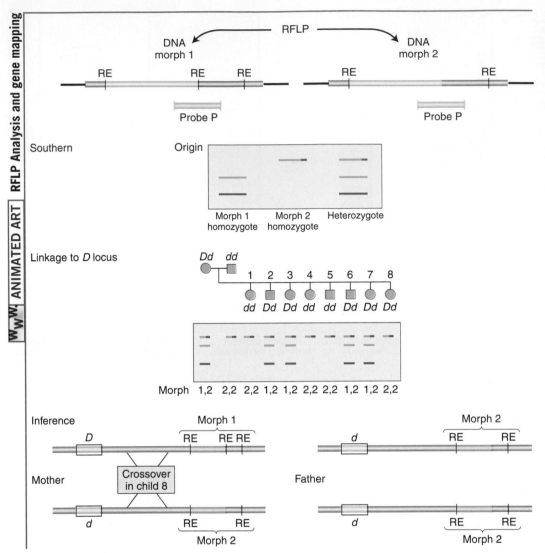

Figure 12-10 The detection and inheritance of a restriction fragment length polymorphism (RFLP). A probe P detects two DNA morphs when the DNA is cut by a certain restriction enzyme (RE). The pedigree of the dominant disease phenotype D shows linkage of the *D* locus to the RFLP locus; only child 8 is recombinant.

SIMPLE-SEQUENCE LENGTH POLYMORPHISMS We learned in Chapter 3 that one category of repetitive DNA has repeats of a very simple short sequence. DNA markers based on variable numbers of short-sequence repeats are collectively called **simple-sequence length polymorphisms (SSLPs).**

SSLPs have two basic advantages over RFLPs. First, an RFLP usually has only 1 or 2 "alleles," or morphs, in a pedigree or population under study. In contrast, SSLPs commonly have multiple alleles, and as many as 15 alleles have been found for an SSLP locus. As a consequence, sometimes 4 alleles (2 from each parent) can be tracked in a pedigree. Two types of SSLPs are now routinely used in genomics—minisatellite and microsatellite markers.

Minisatellite markers **Minisatellite markers** are based on variation of the number of VNTRs (variable number tandem repeats). The VNTR loci in humans are from 1- to 5-kb sequences consisting of variable numbers of a repeating unit from 15 to 100 nucleotides long. VNTRs with the same repeating unit but different numbers of repeats are dispersed throughout the genome.

To find a VNTR, the total genomic DNA is first cut with a restriction enzyme that has no target sites within the VNTR arrays but does cut flanking sites; thus the length of each fragment containing a VNTR will correspond to the number of repeats. If a probe that recognizes one of the VNTRs is available, then a Southern blot will reveal a large number of different-sized fragments that are bound by the probe. In fact, these patterns are sometimes called **DNA fingerprints.** If parents differ for a particular band, then this difference becomes a heterozygous site that can be used in mapping. A simple example is shown in Figure 12-11.

Microsatellite markers **Microsatellite markers** are dispersed regions of the genome composed of variable numbers of dinucleotides repeated in tandem. The most common type is a repeat of CA and its complement GT, as in the following example:

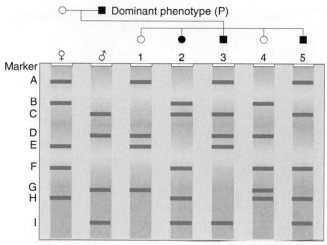

ANALYSIS EXAMPLES

F and H   Always inherited together — linked?
A and B   In progeny, always either A *or* B — "allelic"?
A and D   Four combinations; A and D, A, D, or neither — unlinked?
F, H, and E   Always *either* F and H *or* E — closely linked in trans?
Allele *P*   Possibly linked to I and C.

Figure 12-11 Using DNA fingerprint bands as DNA markers in mapping. Simplified fingerprints are shown for parents and five progeny. Examples illustrate methods of linkage analysis.

5′ C-A-C-A-C-A-C-A-C-A-C-A-C-A-C-A 3′
3′ G-T-G-T- G-T- G-T- G-T- G-T- G-T- G-T 5′

Variant markers are distinguished by PCR with the use of primers for regions surrounding individual microsatellite repeats. Cloned DNA containing the microsatellite and flanking DNA is sequenced and hence primers can be designed from the abutting regions.

Primer 1

(CA)ₙ

(GT)ₙ

Primer 2

An individual primer pair will amplify that microsatellite locus only because the primers are specific. Any variants of microsatellite size will be revealed by gel electrophoresis of the PCR products from different individuals. Thus, all size variations of the microsatellite repeat will be detectable. A high proportion of such PCR analyses reveal several marker "alleles," or different-sized strings of repeats. An example of the microsatellite mapping technique is shown in Figure 12-12. Thousands of microsatellite primer pairs can be made that likewise detect thousands of marker loci.

Figure 12-13 shows how molecular markers can flesh out a linkage map. One centimorgan (one map unit) of human DNA is a huge segment, estimated as 1 megabase (1 Mb = 1 million base pairs, or 1000 kb).

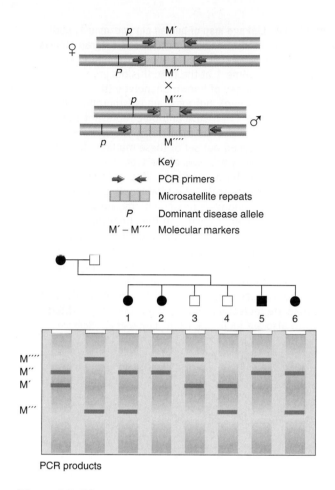

Figure 12-12 Using microsatellite repeats as molecular markers for mapping. A hybridization pattern is shown for a family with six children, and this pattern is interpreted at the top of the illustration with the use of four different-sized microsatellite "alleles," M′ through M′′′′, one of which (M′′) is probably linked in cis configuration to the disease allele *P*.

You can see that the number of mapped molecular markers greatly exceeds the number of mapped phenotypic markers (referred to as genes). Note that, because of their much higher density, SNPs cannot be represented on a whole chromosome map such as this one.

MESSAGE Recombination analysis that uses both the locations of genes with known phenotypic effect and the locations of DNA markers has produced high-density linkage maps.

Placing molecular markers on cytogenetic maps

Recombination-based maps of molecular markers are abstract concepts. It is desirable to correlate such a map with the real structure that it represents, the chromosome. Molecular markers can be placed on cytogenetic

Figure 12-13 Linkage map of human chromosome 1, correlated with chromosome banding pattern. The diagram shows the distribution of all genetic differences that had been mapped to chromosome 1 at the time this diagram was drawn. Some markers are genes of known phenotype (their numbers are represented in green), but most are polymorphic DNA markers (the numbers in mauve and blue represent two different classes of molecular markers). A linkage map displaying a well-spaced-out set of these markers, based on recombinant frequency analyses of the type described in this chapter, is in the center of the figure. Map distances are shown in centimorgans (cM). At a total length of 356 cM, chromosome 1 is the longest human chromosome. Some markers have also been localized on the chromosome 1 cytogenetic map (right-hand map, called an *idiogram*), by using techniques discussed later in this chapter. Having common landmark markers on the different genetic maps permits the locations of other genes and molecular markers to be estimated on each map. Most of the markers shown on the linkage map are molecular, but several genes (highlighted in light green) also are included. [B. R. Jasney et al., *Science*, September 30, 1994.]

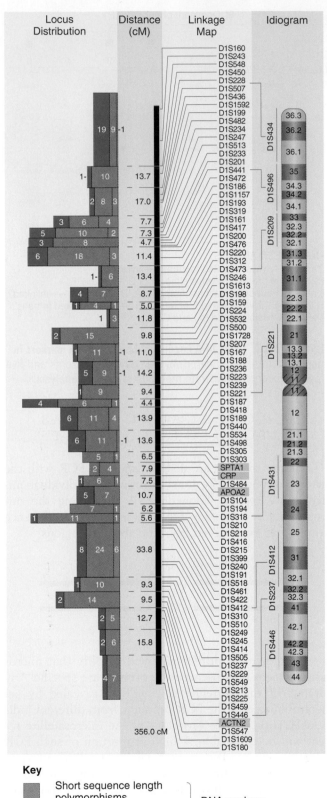

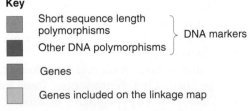

chromosome maps in a variety of ways, by relating the locations of DNA markers to cytogenetic landmarks such as chromosome bands or to chromosome breaks of known position. Let's look at some common techniques.

IN SITU HYBRIDIZATION MAPPING If part of a genome has been cloned, then it can be used to make a labeled probe for hybridization to chromosomes in situ. The logic of this approach is identical with that of any hybridization technique, such as Southern blotting, except that here largely intact chromosomes are the target for probe hybridization (rather than DNA on a membrane). In this technique, cells are broken open and their chromosomes are spread out on microscrope slides. The DNA of the chromosomes is denatured so that their DNA is largely single stranded. Then the labeled probe, also denatured, is added to the preparation; the probe will hybridize to the sites of homologous sequences "in situ" within the chromosomal DNA. In the process called **fluorescent in situ hybridization,** or **FISH,** the probe is labeled with a fluorescent dye and the location of the homologous fragment is revealed by a bright fluorescent spot on the chromosome (Figure 12-14). The probe sequence can be mapped to the approximate position on the chromosome to which it hybridizes by noting its position in relation to banding patterns, centromere, or other cytological features. The location is *approximate* because this technique does not have the resolving power of recombinational mapping. For example, the positions of two genes 5 cM apart in the human genome will be indistinguishable by in situ hybridization mapping.

An extension of FISH is **chromosome painting.** Rather than using visible chromosome features as landmarks, this technique uses a standard control set of probes homologous to known locations to establish the cytogenetic map. The probes are labeled with different fluorescent dyes, which thus "paint" specific regions and make them identifiable under the microscope (Figure 12-15). If a probe consisting of a cloned sequence of unknown location is labeled with yet another dye, then its position can be established in the painted array.

REARRANGEMENT BREAKPOINT MAPPING In Chapter 15, we shall consider chromosomal rearrangements, a class of mutations that result from the breaking of a chromosome at one location and its rejoining with another similarly severed site on the same chromosome or a different one. In some chromosomal rearrangements, there is a swap of DNA at the breakpoints, and so a segment of chromosome is now adjacent to a new neighboring sequence. In other cases, some of the DNA segments are lost from the reglued chromosomes, and so a piece of the genome is missing. A chromosome lacking a segment of its DNA is said to carry a deletion.

Figure 12-14 FISH analysis. Chromosomes probed in situ with a fluorescent probe specific for a gene present in a single copy in each chromosome set—in this case, a muscle protein. Only one locus shows yellow fluorescence, which corresponds to the probe bound to the muscle protein gene. We see four spots corresponding to this one locus because there are two homologous mitotic prophase chromosomes, each of which contains two chromatids. [From P. Lichter et al., "High-Resolution Mapping of Chromosome 11 by In Situ Hybridization with Cosmid Clones," *Science* 247, 1990, 64.]

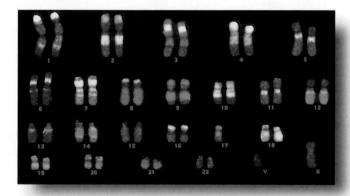

Figure 12-15 Chromosome painting by in situ hybridization with different labeled probes. Each probe fluoresces at a different wavelength. By exposing the slide serially to each of the different wavelengths and then turning each wavelength into a different virtual color on a computer screen, the multicolored images representing different characteristic paint patterns for each chromosome in the karyotype can be generated. [Applied Imaging, Hylton Park, Wessington, Sunderland, U.K.]

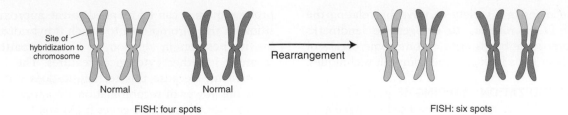

Figure 12-16 Chromosomal rearrangement breakpoints detected by FISH. A probe that spans the breakpoint hybridizes to four spots in the normal karyotype. After rearrangement, two additional spots are visible corresponding to a part of the original sequence now residing on another chromosome.

Why are we interested in such structural rearrangements in the context of maps? The reason is that some breaks cause mutant phenotypes, either because a gene that happens to reside at the breakpoint is deactivated or because the junction fuses two genes together, creating a novel gene. Some deletions also cause mutant phenotypes. Hence a link can be made between a mutation and a chromosomal position. How can we map rearrangement breakpoints relative to molecular markers? One approach is to do a FISH analysis. When a clone that detects some molecular marker spans a breakpoint, the breakpoint is easily detected because, in FISH mapping, there are two sites of labeling instead of one (Figure 12-16).

> **MESSAGE** By the correlation of structural landmarks on chromosomes with the location of cloned probe DNA, molecular markers are introduced onto cytogenetic maps.

Uniting the maps

Now that we know how to insert molecular markers on linkage and cytogenetic maps, how can we connect these maps to our physical and sequence maps? We can do so in several ways (Figure 12-17).

ATTACHING MOLECULAR MARKERS TO PHYSICAL MAPS The probes or PCR primers that were used to

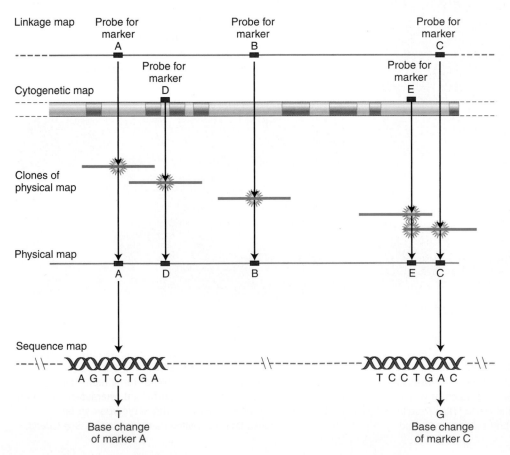

Figure 12-17 Uniting linkage and cytogenetic maps with molecular maps. To connect the different map types, molecular markers from linkage and cytogenetic maps are anchored onto the physical map and the genomic sequence map.

detect molecular markers can be applied to the set of genomic clones that have been arranged into a physical map. This will reveal the locations of the molecular markers on the physical map (Figure 12-17, top).

ATTACHING THE MOLECULAR MARKERS TO GENOMIC SEQUENCE MAPS Typically, probes for mapping molecular markers have already been sequenced. Even if they haven't been, sequencing them is easily accomplished. Once the sequence is known, we can search the sequence map with a computer to identify the location of the identical sequence in the genome (Figure 12-17, bottom). Thus, the locations of the molecular markers can be identified on the sequence map down to single-base-pair resolution.

RELATING MUTATIONS AND REARRANGEMENT BREAKPOINTS TO THE SEQUENCE MAP Everything discussed thus far in this section is prelude. Our real goal is to find a gene of interest. Therefore the task is to be able to delimit a range of base pairs on the genomic sequence within which an interesting gene mutation lies. We now have everything in place to do so. The logic of defining this base pair range is now simple (Figure 12-18).

1. On the linkage map or the cytogenetic map, identify the nearest molecular marker that lies to the left of the mutation or rearrangement breakpoint of interest.

2. Similarly, on the linkage map or the cytogenetic map, identify the nearest molecular marker that lies to the right of the mutation or rearrangement breakpoint of interest.

3. Determine the locations of these markers in the genome sequence map.

4. The mutation or rearrangement breakpoint must lie between these boundary molecular markers.

We now have set limits on the position of the gene of interest, originally identified by a mutant phenotype. However, the geneticist wants to know *exactly* where the mutation is. The next steps depend on how large the delimited molecular region is. If it is relatively small, we might literally sequence the entire region in the mutant individual to identify the exact mutational lesion or rearrangement breakpoint. If it is somewhat larger, we might try the **candidate gene approach.** We make an educated guess about the most likely genes that might be affected in the mutant strain, by comparing the phenotype of the mutant with the known functions of the best candidate genes. We would then focus our sequencing effort specifically on the candidate gene(s) in the delimited region. Alternatively, we might try some indirect approach, such as testing specific segments of the genome for functional complementation. We could divide the delimited region into subregions and apply the complementation test to each, by transforming each subregion

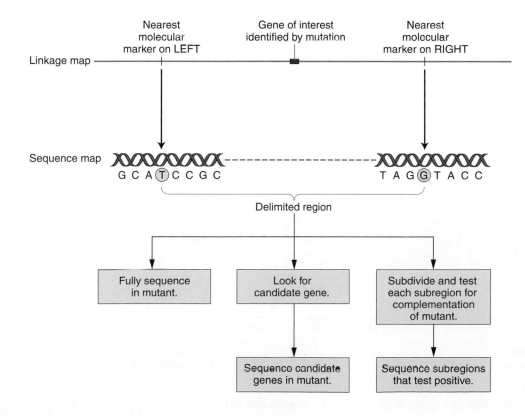

Figure 12-18 Identifying a gene of interest on a sequence map. A mutation identifying a gene of interest is first localized to a delimited region of a sequence map. The boxes at the bottom of the figure indicate three different strategies for identifying the sequence of a mutation and the gene that is mutated.

with a wild-type transgene and seeing if it will revert a recessive mutation. A positive result would be taken as strong evidence that we had identified a DNA fragment that contained the mutant allele or rearrangement breakpoint.

12.5 Bioinformatics: meaning from genomic sequence

The genomic sequence is a highly encrypted code containing the information for building and maintaining a functional organism. The study of the information content of genomes is called **bioinformatics.** We are far from being able to read this information from beginning to end in the way that we would read a book. Even though we know the letters of the alphabet and we know which triplets code for amino acids in the protein-coding segments, much of the information contained in a genome remains indecipherable.

The nature of the information content of DNA

Information is defined literally as "that which is necessary to give form." DNA contains such information, but in what way is it encoded? Conventionally, the information is thought of as the sum of all the gene products, both proteins and RNAs. However, it is more complex than that. Another view of the genome is that it encodes a series of *docking sites* for different proteins and RNAs. Many proteins dock at sites located in the DNA itself, whereas other proteins and RNAs dock at sites located in mRNA (Figure 12-19). The sequence and relative positions of those docking sites permit genes to be transcribed, spliced, and translated properly, at the appropriate time in the appropriate tissue. For example, regulatory-protein-binding sites determine when, where, and at what level a gene will be expressed. For transcription to take place, multiple docking sites on the DNA must be appropriately spaced so that RNA polymerase can simultaneously contact all of them. At the RNA level

in eukaryotes, the locations of docking sites for the RNAs and proteins of spliceosomes will determine the 5' and 3' splice sites where introns are removed. Regardless of whether a docking site actually functions as such in DNA or RNA, the site must be encoded in the DNA. (If we push the docking-site notion to its limit, we might conclude that even codons in the translated region of an mRNA are really docking sites for specific aminoacyl-tRNAs. However, we will not include codons as docking sites.) In summary, the information in the genome can be thought of as the sum of all the sequences that encode proteins and RNAs, plus the docking sites that permit their proper action in time and in tissue.

Deducing the protein-coding genes from genomic sequence

Because the proteins present in a cell largely determine its morphology and physiological properties, one of the first orders of business in genome analysis is to try to determine a list of all of the polypeptides encoded by an organism's genome. This list is termed the organism's **proteome.** It can be considered a "parts list" for the cell. To determine the list of polypeptides, the sequence of each mRNA encoded by the genome must be deduced. Because of intron splicing, this task is particularly challenging in higher eukaryotes, where introns are the norm. In humans, for example, an average gene has about 10 exons. Furthermore, many genes encode alternative exons—that is, some exons are included in some versions of a processed mRNA but are not included in others (Chapter 8). The alternatively processed mRNAs can encode polypeptides having much, but not all, of their amino acid sequence in common. Even though we have a great many examples of completely sequenced genes and mRNAs, it is not yet possible to identify 5' and 3' splice sites merely from DNA sequence with a high degree of accuracy. Therefore, we cannot be certain of which sequences are introns. Predictions of alternatively used exons are even more error prone. For such reasons, deducing the total polypeptides "parts list" in higher eukaryotes is a large problem. Some approaches follow.

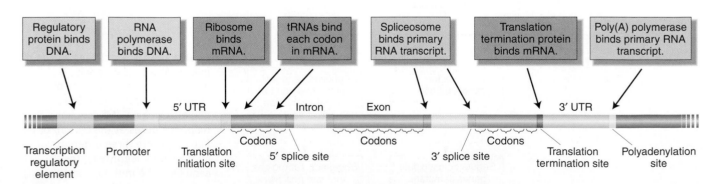

Figure 12-19 A depiction of a gene within DNA as a series of docking sites for proteins and RNAs.

ORF DETECTION The main approach to producing a polypeptide list is to use the computational analysis of the genome sequence to predict mRNA and polypeptide sequences, an important part of bioinformatics. The basic approach is to look for **open reading frames,** or **ORFs.** An ORF is a gene-sized sequence of sense codons with the appropriate 5'- and 3'-end sequences (such as start and stop codons). Possible introns are processed out computationally. Such an ORF seems like a good candidate for being a gene. To find candidate ORFs, the computer scans the DNA sequence on both strands in each reading frame. Because there are three possible reading frames on each strand, there are six possible reading frames in all.

DIRECT EVIDENCE FROM cDNA SEQUENCES cDNA sequences are extremely valuable in identifying the exons of a gene because cDNAs are DNA copies of mRNAs (Figure 12-20). The alignment of cDNAs with their corresponding genomic sequence clearly delineates the exons, and hence introns are revealed as the regions falling between the exons. In the cDNA, the ORF should be continuous from initiation codon through stop codon. Thus, cDNA sequences can greatly assist in identifying the correct reading frame, including the initiation and stop codons. Full-length cDNA evidence is taken as the gold-standard proof for identification of the sequence of a transcription unit, determination of how the transcript is processed, and localization of the ORF that it encodes.

In addition to full-length cDNA sequences, there are large data sets of cDNAs for which only the 5' or the 3' ends or both have been sequenced. These short cDNA sequence reads are called **expressed sequence tags (ESTs).** ESTs can be aligned with genomic DNA and thereby used to determine the 5' and 3' ends of transcripts—in other words, to determine the boundaries of the transcript as shown in Figure 12-20.

PREDICTIONS OF DOCKING SITES As already discussed, a gene consists of a segment of DNA that encodes a transcript, as well as the regulatory signals that determine when, where, and how much of that transcript is made. In turn, that transcript has the signals necessary to determine its splicing into mRNA and the translation of that mRNA into a polypeptide (Figure 12-21). There are now statistical "gene finding" computer programs that search for the predicted sequences of the various docking sites used for transcription start sites, for 3' and 5' splice sites, and for translation initiation codons within genomic DNA. These predictions are based on consensus motifs for such known sequences, but they are by no means perfect.

USING POLYPEPTIDE AND DNA SIMILARITY Candidate ORFs predicted by the above techniques often can be anchored in reality by comparing them with all the other genes that have ever been found. This is done by submitting candidate sequences as "query sequences" to the public databases. This procedure is called a BLAST search (BLAST stands for Basic Local Alignment Search Tool). The sequence can be submitted as a nucleotide sequence (a BLASTn search) or as a translated amino acid sequence (BLASTp). Because organisms share common ancestors, gene sequences are generally very similar between them. Hence a query sequence that represents a real gene will likely have relatives in the genes isolated and sequenced in other organisms, especially in the closely related ones. The computer scans the database and returns a list of full or partial "hits," starting with the closest matches. If the candidate sequence closely resembles that of a gene previously identified from another organism, then this provides a strong indication that the ORF is a real gene. Less close matches are still useful. For example an amino acid identity of only 35 percent, but at identical positions, is a strong indication of common three-dimensional structures.

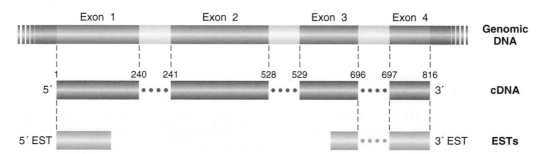

Figure 12-20 Alignment of fully sequenced cDNAs and ESTs with genomic DNA. The dashed lines indicate regions of alignment; for the cDNA, these regions are the exons of the gene. The dots between segments of cDNA or expressed sequence tags (ESTs) indicate regions in the genomic DNA that do not align with cDNA or EST sequences; these regions are the locations of the introns. The numbers above the cDNA line indicate the base coordinates of the cDNA sequence, where base 1 is the 5'-most base and base 816 is the 3'-most base of the cDNA. For the ESTs, only a short sequence read is obtained from each end (5' and 3') of the corresponding cDNA. These sequence reads establish the boundaries of the transcription unit, but they are not informative about the internal structure of the transcript unless the EST sequences cross an intron (as is true for the 3' EST depicted here).

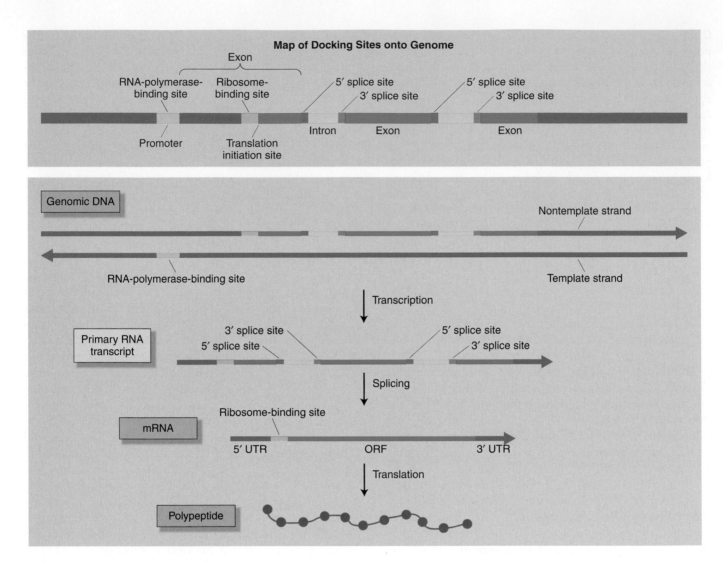

Figure 12-21 Eukaryotic information transfer from gene to polypeptide chain. Note the DNA and RNA "docking sites" that are bound by protein complexes to initiate the events of transcription, splicing, and translation.

BLAST searches are used in many other ways, but always the theme is to find out more about some sequence of interest that the experimenter has identified.

PREDICTIONS BASED ON CODON BIAS Recall from Chapter 9 that the triplet code for amino acids is degenerate; that is, most amino acids are encoded by two or more codons (see Figure 9-8). The multiple codons for a single amino acid are termed *synonymous codons*. In a given species, not all synonymous codons for an amino acid are used with equal frequency. Rather, certain codons are present much more frequently in the ORFs. For example, in *Drosophila melanogaster*, of the two codons for cysteine, UGC is used 73 percent of the time, whereas UGU is used 27 percent. This usage is a diagnostic for *Drosophila* because, in other organisms, this "codon bias" pattern is quite different. Codon biases are thought to be due to the relative abundance of the tRNAs complementary to these various codons in a

given species. If the codon usage of a predicted ORF matches that species' known pattern of codon usage, then this match is supporting evidence that the proposed ORF is genuine.

PUTTING IT ALL TOGETHER A summary of how these different sources of information are combined to create the best possible mRNA and ORF predictions is depicted in Figure 12-22. These different kinds of evidence are complementary and can cross-validate one another. For example, the structure of a gene may be inferred from evidence of protein similarity within a region of genomic DNA bounded by 5' and 3' ESTs. Useful predictions are possible even in the absence of cDNA sequence or evidence from protein similarities. A docking-site prediction program can propose a hypothetical ORF, and proper codon bias would be supporting evidence.

An example of gene predictions for a chromosome from the human genome is shown in Figure 12-23. Such

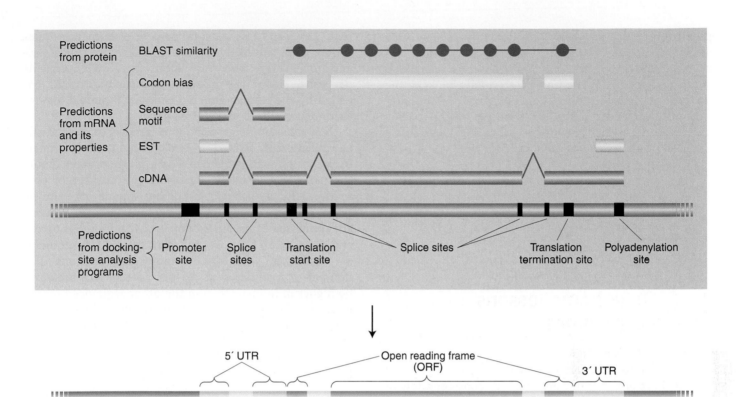

Figure 12-22 Making gene predictions. The different forms of gene-product evidence—cDNAs, ESTs, BLAST similarity hits, codon bias, and motif hits—are integrated to make gene predictions. Where multiple classes of evidence are found to be associated with a particular genomic DNA sequence, there is greater confidence in the likelihood that a gene prediction is accurate.

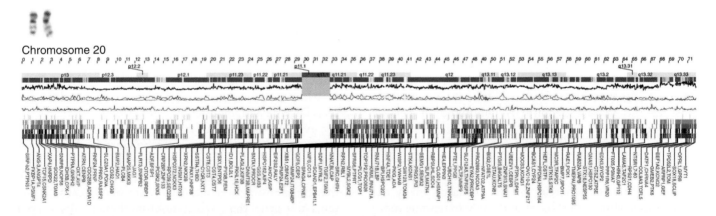

Figure 12-23 The sequence map of human chromosome 20. The recombinational and cytogenetic map coordinates are shown in the top lines of the figure. Various graphics depicting gene density and different DNA properties are shown in the middle sections. The identifiers of the predicted genes are shown at the bottom of the panel. [Courtesy of Jim Kent, Ewan Birney, Darryl Leja, and Francis Collins. Adapted from the International Human Genome Sequencing Consortium, "Initial Sequencing and Analysis of the Human Genome," *Nature* 409, 2001, 860–921.]

predictions are being revised continually as new data and new computer programs become available. The current state of the predictions can be viewed at many Web sites, most notably at the public DNA data banks in the United States and Europe (see Appendix B). These predictions are current best guesses of the protein-coding genes present in the sequenced species and, as such, are works in progress.

> **MESSAGE** Predictions of mRNA and polypeptide structure from genomic DNA sequence depend on an integration of information from cDNA sequence, docking-site predictions, polypeptide similarities, and codon bias.

12.6 Take-home lessons from the genomes

There is now such a flood of genomic information that any attempt to summarize it in a few paragraphs will fall far short of the mark. Instead, let's consider a few of the insights from our first view of the overall genome structures and global parts lists of the species whose genomes have been sequenced. We will view the genome sequences by introspection—that is, we will ask, What can we learn by looking at a single genome by itself? We will use the human genome and yeast genomes as our examples.

The structure of the human genome

In describing the overall structure of the human genome, we wish first to delineate is its repeat structure. A considerable fraction of the human genome, about 45 percent, is repetitive and composed of transposable elements. Indeed, even within the remaining single-copy DNA, a fraction has sequences suggesting that they might be descended from ancient transposable elements that are now immobile and have accumulated random mutations, causing them to diverge in sequence from the ancestral transposable elements. Thus, it appears that most of the human genome is composed of genetic hitchhikers.

Only a small part of the human genome codes for polypeptides; that is, somewhat less than 3 percent of it encodes exons of mRNAs. Exons are typically small (about 150 bases), whereas introns are large, many extending more than 1000 bases and some extending more than 100,000 bases. Transcripts are composed of an average of 10 exons, although many have substantially more. Finally, introns may be spliced out of the same gene in locations that vary. This variation in the location of splice sites generates considerable added diversity in mRNA and polypeptide sequence. On the basis of current cDNA and EST data, 60 percent of human protein-coding genes are likely to have two or more splice variants. On average, there are about three splice variants per gene. Hence the number of distinct proteins encoded by the human genome is about threefold greater than the number of recognized genes.

Proteins can be grouped into families of structurally and functionally related proteins on the basis of similarity in amino acid sequence. For a given protein family that is known in many organisms, the number of proteins in a family is larger in humans than in other invertebrates whose genomes have been sequenced. Proteins are composed of modular domains, with modules mixed and matched to carry out different roles. Many domains are associated with specific biological functions. The number of modular domains per protein also seems to be higher in humans than in the other sequenced organisms. In subsequent chapters in the book, we will encounter numerous protein families and see their roles in biological processes.

The degree of similarity tells us how confidently we can relate the function of a known gene family member to other unknown members. We will see examples of protein family members that are clearly "brothers" or "sisters" in the sense that they have all protein domains in common and act similarly in parallel processes or act in some coordinated or cooperative mode in the same biological process. We will also encounter examples of family members that are "half-brothers" or "half-sisters," meaning that they have some protein domains (and some functional attributes) in common but not others. Finally, in many cases, the degree of protein similarity puts us in a statistical gray zone, where we don't know if we are looking at a significant relationship (perhaps "second or third cousins") or not. At present, it is very difficult to make sense of such cases simply by comparing primary amino acid sequences, but it is hoped that comprehensive studies of the three-dimensional structures of all protein families (a project sometimes called **structural genomics**) will shed light on these weak relationships. The results of these studies will reveal the degree to which weakly related sequences correspond to conserved shape and, by inference, function.

As more refined information on the human genome emerges, additional features can be gleaned. A recent example is the finished sequence map of one of the best-studied human chromosomes—chromosome 7. Initially, this chromosome was intensively studied because it contains the gene that, when nonfunctional, causes cystic fibrosis. The location of the cystic fibrosis gene was identified in the early days of the human genome project by overlaying the linkage map on the physical and sequence maps, as discussed earlier in this chapter. Groups have continued to study this chromosome in detail. About 1700 genes are known or predicted to reside on

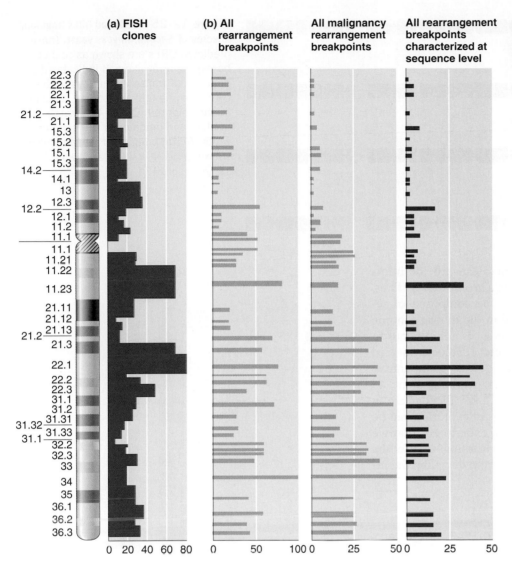

Figure 12-24 Cytogenetic map of human chromosome 7. (a) FISH physical map clones on the cytogenetic map. (b) Distribution of rearrangement breakpoints from patients with genetic disorders. [After W. S. Scherer et al., "Human Chromosome 7: DNA Sequence and Biology," *Science*, May 2, 2003, pp. 769 and 771, Figures 2 and 5.]

chromosome 7. About 800 physical map clones have been FISH mapped to human chromosome 7, creating a high-density cytogenetic map (Figure 12-24a). With the use of these clones, about 1600 rearrangement breakpoints associated with human disease have been FISH mapped (by the method described earlier) and 440 of these breakpoints have been characterized at the sequence level, allowing the association of mutant phenotype and genes found in the DNA sequences (Figure 12-24b).

Deciphering encoded information through comparative genomics

The area of genomics that compares and contrasts genomes from different organisms is called **comparative genomics.** It provides a powerful method of identifying essential sequences. Retention of an identical sequence in many different groups in the years since their divergence is taken as evidence that the sequence is being maintained by natural selection because it plays an essential role. However, such conservation of sequence cannot itself tell us what that essential role is. The results of recent studies of four related species of the yeast genus *Saccharomyces* have demonstrated the power of comparisons among multiple species that have diverged at different times. Figure 12-25 depicts a 50-kb region of the *S. cerevisiae* genome. Even though the genetics of *S. cerevisiae* has been studied extensively for more than half a century and the genome was fully sequenced about a decade ago, the comparative study found evidence of numerous protein-coding genes that had never been previously predicted. Additional sequences were found to be conserved in the regions 5′ to the proposed new transcription units, suggesting that these sequences are important in the regulation of transcription. Finally,

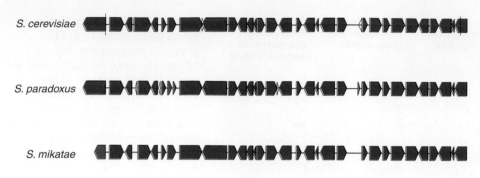

Figure 12-25 Predicted ORFs from four species of *Saccharomyces* yeast. The predicted ORFs are shown as colored boxes; the arrowlike end indicates the direction of transcription. Most of the ORFs are one-to-one matches, indicated by the color red. One-to-two matches are in blue, and unmatched ORFs are in white. A 50-kb region of the genome is shown. [After M. Kellis et al., *Nature*, May 15, 2003, p. 243.]

the presence of conserved sequences can provide information on the evolution of entire chromosomes. Thus, human chromosome 7 appears to be composed of sequences that are largely conserved in the mouse genome (Figure 12-26), but the mouse sequences homologous to human chromosome 7 are split into 19 blocks located on six different chromosomes. Thus comparative genomics reveals that chromosomal rearrangement, to be fully discussed in Chapter 15, is a key mechanism in chromosomal evolution. The process of evolution can be considered the longest-term genetic experiment that we have available to us. We are just learning to use the power of comparative genomics to help decipher the information encrypted in a genome.

The effects of genomics on biological research

Even though genomics as a technology and an approach is not even 20 years old, it has already become an academic household word. The analysis of whole genomes permeates every corner of biological research. From an evolutionary perspective, genomics provides a detailed view of how genomes have diverged and adapted over geological time. In ecological research, ecologists are developing tools for assaying organism distribution based on detecting the presence and concentration of different genomes in natural samples. In basic cell and molecular biology, the analysis of whole genomes is providing the basis for the global analysis of the physiological role of all gene products through the development of the field of *systems biology*. In human genetics, genomics is providing new ways to locate genes that contribute to the many genetic diseases thought to be determined by a complex combination of factors (called multifactorial diseases).

The day when a person's genome sequence is a standard part of his or her medical record may be only from 10 to 15 years away. If these data are not private, what might be the effect of their availability to third parties such as employers or insurers on the individual person?

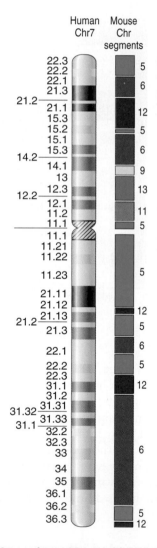

Figure 12-26 Comparison of human chromosome 7 and mouse chromosome. Segments of human chromosome 7 are conserved in the mouse but distributed among six mouse chromosomes. The numbers at the right give the number of the mouse chromosome on which the segment resides. [After W. S. Scherer et al., "Human Chromosome 7: DNA Sequence and Biology," *Science*, May 2, 2003, p. 769, Figure 2.]

Public policies that address privacy and other societal matters will drive the use of genomics in the future.

12.7 Functional genomics

Geneticists have been studying the expression and interactions of gene products for the past 50 years or so. However, these studies were small scale, taking one or a few genes at a time. With the advent of genomics, we have an opportunity to expand these studies to a global level, by using genomic approaches to study most or all gene products simultaneously. This global approach to the study of the expression and interaction of gene products is termed **functional genomics.**

Ome, Sweet Ome

In addition to the genome, there are other global data sets that are of interest. Following the example of the term genome, for which "gene" plus "ome" becomes a word for "all genes," genomics researchers have coined a number of terms to describe other global data sets on which they are working. This -ome wish list includes:

The transcriptome. The sequence and expression patterns of all transcripts (where, when, how much).

The proteome. The sequence and expression patterns of all proteins (where, when, how much).

The interactome. The complete set of physical interactions between proteins and DNA segments, between proteins and RNA segments, and between proteins.

The phenome. The description of the complete set of phenotypes produced by the inactivation of gene function for each gene in the genome.

We will not consider all of these -omes in this section but will focus on some of the global techniques that are beginning to be exploited to harvest these data sets.

The use of DNA microarrays to study the transcriptome and the interactome

DNA chips are samples of DNA laid out as a series of microscopic spots bound to a glass "chip" the size of a microscope cover slip. One chip can contain spots of DNA segments corresponding to all the genes of a complex genome. The set of DNAs so displayed is called a **microarray.** DNA chips have revolutionized genetics by permitting the straightforward assay of all gene products simultaneously in a single experiment.

One protocol for making DNA chips is as follows. Robotic machines with multiple printing tips resembling miniature fountain pen nibs deliver microscopic droplets of DNA solution to specific positions (addresses) on the chip. The DNA is dried and treated so that it will bind to the glass. Many thousands of samples can be applied to one chip. In one approach, the array of DNAs consists of all known cDNAs from the genome. Another type of array is of oligonucleotides chemically synthesized in situ on the chip itself. Such arrays are exposed to a probe, for example, one consisting of the total set of RNA molecules extracted from a particular cell type at a specific stage in development. Fluorescent labels are attached to the probe, and the binding of the probe molecules to the homologous DNA spots on the glass chip is monitored automatically with the use of a laser-beam-illuminated microscope. Typical results are shown in Figure 12-27. In this way, the genes that are active at any stage of development or under any environmental condition can be assayed. The idea is to identify protein networks that are active in the cell at any particular stage of interest. Figure 12-28 shows an example of a developmental expression pattern generated by assaying this kind of chip.

Note that these DNA array methods basically take an approach to genetic dissection that is an alternative to mutational analysis. Under either method, the goal is to define the set of genes or proteins taking part in any specific process under study. Traditional mutational analysis accomplishes this objective by amassing mutations that define the total set of genes active in the process. Chip technology does the same thing directly by detecting the specific mRNAs that are expressed in that process (see Chapter 16).

DNA chips can also be used to detect protein–DNA interactions. For example, a DNA-binding protein can be fluorescently tagged and bound to DNA sequences on a chip to identify specific binding sites within the genome.

The interactome

One way of studying the interactome uses an engineered system in yeast cells called the **two-hybrid test,** which detects physical interactions between two proteins. The basis for the test is the yeast transcriptional activator for the *GAL4* gene. This protein has two domains, a DNA-binding domain that binds to the transcriptional start site and an activation domain that will activate transcription but cannot itself bind to DNA. Thus, both domains must be in close proximity in order for transcription of the *GAL4* gene to take place. In the two-hybrid system, the gene for the GAL4 transcriptional activator is divided between two plasmids so that one contains the part encoding the DNA-binding domain and the other the part encoding the activation domain. On one plasmid, a gene for one protein under investigation is spliced next to the DNA-binding domain, and this fusion protein acts as

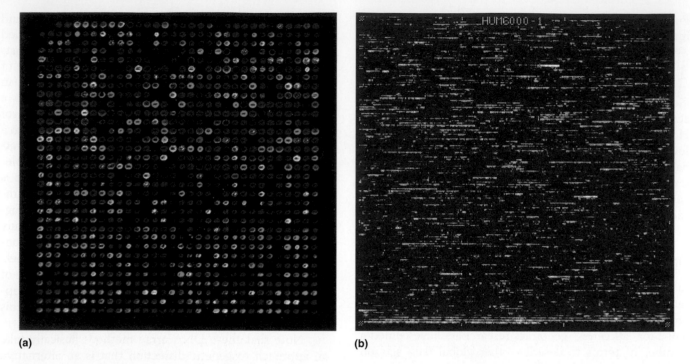

(a)   **(b)**

Figure 12-27 Fluorescence detection of binding to DNA microarrays. (a) Array of 1046 cDNAs probed with fluorescently labeled cDNA made from bone-marrow mRNA. Level of hybridization signal follows the colors of the spectrum, with red highest and blue lowest. (b) Affymetrix GeneChip 65,000 oligonucleotide array representing 1641 genes, probed with tissue-specific cDNAs. [Part a courtesy of Mark Scheria, Stanford University. Image appeared in *Nature Genetics* 16, June 1997, p. 127, Figure 1a. Part b courtesy of Affymetrix Inc. Santa Clara, CA. Image generated by David Lockhart. Affymetrix and GeneChip are U.S. registered trademarks used by Affymetrix. Image appeared in *Nature Genetics* 16, June 1997, p. 127, Figure 1b.]

"bait." On another plasmid, a gene for another protein under investigation is spliced to the activation domain and the resulting fusion protein is said to be the "target" (Figure 12-29). The two hybrid plasmids are then introduced into the same yeast cell—perhaps by mat-

ing haploid cells containing bait and target plasmids. The final step is to look for activation of GAL4 transcription, which would be proof of binding of bait and target. Successful binding can be revealed by use of a **reporter gene** (a gene for an easily detected protein)

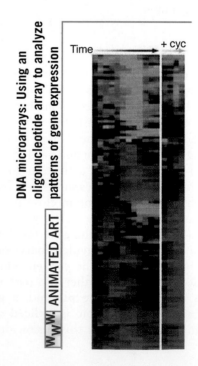

Figure 12-28 Display of gene-expression patterns detected by DNA microarrays. Each row is a different gene and each column a different time point. Red means that transcript levels for the gene are higher than at the initial time point; green means that transcript levels are lower. The four columns labeled +cyc are from cells grown on cycloheximide, meaning that no protein synthesis was taking place in these cells. [Mike Eisen and Vishy Iyer, Stanford University. Image appeared in *Nature Genetics* 18, March 1998, p. 196, Figure 1.]

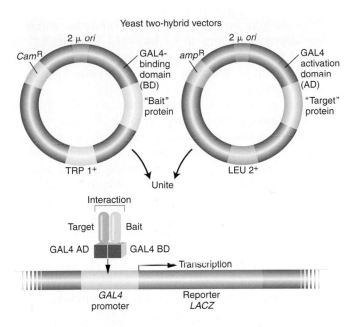

Figure 12-29 The yeast two-hybrid system for detecting gene interaction. The system uses the binding of two proteins under investigation to restore the function of the GAL4 protein, which activates a reporter gene. *Cam*, TRP, and LEU are components of the selection systems for moving the plasmids around between cells. The reporter gene is *LACZ*, which resides on a yeast chromosome (shown in blue).

that is fused downstream of the GAL4 initiation region. The two-hybrid system can be automated to make it possible to hunt for protein interactions throughout the proteome.

Genomics and all the other "omics" have spawned a discipline called **systems biology.** Whereas the approach of genetics has traditionally been reductionist, dissecting an organism with mutations to see what the parts are, systems biology tries to put the parts together to understand the whole as a system. Systems biologists illustrate their approach with the truism that knowing all the parts of an airplane does not tell one how the plane can fly. The flying plane is a system, requiring the interaction of the parts in an integrated whole. A biological system comprises gene-regulation networks, signal-transduction cascades, cell-to-cell communication, and many forms of interactions not only between "genetic" molecules but with all the other molecules of the cell and the environment. Understanding the system (as with the plane) requires not only seeing the parts in action, but, more importantly, formulating the principles by which the system can exist and function. We are said to live in a postgenomic era. With the sequencing work done, we stand on the threshold of an era in which there will be new ways of knowing about life.

KEY QUESTIONS REVISITED

- **What is genomics?**

The study of one or more complete genome sequences.

- **How are sequence maps of genomes produced?**

From sequencing large numbers of clones. Either the clones are mapped first by using molecular markers and then sequenced or the clones are sequenced at random and then aligned by overlaps.

- **How can genomic sequence maps help find a gene of interest?**

If a mutant gene has been mapped on a linkage map, its relative position on the linkage map can be compared with its position on the genome sequence and candidate genes tested—for example, for the presence of mutations.

- **What is the nature of the information in the genome?**

It can be viewed as the sum of functional products (proteins and RNAs) or as an array of docking sites for molecules such as proteins to bind to DNA or RNA.

- **What new questions can be addressed by genome-level analysis?**

Comparative genomics addresses the mechanisms of genome evolution. Functional genomics addresses the total set of genes active during one specific developmental stage. Systems biology tries to understand the genome as a series of interacting components that unite to produce life.

SUMMARY

Genomic analysis takes the approaches of genetic analysis and applies them to the collection of global data sets to fulfill goals such as the mapping and sequencing of whole genomes and the characterization of all transcripts and proteins. Genomic techniques require the rapid processing of large sets of experimental material, all dependent on extensive automation.

The key problem in compiling an accurate sequence of a genome is to take short sequence reads and relate them to one another by sequence identity to build up a consensus sequence of an entire genome. This can be done straightforwardly for prokaryotic genomes by aligning overlapping sequences from different sequence reads to compile the entire genome, because there are few or no DNA segments that are present in more than one copy in prokaryotes. The problem is that complex genomes are replete with such repetitive sequences. These repetitive sequences interfere with accurate sequence contig production. The problem is resolved by either whole genome shotgun (WGS) sequencing with the use of paired-end reads or clone-by-clone sequencing, which treats dispersed repetitive elements such as mobile elements as unique in the context of a clone. Unlike WGS sequencing, clone-by-clone sequencing requires that a physical map of the distribution of ordered and oriented clones be produced first so that appropriate clones forming a minimum tiling path through the genome can be selected for sequencing.

Genomic sequence maps form the basis for making predictions of gene structure. Linkage and cytogenetic maps show the approximate locations of mutations with defined phenotypic effects. Locating polymorphic molecular markers on both the genetic and the sequence maps allows mutations to be localized to delimited regions of the genomic sequence maps. This method has greatly advanced progress in the positional cloning of genes of interest.

Having a genomic sequence map provides the raw, encrypted text of the genome. The job of bioinformatics is to interpret this encrypted information. For the analysis of gene products, interpretation is accomplished by combining available experimental evidence for transcript structures (cDNA sequences), protein similarities, knowledge of characteristic sequence motifs, and comparative genomics.

Functional genomics attempts to understand the working of the genome as a whole system. A key approach is to determine the interactome, the set of interacting gene products and other molecules that together allow a living cell to be produced and to function.

KEY TERMS

bioinformatics (p. 406)

candidate gene approach (p. 405)

chromosome painting (p. 403)

clone contig (p. 396)

comparative genomics (p. 411)

consensus sequence (p. 392)

DNA fingerprint (p. 400)

expressed sequence tag (EST) (p. 407)

fluorescent in situ hybridization (FISH) (p. 403)

functional genomics (p.413)

genome project (p. 391)

genomics (p. 390)

microarray (p. 413)

microsatellite marker (p. 400)

minimum tiling path (p. 397)

minisatellite marker (p. 400)

open reading frame (ORF) (p. 407)

paired-end reads (p. 395)

physical map (p. 396)

primer walking (p. 395)

proteome (p. 406)

reporter gene (p. 414)

scaffold (p. 396)

sequence assembly (p. 392)

sequence contig (p. 395)

simple-sequence length polymorphism (SSLP) (p. 400)

structural genomics (p. 410)

supercontig (p. 396)

systems biology (p. 415)

two-hybrid test (p. 413)

SOLVED PROBLEMS

1. A *Neurospora* geneticist has just isolated a new mutation that causes aluminum insensitivity (*al*) in a wild-type strain called Oak Ridge. She wishes to isolate the gene from the genomic sequence and therefore needs to map it by using RFLP. It so happens that there are many RFLPs between Oak Ridge and another wild type called Mauriceville. For reasons that we do not need to go into, she suspects that it is located near the tip of the right arm of chromosome 4. Luckily, there are three RFLP markers (1, 2, and 3) available in that vicinity, and so the following cross is made:

al (Oak Ridge wild type) × *al*+ (Mauriceville wild type)

One hundred progeny are isolated and tested for *al* and the six RFLP alleles 1^O, 2^O, 3°, 1^M, 2^M, and 3^M. The results were as follows, where O and M represent the RFLP alleles, and *al* and + represent *al* and *al*+, respectively:

RFLP 1	O	M	O	M	O	M
RFLP 2	O	M	M	O	O	M
RFLP 3	O	M	M	O	M	O
al locus	*al*	+	*al*	+	*al*	+
Total of genotype	34	36	6	4	12	8

a. Is the *al* locus in fact in this vicinity?

b. If so, to which RFLP is it closest?

c. How many map units separate the three RFLP loci?

Solution

This is a mapping problem, but with the twist that some of the markers are traditional types (which we have encountered in preceding chapters) and others are molecular markers (in this case, RFLPs). Nevertheless, the principle of mapping is the same as that which we used before; in other words, it is based on recombinant frequency. In any recombination analysis, we must be clear about the genotype of the parents before we can classify progeny into recombinant classes. In this case, we know that the Oak Ridge parent must contain all O alleles, and Mauriceville all M alleles; therefore, the parents are

$$al\ 1^O2^O3^O\ \times\ al^+\ 1^M\ 2^M\ 3^M$$

and knowing this makes determining recombinant classes easy. We see from the data that the parental classes are the two most common (34 and 36). We first of all notice that the al alleles are tightly linked to RFLP 1 (all progeny are $al\ 1^O$ or $+1^M$). Therefore, the al locus is definitely on this part of chromosome 4. There are $6 + 4 = 10$ recombinants between RFLP 1 and 2, and so these loci must be 10 map units apart. There are $12 + 8 = 20$ recombinants between RFLP 2 and 3; that is, they are 20 map units apart. There are $6 + 4 + 12 + 8 = 30$ recombinants between RFLP 1 and 3, showing that these loci must flank RFLP 2. Therefore, the map is

$$\underline{al\ \text{RFLP 1}\quad 10\ \text{m.u.}\quad \text{RFLP 2}\quad 20\ \text{m.u.}\quad \text{RFLP 3}}$$

There are evidently no double recombinants, which would have been of the type M O M and O M O.

Notice that there is no new principle at work in the solution of this problem; the real challenge is to understand the nature of RFLPs and how they are used in genome mapping. If you still don't understand RFLPs, you might ask yourself how we assess which RFLP alleles are present.

2. Duchenne muscular dystrophy (DMD) is an X-linked recessive human disease affecting muscles. Six small boys had DMD, together with various other disorders, and they were found to have small deletions of the X chromosome, as shown here:

a. On the basis of this information, which chromosomal region most likely contains the gene for DMD?

b. Why did the boys show other symptoms in addition to DMD?

c. How would you use DNA samples from these six boys and DNA from unaffected boys to obtain an enriched sample of DNA containing the gene for DMD, as a prelude to cloning the gene?

Solution

a. The only region that all the deletions are lacking is the chromosomal region labeled 5, and so this region presumably contains the gene for DMD.

b. The other symptoms probably result from the deletion of the other regions surrounding the DMD region.

c. If the DNA from all the DMD deletions is denatured (that is, its strands separated) and bound to some kind of filter, the normal DNA can be cut by shearing or by restriction-enzyme treatment, denatured, and passed through the filter containing the deleted DNA. Most DNA will bind to the filter, but the region-5 DNA will pass through. This process can be repeated several times. The filtrate DNA can be cloned and then used in a FISH analysis to see if it binds to the DMD X chromosomes. If not, it becomes a candidate for the DMD-containing sequence.

PROBLEMS

BASIC PROBLEMS

1. The word *contig* is derived from the word *contiguous*. Explain the derivation.

2. Explain the approach that you would apply to sequencing the genome of a newly discovered bacterial species.

3. A FISH analysis with a probe of unknown sequence (the label on the tube fell off) reveals a fluorescent spot at one end of every chromosome. Could this be

a. a unique gene probe?

b. a centromere probe?

c. a telomere probe?

4. Terminal sequencing reads of clone inserts are a routine part of genome sequencing. How is the central part of the clone insert ever obtained?

5. In sequencing a 1-gigabase genome with fivefold coverage by using BACs, how many clones would you be handling?

6. Fingerprints of three cloned inserts labeled P, Q, and R are obtained. Clone P has no bands in common with clone Q but has two bands in common with clone R. Q has three bands in common with R. What is the order of these three inserts in the genome? Show how they overlap.

7. What is the reason for choosing a set of clones that represents a minimal tiling path?

8. How would you subclone a BAC clone? Why is it necessary for genome sequence assembly?

9. What is the difference between a contig and a scaffold?

10. Two particular contigs are suspected to be adjacent, possibly separated by repetitive DNA. In an attempt to link them, end sequences are used as primers to try to bridge the gap. Is this approach reasonable? In what situation will it not work?

11. Individual sperm can be tested for certain markers by using PCR. A man heterozygous for a microsatellite locus (M′ / M″) is also heterozygous for a disease allele caused by a short deletion. Half his sperm containing the deletion are found to contain M′, and half contain M″. Do you think these loci are linked? Draw a diagram to explain your answer.

12. A segment of cloned DNA containing a protein-coding gene is radioactively labeled and used in an in situ hybridization. Radioactivity was observed over five regions on different chromosomes. How is this possible?

13. In an in situ hybridization experiment, a certain clone bound to only the X chromosome in a boy with no disease symptoms. However, in a boy with Duchenne muscular dystrophy (X-linked recessive disease), it bound to the X chromosome and to an autosome. Explain. Could this clone be useful in isolating the DMD gene?

14. In a recombination analysis, a previously obtained *Neurospora* morphological mutant called "stubby" (*stu*), which had abnormal branching, mapped to the left end of chromosome 5. The complete *Neurospora* genomic sequence available in the public database showed three likely candidate genes in that vicinity. How would you determine which (if any) of these candidates corresponds to stubby? (**Hint:** It is quite easy to transform *Neurospora*.)

15. In a genomic analysis looking for a specific gene, one candidate gene was found to have a single-base-pair substitution resulting in a nonsynonymous amino acid change. What would you have to check before breaking out the champagne?

16. In what sense is a codon within an ORF a docking element?

17. Is a bacterial operator a docking element?

18. A certain cDNA of size 2 kb hybridized to eight genomic fragments of total size 30 kb and contained two short ESTs. The ESTs were also found in two of the genomic fragments each of size 2 kb. Sketch a possible explanation for these results.

19. A sequenced fragment of DNA in *Drosophila* was used in a BLAST search. The best (closest) match was to a kinase gene from *Neurospora*. Does this match mean that the *Drosophila* sequence contains a kinase gene?

20. In a two-hybrid test, a certain gene *A* gave positive results with two clones M and N. When M was used, it gave positives with three clones, A, S, and Q. M gave only one positive (with A). Develop a tentative interpretation of these results.

21. A cloned gene from *Arabidopsis* is used as a radioactive probe against DNA samples from cabbage (which is in the same plant family) digested by three different restriction enzymes. For enzyme 1, there were three radioactive bands on the autoradiogram; for enzyme 2, there was one band; and, for enzyme 3, there were two bands. How can these results be explained?

22. Five YAC clones of human DNA (YAC A through YAC E) were tested for sequence-tagged sites (STSs): STS1 through STS7. (An STS is a short unique sequence that is PCR amplified by a specific primer pair.) The results are shown in the following table, in which a plus sign shows that the YAC contains that STS:

YAC	STS						
	1	2	3	4	5	6	7
A	+	−	+	+	−	−	−
B	+	−	−	−	+	−	−
C	−	−	+	+	−	−	+
D	−	+	−	−	+	+	−
E	−	−	+	−	−	−	+

a. Draw a physical map showing the STS order.

b. Align the YACs into a contig.

23. You have two strains of the ascomycete fungus *Aspergillus nidulans* that are from different continents and are likely to have accumulated many different genetic variations in the time since they

became geographically isolated from each other. You detect polymorphisms in this strain by allowing a single sequence to act as primer for PCR. This by chance spans and amplifies various regions of the genome, showing up as bands on a gel. (This is called RAPD analysis, standing for randomly amplified polymorphic DNA.) One primer amplified two bands in haploid strain 1 and no bands in strain 2. These strains were crossed, and seven progeny were analyzed. The results were as follows:

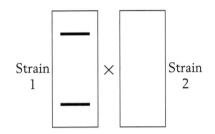

Progeny

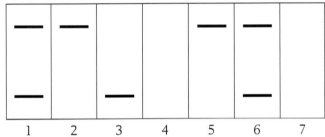

a. Draw diagrams that explain the difference between the parents.

b. Explain the origin of the progeny and their relative frequencies.

c. Draw an example of a single tetrad from this cross, showing the amplified bands.

24. A certain disease is inherited as an autosomal dominant N. It was noted that some patients carry chromosomal rearrangements that have in common the property of always having a chromosomal breakpoint in band 3q3.1 on chromosome 3. Four molecular probes (a through d) are known to hybridize in situ to this band, but their order is not known. In the rearrangements, only probe c hybridizes to one side of the chromosome 3 breakpoint, and probes a, b, and d always hybridize to the other side of the breakpoint.

a. Draw diagrams that illustrate the meaning of these findings.

b. How would you use this information for positional cloning of the normal allele n?

c. Once n has been cloned, how would you use this clone to investigate the nature of the mutations in patients who have the same disease but do not have a chromosomal rearrangement with a breakpoint in 3q3.1?

25. Place the following techniques in the order in which they would be employed in taking a genome-sequencing project from low to highest resolution. (Not all techniques necessarily need to be used.)

a. RFLP mapping

b. Clone contig assembly

c. Microsatellite mapping

d. DNA fingerprint mapping

e. DNA sequencing of BAC clones

f. Mapping of phenotypic markers

g. Clone scaffold assembly

h. Paired-end reads

26. You have the following sequence reads from a genomic clone of the *Drosophila melanogaster* genome:

Read 1: TGGCCGTGATGGGCAGTTCCGGTG
Read 2: TTCCGGTGCCGGAAAGA
Read 3: CTATCCGGGCGAACTTTTGGCCG
Read 4: CGTGATGGGCAGTTCCGGTG
Read 5: TTGGCCGTGATGGGCAGTT
Read 6: CGAACTTTTGGCCGTGATGGGCAGTTCC

Use these six sequence reads to create a sequence contig of this part of the *D. melanogaster* genome.

27. In whole genome shotgun sequencing, paired-end reads are used to join contigs together into scaffolds. You have two contigs, called *contig A* and *contig B*. Contig A is 4833 nucleotides long, and contig B is 3320 nucleotides long. Paired-end reads have been made from two ends of a clone containing a 2000-bp genomic insert. The sequencing read from one end of this clone is 210 bp long, and it aligns with nucleotides 4572–4781 of contig A. The sequencing read from the other end of the clone is 342 nucleotides long and aligns with nucleotides 245–586 of contig B. From this information, draw a map of the scaffold containing contig A and contig B, indicating the overall size of the scaffold and the size of the gap between contig A and contig B.

28. Sometimes, cDNAs turn out to be "monsters"; that is, fusions of DNA copies of two different mRNAs accidentally inserted adjacently to each other in the same clone. You suspect that a cDNA clone from the nematode *Caenorhabditis elegans* is such a monster because the sequence of the cDNA insert predicts a protein with two structural domains not normally observed in the same protein. How would you use the availability of the entire genomic sequence to assess if this cDNA clone is a monster or not?

29. You have sequenced the genome of the bacterium *Salmonella typhimurium*, and you are using BLAST analysis to identify similarities within the *S. typhimurium* genome to known proteins. You find a protein that is 100 percent identical in the bacterium *Escherichia coli*. When you compare nucleotide sequences of the *S. typhimurium* and *E. coli* genes, you find that their nucleotide sequences are only 87 percent identical.

    a. Explain this observation.

    b. What do these observations tell you about the merits of nucleotide versus protein similarity searches in identifying related genes?

Challenging Problems

30. From in situ hybridizations, five different YACs containing genomic fragments were known to hybridize to one specific chromosome band of the human genome. Rare-cutting restriction enzymes (which have 8-bp recognition sequences and cut on average once in every 64,000 bp) were used to digest genomic DNA. Radioactively labeled YACs were each hybridized to blots of the digest. The autoradiogram was as follows:

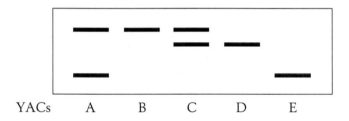

YACs   A   B   C   D   E

    a. Use these results to order the three hybridized restriction fragments.

    b. Show the locations of the YACs in relation to the three genomic restriction fragments in part a.

 **Unpacking Problem 30**

1. State two types of hybridization used in genetics. What types of hybridizations are used in this problem, and what is the molecular basis for such hybridizations? (Draw a rough sketch of what happens at the molecular level during hybridization.)

2. How are in situ hybridizations done in general? How would the specific in situ hybridizations in this problem be done (as in the first sentence of the problem)?

3. What is a YAC?

4. What are chromosome bands, and what procedure is used to produce them? Sketch a chromosome with some bands and show how the in situ hybridizations would look.

5. How would five different YACs have been shown to hybridize to one band?

6. What is a genomic fragment? Would you expect the five YACs to contain the same genomic fragment or different ones? How do you think these genomic fragments were produced (what are some general ways of fragmenting DNA)? Does it matter how the DNA was fragmented?

7. What is a restriction enzyme?

8. What is a rare cutter?

9. Why were the YACs radioactively labeled? (What does it mean to radioactively label something?)

10. What is an autoradiogram?

11. Write a sentence that uses the terms *DNA*, *digestion*, *restriction enzyme*, *blot*, and *autoradiogram*.

12. Explain exactly how the pattern of dark bands shown in the problem was obtained.

13. Approximately how many kilobases of DNA are in a human genome?

14. If human genomic DNA were digested with a restriction enzyme, roughly how many fragments would be produced? Tens? Hundreds? Thousands? Tens of thousands? Hundreds of thousands?

15. Would all these DNA fragments be different? Would most of them be different?

16. If these fragments were separated on an electrophoretic gel, what would you see if you added a DNA stain to the gel?

17. How does your answer to the preceding question compare with the number of autoradiogram bands in the diagram?

18. Part a of the problem mentions "three hybridized restriction fragments." Point to them in the diagram.

19. Would there actually be any restriction fragments on an autoradiogram?

20. Which YACs hybridize to one restriction fragment and which YACs hybridize to two?

21. How is it possible for a YAC to hybridize to two DNA fragments? Suggest two explanations, and decide which makes more sense in this problem. Does the fact that all the YACs in this problem bind to one chromosome band (and apparently to nothing else) help you in deciding? Could a YAC hybridize to more than two fragments?

22. Distinguish the use of the word *band* by cytogeneticists (chromosome microscopists) from the use of the word *band* by molecular geneticists. In what way do these uses come together in this problem?

31. You have the following sequence reads from a genomic clone of the *Homo sapiens* genome:

Read 1: ATGCGATCTGTGAGCCGAGTCTTTA
Read 2: AACAAAAATGTTGTTATTTTTATTTCAGATG
Read 3: TTCAGATGCGATCTGTGAGCCGAG
Read 4: TGTCTGCCATTCTTAAAAACAΛΛΛΛTGT
Read 5: TGTTATTTTTATTTCAGATGCGA
Read 6: AACAAAAATGTTGTTATT

a. Use these six sequence reads to create a sequence contig of this part of the *H. sapiens* genome.

b. Translate the sequence contig in all possible reading frames.

c. Go to the BLAST page of NCBI (http://www.ncbi.nlm.nih.gov/BLAST/ and see Appendix B) and see if you can identify the gene of which this sequence is a part by using each of the reading frames as a query for protein–protein comparison (BLASTp).

32. A *Neurospora* geneticist wanted to clone the gene *cys-1*, which was believed to be near the centromere on chromosome 5. Two RFLP markers (RFLP1 and RFLP2) were available in that vicinity, and so he made the following cross:

Oak Ridge *cys-1* × Mauriceville *cys-1*⁺

(See Solved Problem 1 for an explanation of the Oak Ridge and Mauriceville strains.) Then 100 ascospores were tested for RFLP and *cys-1* genotypes, and the following results were obtained:

RFLP 1	O	M	O	M	O	M
RFLP 2	O	M	M	O	M	O
cys locus	*cys*	+	+	*cys*	*cys*	+
Total of genotype	40	43	2	3	7	5

a. Is *cys-1* in this region of the chromosome?

b. If so, draw a map of the loci in this region, labeled with map units.

c. What would be a suitable next step in cloning the *cys-1* gene?

33. It turns out that some sizable regions of different chromosomes of the human genome are more than 99 percent nucleotide identical with one another. These regions were overlooked in the production of the draft genome sequence of the human genome because of their high level of similarity. Of the mapping techniques discussed in this chapter, which would allow genome researchers to identify the existence of such duplicate regions?

34. A *Caenorhabditis* contig for one region of chromosome 2 is as follows, in which A through H are cosmids:

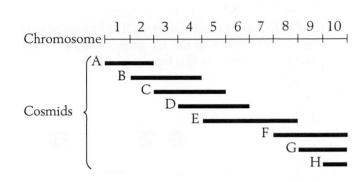

a. A cloned gene *pBR322-x* hybridized to cosmids C, D, and E. What is the approximate location of this gene *x* on the chromosome?

b. A cloned gene *pUC18-y* hybridized only to cosmids E and F. What is its location?

c. Explain exactly how it is possible for both probes to hybridize to cosmid E.

 35. The gene for the autosomal dominant disease shown in the pedigree below is thought to be on chromosome 4, and so five RFLPs (1–5) mapped on chromosome 4 were tested in all family members. The results of the testing are shown below each family member listed in the pedigree. Vertical lines represent the two homologous chromosomes, and the superscripts represent different alleles of the RFLP loci.

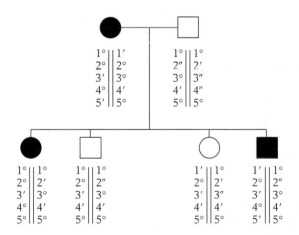

a. Explain how this experiment was carried out.

b. Decide which RFLP locus is closest to the disease gene (explain your logic).

c. How would you use this information to clone the disease gene?

36. Cystic fibrosis is a disease that shows autosomal recessive inheritance. A couple has three children with cystic fibrosis (CF), as shown in the following pedigree. Their oldest son has recently married his second cousin. He has molecular testing done to determine if there is a chance that he may have children

with CF. Three probes detecting RFLPs known to be very closely linked to the *CF* gene were used to assess the genotypes in this family.

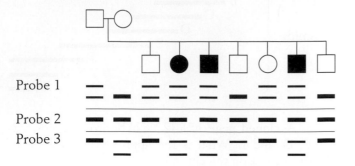

Answer the following questions, describing your logic.

a. Is this man homozygous normal or a carrier?

b. Are his three normal siblings homozygous normal or carriers?

c. From which parent did each carrier inherit the disease allele?

(Problem 36 is from Tamara Western.)

37. Some exons in the human genome are quite small (less than 75 bp long). Identification of such "microexons" is difficult because these distances are too short to reliably use ORF identification or codon bias to determine if small genomic sequences are truly part of an mRNA and a polypeptide. What techniques of "gene finding" can be used to try to assess if a given region of 75 bp constitutes an exon?

EXPLORING GENOMES   A Web-Based Bioinformatics Tutorial

Introduction to Genomic Databases

Where does a researcher turn to find information about a gene? Integrated genetic databases are maintained by a number of private and government organizations. In the first Genomics tutorial at www.whfreeman.com/iga, you will be introduced to the resources available through the National Center for Biotechnology Information (NCBI) in Washington, D.C.

Learning to Use ENTREZ

The ENTREZ program at NCBI is an integrated search tool that links together a variety of databases that have different types of content. In the Genomics tutorial at Web site www.whfreeman.com/iga, we will use it to look up the dystrophin gene associated with muscular dystrophy and find research literature references, the gene sequence as well as conserved domains, the equivalent gene from a variety of organisms other than human, and the chromosome map of its location.

Learning to Use BLAST

To compare one protein sequence with another, we most often use a computer program called BLAST. This program allows us use a protein sequence to search and to find sequences from other organisms that are similar to it. In the Genomics tutorial at www.whfreeman.com/iga, we will run BLAST on a small simple protein, insulin (Chapter 11), and on a large complex one, dystrophin.

Using BLAST to Compare Nucleic Acid Sequences

The BLAST algorithm is also able to search for nucleic acid sequences and compare them. In the Genomics tutorial at www.whfreeman.com/iga, we will find that comparing transfer RNA sequences among species is a good way to explore this capability.

Learning to Use PubMed

PubMed provides a searchable database of the world's scientific literature. In the Genomics tutorial at www.whfreeman.com/iga, you will learn to do a literature search to find the first report of a gene sequence and subsequent papers demonstrating the function of the gene.

THE DYNAMIC GENOME: TRANSPOSABLE ELEMENTS

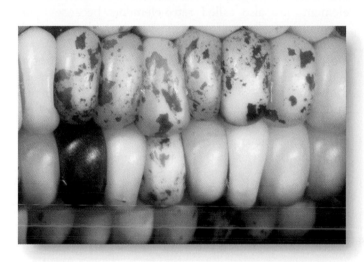

Kernels on an ear of corn. The spotted kernels on this ear of corn result from the interaction of a mobile genetic element (a transposable element) with a corn gene whose product is required for pigmentation. [Cliff Weil and Susan Wessler.]

KEY QUESTIONS

- Why were transposable elements first discovered genetically in maize but first isolated molecularly from *E. coli?*

- How do transposable elements participate in the spread of antibiotic-resistant bacteria?

- Why are transposable elements classified as DNA transposons or RNA transposons?

- How do autonomous and nonautonomous transposable elements differ?

- How can humans survive given that up to 50 percent of the human genome is derived from transposable elements?

- How can the study of yeast retrotransposons lead to improved procedures for human gene therapy?

OUTLINE

13.1 Discovery of transposable elements in maize

13.2 Transposable elements in prokaryotes

13.3 Transposable elements in eukaryotes

13.4 The dynamic genome: more transposable elements than ever imagined

13.5 Host regulation of transposable elements

CHAPTER OVERVIEW

Starting in the 1930s, genetic studies of maize yielded results that greatly upset the classical genetic picture of genes residing only at fixed loci on the main chromosome. The research literature began to carry reports suggesting that certain genetic elements present in the main chromosomes can somehow move from one location to another. These findings were viewed with skepticism for many years, but it is now clear that such mobile elements are widespread in nature.

A variety of colorful names (some of which help to describe their respective properties) have been applied to these genetic elements: controlling elements, jumping genes, mobile genes, mobile genetic elements, and transposons. Here we use the terms transposable elements and transposons, which embrace the entire family of types. Transposable elements can move to new positions within the same chromosome or even to a different chromosome. They have been detected genetically in model organisms such as *E. coli*, maize, yeast, and *Drosophila* through the mutations that they produce when they inactivate genes into which they insert.

DNA sequencing of genomes from a variety of microbes, plants, and animals indicates that transposable elements exist in virtually all organisms. Surprisingly, they are by far the largest component of the human genome, accounting for almost 50 percent of our chromosomes. Despite their abundance, the normal genetic role of these elements is not known with certainty. What is becoming clear is that both plants and animals use epigenetic mechanisms to regulate the proliferation of their transposable elements (Chapter 10).

As shown in Figure 13-1, eukaryotic transposable elements fall into one of two classes, called **class 1** and **class 2 elements,** that are distinguished by whether the DNA element transposes through an RNA copy (class 1) or directly (class 2). Class 1 elements are also called **RNA elements** because the DNA element in the genome is transcribed into an RNA copy. The RNA copy is converted back into a DNA copy that can then insert into another (target) site in the host genome. Class 1 elements are also called **retro-elements** because their movement (called **retrotransposition**) is characterized by the reverse flow of genetic information, from RNA to DNA. The copy number of class 1 elements in a host genome can be enormous. For example, there are almost a million copies of a class 1 element called *Alu* in the human genome. Such high copy numbers are possible because many RNAs can be transcribed from a single class 1 element and each can, theoretically, result in a new DNA insertion in the host genome. In addition, because the RNA copy is the transposition intermediate,

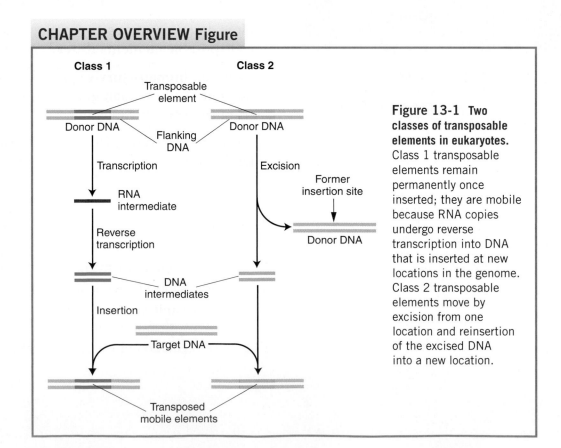

CHAPTER OVERVIEW Figure

Figure 13-1 Two classes of transposable elements in eukaryotes. Class 1 transposable elements remain permanently once inserted; they are mobile because RNA copies undergo reverse transcription into DNA that is inserted at new locations in the genome. Class 2 transposable elements move by excision from one location and reinsertion of the excised DNA into a new location.

insertions of class 1 elements into the genome are essentially permanent. That is, they cannot be excised from the donor site. Nonetheless, they are still considered to be mobile because their copies can insert into new target DNA.

In contrast, class 2 elements are called **DNA elements** because a class 2 element itself moves from one site in the genome to another. Unlike class 1 elements, class 2 elements can excise from the donor site, which means that, if insertion into a gene has created a mutation, excision of the element can lead to the reversion of the original mutation. The first transposable elements discovered genetically in maize were class 2 elements that led to unusual spotted kernels due to their excision from genes responsible for pigment production in the kernel.

In their studies, scientists are able to exploit the ability of transposable elements to insert into new sites in the genome. Transposable elements engineered in the test tube are valuable tools, both in prokaryotes and in eukaryotes, for genetic mapping, creating mutants, cloning genes, and even producing transgenic organisms. Let us reconstruct some of the steps in the evolution of our present understanding of transposable elements. In doing so, we will uncover the principles guiding these fascinating genetic units.

13.1 Discovery of transposable elements in maize

McClintock's experiments: the *Ds* element

In the 1940s, Barbara McClintock made an astonishing discovery while studying the colored kernels of so-called Indian corn (maize; see the Model Organism box on page 428). McClintock was analyzing the breakage of chromosomes in maize, which has 10 chromosomes, numbered from largest (1) to smallest (10). Chromosome breakage happens randomly and infrequently in the lifetime of any organism. However, McClintock found that, in one strain of maize, chromosome 9 broke very frequently and at one particular site (locus; Figure 13-2). Breakage of the chromosome at this locus, she determined, was due to the presence of two genetic factors. One factor that she called *Ds* (for *Dissociation*) was located at the site of the break. Another, unlinked genetic factor was required to "activate" the breakage of chromosome 9 at the *Ds* locus. Thus, McClintock called this second factor *Ac* (for *Activator*).

McClintock began to suspect that *Ac* and *Ds* were actually mobile genetic elements when she found it impossible to map *Ac*. In some plants, it mapped to one position; in other plants of the same line, it mapped to different positions. As if this mapping were not enough of a curiosity, rare kernels with dramatically different

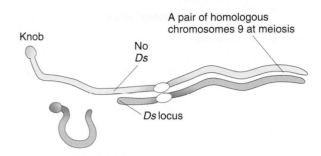

Figure 13-2 Chromosomal breakage at the *Ds* locus in corn. Chromosome 9 breaks at the *Ds* locus, where the *Ds* transposable has inserted.

phenotypes could be derived from the original strain that had frequent breaks in chromosome 9. One such phenotype was a rare colorless kernel containing pigmented spots.

Figure 13-3 compares the phenotype of the chromosome-breaking strain with the phenotype of one of these derivative strains. For the chromosome-breaking strain, breakage of the chromosome at or near *Ds* results in the loss of the wild-type alleles of the *C*, *Sh*, and *Wx* genes. In the example shown in Figure 13-3a, a break occurred in a single cell, which divided mitotically to produce the large sector of mutant tissue (*c sh wx*). Breakage can happen many times in a single kernel, but each sector of tissue will display the loss of expression of all three genes. In contrast, each new derivative affected the expression of only a single gene. One derivative that affected the expression of only the C gene is shown in Figure 13-3b. In this example, pigmented spots appeared on a colorless kernel background. Although C gene expression was altered in this strange way, the expression of *Sh* and *Wx* was normal.

To explain the new derivatives, McClintock hypothesized that *Ds* had moved from a site near the centromere into the C gene located close to the telomeric end. In its new location, *Ds* produced very unusual changes in the expression of the C gene. The spotted kernel is an example of an **unstable phenotype.** McClintock concluded that such unstable phenotypes resulted from the movement or transposition of *Ds* away from the C gene. That is, the kernel begins development with a C gene that has been mutated by the insertion of *Ds*. However, in some cells of the kernel, *Ds* leaves the C gene, allowing the mutant phenotype to revert to wild type (as evidenced by pigment production) in the original cell and in all its mitotic descendants. There are big spots of color when *Ds* leaves the C gene early in kernel development (because there are more mitotic descendants), whereas there are small spots when *Ds* leaves the C gene later in kernel development. Unstable mutant phenotypes that revert to wild type are a clue to the participation of mobile elements.

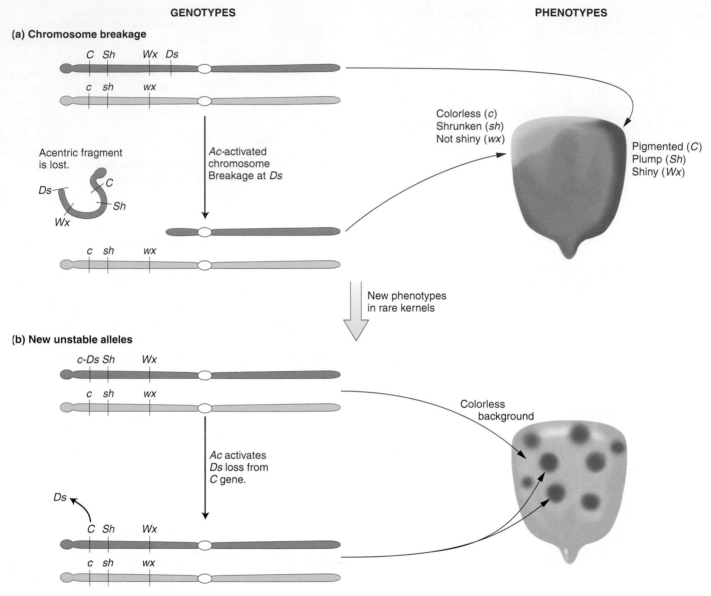

GENOTYPES

(a) Chromosome breakage

PHENOTYPES

Colorless (*c*)
Shrunken (*sh*)
Not shiny (*wx*)

Pigmented (*C*)
Plump (*Sh*)
Shiny (*Wx*)

Acentric fragment
is lost.

Ac-activated
chromosome
Breakage at *Ds*

New phenotypes
in rare kernels

(b) New unstable alleles

Colorless
background

Ac activates
Ds loss from
C gene.

Figure 13-3 New phenotypes in corn produced through the movement of the *Ds* transposable element on chromosome 9. (a) A chromosome fragment is lost through breakage at the *Ds* locus. Recessive alleles on the homologous chromosome are expressed, producing the colorless sector in the kernel. (b) Insertion of *Ds* in the *C* gene (top) creates colorless corn-kernel cells. Excision of *Ds* from the *C* gene through the action of *Ac* in cells and their mitotic descendants allows color to be expressed again, producing the spotted phenotype.

Autonomous and nonautomous elements

What is the relation between *Ac* and *Ds*? How do they interact with genes and chromosomes to produce these interesting and unusual phenotypes? These questions were answered by further genetic analysis. Interactions between *Ds*, *Ac*, and the pigment gene C are used as an example in Figure 13-4. Here *Ds* is shown as a piece of DNA that has inactivated the C gene [the allele is called *c-mutable(Ds)* or *c-m(Ds)* for short] by inserting into its coding region. A strain with *c-m(Ds)* and no *Ac* has col-

orless kernels because *Ds* cannot move; it is stuck in the C gene. A strain with *c-m(Ds)* and *Ac* has spotted kernels because *Ac* activates *Ds*, in some cells, to leave (called **excise** or **transpose**) the C gene, thereby restoring gene function.

Other strains were isolated in which the *Ac* element itself had inserted into the C gene [called *c-m(Ac)*]. Unlike the *c-m(Ds)* allele, which is unstable only when *Ac* is in the genome, *c-m(Ac)* is always unstable. Furthermore, McClintock found that on rare occasions an allele

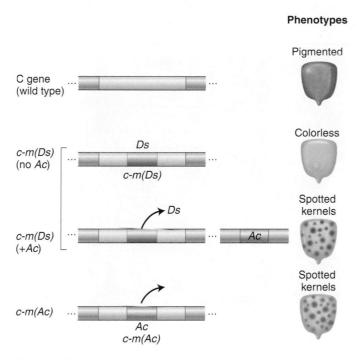

Phenotypes

Pigmented

Colorless

Spotted kernels

Spotted kernels

Figure 13-4 Summary of the main effects of transposable elements in corn. *Ac* and *Ds* are used as examples, acting on the *C* gene controlling pigment.

of the *Ac* type could be transformed into an allele of the *Ds* type. This transformation was due to the spontaneous generation of a *Ds* element from the inserted *Ac* element. In other words, *Ds* is, in all likelihood, an incomplete, mutated version of *Ac* itself.

Several systems like *Ac/Ds* were found by McClintock and other geneticists working with maize. Two other

systems are *Dotted* (*Dt*, discovered by Marcus Rhoades) and *Suppressor/mutator* [(*Spm*), independently discovered by McClintock and Peter Peterson, who called it *Enhancer/ Inhibitor* (*En/In*)]. In addition, as you will see in the sections that follow, elements with similar genetic behavior have been isolated from bacteria, plants, and animals.

The common genetic behavior of these elements led geneticists to propose new categories for all the elements. *Ac* and elements with similar genetic properties are now called **autonomous elements** because they require no other elements for their mobility. Similarly, *Ds* and elements with similar genetic properties are called **nonautonomous elements.** An element *family* is composed of an autonomous element and the nonautonomous members that it can mobilize. Autonomous elements encode the information necessary for their own movement and for the movement of unlinked nonautonomous elements in the genome. Because nonautonomous elements do not encode the functions necessary for their own movement, they cannot move unless an autonomous element from their family is present somewhere else in the genome.

Figures 13-5 and 13-6 show examples of the effects of transposons in maize and similar effects in snapdragon.

MESSAGE Transposable elements in maize can inactivate a gene in which they reside, cause chromosome breaks, and transpose to new locations within the genome. Autonomous elements can perform these functions unaided; nonautonomous elements can transpose only with the help of an autonomous element elsewhere in the genome.

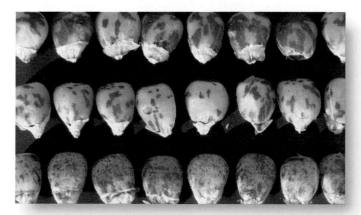

Figure 13-5 Mosaicism through the activity of a transposable element in maize. Insertion of a transposable element disrupts pigment production, resulting in yellow (colorless) kernels. Excision of the transposable element restores pigment production. Shown are kernels in which the transposable element excised at different stages. The larger the spot, the earlier the excisions in kernel development. [Anthony Griffiths.]

Figure 13-6 Mosaicism through the excision of transposable elements in snapdragons *(Antirrhinum).* Insertion of a transposable element disrupts pigment production, resulting in white flowers. Excisions of the transposable element restores pigment production, resulting in red floral tissue sectors. [Photograph from Rosemary Carpenter and Enrico Coen.]

MODEL ORGANISM Maize

Maize, also known as corn, is actually *Zea mays*, a member of the grass family. Grasses—also including rice, wheat, and barley—are the most important source of calories for humanity. Maize was domesticated from the wild grass teosinte by Native Americans in Mexico and Central America and was first introduced to Europe by Columbus on his return from the New World.

In the 1920s, Rollins A. Emerson set up a laboratory at Cornell University to study the genetics of corn traits, including kernel color, which were ideal for genetic analysis. In addition, the physical separation of male and female flowers into the tassel and ear, respectively, made controlled genetic crosses relatively easy to accomplish. Among the outstanding geneticists attracted to the Emerson laboratory were Marcus Rhoades, Barbara McClintock, and George Beadle (see Chapter 6). Before the advent of molecular biology and the rise of microorganisms as model organisms, geneticists performed microscopic analysis of chromosomes and related their behavior to the segregation of traits. The large pachytene chromosomes of maize and the salivary gland chromosomes of *Drosophila* made these the organisms of choice for cytogenetic analyses. The results of these early studies led to an understanding of chromosome behavior during meiosis and mitosis, including such events as recombination, the consequences of chromosome breakage such as inversions, translocations, and duplication, and the ability of knob-like structures to behave like centromeres (called neo-centromeres) in meiosis.

The maize lab of Rollins A. Emerson at Cornell University, 1929. Standing from left to right: Charles Burnham, Marcus Rhoades, R. A. Emerson, and Barbara McClintock. Kneeling is George Beadle. Both McClintock and Beadle (see Chapter 6) went on to win the Nobel Prize. [Courtesy Department of Plant Breeding, Cornell University.]

Maize still serves as a model genetic organism. Molecular biologists continue to exploit its beautiful pachytene chromosomes with new antibody probes (see the photo below) and have used its wealth of genetically well characterized transposable elements as tools to identify and isolate important genes.

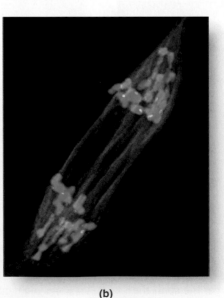

Analysis of maize chromosomes, then and now. Maize chromosomes are large and easily visualized by light microscopy. (a) An image from Marcus Rhoades (1952). (b) This image is comparable to part a except that the spindle is shown in blue (stained with antibodies to tubulin), the centromere is shown in red (stained with antibodies to a centromere-associated protein), and the chromosomes are green. [Part a is from M. M. Rhoades, "Preferential Segregation in Maize," in J. W. Gowen, ed., *Heterosis.* Iowa State College Press, Ames, 1952, pp. 66–80. Part b is from R. K. Dawe, L. Reed, H.-G. Yu, M. G. Muszynski, and E. N. Hiatt, "A Maize Homolog of Mammalian CENPC Is a Constitutive Component of the Inner Kinetochore," *Plant Cell* 11, 1999, 1227–1238.]

(a)   (b)

Transposable elements: only in maize?

Although geneticists readily accepted McClintock's discovery of transposable elements in maize, many were reluctant to consider the possibility that similar elements resided in the genomes of other organisms. Their existence in all organisms would imply that genomes are inherently unstable and dynamic. This view was inconsistent with the fact that the genetic maps of members of the same species were the same. After all, if genes can be genetically mapped to a precise chromosomal location, doesn't this mapping indicate that they are not moving around?

Because McClintock was a highly respected geneticist, her results were explained by saying that maize is not a natural organism; it is a crop plant that is the product of human selection and domestication. This view was held by some until the 1960s when the first transposable elements were isolated from the *E. coli* genome and studied at the DNA sequence level. Transposable elements were subsequently isolated from the genomes of many organisms including *Drosophila* and yeast. When it became apparent that transposable elements are a significant component of the genomes of most and perhaps all organisms, Barbara McClintock was recognized for her seminal discovery by being awarded the 1983 Nobel Prize in Medicine or Physiology.

13.2 Transposable elements in prokaryotes

The genetic discovery of transposable elements led to many questions about what such elements might look like at the DNA sequence level and how they are able to move from one site to another in the genome. Did all organisms have them? Did all elements look alike or were there different classes of transposable elements? If there were many classes of elements, could they coexist in one genome? Did the number of transposable elements in the genome vary from species to species? The molecular nature of transposable genetic elements was first understood in bacteria. Therefore, we shall continue this story by examining the original studies performed with prokaryotes.

Bacterial insertion sequences

Insertion sequences, or **insertion-sequence (IS) elements,** are segments of bacterial DNA that can move from one position on a chromosome to a different position on the same chromosome or on a different chromosome. When IS elements appear in the middle of genes, they interrupt the coding sequence and inactivate the expression of that gene. Owing to their size and in some cases the presence of transcription and translation termination signals, IS elements can also block the expression of other genes in the same operon if those genes are downstream from the promoter of the operon. IS elements were first found in *E. coli* in the *gal* operon—a set of three genes taking part in the metabolism of the sugar galactose.

Physical demonstration of DNA insertion

Recall from Chapter 5 that phage λ inserts next to the *gal* operon and that it is a simple matter to obtain λd*gal* phage particles that have picked up the *gal* region. When the IS mutations in *gal* are incorporated into λd*gal* phage and the buoyant density in a cesium chloride (CsCl) gradient of the phage is compared with that of the normal λd*gal* phage, it is evident that the DNA carrying the IS mutation is longer than the wild-type DNA. This experiment clearly demonstrates that the mutations are caused by the insertion of a significant amount of DNA into the *gal* operon. Figure 13-7 depicts this experiment in more detail.

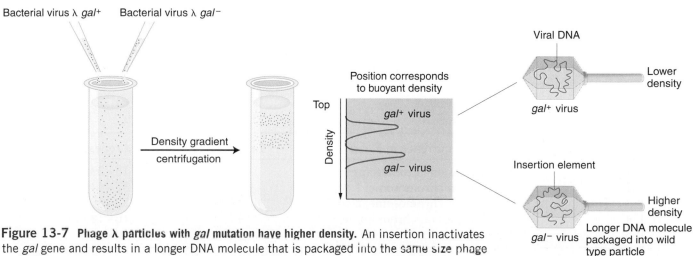

Figure 13-7 Phage λ particles with *gal* mutation have higher density. An insertion inactivates the *gal* gene and results in a longer DNA molecule that is packaged into the same size phage particle as wild type. [From S. M. Cohen and J. A. Shapiro, "Transposable Genetic Elements." Copyright 1980 by Scientific American, Inc. All rights reserved.]

IDENTIFICATION OF DISCRETE IS ELEMENTS Several *E. coli gal⁻* mutants were found to contain large insertions of DNA into the *gal* operon. This finding led naturally to the next question: Are the segments of DNA that insert into genes merely random DNA fragments or are they distinct genetic entities? The answer to this question came from the results of hybridization experiments showing that many different insertion mutations are caused by a small set of insertion sequences. These experiments are performed with the use of λ*dgal* phage that contain the *gal⁻* operon from several independently isolated *gal* mutant strains. Individual phage from the strains are isolated, and their DNA is used to synthesize radioactive RNA in vitro. Certain fragments of this RNA are found to hybridize with the DNA from other *gal⁻* mutations containing large DNA insertions but not with wild-type DNA. These results were interpreted to mean that independently isolated *gal* mutants contain the same extra piece of DNA. These particular RNA fragments also hybridize to DNA from other IS mutants, showing that the same bit of DNA is inserted in different places in the different IS mutants.

On the basis of their patterns of cross-hybridization, the insertion mutants are placed into categories. The first sequence, the 800-bp segment identified in *gal*, is termed IS1. A second sequence, termed IS2, is 1350 bp long. Table 13-1 lists some of the insertion sequences and their sizes. Although IS elements differ in DNA sequence, they have several features in common. For example, all IS elements encode a protein, called a **transposase,** which is an enzyme required for the movement of IS elements from one site in the chromosome to another. In addition, all IS elements begin and end with short inverted repeat sequences that are required for their mobility. The transposition of IS elements and other mobile genetic elements will be considered later in the chapter.

The genome of the standard wild-type *E. coli* is rich in IS elements: it contains eight copies of IS1, five copies of IS2, and copies of other less well studied IS types. Because IS elements are regions of identical sequence, they are sites where crossovers may take place. For example,

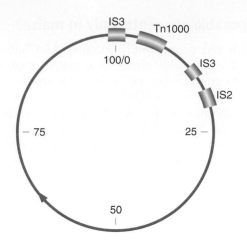

Figure 13-8 Distribution of IS elements in one F factor. The positions of the elements IS2, IS3, and Tn1000 (boxes) are shown relative to the map coordinates from 0 to 100. The arrow indicates the origin and direction of F DNA transfer during conjugation.

recombination between the F factor plasmid and the *E. coli* chromosome to form *Hfr* strains is mediated by a single crossover between an IS element located on the plasmid and an IS element located on the chromosome. Figure 13-8 shows an example of the IS element distribution on one F factor plasmid.

> **MESSAGE** The bacterial genome contains segments of DNA, termed IS elements, that can move from one position on the chromosome to a different position on the same chromosome or on a different chromosome.

Prokaryotic transposons

In Chapter 5, you learned about **R factors,** which are plasmids carrying genes that encode resistance to several antibiotics. These R factors (for resistance) are transferred rapidly on cell conjugation, much like the F factor in *E. coli.*

R factors proved to be just the first of many similar F-like factors to be discovered. R factors have been

Table 13-1 Prokaryotic Insertion Elements

Insertion sequence	Normal occurrence in *E. coli*	Length (bp)	Inverted repeat (bp)*
IS1	5–8 copies on chromosome	768	18–23
IS2	5 copies on chromosome; 1 copy on F	1327	32–41
IS3	5 copies on chromosome; 2 copies on F	1400	32–38
IS4	1 or 2 copies on chromosome	1400	16–18
IS5	Unknown	1250	Short

* The numbers represent the lengths of the 5′ and 3′ copies of the imperfect inverted repeats.
Source: M. P. Calos and J. H. Miller, *Cell* 20, 1980, 579–595

found to carry many different kinds of genes in bacteria. What is the mode of action of these plasmids? How do they acquire their new genetic abilities? How do they carry them from cell to cell? It turns out that the drug-resistance genes reside on a mobile genetic element called a **transposon (Tn)**. There are two types of bacterial transposons. **Composite transposons** contain a variety of genes that reside between two nearly identical IS elements that are oriented in opposite direction (Figure 13-9a) and as such, form what is called an **inverted repeat (IR)** sequence. Transposase encoded by one of the two IS elements is necessary to catalyze the movement of the entire transposon. An example of a composite transposon is Tn10, shown in Figure 13-9a. Tn10 carries a gene that confers resistance to the antibiotic tetracycline and is flanked by two IS10 elements in opposite orientation. The IS elements that make up composite transposons are not capable of transposing on their own because of mutations.

Simple transposons are flanked by IR sequences, but these sequences are short (<50 bp) and do not encode the transposase enzyme that is necessary for transposition. Thus, their mobility is not due to an association with IS elements. Instead, simple transposons encode their own transposase in addition to carrying bacterial genes. An example of a simple transposon is Tn3, shown in Figure 13-9b.

To review, IS elements are short mobile sequences that encode only those proteins necessary for their mobility. Composite transposons and simple transposons contain additional genes that confer new functions to bacterial cells. Whether composite or simple, trans-

Table 13-2 Genetic Determinants Borne by Plasmids

Characteristic	Plasmid examples
Fertility	F, RI, col
Bacteriocin production	Col E1
Heavy-metal resistance	R6
Enterotoxin production	Ent
Metabolism of camphor	Cam

posons are usually just called transposons, and different transposons are designated Tn1, Tn2, Tn505, and so forth.

Transposons are longer than IS elements (usually a few kilobases in length), inasmuch as they contain extra protein-coding genes. Table 13-2 lists some of the genetic determinants that can be borne by plasmids. Although IS elements and transposons as defined here are prokaryotic mobile elements, their properties typify many kinds of mobile elements that are found in eukaryotes as well.

> **MESSAGE** Transposons were originally detected as mobile genetic elements that confer drug resistance. Many of these elements consist of recognizable IS elements flanking a gene that encodes drug resistance. IS elements and transposons are now grouped together under the single term *transposable elements*.

A transposon can jump from a plasmid to a bacterial chromosome or from one plasmid to another plasmid. In this manner, multiple drug-resistant plasmids are generated. Figure 13-10 is a composite diagram of an **R plasmid,** indicating the various places at which transposons can be located. We next consider the question of how such **transposition** or mobilization events occur.

Mechanism of transposition

As already discussed, the movement of a transposable element from one site in the chromosome to another or between a plasmid and the chromosome is mediated by a transposase. In one of the first steps of transposition, the transposase makes a staggered cut in the target-site DNA (not unlike the staggered breaks catalyzed by restriction endonucleases in the sugar-phosphate backbone of DNA). Figure 13-11 shows the steps in the integration of a generic transposon mediated by a transposase that makes a five-base-pair staggered cut. The transposon inserts between the staggered ends, and the single-strand overhangs are used as templates (by the host DNA repair machinery) for creating a second

(a) Composite transposon

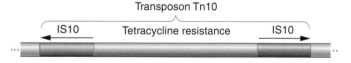

(b) Simple transposon

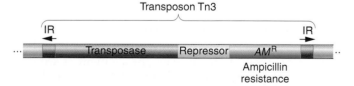

Figure 13-9 Structural features of composite and simple transposons. (a) Tn10, an example of a composite transposon. The IS elements are inserted in opposite orientation and form inverted repeats (IRs). (b) Tn3, an example of a simple transposon. Short inverted repeats contain no transposase. Instead, simple transposons encode their own transposase. The repressor is a protein that regulates the transposase gene.

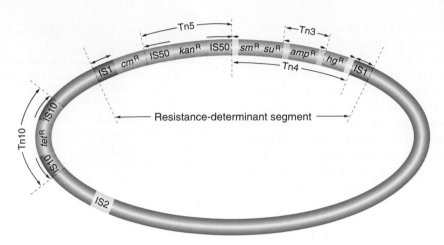

Figure 13-10 A schematic map of a plasmid carrying simple and composite transposon-resistance genes. Genes encoding resistance to the antibiotics tetracycline (*tet*^{R}), kanamycin (*kan*^{R}), streptomycin (*sm*^{R}), sulfonamide (*su*^{R}), and ampicillin (*amp*^{R}) and to mercury (*Hg*^{R}) are shown. The resistant-determinant segment can move as a cluster of resistance genes. Tn3 is within Tn4. Each transposon can be transferred independently. [Simplified from S. N. Cohen and J. A. Shapiro, "Transposable Genetic Elements." Copyright 1980 by Scientific American, Inc. All rights reserved.]

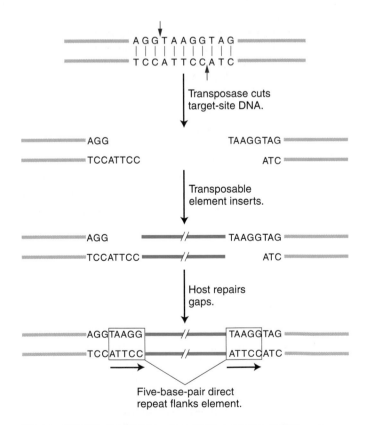

Figure 13-11 Duplication of a short sequence of DNA at the insertion site. The recipient DNA is cleaved at staggered sites (a 5-bp staggered cut is shown), leading to the production of two copies of the five-base-pair sequence flanking the inserted element.

complementary strand. In this example, integration generates a five-base-pair duplication, called a **target-site duplication.** Virtually all transposable elements (in both prokaryotes and eukaryotes) are flanked by a target-site duplication, indicating that all use a mechanism of integration similar to that shown in Figure 13-11. What differs is the length of the duplication; a particular type of transposable element in prokaryotes (and in eukaryotes as well) has a characteristic length for its target-site duplication—as small as two base pairs for some elements.

Most transposable elements in prokaryotes (and in eukaryotes) employ one of two mechanisms of transposition, called **replicative** and **conservative** (nonreplicative), as illustrated in Figure 13-12. In the replicative pathway (as shown for Tn3), a new copy of the transposable element is generated in the transposition event. The results of the transposition are that one copy appears at the new site and one copy remains at the old site. In the conservative pathway (as shown for Tn10), there is no replication. Instead, the element is excised from the chromosome or plasmid and is integrated into the new site. The conservative pathway is also called "**cut and paste.**"

REPLICATIVE TRANSPOSITION Because this mechanism is a bit complicated, it will be described here in more detail. As Figure 13-12 illustrates, one copy of Tn3 is produced from an initial single copy, yielding two copies of Tn3 altogether.

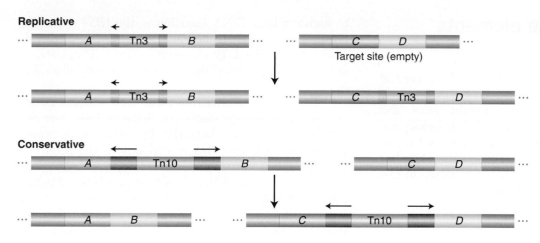

Figure 13-12 Two general modes of mobile-element transposition. See text for details. [Adapted with permission from *Nature Reviews: Genetics* 1, no. 2, p. 138, Figure 3, November 2000, "Mobile Elements and the Human Genome," E. T. Luning Prak and H. H. Kazazian, Jr. Copyright 2000 by Macmillan Magazines Ltd.]

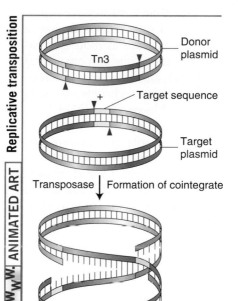

Figure 13-13 Replicative transposition of Tn3 takes place through a cointegrate intermediate. [Adapted from Figure 18.14, in Robert J. Brooker, *Genetics: Analysis and Principles.* Benjamin-Cummings, 1999.]

Figure 13-13 shows the details of the intermediates in the transposition of Tn3 from one plasmid (the donor) to another plasmid (the target). The transposition intermediate consists of a double plasmid with both donor and recipient plasmids being fused together. The formation of this intermediate is catalyzed by Tn3-encoded transposase, which makes single-strand cuts at the two ends of Tn3 and staggered cuts at the target sequence (recall this reaction from Figure 13-11) and joins the free ends together, forming a fused circle called a **cointegrate.** The transposable element is duplicated during the fusion event. The cointegrate then resolves by a recombination-like event that turns a cointegrate into two smaller circles, leaving one copy of the transposable element in each plasmid. The result is that one copy remains at the original location of the element, whereas the other is integrated at a new genomic position.

CONSERVATIVE TRANSPOSITION Some transposons, such as Tn10, excise from the chromosome and integrate into the target DNA. In these cases, DNA replication of the element does not take place, and the element is lost from the site of the original chromosome (see Figure 13-12). Like replicative transposition, this reaction is initiated by the element-encoded transposase, which cuts at the ends of the transposon. However, in contrast with replicative transposition, the transposase cuts the element out of the donor site. It then makes a staggered cut at a target site and inserts the element into the target site. We will revisit this mechanism in more detail in a discussion of the transposition of eukaryotic transposable elements, including the *Ac/Ds* family of maize.

MESSAGE In prokaryotes, transposition occurs by at least two different pathways. Some transposable elements can replicate a copy of the element into a target site, leaving one copy behind at the original site. In other cases, transposition consists of the direct excision of the element and its reinsertion into a new site.

13.3 Transposable elements in eukaryotes

Although transposable elements were first discovered in maize, the first eukaryotic elements to be molecularly characterized were isolated from mutant yeast and *Drosophila* genes. As shown in Figure 13-1, eukaryotic transposable elements fall into two classes: class 1 retrotransposons and class 2 DNA transposons. The first class to be isolated, the retrotransposons, were not at all like the prokaryotic IS elements and transposons.

Class I: retrotransposons

The laboratory of Gerry Fink was among the first to use yeast as a model organism to study eukaryotic gene regulation. Through the years, he and his colleagues isolated thousands of mutations in the *HIS4* gene, which codes for one of the enzymes in the pathway leading to the synthesis of the amino acid histidine.

They isolated more than 1500 spontaneous *HIS4* mutants and found that two of them had an unstable mutant phenotype. The unstable mutants had a reversion frequency [from his⁻ to His⁺ (uppercase letters and a superscript plus sign are used to indicate wild type, whereas lowercase letters and a superscript minus sign or mutation number indicate a mutant)] that was 1000-fold higher than their other *HIS4* mutants. Like the *E. coli gal⁻* mutants, these yeast mutants were found

to harbor a large DNA insertion in the *HIS4* gene. The insertion turned out to be very similar to one of the already characterized **Ty elements** in yeast. There are, in fact, about 35 copies of the inserted element, called *Ty1*, in the yeast genome.

Cloning of the elements from these mutant alleles led to the surprising discovery that the insertions did not look at all like bacterial IS elements or transposons. Instead, they resembled a well-characterized class of animal viruses called retroviruses. A **retrovirus** is a single-stranded RNA virus that employs a double-stranded DNA intermediate for replication. The RNA is copied into DNA by the enzyme **reverse transcriptase**. The double-stranded DNA is integrated into host chromosomes, from which it is transcribed to produce the viral genome and proteins that form new viral particles. The life cycle of a typical retrovirus is shown in Figure 13-14. Some retroviruses, such as mouse mammary tumor virus (MMTV) and Rous sarcoma virus (RSV), are responsible for the induction of cancerous tumors. When integrated into host chromosomes as double-stranded DNA, the double-stranded DNA copy of the retroviral genome is called a **provirus**.

Figure 13-15 shows the similarity in structure and gene content of a retrovirus and the *Ty1* element isolated from the *HIS4* mutants. Both are flanked by **long terminal repeat** sequences (called **LTRs**) that are several hundred base pairs long. Both contain two genes in common, *gag* and *pol*.

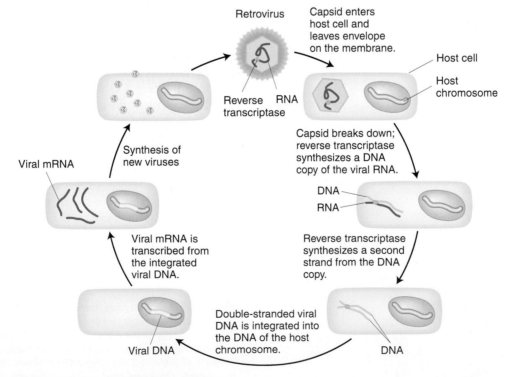

Figure 13-14 Life cycle of a retrovirus.

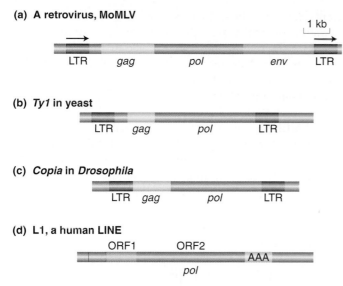

(a) A retrovirus, MoMLV

1 kb

LTR   *gag*   *pol*   *env*   LTR

(b) *Ty1* in yeast

LTR   *gag*   *pol*   LTR

(c) *Copia* in *Drosophila*

LTR   *gag*   *pol*   LTR

(d) L1, a human LINE

ORF1   ORF2

AAA

pol

Figure 13-15 Structural comparison of a retrovirus to retrotransposons found in eukaryotic genomes. (a) A retrovirus, Moloney murine leukemia virus (MoMLV), of mice. LTR = long terminal repeat. (b) A retrotransposon, *Ty1*, in yeast. (c) A retrotransposon, *copia*, in *Drosophila*. (d) A long interspersed nuclear element (LINE) in humans. ORF = open reading frame.

Retroviruses encode at least three proteins that take part in viral replication: the products of the *gag*, *pol*, and *env* genes. The *gag*-encoded protein has a role in the maturation of the RNA genome, *pol* encodes the all-important reverse transcriptase, and *env* encodes the structural protein that surrounds the virus. This protein is necessary for the virus to leave the cell to infect other cells. Interestingly, *Ty1* elements have genes related to *gag* and *pol* but not *env*. These features led to the hypothesis that, like retroviruses, *Ty1* elements are transcribed into RNA transcripts that are copied into double-stranded DNA by the reverse transcriptase. However, unlike retroviruses, *Ty1* elements cannot leave the cell, because they do not encode *env*. Instead, the double-stranded DNA copies are inserted back into the genome of the same cell. These steps are diagrammed in Figure 13-16.

In 1985, Jef Boeke and Gerald Fink showed that, like retroviruses, *Ty* elements do in fact transpose through an RNA intermediate. Figure 13-17 diagrams their experimental design. They began by altering a yeast *Ty1* element, cloned on a plasmid. First, near one end of an element, they inserted a promoter that could be activated by the addition of galactose to the medium. Second, they introduced an intron from another yeast gene into the coding region of the *Ty* transposon.

The addition of galactose greatly increases the frequency of transposition of the altered *Ty* element. This increased frequency suggests the involvement of RNA, because galactose stimulates transcription of *Ty* RNA, beginning at the galactose-sensitive promoter. The key

experimental result, however, is the fate of the transposed *Ty* DNA. Researchers found that the intron had been removed from the *Ty* DNA resulting from transpositions. Because introns are spliced only in the course of RNA processing (see Chapter 8), the transposed *Ty* DNA must have been copied from an RNA intermediate transcribed from the original *Ty* element and spliced before reverse transcription. The DNA copy of the spliced mRNA is then integrated into the yeast chromosome. Transposable elements that utilize reverse transcriptase to transpose through an RNA intermediate are termed **retrotransposons**. They are also known as class 1 transposable elements. Retrotransposons such as *Ty1* that have *long terminal repeats* at their ends are called **LTR-retrotransposons**.

Several spontaneous mutations isolated through the years in *Drosophila* also were shown to contain retrotransposon insertions. The ***copia*-like elements** of *Drosophila* are structurally similar to *Ty1* elements and appear at from 10 to 100 positions in the *Drosophila* genome (see Figure 13-15c). Certain classic *Drosophila* mutations result from the insertion of *copia*-like and other elements. For example, the *white-apricot* (w^a) mutation for eye color is caused by the insertion of an element of the *copia* family into the *white* locus. The insertion of LTR retrotransposons into plant genes (including maize) also has been shown to contribute to spontaneous mutations in this kingdom.

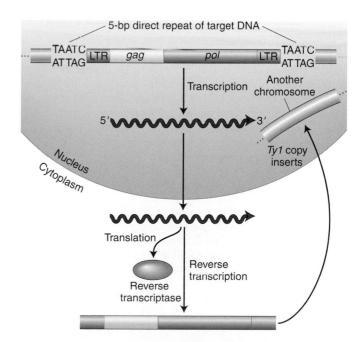

Figure 13-16 Transposition by a retrotransposon.
An RNA transcript from the retrotransposon undergoes reverse transcription into DNA, by a reverse transcriptase encoded by the retrotransposon. The DNA copy is inserted at a new location in the genome.

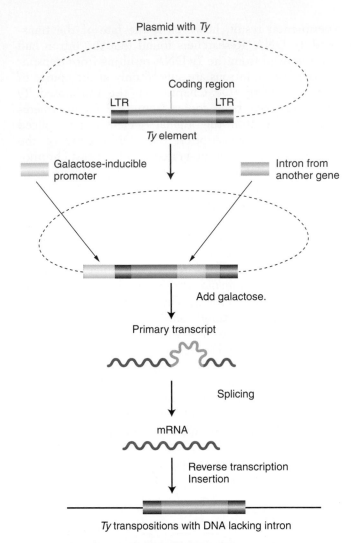

Figure 13-17 Demonstration of transposition through an RNA intermediate. A *Ty* element is altered by adding an intron and a promoter that can be activated by the addition of galactose. The intron sequences are spliced before reverse transcription. [After H. Lodish, D. Baltimore, A. Berk, S. L. Zipursky, P. Matsudaira, and J. Darnell, *Molecular Cell Biology*, 3d ed., p. 332. Copyright 1995 by Scientific American Books.]

> **MESSAGE** Transposable elements that transpose through RNA intermediates are unique to eukaryotes. Retrotransposons, also known as class 1 elements, encode a reverse transcriptase that produces a double-stranded DNA copy (from an RNA intermediate) that is capable of integrating at a new position in the genome.

DNA transposons

Some mobile elements found in eukaryotes appear to transpose by mechanisms similar to those in bacteria. As we saw in Figure 13-12, for IS elements and transposons, the entity that inserts into a new position in the genome is either the element itself or a copy of the element. Ele-

ments that transpose in this manner are called class 2 elements or **DNA transposons.** The first transposable elements discovered by McClintock in maize are now known to be DNA transposons. However, the first DNA transposons to be molecularly characterized were the *P* elements in *Drosophila*.

P **ELEMENTS** Of all the transposable elements in *Drosophila*, the most intriguing and useful to the geneticist are the *P* **elements.** These elements were discovered as a result of studying **hybrid dysgenesis**—a phenomenon that occurs when females from laboratory strains of *Drosophila melanogaster* are mated with males derived from natural populations. In such crosses, the laboratory stocks are said to possess an **M cytotype** (cell type), and the natural stocks are said to possess a **P cytotype.** In a cross of M (female) × P (male), the progeny show a range of surprising phenotypes that are manifested in the germ line, including sterility, a high mutation rate, and a high frequency of chromosomal aberration and nondisjunction (Figure 13-18). These hybrid progeny are *dysgenic*, or biologically deficient (hence, the expression *hybrid dysgenesis*). Interestingly, the reciprocal cross, P (female) × M (male), produces no dysgenic offspring. An important observation is that a large percentage of the dysgenically induced mutations are unstable; that is, they revert to wild type or to other mutant alleles at very high frequencies. This instability is generally restricted to the germ line of an individual fly possessing an M cytotype.

Similarities between the unstable *Drosophila* mutants and the maize mutants characterized by McClintock led to the hypothesis that the dysgenic mutations are caused by the insertion of transposable elements into specific genes, thereby rendering them inactive. According to this view, reversion would usually result from the excision of these inserted sequences. This hypothesis has been critically tested by isolating dysgenically derived unstable mutations at the eye-color locus *white*. Most of the mutations were found to be caused by the insertion of a transposable element into the *white*+ gene. The element, called the *P element*, was found to be present in from 30 to 50 copies per genome in P strains but to be completely absent in M strains. The *P* elements vary in size, ranging from 0.5 to 2.9 kb in length. This size difference is due to the presence of many defective *P* elements from which portions of the middle of the element have been deleted. The full-sized *P* element resembles the simple transposons of bacteria, in that its ends are short (31 bp) inverted repeats and it encodes a transposase. However, this eukaryotic transposase gene contains three introns and four exons (Figure 13-19).

The current explanation of hybrid dysgenesis is based on the proposal that P strains contain *P* elements and a repressor that prevents transposition of the *P* ele-

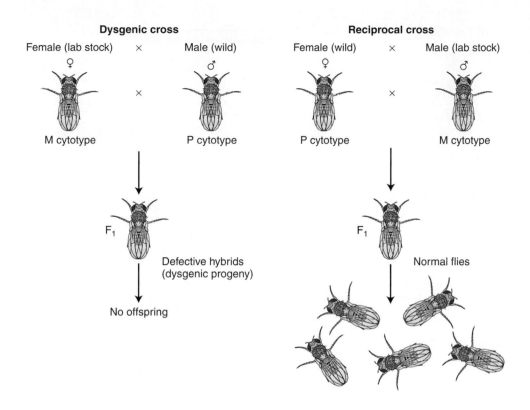

Figure 13-18 Hybrid dysgenesis. See text for details.

ments within the genome. According to this model, which is depicted in Figure 13-20, *P* elements, like bacterial IS and Tn elements, encode a transposase that is responsible for their mobilization. In addition, *P* elements encode a repressor whose job is to prevent transposase production, thereby blocking transposition. For some reason, most laboratory strains have no *P* elements; consequently, there is no *P*-element-encoded repressor in the cytoplasm. In hybrids from the cross M (female, no *P* elements) × P (male, *P* elements), the *P* elements in the newly formed zygote are in a repressor-free environment because the sperm contributes the genome (with the *P* elements) but no cytoplasm (with the repressor). The *P* elements derived from the male genome can now transpose throughout the genome, causing a variety of damage as they insert into genes and cause mutations. These molecular events are expressed as

the various manifestations of hybrid dysgenesis. On the other hand, as noted earlier, P (female) × M (male) crosses do not result in dysgenesis because, in this case, the egg cytoplasm contains *P* repressor.

An intriguing question remains unanswered: Why do laboratory strains lack *P* elements, whereas strains in the wild have *P* elements? One hypothesis is that most of the current laboratory strains descended from the original isolates taken from the wild by Morgan and his students almost a century ago. At some point in time between the capture of those original strains and the present, *P* elements spread through natural populations but not through laboratory strains. This difference was not noticed until wild strains were again captured and mated with laboratory strains.

Although the exact scenario of how *P* elements have spread through wild populations is not clear, what is clear is that transposable elements can spread rapidly from a few individual members of a population. In this regard, the spread of *P* elements resembles the spread of transposons carrying resistance genes to previously susceptible bacterial populations.

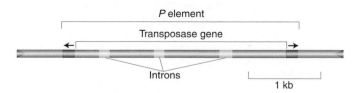

Figure 13-19 *P*-element structure. DNA sequence analysis of the 2.9-kb element reveals a gene, composed of four exons and three introns, that encodes transposase. There is a perfect 31-bp inverted repeat at each terminus. [From G. Robin, in J. A. Shapiro, ed., *Mobile Genetic Elements*, pp. 329–361. Copyright 1983 by Academic Press.]

MAIZE TRANSPOSABLE ELEMENTS REVISITED Although the causative agent responsible for unstable mutants was first shown genetically to be transposable elements in maize, it was almost 50 years before the maize *Ac* and *Ds* elements were isolated and shown to be related to DNA transposons in bacteria and in other eukaryotes. Like the *P* element of *Drosophila*, *Ac* has

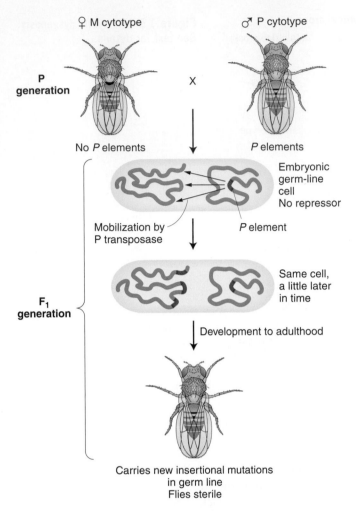

Figure 13-20 Molecular events underlying hybrid dysgenesis. Crosses of male *Drosophila* bearing P transposase with female *Drosophila* that do not have functional *P* elements produce mutations in the germ line of F₁ progeny caused by *P*-element insertions. *P* elements are able to move, causing mutations, because the male sperm do not bring repressors with them.

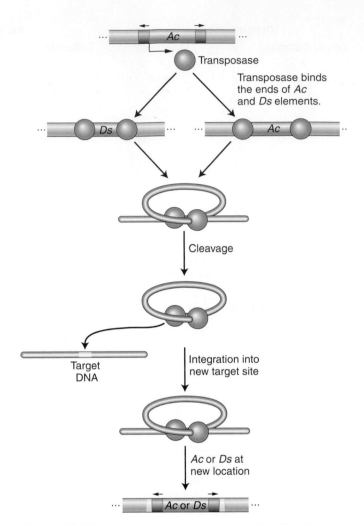

Figure 13-21 Action of the *Ac* element in maize. The *Ac* element encodes a transposase that binds its own ends or those of a *Ds* element, excising the element, cleaving the target site, and allowing the element to insert elsewhere in the genome.

terminal inverted repeats and encodes a single protein, the transposase. The nonautonomous *Ds* element does not encode transposase and thus cannot transpose on its own. When *Ac* is in the genome, its transposase can bind to both ends of *Ac* or *Ds* elements and promote their transposition (Figure 13-21).

As noted earlier in the chapter, *Ac* and *Ds* are members of a single transposon family, and there are other transposable-element families in maize. Each family contains an autonomous element encoding a transposase that can move elements in the same family but cannot move elements in other families, because the transposase can bind only to the ends of family members.

Although some organisms such as yeast have no DNA transposons, elements structurally similar to the *P* and *Ac* elements have been isolated from many plant and animal species. In fact, the pigment gene responsible

for the snapdragon floral mutant pictured in Figure 13-6 is caused by the insertion of an element called *Tam3*, which is very similar to *Ac*. Several copies of *Tam3* normally reside in the snapdragon genome.

> **MESSAGE** The first known transposable elements in maize are DNA transposons that structurally resemble DNA transposons in bacteria and other eukaryotes. DNA transposons encode a transposase that cuts the transposon from the chromosome and catalyzes its reinsertion at other chromosomal locations.

Utility of DNA transposons for gene discovery

Quite apart from their interest as a genetic phenomenon, DNA transposons have become major tools used by geneticists working with a variety of organisms. Their mobility has been exploited to tag genes for cloning and to insert transgenes. The *P* element in *Drosophila* provides one of the best examples of how geneticists exploit the properties of transposable elements in eukaryotes.

USING *P* ELEMENTS TO TAG GENES FOR CLONING

P elements can be used to create mutations by insertion, to mark the position of genes, and to facilitate the cloning of genes. *P* elements inserted into genes in vivo disrupt genes at random, creating mutants with different phenotypes. Fruit flies with interesting mutant phenotypes can be selected for cloning of the mutant gene, which is marked by the presence of the *P* element. The interrupted gene can be cloned with the use of *P*-element segments as a probe, a method termed **transposon tagging**. Fragments from the mutant gene can than be used as a probe to isolate the wild-type gene.

USING *P* ELEMENTS TO INSERT GENES

Gerald Rubin and Allan Spradling showed that *P*-element DNA can be used as an effective vehicle for transferring donor genes into the germ line of a recipient fly. They devised the following experimental procedure (Figure 13-22). The recipient genotype is homozygous for the *rosy* (*ry⁻*) mutation, which confers a characteristic eye color. From this strain, embryos are collected at the completion of about nine nuclear divisions. At this stage, the embryo is one multinucleate cell, and the nuclei destined to form the germ cells are clustered at one end. (*P* elements mobilize only in germ-line cells.) Two types of DNA are injected into embryos of this type. The first is a bacterial plasmid carrying a defective *P* element (that resembles the maize *Ds* element in that it does not encode transposase but still has the ends that bind transposase and allow transposition) into which the *ry⁺* gene has been inserted. This deleted element is not able to transpose, and so, as mentioned earlier, a helper plasmid bearing a complete element also is injected. Flies developing from these embryos are phenotypically still *rosy* mutants, but their offspring include a large proportion of *ry⁺* flies. These *ry⁺* descendants show Mendelian inheritance of the newly acquired *ry⁺* gene, suggesting that it is located on a chromosome. This location was confirmed by in situ hybridization, which shows that the *ry⁺* gene, together with the deleted *P* element, has been inserted into one of

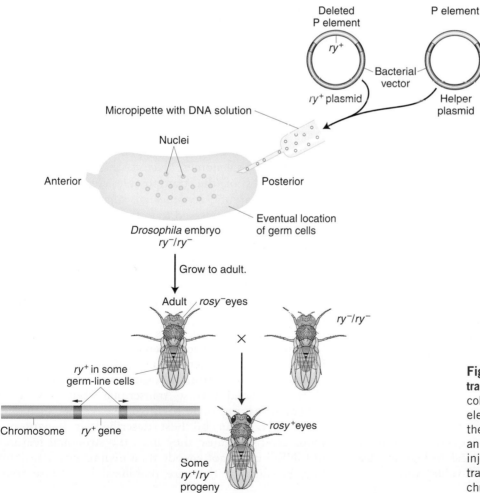

Figure 13-22 *P*-element-mediated gene transfer in *Drosophila*. The *rosy⁺* (*ry⁺*) eye-color gene is engineered into a deleted *P* element carried on a bacterial vector. At the same time, a helper plasmid bearing an intact *P* element is used. Both are injected into an *ry⁻* embryo, where *ry⁺* transposes with the *P* element into the chromosomes of the germ-line cells.

several distinct chromosome locations. None appears exactly at the normal locus of the *rosy* gene. These new *ry*⁺ genes are found to be inherited in a stable fashion.

Because the *P* element can transpose only in *Drosophila*, these applications are restricted in their usage. In contrast, the maize *Ac* element is able to transpose after its introduction into the genomes of plant species including the weed *Arabidopsis*, lettuce, carrot, rice, barley, and many more. Like *P* elements, *Ac* has been engineered by geneticists for use in gene isolation by transposon tagging. In this way, *Ac*, the first transposable element discovered by Barbara McClintock, serves as an important tool of plant geneticists more than 50 years later.

> **MESSAGE** DNA transposons have been modified and used by scientists in two important ways: (1) to make mutants that can be identified molecularly by having a transposon tag and (2) as vectors that can introduce foreign genes into the chromosome.

13.4 The dynamic genome: more transposable elements than ever imagined

As you have seen, transposable elements were first discovered with the use of genetic approaches. In these studies, the elements made their presence known when they transposed into a gene or were sites of chromosome breakage or rearrangement. After the DNA of transposable elements was isolated from unstable mutations, scientists could use that DNA as molecular probes to determine if there were more related copies in the genome. In all cases, there were always at least several copies of the element in the genome and as many as several hundred.

Scientists wondered about the prevalence of transposable elements in genomes. Were there other transposable elements in the genome that remained unknown because they had not caused a mutation that could be studied in the laboratory? Were there transposable elements in the vast majority of organisms that were not amenable to genetic analysis? Asked another way, do organisms without mutations induced by transposable elements nonetheless have transposable elements in their genomes? These questions are reminiscent of the question, If a tree falls in the forest, does it make a sound if no one is listening?

Large genomes are largely transposable elements

Long before the advent of DNA-sequencing projects, scientists using a variety of biochemical techniques discovered that DNA content (called **C-value**) varied dra-

matically in eukaryotes and did not correlate with biological complexity. For example, the genomes of salamanders are 20 times as large as the human genome, whereas the genome of barley is more than 10 times as large as the genome of rice, a related grass. The lack of correlation between genome size and the biological complexity of an organism is known as the **C-value paradox.**

Barley and rice are both cereal grasses and as such their gene content should be similar. However, if genes are a relatively constant component of the genomes of multicellular organisms, what is responsible for the C-value paradox? On the basis of the results of additional experiments, scientists were able to determine that DNA sequences that are repeated thousands, even hundreds of thousands, of times make up a large fraction of eukaryotic genomes and that some genomes contain much more repetitive DNA than others.

Thanks to many recent projects to sequence the genomes of a wide variety of taxa (including *Drosophila*, humans, the mouse, *Arabidopsis*, and rice), we now know that there are many classes of repetitive sequences in the genomes of higher organisms and that some are similar to the DNA transposons and retrotransposons shown to be responsible for mutations in plants, yeast, and insects. Most remarkably, these sequences make up most of the DNA in the genomes of multicellular eukaryotes.

Rather than correlating with gene content, genome size frequently correlates with the amount of DNA in the genome that is derived from transposable elements. Organisms with big genomes have lots of sequences that resemble transposable elements, whereas organisms with small genomes have many fewer. Two examples, one from the human genome and the other from a comparison of the grass genomes, illustrate this point. The structural features of the transposable elements that are found in eukaryotic genomes are summarized in Figure 13-23 and will be referred to in the next section.

Transposable elements in the human genome

Almost half of the human genome is derived from transposable elements. The vast majority of these transposable elements are two types of retrotransposons called **long interspersed nuclear elements** or **LINEs** and **short interspersed nuclear elements** or **SINEs** (see Figure 13-23). LINEs move by retrotransposition with the use of an element-encoded reverse transcriptase but lack some structural features of retrovirus-like elements, including LTRs. SINEs can be best described as nonautonomous LINEs, because they have the structural features of LINEs but do not encode their own reverse transcriptase. Presumably, they are mobilized by reverse tran-

Types of transposable elements in the human genome

Element	Transposition	Structure	Length	Copy number	Fraction of genome
LINEs	Autonomous	ORF1 ORF2 *(pol)* AAA	1–5 kb	20,000–40,000	21%
SINEs	Nonautonomous	AAA	100–300 bp	1,500,000	13%
DNA transposons	Autonomous	← transposase →	2–3 kb	300,000	3%
	Nonautonomous	← →	80–3000 bp		

Figure 13-23 The general classes of transposable elements found in the human genome.
[Reprinted by permission from *Nature* 409, 880 (15 February 2001), "Initial Sequencing and Analysis of the Human Genome," The International Human Genome Sequencing Consortium. Copyright 2001 by Macmillan Magazines Ltd.]

scriptase enzymes that are encoded by LINEs that reside in the genome.

The most abundant SINE in humans is called **Alu,** so named because it contains a target site for the *Alu* restriction enzyme. The human genome contains well over 1 million whole and partial *Alu* sequences, scattered between genes and within introns. These *Alu* sequences make up more than 10 percent of the human genome. The full *Alu* sequence is about 200 nucleotides long and bears remarkable resemblance to 7SL RNA, an RNA that is part of a complex by which newly synthesized polypeptides are secreted through the endoplasmic reticulum. Presumably, the *Alu* sequences originated as reverse transcripts of these RNA molecules.

There is about 20 times more DNA in the human genome derived from transposable elements than there is DNA encoding all human proteins. Figure 13-24 illus-

trates the number and diversity of transposable elements present in the human genome, using as an example the positions of individual *Alu*s, other SINEs, and LINEs in the vicinity of a typical human gene.

The human genome seems to be typical for a multicellular organism in the abundance and distribution of transposable elements. Thus, an obvious question is, How do plants and animals survive and thrive with so many insertions in genes and so much mobile DNA in the genome? First, with regard to gene function, all of the elements shown in Figure 13-24 are inserted into introns. Thus, the mRNA produced by this gene will not include any sequences from transposable elements, because they will have been spliced out of the pre-mRNA with the surrounding intron. Presumably, transposable elements insert into both exons and introns, but only the insertions into introns will remain in

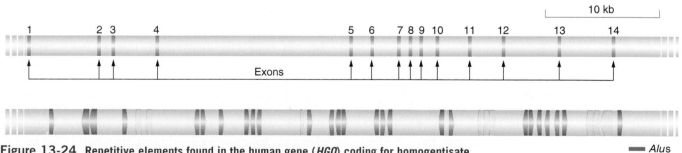

Figure 13-24 Repetitive elements found in the human gene (*HGO*) coding for homogentisate 1,2-dioxygenase, the enzyme whose deficiency causes alkaptonuria. The upper row diagrams the positions of the *HGO* exons. The locations of *Alu*s (blue), other SINEs (purple), and LINEs (yellow) in the *HGO* sequence are indicated in the lower row. [After B. Granadino, D. Beltrán-Valero de Bernabé, J. M. Fernández-Cañón, M. A. Peñalva, and S. Rodríguez de Córdoba, "The Human Homogentisate 1, 2-Dioxygenase (*HGO*) Gene," *Genomics* 43, 1997, 115.]

— Alus
— SINEs
— LINEs

the population because they are less likely to cause a mutation. Insertions into exons are said to be subjected to **negative selection.** Second, humans, as well as all other multicellular organisms, can survive with so much mobile DNA in the genome because the vast majority is inactive and cannot move or increase in copy number. Most transposable-element sequences in a genome are relics that have accumulated inactivating mutations over evolutionary time. Others are still capable of movement but are rendered inactive by host regulatory mechanisms. The epigenetic mechanisms that serve to inactivate transposable elements will be discussed in more detail later in this chapter. There are, however, a few active LINEs and *Alu*s that have managed to escape host control and have inserted into important genes causing several human diseases. Three separate insertions of LINEs have disrupted the factor VIII gene causing hemophilia A. At least 11 *Alu* insertions into human genes have been shown to cause several diseases, including hemophilia B (in the factor IX gene), neurofibromatosis (in the *NF1* gene), and breast cancer (in the *BRCA2* gene).

The overall frequency of spontaneous mutation due to the insertion of class 2 elements in humans is quite low, accounting for less than 0.2 percent (1 in 500) of all characterized spontaneous mutations. Surprisingly, retrotransposon insertions account for about 10 percent of spontaneous mutation in another mammal, the mouse. The approximately 50-fold increase in spontaneous mutations due to retrotransposon insertion in the mouse most likely corresponds to the much higher activity of these elements in the mouse than in the human genome.

> **MESSAGE** Transposable elements compose the largest fraction of the human genome, with LINEs and SINEs being the most abundant. The vast majority of transposable elements are ancient relics that can no longer move or increase their copy number. A few elements remain active and their movement into genes can cause disease.

The grasses: LTR retrotransposons thrive in large genomes

As previously mentioned, the C-value paradox refers to a lack of correlation between genome size and biological complexity. How can organisms have very similar gene content but differ dramatically in the size of their genomes? This situation has been investigated in the cereal grasses. Differences in the genome size of these grasses have been shown to correlate primarily with the number of one class of elements, the LTR retrotransposons. The cereal grasses are evolutionary relatives that have arisen from a common ancestor in the past 70 million years. As such, their genomes are still very similar with respect to gene content and organization (called

synteny; see Chapter 21), and regions can be compared directly. These comparisons reveal that linked genes in the small rice genome are physically closer together than the same genes in the larger maize and barley genomes. In the later genomes, genes are separated by large clusters of retrotransposons (Figure 13-25).

> **MESSAGE** The *C*-value paradox refers to a lack of correlation between genome size and biological complexity. Because genes make up such a small proportion of the genomes of multicellular organisms, genome size usually corresponds to the amount of transposable-element sequences rather than gene content.

Safe havens

The abundance of transposable elements in the genomes of multicellular organisms led some investigators to postulate that successful transposable elements (those that are able to attain very high copy numbers) have evolved mechanisms to prevent harm to their hosts by not inserting into host genes. Instead, successful transposable elements insert into so-called **safe havens** in the genome. For the grasses, a safe haven for new insertions appears to be into other retrotransposons. Another safe haven for the insertion of many classes of transposable elements in both plant and animal species is the centric heterochromatin, where there are very few genes but lots of repetitive DNA.

SAFE HAVENS IN SMALL GENOMES: TARGETED INSERTIONS In contrast with the genomes of multicellular eukaryotes, the genome of unicellular yeast is very compact, with closely spaced genes and very few introns. With almost 70 percent of its genome as exons, there is a high probability that new insertions of transposable elements will disrupt a gene-coding sequence. Yet, as we have seen earlier in this chapter, the yeast genome supports a collection of LTR-retrotransposons called *Ty* elements.

How are transposable elements able to spread to new sites in genomes with few safe havens? Investigators have identified hundreds of *Ty* elements in the sequenced yeast genome and have determined that they are not randomly distributed. Instead, each family of *Ty* elements inserts into a particular genomic region. For example, the *Ty3* family inserts almost exclusively near but not in tRNA genes, at sites where they do not interfere with the production of tRNAs and, presumably, do not harm their hosts. The mechanism that *Ty* elements have evolved to insert into particular regions of the genome entails the specific interaction of *Ty* proteins necessary for integration with yeast proteins bound to genomic DNA. *Ty3* proteins, for example, recognize and bind to subunits of the RNA polymerase complex that have assembled at tRNA promoters (Figure 13-26a).

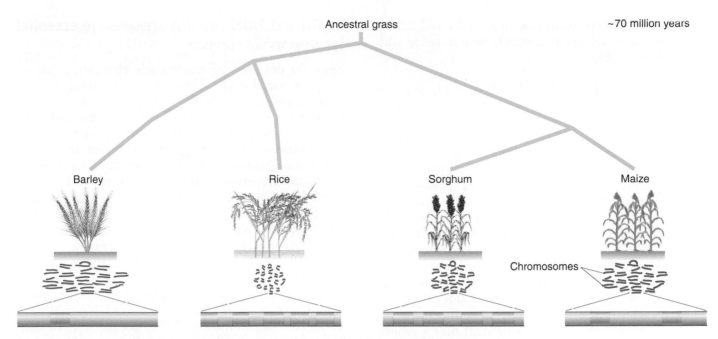

Figure 13-25 Transposable elements in grasses are responsible for genome size differences. The grasses, including barley, rice, sorghum, and maize, arose from a common ancestor about 70 million years ago. Since that time, the transposable elements have accumulated to different levels in each species. Chromosomes are larger in maize and barley, whose genomes contain large amounts of LTR retrotransposons. Green in the partial genome at the bottom represents a cluster of transposons, whereas orange represents genes.

The ability of some transposons to insert preferentially into certain sequences or genomic regions is called **targeting.** A remarkable example of targeting is illustrated by the *R1* and *R2* elements of arthropods, including *Drosophila*. *R1* and *R2* are LINEs (see Figure 13.23) that insert only into the genes that produce ribosomal RNA. In arthropods, several hundred rRNA genes are organized in tandem arrays (Figure 13-26b). With so many genes encoding the same product, the host tolerates insertion into a subset. However, it has

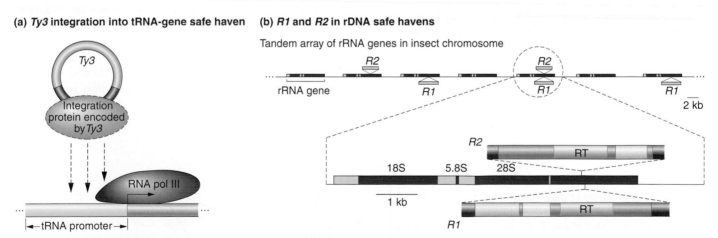

Figure 13-26 Insertion of transposable elements in safe havens. (a) The yeast *Ty3* retrotransposon inserts into the promoter region of tRNA genes. (b) The *Drosophila R1* and *R2* non-LTR retrotrasposons (LINEs) insert into the genes encoding ribosomal RNA that are found in long tandem arrays on the chromosome. Only the reverse transcriptase (RT) genes of *R1* and *R2* are noted. [Part a inspired by D. F. Voytas and J. D. Boeke, "Ty1 and Ty5 of *Saccharomyces cerevisiae,*" in *Mobile DNA II,* Chapter 26, Figure 15, p. 652. ASM Press, 2002. Part b, adapted from T. H. Eickbush, in *Mobile DNA II,* Chapter 34, "R2 and Related Site-Specific Non-Long Terminal Inverted Repeat Retrotransposons," Figure 1, p. 814. ASM Press, 2002.]

been shown that too many insertions of *R1* and *R2* can decrease insect viability, presumably by interfering with ribosome assembly.

GENE THERAPY REVISITED In Chapter 11, you saw that modified retroviruses have been used in gene-therapy trials to deliver transgenes that may correct certain human diseases. One of the first trials included patients with X-linked severe combined human immunodeficiency disease (SCID), a fatal disease if left untreated because it severely compromises the immune system. Bone-marrow cells from each patient were collected and treated with a retrovirus vector containing a good gene for one of the chains of the interleukon-2 receptor (the gene that is mutated in these patients), The transformed cells were then infused back into the patient. The immune systems of most of the patients showed significant improvement. However, the therapy had a very serious side effect: two of the patients developed leukemia. In both patients, the retroviral vector had inserted (integrated) near a cellular gene whose aberrant expression is associated with leukemia. A likely scenario is that insertion of the retroviral vector near the cellular gene altered its expression and, either directly or indirectly, caused the leukemia.

Clearly, this form of gene therapy might be greatly improved if doctors were able to control where the retroviral vector integrates into the human genome. We have already seen that there are many similarities between LTR retrotransposons and retroviruses. It is hoped that, by understanding *Ty* targeting in yeast, we can learn how to construct retroviral vectors that insert themselves and their transgene cargo into safe havens in the human genome.

> **MESSAGE** A successful transposable element increases copy number without harming its host. One way in which an element safely increases copy number is to target new insertions into safe havens, regions of the genome where there are few genes.

HeT-A and *TART* retrotransposons are essential for *Drosophila* survival

Since the discovery of transposable elements, a controversy has raged over whether they are simply genome parasites (so-called **junk DNA**) or whether they are useful to the organism in which they reside. This debate has entered a new phase with the discovery that transposable elements often make up the largest fraction of eukaryotic genomes. It is clear that many transposable elements have evolved strategies to increase their copy number without killing their hosts. In several instances transposable-element sequences have become incorporated into the regulatory regions of cellular genes or even into coding regions of cellular proteins.

Remarkable examples of useful transposable elements are the **_HeT-A_** and **_TART_** elements that function as the telomeres of all *Drosophila* chromosomes. Recall from Chapter 7 that telomeres are the DNA sequences at the ends of chromosomes. They are made up of short repeated sequences that are added after replication by the RNA-containing telomerase enzyme (see Figure 7-25). Telomerase is actually a reverse transcriptase that uses its RNA as a template for DNA synthesis. Without telomerase, chromosomes would become progressively shorter from generation to generation. Scientists were surprised to find that *Drosophila* telomeres are not made up of short repeated sequences but instead contain many copies of the *HeT-A* and *TART* non-LTR retrotransposons (LINEs). *Drosophila* ends are maintained by the repeated transposition of these elements into the ends. This process is shown in Figure 13-27, where the RNA encoded by the retrotransposon can be seen to serve a role similar to that of the RNA of the telomerase. The retrotransposon contains a reverse transcriptase analogous to telomerase that catalyzes the addition of DNA to the 5′ end of the chromosome.

Why *Drosophila* telomeres are lengthened by retrotransposition and how the current situation might have

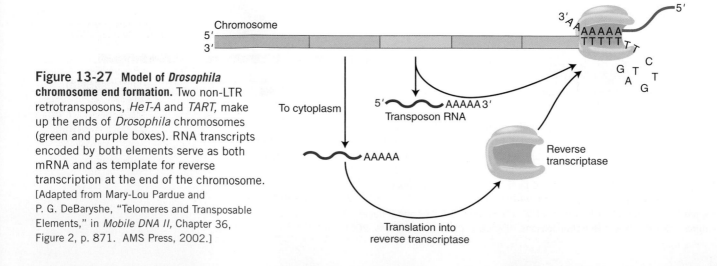

Figure 13-27 Model of *Drosophila* chromosome end formation. Two non-LTR retrotransposons, *HeT-A* and *TART*, make up the ends of *Drosophila* chromosomes (green and purple boxes). RNA transcripts encoded by both elements serve as both mRNA and as template for reverse transcription at the end of the chromosome. [Adapted from Mary-Lou Pardue and P. G. DeBaryshe, "Telomeres and Transposable Elements," in *Mobile DNA II*, Chapter 36, Figure 2, p. 871. AMS Press, 2002.]

evolved are not known. However, it is reasonable to assume that a *Drosophila* ancestor used telomerase like other organisms but that this mechanism was lost in favor of retrotransposition of *HeT-A* and *TART* elements.

13.5 Host regulation of transposable elements

Reversible changes in *Ac* activity

While Barbara McClintock was characterizing the transposable elements of maize, she discovered a very unusual phenomenon. Some autonomous elements, such as *Ac*, would turn off for part of the plant life cycle or even for several generations and then turn back on again. While off, *Ac* could not itself move or promote the movement of *Ds* elements in the genome. She called this reversible change in the activity of *Ac* **change in phase.**

To better understand what is meant by change in phase, let's see how the phenomenon appears in the pigment of corn kernels. Recall from Figure 13-4 that a strain with *Ds* inserted into the C gene [the *c-m(Ds)* allele] will have spotted kernels if *Ac* is in the genome. However, in some of these strains, the kernels are not spotted, but rather all kernels on an ear are colorless (Figure 13-28). In subsequent generations, however, spotted kernels reappear. When confronted with a situation similar to this one, McClintock hypothesized that *Ac* was undergoing phases of *reversible* activation and inactivation. This puzzling phenomenon is reminiscent of two other instances of reversible activation and inactivation: imprinting and X-chromosome inactivation. As discussed in Chapter 10, imprinting and X-chromosome inactivation are now known to be epigenetic phenomena. That is, the genes are rendered inactive **(silenced)** by alterations in chromatin structure rather than by mutations in the DNA. Unlike mutant genes, silenced genes can be reactivated when chromatin becomes more open, making the gene accessible to RNA polymerase II. Re-

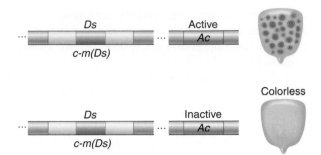

Figure 13-28 Change of phase produces colorless corn kernels. *(top)* An active *Ac* element mobilizes a *Ds* element, excising it from the *C* gene in some cells and producing a spotted phenotype. *(bottom)* An inactive *Ac* element cannot mobilize *Ds* elements, so no pigmented spots are produced.

versible mutations such as the inactive *Ac* element or genes that cannot be expressed, owing to imprinting, are indicative of changes in epigenetic regulation and are now referred to as **epimutations.**

Transgene silencing

We now know that the epigenetic inactivation of *Ac* elements in maize is part of a complex system that has evolved to protect organisms from the mutagenic effects of transposons. But this was not immediately apparent. As is often the case in science, the results of a seemingly unrelated series of experiments were sources of insight into host regulation of transposon activity.

In the late 1980s, Richard Jorgensen and Carolyn Napoli performed a series of experiments in which they inserted different transgenes containing floral pigment genes into mutant petunia plants and into normal purple-blue flowered plants. They did not expect to see any change in floral coloration in the normal petunia strain, which served as a control. Instead, they observed very unusual floral patterns like those shown in Figure 13-29. They concluded that the transgene had, by an

(a)   **(b)**   **(c)**

Figure 13-29 Petunia flowers demonstrating cosuppression. On the left is the wild-type (no transgene) phenotype. At the right and center are so-called cosuppression phenotypes resulting from the transformation of the wild-type petunia shown in (a) with a petunia gene required for pigmentation. In the colorless regions, both the transgene and the chromosomal copy of the same gene have been epigenetically inactivated. [Photos courtesy of Richard A. Jorgensen, Department of Plant Sciences, University of Arizona.]

unknown mechanism, triggered the suppression of both itself and the homologous gene in the petunia chromosome. This phenomenon has come to be known as **cosuppression.**

As the transformation of some plant species with foreign genes became routine, scientists noticed that a variety of transgenes were efficiently silenced in the plant host genome. Because the transgenes could often be reactivated, this silencing was recognized as a form of epigenetic regulation. However, it is highly unlikely that organisms have evolved mechanisms to turn off transgenes introduced by plant scientists. Instead, it was reasoned that transgenes are silenced because they resemble a natural threat to the host, perhaps their own transposable elements or infecting viruses or both. Like transposable elements and viruses, transgenes can insert into new sites in the host genome. What if organisms had defense mechanisms that could recognize these "invaders" and turn them off by silencing their expression, possibly through changes in chromatin structure?

In an attempt to identify host genes contributing to transgene and possibly transposable-element silencing, geneticists sought suppressor strains that had lost the ability to silence transgenes. One approach used a strain of the unicellular green alga *Chlamydomonas rheinhardii* that contained a silenced transgene that normally conferred resistance to the antibiotic spectinomycin. This strain, which could not grow on agar plates containing spectinomycin, was treated with a mutagen and then spread onto plates containing the antibiotic. A cell with a mutation in a gene required to silence the transgene should be able to grow on these plates. Mutant strains were indeed isolated in this way and, as expected, these strains were unable to silence the spectinomycin-resistance gene or other introduced transgenes. In addition, several transposable elements normally inactive in the genome of *Clamydomonas* were reactivated in the mutant strains and shown to insert into new chromosomal locations.

Similar results in plants and animals have made it increasingly clear that epigenetic regulation serves not only as an effective way to silence cellular genes (Chapter 10) but also as a major line of defense against the potentially mutagenic effects of transposon activity.

A genomic battleground?

We have already seen that transposable elements have caused a variety of mutations in plants and animals. Thus, there must be times when host regulation of transposable elements can be overcome and silenced elements are reactivated. Looked at another way, if host regulation were completely successful, transposable elements would no longer exist; they would be silenced, unable to transpose, and would gradually mutate into unrecognizable sequences. Rather, there appears to be a constant battle between the proliferation of transposable elements and host attempts to silence or otherwise inactivate them.

In this regard, some of you may now be concerned that almost 50 percent of your genome is derived from transposable elements. There is really no need to worry. Humans and all other organisms have coevolved with their transposable elements and have worked out a variety of mechanisms so that both are able to coexist. Organisms not able to evolve a satisfactory accommodation with their transposable elements have most likely become extinct.

KEY QUESTIONS REVISITED

• **Why were transposable elements first discovered genetically in maize but first isolated molecularly from *E. coli*?**

The genetic behavior of transposable elements in maize genes produced striking kernel phenotypes that were deciphered by maize geneticists, especially Barbara McClintock. However, the maize genome is huge (about the size of the human genome) and the molecular isolation of maize elements was not possible until decades after their genetic discovery. In contrast, gene isolation was pioneered in *E. coli* (>1000-fold smaller genome than that of maize) where the first elements to be cloned were the IS elements from *E. coli* mutations.

• **How do transposable elements participate in the spread of antibiotic-resistant bacteria?**

Antibiotic-resistance genes are frequently found in the chromosome or on plasmids, where they are flanked on either side by IS elements. The IS elements move themselves and the gene between them to plasmids that can pass to nonresistant cells though bacterial conjugation.

• **Why are transposable elements classified as RNA transposons or DNA transposons?**

RNA transposons, also called class 1, include retrotransposons (LINEs and LTR-retrotransposons) and SINEs (such as the human *Alu*). For all RNA elements, the transposition intermediate is RNA. In contrast, the transposition intermediate of all DNA elements, also called class 2, is DNA.

• **How do autonomous and nonautonomous transposable elements differ?**

Autonomous elements encode all the proteins necessary to mobilize themselves and the nonautonomous elements in their family. Nonautonomous elements rely on autonomous elements for their movement because they do not encode the necessary proteins, including reverse transcriptase (for RNA elements) and transposase (for DNA elements).

- **How can humans survive given that up to 50 percent of the human genome is derived from transposable elements?**

Three major reasons. First, most of the transposable elements' sequences are mutant and no longer capable of transposition. Second, the transposition of the few active elements in the genome is usually prevented by host regulatory mechanisms. Finally, the vast majority of transposable-element sequences in the human genome are in noncoding DNA including telomeres, centromeres, intergenic DNA, and introns.

- **How can the study of retrotransposons in yeast lead to improved procedures for human gene therapy?**

Yeast retrotransposons target their new insertions to so-called safe havens, regions of the genome with few genes. By understanding the underlying mechanisms, scientists may be able to devise new strategies to target genes for gene therapy into safe havens in the human genome.

SUMMARY

Transposable elements were discovered in maize by Barbara McClintock as the cause of several unstable mutations. *Ds* is an example of a nonautonomous element that requires the presence of the autonomous *Ac* element in the genome for it to transpose.

Bacterial insertion sequence (IS) elements were the first transposable elements isolated molecularly. There are many different types of IS elements in *E. coli* strains, and they are usually present in at least several copies. Composite transposons contain IS elements flanking one or more genes, such as genes conferring resistance to antibiotics. Transposons with resistance genes can insert into plasmids and are then transferred by conjugation to nonresistant bacteria.

There are two major groups of transposable elements in eukaryotes: class 1 retroelements and class 2 DNA elements. The *P* element was the first class 2 DNA transposon to be isolated molecularly. It was isolated from unstable mutations in *Drosophila* that were induced by hybrid dysgenesis. *P* elements have been developed into vectors for the introduction of foreign DNA into *Drosophila* germ cells.

Ac, *Ds*, and *P* are examples of DNA transposons, so named because the transposition intermediate is the DNA element itself. Autonomous elements such as *Ac* encode a transposase that binds to the ends of autonomous and nonautonomous elements and catalyzes excision of the element from the donor site and reinsertion into a new target site elsewhere in the genome.

Retrotransposons were first molecularly isolated from yeast mutants and their resemblance to retroviruses was immediately apparent. Retrotransposons are class 1 elements, as are all transposable elements that use RNA as their transposition intermediate.

The active transposable elements isolated from such model organisms as yeast, *Drosophila*, *E. coli*, and maize constitute a very small fraction of all the transposable elements in the genome. DNA sequencing of whole genomes, including the human genome, has led to the remarkable finding that almost half of the human genome is derived from transposable elements. Organisms coexist with their elements largely because epigenetic mechanisms have evolved to suppress the movement of the elements.

KEY TERMS

Activator (*Ac*) (p. 425)

Alu (p. 441)

autonomous element (p. 427)

change in phase (p. 445)

class 1 element (p. 424)

class 2 element (p. 424)

cointegrate (p. 432)

composite transposon (p. 431)

conservative transposition (p. 432)

copia-like element (p. 435)

cosuppression (p. 446)

"cut and paste" (p. 432)

C-value (p. 440)

C-value paradox (p. 440)

Dissociation (*Ds*) (p. 425)

DNA element (p. 425)

DNA transposon (p. 436)

epimutation (p. 445)

excise (p. 426)

HeT-A (p. 444)

hybrid dysgenesis (p. 436)

insertion-sequence (IS) element (p. 429)

inverted repeat (IR) (p. 431)

junk DNA (p. 444)

long interspersed nuclear element (LINE) (p. 440)

long terminal repeat (LTR) (p. 434)

LTR-retrotransposon (p. 435)

M cytotype (p. 436)

negative selection (p. 442)

nonautonomous element (p. 427)

P cytotype (p. 436)

P element (p. 436)

provirus (p. 434)

replicative transposition (p. 432)

retro-element (p. 424)

retrotransposition (p. 424)

retrotransposon (p. 435)

retrovirus (p. 434)

reverse transcriptase (p. 432)

R factor (p. 430)

RNA element (p. 424)

R plasmid (p. 431)

safe haven (p. 442)

short interspersed nuclear element
  (SINE) (p. 440)

silenced (p. 445)

simple transposon (p. 431)

synteny (p. 442)

targeting (p. 443)

target-site duplication (p. 432)

TART (p. 444)

transposase (p. 430)

transpose (p. 426)

transposition (p. 431)

transposon (Tn) (p. 431)

transposon tagging (p. 439)

Ty element (p. 434)

unstable phenotype (p. 425)

SOLVED PROBLEMS

1. In Chapter 10, we studied the operon model. Note that, for the *gal* operon, the order of transcription of the genes in the operon is *E-T-K*. Suppose we have five different mutations in *galT*: *gal-1*, *gal-2*, *gal-3*, *gal-4*, and *gal-5*. The following table shows the expression of *galE* and *galK* in mutants carrying each of these mutations:

galT mutation	Expression of *galE*	Expression of *galK*
gal-1	1	2
gal-2	1	2
gal-3	1	2
gal-4	1	1
gal-5	1	1

In addition, the reversion patterns of these mutations with several mutagens that we studied in Chapter 14 are shown in the following table. Here, a "1" indicates a high rate of reversion in the presence of each mutagen, a "2" depicts no reversion, and a "low" indicates a low rate of reversion.

Mutation	Spontaneous	2-Amino purine	ICR191	UV	EMS
gal-1	2	2	1	1	2
gal-2	2	2	1	2	2
gal-3	Low	Low	Low	Low	Low
gal-4	2	2	2	2	2
gal-5	Low	1	Low	1	1

Reversion

Which mutation is most likely to result from the insertion of a transposable element such as IS1 and why? Can you assign the other mutations to other categories?

Solution

Transposable elements will cause polarity, preventing the expression of genes downstream of the point of insertion but not of upstream genes. Therefore, we would expect the insertion mutation to prevent the expression of the *galK* gene. Three mutations are in this category, *gal-1*, *gal-2*, and *gal-3*. These mutations could be frameshifts, nonsense mutations, or insertions, because each of them can lead to polarity. If we examine the reversion data, however, we can distinguish among these possibilities. Transposable elements revert at low rates spontaneously, and this rate is not stimulated by base analogs, frameshift mutagens, alkylating agents, or UV light. On the basis of these criteria, the *gal-3* mutation is most likely to result from an insertion, because it reverts at a low rate that is not stimulated by any of the mutagens, *gal-1* might be a frameshift, because it does not revert with 2-AP and EMS but does revert with ICR191, a frameshift mutagen, and UV light. (Refer to Chapter 16 for details of each mutagen.) Likewise, *gal-2* is probably a frameshift, because it reverts only with ICR191. The *gal-4* mutation is probably a deletion, because it is not stimulated to revert at all. The *gal-5* mutation appears to be a base substitution, because it reverts with 2-AP, but not above the spontaneous background rate with ICR191.

2. Transposable elements have been referred to as "jumping genes" because they appear to jump from one position to another, leaving the old locus and appearing at a new locus. In light of what we now know concerning the mechanism of transposition, how appropriate is the term "jumping genes" for bacterial transposable elements?

Solution

In bacteria, transposition takes place by two different modes. The conservative mode results in true jumping genes, because, in this case, the transposable element excises from its original position and inserts at a new position. A second mode is termed the *replicative mode*. In this pathway, a transposable element moves to a new location by replicating into the target DNA, leaving behind a copy of the transposable element at the original site. When operating by the replicative mode, transposable elements are not really jumping genes, because a copy does remain at the original site.

PROBLEMS

BASIC PROBLEMS

1. Suppose that you want to determine whether a new mutation in the *gal* region of *E. coli* is the result of an insertion of DNA. Describe a physical experiment that would allow you to demonstrate the presence of an insertion.

2. Explain the difference between the replicative and the conservative modes of transposition. Briefly describe an experiment demonstrating each of these modes in prokaryotes.

3. Describe the generation of multiple drug-resistance plasmids.

4. Briefly describe the experiment that demonstrates that the transposition of the *Ty1* element in yeast takes place through an RNA intermediate.

5. Explain how the properties of *P* elements in *Drosophila* make gene-transfer experiments possible in this organism.

6. Nobel prizes are usually awarded many years after the actual discovery. For example, Watson, Crick, and Wilkens won the Nobel Prize in Medicine or Physiology in 1962, almost a decade after their discovery of the double helical structure of DNA. However, Barbara McClintock was awarded her Nobel Prize in 1983, almost four decades after her discovery of transposable elements in maize. Why do you think it took this long?

CHALLENGING PROBLEMS

7. Prior to the integration of a transposon, its transposase makes a staggered cut in the host target DNA. If the staggered cut occurs at the sites of the arrows below, draw what the sequence of the host DNA will be after the transposon is inserted. You can represent the transposon as a rectangle.

$$\downarrow$$

AATTTGGCCTAGTACTAATTGGTTGG

TTAAACCGGATCATGATTAACCAACC

$$\uparrow$$

8. In *Drosophila*, M. Green found a *singed* allele *(sn)* with some unusual characteristics. Females homozygous for this X-linked allele have singed bristles, but they have numerous patches of sn^+ (wild-type) bristles on their heads, thoraxes, and abdomens. When these flies are mated with *sn* males, some females give only singed progeny, but others give both singed and wild-type progeny in variable proportions. Explain these results.

9. Consider two maize plants:

   a. Genotype C/c^m ; Ac/Ac^+, where c^m is an unstable allele caused by *Ds* insertion

   b. Genotype C/c^m, where c^m is an unstable allele caused by *Ac* insertion.

   What phenotypes would be produced and in what proportions when (1) each plant is crossed with a base-pair-substitution mutant *c/c* and (2) the plant in part a is crossed with the plant in part b? Assume that *Ac* and *c* are unlinked, that the chromosome breakage frequency is negligible, and that mutant *c/C* is Ac^+.

10. You meet your friend, the scientist, at the gym and she begins telling you about a mouse gene she is studying in the lab. The product of this gene is an enzyme required to make the fur brown. The gene is called *FB* and the enzyme is called FB enzyme. When *FB* is mutant and cannot produce the FB enzyme, the fur is white. The scientist tells you that she has isolated the gene from two mice with brown fur and that surprisingly she found that the two genes differ by the presence of a 250 bp SINE (like the human *Alu* element) in the *FB* gene of one mouse but not the gene from the other. She does not understand how this is possible, especially when she determined that both mice make the FB enzyme. Can you help her formulate a hypothesis that explains why the mouse can still produce FB enzyme with a transposable element in its *FB* gene?

11. The yeast genome has class 1 elements (*Ty1*, *Ty2*, etc.) but no class 2 elements. Can you think of a possible reason why DNA elements have not been successful in the yeast genome?

MUTATION, REPAIR, AND RECOMBINATION

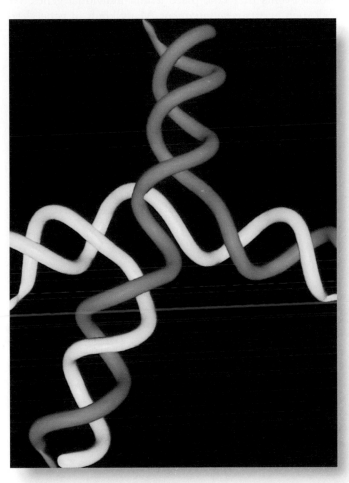

Computer model of a Holliday junction. [Julie Newdol, Computer Graphics Laboratory, University of California, San Francisco. Copyright by Regents, University of California.]

KEY QUESTIONS

- What is the molecular nature of mutations?
- How do certain types of radiation and chemicals cause mutation?
- Are induced mutations different from spontaneous mutations?
- Can a cell repair mutations?
- What is the molecular mechanism of crossing-over?
- Do mutational repair systems participate in crossing-over?

OUTLINE

14.1 Point mutations

14.2 Spontaneous mutation

14.3 Biological repair mechanisms

14.4 The mechanism of meiotic crossing-over

CHAPTER OVERVIEW

Genetic variation among individuals provides the raw material for evolution. Because genetics is the study of inherited differences, genetic analysis would not be possible without *variants*—individuals that show phenotypic differences in one or more particular characters. In previous chapters we performed many analyses of the inheritance of such variants; now we consider their origin. How do genetic variants arise?

Two major processes are responsible for genetic variation, *mutation* and *recombination*. We have seen that mutation is a change in the DNA sequence of a gene. Mutation is the ultimate source of evolutionary change; new alleles arise in all organisms, some spontaneously, others as a result of exposure to radiation and chemicals in the environment. The new alleles produced by mutation become the raw material for a second level of varia-

tion, effected by recombination. As its name suggests, recombination is the outcome of cellular processes that cause alleles of different genes to become grouped in new combinations. To use an analogy, mutation produces new playing cards, and then recombination shuffles them and deals them out as different hands.

In the cellular environment, DNA molecules are not absolutely stable; each base pair in a DNA double helix has a certain probability of mutating. As we shall see, the term *mutation* covers a broad array of different kinds of changes. In the next chapter, we shall consider mutational changes that affect entire chromosomes or large pieces of chromosomes. In the present chapter, we focus on mutational events that take place *within* individual genes. We call such events *gene mutations*. Many kinds of gene alterations can occur within DNA molecules. These events can be as simple as the swapping of one base pair for another. Alternatively, some mutations entail a change in the num-

CHAPTER OVERVIEW Figure

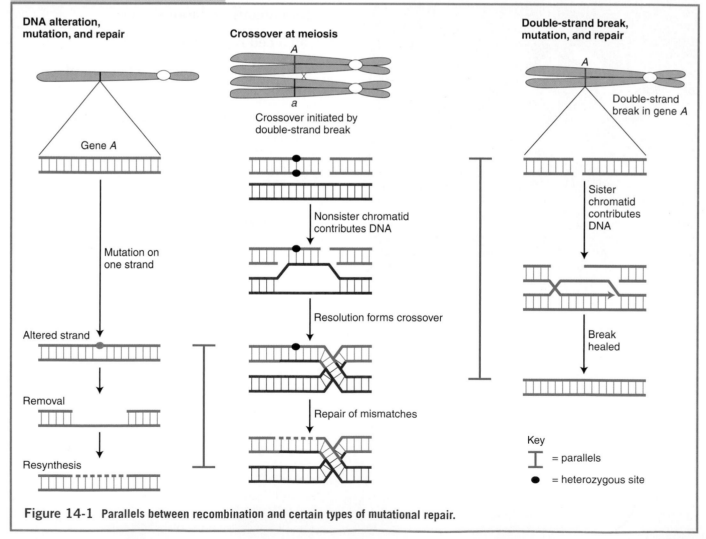

Figure 14-1 Parallels between recombination and certain types of mutational repair.

ber of copies of a trinucleotide repeat sequence [as when $(AGC)_3$ becomes $(AGC)_5$]. Mutations can even be caused by the insertion of a transposable element from elsewhere in the genome (Chapter 13). In this chapter we focus on mutations that do not involve transposable elements.

We can view DNA as being subjected to a dynamic tug of war between the chemical processes that damage DNA and lead to new mutations and the cellular repair processes that constantly monitor DNA for such damage and correct it. Mutations often arise through the action of certain agents, called *mutagens*, that increase the rate at which mutations occur. Alternatively, mutations can occur "spontaneously." Spontaneous mutations are much less frequent (and hence harder to study) than induced mutations, but they are evolutionarily more important. A host of different molecular mechanisms underlie mutation, ranging from the reaction of DNA with highly reactive products of cell metabolism to mistakes in the DNA replication process.

Cells have evolved sophisticated systems to identify and repair damaged DNA, thereby preventing the occurrence of mutations. Most notably, there are a variety of repair systems, and most of them rely on DNA complementarity. That is, they use one DNA strand as a template for the correction of DNA damage. For example, in the type of repair called *excision* repair, damage in one strand is cut out along with adjacent nucleotides, and then the correct sequence is resynthesized using the undamaged complementary strand as template (Figure 14-1, left column).

Finally, we will see that what is potentially the most serious class of DNA damage, a double-strand break, is also an intermediate step in a normal cellular process, recombination via meiotic crossing-over. Thus, we can draw parallels between mutation and recombination at two levels. First, as mentioned above, mutation and recombination are the major sources of variation. Second, mechanisms of DNA repair and recombination share some features, including the use of some of the same proteins. For this reason, we will explore mechanisms of DNA recombination and compare these with mechanisms of DNA repair. Figure 14-1 diagrams the parallels between crossing-over and two kinds of mutational repair (excision and double-strand break repair).

We consider two general classes of gene mutation:

- Mutations affecting single base pairs of DNA
- Mutations altering the number of copies of a small repeated sequence within a gene

14.1 Point mutations

Point mutations typically refer to alterations of single base pairs of DNA or of a small number of adjacent base pairs—that is, mutations that map to a single location, or "point," within a gene. Here we will focus on the point mutations that alter one base pair at a time. Essentially, these "point" mutations are the minimum changes that can be produced—changing only one "letter" in the "book of DNA." The constellation of possible ways in which a point mutation could change a wild-type gene is very large. However, it is always true that such mutations are more likely to reduce or eliminate gene function (thus they are loss-of-function mutations) than to enhance it (gain-of-function mutations). The reason is simple: by randomly changing or removing one of the components of a machine, it is much easier to break it than to alter the way that it works. Conversely, mutations that increase a gene's activity or alter the type of activity or change the location within a multicellular body where the gene is expressed are much rarer.

The origin of point mutations

Newly arising mutations are categorized as *induced* or *spontaneous*. Induced mutations are defined as those that arise after purposeful treatment with mutagens, environmental agents that are known to increase the rate of mutations. Spontaneous mutations are those that arise in the absence of *known* mutagen treatment. They account for the "background rate" of mutation and are presumably the ultimate source of natural genetic variation that is seen in populations.

The frequency at which spontaneous mutations occur is low, generally in the range of one cell in 10^5 to 10^8. Therefore, if a large number of mutants are required for genetic analysis, mutations must be induced. The induction of mutations is accomplished by treating cells with mutagens. The production of mutations through exposure to mutagens is called **mutagenesis,** and the organism is said to be *mutagenized*. The most commonly used mutagens are high-energy radiation or specific chemicals; examples of these mutagens and their efficacy are given in Table 14-1. The greater the dose of mutagen, the greater the number of mutations induced, as shown in Figure 14-2. Note that Figure 14-2 shows a *linear* dose response, which is often observed in the induction of point mutations.

Recognize that the distinction between induced and spontaneous is purely operational. If we are aware that an organism was exposed to a mutagen, then we surmise that the bulk of the mutations that arise afterward were induced by that mutagen. However, this is not true in an absolute sense. The mechanisms that give rise to spontaneous mutations are also acting in this mutagenized organism. In reality, there will always be a subset of mutations recovered after mutagenesis that arose independently of the action of the mutagen. The proportion of mutations that fall into this subset depends on how potent a mutagen is. The higher the rate of induced mutations, the lower the proportion of recovered mutations that are actually "spontaneous" in origin.

Table 14-1   Mutation Frequencies Obtained with Various Mutagens in *Neurospora*

Mutagenic treatment	Exposure time (minutes)	Survival (%)	Number of *ad-3* mutants per 10^6 survivors
No treatment (spontaneous rate)	–	100	~0.4
Amino purine (1–5 mg/ml)	During growth	100	3
Ethylmethanesulfonate (1%)	90	56	25
Nitrous acid (0.05 M)	160	23	128
X rays (2000 r/min)	18	16	259
Methyl methanesulfonate (20 mM)	300	26	350
UV rays (600 erg/mm²/min)	6	18	375
Nitrosoguanidine (25 mM)	240	65	1500
ICR-170 acridine mustard (5 mg/ml)	480	28	2287

Note: The assay measures the frequency of *ad-3* mutants. It so happens that such mutants are red, so they can be detected against a background of white *ad-3⁺* colonies.

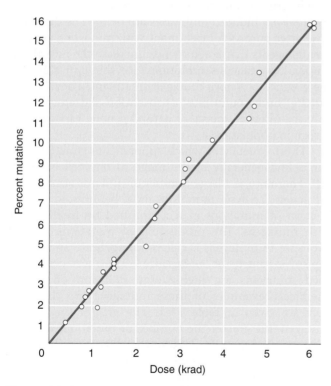

Figure 14-2 Linear relationship between X-ray dose and mutation. The relationship is measured by the induction of sex-linked recessive lethals in *Drosophila*.

Induced and spontaneous mutations arise by generally different mechanisms, and so they will be covered separately. After considering these mechanisms, we shall explore the subject of biological mutation repair. Without these repair mechanisms, the rate of mutation would be so high that cells would accumulate too many mutations to remain viable and capable of reproduction. Thus, the mutational events that do occur are those rare events that have somehow been overlooked or bypassed by the repair processes.

Types of point mutations

Point mutations are classified in molecular terms in Table 14-2, which shows the main types of DNA changes and their effects on protein function when they occur within the protein-coding region of a gene.

The two main types of point mutation in DNA are *base substitutions* and *base additions* or *deletions*. Base substitutions are mutations in which one base pair is replaced by another. Base substitutions also can be divided into two subtypes: transitions and transversions. To describe these subtypes, we consider how a mutation alters the sequence on one DNA strand (the complementary change will take place on the other strand). A **transition** is the replacement of a base by the other base of the same chemical category (purine replaced by purine: either A to G or G to A; pyrimidine replaced by pyrimidine: either C to T or T to C). A **transversion** is the opposite—the replacement of a base of one chemical category by a base of the other (pyrimidine replaced by purine: C to A, C to G, T to A, T to G; purine replaced by pyrimidine: A to C, A to T, G to C, G to T). In describing the same changes at the double-stranded level of DNA, we must represent both members of a base pair in the same relative location. Thus, an example of a transition would be G · C → A · T; that of a transversion would be G · C → T · A.

Addition or deletion mutations are actually additions or deletions of *nucleotide* pairs; nevertheless, the convention is to call them *base*-pair additions or deletions. Collectively, they are termed *indel mutations* (for *insertion-deletion*). The simplest of these mutations are single-base-pair additions or single-base-pair deletions. Mutations sometimes arise through the simultaneous addition or deletion of multiple base pairs at once. As we shall see later in this chapter, mechanisms that selectively produce certain kinds of multiple-base-pair additions or deletions are the cause of certain human genetic diseases.)

Table 14-2 Point Mutations at the Molecular Level

Type of mutation	Result and examples
At DNA level	
Transition	Purine replaced by a different purine, or pyrimidine replaced by a different pyrimidine: A · T ⟶ G · C ⟶ G · C ⟶ A · T C · G ⟶ T · A T · A ⟶ C · G
Transversion	Purine replaced by a pyrimidine, or pyrimidine replaced by a purine: A · T ⟶ C · G A · T ⟶ T · A G · C ⟶ T · A G · C ⟶ C · G T · A ⟶ G · C T · A ⟶ A · T C · G ⟶ A · T C · G ⟶ G · C
Indel	Addition or deletion of one or more base pairs of DNA (inserted or deleted bases are underlined): AAGACTCCT ⟶ AAGA<u>G</u>CTCCT AA<u>G</u>ACTCCT ⟶ AAACTCCT
At protein level	
Synonymous mutation	Codons specify the same amino acid: AGG ⟶ CGG Arg Arg

Missense mutation
  Conservative missense mutation
	Codon specifies a different amino acid
	  Codon specifies chemically similar amino acid:
	
	AAA ⟶ AGA
	Lys Arg
	(basic) (basic)
	
	Does not alter protein function in many cases

Nonconservative missense mutation	Codon specifies chemically dissimilar amino acid: UUU ⟶ UCU Hydrophobic Polar phenylalanine serine
Nonsense mutation	Codon signals chain termination: CAG ⟶ UAG Gln Amber termination codon
Frameshift mutation	One base-pair addition (underlined) AAG ACT CCT ⟶ AAG A<u>G</u>C TCC T... One base-pair deletion (underlined) AA<u>G</u> ACT CCT ⟶ AAA CTC CT...

The molecular consequences of point mutations on gene structure and expression

What are the functional consequences of these different types of point mutations? First, consider what happens when a mutation arises in a polypeptide-coding part of a gene. For single-base substitutions, there are several possible outcomes, but all are direct consequences of two aspects of the genetic code: degeneracy of the code and the existence of translation termination codons (Figure 14-3).

- **Synonymous mutations.** The mutation changes one codon for an amino acid into another codon for that same amino acid. Synonymous mutations are also referred to as *silent* mutations.

- **Missense mutations.** The codon for one amino acid is changed into a codon for another amino acid. Missense mutations are sometimes called *nonsynonymous* mutations.

- **Nonsense mutations.** The codon for one amino acid is changed into a translation termination (stop) codon.

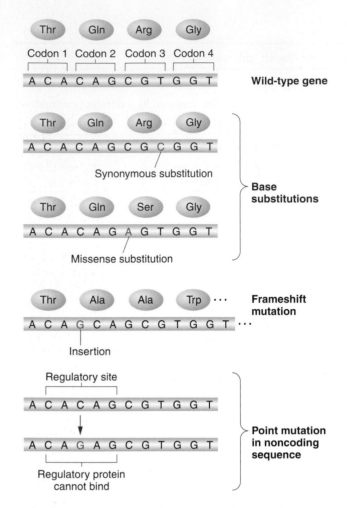

Figure 14-3 Consequences of point mutations within genes. In the top four panels, codons numbered 1–4 are located within the coding region of a gene.

Synonymous substitutions never alter the amino acid sequence of the polypeptide chain. The severity of the effect of missense and nonsense mutations on the polypeptide differs from case to case. For example, if a missense mutation replaces one amino acid with a chemically similar amino acid, referred to as a **conservative substitution,** then the alteration is less likely to affect the protein's structure and function severely. Alternatively, chemically different amino acid substitutions, called **nonconservative substitutions,** are more likely to produce severe changes in protein structure and function. Nonsense mutations will lead to the premature termination of translation. Thus, they have a considerable effect on protein function. The closer a nonsense mutation is to the 3′ end of the open reading frame, the more plausible it is that the resulting protein might possess some biological activity. However, many nonsense mutations produce completely inactive protein products.

Like nonsense mutations, indel mutations (base-pair additions or deletions) have consequences on polypep-

tide sequence that extend far beyond the site of the mutation itself (see Figure 14-3). Recall that the sequence of mRNA is "read" by the translational apparatus in register ("in frame"), three bases (one codon) at a time. The addition or deletion of a single base pair of DNA changes the reading frame for the remainder of the translation process, from the site of the base-pair mutation to the next stop codon in the new reading frame. Hence, these lesions are called **frameshift mutations.** These mutations cause the entire amino acid sequence translationally downstream of the mutant site to bear no relation to the original amino acid sequence. Thus, frameshift mutations typically result in complete loss of normal protein structure and function.

Now let's turn to mutations that occur in regulatory and other noncoding sequences (see Figure 14-3). Those parts of a gene that do not directly encode a protein contain many crucial DNA binding sites for proteins interspersed among sequences that are nonessential to gene expression or gene activity. At the DNA level, the docking sites include the sites to which RNA polymerase and its associated factors bind, as well as sites to which specific transcription-regulating proteins bind. At the RNA level, additional important docking sites include the ribosome-binding sites of bacterial mRNAs, the 5′ and 3′ splice sites for exon-joining in eukaryotic mRNAs, and sites that regulate translation and localize the mRNA to particular areas and compartments within the cell.

It is much harder to predict the ramifications of mutations in parts of a gene other than the polypeptide-coding segments. In general, the functional consequences of any point mutation in such a region depend on whether it disrupts (or creates) a binding site. Mutations that disrupt these sites have the potential to change the expression pattern of a gene by altering the amount of product expressed at a certain time or in a certain tissue or by altering the response to certain environmental cues. Such regulatory mutations will alter the amount of the protein product produced but *not* the structure of the protein. Alternatively, some binding-site mutations might completely obliterate a required step in normal gene expression (such as the binding of RNA polymerase or splicing factors) and hence totally inactivate the gene product or block its formation. Figure 14-4 shows some examples of how different types of mutations affect mRNA and protein.

It is important to keep in mind the distinction between the occurrence of a gene mutation—that is, a change in the DNA sequence of a given gene—and the detection of such an event at the phenotypic level. Many point mutations within noncoding sequences elicit little or no phenotypic change; these mutations fall within sites that, for example, are between DNA binding sites for regulatory proteins. Such sites may be functionally irrelevant, or other sites within the gene may duplicate their function.

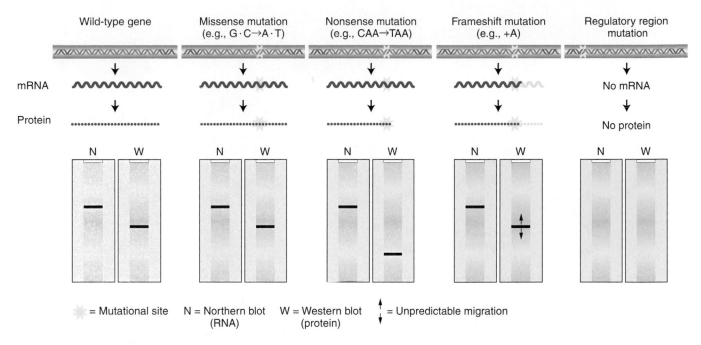

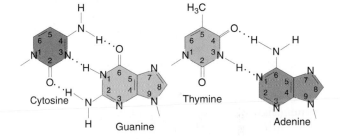

Figure 14-4 **Effects of common mutations.** The effects of some common types of mutations at the RNA and protein levels.

Mechanisms of point-mutation induction

When we examine the array of mutations induced by different mutagens, we see that each mutagen is characterized by a distinct **mutational specificity,** or "preference," both for a certain *type* of mutation (e.g., G · C → A · T transitions) and for certain mutational *sites,* called **hot spots.** Such mutational specificity was first noted at the *rII* locus of the bacteriophage T4.

Mutagens act through at least three different mechanisms. They can *replace* a base in the DNA, *alter* a base so that it specifically mispairs with another base, or *damage* a base so that it can no longer pair with any base under normal conditions.

BASE REPLACEMENT Some chemical compounds are sufficiently similar to the normal nitrogen bases of DNA that they are occasionally incorporated into DNA in place of normal bases; such compounds are called **base analogs.** Many of these analogs have pairing properties unlike those of the normal bases; thus they can produce mutations by causing incorrect nucleotides to be inserted in the course of replication. To understand the action of base analogs, we must first consider the natural tendency of bases to assume different forms.

Each of the bases in DNA can appear in one of several forms, called **tautomers,** which are isomers that differ in the positions of their atoms and in the bonds between the atoms. The forms are in equilibrium. The **keto** form of each base is normally present in DNA (Figure 14-5), whereas the **imino** and **enol** forms of the bases are

Figure 14-5 **Pairing between the normal (keto) forms of the bases.**

rare. The imino or enol tautomer may pair with the wrong base, forming a *mispair.* The ability of such a mispair to cause a mutation in the course of DNA replication was first noted by Watson and Crick when they formulated their model for the structure of DNA (Chapter 7). Figure 14-6 demonstrates some possible mispairs resulting from the change of one tautomer into another, termed a **tautomeric shift.**

Mispairs can arise spontaneously but can also arise when bases become ionized. The mutagen 5-bromouracil (5-BU) is an analog of thymine that has bromine at the carbon 5 position in place of the CH_3 group found in thymine (Figure 14-7a). Its mutagenic action is based on enolization and ionization. In 5-BU, the bromine atom is not in a position in which it can hydrogen-bond during base pairing. Thus the keto form of 5-BU pairs with adenine, as would thymine; this pairing is shown in Figure 14-7a. However, the presence of

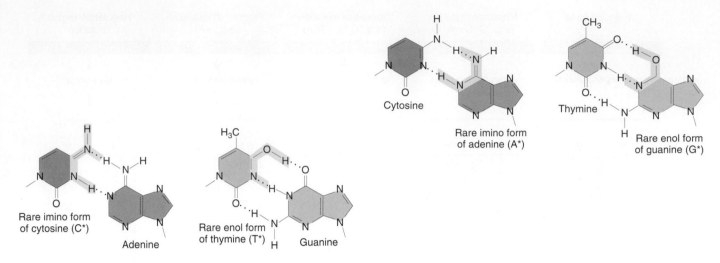

Figure 14-6 Mismatched bases. Rare tautomeric forms of bases result in mismatches.

the bromine atom significantly alters the distribution of electrons in the base ring; so 5-BU can *frequently* change to either the enol form or an ionized form; the enol and ionized forms of 5-BU pair with guanine (Figure 14-7b). 5-BU causes $G \cdot C \rightarrow A \cdot T$ or $A \cdot T \rightarrow G \cdot C$ transitions in the course of replication, depending on whether 5-BU has been enolized or ionized within the DNA molecule or as an incoming base. Hence the action of 5-BU as a mutagen is due to the fact that the molecule spends more of its time in the enol or ion form.

Another widely employed mutagen is 2-amino-purine (2-AP), an analog of adenine that can pair with thymine (Figure 14-8a). When protonated, 2-AP can mispair with cytosine (Figure 14-8b). Therefore, when 2-AP is incorporated into DNA by pairing with thymine, it can generate $A \cdot T \rightarrow G \cdot C$ transitions by mispairing with cytosine in subsequent replications. Or, if 2-AP is incorporated by mispairing with cytosine, then $G \cdot C \rightarrow A \cdot T$ transitions will result when 2-AP pairs with thymine in subsequent replications. Genetic studies have shown that 2-AP, like 5-BU, is highly specific for transitions.

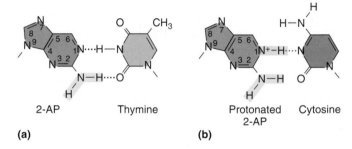

Figure 14-8 Alternative pairings for 2-aminopurine (2-AP). This analog of adenine can pair with cytosine in its protonated state (b).

BASE ALTERATION Some mutagens are not incorporated into the DNA but instead alter a base, causing specific mispairing. Certain alkylating agents commonly used as mutagens, such as ethylmethanesulfonate (EMS) and nitrosoguanidine (NG), operate by this pathway.

Such agents add alkyl groups (an ethyl group in the case of EMS and a methyl group in the case of NG) to many positions on all four bases. However, a mutation is most likely to occur when the alkyl group is added to the oxygen at position 6 of guanine to create an O-6-alkylguanine. This alkylation leads to direct mispairing with thymine, as shown in Figure 14-9, and results in $G \cdot C \rightarrow A \cdot T$ transitions in the next round of replication. Alkylating agents can also modify the bases of incoming nucleotides in the course of DNA synthesis.

The **intercalating agents** are another important class of DNA modifiers. This group of compounds includes **proflavin, acridine orange,** and a class of chemicals termed **ICR compounds** (Figure 14-10a). These agents are flat planar molecules that mimic base pairs and are able to slip themselves in (*intercalate*) between the stacked nitrogen bases at the core of the DNA double

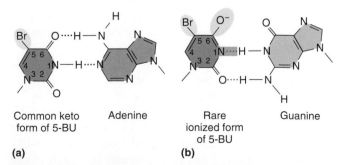

Figure 14-7 Alternative pairings for 5-bromouracil (5-BU). An analog of thymine, 5-BU can be mistakenly incorporated into DNA as a base. The ionized form base pairs with guanine.

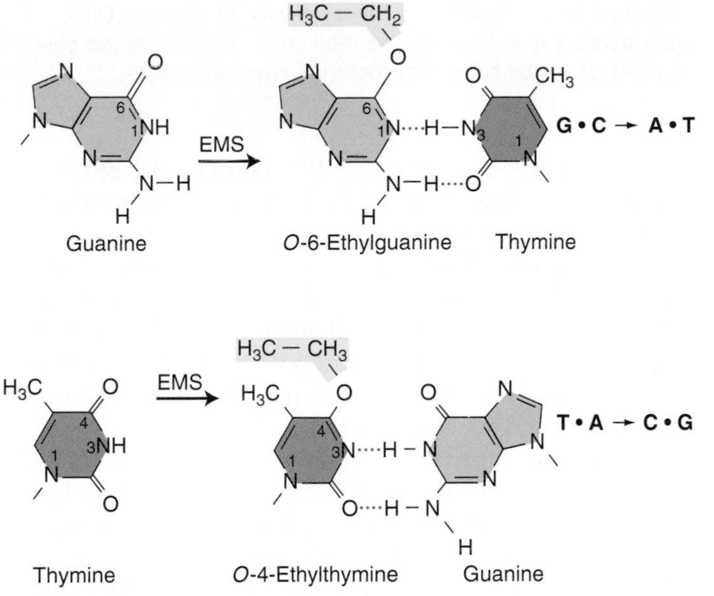

Figure 14-9 Alkylation-induced specific mispairings.
Treatment with EMS alters the structure of guanine and
thymine and leads to mispairings.

helix (Figure 14-10b). In this intercalated position,
an agent can cause single-nucleotide-pair insertions or
deletions.

BASE DAMAGE A large number of mutagens *damage*
one or more bases; so no specific base pairing is possible.
The result is a replication block, because DNA polymerase cannot continue DNA synthesis past such a damaged template base. In both prokaryotes and eukaryotes,
such replication blocks can be *bypassed* by inserting nonspecific bases. In *E. coli*, this process requires the activation of the **SOS system.** SOS and other mechanisms of
biological repair will be described later in this chapter.
However, an overview of this repair mechanism will be
presented in this section because, somewhat ironically,

some repair mechanisms are themselves responsible for
mutating DNA. The name *SOS* comes from the idea that
this system is induced as an emergency response to prevent cell death in the presence of significant DNA damage. As such, SOS induction is a mechanism of last
resort, a form of damage tolerance that allows the cell to
trade death for a certain level of mutagenesis.

It has taken over 30 years to figure how the SOS
system generates mutations while allowing DNA polymerase to bypass lesions at stalled replication forks. As
you will see below, UV light usually causes damage to
nucleotide bases in most organisms. An unusual class of
E. coli mutants that survived UV exposure without sus-
taining additional mutations was isolated in the 1970s.
The fact that such mutants even existed suggested that
some *E. coli* genes function to generate mutations when
exposed to UV light.

UV-induced mutation will not occur if the *DinB*,
UmuC, or *UmuD'* genes are mutated. Recently it was
discovered that these genes encode two error-prone
DNA polymerases: *DinB* encodes DNA polymerase IV,
while *UmuC* and *UmuD'* encode subunits of DNA
polymerase V. These polymerases overcome the block in
replication by adding nucleotides to the strand opposite
the damaged bases. Error-prone polymerases (also called
EP polymerases or *sloppy copiers*) have also been found in
diverse taxa of eukaryotes from yeast to human, where
they contribute to a damage-tolerance mechanism called
translesion DNA synthesis that resembles the SOS bypass
system in *E. coli*. Figure 14-11 shows how these polymerases (called pol τ and pol η in humans) function in
humans.

Whereas error-prone polymerases always appear to
be present in eukaryotic cells, they are induced by UV
exposure in *E. coli*. The first step in the SOS mechanism
occurs when UV induces the synthesis of a protein called
RecA. We will see more of the RecA protein later in the
chapter because it is a key player in many mechanisms of
DNA repair and recombination. When the replicative

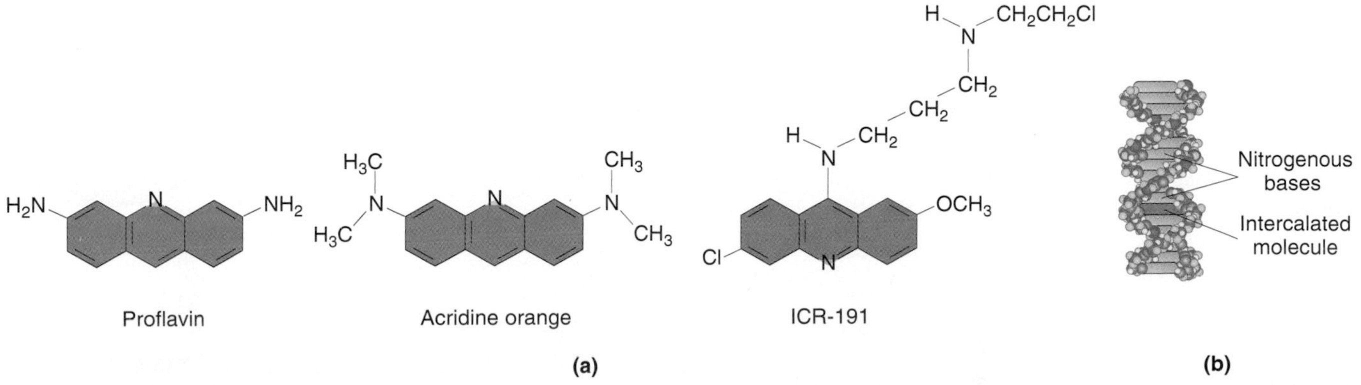

Figure 14-10 Intercalating agents. (a) Structures of common intercalating agents and
(b) their interaction with DNA. [From L. S. Lerman, *Proc. Natl. Acad. Sci. USA* 39, 1963, 94.]

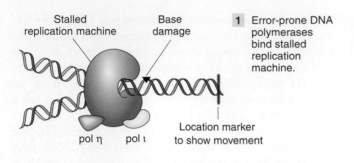

Stalled replication machine

Base damage

1 Error-prone DNA polymerases bind stalled replication machine.

Location marker to show movement

pol η   pol ι

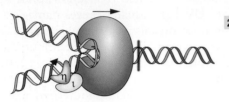

2 Binding promotes conformational change. Replication begins. pol η adds erroneous bases.

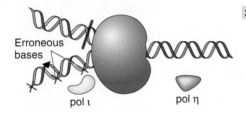

Erroneous bases

pol ι   pol η

3 Pol η disassociates. pol ι continues incorporating erroneous bases on strand opposite damaged base.

Figure 14-11 Translesion DNA synthesis. Error-prone DNA polymerases η and τ allow the replication machine to get past a bulky damaged base, incorporating erroneous bases.

polymerase (DNA polymerase III) stalls at a site of DNA damage, the DNA ahead of the polymerase continues to be unwound, exposing regions of single-stranded DNA that become bound by single-strand-binding protein (SSB). Next, RecA proteins join the SSB and form a protein–DNA filament. The RecA filament is the biologically active form of this protein. In this situation, RecA acts as a signal that leads to the induction of the error-prone polymerase and attracts it to the stalled fork.

Mutagens that create bases unable to form stable base pairs are thus dependent on SOS and similar systems for their mutagenic action, because the incorporation of incorrect nucleotides requires the activation of the SOS system. The category of SOS-dependent mutagens is important because it includes most cancer-causing agents (carcinogens), such as ultraviolet (UV) light and aflatoxin B_1. Indeed, a great deal of work has been done on the relationship of mutagens to carcinogens. The connections between mutation and cancer will be discussed in detail in Chapter 17.

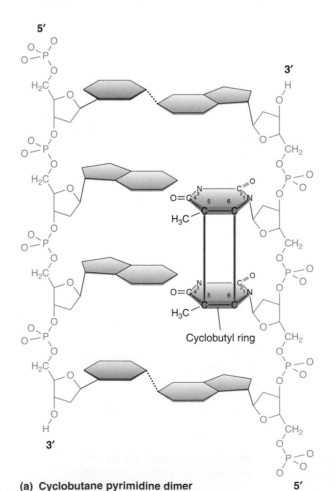

Figure 14-12 UV light–generated photoproducts. Photoproducts that unite adjacent pyrimidines in DNA are strongly correlated with mutagenesis.
[Left panel adapted from E. C. Friedberg, *DNA Repair*. Copyright 1985 by W. H. Freeman and Company. Right panel from J. S. Taylor et al.]

Cyclobutyl ring

(a) Cyclobutane pyrimidine dimer

T (6-4) T

(b) 6-4 Photoproduct

Ultraviolet light generates a number of distinct types of alterations in DNA, called *photoproducts*, from the word *photo* for "light." The most likely to lead to mutations are two different lesions that unite adjacent pyrimidines in the same strand. These lesions are the cyclobutane pyrimidine photodimer and the 6-4 photoproduct. For the cyclobutane pyrimidine dimer, ultraviolet light stimulates the formation of a four-membered cyclobutyl ring (shown in green in Figure 14-12a) between two adjacent pyrimidines on the same DNA strand by acting on 5,6 double bonds. The 6-4 photoproduct structure (Figure 14-12b) forms between the C-6 and C-4 positions of two adjacent pyrimidines, most prevalently 5'-CC-3'and 5'-TC-3'. The UV photoproducts significantly perturb the local structure of the double helix. These lesions interfere with normal base pairing; hence, induction of the SOS system is required for mutagenesis. The incorrect bases are inserted across from UV photoproducts at the 3'position of the dimer. The C → T transition is the most frequent mutation, but UV light also induces other base substitutions (transversions) and frameshifts, as well as larger duplications and deletions.

Aflatoxin B$_1$ (AFB$_1$) is a powerful carcinogen originally isolated from peanuts infected with a fungus. Aflatoxin attaches to guanine at the N-7 position (Figure 14-13). The formation of this addition product leads to the breakage of the bond between the base and the sugar, thereby liberating the base and resulting in an **apurinic site** (Figure 14-14). Studies of apurinic sites generated in vitro have demonstrated that the SOS bypass of these sites often leads to the insertion of an adenine residue across from an apurinic site. Thus, agents that cause depurination at guanine residues should tend to induce G · C → T · A transversions.

> **MESSAGE** Mutagens induce mutations by a variety of mechanisms. Some mutagens mimic normal bases and are incorporated into DNA, where they can mispair. Others damage bases, which then are not correctly recognized by DNA polymerase during replication, resulting in mispairing.

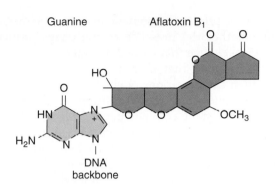

Figure 14-13 **The binding of metabolically activated aflatoxin B$_1$ to DNA.**

14.2 Spontaneous mutation

The origin of spontaneous hereditary change has always been a topic of considerable interest. One of the first questions asked by geneticists was whether spontaneous mutations are induced in response to external stimuli, or whether variants are present at a low frequency in most populations. An ideal experimental system to address this important question was the analysis of mutations in bacteria that confer resistance to specific environmental agents not normally tolerated by wild types.

Luria and Delbrück fluctuation test

One experiment by Salvador Luria and Max Delbrück in 1943 was particularly influential in shaping our understanding of the nature of mutation, not only in bacteria, but in organisms generally. It was known at the time that if *E. coli* bacteria are spread on a plate of nutrient medium in the presence of phage T1, the phage soon infect and kill the bacteria. However, rarely but regularly, colonies were seen that were resistant to phage attack; these colonies were stable and so appeared to be genuine mutants. However, it was not known whether these

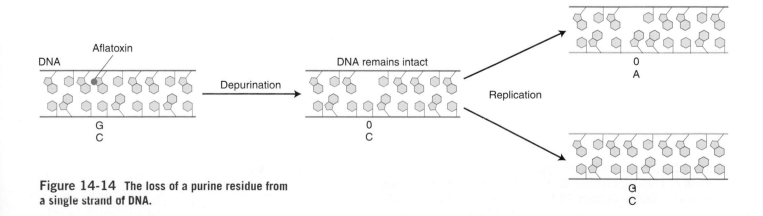

Figure 14-14 **The loss of a purine residue from a single strand of DNA.**

mutants were produced spontaneously but randomly in time or whether the presence of the phage induced a physiological change that caused resistance.

Luria reasoned that if mutations occurred spontaneously, then the mutations might be expected to occur at different times in different cultures; so the resulting numbers of resistant colonies per culture should show high variation (or "fluctuation" in his word). He later claimed that he obtained the idea while watching the fluctuating returns obtained by colleagues gambling on a slot machine at a faculty dance in a local country club; hence the term *"jackpot"* mutation.

Luria and Delbrück designed their **"fluctuation test"** as follows: They inoculated 20 small cultures, each with a few cells, and incubated them until there were 10^8 cells per milliliter. At the same time, a much larger culture also was inoculated and incubated until there were 10^8 cells per milliliter. The 20 individual cultures and 20 samples of the same size from the large culture were plated in the presence of phage. The 20 individual cultures showed high variation in the number of resistant colonies: 11 plates had 0 resistant colonies, and the remainder had 1, 1, 3, 5, 5, 6, 35, 64, and 107 per plate. The 20 samples from the large culture showed much less variation from plate to plate, all in the range of 14 to

26. If the phage were inducing mutations, there was no reason why fluctuation should be higher on the individual cultures, because all were exposed to phage similarly. The best explanation was that mutation was occurring randomly in time: the early mutations gave the higher numbers of resistant cells because they had time to produce many resistant descendants. The later mutations produced fewer resistant cells (Figure 14-15b).

This elegant analysis suggests that the resistant cells are selected by the environmental agent (here, phage) rather than produced by it. Can the existence of mutants in a population before selection be demonstrated directly? This demonstration was made possible by the use of a technique called **replica plating,** developed by Joshua and Esther Lederberg in 1952. A population of bacteria was plated on nonselective medium—that is, medium containing no phage—and from each cell a colony grew. This plate was called the *master plate.* A sterile piece of velvet was pressed down lightly on the surface of the master plate, and the velvet picked up cells wherever there was a colony (Figure 14-16). In this way, the velvet picked up a colony "imprint" from the whole plate. On touching the velvet to replica plates containing selective medium (that is, containing T1 phage), cells clinging to the velvet are inoculated onto the replica plates in the

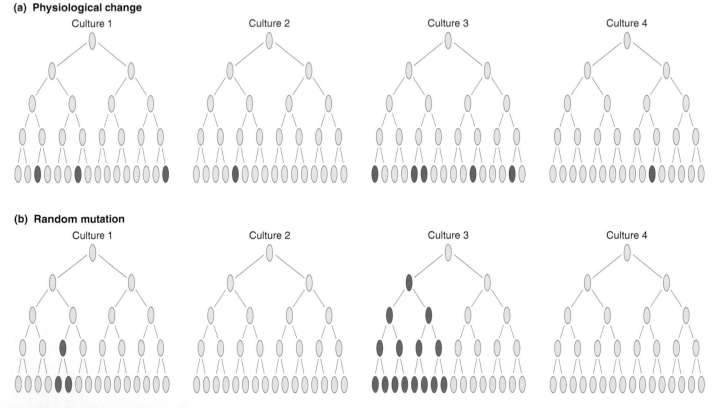

(a) Physiological change

Culture 1 Culture 2 Culture 3 Culture 4

(b) Random mutation

Culture 1 Culture 2 Culture 3 Culture 4

Figure 14-15 "Fluctuation test" hypotheses. These cell pedigrees illustrate the expectations from two contrasting hypotheses about the origin of resistant cells. [From G. S. Stent and R. Calendar, *Molecular Genetics*, 2d ed. W. H. Freeman and Company, 1978.]

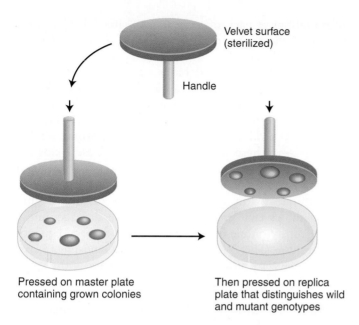

Figure 14-16 Replica plating. Replica plating reveals mutant colonies on a master plate through their behavior on selective replica plates. [From G. S. Stent and R. Calendar, *Molecular Genetics*, 2d ed. W. H. Freeman and Company, 1978.]

same relative positions as those of the colonies on the original master plate. As expected, rare resistant mutant colonies were found on the replica plates, but the multiple replica plates showed identical patterns of resistant colonies (Figure 14-17). If the mutations had occurred

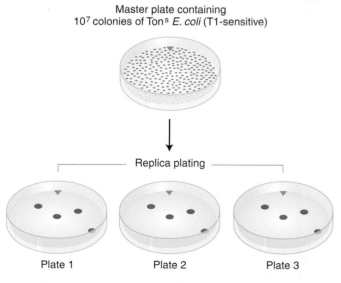

Figure 14-17 Replica plating demonstrates the presence of mutants before selection. The identical patterns on the replicas show that the resistant colonies are from the master. [From G. S. Stent and R. Calendar, *Molecular Genetics*, 2d ed. W. H. Freeman and Company, 1978.]

after exposure to the selective agents, the patterns for each plate would have been as random as the mutations themselves. The mutation events must have occurred before exposure to the selective agent.

> **MESSAGE** Mutation is a random process. Any allele in any cell may mutate at any time.

Replica plating has become an important technique of microbial genetics. It is useful in screening for mutants that *fail* to grow under the selective regime. The position of an absent colony on the replica plate is used to retrieve the mutant from the master. For example, replica plating can be used to screen auxotrophic mutants in precisely this way. In general, replica plating is a way of retaining an original set of strains on a master plate while simultaneously subjecting replicas to various kinds of tests on different media or under different environmental conditions.

Mechanisms of spontaneous mutations

We now know that spontaneous mutations arise from a variety of sources, including errors in DNA replication, spontaneous lesions, and as we saw in Chapter 13, the insertion of transposable elements. Spontaneous mutations are very rare, making it difficult to determine the underlying mechanisms. What sources of insight do we then have into the processes governing spontaneous mutation? Even though these mutations are rare, some selective systems allow spontaneous mutations to be obtained and then characterized at the molecular level — for example, their DNA sequences can be determined. From the nature of the sequence changes, inferences can be made about the processes that have led to the spontaneous mutations.

SPONTANEOUS LESIONS Naturally occurring damage to DNA, called **spontaneous lesions,** can generate mutations. Two of the most frequent spontaneous lesions are depurination and deamination, the former being more common.

We learned earlier that aflatoxin induces **depurination,** the loss of a purine base; however, depurination also occurs spontaneously. A mammalian cell spontaneously loses about 10,000 purines from its DNA in a 20-hour cell-generation period at 37°C. If these lesions were to persist, they would result in significant genetic damage because, during replication, the apurinic sites cannot specify any kind of base, let alone the correct one. However, as mentioned earlier in the chapter, under certain conditions, a base can be inserted across from an apurinic site, frequently resulting in a mutation.

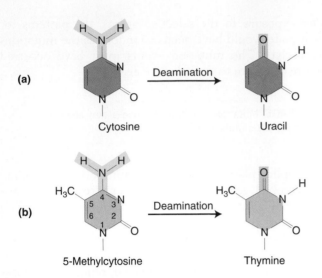

Figure 14-18 Deamination of (a) cytosine and (b) 5-methylcytosine.

The **deamination** of cytosine yields uracil (Figure 14-18a). Unless corrected, uracil residues will pair with adenine in the course of replication, resulting in the conversion of a G · C pair into an A · T pair (a G · C → A · T transition). Deamination of 5-methylcytosine also occurs (Figure 14-18b). (Certain bases in prokaryotes and eukaryotes are normally methylated.) The deamination of 5-methylcytosine generates thymine (5-methyluracil). Thus, C to T transitions generated by deamination are also seen frequently at 5-methylcytosine sites. DNA sequence analysis of hot spots for G · C → A · T transitions in the *lacI* gene has shown that 5-methylcytosine residues are present at the position of each hot spot. Some of the data from this *lacI* study are shown in Figure 14-19. The height of each bar on the graph represents the frequency

of mutations at each of a number of sites. The positions of 5-methylcytosine residues correlate nicely with the most mutable sites.

Oxidatively damaged bases constitute a third type of spontaneous lesion that can lead to mutation. Active oxygen species, such as superoxide radicals ($O_2^{\cdot-}$), hydrogen peroxide (H_2O_2), and hydroxyl radicals ($\cdot OH$), are produced as by-products of normal aerobic metabolism. These oxygen species can cause oxidative damage to DNA, as well as to precursors of DNA (such as GTP), resulting in mutation. Such mutations have been implicated in a number of human diseases. Figure 14-20 shows two products of oxidative damage, one a damaged thymidine residue, the other a damaged guanosine residue. The 8-oxo-7-hydrodeoxyguanosine (8-oxo dG, or "GO") product frequently mispairs with A, resulting in a high level of G → T transversions.

ERRORS IN DNA REPLICATION Mistakes made by the DNA replication apparatus are another source of mutations.

Base substitutions No chemical reaction is perfectly efficient. Accordingly, an error in DNA replication can

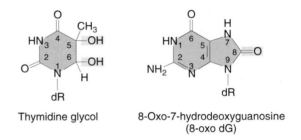

Thymidine glycol   8-Oxo-7-hydrodeoxyguanosine
                                  (8-oxo dG)

Figure 14-20 DNA damage products formed after attack by oxygen radicals.

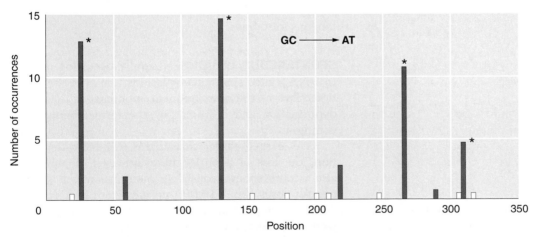

Figure 14-19 5-Methylcytosine hot spots in *E. coli*. Nonsense mutations at 15 different sites in *lacI* were scored. All resulted in G · C → A · T transition. The asterisk (*) marks the positions of 5-methylcytosines, and the white bars mark sites where transitions known to occur were not isolated in this group.[From C. Coulondre, J. H. Miller, P. J. Farabaugh, and W. Gilbert, *Nature* 274, 1978,775.]

Molecular mechanism of mutation

ANIMATED ART

WWW

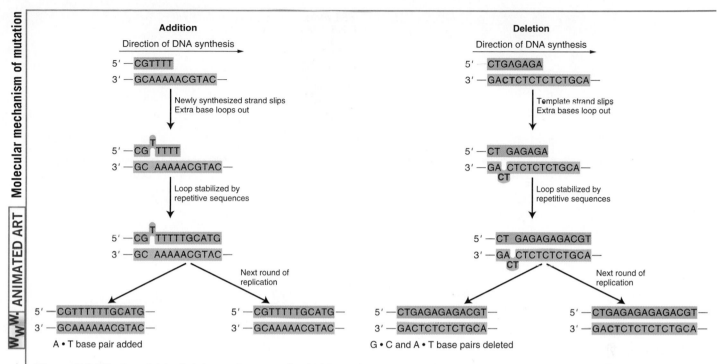

Figure 14-21 A model for indel mutations resulting in frameshifts. dr = deoxyribose.

occur when an illegitimate nucleotide pair (say, A · C) forms in DNA synthesis, leading to a base substitution. Also recall that earlier in the chapter we discussed that there are several forms of each base, called *tautomers*, and that the change of one tautomeric form to another can result in mispairing during DNA replication.

Base insertion and deletion Although some errors in replication produce base-substitution mutations, other kinds of replication errors can lead to **indel mutations**—that is, insertions or deletions of one or more base pairs. When such mutations add or subtract a number of bases not divisible by three, they produce frameshift mutations in the protein-coding regions. The nucleotide sequence at frameshift mutation hot spots was determined in the lysozyme-encoding gene of phage T4. These mutations often occur at repeated bases. The prevailing model (Figure 14-21) proposes that indels arise when loops in single-stranded regions are stabilized by the "slipped mispairing" of repeated sequences in the course of replication. This mechanism is sometimes called *replication slippage*. In the *E. coli lacI* gene, certain hot spots result from repeated sequences, just as predicted by this model. Figure 14-22 depicts the distribution of spontaneous mutations in the *lacI* gene. Note how one site dominates the distribution. In *lacI*, the major indel hot spot is a four-base-pair sequence (CTGG) repeated three times in tandem in the wild type (for simplicity, only one strand of the DNA is shown):

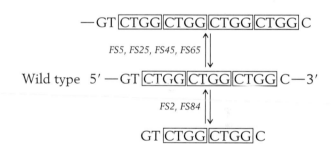

The majority of mutations at this site (represented here by the mutations *FS5*, *FS25*, *FS45*, and *FS65*) result from the addition of one extra set of the four bases CTGG. A minority (represented here by the mutations *FS2* and *FS84*) result from the loss of one set of the four bases CTGG.

How can we explain these observations? The model predicts that the frequency of a particular indel depends on the number of base pairs that can form during the slipped mispairing of repeated sequences. The wild-type sequence shown for the *lacI* gene can slip out one CTGG sequence and stabilize this structure by forming nine base pairs (apply the model in Figure 14-21 to the sequence shown for *lacI*). Whether a deletion or an insertion is generated depends on whether the slippage is on the template or on the newly synthesized strand, respectively.

Larger deletions (more than a few base pairs) constitute a sizable fraction of observed spontaneous mutations, as shown in Figure 14-22. Most, although not all,

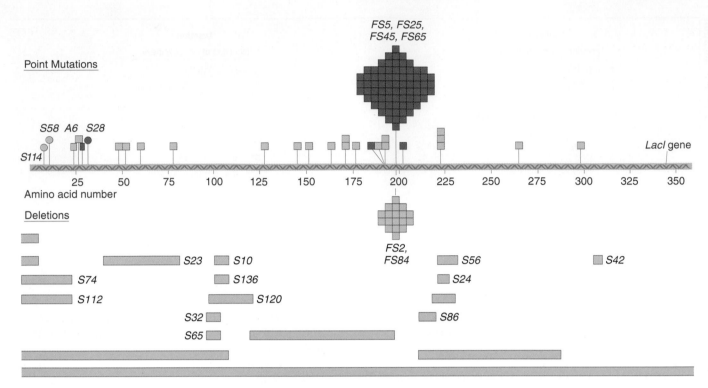

Figure 14-22 The distribution of 140 spontaneous mutations in *lacI*. Boxes indicate position of point mutations, with red designating fast-reverting mutants. Deletions are represented in gold. Circles designate larger deletion and insertion mutants. Allele numbers correspond to mutants that have been sequenced. [From P. J. Farabaugh, U. Schmeissner, M. Hofer, and J. H. Miller, *Journal of Molecular Biology* 126, 1978, 847.]

of the deletions are of repeated sequences. Figure 14-23 shows 9 deletions analyzed at the DNA sequence level in the *lacI* gene of *E. coli*. The results of further studies have shown that the longer repeats constitute hot spots for deletions. Duplications of DNA segments have also been observed in many organisms. Like deletions, they often occur at sequence repeats.

It must be noted that, in addition to their origin by replication slippage, deletions and duplications could be generated by offset homologous recombination between copies of the repeats.

> **MESSAGE** Spontaneous mutations can be generated by several different processes. Spontaneous lesions and replication errors generate most of the base-substitution and indel mutations, respectively.

Spontaneous mutations in humans— trinucleotide repeat diseases

DNA sequence analysis has revealed the gene mutations contributing to numerous human hereditary diseases. Many are of the expected base-substitution or single-base-pair indel type. However, some mutations are more complex. A number of these human disorders are due to duplications of short repeated sequences.

A common mechanism responsible for a number of genetic diseases is the expansion of a three-base-pair repeat. For this reason, they are termed **trinucleotide repeat** diseases. An example is the human disease called *fragile X syndrome*. This disease is the most common form of inherited mental retardation, occurring in close

---

S74, S112   75 bases

CAATTCAGG**GTGGTGAA**TGTGAAACC------CGC**GTGGTGAA**CCAGG

Site (no. of bp)	Sequence repeat	No. of bases deleted	Occurrences	
20 to 95	GTGGTGAA	75	2	S74, S112
146 to 269	GCGGCGAT	123	1	S23
331 to 351	AAGCGGCG	20	2	S10, S136
316 to 338	GTCGA	22	2	S32, S65
694 to 707	CA	13	1	S24
694 to 719	CA	25	1	S56
943 to 956	G	13	1	S42
322 to 393	None	71	1	S120
658 to 685	None	27	1	S86

Figure 14-23 Analysis of deletions in *lacI* in regions containing repeated sequences. For *S74* and *S112*, one of two repeated sequences and all of the intervening sequences are deleted, producing the same final sequence. [From P. J. Farabaugh, U. Schmeissner, M. Hofer, and J. H. Miller, *Journal of Molecular Biology* 126, 1978, 847.]

to 1 of 1500 males and 1 of 2500 females. It is manifested cytologically by a fragile site in the X chromosome that results in breaks in vitro. Fragile X syndrome results from changes in the number of a (CGG)$_n$ repeat in a region of the *FMR-1* gene that is transcribed but not translated (Figure 14-24a).

How does repeat number correlate with the disease phenotype? Humans normally show considerable variation in the number of CGG repeats in the *FMR-1* gene, ranging from 6 to 54, with the most frequent allele containing 29 repeats. Sometimes, unaffected parents and grandparents give rise to several offspring with fragile X syndrome. The offspring with the symptoms of the disease have enormous repeat numbers, ranging from 200 to 1300 (see Figure 14-24b). The unaffected parents and grandparents have also been found to contain increased copy numbers of the repeat, but ranging from only 50 to 200. For this reason, these ancestors have been said to carry *premutations*. The repeats in these premutation alleles are not sufficient to cause the disease phenotype, but they are much more unstable (i.e., readily expanded) than normal alleles, and so they lead to even greater expansion in their offspring. (In general, it appears that the more expanded the repeat number, the greater the instability.)

The proposed mechanism for the generation of these repeats is a slipped mispairing in the course of DNA synthesis, just as discussed previously for the expansion of the repeat at the *lacI* hot spot. However, the extraordinarily high frequency of mutation at the trinucleotide repeats in fragile X syndrome suggests that in human cells, after a threshold level of about 50 repeats, the replication machinery cannot faithfully replicate the correct sequence, and large variations in repeat numbers result.

Other diseases, such as Huntington disease (see Chapter 2), also have been associated with the expansion of trinucleotide repeats in a gene. Several general themes apply to these diseases. In Huntington disease, for example, the wild-type HD gene includes a repeated sequence, often within the protein-coding region, and mutation correlates with a considerable expansion of this repeat region. The severity of the disease correlates with the number of repeat copies.

Huntington disease and Kennedy disease (also called *X-linked spinal and bulbar muscular atrophy*) result from the amplification of a three-base-pair repeat, CAG. Normal persons have an average of 19 to 21 CAG repeats, whereas affected patients have an average of about 46. In Kennedy disease, which is characterized by progressive

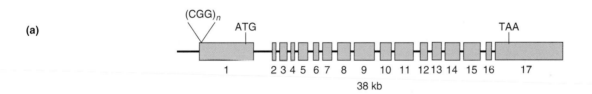

(a)

1   2 3 4 5 6 7 8   9   10 11 12 13 14 15 16   17

38 kb

FMR-1 gene	Phenotype	Transmission	Methylation	Transcription
Normal 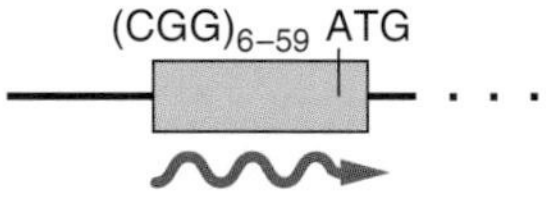 (CGG)$_{6-59}$ ATG	Normal	Stable	No	Yes
Premutation 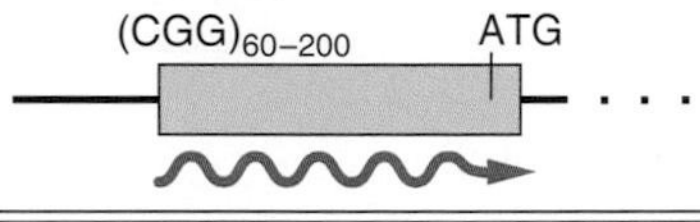 (CGG)$_{60-200}$ ATG	Largely normal	Unstable, prone to expansion	No	Yes
Full mutation 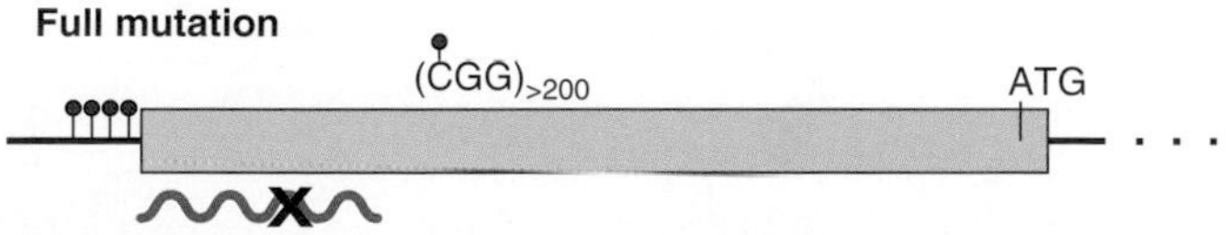 (CGG)$_{>200}$ ATG	Affected	Unstable	Yes	No

Figure 14-24 The *FMR-1* gene involved in fragile X syndrome. (a) Exon structure and upstream CGG repeat. (b) Transcription and methylation in normal, premutation, and full mutation alleles. The red circles are methyl groups. [W. T. O'Donnell and S. T. Warren, *Ann. Rev. Neuroscience* 25, 2002, 315–338, Figure 1.]

muscle weakness and atrophy, the expansion of the trinucleotide repeat occurs in the gene that encodes the androgen receptor.

Properties common to some trinucleotide-repeat diseases suggest a common mechanism by which the abnormal phenotypes are produced. First, many of these diseases seem to include neurodegeneration—that is, cell death within the nervous system. Second, in such diseases the trinucleotide repeats fall within the open reading frames of the transcripts of these genes, leading to expansions or contractions of the number of repeats of a single amino acid in the polypeptide (for example, CAG repeats encode a polyglutamine repeat). Thus, it is no accident that these diseases entail expansions of codon-size three-base-pair units.

But this explanation cannot hold for all trinucleotide-repeat diseases. After all, in fragile X syndrome, the trinucleotide expansion occurs near the 5′ end of the *FMR-1* mRNA, *before* the translation start site. Thus, we cannot ascribe the phenotypic abnormalities of the *FMR-1* mutations to an effect on protein structure. One clue to the problem with the mutant *FMR-1* genes is that they, unlike the normal gene, are hypermethylated, a feature associated with transcriptionally silenced genes (see Figure 14-24b). Based on these findings it is hypothesized that repeat expansion leads to changes in chromatin structure that silence transcription of the mutant gene (see Chapter 10). In support of this model is the finding that the *FMR-1* gene is deleted in some patients with fragile X syndrome. These observations support a loss-of-function mutation.

> **MESSAGE** Trinucleotide-repeat diseases arise through the expansion of the number of copies of a three-base-pair sequence normally present in several copies, often within the coding region of a gene.

14.3 Biological repair mechanisms

Living cells have evolved a series of enzymatic systems that repair DNA damage in a variety of ways. The low rate of spontaneous mutation is indicative of the efficiency of these repair systems. We can think of the spontaneous mutation rate as being at a balance point between the rate at which premutational damage arises and the rate at which repair systems recognize this damage and restore the normal base sequence. Failure of these systems can lead to a higher mutation rate, as we shall see later.

Let's now examine some of the repair pathways, beginning with error-free repair. For this pathway one of two things can happen:

- The repair pathway chemically repairs the damage to the DNA base.

- The repair pathway deletes the damaged DNA and uses an existing complementary sequence as a template to restore the normal sequence.

Direct reversal of damaged DNA

The most straightforward way to repair a lesion is to reverse it directly, thereby regenerating the normal base (Figure 14-25a). Although some types of damage are essentially irreversible, in a few cases lesions can be repaired by direct reversal. One case is a mutagenic photodimer caused by UV light. The cyclobutane pyrimidine photodimer can be repaired by an enzyme called a *photolyase*. The enzyme binds to the photodimer and splits it, in the presence of certain wavelengths of visible light, to regenerate the original bases (Figure 14-26). This repair mechanism is called *light repair* or *photorepair*. The photolyase enzyme cannot operate in the dark, and so other repair pathways are required to remove UV damage in the absence of visible light.

Alkyltransferases also are enzymes that directly reverse lesions. They remove certain alkyl groups that have been added to the O-6 positions of guanine (see Figure 14-9) by such mutagens as nitrosoguanidine and ethylmethanesulfonate. The methyltransferase from *E. coli* has been well studied. This enzyme transfers the methyl

(a) Direct reversal

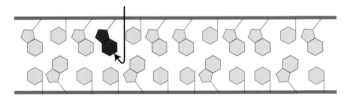

(b) Base excision and replacement

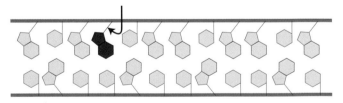

(c) Segment removal and replacement

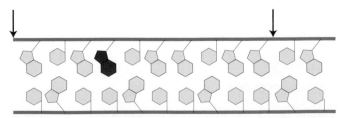

Figure 14-25 Three types of repair of DNA with a damaged base.

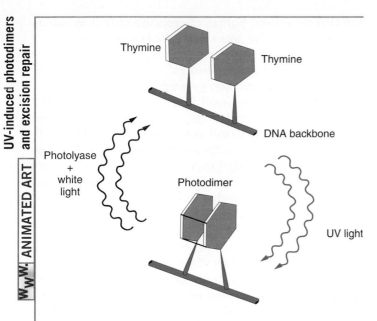

Figure 14-26 Induction and removal of UV-induced pyrimidine photodimer.

group from O-6-methylguanine to a cysteine residue on the protein. However, the transfer inactivates the enzyme, so this repair system can be saturated if the level of alkylation is high enough.

Homology-dependent repair systems

One of the overarching principles guiding cellular genetic systems is the power of nucleotide sequence complementarity. (You will recall that genetic analysis also depends heavily on this principle.) Important repair systems exploit the properties of antiparallel complementarity to restore damaged DNA segments back to their initial, undamaged state. In these systems, a segment of a DNA chain is removed and replaced with a newly synthesized nucleotide segment complementary to the opposite template strand. Because these systems depend on the complementarity, or homology, of the template strand to the strand being repaired, they are called **homology-dependent repair systems.** Because repair takes place through a template, the rules of DNA replication ensure that repair is accomplished with high fidelity—that is, it is *error-free*. There are two major homology-dependent error-free repair systems. One system (excision repair) repairs damage that has been detected before replication. The other (postreplication repair) repairs damage that is detected in the course of the replication process or afterward.

EXCISION-REPAIR PATHWAYS Unlike the examples of reversal of damage described above, excision repair entails the removal and replacement of an entire base.

Base excision repair Base excision repair (see Figure 14-25b) is carried out by **DNA glycosylases** that cleave base–sugar bonds, thereby liberating the altered bases and generating apurinic or apyrimidinic sites. An enzyme called AP endonuclease then cuts the sugar-phosphate backbone around the site lacking a base. A third enzyme, deoxyribophosphodiesterase, cleans up the backbone by removing a stretch of neighboring sugar-phosphate residues so that a DNA polymerase can fill the gap with nucleotides complementary to the other strand. DNA ligase then seals the new nucleotide into the backbone (Figure 14-27).

Numerous DNA glycosylases exist. One, uracil–DNA glycosylase, removes uracil from DNA. Uracil residues, which result from the spontaneous deamination of cytosine (see Figure 14-18), can lead to a C → T transition if unrepaired. One advantage of having thymine (5-methyluracil) rather than uracil as the natural pairing partner of adenine in DNA is that spontaneous cytosine deamination events can be recognized as abnormal and then excised and repaired. If uracil were a normal constituent of DNA, such repair would not be possible.

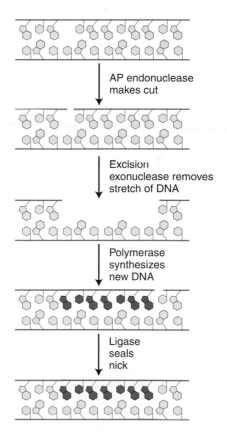

Figure 14-27 Repair of AP (apurinic or apyrimidinic) sites. AP endonucleases recognize AP sites and cut the phosphodiester bond. A stretch of DNA is removed by an exonuclease, and the resulting gap is filled in by DNA polymerase I and DNA ligase, using the complementary strand as template. [After B. Lewin, *Genes.* Copyright 1983 by John Wiley.]

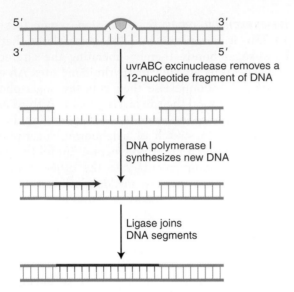

Figure 14-28 Nucleotide excision repair. Repair of a region of DNA containing a thymidine dimer. The thymidine dimer is shown in blue, and the new region of DNA is shown in red.

Nucleotide excision repair Base excision repair can correct only damaged bases that can be removed by a specific DNA glycosylase. However, there are more ways to damage a base than there are glycosylases to remove them. Thus, another system is required to repair the damage that the glycosylases cannot. Rather than recognize a particular damaged base, the **nucleotide excision-repair system** detects distortions in the double helix caused by the presence of an abnormal base (see Figure 14-25c). Mutations that cause such distortions include the pyrimidine dimers caused by UV light and the addition of aflatoxin to guanine residues. Detection of a distortion initiates a multistep repair process involving many proteins. In *E. coli*, a complex consisting of three enzymatic activities encoded by the *uvrABC* genes detects the distortion and cuts the damaged strand at two sites flanking the lesion (Figure 14-28). The uvrABC exinuclease, as it is called, excises precisely 12 nucleotides: 8 from one side of the damage and 4 from the other side. The 12-nucleotide gap is then filled by DNA polymerase I, using the template strand to produce an accurate copy of the original DNA sequence. DNA ligase then seals the new oligonucleotide into place.

> **MESSAGE** Base excision repair and nucleotide excision repair are error-free repair mechanisms that recognize and remove mispaired bases prior to replication and utilize the undamaged DNA complementary strand to guide repair.

Transcription-coupled repair in eukaryotes As in prokaryotes, eukaryotic error-free repair involves distinct base excision- and nucleotide excision-repair systems. DNA repair systems in eukaryotes are highly conserved from yeast to humans, and for this reason, yeast has again proved to be a useful model. As we will see later in

the chapter, several human diseases are caused by mutations in some of the genes that encode repair proteins.

In yeast, nucleotide excision repair is carried out by the multisubunit *repairosome*, a complex made up of more than 20 different polypeptides. The repairosome is able to recognize damaged DNA, excise about 30 nucleotides around the damage, and fill in the gap using the complementary strand as template (Figure 14-29). It has been noted that this system preferentially repairs the template (transcribed) DNA strand. How can the repairosome "know" which strand of a gene is transcribed and which is not? One clue is that seven of the polypeptides of the repairosome are also subunits of the basal transcription apparatus that is able to discriminate between the template and nontemplate strands of DNA (described in Chapter 8). One model is that the presence of DNA damage leads to the dissociation of the basal transcription apparatus and the assembly of the repairosome. Presumably, in this way, a mutant gene will be repaired before it can be transcribed.

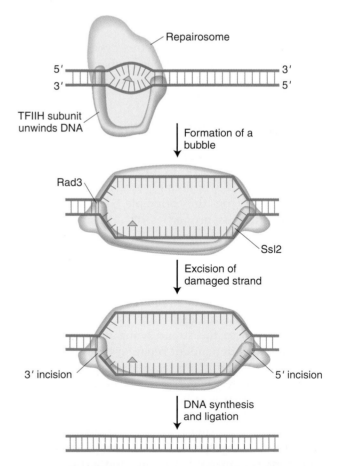

Figure 14-29 Nucleotide excision repair in eukaryotes. In this example a pyrimidine dimer (triangle) causes a bulge, which is recognized by a repairosome. Various proteins in the repairosome make a single-stranded bubble, and one single strand is cut out and resynthesized. [Adapted from *Encyclopedia of Life Sciences*, 2001, E. C. Friedberg, "Nucleotide Excision Repair in Eukaryotes," Figure 1.]

Why is it important to couple transcription and repair? Unlike actively dividing *E. coli* and most other prokaryotes, most of the cells in a multicellular organism are terminally differentiated and no longer dividing, so replication repair is not possible. These cells are not dead, however; their genes are being actively transcribed into mRNAs, which are translated into proteins. DNA that is damaged in these cells must also be repaired, since mutations in certain genes could have devastating consequences to the health of the whole organism. As you will see in Chapter 17, most cancerous tumors develop from somatic cells that have sustained mutations that have not been repaired.

POSTREPLICATION REPAIR Some repair pathways are capable of recognizing errors that usually occur during replication but fail to be corrected by the 3′-to-5′ proofreading function of the replicative polymerase. One such pathway, termed the **mismatch-repair system,** can detect such mismatches. Mismatch-repair systems have to do at least three things:

1. Recognize mismatched base pairs

2. Determine which base in the mismatch is the incorrect one

3. Excise the incorrect base and carry out repair synthesis

The second property is the crucial one of such a system. Unless it is capable of discriminating between the correct and the incorrect bases, the mismatch-repair system cannot determine which base to excise to prevent a mutation from arising. If, for example, a G · T mismatch occurs as a replication error, how can the system determine whether G or T is incorrect? Both are normal bases in DNA. But replication errors produce mismatches on the newly synthesized strand, and so the repair system knows that it is the base on this strand that must be recognized and excised.

The mismatch-repair system is best characterized in bacteria. Recall from Chapter 10 that bacterial DNA is methylated; this methylation normally takes place after replication. To distinguish the old, template strand from the newly synthesized strand, the bacterial repair system takes advantage of a delay in the methylation of the following sequence:

$$5'\text{-}G\text{-}A\text{-}T\text{-}C\text{-}3'$$
$$3'\text{-}C\text{-}T\text{-}A\text{-}G\text{-}5'$$

The methylating enzyme is **adenine methylase,** which creates 6-methyladenine on each strand. However, it takes adenine methylase several minutes to recognize and modify the newly synthesized GATC stretches. During that interval, the mismatch-repair system can operate because it can now distinguish the old strand from the new one by the methylation pattern. Methylating position 6 of adenine does not affect base pair-

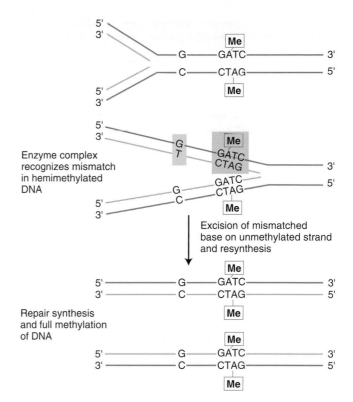

Figure 14-30 Model for mismatch repair in *E. coli*. DNA is methylated at A residues in the sequence GATC. DNA replication yields a hemimethylated duplex that exists until methylase can modify the newly synthesized strand. The mismatch-repair system makes any necessary corrections based on the sequence found on the methylated strand (original template). [After E. C. Friedberg, *DNA Repair*. Copyright 1985 by W. H. Freeman and Company.]

ing, and it provides a convenient tag that can be detected by other enzyme systems. Figure 14-30 shows the replication fork during mismatch correction. Note that only the old strand is methylated at GATC sequences right after replication. After the mismatched site has been identified, the mismatch-repair system corrects the error.

The mismatch-repair system has also been characterized in humans. Figure 14-31 depicts a model of how the human mismatch-repair system carries out the correction. An important target of the human mismatch-repair system is short repeat sequences that can be expanded or deleted during replication by the slipped-mispairing mechanism described previously (see Figure 14-21). Mutations in some of the components of this pathway have been shown to be responsible for several human diseases, especially cancers. There are thousands of short repeats (microsatellites) located all over the genome. Although most are located in noncoding regions (since most of the genome is noncoding), a few are located in genes that are critical for normal growth and development.

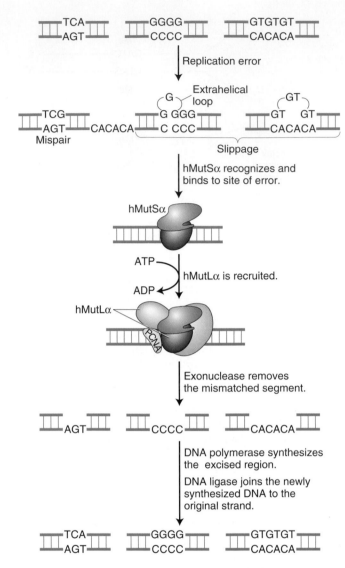

Figure 14-31 **Model for mismatch repair in humans.** Errors arising at replication, such as mispaired regions and loops from replication slippage, can be removed and repolymerized by the proteins shown. [Adapted from *Encyclopedia of Life Sciences*, 2001, P. Karran, "Human Mismatch Repair: Defects and Predisposition to Cancer," Figure 1.]

MESSAGE The mismatch-repair system corrects errors in replication that are not corrected by the proofreading function of the replicative DNA polymerase. Repair is restricted to the newly synthesized strand, which is recognized by the repair machinery in prokaryotes because it lacks a methylation marker.

Repair of double-strand breaks

As we have seen, DNA complementarity is an important resource that is exploited by many error-free correction systems. Such error-free repair is characterized by two stages: (1) removal of damaged and nearby DNA from one strand of the double helix and (2) use of the other strand as a template for the DNA synthesis needed to fill the single-strand gap. However, what would happen if *both* strands of the double helix were damaged in such a way that complementarity could not be exploited? One way this might happen is if both strands of the double helix were to break at sites that were close together. A mutation like this is called a **double-strand break.** If left unrepaired, double-strand breaks can cause a variety of chromosomal aberrations resulting in cell death or a precancerous state.

Interestingly, the ability of double-strand breaks to initiate chromosomal instability is an integral feature of some normal cellular processes that require DNA rearrangements. One example is the generation of the diversity of antibodies in the cells of the mammalian immune system. Another is meiotic recombination, which uses double-strand breaks to generate genetic diversity. As will be seen in the remainder of this chapter, the cell uses many of the same proteins and pathways to repair double-strand breaks and to carryout meiotic recombination. For this reason, we begin by focusing on the molecular mechanisms that repair double-strand breaks before turning our attention to the mechanism of meiotic recombination.

Double-strand breaks can arise spontaneously (for example, in response to reactive oxygen species), or they can be induced by ionizing radiation. Two distinct mechanisms are used to repair these potentially lethal lesions: nonhomologous end joining and homologous recombination.

NONHOMOLOGOUS END-JOINING As mentioned earlier, DNA repair is important to prevent precancerous mutations from occurring in the nondividing cells of multicellular organisms. However, when a double-strand break occurs in cells that have stopped dividing, error-free repair is not possible because neither of the two usual sources of undamaged DNA is available as a template for new DNA synthesis. That is, complementarity cannot be exploited because both strands of the DNA helix are damaged and, in the absence of replication, there is no sister chromatid. However, as was the case for the error-prone translesion synthesis (including the SOS system in *E. coli*), the consequences of imperfect repair may be less harmful to the cell than leaving the lesion unrepaired. In this case, it is better to put the free ends back together so they cannot initiate chromosomal rearrangements, even if this means that some sequence may be lost. Putting the ends back together is accomplished by a mechanism called *nonhomologous end-joining*, which involves the three steps shown in Figure 14-32. These steps include the binding of the broken ends by 3 proteins (KU70, KU80, and a large DNA-dependent protein kinase) followed by the trimming of the ends so that they can be ligated together. In mammals, several of the proteins in this pathway also participate in the end-joining reactions associated with the programmed rearrangements of antibody genes.

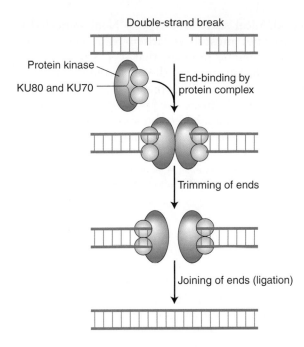

Figure 14-32 Mechanism of nonhomologous end-joining of double-strand breaks. This is an error-prone mechanism.

HOMOLOGOUS RECOMBINATION The mechanism of homologous recombination utilizes the sister chromatid to repair double-strand breaks. For this reason, repair is usually error-free. The mechanism of homologous recombination is shown in Figure 14-33. Key steps are the binding of the broken ends by specialized proteins and enzymes, the trimming of the 5′ ends to expose single-stranded regions, and the coating of these regions with proteins that include the RecA homolog, RAD51. Recall that during the SOS response, RecA monomers associate with regions of single-stranded DNA to form long helical filaments. Similarly, RAD51 forms long filaments as it associates with the exposed single-stranded region. The RAD51–DNA filament then takes part in a remarkable search of the undamaged sister chromatid for the complementary sequence that will be used as a template for DNA synthesis. Once the complementary region is found, a *joint molecule* forms between homologous damaged and undamaged duplex DNAs. The sequences missing from the damaged strand are then copied from the complementary sister chromatid.

> **MESSAGE** Double-strand breaks are extremely dangerous because they can lead to chromosome rearrangements that result in cell death or aberrant growth and development. Nondividing cells efficiently join the ends of double-strand breaks with an error-prone process called *nonhomologous end-joining*. Dividing cells utilize homologous recombination whereby the free ends invade the homologous region of the sister chromatid to initiate DNA synthesis and error-free repair.

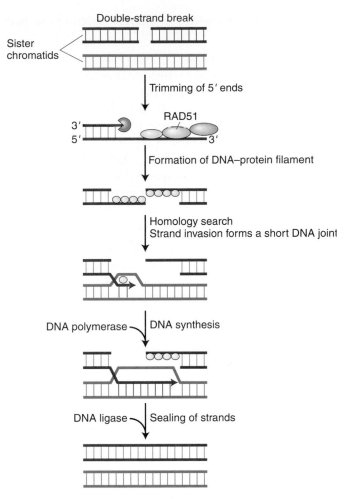

Figure 14-33 Repair of a double-strand break by homologous recombination. A double-strand break induces an enzyme to chew back 5′ ends, leaving 3′ overhangs that are coated with proteins, including RAD51, a RecA homolog. A segment of the sister chromatid (blue) is used as a template to repair the break. [From D. C. van Gent, J. H. J. Hoeijmakers, and R. Kandar, *Nature Reviews: Genetics* 2, 2001, 196–206.]

14.4 The mechanism of meiotic crossing-over

Our discussion of double-strand breaks leads naturally into the topic of crossing-over at meiosis. This is because, according to the current molecular model of crossing-over, a double-strand break initiates the crossover event. The molecular details of the crossover process will seem very familiar from the above discussion of double-strand break repair.

Crossing-over is a remarkably precise process. It takes place between two homologous nonsister chromatids. Some kind of cellular machinery takes these two huge molecular assemblages, breaks them at the same relative position, and then rejoins them in a new arrangement so that no genetic material is lost or gained in either.

The molecular mechanism is thought to comprise two key steps:

1. A double-strand break. One key piece of evidence here was that in yeast transformation the incorporation of a circular plasmid into the yeast genome is stimulated 1000-fold when the plasmid is cut to become linear. Broken DNA ends seem to be *recombinogenic*; that is, they promote recombination.

2. The formation of **heteroduplex DNA.** This is a hybrid type of DNA molecule that is composed of a single DNA strand from a chromatid derived from one parent, and a single strand from a chromatid derived from the other parent.

The first evidence for heteroduplex DNA was also provided by genetics, specifically ascus analysis. Octads are particularly informative in pointing to the existence of heteroduplexes in crossing-over. We saw in Chapter 2 that in fungi a cross $A \times a$ will create a monohybrid meiocyte A/a that is expected to segregate in a $1:1$ ratio in the meiotic products according to the law of equal segregation. Indeed the $1:1$ ratio is found in most fungal meiocytes: we see 4 A and 4 a. However, in rare meiocytes (generally on the order of 0.1 to 1 percent) any one of four types of aberrant ratios can be found, and these give the clues needed to build a heteroduplex crossover model. The aberrant ratios are as follows:

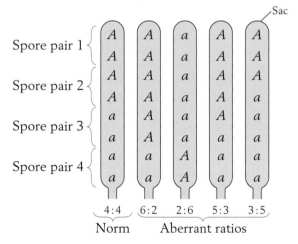

Spore pair 1	A	A	a	A	A
	A	A	a	A	A
Spore pair 2	A	A	a	A	A
	A	A	a	A	a
Spore pair 3	a	A	a	A	a
	a	A	a	a	a
Spore pair 4	a	a	A	a	a
	a	a	A	a	a
	4:4	6:2	2:6	5:3	3:5
	Norm		Aberrant ratios		

These asci all have more than 4 copies of one genotype; this result is unexpected based on Mendel's first law of equal segregation. The one or two "extra" cases are said to have undergone **gene conversion** from wild type to mutant or mutant to wild type. All the aberrant ratios need to be explained, but we will concentrate on two, the $5:3$ and the $6:2$, because their explanation embodies the same elements as all.

The $5:3$ ratio is particularly interesting because in this octad there is a pair of *nonidentical sister spores*. (Recall that the postmeiotic round of mitosis is expected to

produce identical sister genotypes.) Nonidentical sister spore genotypes must have arisen from heteroduplex DNA in the meiotic product, that is, DNA with a segment in which one strand is the nucleotide sequence of the A allele and one strand is the nucleotide sequence of the a allele. After mitosis the two sister cells resulting from division of such a heteroduplex-containing nucleus will be different, one with A and one a.

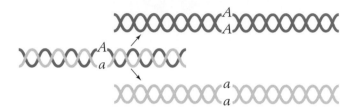

Let's assume that A and a differ by a single base pair; that base pair is G · C in A and A · T in a. Let's further assume that for any particular gene, rarely a single heteroduplex is formed at meiosis such that the A · T in one of the products becomes G · T (we will see the mechanism soon). Then we can represent the products of meiosis as

1. G · C   2. G · C   3. **G · T**   4. A · T

After the postmeiotic mitosis, the resulting octad will be

1. G · C   2. G · C   3. G · C   4. G · C
5. **G · C**   6. A · T   7. A · T   8. A · T

which is the observed $5:3$ ratio.

However, from what we have learned in this chapter, we know that the G · T of the heteroduplex is a candidate for mismatch repair. Let's assume that such a repair system excises the T of the GT heteroduplex and inserts C, so we get GC. In this ascus there will be a $3:1$ ratio of GC:AT and the octad will show a $6:2$ ratio.

In meioses that produce aberrant ratios, it was observed that there is a crossover between flanking genes at much higher frequencies than expected. Hence it seemed likely that heteroduplex formation might be part of the normal crossover process. Furthermore, presumably by chance the heterozygous gene under study rarely happened to be in the middle of the molecular events of a crossover, and it was these molecular events that led to the gene conversion. Putting all these ideas together led to the double-strand break model, one of several heteroduplex models of crossing-over.

The model is shown in Figure 14-34. In one a chromatid a double-strand break occurs, and erosion of the ends results in short regions of single-stranded DNA. The 3' end of one of these strands "invades" an A chromatid. The invader primes synthesis of its missing bases, using the antiparallel strand of the A chromatid as a template. This new synthesis displaces a single-stranded

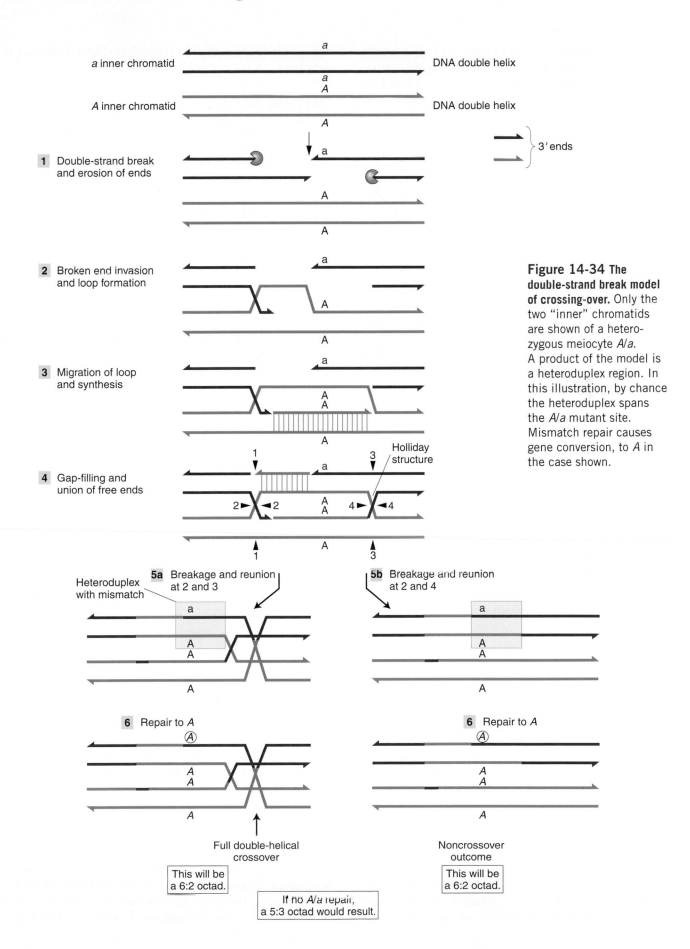

Figure 14-34 The double-strand break model of crossing-over. Only the two "inner" chromatids are shown of a heterozygous meiocyte *A/a*. A product of the model is a heteroduplex region. In this illustration, by chance the heteroduplex spans the *A/a* mutant site. Mismatch repair causes gene conversion, to *A* in the case shown.

loop, which hybridizes to the noninvading *a* single strand, thus forming a small "*Aa*" heteroduplex region, and serves as a template to restore the missing bases on that strand. Filling in of gaps by polymerase activity and joining of DNA ends by ligase result in a peculiar structure that looks like two single-stranded crossovers. Note that this structure also contains the single heteroduplex that we need.

The single-stranded "crossovers" are called *Holliday structures*, after Robin Holliday, who first proposed them in the 1960s. Physical (as opposed to genetic) evidence for Holliday structures has been obtained independently. They are unstable and must be resolved in one of two ways. Simply put, they can each be resolved by either "vertical" or "horizontal" breakage and reunion of single strands, as shown in the figure. One resolution results in a double-stranded reciprocal crossover (shown at left) and the other a noncrossover (right). Hence the association with crossovers is explained. Note that if the heteroduplex formed via a double-strand break on the other chromatid, we could explain the 3 : 5 and 2 : 6 ratios, too.

Overall, note again the use of molecular mechanisms in crossing-over that seem "stolen" from the mutation and repair processes. Although these enzymes undoubt-

edly evolved for mutation repair, they have been put to new uses.

> **MESSAGE** A double-strand break on one chromatid leads to two single-stranded Holliday junctions surrounding a heteroduplex region. This overall molecular structure can be resolved into a single reciprocal double-stranded crossover. Repair or nonrepair of the heteroduplex results in a 6 : 2 or a 5 : 3 ratio.

Recombination between alleles of a gene

In a cross between two mutant alleles, say, $a' \times a''$, the heteroduplex might span one mutant site, or both mutant sites. Such a heteroduplex would be as follows:

$$----a^+---- + ----$$
$$---- + ----a^-----$$

Mismatch repair of at least one site to the wild-type sequence would result in a wholly wild-type strand. These intragenic recombinants can be detected in several ways. In crosses between auxotrophic microorganisms, they would be prototrophs, and thus they can be detected most simply by plating on minimal medium.

KEY QUESTIONS REVISITED

- **What is the molecular nature of mutations?**

A large proportion of mutations are changes to, additions to, or losses of one or a small number of bases in the gene sequence. Such changes may cause a change of amino acid (missense mutations) or produce a new stop codon (nonsense mutations). Additions or losses can cause frameshift mutations.

- **How do certain types of radiation and chemicals cause mutation?**

By acting on and chemically changing the DNA. Common changes are base replacement, base alteration, or base damage.

- **Are induced mutations different from spontaneous mutations?**

Spontaneous mutations are largely the product of errors made by cellular enzymes. A wide spectrum of changes can result. Some mutagens can also produce a wide spectrum of changes, but many produce a preponderance of a certain type such as G · C-to-A · T transition substitutions.

- **Can a cell repair mutations?**

Yes. Some repair systems are very efficient at restoring the original sequence. Others are error-prone: they convert the original chemical event into a permanent mutation.

- **What is the molecular mechanism of crossing-over?**

In the current best model, a double-stranded break in one chromatid initiates formation of a pair of single-stranded Holliday junctions that can be resolved to form one standard double-helical crossover. Heteroduplex DNA is formed in the process, and if it happens to span a heterozygous gene, aberrant ascus ratios can result.

- **Do mutational repair systems participate in crossing-over?**

Yes, certain steps in the double-strand model (such as double-strand breaks, exonuclease activity, mismatch repair, polymerase activity, ligase activity) are very similar to several types of mutational repair. Common enzymes are also involved.

SUMMARY

DNA change within a gene (point mutation) generally involves one or a few base pairs. Single base-pair substitutions can create missense codons or nonsense (tran-

scription termination) codons. A purine replaced by the other purine (or a pyrimidine replaced by the other pyrimidine) is called a *transition*. A purine replaced by a pyrimidine (or vice versa) is called a *transversion*. Single-base-pair additions or deletions (indels) produce frameshift mutations. Certain human genes that contain trinucleotide repeats—especially those that are expressed in neural tissue—become mutated through the expansion of these repeats and can thus cause disease. The formation of monoamino acid repeats within the polypeptides encoded by these genes is responsible for the mutant phenotypes.

Mutations can be either spontaneous or induced by mutagenic radiation or chemicals. Spontaneous changes are generally a range of types. Mutagens often result in a specific type of change because of their chemical specificity. For example, some produce exclusively $G \cdot C \rightarrow A \cdot T$ transitions; others, exclusively frameshifts.

Spontaneous base replacement can be from a type of chemical isomerization called *tautomeric shift*. Some chemicals exacerbate this type of change. Some mutagens alter the structure of a base, leading to new hydrogen-bonding properties. Some agents (such as UV) cause large-scale damage to the base or base removal.

Cellular enzymes participate in mutation in several ways. Errors of enzymes can result in altered DNA. Replication is often needed to fix the new base stably in the DNA. Several enzymes specialize in repair. Some of these provide accurate reversals; others are error-prone and result in mutations. Repair is by direct reversal of damage, excision and resynthesis using existing templates, postreplication repair, transcription-coupled repair, end-joining, or repair by homologous recombination.

The molecular mechanism of crossing-over is thought to involve repairlike processes. In the current best model, a double-stranded break in one chromatid initiates formation of a pair of single-stranded Holliday junctions that can be resolved to result in a one standard double-helical crossover. Heteroduplex DNA is formed in the process, and if this happens to span a heterozygous gene, aberrant ascus ratios can result. Certain steps in the double-strand break model (such as double-strand breaks, exonuclease activity, mismatch repair, polymerase activity, ligase activity) are very similar to those in several types of mutational repair. Common enzymes are involved, and mutants observed for one system often affect the other.

KEY TERMS

acridine orange (p. 458)

adenine methylase (p. 471)

aflatoxin B_1 (AFB$_1$) (p. 461)

apurinic site (p. 461)

base analogs (p. 457)

base excision repair (p. 469)

conservative substitution (p. 456)

deamination (p. 464)

depurination (p. 463)

DNA glycosylases (p. 469)

double-strand break (p. 472)

enol (p. 457)

fluctuation test (p. 462)

frameshift mutations (p. 456)

gene conversion (p. 474)

heteroduplex DNA (p. 474)

homology-dependent repair systems (p. 469)

hot spots (p. 457)

ICR compounds (p. 458)

imino (p. 457)

indel mutations (p. 465)

intercalating agents (p. 458)

keto (p. 457)

mismatch-repair system (p. 471)

missense mutations (p. 455)

mutagenesis (p. 453)

mutational specificity (p. 457)

nonconservative substitution (p. 456)

nonsense mutations (p. 455)

nucleotide excision-repair system (p. 470)

oxidatively damaged bases (p. 464)

proflavin (p. 458)

replica plating (p. 462)

SOS system (p. 459)

spontaneous lesions (p. 463)

synonymous mutations (p. 455)

tautomeric shift (p. 457)

tautomers (p. 457)

transition (p. 454)

transversion (p. 454)

trinucleotide repeat (p. 466)

SOLVED PROBLEMS

1. In Chapter 9, we learned that UAG and UAA codons are two of the chain-terminating nonsense triplets. On the basis of the specificity of aflatoxin B_1 and ethylmethanesulfonate (EMS), describe whether each mutagen would be able to revert these codons to wild type.

Solution

EMS induces primarily $G \cdot C \rightarrow A \cdot T$ transitions. UAG codons could not be reverted to wild type, because only the UAG $\rightarrow$ UAA change would be stimulated by EMS and that generates a nonsense (ochre) codon. UAA codons would not be acted on by EMS. Aflatoxin B_1 induces primarily G · C → T · A transversions. Only the third position of UAG codons would be acted on, resulting in a UAG → UAU change (on the mRNA level), which produces tyrosine. Therefore, if tyrosine were an acceptable amino acid at the corresponding site in the protein, aflatoxin B₁ could revert UAG codons. Aflatoxin B₁ would not revert UAA codons, because no G · C base pairs appear at the corresponding position in the DNA.

2. Explain why mutations induced by acridines in phage T4 or by ICR-191 in bacteria cannot be reverted by 5-bromouracil.

Solution

Acridines and ICR-191 induce mutations by deleting or adding one or more base pairs, which results in a frameshift. However, 5-bromouracil induces mutations

by causing the substitution of one base for another. This substitution cannot compensate for the frameshift resulting from ICR-191 and acridines.

3. A mutant of *E. coli* is highly resistant to mutagenesis by a variety of agents, including ultraviolet light, aflatoxin B₁, and benzo(*a*)pyrene. Explain one possible cause of this mutant phenotype.

Solution

The mutant might lack the SOS system and perhaps carries a defect in the *UmuC* gene. Such strains would not be able to bypass replication-blocking lesions of the type caused by the three mutagens listed. Without the processing of premutational lesions, mutations would not be recovered in viable cells.

PROBLEMS

BASIC PROBLEMS

1. Consider the wild-type and mutant sequences below:

   Wild ....CTTGCAAGCGAATC....
   Mutant ....CTTGC**TAG**CGAATC....

   The substitution shown *seems* to have created a stop codon. What further information do you need to be confident that this is so?

2. What type of mutation is the following (shown as mRNA)?

   Wild type .....5′AAUCCUUACGGA 3′.....
   Mutant .....5′AAUCCUACGGA 3′.......

3. Can a missense mutation of proline to histidine be made with a G · C → A · T transition-causing mutagen? What about a proline-to-serine missense mutation?

4. By base-pair substitution, what are all the synonymous changes that can be made starting with the codon CGG?

5. a. What are all the transversions that can be made starting with the codon CGG?

   b. Which of these will be missense? Can you be sure?

6. Which tautomer of thymine can form the most hydrogen bonds in pairing with other DNA bases?

7. a. If the enol form of thymine is inserted on a single-stranded template during replication, what base-pair substitution will result?

   b. If thymine enolizes while acting as a template during replication, what base-pair substitution will result?

8. a. Acridine orange is an effective mutagen for producing null alleles by mutation. Why do you think this is so?

   b. A certain acridinelike compound generates only single insertions. A mutation induced with this compound is treated with the same compound, and some revertants are produced. How is this possible?

9. A newly discovered SOS bypass system is found to preferentially insert thymine opposite apurinic sites. What type of mutations should be preferentially produced?

10. Draw diagrams that contrast replication slippage and asymmetrical crossovers as possible causes of multiple tandem repeats.

11. In a project in which she is trying to induce mutations using UV radiation, a student notices that on bright sunny days far fewer mutations are obtained. Suggest an explanation.

12. A mutational lesion results in a sequence containing a mismatched base pair:

    5′AGCTGCCTT 3′
    3′ACGATGGAA 5′
    Codon

    If mismatch repair occurs in either direction, which amino acids could be found at this site?

13. What aspect of the double-strand break model is responsible for the finding that gene conversion is often accompanied by a crossover?

14. Normally, 6:2 aberrant asci are more frequent than 5:3 aberrant asci. What might be the explanation for this on the basis of the double-strand break model?

15. In the double-strand break model, list all the stages at which exo- and endonucleases act.

16. Differentiate between the elements of the following pairs:
    a. Transitions and transversions
    b. Synonymous and neutral mutations
    c. Missense and nonsense mutations
    d. Frameshift and nonsense mutations

17. Why are frameshift mutations more likely than missense mutations to result in proteins that lack normal function?

18. Diagram two different mechanisms for deletion formation. What type of information provided from DNA sequencing can distinguish between these possibilities?

19. Describe two spontaneous lesions that can lead to mutations.

20. Compare the mechanism of action of 5-bromouracil (5-BU) with ethylmethanesulfonate (EMS) in causing mutations. Explain the specificity of mutagenesis for each agent in light of the proposed mechanism.

21. Compare the two different systems required for the repair of AP sites and the removal of photodimers.

22. In adult cells that have stopped dividing, what types of repair systems are possible?

23. A certain compound that is an analog of the base cytosine can become incorporated into DNA. It normally hydrogen-bonds just as cytosine does, but it quite often isomerizes to a form that hydrogen-bonds as thymine does. Do you expect this compound to be mutagenic, and if so, what types of changes might it induce at the DNA level?

24. Describe the repair systems that operate after depurination and deamination.

25. Describe the model for indel formation. Show how this model can explain mutational hot spots in the *lacI* gene of *E. coli*.

CHALLENGING PROBLEMS

26. a. Why is it impossible to induce nonsense mutations (represented at the mRNA level by the triplets UAG, UAA, and UGA) by treating wild-type strains with mutagens that cause only $A \cdot T \rightarrow G \cdot C$ transitions in DNA?

    b. Hydroxylamine (HA) causes only $G \cdot C \rightarrow A \cdot T$ transitions in DNA. Will HA produce nonsense mutations in wild-type strains?

    c. Will HA treatment revert nonsense mutations?

27. Several auxotrophic point mutants in *Neurospora* are treated with various agents to see if reversion will take place. The following results were obtained (a plus sign indicates reversion; HA causes only $G \cdot C \rightarrow A \cdot T$ transitions).

Mutant	5-BU	HA	Proflavin	Spontaneous reversion
1	−	−	−	−
2	−	−	+	+
3	+	−	−	+
4	−	−	−	+
5	+	+	−	+

a. For each of the five mutants, describe the nature of the original mutation event (not the reversion) at the molecular level. Be as specific as possible.

b. For each of the five mutants, name a possible mutagen that could have caused the original mutation event. (Spontaneous mutation is not an acceptable answer.)

c. In the reversion experiment for mutant 5, a particularly interesting prototrophic derivative is obtained. When this type is crossed with a standard wild-type strain, the progeny consist of 90 percent prototrophs and 10 percent auxotrophs. Give a full explanation for these results, including a precise reason for the frequencies observed.

28. You are using nitrosoguanidine to "revert" mutant *nic-2* (nicotinamide-requiring) alleles in *Neurospora*. You treat cells, plate them on a medium without nicotinamide, and look for prototrophic colonies. You obtain the following results for two mutant alleles. Explain these results at the molecular level, and indicate how you would test your hypotheses.

a. With *nic-2* allele 1, you obtain no prototrophs at all.

b. With *nic-2* allele 2, you obtain three prototrophic colonies A, B, and C, and you cross each separately with a wild-type strain. From the cross prototroph A × wild type, you obtain 100 progeny, all of which are prototrophic. From the cross prototroph B × wild type, you obtain 100 progeny, of which 78 are prototrophic and 22 are nicotinamide-requiring. From the cross prototroph C × wild type, you obtain 1000 progeny, of which 996 are prototrophic and 4 are nicotinamide-requiring.

29. Fill in the following table, using a plus sign (+) to indicate that the mutagenic lesion (base damage) induces the indicated base change and a minus sign (−) to show that it does not.

Base change	O-6-Methyl G	8-Oxo dG	C–C photodimer
$A \cdot T$ to $G \cdot T$			
$G \cdot C$ to $T \cdot A$			
$G \cdot C$ to $A \cdot T$			

30. You are working with a newly discovered mutagen, and you wish to determine the base change that it introduces into DNA. Thus far you have determined that the mutagen chemically alters a single base in such a way that its base-pairing properties are altered permanently. In order to determine the specificity of the alteration, you examine the amino acid changes that take place after mutagenesis. A sample of what you find is shown below:

Original: Gln-His-Ile-Glu-Lys
Mutant: Gln-His-Met-Glu-Lys

Original: Ala-Val-Asn-Arg
Mutant: Ala-Val-Ser-Arg

Original: Arg-Ser-Leu
Mutant: Arg-Ser-Leu-Trp-Lys-Thr-Phe

What is the base-change specificity of the mutagen?

31. You now find an additional mutant from the experiment in Problem 30:

Original: Ile-Leu-His-Gln
Mutant: Ile-Pro-His-Gln

Could the base-change specificity in your answer to Problem 30 account for this mutation? Why or why not?

32. Strain A of *Neurospora* contains an *ad-3* mutation that reverts spontaneously at a rate of 10^{26}. Strain A is crossed with a newly acquired wild-type isolate, and *ad-3* strains are recovered from the progeny. When 28 different *ad-3* progeny strains are examined, 13 lines are found to revert at the rate of 10^{26}, but the remaining 15 lines revert at the rate of 10^{23}. Formulate a hypothesis to account for these findings, and outline an experimental program to test your hypothesis.

33. For each lesion in parts a–g, indicate which of the following repair systems repairs that lesion:

(1) alkyltransferase

(2) endonuclease

(3) photolyase

(4) MutY glycosylase

(5) MutM glycosylase

(6) uracil DNA glycosylase

(7) general nucleotide excision repair

(8) methyl-directed mismatch repair

a. Deamination of cytosine

b. 8-Oxo dG

c. Aflatoxin B_1 adduct

d. G · T mispair as replication error

e. 5′-CC-3′ dimer

f. AP site

g. O^6-methylguanine

34. Which of the following linear asci shows gene conversion at the *arg-2* locus?

1	2	3	4	5	6
+	+	+	+	+	+
+	+	+	+	+	+
+	arg	+	+	arg	arg
+	arg	arg	arg	arg	arg
arg	arg	arg	+	+	arg
arg	arg	arg	arg	+	arg
arg	+	arg	arg	arg	arg
arg	+	arg	arg	arg	arg

35. Assume you have made a cross in *Neurospora* using a mutant that has three mutant sites in the same gene, called *1*, *2* and *3*, which are spaced evenly through the 2-kb gene:

1 2 3 × *+ + +*

Explain a likely origin of the following two asci:

1 2 3	*1 2 3*
1 2 3	*1 2 3*
+ + +	*1 2 +*
+ + +	*1 2 +*
+ + +	*+ + +*
+ + +	*+ + +*
+ + +	*+ + +*
+ + +	*+ + +*

36. The double-strand break model illustrated in this chapter generated one heteroduplex. Other models generate two identical heteroduplexes in the same meiosis, so that the chromatids are parent 1, heteroduplex, heteroduplex, parent 2. What octad patterns would be produced from the various combinations of repair and nonrepair of these two heteroduplex mismatches?

15

LARGE-SCALE CHROMOSOMAL CHANGES

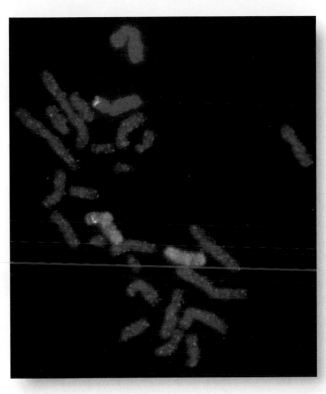

A reciprocal translocation demonstrated by chromosome painting. A suspension of chromosomes from many cells is passed through an electronic device that sorts them by size. DNA is extracted from individual chromosomes, denatured, joined with one of several fluorescent dyes, and then added to partially denatured chromosomes on a slide. The fluorescent DNA "finds" its own chromosome and binds along its length by base complementarity, thus "painting" it. In this preparation, a bright blue and a pink dye have been used to paint different chromosomes. The preparation shows one normal pink chromosome, one normal light blue, and two that have exchanged their tips. [Lawrence Berkeley Laboratory.]

KEY QUESTIONS

- How common are polyploids (organisms with multiple chromosome sets)?

- How do polyploids arise?

- Do polyploids have any special properties?

- Is the polyploid state transmissible to offspring?

- What inheritance patterns are observed in the progeny of polyploids?

- How do aneuploids (variants in which a single chromosome has been gained or lost) arise?

- Do aneuploids show any special properties?

- What inheritance patterns are produced by aneuploids?

- How do large-scale chromosome rearrangements (deletions, duplications, inversions, and translocations) arise?

- Do these rearrangements have any special properties?

- What inheritance patterns are produced by rearrangements?

OUTLINE

15.1 Changes in chromosome number

15.2 Changes in chromosome structure

15.3 Overall incidence of human chromosome mutations

CHAPTER OVERVIEW

A young couple is planning to have children. The husband knows that his grandmother had a child with Down syndrome by a second marriage. Down syndrome is a set of physical and mental disorders caused by the presence of an extra chromosome 21 (Figure 15-1). No records of the birth, which occurred early in the twentieth century, are available, but the couple knows of no other cases of Down syndrome in their families.

The couple has heard that Down syndrome results from a rare chance mistake in egg production and therefore decide that they stand only a low chance of having such a child. They decide to have children. Their first child is unaffected, but the next conception aborts spontaneously (a miscarriage), and their second child is born with Down syndrome. Was this a coincidence, or is it possible that there is a connection between the genetic makeup of the man and that of his grandmother that led to their both having Down syndrome children? Was the spontaneous abortion significant? What tests might be necessary to investigate this situation? The analysis of such questions is the topic of this chapter.

We saw in Chapter 10 that gene mutations are one source of genomic change. However, the genome can also be remodeled on a larger scale by alterations to chromosome structure or by changes in the number of copies of chromosomes in a cell. These large-scale variations are termed **chromosome mutations** to distinguish them from gene mutations. Broadly speaking, gene mutations are defined as changes that take place within a gene, whereas

Figure 15-1 Children with Down syndrome.
[Bob Daemmrich/The Image Works.]

chromosome mutations are changes to a chromosome region encompassing multiple genes. Chromosome mutations can be detected by microscopic examination, genetic analysis, or both. In contrast, gene mutations are never detectable microscopically; a chromosome bearing a gene mutation looks the same under the microscope as one carrying the wild-type allele. Chromosome mutations

CHAPTER OVERVIEW Figure

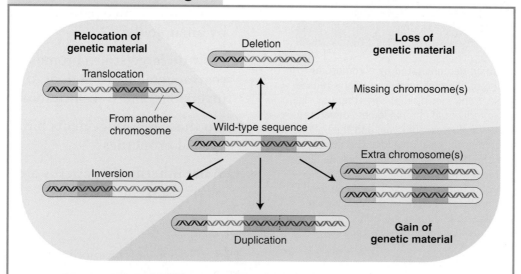

Figure 15-2 Overview of chromosome mutations. The figure has been divided into three colored regions to depict the main types of chromosome mutations that can occur. These involve the loss, gain, or relocation of entire chromosomes or chromosome segments. The wild-type chromosome is shown in the center.

have been best characterized in eukaryotes, and all the examples in this chapter are from that group.

Chromosome mutations are important from several biological perspectives. First, they can be sources of insight into how genes act in concert on a genomic scale. Second, they reveal several important features of meiosis and chromosome architecture. Third, they constitute useful tools for experimental genomic manipulation. Fourth, they are sources of insight into evolutionary processes.

Many chromosome mutations cause abnormalities in cell and organismal function. Most of these abnormalities stem from changes in *gene number* or *gene position*. In some cases, a chromosome mutation results from chromosome breakage. If the break occurs within a gene, the result is functional *disruption* of that gene.

For our purposes, we shall divide chromosome mutations into two groups: changes in chromosome *number* and changes in chromosome *structure*. These two groups represent two fundamentally different kinds of events. Changes in chromosome number are not associated with structural alterations of any of the DNA molecules of the cell. Rather, it is the *number* of these DNA molecules that is changed, and this change in number is the basis of their genetic effects. Changes in chromosome structure, on the other hand, result in novel sequence arrangements within one or more DNA double helices. These two types of chromosome mutations are illustrated in Figure 15-2, which is a summary of the topics of this chapter. We begin by exploring the nature and consequences of changes in chromosome number.

15.1 Changes in chromosome number

In genetics as a whole, few topics impinge on human affairs quite so directly as that of changes in the number of chromosomes present in our cells. Foremost is the fact that a group of common genetic disorders results from the presence of an abnormal number of chromosomes. Although this group of disorders is small, it accounts for a large proportion of the genetically determined health problems that afflict humans. Also of relevance to humans is the role of chromosome mutations in plant breeding: plant breeders have routinely manipulated chromosome number to improve commercially important agricultural crops.

Changes in chromosome number are of two basic types: changes in *whole* chromosome sets, resulting in a condition called *aberrant euploidy*, and changes in *parts* of chromosome sets, resulting in a condition called *aneuploidy*.

Aberrant euploidy

Organisms with multiples of the basic chromosome set (genome) are referred to as **euploid**. We learned in earlier chapters that familiar eukaryotes such as plants, animals, and fungi carry in their cells either one chromosome set (haploidy) or two chromosome sets (diploidy). In these species, the haploid and diploid states are both cases of normal euploidy. Organisms that have more or fewer than the normal number of sets are aberrant euploids. **Polyploids** are individual organisms that have more than two chromosome sets. They can be represented by $3n$ (**triploid**), $4n$ (**tetraploid**), $5n$ (**pentaploid**), $6n$ (**hexaploid**), and so forth. (The number of chromosome sets is called the ploidy or ploidy level.) An individual of a normally diploid species that has only one chromosome set (n) is called a **monoploid** to distinguish it from an individual of a normally haploid species (also n). Examples of these conditions are shown in the first four rows of Table 15-1.

TABLE 15-1 Chromosome Constitutions in a Normally Diploid Organism with Three Chromosomes (Labeled A, B, and C) in the Basic Set

Name	Designation	Constitution	Number of chromosomes
Euploids			
Monoploid	n	A B C	3
Diploid	$2n$	AA BB CC	6
Triploid	$3n$	AAA BBB CCC	9
Tetraploid	$4n$	AAAA BBBB CCCC	12
Aneuploids			
Monosomic	$2n - 1$	A BB CC	5
		AA B CC	5
		AA BB C	5
Trisomic	$2n + 1$	AAA BB CC	7
		AA BBB CC	7
		AA BB CCC	7

MONOPLOIDS Male bees, wasps, and ants are monoploid. In the normal life cycles of these insects, males develop by **parthenogenesis** (the development of a specialized type of unfertilized egg into an embryo without the need for fertilization). In most other species, however, monoploid zygotes fail to develop. The reason is that virtually all individuals in a diploid species carry a number of deleterious recessive mutations, together called a **"genetic load."** The deleterious recessive alleles are masked by wild-type alleles in the diploid condition, but are automatically expressed in a monoploid derived from a diploid. Monoploids that do develop to advanced stages are abnormal. If they survive to adulthood, their germ cells cannot proceed through meiosis normally because the chromosomes have no pairing partners. Thus, monoploids are characteristically sterile. (Male bees, wasps, and ants bypass meiosis; in these groups, gametes are produced by *mitosis*.)

POLYPLOIDS Polyploidy is very common in plants but rarer in animals (for reasons that we will consider later). Indeed, an increase in the number of chromosome sets has

been an important factor in the origin of new plant species. The evidence for this is shown in Figure 15-3, which displays the frequency distribution of haploid chromosome numbers in dicotyledonous plant species. Above a haploid number of about 12, even numbers are much more common than odd numbers. This pattern is a consequence of the polyploid origin of many plant species, because doubling and redoubling of a number can give rise only to even numbers. Animal species do not show such a distribution, owing to the relative rareness of polyploid animals.

In aberrant euploids, there is often a correlation between the number of copies of the chromosome set and the size of the organism. A tetraploid organism, for example, typically looks very similar to its diploid counterpart in its proportions, except that the tetraploid is bigger, both as a whole and in its component parts. The higher the ploidy level, the larger the size of the organism (Figure 15-4).

> **MESSAGE** Polyploid plants are often larger and have larger component parts than their diploid relatives.

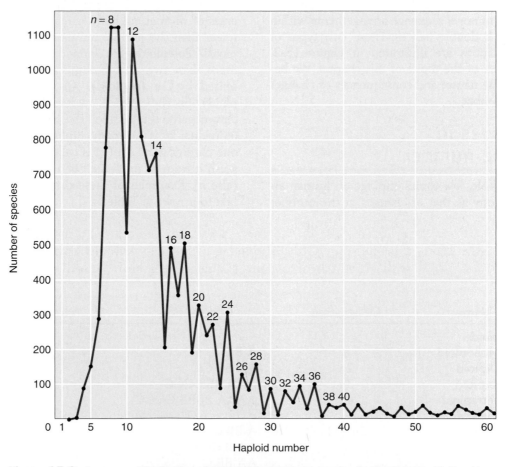

Figure 15-3 Frequency distribution of haploid chromosome number in dicot plants. Notice the excess of even-numbered values in the higher ranges, suggesting ancestral polyploidization.
[Adapted from Verne Grant, *The Origin of Adaptations*. Columbia University Press, 1963.]

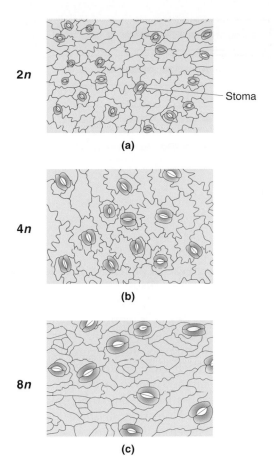

Figure 15-4 Epidermal leaf cells of tobacco plants with increasing ploidy. Cell size increases, particularly evident in stoma size, with an increase in ploidy. (a) Diploid; (b) tetraploid; (c) octoploid. [From W. Williams, *Genetic Principles and Plant Breeding*. Blackwell Scientific Publications, Ltd.]

In the realm of polyploids, we must distinguish between **autopolyploids,** which have multiple chromosome sets originating from within one species, and **allopolyploids,** which have sets from two or more different species. Allopolyploids form only between closely related species; however, the different chromosome sets are only **homeologous** (partially homologous), not fully homologous as they are in autopolyploids.

Autopolyploids Triploids are usually autopolyploids. They arise spontaneously in nature, but they can be constructed by geneticists from the cross of a 4n (tetraploid) and a 2n (diploid). The 2n and the n gametes produced by the tetraploid and the diploid, respectively, unite to form a 3n triploid. Triploids are characteristically sterile. The problem, as in monoploids, lies in pairing at meiosis. The molecular mechanisms for synapsis, or true pairing, dictate that pairing can take place between only two of the three chromosomes of each type (Figure 15-5). Paired homologs **(bivalents)** segregate to opposite poles, but the unpaired homologs **(univalents)** pass to either pole randomly. In the case of a **trivalent,** a paired group of three, the paired centromeres segregate as a bivalent and the unpaired one as a univalent. These segregations take place for every chromosome threesome, so for any chromosomal type, the gamete could receive either one or two chromosomes. It is unlikely that a gamete will receive *two* for *every* chromosomal type, or that it will receive *one* for *every* chromosomal type. Hence the likelihood is that gametes will have chromosome numbers intermediate between the haploid and diploid number; such genomes are of a type called **aneuploid** ("not euploid").

Aneuploid gametes generally do not give rise to viable offspring. In plants, aneuploid pollen grains are generally inviable and hence unable to fertilize the female gamete. In any organism zygotes that might arise from the fusion of a haploid and an aneuploid gamete will themselves be aneuploid, and typically these zygotes also are inviable. We will examine the underlying reason for the inviability of aneuploids when we consider gene balance later in the chapter.

> **MESSAGE** Polyploids with odd numbers of chromosome sets, such as triploids, are sterile or highly infertile because their gametes and offspring are aneuploid.

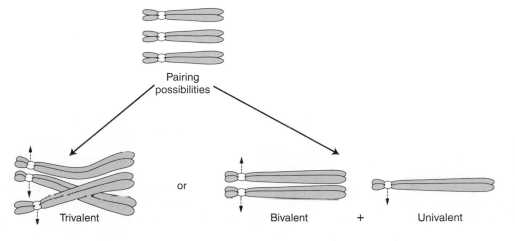

Figure 15-5 Pairing of three homologous chromosomes. The three homologous chromosomes of a triploid may pair in two ways at meiosis, as a trivalent or as a bivalent plus a univalent.

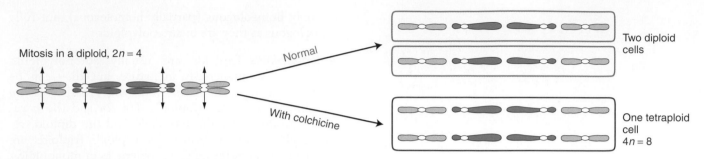

Mitosis in a diploid, 2n = 4

Normal

With colchicine

Two diploid cells

One tetraploid cell
4n = 8

Figure 15-6 The use of colchicine to generate a tetraploid from a diploid. Colchicine added to mitotic cells during metaphase and anaphase disrupts spindle-fiber formation, preventing the migration of chromatids after the centromere is split. A single cell is created that contains pairs of identical chromosomes that are homozygous at all loci.

Autotetraploids arise by the doubling of a 2n complement to 4n. This doubling can occur spontaneously, but it can also be induced artificially by applying chemical agents that disrupt microtubule polymerization. We saw in Chapter 3 that chromosome segregation is powered by spindle fibers, which are polymers of the protein tubulin. Hence disruption of microtubule polymerization blocks chromosome segregation. The chemical treatment is normally applied to somatic tissue during the formation of spindle fibers in cells undergoing division. The resulting polyploid tissue (such as a polyploid branch of a plant) can be detected by examining stained chromosomes from the tissue under a microscope. Such a branch can be removed and used as a cutting to generate a polyploid plant or allowed to produce flowers, which when selfed would produce polyploid offspring. A commonly used antitubulin agent is colchicine, an alkaloid extracted from the autumn crocus. In colchicine-treated cells, the S phase of the cell cycle occurs, but

chromosome segregation or cell division does not. As the treated cell enters telophase, a nuclear membrane forms around the entire doubled set of chromosomes. Thus, treating diploid (2n) cells with colchicine for one cell cycle leads to tetraploids (4n) with exactly four copies of each type of chromosome (Figure 15-6). Treatment for an additional cell cycle produces octoploids (8n), and so forth. This method works in both plant and animal cells, but generally plants seem to be much more tolerant of polyploidy. Note that all alleles in the genotype are doubled. Therefore, if a diploid cell of genotype A/a ; B/b is doubled, the resulting autotetraploid will be of genotype A/A/a/a ; B/B/b/b.

Because 4 is an even number, autotetraploids can have a regular meiosis, although this is by no means always the case. The crucial factor is how the four chromosomes of each set pair and segregate. There are several possibilities, as shown in Figure 15-7. If the chromosomes pair as bivalents or quadrivalents, the chromo-

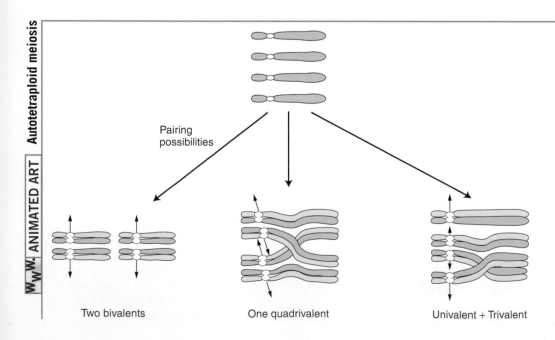

Pairing possibilities

Two bivalents

One quadrivalent

Univalent + Trivalent

Figure 15-7 Three different meiotic pairing possibilities in tetraploids. The four homologous chromosomes may pair as two bivalents or a quadrivalent, and each can yield functional gametes. A third possibility, a trivalent plus a univalent, yields nonfunctional gametes.

somes segregate normally, producing diploid gametes. The fusion of gametes at fertilization regenerates the tetraploid state. If trivalents form, segregation leads to nonfunctional aneuploid gametes, and hence sterility.

What genetic ratios are produced by an autotetraploid? Assume for simplicity that the tetraploid forms only bivalents. If we start with an $A/A/a/a$ tetraploid plant and self it, what proportion of progeny would be $a/a/a/a$? Obviously we first need to deduce the frequency of a/a gametes because this is the only type that can produce a recessive homozygote. The a/a gametes can arise only if the pairings are both A with a, and then the a alleles must both segregate to the same pole. Let's calculate the frequencies of the possible outcomes by means of the following thought experiment. Consider the options from the point of view of one of the a chromosomes faced with the options of pairing with the other a chromosome or with one of the two A chromosomes; if pairing is random there is a $\frac{2}{3}$ chance that it will pair with an A chromosome. If it does, then the pairing of the remaining two chromosomes will necessarily also be A with a because those are the only chromosomes remaining. With these two A-with-a pairings there are two equally likely segregations, and overall $\frac{1}{4}$ of the products will contain both a alleles at one pole. Hence the probability of an a/a gamete will be $\frac{2}{3} \times \frac{1}{4} = \frac{1}{6}$. Hence, if gametes pair randomly, the probability of an $a/a/a/a$ zygote will be $\frac{1}{6} \times \frac{1}{6} = \frac{1}{36}$, and by subtraction the probability of $A/-/-/-$ will be $\frac{35}{36}$. Therefore a $35:1$ phenotypic ratio is expected.

Genomic sequencing has shown that many species that act like "normal" diploids or haploids are in fact descendants of autopolyploids that occurred in past evolutionary times. For example, genomic analysis of the haploid yeast *Saccharomyces cerevisiae* has shown that most chromosomal regions have a duplicate somewhere else in the genome. In fact the ancestral genome of this yeast was probably very similar to that of the filamentous fungus *Ashbya gossypii*, as revealed by comparisons of the size and gene content of their fully sequenced genomes. Hence the smaller *Ashbya* genome presumably doubled at some point in the distant past. Subsequent rearrangement, mutation, and partial loss of segments gave rise to the modern yeast species.

Allopolyploids An allopolyploid is a plant that is a hybrid of two or more species, containing two or more copies of each of the input genomes. The prototypic allopolyploid was an allotetraploid synthesized by G. Karpechenko in 1928. He wanted to make a fertile hybrid that would have the leaves of the cabbage *(Brassica)* and the roots of the radish *(Raphanus)*, because these were the agriculturally important parts of each plant. Each of these two species has 18 chromosomes, so $2n_1 = 2n_2 = 18$, and $n_1 = n_2 = 9$. The species are related closely enough

to allow intercrossing. Fusion of an n_1 and an n_2 gamete produced a viable hybrid progeny individual of constitution $n_1 + n_2 = 18$. However, this hybrid was functionally sterile because the 9 chromosomes from the cabbage parent were different enough from the radish chromosomes that pairs did not synapse and segregate normally at meiosis, and thus the hybrid could not produce functional gametes.

Eventually, one part of the hybrid plant produced some seeds. On planting, these seeds produced fertile individuals with 36 chromosomes. All these individuals were allopolyploids. They had apparently been derived from spontaneous, accidental chromosome doubling to $2n_1 + 2n_2$ in one region of the sterile hybrid, presumably in tissue that eventually became a flower and underwent meiosis to produce gametes. In $2n_1 + 2n_2$ tissue, there is a pairing partner for each chromosome, and functional gametes of the type $n_1 + n_2$ are produced. These gametes fuse to give $2n_1 + 2n_2$ allopolyploid progeny, which also are fertile. This kind of allopolyploid is sometimes called an **amphidiploid,** or doubled diploid (Figure 15-8). Treating a sterile hybrid with colchicine greatly increases the chances that the chromosome sets will double. Amphidiploids are now synthesized routinely in this manner. (Unfortunately for Karpechenko, his amphidiploid had the roots of a cabbage and the leaves of a radish.)

When Karpechenko's allopolyploid was crossed with either parental species—the cabbage or the radish—sterile offspring resulted. The offspring of the cross with cabbage were $2n_1 + n_2$, constituted from an $n_1 + n_2$

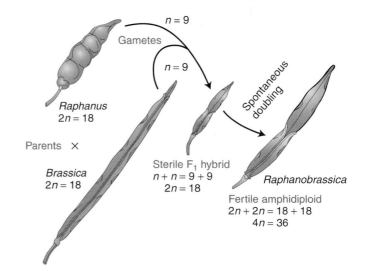

Figure 15-8 The origin of the amphidiploid *(Raphanobrassica)* from cabbage *(Brassica)* and radish *(Raphanus)*. The fertile amphidiploid arose from spontaneous doubling in the $2n = 18$ sterile hybrid. [From A. M. Srb, R. D. Owen, and R. S. Edgar, *General Genetics*, 2d ed. Copyright 1965 by W. H. Freeman. Adapted from G. Karpechenko, *Z. Indukt. Abst. Vererb.* 48, 1928, 27.]

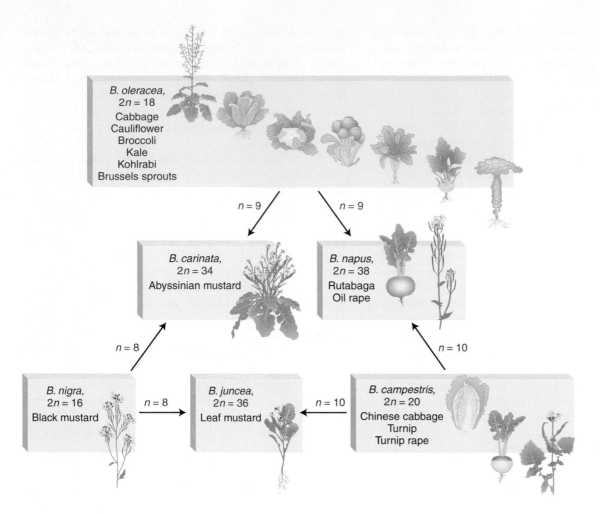

Figure 15-9
Three species of *Brassica* (blue boxes) and their allopolyploids (pink boxes), showing the importance of allopolyploidy in the production of new species.

gamete from the allopolyploid and an n_1 gamete from the cabbage. The n_2 chromosomes had no pairing partners; hence, a normal meiosis could not take place, and the offspring were sterile. Thus, Karpechenko had effectively created a new species, with no possibility of gene exchange with either cabbage or radish. He called his new plant *Raphanobrassica*.

In nature, allopolyploidy seems to have been a major force in the evolution of new plant species. One convincing example is shown by the genus *Brassica*, as illustrated in Figure 15-9. Here three different parent species have hybridized in all possible pair combinations to form new amphidiploid species. Natural polyploidy was once viewed as a somewhat rare occurrence, but recent work has shown that it is a recurrent event in many plant species. The use of DNA markers has made it possible to show that polyploids in any population or area which appear to be the same can have many different parental genotypes as a result of many independent past fusions. It has been estimated that about 50 percent of all angiosperm plants are polyploids, resulting from auto- or allopolyploidy. As a result of multiple polyploidizations, the amount of allelic variation within a polyploid species is much higher than previously thought, perhaps contributing to its potential for adaptation.

A particularly interesting natural allopolyploid is bread wheat, *Triticum aestivum* ($6n = 42$). By studying its wild relatives, geneticists have reconstructed a probable evolutionary history of this plant. Figure 15-10 shows that bread wheat is composed of two sets each of three ancestral genomes. At meiosis, pairing is always between homologs from the same ancestral genome. Hence, in bread wheat meiosis, there are always 21 bivalents.

Allopolyploid plant cells can also be produced artificially by fusing diploid cells from different species. First the walls of two diploid cells are removed by treatment with an enzyme, and the membranes of the two cells fuse and become one. The nuclei often fuse, too, resulting in the polyploid. If the cell is nurtured with the appropriate hormones and nutrients, it divides to become a small allopolyploid plantlet, which can then be transferred to soil.

> **MESSAGE** Allopolyploid plants can be synthesized by crossing related species and doubling the chromosomes of the hybrid or by fusing diploid cells.

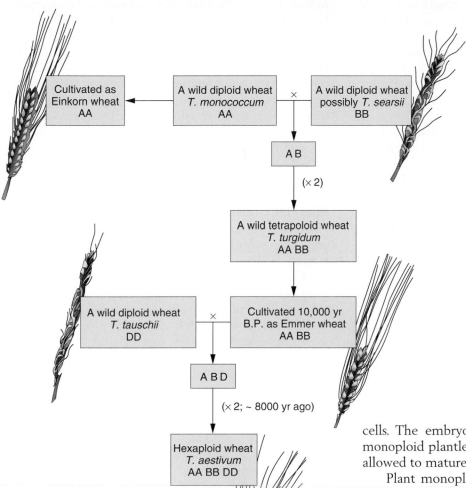

Figure 15-10 The proposed evolutionary history of modern hexaploid wheat. Amphidiploids were produced at two points in this history. A, B, and D are different chromosome sets.

AGRICULTURAL APPLICATIONS

Variations in chromosome number have been exploited to create new plant lines with desirable features. Some examples follow.

Monoploids Diploidy is an inherent nuisance for plant breeders. When they want to induce and select new recessive mutations that are favorable for agricultural purposes, the new mutations cannot be detected unless they are homozygous. Breeders may also want to find favorable new combinations of alleles at different loci, but such favorable allele combinations in heterozygotes will be broken up by recombination at meiosis. Monoploids provide a way around some of these problems.

Monoploids can be artificially derived from the products of meiosis in the plant's anthers. A cell destined to become a pollen grain can instead be induced by cold treatment to grow into an **embryoid,** a small dividing mass of monoploid cells. The embryoid can be grown on agar to form a monoploid plantlet, which can then be potted in soil and allowed to mature (Figure 15-11).

Plant monoploids can be exploited in several ways. In one approach, they are first examined for favorable allelic combinations that have arisen from the recombination of alleles already present in a heterozygous diploid parent. Hence from a parent that is *A/a ; B/b* might come a favorable monoploid combination *a ; b.* The monoploid can then be subjected to chromosome doubling to produce homozygous diploid cells, *a/a ; b/b,* that are capable of normal reproduction.

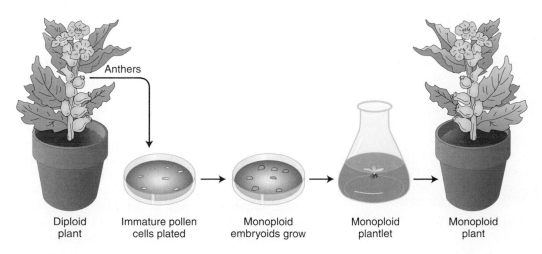

Figure 15-11 Generating monoploid plants by tissue culture.

Diploid plant   Immature pollen cells plated   Monoploid embryoids grow   Monoploid plantlet   Monoploid plant

Another approach is to treat monoploid cells basically as a population of haploid organisms in a mutagenesis-and-selection procedure. A population of monoploid cells is isolated, their walls are removed by enzymatic treatment, and they are exposed to a mutagen. They are then plated on a medium that selects for some desirable phenotype. This approach has been used to select for resistance to toxic compounds produced by a plant parasite as well as to select for resistance to herbicides being used by farmers to kill weeds. Resistant plantlets eventually grow into monoploid plants, whose chromosome number can then be doubled using colchicine. This treatment produces diploid tissue and eventually, by taking a cutting or by selfing a flower, a fully resistant diploid plant. These powerful techniques can circumvent the normally slow process of meiosis-based plant breeding. They have been successfully applied to important crop plants such as soybeans and tobacco.

> **MESSAGE** Geneticists can create new plant lines by producing monoploids with favorable genotypes and then doubling their chromosomes to form fertile, homozygous diploids.

Autotriploids The bananas that are widely available commercially are sterile triploids with 11 chromosomes in each set ($3n = 33$). The most obvious expression of the sterility of bananas is the absence of seeds in the fruit that we eat. (The black specks in bananas are not seeds; banana seeds are rock hard—real tooth-breakers.) Seedless watermelons are another example of the commercial exploitation of triploidy in plants.

Autotetraploids Many autotetraploid plants have been developed as commercial crops to take advantage of their increased size (Figure 15-12). Large fruits and flowers are particularly favored.

Figure 15-12 Diploid (left) and tetraploid (right) grapes.
[Copyright Leonard Lessin/Peter Arnold Inc.]

Allopolyploids Allopolyploidy (formation of polyploids between different species) has been important in the production of modern crop plants. New World cotton is a natural allopolyploid that occurred spontaneously, as is wheat. Allopolyploids also are synthesized artificially to combine the useful features of parental species into one type. Only one synthetic amphidiploid has ever been widely used commercially, a crop known as *Triticale*. This is an amphidiploid between wheat (*Triticum*, $6n = 42$) and rye (*Secale*, $2n = 14$). Hence, for *Triticale*, $2n = 2 \times (21 + 7) = 56$. This novel plant combines the high yields of wheat with the ruggedness of rye.

POLYPLOID ANIMALS Polyploidy is more common in plants than in animals, but there are cases of naturally occurring polyploid animals. Polyploid species of flatworms, leeches, and brine shrimps reproduce by parthenogenesis. Triploid and tetraploid *Drosophila* have been synthesized experimentally. However, examples are not limited to these so-called lower forms. Naturally occurring polyploid amphibians and reptiles are surprisingly common. They have several modes of reproduction: polyploid species of frogs and toads participate in sexual reproduction, whereas polyploid salamanders and lizards are parthenogenetic. The Salmonidae (the family of fishes that includes salmon and trout) provide a familiar example of the numerous animal species that appear to have originated through ancestral polyploidy.

The sterility of triploids has been commercially exploited in animals as well as in plants. Triploid oysters have been developed because they have a commercial advantage over their diploid relatives. The diploids go through a spawning season, when they are unpalatable, but the sterile triploids do not spawn and are palatable year-round.

Aneuploidy

Aneuploidy is the second major category of chromosomal aberrations in which the chromosome number is abnormal. An aneuploid is an individual organism whose chromosome number differs from the wild type by part of a chromosome set. Generally, the aneuploid chromosome set differs from the wild type by only one chromosome or by a small number of chromosomes. An aneuploid can have a chromosome number either greater or smaller than that of the wild type. Aneuploid nomenclature (see Table 15-1) is based on the number of copies of the specific chromosome in the aneuploid state. For autosomes in diploid organisms, the aneuploid $2n + 1$ is **trisomic**, $2n - 1$ is **monosomic**, and $2n - 2$ (the -2 represents the loss of both homologs of a chromosome) is **nullisomic**. In haploids, $n + 1$ is **disomic**. Special notation is used to describe sex chromosome aneuploids because it must deal with the two different chromosomes. The notation merely lists the copies of each sex chromosome, such as XXY, XYY, XXX, or XO (the "O" stands for absence of a chromosome and is included to show that the single X symbol is not a typographical error).

NONDISJUNCTION The cause of most aneuploidy is **nondisjunction** in the course of meiosis or mitosis. *Disjunction* is another word for the normal segregation of homologous chromosomes or chromatids to opposite poles at meiotic or mitotic divisions. Nondisjunction is a failure of this process, in which two chromosomes or chromatids incorrectly go to one pole and none to the other.

Mitotic nondisjunction during development results in aneuploid sections of the body (aneuploid *sectors*). *Meiotic* nondisjunction is more commonly encountered. It results in aneuploid meiotic products, leading to descendants in which the entire organism is aneuploid. In meiotic nondisjunction, the chromosomes may fail to disjoin at either the first or the second meiotic division (Figure 15-13). Either way, $n - 1$ and $n + 1$ gametes are produced. If an $n - 1$ gamete is fertilized by an n gamete,

a monosomic $(2n - 1)$ zygote is produced. The fusion of an $n + 1$ and an n gamete yields a trisomic $2n + 1$.

> **MESSAGE** Aneuploid organisms result mainly from nondisjunction during a parental meiosis.

Nondisjunction occurs spontaneously. Like most gene mutations, it is an example of a chance failure of a basic cellular process. The precise molecular processes that fail are not known, but in experimental systems, the frequency of nondisjunction can be increased by interference with microtubule polymerization action, thereby inhibiting normal chromosome movement. It appears that disjunction is more likely to go awry in meiosis I. This may not be surprising, because normal anaphase I disjunction requires that the homologous chromosomes of the tetrad remain paired during prophase I and metaphase I, and also requires crossovers. In contrast, proper disjunction at anaphase II or at mitosis requires that the centromere split properly but does not require chromosome pairing or crossing-over.

Crossovers are a necessary component of the normal disjunction process. Somehow the formation of a chiasma in a chromosome pair helps to hold the tetrad together and ensures that the members of a pair will go to opposite poles. In most organisms, the amount of crossing-over is sufficient to ensure that all tetrads will have at least one chiasma per meiosis. In *Drosophila*, many of the nondisjunctional chromosomes seen in disomic $(n + 1)$ gametes are nonrecombinant, showing that they arise from meioses in which there is no crossing-over on that chromosome. Similar observations have been made in human trisomies. In addition, in several different experimental organisms, mutations that interfere with recombination have the effect of massively increasing the frequency of meiosis I nondisjunction. All these observations provide evidence for the role of crossing-over in maintaining chromosome associations in the tetrad; in the absence of these associations, chromosomes are vulnerable to anaphase I nondisjunction.

> **MESSAGE** Crossovers are needed to maintain the intact tetrad until anaphase I. If crossing-over fails for some reason, first-division nondisjunction occurs.

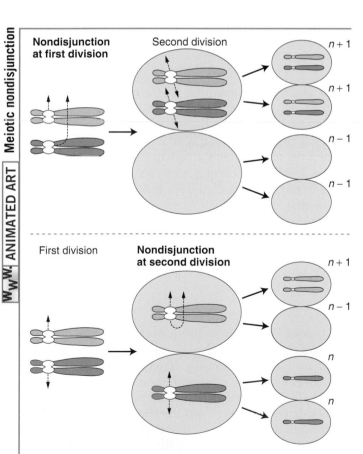

Meiotic nondisjunction
ANIMATED ART *www*

Figure 15-13 The origin of aneuploid gametes by nondisjunction at the first or second meiotic division. Note that all other chromosomes are present in normal number, including cells in which no chromosomes are shown.

MONOSOMICS ($2n - 1$) Monosomics are missing one copy of a chromosome. In most diploid organisms, the absence of one chromosome copy from a pair is deleterious. In humans, monosomics for any of the autosomes die in utero. Many X chromosome monosomics also die in utero, but some are viable. A human chromosome complement of 44 autosomes plus a single X produces a condition known as **Turner syndrome,** represented as

XO. Affected persons have a characteristic phenotype: they are sterile females, short in stature, and often have a web of skin extending between the neck and shoulders (Figure 15-14). Although their intelligence is near normal, some of their specific cognitive functions are defective. About 1 in 5000 female births show Turner syndrome.

Geneticists have used viable plant monosomics to match newly discovered recessive mutant alleles to a specific chromosome. For example, one can make a set of monosomic lines, each known to lack a different chromosome. Homozygotes for the new mutant allele are crossed with each monosomic line, and the progeny of each cross are inspected for the recessive phenotype. The appearance of the recessive phenotype identifies the chromosome that has one copy missing as the one the gene is normally located on. The test works because half the gametes of a fertile monosomic will be $n - 1$, and when an $n - 1$ gamete is fertilized by a gamete bearing a new mutation on the homologous chromosome, the mutant allele will be the only allele of that gene present and hence will be expressed.

As an example, let's assume that a gene A/a is on chromosome 2. Crosses of a/a to monosomics for chromosome 1 and chromosome 2 illustrate the method (chromosome 1 is abbreviated chr1):

chr1/chr1 ; *a/a*   ×   chr1/0 ; *A/A*
Mutant   Chromosome 1 monosomic
                                        genotype *A*

                    ↓

        progeny   all *A/a*

chr1/chr1 ; *a/a*   ×   chr1/chr1 ; *A/0*
Mutant   Chromosome 2 monosomic
                                        genotype A

                    ↓

        progeny $\frac{1}{2}$ *A/a*
                       $\frac{1}{2}$ *a/0*

TRISOMICS (2n + 1) Trisomics contain an extra copy of one chromosome. In diploid organisms generally, the chromosomal imbalance from the trisomic condition can result in abnormality or death. However, there are many examples of viable trisomics. Furthermore, trisomics can be fertile. When cells from some trisomic organisms are observed under the microscope at the time of meiotic chromosome pairing, the trisomic chromosomes are seen to form an associated group of three (a trivalent), whereas the other chromosomes form regular pairs.

What genetic ratios might we expect for genes on the trisomic chromosome? Let us consider a gene A that is close to the centromere on that chromosome, and let

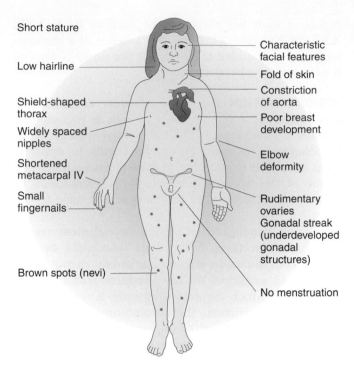

Figure 15-14 Characteristics of Turner syndrome.
The condition results from the presence of a single X chromosome (XO). [Adapted from F. Vogel and A. G. Motulsky, *Human Genetics*. Springer-Verlag, 1982.]

us assume that the genotype is $A/a/a$. Furthermore, let's postulate that at anaphase I the two paired centromeres in the trivalent pass to opposite poles and that the other centromere passes randomly to either pole. Then we can predict the three equally frequent segregations shown in Figure 15-15. These segregations result in an overall gametic ratio as shown in the six compartments of Figure 15-15; that is,

$\frac{1}{6}$ *A*

$\frac{2}{6}$ *a*

$\frac{2}{6}$ *A/a*

$\frac{1}{6}$ *a/a*

If a set of lines is available, each carrying a different trisomic chromosome, then a gene mutation can be located to a chromosome by determining which of the lines gives a trisomic ratio of the above type.

There are several examples of viable human trisomies. Several types of sex chromosome trisomics can live to adulthood. Each of these types is found at a frequency of about 1 in 1000 births of the relevant sex. (In considering human sex chromosome trisomies, recall that mammalian sex is determined by the presence or absence of the Y chromosome.) The combination XXY results in **Klinefelter syndrome.** Persons with this syn-

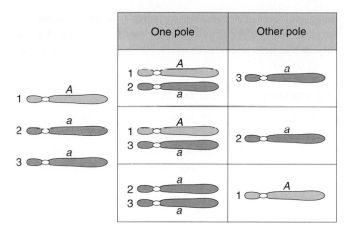

Figure 15-15 Genotypes of the meiotic products of an *A/a/a* trisomic. The three segregations shown are equally likely.

drome are males with lanky builds and a mildly impaired IQ and are sterile (Figure 15-16). Another abnormal combination, XYY, has a controversial history. Attempts have been made to link the XYY condition with a predisposition toward violence. However, it is now clear that an XYY condition in no way guarantees such

behavior. Males with XYY are usually fertile. Meioses show normal pairing of the X with one of the Y's; the other Y does not pair and is not transmitted to gametes. Therefore the gametes contain either X or Y, never YY or XY. Triplo-X trisomics (XXX) are phenotypically normal and fertile females. Meiosis shows pairing of only two X chromosomes; the third does not pair. Hence eggs bear only one X and, as in the case of XYY individuals, the condition is not passed on to progeny.

Of human trisomies, the most familiar type is **Down syndrome** (Figure 15-17), which we discussed briefly at the beginning of the chapter. Down syndrome occurs at a frequency of about 0.15 percent of all live births. Most affected individuals have an extra copy of chromosome 21 caused by nondisjunction of chromosome 21 in a parent who is chromosomally normal. In this *sporadic* type of Down syndrome, there is no family history of aneuploidy. Some rare types of Down syndrome arise from translocations (a type of chromosomal rearrangement discussed later in the chapter); in these cases, as we shall see, Down syndrome recurs in the pedigree because the translocation may be transmitted from parent to child.

The combined phenotypes that make up Down syndrome include mental retardation (with an IQ in the 20-to-50 range); a broad, flat face; eyes with an epicanthic fold; short stature; short hands with a crease across the middle; and a large, wrinkled tongue. Females may be fertile and may produce normal or trisomic progeny, but males do not reproduce. Mean life expectancy is about 17 years, and only 8 percent of persons with Down syndrome survive past age 40.

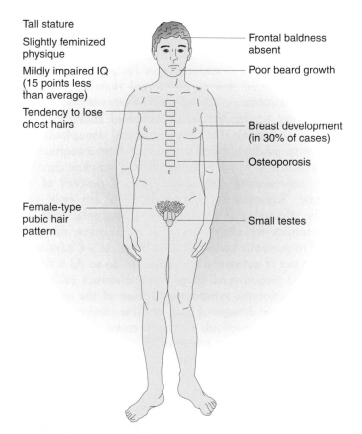

Figure 15-16 Characteristics of Klinefelter syndrome (XXY). [Adapted from F. Vogel and A. G. Motulsky, *Human Genetics.* Springer-Verlag, 1982.]

Tall stature

Slightly feminized physique

Mildly impaired IQ (15 points less than average)

Tendency to lose chest hairs

Female-type pubic hair pattern

Frontal baldness absent

Poor beard growth

Breast development (in 30% of cases)

Osteoporosis

Small testes

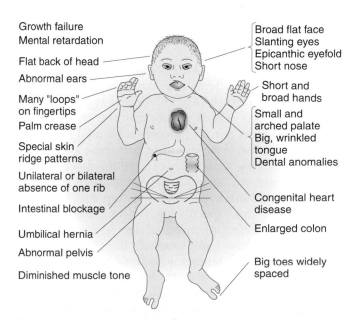

Figure 15-17 Characteristics of Down syndrome (trisomy 21). [Adapted from F. Vogel and A. G. Motulsky, *Human Genetics.* Springer-Verlag, 1982.]

Growth failure
Mental retardation

Flat back of head

Abnormal ears

Many "loops" on fingertips

Palm crease

Special skin ridge patterns

Unilateral or bilateral absence of one rib

Intestinal blockage

Umbilical hernia

Abnormal pelvis

Diminished muscle tone

Broad flat face
Slanting eyes
Epicanthic eyefold
Short nose

Short and broad hands

Small and arched palate
Big, wrinkled tongue
Dental anomalies

Congenital heart disease

Enlarged colon

Big toes widely spaced

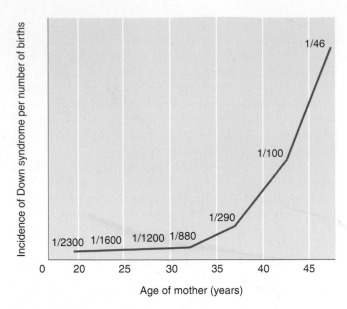

Figure 15-18 Maternal age and the production of offspring with Down syndrome. [From L. S. Penrose and G. F. Smith, *Down's Anomaly*. Little, Brown and Company, 1966.]

The incidence of Down syndrome is related to maternal age; older mothers run a greatly elevated risk of having a child with Down syndrome (Figure 15-18). For this reason, fetal chromosome analysis (by amniocentesis or by chorionic villus sampling) is now recommended for older mothers. A less pronounced paternal-age effect also has been demonstrated.

Even though the maternal-age effect has been known for many years, its cause is still not known. Nonetheless, there are some interesting biological correlations. It is possible that with age it becomes less likely that the chromosome tetrad will keep together during prophase I of meiosis. Meiotic arrest of oocytes (female meiocytes) in late prophase I is a common phenomenon in many animals. In female humans, all oocytes are arrested at diplotene before birth. Meiosis resumes at each menstrual period, which means that the chromosomes in the tetrad must remain properly associated for as long as several decades. If we speculate that these associations have an increasing probability of breaking down by accident over time, we can envision a mechanism contributing to increased maternal nondisjunction with age. Consistent with this speculation, most nondisjunction related to the effect of maternal age is due to nondisjunction at anaphase I, not anaphase II.

The only other human autosomal trisomics to survive to birth are those with trisomy 13 (Patau syndrome) and trisomy 18 (Edwards syndrome). Both show severe physical and mental abnormalities. The phenotypic syndrome of trisomy 13 includes a harelip; a small, malformed head; "rockerbottom" feet; and a mean life expectancy of 130 days. That of trisomy 18 includes "faunlike" ears, a small jaw, a narrow pelvis, and rocker-bottom feet; almost all babies with trisomy 18 die within the first few weeks after birth. All other trisomics die in utero.

The concept of gene balance

In considering aberrant euploidy, we noted that an increase in the number of full chromosome sets correlates with increased organism size, but that the general shape and proportions of the organism remain very much the same. In contrast, autosomal aneuploidy typically alters the organism's shape and proportions in characteristic ways.

Plants tend to be somewhat more tolerant of aneuploidy than are animals. Studies in jimsonweed (*Datura stramonium*) provide a classic example of the effects of aneuploidy and polyploidy. In jimsonweed, the haploid chromosome number is 12. As expected, the polyploid jimsonweed is proportioned like the normal diploid, only larger. In contrast, each of the 12 possible trisomics is disproportionate, but in ways different from one another, as exemplified by changes in the shape of the seed capsule (see Figure 3-3). The 12 different trisomies lead to 12 different and characteristic shape changes in the capsule. Indeed, these and other characteristics of the individual trisomics are so reliable that the phenotypic syndrome can be used to identify plants carrying a particular trisomy. Similarly, the 12 monosomics are themselves different from one another and from each of the trisomics. In general, a monosomic for a particular chromosome is more severely abnormal than is the corresponding trisomic.

We see similar trends in aneuploid animals. In the fruit fly *Drosophila*, the only autosomal aneuploids that survive to adulthood are trisomics and monosomics for chromosome 4, which is the smallest *Drosophila* chromosome, representing only about 1 to 2 percent of the genome. Trisomics for chromosome 4 are only very mildly affected and are much less abnormal than are monosomics for chromosome 4. In humans, no autosomal monosomic survives to birth, but as we have seen, three types of autosomal trisomics can do so. As is true of aneuploid jimsonweed, these three trisomics each show unique phenotypic syndromes because of the special effects of altered dosages of each of these chromosomes.

Why are aneuploids so much more abnormal than polyploids? Why does aneuploidy for each chromosome have its own characteristic phenotypic effects? And why are monosomics typically more severely affected than are the corresponding trisomics? The answers seem certain to be a matter of **gene balance.** In a euploid, the ratio of genes on any one chromosome to the genes on other chromosomes always is 1 : 1, regardless of whether we are considering a monoploid, diploid, triploid, or tetraploid. For example, in a tetraploid, for gene *A* on chromosome 1 and gene B on chromosome 2, the ratio is 4 A:4 B, or 1:1. In contrast, in an aneuploid, the ratio of genes on the aneuploid chromosome to genes on the other chromosomes differs from the wild type by 50 percent; 50 percent for monosomics; 150 percent for trisomics. Using the same example as above, in a trisomic for chromosome 2, the ratio of the A and B genes is 2A:3B. Thus, we can see that the aneuploid genes are out of balance. How does this help us answer the questions raised?

In general, the amount of transcript produced by a gene is directly proportional to the number of copies of that gene in a cell. That is, for a given gene, the rate of transcription is directly related to the number of DNA templates available. Thus, the more copies of the gene, the more transcripts are produced and the more of the corresponding protein product is made. This relationship between the number of copies of a gene and the amount of the gene's product made is called a **gene-dosage effect.**

We can infer that normal physiology in a cell depends on the proper ratio of gene products in the euploid cell. This ratio is the normal gene balance. If the relative dosage of certain genes changes—for example, because of the removal of one of the two copies of a chromosome (or even a segment thereof)—physiological imbalances in cellular pathways can arise.

In some cases, the imbalances of aneuploidy result from the effects of a few "major" genes whose dosage has changed, rather than from changes in the dosage of all the genes on a chromosome. Such genes can be viewed as *haplo-abnormal* (resulting in an abnormal phenotype if present only once) or *triplo-abnormal* (resulting in an abnormal phenotype if present in three copies) or both. They contribute significantly to the aneuploid phenotypic syndromes. For example, the study of persons trisomic for only part of chromosome 21 has made it possible to localize genes contributing to Down syndrome to various regions of chromosome 21; the results hint that some aspects of the phenotype might be due to triplo-abnormality for single major genes in these chromosome regions. In addition to these major gene effects, other aspects of aneuploid syndromes are likely to result from the cumulative effects of aneuploidy for numerous genes whose products are all out of balance. Undoubtedly, the entire aneuploid phenotype results from a combination of the imbalance effects of a few major genes, together with a cumulative imbalance of many minor genes.

However, the concept of gene balance does not tell us why having too few gene products (monosomy) is much worse for an organism than having too many gene products (trisomy). In a parallel manner, we can ask why there are many more haplo-abnormal genes than triplo-abnormal ones. A key to explaining the extreme abnormality of monosomics is that any deleterious recessive alleles present on a monosomic autosome will be automatically expressed.

How do we apply the idea of gene balance to cases of sex chromosome aneuploidy? Gene balance holds for sex chromosomes as well, but we also have to take into account the special properties of the sex chromosomes. In organisms with XY sex determination, the Y chromosome seems to be a degenerate X chromosome in which there are very few functional genes other than some involved in sex determination itself, in sperm production, or in both. The X chromosome, on the other hand, contains many genes involved in basic cellular processes ("housekeeping genes") that just happen to reside on the chromosome that eventually evolved into the X chromosome. XY sex determination mechanisms have probably evolved independently from 10 to 20 times in different taxonomic groups. Thus, there appears to be one sex determination mechanism for all mammals, but it is completely different from the mechanism governing XY sex determination in fruit flies.

In a sense, X chromosomes are naturally aneuploid. In species with an XY sex determination system, females have two X chromosomes, whereas males have only one. Nonetheless, it has been found that the X chromosome's housekeeping genes are expressed to approximately equal extents per cell in females and in males. In other words, there is **dosage compensation.** How is this accomplished? The answer depends on the organism. In fruit flies, the male's X chromosome appears to be hyperactivated, allowing it to be transcribed at twice the rate of either X chromosome in the female. As a result, the XY male *Drosophila* has an X gene dosage equivalent to that of an XX female. In mammals, in contrast, the rule is that no matter how many X chromosomes are present, there is only one transcriptionally active X chromosome in each somatic cell. This rule gives the XX female mammal an X gene dosage equivalent to that of an XY male. Dosage compensation in mammals is achieved by X-chromosome inactivation. A female with two X chromosomes, for example, is a mosaic of two cell types in which one or the other X is active. We examined this phenomenon in Chapter 10. Thus, XY and XX individuals produce the same amounts of X-chromosome housekeeping-gene products. X-chromosome inactivation also explains why triplo-X humans are phenotypically normal—only one of the three X chromosomes is transcriptionally active in a given cell. Similarly, an XXY male is only moderately affected because only one of his two X chromosomes is active in each cell.

Why are XXY individuals abnormal at all, given that triplo-X individuals are phenotypically normal? It turns out that a few genes scattered throughout an "inactive X" are still transcriptionally active. In XXY males, these genes are transcribed at twice the level they are in

XY males. In XXX females, on the other hand, the few transcribed genes are active at only 1.5 times the level that they are in XX females. This lower level of "functional aneuploidy" in XXX than in XXY, plus the fact that the active X genes appear to lead to feminization, may explain the feminized phenotype of XXY individuals. The severity of Turner syndrome (XO) may be due to the deleterious effects of monosomy and to the lower activity of the transcribed genes of the X (compared with XX females). As is usually observed for aneuploids, monosomy for the X chromosome produces a more abnormal phenotype than does having an extra copy of the same chromosome (triplo-X females or XXY males).

Gene dosage is also important in the phenotypes of polyploids. Human polyploid zygotes do arise through various kinds of mistakes in cell division. Most die in utero. Occasionally, triploid babies are born, but none survive. This fact seems to violate the principle that we have been discussing—namely, that polyploids are more normal than aneuploids. The explanation for this contradiction seems to lie with X-chromosome dosage compensation. Part of the rule for gene balance in organisms that have a single active X seems to be that there must be one active X for every two copies of the autosomal chromosome complement. Thus, some cells in triploid

mammals are found to have one active X, whereas others, surprisingly, have two. Neither situation is in balance with autosomal genes.

> **MESSAGE** Aneuploidy is nearly always deleterious because of gene imbalance—the ratio of genes is different from that in euploids, and this difference interferes with the normal function of the genome.

15.2 Changes in chromosome structure

Changes in chromosome structure, called **rearrangements,** encompass several major classes of events. A chromosome segment can be lost, constituting a **deletion,** or doubled, to form a **duplication.** The orientation of a segment within the chromosome can be reversed, constituting an **inversion.** Or a segment can be moved to a different chromosome, constituting a **translocation.** DNA breakage is a major cause of each of these events. Both DNA strands must break at two different locations, followed by a rejoining of the broken ends to produce a new chromosomal arrangement (Figure 15-19). Chro-

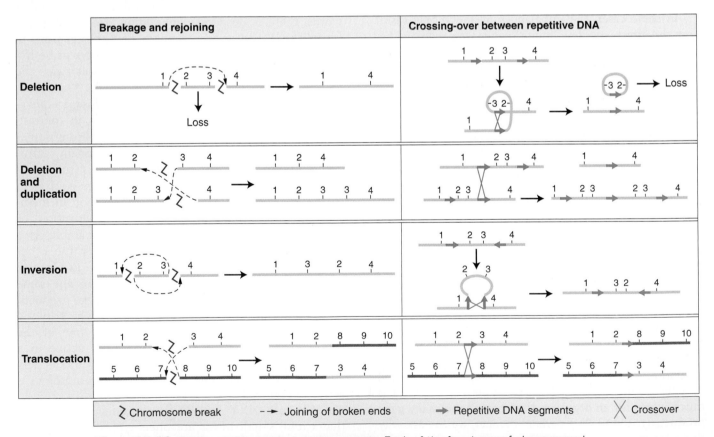

Figure 15-19 Origins of chromosomal rearrangements. Each of the four types of chromosomal rearrangements can be produced by either of two basic mechanisms: chromosome breakage and rejoining or crossing-over between repetitive DNA. Chromosome regions are numbered 1 through 10. Homologous chromosomes are the same color.

mosomal rearrangements by breakage can be induced artificially by using ionizing radiation. This kind of radiation, particularly X rays and gamma rays, is highly energetic and causes numerous double-stranded breaks in DNA.

To understand how chromosomal rearrangements are produced by breakage, several points should be kept in mind:

1. Each chromosome is a single double-stranded DNA molecule.

2. The first event in the production of a chromosomal rearrangement is the generation of two or more double-stranded breaks in the chromosomes of a cell (see Figure 15-19, top row at left).

3. Double-stranded breaks are potentially lethal, unless they are repaired.

4. Repair systems in the cell correct the double-stranded breaks by joining broken ends back together (see Chapter 14 for a detailed discussion of DNA repair).

5. If the two ends of the same break are rejoined, the original DNA order is restored. If the ends of two different breaks are joined together, however, one result is one or another type of chromosomal rearrangement.

6. The only chromosomal rearrangements that survive meiosis are those that produce DNA molecules that have one centromere and two telomeres. If a rearrangement produces a chromosome that lacks a centromere, such an **acentric** chromosome will not be dragged to either pole at anaphase of mitosis or meiosis and will not be incorporated into either progeny nucleus. Therefore acentric chromosomes are not inherited. If a rearrangement produces a chromosome with two centromeres (a **dicentric**), it will often be pulled simultaneously to opposite poles at anaphase, forming an **anaphase bridge**. Anaphase bridge chromosomes typically will not be incorporated into either progeny cell. If a chromosome break produces a chromosome lacking a telomere, that chromosome cannot replicate properly. Recall from Chapter 7 that telomeres are needed to prime proper DNA replication at the ends (see Figure 7-24).

7. If a rearrangement duplicates or deletes a segment of a chromosome, gene balance may be affected. The larger the segment that is lost or duplicated, the more likely it is that gene imbalance will cause phenotypic abnormalities.

Another important cause of rearrangements is crossing-over between repetitive DNA segments. In organisms with repeated short DNA sequences within one chromosome or on different chromosomes there is ambiguity about which of the repeats will pair with each other at meiosis. If sequences pair up that are not in the

same relative positions on the homologs, crossing-over can produce aberrant chromosomes. Deletions, duplications, inversions, and translocations can all be produced by such crossing-over (see Figure 15-19, right side).

There are two general types of rearrangements, balanced and unbalanced. **Balanced rearrangements** change the chromosomal gene order but do not remove or duplicate any DNA. The two simple classes of balanced rearrangements are inversions and reciprocal translocations. An **inversion** is a rearrangement in which an internal segment of a chromosome has been broken twice, flipped 180 degrees, and rejoined.

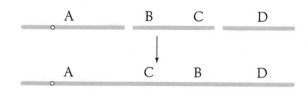

A **reciprocal translocation** is a rearrangement in which two nonhomologous chromosomes are each broken once, creating acentric fragments, which then trade places:

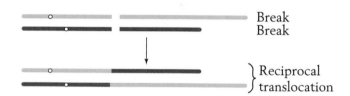

Sometimes the DNA breaks that precede the formation of a rearrangement occur *within* genes. When they do, they disrupt gene function because part of the gene moves to a new location and no complete transcript can be made. In addition, the DNA sequences on either side of the rejoined ends of a rearranged chromosome are ones that are not normally juxtaposed. Sometimes the junction occurs in such a way that fusion produces a nonfunctional hybrid gene composed of parts of two other genes.

Unbalanced rearrangements change the gene dosage of a chromosome segment. As with aneuploidy for whole chromosomes, the loss of one copy of a segment or the addition of an extra copy can disrupt normal gene balance. The two simple classes of unbalanced rearrangements are deletions and duplications. A **deletion** is the loss of a segment within one chromosome arm and the juxtaposition of the two segments on either side of the deleted segment, as in this example, which shows loss of segment C–D:

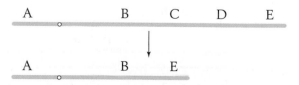

A **duplication** is the repetition of a segment of a chromosome arm. In the simplest type of duplication, the two segments are adjacent to each other (a tandem duplication), as in this duplication of segment C:

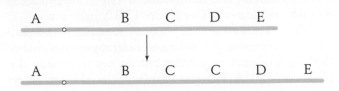

However, the duplicate segment can end up at a different position on the same chromosome, or even on a different chromosome.

The following sections consider the properties of these balanced and unbalanced rearrangements.

Inversions

Inversions are of two basic types. If the centromere is outside the inversion, the inversion is said to be **paracentric**. Inversions spanning the centromere are **pericentric**.

	A	B		C	D	E	F
Normal sequence							
	A	B		C	E	D	F
Paracentric							
	A	D	C		B	E	F
Pericentric							

Because inversions are balanced rearrangements, they do not change the overall amount of genetic material, so they do not result in gene imbalance. Individuals with inversions are generally normal, if there are no breaks within genes. A break that disrupts a gene produces a mutation that may be detectable as an abnormal phenotype. If the gene has an essential function, then the breakpoint acts as a lethal mutation linked to the inversion. In such a case, the inversion cannot be bred to homozygosity. However, many inversions can be made homozygous, and furthermore, inversions can be detected in haploid organisms. In these cases, the breakpoints of the inversion are clearly not in essential regions. Some of the possible consequences of inversion at the DNA level are shown in Figure 15-20.

Breakpoints *between* genes

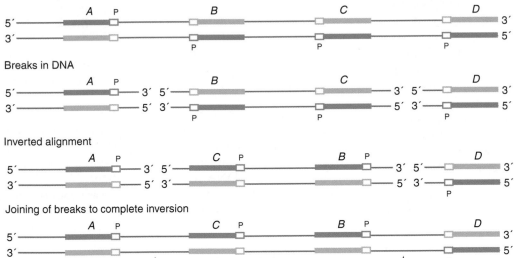

One breakpoint *between* genes
One *within* gene *C* (*C* disrupted)

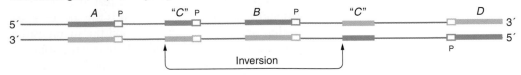

Breakpoints *in* genes *A* and *D*
Creating gene fusions

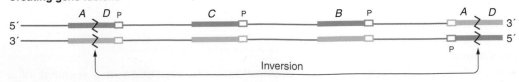

Figure 15-20 Effects of inversions at the DNA level. Genes are represented by *A*, *B*, *C*, and *D*. Template strand is dark green; nontemplate strand is light green; jagged lines indicate where breaks in the DNA produced gene fusions (*A* with *D*) after inversion and rejoining. The letter P stands for promoter; arrows indicate the positions of the breakpoints.

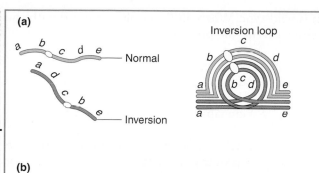
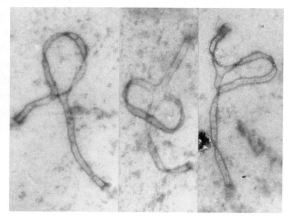

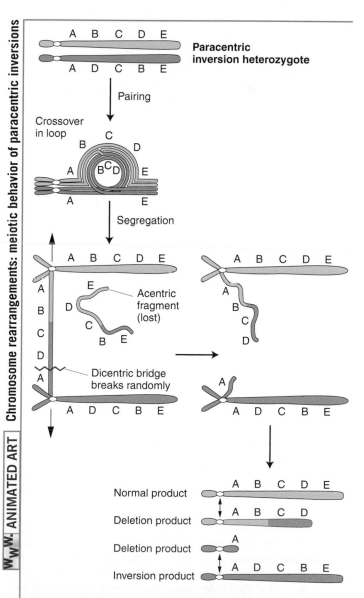

Figure 15-21 **The chromosomes of inversion heterozygotes pair in a loop at meiosis.** (a) Diagrammatic representation. (b) Electron micrographs of synaptonemal complexes at prophase I of meiosis in a mouse heterozygous for a paracentric inversion. Three different meiocytes are shown. [Part b from M. J. Moses, Department of Anatomy, Duke Medical Center.]

Most analyses of inversions are carried out on diploid cells that contain one normal chromosome set plus one set carrying the inversion. This type of cell is called an **inversion heterozygote,** but note that this designation does not imply that any gene locus is heterozygous, but rather that one normal and one abnormal chromosome set are present. The location of the inverted segment often can be detected microscopically. During meiosis, one chromosome twists once at the ends of the inversion to pair with the other untwisted chromosome; in this way the paired homologs form a visible **inversion loop** (Figure 15-21).

In the case of a *paracentric* inversion, crossing-over within the inversion loop at meiosis connects homologous centromeres in a **dicentric bridge** while also producing an **acentric fragment** (Figure 15-22). Then, as the chromosomes separate during anaphase I, the centromeres remain linked by the bridge. The acentric fragment cannot align itself or move, and consequently it is lost. Tension eventually breaks the dicentric bridge, forming two chromosomes with terminal deletions. Either the gametes containing such chromosomes or the zygotes that they eventually form will probably be invi-

able. Hence, a crossover event, which normally generates the recombinant class of meiotic products, is instead lethal to those products. The overall result is a drastically lower frequency of viable recombinants. In fact, for genes within the inversion, the RF is close to zero. (It is not exactly zero because double crossovers involving only two chromatids—which are rare—are viable.) For genes flanking the inversion, the RF is reduced in proportion to the size of the inversion, because for a longer inversion, there is a greater probability of a crossover occurring within it and producing an inviable meiotic product.

In the case of a heterozygous *pericentric* inversion, the net genetic effect is the same as that of a paracentric

Figure 15-22 **Meiotic products resulting from a single crossover within a paracentric inversion loop.** Two nonsister chromatids cross over within the loop.

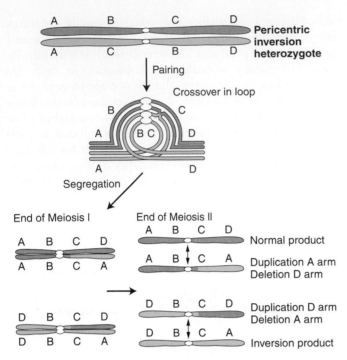

Figure 15-23 Meiotic products resulting from a single crossover within a pericentric inversion loop.

inversion—crossover products are not recovered—but the reasons are different. In a pericentric inversion, the centromeres are contained within the inverted region. Consequently, the chromosomes that have engaged in crossing-over separate in the normal fashion, without the creation of a bridge (Figure 15-23). However, the crossover produces chromatids that contain a duplication and a deletion for different parts of the chromosome. In this case, if a gamete carrying a crossover chromosome is fertilized, the zygote dies because of gene imbalance. Again, the result is that only noncrossover chromatids are present in viable progeny. Hence, the RF value of genes within a pericentric inversion is also zero.

Inversions affect recombination in another way, too. Inversion heterozygotes often have mechanical pairing problems in the region of the inversion. The inversion loop causes a large distortion that can extend beyond the loop itself. This distortion reduces the opportunity for crossing-over in the neighboring regions.

Let us consider an example of the effects of an inversion on recombinant frequency. A wild-type *Drosophila* specimen from a natural population is crossed with a homozygous recessive laboratory stock *dp cn/dp cn*. (The *dp* allele codes for dumpy wings and *cn* codes for cinnabar eyes. The two genes are known to be 45 map units apart on chromosome 2.) The F$_1$ genera-

tion is wild-type. When an F$_1$ female is crossed with the recessive parent, the progeny are

250	wild type	+ +/dp cn
246	dumpy cinnabar	dp cn/dp cn
5	dumpy	dp +/dp cn
7	cinnabar	+ cn/dp cn

In this cross, which is effectively a dihybrid testcross, 45 percent of the progeny are expected to be dumpy or cinnabar (they constitute the crossover classes), but only 12 out of 508, about 2 percent, are obtained. Something is reducing crossing-over in this region, and a likely explanation is an inversion spanning most of the *dp–cn* region. Because the expected RF was based on measurements made on laboratory strains, the wild-type fly from nature was the most likely source of the inverted chromosome. Hence chromosome 2 in the F$_1$ can be represented as follows:

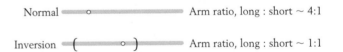

Pericentric inversions also can be detected microscopically through new arm ratios. Consider the following pericentric inversion:

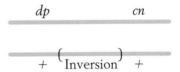

Note that the length ratio of the long arm to the short arm has been changed from about 4:1 to about 1:1 by the inversion. Paracentric inversions do not alter the arm ratio, but they may be detected microscopically by observing changes in banding or other chromosomal landmarks, if available.

> **MESSAGE** The main diagnostic features of heterozygous inversions are inversion loops, reduced recombinant frequency, and reduced fertility because of unbalanced or deleted meiotic products.

In some model experimental systems, notably the fruit fly (*Drosophila*) and the nematode (*Caenorhabditis elegans*), inversions are used as balancers. A **balancer** chromosome contains *multiple* inversions, so that when it is combined with the corresponding wild-type chromosome, there can be no viable crossover products. In some analyses, it is important to keep stock with all the alleles on one chromosome together. Combining such chromosomes with a balancer eliminates crossovers, and only parental combinations survive. For convenience, balancer chromosomes are marked with a dominant morphological mutation. The marker allows the geneticist to track the segregation of the entire balancer or its normal homolog by following the presence or absence of the marker.

Reciprocal translocations

There are several types of translocations, but here we consider only reciprocal translocations, the simplest type. As with other rearrangements, meiosis in heterozygotes having two translocated chromosomes and their normal counterparts produces characteristic configurations. Figure 15-24 illustrates meiosis in an individual that is heterozygous for a reciprocal translocation. Notice that the pairing configuration is cross-shaped. Because the law of independent assortment is still in force, there are two common patterns of segregation. Let us use N_1 and N_2 to represent the normal chromosomes, and T_1 and T_2 the

Figure 15-25 **Photomicrograph of normal and aborted pollen of a semisterile corn plant.** The clear pollen grains contain chromosomally unbalanced meiotic products of a reciprocal translocation heterozygote. The opaque pollen grains, which contain either the complete translocation genotype or normal chromosomes, are functional in fertilization and development. [William Sheridan.]

translocated chromosomes. The segregation of each of the structurally normal chromosomes with one of the translocated ones $(T_1 + N_2$ and $T_2 + N_1)$ is called **adjacent-1 segregation.** Each of the two meiotic products is deficient for a different arm of the cross and has a duplicate of the other. These products are inviable. On the other hand, the two normal chromosomes may segregate together, as will the reciprocal parts of the translocated ones, to produce $N_1 + N_2$ and $T_1 + T_2$ products. This segregation pattern is called **alternate segregation.** These products are both balanced and viable.

Adjacent-1 and alternate segregations are equal in number, so half the overall population of gametes will be nonfunctional, a condition known as **semisterility** or "half sterility." Semisterility is an important diagnostic tool for identifying translocation heterozygotes. However, semisterility is defined differently for plants and animals. In plants, the 50 percent of meiotic products that are from the adjacent-1 segregation generally abort at the gametic stage (Figure 15-25). In animals, these products are viable as gametes but lethal to the zygotes they produce upon fertilization.

Remember that heterozygotes for inversions may also show some reduction in fertility, but by an amount dependent on the size of the affected region. The precise 50 percent reduction in viable gametes or zygotes is usually a reliable diagnostic clue for a translocation.

Genetically, genes on translocated chromosomes act as though they are linked if their loci are close to the translocation breakpoint. Figure 15-26 shows a translocation heterozygote that has been established by crossing an a/a ; b/b individual with a translocation homozygote bearing the wild-type alleles. On testcrossing the heterozygote, the only viable progeny are those bearing the parental genotypes, so linkage is seen between loci

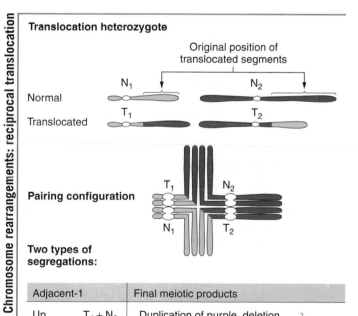

Translocation heterozygote

Original position of translocated segments

Normal N_1 N_2

Translocated T_1 T_2

Pairing configuration

T_1 N_2

N_1 T_2

Two types of segregations:

Adjacent-1		Final meiotic products	
Up	$T_1 + N_2$	Duplication of purple, deletion of orange translocated segment	Often inviable
Down	$N_1 + T_2$	Duplication of orange, deletion of purple translocated segment	
Alternate			
Up	$T_1 + T_2$	Translocation genotype	Both complete and viable
Down	$N_1 + N_2$	Normal	

Figure 15-24 **The two most commonly encountered chromosome segregation patterns in a reciprocal translocation heterozygote.** N_1 and N_2, normal nonhomologous chromosomes; T_1 and T_2, translocated chromosomes. Up and Down designate the opposite poles that homologs migrate to during anaphase I.

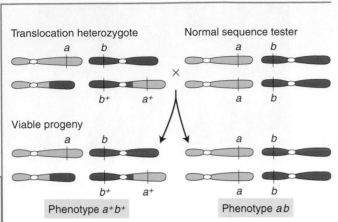

Translocation heterozygote

a   *b*

b⁺   *a⁺*

Viable progeny

a   *b*

b⁺   *a⁺*

Phenotype *a⁺b⁺*

×

Normal sequence tester

a   *b*

a   *b*

a   *b*

a   *b*

Phenotype *ab*

Figure 15-26 Pseudolinkage of genes due to translocation.
When a translocated fragment carries a marker gene, this marker can show linkage to genes on the other chromosome.

that were originally on different chromosomes. Apparent linkage of genes known normally to be on separate nonhomologous chromosomes—sometimes called **pseudolinkage**—is a genetic diagnostic clue to the presence of a translocation.

> **MESSAGE** Heterozygous reciprocal translocations are diagnosed genetically by semisterility and by the apparent linkage of genes whose normal loci are on separate chromosomes.

Applications of inversions and translocations

Inversions and translocations have proved to be useful genetic tools; some examples of their uses follow.

GENE MAPPING Inversions and translocations are useful for the mapping and subsequent isolation of specific genes. The gene for human neurofibromatosis was isolated in this way. The critical information came from people who not only had the disease, but also carried chromosomal translocations. All the translocations had one breakpoint in common, in a band close to the centromere of chromosome 17. Hence it appeared that this band must be the locus of the neurofibromatosis gene, which had been disrupted by the translocation breakpoint. Subsequent analysis showed that the chromosome 17 breakpoints were not at identical positions; however, since they must have been within the gene, the range of their positions revealed the segment of the chromosome that constituted the neurofibromatosis gene. Isolation of DNA fragments from this region eventually led to the recovery of the gene itself.

SYNTHESIZING SPECIFIC DUPLICATIONS OR DELETIONS Translocations and inversions are routinely used to delete or duplicate specific chromosome segments. Recall, for example, that both translocations and pericentric inversions generate products of meiosis that contain a duplication *and* a deletion (see Figures 15-23 and 15-24). If the duplicated or the deleted segment is very small, then the duplication-deletion meiotic products are tantamount to duplications or deletions, respectively. Duplications and deletions are useful for a variety of experimental applications, including the mapping of genes and the varying of gene dosage for the study of regulation, as we shall see in the following sections.

Another approach to creating duplications uses unidirectional *insertional* translocations, in which a segment of one chromosome is removed and inserted into another. In an insertional translocation heterozygote, a duplication results if the chromosome with the insertion segregates along with the normal copy.

POSITION-EFFECT VARIEGATION Gene action can be blocked by proximity to the densely staining chromosome regions called *heterochromatin*, and translocations or inversions can be used to study this effect. The locus for white eye color in *Drosophila* is near the tip of the X chromosome. Consider a translocation in which the tip of an X chromosome carrying w^+ is relocated next to the heterochromatic region of, say, chromosome 4 (Figure 15-27a, top section). **Position-effect variegation** is observed in flies that are heterozygotes for such a translocation. The normal X chromosome in such a heterozygote carries the recessive allele w. The eye phenotype is expected to be red because the wild-type allele is dominant to w. However, in such cases, the observed phenotype is a variegated mixture of red and white eye facets (Figure 15-27b). How can we explain the white areas? The w^+ allele is not always expressed because the heterochromatin boundary is somewhat variable: in some cells it engulfs and inactivates the w^+ gene, thereby allowing the expression of w. If the position of the w^+ and w alleles is exchanged by a crossover, then position-effect variegation is not detected (Figure 15-27a, lower section).

Deletions

A deletion is simply the loss of a part of one chromosome arm. The process of deletion requires two chromosome breaks to cut out the intervening segment. The deleted fragment has no centromere; consequently, it cannot be pulled to a spindle pole in cell division and is lost. The effects of deletions depend on their size. A small deletion *within* a gene, called an **intragenic deletion,** inactivates the gene and has the same effect as other null mutations of that gene. If the homozygous null phenotype is viable (as, for example, in human al-

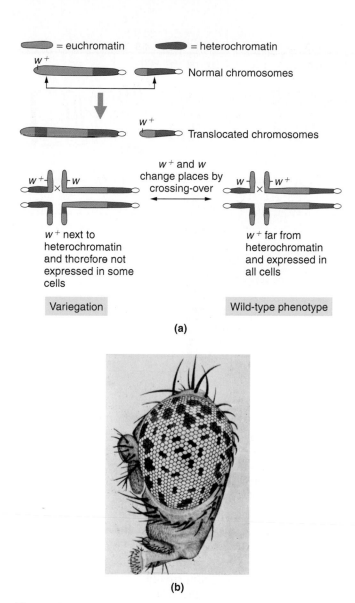

Variegation   Wild-type phenotype

(a)

(b)

Figure 15-27 Position-effect variegation. (a) The translocation of w^+ to a position next to heterochromatin causes the w^+ function to fail in some cells, producing position-effect variegation. (b) A *Drosophila* eye showing position-effect variegation. [Part b from Randy Mottus.]

binism), the homozygous deletion also will be viable. Intragenic deletions can be distinguished from mutations caused by single nucleotide changes because genes with such deletions never revert to wild-type.

For most of this section, we shall be dealing with **multigenic deletions,** in which several to many genes are missing. These have more severe consequences than do intragenic deletions. If such a deletion is made homozygous by inbreeding (that is, if both homologs have the same deletion), the combination is always lethal. This fact suggests that all regions of the chromosomes are essential for normal viability and that complete elimination of any segment from the genome is deleterious. Even an individual organism heterozygous for a multigenic deletion—

that is, having one normal homolog and one that carries the deletion—may not survive. Principally, this lethal outcome is due to disruption of normal gene balance. Alternatively, the deletion may "uncover" deleterious recessive alleles, allowing the single copies to be expressed.

> **MESSAGE** The lethality of large heterozygous deletions can be explained by gene imbalance and the expression of deleterious recessives.

Small deletions are sometimes viable in combination with a normal homolog. Such deletions may be identified by examining meiotic chromosomes under the microscope. The failure of the corresponding segment on the normal homolog to pair creates a visible **deletion loop** (Figure 15-28a). In flies, deletion loops are also detected in the polytene chromosomes, in which the homologs are tightly paired and aligned (Figure 15-28b). A deletion can be assigned to a specific chromosome location by examining polytene chromosomes microscopically and determining the position of the deletion loop.

Another clue to the presence of a deletion is that deletion of a segment on one homolog sometimes unmasks recessive alleles present on the other homolog, leading to their unexpected expression. Consider, for example, the deletion shown in the following diagram:

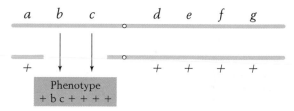

If there is no deletion, none of the seven recessive alleles is expected to be expressed; however, if *b* and *c* are

(a) Meiotic chromosomes   **(b) Polytene chromosomes**

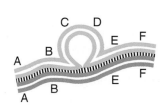

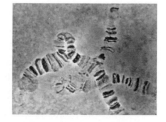

Figure 15-28 Looped configurations in a *Drosophila* deletion heterozygote. (a) In meiotic pairing, the normal homolog forms a loop. The genes in this loop have no alleles with which to synapse. (b) Because polytene chromosomes in *Drosophila* have specific banding patterns, we can infer which bands are missing from the homolog with the deletion by observing which bands appear in the loop of the normal homolog. [Part b from William M. Gelbart.]

Figure 15-29

Mapping mutant alleles using pseudodominance. A *Drosophila* strain heterozygous for deletion and normal chromosomes is used. The red bars show the extent of the deleted segments in 13 deletions. All recessive alleles in the same region that is deleted in a homologous chromosome will be expressed.

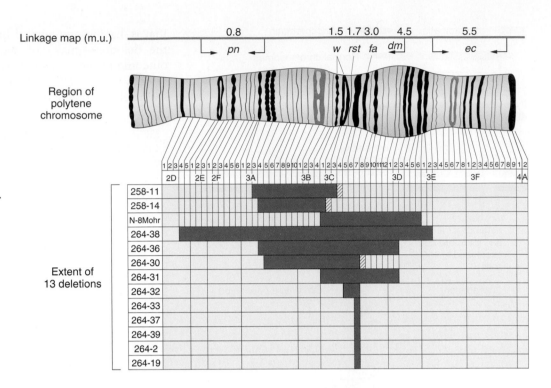

expressed, then a deletion spanning the b^+ and c^+ genes has probably occurred on the other homolog. Because in such cases it seems as if recessive alleles are showing dominance, the effect is called **pseudodominance.**

In the reverse case—if we already know the location of the deletion—we can apply the pseudodominance effect in the opposite direction to map the positions of mutant alleles. This procedure, called **deletion mapping,** pairs mutations against a set of defined overlapping deletions. An example from the fruit fly (*Drosophila*) is shown in Figure 15-29. In this diagram, the recombination map is shown at the top, marked with distances in map units from the left end. The horizontal red bars below the chromosome show the extent of the deletions listed at the left. Each deletion is paired with each mutation under test, and the phenotype is observed to see if the mutation is pseudodominant. The mutation prune (*pn*), for example, shows pseudodominance only with deletion 264-38, and this result determines its location in the 2D-4 to 3A-2 region. However, *fa* shows pseudodominance with all but two deletions (258-11 and 258-14), so its position can be pinpointed to band 3C-7, which is the region that all but two deletions have in common.

> **MESSAGE** Deletions can be recognized by deletion loops and pseudodominance.

Clinicians regularly find deletions in human chromosomes. The deletions are usually relatively small, but they do have an adverse effect, even though heterozy-

gous. Deletions of specific human chromosome regions cause unique syndromes of phenotypic abnormalities. One example is *cri du chat* syndrome, caused by a heterozygous deletion of the tip of the short arm of chromosome 5 (Figure 15-30). The specific bands deleted in *cri du chat* syndrome are 5p15.2 and 5p15.3, the two most distal bands identifiable on 5p. The most characteristic phenotype in the syndrome is the one that gives it its name, the distinctive catlike mewing cries made by affected infants. Other manifestations of the syndrome are microencephaly (abnormally small head) and a moonlike face. Like syndromes caused by other deletions, cri du chat syndrome also includes mental retardation. Fatality rates are low, and many persons with this deletion reach adulthood.

Another instructive example is Williams syndrome. This is an autosomal dominant syndrome characterized by unusual development of the nervous system and certain external features. Williams syndrome is found at a frequency of about 1 in 10,000 people. Patients often have pronounced musical or singing ability. It is almost always caused by a 1.5 Mb deletion on one homolog of chromosome 7. Sequence analysis showed that this segment contains 17 genes of known and unknown function. The abnormal phenotype is thus caused by haploinsufficiency of one or more of these 17 genes. Sequence analysis also reveals the origin of this deletion because the normal sequence is bounded by repeated copies of a gene called *PMS*, which happens to encode a DNA repair protein. As we have seen, repeated sequences can act as substrates for unequal crossing-over. Crossing-over between one of the copies from each end leads to a

Cri du chat syndrome

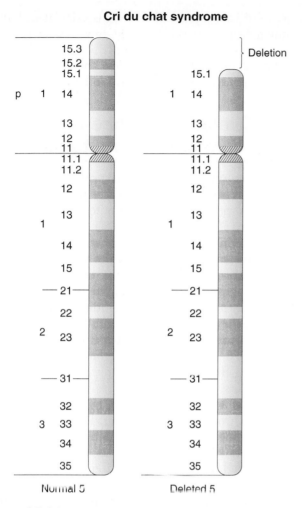

Figure 15-30 **Origin of cri du chat syndrome.** This human syndrome is caused by the loss of the tip of the short arm of one of the homologs of chromosome 5.

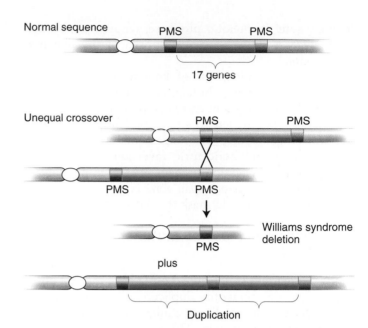

Figure 15-31 **Probable origin of Williams syndrome deletion.** A crossover between left and right repetitive flanking genes results in two reciprocal rearrangements, one of which corresponds to the Williams syndrome deletion.

netic content. In diploid plants, on the other hand, the pollen produced by a deletion heterozygote is of two types: functional pollen carrying the normal chromosome and nonfunctional (aborted) pollen carrying the deficient homolog. Thus, pollen cells seem to be sensitive to changes in the amount of chromosomal material, and this sensitivity might act to weed out deletions. This effect is analogous to the sensitivity of pollen to whole-chromosome aneuploidy, described earlier in this chapter. Unlike animal sperm cells, whose metabolic activity relies on enzymes that have already been deposited in them during their formation, pollen cells must germinate and then produce a long pollen tube that grows to fertilize the ovule. This growth requires that the pollen cell manufacture large amounts of protein, thus making it sensitive to genetic abnormalities in its own nucleus. Plant ovules, in contrast, are quite tolerant of deletions, presumably because they receive their nourishment from the surrounding maternal tissues.

Duplications

The processes of chromosome mutation sometimes produce an extra copy of some chromosome region. The duplicate regions can be located adjacent to each other—called a **tandem duplication**—or the extra copy can be located elsewhere in the genome—called an **insertional duplication**. A diploid cell containing a duplication will have three copies of the chromosome region in question; two in one chromosome set and one in the other. This is an example of a duplication

duplication (not found) and a Williams syndrome deletion, as shown in Figure 15-31.

Most human deletions, such as those that we have just considered, arise spontaneously in the gonads of a normal parent of an affected person; thus usually no signs of the deletions are found in the chromosomes of the parents. Less commonly, deletion-bearing individuals appear among the offspring of an individual having an undetected balanced rearrangement of chromosomes. For example, *cri du chat* syndrome can result from a parent heterozygous for a reciprocal translocation, because adjacent segregation produces deletions. Recombination within a pericentric inversion heterozygote also produces deletions. Animals and plants show differences in the survival of gametes or offspring that bear deletions. A male animal with a deletion in one chromosome produces sperm carrying one or the other of the two chromosomes in approximately equal numbers. These sperm seem to function to some extent regardless of their ge-

heterozygote. In meiotic prophase, tandem duplication heterozygotes show a loop comprised of the unpaired extra region.

Synthetic duplications of known coverage can be used for gene mapping. In haploids, for example, a chromosomally normal strain carrying a new recessive mutation *m* may be crossed to strains bearing a number of duplication-generating rearrangements (for example, translocations and pericentric inversions). In any one cross, if some duplication progeny are "m" in phenotype, the duplication does not span gene *m*, because if it did, its extra segment would mask the recessive *m* allele.

Identifying chromosome mutations by genomics

DNA microarrays (see Figure 12-27) have made it possible to detect and quantify duplications or deletions of a given DNA segment. The technique is called *comparative genomic hybridization*. The total DNA of the wild type and of a mutant are labeled with two different fluorescent dyes that emit distinct wavelengths of light. These labeled DNAs are added to a cDNA microarray together, and they both hybridize to the array. The array is then scanned with a detector tuned to one fluorescent wavelength, and then again for the other wavelength.

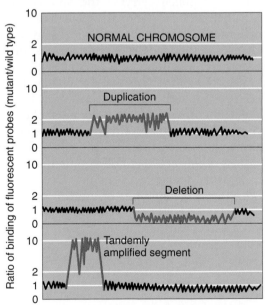

Figure 15-32 **Detecting changes in the amount of chromosomal DNA by comparative genomic hybridization.** Mutant and wild-type genomic DNA is tagged with dyes that fluoresce at different wavelengths. These are added to cDNA clones arranged in chromosomally ordered microarrays, and the ratio of bound fluorescence at each wavelength is calculated for each clone. The expected results for a normal genome and three types of mutants are illustrated.

The ratio of values for each cDNA is calculated. Ratios for mutant/wild type substantially greater than 1 represent regions that have been amplified. A ratio of 2 points to a duplication, and a ratio of less than 1 points to a deletion. Some examples are shown in Figure 15-32.

15.3 Overall incidence of human chromosome mutations

Chromosome mutations arise surprisingly frequently in human sexual reproduction, showing that the relevant cellular processes are prone to a high level of error. Figure 15-33 shows the estimated distribution of chromosome mutations among human conceptions that develop sufficiently to implant in the uterus. Of the estimated 15 percent of conceptions that abort spontaneously (pregnancies that terminate naturally), fully half show chromosomal abnormalities. Some medical geneticists believe that even this high level is an underestimate because many cases are never detected. Among live births, 0.6 percent have chromosomal abnormalities, resulting from both aneuploidy and chromosomal rearrangements.

Now that we have covered a number of analyses bearing on chromosome mutations, let's return to the family with the Down syndrome child, introduced at the beginning of the chapter. It is possible that the birth is indeed a coincidence—after all, coincidences do happen. However, the miscarriage gives a clue that something else might be going on. Recall that a large proportion of spontaneous abortions carry chromosomal abnormalities, so perhaps that is the case in this example. If so, the couple may have had two conceptions with chromosome mutations, which would be very unlikely unless there was a common cause. It is known that a small proportion of Down syndrome cases result from a translocation in one of the parents. We have seen that translocations can produce progeny that have extra material from part of the genome, so a translocation involving chromosome 21 can produce progeny that have extra material from that chromosome. In Down syndrome, the translocation responsible is of a type called a *Robertsonian translocation*. It produces progeny carrying an almost complete extra copy of chromosome 21. The translocation and its segregation are illustrated in Figure 15-34. Note that not only complements causing Down syndrome are produced, but other aberrant chromosome complements, too, most of which abort. In our example, the man may have this translocation, which he may have inherited from his grandmother. To confirm this, his chromosomes would be checked. His unaffected child might have normal chromosomes or might have inherited his translocation.

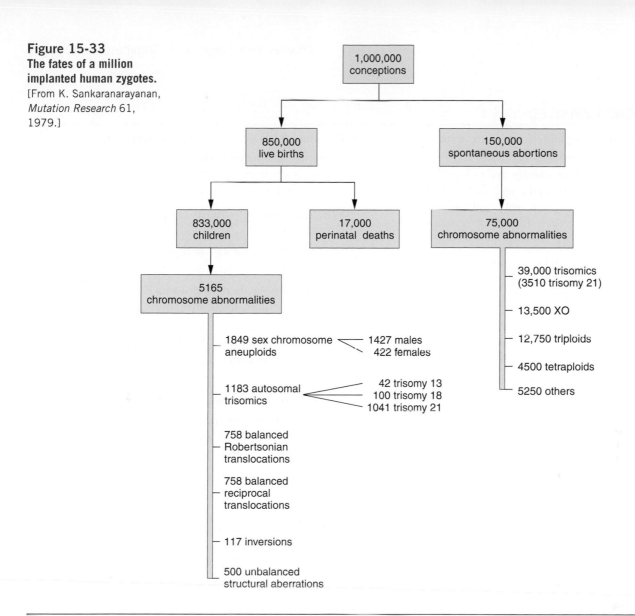

Figure 15-33
The fates of a million implanted human zygotes.
[From K. Sankaranarayanan, *Mutation Research* 61, 1979.]

1,000,000 conceptions

850,000 live births

150,000 spontaneous abortions

833,000 children

17,000 perinatal deaths

75,000 chromosome abnormalities

5165 chromosome abnormalities

1849 sex chromosome aneuploids
— 1427 males
— 422 females

1183 autosomal trisomics
— 42 trisomy 13
— 100 trisomy 18
— 1041 trisomy 21

758 balanced Robertsonian translocations

758 balanced reciprocal translocations

117 inversions

500 unbalanced structural aberrations

39,000 trisomics (3510 trisomy 21)

13,500 XO

12,750 triploids

4500 tetraploids

5250 others

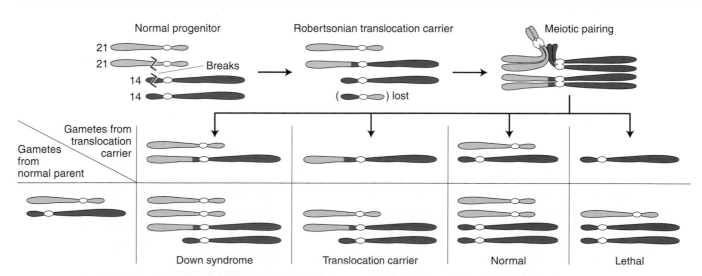

Figure 15-34 One origin of Down syndrome. In a small minority of cases, the origin of Down syndrome is a parent heterozygous for a Robertsonian translocation involving chromosome 21. Meiotic segregation results in some gametes carrying a chromosome with a large additional segment of chromosome 21. In combination with a normal chromosome 21 provided by the gamete from the opposite sex, the symptoms of Down syndrome are produced even though there is not full trisomy 21.

KEY QUESTIONS REVISITED

- **How common are polyploids (organisms with multiple chromosome sets)?**

A large proportion of plant species are polyploids of recent or ancient origin. Animals and fungi, too, show many examples of (mainly ancient) polyploidy. Many food crops are polyploid, created by plant geneticists.

- **How do polyploids arise?**

Either by spontaneous or colchicine-induced chromosome doubling (autopolyploids) or by fusion of gametes of different species, followed by doubling (allopolyploids).

- **Do polyploids have any special properties?**

Often they are bigger, and allopolyploids sometimes show combinations of parental characteristics. Some are sterile because of unpaired chromosomes.

- **Is the polyploid state transmissible to offspring?**

Yes, in even number polyploids the ploidy level is often transmitted to all progeny.

- **What inheritance patterns are observed in the progeny of polyploids?**

The phenotypic ratios are non-Mendelian because the starting point is not two alleles as in diploids, but four or six, etc. For example if an *A/A/a/a* autotetraploid is selfed, a 35 : 1 ratio is produced.

- **How do aneuploids (variants in which a single chromosome has been gained or lost) arise?**

Mainly by nondisjunction at meiosis I or II.

- **Do aneuploids show any special properties?**

The relative proportions of genes on the aneuploid chromosome is changed, generally leading to abnormalities caused by genetic imbalance.

- **What inheritance patterns are produced by aneuploids?**

They are non-Mendelian and depend on the nature of the aneuploid.

- **How do large-scale chromosome rearrangements (deletions, duplications, inversions, and translocations) arise?**

Either by chromosome breakage and reunion or by crossing-over between repetitive segments in different locations.

- **Do these rearrangements have any special properties?**

Yes, each has its own properties, e.g., semisterility for heterozygous reciprocal translocations, and pseudodominance of recessive alleles spanned by a deletion.

- **What inheritance patterns are produced by rearrangements?**

This varies with the rearrangement; for example, heterozygous translocations show linkage of genes on different chromosomes, and heterozygous inversions show lower RF values for genes in or spanning the inversion.

SUMMARY

Polyploidy is an abnormal condition in which there is a larger than normal number of chromosome sets. Polyploids such as triploids (3×) and tetraploids (4×) are common among plants and are represented even among animals. Organisms with an odd number of chromosome sets are sterile because there is not a partner for each chromosome at meiosis. Unpaired chromosomes pass randomly to the poles of the cell during meiosis, leading to unbalanced sets of chromosomes in the resulting gametes. Such unbalanced gametes do not yield viable progeny. In polyploids with an even number of sets, each chromosome has a potential pairing partner and hence can produce balanced gametes and progeny. Polyploidy can result in an organism of larger dimensions; this discovery has permitted important advances in horticulture and in crop breeding.

In plants, allopolyploids (polyploids formed by combining chromosome sets from different species) can be made by crossing two related species and then doubling the progeny chromosomes through the use of colchicine or through somatic cell fusion. These techniques have potential applications in crop breeding because allopolyploids combine the features of the two parental species.

When cellular accidents change parts of chromosome sets, aneuploids result. Aneuploids are important in the engineering of specific crop genotypes, although aneuploidy per se usually results in an unbalanced genotype with an abnormal phenotype. Examples of aneuploids include monosomics ($2n - 1$) and trisomics ($2n + 1$). Down syndrome (trisomy 21), Klinefelter syndrome (XXY), and Turner syndrome (XO) are well-documented examples of aneuploid conditions in humans. The spontaneous level of aneuploidy in humans is quite high and accounts for a large proportion of genetically based ill health in human populations. The phenotype of an aneuploid organism depends very much on the particular chromosome involved. In some cases, such as human trisomy 21, there is a highly characteristic constellation of associated phenotypes.

Most instances of aneuploidy result from accidental chromosome mis-segregation at meiosis (nondisjunction). The error is spontaneous and can occur in any particular meiocyte at the first or second division. In humans there is a maternal-age effect associated with nondisjunction of chromosome 21, resulting in a higher incidence of Down syndrome in the children of older mothers.

The other general category of chromosome mutations is structural rearrangements, which include deletions, duplications, inversions, and translocations. Chromosomal rearrangements are an important cause of ill health in human populations and are useful in engineering special strains of organisms for experimental and applied genetics. In organisms with one normal chromosome set plus a rearranged set (heterozygous rearrangements), there are unusual pairing structures at meiosis resulting from the strong pairing affinity of homologous chromosome regions. For example, heterozygous inversions show loops, and reciprocal translocations show cross-shaped structures. Segregation of these structures results in abnormal meiotic products unique to the rearrangement.

An inversion is a 180-degree turn of a part of a chromosome. In the homozygous state, inversions may cause little problem for an organism unless heterochromatin brings about a position effect or one of the breaks disrupts a gene. On the other hand, inversion heterozygotes show inversion loops at meiosis, and crossing-over within the loop results in inviable products. The crossover products of pericentric inversions, which span the centromere, differ from those of paracentric inversions, which do not, but both show reduced recombinant frequency in the affected region and often result in reduced fertility.

A translocation moves a chromosome segment to another position in the genome. A simple example is a reciprocal translocation, in which parts of nonhomologous chromosomes exchange positions. In the heterozygous state, translocations produce duplication-deletion meiotic products, which can lead to unbalanced zygotes. New gene linkages can be produced by translocations. Random segregation of centromeres in a translocation heterozygote results in 50 percent unbalanced meiotic products and, hence, 50 percent sterility (called *semisterility*).

Deletions are losses of a section of chromosome, either because of chromosome breaks followed by loss of the intervening segment or because of segregation in other heterozygous translocations or inversions. If the region removed in a deletion is essential to life, a homozygous deletion is lethal. Heterozygous deletions may be nonlethal, or they may be lethal because of chromosomal imbalance or because they uncover recessive deleterious alleles. When a deletion in one homolog allows the phenotypic expression of recessive alleles in the other, the unmasking of the recessive alleles is called *pseudodominance*.

Duplications are generally produced from other rearrangements or by aberrant crossing-over. They also unbalance the genetic material, producing a deleterious phenotypic effect or death of the organism. However, duplications can be a source of new material for evolution because function can be maintained in one copy, leaving the other copy free to evolve new functions.

KEY TERMS

acentric (p. 497)

acentric fragment (p. 499)

adjacent-1 segregation (p. 501)

allopolyploids (p. 485)

alternate segregation (p. 501)

amphidiploid (p. 487)

anaphase bridge (p. 497)

aneuploid (p. 485)

autopolyploids (p. 485)

balanced rearrangements (p. 497)

balancer (p. 500)

bivalents (p. 485)

chromosome mutations (p. 482)

deletion (p. 497)

deletion loop (p. 503)

deletion mapping (p. 504)

dicentric (p. 497)

dicentric bridge (p. 499)

disomic (p. 490)

dosage compensation (p. 495)

Down syndrome (p. 493)

duplication (p. 497)

embryoid (p. 489)

euploid (p. 483)

gene balance (p. 494)

gene-dosage effect (p. 495)

genetic load (p. 484)

hexaploid (p. 483)

homeologous (p. 485)

intragenic deletion (p. 502)

insertional duplication (p. 505)

inversion (p. 496)

inversion heterozygote (p. 499)

inversion loop (p. 499)

Klinefelter syndrome (p. 492)

monoploid (p. 483)

monosomic (p. 490)

multigenic deletions (p. 503)

nondisjunction (p. 491)

nullisomic (p. 490)

paracentric (p. 498)

parthenogenesis (p. 484)

pentaploid (p. 483)

pericentric (p. 498)

polyploids (p. 483)

position-effect variegation (p. 502)

pseudodominance (p. 504)

pseudolinkage (p. 502)

rearrangements (p. 496)

reciprocal translocation (p. 497)

semisterility (p. 501)

tandem duplication (p. 505)

tetraploid (p. 483)

translocation (p. 496)

triploid (p. 483)

trisomic (p. 490)

trivalent (p. 485)

Turner syndrome (p. 491)

unbalanced rearrangements (p. 497)

univalents (p. 485)

SOLVED PROBLEMS

1. A corn plant is obtained that is heterozygous for a reciprocal translocation and therefore is semisterile. This plant is crossed to a chromosomally normal strain that is homozygous for the recessive allele brachytic (*b*), located on chromosome 2. A semisterile F_1 plant is then backcrossed to the homozygous brachytic strain. The progeny obtained show the following phenotypes:

Nonbrachytic		Brachytic	
Semisterile	Fertile	Semisterile	Fertile
334	27	42	279

a. What ratio would you expect to result if the chromosome carrying the brachytic allele is not involved in the translocation?

b. Do you think that chromosome 2 is involved in the translocation? Explain your answer, showing the conformation of the relevant chromosomes of the semisterile F_1 and the reason for the specific numbers obtained.

Solution

a. We should start with the methodical approach and simply restate the data in the form of a diagram, where

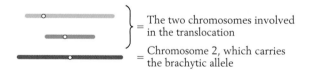

= The two chromosomes involved in the translocation

= Chromosome 2, which carries the brachytic allele

To simplify the diagram, we do not show the chromosomes divided into chromatids (although they would be at this stage of meiosis). We then show the first cross:

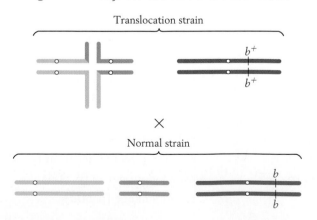

Translocation strain

×

Normal strain

All the progeny from this cross will be heterozygous for the chromosome carrying the brachytic allele, but what about the chromosomes involved in the translocation? In this chapter, we have seen that only alternate-segregation products survive, and that half of these survivors will be chromosomally normal and half will carry the two rearranged chromosomes. The rearranged combination will regenerate a translocation heterozygote when it combines with the chromosomally normal complement from the normal parent. These latter types— the semisterile F1's—are diagrammed as part of the backcross to the parental brachytic strain:

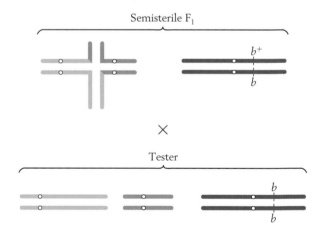

Semisterile F_1

×

Tester

In calculating the expected ratio of phenotypes from this cross, we can treat the behavior of the translocated chromosomes independently of the behavior of chromosome 2. Hence, we can predict that the progeny will be

$\frac{1}{2}$ translocation heterozygotes (semisterile)

⟶ $\frac{1}{2} b^+/b$ ⟶ $\frac{1}{4}$ semisterile nonbrachytic

⟶ $\frac{1}{2} b/b$ ⟶ $\frac{1}{4}$ semisterile brachytic

$\frac{1}{2}$ normal (fertile)

⟶ $\frac{1}{2} b^+/b$ ⟶ $\frac{1}{4}$ fertile nonbrachytic

⟶ $\frac{1}{2} b/b$ ⟶ $\frac{1}{4}$ fertile brachytic

This predicted $1:1:1:1$ ratio is quite different from that obtained in the actual cross.

b. Because we observe a departure from the expected ratio based on the independence of the brachytic phenotype and semisterility, it seems likely that chromosome 2 *is* involved in the translocation. Let's assume that the brachytic locus (*b*) is on the orange chromosome. But where? For the purpose of the diagram, it doesn't matter where we put it, but it does matter genetically because the position of the *b* locus affects the ratios in the progeny. If we assume that the *b* locus is near the tip of the piece that is translocated, we can redraw the pedigree:

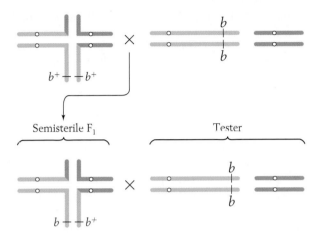

If the chromosomes of the semisterile F_1 segregate as diagrammed here, we could then predict

$\frac{1}{2}$ fertile, brachytic

$\frac{1}{2}$ semisterile, nonbrachytic

Most progeny are certainly of this type, so we must be on the right track. How are the two less frequent types produced? Somehow, we have to get the b^+ allele onto the normal orange chromosome and the *b* allele onto the translocated chromosome. This must be achieved by crossing-over between the translocation breakpoint (the center of the cross-shaped structure) and the brachytic locus.

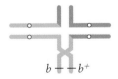

The recombinant chromosomes produce some progeny that are fertile and nonbrachytic and some that are semisterile and brachytic (these two classes together constitute 69 progeny of a total of 682, or a frequency of about 10 percent). We can see that this frequency is really a measure of the map distance (10 m.u.) of the brachytic locus from the breakpoint. (The same basic result would have been obtained if we had drawn the brachytic locus in the part of the chromosome on the other side of the breakpoint.)

2. We have lines of mice that breed true for two alternative behavioral phenotypes that we know are determined by two alleles at a single locus: *v* causes a mouse to move with a "waltzing" gait, whereas *V*

determines a normal gait. After crossing the true-breeding waltzers and normals, we observe that most of the F_1 is normal, but, unexpectedly, there is one waltzer female. We mate the F_1 waltzer with two different waltzer males and note that she produces only waltzer progeny. When we mate her with normal males, she produces normal progeny and no waltzers. We mate three of her normal female progeny with two of their brothers, and these mice produce sixty progeny, all normal. When, however, we mate one of these same three females with a third brother, we get six normals and two waltzers in a litter of eight. By thinking about the parents of the F_1 waltzer, we can consider some possible explanations of these results:

a. A dominant allele may have mutated to a recessive allele in her normal parent.

b. In one parent there may have been a dominant mutation in a second gene to create an epistatic allele that acts to prevent *V*'s expression, leading to waltzing.

c. Meiotic nondisjunction of the chromosome carrying *V* in her normal parent may have given a viable aneuploid.

d. There may have been a viable deletion spanning *V* in the meiocyte from her normal parent.

Which of these explanations are possible, and which are eliminated by the genetic analysis? Explain in detail.

Solution

The best way to answer the question is to take the explanations one at a time and see if each fits the results given.

a. Mutation *V* to *v*

This hypothesis requires that the exceptional waltzer female be homozygous *v/v*. This assumption is compatible with the results of mating her both to waltzer males, which would, if she is *v/v*, produce all waltzer offspring (*v/v*), and to normal males, which would produce all normal offspring (*V/v*). However, brother-sister matings within this normal progeny should then produce a 3 : 1 normal-to-waltzer ratio. Because some of the brother-sister matings actually produced no waltzers, this hypothesis does not explain the data.

b. Epistatic mutation *s* to *S*

Here the parents would be *V/V · s/s* and *v/v · s/s*, and a germinal mutation in one of them would give the F_1 waltzer the genotype *V/v · S/s*. When we crossed her with a waltzer male, who would be of the genotype *v/v · s/s*, we would expect some *V/v · s/s* progeny, which would be phenotypically normal. However, we saw no normal progeny from this cross, so the hypothesis is already overthrown. Linkage could save the hypothesis temporarily if we assumed that the mutation was in the normal parent, giving a gamete *V S*. Then the F_1 waltzer would be *V S/v s*, and if linkage were tight enough, few or no *V s* gametes would be produced, the type that are necessary to combine with the *v s* gamete from the male to give *V s/v s* normals. However, if this were true, the cross with the normal males would be *V S/v s* × *V s/V s*, and this would give a high percentage of *V S/V s* progeny, which would be waltzers, none of which were seen.

c. Nondisjunction in the normal parent

This explanation would give a nullisomic gamete that would combine with *v* to give the F_1 waltzer the hemizygous genotype *v*. The subsequent matings would be

- *v* × *v/v*, which gives *v/v* and *v* progeny, all waltzers. This fits.
- *v* × *V/V*, which gives *V/v* and *V* progeny, all normals. This also fits.
- First intercrosses of normal progeny: *V* × *V*. This gives *V* and *V/V*, which are all normal. This fits.
- Second intercrosses of normal progeny: *V* × *V/v*. This gives 25 percent each of *V/V*, *V/v*, *V* (all normals), and *v* (waltzers). This also fits.

This hypothesis is therefore consistent with the data.

d. Deletion of *V* in normal parent

Let's call the deletion D. The F_1 waltzer would be D/*v*, and the subsequent matings would be:

- D/*v* × *v/v*. This gives *v/v* and D/*v*, all of which are waltzers. This fits.
- D/*v* × *V/V*. This gives *V/v* and D/*V*, all of which are normal. This fits.
- First intercrosses of normal progeny: D/*V* × D/*V*. This gives D/*V* and *V/V*, all normal. This fits.
- Second intercrosses of normal progeny: D/*V* × *V/v*. This gives 25 percent of each of *V/V*, *V/v*, D/*V* (all normals), and D/*v* (waltzers). This also fits.

Once again, the hypothesis fits the data provided, so we are left with two hypotheses that are compatible with the results, and further experiments are necessary to distinguish them. One obvious way of doing this would be to examine the chromosomes of the exceptional female under the microscope; aneuploidy should be relatively easy to distinguish from deletion.

PROBLEMS

BASIC PROBLEMS

1. In keeping with the style of Table 15-1, what would you call organisms that are MM N OO; MM NN OO; MMM NN PP?

2. A large plant arose in a natural population. Qualitatively it looked just the same, except much larger. Is it more likely to be an allopolyploid or an autopolyploid? How would you test that it was a polyploid and not just growing in rich soil?

3. Is a trisomic an aneuploid or a polyploid?

4. In a tetraploid *B/B/b/b*, how many quadrivalent possible pairings are there? Draw them (see Figure 15-7).

5. Someone tells you that cauliflower is an amphidiploid. Do you agree?

6. Why is *Raphanobrassica* fertile, whereas its progenitor wasn't?

7. In the designation of wheat genomes, how many chromosomes are represented by the letter B?

8. How would you "re-create" hexaploid bread wheat from *Triticum tauschii* and Emmer?

9. How would you make a monoploid plantlet starting with a diploid plant?

10. A disomic product of meiosis is obtained. What is its likely origin? What other genotypes would you expect among the products of that meiosis under your hypothesis?

11. Can a trisomic *A/A/a* ever produce a gamete of genotype *a*?

12. Which, if any, of the following sex chromosome aneuploids in humans are fertile: XXX, XXY, XYY, XO?

13. Why are older mothers routinely given amniocentesis or CVS?

14. In an inversion, is a 5′ DNA end ever joined to another 5′ end? Explain.

15. If you observed a dicentric bridge at meiosis, what rearrangement would you predict had occurred?

16. Why do acentric fragments get lost?

17. Draw a case of a translocation arising from repetitive DNA. Repeat for a deletion.

18. From a large stock of *Neurospora* rearrangements available from the fungal genetics stock center, what type would you choose to synthesize a strain that had a duplication of the right arm of chromosome 3 and a deletion for the tip of chromosome 4?

19. You observe a very large pairing loop at meiosis. Is it more likely to be from a heterozygous inversion or heterozygous deletion? Explain.

20. A new recessive mutant allele doesn't show pseudodominance with any of the deletions that span *Drosophila* chromosome 2. What might be the explanation?

21. Compare and contrast the origins of Turner syndrome, Williams syndrome, cri du chat syndrome, and Down syndrome. (Why are they called *syndromes*?)

22. List the diagnostic features (genetic or cytological) that are used to identify these chromosomal alterations:

   a. Deletions

   b. Duplications

   c. Inversions

   d. Reciprocal translocations

23. The normal sequence of nine genes on a certain *Drosophila* chromosome is 123 · 456789, where the dot represents the centromere. Some fruit flies were found to have aberrant chromosomes with the following structures:

   a. 123 · 476589

   b. 123 · 46789

   c. 1654 · 32789

   d. 123 · 4566789

   Name each type of chromosomal rearrangement, and draw diagrams to show how each would synapse with the normal chromosome.

24. The two loci *P* and *Bz* are normally 36 m.u. apart on the same arm of a certain plant chromosome. A paracentric inversion spans about one-fourth of this region but does not include either of the loci. What approximate recombinant frequency between *P* and *Bz* would you predict in plants that are

   a. Heterozygous for the paracentric inversion?

   b. Homozygous for the paracentric inversion?

25. As we saw in Solved Problem 2, certain mice called *waltzers* have a recessive mutation that causes them to execute bizarre steps. W. H. Gates crossed waltzers with homozygous normals and found, among several hundred normal progeny, a single waltzing female mouse. When mated with a waltzing male, she produced all waltzing offspring. When mated with a homozygous normal male, she produced all normal progeny. Some males and females of this normal progeny were intercrossed, and there were no waltzing offspring among their progeny. T. S. Painter examined the chromosomes of waltzing mice that were derived from some of Gates's crosses and that showed a breeding behavior similar to that of the original, unusual waltzing female. He found that these individuals had 40 chromosomes, just as in normal mice or the usual waltzing mice. In the unusual waltzers, however, one member of a chromosome pair was abnormally short. Interpret these observations as completely as possible, both genetically and cytologically.

(Problem 25 is from A. M. Srb, R. D. Owen, and R. S. Edgar, *General Genetics*, 2d ed. W. H. Freeman and Company, 1965.)

26. Six bands in a salivary gland chromosome of *Drosophila* are shown in the following figure, along with the extent of five deletions (Del 1 to Del 5):

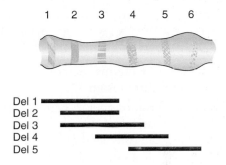

Recessive alleles *a*, *b*, *c*, *d*, *e*, and *f* are known to be in the region, but their order is unknown. When the deletions are combined with each allele, the following results are obtained:

	a	*b*	*c*	*d*	*e*	*f*
Del 1	−	−	−	+	+	+
Del 2	−	+	−	+	+	+
Del 3	−	+	−	+	−	+
Del 4	+	+	−	−	−	+
Del 5	+	+	+	−	−	−

In this table, a minus sign means that the deletion is missing the corresponding wild-type allele (the deletion uncovers the recessive), and a plus sign means that the corresponding wild-type allele is still present. Use these data to infer which salivary band contains each gene.

(Problem 26 is from D. L. Hartl, D. Friefelder, and L. A. Snyder, *Basic Genetics*. Jones and Bartlett, 1988.)

27. A fruit fly was found to be heterozygous for a paracentric inversion. However, it was impossible to obtain flies that were homozygous for the inversion even after many attempts. What is the most likely explanation for this inability to produce a homozygous inversion?

28. Orangutans are an endangered species in their natural environment (the islands of Borneo and Sumatra), so a captive breeding program has been established using orangutans currently held in zoos throughout the world. One component of this program is research into orangutan cytogenetics. This research has shown that all orangutans from Borneo carry one form of chromosome 2, as shown in the accompanying diagram, and all orangutans from Sumatra carry the other form. Before this cytogenetic difference became known, some matings were carried out between animals from different islands, and 14 hybrid progeny are now being raised in captivity.

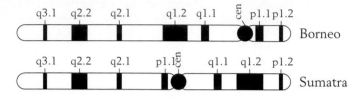

a. What term or terms describe the differences between these chromosomes?

b. Draw the chromosomes 2, paired during the first meiotic prophase, of such a hybrid orangutan. Be sure to show all the landmarks indicated in the accompanying diagram, and label all parts of your drawing.

c. In 30 percent of meioses, there will be a crossover somewhere in the region between bands p1.1 and q1.2. Draw the gamete chromosomes 2 that would result from a meiosis in which a single crossover occurred within band q1.1.

d. What fraction of the gametes produced by a hybrid orangutan will give rise to viable progeny, if these are the only chromosomes that differ between the parents?

(Problem 28 is from Rosemary Redfield.)

29. In corn, the genes for tassel length (alleles *T* and *t*) and rust resistance (alleles *R* and *r*) are known to be on separate chromosomes. In the course of making routine crosses, a breeder noticed that one *T/t* ; *R/r* plant gave unusual results in a testcross with the double-recessive pollen parent *t/t* ; *r/r*. The results were

Progeny: *T/t* ; *R/r*   98
            t/t ; *r/r*   104
            T/t ; *r/r*   3
            t/t ; *R/r*   5
Corncobs: Only about half as many seeds as usual

a. What key features of the data are different from the expected results?

b. State a concise hypothesis that explains the results.

c. Show genotypes of parents and progeny.

d. Draw a diagram showing the arrangement of alleles on the chromosomes.

e. Explain the origin of the two classes of progeny having three and five members.

 UNPACKING PROBLEM 29

1. What do a "gene for tassel length" and a "gene for rust resistance" mean?

2. Does it matter that the precise meaning of the allelic symbols *T*, *t*, *R*, and *r* is not given? Why or why not?

3. How do the terms *gene* and *allele*, as used here, relate to the concepts of locus and gene pair?

4. What prior experimental evidence would give the corn geneticist the idea that the two genes are on separate chromosomes?

5. What do you imagine "routine crosses" are to a corn breeder?

6. What term is used to describe genotypes of the type *T/t* ; *R/r*?

7. What is a "pollen parent"?

8. What are testcrosses, and why do geneticists find them so useful?

9. What progeny types and frequencies might the breeder have been expecting from the testcross?

10. Describe how the observed progeny differ from expectations.

11. What does the approximate equality of the first two progeny classes tell you?

12. What does the approximate equality of the second two progeny classes tell you?

13. What were the gametes from the unusual plant, and what were their proportions?

14. Which gametes were in the majority?

15. Which gametes were in the minority?

16. Which of the progeny types seem to be recombinant?

17. Which allelic combinations appear to be linked in some way?

18. How can there be linkage of genes supposedly on separate chromosomes?

19. What do these majority and minority classes tell us about the genotypes of the parents of the unusual plant?

20. What is a corncob?

21. What does a normal corncob look like? (Sketch one and label it.)

22. What do the corncobs from this cross look like? (Sketch one.)

23. What exactly is a kernel?

24. What effect could lead to the absence of half the kernels?

25. Did half the kernels die? If so, was the female or the male parent the reason for the deaths?

Now try to solve the problem.

30. A yellow body in *Drosophila* is caused by a mutant allele *y* of a gene located at the tip of the X chromosome (the wild-type allele causes a gray body). In a radiation experiment, a wild-type male was irradiated with X rays and then crossed with a yellow-bodied female. Most of the male progeny were yellow, as expected, but the scanning of thousands of flies revealed two gray-bodied (phenotypically wild-type) males. These gray-bodied males were crossed with yellow-bodied females, with the following results:

	Progeny
Gray male 1 × yellow female	Females all yellow
	Males all gray
Gray male 2 × yellow female	$\frac{1}{2}$ females yellow
	$\frac{1}{2}$ females gray
	$\frac{1}{2}$ males yellow
	$\frac{1}{2}$ males gray

a. Explain the origin and crossing behavior of gray male 1.

b. Explain the origin and crossing behavior of gray male 2.

31. In corn, the allele *Pr* stands for green stems, *pr* for purple stems. A corn plant of genotype *pr/pr* that has standard chromosomes is crossed with a *Pr/Pr* plant that is homozygous for a reciprocal translocation between chromosomes 2 and 5. The F₁ is semisterile and phenotypically Pr. A backcross with the parent with standard chromosomes gives 764 semisterile Pr, 145 semisterile pr, 186 normal Pr, and 727 normal pr. What is the map distance between the *Pr* locus and the translocation point?

32. Distinguish between Klinefelter, Down, and Turner syndromes.

33. Show how one could make an allotetraploid between two related diploid plant species, both of which are $2n = 28$.

34. In *Drosophila*, trisomics and monosomics for the tiny chromosome 4 are viable, but nullisomics and tetrasomics are not. The *b* locus is on this chromosome. Deduce the phenotypic proportions in the progeny of the following crosses of trisomics.

a. *b⁺/b/b × b/b*

b. *b⁺/b⁺/b × b/b*

c. *b⁺/b⁺/b × b⁺/b*

35. A woman with Turner syndrome is found to be color-blind (an X-linked recessive phenotype). Both her mother and her father have normal vision.

a. Explain the simultaneous origin of Turner syndrome and color blindness by the abnormal behavior of chromosomes at meiosis.

b. Can your explanation distinguish whether the abnormal chromosome behavior occurred in the father or the mother?

c. Can your explanation distinguish whether the abnormal chromosome behavior occurred at the first or second division of meiosis?

d. Now assume that a color-blind Klinefelter man has parents with normal vision, and answer parts a, b, and c.

36. a. How would you synthesize a pentaploid?

b. How would you synthesize a triploid of genotype *A/a/a*?

c. You have just obtained a rare recessive mutation *a** in a diploid plant, which Mendelian analysis tells you is *A/a**. From this plant, how would you synthesize a tetraploid (4 ×) of genotype *A/A/a*/a**?

d. How would you synthesize a tetraploid of genotype *A/a/a/a*?

37. Suppose you have a line of mice that has cytologically distinct forms of chromosome 4. The tip of the chromosome can have a knob (called 4^K) or a satellite (4^S) or neither (4). Here are sketches of the three types:

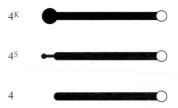

You cross a 4^K/4^S female with a 4/4 male and find that most of the progeny are 4^K/4 or 4^S/4, as expected. However, you occasionally find some rare types as follows (all other chromosomes are normal):

a. 4^K/4^K/4

b. 4^K/4^S/4

c. 4^K

Explain the rare types that you have found. Give, as precisely as possible, the stages at which they originate, and state whether they originate in the male parent, the female parent, or the zygote. (Give reasons briefly.)

38. A cross is made in tomatoes between a female plant that is trisomic for chromosome 6 and a normal diploid male plant that is homozygous for the recessive allele for potato leaf (*p/p*). A trisomic F₁ plant is backcrossed to the potato-leaved male.

a. What is the ratio of normal-leaved plants to potato-leaved plants when you assume that *p* is located on chromosome 6?

b. What is the ratio of normal-leaved to potato-leaved plants when you assume that p is not located on chromosome 6?

39. A tomato geneticist attempts to assign five recessive genes to specific chromosomes by using trisomics. She crosses each homozygous mutant ($2n$) to each of three trisomics, involving chromosomes 1, 7, and 10. From these crosses, the geneticist selects trisomic progeny (which are less vigorous) and backcrosses them to the appropriate homozygous recessive. The *diploid* progeny from these crosses are examined. Her results, in which the ratios are wild type : mutant, are as follows:

Trisomic chromo-some	Gene				
	d	y	c	h	cot
1	48:55	72:29	56:50	53:54	32:28
7	52:56	52:48	52:51	58:56	81:40
10	45:42	36:33	28:32	96:50	20:17

Which of the genes can the geneticist assign to which chromosomes? (Explain your answer fully.)

40. A petunia is heterozygous for the following autosomal homologs:

```
    A   B   C   D   E   F   G   H   I
    ●
    a   b   c   d   h   g   f   e   i
    ●
```

a. Draw the pairing configuration that you would see at metaphase I, and label all parts of your diagram. Number the chromatids sequentially from top to bottom of the page.

b. A three-strand double crossover occurs, with one crossover between the C and D loci on chromatids 1 and 3, and the second crossover between the G and H loci on chromatids 2 and 3. Diagram the results of these recombination events as you would see them at anaphase I, and label your diagram.

c. Draw the chromosome pattern that you would see at anaphase II after the crossovers described in part b.

d. Give the genotypes of the gametes from this meiosis that will lead to the formation of viable progeny. Assume that all gametes are fertilized by pollen that has the gene order A B C D E F G H I.

41. Two groups of geneticists, in California and in Chile, begin work to develop a linkage map of the medfly. They both independently find that the loci for body color (B = black, b = gray) and eye shape (R = round, r = star) are linked 28 m.u. apart. They send strains to each other and perform crosses; a summary of all their findings is shown here:

Cross	F_1	Progeny of F_1 × any b r/b r	
B R/B R (Calif.) × b r/b r (Calif.)	B R/b r	B R/b r	36%
		b r/b r	36
		B r/b r	14
		b R/b r	14
B R/B R (Chile) × b r/b r (Chile)	B R/b r	B R/b r	36
		b r/b r	36
		B r/b r	14
		b R/b r	14
B R/B R (Calif.) × b r/b r (Chile) or b r/b r (Calif.) × B R/B R (Chile)	B R/b r	B R/b r	48
		b r/b r	48
		B r/b r	2
		b R/b r	2

a. Provide a genetic hypothesis that explains the three sets of testcross results.

b. Draw the key chromosomal features of meiosis in the F_1 from a cross of the Californian and Chilean lines.

42. An aberrant corn plant gives the following RF values when testcrossed:

	Interval				
	d–f	f–b	b–x	x–y	y–p
Control	5	18	23	12	6
Aberrant plant	5	2	2	0	6

(The locus order is centromere–d–f–b–x–y–p.) The aberrant plant is a healthy plant, but it produces far fewer normal ovules and pollen than the control plant.

a. Propose a hypothesis to account for the abnormal recombination values and the reduced fertility in the aberrant plant.

b. Use diagrams to explain the origin of the recombinants according to your hypothesis.

43. The following corn loci are on one arm of chromosome 9 in the order indicated (the distances between them are shown in map units):

$$c\mathrm{—}bz\mathrm{—}wx\mathrm{—}sh\mathrm{—}d\mathrm{—centromere}$$
$$\ \ \ 12\ \ \ 8\ \ \ \ 10\ \ \ 20\ \ 10$$

C gives colored aleurone; c, white aleurone.
Bz gives green leaves; *bz*, bronze leaves.
Wx gives starchy seeds; *wx*, waxy seeds.
Sh gives smooth seeds; *sh*, shrunken seeds.
D gives tall plants; *d*, dwarf.

Table for Problem 45

Mating	Embryos (mean number)			Degeneration (%)
	Implanted in the uterine wall	Degenerating after implantation	Normal	
exceptional ♂ × normal ♀	8.7	5.0	3.7	57.5
normal ♂ × normal ♀	9.5	0.6	8.9	6.5

A plant from a standard stock that is homozygous for all five recessive alleles is crossed to a wild-type plant from Mexico that is homozygous for all five dominant alleles. The F_1 plants express all the dominant alleles and, when backcrossed to the recessive parent, give the following progeny phenotypes:

colored, green, starchy, smooth, tall	360
white, bronze, waxy, shrunk, dwarf	355
colored, bronze, waxy, shrunk, dwarf	40
white, green, starchy, smooth, tall	46
colored, green, starchy, smooth, dwarf	85
white, bronze, waxy, shrunk, tall	84
colored, bronze, waxy, shrunk, tall	8
white, green, starchy, smooth, dwarf	9
colored, green, waxy, smooth, tall	7
white, bronze, starchy, shrunk, dwarf	6

Propose a hypothesis to explain these results. Include:

a. A general statement of your hypothesis, with diagrams if necessary

b. Why there are 10 classes

c. An account of the origin of each class, including its frequency

d. At least one test of your hypothesis

44. Chromosomally normal corn plants have a p locus on chromosome 1 and an s locus on chromosome 5.

P gives dark-green leaves; p, pale-green.
S gives large ears; s, shrunken ears.

An original plant of genotype P/p ; S/s has the expected phenotype (dark-green, large ears) but gives unexpected results in crosses as follows:

- On selfing, fertility is normal, but the frequency of p/p ; s/s types is $\frac{1}{4}$ (not $\frac{1}{16}$, as expected).
- When crossed with a normal tester of genotype p/p ; s/s, the F_1 progeny are $\frac{1}{2}$ P/p ; S/s and $\frac{1}{2}$ p/p ; s/s ; fertility is normal.
- When an F_1 P/p ; S/s plant is crossed with a normal p/p ; s/s tester, it proves to be semisterile, but again the progeny are $\frac{1}{2}$ P/p ; S/s and $\frac{1}{2}$ p/p ; s/s.

Explain these results, showing the full genotypes of the original plant, the tester, and the F_1 plants. How would you test your hypothesis?

45. A male rat that is phenotypically normal shows reproductive anomalies when compared with normal male rats, as shown in the table above. Propose a genetic explanation of these unusual results, and indicate how your idea could be tested.

46. A tomato geneticist working on Fr, a dominant mutant allele that causes rapid fruit ripening, decides to find out which chromosome contains this gene by using a set of lines, each of which is trisomic for one chromosome. To do so, she crosses a homozygous diploid mutant to each of the wild-type trisomic lines.

a. A trisomic F_1 plant is crossed to a diploid wild-type plant. What is the ratio of fast- to slow-ripening plants in the *diploid* progeny of this second cross if Fr is on the trisomic chromosome? Use diagrams to explain.

b. What is the ratio of fast- to slow-ripening plants in the *diploid* progeny of this second cross if Fr is not located on the trisomic chromosome? Use diagrams to explain.

c. Here are the results of the crosses. On which chromosome is Fr and why?

Trisomic chromosome	Fast ripening:slow ripening in diploid progeny
1	45:47
2	33:34
3	55:52
4	26:30
5	31:32
6	37:41
7	44:79
8	49:53
9	34:34
10	37:39

(Problem 46 is from Tamara Western.)

CHALLENGING PROBLEMS

47. The *Neurospora un-3* locus is near the centromere on chromosome 1, and crossovers between *un-3* and the centromere are very rare. The *ad-3* locus is on the other side of the centromere of the same chromosome, and crossovers occur between *ad-3* and the centromere in about 20 percent of meioses (no multiple crossovers occur).

a. What types of linear asci (see Chapter 4) do you predict, and in what frequencies, in a normal cross of *un-3 ad-3* × wild type? (Specify genotypes of spores in the asci.)

b. Most of the time such crosses behave predictably, but in one case, a standard *un-3 ad-3* strain was crossed with a wild type isolated from a field of sugarcane in Hawaii. The results follow:

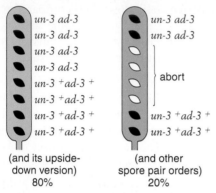

un-3 ad-3
un-3 ad-3
un-3 ad-3
un-3 ad-3
un-3 +ad-3 +
un-3 +ad-3 +
un-3 +ad-3 +
un-3 +ad-3 +

un-3 ad-3
un-3 ad-3
abort
un-3 +ad-3 +
un-3 +ad-3 +

(and its upside-down version)
80%

(and other spore pair orders)
20%

Explain these results, and state how you could test your idea. (**Note:** In *Neurospora*, ascospores with extra chromosomal material survive and are the normal black color, whereas ascospores lacking any chromosome region are white and inviable.)

48. Two mutations in *Neurospora*, *ad-3* and *pan-2*, are located on chromosomes 1 and 6, respectively. An unusual *ad-3* line arises in the laboratory, giving the results shown in the table at the bottom of this page. Explain all three results with the aid of clearly labeled diagrams. (**Note:** In *Neurospora*, ascospores with extra chromosomal material survive and are the normal black color, whereas ascospores lacking any chromosome region are white and inviable.)

49. Deduce the phenotypic proportions in the progeny of the following crosses of autotetraploids in which the a^+/a locus is very close to the centromere. (Assume that the four homologous chromosomes of any one type pair randomly two-by-two and that only one copy of the a^+ allele is necessary for the wild-type phenotype.)

a. $a^+/a^+/a/a$ × $a/a/a/a$
b. $a^+/a/a/a$ × $a/a/a/a$
c. $a^+/a/a/a$ × $a^+/a/a/a$
d. $a^+/a^+/a/a$ × $a^+/a/a/a$

50. The New World cotton species *Gossypium hirsutum* has a $2n$ chromosome number of 52. The Old World species *G. thurberi* and *G. herbaceum* each have a $2n$ number of 26. Hybrids between these species show the following chromosome pairing arrangements at meiosis:

Hybrid	Pairing arrangement
G. hirsutum × *G. thurberi*	13 small bivalents + 13 large univalents
G. hirsutum × *G. herbaceum*	13 large bivalents + 13 small univalents
G. thurberi × *G. herbaceum*	13 large univalents + 13 small univalents

Draw diagrams to interpret these observations phylogenetically, clearly indicating the relationships between the species. How would you go about proving that your interpretation is correct?

(Problem 50 is adapted from A. M. Srb, R. D. Owen, and R. S. Edgar, *General Genetics*, 2d ed. W. H. Freeman and Company, 1965.)

51. There are six main species in the *Brassica* genus: *B. carinata*, *B. campestris*, *B. nigra*, *B. oleracea*, *B. juncea*, and *B. napus*. You can deduce the interrelationships between these six species from the following table:

Species or F₁ hybrid	Chromosome number	Number of bivalents	Number of univalents
B. juncea	36	18	0
B. carinata	34	17	0
B. napus	38	19	0
B. juncea × *B. nigra*	26	8	10
B. napus × *B. campestris*	29	10	9
B. carinata × *B. oleracea*	26	9	8
B. juncea × *B. oleracea*	27	0	27
B. carinata × *B. campestris*	27	0	27
B. napus × *B. nigra*	27	0	27

"Results" for Problem 48	Ascospore appearance	RF between *ad-3* and *pan-2*
1. Normal *ad-3* × normal *pan-2*	All black	50%
2. Abnormal *ad-3* × normal *pan-2*	About ½ black and ½ white (inviable)	1%
3. Of the black spores from cross 2, about half were completely normal, and half repeated the same behavior as the original abnormal *ad-3* strain		

a. Deduce the chromosome number of *B. campestris*, *B. nigra*, and *B. oleracea*.

b. Show clearly any evolutionary relationships between the six species that you can deduce at the chromosomal level.

52. Several kinds of sexual mosaicism are well documented in humans. Suggest how each of the following examples may have arisen by nondisjunction at *mitosis*:

a. XX/XO (that is, there are two cell types in the body, XX and XO)

b. XX/XXYY

c. XO/XXX

d. XX/XY

e. XO/XX/XXX

53. In *Drosophila*, a cross (cross 1) was made between two mutant flies, one homozygous for the recessive mutation bent wing (*b*) and the other homozygous for the recessive mutation eyeless (*e*). The mutations *e* and *b* are alleles of two different genes that are known to be very closely linked on the tiny autosomal chromosome 4. All the progeny had a wild-type phenotype. One of the female progeny was crossed with a male of genotype *b e/b e*; we shall call this *cross 2*. Most of the progeny of cross 2 were of the expected types, but there was also one rare female of wild-type phenotype.

a. Explain what the common progeny are expected to be from cross 2.

b. Could the rare wild-type female have arisen by (1) crossing-over or (2) nondisjunction? Explain.

c. The rare wild-type female was testcrossed to a male of genotype *b e/b e* (cross 3). The progeny were

$\frac{1}{6}$ wild type

$\frac{1}{6}$ bent, eyeless

$\frac{1}{3}$ bent

$\frac{1}{3}$ eyeless

Which of the explanations in part b is compatible with this result? Explain the genotypes and phenotypes of the progeny of cross 3 and their proportions.

 UNPACKING PROBLEM 53

1. Define *homozygous, mutation, allele, closely linked, recessive, wild type, crossing-over, nondisjunction, testcross, phenotype,* and *genotype.*

2. Does this problem concern sex linkage? Explain.

3. How many chromosomes does *Drosophila* have?

4. Draw a clear pedigree summarizing the results of crosses 1, 2, and 3.

5. Draw the gametes produced by both parents in cross 1.

6. Draw the chromosome 4 constitution of the progeny of cross 1.

7. Is it surprising that the progeny of cross 1 are wild-type phenotype? What does this outcome tell you?

8. Draw the chromosome 4 constitution of the male tester used in cross 2 and the gametes that he can produce.

9. With respect to chromosome 4, what gametes can the female parent in cross 2 produce in the absence of nondisjunction? Which would be common and which rare?

10. Draw first- and second-division meiotic nondisjunction in the female parent of cross 2, as well as the resulting gametes.

11. Are any of the gametes from part 10 aneuploid?

12. Would you expect aneuploid gametes to give rise to viable progeny? Would these progeny be nullisomic, monosomic, disomic, or trisomic?

13. What progeny phenotypes would be produced by the various gametes considered in parts 9 and 10?

14. Consider the phenotypic ratio in the progeny of cross 3. Many genetic ratios are based on halves and quarters, but this ratio is based on thirds and sixths. What might this point to?

15. Could there be any significance to the fact that the crosses concern genes on a very small chromosome? When is chromosome size relevant in genetics?

16. Draw the progeny expected from cross 3 under the two hypotheses, and give some idea of relative proportions.

54. In the fungus *Ascobolus* (similar to *Neurospora*), ascospores are normally black. The mutation *f*, producing fawn-colored ascospores, is in a gene just to the right of the centromere on chromosome 6, whereas mutation *b*, producing beige ascospores, is in a gene just to the left of the same centromere. In a cross of fawn and beige parents (+ *f* × *b* +), most octads showed four fawn and four beige ascospores, but three rare exceptional octads were found, as shown in the accompanying illustration. In the sketch, black is the wild-type phenotype, a vertical line is fawn, a horizontal line is beige, and an empty circle represents an aborted (dead) ascospore.

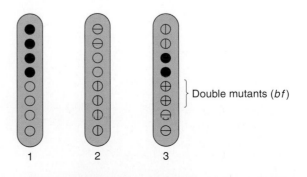

a. Provide reasonable explanations for these three exceptional octads.

b. Diagram the meiosis that gave rise to octad 2.

55. The life cycle of the haploid fungus *Ascobolus* is similar to that of *Neurospora*. A mutational treatment produced two mutant strains, 1 and 2, both of which when crossed to wild type gave unordered tetrads, all of the following type (fawn is a light-brown color; normally, crosses produce all black ascospores):

spore pair 1	black
spore pair 2	black
spore pair 3	fawn
spore pair 4	fawn

a. What does this result show? Explain.

The two mutant strains were crossed. Most of the unordered tetrads were of the following type:

spore pair 1	fawn
spore pair 2	fawn
spore pair 3	fawn
spore pair 4	fawn

b. What does this result suggest? Explain.

When large numbers of unordered tetrads were screened under the microscope, some rare ones that contained black spores were found. Four cases are shown here:

	Case A	Case B	Case C	Case D
spore pair 1	black	black	black	black
spore pair 2	black	fawn	black	abort
spore pair 3	fawn	fawn	abort	fawn
spore pair 4	fawn	fawn	abort	fawn

(**Note:** Ascospores with extra genetic material survive, but those with less than a haploid genome abort.)

c. Propose reasonable genetic explanations for each of these four rare cases.

d. Do you think the mutations in the two original mutant strains were in one single gene? Explain.

16

DISSECTION OF GENE FUNCTION

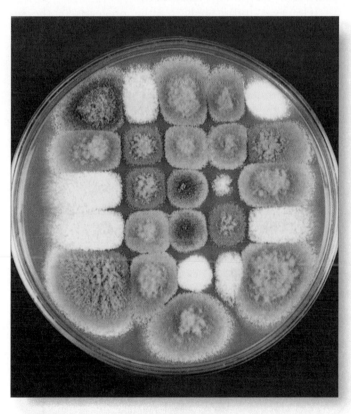

Mutant colonies of the mold *Aspergillus*. The mutations are in genes that control the type and amount of several different pigments synthesized by this fungus, whose normal color is dark green. [Courtesy of J. Peberdy, Department of Life Sciences, University of Nottingham, England.]

KEY QUESTIONS

- What is the purpose of mutational dissection?
- What are the general strategies of mutational dissection?
- How are mutations characterized?
- Are there alternatives to classical mutagens for dissecting gene function?

OUTLINE

16.1 Forward genetics

16.2 Reverse genetics

16.3 Analysis of recovered mutations

16.4 Broader applications of functional dissection

CHAPTER OVERVIEW

Acommon endeavor of genetic analysts is to understand the functions of the genes that affect some particular biological process of interest. The traditional approach of genetics is to do this by first looking at abnormal gene function. The strategy is to disrupt normal gene activity and then analyze the phenotypes of the resulting mutant organisms, which provide clues to the function of the disrupted gene. As we will see, the genetic approach to disrupting gene activity is through mutation. In recent years, alternative techniques have emerged (generally called phenocopying) that disrupt gene expression or activity. Unlike mutations these disruptions are not inherited. This chapter will explore both these approaches. The major focus, however, will be on mutational approaches, since they allow for a wider and deeper exploration of a gene's normal biological contributions.

In some cases, the geneticist wants to survey the genome for all the genes that contribute to a particular biological process—for example, brain development. In such cases, after mutagenizing a large population of genomes, the challenge is to sieve through this collection

of individuals and identify the few with phenotypes suggestive of a mutation affecting a process such as brain development. In other cases, the geneticist already knows of a phenotype produced by mutations in a gene but wants to study a broader range of mutations in that single gene in order to understand its effects fully. In such cases, the collection of mutagenized individuals is sieved to identify those with mutations in the one particular gene of interest. The next step is to clone and sequence that gene and find out what is wrong with it. All these cases can be thought of as **forward genetics,** because the geneticist first analyzes heritable phenotypes at the genetic level before performing the molecular analysis of the mutants discovered.

However, today analysis often proceeds in the reverse direction. A researcher has identified a stretch of DNA, an RNA, or a protein of interest and wants to know how it affects the organism. A key step is find what the phenotype is when the gene encoding this product is mutated. This approach, starting with a molecule and then mutating the gene that encodes it, is called **reverse genetics.** Both approaches are summarized in Figure 16-1.

CHAPTER OVERVIEW Figure

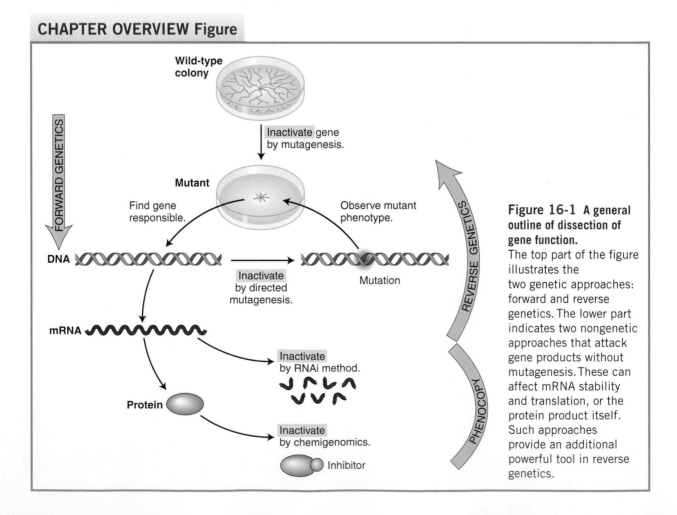

Figure 16-1 A general outline of dissection of gene function. The top part of the figure illustrates the two genetic approaches: forward and reverse genetics. The lower part indicates two nongenetic approaches that attack gene products without mutagenesis. These can affect mRNA stability and translation, or the protein product itself. Such approaches provide an additional powerful tool in reverse genetics.

> **MESSAGE** The "mutations first, molecular analysis second" approach to genetic dissection is called *forward genetics*. The "molecular analysis first, mutations second" approach to genetic dissection is called *reverse genetics*.

16.1 Forward genetics

Typically, forward genetics starts with the wild-type genome, mutates it randomly with a mutagen, and systematically surveys it for mutations that share some common phenotype. This procedure is sometimes called a *mutant hunt*. Ideally, we will identify mutations in literally all the genes in the genome that can be mutated to a state that confers that particular phenotype. Then we can say that we have *saturated* the genome for mutations of that class. In reality, it is very difficult to achieve total saturation—for several reasons. Typically a mutagen has a low, more or less equivalent probability of mutating any region of the genome. However, the sizes of genes within a species can vary considerably. The larger the size of the gene, the greater is the mutagen's target, and the more likely it is that a mutation will be produced in it. Furthermore, because mutations are often pleiotropic (have numerous effects on the phenotype), it is possible that a mutation's more severe effect—such as causing death during development—might mask its milder manifestations, such as altering adult hair color. Thus, a forward genetic analysis may fail to recover mutations if they have pleiotropic effects.

One of the first examples of a mutant hunt was carried out by H. J. Muller in 1927. Specifically he was looking for any mutation on the *Drosophila melanogaster* X chromosome that was both lethal and recessive. Follow his protocol in Figure 16-2. He used X rays to irradiate wild-type X chromosomes in males. The females carried X chromosomes called *ClB*. The C stands for "crossover suppressor"; it is actually an inversion. The *l* stands for a known recessive lethal. The symbol B is the dominant bar-eyed mutation. In the F_1 the only males to survive are the non-*ClB* types that did not inherit the *ClB* chromosomes from their mothers. These are crossed individually to *ClB* F_1 females, who must carry an irradiated X chromosome (indicated by asterisks in the figure) inherited from their fathers. The absence of male progeny in an individual vial shows that the female did carry at least one newly induced lethal on her other X.

Choice of mutagens for forward genetics

Choice of mutagen is an important contributor to the outcome of forward genetics. In such experiments, the geneticist hopes to create a population of individuals that together have mutations in each of the genes in the genome (Figure 16-3a). For such random mutageneses, mutagens that work on any type of DNA sequence are optimal. The mutagen needs to be efficacious enough to raise the mutation rate sufficiently to facilitate mutant recovery. However, dose is also important. Point mutations generally show a linear increase in response to dose. If too many mutations are induced in one genome, there is the possibility of multiple mutations, which would make genetic analysis difficult. Also, it is possible that the cell will not survive. A balance is needed. In mutagenizing the cells of microorganisms, a dose that leads to 50 percent survival is a good rule of thumb for optimal recovery of single-gene mutations.

The specific choice of mutagen varies with the organism. UV radiation is convenient to use on microbes and produces a wide range of types of point mutations. In these organisms, often used chemical mutagens are nitrosoguanidine and nitrous acid. In *Drosophila* the alkylating agents EMS and MMS are convenient because the flies will readily ingest the mutagen when it is added to filter paper in a sugar solution. Generally, a range of mutational types including missense, nonsense, and frameshift mutations is desirable because these alter the gene in a number of different ways, any one of which might produce the desired functional alteration. For large-scale changes, such as intragenic deletions, ionizing radiation such as X or gamma rays is the choice.

Transposable elements, described in Chapter 13, can be very potent and convenient mutagens for forward genetics. Once introduced into a cell, an active transposon is capable of inserting into the nucleotide sequence more or less anywhere in the genome. If it integrates

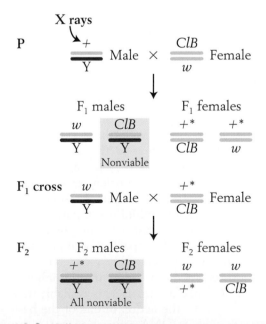

Figure 16-2 Muller's scheme for obtaining recessive lethal mutations anywhere on the X chromosome of *Drosophila*. See the text for a discussion.

Figure 16-3 General versus targeted mutagenesis. (a) General mutagenesis produces a variety of mutations (represented by the different-colored ovals). (b) In contrast, targeted mutagenesis, discussed under reverse genetics, produces mutations (represented by the red ovals) in only the targeted gene.

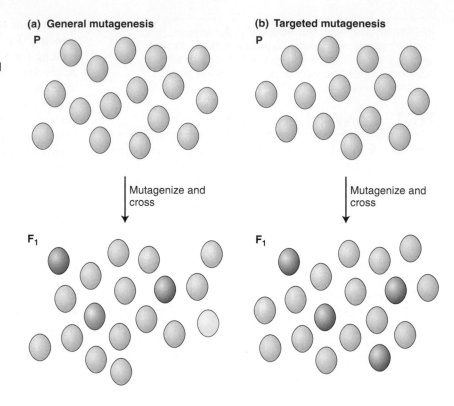

into the sequence of a gene, it may disrupt the integrity of one of its exons, or its splicing pattern, or its regulation, depending on where it lands. As we shall see, transposons have an added level of convenience in that if a mutation is induced, the transposon acts as a tag for the mutant allele. This can lead to easy identification of the inserted gene, for example by PCR. The procedure is known as **transposon tagging.**

> **MESSAGE** Mutagenic chemicals and radiation or transposons can be used to increase the rate of production of mutants in a mutant hunt.

Mutational assay systems for forward genetics

Mutations and mutants are rare even after mutagenesis, often at frequencies of 10^{-5} or less. Therefore an effective assay system is needed to find them. Broadly there are two approaches: selections and screens.

Some mutagenesis protocols are cleverly designed so that only the desired mutant phenotype survives or replicates. Such protocols are called **genetic selections** (Figure 16-4, *left*). An alternative approach is to set up the protocol in such a way that it is easy to identify the desired phenotype among a large number of individuals. These protocols are called **genetic screens** (see Figure 16-4, *right*). What are the advantages of one approach over the other? Selections are effective at obtaining one specific type of preordained mutation.

Screens are harder work but have the advantage of being able to choose a range of phenotypes within a broad general class.

> **MESSAGE** Genetic selections are highly efficient but cannot detect many kinds of mutant phenotypes. Genetic screens are more laborious but far more adaptable to the detection of multiple classes of mutant phenotypes.

Forward genetic selection

Microbes are amenable to selections because they can be grown in very large numbers on plates or in test tubes, on media containing specific substances that act as selective agents. These agents will kill all cells except those with the mutations leading to the desired phenotype. There are many possible selectable phenotypes, such as the ability to use a specific nutrient as a carbon source, the ability to grow in the absence of a specific nutrient, and resistance to various inhibitors or pathogens.

Consider mutations that have the phenotype of survival at toxic levels of a specific inhibitor. Individuals with such mutations are readily recovered by growing the organism on medium containing the inhibitor. Wild types all die. A range of inhibitors is available that specifically interfere with the action of one of the basic cellular processes such as DNA replication, transcription, or translation. For example, *E. coli* mutations can be recovered that are resistant to the antibiotic streptomycin.

Screening and selecting for mutations

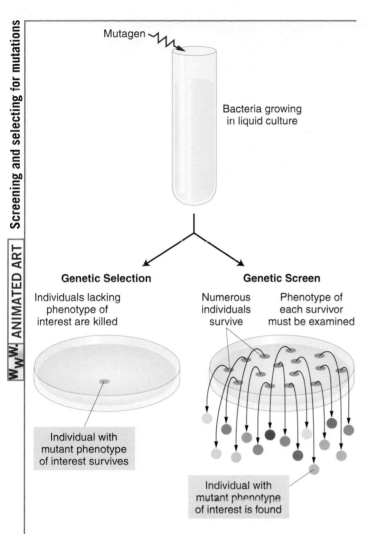

Figure 16-4 Comparison of genetic selections and genetic screens. In selections only the mutant survives. In screens, large populations are visually checked for mutants.

Streptomycin binds to a protein subunit of the *E. coli* ribosome and interferes with translation. Mutant forms of this protein cannot bind to streptomycin, so cells containing them can survive otherwise toxic doses of the antibiotic.

Resistance can also arise by many other types of mutation. In some cases the mutations inactivate proteins that are necessary for the cell to take in the inhibitor. Hence cells plated on a toxic amino acid analog are often found to be defective in the active transport systems that take up amino acids from the medium. A hunt for a range of such mutations identifies the various components of the uptake system. In filamentous fungi, chlorate-resistant mutants are found to have mutations in the gene for the key nitrogen-metabolizing enzyme nitrate reductase, or in some cases defective uptake systems for the inhibitor. Phage resistance in bacteria is generally the result of a mutation in a gene for the protein in the bacterial coat that acts as a phage receptor.

Selection systems for auxotrophic mutations are possible in some microbes. For example, in filamentous fungi the physical nature of the organism can be used to good effect in a procedure called *filtration enrichment*, which is used for selecting auxotrophic mutants (Figure 16-5). On liquid minimal medium prototrophic wild-type cells grow and form fuzzy balls, whereas auxotrophic mutants do not grow. The wild types can be filtered off and discarded, leaving the desired auxotrophic cells, which can be coaxed to grow later with the addition of some specific nutrient they cannot synthesize. Hence the filtration efficiently distinguishes wild and mutant states, and paradoxically lack of growth enables survival. This has been one of the methods used to amass nutritional mutations from which it has been possible to piece together metabolic pathways. The procedure for genotyping auxotrophs is shown in Figure 16-6.

Reversion to wild type often lends itself well to selection: the conditions are such that the mutant dies whereas the wild or partially wild revertant survives. Reversion of auxotrophy to prototrophy is a good example. Reverse selections often produce not only true revertants

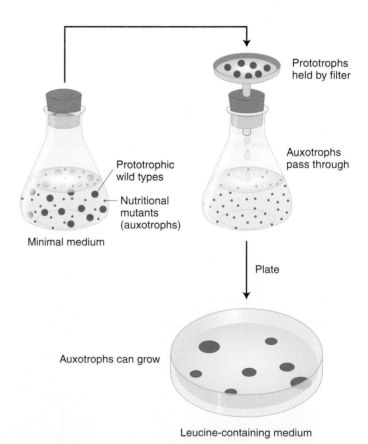

Figure 16-5 Selection of fungal auxotrophs by filter enrichment. Only the desired phenotypes evade the filter. In this case the desired phenotype is a requirement for leucine in the medium.

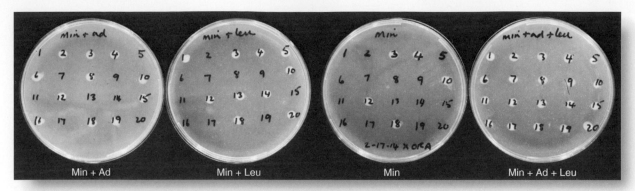

Figure 16-6 Testing strains of *Neurospora crassa* for auxotrophy and prototrophy.
In this example, 20 progeny from a cross *ad · leu⁺* × *ad⁺ · leu* are inoculated on
minimal medium (Min) with either adenine (Ad, *first*), leucine (Leu, *second*), neither
(third), or both *(fourth)*. Growth appears as a small circular colony (white in the
photograph). Any culture growing on minimal medium must be *ad⁺ · leu⁺*, one growing
on adenine and no leucine must be *leu⁺*, and one growing on leucine and no adenine
must be *ad⁺*. For example, culture 8 must be *ad · leu⁺*, 9 must be *ad · leu*,
10 must be *ad⁺ · leu⁺*, and 13 must be *ad⁺ · leu*. [Anthony Griffiths.]

but suppressors that can cancel the effects of the mutation. Suppressors identify genes whose proteins interact
with the original mutation-carrying gene.

Mutants for certain types of animal behavior are
amenable to selections. An example is the light response
of *Drosophila*. Wild-type flies migrate toward light, and
mutants that lack this response can be selected by put-

ting the flies into a T-maze in which one arm of the maze
is illuminated and the other is in the dark (Figure 16-7).

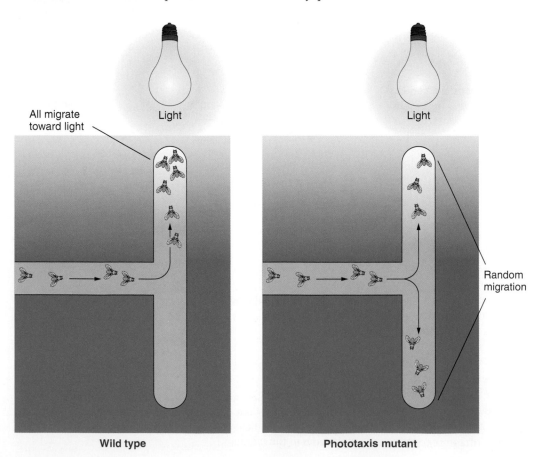

**Figure 16-7 Selection
for behavioral mutants
in *Drosophila*.** A T-maze for
identifying mutants that
are unable to orient
and travel toward light.
Wild-type flies display
positive phototaxis, and
all accumulate on the
illuminated end of the T-
maze. Phototaxis-defective
mutants go to either the
light or the dark end
of the T-maze with equal
probability.

Forward genetic screens

Genetic screens can be used to dissect any biological process. Their effectiveness depends only on the ingenuity of the researcher in coming up with a protocol that reveals the desired class of mutations. Some examples follow.

DISSECTION OF MORPHOGENESIS IN *NEUROSPORA*

Morphogenesis (development of form) in filamentous fungi such as *Neurospora* is merely a reiteration of the processes of hyphal tip growth and branching. If there is a mutation in any one of the genes affecting these processes, the colony takes on an abnormal appearance (Figure 16-8). Hence it is a simple matter to mutagenize a large population of haploid cells, plate them out at high densities, and then visually screen for odd-looking colonies. Hundreds of loci have been identified in such screens, representing the set of genes that together promote tip growth and branching. Among these are genes for actin, dynactin, and dynein, all associated with the cytoskeleton.

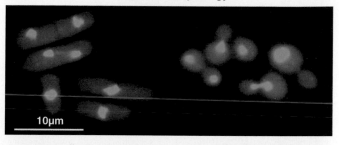

Normal morphology

10μm

Cell-cycle defective mutants

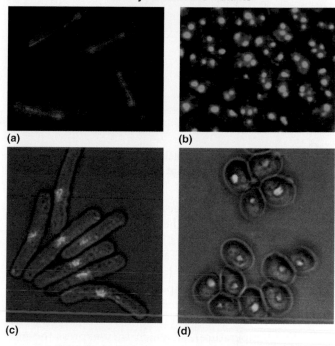

(a)   (b)

(c)   (d)

Figure 16-9 Wild types and cell-cycle mutants of the yeasts *Schizosaccharomyces pombe* (*left*) and *Saccharomyces cerevisiae* (*right*). Mutants are detected on the basis of abnormal cell shape or nuclear position and number (stained). (a) Abnormal mitosis: DNA segregates irregularly along the spindle in mutants. (b) Mutants enter meiosis from the haploid state. (c) Mutants elongate without dividing. (d) Mutants arrest without budding. [Photos courtesy of Susan L. Forsburg, The Salk Institute. "The Art and Design of Genetic Screens: Yeast," *Nature Reviews: Genetics* 2, 2001, 659–668.]

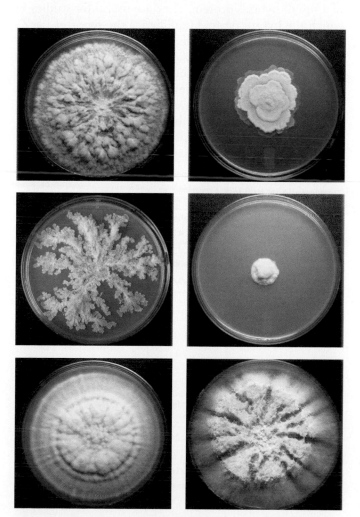

Figure 16-8 Growth mutants of *Neurospora* obtained from a screen seeking abnormal morphology. Each mutant phenotype is caused by a mutation of a different growth gene. [Olivera Gavric and Anthony Griffiths.]

DISSECTION OF CELL CYCLE IN YEAST

The budding yeast *Saccharomyces cerevisiae* and the fission yeast *Schizosaccharomyces pombe* have been at the fore of genetic dissection. Indeed an acronym has been coined in their honor—TAPOYG, "the awesome power of yeast genetics." One relatively straightforward but important screen was used to identify mutants that interfered with the cell-division cycle (*cdc* mutants), for which Leland Hartwell and Paul Nurse received Nobel prizes. Figure 16-9 depicts the kinds of mutations recovered in one screen, looking for mutations that block the mitotic cell

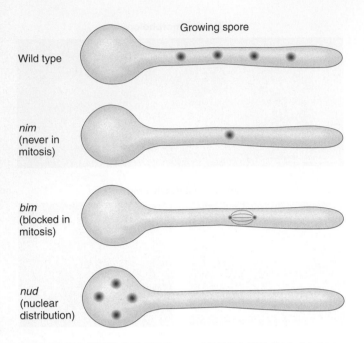

Figure 16-10 A screen for nuclear division defects in *Aspergillus* revealed three mutant classes.

cycle at specific points. Because many such mutations are expected to be lethal, the screen was for conditional heat-sensitive *cdc* mutations, which are wild type at low temperatures but mutant at high. They result from amino acid changes that lead to deleterious protein shape changes at high temperature. These mutants can be propagated at room temperature (the permissive temperature) and then shifted to high temperature (the restrictive temperature), at which they express the mutant phenotype. The set of mutants derived from this type of screen has enabled researchers to define many of the proteins that regulate the highly programmed progression through the cell cycle. Comparative genomics has shown that these same genes are at work in the cell cycle of humans, and that many of these genes are defective in cancers.

DISSECTION OF NUCLEAR DIVISION IN ASPERGILLUS *Aspergillus* is a filamentous fungus that like *Neurospora* has been an important genetic model organism. One interesting screen using *Aspergillus* was a visual screen for mutants with altered nuclear division. The screen revealed three main classes of mutants, *nim* (never in mitosis), *bim* (blocked in mitosis), and *nud* (nuclear distribution), as shown in Figure 16-10. As was also true for the yeast *cdc* mutants, these were heat-sensitive alleles that could be grown at permissive temperature, but shifted to restrictive temperature for study of the phenotype. Subsequent studies showed that NimA is a kinase (phosphorylates other proteins), BimC is a kinesin (a motor; a protein that moves organelles on the cytoskeleton), and NudA is a dynein subunit (another

motor). Thus these proteins are revealed as key players in the cell-division and growth process.

DISSECTION OF SECRETION IN ESCHERICHIA COLI *E. coli* has several systems that make it ideal for genetic dissection. Foremost is the *lacZ* gene. The function of this gene (for the enzyme β galactosidase) can be conveniently assayed by adding a compound called *Xgal* to the medium. The LacZ protein converts Xgal to a blue color (which happens to be the same dye that is used to stain blue jeans). The power of this system is that *lacZ* can be fused to genes of other proteins of interest; then its blue dye production acts as a reporter for that gene. Two types of fusions are possible, transcriptional fusions and translational fusions (Figure 16-11). Transcriptional fusions result in two separate proteins' being made off one transcript. They are useful only for monitoring transcription levels because LacZ is made separately from the other protein. Translational fusions result in translation of a fused "hybrid" protein. They are useful for studies in which the hybrid protein (and hence the reporter LacZ) participate in the usual cellular transactions of the gene of interest. Translational fusions have been useful in the study of protein secretion out of an *E. coli* cell. Various secretory proteins such as membrane targeting proteins (signal sequences), membrane anchoring proteins, and secretory proteins have all been fused to LacZ. For example, when *lacZ* was transcriptionally fused to MalF, the gene for a membrane secretory protein, the accom-

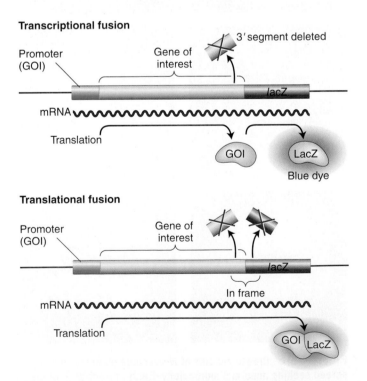

Figure 16-11 LacZ reporter fusions to examine cellular expression of a gene of interest. GOI = gene of interest.

panying βGAL protein was thrust part way through the membrane and could not produce the blue color, and so colonies were white. When this strain is mutagenized, mutants that affect any stage of the secretion process leave the βGAL in the cytoplasm and a blue colony color results. Hence, a screen for blue mutants revealed many mutants of interest. From such studies the various players in the secretion process can be pieced together.

Dissection of development in zebra fish

In the last 10 years, the zebra fish *(Danio rerio)* has emerged as an important genetic model for studying vertebrate development and neurobiology. These fish are small, develop rapidly (for a vertebrate), and produce many offspring. They have transparent embryos that make it easy to observe abnormalities in early development.

In a typical large-scale forward screen for recessive mutations in the zebra fish, a parental generation male is exposed to tank water containing a chemical mutagen (ENU, ethylnitrosourea), which produces mainly base substitutions. Each of the male's sperm potentially carries a different set of base substitution mutations. Each F_1 contains one normal genome (from the mother) and one mutagenized genome (from the father). F_1 males are crossed individually to wild females and descendants allowed to interbreed randomly in the same tank. The inbreeding allows any recessive mutation to come to homozygosity. Some examples of phenotypic mutants that arise in such screens are shown in Figure 16-12.

The above screen can be lengthy and laborious; however, special tricks can speed up the identification of mutations in zebra fish. One of these is the creation of

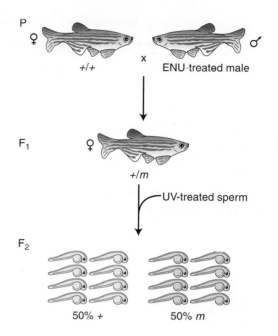

Figure 16-13 Screening using haploid zebra fish. Irradiation of zebra fish sperm results in haploid progeny, half of which will express a newly induced mutant from the female parent.

haploid fish. To create haploid fish, males are treated with large doses of UV light. The exposed sperm nuclei become so heavily mutagenized that they are unable to contribute their genome to the zygote. However, they are still able to penetrate the egg membrane and activate development of the haploid oocyte nucleus. The haploid fish that are produced typically fail to form adults, but they do survive for several days, and these immature fish can be assayed for recessive phenotypes (Figure 16-13).

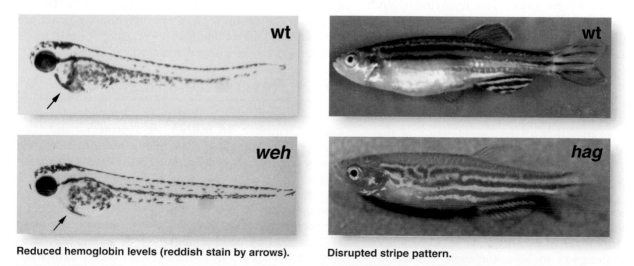

Reduced hemoglobin levels (reddish stain by arrows).   Disrupted stripe pattern.

Figure 16-12 Some phenotypes arising from screens for developmental mutants in zebra fish. *weh = weissherbst*; reduced hemoglobin in blood. *hag = hagoramo;* an abnormal striping. [*Left:* D. F. Ransom et al., OHSU, "Characterization of Zebrafish Mutants with Defects in Embryonic Hematopoiesis," *Development* 123, 1996, 311–319. *Right:* Photos courtesy of Nancy Hopkins. K. Kawakami et al., *Current Biology* 10, 2000, 436–466. Copyright 2000 by Elsevier Science.]

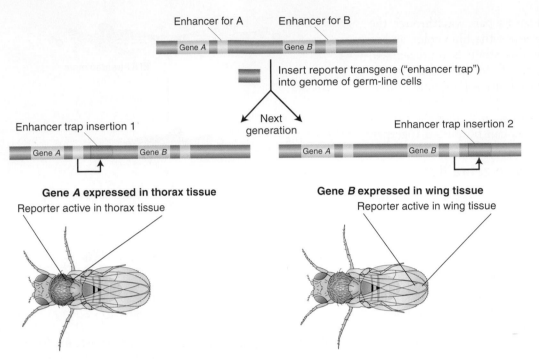

Figure 16-14 Enhancer traps. Different enhancer trap insertions lead to expression of a reporter gene in different tissues. Insertion of the enhancer-trap construct near enhancer elements that drive expression in the thorax *(left)* and wing *(right)* of *Drosophila* are shown.

The geneticist is thus able to focus on those tanks producing interesting mutations, and the mutants can be recovered by inbreeding the parental F₁ female.

Another way to speed up the screen is to use a molecular tag. Earlier we saw that transposons can be used as tags. In zebra fish, retrovirus vectors serve as tags. Recall that retroviruses are RNA viruses that use reverse transcriptase to create a double-stranded DNA copy of their viral genome. The virus is injected into embryos of size 1000–2000 cells. When retroviral DNA is replicated in a cell, the new DNA becomes integrated at random into the host chromosomes. The retroviral DNA causes mutations by gene disruption, and also serves as a convenient molecular tag for the disrupted genes. The resulting fish are subjected to an inbreeding program as before to bring mutations to homozygosity. Once a mutant is identified in a subsequent generation, the gene that it has inserted into can be determined from its sequence, determined with the help of PCR. The viral insert serves as a primer that allows amplification of the sequence into the adjacent region.

Forward genetic screens using enhancer traps

In higher eukaryotes, many DNA regulatory sequences act as enhancers to control transcription (as we saw in Chapter 10). In model organisms such as *Drosophila*, screens can be designed to hunt for these enhancers. These regulatory elements enhance transcription of any gene whose transcription start site is nearby, so the strategy is to randomly insert a transgenic reporter construct designed to respond to any nearby enhancer. The construct will have a transcription start site and a "reporter"

gene such as green fluorescent protein (GFP) or blue dye–producing β galactosidase. The construct is carried on a transposon. Crosses are made that mobilize this reporter construct so that it transposes itself into various sites in the genome, and then the distribution of the reporter protein product is observed. By this means, we can identify the locations of enhancer elements that drive a particular pattern of gene expression (Figure 16-14). The reporter-transgene insertions are called **enhancer traps.**

Suppose that one particular reporter insertion is expressed only in developing *Drosophila* eye tissue. We can infer that it is likely that a gene expressed in the eye resides in the vicinity. Thus, the neighboring genes are candidates for involvement in some aspect of eye development and can be isolated and studied.

16.2 Reverse genetics

Reverse genetic analysis starts with a known molecule—a DNA sequence, an mRNA, or a protein—and then attempts to disrupt this molecule in order to assess the role of the normal gene product in the biology of the organism.

There are several approaches to reverse genetics. One approach is to mutagenize the genome randomly but then home in on the gene of interest by mapping or allelism tests by complementation. A second way is to conduct a targeted mutagenesis that highly favors the production of mutations in the gene of interest. A third way is to create *phenocopies*—effects comparable to mutant phenotypes—by treatment with agents that interfere with mRNA or with the activity of the final protein product.

There are advantages to each of these approaches. Random mutagenesis is the easiest to carry out, but it requires time and effort to sift through all the mutations to find the small proportion that includes the gene of interest. Targeted mutagenesis is also labor-intensive, but once the targeted mutation is obtained, it is more straightforward to characterize. Creating phenocopies can be very efficient, but there are limits to the kinds of phenotypes that can be copied. We will consider examples of each of these approaches.

Reverse genetics through random mutagenesis

Random mutagenesis for reverse genetics employs the same kinds of general mutagens that are used for forward genetics: chemical agents, radiation, or transposable genetic elements. However, instead of screening the genome at large for mutations that exert a particular phenotypic effect, reverse genetics focuses in on the gene in question. This can be done in one of two general ways.

One approach is to focus in on the map location of the gene. Only mutations falling in the region of the genome where the gene is located are retained for further detailed molecular analysis. Thus in this approach the recovered mutations must be mapped. One straightforward way is to combine the new mutants with a known deletion or mutation of the gene of interest. Symbolically the pairings are *new mutant/Δ* or *new mutant/known mutant*. Only the pairings that result in a mutant phenotype (showing lack of complementation) are saved for study.

In another approach, molecular lesions are directly assayed at the DNA level and compared with the gene of interest. For example, if the mutagen causes small deletions, then after PCR amplification, genes from the parental and mutagenized genomes can be compared, looking for a mutagenized genome that contains a PCR fragment that is reduced in size. Similarly, transposable element insertions into the gene of interest can be readily detected because they increase its size. Techniques for recognizing single-base-pair substitutions are also available. In these ways, a set of randomly mutagenized genomes can be effectively screened to identify the small fraction of mutations that are of interest to a researcher.

Reverse genetics by gene-specific mutagenesis

For most of the twentieth century, researchers viewed the ability to direct mutations to a specific gene (refer back to Figure 16-3, *right*) as the unattainable "holy grail" of genetics. However, now several such techniques are available. Once a gene is inactivated in an individual, geneticists can evaluate the phenotype exhibited for clues to the gene's function. We will look at several methods for targeting mutations to specific genes.

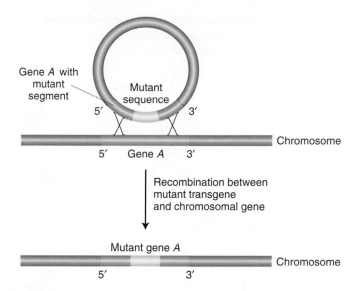

Figure 16-15 The basic molecular event in targeted gene replacement. A transgene containing sequences from two ends of a gene but with a selectable segment of DNA in between is introduced into a cell. Double recombination between the transgene and a normal chromosomal gene produces a recombinant chromosomal gene that has incorporated the abnormal segment.

Gene-specific mutagenesis usually involves replacing a resident wild-type copy of a gene with a transgenic segment containing a mutated version of the gene. The mutated gene inserts into the chromosome by a mechanism resembling homologous recombination, replacing the normal sequence with the mutant (Figure 16-15).

This approach can be used for targeted gene knockout, in which a null allele replaces the wild-type copy.

A finer scale manipulation is **site-directed mutagenesis.** This method can create mutations at any specific site in a gene that has been cloned and sequenced. The mutation must be introduced into a transgene borne on a vector. Then the mutated construct is introduced into a recipient cell. In one protocol, the gene of interest is inserted into a single-stranded bacteriophage vector, such as the phage M13. A synthetic oligonucleotide containing the desired mutation is designed. This oligonucleotide is allowed to hybridize to the complementary site in the gene of interest residing in the vector. The oligonucleotide then serves as a primer for the in vitro synthesis of the complementary strand of the M13 vector (Figure 16-16a). Any desired specific base change can be programmed into the sequence of the synthetic primer [Figure 16-16a(i)]. Although there will be a mispaired base, the synthetic oligonucleotide still hybridizes with the complementary sequence on the M13 vector. After replication in *E. coli*, many of the resulting phages will carry the desired mutant. Oligonucleotides with insertions or deletions will also cause similar

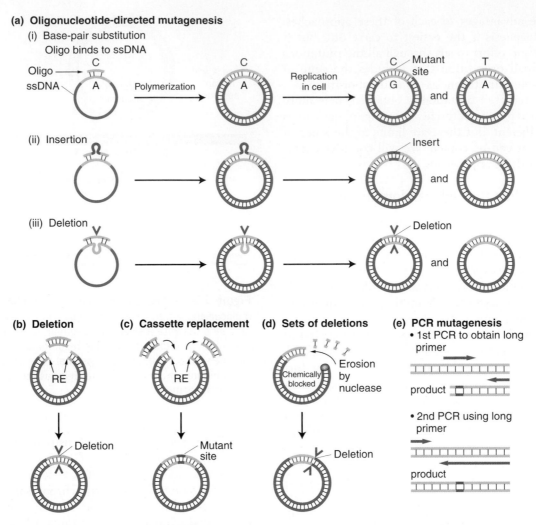

Figure 16-16 Site-directed mutagenesis. (Oligo = oligonucleotide; PCR = polymerase chain reaction; RE = restriction enzyme; ssDNA = single-stranded DNA.) See the text for a discussion.

mutations in the resident gene [Figure 16-16a(ii) and (iii)]. The oligonucleotide-directed method can also be used on genes cloned in double-stranded vectors if their DNA is first denatured.

A knowledge of restriction sites is also useful in directed mutation of a transgene. For example, a small deletion can be made by removing the fragment that is liberated by cutting at two restriction sites (Figure 16-16b). With a similar double cut, a fragment, or *cassette*, can be inserted at a single restriction cut to create a duplication or other modification (Figure 16-16c). Another approach is to erode enzymatically a cut end created by a restriction enzyme to create deletions of various lengths (Figure 16-16d). PCR also can be used to generate a DNA fragment containing a specific mutation, for eventual introduction as a transgene (Figure 16-16e).

In the fungus *Neurospora*, the routine manufacture of transgenic cells revealed a previously unimagined mu-

tational mechanism that acts spontaneously in the organism. The mechanism is called **RIP,** or **repeat-induced point mutation,** and is very useful in targeting mutations to one specific gene. When a transgene is introduced into haploid cells, it does not replace the resident gene but inserts at random locations (that is, ectopically). Hence the cell contains two copies of the gene, the resident copy plus the transgene. However, when such a strain is crossed, both copies of the gene from the transgenic strain emerge from the cross with multiple GC · AT transitions, which inactivate them both. The mutagenic system, which acts just before meiosis, seems to be a defense against pesky transposons or viruses because *Neurospora* is remarkably free of repetitive DNA. Geneticists can capitalize on RIP because any cloned wild-type copy of a gene can be introduced as a transgene, and then it will be rendered null by the RIP process (Figure 16-17).

Introduce wild transgene of interest into haploid cell.

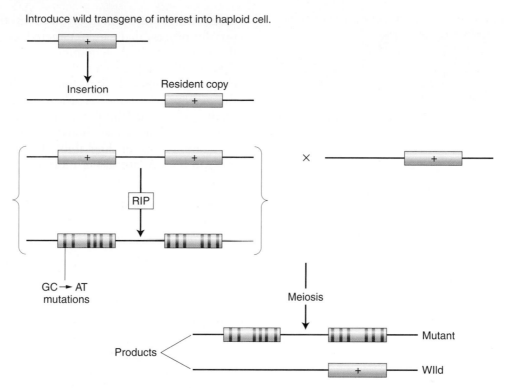

Figure 16-17 Using the RIP process to obtain targeted mutations in *Neurospora.*

> **MESSAGE** Targeted gene replacement and RIP are two ways of inactivating specific genes for the purpose of deducing their function on the basis of mutant phenotype.

Reverse genetics by phenocopying

The advantage of inactivating a gene itself is that mutations will be passed on from one generation to the next, so that once obtained a line of mutants is always available for future study. However, only organisms well developed as molecular genetic models can be used for such manipulations. On the other hand, phenocopying can be applied to a great many organisms regardless of how well developed the genetic technology is for a given species. The next two sections describe two phenocopying techniques.

RNA INTERFERENCE An exciting finding of the past ten years has been the discovery of another widespread mechanism whose natural function seems to be to protect the cell from foreign DNA. This mechanism is called **RNA interference,** or **RNAi.** As with RIP, researchers have capitalized on this cellular mechanism to make a powerful method for inactivating specific genes. The inactivation is achieved as follows. A double-stranded RNA is made with sequences homologous to part of the gene, and this is introduced into a cell (Figure 16-18). The net result is a considerable reduction of mRNA levels that lasts for hours or days, thereby nullifying expression of that gene. The technique has been applied successfully in several model systems, including *Caenorhabditis elegans, Drosophila,* and several plant species.

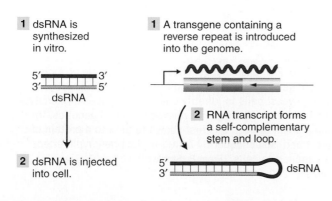

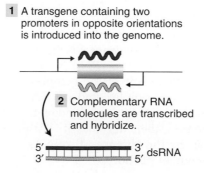

Figure 16-18 Three ways to create and introduce double-stranded RNA into a cell.
The dsRNA will then stimulate RNAi.
[Reprinted with permission from S. Hammond, A. Caudy, and G. Hannon, *Nature Reviews: Genetics* 2, 2001, 116.]

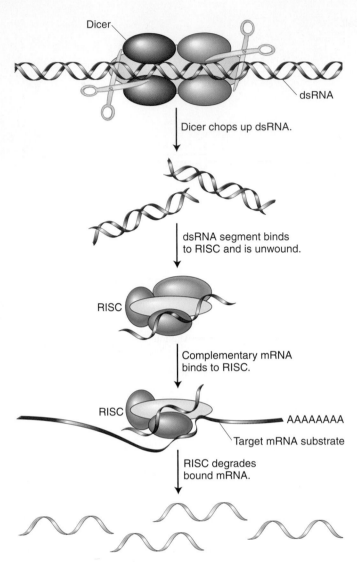

Figure 16-19 Mechanism of action of RNAi. In this mechanism, the dsRNA specifically interacts with the dicer complex, which chops it up. The RISC complex uses the small dsRNAs to find and destroy homologous mRNA transcribed from the target DNA, thereby canceling gene expression. [Modified from S. Hammond, A. Caudy, and G. Hannon, *Nature Reviews: Genetics* 2, 2001, 115.]

How does the introduction of double-stranded RNA inactivate the homologous gene? The mechanism of action of RNAi has been worked out in *C. elegans* (Figure 16-19). The introduced double-stranded RNA is chopped into segments 22 nucleotides long by a molecular complex called *dicer*. The pieces, called *interfering RNAs*, are then bound to a complex called the *RISC (RNA-induced silencing complex)*. An RNA component of the complex, called the *guide RNA*, helps the complex find mRNAs complementary to the interfering RNAs. After binding to the target mRNA, the complex then degrades it. With translation brought to a halt by the absence of mRNAs, the organism expresses the mutant phenotype.

CHEMIGENOMICS Another stage in the information-transfer process that can be targeted for phenocopying is the protein itself. A genomic-scale technique has been developed, which goes by the names **chemigenomics** or **chemical genetics**. This technique is based on reducing the activity of a target gene's protein product through binding of a small inhibitory molecule (Figure 16-20). Robots test libraries of thousands of related small synthetic molecules for their ability to bind tightly to a specific protein. A promising molecule is then introduced

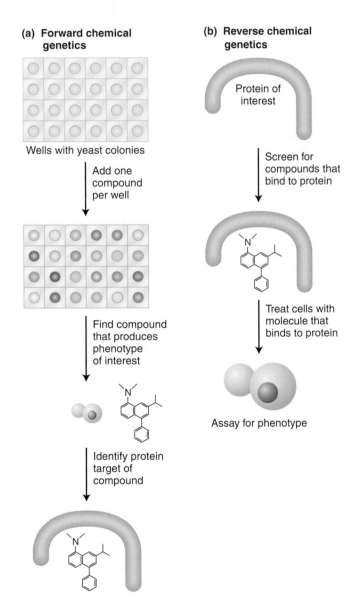

Figure 16-20 Chemical genetics. (a) An example of *forward* chemical genetics, in which small molecules are directly tested on yeast cells to identify one that produces a phenotype of interest. (b) An example of *reverse* chemical genetics, in which a small molecule is first shown to bind to a protein of interest and is subsequently tested for its phenotypic effect when applied to cells. [From B. Stockwell, *Nature Reviews: Genetics* 1, 2000, 117.]

into a cell and tested for its ability to inhibit the protein's activity. If it does inhibit protein activity sufficiently, then a cell or an organism may be treated with that chemical compound to achieve a phenocopy of the mutant phenotype for the target gene.

Despite its names, chemigenomics or chemical genetics is not a genetic technique, because it does not involve inheritance. Rather, it is a systematic extension of the long-standing use of inhibitory drugs (a form of phenocopying) to inactivate a protein involved in a specific biochemical process in the cell. The problem with most inhibitory drugs is that they are not 100 percent specific to a single protein, and so, inadvertently, they inhibit multiple proteins and multiple biochemical processes in an organism, causing ambiguities in interpreting results. Through the use of chemical libraries and robotic tests for specificity, chemigenomics holds the promise of delivering much greater specificity than the traditional inhibitory compounds.

MESSAGE RNAi and chemigenomics provide ways of experimentally interfering with the function of a specific gene, without changing its DNA sequence (generally called *phenocopying*). This represents an alternative approach to reverse genetics.

16.3 Analysis of recovered mutations

Once mutations have been detected and isolated, their properties can be used to infer how the normal gene products work and interact. Chapters 17 and 18 provide detailed examples of how to analyze a series of mutations in order to understand a biological process. Here we discuss general aspects of the analysis.

Counting the genes in a biological process

A typical mutational screen or selection recovers a large number of mutations that represent multiple "hits" in a smaller number of genes. How many genes are represented by this mutant collection? The answer is obtained by mapping the mutations and applying the complementation test (Figure 16-21). Recall that two mutations that are separable by recombination must be mutations in different genes. Mutations that map in the same region of the genome may represent multiple hits in the same gene or may represent mutations in a cluster of genes. The complementation test can resolve these two possibilities. We saw in Chapter 6 that if two *recessive* mutations *m1* and *m2* complement one another (that is, if *m1/m2* is wild type in phenotype), then we infer that the two mutations represent different genes (hence the genotype must be *m1 +/+ m2*). On the other hand, if two recessive mutations fail to complement (that is, if *m1/m2* is

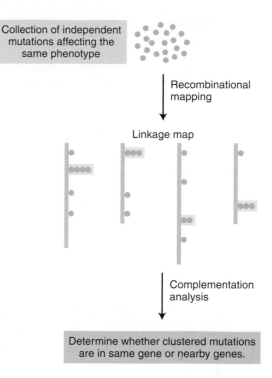

Figure 16-21 Counting genes that contribute to a phenotype. Recombinational mapping and complementation analysis can give the number of genes mutated in a given collection of mutations that share a common phenotype. Recombinational analysis helps us know where mutations map, and clusters of mutations (light-blue boxes) must be tested by complementation analysis to determine whether mutations in the same cluster are in the same gene.

mutant in phenotype), then we infer that the two mutants are allelic mutations in the same gene. (Remember: this test doesn't work with dominant mutations—by the definition of dominance, their phenotype will be mutant regardless of the mutant state of the other allele.)

MESSAGE New mutations with the same or closely related phenotypes can be grouped by recombination mapping; then, closely linked recessive mutations can be resolved by complementation testing.

Distinguishing loss of function versus gain of function

We have been following the idea that geneticists gain insight into a process by disrupting the process through mutagenesis or other means, observing the consequences, and using this information to work out the steps of the normal process. Such analysis is furthered by an understanding of what has gone wrong with the mutant gene because then better sense can be made of a specific functional change. One important piece of information

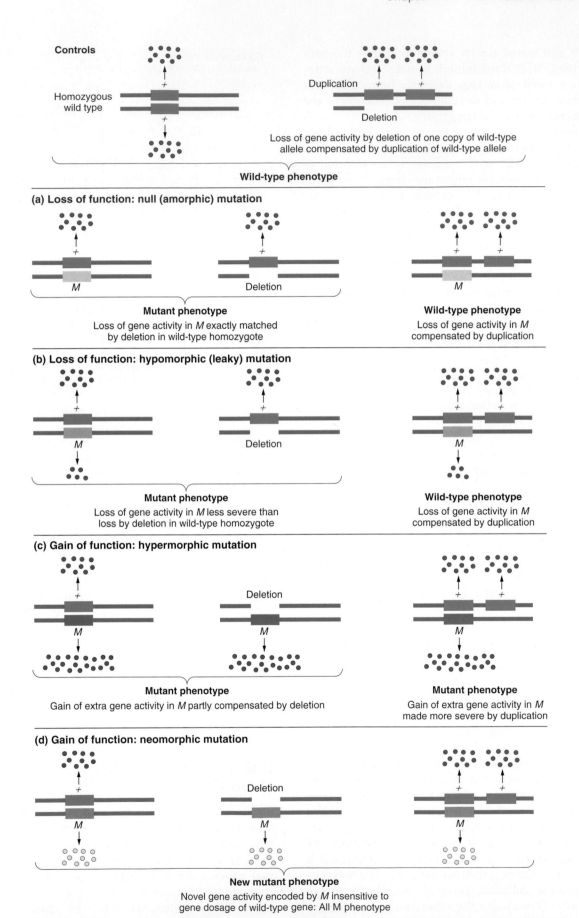

Figure 16-22 Distinguishing various functional causes of dominant mutations. See the text for discussion.

is whether the mutation represents a *loss of function* or a *gain of function*. Recessive mutations are generally loss-of-function alleles of a haplosufficient gene. Indeed most mutations are of this type. Dominant mutations are more diverse but often more interesting, and special tests are needed to distinguish dominant mutations that represent a loss of function from those that represent a gain of function.

Let's consider dominant mutations (Figure 16-22). Consider a dominant null mutation. In that case, a single copy of the wild-type allele does not make enough gene product to generate a wild-type phenotype. Hence if a gene generates a loss-of-function dominant mutation, we can infer that it falls into the class called *haploinsufficient*. We can recognize such a mutation by comparing the $+/M$ phenotype first with the phenotype produced by a deletion $+/\Delta$, if available, and then with the phenotype produced by a duplication Dup/M (see Figure 16-22a). First, the phenotype of a mutant gene paired with a wild-type gene should be the same as the phenotype shown by a deletion. Second, the dominant mutant phenotype should be "cured" by adding a duplicate copy of the wild type.

Loss of function is not always 100 percent; it can be at some intermediate level. Some loss-of-function mutations completely eliminate the activity of the gene product (null mutations; Figure 16-22a). Others merely decrease the activity of the gene product; these are called **hypomorphic**, or **leaky, mutations** (Figure 16-22b). Gene dosage (deletions and duplications) also can be used to distinguish between levels of loss of function. In the preceding paragraph describing a null mutation, we saw that the mutant heterozygote with wild type would have a phenotype identical with that of the deletion heterozygote with wild type. We represented this symbolically as $M/+ = \Delta/+$. For a hypomorphic mutation of a haploinsufficient gene, the mutant heterozygote with wild type should have *more* normal activity than the deletion heterozygote (see Figure 16-22b). Symbolically, $M/+ > \Delta/+$ (where ">" means "more normal than").

Let's now compare these predictions with those for dominant *gain-of-function* mutations. Two examples of dominant gain-of-function mutations are hypermorphs and neomorphs. A **hypermorph** is a mutation that produces *more* gene activity per gene dose than wild type, but in all other respects the gene product is normal (see Figure 16-22c). If the hypermorph mutation is dominant, the extra gene activity of the mutant allele M produces a novel phenotype in the $M/+$ heterozygote. If we remove the $+$ allele with a deletion (M/Δ), we reduce the combined activity of the two alleles and thus the phenotype should become more normal—symbolically, $M/\Delta > M/+$. On the other hand, if we increase the dosage of wild type with a duplication, then the phenotype should become more mutant; that is, $M/+ > M/\text{Dup}$.

A **neomorph** is a mutation that produces novel gene activity that is not characteristic of the wild type. For example, if the coding sequences of two genes are fused in-frame, a novel protein can be produced that may have cellular activities different from those of either parental protein. Alternatively, by fusion with a different promoter a wild-type protein can be misregulated, so that it is expressed in a tissue in which the wild-type gene product is normally never expressed. This protein may then alter the biochemical pathways in this tissue's cells and thereby produce a completely unpredictable and novel phenotype. How is a neomorphic mutation identified? Because the gene product or site of action is novel, a neomorphic mutation is insensitive to dosage of the wild-type allele (see Figure 16-22d). Having zero, one, or two copies of the wild-type allele in a genotype with the corresponding neomorphic mutation produces the same dominant mutant phenotype. Symbolically, for a neomorphic mutation, $M/+ = M/\Delta = M/\text{Dup}$.

> **MESSAGE** Dominant mutations can arise from several types of changes at the functional level, including complete or incomplete loss of function in a haploinsufficient gene, production of more wild-type product, or production of a novel product. These can be distinguished by combining the new allele with varying doses of the wild-type allele.

What other information can be gleaned from a set of mutations? Important steps are the molecular identification of genes and their mRNA and protein products. This generally requires cloning, sequencing, and functional characterization of the gene and its products, as described in Chapter 11. After each gene has been studied at the molecular and cell levels, the next task is to integrate the gene functions that are relevant to the particular biological process under investigation. This can be a long and difficult process requiring many years and many people. The next two chapters will look at how these questions were answered for pathways underlying the regulation of cell number and the regulation of development.

16.4 Broader applications of functional dissection

Much of our discussion of mutational dissection and phenocopying has focused on genetic model organisms. One of the next challenges is to apply these systems more broadly, including to species that have negative impacts on human society, such as parasites, disease carriers, or agricultural pests. Classical genetic techniques are not readily applicable to most of these species, but functional dissection can be carried out by either transgenesis or phenocopying.

The first approach—functional dissection of non-model organisms by creating transgenics—is shown in

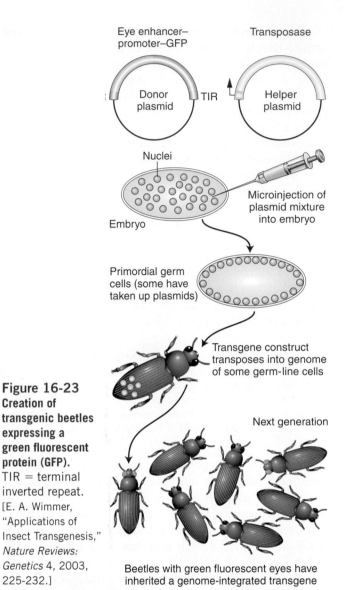

**Figure 16-23
Creation of
transgenic beetles
expressing a
green fluorescent
protein (GFP).**
TIR = terminal
inverted repeat.
[E. A. Wimmer,
"Applications of
Insect Transgenesis,"
*Nature Reviews:
Genetics* 4, 2003,
225-232.]

Figure 16-23. This example concerns beetles, many of which are agricultural pests. Transgenic beetles can be produced using the same methodology used to produce transgenic *Drosophila*. However, one problem is that for most beetle species no recessive markers have been identified that could be used in recipient animals for identifying successful transgenesis. Therefore the technique depends instead on using transgenic constructs carrying dominant mutant phenotypes that can be expressed in a wild-type recipient. The green fluorescent protein (GFP), originally isolated from a jellyfish, is a useful reporter for this application. As in *Drosophila*, transgene insertions can be mobilized using transposons, and these then serve as insertional mutagens throughout the beetle genome. Figure 16-23 shows the use of GFP transgenes driven by an enhancer element that determines expression in the insect eye. This method has also been effectively used to create GFP-expressing transgenes in the mosquito that carries yellow fever and dengue fever *(Aedes aegypti)*, a flour beetle *(Tribolium castaneum)*, and the silkworm moth *(Bombyx mori)* (Figure 16-24).

Phenocopying techniques are also widely applicable in non-model organisms. Target genes can be identified by comparative genomics. Then RNAi sequences can be produced to target the inhibition of the specific target genes. This technique has already been applied to a mosquito that carries malaria *(Anopheles gambiae)*. Using these techniques, scientists can better understand the biological mechanisms relating to the medical or economic impact of these species. For example, the genes that control the complicated life cycle of the malaria parasite, partly inside of a mosquito host and partly in the human body, can be better understood, revealing new ways to control the single most common infectious disease in the world. Many other applications can be envisioned.

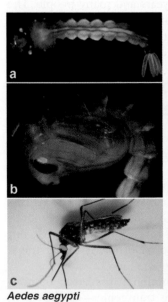

Aedes aegypti

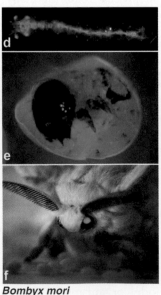

Bombyx mori

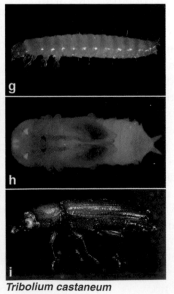

Tribolium castaneum

**Figure 16-24 Examples of a
transgenic green fluorescent protein
reporter expressed in the eyes of
some non-model insects.**
Expression is driven from one
single promoter active in the
eye. The insects are mosquito
(Aedes aegypti), silkworm moth
(Bombyx mori), and beetle
(Tribolium castaneum).
[*Left:* Courtesy of V. A. Kooks and
Alexander S. Raikhel. *Middle:* (*top*)
J. L. Thomas et al. Copyright 2002 by
Elsevier Science; (*middle and bottom*)
courtesy of Marek Jindra. *Right:*
Copyright 2000 by Elsevier Science.]

KEY QUESTIONS REVISITED

• **What is the purpose of mutational dissection?**

To find and characterize all or most of the genes that contribute to some biological process of interest.

• **What are the general strategies of mutational dissection?**

Start with a collection of mutants obtained using selections or screens and then characterize their wild functions at the molecular level (forward genetics); alternatively, start with a set of specific DNA sequences, mutate them, and observe the phenotypes (reverse genetics).

• **How are mutations characterized?**

They can be classified first as gain of function, loss of function, or new function; then their sequence and cell-expression patterns can be characterized.

• **Are there alternatives to classical mutagens for dissecting gene function?**

Yes, there are ways to mimic mutant phenotypes (make phenocopies). This can be achieved by interference with mRNA (RNAi) or by binding highly specific inhibitors to proteins (chemigenomics).

SUMMARY

Mutational dissection is used for several purposes. One of the major reasons is to identify the genes (and, therefore, the gene products) that contribute to a specific biological process. For example, by identifying mutations that block the yeast cell-division cycle at discrete points, the pathway of genes that regulate cell division can be discovered. This is an example of forward genetic analysis. Another reason to carry out a mutational dissection is to understand the phenotypic role of genes that have been characterized at the molecular level but not genetically. This type of mutational dissection is a reverse genetic analysis. In these cases, molecular information is used to focus mutagenesis on a specific small region of the genome. Mutations are obtained in the gene of interest as expeditiously as possible, either through random or targeted mutagenesis, or by use of phenocopying technologies.

There are three key parts to developing a plan for mutational dissection: selection of the mutagenic agent, identification of the plan for phenotypic selection or screening, and determining the breeding scheme most appropriate for a specific species. Mutations are identified on the basis of phenotype. Some phenotypes are directly selectable. This is the most effective type of identification, since selection permits the researcher to sift through enormous numbers of mutagenized genomes. Screens are more labor-intensive, but also more versatile, since many phenotypes don't lend themselves to selections. Phenocopying offers powerful alternatives to classical techniques for nullifying gene function. We explored two examples of phenocopying. In RNAi, a cellular mechanism that degrades any RNA sequences corresponding to fragments of double-stranded RNA is exploited to target specific mRNAs for degradation. In chemigenomics, libraries of small molecules are screened for their ability to bind to proteins of interest. Binding ligands are then screened for their ability to inhibit or activate the target proteins. Because these techniques do not require genetic manipulations, they are amenable to automation.

The follow-up analysis includes molecular and phenotypic characterization. It is important to analyze the effect of the mutation on gene function—whether the mutation causes a gain of function or loss of function. By combining a dominant mutant with known deletions or duplications of the gene, it is possible to infer whether the mutation is acting to increase, reduce, or eliminate the normal gene activity or to generate a novel function.

KEY TERMS

chemical genetics (p. 534)

chemigenomics (p. 534)

enhancer traps (p. 530)

forward genetics (p. 522)

genetic screens (p. 524)

genetic selections (p. 524)

hypermorph (p. 537)

hypomorph (p. 537)

hypomorphic, or leaky, mutations (p. 537)

neomorph (p. 537)

repeat-induced point mutation (RIP) (p. 532)

reverse genetics (p.522)

RNA interference (RNAi) (p. 533)

site-directed mutagenesis (p. 531)

transposon tagging (p. 524)

SOLVED PROBLEM

You want to study the development of the olfactory (smell-reception) system in the mouse. You know that the cells that sense specific chemical odors (odorants) are located in the lining of the nasal passages of the mouse. Describe some approaches for doing both forward and reverse genetics to study olfaction.

Solution

There are many approaches that can be imagined.

a. For forward genetics, the first trick is developing an assay system. One such assay system would be behavioral. For example, we could identify odorants that either attracted or repelled wild-type mice. We could then perform a genetic screen after mutagenesis, looking for mice that do not recognize the presence of this odorant, by using a maze in which a stream of air containing the odorant is directed at the mice from one end of the maze. This assay has the advantage of allowing us to process lots of mutagenized mice without any anatomical analysis.

As a first step, you might want to identify mutations that act dominantly to affect odorant reception, simply because these mutations could be identified in an F_1 screen. If this didn't work, you might need to resort to an F_2 screen for X-linked recessives or an F_3 screen for autosomal recessive mutations. Given that you don't

know anything about the number or target size of genes that might produce this phenotype, you would want to use a highly mutagenic agent, such as a base-substitution mutagen, that would be able to generate mutations very efficiently in most protein-coding genes.

b. For reverse genetics, we would want to identify candidate genes that are expressed in the lining of the nasal passages. Given the techniques of functional genomics (Chapter 9), this could be accomplished by purifying RNA from isolated nasal-passage-lining cells and using this RNA as a probe of DNA chips containing sequences that correspond to all known mRNAs in the mouse. For example, you may choose first to examine mRNAs that are expressed in the nasal passage lining but nowhere else in the mouse as important candidates for a specific role in olfaction. (Many of the important molecules may also have other jobs elsewhere in the body, but you have to start somewhere.) Alternatively, you may choose to start with those genes whose protein products are located in the cell membrane as candidate proteins for binding the odorants themselves. Regardless of your choice, the next step would be to engineer a targeted knockout of the gene that encodes each mRNA or protein of interest, or to use antisense or dsRNA injection to attempt to phenocopy the loss-of-function phenotype of each of the candidate genes.

PROBLEMS

BASIC PROBLEMS

1. One way to look for auxotrophic mutants in microbes is to inoculate thousands of mutagenized colonies in specific positions on plates of fully supplemented medium, then to use a pad of sterile felt to transfer cells from these colonies into similar positions on plates of minimal medium ("replica-plating"). Auxotrophs do not grow on minimal medium. Would you call this a selection or a screen for auxotrophs? Explain.

2. You wish to find mutants of the haploid alga *Chlamydomonas* that are defective in flagellar function. Devise a selection and a screen for such mutants.

3. In *Neurospora*, *ad-3* colonies are purple. A heterokaryon is available that is *ad-3*$^+$/Δ (Δ = deletion of *ad-3*). Using this information, devise a selection and a screen for *ad-3* mutants.

4. The haploid fungus *Ustilago hordei* is a pathogen on barley, but will also grow like *Neurospora* or *Aspergillus* in the lab. On medium containing low concentrations of red food coloring (used for cake decoration) *Ustilago* colonies take up and concentrate the

dye and become blood-red. Use this system to devise a genetic dissection that explores fungal nutrient uptake in pathogenicity.

5. In the tests in Figure 16-6, is the Min + Ad + Leu plate really necessary?

6. In the experiment in Figure 16-7, what types of mutated gene functions might send flies down the dark pathway?

7. Why are yeast *cdc* mutants relevant to human cancer?

8. In *Aspergillus* you make the double mutants *nim · bim*, *nim · nud*, and *bim · nud*. Which mutant would you expect to be epistatic in each case?

9. You wish to use *lacZ* fusions to study secretion of proteins into the medium by *E. coli*. Would you use a transcriptional or a translational fusion? How would you design the experiment?

10. Your 10-year old nephew looks at your genetics homework and sees the terms "your favorite gene" and "gene of interest" and wants an explanation of these peculiar phrases. What would you say?

11. In a haploid zebra fish screen you find a very interesting female that generates exactly the right developmental phenotype you seek in her haploid progeny. How would you go about producing a diploid line homozygous for that mutation (assume it is viable)?

12. You are interested in the development of the *Drosophila* head. How would you use a *lacZ* enhancer trap to find "head genes"?

13. In *Neurospora* you are interested in the gene for phospholipase C (PLC). You have its sequence, and two clones of it. One clone is wild type, and the other has a hygromycin-resistance gene spliced into the middle of it. Show how you might knock out PLC by replacement by RIP.

14. In PCR mutagenesis, why is the "long PCR primer" necessary?

15. In RNAi, distinguish between the different roles of RISC and dicer complexes.

16. To inactivate a gene by RNAi, what information do you need? Do you need the map position of the target gene?

17. In a haploid fungus, you have performed a forward mutation hunt for growth mutants and have nine very similar looking mutant strains. They map to two positions; five are on chromosome 1 and four are on chromosome 5. The ones on chromosome 1 never complement one another. The ones on chromosome 5 show the following complementation pattern (+ = complementation):

	1	2	3	4
1	−	+	+	−
2		−	−	+
3			−	+
4				−

How many genes have you discovered?

18. In most organisms the active site of the enzyme you have been studying in a model organism (say, a fungus) is between nucleotide positions 50 and 70. How would you genetically test that this is the active site in your organism?

19. A new mutant when paired with a wild type gives wild-type phenotype. How would you classify the gene involved?

20. A new *Drosophila* mutation, when in combination with a wild-type allele, gives an unusual black pigment throughout the body. A mutant paired with a deletion and with a duplication shows the same amount of black. A homozygote is extremely black. How would you classify this mutation?

21. A new mutant when combined with a wild-type allele gives a wild-type phenotype. When homozygous, a mutant phenotype appears. When the mutant allele is paired with a deletion, the combination is lethal. Provide a model that accounts for these results.

22. A certain species of plant produces flowers with petals that are normally blue. Plants with the mutation *w* produce white petals. In a plant of genotype *w/w*, one *w* allele reverts during the development of a petal. What detectable outcome would this reversion produce in the resulting petal? Would this mutation be inherited?

23. How would you select revertants of the yeast allele *pro-1*? This allele confers an inability to synthesize the amino acid proline, which can be synthesized by wild-type yeast and which is necessary for growth.

24. Suppose that you want to determine whether caffeine induces mutations in higher organisms such as humans. Describe how you might do so (include control tests).

25. One of the jobs of the Hiroshima-Nagasaki Atomic Bomb Casualty Commission was to assess the genetic consequences of the blast. One of the first things they studied was the sex ratio in the offspring of the survivors. Why do you suppose they did this?

26. Cells of a haploid wild-type *Neurospora* strain were mutagenized with EMS. Large numbers of these cells were plated and grown into colonies at 25°C on complete medium (containing all possible nutrients). These strains were tested on minimal medium and complete medium at both 25°C and 37°C. There were several mutant phenotypes, as shown in the accompanying figure. The large circles represent luxuriant growth, the spidery symbols represent weak growth, and a blank means no growth at all. How would you categorize the types of mutants represented by isolates 1 through 5?

Strain	Minimal 25°C	Minimal 37°C	Complete 25°C	Complete 37°C
Mutant 1			○	○
Mutant 2	○		○	
Mutant 3	❋	❋	○	○
Mutant 4	○	❋	○	❋
Mutant 5	○		○	○
Wild type (control)	○	○	○	○

27. Devise imaginative screening procedures for detecting the following:

a. Nerve mutants in *Drosophila*

b. Mutants lacking flagella in a haploid unicellular alga

c. Supercolossal-size mutants in bacteria

d. Mutants that overproduce the black compound melanin in normally white haploid fungus cultures

e. Individual humans (in large populations) whose eyes polarize incoming light

f. Negatively phototrophic *Drosophila* or unicellular algae

g. UV-sensitive mutants in haploid yeast

28. A man and a woman with no record of genetic disease in their families have one child with neurofibromatosis (autosomal dominant) and another child who is unaffected. The penetrance of neurofibromatosis is close to 100 percent.

a. Explain the birth of the affected child.

b. How would you counsel the parents if they contemplated having another child?

29. Describe three different methods used to generate phenocopies. What is the purpose of generating a phenocopy?

30. What is the benefit of using a balancer chromosome?

31. What is the difference between forward and reverse mutation?

32. What are the advantages and disadvantages of genetic screens versus selections?

CHALLENGING PROBLEMS

33. A haploid strain of *Aspergillus nidulans* carried an auxotrophic *met-8* mutation conferring a requirement for methionine. Several million asexual spores were plated on minimal medium, and two prototrophic colonies grew and were isolated. These prototrophs were crossed sexually with two different strains, with the progeny shown in the body of the following table, where *met*⁺ means that methionine is not required for growth and *met*⁻ means that methionine is required for growth.

	Crossed with a wild-type strain	Crossed with a strain carrying the original *met-8* allele
Prototroph 1	All *met*⁺	1/2 *met*⁺ 1/2 *met*⁻
Prototroph 2	3/4 *met*⁺ 1/4 *met*⁻	1/2 *met*⁺ 1/2 *met*⁻

a. Explain the origin of both of the original prototrophic colonies.

b. Explain the results of all four crosses, using clearly defined gene symbols.

UNPACKING PROBLEM 33

Before you try to solve this problem, follow the instructions and answer the questions that pertain to the experimental system.

1. Draw a labeled diagram that shows how this experiment was done. Show test tubes, plates, and so forth.

2. Define all the genetic terms in this problem.

3. Many problems show a number next to the auxotrophic mutation's symbol—here the number *8* next to *met*. What does the number mean? Is it necessary to know in order to solve the problem?

4. How many crosses were actually made? What were they?

5. Represent the crosses by using genetic symbols.

6. Is the question about somatic mutation or germinal mutation?

7. Is the question about forward mutation or reversion?

8. Why were such a small number of prototrophic colonies (two) found on the plate?

9. Why didn't the several million asexual spores grow?

10. Do you think any of the millions of spores that did not grow were mutant? Dead?

11. Do you think the wild type used in the crosses was prototrophic or auxotrophic? Explain.

12. If you had the two prototrophs from the plates and the wild-type strain in three different culture tubes, could you tell them apart just by looking at them?

13. How do you think the *met-8* mutation was obtained in the first place? (Show with a simple diagram. **Note:** *Aspergillus* is a filamentous fungus.)

14. Is the concept of recombination relevant to any part of this problem?

15. What progeny do you predict from crossing *met-8* with a wild type? What about *met-8* × *met-8*? What about wild type × wild type?

16. Do you think this is a random meiotic progeny analysis or a tetrad analysis?

17. Draw a simple life-cycle diagram of a haploid organism showing where meiosis takes place in the cycle.

18. Consider the 3/4:1/4 ratio. In haploids, crosses heterozygous for one gene generally give progeny ratios based on halves. How can this idea be extended to give ratios based on quarters?

34. Every mutagen produces a certain characteristic type of mutational event. Explain your answers to the following questions:

   a. Would you expect hypomorphic mutations to be more frequent among all mutations produced by base-substitution mutagens or by frameshift mutagens?

   b. Would you expect null mutations to be more frequent among all mutations produced by base-substitution mutagens or by frameshift mutagens?

35. Neomorphic mutations can be reverted by treatment with standard mutagens. When the revertants are examined, they typically turn out to be recessive loss-of-function mutations. Explain this observation.

36. You are trying to identify all mutations that affect development of the dorsal fin in the zebra fish. You do an F_3 mutagenesis analysis for recessive mutations that cause loss of the dorsal fin. By recombination and complementation analysis, you find that 40 mutations you've isolated represent mutations in five genes, with 12 mutations in one gene, 10 mutations in each of two others, 7 mutations in a fourth gene, and only 1 mutation in a fifth.

   a. Is it surprising that you recovered so many mutations in each of four of the genes and only one in the fifth? Justify your answer.

   b. Would you expect that this screen has identified all genes that contribute to dorsal fin development? Why or why not? If you think there should be other classes of mutations, propose some experiments to identify them.

37. You are studying proteins involved in translation in the mouse. By BLAST analysis of the predicted proteins of the mouse genome, you identify a set of mouse genes that encode proteins with sequences similar to those of known eukaryotic translation-initiation factors. You are interested in determining the phenotypes associated with loss-of-function mutations of these genes.

   a. Would you use forward or reverse genetics approaches to identify these mutations?

   b. Briefly outline two different approaches you might use to look for loss-of-function phenotypes in one of these genes.

38. Normal ("tight") auxotrophic mutants will not grow at all in the absence of the appropriate supplementation to the medium. However, in mutant hunts for auxotrophic mutants, it is common to find some mutants (called *leaky*) that grow very slowly in the absence of the appropriate supplement but normally in the presence of the supplement. Propose an explanation for the molecular action of the leaky mu-

tants in a biochemical pathway, and say how to test your idea.

39. A botanist interested in the chemical reactions whereby plants capture light energy from the sun decided to dissect this process genetically. She decided that leaf fluorescence would be a useful mutant phenotype to select because it would show something wrong with the process whereby electrons normally are transferred from chlorophyll. Therefore four fluorescent (*fl*) mutants were obtained in the plant *Arabidopsis* after mutagenesis. All were found to be recessive in the simple Mendelian manner. Homozygous stocks of the mutants were intercrossed, and the F_1s were each testcrossed to a strain that was homozygous recessive for all the genes involved in that cross. The results are shown in the following table.

Cross	Percentage of wild types in F_1	Percentage of wild types in progeny of testcross of F_1
1 × 2	100	25
1 × 3	100	25
1 × 4	0	0
2 × 3	100	10
2 × 4	100	25
3 × 4	100	25

   a. How many genes are represented by these mutants?

   b. What can you deduce about the chromosomal location of the genes?

   c. Use your own gene symbols to explain the F_1 and testcross results.

40. The entire genome of the yeast *Saccharomyces cerevisiae* has been sequenced. This sequencing has led to the identification of all the open reading frames (ORFs) in the genome (gene-sized sequences with appropriate translational initiation and termination signals). Some of these ORFs are previously known genes with established functions; however, the remainder are unassigned reading frames (URFs). To deduce the possible functions of the URFs, they are being systematically, one at a time, converted into null alleles by in vitro knockout techniques. The results are as follows:

15 percent are lethal when knocked out.
25 percent show some mutant phenotype (altered morphology, altered nutrition, and so forth).
60 percent show no detectable mutant phenotype at all and resemble wild type.

Explain the possible molecular-genetic basis of these three mutant categories, inventing examples where possible.

41. In *Drosophila*, the genes for ebony body (*e*) and stubby bristles (*s*) are linked on the same arm of chromosome 2. Flies of genotype + *s/e* + develop predominantly as wild type but occasionally show two different kinds of unexpected abnormalities on their bodies. The first abnormality is the presence of pairs of adjacent patches, one with stubby bristles and the other with ebony color. The second abnormality is the presence of solitary patches of ebony color.

a. Draw diagrams to show the likely origin of these two types of unexpected abnormalities.

b. Explain why there are no single patches that are stubby. (**Hint:** What would be the outcome of a crossover between homologs that accidentally pair at *mitosis?*)

42. A strain of *Aspergillus* was subjected to mutagenesis by X rays, and two tryptophan-requiring mutants (A and B) were isolated. These tryptophan-requiring strains were plated in large numbers to obtain revertants to wild type. You failed to recover any revertants from mutant A and recovered one revertant from mutant B. This revertant was crossed with a normal wild-type strain.

a. What proportion of the progeny from this cross would be wild type if the reversion precisely reversed the original change that produced the *trp⁻* mutant allele?

b. What proportion of the progeny from this cross would be wild type if the revertant phenotype was produced by a mutation in a second gene located on a different chromosome (the new mutation suppresses *trp⁻*)?

c. Propose an explanation of why no revertants from mutant A were recovered.

(Dana Burns-Pizer.)

17

GENETIC REGULATION OF CELL NUMBER: NORMAL AND CANCER CELLS

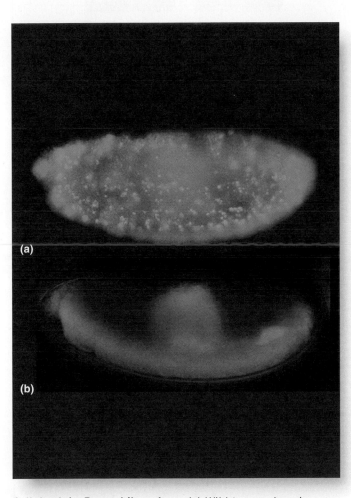

Cell death in *Drosophila* embryo. (a) Wild-type embryo in which the bright spots are cells carrying out a genetic program to die (apoptosis). (b) Mutant embryo in which this genetic program does not proceed. [Kristin White, Massachusetts General Hospital and Harvard Medical School.]

KEY QUESTIONS

- Why is it important for multicellular organisms to have mechanisms for regulating cell numbers?

- Why is cell-cycle progression regulated?

- Why is programmed cell death necessary?

- How do neighboring cells influence cell proliferation and cell death?

- Why is cancer considered a genetic disease of somatic cells?

- How do mutations promote tumors?

- How can genomic methodologies be applied to cancer research and medicine?

OUTLINE

17.1 The balance between cell loss and cell proliferation

17.2 The cell-proliferation machinery of the cell cycle

17.3 The machinery of programmed cell death

17.4 Extracellular signals

17.5 Cancer: the genetics of aberrant cell number regulation

17.6 Applying genomic approaches to cancer research, diagnosis, and therapies

CHAPTER OVERVIEW

The regulation of somatic-cell number is a splendid example of homeostasis—the sophisticated mechanisms that maintain an organism's physiology within normal limits. Cell proliferation is shut down and cells die when too many cells of a given type are present, and cell proliferation accelerates and cell death is inhibited when there is a deficiency of cells of a particular type. If cell proliferation and cell death are carefully balanced normally, what happens when these homeostatic mechanisms go awry because of mutations in the genes governing these processes? The consequences in many organisms, including humans, are dramatic: the accumulation of multiple mutations accelerating proliferation and blocking cell death in the same somatic cell is the underlying cause of cancer.

This chapter will explore the regulation of cell number and the underlying causes of misregulation that lead to tumor formation and cancer (Figure 17-1). In its sim-

CHAPTER OVERVIEW Figure

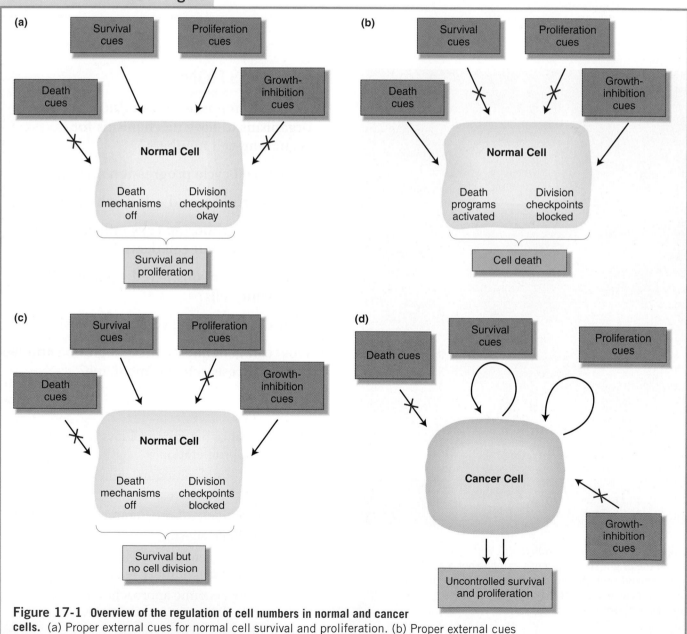

Figure 17-1 Overview of the regulation of cell numbers in normal and cancer cells. (a) Proper external cues for normal cell survival and proliferation. (b) Proper external cues for normal cell death or inhibition of proliferation. (c) Proper external cues for normal cell survival without proliferation. (d) Self-generated survival and proliferation signals in cancer cells.

plest form, the normal regulation of cell numbers can be regarded as the interplay between mechanisms that control cell proliferation and those that control cell death. Cell proliferation is controlled by the mitotic cell cycle, whereas cell death is achieved through a mechanism called programmed cell death, or *apoptosis*. In both cases, sequential biochemical events depend on the successful occurrence of prior events. In the cell cycle, fail-safe mechanisms—called *checkpoints*—prevent the cell cycle from progressing until prior events have been successfully completed. Similarly, survival factors block advancement of the death pathways. In multicellular animals, there is community input into the decision-making process for these pathways. Sensor molecules (receptor proteins) interact with chemical signaling proteins present in the immediate external milieu of the cell. These protein signals and their sensors are wired into the machinery of the cell cycle or apoptosis to act as either accelerators or brakes on one of these two pathways.

The genetic basis of cancer is the disruption of these homeostatic mechanisms through mutations. Cancers are *malignant* tumors, exhibiting uncontrolled growth and, as they become more advanced, acquiring the ability to *metastasize*—spread throughout the body.

When we say that cancer has a "genetic basis," we mean something different from the way in which we use the term in standard genetic analysis. In standard genetic analysis, we are referring to the transmission of distinct alleles of a gene from parent to offspring. Although some cancers have a heritable form, cancer occurrence is sporadic in most cases; that is, a particular type of cancer occurs in one member of a family but not in any of his or her relatives. In these cases, the underlying mutations have arisen within a lineage of somatic cells in that member, after the time early in development when the germ line and the soma separated from each other. Over time, several somatic mutations accumulate within the same somatic-cell lineage, altering or knocking out the function of several genes, until finally a cancerous cell is produced. The progeny of this cancerous cell expand into a clone of cancerous cells—the primary tumor. If left unchecked, this tumor will continue to grow and eventually invade other tissues and organs in the body.

Given that most cancers are somatic, not germ-line mutational events, we cannot study them by standard genetic analysis. Rather, we need to use other approaches to discover the mutational alterations and to understand how they affect their cellular pathways. We will see that there are many pathways to cancer formation, but all ultimately lead to the creation of cells that are insensitive to normal regulation of the cell cycle and apoptosis.

In our exploration of these topics, we will see again the recurrent theme that cells modulate the activity of key target proteins by relatively minor modifications in these proteins, such as the formation of complexes of the proteins with allosteric effectors. Much of genetics, indeed much of the biology of a cell, depends on such modulations, in which key proteins are toggled between active and inactive states.

17.1 The balance between cell loss and cell proliferation

Adult tissues are composed mostly of **differentiated cells**—nondividing cells that fulfill specialized physiological roles within the tissue. Typically, there is constant low-level turnover of these differentiated cell populations. That is, occasionally cells die and, when they do, they are replaced. Some cell loss is accidental—for example, from a burn or a wound. Other cell loss is through programmed cell death (apoptosis) of a cell that is abnormal in some way (perhaps dividing too rapidly or replicating an infecting virus).

Abnormal cells can do considerable harm. A cell that is dividing too rapidly may be on its way to becoming cancerous. A cell infected by a virus may shed virus particles that will circulate through the body, spreading the infection. The loss of such a cell would be negligible, but its continued presence could be catastrophic to the organism as a whole. Because we are concerned with regulation of *somatic* cell numbers (not germ-line cells), cell loss is not a problem for the organism as long as a cell population can be replenished. Thus, one of the roles of the apoptosis machinery is to survey for cellular abnormalities and execute a self-destruct mechanism when they are detected.

If there were only cell loss and no proliferation, then ultimately we would all run out of cells. We do not do so, because most mature cell populations have **stem cells** in reserve. Stem cells are undifferentiated cells that can divide in a way that gives rise to a variety of differentiated cell types—skin cells or hair follicle cells in the epidermis, T-cells, B-cells, or mast cells in the immune system, and so forth. Each tissue has its own kind of stem cells. When surveillance mechanisms detect an underrepresentation of a specific cell type, nearby stem cells are induced to undergo mitosis. Stem-cell mitoses are asymmetrical, giving rise to two progeny cells of different size. The larger of the two cells will be like its parent—a stem cell—and the other, smaller cell will differentiate into a specialized cell of the type that had become underpopulated in the tissue. Because a stem-cell division always produces another stem cell, there is an essentially limitless supply of replacement cells in reserve for tissue repair.

How can a stem cell tell that it needs to divide to replenish a particular cell type? It receives commands from neighboring cells to engage the cell-division machinery. How does the cell-death program begin? Often, the cell itself senses damage and initiates its own death. In other cases, the cell-death program is initiated by signals from neighboring cells that tell the cell to die. These signals are in the form of proteins or other molecules that attach to receptors on the surface of the stem cell or the dying cell. Eukaryotic cells have elaborate intercellular signaling pathways as status indicators of the environment. Some signals, called *mitogens*, stimulate proliferation, whereas others inhibit it. Death signals can activate apoptosis, whereas survival signals block activation.

The cell-proliferation and cell-death machinery must be wired into the cellular systems that are surveying the cell's external environment for the presence of signaling molecules. Thus intercellular signaling pathways typically consist of several components: the signals themselves, the receptors that receive the signals, and the transduction systems that relay the signal to various parts of the cell. The various components of the intercellular signaling systems are modified—by protein phosphorylation, allosteric interactions between proteins and small molecules, and interaction between protein subunits—to control the activity of these pathways.

17.2 The cell-proliferation machinery of the cell cycle

Before we consider the regulatory systems that control cell proliferation, it is first important to understand the basic events in the cell cycle. The cell cycle has four main parts: M phase—mitosis, the nuclear division process described in detail in Chapter 3—and three parts that are components of interphase—G1, the period between the end of mitosis and the start of DNA replication; S, the period of DNA synthesis; and G2, the period that follows DNA replication and precedes the mitotic prophase. S, G2, and M phases are normally fixed in duration. G1, however, can be quite variable because the cell cycle can enter an optional G0 resting phase.

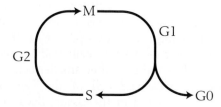

Some cells, such as the cells of the newly fertilized embryo, are rapidly dividing and have no G0 phase. At the other extreme, differentiated cells are in G0 for the remainder of their natural lives. Stem cells fluctuate between G0 and the cell-division cycle. This section will consider the molecular signals that drive the cell cycle. A

later section will consider how these molecules are integrated into the overall biology of the cell.

Cyclins and cyclin-dependent protein kinases

The engines that drive the cell cycle from one step to the next are a series of protein complexes called **CDK-cyclin complexes.** Ultimately, going to the next step requires the activation of genes whose protein products are necessary for the next phase of the cell cycle. This activation occurs through the turning on of transcription factors by the CDK-cyclin complexes (Figure 17-2). Consider, for example, the CDK-cyclin complex active during G1, which takes the cell cycle into the S phase, when DNA is synthesized. The G1 CDK-cyclin complex activates multiple cellular components:

- A transcription factor that turns on the genes encoding DNA polymerase subunits
- The genes for the enzymes that produce deoxyribonucleotides
- Other proteins with roles in the duplication of the chromosomes
- The gene for one of the subunits of the next required CDK-cyclin complex

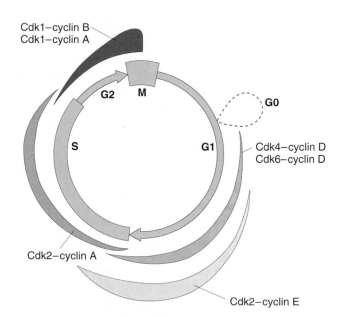

Figure 17-2 Variations in CDK-cyclin activities throughout the cell cycle of a mammalian cell. The widths of the bands indicate the relative kinase activities of the various CDK-cyclin complexes. Note that different CDKs and cyclins can bind to one another to form different complexes, increasing the array of combinations of CDK-cyclin complexes that can form in the course of the cell cycle. [After H. Lodish, A. Berk, S. L. Zipursky, P. Matsudaira, D. Baltimore, and J. Darnell, *Molecular Cell Biology*, 4th ed. Copyright 2000 by W. H. Freeman and Company.]

As different CDK-cyclin complexes appear and disappear, the proteins necessary for carrying out the different phases of the cell-division cycle are transcribed and translated like a well-choreographed ballet. The choreography requires that each active CDK-cyclin complex be "on stage" only for a limited part of the cell cycle. If an active CDK-cyclin complex is present at the wrong time, it will cause wholly inappropriate sets of genes to be transcribed or shut off.

The CDK-cyclin protein complexes are composed of two subunits: a **cyclin** and a **cyclin-dependent protein kinase** (abbreviated **CDK**).

- *Cyclins.* Every eukaryote has a family of structurally and functionally related cyclin proteins. Cyclins are so named because each is present in the cell only during one or more defined segments of the cell cycle. The appearance of a specific cyclin is the result of the activity of the preceding CDK-cyclin complex, which leads to the activation of a transcription factor for the new cyclin.

- *CDKs.* Cyclin-dependent protein kinases are another family of structurally and functionally related proteins. **Kinases** are enzymes that add phosphate groups to target substrates; for protein kinases, such as CDKs, the substrates are the side groups of specific amino acids on specific proteins. Each CDK catalyzes the phosphorylation of specific serine and threonine residues belonging to one or more unique target proteins. Becoming phosphorylated changes the activity of the target protein.

CDKs are named "cyclin dependent" because each CDK must be attached to a cyclin to function. The cyclin tethers the target protein so that the CDK can phosphorylate it (Figure 17-3). The CDKs are present throughout the cell cycle, and so which complex is active is a function of which cyclin is present. Because different cyclins are present at different phases of the cell cycle (see Figure 17-2), each phase is characterized by the phosphorylation of different target proteins.

As the cell cycle proceeds from one stage to the next, it is just as important that the currently active CDK-cyclin complexes be deactivated as that new ones be activated. Both the cyclin mRNA and the cyclin itself are highly unstable, and so the existing pool of cyclin and its mRNA will be quickly eliminated when the transcription factor for the cylin's gene is inactivated. The phosphorylation events are likewise transient and reversible. When a CDK-cyclin complex disappears, its phosphorylated target proteins are rapidly dephosphorylated by other enzymes always present in the cell.

CDK targets

How does the phosphorylation of some target proteins control the cell cycle? Phosphorylation initiates a chain of events that culminates in the activation of certain transcription factors. These transcription factors promote the transcription of certain genes whose products are required for the next stage of the cell cycle.

Much of our knowledge of the cell cycle comes from genetic studies of yeast (see the Model Organism box, Yeast) or from biochemical studies of cultured mammalian cells. Indeed, work on the cell cycle was recognized with the Nobel Prize for Medicine or Physiology in 2001. A well-understood example of the path connecting the appearance of cyclin to gene transcription is the Rb-E2F pathway in mammalian cells. Rb is the target protein of a CDK-cyclin complex called Cdk2–cyclin A, and E2F is the transcription factor that Rb regulates (Figure 17-4). From late M phase through the middle of G1, the Rb and E2F proteins are combined in a protein complex that does not promote transcription. In late G1, however, the active Cdk2–cyclin A complex is produced, which phosphorylates the Rb protein. This phosphorylation produces

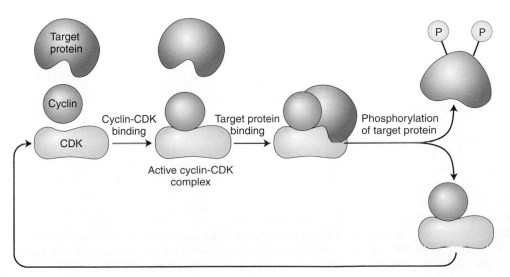

Figure 17-3 Phosphorylation of target proteins by the CDK-cyclin complex. The target protein binds to the cyclin part of an active CDK-cyclin complex, placing the target phosphorylation sites close to the active site of CDK. Once phosphorylated, the target protein is no longer able to bind to cyclin and is released from the complex.

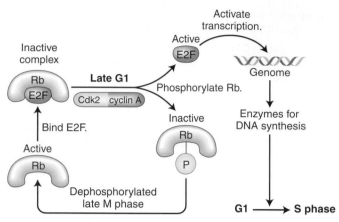

Figure 17-4 Contributions of the Rb and E2F proteins in the regulation of the G1-to-S-phase transition in a mammalian cell. [After H. Lodish, D. Baltimore, A. Berk, S. L. Zipursky, P. Matsudaira, and J. Darnell, *Molecular Cell Biology*, 3d ed. Copyright 1995 by Scientific American Books.]

a change in the shape of Rb so that it can no longer bind to the E2F protein. The unattached E2F protein is then free to promote the transcription of certain genes that encode the enzymes vital for DNA synthesis and other aspects of chromosome replication. It also activates gene expression of the next cyclin to be expressed, cyclin B. With cyclin B's appearance, the next phase of the cell cycle—S phase—can proceed.

Rb and E2F are in fact representatives of two families of related proteins. In mammals, different CDK-cyclin complexes selectively phosphorylate different proteins of the Rb family, each of which in turn releases a specific E2F family member to which it is bound. The different E2F transcription factors then promote the transcription of different genes that execute different aspects of the cell cycle.

> **MESSAGE** Sequential activation of different CDK-cyclin complexes ultimately controls progression of the cell cycle.

Checkpoints as brakes on cell-cycle progression

Proper choreography of the cell cycle's events is crucial to the production of progeny cells with the proper number of intact chromosomes. For example, attempting to condense chromosomes and move them to metaphase (recall the stages of mitosis, Figure 3-28) before their DNA molecules have completed replication could lead to the production of chromosome fragments or other kinds of damage. Attempting to separate the chromosomes to opposite poles at anaphase before all kinetochores are attached to spindle fibers could lead to nondisjunction and aneuploidy of the progeny cells. These potentially disastrous problems are prevented by having a series of **checkpoints**—circuitry attached to the cell-cycle system

that prevents progression to the next stage until the preceding stage has been successfully fulfilled.

How do checkpoints act as brakes on the cell cycle? They activate proteins that can inhibit the protein kinase activity of one of the CDK-cyclin complexes. In this way, the cell cycle can be held in check until the checkpoint monitoring mechanisms give a "green light," indicating that the cell is properly prepared to proceed to the next phase of the cycle.

One example of how this checkpoint system operates begins with damaged DNA (Figure 17-5). When DNA is damaged during G1 (for example, by X-irradiation), CDK-cyclin complexes stop phosphorylating their target proteins. A protein called p53 recognizes certain kinds of DNA mismatches and then proceeds to activate another protein, p21. When its levels are high, p21 binds to the CDK-cyclin complex and inhibits its protein kinase activity. Thus CDK cannot phosphorylate its target proteins, and the cell cycle is unable to progress from G1 to S. The inhibiting processes are reversed by a drop in p53 levels after DNA repair is complete. In the absence of p53, CDK-cyclin protein kinase activity is no longer inhibited, and the G1-to-S checkpoint block is removed. A checkpoint is thus a regulatory protein, such as p53, that can act as a brake on the protein kinase activity of the CDK-cyclin complexes.

> **MESSAGE** Fail-safe systems called checkpoints ensure that the cell cycle does not progress until the cell has completed all prior events necessary to assure its survival through the next steps.

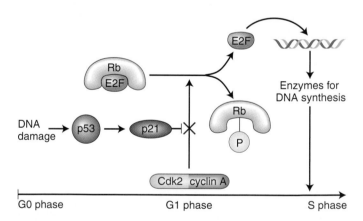

Figure 17-5 Inhibitory control of the progression of the cell cycle in mammals. In the presence of damaged DNA, p53 protein is induced, which in turn induces p21 protein. Elevated levels of p21 inhibit the protein kinase activity of the Cdk2−cyclin A complex. E2F remains complexed with Rb, and the cell does not progress to S phase. After the damaged DNA has been repaired, p53 and then p21 levels drop. The inhibition of the Cdk2-cyclin protein kinase activity is relieved, permitting the cell to enter S phase. [After C. J. Sherr and J. M. Roberts, *Genes and Development* 9, 1995, 1150.]

MODEL ORGANISM Yeast

Yeast As a Model for the Cell Cycle

Work initiated by Leland Hartwell and his associates on the cell-cycle genetics of the budding yeast *Saccharomyces cerevisiae* revealed a large array of genetic functions that maintain the proper cell cycle. These functions were identified as a specific subset of temperature-sensitive (ts) mutations called *cdc* (cell division cycle) mutations. When grown at low temperature, yeasts with these *cdc* mutations grew normally. When shifted to higher, restrictive temperatures, the yeasts stopped growing. What made these *cdc* mutations novel among the more general class of ts mutations was that a particular *cdc* mutant would stop growing at a specific time in the cell cycle, and all the yeast cells would look alike. Consider some examples of *S. cerevisiae*, a yeast that divides through budding, a process in which a mother cell develops a small outpocketing, a "bud." The bud grows and mitosis takes

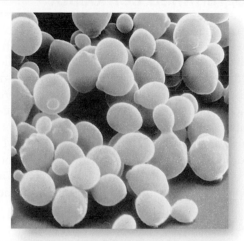

Scanning electron micrograph of *S. cerevisiae* cells at different points in the cell cycle, as indicated by different bud sizes. [Courtesy of E. Schachtbach and I. Herskowitz.]

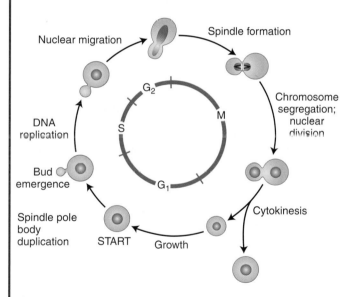

Cell cycle of *S. cerevisiae*. [From H. Lodish, D. Baltimore, A. Berk, S. L. Zipursky, P. Matsudaira, and J. Darnell, *Molecular Cell Biology*, 3d ed. Copyright 1995 by Scientific American Books.]

place in such a way that one spindle pole is in the mother cell and the other in the bud. The bud continues to grow until it is as big as the mother cell, and the mother cell and bud then separate into two daughter cells. Any run-of-the-mill ts mutations in *S. cerevisiae*, when shifted to restrictive temperature, stop growing at various times in the cycle of bud formation and growth. In contrast, after a shift to restrictive temperature, one *S. cerevisiae cdc* mutation produces yeast cells with only tiny buds, whereas another produces yeast cells exclusively with larger buds, half the size of the mother cell. Such different Cdc phenotypes indicate different defects in the machinery required to execute specific events in the cell cycle.

With the sequencing of the *S. cerevisiae* genome now complete, we are able to identify the entire array of proteins of the cyclin and CDK families (22 and 5 proteins, respectively). Their genes are now being systematically mutagenized and genetically characterized to reveal how each contributes to the cell cycle.

It is necessary not only to release the cell-cycle "brake" but also to engage the "transmission" and the "engine" to advance the cell cycle. In regard to the CDK-cyclin complexes, it is not enough to remove any inhibition and add the correct cyclin; the complexes must also be activated by phosphorylation. Once the brake has been released, independent signals from within or outside the cell induce a cascade of protein kinases that phosphorylate the appropriate CDK-cyclin complex. The activated complex is now able to phosphorylate its target proteins. Just as an automobile doesn't perform unless it has a functional engine, transmission, steering, acceleration, and braking system, the cell-cycle machinery is ineffective unless all the necessary components for governing its progression are present.

17.3 The machinery of programmed cell death

In multicellular organisms, systems have evolved to eliminate damaged (and hence potentially harmful) cells by a self-destruct and disposal mechanism called **programmed cell death** or **apoptosis**. This self-destruct mechanism can be activated under many different circumstances, such as cells that are no longer needed for development. In all cases, however, the events in apoptosis seem to be the same (Figure 17-6). First, the DNA of the chromosomes is fragmented, organelle structure is disrupted, and the cell loses its normal shape and becomes spherical. Then, the cell breaks up into small fragments called *apoptotic bodies*, which are phagocytosed (literally, eaten up) by motile scavenger cells.

The engines of self-destruction are a series of enzymes called **caspases,** which is short for *cysteine-containing aspartate-specific proteases* (see the Model Organism box, *Caenorhabditis elegans*). Proteases are enzymes that cleave other proteins. The general term for such cleavage of proteins is *proteolysis*. Each caspase is a protease rich in cysteines: when activated, it cleaves certain target proteins at a specific aspartate. These target proteins initiate fragmentation of DNA, disruption of organelles, and other events that characterize apoptosis.

Every multicellular animal has a family of caspase proteins, related to one another by polypeptide sequence; in humans, for example, 14 caspases have been identified so far. In normal cells, each caspase is in an inactive state, called the **zymogen** form. In general, a zymogen is an inactive precursor form of an enzyme having a longer polypeptide chain than that of the final active enzyme. To turn the zymogen form of an enzyme into the active form, a part of the polypeptide is removed by enzyme cleavage.

There are two classes of caspases: *initiators* and *executioners*. The initiator caspases are cleaved in response to activation signals coming from other classes of proteins. They in turn cleave one of the executioner caspases, which in turn cleaves another, and so forth, until all executioner caspases are active. Unlike the cell *cycle*, which as its name implies can be activated repeatedly, a given cell can only die once. Thus, instead of being linked to one another in a circular pathway, as in the cell cycle, the events of programmed cell death need only proceed unidirectionally. The logic of the programmed steps in each of these systems fits well with its need to proceed cyclically or reach a final conclusion.

> **MESSAGE** Programmed cell death is mediated by a sequential cascade of proteolysis events that activate enzymes that destroy several key targeted cellular components.

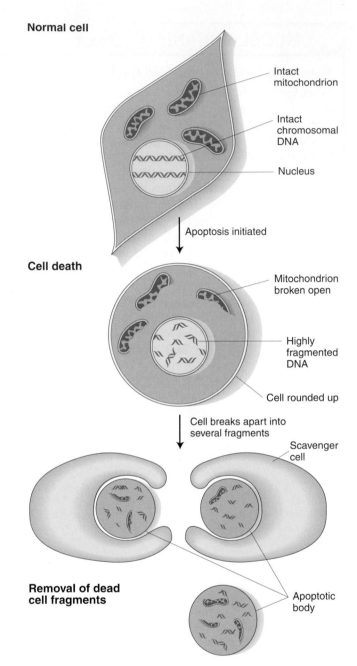

Figure 17-6 Sequence of events in apoptosis. First, the membranes of organelles such as mitochondria are disrupted and their contents leak into the cytoplasm, chromosomal DNA breaks into small pieces, and the cell loses its normal shape. Then the cell breaks apart into small fragments, which are disposed of by phagocytotic scavenger cells.

How do the executioner caspases carry out the sentence of death? In addition to activating other caspases, executioner caspases enzymatically cleave other target proteins in the damaged cell (Figure 17-7). One target is a "sequestering" protein that forms a complex with a DNA endonuclease, thereby holding (sequestering) the endonuclease in the cytoplasm. When the sequestering protein has been cleaved, the endonuclease is free to

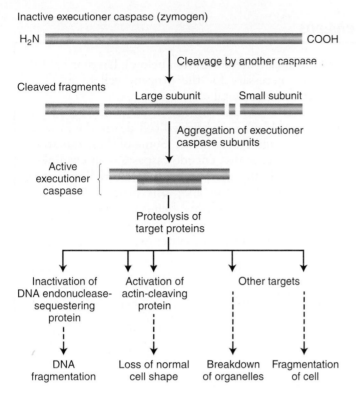

Inactive executioner caspase (zymogen)

H₂N ———————————————— COOH

Cleavage by another caspase

Cleaved fragments

Large subunit   Small subunit

Aggregation of executioner caspase subunits

Active executioner caspase

Proteolysis of target proteins

Inactivation of DNA endonuclease-sequestering protein

Activation of actin-cleaving protein

Other targets

DNA fragmentation

Loss of normal cell shape

Breakdown of organelles

Fragmentation of cell

Figure 17-7 Role of executioner caspases in apoptosis. Cleavage of the zymogen precursor leads to the formation of enzymatically active executioner caspases. As the executioner caspases cleave target proteins, the various cellular breakdown events take place, leading to cell death and removal.

enter the nucleus and chop up the cell's DNA. Another target is a protein that, when cut by the caspases, cleaves actin, a major component of the cytoskeleton. The disruption of actin filaments leads to the loss of normal cell shape. In similar fashion, caspase-activated proteases are thought to initiate all other events of apoptosis.

What triggers caspase activation? It has been known for several years that, in some manner, many forms of cellular damage trigger leakage of material from mitochondria and that this leakage somehow induces the apoptotic response. Indeed, it now appears that one of the switches for apoptosis activation is cytochrome *c*, a mitochondrial protein that normally takes part in cell respiration. Cytochrome *c* that has leaked into the cytoplasm is thought to bind to another protein called Apaf (apoptotic protease-activating factor). The cytochrome *c*-Apaf complex then binds to the initiator caspase and activates it.

Because cell death is irreversible, it is crucial that healthy cells not be destroyed by apoptosis. The need to preserve healthy cells has probably been the compelling factor in the evolution of systems to make sure that the apoptosis pathway remains "off" under normal conditions. Proteins such as Bcl-2 and Bcl-x in mammals act as roadblocks to prevent apoptosis from being triggered. Among the possible actions of these Bcl proteins is

blocking the release of cytochrome *c* from mitochondria (possibly by making it more difficult for mitochondria to burst) and binding to Apaf, preventing its interaction with the initiator caspase.

17.4 Extracellular signals

A cell in a multicellular organism continually assesses its own internal status to determine whether conditions are appropriate to initiate the cell cycle or commit cellular suicide. Ultimately, however, the fate of a cell must be subservient to the needs of the population of cells of which it is a member (such as the entire early embryo, a tissue, or a body part such as a limb or an organ). Hence, when cell numbers are depleted, cells in concert determine that their normal neighbors are missing and pass an extracellular signal to the stem cells that they need to divide. To consider how these signals are wired into the proliferation and apoptosis pathways, we first need to look at the basic circuitry that mediates cross talk between cells, often called **cell–cell communication.**

Mechanisms for cell–cell communication

Many kinds of signals are transmitted between cells to coordinate virtually all aspects of a multicellular organism's development and physiology. All systems for intercellular communication have multiple components. To initiate communication, a signaling cell secretes a molecule called a **ligand** (Figure 17-8). Some ligands, called hormones, are long-range **endocrine signals** that are

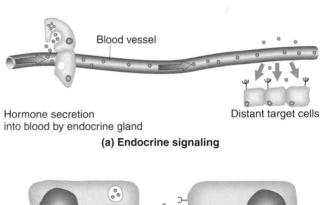

Blood vessel

Hormone secretion into blood by endocrine gland

Distant target cells

(a) Endocrine signaling

Secretory cell

Adjacent target cell

Figure 17-8 Modes of intercellular signaling. (a) Endocrine signals enter the circulatory system and can be received by distant target cells. (b) Paracrine signals act locally and are received by nearby target cells. [From H. Lodish, D. Baltimore, A. Berk, S. L. Zipursky, P. Matsudaira, and J. Darnell, *Molecular Cell Biology*, 3d ed. Copyright 1995 by Scientific American Books.]

MODEL ORGANISM *Caenorhabditis elegans*

The Nematode *Caenorhabditis elegans* As a Model for Programmed Cell Death

In the past 15 years, genetic studies by Robert Horvitz and his associates of the nematode (roundworm) *Caenorhabditis elegans* have advanced our understanding of programmed cell death. Researchers have mapped the entire series of somatic-cell divisions that produce the 1000 or so cells of the adult worm. For some of the embryonic and larval cell divisions, particularly those that will contribute to the worm's nervous system, a progenitor cell gives rise to two progeny cells, one of which then undergoes programmed cell death (see the illustration below). Divisions of this type are necessary for the progeny cell to fulfill its proper developmental role.

A set of mutations has been identified in the worm that block programmed cell death; the progeny cell that would otherwise die. Some of these mutations knock out genes that encode caspases—an example is *ced-3* (cell-death gene number 3)—clearly implicating these caspases in the apoptosis process. Other key players in this process are being uncovered through the analysis of other genes with mutant cell-death phenotypes in worms and other experimental systems.

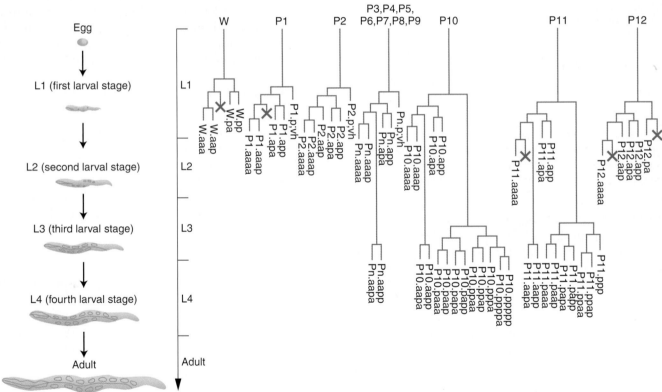

The life cycle of the nematode *C. elegans.*

Examples of programmed cell death in the development of *C. elegans.* A symbolic representation of the cell lineages of 13 cells (the W cell, the P1 cell, and so forth) produced during embryogenesis. The vertical axis is developmental time, beginning with the hatching of the egg into the first larval stage (L1). In each lineage, a vertical blue bar connects the various division events (horizontal blue bars). The names of the final cells are shown, such as W.aaa or P1.apa. In several cases, a cell division produces one viable cell and one cell that undergoes apoptosis. A cell that undergoes programmed cell death is indicated with a blue × at the end of a branch of a lineage. In homozygotes for mutations such as *ced-3,* these cell deaths do not occur.

released from endocrine organs into the circulatory system, which transmits them throughout the body. Hormones can act as master control switches for many different tissues, which can then respond in a coordinated fashion. Other secreted ligands act as **paracrine signals;** that is, they do not enter the circulatory system but act only locally, in some cases only on adjacent cells. Most (but not all) endocrine signals are small molecules, such as the mammalian steroid hormones androgen or estrogen that are responsible for male or female sex-specific phenotypes. In contrast, most paracrine signals are proteins. Because the signals initiating cell proliferation and cell death emanate from nearby, this discussion will focus on paracrine signaling through protein ligands.

Protein ligands and transmembrane receptors

Protein ligands act as signals by binding to transmembrane receptors, proteins that are embedded in the plasma membrane at the surface of the cell. These ligand–receptor complexes release chemical signals into the cytoplasm just inside the plasma membrane. Such signals are passed through a series of intermediary molecules until they finally alter the structure of transcription factors in the nucleus, activating the transcription of some genes and the repression of others.

Transmembrane receptors have one part (the extracellular domain) outside of the cell, a middle part that passes once or several times through the plasma membrane, and another part (the cytoplasmic domain) inside the cell (Figure 17-9). The ligand binds to the extracellular domain of the receptor. The binding of the ligand changes the structure of the receptor in some way that initiates signaling activity. Many polypeptide ligands are dimers and can simultaneously bind two receptor monomers. This simultaneous binding brings the cytoplasmic domains of the two receptor subunits into close proximity and activates their signaling activity. Some receptors for polypeptide ligands are receptor tyrosine kinases (RTKs, see Figure 17-9b). Their cytoplasmic domains, when activated, have the ability to phosphorylate certain tyrosine residues on target proteins. Other receptors are serine/threonine kinases. Still others have no enzymatic activity but, when bound to ligands, they undergo conformational changes that in turn cause conformational changes in (and activation of) proteins bound to their cytoplasmic domains.

Sequentially, modification of the conformation of one protein leads to modification of the conformation of another, ending with the modification of transcriptional activators and repressors that alter the activities of many genes in the target cell. Each altered protein acts as one signal in a chain. The chain of signals leading from the ligand–receptor complex to the altered gene activity is called a **signal-transduction cascade.**

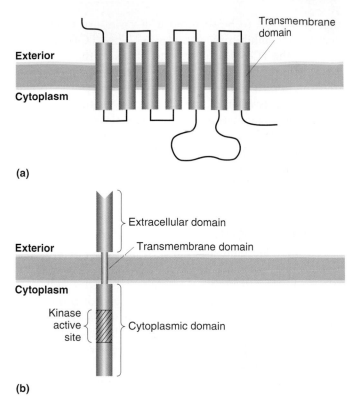

Figure 17-9 Examples of transmembrane receptors.
(a) A receptor that passes through the cell membrane seven times. (b) Receptor tyrosine kinase, which has a single transmembrane domain. The extracellular domain binds to ligand. The active site of the tyrosine kinase is in the cytoplasmic domain. [After H. Lodish, D. Baltimore, A. Berk, S. L. Zipursky, P. Matsudaira, and J. Darnell, *Molecular Cell Biology*, 3d ed. Copyright 1995 by Scientific American Books.]

Protein ligands released in a tissue may act as signals that regulate the cell-division cycle or apoptosis. The general theme is that there are pathways for controlling cell proliferation and self-destruction and that activation of these pathways requires the correct array of positive inputs and the absence of negative, or inhibitory, inputs.

THE CELL CYCLE: POSITIVE EXTRACELLULAR CONTROLS Cell division is promoted by the action of **mitogens,** protein ligands usually released from a paracrine (nearby) source. Mitogens are also called **growth factors.** Many mitogens activate a receptor tyrosine kinase and initiate a signal-transduction pathway that eventually leads to the expression of the G1 cyclin D genes. Mitogens include epidermal growth factor (EGF), which signals by binding to a receptor tyrosine kinase called the EGF receptor (EGFR).

THE CELL CYCLE: NEGATIVE EXTRACELLULAR CONTROLS Certain secreted protein ligands are known to inhibit cell division in completely intact tissue, when growth is not necessary. One example is TGF-β, a ligand that is thought to be secreted in a variety of tissues under growth-inhibitory conditions. The TGF-β ligand binds to the TGF-β receptor and activates the receptor's serine/threonine kinase activity. The ensuing signal-transduction cascade eventually blocks the phosphorylation and inactivation of the Rb protein. Recall, from earlier in the chapter, the role that Rb plays in regulating the cell cycle by preventing activation of the E2F transcription factor. Blocking Rb inactivation thus keeps E2F expression turned off. Consequently, the ingredients necessary for DNA replication are not present and the cell cycle will progress no farther.

APOPTOSIS: POSITIVE EXTRACELLULAR CONTROLS

Often, the command for self-destruction comes from a neighboring cell. For example, within the immune system, only a small percentage of B cells and T cells mature to make functional antibody or receptor protein, respectively. The nonfunctional immature B cells and T cells must be eliminated by induced self-destruction; otherwise, the numbers of these unnecessary cells would clog the immune system. The self-destruction signal is activated through the Fas system (Figure 17-10) and other death receptors. A membrane-bound protein called FasL (Fas ligand) protrudes from the surface of the cell; its presence is the signal to an adjacent cell to self-destruct. The FasL protein binds to transmembrane Fas receptors on the surface of the adjacent cell. This binding induces three receptors to join to form a trimer bound to the single FasL protein. The trimerization of their cytoplasmic domains in turn indirectly activates a molecule such as Apaf. As described earlier in this chapter, Apaf activates an initiator caspase, which sets off the cascade leading to cell death.

APOPTOSIS: NEGATIVE EXTRACELLULAR CONTROLS

It appears that self-destruction may in some ways be a default state and that cells receive constant signals instructing them to stay alive. The advantage of such a system may be that it is able to eliminate abnormal cells rapidly to ward off the dangers that they pose to the organism as a whole. The signals that can block activation of the apoptosis pathway are secreted factors called **survival factors**. Some survival factors do double duty as mitogens, such as EGF. The action of EGF and the EGFR on survival is largely mediated by a different signal-transduction cascade from the usual receptor tyrosine kinase pathway. This alternative cascade eventually leads directly or indirectly to activation of proteins of the Bcl-2 family.

MESSAGE Intercellular signaling systems communicate instructions to activate or to arrest the cell cycle and to initiate or postpone self-destruction.

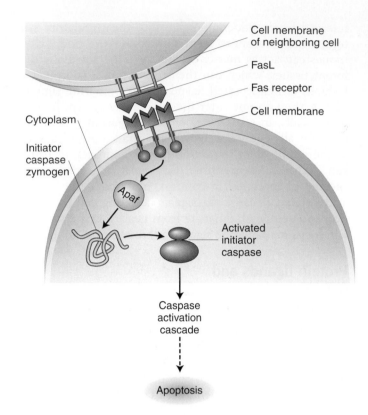

Figure 17-10 Positive extracellular control of apoptosis. Receptor–ligand interaction leads to activation of a molecule (Apaf), which in turn causes proteolysis and activation of the initiator caspase. A series of caspases are then proteolysed and activated in turn, ultimately leading to apoptosis of the cell. [After S. Nagata, *Cell* 88, 1997, 357.]

Signal-transduction cascades

Perhaps the best understood of the receptors for polypeptide ligands are the receptor tyrosine kinases (Figure 17-11). A receptor tyrosine kinase, or RTK, is a single unit, or monomer, "floating" in the plasma membrane. When ligand binds to two neighboring RTK units, the two monomers bind to form a dimer. Thus a ligand–RTK complex consists of a single ligand molecule bound to a dimer of two RTK molecules. RTK dimerization activates the protein kinase activity of the receptor's cytoplasmic domain. The first phosphorylation targets of the kinase are several tyrosines in the cytoplasmic domain of the RTK itself; this process is called *autophosphorylation*, because the kinase acts on itself. The phosphorylated RTK is conformationally changed so that its tyrosine kinase active site phosphorylates other target proteins.

Quite often, the next step in propagating the signal is to activate a G-protein. A G-protein cycles between being bound by GDP in the inactive state and being bound by GTP in the activated state. The autophosphorylated RTK activates a protein that binds to the inactive

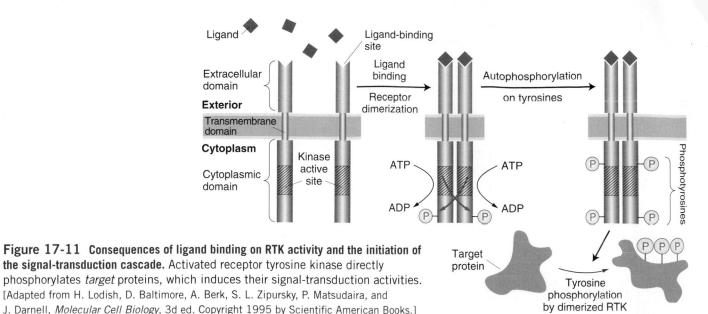

Figure 17-11 **Consequences of ligand binding on RTK activity and the initiation of the signal-transduction cascade.** Activated receptor tyrosine kinase directly phosphorylates *target* proteins, which induces their signal-transduction activities. [Adapted from H. Lodish, D. Baltimore, A. Berk, S. L. Zipursky, P. Matsudaira, and J. Darnell, *Molecular Cell Biology*, 3d ed. Copyright 1995 by Scientific American Books.]

G-protein, changing its conformation so that it binds to a molecule of GTP and becomes activated (Figure 17-12). The specific G-protein called Ras is especially important in carcinogenesis, as will be explained later.

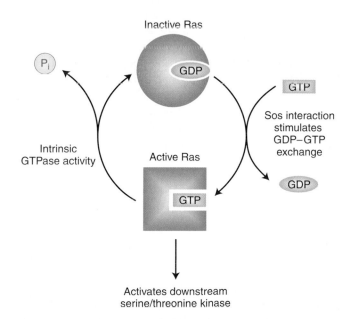

Figure 17-12 **Ras as an example of the G-protein activity cycle.** When bound to GDP, Ras does not signal. Interaction with a second protein, Sos, promotes the exchange of GDP for GTP. The active Ras–GTP complex interacts with a cytoplasmic serine/threonine kinase, which transmits the signal to the next step in the signal-transduction pathway. When Ras–GTP is released from Sos, it hydrolyzes GTP to GDP and reassumes the inactive Ras–GDP state. [After J. D. Watson, M. Gilman, J. Witkowski, and M. Zoller, *Recombinant DNA*, 2d ed. Copyright 1992 by James D. Watson, Michael Gilman, Jan Witkowski, and Mark Zoller.]

The activated GTP-bound G-protein then binds to a cytoplasmic protein kinase, in turn changing its conformation and activating its kinase activity. This protein kinase then phosphorylates other proteins, including other protein kinases. (In the example in Figure 17-13, the protein kinases farther down the cascade are called Raf, MEK, and MAP kinase.) The targets of some of these protein kinases are transcriptional activators and repressors. The phosphorylated transcription factors alter their conformations, leading to the transcription of some genes and the repression of others (see Figure 17-13).

The steps in ligand–receptor binding and in signaling within the cell depend on conformational changes. For example, the conformational changes caused by the binding of ligands to receptors activate the signaling pathways. Likewise, conformational changes in protein kinases enable them to phosphorylate specific amino acids on specific proteins, and other proteins undergo conformational changes when they bind to GTP. Not only do these conformational changes permit rapid response to an initial signal, but they are readily reversible, enabling signals to be shut down rapidly. Thus the components of the signaling system can be recycled so that they are ready to receive further signals.

An integrated view of the control of cell numbers

We are learning that there is intimate coupling between cell-proliferation and cell-death decisions. We saw the example of the EGF–EGFR signaling system, which acts both as a mitogenic stimulus and as a survival factor, inhibiting apoptosis. Similarly, it turns out that active p53 protein, which is a key intermediary in

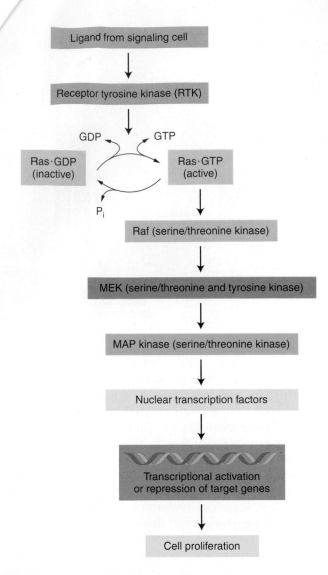

Figure 17-13 One pathway for RTK signaling. Raf, MEK, and MAP kinase are three cytoplasmic protein kinases that are sequentially activated in the signal-transduction cascade. [After H. Lodish, D. Baltimore, A. Berk, S. L. Zipursky, P. Matsudaira, and J. Darnell, *Molecular Cell Biology*, 3d ed. Copyright 1995 by Scientific American Books.]

the S to G2 checkpoint, promotes apoptosis by attacking the integrity of the mitochondrion as well as by inhibiting the expression of genes whose products contribute to survival. Why does it make sense that the cell cycle and apoptosis are linked? Simply put, the conditions that encourage proliferation also demand that the cell survives to duplicate itself. On the other hand, a cell with damage to its DNA poses a danger to the organism itself. Thus, it is in the organism's interest to have evolved mechanisms that not only block cell-cycle progression but also remove a mutated cell that might be on its way to becoming abnormal.

17.5 Cancer: the genetics of aberrant cell number regulation

A basic article of faith in genetic analysis is that we learn a great deal about biology, both normal and diseased, by studying the properties of mutations that disrupt normal processes. It has certainly been true in regard to cancer. It has become clear that virtually all cancers of somatic cells arise owing to a series of special mutations that accumulate in a cell. Some of these mutations alter the activity of a gene; others simply eliminate the gene's activity. Cancer-promoting mutations fall into a few major categories: those that increase the ability of a cell to proliferate, those that decrease the susceptibility of a cell to apoptosis, or those that increase the general mutation rate of the cell or its longevity, so that all mutations, including those that encourage proliferation or apoptosis, are more likely to occur. Improvements in diagnosis and treatment of cancers have been dramatic, but, though some battles have been won, we are a very long way from declaring victory in the war against cancer. The genetic and genomic analysis of cancer offers important new avenues to explore.

How cancer cells differ from normal cells

Malignant tumors, or **cancers,** are aggregates of cells, all descended from an initial aberrant founder cell. In other words, the malignant cells are all members of a single clone. This is true even in advanced cancers having multiple tumors at many sites in the body. Cancer cells typically differ from their normal neighbors by a host of phenotypic characters, such as rapid division rate, ability to invade new cellular territories, high metabolic rate, and abnormal shape. For example, when cells from normal epithelial cell sheets are placed in cell culture, they can grow only when anchored to the culture dish itself. In addition, normal epithelial cells in culture divide only until they form a continuous monolayer (Figure 17-14a). At that point, they somehow recognize that they have formed a single epithelial sheet and stop dividing. In contrast, malignant cells derived from epithelial tissue continue to proliferate, piling up on one another (Figure 17-14b).

Clearly, the factors regulating normal cellular physiology have been altered. What, then, is the underlying cause of cancer? Many different cell types can be converted into a malignant state. Is there a common theme? or does each arise in a quite different way? We can think about cancer in a general way: as being due to the accumulation of multiple mutations in a single cell that cause it to proliferate out of control. Some of those mutations may be transmitted from the parents through the germ line. But most arise de novo in the somatic-cell lineage of a particular cell.

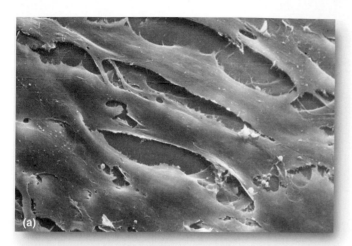 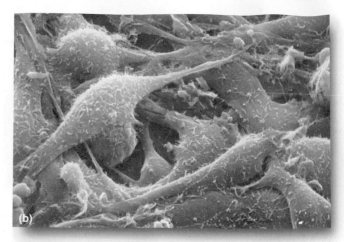

Figure 17-14 Normal cells and cell transformed by an oncogene. Scanning electron micrographs of (a) normal cells and (b) cells transformed by Rous sarcoma virus, which infects cells with the *src* oncogene. (a) A normal cell line called 3T3. Note the organized monolayer structure of the cells. (b) A transformed derivative of 3T3. Note how the cells are rounder and piled up on one another. [From H. Lodish, A. Berk, S. L. Zipursky, P. Matsudaira, D. Baltimore, and J. Darnell, *Molecular Cell Biology*, 4th ed. Copyright 2000 by W. H. Freeman and Company. Courtesy of L.-B. Chen.]

The multicellular solution to life that has evolved in higher animals relies on the collaboration and communication exhibited by the organized cells of the tissues and organ systems. In a sense, cancer cells have reverted to an antisocial, isolated state where they operate without external constraint. Cancerous cells are "deaf" to signals from neighboring cells to stop dividing or to activate their self-destruct systems.

Mutations in cancer cells

Several lines of evidence point to a genetic origin of the transformation of cells from the benign into the cancerous state.

1. Most carcinogenic agents (chemicals and radiation) are also mutagenic, suggesting that they produce cancer by introducing mutations into genes.

2. In the past few years, several alleles that increase susceptibility to cancer have been cloned and mapped.

3. Mutations that are frequently associated with particular kinds of cancers have been identified. In some cases, investigators have created transgenic or knockout mimics of these naturally occurring mutations in cell lines in culture or in intact experimental models. These artificially created mutations produce cancerous or cancerlike phenotypes.

MESSAGE Tumors arise from a sequence of mutational events that lead to uncontrolled proliferation and cellular immortality.

A tumor does not arise as a result of a single genetic event but rather as the result of multiple hits; that is, several mutations must arise within a single cell for it to become cancerous. Occasionally, a single mutation is powerful enough to guarantee that a cancer will form. For example, certain cancers are inherited as highly penetrant single Mendelian factors; an example is familial retinoblastoma, a cancer of the retina further discussed on page 564. Perhaps more common are less penetrant alleles that increase the probability of developing a particular type of cancer. In some of the best-studied cases, the progression of colon cancer and astrocytoma (a brain cancer) has been shown to entail the accumulation of several different mutations in the malignant cells (Figure 17-15).

In regard to how mutations act within the cancer cell, two general kinds are associated with tumors: oncogene mutations and mutations in tumor-suppressor genes.

- **Oncogene** mutations act in the cancer cell as gain-of-function dominant mutations (see Chapter 16 for a discussion of gain-of-function dominant mutations). That is, the mutation need be present as only one allele to contribute to tumor formation. When the mutation is present in protein-coding DNA, the oncogene causes a structural change in the encoded protein. When the mutation is present in a regulatory element, the oncogene causes a structurally normal protein to be misregulated. The gene in its normal, unmutated form is called a **proto-oncogene.**

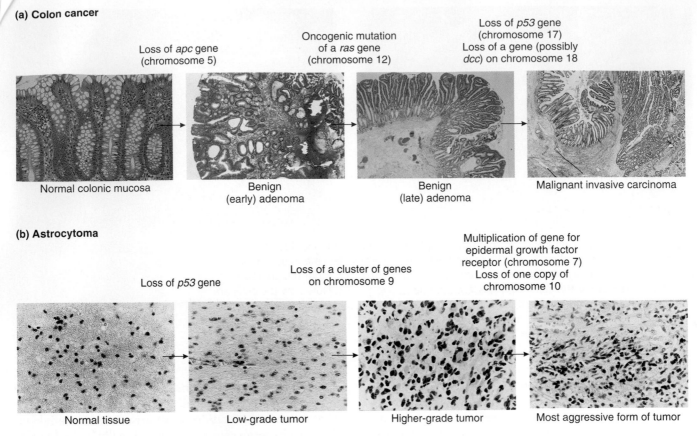

Figure 17-15 Multistep progression to malignancy in cancers of the colon and brain. Several histologically distinct stages can be distinguished in the progression of these tissues from the normal state to benign tumors to a malignant cancer. (a) A common sequence of mutational events in the progression to colon cancer. Note that the tissue becomes more disorganized as the tumor progresses toward malignancy. (b) A different characteristic series of mutations mark the progression toward a malignant astrocytoma, a form of brain cancer. [Micrographs by E. R. Fearon and K. Cho. From W. K. Cavanee and R. L. White, *Scientific American*, March 1995, pp. 78–79.]

- Mutations in **tumor-suppressor genes** that promote tumor formation are loss-of-function recessive mutations. That is, for cancer to occur, both alleles of the gene must encode gene products having reduced or no residual activity (that is, they are null mutations).

> **MESSAGE** The proteins that oncogenes encode are *activated* in tumor cells, whereas the proteins that tumor-suppressor genes encode are *inactivated*.

Many proteins that are altered by cancer-producing mutations take part in intercellular communication and the regulation of the cell cycle and apoptosis (Table 17-1). Genes that become oncogenes are genes encoding proteins that positively control (turn on) the cell cycle or negatively control (block) apoptosis. The mutant proteins are now active even in the absence of the appropriate activation signals. As a consequence, oncogenes act either to increase the rate of cell proliferation or to prevent apoptosis. On the other hand, tumor-suppressor genes encode proteins that arrest the cell cycle or induce apoptosis; in these cases, the cell loses a brake that can stop cell proliferation.

It is obvious why mutations that increase the rate of cell proliferation cause tumors. It is less obvious why mutations that decrease the chances that a cell will undergo apoptosis cause tumors. The reason seems to be twofold: (1) a cell that cannot undergo apoptosis has a much longer lifetime in which to accumulate proliferation-promoting mutations and (2) the sorts of damage and unusual physiological changes that take place inside a tumor cell would otherwise induce the self-destruction pathway. Thus tumor cells will not be able to survive unless they acquire mutations that prevent the triggering of apoptosis.

How have tumor-promoting mutations been identified? Several approaches have been used. It is well known that certain types of cancer can "run in families."

TABLE 17-1 Functions of Wild-Type Proteins and Properties of Tumor-Promoting Mutations in the Corresponding Genes

Wild-type protein function	Properties of tumor-promoting mutations
Promotes cell-cycle progression	Oncogene (gain-of-function)
Inhibits cell-cycle progression	Tumor-suppressor mutation (loss-of-function)
Promotes apoptosis	Tumor-suppressor mutation (loss-of-function)
Inhibits apoptosis	Oncogene (gain-of-function)
Promotes DNA repair	Tumor-suppressor mutation (loss of function)

With modern techniques for pedigree analysis, familial tendencies toward certain kinds of cancer can be matched with molecular markers such as microsatellites; in several cases, this process has led to the successful identification of a mutated gene. Alternatively, many types of tumors have characteristic chromosomal translocations or deletions of particular chromosomal regions. In some cases, these chromosomal rearrangements are so reliably a part of a particular cancer that they can be used for diagnosis. For example, 95 percent of patients with chronic myelogenous leukemia (CML) have a characteristic translocation between chromosomes 9 and 22. This translocation, called the *Philadelphia chromosome* after the city where the translocation was first described, is a critical part of the CML diagnosis. However, not all tumor-promoting mutations are specific to a given type of cancer. Rather, the same mutations seem to promote tumors in a variety of cell types and thus are seen in many different cancers.

MESSAGE Tumor-promoting mutations can be identified in various ways. When located, they can be cloned and studied to learn how they contribute to the malignant state.

Classes of oncogenes

Roughly a hundred different oncogenes have been identified (some examples are listed in Table 17-2). How do their normal counterparts, proto-oncogenes, function? Proto-oncogenes generally encode a class of proteins that are active only when the proper regulatory signals allow them to be activated. As already mentioned, many proto-oncogene products are elements in pathways that induce (positively control) the cell cycle. These products include growth-factor receptors, signal-transduction proteins, and transcriptional regulators. Other proto-oncogene products act to inhibit (negatively control) the apoptotic pathway. In both types of oncogene mutation, the activity of the mutant protein

Table 17-2 Some Well-Characterized Oncogenes and Functions of the Corresponding Proteins

Oncogene	Location	Function
Nuclear transcription regulators		
jun	Nucleus	Transcription factor
fos	Nucleus	Transcription factor
erbA	Nucleus	Member of steroid-receptor family
Intracellular signal transducers		
abl	Cytoplasm	Protein tyrosine kinase
raf	Cytoplasm	Protein serine kinase
gsp	Cytoplasm	G-protein α subunit
ras	Cytoplasm	GTP/GDP-binding protein
Mitogen		
sis	Extracellular	Secreted growth factor
Mitogen receptors		
erbB	Transmembrane	Receptor tyrosine kinase
fms	Transmembrane	Receptor tyrosine kinase
Apoptosis inhibitor		
bcl2	Cytoplasm	Upstream inhibitor of caspase cascade

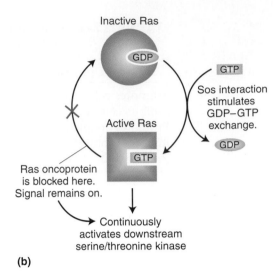

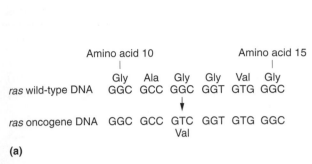

(a)

Figure 17-16 Ras oncoprotein. (a) The *ras* oncogene differs from the wild type by a single base pair, producing a Ras oncoprotein that differs from the wild type in one amino acid, at position 12 in the *ras* open reading frame. (b) The Ras oncoprotein that cannot hydrolyze GTP to GDP. Because of this defect, the Ras oncoprotein remains in the active Ras–GTP complex and continuously activates the downstream serine/threonine kinase (see Figure 17-13).

(b)

has been uncoupled from its normal regulatory pathway, leading to its continuous unregulated expression. The continuously expressed protein product of an oncogene is called an **oncoprotein.** Several categories of oncogenes have been identified according to the different ways in which the regulatory functions have been uncoupled.

> **MESSAGE** The wild-type counterparts of oncogenes act to positively control the cell cycle or negatively control apoptosis.

The following examples illustrate some of the ways in which oncogenes are mutated and act to produce a cancerous cell.

POINT MUTATION OF AN INTRACELLULAR SIGNAL TRANSDUCER The *ras* oncogene illustrates a tumor-promoting mutation of a molecule in a signal-transduction pathway. As is often the case, the change from normal protein to oncoprotein entails structural modifications of the protein itself, in this case caused by a simple point mutation. A single base-pair substitution that converts glycine into valine at amino acid number 12 of the Ras protein, for example, creates the oncoprotein found in human bladder cancer (Figure 17-16a). Recall that the normal Ras protein is a G-protein subunit that takes part in signal transduction. It normally functions by cycling between the active GTP-bound state and the inactive GDP-bound state (see Figure 17-12). The missense mutation in the *ras* oncogene produces an oncoprotein that always binds GTP (Figure 17-16b), even in the absence of normal signals. Thus, the Ras oncoprotein continuously propagates a signal that promotes cell proliferation.

PROTEIN DOMAIN DELETION OF A MITOGEN RE-CEPTOR TYROSINE KINASE A growth factor is not necessary to initiate the cell proliferation pathway if its receptor has been altered appropriately. The v-*erbB* oncogene is a mutated gene in a tumor-producing virus that infects birds. It encodes a mutated form of a receptor tyrosine kinase known as the EGFR, a receptor for the epidermal growth-factor ligand (Figure 17-17). Sev-

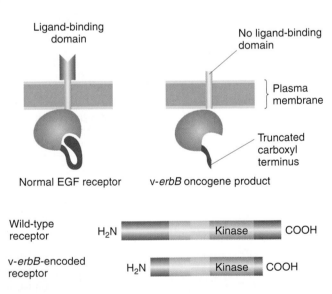

Figure 17-17 Oncogenic mutation that affects signaling between cells. Normally, the kinase activity of the EGFR is activated only when EGF binds to the ligand-binding domain. The erythroblastosis tumor virus carries the v-*erbB* oncogene, which encodes a mutant form of the EGFR. The mutant protein dimerizes constitutively, leading to continuous autophosphorylation and continuous transduction of a signal from the receptor. [After J. D. Watson, M. Gilman, J. Witkowski, and M. Zoller, *Recombinant DNA*, 2d ed. Copyright 1992 by James D. Watson, Michael Gilman, Jan Witkowski, and Mark Zoller.]

eral parts of the normal EGFR are missing in the mutated form. The mutated form of the EGFR lacks the extracellular, ligand-binding domain as well as some regulatory components of the cytoplasmic domain. As a consequence, the truncated EGFR oncoprotein is able to dimerize even in the absence of the EGF ligand. The EGFR oncoprotein dimer is always autophosphorylated through its tyrosine kinase activity and so continuously initiates a signal-transduction cascade that promotes cell proliferation.

A PROTEIN FUSION INVOLVING AN INTRACELLULAR SIGNAL TRANSDUCER Perhaps the most remarkable type of structurally altered oncoprotein is caused by a gene fusion. The classic example emerged from the results of studies of the Philadelphia chromosome, which, as already mentioned, is a translocation between chromosomes 9 and 22 that is a diagnostic feature of chronic myelogenous leukemia. Recombinant DNA methods have shown that the breakpoints of the Philadelphia chromosome translocation in different CML patients are quite similar and cause the fusion of two genes, *bcr1* and *abl* (Figure 17-18). The *abl* (*Abelson*) proto-oncogene encodes a tyrosine-specific protein kinase present in the cytoplasm that participates in a signal-transduction pathway—specifically, a signal-transduction pathway initiated by a growth factor that leads to cell proliferation. The Bcr1–Abl fusion oncoprotein has a permanent

protein kinase activity, which is responsible for its oncogenic effect. The oncoprotein continually propagates its growth signal downstream in the pathway, regardless of whether an upstream signal is present.

GENE FUSION CAUSING MISEXPRESSION OF AN APOPTOSIS INHIBITOR Some oncogenes produce oncoproteins that are identical in structure with the normal proteins. In these cases, the mutation induces misexpression of the protein; that is, the protein is expressed in cell types from which it is ordinarily absent. Here we see how, through gene fusion, an oncogene causes a protein that inhibits apoptosis to be overexpressed in an immune system cell.

Several oncogenes that cause misexpression are associated with chromosomal translocations diagnostic of various B-lymphocyte tumors. B lymphocytes and their descendants, plasma cells, are the cells that synthesize antibodies, also called immunoglobulins. In the translocations observed in B-cell oncogenes, no protein fusion results; rather, the chromosomal rearrangement causes a gene near one breakpoint to be turned on in the wrong tissue. In follicular lymphoma, 85 percent of patients have a translocation between chromosomes 14 and 18 (Figure 17-19). Near the chromosome 14 breakpoint is a transcriptional enhancer from one of the immunoglobulin genes. This translocated enhancer element is fused with the *bcl-2* gene, which is a negative regulator

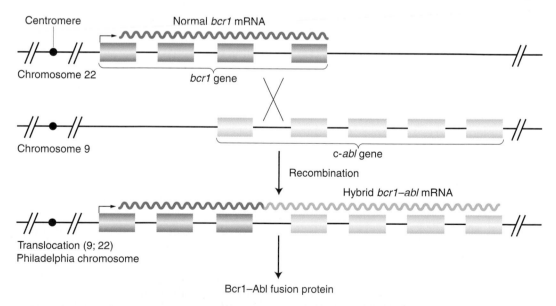

Figure 17-18 Chromosome rearrangement in CML. The Philadelphia chromosome, which is diagnostic of chronic myelogenous leukemia, is a translocation between chromosomes 9 and 22. The translocation produces a hybrid Bcr1–Abl protein that lacks the normal controls for repressing *c-abl*-encoded protein tyrosine kinase activity. Only one of the two rearranged chromosomes of the reciprocal translocation is shown. [After J. D. Watson, M. Gilman, J. Witkowski, and M. Zoller, *Recombinant DNA*, 2d ed. Copyright 1992 by James D. Watson, Michael Gilman, Jan Witkowski, and Mark Zoller.]

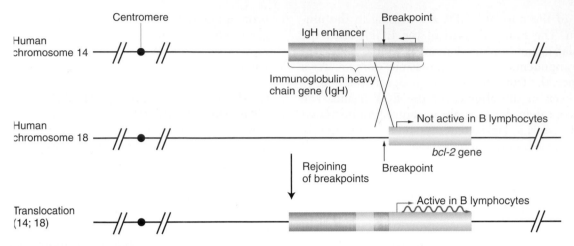

Figure 17-19 Chromosomal rearrangement in follicular lymphoma. The translocation fuses the transcriptional enhancer element of an antibody-coding gene, on chromosome 14, with the transcription unit of a gene on chromosome 18 that encodes Bcl-2, a negative regulator of apoptosis. This gene fusion causes the Bcl-2 protein to be produced in antibody-producing cells, thus preventing any self-destruction signals from inducing apoptosis in those cells.

of apoptosis. This enhancer–*bcl-2* fusion causes large amounts of Bcl-2 protein to be expressed in B lymphocytes. The plentiful Bcl-2 protein effectively blocks apoptosis in these lymphocytes, giving them an unusually long lifetime in which to accumulate mutations that promote cell proliferation.

> **MESSAGE** Dominant oncogenes contribute to the oncogenic state by causing a protein to be expressed in an activated form or in the wrong cells.

Classes of tumor-suppressor genes

The normal functions of tumor-suppressor genes fall into categories complementary to those of proto-oncogenes (see Table 17-1). Some tumor-suppressor genes encode negative regulators of the cell cycle, such as the Rb protein or elements of the TGF-β signaling pathway. Others encode positive regulators of apoptosis (at least part of the function of *p53* falls into this category). Still others are indirect players in cancer, with a normal role in the repair of damaged DNA or in controlling cellular longevity. We shall consider two examples here.

THE KNOCKOUT OF A PROTEIN THAT INHIBITS CELL PROLIFERATION

Retinoblastoma is a cancer of the retina that typically affects young children. In retinoblastoma, the gene that encodes the Rb protein has mutated, and retinal cells that lack a functional *RB* gene proliferate out of control. The cancer is a recessive trait at the cellular level: both alleles of the gene coding

for the Rb protein must be inactivated, either by the same mutation or by a different mutation in each. Most patients have one or a few tumors localized to one site in one eye, and the condition is sporadic—in other words, there is no history of retinoblastoma in the family and the affected person does not transmit it to his or her offspring. In these cases, the *rb* mutations arise in a somatic cell whose descendants populate the retina (Figure 17-20). Presumably, the mutations arise by chance at different times in the course of development in the same cell lineage.

A few patients, however, have an inherited form of the disease, called hereditary binocular retinoblastoma (HBR). Such patients have many tumors, in the retinas of both eyes. Paradoxically, even though *rb* is a recessive allele at the cellular level, standard genetic analysis would describe HBR as transmitted as an autosomal dominant (see Figure 17-20b). How do we resolve this paradox? A germ-line mutation knocks out one of the two copies of the *RB* gene in all cells of the retina in both eyes. It is virtually certain that the single remaining normal *RB* gene will acquire an *rb* mutation in at least some of the retinal cells. These cells will produce no functional Rb protein.

Why does the absence of Rb protein promote tumor growth? Recall from our consideration of the cell cycle that Rb protein functions by binding the E2F transcription factor. Bound Rb prevents E2F from promoting the transcription of genes whose products are needed for DNA replication and other S-phase functions. An inactive Rb is unable to bind E2F, and so E2F can promote the transcription of S-phase genes. In homozygous null *rb* cells, Rb protein is permanently inactive. Thus, E2F is

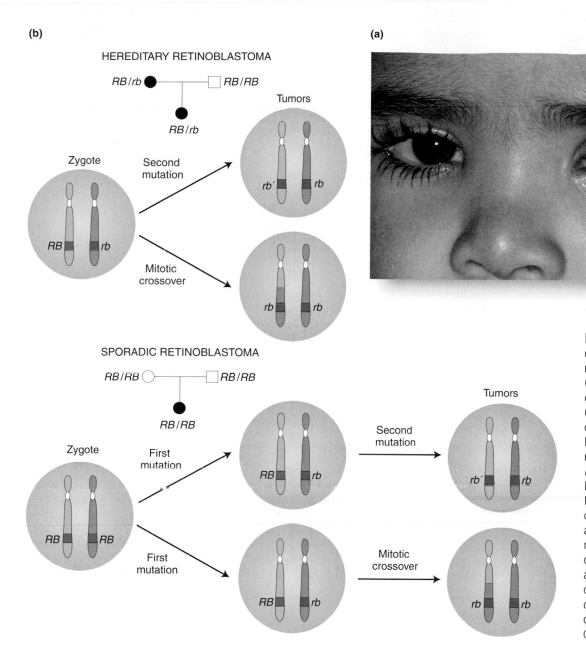

Figure 17-20 Origin of the two types of retinoblastoma.
(a) Retinoblastoma, a cancer of the retina.
(b) The mutational origin of retinal tumors in hereditary and sporadic retinoblastoma. Recessive *rb* alleles of the *RB* gene lead to tumor development. Note that, if one chromosome in a cell already has an *rb* mutation, the homologous chromosome may also acquire it through either chance mutation or crossing-over in the course of mitosis. [Part a from Custom Medical Stock.]

always able to promote S phase, and the arrest of normal cells in late G1 does not occur in retinoblastoma cells.

As already noted, the germ-line *rb* mutation is only one event leading to the loss-of-function rb phenotype. The other allele has to be knocked out as well, which can occur through mutation or through deletion of the other allele or through a mitotic abnormality (mitotic crossing-over or mitotic nondisjunction). Of these, deletion and mitotic crossing-over or nondisjunction would lead to the loss of one of the two alleles of the gene, a condition called **loss-of-heterozygosity,** or **LOH** for short. There are now many ways to identify loss-of-heterozygosity in tumor cells by comparing their DNA content with their wild-type neighbors, by using molecular polymorphisms such as SNPs, SSLPs,

or RFLPs (see Chapter 12). To think about this from the opposite direction, if we were to identify a region of the genome that consistently exhibited LOH in a particular type of tumor, we would be highly suspicious that a tumor-suppressor gene normally resided in that region of the genome.

THE KNOCKOUT OF A PROTEIN THAT INHIBITS CELL PROLIFERATION AND PROMOTES APOPTOSIS

The *p53* gene has been identified as another tumor-suppressor gene. Mutations in *p53* are associated with many types of tumors, and estimates are that 50 percent of human tumors lack a functional *p53* gene. The active p53 protein is a transcriptional regulator that is activated in response to DNA damage. Activated wild-type p53 serves double duty: it prevents progression of the cell cycle until the DNA damage is repaired, and, under some circumstances, it induces apoptosis. If no functional *p53* gene is present, the cell cycle progresses even if damaged DNA has not been repaired. The progression of the cell cycle into mitosis elevates the overall frequency of mutations, chromosomal rearrangements, and aneuploidy and thus increases the chances that other mutations that promote cell proliferation or block apoptosis will arise.

It is now clear that null mutations able to elevate the mutation rate are important contributors to the progression of tumors in humans. These null mutations are recessive mutations in tumor-suppressor genes that normally function in DNA repair pathways and thus interfere with DNA repair. They promote tumor growth indirectly by elevating the mutation rate, which makes it much more likely that a series of oncogene and tumor-suppressor mutations will arise, corrupting the normal regulation of the cell cycle and programmed cell death. Large numbers of such tumor-suppressor-gene mutations have been identified, including some associated with heritable forms of cancer in specific tissues. Examples are the *BRCA1* and *BRCA2* mutations and breast cancer.

> **MESSAGE** Mutations in tumor-suppressor genes, like mutations in oncogenes, act directly or indirectly to promote the cell cycle or block apoptosis.

17.6 Applying genomic approaches to cancer research, diagnosis, and therapies

At this point, we have a thematic understanding of the sorts of mutational lesions that arise in cancers. Most of this information historically has come from the results of studies directed at specific genes, but there is now a great deal of emphasis on applying genomewide analytical methods to a variety of subjects of inquiry in cancer research and medicine. Let's briefly consider some of these approaches.

Genomic approaches to identifying tumor-promoting mutations

Although a couple of hundred genes have already been demonstrated to mutate to a tumor-promoting state under some circumstances, the hunt for such genes is by no means over. As one example of a genome-scale approach to finding cancer genes, let's consider how loss-of-heterozygosity could be exploited. Recall that mutations in tumor-suppressor genes act as recessive alleles in tumor cells. Often, the knockout of one allele occurs

through its removal from the genome, and the location of the absent allele can be determined by looking for polymorphic markers that also show LOH. Such a polymorphism can be assumed to map to the vicinity of the tumor-suppressor gene on the sequence map of the human genome.

A recent study of prostate cancer surveyed LOH in 11 patients, using approximately 1500 SNPs across the genome that could be detected by differential hybridization on DNA oligonucleotide chips (SNP-chips). The idea of this approach is to identify heterozygous SNPs in normal cells from these patients (only a subset of patients will be heterozygous for a given SNP) and then see if both variants of these SNPs are present in the prostate tumor cells. If only one SNP variant is present, then LOH has occurred. Figure 17-21 shows the distribution of 32 sites in the genome that exhibited LOH, superimposed on the human cytogenetic map. Each of these regions could then be subjected to more intensive screening for SNPs or other heterozygous polymorphic markers in the vicinity of the detected LOH sites to determine the size of the lost region. After the sizes of these regions have been determined, they can be compared with the known genes on the sequence map that fall into these regions. The comparison will allow researchers to focus in on candidate tumor-suppressor genes and to determine which of them are actually knocked out in these tumors.

Genomic approaches to cancer diagnostics and therapies

A goal of cancer diagnostics is to properly classify the type of cancer that a patient has contracted, often by inspecting tumor cells under the microscope. However, genomic approaches have revealed that tumors that look the same under a microscope may have very different molecular bases. Let's examine how genomics can be used to address this diagnostic problem. As an example, we will consider a form of lymphoma.

Lymphomas and leukemias are cancers of the white blood cells—the cells that make up the immune system. In these diseases, certain white blood cells massively overproliferate, leading to an imbalance in immune cells and, ultimately, a failure of the immune system. One class of lymphoma is called diffuse large B-cell lymphoma (DLBCL), the most common form of non-Hodgkin's lymphoma. In the United States alone, about 25,000 new cases of DLBCL are diagnosed every year. The diagnosis is based on finding a characteristic set of symptoms and on the histology (microscopic examination of cell and tissue morphology) of affected lymph nodes (Figure 17-22). Standard chemotherapy cures about 40 percent of DLBCL patients, but other patients do not respond to this regimen. Why is there this dichotomy among the patient population? It could

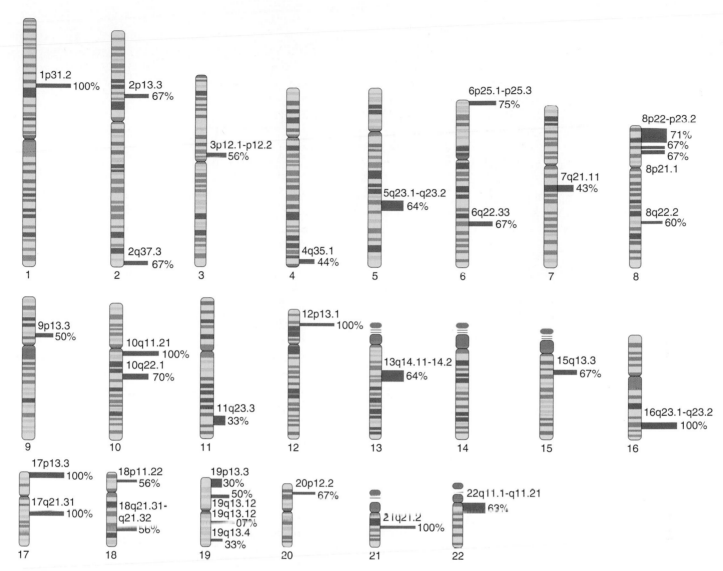

Figure 17-21 Genomic SNP analysis of LOH in 11 prostate cancer patients. The blocks indicate the location of the LOH regions from the sequence map overlaid on the cytogenetic map. The horizonal bars represent the proportion (in percentage) of prostate cancer patients showing LOH at the indicated loci. [From C. I. Dumur et al., "Genomic-wide Detection of LOH in Prostate Cancer Using Human SNP Microarray Technology," *Genomics* 81, 2003, 265.]

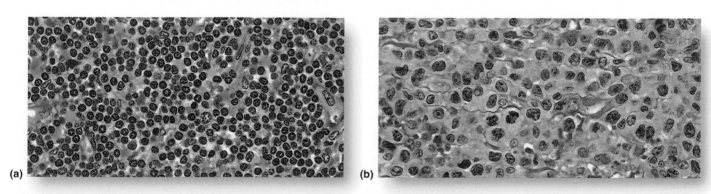

Figure 17-22 Histology of (a) a normal lymph node and (b) diffuse large B-cell lymphoma (DLBCL). The darkly stained cells are lymphoid cells. Note that the lymphocytes in the normal lymph node are small and uniformly shaped, but the lymphoma cells are large and more varied in shape. [G. W. Willis, M.D./Visuals Unlimited.]

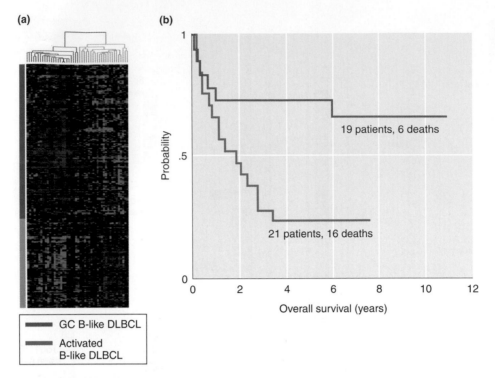

Figure 17-23 Functional genomic analysis of DLBCL. (a) Microarray analysis of malignant cells indicates that patients possess one of two basic RNA-expression profiles, one termed GC B-like and the other activated B-like. (b) Higher patient survival correlates with the GC B-like RNA profile. [Reprinted by permission from L. Liotta and E. Petricoin, *Nature Reviews: Genetics* 1, 2000, 53. Macmillan Magazines Ltd.]

result from differences in genetic or environmental factors, differences in disease stage, or any of many other factors.

A breakthrough was achieved from profiling gene transcription in malignant cells from 40 patients, with the use of microarray technology (Figure 17-23). In this study, two different patterns of gene expression were observed. One pattern, called the germinal-center B-cell-like pattern (GCB), correlates with a much higher probability of patient survival after standard chemotherapy than does the other pattern, called activated B-cell-like DLBCL. (These patterns are named for their similarities in gene expression to certain normal B-lymphocyte cell types.) This finding suggests that even though DLBCL is diagnosed histologically as one disease, at a molecular level it is more like two different diseases. This result raises the hope of more accurate diagnosis and of treatment that can be focused specifically on each disease. Further, such microarray experiments are beginning to reveal unexpected RNAs that differ in level either between normal and tumor cells or between different stages of malignancy; the protein products of these RNAs are candidate targets for drug therapy.

A challenge: These examples just barely touch the surface of how genomic technologies are being applied to cancer research and medicine. On the basis of techniques discussed earlier in the text, think of other ways that you would apply genomic techniques for understanding the molecular basis of cancer, for diagnostics and early detection of cancer, and for developing new cancer therapies.

The complexities of cancer

We have seen that numerous mutations that promote tumor growth can arise. These mutations alter the normal processes that govern proliferation and apoptosis (Figure 17-24). The story does not end here, however. There is evidence that other modes of gene inactivation, such as epigenetic imprinting, can produce tumor-promoting lesions. There is also evidence that the overexpression of telomerase is another condition required for cell immortality, a feature of cancer cells. (Normal human somatic cells can undergo only a relatively small number of divisions before their telomeres are reduced in size to the point at which the cells can no longer grow. In human tumor cells, however, the length of the telomeres appears to be substantially extended, proba-

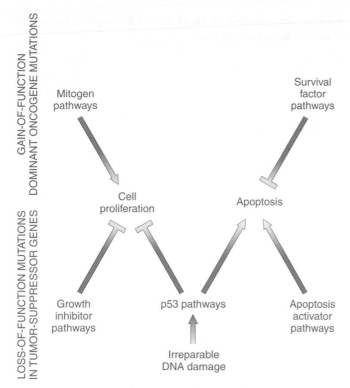

Figure 17-24 Main events contributing to tumor formation: increased cell proliferation and cell survival (decreased apoptosis). The pathways in red are susceptible to gain-of-function oncogene mutation. The pathways in blue are susceptible to loss-of-function tumor-suppressor-gene mutation.

bly through the overexpression of telomerase.) Even malignant tumors differ in their rates of proliferation and their ability to invade other tissues, or metastasize. Undoubtedly, even after a malignant state is achieved, many more mutations accumulate in the tumor cell that further promote its proliferation and invasiveness. Thus, there is a considerable way to go before we have a truly comprehensive view of how tumors arise and progress.

Even so, there is light at the end of the tunnel. In the early 1990s, researchers were able to correlate overexpression of the Abelson tyrosine kinase, through the Philadelphia chromosome gene fusion, with chronic myelogenous leukemia. Chemists then developed a compound, called ST1571, that bound to the ATP-binding site of the Abelson tyrosine kinase and thereby inhibited its ability to phosphorylate tyrosines on target proteins. ST1571 was shown to inhibit cell proliferation induced by *bcr–abl* in cell lines and was then tested as the drug Gleevec in clinical trials. The treatment had dramatic effects; more than 90 percent of treated patients have recovered normal blood counts. Even more striking, more than half show no evidence of cells containing the Philadelphia chromosome in their bloodstreams. No other treatment for CML has been this effective. This is the first clear case in which an understanding of the molecular biology of a malignancy has been translated into a highly effective, targeted treatment.

KEY QUESTIONS REVISITED

- **Why is it important for multicellular organisms to have mechanisms for regulating cell numbers?**

Multicellular organisms depend heavily on a division of labor among the various cell types in a tissue. To maintain the proper proportion and number of cells of the various types within a tissue, mechanisms are needed to replace accidentally lost cells and to eliminate abnormal cells that present a danger to the organism as a whole.

- **Why is cell-cycle progression regulated?**

First, regulation at checkpoints ensures that the cell cycle progresses to the next stage only when the proper preconditions have been fulfilled. Second, because the cell cycle is wired into the intercellular signaling system, proliferation can be halted until the proper signals for it are received.

- **Why is programmed cell death necessary?**

Apoptosis is a normal self-destruction mechanism that eliminates damaged and potentially harmful cells as well as cells that have only temporary functions in development. Cell survival requires direct survival signals that inhibit activation of the self-destruction program. The cost of replacing discarded cells is far less than the cost of compromising the health of the entire organism.

- **How do neighboring cells influence cell proliferation and cell death?**

Proteins secreted or presented by nearby cells act as signals. These proteins bind to transmembrane receptor proteins on the cells receiving the signal, activating a signal-transduction pathway that ultimately changes the expression of key genes in the cell-cycle or apoptosis pathways.

- **Why is cancer considered a genetic disease of somatic cells?**

Cancers are caused by the accumulation of a set of mutations in a single somatic cell. These mutations undermine the basic control mechanisms that keep a cell from proliferating out of control.

• **How do mutations promote tumors?**

Tumor-producing mutations either release normal inhibitory controls on the cell cycle or block cell death. In addition, mutations that increase the mutation rate of the cell are tumor promoting, presumably because mutations increase in the cell-proliferation or apoptosis pathways.

• **How can genomic methodologies be applied to cancer research and medicine?**

Microarray technology can be used to search for mutations in tumor-suppressor genes and to more precisely diagnose the type of cancer in patients presenting the same histological tumor phenotype.

SUMMARY

Somatic cells of higher eukaryotes are integrated into a variety of tissues and may be of many different types, each with a specialized role in the physiology of an organism. Regulation of cell numbers is essential for maintaining proper physiological balance among the various tissues and cell types. Higher eukaryotic organisms have evolved mechanisms that control cell survival and the ability to proliferate.

Normal cell proliferation is controlled by regulation of the cell cycle. This regulation achieves multiple ends. First, regulation at checkpoints of the cell cycle ensures that the cell cycle can progress to the next stage only when a set of preconditions that ensure proper chromosome replication and segregation are fulfilled. Second, because the cell cycle is wired into the intercellular signaling system, proliferation can be halted until the proper signals for it are received.

Apoptosis is a normal self-destruction mechanism that eliminates damaged and potentially harmful cells as well as cells that have only temporary functions in development. We now know that cell survival requires direct survival signals that inhibit activation of the execution program and the caspase cascade, which lead to self-destruction.

Cell proliferation and apoptosis are coordinated in a population of cells through intercellular signaling systems. Signaling consists of the secretion of a signal from sending cells, receipt of that signal by surface receptors on target cells, and transduction of that signal from the transmembrane receptor to the interior (usually the nucleus) of the target cells.

Cancer is a genetic disease of somatic cells. In cancer, cells proliferate out of control and avoid self-destruct mechanisms through the accumulation of a series of tumor-promoting mutations in the same somatic cell. Many of the genes that promote cancers when they are mutated contribute directly or indirectly to growth control and differentiation. Genes that normally accelerate survival or proliferation may be mutated to become oncogenes. Oncogene mutations behave in the cancer cell as gain-of-function dominant mutations, in which their tumor-promoting effect is due to altered activity of the mutant gene product. Genes that normally accelerate apoptosis or inhibit mitosis are tumor-suppressor genes. Mutations in tumor-suppressor genes behave in the cancer cell as recessive loss-of-function mutations. In other words, the abolition of activity of these gene products promotes tumor growth. In addition, mutants in repair pathways, as well as those that increase the general mutation rate, increase the likelihood of tumor-promoting mutations and hence indirectly promote tumor development.

Functional genomics is being used in diagnostic tests and to identify drug targets in studies to improve the detection and treatment of cancer. Instead of applying the former method of screening for specific genetic contributions to tumor development, we are now able to survey virtually the entire genome for changes in gene activity associated with tumor development. Even if we don't understand the nature of all the changes, functional genomics reveals gene-expression "fingerprints" that allow different disease states to be identified.

KEY TERMS

apoptosis (p. 552)

cancer (p. 558)

caspase (p. 552)

CDK-cyclin complexes (p. 548)

cell–cell communication (p. 553)

checkpoint (p. 550)

cyclin (p. 549)

cyclin-dependent protein kinase (CDK) (p. 549)

differentiated cell (p. 547)

endocrine signal (p. 553)

growth factor (p. 555)

kinase (p. 549)

ligand (p. 553)

loss-of-heterozygosity (LOH) (p. 565)

mitogen (p. 555)

oncogene (p. 559)

oncoprotein (p. 562)

paracrine signal (p. 555)

programmed cell death (p. 552)

proto-oncogene (p. 559)

signal-transduction cascade (p. 555)

stem cell (p. 547)

survival factor (p. 556)

tumor-suppressor gene (p. 560)

zymogen (p. 552)

SOLVED PROBLEM

In the inherited form of retinoblastoma, an affected child is heterozygous for an *rb* mutation, which is either passed on from a parent or has newly arisen in the sperm or the oocyte nucleus that gave rise to the child. The heterozygous *RB/rb* cells are nonmalignant, however. The *RB* allele of the heterozygote must be knocked out in the developing retinal tissue to create a tumorous cell. Such a knockout can occur through an independent mutation of the *RB* allele or by mitotic crossing-over such that the original *rb* mutation is now homozygous.

a. If retinoblastoma is passed on to other siblings as well, could we determine whether the original mutation was derived from the mother or the father? How?

b. Could we determine whether the *rb* mutation was maternally or paternally derived if it arose de novo in a germ cell of one parent?

Solution

a. If the trait is inherited, we can determine from which parent it came. The most straightforward approach is to identify DNA polymorphisms, such as restriction fragment length polymorphisms (RFLPs), that map within or near the *RB* gene. *RB* has the curious property of being inherited as an autosomal dominant, even though it is recessive on a cellular basis. Given the dominant pattern of inheritance, by finding DNA differences in the parental genomes that map near each of the four parental alleles, we should be able to determine which

allele has been passed on to all affected offspring. That allele is the mutant one.

b. If the trait arose de novo in the sperm or the oocyte, we can still possibly determine from which parent it came, but with considerable difficulty. One way to do so would be to use recombinant DNA cloning techniques to isolate each of the two copies of the *RB* gene from normal cells. Only one of these alleles should be mutant. When the gene has been cloned, by DNA sequencing of the two alleles we may be able to identify the mutation that inactivates *RB*. If it arose de novo in the sperm or the oocyte, this mutation would not be present in the somatic cells of the parents. If sequencing also reveals some polymorphisms (for example, in restriction-enzyme recognition sites) that distinguish the alleles, we should then be able to go back to the parents' DNA to find out whether the mother or the father carries the polymorphisms that were found in the cloned mutant allele. (Whether this approach will work depends on the exact nature of the parental alleles and the mutation.)

If the mutation arose by mitotic crossing-over, additional tools are available. In this case, the entire region around the *rb* gene will be homozygous for the mutant chromosome. By examining DNA polymorphisms known to map in this region, we should be able to determine whether this chromosome derived from the mother or from the father. This approach becomes a standard exercise in DNA fingerprinting similar to that described in part a.

PROBLEMS

BASIC PROBLEMS

1. Describe three lines of evidence that cancer is a genetic disease.

2. Describe three mechanisms used to control the activities of the proteins in the cell-cycle and cell-death pathways.

3. What are the two roles of cyclins? How are the levels of cyclins regulated?

4. What are three major categories of mutations that lead to the development of cancer?

5. What are the roles of Apaf and Bcl proteins in apoptosis?

6. What are two mechanisms by which translocations can lead to oncogene formation?

7. How is the activity of caspases controlled?

8. Give an example of an oncogene. Why is the mutation dominant?

9. Give an example of a tumor-suppressor gene. Why is the mutation recessive?

10. How do the following mutations lead to cancer?

    a. v-*erbB*

    b. *ras* oncogene

    c. Philadelphia chromosome

11. For each of the following proteins, describe how a mutation could lead to the formation of an oncogene.

    a. growth-factor receptor

    b. transcriptional regulator

    c. G-protein

CHALLENGING PROBLEMS

12. Cancer is thought to be caused by the accumulation of two or more "hits"—that is, two or more mutations that affect cell proliferation and survival in the same cell. Many of these oncogenic mutations are dominant: one mutant copy of the gene is sufficient to change the proliferative properties of a cell. Which of the following general types of mutations have the potential to create dominant oncogenes? Justify each answer.

    a. a mutation that increases the number of copies of a transcriptional activator of cyclin A

    b. a nonsense mutation located shortly after the beginning of translation in a gene that encodes a growth-factor receptor

    c. a mutation that increases the level of FasL

    d. a mutation that disrupts the active site of a cytoplasmic tyrosine-specific protein kinase

    e. a translocation that joins a gene encoding an inhibitor of apoptosis to an enhancer element for gene expression in the liver

13. Many of the proteins that participate in the progression pathway of the cell cycle are reversibly modified, whereas, in the apoptosis pathway, the modification events are irreversible. Rationalize these observations in relation to the nature and end result of the two pathways.

14. Tumor-promoting mutations are described as being either gain-of-function or loss-of-function. Describe the effects of both classes of tumor-promoting mutations on the cell cycle and apoptosis.

15. Normally, FasL is present on cells only when a message needs to be sent to neighboring cells instructing them to undergo apoptosis. Suppose that you have a mutation that produces FasL on the surfaces of all liver cells.

    a. If the mutation were present in the germ line, would you predict it to be dominant or recessive?

    b. If such a mutant arose in somatic tissues, would you expect it to be tumor promoting? Why or why not?

16. Some genes can be mutated to oncogenes by increasing the copy number of the gene which, for example, is true of the Myc transcription factor. On the other hand, oncogenic mutations of *ras* are always point mutations that alter the protein structure of Ras. Rationalize these observations in relation to the roles of normal and oncogenic versions of Ras and Myc.

17. We now understand that mutations that inhibit apoptosis are found in tumors. Because proliferation itself is not induced by the inhibition of apoptosis, explain how this inhibition might contribute to tumor formation.

18. Suppose that you had the ability to introduce normal copies of a gene into a tumor cell in which mutations in the gene caused it to promote tumor growth.

    a. If the mutations were in a tumor-suppressor gene, would you expect these normal transgenes to block the tumor-promoting activity of the mutations? Why or why not?

    b. If the mutations were of the oncogene type, would you expect the normal transgenes to block their tumor-promoting activity? Why or why not?

19. Insulin is a protein that is secreted by the pancreas (an endocrine organ) when the level of blood sugar is high. Insulin acts on many distant tissues by binding and activating a receptor tyrosine kinase, leading to a reduction in blood sugar by appropriate storage of the products of sugar metabolism. Diabetes is a disease in which the level of blood sugar remains high because some part of the insulin pathway is defective. One kind of diabetes (call it type A) can be treated by giving the patient insulin. Another kind of diabetes (call it type B) does not respond to insulin treatment.

    a. Which type of diabetes is likely to be due to a defect in the pancreas, and which type is likely to be due to a defect in the target cells? Justify your answer.

    b. Type B diabetes can be due to mutations in any of several different genes. Explain this observation.

20. Irreparable DNA damage can have consequences for both the cell cycle and apoptosis. Explain what the consequences are, as well as the pathways by which the cell implements them.

21. Retinoblastomas arise through mutations in the *RB* gene. In HBR (hereditary binocular retinoblastoma), the gene is inherited from parent to offspring as a simple autosomal dominant. Nonetheless, *RB* is thought to be a recessive tumor-suppressor gene. In HBR, there are typically multiple tumors in both eyes. In contrast, in the form of retinoblastoma that is not inherited (called sporadic retinoblastoma), there are fewer tumors and they are usually restricted to one eye. Account for all of these observations in relation to your understanding of how tumor-suppressor genes mutate to cause neoplasia.

22. The wild-type Rb protein functions to sequester E2F in the cytoplasm. At the appropriate time in the cell cycle, E2F is released so that it can act as a functional transcription factor.

   a. In cells homozygous for a loss-of-function *rb* mutation, where in the cell do you expect to find E2F?

   b. If a cell were doubly homozygous for loss-of-function mutations in both the *RB* and *E2F* genes, would you expect tumor growth? Explain.

   c. How would a mutation in the RB protein that irreversibly binds E2F affect the cell cycle?

EXPLORING GENOMES   A Web-Based Bioinformatics Tutorial

Exploring the Cancer Anatomy Project

The Cancer Anatomy project at NCBI is a specialized database focused on a particular subset of data. It brings together, in a graphical and searchable format, all known information about individual genes, map positions, and chromosomal location for genes associated with cancer. In the Genomics tutorial at www.whfreeman.com/mga, we explore gene-expression data, gene mutations, and chromosomal aberrations associated with the various tumor types.

18

THE GENETIC BASIS OF DEVELOPMENT

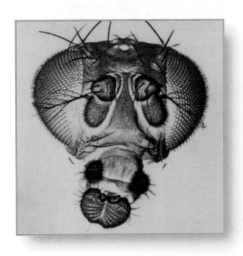

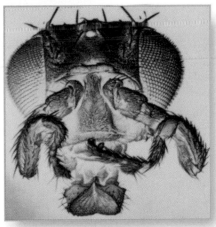

A homeotic mutation that alters the basic *Drosophila* body plan. Homeosis is the replacement of one body part by another. In place of the normal antennae (shown at the top), an *Antennapedia* mutation causes the antennal precursor cells to develop into a leg. [F. R. Turner/BPS.]

KEY QUESTIONS

- What sequence of events produces the basic body plan of an animal?

- How are polarities that give rise to the main body axes created?

- How do cells recognize their locations along the developing body axes?

- Do cells participate in single or multiple decision-making processes?

- What role does cell-to-cell communication play in building the basic body plan?

- Are the pathways for building biological pattern conserved among distant species?

OUTLINE

18.1 The logic of building the body plan

18.2 Binary fate decisions: the germ line versus the soma

18.3 Forming complex pattern: the logic of the decision-making process

18.4 Forming complex pattern: establishing positional information

18.5 Forming complex pattern: utilizing positional information to establish cell fates

18.6 Refining the pattern

18.7 The many parallels in vertebrate and insect pattern formation

18.8 The genetics of sex determination in humans

18.9 Do the lessons of animal development apply to plants?

18.10 Genomic approaches to understanding pattern formation

CHAPTER OVERVIEW

The general body plan of an animal is a specific pattern of structures common to all members of its species. Such body plans may be common to many very different species: all mammalian species have four limbs, whereas all insects have six. However, all mammals and insects must, in the course of their development, differentiate the anterior from the posterior end and the dorsal from the ventral side. Eyes and legs always appear in the appropriate places. The basic body plan of a species appears to be quite robust—that is, the internal genetic program produces the same body plan within a broad range of environmental conditions. We should not forget, however, that the study of the genetic determination of these basic developmental processes does not provide an explanation of the phenotypic differences between *individual* members of a species. This chapter focuses on the processes that underlie pattern formation, the construction of complex biological form. These processes are dictated by a genetic developmental program for the basic body plan.

This chapter will show that the production of the body plan is a step-by-step process in which the outline of the major body subdivisions is first painted with broad strokes and then refined until fine-grained pattern is finally laid down. The general types of steps in the development of a *Drosophila* embryo are summarized in Figure 18-1. Substances are laid down at the poles of the egg by the maternal parent, and these substances give rise to gradients in both the anterior–posterior and the dorsal–ventral axes. Interaction of these gradients leads to broad fields of gene expression. These patterns are remembered through cell division, and so all descendant cells are programmed to that developmental fate. Subsequent gene signaling, both within and between cells, sets up finer regions of gene expression.

This approach to building pattern is sensible, because it is well set up to compensate for errors. Any adjustments needed to correct errors are carried out as the system progresses. First, a few different populations of cells are made, and each population is subdivided in step-by-step fashion into a larger and larger set of distinct cell types, eventually forming functional tissues and organs. Often, cell proliferation is going on concomitantly. At each step in the process, there is an opportunity to adjust cell types and cell numbers if the preceding step didn't work quite right. Only after midcourse evaluations and corrections have been executed can the system proceed to the next step.

Understanding the development of a pattern requires understanding all the alternative routes through a network of developmental pathways, but with particular focus on the key points—the switches that allow direction to specific routes through the network. The devel-

CHAPTER OVERVIEW Figure

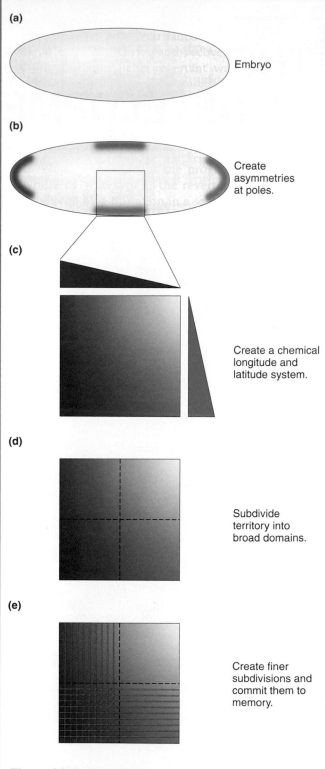

(a) Embryo

(b) Create asymmetries at poles.

(c) Create a chemical longitude and latitude system.

(d) Subdivide territory into broad domains.

(e) Create finer subdivisions and commit them to memory.

Figure 18-1 An overview of developmental strategies. The mother's tissue deposits material that creates asymmetries, which generates gradients that define several broad areas. Subsequent gene interaction subdivides these areas. These commitments are "remembered."

opment of a pattern includes points at which choices must be made: some are like toggle switches and others are analogous to the preset buttons on a radio. Toggle switches have only two states: ON or OFF. Preset buttons allow the selection of a single option among many alternative states. All of the molecular elements that constitute choice points have been discussed in earlier chapters. The challenge of pattern formation is to understand how these molecular elements are interconnected in a coherent and reproducible way to build the same basic body plan in every member of a species.

In genetic circuits, the ultimate effect of these choice points is to modulate the activity of a constellation of genes that together give a cell its defining developmental characteristics—the kinds of cell types to which it can give rise. A body plan is produced by wiring the gene transcriptional apparatus in a cell in such a way that it communicates its decisions to adjacent cells. In this way, each cell can share with its neighbors information concerning the status of its developmental decisions. As a result of this information sharing, the entire cell population can coordinate a community effort, to ensure that all of the necessary cell types are represented in the proper spatial deployment for building the tissues, organs, and appendages of the mature body plan. Thus, this chapter will not present new types of molecular functions. Rather, we will encounter the same cast of characters that take part in gene regulation and intercellular signaling but in more highly integrated and coordinated contexts. This should not be surprising, because one of the basic themes of biology is that natural selection mixes and matches a limited set of existing tools to solve new problems, such as the construction of complex body plans.

18.1 The logic of building the body plan

In all higher organisms, life begins as a single cell, the newly fertilized egg. It reaches maturity as a population of thousands, millions, or even trillions of cells combine into a complex organism with many integrated organ systems. The goal of developmental biology is to unravel the fascinating and mysterious processes that achieve the transfiguration of egg into adult. Because we know more about development in animal systems than in plants or multicellular fungi, we shall focus our attention initially on animals.

Each cell has a protein profile consisting of the types and amounts of proteins that it contains. In a multicellular organism, a cell's protein profile is the end result of a series of decisions that determine the "when, where, and how much" of gene expression. Thus, for a particular gene, a geneticist wants to know in which tissues and at what times in the course of development the gene is transcribed and how much of the gene product is synthesized. From that point of view, all developmental programming is determined by the regulatory information encoded in DNA. We can look at the genome as a parts list of all the gene products (RNAs and polypeptides) that can possibly be produced and as an instruction manual of when, where, and how much of these products are to be expressed. Thus, one aspect of developmental genetics is to understand how this instruction manual operates to send cells down different developmental pathways, ultimately producing a large constellation of characteristic cell types (refer to Figure 18-1).

This is not all that there is to the production of cellular diversity during development, however. How are these different cell types deployed coherently and constructively? In other words,

- how do they become organized into organs and tissues?
- how are those organ systems and tissues organized into an integrated, coherently functioning individual organism?

We shall explore these questions by examining the formation of the basic animal body plan.

In the elaboration of the body plan, at some stage a cell must commit to a specific **cell fate**—that is, differentiate to become one particular kind of cell. The commitment has to make sense in light of the cell's location; it would be neither useful for a cell at the body's tail end to become a brain cell, for example, nor helpful to have lens cells develop in the location of the retina. Because all organs and tissues are made up of many cells, the structure of an organ or tissue requires a cooperative division of labor among the participating cells. Thus, somehow, cell position must be identified, and fate assignments must be parceled out within a cooperating group of cells; such a group of interacting cells is called a **developmental field.**

A cell's position is generally established through protein signals that emanate from a specific position within the initial one-cell zygote or from one or more cells clustered within a developmental field. Just as we need longitudes and latitudes to navigate on earth, a cell needs information specifying its location within a developmental field. This information is called **positional information.** By using this information, a cell can execute the developmental program appropriate to that location. When that positional information has been received, generally a few different precursor cell types are created within a field. Through further processes of cell division and decision making, a population of cells with the necessary final diversity and spatial distribution of fates will be established.

These further processes—**fate refinement**—can be of two types. Sometimes, through asymmetrical divisions of one of the precursor types of cells, descendants are created that have received different regulatory instructions and therefore become committed to different fates. This mechanism for partitioning fates can be thought of as *cell lineage dependent.* Alternatively, such fate decisions are made by "committee"; that is, the fate of a cell depends on input from neighboring cells and feed-back to them. This mechanism for partitioning fates can be thought of as *neighborhood dependent.* The cell-neighborhood-dependent mechanisms provide for a certain flexibility so that an organism can compensate for unscripted events such as the accidental death of a cell. If some cells are lost by accident, the cell-to-cell communication system is used to reprogram the surviving neighbors to divide and divert a subset of their descendants to adopt the fates of the deceased cells. Indeed, the complete regeneration of severed limbs, as occurs in some animals such as starfish or amphibians, depends on neighborhood-dependent mechanisms.

> **MESSAGE** Cells within a developmental field must be able to identify their spatial locations and make developmental decisions in the context of the decisions being made by their neighbors.

Many developmental decisions have two distinguishable phases: (1) establishing the decision and (2) maintaining it. Establishing the decision, as previously stated, is analogous to flipping a toggle switch or pressing a button on a radio. Maintenance entails setting up a memory system that permanently locks a cell *and its descendants* into the "on" or "off" position with regard to a specific decision. Making a developmental decision and subsequently remembering that decision through subsequent cell divisions are keys to the commitment of cells to their fates.

The process of commitment to a particular fate is a gradual one. A cell does not go in one step from being totally uncommitted, or **totipotent,** to becoming earmarked for a single fate. Each major patterning decision is a series of events in which cells are committed to their fates step by step. Thus, if we examine a cell lineage—that is, the family tree of a somatic cell and its descendants—we see that parental cells in the tree are less committed than their descendants.

> **MESSAGE** As cells proliferate in the developing organism, decisions are made that specify more and more precisely the options for the fate of cells of a given lineage.

18.2 Binary fate decisions: the germ line versus the soma

The toggle-switch type of decision making results in one of two outcomes; hence it is called a **binary fate decision.** How do toggle switches work in cells? We are interested in the processes that subdivide the embryo into different cellular populations. One key binary decision is between becoming either part of the germ line (the gamete-forming cells) or the soma (everything else). Once this separation occurs, it is irreversible. Germ cells do not contribute to somatic structures. Somatic cells cannot form gametes, and thus their descendants never contribute genetic material to the next generation. Let's examine the toggle switch that creates this decision.

In the cases that are best understood, the germ line versus soma decision takes advantage of underlying cellular asymmetries. From the time that an oocyte is fertilized, special regulatory molecules are tethered to one end of the cell. Hence, after division, only a subset of the progeny cells acquire these regulatory molecules in their cytoplasms. That subset of cells becomes the germ line. Cells not acquiring these regulatory molecules develop as soma. In other words, for this toggle switch, we can think of soma as the default, or OFF, state, and the germ line as the shunted, or ON, state of the toggle switch. Thus, we need to consider three questions:

- What is the nature of the underlying asymmetrical cellular structures?
- How do regulatory molecules accumulate at one end of these structures?
- What are these regulatory molecules and how do they work?

Let's begin by considering the underlying cellular asymmetries that are exploited by the germ-line-determination system—the asymmetrical components of the **cytoskeleton,** the "roadways" and "girders" that traverse the cell's cytoplasm and give the cell its shape and structure.

The cytoskeleton of the cell

The cytoskeleton consists of several networks of very highly organized structural rods that run through each cell. Each network is made up of one of three types of rod: intermediate filaments, microfilaments, and microtubules (Figure 18-2). Each type of rod has its own architecture formed of unique protein subunits that act to promote polymerization or disassembly of the rods. We will focus on the last two types. **Microfilaments** are linear polymers of the structural protein actin, whereas **microtubules** are linear polymers of the two closely related

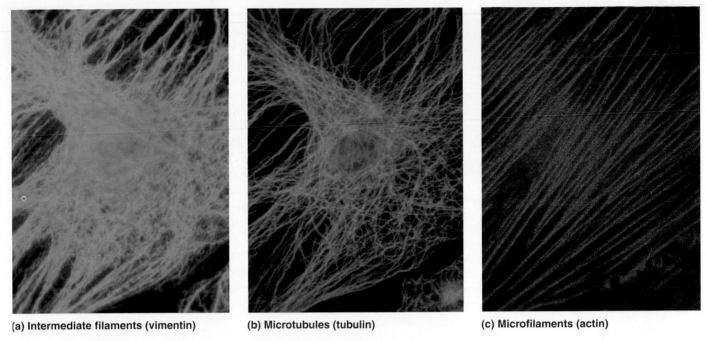

(a) Intermediate filaments (vimentin)   **(b) Microtubules (tubulin)**   **(c) Microfilaments (actin)**

Figure 18-2 Different cytoskeletal systems in the same cell. The distributions of
(a) the intermediate filament protein vimentin, (b) the microtubulin protein tubulin, and
(c) the microfilament protein actin are shown. [Courtesy of V. Small. Reprinted from H. Lodish,
D. Baltimore, A. Berk, S. L. Zipursky, P. Matsudaira, and J. Darnell, *Molecular Cell Biology*,
3d ed. Copyright Scientific American Books, 1995.]

structural proteins, α-tubulin and β-tubulin. Each type of rod participates in higher-order networks by forming reversible cross-links with neighboring rods of its type. The different cytoskeletal systems often play different biological roles. For example, actin microfilaments often play a primary role in cellular motility, such as muscle contraction or the "crawling" movement of amebae, whereas microtubules are the cytoskeletal elements within the mitotic and meiotic spindles of cell division.

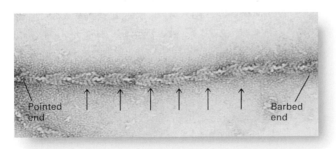

Figure 18-3 Polarity of subunits in an actin microfilament. An actin microfilament does not ordinarily have this appearance, but the microfilament has been coated with a protein that binds in a fashion that reveals the underlying polarity of the actin microfilament itself. [Courtesy of R. Craig. Reprinted from H. Lodish, D. Baltimore, A. Berk, S. L. Zipursky, P. Matsudaira, and J. Darnell, *Molecular Cell Biology*, 3d ed. Copyright Scientific American Books, 1995.]

The cytoskeleton performs several tasks that are important to the formation of asymmetries: control of the location of the mitotic cleavage plane within the cell, control of cell shape, and directed transport of molecules and organelles within the cell. All of these tasks depend on the fact that the cytoskeletal rods are polar structures exhibiting directionality (Figure 18-3). That is, one end of a cytoskeletal rod can be chemically distinguished from the other.

Exploiting the polarity of microfilaments and microtubules

Microfilaments and microtubules are intracellular highways, along which other molecules move up and down throughout the cell. They are multilane two-way highways in the sense that different vehicles can travel on them in each of the two possible directions.

To understand how directional travel along cytoskeletal rods can occur, let's consider microtubules. Near the center of most cell types are found all the "−" (minus) ends of the microtubules (Figure 18-4). This location is called the microtubule organizing center (MTOC). The "+" (plus) ends of microtubules are located at the periphery of the cell. Very much as an automobile uses the combustion of gasoline to create energy that is then transduced into motion, special "motor"

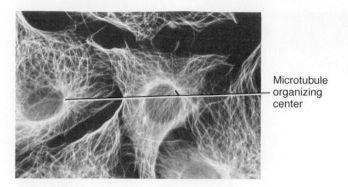

Figure 18-4 Fluorescence micrograph showing the distribution of tubulin in an interphase animal fibroblast. Notice that the microtubules radiate out from a microtubule organizing center. The negative (minus) ends of the microtubules are in the center, and the positive (plus) ends are at the periphery of the cell. [Courtesy of M. Osborn. Reprinted from H. Lodish, D. Baltimore, A. Berk, S. L. Zipursky, P. Matsudaira, and J. Darnell, *Molecular Cell Biology,* 3d ed. Copyright Scientific American Books, 1995.]

proteins hydrolyze ATP for energy that is utilized to propel movement along a microtubule. For example, a protein called kinesin is able to move in a − to + direction along microtubules, carrying cargos such as vesicles from the center of the cell to its periphery (Figure 18-5a and b). The motor—the part of the kinesin protein that directly interacts with the microtubule rod—is contained in the globular heads of the protein (Figure 18-5c). The tail of kinesin is thought to be where the cargo is attached. These cargos might be individual molecules, organelles, or other subcellular particles to be towed from one part of the cell to another. (Other motors exist that travel in the + to − direction on microtubules, and still other motors can travel in each of the two directions on actin microfilaments.) What is the value of having multiple independent trafficking systems? A part of the answer is division of labor. Different cell components must be moved to different places and at different rates.

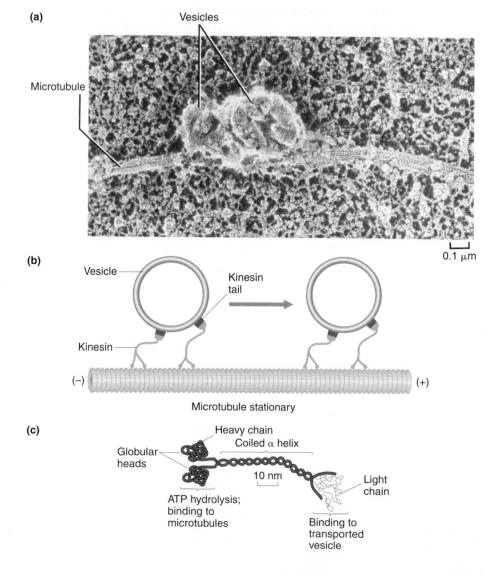

Figure 18-5 Movement of vesicles along microtubules. (a) A scanning electron micrograph of two small vesicles attached to a microtubule. (b) A diagram of how kinesin is thought to attach to cellular cargos such as vesicles and to move the cargos along the microtubule in the minus-to-plus direction by using the motor domain in the kinesin heads. (c) A diagram of the kinesin protein showing the functions associated with various parts of the molecule. [Part a from B. J. Schnapp et al., *Cell* 40, 1985, 455. Courtesy of B. J. Schnapp, R. D. Valle, M. P. Sheetz, and T. S. Reese. All parts: Reprinted from H. Lodish, D. Baltimore, A. Berk, S. L. Zipursky, P. Matsudaira, and J. Darnell, *Molecular Cell Biology,* 3d ed. Copyright Scientific American Books, 1995.]

> **MESSAGE** The cytoskeleton serves as a highway system for the directed movement of subcellular particles and organelles.

Determining germ-line cells through cytoskeletal asymmetries

In many organisms, a visible particle is asymmetrically distributed to the cells that will form the germ line. These particles are called *P granules* in *Caenorhabditis elegans*, *polar granules* in *Drosophila*, and *nuage* in frogs. They are thought to be transport vehicles that ride on specific cytoskeletal highways to deliver attached regulatory molecules, which are the germ-cell determinants, to future germ-line cells. In *C. elegans* and *Drosophila*, the evidence relating germ-line determination to mechanisms that depend on the cytoskeleton is particularly strong. Let's consider each of these cases in turn and then review the features that they have in common.

DETERMINING THE *C. ELEGANS* GERM LINE The early cell divisions of the *C. elegans* zygote provide an example of how cytoskeletal asymmetries help form the germ line. One of the favorable properties of *C. elegans* as an experimental system is that the same pattern of cell divisions is present from one worm to another—and this pattern can be readily followed under the microscope. A lineage tree can then be constructed that traces the descent of each of the thousand or so somatic cells of the worm.

Because they are traceable, every somatic cell in the worm is given a name. Fertilization in *C. elegans* initially produces a one-cell zygote called the P_0 cell. It is an ellipsoidal cell that divides asymmetrically across its long axis to produce a larger, anterior AB cell and a smaller, posterior P_1 cell (Figure 18-6). This division is very important in that it sets up specialized roles for the descendants of these first two cells. Descendants of the AB cell will form most of the skin of the worm (the hypoderm) and most of the nervous system, whereas descendants of the P_1 cell will give rise to most of the muscles, all of the digestive system, and the entire germ line.

The key to the formation of the germ line is the presence of fluorescent particles called P granules. Before fertilization, P granules are distributed uniformly throughout the oocyte cytoplasm. The event of sperm entry itself marks the anterior side of the future embryo, causing a reorganization of the actin microfilament network. This reorganized network delivers the P granules to the posterior side of the cytoplasm of the newly formed zygote. This phenomenon, in which fertilization causes a reorganization of the actin cytoskeletal network in a manner of a few seconds, occurs in many animals including some vertebrates such as amphibians. It is a very powerful way

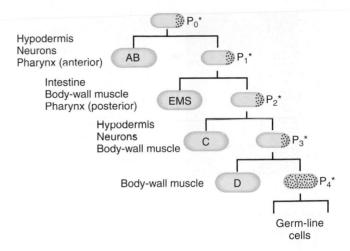

Figure 18-6 Early divisions of the *C. elegans* zygote. The mature cell types that arise from the various daughter cells of the early divisions are indicated. Note that the entire germ line comes from the P_4 cell. Each of the P cell divisions indicated by an asterisk is asymmetrical, and each of the posterior daughter cells inherits all the P granules, which are thought to be germ-line determinants in the worm. The letters (for example, AB, EMS) are symbols for names of daughter cells.

to take advantage of an asymmetrical mark (the point of fertilization) to begin to create pattern.

The P granules, having been delivered to the zygote's posterior side, end up in the P_1 cell. That cell divides, also asymmetrically, again producing a larger, anterior cell and a smaller, posterior P cell, called P_2. This pattern of asymmetrical anterior–posterior cell divisions continues, producing a series of P cells as the more posterior of the two progeny cells produced at each successive cell division. The posterior descendants of P_0 (P_1, P_2, and so forth) each acquire , the P granules. The P_4 cell becomes the germ line of the worm—all other cells are somatic. P granules depend on microfilaments to carry them to their original location in the P_0 cells. When microfilaments in the P_0 cell are disrupted by drugs, P granules end up distributed evenly between the two progeny cells. (Presumably because other fate determinants also are abnormally distributed owing to the actin disruption, the resulting embryos are quite "confused" and die as masses of cells that look nothing like normal worms.)

DETERMINING THE *D. MELANOGASTER* GERM LINE The cytoskeleton is also exploited in germ-line determination in early *Drosophila* development. Here, the regulatory molecules are the *polar granules*, which appear under a light microscope as dense particles. During oogenesis in the ovary of the mother, the polar granules become tethered to the posterior pole of the oocyte. They remain in this location throughout early embryogenesis. An unusual feature of early *Drosophila* development is that the first 13 mitoses are nuclear divisions without

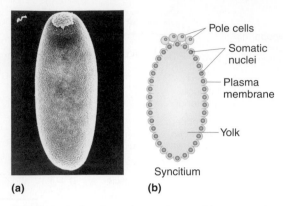

(a)   **(b)**

Figure 18-7 **Germ-line formation at the syncitial stage of the early *Drosophila* embryo.** (a) A scanning electron micrograph of a *Drosophila* embryo with the egg shell removed. Note that the pole cells (the cap of cells on top of the embryo) lie outside the somatic syncitium. (b) A diagram of a longitudinal section through the embryo in part a, showing that the germ-line cells—the pole cells—have formed, whereas the soma still consists of syncitial nuclei. [Modified from F. R. Turner and A. P. Mahowald, *Developmental Biology* 50, 1976, 95.]

any cell division, making the early embryo a syncitium—a multinucleate cell. At nuclear division 9, a few nuclei migrate to the posterior pole where the polar granules are tethered. The plasma membrane of the syncitial egg then moves inward at the posterior pole to surround each nucleus, incorporating some of the polar granules into the cytoplasm that surrounds the nucleus. This action creates the **pole cells,** the first mononucleate cells formed in the embryo. The pole cells will uniquely form the fly's germ line (Figure 18-7).

How do the polar granules get tethered to the posterior pole of the oocyte? Again, one of the cytoskeletal networks carries the granules to the posterior pole and provides the rods to which the granules attach. In contrast with *C. elegans*, in which the P granules attach to actin-based microfilaments, here the tubulin-based microtubules provide the essential trafficking system.

18.3 Forming complex pattern: the logic of the decision-making process

The major pattern-formation decisions within the soma typically require a choreographed set of steps that lead to the establishment of all the necessary cell types with the appropriate spatial distribution. In the newly fertilized animal embryo, this set of steps is used to establish the overall body plan. Later on in development, nests of cells are set aside to become the various organs and appendages of the body. When first set aside, each of these nests of cells undergoes a similar choreography to subdivide the initial population into its final pattern of cell

types. In other words, once metazoan evolution "invented" a basic "dance" to form complex pattern, it applied that "innovation" as a solution to a great many problems in embryogenesis and organogenesis.

What are the choreographed steps of this process? In outline form (Figure 18-8), they are:

- Create a population of developmentally identical cells (part a in Figure 18-8).
- Create an asymmetry within that population (part b).
- Exploit the asymmetry to set up a chemical concentration gradient within the population of cells (part c).

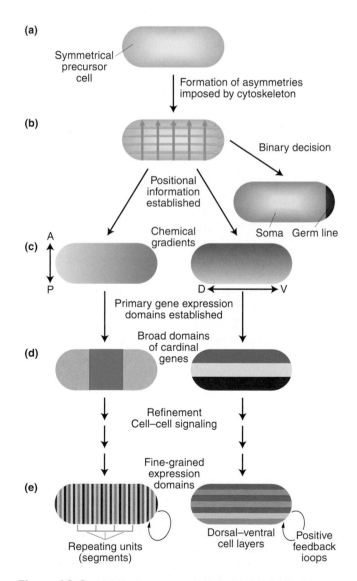

Figure 18-8 **Choreographed steps of pattern formation in *Drosophila*.** (a) The cytoskeleton imposes an asymmetry on the egg. (b) A germ line is separated off by cell division. (c) Gradients of soluble chemicals are laid down on the A–P and D–V axes, stimulating gene action and (d) leading to broad bands. (e) Within these bands more complex gene interactions subdivide the broad regions.

- Create a system that can mount a differential response to that local concentration, thereby creating multiple cell states (part d).

- Create mechanisms for cross talk between the cells of different states, such that cells can take on additional cell states, refine their positions, and adjust their numbers (part e).

- As key decisions are made, develop systems that allow these decisions to become fixed in the molecular memory banks of a cell and its descendants.

We will focus largely on the construction of the basic *Drosophila* body plan in the embryo as a well-understood application of the logic of this choreography. Specifically, we will look at the establishment of the two major body axes of the embryo: anterior–posterior (head to tail) and dorsal–ventral (back to front). Development of these two axes leads to the following events:

- Subdivision of the embryonic anterior–posterior axis into a series of distinct units called **segments,** or metameres, and assignment of distinct roles to each segment according to its location in the developing animal. Here, a developing cell can adopt one of many possible cell states, corresponding to the 14 different segments, and, even within each segment, multiple cell states will be established. Thus, this decision is much more complex than simply the yes–no decision of germ line versus soma.

- Subdivision of the embryonic dorsal–ventral axis into the outer, middle, and inner sheets of cells, called the **germ layers** (not to be confused with germ lines), and assignment of distinct roles to each of these layers. Here again, we are looking at the establishment of a wide variety of cell types within each of the three primary germ layers: the cells of the epidermis, nervous, circulatory, immune, musculature, respiratory, digestive, extractory, and reproductive systems, and so forth.

18.4 Forming complex pattern: establishing positional information

The first steps in the process of pattern formation take place in the oocyte while it is developing in the ovary of the mother. A genetic circuit is activated that leads to the **maternal expression** of the genes responsible for creating the asymmetries and the chemical concentration gradients of positional information that the embryo uses for its spatial coordinates. The existence of the maternal effect (and many other aspects of *Drosophila* development) became known from studies on mutants. Specifically, geneticists discovered mutations in **maternal-effect genes,** which produce anterior–posterior axis abnormalities only when expressed in the mother. Hence, if we label a reces-

sive maternal-effect mutation "*m*," in a cross of an *m/m* female with a +/+ male, all the embryos are abnormal even though they are genetically normal (+/*m*). See the Model Organism box, *Drosophila*, for a description of how these mutants were recovered and analyzed.

Cytoskeletal asymmetries and the *Drosophila* anterior–posterior axis

We have seen that the *Drosophila* germ line is established when specific regulatory molecules, anchored to microtubules, are localized to one area of the embryo. The same is also true for the determination of cell type from the front to the back of the organism (the anterior–posterior, or A–P, axis). Positional information along the A–P axis of the syncitial *Drosophila* embryo is initially established through the creation of concentration gradients of two transcription factors: the BCD and HB-M proteins, which are products of two maternal-effect genes. These factors interact to generate different patterns of gene expression along the axis. The BCD protein, encoded by the *bicoid (bcd)* gene, is distributed in a steeper gradient in the early embryo, whereas the HB-M protein, encoded by one of the mRNAs of the *hunchback (hb)* gene, is distributed in a shallower but longer gradient (Figure 18-9). Both concentration gradients

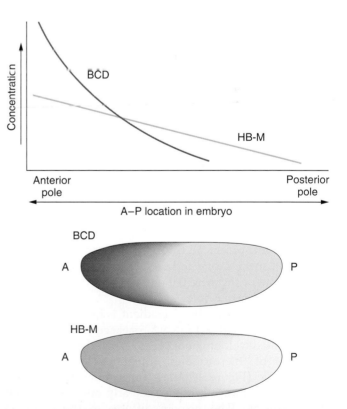

Figure 18-9 Concentration gradients of BCD and HB-M proteins. The BCD gradient is steeper, and the BCD protein is not detectable in the posterior half of the early *Drosophila* embryo. The HB-M gradient is shallower, and the HB-M protein can be detected well into the posterior half of the embryo.

MODEL ORGANISM *Drosophila*

Mutational Analysis of Early *Drosophila* Development

The initial insights into the genetic control of pattern formation emerged from studies of the fruit fly *Drosophila melanogaster*. *Drosophila* development has proved to be a gold mine to researchers because developmental problems can be approached by the use of genetic and molecular techniques simultaneously. Let's consider the basic genetic and molecular techniques that are employed.

The *Drosophila* embryo has been especially important in understanding the formation of the basic animal body plan. One important reason is that an abnormality in the body plan of a mutant is easily identified in the larval exoskeleton in the *Drosophila* embryo. The larval exoskeleton is a noncellular structure, made of a polysaccharide polymer, called chitin, that is produced as a secretion of the epidermal cells of the embryo. Each structure of the exoskeleton is formed from epidermal cells or cells immediately underlying that structure. With its intricate pattern of hairs, indentations, and other structures, the exoskeleton provides numerous landmarks to serve as indicators of the fates assigned to the many epidermal cells. In particular, there are many distinct anatomical structures along the anterior–posterior (A–P) and dorsal–ventral (D–V) axes. Furthermore, because all the nutrients necessary to develop to the larval stage are prepackaged in the egg, mutant embryos in which the A–P or D–V cell fates are drastically altered can nonetheless develop to the end of embryogenesis and produce a mutant larva. The exo-

skeleton of such a mutant larva mirrors the mutant fates assigned to subsets of the epidermal cells and can thus identify genes worthy of detailed analysis.

Researchers, most notably Christiane Nüsslein-Volhard, Eric Wieschaus, and their colleagues, have performed extensive mutational screens, essentially saturating the genome for mutations that alter the A–P or D–V patterns of the larval exoskeleton. These mutational screens identified two broad classes of genes that affect the basic body plan: zygotically acting genes and maternal-effect genes (see diagrams). The zygotically acting genes are those genes whose expression in the zygote is the sole source of their gene products. They are part of the DNA of the zygote itself and are the "standard" sorts of genes that we are

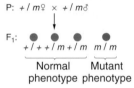

Zygotically acting genes

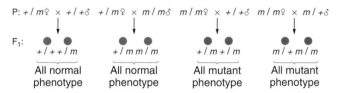

The genetic distinction between recessive zygotically acting and maternal-effect mutations.

---

have their high points at the anterior pole. In somewhat different ways, the gradients of both these proteins depend on the diffusion of protein from a localized origin: each protein is produced through localized translation of an mRNA species, one tethered to microtubules at the anterior pole of the syncitial embryo, and the other tethered at the posterior pole.

The origin of the BCD gradient is quite straightforward. The maternal *bcd* mRNA, packaged during oogenesis into the developing oocyte, is tethered to the − (minus) ends of microtubules, which are located at the anterior pole (Figure 18-10a). Translation of BCD protein begins midway through the early nuclear divisions of the embryo. The protein diffuses in the common cytoplasm of the syncitium. Because the protein is a transcription factor, it contains signals that direct it to the nuclei. Those nuclei nearer to the anterior pole

incorporate a higher concentration of the diffusing BCD protein than do those farther away; this difference results in the steep BCD protein gradient (Figure 18-10b).

The origin of the HB-M protein gradient is more complex. Unlike the *bcd* mRNA, the *hb-m* RNA is

---

Figure 18-10 Photomicrographs showing the expression of localized A–P determinants in the embryo. (a) By in situ hybridization to RNA, the localization of *bcd (bicoid)* mRNA to the anterior (left) tip of the embryo can be seen. (b) By antibody staining, a gradient of Bicoid protein (brown stain) can be visualized, with its highest concentration at the anterior tip. (c) Similarly, nanos (*nos*) mRNA localizes to the posterior (right) tip of the embryo, and (d) NOS protein is in a gradient with a high point at the posterior tip. [Parts a and b from C. Nüsslein-Volhard, *Development*, Suppl. 1, 1991, 1. Parts c and d provided by E. R. Gavis, L. K. Dickinson, and R. Lehmann, then of Massachusetts Institute of Technology.]

used to thinking about. Recessive mutations in zygotically acting genes elicit mutant phenotypes only in homozygous mutant animals. The alternative category—the maternal-effect genes—has already produced its gene products in the ovary of the mother. Thus, the maternal-effect proteins are already present in the oocyte at fertilization. In maternal-effect mutations, the phenotype of the offspring depends on the genotype of the mother, not of the offspring, because the source of the gene products is the mother's genes. A recessive maternal-effect mutation will produce mutant animals only when the mother is a mutant homozygote.

Genes that contribute to the *Drosophila* body plan can be cloned and characterized at the molecular level with relative ease. Any *Drosophila* gene with a well-established chromosomal map location can be cloned by using recombinant DNA techniques such as those described in Chapter 11. The analysis of the cloned genes often provides valuable information on the function of the protein product—usually by identifying close relatives in amino acid sequence of the encoded polypeptide through comparisons with all the protein sequences stored in public databases. In addition, one can investigate the spatial and temporal pattern of expression of (1) an mRNA, by using histochemically tagged single-stranded DNA sequences complementary to the mRNA to perform RNA in situ hybridization, or (2) a protein, by using histochemically tagged antibodies that bind specifically to that protein.

Extensive use is made of in vitro mutagenesis techniques, as well. P elements are used for germ-line transformation in *Drosophila* (see Chapter 11). A cloned pattern-formation gene is mutated in a test tube and put back into the fly. The mutated gene is then analyzed to see how the mutation alters the gene's function.

Using Knowledge from One Model Organism to Fasttrack Pattern-Formation Research in Others

With the discovery that there are numerous homeobox genes within the *Drosophila* genome, similarities among the DNA sequences of these genes could be exploited in treasure hunts for other members of the homeotic gene family. These hunts depend on DNA base-pair complementarity. For this purpose, DNA hybridizations were carried out under *moderate stringency conditions*, in which there could be some mismatch of bases between the hybridizing strands without disrupting the proper hydrogen bonding of nearby base pairs. Some of these treasure hunts were carried out in the *Drosophila* genome itself, in looking for more family members. Others searched for homeobox genes in other animals, by means of *zoo blots* (Southern blots of restriction-enzyme-digested DNA from different animals), by using radioactive *Drosophila* homeobox DNA as the probe. This approach led to the discovery of homologous homeobox sequences in many different animals, including humans and mice. (Indeed, it is a very powerful approach for "fishing" for relatives of almost any gene in your favorite organism.) Some of these mammalian homeobox genes are very similar in sequence to the *Drosophila* genes.

(a) *Bicoid* mRNA

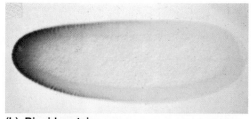

(b) Bicoid protein

(c) *nos* mRNA

(d) NOS protein

uniformly distributed throughout the oocyte and the syncitial embryo. However, translation of *hb-m* mRNA is blocked by a repressor protein—the NOS protein product. This protein is encoded by the *nanos (nos)* gene. In the mother, *nos* mRNA is deposited at the posterior pole in the oocyte, through its association with the + (plus) ends of microtubules (Figure 18-10c). At the time of its translation, NOS protein becomes distributed by diffusion in a gradient opposite that of BCD. The NOS gradient has a high point at the posterior pole and drops down to background levels in the middle of the A–P axis of the embryo (Figure 18-10d). NOS protein inhibits translation of *hb-m* mRNA, and the level of inhibition is proportional to the concentration of NOS protein, producing the shallow gradient of HB-M protein from the anterior to the posterior of the embryo.

> **MESSAGE** Localization of mRNAs within a cell is accomplished by anchoring the mRNAs to one end of polarized cytoskeleton chains.

How do the maternal *bcd* and *nos* mRNAs get tethered to opposite ends of the polarized microtubules of the oocyte and early embryo? The answer is that there are specific microtubule-association sequences located within the untranslated regions of the mRNA 3' to the translation-termination codon (3' UTRs). The 3' UTR localization sequences of *bcd* mRNA are bound by a protein that can also bind the minus ends of the microtubules. In contrast, the 3' UTR localization sequences of *nos* mRNA are bound by a protein that can also bind the plus ends of microtubules.

How can we demonstrate that the 3' UTRs of the mRNAs are where the localization sequences reside? This has been determined in part by "swapping" experiments. A synthetic transgene is constructed that produces an mRNA with the 5' UTR and protein-coding regions of the normal *nos* mRNA glued to the 3' UTR of the normal *bcd* mRNA. When inserted into the fly genome, this fused *nos–bcd* mRNA will be localized at the anterior pole of the oocyte. This localization causes a double gradient of NOS: one from anterior to posterior (due to the transgene's mRNA) and one from posterior to anterior (due to the normal *nos* gene's mRNA). This procedure produces a very weird embryo, with two mirror-image posteriors and no anterior (Figure 18-11). This double-abdomen embryo arises because NOS protein is now present throughout the embryo and represses the translation of *hb-m* mRNA (it also represses translation of *bcd* mRNA, although it is not clear that repression of *bcd* mRNA is its normal function in wild-type animals).

(a)

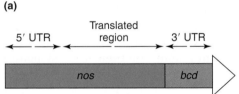

(b)

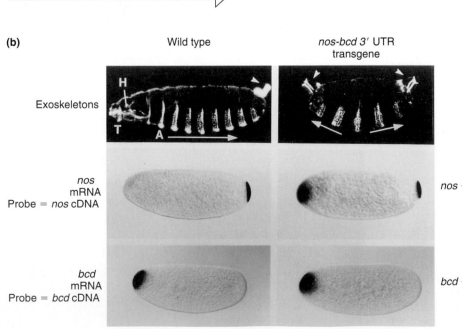

Figure 18-11 Effect of replacement of the 3' UTR of the *nanos* mRNA with the 3' UTR of the *bicoid* mRNA on mRNA localization and embryonic phenotypes. (a) The structure of the *nos–bcd* 3' UTR transgene. (b) The effects on embryonic development of the transgene. The embryos and larva in the left-hand column are derived from wild-type mothers. Those in the right-hand column are derived from transgenic mothers. All embryos are shown with anterior to the left and posterior to the right. The exoskeletons are shown for comparison. The transgene causes a perfect mirror-image double abdomen. In the embryo from a transgenic mother, mRNAs coding for NOS protein are now present at both poles of the embryo. NOS protein will inhibit translation of *hb-m* mRNA (and, actually, of *bcd* mRNA as well). [From E. R. Gavis and R. Lehmann, *Cell* 71, 1992, 303.]

Studying the BCD gradient

How do we know that molecules such as BCD and HB-M contribute A–P positional information? Let's consider the example of BCD in detail.

1. Genetic changes in the *bcd* gene alter anterior fates. Embryos derived from mothers that are homozygous for *bcd* null mutants lack anterior segments (Figure 18-12). If, on the other hand, we overexpress *bcd* in the mother, by increasing the number of copies of the *bcd⁺* gene from the normal two to three, four, or more, we "push" fates that ordinarily appear in anterior positions into increasingly posterior zones of the resulting embryos, as shown in Figure 18-13. The position of the cephalic furrow (a normal feature of the embryo) serves as an example in Figure 18-13. These observations suggest that BCD protein exerts global control of anterior positional information.

2. The *bcd* mRNA can completely substitute for the anterior determinant activity of the anterior cytoplasm (Figure 18-14). If the anterior cytoplasm

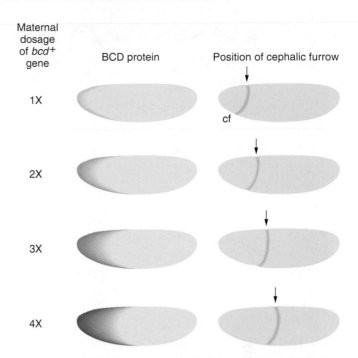

Figure 18-13 Concentration of BCD protein affects A–P cell fates. The amount of BCD protein can be changed by varying the number of copies of the *bcd⁺* gene in the mother. Embryos derived from mothers carrying from one to four copies of *bcd⁺* have increasing amounts of BCD protein. (Two copies is the normal diploid gene dosage.) Cells that invaginate to form the cephalic furrow (cf) are determined by a specific concentration of BCD protein. In the progression from one maternal copy to four copies of *bcd⁺*, this specific concentration is present more and more posteriorly in the embryo. Thus, the position of the cephalic furrow (marked dorsally by the arrow) arises farther toward the posterior according to *bcd⁺* gene dosage. [After W. Driever and C. Nüsslein-Volhard, *Cell* 54, 1988, 100.]

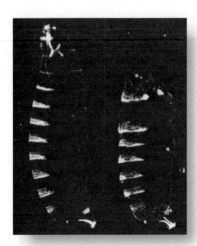

Figure 18-12 Photomicrographs of the exoskeletons of larvae derived from wild-type and *bcd* maternal-effect-lethal mutant mothers. These photomicrographs are dark-field, and so dense structures appear white, as in a photographic negative. Note the bright, segmentally repeated denticle bands present on the ventral side of the embryo. Maternal genotypes and larval phenotypes (and class of mutation) are as follows: *(left)* wild type, normal phenotype; *(right) bcd (bicoid),* anterior head and thoracic structures missing (anterior). [From C. H. Nüsslein-Volhard, G. Frohnhöfer, and R. Lehmann, *Science* 238, 1987, 1678.]

is removed from a punctured syncitial embryo, anterior segments (head and thorax) are lost (not shown). Injection of anterior cytoplasm from another embryo into the anterior region of the anterior-cytoplasm-depleted embryo restores normal anterior segment formation, and a normal larva is produced. Similarly, synthetic *bcd* mRNA can be made in a test tube and injected into the anterior region of an anterior-cytoplasm-depleted embryo. Again, a normal larva is produced. In a control experiment, transplantation of cytoplasm from middle or posterior regions of a syncitial embryo does not restore normal anterior formation. Thus, the anterior determinant should be located *only* at the anterior end of the egg. As already discussed, this is exactly where *bcd* mRNA is found.

3. As has already been described (see Figure 18-10b), the BCD protein shows the predicted asymmetrical and graded distribution required to fulfill its role of establishing positional information.

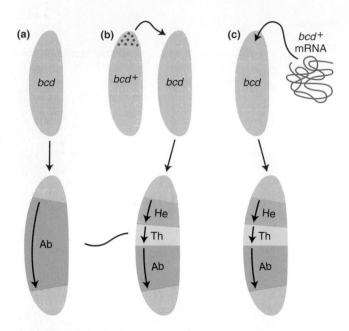

Figure 18-14 The *bcd* "anteriorless" mutant phenotype can be rescued by wild-type cytoplasm or purified *bcd*⁺ mRNA. (a) In embryos derived from *bcd* mothers, the anterior (head and thoracic) segments do not form, producing the anteriorless phenotype. (b) If anterior cytoplasm from an early wild-type donor embryo is injected into the anterior of a recipient embryo derived from a *bcd* mutant mother, a fully normal embryo and larva are produced. Cytoplasm from any other part of the donor embryo does not rescue. (c) Injection of *bcd*⁺ mRNA into the anterior of an embryo derived from a *bcd* mutant mother also rescues the wild-type segmentation pattern. (Ab = abdomen; He = head; Th = thorax.) [After C. H. Nüsslein-Volhard, G. Frohnhöfer, and R. Lehmann, *Science* 238, 1987, 1678.]

Cell-to-cell signaling and the *Drosophila* dorsal–ventral axis

In the examples considered thus far, products *inside the cell* determined the position of future body parts: these products were mRNAs or larger macromolecular assemblies packaged into the oocyte. In many circumstances, though, positional information depends on proteins *outside the cell* that have been secreted from a localized subset of cells within a developing field. These secreted proteins diffuse in the extracellular space to form a concentration gradient of ligand that binds to receptors on target cells. The ligand thus activates the target cells in a concentration-dependent fashion through a receptor–signal-transduction system. The level of signal outside a given target cell determines the level of signal transduction and response to the signal inside the cell. The level of response inside the cell ultimately determines the combination of genes whose transcription rates are elevated or lowered in the target cell.

An example of such a mechanism for establishing position is the establishment of the dorsal–ventral

(D–V) axis in the early *Drosophila* embryo by the DL protein. The immediate effect of the D–V positional information is to create a gradient of DL protein activity in cells along the D–V axis. The DL protein is a transcription factor encoded by the *dorsal (dl)* gene. It exists in two forms: (1) active transcription factor located in the nucleus and (2) inactive protein located in the cytoplasm, where it is sequestered in a complex bound to the CACT protein encoded by the *cactus (cact)* gene. A concentration gradient of active DL protein determines cell fate along the D–V axis. Both *dl* mRNA and inactive DL protein are distributed uniformly in the oocyte and the very early embryo. However, late in the syncitial embryo stage, there develops a gradient of active DL protein, with its high point in cells at the ventral midline of the embryo (Figure 18-15).

How is the gradient of active DL protein generated? By a rather complex set of gene interactions. The key events take place in the developing egg (Figure 18-16a), well before fertilization. At this time an interaction takes place between the oocyte itself and the layer of surrounding somatic cells—the follicle cells, which also make the eggshell (Figure 18-16b). The follicle cells on the ventral side of the oocyte secrete some proteins that activate a secreted precursor of the SPZ ligand, encoded by the *spaetzle (spz)* gene. As a consequence, the activated SPZ ligands are concentrated on the ventral side of the oocyte, where they form a gradient that is highest at the ventral midline. On the inner boundary of the eggshell is the vitelline membrane (Figure 18-17a). The SPZ ligand is sequestered in the vitelline membrane until near the end of the syncitial stage of early embryogenesis, when it is released. Active SPZ ligand (with its highest concentration at the ven-

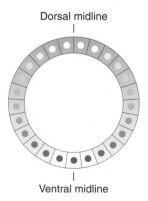

Figure 18-15 Distribution of DL shown in a cross section of the cellular blastoderm-stage *Drosophila* embryo. The dorsal midline is at the top and the ventral midline is at the bottom. Note that the DL protein is in the nucleus (where it is active) ventrally, throughout the cell laterally, and in the cytoplasm (where it is inactive) dorsally. [After S. Roth, D. Stein, and C. Nüsslein-Volhard, *Cell* 59, 1989, 1196.]

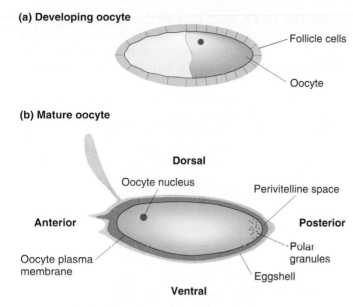

(a) Developing oocyte

Follicle cells

Oocyte

(b) Mature oocyte

Dorsal

Oocyte nucleus

Perivitelline space

Anterior

Posterior

Oocyte plasma membrane

Polar granules

Eggshell

Ventral

Figure 18-16 Developing (a) and mature (b) oocyte. The follicle cells that built and surrounded the eggshell in part a have been discarded in the mature oocyte, and the plasma membrane of the oocyte is interior to the eggshell. Note the polar granules at the posterior end of the oocyte. [After C. Nüsslein-Volhard, *Development*, Suppl. 1, 1991, 1.]

tral midline) then binds to the TOLL transmembrane receptor, encoded by the *Toll* gene, present uniformly in the oocyte plasma membrane (see Figure 18-17a). In a concentration-dependent manner, the SPZ–TOLL complex triggers a signal-transduction pathway that ends up phosphorylating the inactive DL and CACT cytoplasmic proteins of the DL–CACT complex (Figure 18-17b). Phosphorylation of DL and CACT causes conformational changes that break apart the cytoplasmic complex. The free phosphorylated DL protein is then able to migrate into the nucleus, where it serves as a transcription factor that activates genes necessary for establishing the ventral fates. The crucial arrangement is the positioning of the SPZ ligand near the ventral midline, which ensures that only the DL on the ventral side of the embryo is able to activate the genes that create structures for the ventral part of the body.

> **MESSAGE** Positional information can be established by cell-to-cell signaling through a concentration gradient of a secreted signaling molecule.

Other roles of the cytoskeleton in establishing the body axes

The cytoskeleton can play other key roles in establishing the body axes. We will consider one example: establishing the poles in the early *Drosophila* oocyte.

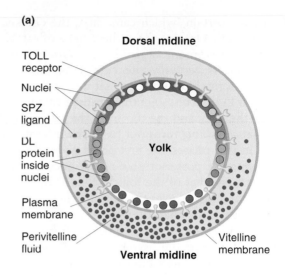

(a)

Dorsal midline

TOLL receptor

Nuclei

SPZ ligand

DL protein inside nuclei

Plasma membrane

Perivitelline fluid

Ventral midline

Yolk

Vitelline membrane

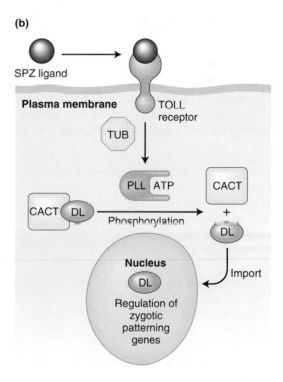

(b)

SPZ ligand

Plasma membrane

TOLL receptor

TUB

PLL ATP

CACT

CACT DL

Phosphorylation

+

DL

Nucleus

DL

Import

Regulation of zygotic patterning genes

Figure 18-17 Signaling pathway leading to the gradient of nuclear versus cytoplasmic localization of DL protein shown in Figure 18-15. (a) A cross section of a *Drosophila* embryo showing the blastoderm cells inside the plasma membrane and the space (perivitelline space) between the inside boundary of the eggshell (vitelline membrane) and the plasma membrane, where the active SPZ ligand is produced on the ventral side of the embryo. (b) The SPZ ligand binds to the TOLL receptor, activating a signal-transduction cascade through two proteins called TUB and PLL, leading to the phosphorylation of DL and its release from CACT. DL then is able to migrate into the nucleus, where it serves as a transcription factor for D–V cardinal genes. [After H. Lodish, D. Baltimore, A. Berk, S. L. Zipursky, P. Matsudaira, and J. Darnell, *Molecular Cell Biology*, 3d ed. Copyright Scientific American Books, 1995.]

There is a certain "which came first, the chicken or the egg?" aspect to tracing the earliest steps of establishing the body axes in any organism. For *Drosophila* axis formation, this is certainly the case. Even before the events just described take place to set up gradients along the A–P and D–V axes, something must distinguish the front from the back and the top from the bottom of the oocyte. That *something* turns out to be the microtubules, which act in two phases: the first phase gives the follicle cells overlying the posterior pole of the oocyte a different identity from that of the follicle cells at the anterior pole; the second phase gives the dorsal follicle cells a different identity from that of their ventral and lateral counterparts. Later on, instructive signals from these follicle cells will create the asymmetries that have been outlined in the preceding sections.

The early oocyte transcribes very few genes, but one of these few is a gene that encodes an epidermal growth factor-like (EGF-like) ligand that acts by binding to a receptor tyrosine kinase called the EGF receptor (EGFR) and activating it. The EGF-like ligand is secreted to the nearest cell surface through the channels of the endoplasmic reticulum, which are located near the nucleus. Thus, if the nucleus is nearer to one side of the oocyte, that side will secrete a higher concentration of EGF-like

ligand. The EGFR, the target of the EGF-like ligand, is on the surface of all follicle cells. However, the EGFR will have its tyrosine kinase activated *only* where EGF-like ligand is concentrated—namely, in the vicinity of the nucleus.

So, the name of the game is to strategically move the nucleus around the oocyte. Briefly, this movement is accomplished in a two-phase process (Figure 18-18). In the first phase, the microtubule network orients with its plus ends at the prospective posterior pole of the early oocyte, dragging the oocyte nucleus to that pole. Then, after EGFR activation in the posterior follicle cells, instructing them to be different from their anterior counterparts, these follicle cells send a signal back to the oocyte that causes the microtubule network to be reoriented perpendicularly, so that the plus ends of the network are at the prospective dorsal side of the oocyte. The nucleus is dragged along and, because it is attached to the nucleus, so is the rough endoplasmic reticulum and hence the EGF-like signal. This secreted signal then activates the EGFR in the follicle cells overlying the nucleus, instructing them to become dorsal follicle cells.

Figure 18-19 summarizes the two types of positional information that we have covered.

Gene interactions in *Drosophila* embryogenesis: Cell signaling in dorsal–ventral axis formation

(a)

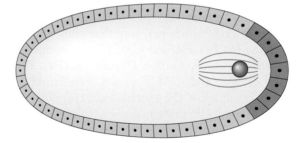

(b) A–P commitment

(c) D–V commitment

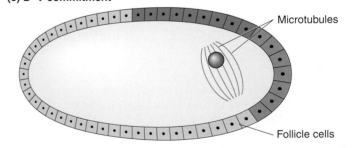

Figure 18-18 Determination of posterior and anterior ends in the *Drosophila* embryo. (a) The cytoskeleton is reorganized to create populations of posterior and dorsal follicle cells through EGF-like ligand and EGFR signaling between the oocyte and the overlying follicle cells. (The EGF-like ligand is called gurken (GRK) because mutations in the *gurken* gene produce a phenotype like little pickles.) (b) Microtubule reorganization delivers the nucleus to the posterior pole, thereby delivering the EGF-like signal to the posterior pole and creating the posterior polar follicle cells. (c) These posterior polar follicle cells send a signal back to the oocyte that causes the microtubules to reorganize again, reversing their polarity and bringing the nucleus and GRK to the dorsal anterior sector of the oocyte, leading to the establishment of the dorsal follicle cells. [After A. Gonzalez-Reyes, H. Elliott, and D. St. Johnston, *Nature* 375, 1995, 657.]

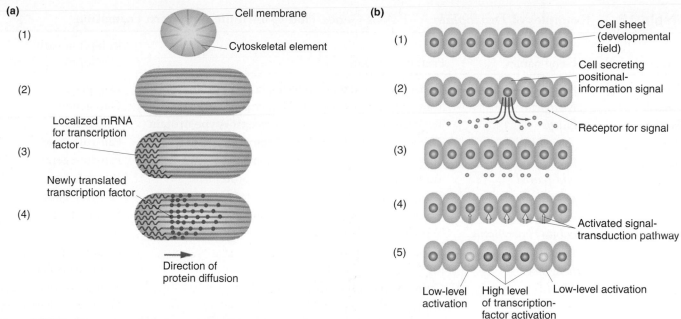

Figure 18-19 The two general classes of positional information. (a) Asymmetrical organization of the cytoskeletal system permits localization of mRNA encoding a transcription factor that will provide positional information. Translation of the anchored mRNA will lead to diffusion of the newly translated transcription factor and the formation of a transcription-factor gradient with a high point near the site of the mRNA. (b) Secretion of a positional-information signaling molecule from a localized source (cell) activates the signal-transduction apparatus and the target transcription factors according to the level of signaling molecule that binds to its transmembrane receptor.

18.5 Forming complex pattern: utilizing positional information to establish cell fates

To develop correctly in a specific position, a cell needs to assess and respond to the local positional cues. To use a geographical analogy, having a system of longitudes and latitudes is not sufficient; we also need equipment that can receive longitude and latitude information— special instruments to read the positions of stars or receivers that can triangulate signals transmitted from radio beacons. In the same way, the developmental positional-information system requires that elements within the cell be able to interpret transmitted signals.

The initial interpretation of positional information

As described earlier, two very different kinds of positional signal can be produced, one kind generated from within the cell and the other kind diffusing from outside the cell. However, both lead to the same outcome: a gradient in the concentration of one or more specific transcription factors within the cells of the developmental field. What elements in the cell interpret the information in the gradient?

Given that positional information is given by a gradient of active transcription factor, we might expect that the receivers are regulatory elements (enhancers and silencers) of special genes. Then the protein products of these special genes can begin the gradual process of specifying cell fate. This is exactly what we see. The genes targeted by the A–P and D–V transcription factors are genes expressed in the zygote and collectively known as the **cardinal genes** because they are the first genes to respond to the positional information supplied by the oocyte. The cardinal genes of the A–P axis are listed in (Table 18-1).

To see how these genes work, we need to review a bit of *Drosophila* embryology. After 2–3 hours of development, all somatic nuclei migrate to the surface of the egg, and the egg's plasma membrane invaginates around each nucleus and its cytoplasm (Figures 18-20 and 18-21a). The result is in an embryonic stage called the *cellular blastoderm*, a hollow spheroid with a wall that is one cell thick. A few hours later, this wall of cells has undergone numerous foldings, outpocketings, and other movements and now manifests the first signs of segmentation. At the end of 10 hours of development, the embryo

Table 18-1   Examples of *Drosophila* A–P Axis Genes That Contribute to Pattern Formation

Gene symbol	Gene name	Protein function	Role(s) in early development
hb-z	*hunchback-zygotic*	Transcription factor—zinc-finger protein	Gap gene
Kr	*Krüppel*	Transcription factor—zinc-finger protein	Gap gene
kni	*knirps*	Transcription factor—steroid receptor-type protein	Gap gene
eve	*even-skipped*	Transcription factor—homeodomain protein	Pair-rule gene
ftz	*fushi tarazu*	Transcription factor—homeodomain protein	Pair-rule gene
opa	*odd-paired*	Transcription factor—zinc-finger protein	Pair-rule gene
prd	*paired*	Transcription factor—PHOX protein	Pair-rule gene
en	*engrailed*	Transcription factor—homeodomain protein	Segment-polarity gene
ci	*cubitus-interruptus*	Transcription factor—zinc-finger protein	Segment-polarity gene
wg	*wingless*	Signaling WG protein	Segment-polarity gene
hh	*hedgehog*	Signaling HH protein	Segment-polarity gene
fu	*fused*	Cytoplasmic serine/threonine kinase	Segment-polarity gene
ptc	*patched*	Transmembrane protein	Segment-polarity gene
arm	*armadillo*	Cell-to-cell junction protein	Segment-polarity gene
lab	*labial*	Transcription factor—homeodomain protein	Segment-identity gene
Dfd	*Deformed*	Transcription factor—homeodomain protein	Segment-identity gene
Antp	*Antennapedia*	Transcription factor—homeodomain protein	Segment-identity gene
Ubx	*Ultrabithorax*	Transcription factor—homeodomain protein	Segment-identity gene

is already externally divided into 14 segments from anterior to posterior—3 head, 3 thoracic, and 8 abdominal segments (Figure 18-21b). By this time, each segment has developed a unique set of anatomical structures, corresponding to its special identity and role in the biology of the animal. At the end of 12 hours, distinct organs first form. At 15 hours, the exoskeleton of the larva begins to form, with its specialized hairs and other external structures. Only 24 hours after development began at fertilization, a fully formed larva hatches out of the eggshell (Figure 18-21c). The segmental arrangement of spikes, hairs, and other sensory structures on the larval exoskeleton makes each segment distinct and recognizable under the microscope. Now let's return to the A–P cardinal genes.

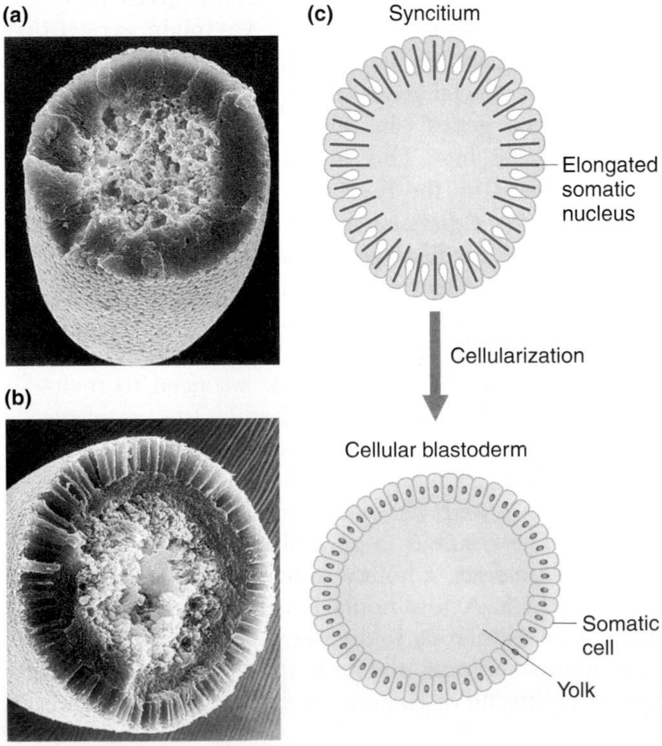

(a)

(b)

(c) Syncitium

— Elongated somatic nucleus

Cellularization

Cellular blastoderm

— Somatic cell

Yolk

Figure 18-20 Cellularization of the *Drosophila* embryo. The embryo begins as a syncitium; cellularization is not completed until there are about 6000 nuclei. (a) and (b) Scanning electron micrographs of embryos removed from the eggshell. (a) A syncitium-stage embryo, fractured to reveal the common cytoplasm toward the periphery and the central yolk-filled region. The bumps on the outside of the embryo are the beginning of cellularization, in which the plasma membrane of the egg folds inward from the outside. (b) A cellular blastoderm embryo, fractured to reveal the columnar cells that have formed by cell membranes being drawn down between the elongated nuclei to create some 6000 mononucleate somatic cells. (c) Diagrams of the changes taking place during cellularization. [After F. R. Turner and A. P. Mahowald, *Developmental Biology* 50, 1976, 95.]

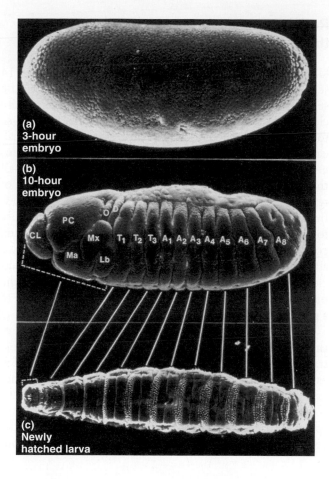

Figure 18-21 Scanning electron micrographs of (a) a 3-hour *Drosophila* embryo, (b) a 10-hour embryo, and (c) a newly hatched larva. Note that outlines of individual cells are visible by 3 hours; by 10 hours, the segmentation of the embryo is obvious. Lines are drawn to indicate that the segmental identity of cells along the A–P axis is already fated early in development. The abbreviations refer to different segments of the head (CL, PC, O, D, Mx, Ma, Lb), thorax (T_1 to T_3), and abdomen (A_1 to A_8). [T. C. Kaufman and F. R. Turner, Indiana University.]

The A–P cardinal genes are also known as **gap genes,** because flies that have mutations in these genes do not have the proper sequential set of larval segments: there is a gap in the normal segmentation pattern. (For examples, look at the phenotypes of mutations in two gap genes, *Krüppel* and *knirps* in Figure 18-22). Through

Figure 18-22 Types of *Drosophila* mutants. These diagrams depict representative mutants in each class of mutant larval phenotypes due to mutations in the three classes of zygotically acting genes controlling segment number in *Drosophila*. The red trapezoids are segmentally repeating swatches of dense projections on the ventral surface of the larval exoskeleton. The boundary of each segment is indicated by a dotted line. The left-hand diagram of each pair depicts a wild-type larva, and the right-hand diagram shows the indicated A–P mutant. The pink regions on the wild-type diagram indicate the A–P domains of the larva that are missing in the mutant.

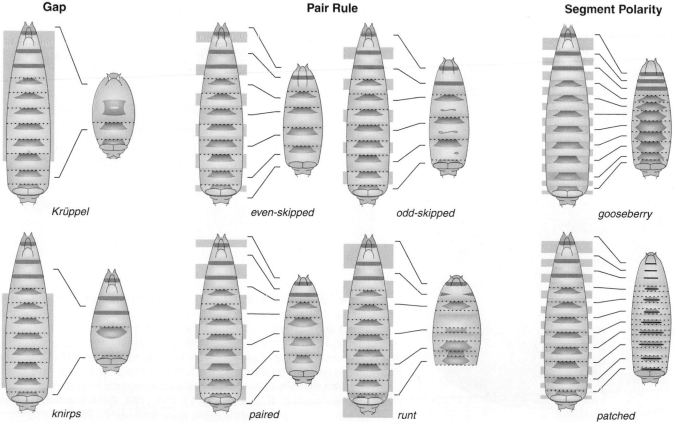

Gap · Pair Rule · Segment Polarity

Krüppel · *even-skipped* · *odd-skipped* · *gooseberry*

knirps · *paired* · *runt* · *patched*

its expression limited to a restricted region of the A–P axis, each gap gene organizes the formation of a spatially localized set of segments. How is expression confined to just the proper location? BCD or HB-M protein or both bind to enhancer elements of the promoters of the gap genes, thereby regulating their transcription. For example, transcription of one gene, *Krüppel (Kr)*, is repressed by high levels of the BCD transcription factor but is activated by low levels of BCD and HB-M. In contrast, the *knirps (kni)* gene is repressed by the presence of any BCD protein but requires low levels of the HB-M transcription factor for its expression. Thus the gap-gene regulatory sequences have been exquisitely "tuned" by evolution each to have a unique sensitivity to the concentrations of the transcription factors of the A–P positional-information system. The result of this regulation is that the *kni* gene is expressed more posteriorly than is *Kr* (Figure 18-23a). In general, through differences in their regulatory element properties, the gap genes are expressed in a series of distinct domains. These domains then become different developmental fields, and the cells in the different domains become committed to different A–P fates. Each gap gene itself encodes a unique transcription factor and hence regulates a different set of downstream target genes necessary to refine the fate of the segments arising in that field.

Refining fate assignments through transcription-factor interactions

Ultimately, every one of the cell rows in each of the 14 segments will establish its own A–P identity during the segmentation process. The number of gap genes are clearly too few and their expression patterns too coarse to account for all of the fates necessary to distinguish each row in each segment. Thus, these gap-gene domains need to be refined to produce the detailed anatomy of the 14 segments of the fully formed larva. The process of refinement begins at about the time when the embryo is transformed from a mass of cytoplasm into a structure formed of separate cells. The cytoplasm of each cell then contains a particular concentration of one or perhaps two adjacent gap-gene-encoded proteins, which enter its nucleus. Essentially all further decisions are driven by the particular A–P gap proteins "captured" in the nucleus of a given blastoderm cell at the time of cellularization.

The A–P developmental pathway downstream of the gap genes bifurcates. Each of the two branches illustrates strategies for refining the pattern. One branch establishes the correct number of segments. The other assigns the proper identity to each segment. (These different identities are manifest in the unique patterns of spikes and hairs on each segment of the larva, as described earlier.) The existence of two branches means

(a) Gap genes

hb

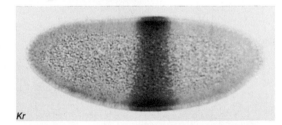

Kr

kni

(b) Pair-rule genes

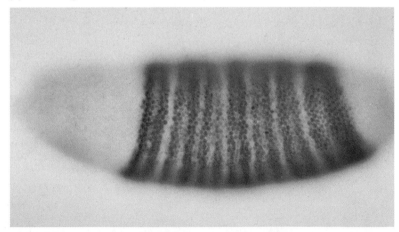

Figure 18-23 Photomicrographs showing the early embryonic expression patterns of gap and pair-rule genes. All embryos are shown with anterior to the left and posterior to the right. (a) Early blastoderm expression patterns of proteins from three gap genes: *hb-z*, *Kr*, and *kni*. (b) Late blastoderm expression patterns of proteins from two pair-rule genes: *ftz* (stained gray) and *eve* (stained brown). Note the localized gap-gene expression patterns compared with the reiterated pair-rule-gene expression pattern. [Part a from M. Hulskamp and D. Tautz, *BioEssays* 13, 1991, 261. Part b from Peter Lawrence, *The Making of a Fly.* Copyright 1992 by Blackwell Scientific Publications.]

that the transcription factors encoded by the gap genes regulate two different sets of target genes.

First, let's briefly consider the formation of segment number. (Refer to Figure 18-22 for a description of the mutant phenotypes produced by the different classes of segment-number genes.) The gap genes activate a set of A–P patterning genes called the **pair-rule genes,** each of which encodes a transcription factor, expressed in a repeating pattern of seven stripes (see Figure 18-23b). The different pair-rule genes produce a slightly offset pattern of stripes, as shown by the gray and reddish bands in Figure 18-23b. Within a given cell, several different pair-rule proteins are expressed.

There is a hierarchy within the pair-rule-gene class. Some of them, called *primary pair-rule genes,* are regulated directly by the gap genes. How do the asymmetrically distributed gap genes produce the repeating seven stripe expression pattern of the primary pair-rule genes? The gap proteins at work here are expressed in a very different, asymmetrical pattern. The key is that the regulatory elements for certain pair-rule genes are quite complex. One primary pair-rule gene is *eve (even-skipped).* The *eve* gene contains separate enhancer elements, each of which interacts with different combinations of gap transcription factors to produce the seven *eve* stripes. For example, the enhancer for the first *eve* stripe is activated by high levels of the HB-Z gap transcription factor, the enhancer for the second *eve* stripe by low levels of HB-Z but high levels of the KR gap transcription factor, and so forth.

> **MESSAGE** The complexity of regulatory elements of the primary pair-rule genes turns an asymmetrical (gap-gene) expression pattern into a repeating one.

Once the primary pair-rule genes are activated, they in turn activate the expression of the other pair-rule genes (also encoding transcription factors) to produce the full seven-stripe pattern. No two pair-rule genes yield exactly the same pattern of stripes. The products of the pair-rule genes then act combinatorially to regulate the transcription of *segment-polarity genes,* which are expressed in offset patterns of 14 stripes. Thus, the hierarchy of transcription-factor regulation extends all the way from the positional-information system to the repeating pattern of segment-polarity-gene expression. The products of the segment-polarity genes permit the 14 segments to form and define the individual A–P rows of cells within each segment.

> **MESSAGE** Through a hierarchy of transcription-factor regulation patterns, positional information leads to the formation of the correct number of segments. In the readout of positional information, transcription factors act combinatorially to create the proper segment-number fates.

Next, let's briefly consider the establishment of segmental identity. The gap genes also target a clustered group of genes known as the **homeotic gene complexes.** They are called gene complexes because several of the genes are clustered together on the DNA. *Drosophila* has two homeotic gene clusters. The ANT-C (*Antennapedia* complex) is largely responsible for segmental identity in the head and anterior thorax, whereas the BX-C (*Bithorax* complex) is responsible for segmental identity in the posterior thorax and abdomen.

Homeosis, or homeotic transformation, is the development of one body part with the phenotype of another. Three examples of body-part-conversion phenotypes due to homeotic gene mutations are:

1. the loss-of-function *bithorax* class of mutations that cause the entire third thoracic segment (T3) to be transformed into a second thoracic segment (T2), giving rise to flies with four wings instead of the normal two (Figure 18-24);

2. the gain-of-function dominant *Tab* mutation, described in Chapter 10 (see Figure 10-28), that transforms part of the adult T2 segment into the sixth abdominal segment (A6); and

3. the gain-of-function dominant *Antennapedia (Antp)* mutation that transforms antenna into leg (see the photograph on the first page of this chapter).

Note that, in each of these cases, the number of segments in the animal remains the same; the only change is in the identity of the segments. The results of studies of these homeotic mutations have revealed much about how segment identity is established.

The various gap proteins activate the target homeotic genes to initially produce a series of overlapping domains as shown in Figure 18-25, in which *Scr* and *Antp* are both genes in the *Antennapedia* complex, and *Ubx* and *Abd-B* are genes in the *Bithorax* complex. The homeotic genes encode transcription factors of a class called homeodomain proteins. **Homeodomain proteins** interact with the regulatory elements of other genes in a specific combination so that the expression pattern in each of the different segments is unique. (We shall consider the relations of structure and function within the homeotic gene complexes later in this chapter, in the context of the evolution of developmental mechanisms.)

Summary of the cascade of regulatory events

As we have seen, A–P patterning of the *Drosophila* embryo is accomplished through a sequential triggering of regulatory events. Positional information establishes different concentrations of transcription factors along the A–P axis, and target regulatory genes are then deployed

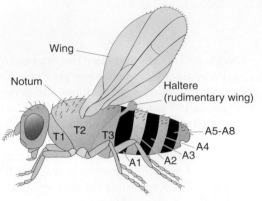

(a) Segment names

(b) Wild type

(c) Bithorax

Figure 18-24 Homeotic transformation of the third thoracic segment (T3) of *Drosophila* into an extra second thoracic segment (T2). (a) Diagram showing the normal thoracic and abdominal segments; note the rudimentary wing structure (haltere) normally derived from T3. Most of the thorax of the fly, including the wings and the dorsal part of the thorax, comes from T2. (b) A wild-type fly with one copy of T2 and one of T3. (c) A *bithorax* triple mutant homozygote completely transforms T3 into a second copy of T2. Note the second dorsal thorax and second pair of wings (T2 structures) and the absence of the halteres (T3 structures). [From E. B. Lewis, *Nature* 276, 1978, 565. Photographs courtesy of E. B. Lewis. Reprinted by permission of Nature. Copyright 1978 by Macmillan Journals Ltd.]

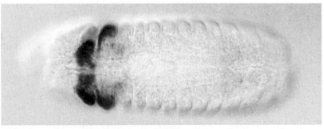

(a) SCR

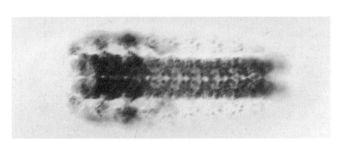

(b) ANTP

(c) UBX

(d) ABD-B

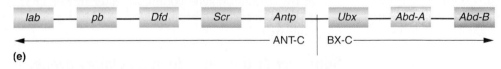

(e)

Figure 18-25 Homeotic gene expression in *Drosophila*. Photomicrographs of embryos that exhibit protein expression patterns encoded by homeotic genes in *Drosophila*. (a–d) The anterior boundary of homeotic gene expression is ordered from SCR (most anterior) to ANTP, UBX, and ABD-B (most posterior). (e) This order is matched by the linear arrangement of the corresponding genes along chromosome 3. [Parts a and b from T. C. Kaufman, Indiana University. Parts c and d from S. Celniker and E. B. Lewis, California Institute of Technology.]

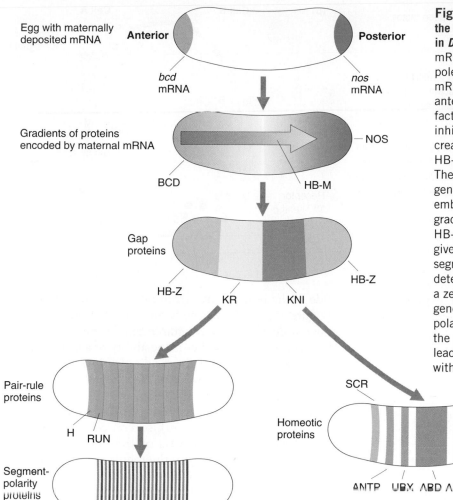

Egg with maternally deposited mRNA

Anterior

Posterior

bcd mRNA

nos mRNA

Gradients of proteins encoded by maternal mRNA

NOS

BCD

HB-M

Gap proteins

HB-Z

HB-Z

KR

KNI

Pair-rule proteins

SCR

H   RUN

Homeotic proteins

Segment-polarity proteins

WG   EN

ANTP   UBX   ABD A   ABD B

Segment Number

Segment Identity

Figure 18-26 Hierarchical cascade that activates the elements forming the A–P segmentation pattern in *Drosophila*. The maternally derived *bcd* and *nos* mRNAs are located at the anterior and posterior poles, respectively. Early in embryogenesis, these mRNAs are translated to produce a steep anterior–posterior gradient of BCD transcription factor. The posterior–anterior gradient of NOS inhibits translation of *hb-m* mRNA, thereby creating a shallow anterior–posterior gradient of HB-M transcription factor (shown as an arrow). The gap genes, which are the A–P cardinal genes, are activated in different parts of the embryo in response to the anterior–posterior gradients of the two factors BCD and HB-M. (The posterior band of HB-Z expression gives rise to certain internal organs, not to segments.) The correct number of segments is determined by activation of the pair-rule genes in a zebra-stripe pattern in response to the gap-gene-encoded transcription factors. The segment-polarity genes are then activated in response to the activities of the several pair-rule proteins, leading to further refinement of the organization within each segment. The correct identities of the segments are determined by expression of the homeotic genes due to direct regulation by the transcription factors encoded by the gap genes. [After J. D. Watson, M. Gilman, J. Witkowski, and M. Zoller, *Recombinant DNA*, 2d ed. Copyright 1992 by James D. Watson, Michael Gilman, Jan Witkowski, and Mark Zoller.]

accordingly to execute the increasingly finer subdivisions of the embryo, establishing both segment number and segment identity (Figure 18-26).

18.6 Refining the pattern

The principles delineated in the preceding sections lay out initial fates, but additional mechanisms must be in place to ensure that all aspects of patterning are elaborated. Some of these mechanisms are considered in this section.

Memory systems for remembering cell fate

Patterning decisions frequently need to be maintained in a cell lineage for the lifetime of the organism. This requirement is certainly true of the segment-polarity and homeotic gene expression patterns that are set up by the A–P patterning system. The key is that patterning decisions are maintained through positive-feedback loops.

Such positive-feedback loops may operate entirely within a cell. In several tissues, positive-feedback loops are established in which the homeodomain protein that is expressed binds to enhancer elements in its own gene, ensuring that more of that homeodomain protein will continue to be produced (Figure 18-27a).

In other cases, the memory system requires cell-to-cell interactions (Figure 18-27b). For example, among the segment-polarity genes, adjacent cells express the WG (wingless) and EN (engrailed) proteins. The EN protein is a transcription factor that activates the gene encoding the molecule HH (hedgehog) in the same cells. HH is a protein-signaling molecule that is secreted from the cell and binds to a receptor on the surface of the WG-expressing cell. It thereby induces a signal-transduction cascade in the WG-expressing cell, activating *wingless* gene expression and causing more WG protein to be expressed. Similarly, WG is a

(a)

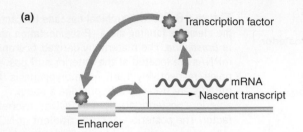

(b)

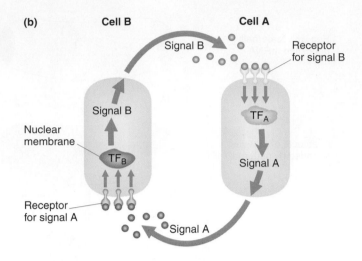

Figure 18-27 Two types of positive-feedback loops that maintain the level of activity of transcription factors that determine cell fate. (a) The transcription factor binds to an enhancer of its own gene, maintaining its transcription. (b) Each adjacent cell sends out a signal (different signals from each cell) that activates receptors, signal-transduction pathways, and transcription-factor (TF) expression in the other cell. This activation leads to a mutual positive-feedback loop between the cells.

secreted protein that activates *engrailed* expression in the adjacent cell, inducing more EN protein in that cell.

> **MESSAGE** When the fate of a cell lineage has been established, it must be remembered. This is accomplished by intracellular or intercellular positive-feedback loops.

Ensuring that all fates are allocated: decisions by committee

Ultimately, for a developmental field to mature into a functional organ or tissue, cells must be committed in appropriate numbers and locations to the full range of functions that are needed. Cell-to-cell interactions ensure that these proper allocations are made. Here, we focus on two types of interactions, both of which operate in the development of the vulva, the opening to the out-

side of the reproductive tract of the nematode *C. elegans* (Figure 18-28). One type is the ability of one cell to induce a developmental commitment in only one neighboring cell, and the other is the ability of a cell to inhibit its immediate neighbors from adopting its fate.

Vulva development has been studied in detail through the analysis of *C. elegans* mutants that have either no vulva or more than one. Within the hypodermis (the body wall of the worm), several cells have the potential to build certain parts of the vulva. Initially, all these cells can adopt any of the required roles and so are called an **equivalence group**—in essence, a developmental field. To make an intact vulva, one of the cells must become the primary vulva cell, and two others must become secondary vulva cells; yet others become tertiary cells that contribute to the surrounding hypodermis (Figure 18-29a and b).

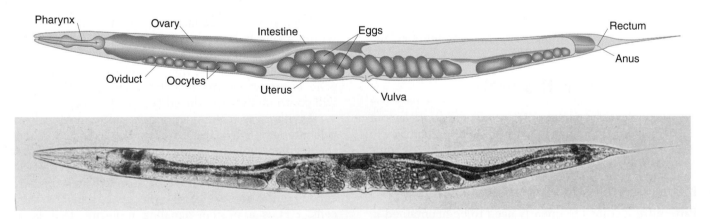

Figure 18-28 Adult *Caenorhabditis elegans.* Photomicrograph and drawing of an adult hermaphrodite, showing various organs readily identified by their location. Note the position of the vulva midway along the anterior–posterior axis of the worm. [From J. E. Sulston and H. R. Horvitz, *Developmental Biology* 56, 1977, 111.]

Figure 18-29 Production of the *C. elegans* vulva by cell-to-cell interactions. (a) The parts of the vulva anatomy occupied by the descendants of primary, secondary, and tertiary cells. (b) The primary, secondary, and tertiary cell types are distinguished by the cell-division patterns that they undergo. (c) Early in development, there is no signal from the anchor cell, and all the cells are in the default tertiary cell state. (d) Later in development, the anchor cell sends a signal that activates a receptor tyrosine kinase signal-transduction cascade. The cell nearest the anchor cell receives the strongest signal and becomes the primary vulva cell. It then sends out lateral inhibition signals to its neighbors, preventing them from also becoming primary vulva cells and shunting them into the secondary vulva cell pathway. [Part b from R. Horvitz and P. Sternberg, *Nature* 351, 1991, 357. Parts a, c, and d after I. Greenwald, *Trends in Genetics* 7, 1991, 366.]

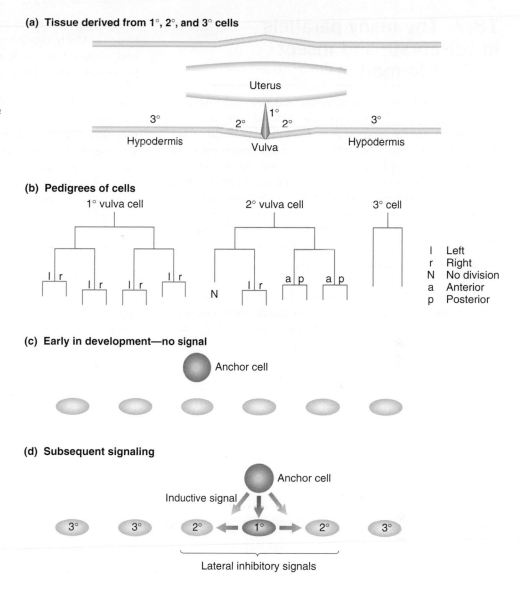

(a) **Tissue derived from 1°, 2°, and 3° cells**

Uterus

3°   2° 1° 2°   3°

Hypodermis   Vulva   Hypodermis

(b) **Pedigrees of cells**

1° vulva cell   2° vulva cell   3° cell

l   Left
r   Right
N   No division
a   Anterior
p   Posterior

(c) **Early in development—no signal**

Anchor cell

(d) **Subsequent signaling**

Anchor cell

Inductive signal

3°   3°   2° 1° 2°   3°

Lateral inhibitory signals

The key to allocating the different roles to these cells is another single cell, called the **anchor cell,** which lies underneath the cells of the equivalence group (Figure 18-29c). The anchor cell secretes a polypeptide ligand that binds to a receptor tyrosine kinase (RTK) present on all the cells of the equivalence group. Only the cell that receives the highest level of this signal (the equivalence-group cell nearest the anchor cell) becomes a primary vulva cell. Only its signal-transduction pathway is sufficiently triggered so that it in turn can activate the transcription factors necessary to become a primary cell (Figure 18-29d). Thus we can say that the anchor cell operates through an **inductive interaction** to commit a cell to the primary vulva fate.

Having acquired its fate, the primary vulva cell sends out a different signal to its immediate neighbors in the equivalence group. That signal inhibits those cells from similarly interpreting the anchor-cell signal, thus preventing them from also adopting the primary role. This process of **lateral inhibition** leads these neighboring cells to adopt the secondary fate. The remaining cells of the equivalence group develop as tertiary cells and contribute to the hypoderm surrounding the vulva. For each of the three cell types into which the equivalence group develops, a specific constellation of activated transcription factors typifies the state of the cell: primary, secondary, or tertiary. Thus, through a series of intercellular signals, a group of equivalent cells can develop into the three necessary cell types.

MESSAGE Fate allocation can be made through a combination of inductive and lateral inhibitory interactions between cells.

18.7 The many parallels in vertebrate and insect pattern formation

How universal are the developmental principles uncovered in *Drosophila*? Even now, the type of genetic analysis possible in *Drosophila* is not feasible in most other organisms, at least not without a huge investment to develop comparable genetic tools and to breed and maintain colonies of larger animals. However, in the past two decades, recombinant DNA technology has provided the necessary tools. One important approach is simply to use the power of DNA-DNA hybridization to fish out cognate genes from a different organism. If genomes have been sequenced, then it is a simple matter to use the computer to find the gene in question. Some of the most spectacular and unexpected parallels have come from comparing early fly and mouse development, given that the evolutionary distance between fly and mammal is so great.

Perhaps the most striking finding is the similarity between certain clusters of homeotic genes in mammals and *Drosophila*. In humans, the clusters of homeotic genes are called Hox complexes. These clusters closely resemble the insect ANT-C and BX-C homeotic gene clusters, collectively called the HOM-C (homeotic gene complex) (Figure 18-30). The ANT-C and BX-C clusters, which are far apart on chromosome 3 of *Drosophila*, are together in one cluster in more primitive insects such as the flour beetle *Tribolium castaneum*. This indicates that there is only one homeotic gene cluster— HOM-C—in insects and that, in the evolution of the *Drosophila* lineage, it was separated into two clusters. Moreover, as noted in Figure 18-25e, the genes of the HOM-C cluster are arranged on the chromosome in an order that is colinear with their spatial pattern of expression: the genes at the left-hand end of the complex are transcribed near the anterior end of the embryo; rightward along the chromosome, the genes are transcribed progressively more toward the posterior (compare Figure 18-30a and b).

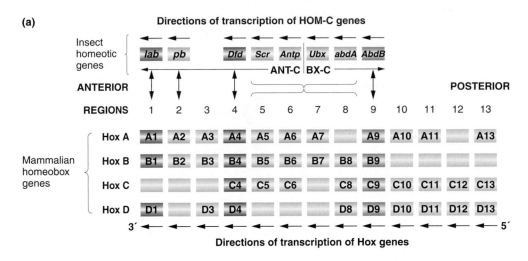

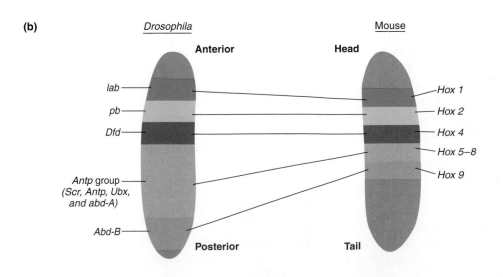

Figure 18-30 Comparison of the structures and functions of the insect and mammalian homeotic genes. (a) The comparative anatomy of the HOM-C (insect) and Hox (mammalian) gene clusters. The genes of the HOM-C are shown at the top. Each of the four paralogous (see text) Hox clusters maps on a different chromosome. Genes shown in the same color are most closely related to one another in structure and function. (b) The expression domains and regions of the *Drosophila* and mouse embryos that require the various HOM-C and Hox genes. The color scheme parallels that in part a. Note that the order of domains in the two embryos is the same. [After H. Lodish, D. Baltimore, A. Berk, S. L. Zipursky, P. Matsudaira, and J. Darnell, *Molecular Cell Biology*, 3d ed. Copyright Scientific American Books, 1995.]

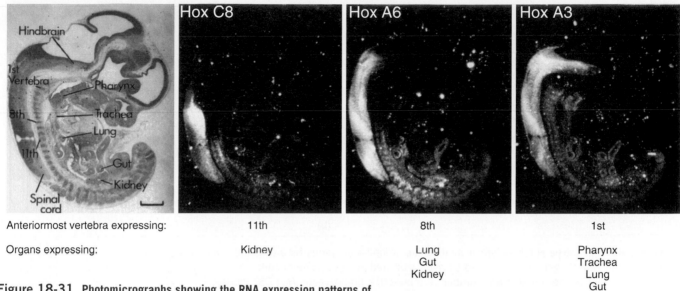

| Anteriormost vertebra expressing: | 11th | 8th | 1st |
| Organs expressing: | Kidney | Lung
Gut
Kidney | Pharynx
Trachea
Lung
Gut
Kidney |

Figure 18-31 Photomicrographs showing the RNA expression patterns of three mouse Hox genes in the vertebral column of a sectioned 12.5-day-old mouse embryo. Note that the anterior limit of each of the expression patterns is different. [From S. J. Gaunt and P. B. Singh, *Trends in Genetics* 6, 1990, 208.]

We still do not know why the insect genes are clustered or organized in this colinear fashion, but, regardless of the roles of these features, the same structural organization—clustering and colinearity—is seen for the equivalent genes in mammals, which are organized into the Hox clusters (see Figure 18-30a). The major difference between flies and mammals is that there is only one HOM-C cluster in the insect genome, whereas there are four Hox clusters, each located on a different chromosome, in mammals. These four Hox clusters are **paralogous,** meaning that the order of genes in each cluster is very similar, as if the entire cluster had been quadruplicated in the course of vertebrate evolution. Each of the genes near the left end of each Hox cluster is quite similar not only to the others, but also to one of the insect HOM-C genes at the left end of the cluster. Similar relations hold throughout the clusters. Finally, and most notably, the Hox genes are expressed so as to define segments in the developing somites (the segmental units of the developing spinal column) and central nervous system of the mouse and presumably the human embryos. Each Hox gene is expressed in a continuous block beginning at a specific anterior limit and running posteriorly to the end of the developing vertebral column (Figure 18-31). The anterior limit differs for different Hox genes. Within each Hox cluster, the leftmost genes have the most anterior limits. These limits proceed more and more posteriorly in the rightward direction in each Hox cluster. Overall, the Hox gene clusters appear to be arranged and expressed in an order that is strikingly similar to that of the insect HOM-C genes (see Figure 18-30b).

The correlations between structure and expression pattern are further strengthened by consideration of mutant phenotypes. In vitro mutagenesis techniques permit efficient gene knockouts in the mouse. Many of the Hox genes have now been knocked out, and the striking result is that the phenotypes of the homozygous knockout mice are thematically parallel to the phenotypes of homozygous null HOM-C flies. For example, the *Hox-C8* knockout causes ribs to be produced on the first lumbar vertebra, L1, which is ordinarily the first nonribbed vertebra behind those vertebrae-bearing ribs (Figure 18-32). Thus, when *Hox C8* is knocked out, the L1 vertebra is homeotically transformed into the segmental identity of a more anterior vertebra. To use geneticists' jargon, *Hox C8*[-] has caused a fate shift toward anterior. Clearly, this Hox gene seems to control segmental fate in a manner quite similar to that of the HOM-C genes, because, for example, a null allele of the *Drosophila Ubx* gene also causes a fate shift toward anterior in which T3 and A1 are transformed into T2.

How can such disparate organisms—fly, mouse, human (and *C. elegans*)—have such similar gene sequences? The simplest interpretation is that the Hox and HOM-C genes are the vertebrate and insect descendants of a homeobox gene cluster present in a common ancestor some 600 million years ago. The evolutionary conservation of the HOM-C and Hox genes is not an unusual occurrence. Indeed, as we are beginning to compare whole genomes, we are finding that such evolutionary and functional conservation seems to be the norm rather than the exception. For example, 60 percent of human genes associated with a heritable disease have related genes in *Drosophila*.

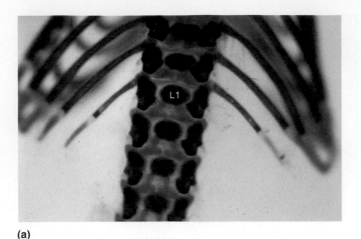

(a)

(b)

Figure 18-32 Phenotype of a homeotic mutant mouse. Mice homozygous for a targeted knockout of the *Hox C8* gene were created by using cultured embryonic stem cells. (a) An enlargement of the thoracic and lumbar vertebrae of a homozygous *Hox C8⁻* mouse. Note the ribs coming from L1, the first lumbar vertebra. L1 in wild-type mice had no ribs. (b) An unexpected second phenotype of the *Hox C8⁻* knockout: the homozygous mutant mouse on the right has clenched fingers, whereas the wild-type mouse on the left has normal fingers. [From H. Le Mouellic, Y. Lallemand, and P. Brulet, *Cell* 69, 1992, 251.]

MESSAGE Developmental strategies in animals are quite ancient and highly conserved. In essence, a mammal, a worm, and a fly are put together with the same basic building blocks and regulatory devices. *Plus ça change, plus c'est la même chose!*

18.8 The genetics of sex determination in humans

An important part of development is the development of sex. Most animals and many plants show sexual dimorphism and, in most of these cases, sex determination is genetically "hardwired." We will look at humans as an example, while noting that there is a range of determination mechanisms quite different from that which follows. However, in human sex determination, we will see some developmental players that by now should be familiar—notably, a toggle switch, wide-acting transcription factors, and cell-to-cell communication.

Follow this story by referring to Figure 18-33, which shows sexual development in men. It is clear that the key toggle switch is the gene on the short arm of the Y chromosome called *SRY* (*s*ex *r*egulation on the *Y*). The presence of *SRY* leads to maleness, whereas its absence leads to femaleness. Its importance is witnessed by the observation that if *SRY* is deleted or has null function, then an XY female results, and, conversely, if *SRY* is translocated to another chromosome, then XX males are

possible. How does the SRY product work? It is a DNA-binding transcription factor that acts on genes in the undifferentiated gonad, transforming it into a testis. If there is no SRY product (as in normal XX females), then the undifferentiated gonad develops into an ovary. Once a testis has developed, the genes for testosterone synthesis are activated. Testosterone (androgen) is a steroid hormone that is responsible for male secondary sexual characteristics such as body shape and body hair patterns—the male phenotype.

Leaving the testis, testosterone enters the bloodstream, which transports it to target cells that will produce the male characteristics. The hormone is lipid soluble, and so it passes straight through the membranes of its target cells and enters the cytoplasm. It binds to a proteinaceous receptor called the androgen receptor, encoded by a gene called *AR*, which resides on the X chromosome. Together, the testosterone and its bound receptor enter the cell's nucleus and act as a transcription factor that turns on maleness genes. The AR protein is crucial to male sex determination. If *AR* is deleted or null in function, then the testosterone cannot act and no maleness results. Recall from Chapter 2 the X-linked recessive human variant called androgen insensitivity syndrome, this syndrome results from mutation of the *AR* gene.

Hence we see two key genes that code for transcription factors and a range of downstream "target genes" both in the testis (primary sex) and in somatic cells that will undergo secondary sexual differentiation.

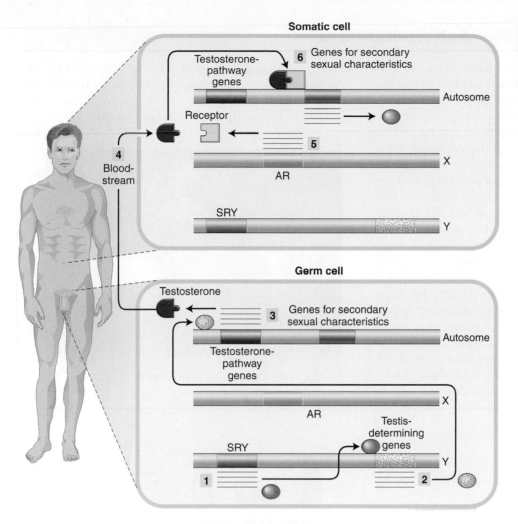

Somatic cell

Testosterone-pathway genes

6 Genes for secondary sexual characteristics

Autosome

Receptor

AR

X

4 Blood-stream

5

SRY

Y

Germ cell

Testosterone

3 Genes for secondary sexual characteristics

Autosome

Testosterone-pathway genes

X

AR

Testis-determining genes

SRY

Y

1

2

Figure 18-33 Gene interaction in sex determination in the human male. The sequence of gene action is numbered, starting with synthesis of the SRY protein in the gonad.

18.9 Do the lessons of animal development apply to plants?

The evidence emerging from the results of comparative studies of pattern formation in a variety of animals indicates that many important developmental pathways are ancient inventions conserved and maintained in many, if not all, animal species. The life history, cell biology, and evolutionary origins of plants would, in contrast, argue against the appearance of these same sets of pathways in the regulation of plant development. Plants have very different organ systems from those of animals, depend on inflexible cell walls for structural rigidity, separate germ line from soma very late in development, and are very dependent on light intensity and duration to trigger various developmental events. Certainly, plants use hormones to regulate gene activity, to signal locally between cells by as yet unknown signals, and to create cell-fate differences by means of transcription factors. The general themes for establishing cell fates in animals are likely to be seen in plants as well, but the participating molecules in these developmental pathways are likely to

be considerably different from those encountered in animal development.

An active area of plant developmental genetics research utilizes a small flowering plant called *Arabidopsis thaliana* as a model genetic system (see the Model Organism box on the next page). The most intensively studied developmental process in *Arabidopsis* is flower pattern formation. Just as the homeotic gene cluster controls segmental identity in animal development, a series of transcriptional regulators determine the fate of the four layers (whorls) of the flower. The outermost whorl of the flower normally develops into the sepals; the next whorl, the petals; the next, the stamens; and the innermost develops into the carpels (Figure 18-34). Several genes have been identified that, when knocked out or ectopically expressed, transform one or more of these whorls into another. For example, the gene *AP1 (Apetala-1)* causes the homeotic transformation of the outer two whorls into the inner two. Analogously to the homeotic mutants in animals, the number of whorls remains the same (four), but their identities are transformed. The study of the spatial expression patterns and mutant phenotypes of the

MODEL ORGANISM   *Arabidopsis thaliana*

More than any other genetic model organism discussed in this text, *Arabidopsis* is a product of the genomics era. It has a genome of only 120 megabase pairs of DNA that is organized into a haploid complement of five chromosomes (the diploid number is 10). Thus, *Arabidopsis* genome size and complexity compare to those of the fruit fly, *Drosophila melanogaster.* In contrast, the genome of corn, a genetic model organism of long standing (see the Model Organism box in Chapter 13), is about 2500 Mb, almost the same as human. Another feature that makes *Arabidopsis* attractive to geneticists is its rapid life cycle: it takes only about 6 weeks for a planted seed to produce a new crop of seed. *Arabidopsis* is also very small (see the photo), growing to less than 10 cm. Its short stature makes it easy to grow in the laboratory in culture tubes or on petri plates, and, because it is a self-fertilizing plant, F_2 mutagenesis screens can be performed in a straightforward manner. Thus, geneticists have obtained many mutations with interesting phenotypes that affect a variety of developmental events

and biochemical pathways. Finally, the genome of *Arabidopsis* has been sequenced, and many tools, such as insertional mutagenesis, exist for overlaying the genetic and transcriptional maps of this plant.

A single *Arabidopsis* plant. [Dan Tenaglia, www.missouriplants.com.]

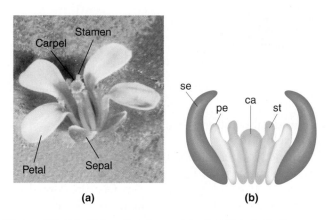

Figure 18-34 Flower development in *Arabidopsis thaliana.*
(a) The mature products of the four whorls of a flower.
(b) A cross-sectional diagram of the developing flower, with the normal fates of the four whorls indicated. From outermost to innermost, they are sepal (se), petal (pe), stamen (st), and carpel (ca). [Photograph courtesy of Vivian F. Irish; from V. F. Irish, "Patterning the Flower," *Developmental Biology* 209, 1999, 211–222.]

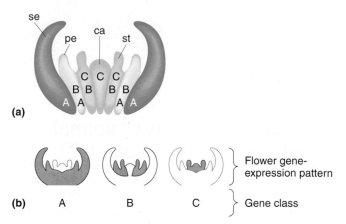

Figure 18-35 Flower-identity gene expression and the establishment of whorl fate. (a) The patterns of gene expression corresponding to the different whorl fates. (b) The shaded regions of the cross-sectional diagrams of the developing flower indicate the gene expression patterns for genes of the A, B, and C classes. Refer to Figure 18-34 for the normal anatomy of the developing flower. [From V. F. Irish, "Patterning the Flower," *Developmental Biology* 209, 1999, 211–222.]

various flower-identity genes has produced a model in which whorl fate is established through the combinatorial action of multiple transcription factors (Figure 18-35). Thus, sepal (outermost whorl fate) is established through the action of transcription factors expressed by genes of the class A type only. Petal fate is established through the action of transcription factors produced by simultaneous

expression of the class A and class B genes. Stamen fate is established through the action of transcription factors produced by simultaneous expression of class B and class C genes. Finally, carpel fate is established through the action of transcription factors expressed by class C genes. Just as the homeotic segment-identity genes in animals encode a series of structurally related (homeodomain-containing) transcription factors, the flower-identity genes encode a series of structurally related transcription factors called MADS transcription factors, found in all eukaryotic kingdoms. (The word *MADS* is composed of the first letters of the names of four prototypic member genes of this family.) Thus, although different in detail, the overall strategy of differentially expressed transcription factors is one of the approaches by which plant cell fate is established. With its combination of sophisticated genetics and genomics, findings from studies of *Arabidopsis* pattern formation should reveal much about the ways in which plants develop.

18.10 Genomic approaches to understanding pattern formation

The key to the study of a pathway or network such as the formation of one of the body axes is to know all of the participating elements. Mutational analysis, though it has proved to be very powerful as a way of first un-

covering these elements, is only one approach. Many genes, for one reason or another, participate in processes of interest to us but cannot be detected by mutational inactivation. For example, we now know that many gene functions are redundant, with more than one gene in the genome contributing to that function. Knock out one of the genes and the product of the other is sufficient to produce a normal phenotype. How then can we identify these pieces of the puzzle?

One way is by detecting expression in interesting patterns along one of the body axes. However, the gene-by-gene approach that a geneticist might use is quite tedious, given that the fly genome (for example) has about 15,000 genes. Instead, with some genetic tricks, we can use transcriptional profiling with expression microarrays to efficiently screen for such mutations. Here's a recent example. A *Drosophila* research group was interested in studying genes that might be positively or negatively regulated by the DL protein, the protein that is distributed in a gradient along the D–V axis of the *Drosophila* embryo. Embryos were studied from mothers who were mutant

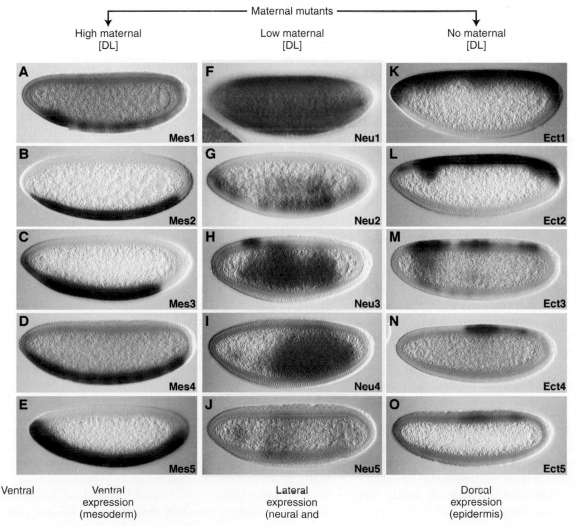

Figure 18-36
Examples of early embryonic expression patterns of genes found in a microarray transcriptional profiling experiment searching for new DL-regulated genes. Maternal mutants showing high, low, or no DL concentrations revealed three different sets of target genes with either ventral, lateral, or dorsal expression. [From A. Stathopoulo, M. Van Drenth, A. Erives, M. Markstein, and M. Levine, *Cell* 111, 2002, 694. Copyright 2002 by Cell Press.]

for one of three different genes that resulted in embryos that had high, low, or no concentrations of DL protein.

Preparations of mRNA from each of these three maternal genotypes were hybridized to microarrays containing every gene in the *Drosophila* genome, and hybridization patterns were compared. A total of 40 new genes were found that were activated or repressed by the DL protein, increasing the number of known DL target genes by 500 percent. In follow-up experiments, the research group was able to show that, indeed, these genes were expressed in the early embryo in domains corresponding to the germ layers along the D–V axis (Figure 18-36). In another approach, high throughput two-hybrid analysis has been used to make protein interaction maps that expand and refine the *Drosophila* body plan networks obtained by mutational analysis.

This is just one of many applications of high-throughput techniques for systematically screening an entire genome for genes of interest that can then be intensively analyzed to help tease apart an entire developmental pathway. This approach does not invalidate the utility of direct mutational screens; rather it is a powerful complementary approach. Neither approach alone will identify all of the relevant genes, but, together, they will contribute to a much more complete picture of developmental processes.

KEY QUESTIONS REVISITED

• **What sequence of events produces the basic body plan of an animal?**

Starting with a totipotent cell, genetic mechanisms produce descendent cells with different genetic fates. These fates are progressively subdivided so that the normal set of organs results, each with its own pattern of gene expression.

• **How are polarities that give rise to the main body axes created?**

Both anterior–posterior and dorsal–ventral axes are established by gradients of molecules laid down in or around the egg by the organism's mother.

• **How do cells recognize their locations along the developing body axes?**

Position is manifest as a specific pattern of transcription factors that activate different sets of cardinal genes that define developmental domains.

• **Do cells participate in single or multiple decision-making processes?**

Multiple: for example, a cell must deal with its position on the A–P and D–V axes at the same time.

• **What role does cell-to-cell communication play in building the basic body plan?**

It further refines cell fate within the general broad developmental fields.

• **Are the pathways for building biological pattern conserved among distant species?**

Yes, to a greater or lesser degree, depending on the organism. The HOM and Hox genes of flies and mammals are homologous and act in similar ways to determine body plan. Transcription-factor gradients are at work in both plants and animals. All organisms use genetic switches of various kinds.

SUMMARY

A programmed set of instructions in the genome of a higher organism establishes the developmental fates of cells with respect to the major features of the basic body plan. These instructions eventually produce a fine-grained mosaic of different cell types deployed in the proper spatial pattern.

The zygote is totipotent, giving rise to every adult cell type. As development proceeds, successive decisions restrict each cell and its descendants (a lineage) to its particular fate. The first developmental decisions are very coarse. Gradients of maternally derived regulatory proteins establish polarity along the major body axes. In all cases that have been well described, the intrinsic polarity of the cytoskeletal system underlies the establishment of this primary positional information within the embryo. Ultimately, the positional information that is established along each of the major body axes leads to differential expression of transcription factors along each axis. The targets of these transcription factors are regulatory elements of cardinal genes. Typically, these cardinal genes are themselves transcription factors or they are another class of molecule that activates other transcription factors. The activation of cardinal genes begins the process of subdividing the animal into a series of coarsely defined developmental domains.

These coarse patterns are refined by a multistep process. Cells communicate patterning information among themselves by using intercellular signaling systems to ensure that the developing structure (embryo, tissue, organ) operates coherently.

The same basic set of genes identified in *Drosophila* and the regulatory proteins that they encode are conserved in mammals and appear to govern major developmental events in many—perhaps all—higher animals. It is fair to say that the majority of genes in the metazoan genome are common to most members of the animal kingdom. The take-home message is that the basic pathways that underlie pattern formation are ancient and are exploited in many different ways to produce animals that are superficially very different from one another. However, there are limits to the generality of the detailed mechanisms. The underlying molecules that regulate development in plants are different from those in animals, but many of the same themes are seen in plant and animal development. In both plants and animals, transcription factors and signaling systems are exploited to create pattern. However, because of the ancient phylogenetic split between plants and animals and because the life strategies of plants and animals are dramatically different, it is not surprising that different molecules fulfill the parallel roles.

KEY TERMS

anchor cell (p. 599)	germ layer (p. 583)	microtubule (p. 578)
binary fate decision (p. 578)	homeodomain protein (p. 595)	pair-rule gene (p. 595)
cardinal gene (p. 591)	homeosis (p. 595)	paralogous (p. 601)
cell fate (p. 577)	homeotic gene complex (p. 595)	pole cell (p. 582)
cytoskeleton (p. 578)	inductive interaction (p. 599)	positional information (p. 577)
developmental field (p. 577)	lateral inhibition (p. 599)	segment (p. 583)
equivalence group (p. 598)	maternal-effect gene (p. 583)	totipotent (p. 578)
fate refinement (p. 578)	maternal expression (p. 583)	
gap gene (p. 593)	microfilament (p. 578)	

SOLVED PROBLEMS

1. The anterior determinant in the *Drosophila* egg is *bcd*. A mother heterozygous for a *bcd* deletion has only one copy of the *bcd*⁺ gene. With the use of P elements to insert copies of the cloned *bcd*⁺ gene into the genome by transformation, it is possible to produce mothers with extra copies of the gene. Shortly after the blastoderm has formed, the *Drosophila* embryo develops an indentation called the cephalic furrow that is more or less perpendicular to the longitudinal body axis. In the progeny of *bcd*⁺ monosomics, this furrow is very close to the anterior tip, lying at a position one-sixth of the distance from the anterior to the posterior tip. In the progeny of standard wild-type diploids (disomic for *bcd*⁺), the cephalic furrow arises more posteriorly, at a position one-fifth of the distance from the anterior to the posterior tip of the embryo. In the progeny of *bcd*⁺ trisomics, it is even more posterior. As additional gene doses are added, the cephalic furrow moves more and more posteriorly, until, in the progeny of hexasomics, the it is midway along the A–P axis of the embryo.

a. Explain the gene-dosage effect of *bcd*⁺ on the formation of the cephalic furrow in relation to the contribution that *bcd* makes to A–P pattern formation.

b. Diagram the relative expression patterns of mRNAs from the gap genes *Kr* and *kni* in blastoderm embryos derived from *bcd* monosomic, trisomic, and hexasomic mothers.

Solution

a. The determination of anterior–posterior parts of the embryo is governed by a concentration gradient of BCD. The furrow develops at a critical concentration of *bcd*. As *bcd*⁺ gene dosage (and, therefore, BCD concentration) decreases, the furrow shifts anteriorly; as the gene dosage increases, the furrow shifts posteriorly.

b.

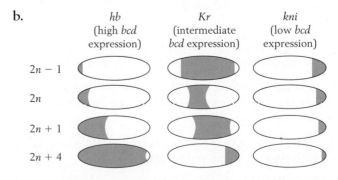

2. In developmental pathways, the crucial events seem to be the activation of master switches that set in motion a programmed cascade of regulatory responses.

Identify the master switches and explain how they operate in sex determination in mammals.

Solution

In mammalian sex determination, the master switch is the presence or absence of the *SRY* gene, which is ordinarily located on the Y chromosome. In the presence of the protein product of this gene, which acts as a DNA-binding protein, certain cells of the gonad (Leydig cells) synthesize androgens, male-inducing steroid hormones. These hormones are secreted into the bloodstream and act on target tissues to induce the transcription-factor activity of the androgen receptors. In the absence of androgen-receptor activation, development proceeds along the default pathway leading to female development. The factors that activate *SRY* expression in the testis are not understood. Because the master switch here is the actual presence or absence of the *SRY* gene itself, it is likely that the regulatory molecules that activate *SRY* are present in the indifferent gonad early in development.

3. In the embryogenesis of mammals, the inner cell mass, or ICM (the prospective fetus), quickly separates from the cells that will serve as enclosing membranes and respiratory, nutritive, and excretory channels between the mother and the fetus.

   a. Design experiments using mosaics in mice to determine when the two fates are decided.

b. How would you trace the formation of different fetal membranes?

Solution

a. We must have markers that enable us to distinguish different cell lineages, which can be done with mice by using strains that differ in chromosomal or biochemical markers. (Other ways would be to use differences in sex chromosomes of XX and XY cells or to induce chromosome loss or aberrations by irradiating embryos.)

When you have decided on the marker difference to be used, one way to approach the design is to inject a single cell from one of two strains into embryos of the other strain at various developmental stages. Another approach is to fuse embryos of defined cell numbers from the two strains. In either case, you would inspect the embryos when the ICM and membranes are distinct and recognizable. When cell insertion or fusion results in membranes and an ICM that are exclusively made up of one cell type and never a mosaic of the two, the two developmental fates have been set.

b. Carry out the same injection or fusion experiment on early embryos. Now look for the pattern of mosaicism. Correlate the presence of cells of similar genotype in different membranes. It should be possible to determine the lineage of cells in each set of membranes.

PROBLEMS

BASIC PROBLEMS

1. In what ways does the cytoskeleton resemble and not resemble a body skeleton?

2. In what sense is a microfilament polar?

3. If you were describing to a nonbiologist the transport of a cargo along a microtubule, what everyday analogy would be most suitable? ("It is like . . .")

4. In *C. elegans*, how many cell divisions of the zygote are necessary to give rise to the cell that will become the precursor to the germ line? Why this number?

5. A rough translation of the word syncitium is "cells together." Is it a good word to use to describe the *Drosophila* syncitium?

6. Draw a graph that represents a *Drosophila* syncitium with two opposing pole-high gradients.

7. Describe the practical details of an experiment in which you would create and demonstrate a gradient of a molecule in solution.

8. Describe briefly the experimental way that the *bicoid* gradient in Figure 18-10a was demonstrated.

9. If you swapped the 5′ UTRs of *bicoid* and *nanos*, what might you predict would be the effect on their gradients?

10. For what purpose is the cephalic furrow being used in the experiment in Figure 18-13?

11. In which part of the cells is the DL protein found in the dorsal region of the *Drosophila* blastoderm?

12. What is the prerequisite for the DL protein to enter a nucleus?

13. *Gooseberry, runt, knirps,* and *antennapedia*. To a *Drosophila* geneticist what are they? How do they differ?

14. Describe the expression pattern of the *Drosophila* gene *eve* in the late blastoderm.

15. Contrast the function of homeotic genes with pair-rule genes.

16. What do geneticists mean when they say that cells "communicate"?

17. What is the difference between the cell communication action of the anchor cell and that of the primary cell in the development of the *C. elegans* vulva?

18. What is the "ancestor" of the Hox gene B6 in the *Drosophila* HOM set?

19. In the human maleness-determining system, what do you think regulates transcription of *SRY* (if anything)?

20. In *Arabidopsis* flowers, what would be the flower whorls in mutants that had no B gene transcripts at all?

21. XYY humans are fertile males. XXX humans are fertile females. What do these observations reveal about the mechanisms of sex determination and dosage compensation?

22. Occasionally, there are humans who are mosaics of XX and XY tissue. They generally exhibit a uniform sexual phenotype. Some of them are phenotypically female, others male. Explain these observations in regard to the mechanism of sex determination in mammals.

23. How are the gradients of BCD and HB-M established during early embryogenesis in *Drosophila*? What is the role of the cytoskeleton in this process?

24. What are the similarities between DL/CACT in *Drosophila* and Rb/E2F (see Chapter 17)?

25. In *C. elegans* vulva development, one anchor cell in the gonad interacts with six equivalence-group cells (cells with the potential to become parts of the vulva). The six equivalence-group cells have three distinct phenotypic fates: primary, secondary, and tertiary. The equivalence-group cell closest to the anchor cell develops the primary vulva phenotype. If the anchor cell is ablated, all six equivalence-group lineages differentiate into the tertiary state.

    a. Set up a model to explain these results.

    b. The anchor cell and the six equivalence-group cells can be isolated and grown in vitro; design an experiment to test your model.

26. There are two types of muscle cell in *C. elegans*: pharyngeal muscles and body-wall muscles. They can be distinguished from each other, even as single cells. In the following diagram, letters designate particular muscle precursor cells, black triangles (▲) are pharyngeal muscles, and white triangles (△) are body-wall muscles.

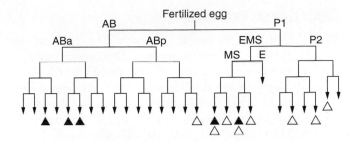

a. It is possible to move these cells around during development. When the positions of ABa and ABp are physically interchanged, the cells develop according to their new position. In other words, the cell that was originally in the ABa position now gives rise to one body-wall muscle cell, whereas the cell that was originally in the ABp position now gives rise to three pharyngeal muscle cells. What does that tell us about the developmental processes controlling the fates of ABa and ABp?

b. If EMS is ablated (by heat inactivation with a laser beam aimed through a microscope lens), no AB descendants make muscles. What does that suggest?

c. If P2 is ablated, all AB descendants turn into muscle cells. What does that suggest?

(Diagram from J. Priess and N. Thomson, *Cell* 48, 1987, 241.)

27. When an embryo is homozygous mutant for the gap gene *Kr*, the fourth and fifth stripes of the pair-rule gene *ftz* (counting from the anterior end) do not form normally. When the gap gene *knl* is mutant, the fifth and sixth *ftz* stripes do not form normally. Explain these results in regard to how segment number is established in the embryo.

28. The *Drosophila* embryo has a polarity that is developed through the action of maternal genes that are expressed in the developing egg follicle in the ovary. Mothers homozygous for a *nanos* mutation produce embryos that lack posterior segments. However, these embryos do not display a mirror-image anterior (bicephalic) pattern. In contrast, mothers homozygous for a *bicoid* mutation produce embryos that not only lack anterior segments but also display a mirror-image posterior (bicaudal) pattern. Explain these observations in relation to the roles of *nanos*, *bicoid*, and *hunchback* in anterior–posterior axis formation.

29. For many of the mammalian Hox genes, some of them have been shown to be more similar to one of the insect HOM-C genes than to the others. Describe an experimental approach by using the tools of molecular biology that would enable you to demonstrate this.

CHALLENGING PROBLEMS

30. **a.** When you remove the anterior 20 percent of the cytoplasm of a newly formed *Drosophila* embryo, you can cause a bicaudal phenotype, in which there is a mirror-image duplication of the abdominal segments. In other words, from the anterior tip of the embryo to the posterior tip, the order of segments is A8-A7-A6-A5-A4-A4-A5-A6-A7-A8. Explain this phenotype in regard to the action of the anterior and posterior determinants and how they affect gap-gene expression.

b. Females homozygous for the maternally acting mutation *nanos (nos)* produce embryos in which the abdominal segments are absent and in which the head and thoracic segments are broad. In regard to the action of the anterior and posterior determinants and gap-gene action, explain how *nos* produces this mutant phenotype. In your answer, explain why there is a loss of segments rather than a mirror-image duplication of anterior segments.

31. The three homeodomain proteins ABD-B, ABD-A, and UBX are encoded by genes within the BX-C of *Drosophila*. In wild-type embryos, the *Abd-B* gene is expressed in the posterior abdominal segments, *Abd-A* in the middle abdominal segments, and *Ubx* in the anterior abdominal and posterior thoracic segments. When the *Abd-B* gene is deleted, *Abd-A* is expressed in both the middle and the posterior abdominal segments. When *Abd-A* is deleted, *Ubx* is expressed in the posterior thorax and in the anterior and middle abdominal segments. When *Ubx* is deleted, the patterns of *Abd-A* and *Abd-B* expression are unchanged from wild type. When both *Abd-A* and *Abd-B* are deleted, *Ubx* is expressed in all segments from the posterior thorax to the posterior end of the embryo. Explain these observations, taking into consideration the fact that the gap genes control the initial expression patterns of the homeotic genes.

32. In considering the formation of the A–P and D–V axes in *Drosophila*, we noted that, for mutations such as *bcd*, homozygous mutant mothers uniformly produce mutant offspring with segmentation defects. This outcome is always true regardless of whether the offspring themselves are *bcd*/*bcd* or *bcd*/*bcd*. Some other maternal-effect lethal mutations are different, in that the mutant phenotype can be "rescued" by introducing a wild-type allele of the gene from the father. In other words, for such rescuable maternal-effect lethals, *mut⁺*/*mut* animals are normal, whereas *mut*/*mut* animals have the mutant defect. Rationalize the difference between rescuable and nonrescuable maternal-effect lethal mutations.

33. In the *Drosophila* embryo, the 3′ untranslated regions (3′ UTRs) of the mRNAs [the regions between the translation-termination codons and the poly(A) tails] are responsible for localizing *bcd* and *nos* to the anterior and posterior poles, respectively. Experiments have been done in which the 3′ UTRs of *bcd* and *nos* have been swapped. Suppose that we make P-element transformation constructs with both swaps (*nos* mRNA with *bcd* 3′ UTR and *bcd* mRNA with *nos* 3′ UTR) and transform them into the *Drosophila* genome. We then make a female that is homozygous mutant for *bcd* and *nos* and carries both swap constructs. What phenotype would you expect for her embryos in regard to A–P axis development?

34. **a.** If you had a mutation affecting anterior–posterior patterning of the *Drosophila* embryo in which every other segment of the developing mutant larva was missing, would you consider it a mutation in a gap gene, a pair-rule gene, a segment-polarity gene, or a segment-identity gene?

b. You have cloned a piece of DNA that contains four genes. How could you use the spatial-expression pattern of their mRNA in a wild-type embryo to identify which represents a candidate gene for the mutation described in part a?

c. Assume that you have identified the candidate gene. If you now examine the wild-type spatial-expression pattern of its mRNA in an embryo homozygous mutant for the gap gene *Krüppel*, would you expect to see a normal expression pattern? Explain.

35. You have in your possession wild-type and *bicoid* mutant strains. You also have cloned cDNAs for *nanos* and *bicoid*. These plasmids are as follows:

plasmid 1	full-length *nanos* cDNA
plasmid 2	full-length *bicoid* cDNA
plasmid 3	*bicoid* 5′ UTR–*bicoid* ORF–*nanos* 3′ UTR
plasmid 4	*nanos* 5′ UTR-*bicoid* ORF–*bicoid* 3′ UTR
plasmid 5	*nanos* 5′ UTR-*bicoid* ORF–*nanos* 3′ UTR

where UTR = untranslated region and ORF = open reading frame (protein-coding region).

The plasmids are constructed so that you can generate a synthetic mRNA corresponding to the sequences described in each cDNA. Describe how you could use these mRNAs to determine that *bicoid* mRNA localization is due to its 3′ UTR.

36. What *Arabidopsis* genotypes would have
a. only carpels?
b. no carpels?

19

POPULATION GENETICS

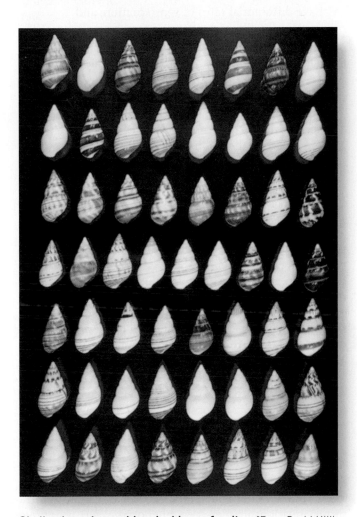

Shell-color polymorphism in *Liguus fascitus*. [From David Hillis, *Journal of Heredity*, July–August 1991.]

KEY QUESTIONS

- How much genetic variation is there in natural populations of organisms?

- What are the effects of patterns of mating on genetic variation?

- What are the sources of the genetic variation that is observed in populations?

- What are the processes that cause changes in the kind and amount of genetic variation in populations?

OUTLINE

19.1 Variation and its modulation
19.2 Effect of sexual reproduction on variation
19.3 Sources of variation
19.4 Selection
19.5 Balanced polymorphism
19.6 Random events

CHAPTER OVERVIEW

So far in our investigation of genetics, we have been concerned with processes that take place in individual organisms and cells. How does the cell copy DNA, and what causes mutations? How do the mechanisms of segregation and recombination affect the kinds and proportions of gametes produced by an individual organism? How is the development of an organism affected by the interactions between its DNA, the cell machinery of protein synthesis, cellular metabolic processes, and the external environment? But organisms do not live only as isolated individuals. They interact with one another in groups, **populations,** and there are questions about the genetic composition of those populations that cannot be answered only from a knowledge of the basic individual-level genetic processes. Why are the alleles of the protein Factor VIII and Factor IX genes that cause a failure of normal blood clotting, hemophilia, so rare in all human populations, whereas the allele of the hemoglobin β gene that causes sickle-cell anemia is very common in some parts of Africa? What changes in the frequency of sickle-anemia are to be expected in the descendants of Africans in North America as a consequence of the change in environment and of the interbreeding between Africans and Europeans and Native Americans? What genetic changes occur in a population of insects subject to insecticides generation after generation? What is the consequence of an increase or decrease in the rate of mating between close relatives? All are questions of what determines the genetic composition of populations and how that composition may be expected to change in time. These questions are the domain of **population genetics.**

> **MESSAGE** Population genetics relates the processes of individual heredity and development to the genetic composition of populations and to changes in that composition over time and space.

The genetic composition of a population is the collection of frequencies of different genotypes in the population. These frequencies are the consequence of processes that act at the level of individual organisms to increase or decrease the number of organisms of each genotype. To relate the basic individual-level genetic processes to population genetic composition, we must investigate the following phenomena (Figure 19-1):

1. The effect of *mating patterns* on different genotypes in the population. Individuals may mate at random, or they may mate preferentially with close relatives (*inbreeding*) or preferentially with individuals of similar or dissimilar genotype or phenotype (*assortative mating*).

2. The changes in population composition due to *migration* of individuals between populations.

3. The rate of introduction of new genetic variation into the population by *mutation*, which brings in new alleles at loci.

4. The production of new combinations of characters by *recombination*, which reassorts combinations of alleles at different loci.

5. The changes in population composition due to the effect of *natural selection*. Different genotypes may have differential rates of reproduction, and genetically different offspring may have differential chances of survival.

6. The consequences of *random fluctuations* in the actual reproductive rates of different genotypes. Because any given individual has only a few offspring and the total population size is limited, genetic ratios from meiosis are never exactly as predicted by theory in real families and real populations. This random fluctuation causes *genetic drift* in allele frequencies from generation to generation.

19.1 Variation and its modulation

Population genetics is both an experimental and a theoretical science. On the experimental side, it provides descriptions of the actual patterns of genetic variation among individuals in populations and estimates the rates of the processes of mating, mutation, recombination, natural selection, and random variation in reproductive rates. On the theoretical side, it makes predictions of how the genetic composition of populations can be expected to change as a consequence of the various forces operating on them.

Observations of variation

Studies of population genetics have been able to explore only limited sets of characters, because of the need for a simple relationship between phenotypic and genotypic variation. The relation between phenotype and genotype varies in simplicity depending on the character that is observed. At one extreme, the phenotype of interest may be the mRNA or polypeptide encoded by a stretch of the genome. At the other extreme lie the bulk of characters of interest to plant and animal breeders and to most evolutionists—the variations in yield, growth rate, body shape, metabolic rate, and behavior that constitute the obvious differences between varieties and species. These characters have a very complex relation to genotype. There is no allele for being 5'8" or being 5'4" tall. Such differences, if they are a consequence of

CHAPTER OVERVIEW Figure

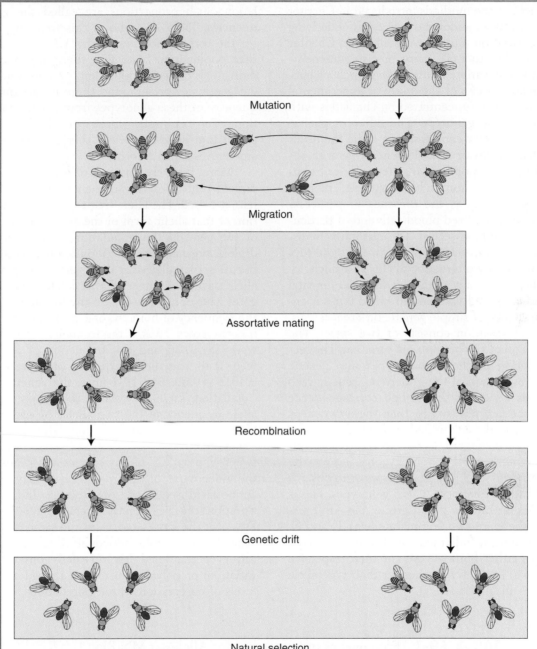

Mutation

Migration

Assortative mating

Recombination

Genetic drift

Natural selection

Figure 19-1 Overview of the phenomena that cause genetic change in populations. We begin with two phenotypically uniform populations, into which variation is introduced by mutation. Migration between populations then introduces these separate mutations into both populations. Assortative mating may occur between individuals of different phenotypes or the same phenotype. Recombination in the offspring of these matings leads to new combinations of characters that were previously found in different individuals. Genetic drift due to random sampling of gametes changes the frequencies of the genotypes and causes some divergence between the two populations in the frequencies of their genotypes and phenotypes. Natural selection then causes the populations to diverge even more.

genetic variation, will be affected by several or many genes and by environmental variation as well. For their analysis we must use the methods introduced in Chapter 22 to say anything at all about the genotypes underlying the phenotypic variation. But, as we shall see in Chapter 22, it is not possible to make very precise statements about the genotypic variation underlying such characters. For that reason, most of the study of experimental population genetics has concentrated on characters with simple relations to the genotype. The different phenotypes for such a character can be shown to be the result of different allelic forms of a single gene. A favorite object of study for human population geneticists, for example, has been the various human blood groups. The phenotype of a blood group is the presence of particular antigens on the surface of red blood cells and of particular antibodies in the blood serum. The qualitatively distinct phenotypes of a given blood group—say, the MN group—are encoded by alternative alleles at a single locus, and the phenotypes are insensitive to environmental variations. Thus, observed variation in blood types is entirely the consequence of simple genetic differences.

The study of variation consists of two stages. The first is a description of the phenotypic variation. The second is a translation of these phenotypes into genetic terms and the description of the underlying genetic variation. If there is a perfect one-to-one correspondence between genotype and phenotype, then these two steps merge into one, as in the MN blood group. If the relation is more complex—for example, as the result of dominance, heterozygotes resemble homozygotes—it may be necessary to carry out experimental crosses or to observe pedigrees to translate phenotypes into genotypes. This is the case for another human blood group, the ABO system, for which there are two dominant alleles, I^A and I^B, and a recessive allele, i. Individuals with type A or type B blood may be either homozygous for their respective alleles ($I^A I^A$ or $I^B I^B$) or heterozygous for their type allele and the recessive allele ($I^A i$ or $I^B i$).

The simplest description of single-gene variation is the list of observed proportions of genotypes in a population. Such proportions are called the **genotype frequencies.** Table 19-1 shows this frequency distribution for the three genotypes of the *MN* gene in several human populations. Note that there is variation between individuals in each population, because there are different genotypes present, and there is variation in the frequencies of these genotypes from population to population. For example, most people in the Eskimo population are MM, while this genotype is quite rare among Australian Aborigines.

More typically, instead of the frequencies of the diploid genotypes, the frequencies of the alternative alleles are used. The allele frequency is simply the proportion of that allelic form of the gene among all the copies of the gene in the population, where each individual diploid organism in the population is counted as contributing two alleles for each gene. Homozygotes for an allele have two copies of that allele, whereas heterozygotes have only one copy. So the frequency of an allele is the frequency of homozygotes plus half the frequency of heterozygotes. Thus, if the frequency of *A/A* individuals were, say, 0.36 and the frequency of *A/a* individuals were 0.48, the allele frequency of *A* would be 0.36 + 0.48/2 = 0.60. Box 19-1 gives the general form of this calculation. Table 19-1 shows the values of *p* and *q*, the **allele frequency** of the two alleles M and N of the MN blood group in the different populations.

Simple variation can be observed within and between populations of any species at various levels, from the phenotype of external morphology down to the amino acid sequences of proteins. Indeed, genotypic variation can be directly characterized by sequencing DNA for the same gene or for the same regions between genes from multiple individuals. Every species of organism ever examined has revealed considerable genetic variation, or **polymorphism,** manifested at one or more levels of observation within populations, between popu-

TABLE 19-1   Frequencies of Genotypes for Alleles at MN Blood Group Locus in Various Human Populations

Population	Genotype			Allele frequencies	
	M/M	*M/N*	*N/N*	*p(M)*	*q(N)*
Eskimo	0.835	0.156	0.009	0.913	0.087
Australian Aborigine	0.024	0.304	0.672	0.176	0.824
Egyptian	0.278	0.489	0.233	0.523	0.477
German	0.297	0.507	0.196	0.550	0.450
Chinese	0.332	0.486	0.182	0.575	0.425
Nigerian	0.301	0.495	0.204	0.548	0.452

Source: W. C. Boyd, *Genetics and the Races of Man.* D. C. Heath, 1950.

BOX 19-1   Calculation of Allele Frequencies

If $f_{A/A}$, $f_{A/a}$, and $f_{a/a}$ are the frequencies of the three genotypes at a locus with two alleles, then the frequency p of the A allele and the frequency q of the a allele are obtained by counting alleles. Because each homozygote A/A consists only of A alleles, and because half the alleles of each heterozygote A/a are A alleles, the total frequency p of A alleles in the population is calculated as

$$p = f_{A/A} + \tfrac{1}{2} f_{A/a} = \text{frequency of } A$$

Similarly, the frequency q of the a allele is given by

$$q = f_{a/a} + \tfrac{1}{2} f_{A/a} = \text{frequency of } a$$

Therefore

$$p + q = f_{A/A} + f_{a/a} + f_{A/a} = 1.00$$

and

$$q = 1 - p$$

If there are more than two different allelic forms, the frequency for each allele is simply the frequency of its homozygote plus half the sum of the frequencies for all the heterozygotes in which it appears.

---

lations, or both. A gene or a phenotypic trait is said to be *polymorphic* if there is more than one form of the gene or more than one phenotype for that character in a population. In some cases nearly the entire population is characterized by one form of the gene or character, with rare exceptional individuals carrying an unusual variant. That extremely common form is called the **wild type,** in contrast to the rare mutants. In other cases two or more forms are common, and it is not possible to pick out one that is the wild type. Genetic variation that might be the basis for evolutionary change is ubiquitous.

It is impossible in this text to provide an adequate picture of the immense richness of even simple genetic variation that exists in species. We can consider only a few examples of the different kinds to gain a sense of the genetic diversity within species. Each of these examples can be multiplied many times over in other species and with other characters.

Protein polymorphisms

IMMUNOLOGIC POLYMORPHISM A number of loci in vertebrates encode antigenic specificities such as the ABO blood types. More than 40 different specificities for antigens on human red cells are known, and several hundred are known in domesticated cattle. Another major polymorphism in humans is the HLA system of cellular antigens, which are implicated in tissue graft compatibility. Table 19-2 gives the allelic frequencies for the ABO blood group locus in some very different human populations. The polymorphism for the HLA system is vastly greater. There appear to be two main loci, each with five distinguishable alleles. Thus, there are $5^2 = 25$ different possible gametic types, making 25 different homozygous forms and $(25)(24)/2 = 300$ different heterozygotes. Not all genotypes are phenotypically distin-

guishable, however, so only 121 phenotypic classes can be seen. Remarkably, in one study of a sample of only 100 Europeans, 53 of the 121 possible phenotypes were actually observed.

AMINO ACID SEQUENCE POLYMORPHISM Studies of genetic polymorphism have been carried down to the level of the polypeptides encoded by the coding regions of the genes themselves. If there is a nonsynonymous codon change in a gene (say, GGU to GAU), the result is an amino acid substitution in the polypeptide produced at translation (in this case, aspartic acid is substituted for glycine). Variation in amino acid sequence of a protein can be detected by sequencing the DNA that codes for the protein from a large number of individuals. This is the method that would be used if one wished to know exactly which amino acids in the protein sequence were varying, but it is extremely time-consuming and expensive to carry out such DNA sequencing projects for many different protein coding genes. There is, however, a

TABLE 19-2 Frequencies of the Alleles I^A, I^B, and i at the ABO Blood Group Locus in Various Human Populations

Population	I^A	I^B	i
Eskimo	0.333	0.026	0.641
Sioux	0.035	0.010	0.955
Belgian	0.257	0.058	0.684
Japanese	0.279	0.172	0.549
Pygmy	0.227	0.219	0.554

Source: W. C. Boyd, *Genetics and the Races of Man.* D. C. Heath, 1950.

practical substitute for DNA sequencing that can be used if one is interested only in detecting variant forms of a protein without knowing the particular amino acid changes involved. This method makes use of the change in the physical properties of a protein when one amino acid is substituted for another. Proteins carry a net charge that is the result of the ionization of side chains on five amino acids (glutamic acid, aspartic acid, arginine, lysine, and histidine). Amino acid substitutions may directly replace one of these charged amino acids, or a noncharged substitution near a charged amino acid may alter the degree of ionization of the charged amino acid, or a substitution at the joining between two α helices may cause a slight shift in the three-dimensional packing of the folded polypeptide. In all these cases, the net charge on the polypeptide will be altered.

The change in net charge on the protein can be detected by gel electrophoresis. Figure 19-2 shows the outcome of such an electrophoretic separation. The tracks represent variants of an esterase enzyme in *Drosophila pseudoobscura*, where each track is the protein from a different individual fly. Figure 19-3 shows a similar gel for variant human hemoglobins. In this case, most individuals are heterozygous for a variant and normal hemoglobin A. Table 19-3 shows the frequencies of different alleles for three enzyme-encoding genes in *D. pseudoobscura* in several populations: a nearly monomorphic locus (malic dehydrogenase), a moderately polymorphic locus (α-amylase), and a highly polymorphic locus (xanthine dehydrogenase).

The technique of gel electrophoresis of proteins (as well as DNA sequencing) differs fundamentally from other methods of genetic analysis in allowing the study of genes that are not actually varying in a population,

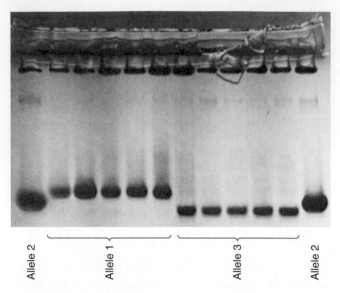

Figure 19-2 **Electrophoretic gel of the proteins encoded by homozygotes for three different alleles at the *esterase-5* locus in *Drosophila pseudoobscura*.** Each lane is the protein from a different individual fly. Repeated samples of proteins encoded by the same allele are identical, but there are repeatable differences between alleles.

because the presence of a protein is prima facie evidence of the DNA sequence encoding the protein. Thus, it has been possible to ask what proportion of all structural genes in the genome of a species are polymorphic and what average fraction of an individual's genome is in heterozygous state, the *heterozygosity*, in a population. Very large numbers of species have been sampled by this method, including bacteria, fungi, higher plants, verte-

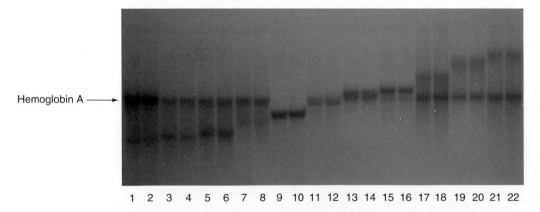

Figure 19-3 **Electrophoretic gel showing normal hemoglobin A and a number of variant hemoglobin alleles.** Each lane represents a different individual. One of the dark-staining bands is marked as normal hemoglobin A. The other dark-staining band seen in most of the lanes (most clearly in lanes 3 and 4) represents any of several variant hemoglobins derived from the second allele of a heterozygote. Hemoglobin A is missing from lanes 9 and 10 because the individuals are homozygotes for a variant allele [Richard C. Lewontin.]

TABLE 19-3 Frequencies of Various Alleles at Three Enzyme-Encoding Loci in Four Populations of *Drosophila pseudoobscura*

Locus (enzyme-encoding)	Allele	Population			
		Berkeley	Mesa Verde	Austin	Bogotá
Malic dehydrogenase	*A*	0.969	0.948	0.957	1.00
	B	0.031	0.052	0.043	0.00
α-Amylase	*A*	0.030	0.000	0.000	0.00
	B	0.290	0.211	0.125	1.00
	C	0.680	0.789	0.875	0.00
Xanthine dehydrogenase	*A*	0.053	0.016	0.018	0.00
	B	0.074	0.073	0.036	0.00
	C	0.263	0.300	0.232	0.00
	D	0.600	0.581	0.661	1.00
	E	0.010	0.030	0.053	0.00

Source: R. C. Lewontin, *The Genetic Basis of Evolutionary Change.* Columbia University Press, 1974.

brates, and invertebrates. The results are remarkably consistent over species. About one-third of genes coding for proteins are detectably polymorphic at the protein level, and the average heterozygosity in a population over all loci sampled is about 10 percent. This means that scanning the genome in virtually any species would show that about 1 in every 10 genes in an individual is in heterozygous condition for genetic variations that are reflected in the amino acid sequence of the proteins, and about one-third of all genes have two or more such alleles segregating in any population. Thus the potential of variation for evolution is immense. The disadvantage of the electrophoretic technique is that it detects variation only in protein-coding regions of genes and misses the important changes in regulatory elements that underlie much of evolution of form and function.

DNA structure and sequence polymorphism

DNA analysis makes it possible to examine variation in genome structure among individuals and between species. There are three levels at which such studies can be done. Examining variation in chromosome number and morphology provides a large-scale view of reorganizations of the genome. Studying variation in the sites recognized by restriction enzymes provides a coarse view of base-pair variation. At the finest level, methods of DNA sequencing allow variation to be observed base pair by base pair.

CHROMOSOMAL POLYMORPHISM Although the karyotype is often regarded as a distinctive characteristic of a species, in fact, numerous species are polymorphic for chromosome number and morphology. Extra chromosomes (supernumeraries), reciprocal translocations, and

inversions are observed in many populations of plants, insects, and even mammals. Figure 19-4 shows a variety of inversion loops found in natural populations of *Drosophila pseudoobscura*. Each such loop is the result of carrying two homologous chromosomes, one of which contains a section that is in reverse linear order with respect to the other chromosome (see Chapter 15). Table 19-4 gives the frequencies of supernumerary chromosomes and translocation heterozygotes in a population of the plant *Clarkia elegans* from California. The "typical" species karyotype would be hard to identify in this plant in which only 56 percent of the individuals lack supernumerary chromosomes and translocations.

RESTRICTION-SITE VARIATION An inexpensive and rapid way to observe overall levels of variation in DNA sequences is to digest that DNA using restriction enzymes (see Chapter 11). There are many different restriction enzymes, each of which will recognize a different base sequence and cut the DNA at the site of that sequence. The result will be two DNA fragments whose lengths are determined by the location of the restriction site in the original uncut molecule. A restriction enzyme that recognizes six-base sequences (a "six cutter") will recognize an appropriate sequence approximately once every $4^6 = 4096$ base pairs along a DNA molecule [determined from the probability that a specific base (of which there are four) will be found at each of the six positions]. If there is polymorphism in the population for one of the six bases at the recognition site, then the enzyme will recognize and cut the DNA in one variant and not in the other (see Chapter 11). Thus there will be a restriction fragment length polymorphism (RFLP) in the population. A panel of, say, eight different six-cutter enzymes will then sample every $4096/8 \cong 500$

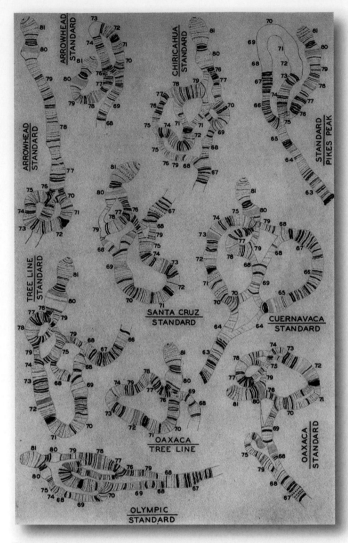

Figure 19-4 Inversion polymorphism of chromosome 3 in natural populations of *Drosophila pseudoobscura*. The inversions are named for the locality at which they were first observed. Pairings of different gene orders show loops in the polytene chromosomes, revealing the location of the breakpoints of the inversions. [T. Dobzhansky, *Chromosomal Races in Drosophila pseudoobscura and Drosophila persimilis*. Carnegie Institution of Washington, 47–144, 1944.]

base pairs for such polymorphisms. However, when one is found, we do not know which of the six base pairs at the recognition site is polymorphic.

If we use enzymes that recognize four-base sequences ("four cutters"), there is a recognition site every $4^4 = 256$ base pairs, so a panel of eight different enzymes can sample about once every $256/8 = 32$ base pairs along the DNA sequence. In addition to single base-pair changes that destroy restriction-enzyme recognition sites, there are insertions and deletions of stretches of DNA that occur along the DNA strand between the locations of restriction sites, and these will also cause restriction fragment lengths to vary.

A variety of different restriction-enzyme studies has been performed for different regions of the X chromosome and the two large autosomes of *Drosophila melanogaster*. These have found between 0.1 and 1.0 percent heterozygosity per nucleotide site, with an average of 0.4 percent. The result of one such study on the xanthine dehydrogenase gene in *Drosophila pseudoobscura* is shown in Figure 19-5. The figure shows, symbolically, the restriction pattern of 58 chromosomes sampled from nature, polymorphic at 78 restriction sites along a sequence 4.5 kb in length. Remarkably, among the 58 patterns there are 53 different ones. (Try to find the identical pairs.)

TANDEM REPEATS Restriction fragment surveys can reveal another form of DNA sequence variation, which arises from the occurrence of multiply repeated DNA sequences. In the human genome, there are a variety of different short DNA sequences dispersed throughout the genome, each one of which is multiply repeated in a row (in *tandem*). The number of repeats may vary from a dozen to more than 100 in different individual genomes. Such sequences are known as **variable number tandem repeats (VNTRs).** If the restriction enzymes cut sequences that flank either side of such a tandem array, a fragment will be produced whose size is proportional to the number of repeated elements. The different-sized

TABLE 19-4 Frequencies of Plants with Supernumerary Chromosomes and of Translocation Heterozygotes in a Population of *Clarkia elegans* from California

No supernumeraries or translocations	Translocations	Supernumeraries	Both translocations and supernumeraries
0.560	0.133	0.265	0.042

Source: H. Lewis, *Evolution* 5, 1951, 142–157.

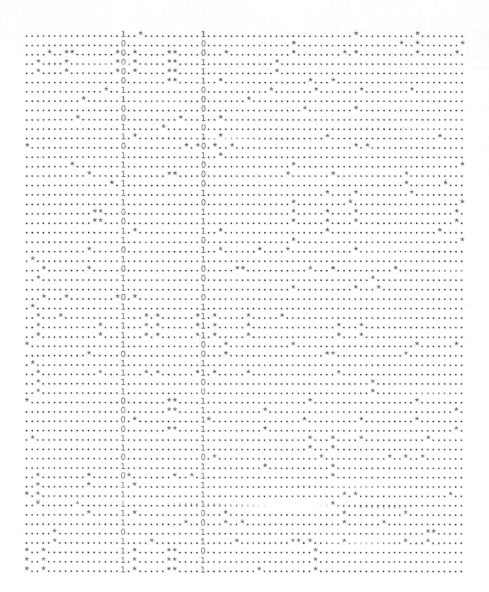

Figure 19-5 The result of a four-cutter survey of 58 chromosomes, probed for the xanthine dehydrogenase gene in _Drosophila pseudoobscura._ Each line is a chromosome (haplotype) sampled from a natural population. Each position along the line is a polymorphic restriction rate along the 4.5-kb sequence studied. Where an asterisk appears, the haplotype differs from the majority, either cutting where most haplotypes are not cut or not cutting where most haplotypes are cut. At two sites, there is no clear majority type, so a 0 or 1 is used to show whether the site is absent or present.

fragments will migrate at different rates in an electrophoretic gel. The individual copies of the repeated sequence elements are too short to allow distinguishing between, say, 64 and 68 repeats, but size classes that include several repeat numbers (_bins_) can be established, and a population can be assayed for the frequencies of the different classes. Table 19-5 shows the data for two different VNTRs sampled in two American Indian groups from Brazil. In one case, D14S1, the Karitiana are nearly homozygous, whereas the Surui are very variable; in the other case, D14S13, both populations are variable but with different frequency patterns.

COMPLETE SEQUENCE VARIATION A ubiquitous form of genetic variation is variation in the nucleotide at a single position, called a **single-nucleotide polymorphism (SNP).** Studies of variation at the level of single base

pairs by DNA sequencing can provide information of two kinds. First, DNA sequence variation can be studied in the protein-coding regions of genes. The sequences of coding regions can be translated to reveal the exact amino acid sequence differences in proteins from different individuals in a population or from different species. DNA sequencing is superior in precision to electrophoretic studies of a protein from different individuals, which can show that there is variation in amino acid sequences but cannot identify how many or which amino acids differ between individuals. So, when DNA sequences were obtained for the various electrophoretic variants of esterase-5 in _Drosophila pseudoobscura_ (see Figure 19-2), electrophoretic variants were found to differ from one another by an average of 8 amino acids, and the 20 different kinds of amino acids were involved in polymorphisms at about the frequency that they were

TABLE 19-5   Size Class Frequencies for Two Different VNTR Sequences, D14S1 and D14S13, in the Karitiana and Surui of Brazil

Size class	D14S1		D14S13	
	Karitiana	Surui	Karitiana	Surui
3–4	105	41	0	0
4–5	0	3	3	14
5–6	0	11	1	4
6–7	0	2	1	2
7–8	0	1	1	2
8–9	3	3	8	16
9–10	0	11	28	9
10–11	0	2	22	0
11–12	0	4	18	8
12–13	0	0	13	18
13–14	0	0	13	3
> 14	0	0	0	2
	108	78	108	78

Source: Data from J. Kidd and K. Kidd, *American Journal of Physical Anthropology* 81, 1992, 249.

present in the protein. Such studies also show that different regions of the same protein have different amounts of polymorphism. For the esterase-5 protein, consisting of 545 amino acids, 7 percent of amino acid positions are polymorphic, but the last amino acids at the carboxyl terminus of the protein are totally invariant between individuals, probably because these amino acids are needed for the protein to function properly.

Second, DNA sequence variation can also be studied in those base pairs that do *not* determine or change the protein sequence. Such base-pair variation can be found in DNA in 5′ flanking sequences that may be regulatory. The importance of studying variation in regulatory sequences cannot be overemphasized. It has been suggested that most of the evolution of shape, physiology, and behavior rests on changes in regulatory sequences. If that is true, then much of the sequence variation in coding regions and in the amino acid sequences for which they code is beside the point. There is also variation in introns, in nontranscribed DNA 3′ to the gene, and in those nucleotide positions in codons (usually third positions) whose variation does not result in amino acid substitutions. These so-called *silent* or *synonymous* base-pair polymorphisms are much more common than are changes that result in amino acid polymorphism, presumably because many amino acid changes interfere with normal function of the protein and are eliminated by natural selection.

An examination of the codon translation table (see Figure 9-8) shows that approximately 25 percent of all random base-pair changes would be synonymous, giving an alternative codon for the same amino acid, whereas 75 percent of random changes would change the amino acid coded. For example, a change from AAT to AAC still encodes asparagine, but a change to ATT, ACT, AAA, AAG, AGT, TAT, CAT, or GAT, all single-base-pair changes from AAT, changes the amino acid encoded. So, if mutations of base pairs are at random and if the substitution of an amino acid made no difference to function, we would expect a 3:1 ratio of amino acid replacement to silent polymorphisms. The actual ratios found in *Drosophila* vary from 2:1 to 1:10. Clearly, there is a great excess of synonymous polymorphism, showing that most amino acid changes make a difference to function and therefore are subject to natural selection. It should not be assumed, however, that silent sites in coding sequences are entirely free from constraints. Different alternative triplet codings for the same amino acid may differ in speed and accuracy of transcription, and the mRNA corresponding to different alternative triplets may have different accuracy and speed of translation because of limitations on the pool of tRNAs available. Evidence for the latter effect is that alternative synonymous triplets for an amino acid are not used equally, and the inequality of use is much more pronounced for genes that are transcribed at a very high rate.

There are also constraints on 5′ and 3′ noncoding sequences and on intron sequences. Both 5′ and 3′ noncoding DNA sequences contain signals for transcription, and introns may contain enhancers of transcription (see Chapter 10).

19.2 Effect of sexual reproduction on variation

Meiotic segregation and genetic equilibrium

If inheritance were based on a continuous substance like blood, then the mating of individuals with different phenotypes would produce offspring that were intermediate in phenotype. When these intermediate types mated with each other, their offspring would again be intermediate. A population in which individuals mated at random would slowly lose all its variation, and eventually every member of the population would have the same phenotype.

The particulate nature of inheritance changes this picture completely. Because of the discrete nature of genes and the segregation of alleles at meiosis, a cross of intermediate with intermediate individuals does *not* result in all intermediate offspring. On the contrary, some of the offspring will be of extreme types—those that are homozygotes. Consider a population in which males and females mate with one another at random with respect to some gene locus A; that is, the genotype at that locus is not a factor in choosing a mate. Such random mating is equivalent to mixing all the sperm and all the eggs in the population together and then matching randomly drawn sperm with randomly drawn eggs.

The outcome of such a random pairing of sperm and eggs is easy to calculate. If, in some population, the allele frequency of A is 0.60 in both sperm and eggs, then the chance that a randomly chosen sperm and a randomly chosen egg are both A is 0.60 × 0.60 = 0.36. Thus, in a random-mating population with this allele frequency, 36 percent of offspring will be A/A. In the same way, the frequency of a/a offspring will be 0.40 × 0.40 = 0.16. Heterozygotes will be produced by the fusion either of an A sperm with an a egg or of an a sperm with an A egg. If gametes pair at random, then the chance of an A sperm and an a egg is 0.60 × 0.40, and the reverse combination has the same probability, so the frequency of heterozygous offspring is 2 × 0.60 × 0.40 = 0.48.

We can now understand why variation is retained in a population. The process of random mating has done nothing to change *allele* frequencies, as can be easily checked by calculating the frequencies of the alleles A and a among the offspring in this example, using the method described in Box 19-1. So the proportions of homozygotes and heterozygotes in each successive generation will remain the same. These constant frequencies form the **equilibrium distribution**. Box 19-2 gives a general form of this equilibrium result.

The equilibrium distribution can be calculated according to the formula

$$\begin{array}{ccc} A/A & A/a & a/a \\ p^2 & 2pq & q^2 \end{array}$$

where p is the frequency of the A allele, q is the frequency of the a allele, and $p + q = 1$.

This distribution is called the **Hardy-Weinberg equilibrium** after G. H. Hardy and W. Weinberg, the two people who independently discovered it. (A third independent discovery was made by the Russian geneticist Sergei Chetverikov.)

The Hardy-Weinberg equilibrium means that sexual reproduction does not cause a constant reduction in genetic variation in each generation; on the contrary, the amount of variation remains constant generation after generation, in the absence of other disturbing forces. The equilibrium is the direct consequence of the segregation of alleles at meiosis in heterozygotes.

Numerically, the equilibrium shows that, irrespective of the particular mixture of genotypes in the parental generation, the genotypic distribution after one round of random mating is completely specified by the allelic frequency p. For example, consider three hypothetical populations that have arisen from the mixing of migrants from different sources:

	$f(A/A)$	$f(A/a)$	$f(a/a)$
I	0.3	0.0	0.7
II	0.2	0.2	0.6
III	0.1	0.4	0.5

The allele frequency p of A in the three populations is

$$\begin{array}{ll} \text{I} & p = f(A/A) + f(A/a) = 0.3 + 1/2(0) \quad = 0.3 \\ \text{II} & p = 0.2 \qquad\qquad\qquad\qquad\quad + 1/2(0.2) = 0.3 \\ \text{III} & p = 0.1 \qquad\qquad\qquad\qquad\quad + 1/2(0.4) = 0.3 \end{array}$$

So, despite their very different genotypic compositions, they have the same allele frequency. After one generation

BOX 19-2   The Hardy-Weinberg Equilibrium

If the frequency of allele A is p in both the sperm and the eggs, and the frequency of allele a is $q = 1 - p$, then the consequences of random unions of sperm and eggs are as shown in the adjoining diagram. The probability that both the sperm and the egg in any mating will carry A is

$$p \times p = p^2$$

so this will be the frequency of A/A homozygotes in the next generation. Likewise, the chance of heterozygotes A/a will be

$$(p \times q) + (q \times p) = 2pq$$

and the chance of homozygotes a/a will be

$$q \times q = q^2$$

The three genotypes, after a generation of random mating, will be in the frequencies

$$p^2 : 2pq : q^2$$

The frequency of A in the F$_1$ will not change (it will still be p), because as the diagram shows, the frequency of A in the zygotes is the frequency of A/A plus half the frequency of A/a, or

$$p^2 + pq = p(p + q) = p$$

Therefore, in the second generation, the frequencies of the three genotypes will again be

$$p^2 : 2pq : q^2$$

and so on, forever. These are the Hardy-Weinberg equilibrium frequencies.

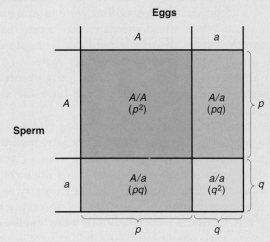

The Hardy-Weinberg equilibrium frequencies that result from random mating.

of random mating within each population, however, each of them will have the same genotypic frequencies:

$$\frac{A/A}{p^2 = (0.3)^2 = 0.09}$$

$$\frac{A/a}{2pq = 2(0.3)(0.7) = 0.42}$$

$$\frac{a/a}{q^2 = (0.7)^2 = 0.49}$$

and they will remain so indefinitely.

One consequence of the Hardy-Weinberg proportions is that rare alleles are virtually never in homozygous condition. An allele with a frequency of 0.001 is present in homozygotes at a frequency of only 1 in a million; most copies of such rare alleles are found in heterozygotes. In general, two copies of an allele are in homozygotes but only one copy of that allele is in each heterozygote, so the relative frequency of the allele in heterozygotes (in contrast with homozygotes) is, from the Hardy-Weinberg equilibrium frequencies,

$$\frac{2pq}{2q^2} = \frac{p}{q}$$

which for $q = 0.001$ is a ratio of 999:1. For example, a recessive mutation causes the potentially lethal disease phenylketonuria (PKU) when it is homozygous. The mutant allele has a frequency of approximately .01 in European and American populations. The frequency of the disease, however, is only about 1 in 10,000 newborns. The general relation between homozygote and heterozygote frequencies as a function of allele frequencies is shown in Figure 19-6.

In our derivation of the equilibrium, we assumed that the allelic frequency p is the same in sperm and eggs. The Hardy-Weinberg equilibrium theorem does not apply to sex-linked genes if males and females start with unequal gene frequencies.

The principle of Hardy-Weinberg equilibrium can be generalized to include cases where there are more than two alleles in the population. In general, no matter how many allelic types are present in the population, the frequency of homozygotes for a particular allele is

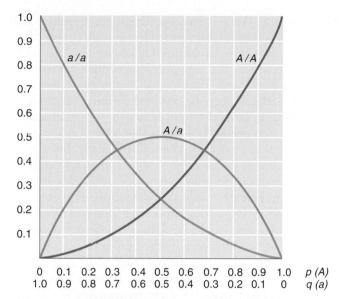

Figure 19-6 Homozygote and heterozygote frequencies as a function of allele frequencies. Curves showing the proportions of homozygotes A/A (blue line), homozygotes a/a (orange line), and heterozygotes A/a (green line) in populations with different allele frequencies if the populations are at Hardy-Weinberg equilibrium.

Heterozygosity

A measure of genetic variation (in contrast with its *description* by allele frequencies) is the amount of **heterozygosity** for a gene in a population, which is given by the total frequency of heterozygotes for the gene. This heterozygosity either can be directly observed by counting heterozygotes or it can be calculated from the allele frequencies, using the Hardy-Weinberg equilibrium proportions. If one allele is in very high frequency and all others are near zero, then there will be very little heterozygosity because most individuals will be homozygous for the common allele. We expect heterozygosity to be greatest when there are many alleles of a gene, all at equal frequency. In Table 19-1 the heterozygosity is simply equal to the frequency of the *M/N* genotype in each population.

When more than one locus is considered, there are two possible ways of calculating heterozygosity. The *S* gene (which encodes the secretor factor, determining whether the M and N proteins are also contained in the saliva) is closely linked to *M/N* in humans. Table 19-6 shows the frequencies of the four combinations of the two alleles for the two genes (*M S*, *M s*, *N S*, and *N s*) in various populations. For the first way of measuring heterozygosity, we can calculate the frequency of heterozygotes at each locus separately (allelic heterozygosity). For the second way, we can consider whether an individual's homologous chromosomes carry the same combination of alleles. The combination of alleles of different genes on the same chromosomal homolog is called a **haplotype**. To determine whether two alleles of different genes are associated on the same chromosomal homolog it is necessary either to sequence the DNA from the individuals or to have information about their parents or offspring. Once we have that information, we consider each haplotype as a unit, as in Table 19-6, and calculate the proportion of all individuals who carry two different haplotypic or gametic forms. This form of heterozygosity is also referred to as *haplotype diversity* or *gametic*

equal to the square of the frequency of the allele. The frequency of heterozygotes for a particular pair of alleles is twice the product of the frequency of those two alleles. For example, suppose there are three alleles, A_1, A_2, and A_3, whose frequencies are .5, .3, and .2, respectively. Then the Hardy-Weinberg equilibrium frequencies of the homozygotes would be

$$
\begin{array}{ccc}
A_1A_1 & A_2A_2 & A_3A_3 \\
(.5)^2 = .25 & (.3)^2 = .09 & (.2)^2 = .04
\end{array}
$$

and the frequencies of the heterozygotes would be

$$
\begin{array}{ccc}
A_1A_2 & A_1A_3 & A_2A_3 \\
2(.5)(.3) = .30 & 2(.5)(.2) = .20 & 2(.3)(.2) = .12
\end{array}
$$

TABLE 19-6 Frequencies of Gametic Types for MNS System in Various Human Populations

| Population | Gametic type | | | | Heterozygosity (H) | |
	M S	*M s*	*N S*	*N s*	From gametes	From alleles
Ainu	0.024	0.381	0.247	0.348	0.672	0.438
Ugandan	0.134	0.357	0.071	0.438	0.658	0.412
Pakistani	0.177	0.405	0.127	0.291	0.704	0.455
English	0.247	0.283	0.080	0.290	0.700	0.469
Navaho	0.185	0.702	0.062	0.051	0.467	0.286

Source: A. E. Mourant, *The Distribution of the Human Blood Groups.* Blackwell Scientific, 1954.

diversity. The results of both calculations are given in Table 19-6. Note that the haplotype diversity is always greater than the average heterozygosity of the separate loci, because an individual is a haplotypic heterozygote if *either* of its genes is heterozygous.

Random mating

The Hardy-Weinberg equilibrium was derived on the assumption of "random mating," but we must carefully distinguish two meanings of that process. First, we may mean that individuals do not choose their mates on the basis of some heritable character. Human beings are randomly mating with respect to blood groups in this first sense, because they generally do not know the blood type of their prospective mates, and even if they did, it is unlikely that they would use blood type as a criterion for choice. In this first sense, random mating will occur with respect to genes that have no effect on appearance, behavior, smell, or other characteristics that directly influence mate choice.

The second sense of random mating is relevant when there is any division of a species into subgroups. If there is genetic differentiation between subgroups so that the frequencies of alleles differ from group to group and if individuals tend to mate within their own subgroup **(endogamy)**, then with respect to the species as a whole, mating is not at random and frequencies of genotypes will depart more or less from Hardy-Weinberg frequencies. In this sense, human beings are not random mating, because ethnic and racial groups and geographically separated populations differ from one another in gene frequencies, and people show high rates of endogamy not only within major geographical races, but also within local ethnic groups. Spaniards and Russians differ in their ABO blood group frequencies, Spaniards usually marry Spaniards and Russians usually marry Russians, so there is unintentional nonrandom mating with respect to ABO blood groups.

Table 19-7 shows random mating in the first sense and nonrandom mating in the second sense for the MN blood group. Within Eskimo, Egyptian, Chinese, and Australian subpopulations, females do not choose their mates by MN type, and thus Hardy-Weinberg equilibrium exists *within* the subpopulations. But Egyptians do not often mate with Eskimos or Australian Aborigines, so the nonrandom associations in the human species *as a whole* result in large differences in genotype frequencies from group to group. It follows that if we took the human species as a whole and could calculate the average allelic frequency in the entire species, we would observe a departure from Hardy-Weinberg equilibrium for the species. To perform such a calculation, however, we would need to know the population size and allele frequencies of every local population. To illustrate the effect, suppose we formed a merged group made up of an equal number of Eskimos and Australian Aborigines. From the observed genotype frequencies in Table 19-7 we can calculate that the allele frequencies in the two subgroups and the merged group are

	$p(M)$	$q(m)$
Eskimos	0.915	0.085
Australian Aborigines	0.178	0.822
Merged average	0.546	0.454

If the merged group were really a single random mating population, we would expect to find the Hardy-Weinberg proportions given by the average allele frequencies,

p^2 (M/M)	$2pq$ (M/N)	q^2 (N/N)
0.298	0.496	0.206

whereas what we actually find is the averaged proportion of homozygotes and heterozygotes from the two original parental populations

(M/M)	(M/N)	(N/N)
0.430	0.230	0.340

TABLE 19-7 Comparison Between Observed Frequencies of Genotypes for the MN Blood Group Locus and the Frequencies Expected from Random Mating

	Observed			Expected		
Population	M/M	M/N	N/N	M/M	M/N	N/N
Eskimo	0.835	0.156	0.009	0.834	0.159	0.008
Egyptian	0.278	0.489	0.233	0.274	0.499	0.228
Chinese	0.332	0.486	0.182	0.331	0.488	0.181
Australian Aborigine	0.024	0.304	0.672	0.031	0.290	0.679

Note: The expected frequencies are computed according to the Hardy-Weinberg equilibrium, using the values of p and q computed from the observed frequencies.

Inbreeding and assortative mating

Random mating with respect to a locus is common within populations, but it is not universal. Two kinds of deviation from random mating must be distinguished. First, individuals may mate with others with whom they share some degree of common ancestry, that is, some degree of genetic relationship. If mating between relatives occurs more commonly than would occur by pure chance, then the population is **inbreeding.** If mating between relatives is less common than would occur by chance, then the population is said to be undergoing **enforced outbreeding,** or **negative inbreeding.**

Second, individuals may tend to choose each other as mates, not because they are related but because of their resemblance to each other in some trait. Bias toward mating of like with like is called **positive assortative mating.** Mating with unlike partners is called **negative assortative mating.** Assortative mating is never complete, so that in any population some matings will be at random and some the result of assortative mating.

Inbreeding and assortative mating are not the same. Close relatives resemble each other more than unrelated individuals on the average but not necessarily for any particular phenotypic trait in particular individuals. So inbreeding can result in the mating of quite dissimilar individuals. On the other hand, individuals who resemble each other for some character may do so because they are relatives, but unrelated individuals also may have specific resemblances. Brothers and sisters do not all have the same eye color, and blue-eyed people are not all related to one another.

Assortative mating for some traits is common. In humans, there is a positive assortative mating bias for skin color and height, for example. An important difference between assortative mating and inbreeding is that the former is specific to a particular phenotype, whereas the latter applies to the entire genome. Individuals may mate assortatively with respect to height but at random with respect to blood group. Cousins, on the other hand, resemble each other genetically on the average to the same degree at all loci.

For both positive assortative mating and inbreeding, the consequence to population structure is the same: there is an increase in homozygosity above the level predicted by the Hardy-Weinberg equilibrium. If two individuals are related, they have at least one common ancestor. Thus, there is some chance that an allele carried by one of them and an allele carried by the other are both descended from the identical DNA molecule. The result is that there is an extra chance of this **homozygosity by descent,** to be added to the chance of homozygosity ($p^2 + q^2$) that arises from the random mating of unrelated individuals. The probability of this extra homozygosity by descent is called the **inbreeding coeffi-**

cient (*F*). Figure 19-7 illustrates the calculation of the probability of homozygosity by descent. Individuals I and II are full sibs because they share both parents. We label each allele in the parents uniquely to keep track of them. Individuals I and II mate to produce individual III. Suppose individual I is A_1/A_3 and the gamete that it contributes to III contains the allele A_1; then we would like to calculate the probability that the gamete produced by II is also A_1. The chance is 1/2 that II will receive A_1 from its father, and, if it does, the chance is 1/2 that II will pass A_1 on to the gamete in question. Thus, the probability that III will receive an A_1 from II is $1/2 \times 1/2 = 1/4$, and this is the chance that III—the product of a full-sib mating—will be homozygous A_1/A_1 by descent from the original ancestor.

Such close inbreeding can have deleterious consequences. Let's consider a rare deleterious allele *a* that, when homozygous, causes a metabolic disorder. If the frequency of the allele in the population is *p*, then the probability that a random couple will produce a homozygous offspring is only p^2 (from the Hardy-Weinberg equilibrium). Thus, if *p* is, say, 1/1000, the frequency of homozygotes will be 1 in 1,000,000. Now suppose that the couple are brother and sister. If one of their common parents is a heterozygote for the disease, they may both receive the deleterious allele and may both pass it on to their offspring. As the calculation shows, there is a 1/4 chance that an offspring of a brother-sister mating will be homozygous for one of the alleles carried by its grandparents. Suppose that among the four copies of the gene possessed by the grandparents, one was a deleterious mutation. Therefore the chance that an offspring of

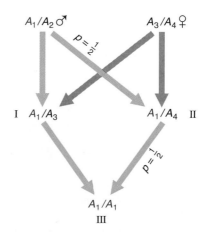

Figure 19-7 Calculation of homozygosity by descent for an offspring (III) of a brother-sister (I-II) mating. Assume that individual III has received one copy of A_1 from its grandfather through individual I. The probability that II will receive A_1 from its father is $\frac{1}{2}$; if it has, the probability that II will pass A_1 on to III is $\frac{1}{2}$. Thus, the probability that III will receive an A_1 from II is $1/2 \times 1/2 = 1/4$.

the brother-sister mating will be homozygous for the deleterious allele is $1/4 \times 1/4 = 1/16$. There are so many rare deleterious alleles for different genes in the human populations that each one of us is a heterozygote for many such rare alleles. Thus the chances are very high that an offspring of a brother-sister mating will be homozygous for at least one of them.

Systematic inbreeding between close relatives eventually leads to complete homozygosity of the population but at different rates, depending on the degree of relationship. If two alleles are present in an inbreeding population, one will eventually be lost and the other will have a frequency of 1.0—in other words, it will become **fixed.** For alleles that are not deleterious, which allele will become fixed is a matter of chance. Suppose, for example, that several groups of individuals are taken from a population and subjected to inbreeding. If, in the original population from which the inbred lines are taken, allele A has frequency p and allele a has frequency $q = 1 - p$, then a proportion p of the homozygous lines established by inbreeding will be homozygous A/A and a proportion q of the lines will be a/a. Inbreeding takes the genetic variation present *within* the original population and converts it into variation *between* homozygous inbred lines sampled from the population (Figure 19-8).

Let's consider how inbreeding leads to a loss of variation. Suppose that a population is founded by some small number of individuals who mate at random to produce the next generation. Assume that no further immigration into the population ever occurs again. (For

example, the rabbits now in Australia probably have descended from a single introduction of a few animals in the nineteenth century.) Even though mating is at random within the population, in later generations everyone is related to everyone else, because their family trees have common ancestors here and there in their pedigrees. Such a population is then inbred, in the sense that there is some probability of a gene's being homozygous by descent. Because the population is, of necessity, finite in size, some of the originally introduced family lines will become extinct in every generation, just as family names disappear in a human population that never receives any migrants, because, by chance, no male offspring are left. As original family lines disappear, the population comes to be made up of descendants of fewer and fewer of the original founder individuals, and all the members of the population become more and more likely to carry the same alleles by descent. In other words, the inbreeding coefficient F increases, and the heterozygosity decreases over time until finally F reaches 1.00 and heterozygosity reaches 0.

The rate of loss of heterozygosity per generation in such a closed, finite, randomly breeding population is inversely proportional to the total number ($2N$) of haploid genomes, where N is the number of diploid individuals in the population. In each generation, $\frac{1}{2N}$ of the remaining heterozygosity is lost.

19.3 Sources of variation

For a given population, there are three sources of variation: mutation, recombination, and immigration of genes. However, recombination between genes by itself does not produce variation unless there is already allelic variation segregating at the different loci. Otherwise there is nothing to recombine. Similarly, migration cannot provide variation if the entire species is homozygous for the same allele. Ultimately, the source of all variation must be mutation.

Variation from mutation

Mutations are the *source* of variation, but the *process* of mutation does not itself drive genetic change in populations. The rate of change in gene frequency from the mutation process is very low because spontaneous mutation rates are low (Table 19-8). The mutation rate is defined as the probability that a copy of an allele changes to some other allelic form in one generation. Thus, the increase in the frequency of a mutant allele will be the product of the mutation rate times the frequency of the nonmutant allele. Suppose that a population were completely homozygous A and mutations to a occurred at the rate of 1/100,000 per newly formed gamete. Then in the next generation, the frequency of a alleles would be

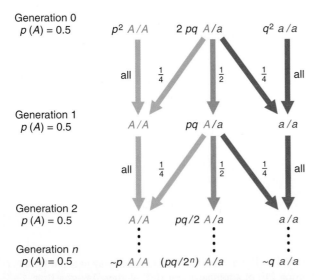

Figure 19-8 Repeated generations of inbreeding (or self-fertilization) will eventually split a heterozygous population into a series of completely homozygous lines. The frequency of A/A lines among the homozygous lines will be equal to the frequency (p) of allele A in the original heterozygous population, while the frequency of a/a lines will be equal to the original frequency of a (q).

Table 19-8 Point-Mutation Rates in Different Organisms

Organism	Gene	Mutation rate per generation
Bacteriophage	Host range	2.5×10^{-9}
Escherichia coli	Phage resistance	2×10^{-8}
Zea mays (corn)	R (color factor)	2.9×10^{-4}
	Y (yellow seeds)	2×10^{-6}
Drosophila melanogaster	Average lethal	2.6×10^{-5}

Source: T. Dobzhansky, *Genetics and the Origin of Species*, 3d ed., rev. Columbia University Press, 1951.

only $1.0 \times 1/100,000 = 0.00001$, and the frequency of A alleles would be 0.99999. After yet another generation of mutation, the frequency of a would be increased by $0.99999 \times 1/100,000 = 0.000009$ to a new frequency of 0.000019, whereas the original allele would be reduced in frequency to 0.999981. It is obvious that the rate of increase of the new allele is extremely slow and that *it gets slower every generation* because there are

fewer copies of the old allele still left to mutate. A general formula for the change in allele frequency under mutation is given in Box 19-3.

> **MESSAGE** Mutation rates are so slow that mutation alone cannot account for rapid genetic changes of populations and species.

BOX 19-3 **The Effect of Mutation on Allele Frequency**

Let μ be the **mutation rate** from allele A to some other allele a (the probability that a copy of gene A will become a during the DNA replication preceding meiosis). If p_t is the frequency of the A allele in generation t, if $q_t = 1 - p_t$ is the frequency of the a allele in generation t, and if there are no other causes of gene frequency change (no natural selection, for example), then the change in allele frequency in one generation is

$$\Delta p = p_t - p_{t-1} = (p_{t-1} - \mu p_{t-1}) - p_{t-1} = -\mu p_{t-1}$$

where p_{t-1} is the frequency in the preceding generation. This tells us that the frequency of A decreases (and the frequency of a increases) by an amount that is proportional to the mutation rate μ and to the proportion p of all the genes that are still available to mutate. Thus Δp gets smaller as the frequency of p itself decreases, because there are fewer and fewer A alleles to mutate into a alleles. We can make an approximation that, after n generations of mutation,

$$p_n = p_0 e^{-n\mu}$$

where e is the base of the natural logarithms.

This relation of allele frequency to number of generations is shown in the adjoining figure for $\mu = 10^{-5}$. After 10,000 generations of continued mutation of A to a,

$$p = p_0 e^{-(10^4) \times (10^{-5})} = p_0 e^{-0.1} = 0.904 p_0$$

If the population starts with only A alleles ($p_0 = 1.0$), it would still have only 10 percent a alleles after 10,000 generations at this rather high mutation rate and would require 60,000 additional generations to reduce p to 0.5.

Even if mutation rates were doubled (say, by environmental mutagens), the rate of change would be very slow. For example, radiation levels of sufficient intensity to double the mutation rate over the reproductive lifetime of an individual human are at the limit of occupational safety regulations, and a dose of radiation sufficient to increase mutation rates by an order of magnitude would be lethal, so rapid genetic change in the species would not be one of the effects of increased radiation. Although we have many things to fear from environmental radiation pollution, turning into a species of monsters is not one of them.

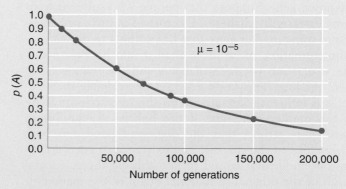

The change over generations in the frequency of a gene A due to mutation from A to a at a constant mutation rate (μ) of 10^{-5}.

Most mutation rates that have been determined are the sum of all mutations of A to *any* mutant form with a detectable effect. The mutation process is even slower if we consider the increase of a *particular* new allelic type. Any *specific* base substitution is likely to be at least two orders of magnitude lower in frequency than the sum of all changes.

Variation from recombination

When a new mutation of a gene arises in a population, it occurs as a single event on a particular copy of a chromosome carried by some individual. But that chromosome copy has a particular allelic composition for all the other polymorphic genes on the chromosome. So, if the mutant allele a arose at the A locus on a chromosome copy that already had the allele b at the B locus, then without recombination all gametes carrying the a allele would also carry the b allele in future generations. The population would then contain only the original $A\ B$ haplotype and the new $a\ b$ haplotype that arose from the mutation. Recombination between the A gene and the B gene in the double heterozygote $A\ B/a\ b$, however, would produce two new haplotypes $A\ b$ and $a\ B$.

The consequence of repeated recombination between genes is to randomize combinations of alleles of different genes. If the allele frequency of a at locus A is, say, .2, and the frequency of allele b at locus B is, say, .4, then the frequency of $a\ b$ would be $(.2)(.4) = .08$ if the combinations were randomized. This randomized condition is **linkage equilibrium.**

Recombination between genes on the same chromosome will not produce linkage equilibrium in a single generation if the alleles at the different genes began in nonrandom association with each other. This original association, **linkage disequilibrium,** decays only slowly from generation to generation at a rate that is proportional to the amount of recombination between the genes. This fact can be used to find the location of unknown genes on chromosomes and to provide evidence that some phenotypic variant is, in fact, influenced by an unknown gene. Suppose that people who suffered from some disease, say, diabetes, also turned out to carry an allele of some marker gene that has nothing to do with insulin formation, more often than would be expected if the association between diabetes and the marker allele were random. This finding would be evidence that diabetes is influenced by a gene on the same chromosome as the marker gene, and if the linkage disequilibrium were quite strong, that the diabetes-related gene was fairly close to the marker. The existence of such a linkage disequilibrium would presumably be the accidental result of the original mutational origin of the marker allele on the same chromosome copy as the allele associated with diabetes.

The creation of genetic variation by recombination can be a much faster process than its creation by mutation. This high rate of the production of variation is simply a consequence of the very large number of different recombinant chromosomes that can be produced even if we take into account only single crossovers. If a pair of homologous chromosomes is heterozygous at n loci, then a crossover can take place in any one of the $n - 1$ intervals between them, and because each recombination produces two recombinant products, there are $2(n - 1)$ new unique gametic types from a single generation of crossing-over, even considering only single crossovers. If the heterozygous loci are well spread out along the chromosome, these new gametic types will be frequent and considerable variation will be generated. Asexual organisms or organisms such as bacteria that very seldom undergo sexual recombination do not have this source of variation, so new mutations are the only way in which a change in gene combinations can be achieved. As a result, populations of asexual organisms may change more slowly than sexual organisms.

Variation from migration

A further source of variation is migration into a population from other populations with different gene frequencies. The resulting mixed population will have an allele frequency that is somewhere intermediate between its original value and the frequency in the donor population.

Suppose a population receives a group of migrants and the number of immigrants is equal to, say, 10 percent of the native population size. Then the newly formed mixed population will have an allele frequency that is a 0.90:0.10 mixture between its original allele frequency and the allele frequency of the donor population. If its original allele frequency of A were, say, 0.70 whereas the donor population had an allele frequency of A that was only, say, 0.40, the new mixed population would have an allele frequency of A that was $0.70 \times 0.90 + 0.40 \times 0.10 = 0.67$. Box 19-4 derives the general result. As shown in Box 19-4, the change in gene frequency is proportional to the difference in frequency between the recipient population and the average of the donor populations. Unlike the mutation rate, the migration rate (m) can be large, so if the difference in allele frequency between the donor and recipient population is large, the change in frequency may be substantial.

We must understand *migration* as meaning any form of the introduction of genes from one population into another. So, for example, genes from Europeans have "migrated" into the population of African origin in North America steadily since the Africans were introduced as slaves. We can determine the amount of this

---

BOX 19-4   The Effect of Migration on Allele Frequency

If p_t is the frequency of an allele in a recipient population in generation t, P is the frequency of that allele in a donor population (or the average over several donor populations), and m is the proportion of the recipient population that is made up of new migrants from the donor population, then the allele frequency in the recipient population in the next generation, p_{t+1}, is the result of mixing $1 - m$ genes from the recipient population with m genes from the donor population. Thus

$$p_{t+1} = (1 - m)p_t + mP = p_t + m(P - p_t)$$

and

$$\Delta p = p_{t+1} - p_t = m(P - p_t)$$

---

migration by looking at the frequency of an allele that is found only in Europeans and not in Africans and comparing its frequency among blacks in North America. We can use the formula for the change in gene frequency from migration if we modify it slightly to account for the fact that several generations of admixture have taken place. If the rate of admixture has not been too great, then (to a close order of approximation) the sum of the single-generation migration rates over several generations (let's call this M) will be related to the total change in the recipient population after these several generations by the same expression as the one used in Box 19-4 for changes due to a single generation of migration. If, as before, P is the allelic frequency in the donor population and p_0 is the original frequency among the recipients, then

$$\Delta p_{\text{total}} = M(P - p_0)$$

so

$$M = \frac{\Delta p_{\text{total}}}{P - p_0}$$

For example, the Duffy blood group allele Fy^a is absent in Africa but has a frequency of 0.42 in whites from the state of Georgia. Among blacks from Georgia, the Fy^a frequency is 0.046. Therefore, the total migration of genes from whites into the black population since the introduction of slaves in the eighteenth century is

$$M = \frac{\Delta p_{\text{total}}}{P - p_0} = \frac{(0.046 - 0.0)}{(0.42 - 0.0)} = .1095$$

That is, on the average over all Americans of African ancestry in Georgia, about 11 percent of their gene alleles have been derived from a European ancestor. This is only an average, however, and different individuals have different proportions of European and African ancestry. When the same analysis is carried out on American blacks from Oakland (California) and Detroit, M is 0.22 and 0.26, respectively, showing either greater admixture

rates in these cities than in Georgia or differential movement into these cities by American blacks who have more European ancestry. In any case, the genetic variation at the Fy locus has been increased by this admixture. At the same time the frequency of the sickle-cell mutation Hb-S has been decreased in African Americans by between 10 and 20 percent of its value in their ancestral African populations as a result of admixture.

19.4   Selection

So far in this chapter, we have considered changes in a population arising from forces of mutation, migration, recombination, and breeding structure. But these changes cannot explain why organisms seem so well adapted to their environments, because they are random with respect to the way in which organisms make a living in the environments in which they live. Changes in a species in response to a changing environment occur because the different genotypes produced by mutation and recombination have different abilities to survive and reproduce. The differential rates of survival and reproduction are what is meant by **selection,** and the process of selection alters the frequencies of the various genotypes in the population. Darwin called the process of differential survival and reproduction of different types **natural selection** by analogy with the **artificial selection** carried out by animal and plant breeders when they deliberately select some individuals of a preferred type.

The relative probability of survival and rate of reproduction of a phenotype or genotype is now called its **Darwinian fitness.** Although geneticists sometimes speak loosely of the fitness of an individual, the concept of fitness really applies to the average probability of survival and average reproductive rate of individuals in a phenotypic or genotypic class. Because of chance events in the life histories of individuals, even two organisms with identical genotypes, living in identical environments, will not live to the same age or leave the same number of offspring. It is the fitness of a genotype on average over all its possessors that matters.

Fitness is a consequence of the relation between the phenotype of the organism and the environment in which the organism lives, so the same genotype will have different fitnesses in different environments. One reason is that even genetically identical organisms may develop different phenotypes if exposed to different environments during development. But, even if the phenotype is the same, the success of the organism depends on the environment. Having webbed feet is fine for paddling in water but a positive disadvantage for walking on land, as a few moments spent observing how a duck walks will reveal. No genotype is unconditionally superior in fitness to all others in all environments.

Reproductive fitness is not to be confused with "physical fitness" in the everyday sense of the term, although they may be related. No matter how strong, healthy, and mentally alert the possessor of a genotype may be, that genotype has a fitness of zero if for some reason its possessors leave no offspring. The fitness of a genotype is a consequence of all the phenotypic effects of the genes involved. Thus, an allele that doubles the fecundity of its carriers but at the same time reduces the average lifetime of its possessors by 10 percent will be more fit than its alternatives, despite its life-shortening property. The most common example is parental care. An adult bird that expends a great deal of its energy gathering food for its young will have a lower probability of survival than one that keeps all the food for itself. But a totally selfish bird will leave no offspring, because its young cannot fend for themselves. As a consequence, parental care is favored by natural selection.

Two forms of selection

Because the differences in reproduction and survival between genotypes depend on the environment in which the genotypes live and develop and because organisms may alter their own environments, there are two fundamentally different forms of selection. In the simple case, the fitness of an individual does not depend on the composition of the population; rather it is a fixed property of the individual's phenotype and the external physical environment. For example, the relative ability of two plants that live at the edge of the desert to get sufficient water will depend on how deep their roots grow and how much water they lose through their leaf surfaces. These characteristics are a consequence of their developmental patterns and are not sensitive to the composition of the population in which they live. The fitness of a genotype in such a case does not depend on how rare or how frequent it is in the population. Fitness is then **frequency-independent.**

In contrast, consider organisms that are competing to catch prey or to avoid being captured by a predator. Then the relative abundances of two different genotypes

will affect their relative fitnesses. An example is Müllerian mimicry in butterflies. Some species of brightly colored butterflies (such as monarchs and viceroys) are distasteful to birds, which learn, after a few trials, to avoid attacking butterflies with a pattern that they associate with distastefulness. Within a species that has more than one pattern, rarer patterns will be selected against. The rarer the pattern, the greater is the selective disadvantage, because birds will be unlikely to have had a prior experience of a low-frequency pattern and therefore will not avoid it. This selection to blend in with the crowd is an example of **frequency-dependent fitness,** because the fitness of a type changes as it becomes more or less frequent in the population.

For reasons of mathematical convenience, most models of natural selection are based on frequency-independent fitness. In fact, however, a very large number of selective processes (perhaps most) are frequency-dependent. The kinetics of the change in allele frequency depends on the exact form of frequency dependence, and for that reason alone, it is difficult to make any generalizations. For the sake of simplicity and as an illustration of the main qualitative features of selection, we deal only with models of frequency-independent selection in this chapter, but convenience should not be confused with reality.

Measuring fitness differences

For the most part, we can measure the differential fitness of different genotypes most easily when the genotypes differ at many loci. In very few cases, such as laboratory mutants, horticultural varieties, and major metabolic disorders, does an allelic substitution at a single locus make enough difference to the phenotype to measurably alter fitness. Figure 19-9 shows the probability of survival from egg to adult—that is, the **viability**—at three different temperatures of a number of different lines made homozygous for the second chromosomes of *D. pseudoobscura*. These chromosomes were sampled from a natural population and carried a variety of different alleles at different loci, as we expect from the very large amount of nucleotide variation present in nature (see pages 619–620). As is generally the case, the fitness (in this case, a component of the total fitness, viability) is different in different environments. The homozygous state is lethal or nearly so in a few cases at all three temperatures, whereas a few cases have consistently high viability. Most genotypes, however, are not consistent in viability between temperatures, and no genotype is unconditionally the most fit at all temperatures.

There are cases in which single-gene substitutions lead to clear-cut fitness differences. Examples are the many "inborn errors of metabolism," where a recessive allele interferes with a metabolic pathway and is lethal

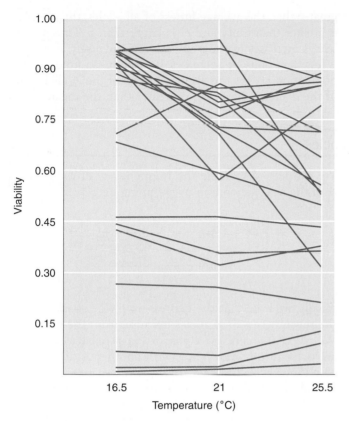

Figure 19-9 Viabilities of various chromosomal homozygotes of *Drosophila pseudoobscura* at three different temperatures.

in homozygotes. One example is sickle-cell anemia. Individuals with this disorder are homozygous for the allele that codes for hemoglobin-S instead of the normal hemoglobin. They die from a severe anemia because hemoglobin-S crystallizes at low oxygen concentrations, causing the red blood cells to become sickle-shaped and then to rupture (Figure 19-10).

As we saw in Chapter 6, another example in humans is phenylketonuria, where tissue degenerates as the result of the accumulation of a toxic intermediate in the pathway of tyrosine metabolism. This case also illustrates how fitness is altered by changes in the environment. People born with PKU will survive if they observe a strict diet that contains no tyrosine.

How selection works

Selection acts by altering allele frequencies in a population. The simplest way to see the effect of selection is to consider an allele *a* that is completely lethal before reproductive age in homozygous condition, such as the allele that leads to Tay-Sachs disease. Suppose that in some generation the allele frequency of this gene is 0.10. Then, in a random-mating population, the proportions of the three genotypes after fertilization are

A/A	*A/a*	*a/a*
0.81	0.18	0.01

At reproductive age, however, the homozygotes *a/a* will have already died, leaving the genotypes at this stage as

A/A	*A/a*	*a/a*
0.81	0.18	0.00

But these proportions add up to only 0.99 because only 99 percent of the population is still surviving. Among the actual surviving reproducing population, the proportions must be recalculated by dividing by 0.99, so that the total proportions add up to 1.00. After this readjustment, we have

A/A	*A/a*	*a/a*
0.818	0.182	0.00

The frequency of the lethal *a* allele among the gametes produced by these survivors is then

$$0.00 + 0.182/2 = 0.091$$

and the change in allelic frequency of the lethal in one generation, expressed as the new value minus the old one, has been $0.091 - 0.100 = -0.019$. Conversely, the change in the frequency of the normal allele has been $+0.019$. We can repeat this calculation in each successive generation to obtain the predicted frequencies of the lethal and normal alleles in a succession of future generations.

The same kind of calculation can be carried out if genotypes are not simply lethal or normal, but if each

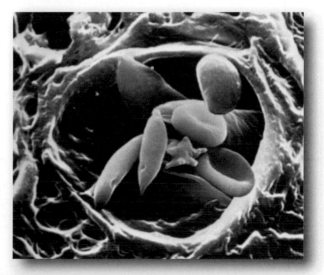

Figure 19-10 Red blood cells of a person with sickle-cell anemia. A few normal disk-shaped red blood cells are surrounded by distorted sickle-shaped cells.

genotype has some relative probability of survival. This general calculation is shown in Box 19-5. After one generation of selection, the new value of the frequency of A is equal to the old value (p) multiplied by the ratio of the mean fitness of A alleles, $\overline{W}_A$, to the mean fitness of the whole population, $\overline{W}$. If the fitness of A alleles is greater than the average fitness of all alleles, then $\overline{W}_A/\overline{W}$ is greater than unity and p' is larger than p. Thus, the allele A increases in the population. Conversely, if $\overline{W}_A/\overline{W}$ is less than unity, A decreases. But the mean fitness of the population ($\overline{W}$) is the average fitness of the A alleles and of the a alleles. So if $\overline{W}_A$ is greater than the mean fitness

---

BOX 19-5   The Effect of Selection on Allele Frequencies

Suppose that a population is mating at random with respect to a given locus with two alleles and that the population is so large that (for the moment) we can ignore inbreeding. Just after the eggs have been fertilized, the genotypes of the zygotes will be in Hardy-Weinberg equilibrium:

Genotype	A/A	A/a	a/a
Frequency	p^2	$2pq$	q^2

and

$$p^2 + 2pq + q^2 = (p + q)^2 = 1.0$$

where p is the frequency of A.

Further suppose that the three genotypes have the relative probabilities of survival to adulthood (viabilities) of $W_{A/A}$, $W_{A/a}$, and $W_{a/a}$. For simplicity, let us also assume that all fitness differences are differences in survivorship between the fertilized egg and the adult stage. (Differences in fertility give rise to much more complex mathematical formulations.) Among the progeny, once they have reached adulthood, the frequencies will be

Genotype	A/A	A/a	a/a
Frequency	$p^2 W_{A/A}$	$2pq W_{A/a}$	$q^2 W_{a/a}$

These adjusted frequencies do not add up to unity, because the W's are all fractions smaller than 1. However, we can readjust them so that they do, without changing their relation to one another, by dividing each frequency by the sum of the frequencies after selection ($\overline{W}$):

$$\overline{W} = p^2 W_{A/A} + 2pq W_{A/a} + q^2 W_{a/a}$$

So defined, $\overline{W}$ is called the **mean fitness** of the population because it is, indeed, the mean of the fitnesses of all individuals in the population. After this adjustment, we have

Genotype	A/A	A/a	a/a
Frequency	$p^2 \dfrac{W_{A/A}}{\overline{W}}$	$2pq \dfrac{W_{A/a}}{\overline{W}}$	$q^2 \dfrac{W_{a/a}}{\overline{W}}$

We can now determine the frequency p' of the allele A in the next generation by summing up genes:

$$p' = A/A + \tfrac{1}{2}A/a = p^2 \frac{W_{A/A}}{\overline{W}} + pq \frac{W_{A/a}}{\overline{W}}$$
$$= p \frac{pW_{A/A} + qW_{A/a}}{\overline{W}}$$

Finally, we note that the expression $pW_{A/A} + qW_{A/a}$ is the mean fitness of A alleles because, from the Hardy-Weinberg frequencies, a proportion p of all A alleles are present in homozygotes with another A, in which case they have a fitness of $W_{A/A}$, whereas a proportion q of all the A alleles are present in heterozygotes with a and have a fitness of $W_{A/a}$. Using $\overline{W}_A$ to denote $pW_{A/A} + qW_{A/a}$, the mean fitness of the allele A yields the final new allele frequency

$$p' = p \frac{\overline{W}_A}{\overline{W}}$$

An alternative way to look at the process of selection is to solve for the *change* in allele frequencies in one generation:

$$\Delta p = p' - p = p \frac{\overline{W}_A}{\overline{W}} - p$$
$$= \frac{p(\overline{W}_A - \overline{W})}{\overline{W}}$$

But $\overline{W}$, the mean fitness of the population, is the average of the allele fitnesses $\overline{W}_A$ and $\overline{W}_a$, so

$$\overline{W} = p\overline{W}_A + q\overline{W}_a$$

where $\overline{W}_a$ is the mean fitness of a alleles. Substituting this expression for $\overline{W}$ in the formula for Δp and remembering that $q = 1 - p$, we obtain (after some algebraic manipulation)

$$\Delta p = \frac{pq(\overline{W}_A - \overline{W}_a)}{\overline{W}}$$

of the population, it must be greater than $\overline{W}_a$, the mean fitness of a alleles. Thus the allele with the higher mean fitness increases in frequency.

It should be noted that the fitnesses $W_{A/A}$, $W_{A/a}$, and $W_{a/a}$ may be expressed as absolute probabilities of survival and absolute reproduction rates, or they may all be rescaled relative to one of the fitnesses, which is given the standard value of 1.0. This rescaling has absolutely no effect on the formula for p', because it cancels out in the numerator and denominator.

> **MESSAGE** As a result of selection, the allele with the higher mean fitness relative to the mean fitnesses of other alleles increases in frequency in the population.

An increase in the allele with the higher fitness means that the average fitness of the population as a whole increases, so selection can also be described as a process that *increases mean fitness*. This rule is strictly true only for frequency-independent genotypic fitnesses, but it is close enough to a general rule to be used as a fruitful generalization. This maximization of fitness does not necessarily lead to any optimal property for the species as a whole, because fitnesses are defined only relative to one another within a population. It is relative (not absolute) fitness that is increased by selection. The population does not necessarily become larger or grow faster, nor is it less likely to become extinct. For example suppose that an allele causes its carriers to lay more eggs than do other genotypes in the population. This higher fecundity allele will increase in the population. But the population size at the adult stage may depend on the total food supply available to the immature stages, so there will not be an increase in the total population size, but only an increase in the number of immature individuals that starve to death before adulthood.

Rate of change in gene frequency

The general expression for the change in allele frequency derived in Box 19-5 is particularly illuminating. It says that Δp will be positive (A will increase) if the mean fitness of A alleles is greater than the mean fitness of a alleles, as we saw before. But it also shows that the speed of the change depends not only on the difference in fitness between the alleles, but also on the factor pq, which is proportional to the frequency of heterozygotes ($2pq$). For a given difference in fitness of alleles, their frequency will change most rapidly when the alleles A and a are in intermediate frequency, so pq is large. If p is near 0 or 1 (that is, if A or a is nearly fixed at frequency 0 or 1), then pq is nearly 0 and selection will proceed very slowly.

The S-shaped curve in Figure 19-11 represents the course of selection of a new favorable allele A that has

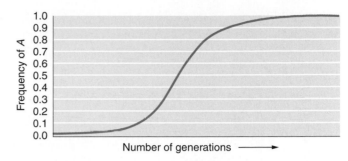

Figure 19-11 The time pattern of increasing frequency of a new favorable allele A that has entered a population of a/a homozygotes.

recently entered a population of homozygotes a/a. At first, the change in frequency is very small because p is still close to 0. Then it accelerates as A becomes more frequent, but it slows down again as A takes over and a becomes very rare (q gets close to 0). This is precisely what is expected from a selection process. When most of the population is of one type, there is nothing to select. For change by natural selection to occur, there must be genetic variation; the more variation, the faster the process.

One consequence of the dynamics shown in Figure 19-11 is that it is extremely difficult to significantly reduce the frequency of an allele that is already rare in a population. Thus, eugenics programs designed to eliminate deleterious recessive alleles from human populations by preventing the reproduction of affected persons do not work. Of course, if all heterozygotes could be prevented from reproducing, the allele could be eliminated (except for new mutations) in a single generation. Because every human being is heterozygous for a number of different deleterious genes, however, no one would be allowed to reproduce.

When alternative alleles are not rare, selection can cause quite rapid changes in allelic frequency. Figure 19-12 shows the course of elimination of a malic dehydrogenase allele that had an initial frequency of 0.5 in a laboratory population of *D. melanogaster*. The fitnesses in this case are

$$W_{A/A} = 1.0 \qquad W_{A/a} = 0.75 \qquad W_{a/a} = 0.40$$

The frequency of a declines rapidly but is not reduced to 0, and to continue reducing its frequency would require longer and longer times, as shown in the negative eugenics case.

> **MESSAGE** Unless alternative alleles are present in intermediate frequencies, selection (especially against recessives) is quite slow. Selection is dependent on genetic variation.

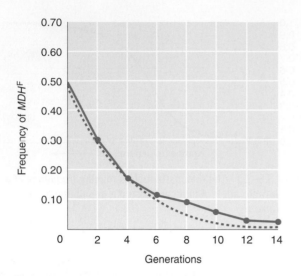

Figure 19-12 The loss of an allele *MDH*^F at the malic dehydrogenase locus due to selection in a laboratory population of *Drosophila melanogaster*. The red dashed line shows the theoretical curve of change computed for fitnesses $W_{A/A}$ = 1.0, $W_{A/a}$ = 0.75, $W_{a/a}$ = 0.4. [From R. C. Lewontin, *The Genetic Basis of Evolutionary Change.* Copyright 1974 by Columbia University Press. Data courtesy of E. Berger.]

19.5 Balanced polymorphism

Thus far we have considered the changes in allelic frequency that occur when the homozygous carriers of one allele, say, *A/A*, are more fit than homozygous carriers for another allele, say, *a/a*, while the heterozygotes, *A/a*, are somewhere between *A/A* and *a/a* in their fitness. But there are other possibilities.

Overdominance and underdominance

First, the heterozygote might be *more* fit than either homozygote, a condition termed **overdominance** in fitness. When one of the alleles, say, *A*, is in low frequency, there are virtually no homozygotes *A/A* and the allele occurs almost entirely in heterozygous condition. Since heterozygotes are more fit than homozygotes, *A* alleles are almost all carried by the most fit genotype, so A will increase in frequency, while *a* decreases in frequency. On the other hand when *a* is in very low frequency, it occurs almost entirely in heterozygous condition. In this case, *a* alleles are almost all carried by the most fit genotype, and they will increase in frequency at the expense of *A* alleles. The net effect of these two pressures, one increasing the frequency of *A* alleles when they are rare, and the other increasing *a* alleles when *they* are rare, is to bring the allele frequencies to a stable equilibrium that is intermediate between a composition consisting of all of one allele or all of the other. Any chance deviation of the allele frequencies on one side or the other of the

equilibrium will be counteracted by the force of selection. We can symbolize the fitnesses of the three genotypes by

$$W_{A/A} \qquad W_{A/a} \qquad W_{a/a}$$
$$1 - t \qquad 1 \qquad 1 - s$$

where t and s are the selective disadvantages of the two homozygotes. Then the equilibrium frequency of the allele *A* is simply the ratio

$$p(A) = s/(s + t)$$

(As an advanced exercise, the reader can derive this result by noting that at equilibrium the average fitness of the *A* allele, $\overline{W}_A$, is equal to the average fitness of the *a* allele, $\overline{W}_a$. Setting these equal to each other and solving for $p(A)$ give the result.) This equilibrium explains the high frequency of the sickle-cell condition in West Africa. Homozygotes for the abnormal allele *Hb-S* die prematurely from anemia. But there is a high mortality in West Africa from falciparum malaria, which kills many of the homozygotes for the normal allele, *Hb-A*. Heterozygotes, *Hb-A/Hb-S*, suffer only a mild, nonfatal anemia, and they are protected against falciparum malaria by the presence of the abnormal hemoglobin in their red blood cells. This overdominance in fitness was lost, however, when slaves were brought to the new world because the falciparum form of malaria does not exist in the western hemisphere. As a consequence, among slaves and their descendants, selection was only against the homozygous *Hb-S/Hb-S*, leading to a reduction in the frequency of this allele through mortality. Recently, sickle-cell anemia has received sufficient medical attention that it is no longer a significant source of mortality, so selection against the *Hb-S* allele is no longer so powerful. Further reductions in the frequency of the allele will then be mostly the consequence of continued admixture with populations of non-African ancestry.

Another possible fitness relation among alleles is that the heterozygote is *less* fit than either homozygote (**underdominance** in fitness) In this case selection favors an allele when it is common, not when it is rare, so an intermediate allele frequency is unstable, and the population should become fixed for either the *A* allele or the *a* allele. Polymorphism resulting from a mixture of an *A/A* population with an *a/a* population should be rapidly lost. A well-known, but mysterious example of underdominance in fitness is Rh incompatibility in humans. Rh-positive children born to Rh-negative mothers often suffer hemolytic anemia as newborns because their mothers produce antibodies against the blood cells of the fetus. Rh-negative mothers are homozygotes, *Rh⁻/Rh⁻*, so their Rh-positive offspring who die from anemia must be heterozygotes, *Rh⁻/Rh⁺*. The mystery is that all human populations are polymorphic for the two Rh alleles. Thus, this human polymorphism must be very old, antedating the origin of modern geographical races. Yet the simple theoretical prediction is that such a polymorphism is unstable and should have disappeared.

Balance between mutation and selection

The overdominant balance of selective forces is not the only situation in which a stable equilibrium of allelic frequencies may arise. Allele frequencies may also reach equilibrium in populations when the introduction of new alleles by repeated mutation is balanced by their removal by natural selection. This balance probably explains the persistence of genetic diseases as low-level polymorphisms in human populations. New deleterious mutations are constantly arising spontaneously or as the result of the action of mutagens. These mutations may be completely recessive or partly dominant. Selection removes them from the population, but there will be an equilibrium between their appearance and removal.

The general equation for this equilibrium is given in detail in Box 19-6. It shows that the frequency of the deleterious allele at equilibrium depends on the ratio μ/s, where μ is the probability of a mutation's occurring in a newly formed gamete (mutation rate) and s is the intensity of selection against the deleterious genotype. For a completely recessive deleterious allele whose fitness in homozygous state is $1 - s$, the equilibrium frequency is

$$q = \sqrt{\frac{\mu}{s}}$$

So, for example, a recessive lethal ($s = 1$) mutating at the rate of $\mu = 10^{-6}$ will have an equilibrium frequency of 10^{-3}. Indeed, if we knew that an allele was a recessive lethal and had no heterozygous effects, we could estimate its mutation rate as the square of its frequency. But the biological basis for the assumptions behind such calculations must be firm. Sickle-cell anemia was once thought to be a recessive lethal with no heterozygous effects, which led to an estimated mutation rate in Africa of 0.1 for this locus, but now we know that its equilibrium is a result of higher fitness of heterozygotes.

A similar result can be obtained for a deleterious allele that has some effect in heterozygotes. If we let the fitnesses be $W_{A/A} = 1.0$, $W_{A/a} = 1 - hs$, and $W_{a/a} = 1 - s$ for a partly dominant allele a, where h is the degree of dominance of the deleterious allele, then a calculation similar to the one in Box 19-6 gives us

$$\hat{q} = \frac{\mu}{hs}$$

Thus, if $\mu = 10^{-6}$ and the lethal is not totally recessive but has a 5 percent deleterious effect in heterozygotes ($s = 1.0$, $h = 0.05$), then

$$\hat{q} = \frac{10^{-6}}{5 \times 10^{-2}} = 2 \times 10^{-5}$$

which is smaller by two orders of magnitude than the equilibrium frequency for the purely recessive case. In general, then, we can expect deleterious, completely

---

BOX 19-6 The Balance Between Selection and Mutation

If we let q be the frequency of the deleterious allele a and $p = 1 - q$ be the frequency of the normal allele A, then the change in allele frequency due to the mutation rate μ is

$$\Delta q_{\text{mut}} = \mu p$$

A simple way to express the fitnesses of the genotypes in the case of a recessive deleterious allele a is $W_{A/A} = W_{A/a} = 1.0$, and $W_{a/a} = 1 - s$, where s is the loss of fitness in the recessive homozygotes. We now can substitute these fitnesses in our general expression for allele frequency change (see Box 19-5) and obtain

$$\Delta q_{\text{sel}} = \frac{-pq(sq)}{1 - sq^2} = \frac{-spq^2}{1 - sq^2}$$

Equilibrium means that the increase in the allele frequency due to mutation must exactly balance the decrease in the allele frequency due to selection, so

$$\Delta \hat{q}_{\text{mut}} + \Delta \hat{q}_{\text{sel}} = 0$$

Remembering that $\hat{q}$ at equilibrium will be quite small, so $1 - s\hat{q}^2 \approx 1$, and substituting the terms for $\Delta \hat{q}_{\text{mut}}$ and $\Delta \hat{q}_{\text{sel}}$ in the preceding formula, we have

$$\mu\hat{p} - \frac{s\hat{p}\hat{q}^2}{1 - s\hat{q}^2} \approx \mu\hat{p} - s\hat{p}\hat{q}^2 = 0$$

or

$$\hat{q}^2 = \frac{\mu}{s} \quad \text{and} \quad \hat{q} = \sqrt{\frac{\mu}{s}}$$

recessive alleles to have frequencies much higher than those of partly dominant alleles, because the recessive alleles are protected in heterozygotes.

19.6 Random events

If a population consists of a finite number of individuals (as all real populations do) and if a given pair of parents has only a small number of offspring, then even in the absence of all selective forces, the frequency of a gene will not be exactly reproduced in the next generation, because of sampling error. If, in a population of 1000 individuals, the frequency of a is 0.5 in one generation, then it may by chance be 0.493 or 0.505 in the next generation because of the chance production of slightly more or slightly fewer progeny of each genotype. In the second generation, there is another sampling error based on the new gene frequency, so the frequency of a may go from 0.505 to 0.511 or back to 0.498. This process of random fluctuation continues generation after generation, with no force pushing the frequency back to its initial state, because the population has no "genetic memory" of its state many generations ago. Each generation is an independent event. This random change in allele frequencies is known as **genetic drift.**

The final result of genetic drift is that the population eventually drifts to $p = 1$ or $p = 0$. After this point, no further change is possible; the population has become homozygous. A different population, isolated from the first, also undergoes this **random genetic drift,** but it may become homozygous for allele A, whereas the first population has become homozygous for allele a. As time goes on, isolated populations diverge from one another, each losing heterozygosity. The variation originally present *within* populations now appears as variation *between* populations.

One form of genetic drift occurs when a small group breaks off from a larger population to found a new colony. This "acute drift," called the **founder effect,** results from a single generation of sampling of a small number of colonizers from the original large population, followed by several generations during which the new colony remains small in number. Even if the population grew large after some time, it would continue to drift, but at a slower rate. The founder effect is probably responsible for the virtually complete lack of blood group B in Native Americans, whose ancestors arrived in very small numbers across the Bering Strait at the end of the last ice age, about 20,000 years ago, but whose ancestral population in northeastern Asia had an intermediate frequency of group B.

The process of genetic drift should sound familiar. It is, in fact, another way of looking at the inbreeding effect in small populations discussed earlier. Populations that are the descendants of a very small number of an-

cestral individuals have a high probability that all the copies of a particular allele are identical by descent from a single common ancestor (see Figure 19-7). Whether regarded as inbreeding or as random sampling of genes, the effect is the same. Populations do not exactly reproduce their genetic constitutions; there is a random component of gene frequency change.

One result of random sampling is that most new mutations, even if they are not selected against, never succeed in becoming part of the long-term genetic composition of the population. Suppose that a single individual is heterozygous for a new mutation. There is some chance that the individual in question will have no offspring at all. Even if it has one offspring, there is a chance of 1/2 that the new mutation will not be transmitted to that offspring. If the individual has two offspring, the probability that neither offspring will carry the new mutation is 1/4, and so forth. Suppose that the new mutation is successfully transmitted to an offspring. Then the lottery is repeated in the next generation, and again the allele may be lost. In fact, if a population is of size N, the chance that a new mutation is eventually lost by chance is $(2N - 1)/2N$. (For a derivation of this result, which is beyond the scope of this book, see Chapters 2 and 3 of Hartl and Clark, *Principles of Population Genetics*, 3d ed. Sinauer Associates, 1997.) But, if the new mutation is not lost, then the only thing that can happen to it in a finite population is that eventually it

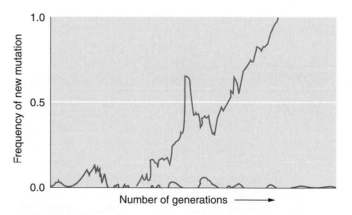

Figure 19-13 The appearance, loss, and eventual incorporation of new mutations in the life of a population. If random genetic drift does not cause the loss of a new mutation, then it must eventually cause the entire population to become homozygous for the mutation (in the absence of selection). In the figure, 10 mutations arise, of which 9 (light red at bottom of graph) increase slightly in frequency and then die out. Only the fourth mutation to occur (blue line) eventually spreads into the population. [After J. Crow and M. Kimura, *An Introduction to the Population Genetics Theory.* Copyright 1970 by Harper & Row.]

will sweep through the population and become fixed. This event has the probability of $1/2N$. In the absence of selection, then, the history of a population looks like Figure 19-13. For some period of time, it is homozygous; then a new mutation appears. In most cases, the new mutant allele will be lost immediately or very soon after it appears. Occasionally, however, a new mutant allele drifts through the population, and the population becomes homozygous for the new allele. The process then begins again.

A striking example of the effect of genetic drift in human populations is the variation in frequencies of the VNTR repeat length variants among populations of South American Indians that we saw illustrated in Table 19-5. For one VNTR, D14S1, the Surui are very variable, but the Karitiana, living several hundred miles away in the Brazilian rain forest, are nearly homozygous for one allele, presumably because of genetic drift in these very small isolated populations. For the other VNTR, D14S13, neither population has become homozygous, but the pattern of frequencies of the alleles is very different in the two.

Even a new mutation that is slightly favorable selectively will usually be lost in the first few generations after it appears in the population, a victim of genetic drift. If a new mutation has a selective advantage of s in the heterozygote in which it appears, then the chance is only $2s$ that the mutation will ever succeed in taking over the population. So a mutation that is 1 percent better in fitness than the standard allele in the population will be lost 98 percent of the time by genetic drift. It is even possible for a very slightly deleterious mutation to rise in frequency and become fixed in a population by drift.

> **MESSAGE** New mutations can become established in a population even though they are not favored by natural selection simply by a process of random genetic drift. Even new favorable mutations are often lost, and occasionally a slightly deleterious mutation can take over a population by drift.

KEY QUESTIONS REVISITED

- **How much genetic variation is there in natural populations of organisms?**

Genetic variation among individuals in populations is extremely common. In many species there is polymorphism within populations for chromosomal rearrangements such as inversions and translocations. Variation at the DNA sequence level is present in all species. Typically a population is polymorphic for 25 to 33 percent of its protein coding genes, where *polymorphism* is defined as having two or more alleles at frequencies of 1 percent or greater in the population, and an average individual is heterozygous for about 10 percent of the nonsynonymous nucleotides in protein coding sequences. On the average individuals within a population differ by between one-tenth and one-half of all their nucleotides in their genomes. Any two humans differ by about 3 million nucleotides.

- **What are the effects of patterns of mating on genetic variation?**

If mating is at random with respect to genotype, then the proportion of heterozygotes and homozygotes for a genetically variable gene is at an equilibrium that depends only on the frequency of the alternative alleles (the Hardy-Weinberg equilibrium). If there is preferential mating between relatives (inbreeding) or between individuals who are more alike (positive assortative mating), the proportion of homozygotes is greater than predicted by the Hardy-Weinberg equilibrium. If mating is solely between relatives, then eventually the population becomes completely homozygous.

- **What are the sources of the genetic variation that is observed in populations?**

The ultimate source of all genetic variation is mutation. Variation for a gene in a given population will be increased if migration from other populations brings into the recipient population gene alleles that are not already present or are in lower frequency than in the donor population. Variation for the genome as a whole is increased by recombination, which brings together new combinations of alleles at different loci.

- **What are the processes that cause changes in the kind and amount of genetic variation in populations?**

Changes in the amount and pattern of variation in a population are the result of (a) recurrent mutation's putting a constant flow of mutations into the population, (b) migration from populations with allele frequencies different from those of the recipient population, (c) recombination of genotypes within the population, (d) natural selection that increases or decreases the frequency of particular genotypes as a result of their differential rates of survival and reproduction, (e) random genetic drift, which causes random changes in the frequencies of genotypes as a result of the sampling of gametes that occurs in successive generations because the population is finite in size.

SUMMARY

The study of changes within a population, or population genetics, relates the heritable changes in populations or organisms to the underlying individual processes of inheritance and development. Population genetics is the study of inherited variation and its modification in time and space.

Identifiable inherited variation within a population can be studied by examining the differences in specific amino acid sequences of proteins, or even examining, most recently, the differences in nucleotide sequences within the DNA. These kinds of observations have revealed that there is considerable polymorphism at many loci within a population. A measure of this variation is the amount of heterozygosity in a population. Population studies have shown that in general the genetic differences between individuals within human races are much greater than the average differences between races.

The ultimate source of all variation is mutation. However, within a population, the quantitative frequency of specific genotypes can be changed by recombination, immigration of genes, continued mutational events, and chance.

One property of Mendelian segregation is that random mating results in an equilibrium distribution of genotypes after one generation. However, if there is inbreeding, the genetic variation within a population is converted into differences between populations by making each separate population homozygous for a randomly chosen allele. On the other hand, for most populations, a balance is reached between inbreeding, mutation from one allele to another, and immigration.

An allele may go up or down in frequency within a population through the natural selection of genotypes with higher probabilities of survival and reproduction. In many cases, such changes lead to homozygosity at a particular locus. On the other hand, the heterozygote may be more fit than either of the homozygotes, leading to a balanced polymorphism.

In general, genetic variation is the result of the interaction of forces. For instance, a deleterious mutant may never be totally eliminated from a population, because mutation will continue to reintroduce it into the population. Immigration may also reintroduce alleles that have been eliminated by natural selection.

Unless alternative alleles are intermediate in frequency, selection (especially against recessives) is very slow, requiring many generations. In many populations, especially those of small size, new mutations can become established even though they are not favored by natural selection, or they may become eliminated even though they are favored, simply by a process of random genetic drift.

KEY TERMS

allele frequency (p. 614)

artificial selection (p. 629)

Darwinian fitness (p. 629)

endogamy (p. 624)

enforced outbreeding (p. 625)

equilibrium distribution (p. 621)

fixed (p. 626)

founder effect (p. 636)

frequency-dependent fitness (p. 630)

frequency-independent (p. 630)

genetic drift (p. 636)

genotype frequency (p. 614)

haplotype (p. 623)

Hardy-Weinberg equilibrium (p. 621)

heterozygosity (p. 623)

homozygosity by descent (p. 625)

inbreeding (p. 625)

inbreeding coefficient (p. 625)

linkage disequilibrium (p. 628)

linkage equilibrium (p. 628)

mean fitness (p. 632)

mutation (p. 627)

natural selection (p. 629)

negative assortative mating (p. 625)

negative inbreeding (p. 625)

overdominance (p. 634)

polymorphism (p. 614)

population (p. 612)

population genetics (p. 612)

positive assortative mating (p. 625)

random genetic drift (p. 636)

selection (p. 629)

single-nucleotide polymorphism (SNP) (p. 619)

underdominance (p. 634)

variable number tandem repeat (VNTR) (p. 618)

viability (p. 630)

wild type (p. 615)

SOLVED PROBLEMS

1. The polymorphisms for shell color (yellow or pink) and for the presence or absence of shell banding in the snail *Cepaea nemoralis* are each the result of a pair of segregating alleles at a separate locus. Design an experimental program that would reveal the forces that determine the frequency and geographical distribution of these polymorphisms.

Solution

a. Describe the frequencies of the different morphs for samples of snails from a large number of populations covering the geographical and ecological range of the species. Each snail must be scored for *both* polymorphisms. At the same time, record a description of the habitat of each population. In addition, estimate the number of snails in each population.

b. Measure migration distances by marking a sample of snails with a spot of paint on the shell, replacing them in the population, and then resampling at a later date.

c. Raise broods from eggs laid by individual snails so that the genotype of male parents can be inferred and nonrandom mating patterns can be observed. The segregation frequencies *within* each family will reveal differences between genotypes in probability of survivorship of early developmental stages.

d. Seek further evidence of selection from (1) geographical patterns in the frequencies of the alleles, (2) correlation between allele frequencies and ecological variables, including population density, (3) correlation between the frequencies of the two different polymorphisms (are populations with, say, high frequencies of pink shells also characterized by, say, high frequencies of banded shells), and (4) nonrandom associations *within* populations of the alleles at the two loci, indicating that certain combinations may have a higher fitness.

e. Seek evidence of the importance of random genetic drift by comparing the variation in allele frequencies among small populations with the variation among large populations. If small populations vary more from each other than do large ones, random drift is implicated.

2. About 70 percent of all white North Americans can taste the chemical phenylthiocarbamide, and the remainder cannot. The ability to taste this chemical is determined by the dominant allele T, and the inability to taste is determined by the recessive allele t. If the population is assumed to be in Hardy-Weinberg equilibrium, what are the genotype and allele frequencies in this population?

Solution

Because 70 percent are tasters (T/T and T/t), 30 percent must be nontasters (t/t). This homozygous recessive frequency is equal to q^2; so to obtain q, we simply take the square root of 0.30:

$$q = \sqrt{0.30} = 0.55$$

Because $p + q = 1$, we can write

$$p = 1 - q = 1 - 0.55 = 0.45$$

Now we can calculate

$$p^2 = (0.45)^2 = 0.20, \text{ the frequency of } T/T$$
$$2pq = 2 \times 0.45 \times 0.55 = 0.50, \text{ the frequency of } T/t$$
$$q^2 = 0.3, \text{ the frequency of } t/t$$

3. In a large natural population of *Mimulus guttatus*, one leaf was sampled from each of a large number of plants. The leaves were crushed and subjected to gel electrophoresis. The gel was then stained for a specific enzyme, X. Six different banding patterns were observed, as shown in the accompanying diagram.

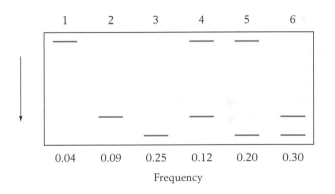

a. Assuming that these patterns are produced by a single locus, propose a genetic explanation for the six types.

b. How can you test your hypothesis?

c. What are the allele frequencies in this population?

d. Is the population in Hardy-Weinberg equilibrium?

Solution

a. Inspection of the gel reveals that there are only three band positions: we shall call them slow, intermediate, and fast, according to how far each has migrated in the gel. Furthermore, any individual can show either one band or two. The simplest explanation is that there are three alleles of one locus (let's call them S, I, and F) and that the individuals with two bands are heterozygotes. Hence, lane 1 = S/S, 2 = I/I, 3 = F/F, 4 = S/I, 5 = S/F, and 6 = I/F.

b. The hypothesis can be tested by making controlled crosses. For example, from a self of type 5, we can predict 1/4 S/S, 1/2 S/F, and 1/4 F/F.

c. The frequencies can be calculated by a simple generalization from the two-allele formulas. Hence:

$$f_S = 0.04 + 1/2(0.12) + 1/2(0.20) = 0.20 = p$$
$$f_I = 0.09 + 1/2(0.12) + 1/2(0.30) = 0.30 = q$$
$$f_F = 0.25 + 1/2(0.20) + 1/2(0.30) = 0.50 = r$$

d. The Hardy-Weinberg genotypic frequencies are

$$(p + q + r)^2 = p^2 + q^2 + r^2 + 2pq + 2pr + 2qr$$
$$= 0.04 + 0.09 + 0.25$$
$$+ 0.12 + 0.20 + 0.30$$

which are precisely the observed frequencies. So it appears that the population is in equilibrium.

4. In a large experimental *Drosophila* population, the fitness of a recessive phenotype is calculated to be 0.90, and the mutation rate to the recessive allele is 5×10^{-5}. If the population is allowed to come to equilibrium, what allele frequencies can be predicted?

Solution

Here mutation and selection are working in opposite directions, so an equilibrium is predicted. Such an equilibrium is described by the formula

$$\hat{q} = \sqrt{\frac{\mu}{s}}$$

In the present question, $\mu = 5 \times 10^{-5}$ and $s = 1 - W = 1 - 0.9 = 0.1$. Hence

$$\hat{q} = \sqrt{\frac{5 \times 10^{-5}}{10^{-1}}} = 2.2 \times 10^{-2} = 0.022$$

$$\hat{p} = 1 - 0.022 = 0.978$$

PROBLEMS

BASIC PROBLEMS

1. What are the forces that can change the frequency of an allele in a population?

2. In a population of mice, there are two alleles of the A locus (A_1 and A_2). Tests showed that in this population there are 384 mice of genotype A_1/A_1, 210 of A_1/A_2, and 260 of A_2/A_2. What are the frequencies of the two alleles in the population?

3. In a randomly mating laboratory population of *Drosophila*, 4 percent of the flies have black bodies (encoded by the autosomal recessive b), and 96 percent have brown bodies (the wild type, encoded by B). If this population is assumed to be in Hardy-Weinberg equilibrium, what are the allele frequencies of B and b and the genotypic frequencies of B/B and B/b?

4. In a wild population of beetles of species X, you notice that there is a 3:1 ratio of shiny to dull wing covers. Does this ratio prove that the *shiny* allele is dominant? (Assume that the two states are caused by two alleles of one gene.) If not, what does it prove? How would you elucidate the situation?

5. The fitnesses of three genotypes are $W_{A/A} = 0.9$, $W_{A/a} = 1.0$, and $W_{a/a} = 0.7$.

   a. If the population starts at the allele frequency $p = 0.5$, what is the value of p in the next generation?

   b. What is the predicted equilibrium allele frequency?

6. *A/A* and *A/a* individuals are equally fertile. If 0.1 percent of the population is *a/a*, what selection pressure exists against *a/a* if the $A - a$ mutation rate is 10^{-5}?

7. In a survey of Native American tribes in Arizona and New Mexico, albinos were completely absent or very rare in most tribes (there is 1 albino per 20,000 North American Caucasians). However, in three Native American populations, albino frequencies are exceptionally high: 1 per 277 Native Americans in Arizona; 1 per 140 Jemez in New Mexico; and 1 per 247 Zuni in New Mexico. All three of these populations are culturally but not linguistically related. What possible factors might explain the high incidence of albinos in these three tribes?

CHALLENGING PROBLEMS

8. In a population, the $D \rightarrow d$ mutation rate is 4×10^{-6}. If $p = 0.8$ today, what will p be after 50,000 generations?

9. You are studying protein polymorphism in a natural population of a certain species of a sexually reproducing haploid organism. You isolate many strains from various parts of the test area and run extracts from each strain on electrophoretic gels. You stain the gels with a reagent specific for enzyme X and find that in the population there are a total of five electrophoretic variants of enzyme X. You speculate that these variants represent various alleles of the structural gene for enzyme X.

   a. How could you demonstrate that your speculation is correct, both genetically and biochemically? (You can make crosses, make diploids, run gels, test enzyme activities, test amino acid sequences, and so forth.) Outline the steps and conclusions precisely.

   b. Name at least one other possible way of generating the different electrophoretic variants, and explain how you would distinguish this possibility from your speculation above.

10. A study made in 1958 in the mining town of Ashibetsu in Hokkaido, Japan, revealed the frequencies of MN blood type genotypes (for individuals and for married couples) shown in the following table:

Genotype	Number of individuals or couples
Individuals	
L^M/L^M	406
L^M/L^N	744
L^N/L^N	332
Total	1482
Couples	
$L^M/L^M \times L^M/L^M$	58
$L^M/L^M \times L^M/L^N$	202
$L^M/L^N \times L^M/L^N$	190
$L^M/L^M \times L^N/L^N$	88
$L^M/L^N \times L^N/L^N$	162
$L^N/L^N \times L^N/L^N$	41
Total	741

a. Show whether the population is in Hardy-Weinberg equilibrium with respect to MN blood types.

b. Show whether mating is random with respect to MN blood types.

(Problem 10 is from J. Kuspira and G. W. Walker, *Genetics: Questions and Problems*. Copyright 1973 by McGraw-Hill.)

11. Consider the populations that have the genotypes shown in the following table:

Population	*A/A*	*A/a*	*a/a*
1	1.0	0.0	0.0
2	0.0	1.0	0.0
3	0.0	0.0	1.0
4	0.50	0.25	0.25
5	0.25	0.25	0.50
6	0.25	0.50	0.25
7	0.33	0.33	0.33
8	0.04	0.32	0.64
9	0.64	0.32	0.04
10	0.986049	0.013902	0.000049

a. Which of the populations are in Hardy-Weinberg equilibrium?

b. What are p and q in each population?

c. In population 10, it is discovered that the $A \rightarrow a$ mutation rate is 5×10^{-6} and that reverse mutation is negligible. What must be the fitness of the *a/a* phenotype?

d. In population 6, the *a* allele is deleterious; furthermore, the *A* allele is incompletely dominant, so that *A/A* is perfectly fit, *A/a* has a fitness of 0.8, and *a/a* has a fitness of 0.6. If there is no mutation, what will p and q be in the next generation?

12. Color blindness results from a sex-linked recessive allele. One in every ten males is color-blind.

a. What proportion of all women are color-blind?

b. By what factor is color blindness more common in men (or, how many color-blind men are there for each color-blind woman)?

c. In what proportion of marriages would color blindness affect half the children of each sex?

d. In what proportion of marriages would all children be normal?

e. In a population that is not in equilibrium, the frequency of the allele for color blindness is 0.2 in women and 0.6 in men. After one generation of random mating, what proportion of the female progeny will be color-blind? What proportion of the male progeny?

f. What will the allele frequencies be in the male and in the female progeny in part e?

(Problem 12 courtesy of Clayton Person.)

13. It seems clear that most new mutations are deleterious. Why?

14. Most mutations are recessive to the wild type. Of those rare mutations that are dominant in *Drosophila*, for example, the majority turn out either to be chromosomal mutations or to be inseparable from chromosomal mutations. Explain why the wild type is usually dominant.

15. Ten percent of the males of a large and randomly mating population are color-blind. A representative group of 1000 people from this population migrates to a South Pacific island, where there are already 1000 inhabitants and where 30 percent of the males are color-blind. Assuming that Hardy-Weinberg equilibrium applies throughout (in the two original populations before the migration and in the mixed population immediately after the migration), what fraction of males and females can be expected to be color-blind in the generation immediately after the arrival of the migrants?

16. Using pedigree diagrams, find the probability of homozygosity by descent of the offspring of (a) parent-offspring matings; (b) first-cousin matings; (c) aunt-nephew or uncle-niece matings.

17. In an animal population, 20 percent of the individuals are *A/A*, 60 percent are *A/a*, and 20 percent are *a/a*.

a. What are the allele frequencies in this population?

b. In this population, mating is always with *like phenotype* but is random within phenotype. What genotype and allele frequencies will prevail in the next generation?

c. Another type of assortative mating takes place only between *unlike* phenotypes. Answer the preceding question with this restriction imposed.

d. What will the end result be after many generations of mating of each type?

18. A *Drosophila* stock isolated from nature has an average of 36 abdominal bristles. By the selective breeding of only those flies with the most bristles, the mean is raised to 56 bristles in 20 generations.

a. What is the source of this genetic flexibility?

b. The 56-bristle stock is infertile, so selection is relaxed for several generations and the bristle number drops to about 45. Why doesn't it drop back to 36?

c. When selection is reapplied, 56 bristles are soon attained, but this time the stock is *not* infertile. How can this situation arise?

19. Allele B is a deleterious autosomal dominant. The frequency of affected individuals is 4.0×10^{-6}. The reproductive capacity of these individuals is about 30 percent that of normal individuals. Estimate μ, the rate at which b mutates to its deleterious allele B.

20. Of 31 children born of father-daughter matings, 6 died in infancy, 12 were very abnormal and died in childhood, and 13 were normal. From this information, calculate roughly how many recessive lethal genes we have, on average, in our human genomes. (**Hint:** If the answer were 1, then a daughter would stand a 50 percent chance of carrying the lethal allele, and the probability of the union's producing a lethal combination would be $1/2 \times 1/4 = 1/8$. So 1 is not the answer.) Consider also the possibility of undetected fatalities in utero in such matings. How would they affect your result?

21. If we define the *total selection cost* to a population of deleterious recessive genes as the loss of fitness per individual affected (s) multiplied by the frequency of affected individuals (q^2), then

$$\text{selection cost} = sq^2$$

a. Suppose that a population is at equilibrium between mutation and selection for a deleterious recessive allele, where $s = 0.5$ and $\mu = 10^{-5}$. What is the equilibrium frequency of the allele? What is the selection cost?

b. Suppose that we start irradiating individual members of the population, so that the mutation rate doubles. What is the new equilibrium frequency of the allele? What is the new selection cost?

c. If we do not change the mutation rate, but we lower the selection intensity to 0.3 instead, what happens to the equilibrium frequency and the selection cost?

EXPLORING GENOMES   Interactive Genetics MegaManual CD-ROM Tutorial

Population Genetics

This activity on the Interactive Genetics CD-ROM includes five interactive problems designed to improve your understanding of what we can learn by looking at the distribution of genes in populations.

20

QUANTITATIVE GENETICS

The composite flowers of *Gaillardia pulchella.* Quantitative variation in flower color, flower diameter, and number of flower parts. [J. Heywood, *Journal of Heredity,* May/June 1986.]

KEY QUESTIONS

- For a particular character, how do we answer the question, Is the observed variation in the character influenced *at all* by genetic variation? Are there alleles segregating in the population that produce some differential effect on the character or is all the variation simply the result of environmental variation and developmental noise (see Chapter 1)?

- If there is genetic variation, what are the norms of reaction of the various genotypes?

- For a particular character, how important is genetic variation as a source of total phenotypic variation? Are the norms of reaction and the environments such that nearly all the variation is a consequence of environmental difference and developmental instabilities or does genetic variation predominate?

- Do many loci (or only a few) vary with respect to a particular character? How are they distributed throughout the genome?

OUTLINE

20.1 Genes and quantitative traits

20.2 Some basic statistical notions

20.3 Genotypes and phenotypic distribution

20.4 Norm of reaction and phenotypic distribution

20.5 Determining norms of reaction

20.6 The heritability of a quantitative character

20.7 Quantifying heritability

20.8 Locating genes

CHAPTER OVERVIEW

Ultimately, the goal of genetics is the analysis of the genotypes of organisms. But a genotype can be identified—and therefore studied—only through its effect on the phenotype. We recognize two genotypes as different from each other because the phenotypes of their carriers are different. Basic genetic experiments, then, depend on the existence of a simple relation between genotype and phenotype. That is why studies of DNA sequences are so important, because we can read off the genotype directly.

In general, we hope to find a uniquely distinguishable phenotype for each genotype and only a simple genotype for each phenotype. At worst, when one allele is completely dominant, it may be necessary to perform a simple genetic cross to distinguish the heterozygote from the homozygote. Where possible, geneticists avoid studying genes that have only partial penetrance and incomplete expressivity (see Chapter 6) because of the difficulty of making genetic inferences from such traits. Imagine how difficult (if not impossible) it would have been for Benzer to study mutations within the *rII* gene in phage, if the only effect of the *rII* mutants was a 5 percent reduction from wild type in their ability to grow on *E. coli* K. For the most part, then, the study of genetics presented in the preceding chapters has been the study of allelic substitutions that cause *qualitative* differences in phenotype—clear-cut differences such as purple flowers versus white flowers.

However, most actual variation between organisms is quantitative, not qualitative. Wheat plants in a cultivated field or wild asters at the side of the road are not neatly sorted into categories of "tall" and "short," any more than humans are neatly sorted into categories of "black" and "white." Height, weight, shape, color, metabolic activity, reproductive rate, and behavior are characteristics that vary more or less continuously over a range (Figure 20-1). Even when the character is intrinsically countable (such as eye facet or bristle number in *Drosophila*), the number of distinguishable classes may be so large that the variation is nearly continuous. If we consider extreme individuals—say, a corn plant 8 feet tall and another one 3 feet tall—a cross between them will not produce a Mendelian result. Such a corn cross will produce plants about 6 feet tall, with some clear variation among siblings. The F_2 from selfing the F_1 will not fall into two or three discrete height classes in ratios of $3:1$ or $1:2:1$. Instead, the F_2 will be continuously distributed in height from one parental extreme to the other.

How do we study quantitative traits when they show such a complex relation between genotype and phenotype? The analysis of a continuously varying character can be carried out by an array of investigations, shown schematically in Figure 20-2:

- Norm of reaction studies, in which different genotypes are allowed to develop in an array of different environments to determine the interaction of genotype and environment in the development of the character.

- Selection studies, in which successive generations are produced from the extreme individuals in the

Figure 20-1 Quantitative inheritance of bract color in Indian paintbrush *(Castilleja hispida)*. The photograph on the left shows the extremes of the color range, and the one on the right shows examples from throughout the phenotypic range.

CHAPTER OVERVIEW Figure

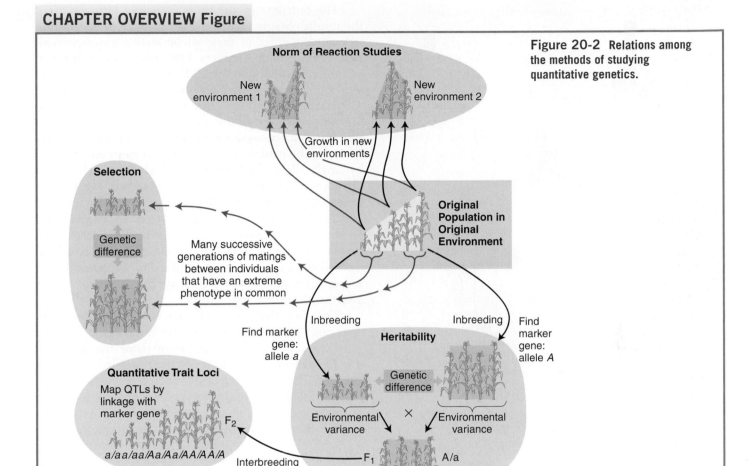

Norm of Reaction Studies

New environment 1

New environment 2

Growth in new environments

Original Population in Original Environment

Selection

Genetic difference

Many successive generations of matings between individuals that have an extreme phenotype in common

Find marker gene: allele a

Inbreeding

Inbreeding

Find marker gene: allele A

Heritability

Genetic difference

Quantitative Trait Loci

Map QTLs by linkage with marker gene

F_2

Environmental variance

$\times$

Environmental variance

$a/aa/aa/Aa/Aa/AA/AA/A$

Interbreeding of F_1

F_1

A/a

Environmental variance

Figure 20-2 Relations among the methods of studying quantitative genetics.

preceding generation. For example we might establish one population from the cross of the two shortest plants and another population from the cross of the two tallest corn plants in the preceding example. Then, in each successive generation, the "short" population and the "tall" population would be bred from the most extreme individuals in each. If, after repeated generations of selection, the populations diverge, then the divergent populations must differ genetically at one or more loci influencing the character.

- Heritability studies, in which the variation in the progeny of crosses is analyzed statistically to estimate the proportion of the variation in the original population that is a consequence of genetic differences and the proportion that is a consequence of environmental differences.

- Quantitative trait locus (QTL) studies, which associate phenotypic differences with alleles of a marker gene of known chromosomal location. Such an association with the marker gene reveals the

approximate location of a gene affecting the quantitative character.

20.1 Genes and quantitative traits

A classic example of the outcome of crosses between strains that differ in a quantitative character is the experiment shown in Figure 20-3. The length of the corolla (flower tube) was measured in a number of individual plants from two true-breeding lines of *Nicotiana longiflora*, a relative of tobacco. The distribution of corolla lengths of the two parental lines is shown in the top panel of Figure 20-3. The difference between the two lines is genetic, but the variation among individual plants within each line is a result of uncontrolled environmental variation and developmental noise. The F_1 plants, whose mean corolla length is very close to halfway between the two parental lines, also vary from one another because of environmental and developmental variation. In the F_2, the mean corolla length remains essentially unchanged from that of the F_1, but there is a

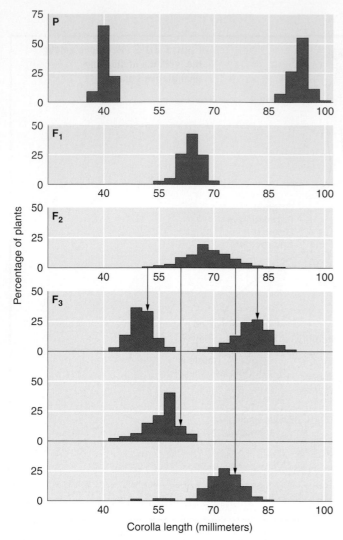

Figure 20-3 Results of crosses between strains of *Nicotiana longiflora* that differ in corolla length. The graphs show *(from top to bottom)* the frequency distribution of corolla lengths in the two parental strains (P); the frequency distribution of corolla lengths in the F_1; the frequency distribution in the F_2; and the frequency distributions in four F_3 crosses made by taking parents from the four indicated parts of the F_2 distribution. [Adapted from K. Mather, *Biometrical Genetics.* Methuen, 1959. Data from E. M. East, *Genetics* 1, 1916, 164–176.]

large increase in the variation because there is now segregation of the genetic differences that were introduced from the two original parental lines. A demonstration that at least part of this variation is the result of genetic differences among the F_2 plants is seen in the F_3. Different pairs of parents were chosen from four different parts of the F_2 distribution and crossed to produce the next, F_3, generation. In each case, the F_3 mean is close to the value of that part of the F_2 distribution from which its parents were sampled.

The outcome of the cross is clearly different from the results obtained when a cross is made between individuals that differ in their allelic state for a gene with clear-cut phenotypic effects. Offspring do not sort into neat Mendelian ratios of $1:2:1$, and there is much more individual variation within each generation of offspring. This behavior of crosses is not an exception; it is the rule for most characters in most species. Mendel obtained his simple results because he worked with horticultural varieties of the garden pea that differed from one another by single allelic differences that had drastic effects on phenotypes. The behavior of crosses seen in Figure 20-3 is not an exception; it is the rule for most characters in most species. Had Mendel conducted his experiments on the natural variation of the weeds in his garden, instead of on abnormal pea varieties, he would never have discovered any of his laws of heredity. In general, size, shape, color, physiological activity, and behavior do not assort in a simple way in crosses.

The fact that most characters vary continuously does not mean that their variation is the result of some genetic mechanisms different from those that apply to the Mendelian genes that we have studied in earlier chapters. The continuity of phenotype is a result of two phenomena. First, each genotype does not have a single phenotypic expression but rather a norm of reaction (see Chapter 1) that covers a wide phenotypic range. As a result, the phenotypic differences between genotypic classes become blurred, and we are not able to assign a particular phenotype unambiguously to a particular genotype.

Second, there may be many segregating loci having alleles that make a difference in the phenotype under observation. Suppose, for example, that five equally important loci affect the number of flowers that will develop in an annual plant and that each locus has two alleles (call them $+$ and $-$). For simplicity, also suppose that there is no dominance and that a $+$ allele adds one flower, whereas a $-$ allele adds nothing. Thus, there are $3^5 = 243$ different possible genotypes [three possible genotypes $(+/+, +/-,$ and $-/-)$ at each of five loci], ranging from

$$\frac{+\ +\ +\ +\ +}{+\ +\ +\ +\ +} \quad \text{through} \quad \frac{+\ +\ +\ +\ +}{-\ -\ -\ -\ -} \quad \text{to} \quad \frac{-\ -\ -\ -\ -}{-\ -\ -\ -\ -}$$

but there are only 11 phenotypic classes (10, 9, 8, . . . , 0) because many of the genotypes will have the same numbers of $+$ and $-$ alleles. For example, although there is only one genotype with 10 $+$ alleles and therefore an average phenotypic value of 10, there are 51 different genotypes with 5 $+$ alleles and 5 $-$ alleles; for example,

$$\frac{+\ +\ +\ +\ -}{+\ -\ -\ -\ -} \quad \text{and} \quad \frac{+\ +\ -\ +\ -}{+\ +\ -\ -\ -}$$

Thus, many different genotypes may have the same average phenotype. At the same time, because of environmental variation, two individuals of the same genotype may not have the same phenotype. This lack of a one-to-one correspondence between genotype and phenotype obscures the underlying Mendelian mechanism.

If we cannot study the behavior of the Mendelian factors controlling such traits directly, then what can we learn about their genetics? Clearly, the methods used to analyze qualitative traits—such as examining the ratios of offspring in a genetic cross—will not work for quantitative traits. Instead, we have to use statistical methods to make predictions about the inheritance of phenotypes in the absence of knowledge about underlying genotypes. This approach is known as quantitative genetics. **Quantitative genetics**—the study of the genetics of continuously varying characters—is concerned with answering the following questions:

1. Is the observed variation in a character influenced *at all* by genetic variation? Is all the variation simply the result of environmental variation and developmental noise (see Chapter 1)? Or, are there alleles segregating in the population that produce some differential effect on the character?

2. If there is genetic variation, what are the norms of reaction of the various genotypes?

3. How important is genetic variation as a source of total phenotypic variation? Is nearly all the variation a consequence of environmental difference and developmental instabilities or does genetic variation predominate?

4. Do many loci (or only a few) contribute to the variation in the character? How are they distributed throughout the genome?

In the end, the purpose of answering these questions is to be able to predict what kinds of offspring will be produced by crosses of different phenotypes.

The precision with which these questions can be framed and answered varies greatly. In experimental organisms, on the one hand, it is relatively simple to determine whether there is any genetic influence at all, but extremely laborious experiments are required to localize the genes (even approximately). In humans, on the other hand, it is extremely difficult to answer even the question of the presence of genetic influence for most traits, because it is almost impossible to separate environmental from genetic effects in an organism that cannot be manipulated experimentally. As a consequence, we know a lot about the genetics of bristle number in *Drosophila* but virtually nothing about the genetics of complex human traits; a few (such as skin color) clearly are influenced by genes, whereas others (such as the spe-

cific language spoken) clearly are not. In this chapter, we will develop the basic statistical and genetic concepts needed to answer these questions and provide some examples of the applications of these concepts to particular characters in particular species.

20.2 Some basic statistical notions

To consider the answers to these questions about the most common kinds of phenotypic variation, quantitative variation, we must first examine a number of statistical tools that are essential in the study of quantitative variation.

Statistical distributions

For a simple variation that depends only on the allelic differences at a single locus, the offspring of a cross will fall into several distinct phenotypic classes. For example, a cross between a red-flowered plant and a white-flowered plant might be expected to yield all red-flowered plants or, if it were a backcross of an F_1 plant to the white-flowered parent, 1/2 red-flowered plants and 1/2 white-flowered plants. However, we require a different mode of description for quantitative characters. If the heights of a large number of male undergraduates are measured to the nearest 5 centimeters (cm), they will vary (say, between 145 and 195 cm), but many more of these undergraduates will fall into the middle measurement classes (say, 170, 175, and 180 cm) than into the classes at the two extremes. Such a description of a set of quantitative measurements is known as a **statistical distribution**.

We can graph such measurements by representing each measurement class as a bar, with its height proportional to the number of individuals in that class, as shown in Figure 20-4a. Such a graph of numbers of individuals observed against measurement class is called a **frequency histogram**. Now suppose that we measure five times as many individuals, each to the nearest centimeter, so that we divide them into even smaller measurement classes, producing a histogram like the one shown in Figure 20-4b. If we continue this process, making each measurement finer but proportionately increasing the number of individuals measured, the histogram eventually takes on the continuous appearance of Figure 20-4c. Such a continuous curve is called the **distribution function** of the measure in the population.

The distribution function is an idealization of the actual frequency distribution of a measurement in any real population, because no measurement can be taken with infinite accuracy or on an unlimited number of individuals. Moreover, the measured character itself may

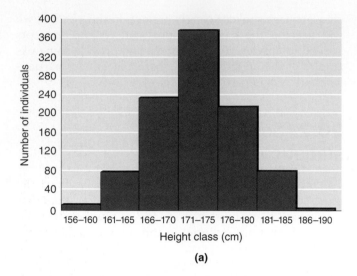

(a)

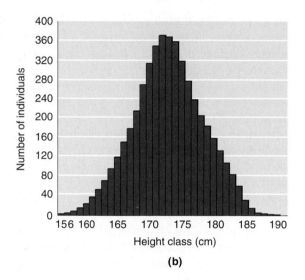

(b)

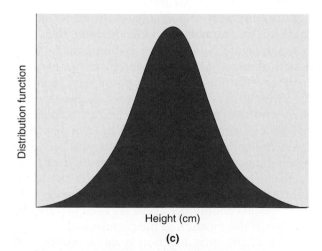

(c)

Figure 20-4 Frequency distributions for height of male undergraduates. (a) A frequency histogram with 5-cm class intervals; (b) a frequency histogram with 1-cm class intervals; (c) the limiting continuous distribution.

be intrinsically discontinuous because it is the count of some number of discrete objects, such as eye facets or bristles. It is sometimes convenient, however, to develop concepts by using this slightly idealized curve as shorthand for the more cumbersome observed frequency histogram.

Statistical measures

Although a statistical distribution contains all of the information that we need about a set of measurements, it is often useful to distill this information into a few characteristic numbers that convey the necessary information about the distribution without giving it in detail. There are several questions about the height distribution for male undergraduates, for example, that we might like to answer:

1. Where is the distribution located along the range of possible values? Are our observed values of height, for example, closer to 100 or to 200 cm? This question can be answered with a measure of **central tendency.**

2. How much variation is there among the individual measurements? Are they all concentrated around the central measurement or do they vary widely across a large range? This question can be answered with a measure of **dispersion.**

3. If we are considering more than one measured quantity, how are the values of the different quantities related? Do taller parents, for example, have taller sons? If they do, we would regard it as evidence that genes influence height. Thus, we need measures of **relation** between measurements.

Among the most commonly used measures of central tendency are the **mode,** which is the most frequent observation, and the **mean,** which is the arithmetic average of the observations. The dispersion of a distribution is almost always measured by the **variance,** which is the average squared deviation of the observations from their mean. The relation between different variables is measured by their **correlation,** which is the average product of the deviation of one variable from its own mean times the deviation of the other variable from its own mean. These common measures are discussed in detail in the Statistical Appendix on statistical analysis at the end of this chapter. The detailed discussion of these statistical concepts is placed in a separate section so as not to interrupt the flow of logic as we consider quantitative genetics. It should not be assumed, however, that an understanding of these statistical concepts is somehow secondary. A proper understanding of quantitative genetics requires a grasp of the basics of statistical analysis.

20.3 Genotypes and phenotypic distribution

The critical difference between quantitative and Mendelian traits

Using the concepts of distribution, mean, and variance, we can understand the difference between quantitative and Mendelian genetic traits.

Suppose that a population of plants contains three genotypes, each of which has some differential effect on growth rate. Furthermore, assume that there is some environmental variation from plant to plant because the soil in which the population is growing is not homogeneous and that there is some developmental noise (see Chapter 1). For each genotype, there will be a separate distribution of phenotypes with a mean and a variance that depend on the genotype and the set of environments. Suppose that these distributions look like the three height distributions in Figure 20-5a. The three distributions are concentrated at three different places on the scale of plant height indicating differences in mean height. The three distributions also have different amounts of spread, which results in their having different variances. Finally, assume that the population consists of a mixture of the three genotypes but in the unequal proportions $1:2:3$ ($a/a:A/a:A/A$).

Under these circumstances, the phenotypic distribution of individual plants in the population as a whole will look like the black line in Figure 20-5b, which is the result of summing the three underlying separate genotypic distributions, weighted by their frequencies in the population. This weighting by frequency is indicated in Figure 20-5b by the different heights of the component distributions. The mean of this total distribution is the average of the three genotypic means, again weighted by the frequencies of the genotypes in the population. The variance of the total distribution is produced partly by the environmental variation within each genotype and partly by the slightly different means of the three genotypes.

Two features of the total distribution are noteworthy. First, there is only a single mode, the most frequent observation represented by the location on the height axis of the peak of the curve. Despite the existence of three separate genotypic distributions underlying it, the population distribution as a whole does not reveal the separate modes. Second, any individual plant whose height lies between the two arrows could have any one of the three genotypes, because the phenotypes of those three genotypes overlap so much. The result is that we cannot carry out a simple Mendelian analysis to determine the genotype of an individual plant. For example, suppose that the three genotypes are the two homozygotes and the heterozygote for a pair of alleles at a locus. Let a/a be the short homozygote and A/A be the tall one, with the heterozygote being of intermediate height. Because the phenotypic distributions overlap so much, we cannot know to which genotype a given individual plant belongs. Conversely, if we cross a homozygote a/a and a heterozygote A/a, the offspring will not fall into two discrete classes, A/a and a/a, in a $1:1$ ratio but will cover almost the entire range of phenotypes smoothly. Thus, we cannot know from looking at the offspring that the cross is in fact $a/a \times A/a$ and not $a/a \times A/A$ or $A/a \times A/a$.

Suppose we grew the hypothetical plants in Figure 20-5 in an environment that exaggerated the differences between genotypes—for example, by doubling the growth rate of all genotypes. At the same time, we were very careful to provide all plants with exactly the same environment. Then, the phenotypic variance of each separate genotype would be reduced because all the plants were grown under identical conditions; at the same time, the phenotypic differences between genotypes would be exaggerated by the more rapid growth. (Figure 20-6a). The result (Figure 20-6b) would be a separation of the population as a whole into three nonoverlapping phenotypic distributions, each characteristic of one genotype. We could now carry out a perfectly conventional Mendelian analysis of plant height. A

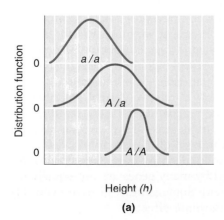

(a)

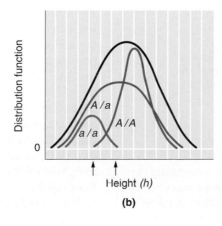

(b)

Figure 20-5 Genotype distribution. (a) Phenotypic distributions of three plant genotypes. (b) A phenotypic distribution for the total population (black line) can be obtained by summing the three genotypic distributions in a proportion $1:2:3$ ($a/a:A/a:A/A$).

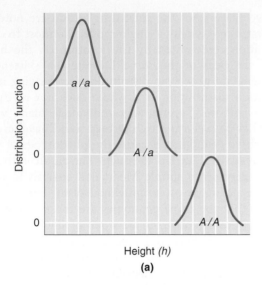

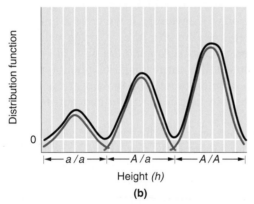

Figure 20-6 **Phenotypic distributions of the same three plant genotypes shown in Figure 20-5 when grown in carefully controlled environments.** The result is a smaller phenotypic variation within each genotype and a greater difference between genotypes. The heights of the individual distributions in part b are proportional to the frequencies of the genotypes in the population.

"quantitative" character has been converted into a "qualitative" one. This conversion has been accomplished by finding a way to make the difference between the means of the genotypes large compared with the variation within genotypes.

> **MESSAGE** A quantitative character is one for which the average phenotypic differences between genotypes are small compared with the variation between individuals within genotypes.

Gene number and quantitative traits

Continuous variation in a character is sometimes assumed to be necessarily caused by a large number of segregating genes, and so continuous variation is taken as

evidence that a character is controlled by many genes. This **multiple-factor hypothesis** (that large numbers of genes, each with a small effect, are segregating to produce quantitative variation) has long been the basic model of quantitative genetics, but, as we have just shown, this hypothesis is not necessarily true. If the difference between genotypic means is small compared with the environmental variance, then even a simple one-gene–two-allele case can result in continuous phenotypic variation.

If the range of a character is limited and if many segregating loci influence it, then we expect the character to show continuous variation, because each allelic substitution must account for only a small difference in the trait. It is important to remember, however, that the *number* of segregating loci that influence a trait is not what separates quantitative and qualitative characters. Even in the absence of large environmental variation, it takes only a few genetically varying loci to produce variation that is indistinguishable from the effect of many loci of small effect. As an example, we can consider one of the earliest experiments in quantitative genetics, that of Wilhelm Johannsen on pure lines. By **inbreeding** (mating close relatives), Johannsen produced 19 homozygous lines of bean plants from an originally genetically heterogeneous population. Each line had a characteristic average seed weight. These weights ranged widely from 0.64 g per seed for the heaviest line to 0.35 g per seed for the lightest line. Suppose all these lines *were* genetically different. In that case, Johannsen's results would be incompatible with a simple one-locus–two-allele model of gene action. If the original population were segregating for the two alleles A and a, all inbred lines derived from that population would have to fall into one of two classes: A/A or a/a. If, in contrast, there were, say, 100 loci, each of small effect, segregating in the original population, then a vast number of different inbred lines could be produced, each with a different combination of homozygotes at different loci.

However, we do not need such a large number of loci to obtain the results observed by Johannsen. If there were only five loci, each with three alleles, then $3^5 = 243$ different kinds of homozygotes could be produced from the inbreeding process. If we make 19 inbred lines at random, there is a good chance (about 50 percent) that each of the 19 lines will belong to a different one of the 243 classes. So Johannsen's experimental results can be easily explained by a relatively small number of genes. Thus, there is no real dividing line between multigenic traits and other traits. It is safe to say that no phenotypic trait above the level of the amino acid sequence in a polypeptide is influenced by only one gene. Moreover, traits influenced by many genes are not equally influenced by all of them. Some genes will have major effects on a trait; others, minor effects.

> **MESSAGE** The critical difference between Mendelian and quantitative traits is not the number of segregating loci but the size of phenotypic differences between genotypes compared with the individual variation within genotypic classes.

20.4 Norm of reaction and phenotypic distribution

The phenotype of an organism depends not only on its genotype but also on the environment that it has experienced at various critical stages in its development. For a given genotype, different phenotypes will develop in different environments. The relation between environment and phenotype for a given genotype is called the genotype's **norm of reaction.** The norm of reaction of a genotype with respect to some environmental variable—say, temperature—can be visualized by a graph showing phenotype as a function of that variable, as exemplified in Figure 20-7, the norms of reaction of abdominal bristle number for different genotypes of *Drosophila*.

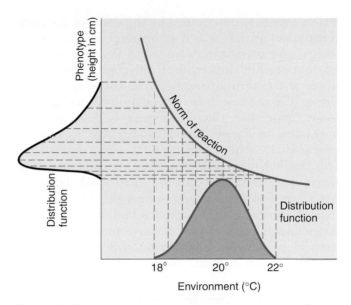

Figure 20-8 Outcomes of norm of reaction study. The distribution of environments on the horizontal axis is converted into the distribution of phenotypes on the vertical axis by the norm of reaction of a genotype.

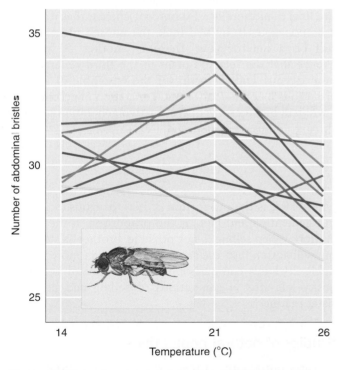

Figure 20-7 Norms of reaction for *Drosophila* bristle number. The number of abdominal bristles in different homozygous genotypes of *Drosophila pseudoobscura* at three different temperatures. Each colored line represents a different genotype. [Data courtesy of A. P. Gupta. Image: Plate IV, University of Texas Publication 4313, *Studies in the Genetics of* Drosophila *III: The Drosophilidae of the Southwest*, by J. T. Patterson. Courtesy of the Life Sciences Library, University of Texas, Austin.]

The phenotypic distribution of a character, as we have seen, is a function of the average phenotypic differences between genotypes and of the phenotypic variation among genotypically identical individuals. But, as the norms of reaction in Figure 20-7 show, both are functions of the environments in which the organisms develop and live. For a given genotype, each environment will result in a given phenotype (for the moment, ignoring developmental noise). Thus, for any given genotype, a *distribution of environments* will result in a *distribution of phenotypes*.

How different environments affect the phenotype of an organism depends on the norm of reaction, as shown in Figure 20-8, in which the horizontal axis represents environment (say, temperature) and the vertical axis represents phenotype (say, plant height). The norm of reaction curve for a genotype shows how each particular temperature results in a particular plant height. This norm of reaction converts a distribution of environments into a distribution of phenotypes. Thus, for example, the dashed line from the 18°C point on the horizontal environment axis is reflected off the norm of reaction curve to a corresponding plant height on the vertical phenotype axis, and so forth for each temperature. If a large number of plants develop at, say, 20°C, then a large number of plants will have the phenotype that corresponds to 20°C, as shown by the dashed line from the 20°C point; if only small numbers develop at 18°C, few plants will have the corresponding plant height. In other words, the frequency distribution of developmental environments will be reflected as a frequency distribution of phenotypes as determined by the shape of the norm of reaction curve. It is as if an observer, standing at the vertical phenotype axis, were seeing the environmental distribution, not directly, but reflected in the curved mirror of the norm of reaction. The shape of its curvature will determine how the environmental distribution is distorted on the phenotype axis. Thus, the norm of reaction in Figure 20-8 falls very rapidly at lower temperatures (the phenotype changes dramatically with small changes in temperature) but flattens out at higher temperatures, showing that plant height is much less sensitive to temperature differences at the higher temperatures. The result is that the symmetrical environmental distribution is converted into an asymmetrical phenotypic distribution with a long tail at the larger plant heights, corresponding to the lower temperatures.

> **MESSAGE** A distribution of environments is reflected biologically as a distribution of phenotypes. The transformation of environmental distribution into phenotypic distribution is determined by the norm of reaction.

20.5 Determining norms of reaction

Remarkably little is known about the norms of reaction for any quantitative traits in any species—partly because determining a norm of reaction requires testing many individual members of identical (or near identical) genotype. In many plants, it is possible to produce identical clones by the simple method of cutting a single plant into many pieces and growing each piece into a complete plant. This method was used to produce the norms of reactions for *Achillea millefolium* shown in Figure 1-21. Animals are not easily clonable, however. For this reason, for example, we do not have a norm of reaction for any genotype for any human quantitative trait.

Domesticated plants and animals

To determine a norm of reaction, we must first create a group of genetically identical individuals—a homozygous line. These genetically identical individuals can then be allowed to develop in different environments to determine a norm of reaction. Alternatively, two different homozygous lines can be crossed and the heterozygous F_1 offspring, all genetically identical with one another, can be tested in different environments.

A few norm of reaction studies have been carried out with plants that can be clonally propagated. The results of one of these experiments are presented in Chap-

ter 1. It is possible to replicate genotypes in sexually reproduced organisms by the technique of mating close relatives, or inbreeding. By selfing (where possible) or by mating brother and sister repeatedly generation after generation, a **segregating line** (one that contains both homozygotes and heterozygotes at a locus) can be made homozygous.

Ideally for a norm of reaction study, all the individuals should be absolutely identical genetically, but the process of inbreeding increases the homozygosity of the group slowly, generation after generation, depending on the closeness of the relatives that are mated. In corn, for example, a single individual plant is chosen and self-pollinated. Then, in the next generation, a single one of its offspring is chosen and self-pollinated. In the third generation, a single one of *its* offspring is chosen and self-pollinated, and so forth. Suppose that the original plant in the first generation is already a homozygote at some locus. Then all of its offspring from self-pollination also will be homozygous and identical at the locus. Future generations of self-pollination will simply preserve the homozygosity. If, on the other hand, the original plant is a heterozygote, then the selfing $A/a \times A/a$ will produce offspring that are $\frac{1}{4}$ A/A homozygotes and $\frac{1}{4}$ a/a homozygotes. If a single offspring is chosen to propagate the line, then there is a 50 percent chance that it is now a homozygote. If, by bad luck, the chosen plant should still be a heterozygote, there is another 50 percent chance that the selected plant in the third generation is homozygous, and so forth. Of the ensemble of all heterozygous loci, then, after one generation of selfing, only $\frac{1}{2}$ will still be heterozygous; after two generations, $\frac{1}{4}$; after three, $\frac{1}{8}$. In the nth generation,

$$\mathrm{Het}_n = \frac{1}{2^n}\mathrm{Het}_0$$

where Het_n is the proportion of heterozygous loci in the nth generation and Het_0 is the proportion in the 0 generation. When selfing is not possible, brother–sister mating will accomplish the same end, although more slowly. Table 20-1 is a comparison of the amount of heterozygosity left after n generations of selfing and brother–sister mating.

Studies of natural populations

To carry out a norm of reaction study of a natural population, a large number of lines are sampled from the population and inbred for a sufficient number of generations to guarantee that each line is virtually homozygous at all its loci. Each line is then homozygous at each locus for a randomly selected allele present in the original population. The inbred lines themselves cannot be used to characterize norms of reaction in the natural popula-

TABLE 20-1 Heterozygosity Remaining After Various Generations of Inbreeding for Two Systems of Mating

Generation	Remaining Heterozygosity	
	Selfing	Brother–sister mating
0	1.000	1.000
1	0.500	0.750
2	0.250	0.625
3	0.125	0.500
4	0.0625	0.406
5	0.03125	0.338
10	0.000977	0.114
20	1.05×10^{-6}	0.014
n	$\mathrm{Het}_n = \frac{1}{2}\mathrm{Het}_{n-1}$	$\mathrm{Het}_n = \frac{1}{2}\mathrm{Het}_{n-1} + \frac{1}{4}\mathrm{Het}_{n-2}$

tion, because such totally homozygous genotypes do not exist in the original population. Each inbred line can be crossed to every other inbred line to produce heterozygotes that reconstitute the original population, and an arbitrary number of individuals from each cross can be produced. If inbred line 1 has the genetic constitution $A/A \cdot B/B \cdot c/c \cdot d/d \cdot E/E \ldots$ and inbred line 2 is $a/a \cdot B/B \cdot C/C \cdot d/d \cdot e/e \ldots$, then a cross between them will produce a large number of offspring, all of whom are identically $A/a \cdot B/B \cdot C/c \cdot d/d \cdot E/e \ldots$ and can be raised in different environments.

Results of norm of reaction studies

Very few norm of reaction studies have been carried out for quantitative characters found in natural populations, but many have been carried out for domesticated species such as corn, which can be self-pollinated, or strawberries, which can be clonally propagated. The outcomes of such studies resemble those given in Figure 20-7. No genotype consistently produces a phenotypic value above or below that of the others under all environmental conditions. Instead, there are small differences between genotypes, and the direction of these differences varies over a wide range of environments.

These features of norms of reaction have important consequences. One consequence is that selection for "superior" genotypes in domesticated animals and cultivated plants will result in varieties adapted to very specific conditions, which may not show their superior properties in other environments. To some extent, this problem can be overcome by deliberately testing genotypes in a range of environments (for example, over several years and in several locations). It would be even better, however, if plant breeders could test their selections in a variety of controlled environments in which different environmental factors could be separately manipulated.

The consequences of actual plant-breeding practices can be seen in Figure 20-9, in which the yields of two varieties of corn are shown as a function of different farm environments. Variety 1 is an older variety of hybrid corn; variety 2 is a later "improved" hybrid. Their performances are compared at a low planting density, which was usual when variety 1 was developed, and at a high planting density, characteristic of farming practice when variety 2 was created. At the high planting density, variety 2 is clearly superior to variety 1 in all environments (Figure 20-9a). At the low planting density (Figure 20-9b), however, the situation is quite different. First, note that the new variety is less sensitive to environmental variation than the older hybrid, as evidenced by its flatter norm of reaction. Second, the new "improved" variety actually performs more poorly than the older variety under the best farm conditions. Third, the yield improvement of the new variety does not occur under the low planting densities characteristic of earlier agricultural practice.

The nature of norms of reaction also has implications for human social relations and policy. Even if it should turn out that there is genetic variation for various mental and emotional traits in the human species—which is by no means clear—this variation is unlikely to favor one genotype over another across a range of environments. We must beware of hypothetical norms of reaction for human cognitive traits that show one genotype being unconditionally superior to another. Even putting aside all questions of moral and political judgment, there is simply no basis for describing different human genotypes as "better" or "worse" on any scale, unless the investigator is able to make a very exact specification of environment.

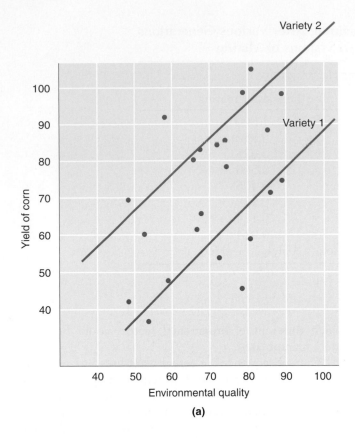

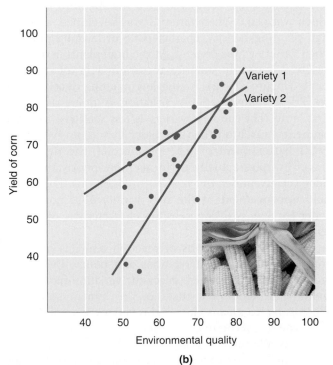

Figure 20-9 Environment and grain yield. Yields of grain of two varieties of corn in different environments: (a) at a high planting density; (b) at a low planting density. [Data courtesy of W. A. Russell, *Proceedings of the 29th Annual Corn and Sorghum Research Conference*, 1974. Photograph copyright by Bonnie Sue/Photo Researchers.]

> **MESSAGE** Norm of reaction studies show that, within a single environment, there are only small phenotypic differences between most genotypes in natural populations and that these differences are not consistent over a wide range of environments. Thus, "superior" genotypes in domesticated animals and cultivated plants may be superior only in certain environments. As with physical traits, if it should turn out that humans exhibit genetic variation for various mental and emotional traits, no one genotype is likely to outperform another across a range of environments.

20.6 The heritability of a quantitative character

The most basic question that we can ask about a quantitative character is whether the observed variation in that character is influenced by genes at all. It is important to note that this question is not the same as asking whether genes play any role in the character's development. Gene-mediated developmental processes lie at the base of every character, but *variation* in a character from individual to individual is not necessarily the result of *genetic variation*. For example, the ability to speak any language at all depends critically on the structures of the central nervous system as well as on the vocal cords, tongue, mouth, and ears, which depend in turn on many genes in the human genome. There is no environment in which cows will speak. But, although the particular language that is spoken by humans varies from nation to nation, this variation is not genetic. A character is said to be **heritable** only if there is genetic variation in that character.

> **MESSAGE** The question "Is a trait is heritable?" is a question about the role that differences in genes play in the phenotypic differences between individuals or groups.

Familiality and heritability

In principle, it is easy to determine whether any genetic variation influences the phenotypic variation in a particular trait. If genes play a role, then (on average) biological relatives should resemble one another more than unrelated individuals do. This resemblance would be seen as a positive correlation in the values of a trait between parents and offspring or between siblings (offspring of the same parents). Parents who are larger than the average, for example, would have offspring who are larger than the average; the more seeds that a plant produces, the more seeds that its siblings would produce. Such correlations between relatives, however, are evidence for genetic variation *only if the relatives do not share common environments more than nonrelatives do*. It is absolutely fundamental to distinguish *familiality* from *heritability*. Character states are **familial** if members of the same family have them in common, for whatever reason. They are heritable only if the similarity arises from having genotypes in common.

There are two general methods for establishing the heritability of a trait as distinct from its familiality. The first depends on *phenotypic similarity* between relatives. For most of the history of genetics, this method has been the only one available, and so nearly all the evidence about heritability for most traits in experimental organisms and in humans has been established by using this approach. The second method, using *marker-gene segregation*, depends on showing that genotypes carrying different alleles of certain marker genes also differ in their average phenotype for a quantitative character. If the marker genes (which have nothing to do with the character under study) are seen to vary in relation to the character, then presumably they are linked to genes that *do* influence the character and its variation. Thus, heritability is demonstrated even if the actual genes causing the variation in the character are not known. This method requires that the organism being studied have large numbers of detectable, genetically variable marker loci spread throughout its genome. Such marker loci can be observed through variants in DNA sequence, electrophoretic studies of protein variation or, in vertebrates, immunological studies of blood-group proteins. Within flocks, for example, chickens with different blood groups show some difference in egg weight, but, as far as is known, the blood-group antigens and antibodies do not themselves cause the difference in egg size. Presumably, genes that do influence egg weight are linked to the loci determining blood group.

Since the introduction of molecular methods for studying DNA sequences, a great deal of genetic variation has been discovered in a great variety of organisms. This variation consists either of substitutions at single nucleotide positions or of variable numbers of insertions or repeats of short sections of DNA. These variations are usually detected by the gain or loss of recognition sites for restriction enzymes or by length variation in DNA sequences between two fixed restriction sites, both of which are forms of restriction fragment length polymorphism (RFLP; see Chapter 19). In tomatoes, for example, strains carrying different RFLP variants differ in fruit characteristics. It is assumed that the DNA sequences in these RFLPs do not themselves influence fruit characteristics; rather, they are landmarks located near genes that do and therefore show high levels of cosegregation for these characteristics.

Because so much of what is known or claimed about heritability still depends on phenotypic similarity between relatives, however, especially in human genetics,

we shall begin our examination of the problem of heritability by analyzing phenotypic similarity.

Phenotypic similarity between relatives

In experimental organisms, there is no problem in separating environmental from genetic similarities. The offspring of a cow producing milk at a high rate and the offspring of a cow producing milk at a low rate can be raised together in the same environment to see whether, despite the environmental similarity, each resembles its own parent. In natural populations, however, and especially in humans, this kind of study is difficult to perform. Because of the nature of human societies, members of the same family have not only genes in common, but also similar environments. Thus, the observation of simple familial similarity of phenotype is genetically uninterpretable. In general, people who speak Hungarian have Hungarian-speaking parents and people who speak Japanese have Japanese-speaking parents. Yet the massive experience of immigration to North America has demonstrated that these linguistic differences, although familial, are nongenetic. The highest correlations between parents and offspring for any social trait in the United States are those for political party and religious affiliation, but these traits are not heritable. The distinction between familiality and heredity is not always so obvious, however. The U.S. Public Health Commission, when it studied the vitamin-deficiency disease pellagra in the southern United States in 1910, came to the conclusion that it was genetic because it ran in families. However, pellagra is now well understood to have been prevalent in southern U.S. populations because of poor diet.

To determine whether a human trait is heritable, we must use studies of certain adopted persons to avoid the usual environmental similarity between biological relatives. The ideal experimental subjects are monozygotic (identical) twins reared apart because they are genetically identical but experience different environments. Such adoption studies must be so contrived that there is no correlation between the social environments of the adopting families and those of the biological families, or else the similarities between the twins' environments will not have been eliminated by the adoption. These requirements are exceedingly difficult to meet; therefore, in practice, we know very little about whether human quantitative characters that are familial are also heritable.

Skin color is clearly heritable, as is adult height—but even for characters such as these we must be very careful. We know that skin color is affected by genes, both from studies of cross-racial adoptions and from observations that the offspring of black African slaves were black even when they were born and reared in North America. But are the differences in height between Japanese and Europeans affected by genes? The children of Japanese immigrants who are born and reared in North America are taller than their parents but shorter than the North American average, and so we might conclude that there is some influence of genetic difference. However, second-generation Japanese Americans are even taller than their American-born parents. It appears that some environmental-cultural influence, possibly nutritional or perhaps an effect of maternal inheritance, is still felt in the first generation of births in North America. We cannot yet say anything definitive about genetic differences that might contribute to the height differences between North Americans of, say, Japanese and Swedish ancestry.

Personality traits, temperament, cognitive performance (including IQ scores), and a whole variety of behaviors such as alcoholism and mental disorders such as schizophrenia have been the subject of heritability studies in human populations. Many of these traits show familiality (that is, familial similarity). There is a positive correlation, for example, between the IQ scores of parents and the scores of their children (the correlation is about 0.5 in white American families), but this correlation does not distinguish familiality from heritability. To make that distinction requires that the environmental correlation between parents and children be broken, and so studies on adopted children are common. Because it is difficult to randomize environments, even in cases of adoption, evidence of heritability for human personality and behavior traits remains equivocal despite the very large number of studies that exist. Prejudices about the causes of human differences are widespread and deep, and, as a result, the canons of evidence adhered to in studies of the heritability of IQ, for example, have been much more lax than in studies of milk yield in cows.

Figure 20-10 summarizes the usual method of testing for heritability in experimental organisms. Individuals from both extremes of the phenotypic distribution are mated with other individuals of their own extreme group, and the offspring are raised in a common controlled environment. If there is an average difference between the two offspring groups, the phenotypic difference is heritable. Most morphological characters in *Drosophila*, for example, turn out to be heritable—but not all of them. If flies with right wings that are slightly longer than their left wings are mated with each other, their offspring have no greater tendency to be "right winged" than do the offspring of "left winged" flies. As we shall see, this method can also be used to obtain quantitative information about heritability.

> **MESSAGE** In experimental organisms, environmental similarity can often be readily distinguished from genetic similarity (or heritability). In humans, however, it is very difficult to determine whether a particular trait is heritable.

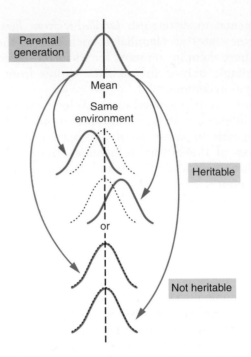

Figure 20-10　Standard method for testing for heritability in experimental organisms. Crosses are performed within two populations of individuals selected from the extremes of the phenotypic distribution in the parental generation. If the phenotypic distributions of the two groups of offspring are significantly different from each other (red curves), then the character difference is heritable. If both offspring distributions resemble the distribution for the parental generation (blue curves), then the phenotypic difference is not heritable.

20.7　Quantifying heritability

If a character is shown to be heritable in a population, then it is possible to quantify the degree of heritability. In Figures 20-5 and 20-6, we saw that the variation between phenotypes in a population arises from two sources. First, there are average differences between the genotypes; second, each genotype exhibits phenotypic variation because of environmental variation. The total phenotypic variance of the population (s_p^2) can thus be broken into two parts: the variance between genotypic means (s_g^2) and the remaining variance (s_e^2). The former is called the **genetic variance,** and the latter is called the **environmental variance;** however, as we shall see, these names are quite misleading. Moreover, the breakdown of the phenotypic variance into environmental and genetic variances leaves out the possibility of some covariance between genotype and environment. For example, suppose it were true (we do not know this) that there were genes that influence musical ability in humans. Parents with such genes might themselves be musicians, who would create a more musical environment for their children, who would then have both the genes and the environment promoting musical performance. The result would be a greater variance among people in musical ability than would be the case if there were no effect of the parental environment on children. If the phenotype is the sum of a genetic and an environmental effect, $p = g + e$, then the variance of the phenotype is, according to the formula on page 672, the sum of the genetic variance, the environmental variance, and twice the covariance between the genotypic and environmental effects:

$$s_p^2 = s_g^2 + s_e^2 + 2 \operatorname{cov} ge$$

If genotypes are not distributed randomly across environments but this is not taken into account, there will be some covariance between genotypic and environmental values, and that covariance will be hidden in the genetic and environmental variances.

The quantitative measure of heritability of a character is that part of the total phenotypic variance that is due to genetic variance:

$$H^2 = \frac{s_g^2}{s_p^2} = \frac{s_g^2}{s_g^2 + s_e^2}$$

H^2, so defined, is called the **broad heritability** of the character.

It must be stressed that this measure of "genetic influence" tells us what part of the population's *variation* in phenotype can be attributed to *variation* in genotype. It does not tell us what parts of an *individual's* phenotype can be ascribed to its genotype and to its environment. This latter distinction is not a reasonable one. An individual's phenotype is a consequence of the interaction between its genes and the sequence of environments that it experiences as it develops. It would be silly to say that 60 inches of your height were produced by your genes and 10 inches were then added by your environment. All measures of the "importance" of genes are framed in terms of the proportion of phenotypic variance ascribable to their variation. This approach is a special application of the more general technique of **analysis of variance,** used for apportioning relative weight to contributing causes. The technique was, in fact, invented originally to deal with experiments in which different environmental and genetic factors were influencing the growth of plants. (For a sophisticated but accessible treatment of the analysis of variance written for biologists, see R. Sokal and J. Rohlf, *Biometry*, 3d ed., W. H. Freeman and Company, 1995.)

Methods of estimating H^2

Heritability in a population can be estimated in several ways. Most directly, we can obtain an estimate of the environmental variance in the population, s_e^2, by making a number of homozygous lines, crossing them in pairs to make heterozygotes typical of the population, and measuring the phenotypic variance *within* each heterozygous genotype. Because all individuals within each group have the same genotype and therefore there is no genetic variance within groups, these variances will (when averaged) provide an estimate of s_e^2. This value can then be subtracted from the value of s_p^2 in the original population to give s_g^2. With the use of this method, any covariance between genotype and environment in the original population will be hidden in the estimate of genetic variance and will inflate it. So, for example, if individuals with genotypes that would make them taller on average over random environments were also given better nutrition than individuals with genotypes that would make them shorter, then the observed difference in heights between the two genotypic groups would be exaggerated.

Other estimates of genetic variance can be obtained by considering the genetic similarities between relatives. By using simple Mendelian principles, we can see that half the genes of two full siblings will (on average) be identical. For simplicity, we can label the alleles at a locus carried by the parents uniquely—say, as A_1/A_2 and A_3/A_4. The older sibling has a probability of 1/2 of getting A_1 from its father, as does the younger sibling, and so the two siblings have a chance of $1/2 \times 1/2 = 1/4$ of both carrying A_1. On the other hand, they might both receive A_2 from their father; so, again, they have a probability of 1/4 of carrying that allele. Thus, the chance is $1/4 + 1/4 = 1/2$ that both siblings will inherit the same allele (either A_1 or A_2) from their father. The other half of the time, one sibling will inherit an A_1 and the other will inherit an A_2. So, as far as paternally inherited genes are concerned, full siblings have a 50 percent chance of carrying the same allele. But the same reasoning applies to their maternally inherited allele. Averaging over their paternally and maternally inherited genes $[(1/2 + 1/2)/2 = 1/2]$, half the genes of full siblings will be identical between them. Their **genetic correlation,** which is equal to the chance that they carry the same allele, will be 1/2, or 0.5.

If we apply this reasoning to half-siblings, say, with a common father but with different mothers, we get a different result. Again, the two siblings have a 50 percent chance of inheriting an identical gene from their father, but this time they have no way of inheriting the same gene from their mothers because they have two different mothers. Averaging the maternally inherited and paternally inherited genes thus gives a probability of $(1/2 + 0)/2 = 1/4$ that these half-siblings will carry the same gene.

We might be tempted to use this theoretical correlation between relatives to estimate H^2. If the observed phenotypic correlation between siblings were, for example, 0.4, and we expected, on purely genetic grounds, a correlation of 0.5, then our estimate of heritability would be 0.4/0.5 = 0.8. But such an estimate fails to take into account the fact that the environments of siblings also may be correlated. Unless we are careful to raise the siblings in independent environments, our estimate of H^2 will be too large and could even exceed 1 if the observed phenotypic correlation were greater than 0.5.

To get around this problem, we use the *differences* between phenotypic correlations of different relatives. For example, the difference in genetic correlation between full and half-siblings is $1/2 - 1/4 = 1/4$. Let's contrast this with their **phenotypic correlations**. If the environmental similarity is the same for half- and full siblings—a very important condition for estimating heritability—then environmental similarities will cancel out if we take the difference in correlation between the two kinds of siblings. This difference in phenotypic correlation will then be proportional to how much of the variance is genetic. Thus:

$$\begin{pmatrix} \text{genetic correlation} \\ \text{of full siblings} \end{pmatrix} - \begin{pmatrix} \text{genetic correlation} \\ \text{of half-siblings} \end{pmatrix} = \tfrac{1}{4}$$

but

$$\begin{pmatrix} \text{phenotypic} \\ \text{correlation} \\ \text{of full siblings} \end{pmatrix} - \begin{pmatrix} \text{phenotypic} \\ \text{correlation} \\ \text{of half-siblings} \end{pmatrix} = H^2 \times \tfrac{1}{4}$$

and so an estimate of H^2 is:

$$H^2 = 4\left[\begin{pmatrix} \text{correlation} \\ \text{of full siblings} \end{pmatrix} - \begin{pmatrix} \text{correlation} \\ \text{of half-siblings} \end{pmatrix} \right]$$

where the correlation here is the *phenotypic* correlation.

This estimate, as well as others based on correlations between relatives, depends *critically* on the assumption that environmental correlations between individuals are the same for all degrees of relationship—which is unlikely to be the case. Full sibs, for example, are usually raised by the same pair of parents, whereas half-sibs are likely to be raised in circumstances with only one parent in common. If closer relatives have more similar environments, as they do among humans, these estimates of heritability will be biased. It is reasonable to assume that most environmental correlations between relatives are positive, in which case the heritabilities would be overestimated. But negative environmental correlations also can exist. For example, if the members of a litter must compete for food that is in short supply, there could be negative correlations in growth rates among siblings.

The difference in phenotypic correlation between monozygotic and dizygotic twins is commonly used in human genetics to estimate H^2 for cognitive or personality traits. Here, the problem of degree of environmental similarity is very severe. Monozygotic (identical) twins are generally treated more similarly to each other than are dizygotic (fraternal) twins. Parents often give their identical twins names that are similar, dress them alike, treat them in the same way, and, in general, accentuate their similarities. As a result, heritability is overestimated.

The meaning of H^2

Attention to the problems of estimating broad heritability distracts from the deeper questions about the meaning of the ratio even when it can be estimated. Despite its widespread use as a measure of how "important" genes are in influencing a character, H^2 actually has a special and limited meaning.

Two alternative conclusions can be drawn from the results of a properly designed heritability study. First, if there is a nonzero heritability, we can conclude that, in the population measured and in the environments in which the organisms have developed, genetic differences have influenced the phenotypic variation among individuals, and so genetic differences do matter to the trait. This finding is not trivial, and it is a first step in a more detailed investigation of the role of genes.

It is important to notice that the reverse is not true. If zero heritability for the trait is found, this finding is not a demonstration that genes are irrelevant to the trait; rather, it demonstrates only that, in the particular population and environment studied, either there is no genetic variation at the relevant loci or different genotypes have the same phenotype. In other populations or other environments, the character might be heritable.

> **MESSAGE** The heritability of a character difference is different in each population and in each set of environments; it cannot be extrapolated from one population and set of environments to another.

Moreover, we must distinguish between *genes* contributing to a trait and *genetic differences* contributing to *differences* in a trait. The natural experiment of immigration to North America has proved that the ability to pronounce the sounds of North American English, rather than French, Swedish, or Russian, is not a consequence of genetic differences between our immigrant ancestors. But, without the appropriate genes, we could not speak any language at all.

Second, the value of H^2 provides a limited prediction of how much a character can be modified by changing the environment. If all the relevant environmental variation is eliminated *and the new constant environment is the same as the mean environment in the original population*, then H^2 estimates how much phenotypic variation will still be present. So, if the heritability of performance on an IQ test were found to be, say, 0.4, then we could predict that, if all children had the same developmental and social environment as the "average child," about 60 percent of the variation in IQ test performance would disappear and 40 percent would remain.

The requirement that the new constant environment be at the mean of the old environmental distribution is absolutely essential to this prediction. If the environment is shifted toward one end or the other of the environmental distribution present in the population used to determine H^2 or if a new environment is introduced, nothing at all can be predicted. In the example of IQ test performance, the heritability gives us no information at all about how variable performance would be if the developmental and social environments of all children were enriched. To understand why this is so, we must return to the concept of the norm of reaction.

The separation of phenotypic variance into genetic and environmental components, s_g^2 and s_e^2, does not really separate the genetic and environmental causes of variation. Consider the results presented in Figure 20-9b. When the environment is poor (an environmental quality of 50), corn variety 2 has a much higher yield than variety 1, and so a population made up of a mixture of the two varieties would have a lot of genetic variance for yield in that environment. But, in a richer environment (scoring 75), there is no difference in yield between varieties 1 and 2, and so a mixed population would have no genetic variance at all for yield in that environment. Thus, *genetic* variance has been changed by changing the *environment*. On the other hand, variety 2 is less sensitive to environment than variety 1, as shown by the slopes of the two lines. So a population made up mostly of variety 2 would have a lower environmental variance than one made up mostly of variety 1. So, *environmental* variance in the population is changed by changing the proportion of *genotypes*.

As a consequence of the argument just given, we cannot predict just from knowing the heritability of a character difference how the distribution of variation in the character will change if either genotypic frequencies or environmental factors change markedly. So, for example, in regard to IQ test performance, knowing that the heritability is 0.4 in one environment does not allow us to predict how IQ test performance will vary among children in a different environment.

> **MESSAGE** A high heritability does not mean that a character is unaffected by the environment. Because genotype and environment interact to produce phenotype, no partition of variation into its genetic and environmental components can actually separate causes of variation.

All that high heritability means is that, for the particular population developing in the particular distribution of environments in which the heritability was measured, average differences between genotypes are large compared with environmental variation within genotypes. If the environment is changed, there may be large differences in phenotype.

Perhaps the best-known example of the erroneous use of heritability arguments to make claims about the changeability of a trait is that of human IQ performance and social success. Many studies have been made of the heritability of IQ performance in the belief that, if heritability is high, then various programs of education designed to increase intellectual performance are a waste of time. The argument is that, if a trait is highly heritable, then it cannot be changed much by environmental changes. But, irrespective of the value of H^2 for IQ test performance, the real error of the argument lies in equating high heritability with unchangeability. In fact, the heritability of IQ is *irrelevant* to the question of how changeable it is.

To see why this is so, let us consider the usual results of IQ studies on children who have been separated from their biological parents in infancy and reared by adoptive parents. Although these results vary quantitatively from study to study, they have three characteristics in common. First, because adoptive parents usually come from a better-educated population than do the biological parents who offer their children for adoption, they generally have higher IQ scores than those of the biological parents. Second, the adopted children have higher IQ scores than those of their biological parents. Third, the adopted children show a higher correlation of IQ scores with their biological parents than with their adoptive parents. The following table is a hypothetical data set that shows all these characteristics, in idealized form, to illustrate these concepts. The scores given for parents are meant to be the average of mother and father.

	Children	Biological parents	Adoptive parents
	110	90	118
	112	92	114
	114	94	110
	116	96	120
	118	98	112
	120	100	116
Mean	115	95	115

First, we can see that the scores of the children have a high correlation with those of their biological parents but a low correlation with those of their adoptive parents. In fact, in our hypothetical example, the correlation of children with biological parents is $r = 1.00$, but with adoptive parents it is $r = 0$. (Remember that a correlation between two sets of numbers does not mean that the two sets are identical, but that, for each unit of increase in one set, there is a constant proportional increase in the other set—see the Statistical Appendix on statistical analysis at the end of this chapter.) This perfect correlation with biological parents and zero correlation with adoptive parents means that $H^2 = 1$, given the arguments just developed. All the variation in IQ score among the children is explained by the variation in IQ score among the biological parents, who have had no chance to influence the environments of their children.

Second, however, we notice that the IQ score of each child is 20 points higher than that of its biological parents and that the mean IQ of the children is equal to the mean IQ of the adoptive parents. Thus, adoption has raised the average IQ of the children 20 points above the average IQ of their biological parents, and so, as a *group*, the children resemble their adoptive parents. So we have perfect heritability, yet high plasticity in response to environmental modification.

An investigator who is seriously interested in knowing how genes might constrain or influence the course of development of any character in any organism must study directly the norms of reaction of the various genotypes in the population over the range of projected environments. No less detailed information will do. Summary measures such as H^2 are not valuable in themselves.

> **MESSAGE** Heritability is not the opposite of phenotypic plasticity. A character may have perfect heritability in a population and still be subject to great changes resulting from environmental variation.

Narrow heritability

Knowledge of the broad heritability (H^2) of a character in a population is not very useful in itself, but a finer subdivision of phenotypic variance can provide important information for plant and animal breeders. The genetic variance can itself be subdivided into two components to provide information about gene action and the possibility of shaping the genetic composition of a population.

Our previous consideration of gene action suggests that the phenotypes of homozygotes and heterozygotes ought to have a simple relation. If one of the alleles encoded a less active gene product or one with no activity at all and if one unit of gene product were sufficient to allow full physiological activity of the organism, then we would expect complete dominance of one allele over the other, as Mendel observed for flower color in peas. If, on the other hand, physiological activity were proportional

to the amount of active gene product, we would expect the heterozygote phenotype to be exactly intermediate between the homozygotes (show no dominance).

For many quantitative traits, however, neither of these simple cases is the rule. In general, heterozygotes are not exactly intermediate between the two homozygotes but are closer to one or the other (show partial dominance), even though there is an equal mixture of the primary products of the two alleles in the heterozygote. Suppose that two alleles, *a* and *A*, segregate at a locus influencing height. In the environments encountered by the population, the mean phenotypes (heights) and frequencies of the three genotypes might be:

	a/a	*A/a*	*A/A*
Phenotype	10	18	20
Frequency	0.36	0.48	0.16

There is genetic variance in the population; the phenotypic means of the three genotypic classes are different. Some of the variance arises because there is an average effect on phenotype of substituting an allele *A* for an allele *a*; that is, the average height of all individuals with *A* alleles is greater than that of all individuals with *a* alleles. By defining the average effect of an allele as the average phenotype of all individuals that carry it, we necessarily make the average effect of the allele depend on the frequencies of the genotypes.

The average effect is calculated by simply counting the *a* and *A* alleles and multiplying them by the heights of the individuals in which they appear. Thus, 0.36 of all the individuals are homozygous *a/a*, each *a/a* individual has two *a* alleles, and the average height of *a/a* individuals is 10 cm. Heterozygotes make up 0.48 of the population, each has only one *a* allele, and the average phenotypic measurement of *A/a* individuals is 18 cm. The total "number" of *a* alleles is $2(0.36) + 1(0.48)$. Thus, the average effect of all the *a* alleles is:

$$\bar{a} = \text{average effect of } a = \frac{2(0.36)(10) + 1(0.48)(18)}{2(0.36) + 1(0.48)}$$

$$= 13.20 \text{ cm}$$

and, by a similar argument,

$$\bar{A} = \text{average effect of } A = \frac{2(0.16)(20) + 1(0.48)(18)}{2(0.16) + 1(0.48)}$$

$$= 18.80 \text{ cm}$$

This average difference in effect between *A* and *a* alleles, the **additive effect,** of 5.60 cm accounts for some of the variance in phenotype—but not for all of it. The heterozygote is not exactly intermediate between the homozygotes; there is some dominance.

We would like to separate the so-called additive effect caused by substituting *a* alleles for *A* alleles from the variation caused by dominance. The reason is that the effect of selective breeding depends on the additive variation and not on the variation caused by dominance. Thus, for purposes of plant and animal breeding or for making predictions about evolution by natural selection, we must determine the additive variation. An extreme example will illustrate the principle. Suppose that plant height is influenced by variation in a gene and that the phenotypic means and frequencies of three genotypes are:

	A/A	*A/a*	*a/a*
Phenotype	10	12	10
Frequency	0.25	0.50	0.25

It is apparent (and a calculation like the preceding one will confirm) that there is no average difference between the *a* and *A* alleles, because each has an effect of 11 units. So there is no *additive* variation, although there is obviously genetic variation because there is variation in phenotype between the genotypes. The tallest plants are heterozygotes. If a breeder attempts to increase height in this population by selective breeding, mating these heterozygotes together will simply reconstitute the original population. Selection will be totally ineffective. This example illustrates the general law that the effect of selection depends on the *additive* genetic variation and not on genetic variation in general.

The total genetic variance in a population can be subdivided into two components: **additive genetic variation** (s_a^2), the variance that arises because there is an average difference between the carriers of *a* alleles and the carriers of *A* alleles, and **dominance variance** (s_d^2), the variance that results from the fact that heterozygotes are not exactly intermediate between the monozygotes. Thus:

$$s_g^2 = s_a^2 + s_d^2$$

The total phenotypic variance can now be written as

$$s_p^2 = s_g^2 + s_e^2 = s_a^2 + s_d^2 + s_e^2$$

We define a new kind of heritability, the **heritability in the narrow sense (h^2)**, as

$$h^2 = \frac{s_a^2}{s_p^2} = \frac{s_a^2}{s_a^2 + s_d^2 + s_e^2}$$

It is this heritability, not to be confused with H^2, that is useful in determining whether a program of selective breeding will succeed in changing the population. The greater the h^2 is, the greater the fraction of the difference between selected parents and the population as a whole that will be preserved in the offspring of the selected parents.

> **MESSAGE** The effect of selection depends on the amount of *additive* genetic variance and not on the genetic variance in general. Therefore, it is the narrow heritability, h^2, not the broad heritability, H^2, that predicts response to selection.

Estimating the components of genetic variance

The different components of genetic variance can be estimated from covariance between relatives—the degree to which the phenotypes of pairs of relatives are correlated with each other—but the derivation of these estimates is beyond the scope of this text. There is, however, another way to estimate narrow heritability h^2 that reveals its real meaning. If, in two generations of a population, we plot the phenotype—say, height—of offspring against the average phenotype of their two parents (the **midparent value**), we may observe a relation like the one illustrated by the red line in Figure 20-11. The regression line will pass through the mean height of all the parents and the mean height of all the offspring, which will be equal to each other because no change has taken place in the population between generations. Moreover, taller parents have taller children and shorter parents have shorter children, and so the slope of the line is positive. But the slope is not unity; very short

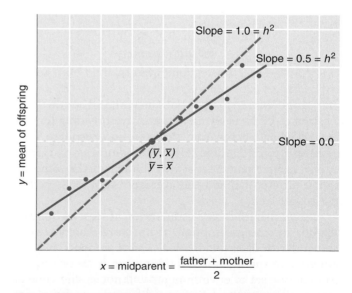

$$x = \text{midparent} = \frac{\text{father} + \text{mother}}{2}$$

Figure 20-11 The regression (red line) of offspring measurements *(y)* on midparent values *(x)* for a trait with narrow heritability *(h²)* of 0.5. The blue line would be the regression slope if the trait were perfectly heritable.

parents on average have children who are somewhat taller, and very tall parents on average have children who are somewhat shorter, than they themselves are. This slope of less than unity arises because heritability is less than perfect. If the phenotype were additively inherited with complete fidelity, then the heights of the offspring would be identical with the midparent values and the slope of the regression line would be 1. On the other hand, if the offspring had no heritable similarity to their parents, all parents would have offspring of the same average height and the slope of the line would be 0. This reasoning suggests that the slope of the regression line relating offspring value to the midparent value provides an estimate of additive heritability. In fact, the slope of the line can be shown mathematically to be a correct estimate of h^2.

The fact that the slope of the regression line estimates additive heritability allows us to use h^2 to predict the effects of artificial selection. Suppose that we select parents for the next generation who are, on average, 2 units above the general mean of the population from which they were chosen. If $h^2 = 0.5$, then the offspring of those selected parents will lie $0.5(2.0) = 1.0$ unit above the mean of the parental population, because the slope of the regression line predicts how much increase in y will result from a unit increase in x. We can define the **selection differential** as the difference between the selected parents and the mean of the entire population in their generation, and the **selection response** as the difference between the offspring of the selected parents and the mean of the parental generation. Thus

$$\text{selection response} = h^2 \times \text{selection differential}$$

or

$$h^2 = \frac{\text{selection response}}{\text{selection differential}}$$

The second expression provides us with yet another way to estimate h^2: by carrying out selective breeding for one generation and comparing the selection response with the selection differential. Usually this process is carried out for several generations with the use of the same selection differential, and the average response is used as an estimate of h^2.

Remember that any estimate of h^2, just as for H^2, depends on the assumption of no correlation between the similarity of the individuals' environments and the similarity of their genotypes. Moreover, h^2 in one population in one set of environments will not be the same as h^2 in a different population in a different set of environments. To illustrate this principle, Figure 20-12 shows the range of narrow-sense heritabilities reported in various studies for a number of characters in chickens. For

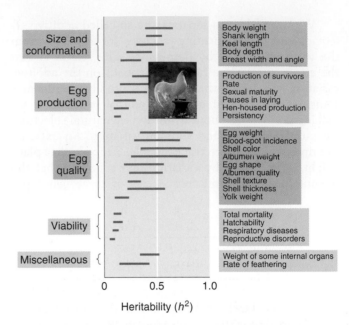

Figure 20-12 Ranges of heritabilities (h^2) reported for a variety of characters in chickens. [From I. M. Lerner and W. J. Libby, *Heredity, Evolution, and Society.* Copyright 1976 by W. H. Freeman and Company. Photograph copyright by Kenneth Thomas/Photo Researchers.]

most traits for which a substantial heritability has been reported, there are big differences from study to study, presumably because different populations have different amounts of genetic variation and because the different studies were carried out in different environments. Thus, breeders who want to know whether selection will be effective in changing some character in their chickens cannot count on the heritabilities found in earlier studies but must estimate the heritability in the particular population and particular environment in which the selection program is to be carried out.

Artificial selection

A vast record demonstrates the effectiveness of artificial selection in changing phenotypes within a population. Animal and plant breeding has, for example, increased milk production in cows and rust resistance in wheat. Selection experiments in the laboratory have made large changes in the physiology and morphology of many organisms including microorganisms, plants, and animals. No analysis of these experiments in terms of allelic frequencies is possible, because individual loci have not been identified and followed. Nevertheless, it is clear that genetic changes have taken place because the populations maintain their characteristics even after the selection has been terminated. Figure 20-13 shows, as an example, that a selection experiment achieved large changes in average bristle number in a population of *D.*

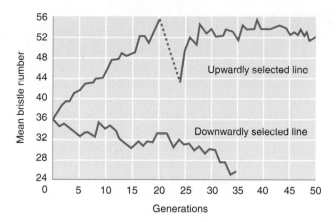

Figure 20-13 Effect of selection on bristle number. Changes in average bristle number obtained in two laboratory populations of *Drosophila melanogaster* through artificial selection for high bristle number in one population and low bristle number in the other. The dashed line shows five generations during which no selection was practiced. [From K. Mather and B. J. Harrison, "The Manifold Effects of Selection," *Heredity* 3, 1949, 1.]

melanogaster. Figure 20-14 shows the increase in the number of eggs laid per chicken as a consequence of 30 years of selection.

The usual method of selection for a continuously varying trait is **truncation selection.** The individuals in a given generation are pooled (irrespective of their families), a sample is measured, and only those individuals above (or below) a given phenotypic value (the truncation point) are chosen as parents for the next generation.

A common experience in artificial selection programs is that, as the population becomes more and more extreme, its viability and fertility decrease. As a result, eventually no further progress under selection is possible, despite the presence of additive genetic variance for

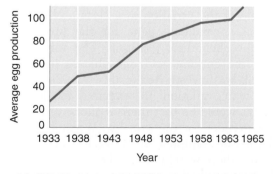

Figure 20-14 Effect of selection on egg production. Changes in average egg production in a chicken population selected for its increase in egg-laying rate over a period of 30 years. [From I. M. Lerner and W. J. Libby, *Heredity, Evolution, and Society*, 2d ed. Copyright 1976 by W. H. Freeman and Company. Data courtesy of D. C. Lowry.]

the character, because the selected individuals do not reproduce. The loss of fitness may be a direct phenotypic effect of the genes for the selected character, in which case nothing much can be done to improve the population further. Often, however, the loss of fitness is tied not to the genes that are under selection but to linked sterility genes that are carried along with them. In such cases, a number of generations are allowed to breed without selection until recombinants form by chance, freeing the genes under selection from their association with the sterility. Selection can then be continued, as in the upwardly selected line in Figure 20-13.

We must be very careful in our interpretation of long-term agricultural selection programs. In the real world of agriculture, changes in cultivation methods, machinery, fertilizers, insecticides, herbicides, and so forth, are taking place along with the production of genetically improved varieties. Increases in average yields are consequences of all of these changes. For example, the average yield of corn in the United States increased from 40 bushels to 80 bushels per acre between 1940 and 1970. But experiments comparing old and new varieties of corn in common environments show that only about half this increase is a direct result of new corn varieties (the other half being a result of improved farming techniques). Furthermore, the new varieties are superior to the old ones only at the high densities of modern planting for which they were selected.

The use of h^2 in breeding

Even though h^2 is a number that applies only to a particular population and a given set of environments, it is still of great practical importance to breeders. A poultry geneticist interested in increasing, say, the growth rate of chickens is not concerned with the genetic variance over all possible flocks and all environmental distributions. Given a particular flock (or a choice between a few particular flocks) under the environmental conditions approximating present husbandry practice, the question becomes, Can a selection scheme be devised to increase growth rate and, if so, how rapidly can it be increased? If one flock has a lot of genetic variance for growth rate and another only a little, the breeder will choose the former flock to carry out selection. If the heritability in the chosen flock is very high, then the mean of the population will respond quickly to the selection imposed, because most of the superiority of the selected parents will appear in the offspring. The higher the h^2 is, the higher the parent–offspring correlation is. If, on the other hand, h^2 is low, then only a small fraction of the superiority of the selected parents will appear in the next generation.

If h^2 is very low, some alternative scheme of selection or husbandry may be needed. In this case, H^2

together with h^2 can be of use to the breeder. Suppose that h^2 and H^2 are both low, which means that there is a large proportion of environmental variance compared with genetic variance. Some scheme of reducing s_e^2 must be used. One method is to change the husbandry conditions so that environmental variance is lowered. Another is to use **family selection.** Rather than selecting the best individuals, the breeder allows pairs to produce several trial progeny, and parental pairs are selected to produce the next generation on the basis of the average performance of those progeny. Averaging over progeny allows uncontrolled environmental variation and developmental noise to be canceled out, and a better estimate of the genotypic difference between pairs can be made so that the best pairs can be chosen as parents of the next generation.

If, on the other hand, h^2 is low but H^2 is high, then there is not much environmental variance. The low h^2 is the result of a small proportion of additive genetic variance compared with dominance variance. Such a situation calls for special breeding schemes that make use of nonadditive variance. One such scheme is the **hybrid–inbred method,** which is used almost universally for corn. A large number of inbred lines are created by selfing. These inbred lines are then crossed in many different combinations (all possible combinations, if it is economically feasible), and the cross that gives the best hybrid is chosen. Then new inbred lines are developed from this best hybrid, and again crosses are made to find the best hybrid cross. This process is continued cycle after cycle. This scheme selects not only for additive effects but also for dominance effects, because it selects the best heterozygotes as parents for the next cycle; it has been the basis of major genetic advances in hybrid maize yield in North America since 1930. Yield in corn does not appear to have large amounts of nonadditive genetic variance, however, and so it is debatable whether this technique *ultimately* produces higher-yielding varieties than those that would have resulted from years of simple selection techniques based on additive variance.

The hybrid–inbred method has been introduced into the breeding of all kinds of plants and animals. Tomatoes and chickens, for example, are now almost exclusively hybrids. Attempts also have been made to breed hybrid wheat, but thus far the wheat hybrids obtained do not yield consistently better than do the nonhybrid varieties now used.

> **MESSAGE** The subdivision of genetic variation and environmental variation provides important information about gene action that can be used in plant and animal breeding.

20.8 Locating genes

It is not possible to identify all the genes that influence the development of a given character by using purely genetic techniques. In a given population, only a subset of the genes that contribute to the development of any given character will be genetically variable. Hence, only some of the possible variation will be observed. This is true even for genes that determine simple qualitative traits—for example, the genes that determine the total antigenic configuration of the membrane of the human red blood cell. About 40 loci determining human blood groups are known at present; each has been discovered by finding at least one person with an immunological specificity that differs from the specificities of other people. Many other loci that determine red blood cell membrane structure may remain undiscovered because all the people studied are genetically identical. *Genetic* analysis detects genes only when there is some allelic variation. In contrast, *molecular* analysis deals directly with DNA and its translated information and so can identify genes as stretches of DNA coding for certain products, even when they do not vary—provided that the gene products can be identified.

Even though a character may show continuous phenotypic variation, the genetic basis for the differences may be allelic variation at a single locus. Most of the classic mutations in *Drosophila* are phenotypically variable in their expression, and in many cases the mutant class differs little from wild type, and so many flies that carry the mutation are indistinguishable from normal flies. Even the genes of the *bithorax* complex, which have dramatic homeotic mutations that turn halteres into wings (see Figure 18-24), also have weak alleles that increase the size of the halteres only slightly on average, and so flies of the mutant genotype may appear to be wild type.

It is sometimes possible to use prior knowledge of the biochemistry and development of an organism to guess that variation at a known locus is responsible for at least some of the variation in a certain character. Such a locus is a **candidate gene** for the investigation of continuous phenotypic variation. The variation in activity of the enzyme acid phosphatase in human red blood cells was investigated in this way. Because we are dealing with variation in enzyme activity, a good hypothesis would be that there is allelic variation at the locus that codes for this enzyme. When H. Harris and D. Hopkinson sampled an English population, they found that there were, indeed, three allelic forms, *A*, *B*, and *C*, that resulted in enzymes with different activity levels. Table 20-2 shows the mean activity, the variance in activity, and the population frequency of the six genotypes. Figure 20-15 shows the distribution of activity for the entire population and how it is composed of the

TABLE 20-2 Red Blood Cell Activity of Different Genotypes of Red-Cell Acid Phosphatase in the English Population

Genotype	Mean activity	Variance of activity	Frequency in population
A/A	122.4	282.4	0.13
A/B	153.9	229.3	0.43
B/B	188.3	380.3	0.36
A/C	183.8	392.0	0.03
B/C	212.3	533.6	0.05
C/C	240	—	0.002
Grand average	166.0	310.7	
Total distribution	166.0	607.8	

Note: Averages are weighted by frequency in population.
Source: H. Harris, *The Principles of Human Biochemical Genetics*, 3d ed. North-Holland, 1980.

distributions of the different genotypes. As Table 20-2 shows, of the variance in activity in the total distribution (607.8), about half is explained by the average variance within genotypes (310.7); so half (607.8 − 310.7 = 297.1) must be accounted for by the variance between the means of the six genotypes. Although so much of the variation in activity is explained by the mean differences between the genotypes, there remains variation within each genotype that may be the result of environmental influences or of the segregation of other, as yet unidentified genes.

By using the candidate-gene method, one often finds that part of the variation in a population is attributable to different alleles at a single locus, but the proportion of variance associated with the single locus is usually less than what was found for acid phosphatase activity. For example, the three common alleles for the gene *apoE*, which encodes the protein apolipoprotein E, account for only about 16 percent of the variance in blood levels of the low-density lipoproteins that carry cholesterol and are implicated in excess cholesterol levels. The remaining variance is a consequence of some unknown combination of genetic variation at other loci and environmental variation.

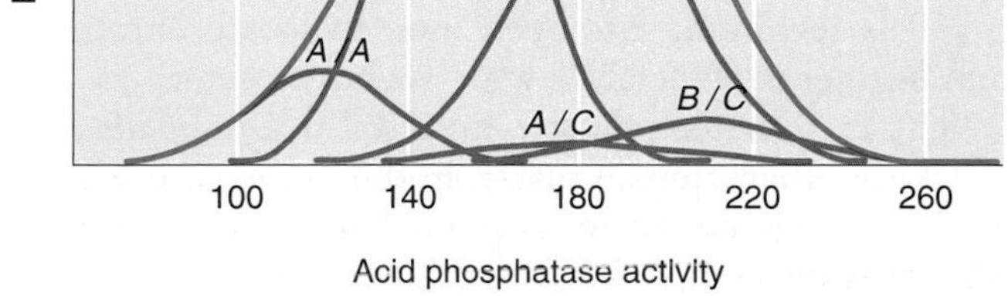

Marker-gene segregation

The genes segregating for a quantitative trait—so-called **quantitative trait loci**, or **QTLs**—cannot be individually identified in most cases. It is possible, however, to locate regions of the genome in which the relevant loci lie and to estimate how much of the total variation is accounted for by QTL variation in each region. This analysis can be done in experimental organisms by crossing two lines that differ markedly in the quantitative trait and that also differ in alleles at well-known loci, called **marker genes**. The marker genes used for such analyses are ones for which the different genotypes can be distinguished by some visible phenotype that cannot be confused with the quantitative trait (for example, eye color in *Drosophila*) or by the electrophoretic mobility of the proteins that they encode or by the DNA sequence of the genes themselves. A typical experiment entails crossing two lines that differ markedly in the quantitative character and that also differ in marker alleles. The F_1 resulting from the cross between the two lines may then be crossed with itself to make a segregating F_2, or it may be backcrossed to one of the parental lines. If there are QTLs closely linked to a marker gene, then the different marker genotypes and the QTLs will be inherited

Figure 20-15 Distribution of enzyme activity. Acid phosphatase activity in red cells for different genotypes (red curves) and the distribution of activity in an English population made up of a mixture of these genotypes (green curve). [H. Harris, *The Principles of Human Biochemical Genetics*, 3d ed. Copyright 1970 by North-Holland.]

together, and the different marker genotypes in the F_2 or backcross will have different average phenotypes for the quantitative character.

Quantitative linkage analysis

The localization of QTLs to small regions within chromosomes requires the presence of closely spaced marker loci along the chromosome. Moreover, it must be possible to create parental lines that differ from each other in the alleles carried at these marker loci. With the advent of molecular techniques that can detect genetic polymorphism at the DNA level, a very high density of variant loci has been discovered along the chromosomes of all species. Especially useful are restriction fragment length polymorphisms (RFLPs), tandem repeats, and single nucleotide polymorphisms (SNPs) in DNA. Such polymorphisms are so common that any two lines selected for a difference in quantitative traits are also sure to differ from each other at some known molecular marker loci spaced a few crossover units from each other along each chromosome.

An experimental protocol for localizing genes, shown in Figure 20-16, uses groups of individuals that differ markedly in the quantitative character of interest as well as at marker loci. These groups may be created by several generations of divergent selection to create extreme lines, or advantage may be taken of existing va-

rieties or family groups that differ markedly in the trait. These lines must then be surveyed for marker loci that differ between them. A cross is made between the two lines, and the F_1 is then crossed with itself to produce a segregating F_2 or is crossed with one of the parental lines to produce a segregating backcross. A large number of offspring from the segregating generation are then measured for the quantitative phenotype and characterized for their genotype at the marker loci. A marker locus that is unlinked or very loosely linked to any QTLs affecting the quantitative trait of interest will have the same average value of the quantitative trait for all its genotypes, whereas one that is closely linked to some QTLs will differ in its mean quantitative phenotype from one marker genotype to another.

How much difference there is in the mean quantitative phenotype between the different marker genotypes depends both on the strength of the effect of the QTL and on the tightness of linkage between the QTL and the marker locus. Suppose, for example, that two selected lines differ by a total of 100 units in some quantitative character. The line with the high value is homozygous $+/+$ at a particular QTL, whereas the line with the low value is homozygous $-/-$, and each $+$allele at this QTL accounts for 5 units of the total difference between the lines. Further, suppose that the high line is M/M and the low line is m/m at a marker locus 10 crossover units away from the QTL. Then, as shown in Figure 20-16, there are 4 units of difference between the average gamete carrying an M allele and the average gamete carrying an m allele in the segregating F_2. We can therefore calculate that 8 units of the difference between an M/M homozygote and an m/m homozygote are attributable to that QTL. Thus, we have accounted for 8 percent of the average difference between the original selected lines. The QTL actually accounts for 10 percent of the difference; the discrepancy comes from the recombination between the marker gene and the QTL. We could then repeat this process by using marker loci at other locations along the chromosome and on different chromosomes to account for yet further fractions of the quantitative difference between the original selected lines.

This technique has been used to locate chromosomal segments associated with such characters as fruit weight in tomatoes, bristle number in *Drosophila*, and vegetative characters in maize. In the maize case, 82 vegetative characters were examined in a cross between lines that differed in 20 DNA markers. On the average, each character was significantly associated with 14 different markers, but the proportion of the character difference between the two lines that was associated with any particular marker was usually very small. Figure 20-17 shows the proportion of the statistically significant marker–character associations (on the y-axis) that accounted for different proportions of character differ-

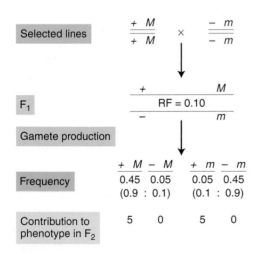

Average phenotypic effect of *M* class = 5 (0.9) + 0 (0.1) = 4.5
Average phenotypic effect of *m* class = 5 (0.1) + 0 (0.9) = 0.5
Difference between *M*-carrying gametes and *m*-carrying gametes =
4.5−0.5 = 4
Difference between average F_2 *M*/*M* homozygotes and average
F_2 *m*/*m* homozygotes = 8

Figure 20-16 Results of a cross between two selected lines that differ at a QTL and at a molecular marker locus 10 crossover units away from the QTL. The QTL $+$allele adds 5 units of difference to the phenotype.

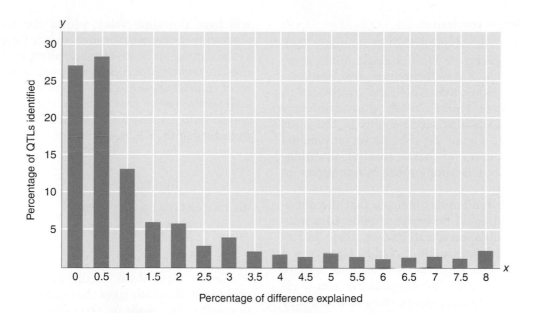

Figure 20-17 Distribution of associations of the trait differences between two lines of maize with an array of DNA markers. The x-axis shows the percentage of the difference explained between the two lines in a given trait that could be associated with any marker gene. The y-axis shows the proportion of all the identified QTLs that had the corresponding percentage of its difference explained. Note that 55 percent of all the associations (first two columns) account for less than 1 percent of their trait differences. [After M. Lynch and B. Walsh, *Genetics and Analysis of Quantitative Traits.* Sinauer Associates, 1998. Data from M. D. Edwards, C. W. Stuber, and J. F. Wendel, *Genetics* 116, 1987, 113-125.]

ence between the lines. As Figure 20-17 shows, most associations accounted for less than 1 percent of the character difference. Unfortunately, in human genetics, although marker-gene segregation can be used to localize single-gene disorders, the small size of human pedigree groups makes the marker segregation technique inapplicable for quantitative trait loci because there are too few progeny from any particular marker cross to provide any accuracy.

For many organisms (for example, humans), it is not possible to make homozygous lines differing in some trait and then cross them to produce a segregating generation. For such organisms, one can use the differences among sibs carrying different marker alleles from heterozygous parents. This method has much less power to find QTLs especially when the number of sibs in any family is small, as it is in human families. As a consequence, the attempts to map QTLs for human traits have not been very successful, although the marker segregation technique has been a success in finding loci whose mutations are responsible for single-gene disorders or for quantitative characters whose variation is strongly influenced by variation at one locus. For example, people vary in their ability to taste the substance phenylthiocarbamate (PTC). Some can detect quite low concentrations, whereas others can detect only high concentrations or are unable to taste PTC at all. Linkage analysis using single nucleotide polymorphisms located a region on human chromsome 7q that accounted for about 75 percent of the variation in taste sensitivity. This chromosomal region was already known to contain several genes coding for bitter taste receptor proteins. When the DNA of these genes was sequenced, three amino acid polymorphisms in one of the genes were found to be strongly associated with the difference between the taster and the nontaster phenotypes.

STATISTICAL APPENDIX

Complete information about the distribution of a phenotype in a population can be given only by specifying the frequency of each measured class, but a great deal of information can be summarized in just two statistics. First, we need some measure of the location of the distribution along the axis of measurement. (For example, do the individual measurements of height for male graduates tend to cluster around 100 cm or 200 cm?) Second, we need some measure of the amount of variation within the distribution. (For example, are the heights of the male undergraduates all concentrated around the central measurement or do they vary widely across a large range?)

Measures of central tendency

The mode   Most distributions of phenotypes look roughly like those in Figure 20-3: a single mode is located near the middle of the distribution, with frequencies decreasing on either side. There are exceptions to this pattern, however. Figure 20-18a shows the very asymmetrical distribution of seed weights in the plant *Crinum longifolium*. Figure 20-18b shows a **bimodal** (two-mode) **distribution** of larval survival probabilities for different second-chromosome homozygotes in *Drosophila willistoni*.

A bimodal distribution may indicate that the population being studied could be better considered a mixture of two populations, each with its own mode. In Figure 20-18b, the left-hand mode probably represents a subpopulation of severe single-locus mutations that are extremely deleterious when homozygous but whose effects are not felt in the heterozygous state in which they usually exist in natural populations. The right-hand mode is part of the distribution of "normal" viability modifiers of small effect.

The mean   A more common measure of central tendency is the arithmetic average, or the **mean**. The mean of the measurement ($\bar{x}$) is simply the sum of all the individual measurements (x_i) divided by the number of measurements in the sample (N):

$$\text{mean} = \bar{x} = \frac{x_1 + x_2 + x_3 + \cdots + x_N}{N} = \frac{1}{N} \Sigma x_i$$

where Σ represents the operation of summing over all values of i from 1 to N, and x_i is the ith measurement.

In a typical large sample, the same measured value will appear more than once, because several individuals will have the same value within the accuracy of the measuring instrument. In such a case, $\bar{x}$ can be rewritten as the sum of all measurement values, each weighted by how frequently it occurs in the population. From a total of N individuals measured, suppose that n_1 fall in the class with value x_1, that n_2 fall in the class with value x_2, and so forth, so that $\Sigma n_i = N$. If we let f_i be the **relative frequency** of the ith measurement class, so that

$$f_i = \frac{n_i}{N},$$

then we can rewrite the mean as

$$\bar{x} = f_1 x_1 + f_2 x_2 + \cdots + f_k x_k = \Sigma f_i x_i$$

where x_i equals the value of the ith measurement class.

Let us apply these calculation methods to the data of Table 20-3, which gives the numbers of toothlike bristles in the sex combs on the right (x) and left (y) front legs and on both legs ($T = x + y$) of 20 *Drosophila*. Looking for the moment only at the sum of the two legs T, we find the mean number of sex comb teeth $\bar{T}$ to be:

$$\bar{T} = \frac{11 + 12 + 12 + 12 + 13 + \cdots + 15 + 16 + 16}{20}$$
$$= \frac{274}{20}$$
$$= 13.7$$

Alternatively, by using the relative frequencies of the different measurement values, we find that

$$\bar{T} = 0.05(11) + 0.15(12) + 0.20(13) + 0.35(14)$$
$$\qquad + 0.15(15) + 0.10(16)$$
$$= 13.7$$

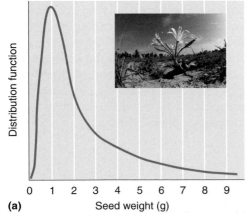

(a)   Seed weight (g)

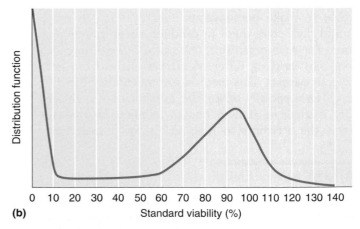

(b)   Standard viability (%)

Figure 20-18 Asymmetrical distribution functions. (a) Asymmetrical distribution of seed weight in *Crinum longifolium*; (b) bimodal distribution of survival of *Drosophila willistoni* expressed as a percentage of standard survival. [After S. Wright, *Evolution and the Genetics of Populations*, vol. 1. Copyright 1968 by University of Chicago Press. Photograph: Earth Scenes/Copyright Thompson GOSF.]

TABLE 20-3 Number of Teeth in the Sex Comb on the Right (x) and Left (y) Legs and the Sum of the Two (T) for 20 *Drosophila* Males

x	y	T	n_i	$f_i = n_i/N$
6	5	11	1	$\frac{1}{20} = 0.05$
6	6	12 ⎫		
5	7	12 ⎬	3	$\frac{3}{20} = 0.15$
6	6	12 ⎭		
7	6	13 ⎫		
5	8	13 ⎪	4	$\frac{4}{20} = 0.20$
6	7	13 ⎪		
7	6	13 ⎭		
8	6	14 ⎫		
6	8	14 ⎪		
7	7	14 ⎪		
7	7	14 ⎬	7	$\frac{7}{20} = 0.35$
7	7	14 ⎪		
6	8	14 ⎪		
8	6	14 ⎭		
8	7	15 ⎫		
7	8	15 ⎬	3	$\frac{3}{20} = 0.15$
6	9	15 ⎭		
8	8	16 ⎫	2	$\frac{2}{20} = 0.10$
7	9	16 ⎭		

$N = 20$ $\qquad$ $s_x^2 = 0.8275$ $\qquad$ $s_x = 0.9096$

$\bar{x} = 6.25$ $\qquad$ $s_y^2 = 1.1475$ $\qquad$ $s_y = 1.0722$

$\bar{y} = 7.05$ $\qquad$ $s_T^2 = 1.71$ $\qquad$ $s_T = 1.308$

$\overline{T} = 13.70$ $\qquad$ cov $xy = -0.1325$

$\qquad\qquad\qquad$ $r_{xy} = -0.1360$

Measures of dispersion: the variance

A second characteristic of a distribution is the width of its spread around the central class. Two distributions with the same mean might differ very much in how closely the measurements are concentrated around the mean. The most common measure of variation around the center is the **variance,** which is defined as the average squared deviation of the observations from the mean, or

$$\text{variance} = s^2$$
$$= \frac{(x_1 - \bar{x})^2(x_2 - \bar{x})^2 + \cdots + (x_N - \bar{x})^2}{N}$$
$$= \frac{1}{N} \Sigma\, (x_i - \bar{x})^2$$

When more than one individual has the same measured value, the variance can be written as

$$s^2 = f_1(x_1 - \bar{x})^2 + f_2(x_2 - \bar{x})^2 + \cdots + f_k(x_k - \bar{x})^2$$
$$= \Sigma f_i(x_i - \bar{x})^2$$

To avoid subtracting every value of x separately from the mean, we can use an alternative computing formula that is algebraically identical with the preceding equation:

$$s^2 = \left(\frac{1}{N}\Sigma x_i^2\right) - \bar{x}^2$$

Because the variance is in squared units (square centimeters, for example), it is common to take the square root of the variance, which then has the same units as the measurement itself. This square-root measure of variation is called the **standard deviation** of the distribution:

$$\text{standard deviation } = s = \sqrt{\text{variance}} = \sqrt{s^2}$$

The data for sex-comb teeth in Table 20-3 can be used to exemplify these calculations:

$$s_T^2 = \frac{(11 - 13.7)^2 + (12 - 13.7)^2 + (12 - 13.7)^2}{20}$$
$$\frac{+ \cdots + (15 - 13.7)^2 + (16 - 13.7)^2}{20}$$
$$= \frac{34.20}{20} = 1.71$$

We can also use the computing formula that avoids taking individual deviations:

$$s_T^2 = \frac{1}{N} \Sigma T_i^2 - \overline{T}^2 = \frac{3788}{20} - 187.69 = 1.71$$

and

$$s = \sqrt{1.71} = 1.308$$

Figure 20-19 shows two distributions having the same mean but different standard deviations (curves A and B) and two distributions having the same standard deviation but different means (curves A and C).

The mean and the variance of a distribution do not describe it completely. They do not distinguish a symmetrical distribution from an asymmetrical one, for example. There are even symmetrical distributions that have the same mean and variance but still have somewhat different shapes. Nevertheless, for the purposes of dealing with most quantitative genetic problems, the mean and variance suffice to characterize a distribution.

Measures of relationship

Covariance and correlation   Another statistical notion that is of use in the study of quantitative genetics is the association, or **correlation,** between variables. As a result of complex paths of causation, many variables in nature vary together but in an imperfect or approximate way. Figure 20-20a provides an example, showing the lengths of two particular teeth in several individual specimens of a fossil mammal, *Phenacodus primaevis*. There is a rough trend such that individuals with longer first molars tend to have longer second molars, but there is considerable scatter of data points around this trend. In contrast, Figure 20-20b shows that the body length and tail length in individual snakes *(Lampropeltis polyzona)* are quite closely related to each other, with all the points falling close to a straight line that could be drawn through them from the lower left of the graph to the upper right.

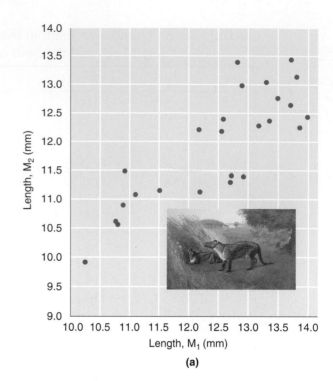

(a)

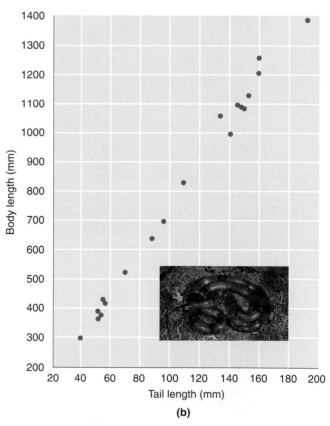

(b)

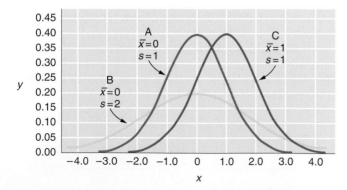

Figure 20-19   Three distribution functions, two of which have the same mean (A and B) and two of which have the same standard deviation (B and C).

Figure 20-20   Scatter diagrams of relations between pairs of variables. (a) Relation between the lengths of the first and second lower molars (M$_1$ and M$_2$) in the extinct mammal *Phenacodus primaevis*. Each point gives the M$_1$ and M$_2$ measurements for one individual. (b) Tail length and body length of 18 *Lampropeltis polyzona* snakes. [Image: Negative no. 2430, *Phenacodus*, painting by Charles Knight; courtesy of Department of Library Services, American Museum of Natural History. Photograph: Animals Animals/Copyright Zig Leszczynski.]

The usual measure of the precision of a relation between two variables x and y is the **correlation coefficient** (r_{xy}). It is calculated in part from the product of the deviation of each observation of x from the mean of the x values and the deviation of each observation of y from the mean of the y values—a quantity called the **covariance** of x and y (cov xy):

$$\text{cov } xy = \frac{(x_1 - \bar{x})(y_1 - \bar{y}) + (x_2 - \bar{x})(y_2 - \bar{y}) + \cdots}{N}$$

$$+ \frac{(x_N - \bar{x})(y_N - \bar{y})}{N}$$

$$= \frac{1}{N} \Sigma (x_i - \bar{x})(y_i - \bar{y})$$

A formula that is exactly algebraically equivalent but that makes computation easier is:

$$\text{cov } xy = \left(\frac{1}{N} \Sigma x_i y_i\right) - \overline{xy}$$

By using this formula, we can calculate the covariance between the right (x) and the left (y) leg counts in Table 20-3.

$$\text{cov } xy = \left(\frac{1}{N} \Sigma xy\right) - \overline{xy}$$

$$= \frac{(6)(5) + (6)(6) + \cdots + (8)(8) + (7)(9)}{20}$$

$$- (6.65)(7.05)$$

$$= -0.1325$$

The correlation, r_{xy}, is defined as:

$$\text{correlation} = r_{xy} = \frac{\text{cov } xy}{s_x s_y}$$

In the formula for correlation, the products of the deviations are divided by the product of the standard deviations of x and y (s_x and s_y). This normalization by the standard deviations has the effect of making r_{xy} a dimensionless number that is independent of the units in which x and y are measured. So defined, r_{xy} will vary from -1, which signifies a perfectly linear negative relation between x and y, to $+1$, which indicates a perfectly linear positive relation between x and y. If $r_{xy} = 0$, there is no linear relation between the variables. Intermediate values between 0 and $+1$ or -1 indicate intermediate degrees of relation between the variables. The data in Figure 20-20a and b have r_{xy} values of 0.82 and 0.99, respectively. In the example of the sex-comb teeth of

Table 20-3, the correlation between left and right legs is:

$$r_{xy} = \frac{\text{cov } xy}{\sqrt{s_x^2 s_y^2}} = \frac{-0.1325}{\sqrt{(0.8275)(1.1475)}} = -0.1360$$

a very small value. It is important to notice, however, that sometimes when there is no *linear* relation between two variables, but there is a regular *nonlinear* relation between them, one variable may be perfectly predicted from the other. Consider, for example, the parabola shown in Figure 20-21. The values of y are perfectly predictable from the values of x; yet $r_{xy} = 0$, because, on average over the whole range of x values, larger x values are not associated with either larger or smaller y values.

Correlation and equality It is important to notice that correlation between two sets of numbers is not the same as numerical identity. For example, two sets of values can be perfectly *correlated*, even though the values in one set are very much larger than the values in the other set. Consider the following pairs of values:

x	y
1	22
2	24
3	26

The variables x and y in the pairs are perfectly correlated ($r = 1.0$), although each value of y is about 20 units greater than the corresponding value of x. Two variables are perfectly correlated if, for a unit increase in one, there is a constant increase in the other (or a constant decrease if r is negative).

The importance of the difference between correlation and identity arises when we consider the effect of environment on heritable characters. Parents and offspring might be perfectly correlated in some character

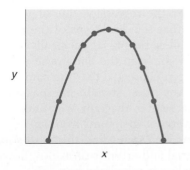

Figure 20-21 A parabola. Each value of y is perfectly predictable from the value of x, but there is no linear correlation.

such as height, yet, because of an environmental difference between generations, every child might be taller than its parents. This phenomenon appears in adoption studies, in which children may be correlated with their biological parents but, on the average, may be quite different from the parents as a result of a change in their social situation.

Covariance and the variance of a sum In Table 20-3, the variances of the left and right legs are 0.8275 and 1.1475, which adds up to 1.975, but the variance of the sum of the two legs T is only 1.71. That is, the variance of the whole is less than the sum of the variances of the parts. This discrepancy is a consequence of the negative correlation between left and right sides. Larger left sides are associated with smaller right sides and vice versa, and so the sum of the two sides varies less than each side separately. If, on the other hand, there were a positive correlation between sides, then larger left and right sides would go together and the variation of the sum of the two sides would be larger than the sum of the two separate variances. In general, if $x + y = T$, then

$$s_T^2 = s_x^2 + s_y^2 + 2 \text{ cov } xy$$

For the data of Table 20-3,

$$s_T^2 = 1.71 = 0.8275 + 1.1475 - 2(0.1325)$$

Regression The correlation coefficient provides us with only an estimate of the *precision* of relation between two variables. A related problem is predicting the value of one variable given the value of the other. If x increases by two units, by how much will y increase? If the two variables are linearly related, then that relation can be expressed as

$$y = bx + a$$

where b is the slope of the line relating y to x and a is the y intercept of that line.

Figure 20-22 shows a scatter diagram of points for two variables, y and x, together with a straight line expressing the general linear trend of y with increasing x. This line, called the **regression line of y on x**, has been positioned so that the deviations of the points from the line are as small as possible. Specifically, if Δy is the distance of any point from the line in the y direction, then the line has been chosen so that the sum of $(\Delta y)^2$ equals a minimum. Any other straight line passed through the points on the scatter diagram will have a larger total squared deviation of the points from it.

Obviously, we cannot find this **least-squares regression line** by trial and error. It turns out, however, that, if slope b of the line is calculated by

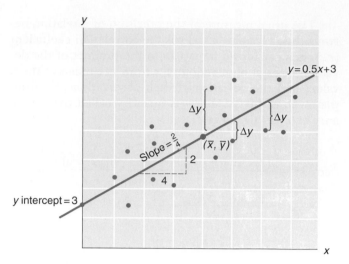

Figure 20-22 A scatter diagram showing the relation between two variables, x and y, with the regression line of y on x. This line, with a slope of $\frac{2}{4}$, minimizes the squares of the deviations (Δy).

$$b = \frac{\text{cov } xy}{s_x^2}$$

and if a is then calculated from

$$a = \bar{y} - b\bar{x}$$

so that the line passes through the point $\bar{x}, \bar{y}$, then these values of b and a will yield the linear equation of the least-squares regression line.

Note that the preceding prediction equation cannot predict y exactly for a given x, because there is scatter around the least-squares regression line. The equation predicts the *average y* for a given x if large enough samples are taken.

Samples and populations The preceding sections have described the distributions of, and some statistics for, particular assemblages of individuals that have been collected in some experiments or sets of observations. For some purposes, however, we are not really interested in the particular 100 undergraduates or 18 snakes that have been measured. Instead, we are interested in the wider world of phenomena of which those particular individuals are representative. For example, we might want to know the average height, in general, of male undergraduates in the United States. That is, we are interested in the characteristics of a **universe**, of which our small collection of observations is only a **sample**. The characteristics of any particular sample are not identical with those of the universe but vary from sample to sample.

We can use the sample mean to estimate the true mean of the universe, but the sample variance and covariance will be on the average a little smaller than the true value in the universe. That is because the deviations from the sample mean are not all independent of one another. In fact, by definition, the sum of all the deviations of the observations from the mean is 0! (Try to prove it as an exercise.) Therefore, if we are told $N - 1$ of the deviations from the mean in a sample of N observations, we can calculate the missing deviation, because all the deviations must add up to zero.

It is simple to correct for this bias in our estimate of variance. Whenever we are interested in the variance of a set of measurements—not as a characteristic of the particular sample but as an estimate of a universe that the sample represents—then the appropriate quantity to use, rather than s^2 itself, is $[N/(N - 1)]s^2$. Note that this new quantity is equivalent to dividing the sum of

squared deviations by $N - 1$ instead of N in the first place, so

$$\left(\frac{N}{N-1}\right)s^2 = \left(\frac{N}{N-1}\right)\frac{1}{N}\Sigma(x_i - \bar{x})^2$$
$$= \frac{1}{N-1}\Sigma(x_i - \bar{x})^2$$

All these considerations about bias also apply to the sample covariance. In the preceding formula for the correlation coefficient, however, the factor $N/(N - 1)$ would appear in both the numerator and the denominator and therefore cancel out, and so we can ignore it for the purposes of computation.

KEY QUESTIONS REVISITED

- **Is the observed variation in a character influenced *at all* by genetic variation? Are there alleles segregating in the population that produce some differential effect on the character or is all the variation simply the result of environmental variation and developmental noise (see Chapter 1)?**

One form of evidence that the observed phenotypic variation is influenced by genotype comes from the results of studies of related individuals. If close relatives resemble one another more than distant relatives or than unrelated individuals this resemblance is evidence of genetic effects on variation only if the groups being compared have developed and are living in the same environment. Otherwise, it is not possible to distinguish genetic similarity from environmental similarity. Another form of evidence can be obtained by comparing individuals with known alternative genotypes for one or more well-defined genes. If there is an association between the marker genotype and the phenotypic character, then there is evidence for a genetic component to the phenotypic variation. The marker gene may be one that is thought to take part in the development of the character or it may simply be a functionally unrelated marker that happens to lie near a relevant functional gene on the chromosome. Again, the environments must be the same for the different genotypes in order to distinguish genetic from environmental causes of similarity.

- **If there is genetic variation, what are the norms of reaction of the various genotypes?**

To carry out norm of reaction studies, it is necessary to have many individuals of the same genotype. They can be produced by a long succession of matings between close relatives (inbreeding), by crosses involving genetic markers that allow the production of homozy-

gous lines of individuals in a few generations, or by cloning techniques. The genetically identical individuals are then allowed to develop in a set of controlled environments differing by some identifiable environmental variable, and the phenotype in each environment is measured. These measurements, when plotted against the environmental variable, give the norm of reaction of the genotype.

- **How important is genetic variation as a source of total phenotypic variation? Are the norms of reaction and the environments such that nearly all the variation is a consequence of environmental difference and developmental instabilities or does genetic variation predominate?**

To obtain a quantitative estimate of the amount of variation in a population that is associated with genetic differences, environmental differences, and developmental instabilities, it is necessary to carry out a heritability study. One form of such a study is to measure the character in groups of individuals that differ by a known amount in their degree of genetic relationship, as, for example, comparing the similarity of identical twin and nonidentical twins or sibs and half-sibs. The heritability can then be estimated by comparing the observed difference in similarity between individuals with the amount predicted from their degree of relationship. This comparison is only valid if the different relationship groups have developed in the same environment. A second method is to determine the realized heritability of a characteristic by a selection experiment. A population is selected to change the character measurement by producing offspring from a group of selected parents. The amount of measured difference between the selected parents and the unselected population (selection differential) is then compared with the amount of difference between the offspring and the unselected parents (realized selection progress). If there is no selection progress, the heritability is zero. If the selection progress is equal to the selection differential, the heritability is unity.

- **Do many loci (or only a few) vary with respect to the character? How are they distributed over the genome?**

An estimate of the number of loci that influence the observed variation in a character is made from the results of linkage studies with known genetic markers distributed across the genome. If a detectable proportion of the differences between individuals in the quantitative character segregates together with the allelic differences at a marker gene, then a quantitative trait locus (QTL) has been detected near the marker gene.

SUMMARY

Many—perhaps most—of the phenotypic traits that we observe in organisms vary continuously. In many cases, the variation of the trait is determined by more than a single segregating locus. Each of these loci may contribute equally to a particular phenotype, but it is more likely that they contribute unequally. The measurement of these phenotypes and the determination of the contributions of specific alleles to the distribution must be made on a statistical basis in these cases. Some of these variations of phenotype (such as height in some plants) may show a normal distribution around a mean value; others (such as seed weight in some plants) will illustrate a skewed distribution around a mean value.

A quantitative character is one for which the average phenotypic differences between genotypes are small compared with the variation between the individuals within the genotypes. This situation may be true even for characters that are influenced by alleles at one locus. The distribution of environments is reflected biologically as a distribution of phenotypes. The transformation of environmental distribution into phenotypic distribution is determined by the norm of reaction. Norms of reaction can be characterized in organisms in which large numbers of genetically identical individuals can be produced. Traits are familial if they are common to members of the same family, for whatever reason. Traits are heritable, however, only if the similarity arises from common genotypes. In experimental organisms, environmental similarities may be readily distinguished from genetic similarities, or heritability. In humans, however, it is

very difficult to determine whether a particular trait is heritable. Norm of reaction studies show only small differences between genotypes, and these differences are not consistent over a wide range of environments. Thus, "superior" genotypes in domesticated animals and cultivated plants may be superior only in certain environments. If it should turn out that humans exhibit genetic variation for various mental and emotional traits, this variation is unlikely to favor one genotype over another across a range of environments.

The attempt to quantify the influence of genes on a particular trait has led to the determination of heritability in the broad sense (H^2). In general, the heritability of a trait is different in each population and each set of environments, and so heritability cannot be extrapolated from one population and set of environments to another. Because H^2 characterizes present populations in present environments only, it is fundamentally flawed as a predictive device. Heritability in the narrow sense (h^2) measures the proportion of phenotypic variation that results from substituting one allele for another. This quantity, if large, predicts that selection for a trait will succeed rapidly. If h^2 is small, special forms of selection are required.

With the use of genetically marked chromosomes, it is possible to determine the relative contributions of different chromosomes to variation in a quantitative trait, to observe dominance and epistasis from whole chromosomes, and, in some cases, to map genes that are segregating for a trait.

KEY TERMS

additive effect (p. 660)

additive genetic
variation (p. 661)

analysis of variance (p. 657)

bimodal distribution (p. 668)

broad heritability (H^2) (p. 657)

candidate gene (p. 664)

central tendency (p. 648)

correlation (p. 648, 670)

correlation coefficient (p. 671)

covariance (p. 671)

dispersion (p. 648)

distribution function (p. 647)

dominance variance (p. 661)

environmental variance (p. 656)

familial (p. 655)

family selection (p. 664)

frequency histogram (p. 647)

genetic correlation (p. 657)

genetic variance (p. 656)

heritability in the narrow
  sense (h^2) (p. 661)

heritable (p. 654)

hybrid–inbred method (p. 664)

inbreeding (p. 650)

least-squares regression line (p. 672)

marker gene (p. 664)

mean (p. 648, 668)

midparent value (p. 661)

mode (p. 648)

multiple-factor hypothesis (p. 650)

norm of reaction (p. 651)

phenotypic correlation (p. 658)

quantitative genetics (p. 647)

quantitative trait locus
  (QTL) (p. 665)

regression line of y on x (p. 672)

relation (p. 648)

relative frequency (p. 668)

sample (p. 672)

segregating line (p. 652)

selection differential (p. 662)

selection response (p. 662)

standard deviation (p. 669)

statistical distribution (p. 647)

truncation selection (p. 663)

universe (p. 672)

variance (p. 648, also 669)

SOLVED PROBLEMS

1. In some species of songbirds, populations living in different geographical regions sing different "local dialects" of the species song. Some people believe that this difference in dialect is the result of genetic differences between populations, whereas others believe that these differences arose from purely individual idiosyncracies in the founders of these populations and have been passed on from generation to generation by learning. Outline an experimental program that would determine the importance of genetic and nongenetic factors and their interaction in this dialect variation. If there is evidence of genetic difference, what experiments could be done to provide a detailed description of the genetic system, including the number of segregating genes, their linkage relations, and their additive and nonadditive phenotypic effects?

Solution

This example has been chosen because it illustrates the very considerable experimental difficulties that arise when we try to examine claims that observed differences in quantitative characters in some species have a genetic basis. To be able to say anything at all about the roles of genes and developmental environment requires, at minimum, that the organisms can be raised from fertilized eggs in a controlled laboratory environment. To be able to make more detailed statements about the genotypes underlying variation in the character requires, further, that the results of crosses between parents of known phenotype and known ancestry be observable and that the offspring of some of those crosses be, in turn, crossed with other individuals of known phenotype and ancestry. Very few animal species can satisfy this requirement, although it is much easier to carry out controlled crosses in plants. We will assume that the songbird species in question can indeed be raised and crossed in captivity, but that is a big assumption.

a. To determine whether there is any genetic difference underlying the observed phenotypic difference in dialect between the populations, we need to raise birds of each population, from the egg, in the absence of auditory input from their own ancestors and in various combinations of auditory environments of other populations. We can do so by raising birds from the egg that have been grouped as follows:

(1) In isolation

(2) Surrounded by hatchlings consisting only of birds derived from the same population

(3) Surrounded by hatchlings consisting of birds derived from other populations

(4) In the presence of singing adults from other populations

(5) In the presence of singing adults from their own population (as a control on the rearing conditions)

If there are no genotypic differences and all dialect differences are learned, then birds from group 5 will sing their population dialect and those from group 4 will sing the foreign dialect. Groups 1, 2, and 3 may not sing at all; they may sing a generalized song not corresponding to any of the dialects; or they may all sing the same song dialect—this dialect would then represent the "intrinsic" developmental program unmodified by learning.

If dialect differences are totally determined by genetic differences, birds from groups 4 and 5 will sing the same dialect, that of their parents. Birds from groups 1, 2, and 3, if they sing at all, will each sing the song dialect of their parent population, irrespective of the other birds in their group. There are then the possibilities of less-clear-cut results, indicating that both genetic and learned differences influence the trait. For example, birds in group 4 might sing a song with both population elements. Note that, if the birds in the control group 5 do not sing their normal dialect, the rest of the results are uninterpretable, because the conditions of artificial rearing are interfering with the normal developmental program.

b. If the results of the first experiments show some heritability in the broad sense, then a further analysis is possible. This analysis requires a genetically segregating population, made from a cross between two dialect populations—say, A and B. A cross between males from population A and females from population B and the reciprocal cross will give an estimate of the average degree of dominance of genes influencing the trait and whether there is any sex linkage. (Remember that, in birds, the female is the heterogametic sex.) The offspring of this cross and all subsequent crosses *must* be raised in conditions that do not confuse the learned and the genetic components of the differences, as revealed in the experiments in part a. If learned effects cannot be separated out, this further genetic analysis is impossible.

c. To localize genes influencing dialect differences would require a large number of segregating genetic markers. These markers could be morphological mutants or molecular variants such as restriction-site polymorphisms. Families segregating for the quantitative trait differences would be examined to see if there were cosegregation of any of the marker loci with the quantitative trait. These cosegregated loci would then be candidates for loci linked to the quantitative trait loci. Further crosses between individuals with and without mutant markers and measure of the quantitative trait values in F_2 individuals would establish whether there was actual linkage between the marker and the quantitative trait loci. In practice, it is very unlikely that such experiments could be carried out on a songbird species, because of the immense time and effort required to establish lines carrying the large number of different marker genes and molecular polymorphisms.

2. Two inbred lines of beans are intercrossed. In the F_1, the variance in bean weight is measured at 1.5. The F_1 is selfed; in the F_2, the variance in bean weight is

6.1. Estimate the broad heritability of bean weight in the F_2 population of this experiment.

Solution
The key here is to recognize that all the variance in the F_1 population must be environmental because all individuals must be of identical genotype. Furthermore, the F_2 variance must be a combination of environmental and genetic components, because all the genes that are heterozygous in the F_1 will segregate in the F_2 to give an array of different genotypes that relate to bean weight. Hence, we can estimate

$$s_e^2 = 1.5$$

$$s_e^2 + s_g^2 = 6.1$$

Therefore

$$s_g^2 = 6.1 - 1.5 = 4.6$$

and broad heritability is

$$H^2 = \frac{4.6}{6.1} = 0.75 \ (75\%)$$

3. In an experimental population of *Tribolium* (flour beetles), the body length shows a continuous distribution with a mean of 6 mm. A group of males and females with body lengths of 9 mm are removed and interbred. The body lengths of their offspring average 7.2 mm. From these data, calculate the heritability in the narrow sense for body length in this population.

Solution
The selection differential is $9 - 6 = 3$ mm, and the selection response is $7.2 - 6 = 1.2$ mm. Therefore, the heritability in the narrow sense is:

$$h^2 = \frac{1.2}{3} = 0.4 \ (40\%)$$

PROBLEMS

BASIC PROBLEMS

1. Distinguish between continuous and discontinuous variation in a population, and give some examples of each.

2. The table at the right shows a distribution of bristle number in *Drosophila*. Calculate the mean, variance, and standard deviation of this distribution.

Bristle number	Number of individuals
1	1
2	4
3	7
4	31
5	56
6	17
7	4

3. A book on the problem of heritability of IQ makes the following three statements. Discuss the validity of each statement and its implications about the authors' understanding of h^2 and H^2.

   a. "The interesting question then is . . . 'How heritable?' The answer [0.01] has a very different theoretical and practical application from the answer [0.99]." (The authors are talking about H^2.)

   b. "As a rule of thumb, when education is at issue, H^2 is usually the more relevant coefficient, and, when eugenics and dysgenics (reproduction of selected individuals) are being discussed, h^2 is ordinarily what is called for."

   c. "But whether the different ability patterns derive from differences in genes . . . is not relevant to assessing discrimination in hiring. Where it could be relevant is in deciding what, in the long run, might be done to change the situation."

   (From J. C. Loehlin, G. Lindzey, and J. N. Spuhler, *Race Differences in Intelligence*. Copyright 1975 by W. H. Freeman and Company.)

4. Using the concepts of norms of reaction, environmental distribution, genotypic distribution, and phenotypic distribution, try to restate the following statement in more exact terms: "80 percent of the difference in IQ performance between the two groups is genetic." What would it mean to talk about the heritability of a difference between two groups?

CHALLENGING PROBLEMS

5. In a large herd of cattle, three different characters showing continuous distribution are measured, and the variances in the following table are calculated:

Characters

Variance	Shank length	Neck length	Fat content
Phenotypic	310.2	730.4	106.0
Environmental	248.1	292.2	53.0
Additive genetic	46.5	73.0	42.4
Dominance genetic	15.6	365.2	10.6

   a. Calculate the broad- *and* narrow-sense heritabilities for each character.

   b. In the population of animals studied, which character would respond best to selection? Why?

   c. A project is undertaken to decrease mean fat content in the herd. The mean fat content is currently 10.5 percent. Animals of 6.5 percent fat content are interbred as parents of the next generation. What mean fat content can be expected in the descendants of these animals?

6. Suppose that two triple heterozygotes A/a ; B/b ; C/c are crossed. Assume that the three loci are in different chromosomes.

   a. What proportions of the offspring are homozygous at one, two, and three loci, respectively?

   b. What proportions of the offspring carry 0, 1, 2, 3, 4, 5, and 6 alleles (represented by capital letters), respectively?

7. In Problem 6, suppose that the average phenotypic effect of the three genotypes at the A locus is $A/A = 4$, $A/a = 3$, and $a/a = 1$ and that similar effects exist for the B and C loci. Moreover, suppose that the effects of loci add to each other. Calculate and graph the distribution of phenotypes in the population (assuming no environmental variance).

8. In Problem 7, suppose that there is a threshold in the phenotypic character so that, when the phenotypic value is above 9, an individual *Drosophila* has three bristles; when it is between 5 and 9, the individual has two bristles; and when the value is 4 or less, the individual has one bristle. Describe the outcome of crosses within and between bristle classes. Given the result, could you infer the underlying genetic situation?

9. Suppose that the general form of a distribution of a trait for a given genotype is:

$$f = 1 - \frac{(x - \bar{x})^2}{s_e^2}$$

   over the range of x where f is positive.

   a. On the same scale, plot the distributions for three genotypes with the following means and environmental variances:

Genotype	$\bar{x}$	s_e^2	Approximate range of phenotype
1	0.20	0.3	$x = 0.03$ to $x = 0.37$
2	0.22	0.1	$x = 0.12$ to $x = 0.24$
3	0.24	0.2	$x = 0.10$ to $x = 0.38$

   b. Plot the phenotypic distribution that would result if the three genotypes were equally frequent in a population. Can you see distinct modes? If so, what are they?

10. The following sets of hypothetical data represent paired observations on two variables (x, y). Plot each set of data pairs as a scatter diagram. Look at the plot of the points, and make an intuitive guess about

the correlation between x and y. Then calculate the correlation coefficient for each set of data pairs, and compare this value with your estimate.

a. (1, 1); (2, 2); (3, 3); (4, 4); (5, 5); (6, 6).

b. (1, 2); (2, 1); (3, 4); (4, 3); (5, 6); (6, 5).

c. (1, 3); (2, 1); (3, 2); (4, 6); (5, 4); (6, 5).

d. (1, 5); (2, 3); (3, 1); (4, 6); (5, 4); (6, 2).

11. Describe an experimental protocol for studies of relatives that could estimate the broad heritability of alcoholism. Remember that you must make an adequate observational definition of the trait itself.

12. A line selected for high bristle number in *Drosophila* has a mean of 25 sternopleural bristles, whereas a low-selected line has a mean of only 2. Marker stocks involving the two large autosomes II and III are used to create stocks with various mixtures of chromosomes from the high (h) and low (l) lines. The mean number of bristles for each chromosomal combination is as follows:

$$\frac{h}{h}\frac{h}{h}\ 25.1 \qquad \frac{h}{l}\frac{h}{h}\ 22.2 \qquad \frac{l}{l}\frac{h}{h}\ 19.0$$

$$\frac{h}{h}\frac{h}{l}\ 23.0 \qquad \frac{h}{l}\frac{h}{l}\ 19.9 \qquad \frac{l}{l}\frac{h}{l}\ 14.7$$

$$\frac{h}{h}\frac{l}{l}\ 11.8 \qquad \frac{h}{l}\frac{l}{l}\ 9.1 \qquad \frac{l}{l}\frac{l}{l}\ 2.3$$

What conclusions can you reach about the distribution of genetic factors and their actions from these data?

13. Suppose that number of eye facets is measured in a population of *Drosophila* under various temperature conditions. Further suppose that it is possible to estimate total genetic variance (s_g^2) as well as the phenotypic distribution. Finally, suppose that there are only two genotypes in the population. Draw pairs of norms of reaction that would lead to the following results:

a. An increase in mean temperature decreases the phenotypic variance.

b. An increase in mean temperature increases H^2.

c. An increase in mean temperature increases s_g^2 but decreases H^2.

d. An increase in temperature *variance* changes a unimodal into a bimodal phenotypic distribution (one norm of reaction is sufficient here).

14. Francis Galton compared the heights of male undergraduates with the heights of their fathers, with the results shown in the following graph.

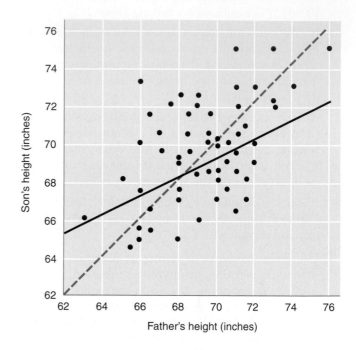

The average height of all fathers is the same as the average height of all sons, but the individual height classes are not equal across generations. The very tall fathers had somewhat shorter sons, whereas the very short fathers had somewhat taller sons. As a result, the best line that can be drawn through the points on the scatter diagram has a slope of about 0.67 *(solid line)* rather than 1.00 *(dashed line)*. Galton used the term *regression* to describe this tendency for the phenotype of the sons to be closer than the phenotype of their fathers to the population mean.

a. Propose an explanation for this regression.

b. How are regression and heritability related here?

(Graph after W. F. Bodmer and L. L. Cavalli-Sforza, *Genetics, Evolution, and Man.* Copyright 1976 by W. H. Freeman and Company.)

21

EVOLUTIONARY GENETICS

Charles Darwin. [Corbis/Bettmann.]

KEY QUESTIONS

- What are the basic principles of the Darwinian mechanism of evolution?

- What are the roles of natural selection and other processes in evolution and how do they interact with one another?

- How do different species arise?

- How different are the genomes of different kinds of organisms?

- How do evolutionary novelties arise?

OUTLINE

21.1 A synthesis of forces: variation and divergence of populations

21.2 Multiple adaptive peaks

21.3 Heritability of variation

21.4 Observed variation within and between populations

21.5 The process of speciation

21.6 Origin of new genes

21.7 Rate of molecular evolution

21.8 Genetic evidence of common ancestry in evolution

21.9 Comparative genomics and proteomics

CHAPTER OVERVIEW

The modern theory of evolution is so completely identified with the name of Charles Darwin (1809–1882) that many people think that it was Darwin who first proposed the concept that organisms have

evolved, but that is certainly not the case. Most scholars had abandoned the notion of fixed species, unchanged since their origin in a grand creation of life, long before publication of Darwin's *Origin of Species* in 1859. By that time, most biologists agreed that new species arise through some process of evolution from older species;

CHAPTER OVERVIEW Figure

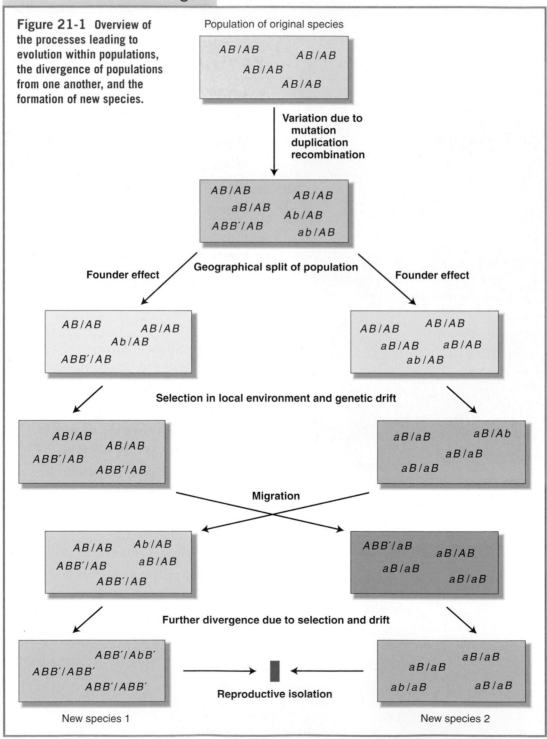

Figure 21-1 Overview of the processes leading to evolution within populations, the divergence of populations from one another, and the formation of new species.

the problem was to explain *how* this evolution could occur.

Darwin provided a detailed explanation of the mechanism of the evolutionary process. Darwin's theory of the mechanism of evolution begins with the variation that exists among organisms within a species. Individuals of one generation are qualitatively different from one another. Evolution of the species as a whole results from the fact that the various types differ in their rates of survival and reproduction, and so the relative frequencies of the types change over time. Evolution, in this view, is a sorting process.

For Darwin, evolution of the group resulted from the differential survival and reproduction of individual variants *already existing* in the group—variants arising in a way unrelated to the environment but whose survival and reproduction do depend on the environment (Figure 21-1).

> **MESSAGE** Darwin proposed a new explanation to account for the accepted phenomenon of evolution. He argued that the population of a given species at a given time includes individuals of varying characteristics. The population of the next generation will contain a higher frequency of those types that most successfully survive and reproduce under the existing environmental conditions. Thus, the frequencies of various types within the species will change over time.

There is an obvious similarity between the process of evolution as Darwin described it and the process by which the plant or animal breeder improves a domestic stock. The plant breeder selects the highest-yielding plants from the current population and (as far as possible) uses them as the parents of the next generation. If the characteristics causing the higher yield are heritable, then the next generation should produce a higher yield. It was no accident that Darwin chose the term **natural selection** to describe his model of evolution through differential rates of reproduction of different variants in the population. As a model for this evolutionary process, he had in mind the selection that breeders exercise on successive generations of domestic plants and animals.

We can summarize Darwin's theory of evolution through natural selection in three principles:

1. *Principle of variation.* Among individuals within any population, there is variation in morphology, physiology, and behavior.
2. *Principle of heredity.* Offspring resemble their parents more than they resemble unrelated individuals.
3. *Principle of selection.* Some forms are more successful at surviving and reproducing than other forms in a given environment.

Clearly, a selective process can produce change in the population composition only if there are some variations among which to select. If all individuals are identical, no amount of differential reproduction of individuals will alter the composition of the population. Furthermore, the variation must be in some part heritable if differential reproduction is to alter the population's genetic composition. If large animals within a population have more offspring than do small ones but their offspring are no larger on average than those of small animals, then there will be no change in population composition from one generation to another. Finally, if all variant types leave, on average, the same number of offspring, then we can expect the population to remain unchanged.

> **MESSAGE** Darwin's principles of variation, heredity, and selection must hold true if there is to be evolution by a variational mechanism.

The Darwinian explanation of evolution must be able to account for two different aspects of the history of life. One is the successive change of form and function that occurs in a single continuous line of descent, **phyletic evolution.** Figure 21-2 shows such a continuous change over a period of 40 million years in the size and curvature of the left shell of the oyster, *Gryphea.* The other is the **diversification** that occurs among species: in the history of life on earth, there have existed many different contemporaneous species having quite different forms and living in different ways. Figure 21-3 shows some of the variety of bivalve mollusc forms that existed at various times in the past 300 million years. Every species eventually becomes extinct and more than 99.9 percent of all the species that have ever existed are already extinct, yet the number of species and the diversity of their forms and functions have increased in the past billion years. Thus species not only must be changing, but must give rise to new and different species in the course of evolution.

Both these processes—phyletic evolution and diversification—are the consequences of heritable variation within populations. Heritable variation provides the raw material for successive changes within a species and for the multiplication of new species. The basic mechanisms of those changes (as discussed in Chapter 19) are the origin of new variation by mutation and chromosomal rearrangements, the change in frequency of alleles within populations by selective and random processes, the divergence of different populations because the selective forces are different or because of random drift, and the reduction of variation between populations by migration. From those basic mechanisms, a set of

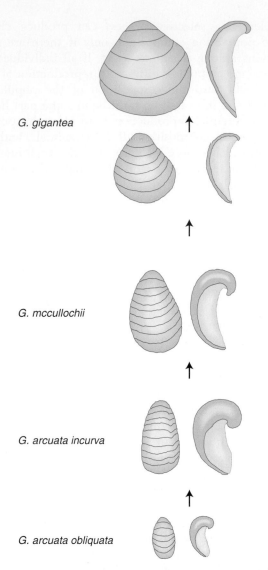

Figure 21-2 Changes in shell size and curvature in the bivalve mollusc *Gryphaea* in the course of its phyletic evolution in the early Jurassic. Only the left shell is shown. In each case, the shell back and a longitudinal section through it are illustrated. [After A. Hallam, "Morphology, Palaeoecology and Evolution of the Genus *Gryphaea* in the British Lias," *Philosophical Transactions of the Royal Society of London Series B* 254, 1968, 124.]

principles governing changes in the genetic composition of populations can be derived. The application of these principles of population genetics provides a detailed genetic theory of evolution.

> **MESSAGE** Evolution, under the Darwinian scheme, is the conversion of heritable variation between individuals within populations into heritable differences between populations in time and in space, by population genetic mechanisms.

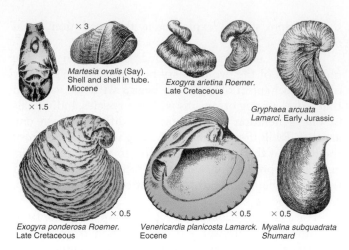

Figure 21-3 A variety of bivalve mollusc shell forms that have appeared in the past 300 million years of evolution. [After C. L. Fenton and M. A. Fenton, *The Fossil Book.* Doubleday, 1958.]

21.1 A synthesis of forces: variation and divergence of populations

When Darwin arrived in the Galapagos Islands in 1835, he found a remarkable group of finchlike birds that provided a very suggestive case for the development of his theory of evolution. The Galapagos archipelago is a cluster of 29 islands and islets of different sizes lying on the equator about 600 miles off the coast of Ecuador. Finches are generally ground-feeding seed eaters with stout bills for cracking the tough outer coats of the seeds. Figure 21-4 shows the 13 Galapagos finch species. The Galapagos species, though clearly finches, exhibit an immense variation in feeding behavior and in the bill shape that corresponds to their food sources. For example, the vegetarian tree finch uses its heavy bill to eat fruits and leaves, the insectivorous finch has a bill with a biting tip for eating large insects, and, most remarkable of all, the woodpecker finch grasps a twig in its bill and uses it to obtain insect prey by probing holes in trees. This diversity of species arose from an original population of a seed-eating finch that arrived in the Galapagos from the mainland of South America and populated the islands. The descendants of the original colonizers spread to the different islands and to different parts of large islands and formed local populations that diverged from one another and eventually formed different species. The finches illustrate the two aspects of evolution that need to be explained. How does one original species with a particular set of characteristics give rise to a diversity of species, each with its own form and function? How do the characteristics of species come to be so

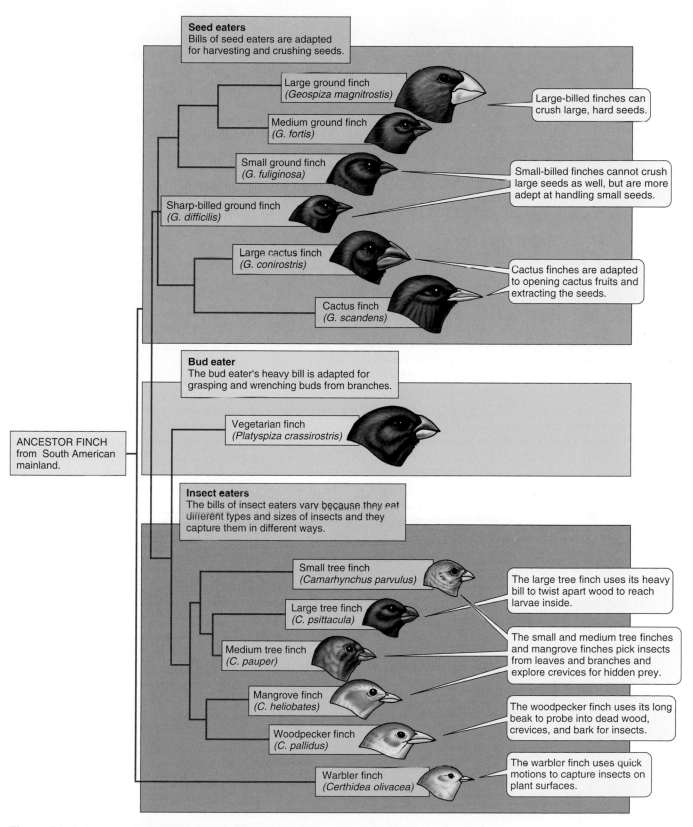

Figure 21-4 The thirteen species of finches found in the Galapagos Islands. [After W. K. Purves, G. H. Orians, and H. C. Heller, *Life: The Science of Biology,* 4th ed. Sinauer Associates/ W. H. Freeman and Company, 1995, Figure 20.3, p. 450.]

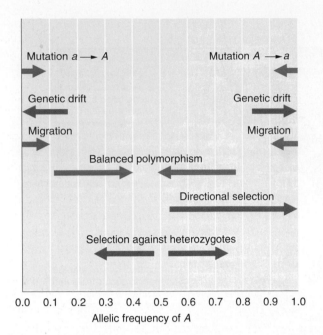

Figure 21-5 The effects on gene frequency of various forces of evolution. The blue arrows show a tendency toward increased variation within the population; the red arrows, decreased variation.

suited to the environments in which the species live? These are the problems of the origin of *diversity* and the origin of *adaptation*.

In evolution, the various forces of breeding structure, mutation, migration, and selection are all acting simultaneously in populations. We need to consider how these forces, operating together, mold the genetic composition of populations to produce both variation within local populations and differences between them.

The genetic variation within and between populations is a result of the interplay of the various evolutionary forces just listed (Figure 21-5). Generally, as Table 21-1 shows, forces that increase or maintain variation

within populations prevent populations from diverging from one another, whereas forces that make each population homozygous cause populations to diverge. Thus, random drift (or inbreeding) produces homozygosity while causing different populations to diverge. This trend toward divergence and homozygosity is counteracted by the constant flux of mutation and the migration of individuals between populations, both of which introduce variation into the populations, making them more alike.

Consider the situation at a genetically variable locus with two alleles, say A and a, in frequencies p and $1 - p$, respectively in a large population. Suppose that a group of isolated island populations were founded by migrants from that single population. The original founders of each population are small samples from the donor population and so differ from one another in allele frequencies because of a random sampling effect. This initial variation is called the **founder effect.** In succeeding generations, random genetic drift further alters allelic frequencies within each population. The frequency of each of the alleles will move toward either 1 or 0 in each population, but average allelic frequency over all the populations remains constant. As time goes on, the gene frequencies among the populations diverge and some become fixed for one of the alleles. By the time $4N$ generations have gone by, 80 percent of the populations are fixed, a proportion p being homozygous A/A and proportion $1 - p$ being homozygous a/a. Eventually, all populations would become fixed at A/A or a/a in these proportions.

The process of differentiation by inbreeding in island populations is slow, but not on an evolutionary or geological time scale. If an island can support, say, 10,000 individuals of a rodent species, then, after 20,000 generations (about 7000 years, assuming 3 generations per year), the population will be homozygous for about half of all the loci that were initially at the maximum of

TABLE 21-1 How the Forces of Evolution Increase (+) or Decrease (−) Variation Within and Between Populations

Force	Variation within populations	Variation between populations
Inbreeding or genetic drift	−	+
Mutation	+	−
Migration	+	−
Directional selection	−	+/−
Balancing	+	−
Incompatible	−	+

heterozygosity. Moreover, the island will be differentiated from other similar islands in two ways: (1) loci that are fixed on that island will either still be segregating on many of the other islands or be fixed at a different allele and (2) loci that are still segregating in all the islands will vary in allele frequency from island to island.

Every population of every species is finite in size, and so all populations should eventually become homozygous and differentiated from one another as a result of inbreeding. All variation would be eliminated and evolution would cease. In nature, however, new variation is always being introduced into populations by mutation and by some migration between localities. Thus, the actual variation available for natural selection is a balance between the introduction of new variation and its loss through local inbreeding. Recall from Chapter 19 that the rate of loss of heterozygosity in a closed population is $1/(2N)$ per generation, and so any effective differentiation between populations that occurs because of drift will be negated if new variation is introduced at this rate or a higher rate. If m is the migration rate into a given population and μ is the rate of mutation to new alleles per generation, then roughly (to an order of magnitude) a population will retain most of its heterozygosity and will not differentiate much from other populations by local inbreeding if

$$m \geq \frac{1}{N} \quad \text{or} \quad \mu \geq \frac{1}{N}$$

or, in other words, if

$$Nm \geq 1 \quad \text{or} \quad N\mu \geq 1$$

For populations of intermediate and even fairly large size, it is unlikely that $N\mu \geq 1$. For example, if the population size is 100,000, then, to prevent loss of variation, the mutation rate must exceed 10^{-5}, which is somewhat on the high side for known mutation rates, although it is not an unknown rate. On the other hand, a migration rate of 10^{-5} per generation is not unreasonably large. In fact

$$m = \frac{\text{number of migrants}}{\text{total population size}} = \frac{\text{number of migrants}}{N}$$

Thus, the requirement that $Nm \geq 1$ is equivalent to the requirement that

$$Nm = N \times \frac{\text{number of migrants}}{N} \geq 1$$

or that

$$\text{number of migrant individuals} \geq 1$$

irrespective of population size. For many populations, more than a single migrant individual per generation is quite likely. Human populations (even isolated tribal populations) have a higher migration rate than this minimal value, and, as a result, no locus is known in humans for which one allele is fixed in some populations and an alternative allele is fixed in others.

The effects of selection are more variable than those of random genetic drift because selection may or may not push a population toward homozygosity. **Directional selection** pushes a population toward homozygosity, rejecting most new mutations as they are introduced but occasionally (if the mutation is advantageous) spreading a new allele through the population to create a new homozygous state. Whether such directional selection promotes differentiation of populations depends on the environment and on chance events. Two populations living in very similar environments may be kept genetically similar by directional selection, but, if there are environmental differences, selection may direct the populations toward different compositions.

Selection favoring heterozygotes **(balancing selection)** will, for the most part, maintain more or less similar polymorphisms in different populations. However, again, if the environments are different enough, then the populations will show some divergence. The opposite of balancing selection is selection against heterozygotes, which produces unstable equilibria. Such selection will cause homozygosity and divergence between populations.

21.2 Multiple adaptive peaks

We must avoid taking an overly simplified view of the consequences of selection. At the level of the gene—or even at the level of the partial phenotype—there is more than one possible outcome of selection for a trait in a given environment. Selection to alter a trait (say, to increase size) may be successful in a number of ways. In 1952, F. Robertson and E. Reeve successfully selected to change wing size in two different populations of *Drosophila*. However, in one case, the *number* of cells in the wing changed, whereas, in the other case, the *size* of the wing cells changed. Two different genotypes had been selected, both causing a change in wing size. The initial state of the population at the outset of selection determined which of these selections occurred.

A simple hypothetical case illustrates how the same selection can lead to different outcomes. Suppose that the variation of two loci (there will usually be many more) influences a character and that (in a particular environment) intermediate phenotypes have the highest fitness. (For example, newborn babies have a higher chance of surviving birth if they are neither too big nor too small.) If the alleles act in a simple way in influencing the phenotype, then the three genetic constitutions

AB/ab, *Ab/Ab*, and *aB/aB* will produce a high fitness because they will all be intermediate in phenotype. On the other hand, very low fitness will characterize the double homozygotes *AB/AB* and *ab/ab*. What will the result of selection be? We can predict the result by using the mean fitness $\overline{W}$ of a population. As previously discussed, selection acts in most simple cases to increase $\overline{W}$. Therefore, if we calculate $\overline{W}$ for every possible combination of gene frequencies at the two loci, we can determine which combinations yield high values of $\overline{W}$. Then we should be able to predict the course of selection by following a curve of increasing $\overline{W}$.

The surface of mean fitness for all possible combinations of allelic frequency is called an **adaptive surface** or an **adaptive landscape** (Figure 21-6). The figure is like a topographic map. The frequency of allele *A* at one locus is plotted on one axis, and the frequency of allele *B* at the other locus is plotted on the other axis. The height above the plane (represented by topographic lines) is the value of $\overline{W}$ that the population would have for a particular combination of frequencies of *A* and *B*. According to the rule of increasing fitness, selection should carry the population from a low-fitness "valley" to a high-fitness "peak." However, Figure 21-6 shows that there are two adaptive peaks, corresponding to a fixed population of *Ab/Ab* and a fixed population of *aB/aB*, with an adaptive valley between them. Which peak the population will ascend—and therefore what its final genetic composition will be—depends on whether the initial genetic composition of the population is on one side or the other of the dashed "fall line" shown in the figure.

> **MESSAGE** Under identical conditions of natural selection, two populations may arrive at two different genetic compositions as a direct result of natural selection.

It is important to note that nothing in the theory of selection requires that the different adaptive peaks be of the same height. The kinetics of selection dictate only that $\overline{W}$ increases, not that it necessarily reaches the highest possible peak in the field of gene frequencies. Suppose, for example, that a population is near the peak *aB/aB* in Figure 21-6 and that this peak is lower than the *Ab/Ab* peak. Selection alone cannot carry the population to *Ab/Ab*, because that would require a temporary decrease in $\overline{W}$ as the population descended the *aB/aB* slope, crossed the saddle, and ascended the other slope. Thus, the force of selection is myopic. It drives the population to a *local* maximum of $\overline{W}$ in the field of gene frequencies—not to a *global* one.

The existence of multiple adaptive peaks for a selective process means that some differences between species are the result of history and not of environmental differences. For example, African rhinoceroses have two horns on their noses, whereas Indian rhinoceroses have one. We need not invent a special story to explain

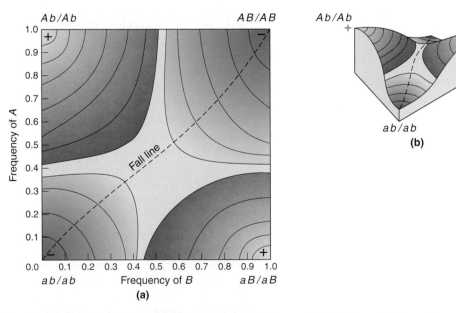

Figure 21-6 An adaptive landscape with two adaptive peaks (red), two adaptive valleys (blue), and a topographic saddle in the center of the landscape. The topographic lines are lines of equal mean fitness. If the genetic composition of a population always changes in such a way as to move the population "uphill" in the landscape (to increasing fitness), then the final composition will depend on where the population began with respect to the fall (dashed) line. (a) Topographic map of the adaptive landscape. (b) A perspective sketch of the surface shown in the map.

why it is better to have two horns on the African plains and one in India. It is much more plausible that the trait of having horns was selected but that two long, slender horns and one short, stout horn are simply alternative adaptive features, and the differences between them are a result of historical accident. Explanations of adaptations by natural selection do not require that every difference between species be an adaptive difference.

Exploration of adaptive peaks

Random and selective forces should not be thought of as simple antagonists. Random drift may counteract the force of selection, but it can enhance it as well. The outcome of the evolutionary process is a result of the simultaneous operation of these two forces. Figure 21-7 illustrates these possibilities. Note that there are multiple adaptive peaks in this landscape. Because of random drift, a population under selection does not ascend an adaptive peak smoothly. Instead, it takes an erratic course in the field of gene frequencies, like an oxygen-starved mountain climber. Pathway I shows a population history where adaptation has failed. The random fluctuations of gene frequency were sufficiently great that the population by chance became fixed at an unfit genotype. In any population, some proportion of loci are fixed at a selectively unfavorable allele because the intensity of selection is insufficient to overcome the random drift to fixation. The existence of multiple adaptive peaks and

the random fixation of less fit alleles are integral features of the evolutionary process. Natural selection cannot be relied on to produce the best of all possible worlds.

Pathway II in Figure 21-7, on the other hand, shows how random drift may improve adaptation. The population was originally in the sphere of influence of the lower adaptive peak; however, by random fluctuation in gene frequency, its composition passed over the adaptive saddle, and the population was captured by the higher, steeper adaptive peak. This passage from a lower to a higher adaptive stable state could never have occurred by selection in an infinite population, because, by selection alone, $\overline{W}$ could never decrease temporarily to cross from one slope to another.

Another important source of indeterminacy in the outcome of a long selective process is the randomness of the mutational process. After the initial genetic variation is exhausted by the selective and random fixation of alleles, new variation arising from mutation can be the source of yet further evolutionary change. The particular direction of this further evolution depends on the particular mutations that occur and the time order in which they take place. A very clear illustration of this historical contingency of the evolutionary process is a selection experiment carried out by H. Wichman and her colleagues. They forced the bacteriophage ΦX174 to reproduce at high temperatures and on the host *Salmonella typhimurium* instead of its normal host *Escherichia coli*. Two independent selection lines were established, labeled TX and ID, and both evolved the ability to reproduce at high temperatures in the new host. In one of the two lines, the ability to reproduce on *E. coli* still existed, but, in the other line, the ability was lost. The bacteriophage has only 11 genes, and so the experimenters were able to record the successive changes in the DNA for all these genes and in the proteins encoded by them during the selection process. There were 15 DNA changes in strain TX, located in 6 different genes; in strain ID, there were 14 changes located in 4 different genes. In only 7 cases were the changes to the two strains identical, including a large deletion, but even these identical changes appeared in each line in a different order. So, for example, the change at DNA site 1533, causing a substitution of isoleucine for threonine, was the third change in the ID strain, but the fourteenth change in the TX strain.

21.3 Heritability of variation

The first rule of any reconstruction or prediction of evolutionary change is that the phenotypic variation must be heritable. It is easy to construct stories of the possible selective advantage of one form of a trait over another, but it is a matter of considerable experimental difficulty to show that the variation in the trait corresponds to genotypic differences (see Chapter 20).

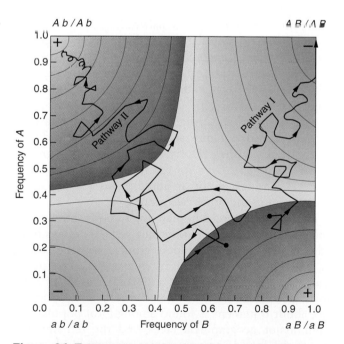

Figure 21-7 Interaction of selection and random drift. Selection and random drift can interact to produce different changes in gene frequency in an adaptive landscape. Without random drift, both populations would have moved toward *aB / aB* as a result of selection alone.

It should not be supposed that all variable traits are heritable. Certain metabolic traits (such as resistance to high salt concentrations in *Drosophila*) show individual variation but no heritability. In general, behavioral traits have lower heritabilities than morphological traits, especially in organisms with more complex nervous systems that exhibit immense individual flexibility in central nervous states. Before any prediction is made about the evolution of a particular quantitative trait, it is essential to determine if there is genetic variance for it in the population whose evolution is to be predicted. Thus, suggestions that such traits in the human species as performance on IQ tests, temperament, and social organization are in the process of evolving or have evolved at particular epochs in human history depend critically on evidence about whether there is genetic variation for these traits. Reciprocally, traits that appear to be completely phenotypically invariant in a species may nevertheless evolve.

One of the most important findings in evolutionary genetics has been the discovery that substantial genetic variation may underly characters that show no morphological variation. They are called **canalized characters,** because the final outcome of their development is held within narrow bounds despite disturbing forces. Different genotypes for canalized characters have the same constant phenotype over the range of environments that is usual for the species. The genetic differences are revealed if the organisms are put in a stressful environment or if a severe mutation stresses the developmental system. For example, all wild-type *Drosophila* have exactly four scutellar bristles (Figure 21-8). If the recessive mutant *scute* is present, the number of bristles is re-

duced, but, in addition, there is variation from fly to fly. This variation is heritable, and lines with zero or one bristle and lines with three or four bristles can be obtained by selection of flies carrying the *scute* mutation. When the mutation is removed, these lines now have two and six bristles, respectively. Similar experiments with similar results have been performed by using extremely stressful environments in place of mutants. A consequence of such hidden genetic variation is that a character that is phenotypically uniform in a species may nevertheless undergo rapid evolution if a stressful environment uncovers the genetic variation.

21.4 Observed variation within and between populations

In Chapter 19, we saw that genetic variation exists within populations at the levels of morphology, karyotype, proteins, and DNA. The general conclusion is that about one-third of all protein-encoding loci are polymorphic and that all classes of DNA, including exons, introns, regulatory sequences, and flanking sequences, show nucleotide diversity among individuals within populations. Several of these examples also documented some differences in genotype frequencies between populations (see Tables 19-1 through 19-3 and 19-5). The relative amounts of variation within and between populations vary from species to species, depending on history and environment. In humans, some gene frequencies (for example, those for skin color or hair form) are well differentiated between populations and major geographical groups (so-called geographical races). If, however, we look at single structural genes identified immunologically or by electrophoresis rather than by these outward phenotypic characters, the situation is rather different. Table 21-2 shows the three loci for which Caucasians, Negroids, and Mongoloids are known to be most different from one another (Duffy and Rhesus blood groups and the P antigen). Even for the most divergent loci, no major geographical group is homozygous for one allele that is absent in the other two groups.

In general, different human populations show rather similar frequencies for polymorphic genes. Findings from studies of polymorphic blood groups, enzyme-coding loci, and DNA polymorphisms in a variety of human populations have shown that about 85 percent of total human genetic diversity is found within local populations, about 6 percent is found among local populations within major geographical races, and the remaining 9 percent is found among major geographical races. Clearly, the genes influencing skin color, hair form, and facial form that are well differentiated among "races" are not a random sample of structural gene loci.

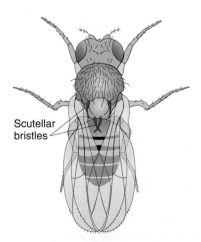

Scutellar
bristles

Figure 21-8   The scutellar bristles of the adult *Drosophila*, shown in blue. This illustration shows an example of a canalized character; all wild-type *Drosophila* have four scutellar bristles in a very wide range of environments.

TABLE 21-2 Examples of Extreme Differentiation in Blood-Group Allelic Frequencies in Three Major Geographical Groups

Gene	Allele	Geographical group		
		Caucasoid	Negroid	Mongoloid
Duffy	Fy	0.0300	0.9393	0.0985
	Fy^a	0.4208	0.0000	0.9015
	Fy^b	0.5492	0.0607	0.0000
Rhesus	R_0	0.0186	0.7395	0.0409
	R_1	0.4036	0.0256	0.7591
	R_2	0.1670	0.0427	0.1951
	r	0.3820	0.1184	0.0049
	r'	0.0049	0.0707	0.0000
	Others	0.0239	0.0021	0.0000
P antigen	P_1	0.5161	0.8911	0.1677
	P_2	0.4839	0.1089	0.8323

Source: A summary is provided in L. L. Cavalli-Sforza and W. F. Bodmer, *The Genetics of Human Populations* (W. H. Freeman and Company, 1971), pp. 724–731. See L. L. Cavalli-Sforza, P. Menozzi, and A. Piazza, *The History and Geography of Human Genes* (Princeton University Press, 1994), for detailed data.

21.5 The process of speciation

When we examine the living world, we see that individual organisms are usually clustered into collections that resemble one another more or less closely and are clearly distinct from other clusters. A close examination of a sibship of *Drosophila* will show differences in bristle number, eye size, and details of color pattern from fly to fly, but an entomologist has no difficulty whatsoever in distinguishing *Drosophila melanogaster* from, say, *Drosophila pseudoobscura*. One never sees a fly that is halfway between these two kinds. Clearly, in nature at least, there is no effective interbreeding between these two forms. A group of organisms that exchanges genes within the group but cannot do so with other groups is what is meant by a **species.** Within a species there may exist local populations that are also easily distinguished from one another by some phenotypic characters, but it is also the case that genes can easily be exchanged between them. For example, no one has any difficulty distinguishing a "typical" Senegalese from a "typical" Swede, but such people are able to mate with each other and produce progeny. In fact, there have been many such matings in North America in the past 300 years, creating an immense number of people of every degree of intermediacy between these local geographical types. They are not separate species. In general, there is some difference in the frequency of various genes in different geographical populations of any species; so the marking out of a particular population as a distinct race is arbitrary and, as a consequence, the concept of race is no longer much used in biology.

MESSAGE A species is a group of organisms that can exchange genes among themselves but are genetically unable to exchange genes in nature with individuals in other such groups. A geographical race is a phenotypically distinguishable local population within a species that is capable of exchanging genes with other races within that species. Because nearly all geographical populations are different from others in the frequencies of some genes, race is a concept that makes no clear biological distinction.

All the species now existing are related to each other, having had a common ancestor at some time in the evolutionary past. That means that each of these species has separated out from a previously existing species and has become genetically distinct and genetically isolated from its ancestral line. In extraordinary circumstances, a single mutation might be enough to found such a genetically isolated group, but the carrier of that mutation would need to be capable of self-fertilization or vegetative reproduction. Moreover, that mutation would have to cause complete mating incompatibility between its carrier and the original species and to allow the new line to compete successfully with the previously established group. Although not impossible, such events must be rare.

More commonly, new species form as a result of geographical isolation. We have already seen how populations that are geographically separated will diverge from one another genetically as a consequence of unique mutations, selection, and genetic drift. Migration between populations will prevent them from diverging too far, however. As shown on page 685, even a single migrant per generation is sufficient to prevent populations from fixing at alternative alleles by genetic drift alone, and even selection toward different adaptive peaks will not succeed in causing complete divergence unless it is extremely strong. As a consequence, populations that diverge enough to become new, reproductively isolated species must first be virtually totally isolated from one another by some mechanical barrier. This isolation almost always requires some spatial separation, and the separation must be great enough or the natural barriers to the passage of migrants must be strong enough to prevent any effective migration. Such spatially isolated populations are referred to as **allopatric**. The isolating barrier might be, for example, the extending tongue of a continental glacier during glacial epochs that forces apart a previously continuously distributed population or the drifting apart of continents that become separated by water or the infrequent colonization of islands that are far from shore. The critical point is that these barriers must make further migration between the separated populations a very rare event. If so, then the populations are now genetically independent and will continue to diverge by mutation, selection, and genetic drift. Eventually, the genetic differentiation between the populations becomes so great that the formation of hybrids between them would be physiologically, developmentally, or behaviorally impossible even if the geographical separation were abolished. These *biologically* isolated populations are now new species, formed by the process of **allopatric speciation**.

> **MESSAGE** Allopatric speciation occurs through an initial geographical and mechanical isolation of populations that prevents any gene flow between them, followed by genetic divergence of the isolated populations sufficient to make it biologically impossible for them to exchange genes in the future.

There are two main biological isolating mechanisms: prezygotic isolating mechanisms and postzygotic mechanisms. **Prezygotic isolation** occurs when there is failure to form zygotes. The cause of this failure may be that the different species mate at different seasons or in different habitats. It may also be that the species are not sexually attractive to each other or their genitalia do not match or male gametes are physiologically incompatable with the female.

Examples of prezygotic isolating mechanisms are well known in plants and animals. The two species of pine growing on the Monterey peninsula, *Pinus radiata* and *P. muricata*, shed their pollen in February and April and so do not exchange genes. The light signals that are emitted by male fireflies and attract females differ in intensity and timing between species. In the tsetse fly, *Glossina*, mechanical incompatibilities cause severe injury and even death if males of one species mate with females of another. The pollen of different species of *Nicotiana*, the genus to which tobacco belongs, either fails to germinate or cannot grow down the style of other species. **Postzygotic isolation** results from the failure of fertilized zygotes to contribute gametes to future generations. Hybrids may fail to develop or have a lower probability of survival than that of the parental species or the hybrids may be partly or completely sterile. Postzygotic isolation is more common in animals than in plants, apparently because the development of many plants is much more tolerant to genetic incompatibilities and chromosomal variations. When the eggs of the leopard frog, *Rana pipiens*, are fertilized by sperm of the wood frog, *R. sylvatica*, the embryos do not succeed in developing. Horses and asses can be easily crossed to produce mules, but, as is well known, these hybrids are sterile.

Genetics of species isolation

Usually, it is not possible to carry out any genetic analysis of the isolating mechanisms between two species for the simple reason that, by definition, they cannot be crossed with each other. It is possible, however, to make use of very closely related species in which the isolating mechanism has not produced complete hybrid sterility and hybrid breakdown. These species can be crossed and the segregating progeny of hybrid F_2 or backcross generations can be analyzed by using genetic markers and the technique of locating quantitative trait loci (QTLs) discussed in Chapter 20. When such marker experiments have been performed on other species, mostly in the genus *Drosophila*, the general conclusions are that gene differences responsible for hybrid inviability are on all the chromosomes more or less equally and that, for hybrid sterility, there is some added effect of the X chromosome. For behavioral sexual isolation, the results are variable. In *Drosophila*, all the chromosomes are involved, but, in Lepidoptera, the genes are much more localized, apparently because of the involvement of specific pheromones whose scent is important in species recognition. The sex chromosome has a very strong effect in butterflies; in the European corn borer, for example, only three loci, one of which is on the sex chromosome, account for the entire isolation between pheromonal types within the species.

21.6 Origin of new genes

It is clear that evolution consists of more than the substitution of one allele for another at loci of defined function. New functions have arisen that have resulted in major new ways of making a living. Many of these new functions—for example, the development of the mammalian inner ear from a transformation of the reptilian jaw bones—result from continuous transformations of shape and do not require totally new genes and proteins. But new genes and proteins are necessary to produce qualitative novelties, such as photosynthesis and cell walls in plants, contractile proteins, new cell and tissue types, oxygenation molecules such as hemoglobin, the immune system, chemical detoxification cycles, and digestive enzymes. Older metabolic functions must have been maintained while new ones were being developed, which in turn means that old genes had to be preserved while new genes with new functions had to evolve. Where does the DNA for new genes come from?

Polyploidy

One process for the provision of new DNA is the duplication of the entire genome by polyploidization, which is much more common in plants than in animals (Chapter 15). Evidence that polyploids have played a major role in the evolution of plant species is presented in Figure 21-9, which shows the frequency distribution of haploid chromosome numbers among dicotyledonous plant species. Above a chromosome number of about 12, even numbers are much more common than odd numbers—a consequence of frequent polyploidy.

Duplications

A second way to increase DNA is by the duplication of small sections of the genome. Such duplication may be a consequence of misreplication of DNA. Alternatively, a transposable element may insert a copy of one part of the genome into another location (see Chapter 11). After a duplicated segment has arisen, one of three

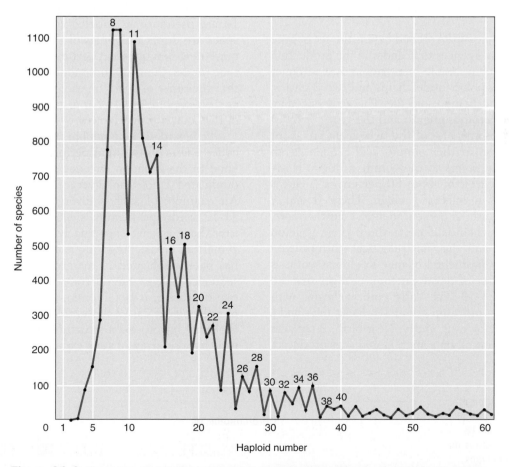

Figure 21-9 **Frequency distribution of haploid chromosome numbers in dicotyledonous plants.**
[After Verne Grant, *The Origin of Adaptations*. Columbia University Press, 1963.]

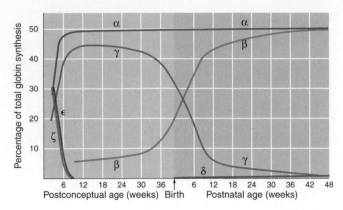

Figure 21-10 Developmental changes in the synthesis of the α-like and β-like globins that make up human hemoglobin.

α	ζ	β	γ	ϵ
α	58	42	39	37
ζ		34	38	37
β			73	75
γ				80

things can happen: (1) the production of the polypeptide may simply increase; (2) the general function of the original sequence is maintained in the new DNA, but there is some differentiation of the sequences by accumulated mutations so that variations on the same protein theme are produced, allowing a somewhat more complex molecular structure; or (3) the new segment may diverge more dramatically and take a whole new function.

A classic example of the second case is the set of gene duplications and divergences that underlie the production of human hemoglobin. Adult hemoglobin is a tetramer consisting of two α polypeptide chains and two β chains, each with its bound heme molecule. The gene encoding the α chain is on chromosome 16 and the gene for the β chain is on chromosome 11, but the two chains are about 49 percent identical in their amino acid sequences, an identity that clearly points to the common origin. However, in fetuses, until birth, about 80 percent of β chains are substituted by a related γ chain. These β and γ polypeptide chains are 75 percent identical. Furthermore, the gene for the γ chain is close to the β-chain gene on chromosome 11 and has an identical intron–exon structure. This developmental change in globin synthesis is part of a larger set of developmental changes that are shown in Figure 21-10. The early embryo begins with α, γ, ϵ, and ζ chains and, after about 10 weeks, the ϵ and ζ chains are replaced by α, β, and γ. Near birth, β replaces γ and a small amount of yet a sixth globin, δ, is produced.

Table 21-3 shows the percentage of amino acid identity among these chains, and Figure 21-11 shows the chromosomal locations and intron—exon structures of the genes encoding them. The story is remarkably consistent. The β, δ, γ, and ϵ chains all belong to a "β-like" group; they have very similar amino acid sequences and are encoded by genes of identical intron—exon structure that are all contained in a 60-kb stretch of DNA on chromosome 11. The α and ζ chains belong to an "α-like" group and are encoded by genes contained in a 40-kb region on chromosome 16. In addition, Figure 21-11 shows that on both chromosome 11 and chromosome 16 are pseudogenes, labeled ψ_α and ψ_β. These pseudogenes are duplicate copies of the genes that did not acquire new functions but accumulated random mutations that render them nonfunctional. What is remarkable is that the order of genes on each chromosome is the same as the temporal order of appearance of the globin chains in the course of development.

In regard to hemoglobin, the duplicated DNA encodes a new protein that performs a function closely related to the function encoded by the original gene. But duplicated DNA can diverge dramatically in function. An example of such a divergence is shown in Figure 21-12. Birds and mammals, like other eukaryotic organisms, have a gene encoding lysozyme, a protective enzyme that breaks down the bacterial cell wall. This gene has been duplicated in mammals to produce a second sequence that encodes a completely different, nonenzymatic protein, α-lactalbumin, a nutritional component of milk. Figure 21-12 shows that the duplicated gene has the same intron—exon structure as that of the lysozyme gene, whose array of four exons and three introns itself

Figure 21-11
Chromosomal distribution of the genes for the α family of globins on chromosome 16 and the β family of globins on chromosome 11 in humans. Gene structure is shown by black bars (exons) and colored bars (introns).

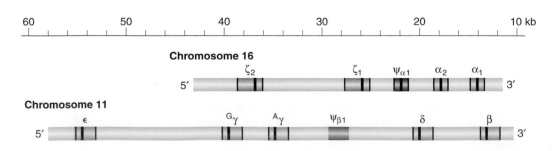

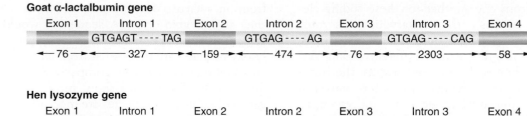

Figure 21-12 Structural homology of the gene for hen lysozyme and mammalian α-lactalbumin. Exons and introns are indicated by dark green bars and light green bars, respectively. Nucleotide sequences at the beginning and end of each intron are indicated, and the numbers refer to the nucleotide lengths of each segment. [After I. Kumagai, S. Takeda, and K.-I. Miura, "Functional Conversion of the Homologous Proteins α-Lactalbumin and Lysozyme by Exon Exchange," *Proceedings of the National Academy of Sciences USA* 89, 1992, 5887–5891.]

suggests an earlier multiple duplication event in the origin of lysozyme.

Imported DNA

DNA duplications are not the only source of new DNA that is the basis of new functions; it can also be imported. Repeatedly in evolution, extra DNA has been imported into the genome from outside sources by mechanisms other than normal sexual reproduction. DNA can be inserted into chromosomes from other chromosomal locations and even from other species. In some cases, genes from totally unrelated organisms can become incorporated into cells to become a functional part of the recipient cell's genome.

CELLULAR ORGANELLES Eukaryotic cells have obtained some of their organelles in this way. Both the chloroplasts of photosynthetic organisms and mitochondria are the descendants of prokaryotes that entered the eukaryotic cells either as infections or by being ingested. These prokaryotes became symbionts, transferring much of their genomes to the nuclei of their eukaryotic hosts but retaining genes that are essential to cellular functions. Mitochondria have retained about three dozen genes concerned with cellular respiration as well as some tRNA genes, whereas chloroplast genomes have about 130 genes encoding enzymes of the photosynthetic cycle as well as ribosomal proteins and tRNAs.

Important evidence for the extracellular origin of mitochondria is to be found in their genetic code. The "universal" DNA–RNA code of nuclear genes is not, in fact, universal and differs in some respects from that in mitochondria. Table 21-4 shows that, for 5 of the 64 RNA triplets, mitochondria differ in their coding from the nuclear genome. Moreover, mitochondria in different

TABLE 21-4 Comparison of the Universal Nuclear DNA Code with Several Mitochondrial Codes for Five Triplets in Which They Differ

	Triplet code				
	TGA	ATA	AGA	AGG	AAA
Nuclear	Stop	Ile	Arg	Arg	Lys
Mitochondrial					
Mammalia	Trp	Met	Stop	Stop	Lys
Aves	Trp	Met	Stop	Stop	Lys
Amphibia	Trp	Met	Stop	Stop	Lys
Echinoderms	Trp	Ile	Ser	Ser	Asn
Insecta	Trp	Met	Ser	Stop	Lys
Nematodes	Trp	Met	Ser	Ser	Lys
Platyhelminth	Trp	Met	Ser	Ser	Asn
Cnidaria	Trp	Ile	Arg	Arg	Lys

organisms differ from one another for these coding elements, providing evidence that eukaryotic cells must have been invaded by prokaryotes at least five times, each time by a prokaryote with a different coding system. For the vertebrates, worms, and insects, the mitochondrial code is more regular than the universal nuclear code. In the nuclear genome, for example, isoleucine is the only amino acid redundantly encoded by precisely three triplets: ATT, ATC, and ATA. The transition of the third base from A to G yields the fourth member of this codon group, ATG, but it codes for methionine. In contrast, in mitochondria, this codon group contains two codons for methionine and two for isoleucine, separated by a transversion.

HORIZONTAL TRANSFER It is now clear that the nuclear genome is open to the insertion of DNA both from other parts of the same genome and from outside. *Within* a genome, DNA can be transferred through the action of transposable elements (see Chapter 13). The chromosomes of an individual *Drosophila*, for example, contain a large variety of families of transposable elements with multiple copies of each distributed throughout the genome. As much as 25 percent of the DNA of *Drosophila* may be of transposable origin. What role this mobile DNA plays in functional evolution is not clear. When transposable elements are introduced into zygotes at mating, such as the P elements of *Drosophila* (see Chapter 11, page 371), the result is an explosive proliferation of the elements in the recipient genome. When a mobile element is inserted into a gene, the effect on the organism is usually drastic and deleterious, but this effect may be an artifact of the methods used to detect the presence of such elements. The results of laboratory selection experiments on quantitative characters have shown that transposition can act as an added source of selectable variation. There is also the possibility that genes are transferred from the nuclear genome of one species to the nuclear genome of another by retroviruses (see Chapter 13). Retroviruses can be carried between very distantly related species by common disease vectors such as insects or by bacterial infections; so any foreign genetic material carried by a retrovirus could be a powerful source of new functions.

Relation of genetic to functional change

There is no simple relation between the amount of change in a gene's DNA and the amount of change in the encoded protein's function. At one extreme, almost the entire amino acid sequence of a protein can be replaced while maintaining the original function. Eukaryotes, from yeast to humans, produce lysozyme, an enzyme that breaks down bacterial cell walls, as mentioned earlier. Virtually every amino acid in this protein has been replaced since yeast and vertebrate lines diverged

from an ancient common ancestor; so an alignment of their two protein or DNA sequences would not reveal any similarity. The evidence that yeast and human lysozyme genes are descended from an original common ancestral gene comes from comparisons of evolutionarily intermediate forms that show more and more divergence of sequence as species are more divergent. The enzyme has maintained its function despite the replacement of the amino acids because just the right amino acids were substituted to maintain the enzyme's three-dimensional structure.

In contrast, it is possible to change the function of an enzyme by a single amino acid substitution. The sheep blow fly, *Lucilia cuprina*, has developed resistance to organophosphate insecticides used widely to control it. R. Newcombe, P. Campbell, and their colleagues showed that this resistance is the consequence of a single substitution of an aspartic acid for a glycine residue in the active site of an enzyme that is ordinarily a carboxylesterase. The mutation causes complete loss of the carboxylesterase activity and its replacement by esterase specificity. Three-dimensional modeling of the molecule indicates that the substituted protein gains the ability to bind a water molecule close to the site of attachment of the organophosphate, which is then hydrolyzed by the water.

> **MESSAGE** There is no regular relation between how much DNA change takes place in evolution and how much functional change results.

When more than one mutation is required for a new function to arise, the order in which these mutations occur in the evolution of the molecule may be critical. B. Hall has experimentally changed a gene to a new function in *E. coli* by a succession of mutations and selection. In addition to the *lacZ* genes specifying the usual lactose-fermenting activity in *E. coli*, another structural gene locus, *ebg*, specifies another β-galactosidase that does not ferment lactose, although it is induced by lactose. The natural function of this second gene is unknown. Hall was able to select mutations of this extra gene to enable *E. coli* to live, without any lactose, on a wholly new substrate, galactobionate. To do so, he first had to mutate the regulatory sequence of *ebg* so that it became constitutive and no longer required lactose to induce its translation. Next, he tried to select mutants that would ferment lactobionate, but he failed. First, it was necessary to select a form that would ferment a related substrate, lactulose, and then he could mutagenize the lactulose fermenters and select from among the mutants those able to operate on lactobionate. Moreover, only some of the independent mutants from lactose fermentation to lactulose utilization could be further mutated and selected to operate on lactobionate. The others were dead ends. Thus, the sequence of evolution had to be (1) from an inducible to a constitutive enzyme, followed by (2) just the right mutation from lactose to lactulose fermentation, followed by (3) a mutation to ferment lactobionate.

> **MESSAGE** In the evolution of new functions by mutation and selection, particular pathways through the array of mutations must be followed. Other pathways come to dead ends that do not allow further evolution.

21.7 Rate of molecular evolution

There is no simple relation between the number of mutations in DNA or substitutions of amino acids in proteins and the amount of functional change in those proteins. Although it is possible that only one or a few mutations can lead to a major change in the function of a protein, the more usual situation is that DNA accumulates substitutions over long periods of evolution without any qualitative change to the functional properties of the encoded proteins. Some of the substitutions may, however, have smaller effects, influencing the kinetic properties, timing of production, or quantities of the encoded proteins that, in turn, will affect the fitness of the organism that carries them. Mutations of DNA can have three effects on fitness. First, they may be deleterious, reducing the probability of survival and reproduction of their carriers. All of the laboratory mutants used by the experimental geneticist have some deleterious effect on fitness. Second, they may increase fitness by increasing efficiency or by expanding the range of environmental conditions in which the species can make a living or by enabling the organism to track changes in the environment. Third, they may have no effect on fitness, leaving the probability of survival and reproduction unchanged; they are the so-called neutral mutations. For the purposes of understanding the rate of molecular evolution, however, we need to make a slightly different distinction—that between *effectively neutral* and *effectively selected* mutations. It is possible to prove that, in a finite population of N individuals, the process of random genetic drift will not be materially altered if the intensity of selection, s, on an allele is of lower order than $1/N$. Thus the class of evolutionarily neutral mutations includes both those that have absolutely no effect on fitness and those whose effects on fitness are less than the reciprocal of population size, so small as to be effectively neutral. On the other hand, if the intensity of selection, s, is of a greater order than $1/N$, then the mutation will be effectively selected.

We would like to know how much of molecular evolution is a consequence of new, favorable adaptive mutations sweeping through a species, the picture presented by a simplistic Darwinian view of evolution, and how much is simply the accumulation of effectively neutral mutations by random fixation. Mutations that are effectively deleterious need not be considered, because they will be kept at low frequencies in populations and will not contribute to evolutionary change. If a newly arisen mutation is effectively neutral, then, as pointed out in Chapter 19, there is a probability of $1/(2N)$ that it will replace the previous allele because of random genetic drift. If μ is the rate of appearance of new effectively neutral mutations at a locus per gene copy per generation, then the absolute number of new mutational copies that will appear in a population of N diploid individuals is $2N\mu$. Each one of these new copies has a probability of $1/(2N)$ of eventually taking over the population. Thus, the absolute rate of replacement of old alleles by new ones at a locus per generation is their rate of appearance multiplied by the probability that any one of them will eventually take over by drift:

$$\text{rate of neutral replacement} = 2N\mu \times 1/(2N) = \mu$$

That is, we expect that in every generation there will be μ substitutions of a new allele for an old one at each locus in the population, purely from genetic drift of effectively neutral mutations.

> **MESSAGE** The rate of replacement in evolution resulting from the random genetic drift of effectively neutral mutations is equal to the mutation rate to such alleles, μ.

The constant rate of neutral substitution predicts that, if the number of nucleotide differences between two species is plotted against the time since their divergence from a common ancestor, the result should be a straight line with slope equal to μ. That is, evolution should proceed according to a **molecular clock** that is ticking at the rate μ. Figure 21-13 shows such a plot for

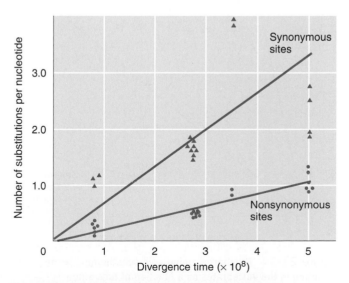

Figure 21-13 The amount of nucleotide divergence at synonymous and nonsynonymous sites of the β-globin gene as a function of time since divergence.

the β-globin gene. The results are quite consistent with the claim that nucleotide substitutions have been effectively neutral in the past 500 million years. Two sorts of nucleotide substitutions are plotted: **synonymous substitutions** that are from one alternative codon to another, making no change in the amino acid, and **nonsynonymous substitutions** that result in an amino acid change. Figure 21-13 shows a much lower slope for nonsynonymous substitutions than that for synonymous changes, which means that the mutation rate to selectively neutral nonsynonymous substitutions is much lower than that to synonymous ones.

This is precisely what we expect. Mutations that cause an amino acid substitution should have a deleterious effect above the threshold for neutral evolution more often than synonymous substitutions that do not change the protein. It is important to note that these observations do not show that synonymous substitutions have *no* selective constraints on them; rather they show that these constraints are, on the average, not as strong as those for mutations that change amino acids.

Another prediction of neutral evolution is that different proteins will have different clock rates, because the metabolic functions of some proteins will be much more sensitive to changes in their amino acid sequences. Proteins in which every amino acid makes a difference will have a smaller effectively neutral mutation rate because a smaller proportion of their mutations will be neutral compared with proteins that are more tolerant of substitution. Figure 21-14 shows a comparison of the clocks for fibrinopeptides, hemoglobin, and cytochrome *c*. That fibrinopeptides have a much higher proportion of neutral mutations is reasonable because these

TABLE 21-5 Synonymous and Nonsynonymous Polymorphisms and Species Differences for Alcohol Dehydrogenase in Three Species of *Drosophila*

	Species differences	Polymorphisms
Nonsynonymous	7	2
Synonymous	17	42
Ratio	0.29 : 0.71	0.05 : 0.95

Source: J. McDonald and M. Kreitman, *Nature* 351, 1991, 652–654.

peptides are merely a nonmetabolic safety catch, cut out of fibrinogen to activate the blood-clotting reaction. From a priori considerations, why hemoglobins are less sensitive to amino acid changes than is cytochrome *c* is less obvious.

> **MESSAGE** The rate of neutral evolution for the amino acid sequence of a protein depends on the sensitivity of a protein's function to amino acid changes.

The demonstration of the molecular clock argues that most nucleotide substitutions that have occurred in evolution were neutral, but it does not tell us how much of molecular evolution has been adaptive. One way of detecting adaptive evolution of a protein is by comparing the synonymous and nonsynonymous polymorphisms within species with the synonymous and nonsynonymous changes between species. Under the operation of neutral evolution by random genetic drift, polymorphism within a species is simply a stage in the eventual fixation of a new allele; so, if all mutations are neutral, the ratio of nonsynonymous to synonymous polymorphisms within a species should be the same as the ratio of nonsynonymous to synonymous substitutions between species. On the other hand, if the amino acid changes between species have been driven by a positive adaptive selection, there ought to be an excess of nonsynonymous changes between species. Table 21-5 shows an application of this principle by J. MacDonald and M. Kreitman to the alcohol dehydrogenase gene in three closely related species of *Drosophila*. Clearly, there is an excess of amino acid replacements between species over what is expected from the polymorphisms.

21.8 Genetic evidence of common ancestry in evolution

When we think of evolution, we think of change. The species living at any particular time are different from their ancestors, having changed in form and function by

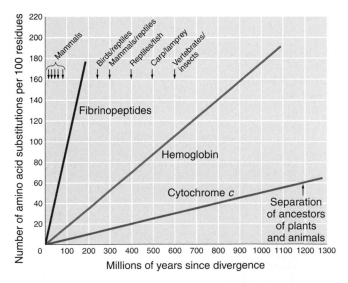

Figure 21-14 Number of amino acid substitutions in the evolution of the vertebrates as a function of time since divergence. The three proteins—fibrinopeptides, hemoglobin, and cytochrome *c*—differ in rate because different proportions of their amino acid substitutions are selectively neutral.

the mechanisms reviewed up to now in the discussion of the genetics of the evolutionary process. But there is a second feature of the diversity of life, one that Darwin took as an important argument for the reality of evolution. Not only have present organisms descended from previous, different organisms; but, if we go back in time, organisms that are currently very different are descended from a single ancestral form. Indeed, if we go back far enough in time to the origin of life, all the organisms on earth are descended from a single common ancestor. Thus we expect to find that apparently different species have underlying similarities, attributes of their common ancestor that have been conserved through evolutionary time despite all the changes that have taken place.

Before the tools of modern biochemistry and genetics were available, the chief evidence of underlying similarity of apparently different structures in different species was taken from anatomical observations of adult and embryonic forms. So, the similar bone structures of the wings of bats and the forelimbs of running mammals make it evident that these structures were derived evolutionarily from a common mammalian ancestor. Moreover, the anatomy of the wings of birds points to the common ancestry of mammals and birds (Figure 21-15). It is even argued that the basic segmentation of the bodies of insects and of vertebrates are evolutionary variants on a common ancestral pattern derived from the common ancestor of invertebrates and vertebrates. Although this argument may seem to push the claim of evolutionary conservation too far, it turns out, as we have seen in the discussion of the Hox and HOM-C genes in Chapter 18, that genetic analysis of patterns of development provides a powerful demonstration of the common ancestry of animals as different as insects and mammals.

We saw in Chapter 18 that such disparate organisms as the fly, the mouse, and human beings have similar sequences for the genes controlling the development of body form. (The same is true for the worm C. *elegans*.) The simplest explanation is that the Hox and HOM-C genes are the vertebrate and insect descendants of a homeobox gene cluster present in a common ancestor some 600 million years ago. The evolutionary conservation of the HOM-C and Hox genes is not a singular occurrence. Many examples have been uncovered of strongly conserved genes and even entire pathways that are similar in function. For example, the pathways for activating the *Drosophila* DL and mammalian $NF_{\kappa}B$ transcription factors are essentially completely conserved from a common ancestral pathway (Figure 21-16). The *Drosophila* protein at any step in the DL activation pathway is similar in amino acid sequence to its counterpart in the mammalian $NF_{\kappa}B$ activation pathway. (Don't worry about what the particular proteins do; just appreciate the incredible conservation of cellular and developmental pathways as demonstrated by the simi-

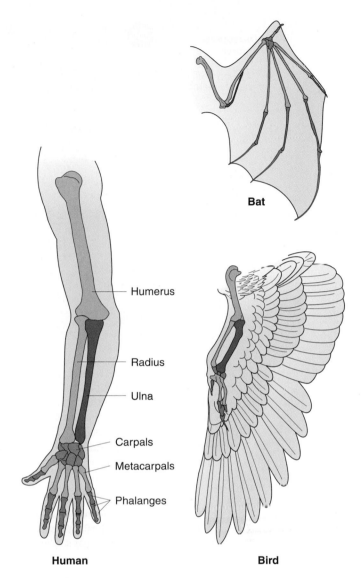

Bat

Humerus

Radius

Ulna

Carpals

Metacarpals

Phalanges

Human   **Bird**

Figure 21-15 The bone structures of a bat wing, a bird wing, and a human arm and hand. These bone structures show the underlying anatomical similarity between them and the way in which different bones have become relatively enlarged or diminished to produce these different structures. [After W. T. Keeton and J. L. Gould, *Biological Science*. W. W. Norton & Company, 1986.]

larity between the corresponding components of the two pathways indicated by similarly shaped objects in the diagrams. We do indeed know that DL and $NF_{\kappa}B$ participate in some equivalent developmental decisions.) Indeed, as can be seen from a selection from the known examples, such evolutionary and functional conservation seems to be the norm rather than the exception. What has made developmental genetics into an extraordinarily exciting field of biological inquiry is the demonstration, by means of genetic analysis, that basic developmental pathways and their genetic basis have been conserved over hundreds of millions of years of evolution.

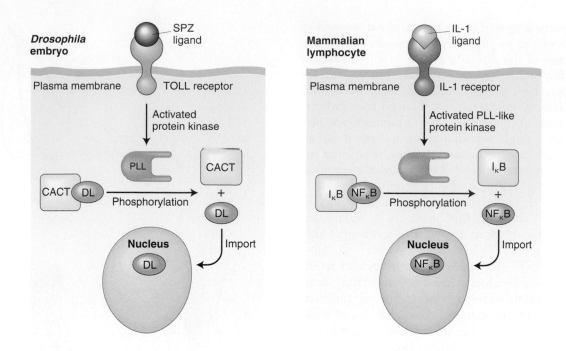

Figure 21-16 Two parallel signaling pathways. The signaling pathway for activation of the *Drosophila* DL morphogen parallels a mammalian signaling pathway for activation of NF$_\kappa$B, the transcription factor that activates the transcription of genes encoding antibody subunits. There are structural protein similarities between SPZ and IL-1, TOLL and IL-1R, CACT and I$_\kappa$B, and DL and NF$_\kappa$B. [After H. Lodish, D. Baltimore, A. Berk, S. L. Zipursky, P. Matsudaira, and J. Darnell, *Molecular Cell Biology*, 3d ed. Copyright Scientific American Books, 1995.]

MESSAGE Developmental strategies in animals are quite ancient and highly conserved. In essence, a mammal, a worm, and a fly are put together with the same basic genetic building blocks and regulatory devices. *Plus ça change, plus c'est le même chose!*

At a second, deeper level, we can observe the common evolutionary origin of organisms in the structure of their proteins and of their genomes. The advantage of direct observation of the protein and DNA sequences is that we do not have to depend on observing similarity of function among the proteins or anatomical structures that result from the possession of particular genes. We have already seen that replacing a single amino acid can change the function of a protein from an esterase to an acid phosphatase. Yet, despite this change in function, we have no difficulty in determining that the two enzymes are produced by reading genes that are virtually identical, one of which was derived by a single mutational step from the other as resistance to insecticides evolved by natural selection.

Over evolutionary time, genes that have descended from a common ancestor will diverge in DNA sequence and in their physical position in the genome, as a result of mutations and chromosomal rearrangements. If enough time elapsed and there were no counteracting force of natural selection, this divergence would finally result in

the loss of any observable similarity in genes or proteins between different species, even if they were descended from a common ancestor. In fact, even the time since the common ancestor of present-day vertebrates and invertebrates has not erased the similarity of DNA and amino acid sequences between *Drosophila* and mice. Not only are mutation rates not high enough to cause complete loss of similarity even over hundreds of millions of years, but also most new mutations are not preserved, because they cause a deleterious loss or change in function of a protein or in the control of the time and place of protein production. Thus the amount of divergence that has been preserved in evolution has been limited.

21.9 Comparative genomics and proteomics

As we saw in Chapter 12, a major effort of molecular genetics is directed toward determining the complete DNA sequence of a variety of different species. At the time at which this paragraph was written, the genomes had been sequenced from more than forty species of bacteria; two species of yeast; the fungus *Neurospora crassa*; the nematode, *Caenorhabditis elegans*; two species of *Drosophila*; two plants, *Arabidopsis* and rice; the mouse; and humans. By the time you read these lines, many more genomes of many more species will have been sequenced. The availability of such data makes it possible to reconstruct the evolution of the genomes of widely diverse species from their common ancestors. Moreover, it is now possible to infer the similarities and differences in the proteomes of these species by comparing the gene sequences in various species with gene sequences that code for the amino acid sequences of proteins with known function.

Comparing the proteomes among distant species

With our current state of knowledge, we can suggest functions for about half the proteins in the proteome of each of the eukaryotes whose genomes have been sequenced, by using the similarity of their sequences with proteins of known function. Figure 21-17 depicts the distribution of this half of each proteome into general functional categories. Strikingly, the group of proteins engaged in defense and immunity have expanded greatly in humans compared with the other species. For other functional categories, though there are greater numbers of proteins in the human lineage, there is no case in which the differences between humans and all other eukaryotes are as pronounced. As discussed in Chapter 10, gene expression is often controlled through regulation of transcription by proteins called transcription factors. Per-

haps as a manifestation of the many cell types that differentiate in humans, the size and distribution of the families of specific transcription factors in humans far exceed the numbers for the other sequenced eukaryotes, with the exception of mustard weed (*Arabidopsis thaliana*, see Figure 21-17).

The distribution of proteins described in the preceding paragraph is a description of only half of each proteome. What about the other half? It can be broken down into two components. One component, comprising about 30 percent of each proteome, consists of proteins that have relatives among the different genomes, but none have had a function ascribed to them. The other component, comprising the remaining 20 percent or so of each proteome, consists of proteins that are unrelated by amino acid sequence to any protein known in another branch of the eukaryotic evolutionary tree. We can imagine two possible explanations for these novel polypeptides. One possibility is that some of these polypeptides first evolved after the sequenced species having a common ancestor diverged from one another. Because none of these species are evolutionarily closer than a few hundred million years, it is perhaps not surprising to find this frequency of newly evolved proteins. The other possibility is that some of these proteins are very rapidly evolving, and so their ancestry has been essentially erased by the overlay of new mutations that have accumulated. It is almost

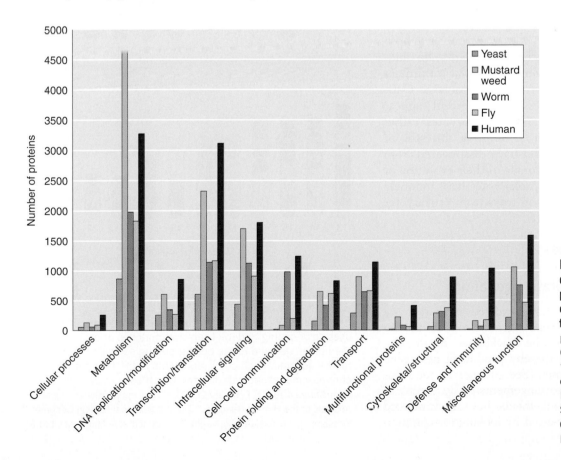

Figure 21-17 The distribution of eukaryotic proteins according to broad categories of biological function. [Reprinted by permission from *Nature* 409 (15 February 2001), 902, "Initial Sequencing and Analysis of the Human Genome," The International Human Genome Sequencing Consortium. Copyright 2001 Macmillan Magazines Ltd.]

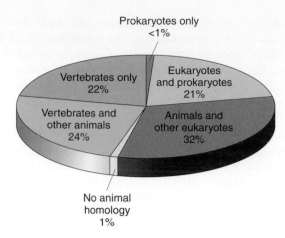

Figure 21-18 The distribution of human proteins according to the identification of significantly related proteins in other species. Note that about one-fifth of human proteins have been identified only within the vertebrate lineage, whereas, at the other extreme, one-fifth have been identified in all of the major branches of the evolutionary tree. [Reprinted by permission from *Nature* 409 (15 February 2001), 902, "Initial Sequencing and Analysis of the Human Genome," The International Human Genome Sequencing Consortium. Copyright 2001 Macmillan Magazines Ltd.]

certain that both possibilities are correct for a subset of these novel polypeptides.

Finally, we can ask, Where do the protein-coding genes in the human genome come from? Figure 21-18 depicts the distribution of human genes in other species. About a fifth of the known human genes have been found only in vertebrates. Another fifth seem to be ubiquitous in eukaryotes and prokaryotes. About a third are found throughout eukaryotes but not in bacteria. Curiously, a few hundred genes (less than 1 percent) appear to be found only in humans and in prokaryotes. Either these genes were present in a common ancestor of prokaryotes and eukaryotes and have disappeared from most other eukaryotes in the course of their evolution or else these genes that we and prokaryotes have uniquely in common have been passed on to us from prokaryotes through horizontal gene transfer.

Comparing the genomes among near neighbors: human–mouse comparative genomics

Genomes are thought to have evolved in part by a process of chromosome rearrangement—that is, the breaking and rejoining of the backbones of double-stranded DNA molecules, thereby producing new gene orders and new chromosomes. (See Chapter 15 for a discussion of chromosome rearrangements.) The extent to which chromosome rearrangements have accumulated during evolution can be assessed by looking for common

gene orders between diverged species. As an example of this approach, we will compare the genomes of the human and the mouse, two species that diverged from a common ancestor about 50 million years ago. The mouse genome has now been sufficiently sequenced that relative gene orders can be determined. It turns out that large blocks of conserved gene order are easily recognized. Through systematic comparisons of this type, one can make **synteny maps,** which show the chromosomal origin of one species essentially painted onto the karyotype of the other. Figure 21-19 depicts a color representation of the syntenic mouse–human genome. In this figure, 21 dif-

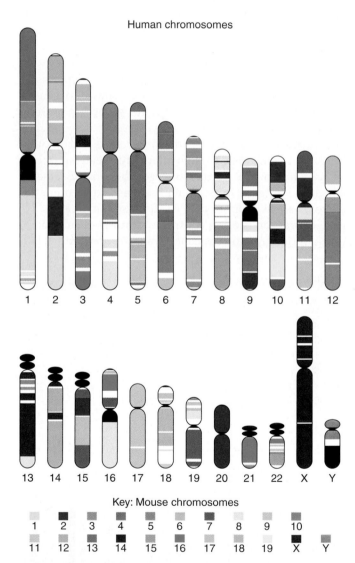

Figure 21-19 A synteny map of the human genome. The map uses color-coding to depict regional matches of each bloc of the human genome to the corresponding sections of the mouse genome. Each color represents a different mouse chromosome, as indicated in the key. [Reprinted by permission from *Nature* 409 (15 February 2001), 910, "Initial Sequencing and Analysis of the Human Genome," The International Human Genome Sequencing Consortium. Copyright 2001 Macmillan Magazines Ltd.]

ferent colors represent the mouse X and Y sex chromosomes and the 19 autosomes. For example, most of mouse chromosome 14 can be found in three blocks on human chromosome 13, but, in addition, small segments can be found on human chromosomes 3, 8, 10, and 14. Similar block-by-block distributions in the human genome are observed for each of the other mouse chromosomes. Thus, we can conclude that many chromosomal re-arrangements have occurred between human beings and the mouse, but not enough to have completely scrambled the two genomes relative to one another.

> **MESSAGE** Comparative genomics is a source of insight into gene-level and chromosome-level changes that occur in the process of evolution.

KEY QUESTIONS REVISITED

- **What are the basic principles of the Darwinian mechanism of evolution?**

The Darwinian explanation of organic evolution is based on a population variation model in which individuals in a population vary from one another and some variants increase in number while others decrease. A variational mechanism for evolution is based on three principles: (1) the principle of variation, that among individuals within any population there is variation in morphology, physiology, and behavior; (2) the principle of heredity, that offspring resemble their parents more than they resemble unrelated individuals; (3) the principle that some variants are more successful at surviving and leaving offspring than other variants in a given environment. Taken together, these three principles provide a mechanism for changes over time in the properties of a population and for the divergence of different populations from one another.

- **What are the roles of natural selection and other processes in evolution, and how do they interact with one another?**

Evolution is a consequence of several processes operating within and between populations and interacting with one another. The heritable variation required by the Darwinian theory arises by mutation and chromosomal changes and by the introduction of new DNA into the genome by duplication of DNA already present or by transposition of DNA from other organisms. In the absence of natural selection, the frequencies of different variants change erratically, and eventually populations will differentiate from one another by this process because some populations become fixed for a new mutation, whereas in others the mutation may never have occurred or may be lost by chance. Migration of individuals between populations increases the variation within populations and decreases the differences between populations. One of the effects of migration is to introduce new mutations into a population and so make it possible for more rapid evolution than would occur if populations were totally isolated and had to wait for the chance occurrence of a new favorable mutation. If the carriers of a new mutation have a greater survival or reproductive rate than that of the genotypes that make up the bulk of the population, these new genotypes will increase in frequency and may eventually become the characteristic type in the population. Even if a new genotype is selectively advantageous, it may not spread if the population is so small that random genetic drift causes the chance loss of the new type.

- **How do different species arise?**

Different species arise from populations that are separated by some geographical barrier that prevents the exchange of genes between them. When such a separation exists, then each population will acquire mutations not shared by the other populations, random genetic drift will cause the fixation of mutations in one population and their loss in others, and ecological differences between the geographical localities may result in some genotypes being favored by natural selection in some populations but not in others. All these forces result in an increasing genetic difference between the isolated populations. Eventually, the populations may become so genetically different from one another that no offspring can be produced by matings between indiviual members of the different populations even when the geographical separation disappears. Such reproductively isolated populations are new species.

- **How different are the genomes of different kinds of organisms?**

The genomes of different kinds of organisms may differ greatly in the total amount of DNA and in the organization of the genes on the chromosomes, but there is a remarkable similarity in the proteins that are encoded by genomes of very divergent forms. Among eukaryotes, which include organisms as different as yeasts and humans, about half the proteins are similar enough to ascribe a similar function to them. Another 30 percent have DNA coding sequences that are similar enough among different organisms to recognize that they are derived from some common ancestral state even though no function has yet been assigned to the proteins for which they code. The remaining 20 percent of the coding DNA has no detectable similarity among organisms and encodes proteins of unknown function.

• **How do evolutionary novelties arise?**

Evolutionary novelties arise from three kinds of change in the genome. First, mutations in already existing coding DNA may result in the substitution of amino acids that completely alter the function of the protein. Second, changes may occur in regulatory DNA sequences such that the genes that they regulate are transcribed at new rates, times, or places in the development of an organism. The result may be novel changes in shape and function of the parts of the organism, as when legs in reptiles evolved into wings in birds. Third, novelties may arise from the evolution of extra DNA that has been added to the genome by duplication or transposition. The extra DNA is free to evolve to new functions while the old functions are still served by the original genes.

SUMMARY

The Darwinian theory of evolution explains the changes that take place in populations of organisms as being the result of changes in the relative frequencies of different variants in the population. If there is no variation within a species for some trait, there can be no evolution. Moreover, that variation must be influenced by genetic differences. If differences are not heritable, they cannot evolve, because the differential reproduction of the different variants will not carry across generational lines. Thus, all hypothetical evolutionary reconstructions depend critically on whether the traits in question are, in fact, heritable. The processes that give rise to the variation within the population are causally independent of the processes that are responsible for the differential reproduction of the various types. It is this independence that is meant when it is said that mutations are "random." The process of mutation supplies undirected variation, whereas the process of natural selection culls this variation, increasing the frequency of those variants that by chance are better able to survive and reproduce. Many are called, but few are chosen.

The evolutionary divergence of populations in space and time is not only a consequence of natural selection. Natural selection is not a globally optimizing process that finds the "best" organisms for a particular environment. Instead, it finds one of a set of alternative "good" solutions to adaptive problems, and the particular outcome of selective evolution in a particular case is subject to chance historical events. Random factors such as genetic drift and the chance occurrence or loss of new mutations may result in radically different outcomes of an evolutionary process even when the force of natural selection is the same. The metaphor usually employed is that there is an "adaptive landscape" of genetic combinations and that natural selection leads the population to a "peak" in that landscape, but only to one of several alternative local peaks.

The vast diversity of different living forms that have existed is a consequence of independent evolutionary histories that have occurred in separate populations. For different populations to diverge from one another, they must not exchange genes; so the independent evolution of large numbers of different species requires that these species be reproductively isolated from one another. Indeed, we define a species as a population of organisms that exchange genes among themselves and are reproductively isolated from other populations. The mechanisms of reproductive isolation may be prezygotic or postzygotic. Prezygotic isolating mechanisms are those that prevent the union of gametes of two species. These mechanisms may be behavioral incompatibility of the males and females of the different species, differences in timing or place of their sexual activity, anatomical differences that make mating mechanically impossible, or physiological incompatibility of the gametes themselves. Postzygotic isolating mechanisms include the inability of hybrid embryos to develop to adulthood, the sterility of hybrid adults, and the breakdown of later generations of recombinant genotypes. For the most part, the genetic differences responsible for the isolation between closely related species are spread throughout all the chromosomes, although in species with chromosomal sex determination there may be a concentration of incompatibility genes on the sex chromosome.

If new functions are to arise in evolution without the sacrifice of previously existing functions, new DNA must be made available for the evolution of added genes. This new DNA may arise by duplication of the entire genome (polyploidy) followed by a slow evolutionary divergence of the extra chromosomal set, which has been a frequent occurrence in plants. An alternative is the duplication of single genes followed by selection for differentiation. Yet another source of DNA, recently discovered, is the entry into the genome of DNA from totally unrelated organisms by infection followed by integration of the foreign DNA into the nuclear genome or by the formation of extranuclear cell organelles with their own genomes. Mitochondria and chloroplasts in higher organisms have arisen by this route.

Not all of evolution is impelled by natural selective forces. If the selective difference between two genetic variants is small enough, less than the reciprocal of population size, there may be a replacement of one allele by another purely by genetic drift. A great deal of molecular evolution seems to be the replacement of one protein sequence by another one of equivalent function. The evidence for this neutral evolution is that the number of amino acid differences between two different species in some molecule—for example, hemoglobin—is directly proportional to the number of generations since their divergence from a common ancestor in the evolutionary past. Such a "molecular clock" with a constant rate of change would not be expected if the selection of differences were dependent on particular changes in the environment. Moreover, we expect the clock to run faster for proteins such as fibrinopeptides, in which the amino acid composition is not critical for the function, and this difference in clock rate is, in fact, observed. Thus we cannot assume without evidence that evolutionary changes are the result of adaptive natural selection.

Overall, genetic evolution is a historical process that is subject to historical contingency and chance, but it is constrained by the necessity of organisms to survive and reproduce in a constantly changing world.

KEY TERMS

adaptive landscape (p. 686)

adaptive surface (p. 686)

allopatric (p. 690)

allopatric speciation (p. 690)

balancing selection (p. 685)

canalized characters (p. 688)

directional selection (p. 685)

diversification (p. 681)

founder effect (p. 684)

molecular clock (p. 695)

natural selection (p. 681)

nonsynonymous substitution (p. 696)

phyletic evolution (p. 681)

postzygotic isolation (p. 690)

prezygotic isolation (p. 690)

species (p. 689)

synonymous substitution (p. 696)

synteny map (p. 700)

SOLVED PROBLEMS

1. An entomologist who studies insects that feed on rotting vegetation has discovered an interesting case of diversification of fungus gnats on several islands in an archipelago. Each island has a gnat population that is extremely similar in morphology, although not identical, to those on the other islands, but each lives on a different kind of rotting vegetation that is not present on the other islands. The entomologist postulates that these populations are closely related species that have diverged by adapting to feeding on slightly different rot conditions.

To support this hypothesis, he carries out an electrophoretic study of the alcohol dehydrogenase enzyme in the different populations. He discovers that each population is characterized by a different electrophoretic form of the alcohol dehydrogenase, and he then reasons that each of these alcohol dehydrogenase forms is specifically adapted to the particular alcohols that are produced in the fermentation of the vegetation characteristic of a particular island. There is, in addition, some polymorphism of alcohol dehydrogenase within each island, but the frequency of variant alleles is low on each island and can be easily explained as the result of an occasional mutation or rare migrant from another island. These fungus gnats then become a textbook example of how species diversity can come about by natural selection adapting each newly forming species to a different environment.

A skeptical population geneticist reads about the case in a textbook and she immediately has some doubts. It seems to her that, given the evidence, an equally plausible explanation is that these populations of gnats are not species at all but just local geographical races that have become slightly differentiated morphologically by random genetic drift. Moreover, the different electrophoretic forms of the alcohol dehydrogenase protein may be physiologically equivalent variants of a gene undergoing neutral molecular evolution in isolated populations.

Outline a program of investigation that could distinguish between these alternative explanations. How could you test whether the different populations are indeed different species? How could you test the hypothesis that the different forms of the alcohol dehydrogenase have diverged selectively?

Solution

To test the species distinctness of the different gnats, it is necessary to be able to manipulate and culture them in captivity. If they cannot be cultured in the laboratory or greenhouse, then their species distinctness cannot be established. The mating-behavior compatibility of the different forms can be tested by placing a mixture of males of two different populations with females of one of the forms to see whether there are any female mating preferences. The same experiment can then be repeated with mixed females and males of one form and with mixtures of males and females of both forms. From such experiments, patterns of mating preference can be observed. Even if there is some small amount of mating of different forms, it may occur only because of the unnatural conditions in which the test is being carried out. On the other hand, no mating of any kind may occur, even between the same forms, because the necessary cues for mating are missing, in which case nothing can be concluded.

If matings between different forms do occur, the survivorship of the interpopulation hybrids can be compared with that of the intrapopulation matings. If hybrids survive, their fertility can be tested by attempting to backcross them to the two different parental strains. As with the mating tests, under the unnatural conditions of the laboratory or greenhouse, some survivorship or fertility of species hybrids is possible even though the isolation in nature is complete. Any clear reduction in observed survivorship or fertility of the hybrids is strong presumptive evidence that they belong to different species.

To test whether the different amino acid sequences underlying the electrophoretic mobility differences are the result of selective divergence, a program of DNA sequencing of the alcohol dehydrogenase locus is necessary. Replicated samples of *Adh* sequences from each of the island populations must be obtained. The number of such sequences needed from each population depends on the degree of nucleotide polymorphism that is present in the populations, but results from many loci in many species suggest that, as a rule of thumb, at least 10 sequences should be obtained from each population. The polymorphic sites within populations are classified into nonsynonymous *(a)* and synonymous *(b)* sites. The fixed nucleotide differences between populations are classified into nonsynonymous *(c)* and synonymous *(d)* differences. If the divergence between the populations is purely the result of random genetic drift, then we expect *a/b* to be equal to *c/d*. If, on the other hand, there has been selective divergence, there should be an excess of fixed nonsynonymous differences, and so *a/b* should be less than *c/d*. The equality of these ratios can be tested by a 2 × 2 contingency χ^2 test of the form

	Polymorphisms	
	Nonsynonymous	Synonymous
Population		
Differences	*a*	*b*
	c	*d*

$$\chi^2 = \frac{(a + b + c + d)\,(ad - bc)^2}{(a + c)\,(b + d)\,(a + b)\,(c + d)}$$

2. Two closely related species are found to be fixed for two different electrophoretically detected alleles at a locus encoding an enzyme. How could you

demonstrate that this divergence is a result of natural selection rather than neutral evolution?

Solution

a. Obtain DNA sequences of the gene from a number of separate individuals or strains from each of the two species. Ten or more sequences from each species would be desirable.

b. Tabulate the nucleotide differences among individuals within each species *(polymorphisms)*, and classify these differences as either those that result in amino acid changes (replacement polymorphisms) or those that do not change the amino acid (synonymous polymorphisms).

c. Make the same tabulation of replacement and synonymous changes for the differences between the species, counting only those differences that completely differentiate the species. That is, do not count a polymorphism in one species that includes a variant that is seen in the other species.

d. If the ratio of replacement differences between the species to synonymous differences between the species is greater than the ratio of replacement polymorphisms to synonymous polymorphisms, then select for amino acid change.

e. Test the statistical significance of the observed greater ratio by a 2 × 2 χ^2 test of the following table:

		Polymorphisms	
		Replacement	Synonymous
Species	Replacement	*a*	*b*
Differences	Synonymous	*c*	*d*

$$\chi^2 = \frac{(a + b + c + d)\,(ad - bc)^2}{(a + c)\,(b + d)\,(a + b)\,(c + d)}$$

3. How could the molecular evolution of a set of different proteins be used to provide evidence of the relative importance of exact amino acid sequence to the function of each protein?

Solution

Obtain DNA sequences from the genes for each protein from a wide variety of very divergent species whose approximate time to a common ancestor is known from the fossil record. Translate the DNA sequences into amino acid sequences. For each protein, plot the observed amino acid difference for each pair of species against the estimated time of divergence for those species. The line for each protein will have a slope that is proportional to the amount of functional constraint on amino acid substitution in that protein. Highly constrained proteins will have very low rates of substitution, whereas more tolerant proteins will have higher slopes.

PROBLEMS

BASIC PROBLEMS

1. What is the difference between a transformational and a variational scheme of evolution? Give an example of each (not including the Darwinian theory of organic evolution).

2. What are the three principles of Darwin's theory of variational evolution?

3. Why is the Mendelian explanation of inheritance essential to Darwin's variational mechanism for evolution? What would the consequences for evolution be if inheritance were by the mixing of blood? What would the consequence for evolution be if heterozygotes did not segregate exactly 50 percent of each of the two alleles at a locus but were consistently biased toward one or the other allele?

4. What is a geographical race? What is the difference between a geographical race and a separate species? Under what conditions will geographical races of a species become new species?

CHALLENGING PROBLEMS

5. If the mutation rate to a new allele is 10^{-5}, how large must isolated populations be to prevent chance differentiation among them to develop in the frequency of this allele?

6. Suppose that a number of local populations of a species are each about 10,000 individuals in size and that there is no migration between them. Suppose, further, that they were originally established from a large population with the frequency of an allele A at some locus equal to 0.4. Show by approximate sketches what the distribution of allele frequencies among the local populations would be after 100, 1000, 5000, 10,000, and 100,000 generations of isolation.

7. Show the results for the populations described in Problem 6 if there were an exchange of migrants among the populations at the rate of **(a)** one migrant individual per population every 10 generations; **(b)** one migrant individual per population every generation.

8. Suppose that a population is segregating for two alleles at each of two loci and that the relative probabilities of survival to sexual maturity of zygotes of the nine genotypes are as follows:

	A/A	A/a	a/a
B/B	0.95	0.90	0.80
B/b	0.90	0.85	0.70
b/b	0.90	0.80	0.65

Calculate the mean fitness, W, of the population if the allele frequencies are $p(A) = 0.8$ and $p(B) = 0.9$. What direction of change do you expect in allele frequencies in the next generation? Make the same calculation and prediction for the allele frequencies $p(A) = 0.2$ and $p(B) = 0.2$. From inspection of the genotypic fitnesses, how many adaptive peaks are there? What are the allele frequencies at the peak(s)?

9. Suppose the genotypic fitnesses in Problem 8 were:

	A/A	A/a	a/a
B/B	0.9	0.8	0.9
B/b	0.7	0.9	0.7
b/b	0.9	0.8	0.9

Calculate the mean fitness, W, for allelic frequencies $p(A) = 0.5$ and $p(B) = 0.5$. What direction of change do you expect for the allele frequencies in the next generation? Repeat the calculation and prediction for $p(A) = 0.1$ and $p(B) = 0.1$. From inspection of the genotype fitnesses, how many adaptive peaks are there and where are they located?

10. What is the evidence that polyploid formation has been important in plant evolution?

11. What is the evidence that gene duplication has been the source of the α and β gene families in human hemoglobin?

12. The human blood group allele I^B has a frequency of about 0.10 in European and Asian populations but is almost entirely absent in Native American populations. What explanations can account for this difference?

13. *Drosophila pseudoobscura* and *D. persimilis* are now considered separate species, but originally they were classified as Race A and Race B of a single species. They are morphologically indistinguishable from each other, except for a small difference in the genitalia of the males. When crossed in the laboratory, abundant adult F$_1$ progeny of both sexes are produced. Outline the program of observations and experiments that you would undertake to test the claim that the two forms are different species.

14. Using the data on amino acid similarity of the α-, β-, γ-, ζ-, and ϵ-globin chains given in Table 21-3, draw a branching tree of the evolution of these chains from an original ancestral sequence in which the order of branching in time is as consistent as possible with the observed amino acid similarity on the assumption of a molecular clock.

15. DNA-sequencing studies for a gene in two closely related species produce the following numbers of sites that vary:

Synonymous polymorphisms	50
Nonsynonymous species differences	2
Synonymous species differences	18
Nonsynonymous polymorphisms	20

Does this result support a neutral evolution of the gene? Does it support an adaptive replacement of amino acids? What explanation would you offer for the observations?

EXPLORING GENOMES A Web-Based Bioinformatics Tutorial

Measuring Phylogenetic Distance

Sequence data allow us to estimate the evolutionary distance among organisms on the basis of the extent of sequence divergence. In the Genomics tutorial at www.whfreeman.com/mga, we will use sequence comparisons to generate or support our conclusions regarding the structure of the evolutionary tree.

A Brief Guide to Model Organisms

Escherichia coli • *Saccharomyces cerevisiae* • *Neurospora crassa* • *Arabidopsis thaliana*
Caenorhabditis elegans • *Drosophila melanogaster* • *Mus musculus*

This brief guide collects in one place the main features of model organisms as they relate to genetics. Each of seven model organisms is given its own two-page spread using a consistent format that allows readers to compare and contrast the features of model organisms. Each treatment focuses on the special features of an organism that have made it useful as a model; the special techniques that have been developed for studying the organism; and the main contributions that studies of the organism have made to our understanding of genetics. Although many differences will be apparent, the general approaches of genetic analysis are similar but have to be tailored to take account of the individual life cycle, ploidy level, size and shape, and genomic properties, such as the presence of natural plasmids and transposons.

Model organisms have always been at the forefront of genetics. Initially in the historical development of a model organism, a researcher selects the organism because of some feature that lends itself particularly well to the study of a genetic process that the researcher is interested in. The advice of the last hundred years has been "choose your organism well." For example, the ascomycete fungi, such as *Saccharomyces cerevisiae* and *Neurospora crassa*, are well suited to the study of meiotic processes, such as crossing-over, because their unique feature, the ascus, holds together the products of a single meiosis.

Different species tend to show remarkably similar processes, even across the members of large groups, such as the eukaryotes. Hence, we can reasonably expect that what is learned in one species can be at least partially applied to others. In particular, geneticists have kept an eye open for new research findings that may apply to our own species. Humans are relatively difficult to study at the genetic level, so advances in human genetics owe a great deal to more than a century of work on model organisms.

All model organisms have far more than one useful feature for genetic or other biological study. Hence, once a model organism is developed by a few people with specific interests, it then acts as a nucleus for the development of a research community—a group of researchers with an interest in various features of one particular model organism. There are organized research communities for all the model organisms mentioned in this summary. The people in these communities are in touch with each other regularly, share their mutant strains, and often meet at least annually at conferences that may attract thousands of people. Such a community makes possible the provision of important services, such as databases of research information, techniques, genetic stocks, clones, DNA libraries, and genomic sequences.

Another advantage to an individual researcher in belonging to such a community is that he or she may develop "a feeling for the organism" (a phrase of maize geneticist and Nobel Laureate Barbara McClintock). This idea is difficult to convey, but it implies understanding of the general ways of an organism. No living process occurs in isolation, so knowing the general ways of an organism is often beneficial in trying to understand one process and to interpret it in its proper context.

As the database for each model organism expands (and currently this is occurring at a great pace thanks to genomics), geneticists are more and more able to take a holistic view, encompassing the integrated workings of all parts of the organism's makeup. In this way model organisms become not only models for isolated processes but also models of integrated life processes. The term *systems biology* is used to describe this holistic approach.

Escherichia coli

Key organism for studying:

- Transcription, translation, replication, recombination
- Mutation
- Gene regulation
- Recombinant DNA technology

Genetic "Vital Statistics"	
Genome size:	4.6 Mb
Chromosomes:	1, circular
Number of genes:	4000
Percentage with human homologs:	8%
Average gene size:	1 kb, no introns
Transposons:	Strain specific, ~ 60 copies per genome
Genome sequenced in:	1997

The unicellular bacterium *Escherichia coli* is widely known as a disease-causing pathogen, the source of food poisoning and intestinal disease. However, this negative reputation is undeserved. Although some strains of *E. coli* are harmful, others are natural and essential residents of the human gut. As model organisms, strains of *E. coli* play an indispensable role in genetic analyses. In the 1940s, several groups began investigating the genetics of *E. coli*. The need was for a simple organism that could be cultured inexpensively to produce large numbers of individual bacteria to be able to find and analyze rare genetic events. Because *E. coli* can be obtained from the human gut and is small and easy to culture, it was a natural choice. Work on *E. coli* defined the beginning of "black box" reasoning in genetics: through the selection and analysis of mutants, the workings of cellular processes could be deduced even though an individual cell was too small to be seen.

E. coli **genome.** Electron micrograph of the genome of the bacterium *E. coli*, released from the cell by osmotic shock. [Dr. Gopal Murti/Science Photo Library/Photo Researchers.]

Special features

Much of *E. coli*'s success as a model organism can be attributed to two statistics: its 1-micrometer cell size and a 20-minute generation time. (Replication of the chromosome takes 40 minutes, but multiple replication forks allow the cell to divide in 20 minutes). Consequently, this prokaryote can be grown in staggering numbers—a feature that allows geneticists to identify mutations and other rare genetic events such as intragenic recombinants. *E. coli* is also remarkably easy to culture. When cells are spread on plates of nutrient medium, each cell divides in situ and forms a visible colony. Alternatively, batches of cells can be grown in liquid shake culture. Phenotypes such as colony size, drug resistance, ability to obtain energy from particular carbon sources, and colored dye production take the place of the morphological phenotypes of eukaryotic genetics.

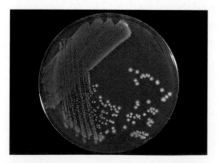

Bacterial colonies. [Biophoto Associates/ Science Source/Photo Researchers.]

LIFE CYCLE

E. coli reproduces asexually by simple cell fission; its haploid genome replicates and partitions with the dividing cell. In the 1940s, Joshua Lederberg and Edward Tatum discovered that *E. coli* also has a type of sexual cycle in which cells of genetically differentiated "sexes" fuse and exchange some or all of their genomes, sometimes leading to re-

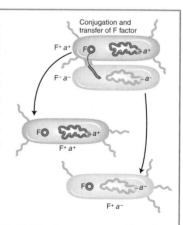

combination (see Chapter 5). "Males" can convert "females" into males by transmission of a particular plasmid. This circular extragenomic 100-kb DNA plasmid, called F, determines a type of "maleness." F⁺ cells acting as male "donors" transmit a copy of the F plasmid to a recipient cell. The F plasmid can integrate into the chromosome to form an Hfr cell type, which transmits the chromosome linearly into F⁻ recipients. Other plasmids are found in *E. coli* in nature. Some carry genes whose functions equip the cell for life in specific environments; R plasmids that carry drug-resistance genes are examples.

Length of life cycle: 20 minutes

Geneticists have also taken advantage of some unique genetic elements associated with *E. coli*. Bacterial plasmids and phages are used as vectors to clone the genes of other organisms within *E. coli*. Transposable elements from *E. coli* are harnessed to disrupt genes in cloned eukaryotic DNA. Such bacterial elements are key players in recombinant DNA technology.

Genetic analysis

Spontaneous *E. coli* mutants show a variety of DNA changes, ranging from simple base substitutions to the insertion of transposable elements. The study of rare spontaneous mutations in *E. coli* is feasible because large populations can be screened. However, mutagens also are used to increase mutation frequencies.

To obtain specific mutant phenotypes that might represent defects in a process under study, screens or selections must be designed (see Chapter 16). For example, nutritional mutations and mutations conferring resistance to drugs or phages can be obtained on plates supplemented with specific chemicals, drugs, or phages. Null mutations of any essential gene will result in no growth; these mutations can be selected by adding penicillin (an antibacterial drug isolated from a fungus), which kills dividing cells but not the nongrowing mutants. For conditional lethal mutations, replica plating can be used: mutated colonies on a master plate are transferred with a felt pad to other plates that are then subjected to some toxic environment. Mutations affecting expression of one specific gene of interest can be screened by fusing it to a reporter gene such as the *lacZ* gene, whose protein product makes a blue dye, or the *GFP* gene, whose product fluoresces when exposed to light of a particular wavelength.

After a set of mutants has been obtained affecting the process of interest, the mutations are sorted into their genes by recombination and complementation. These genes are cloned and sequenced to obtain clues to function. Site-directed mutagenesis can be used to tailor mutational changes at specific protein positions (see p. 531).

In *E. coli*, crosses are used to map mutations and to produce specific cell genotypes (see Chapter 5). Recombinants are made by mixing Hfr cells (having an integrated F plasmid) and F⁻ cells. Generally an Hfr donor transmits part of the bacterial genome, forming a temporary merozygote in which recombination takes place. Hfr crosses can be used to perform mapping by time-of-marker entry or by recombinant frequency. By transfer of F′ derivatives carrying donor genes to F⁻, it is possible to make stable partial diploids to study gene interaction or dominance.

Techniques of Genetic Modification

Standard mutagenesis:

Chemicals and radiation	Random somatic mutations
Transposons	Random somatic insertions

Transgenesis:

On plasmid vector	Free or integrated
On phage vector	Free or integrated
Transformation	Integrated

Targeted gene knockouts:

Null allele on vector	Gene replacement by recombination
Engineered allele on vector	Site-directed mutagenesis by gene replacement

---

Genetic engineering

Transgenesis. *E. coli* plays a key role in introducing transgenes to other organisms (see Chapter 11). It is the standard organism used for cloning genes of any organism. *E. coli* plasmids or bacteriophages are used as vectors, carrying the DNA sequence to be cloned. These vectors are introduced into a bacterial cell by transformation, if a plasmid, or by transduction, if a phage, where they replicate in the cytoplasm. Vectors are specially modified to include unique cloning sites that can be cut by a variety of restriction enzymes. Other "shuttle" vectors are designed to move DNA fragments from yeast ("the eukaryotic *E. coli*") into *E. coli*, for its greater ease of genetic manipulation, and then back into yeast for phenotypic assessment.

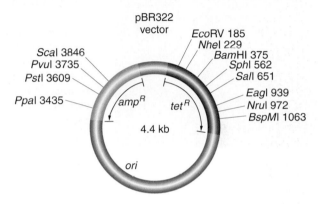

A plasmid designed as a vector for DNA cloning. Successful insertion of a foreign gene into the plasmid is detected by inactivation of either drug-resistance gene (*tet*ᴿ or *amp*ᴿ). Restriction sites are identified.

Targeted gene knockouts. A complete set of gene knockouts is being accumulated. In one procedure, a kanamycin-resistance transposon is introduced into a cloned gene in vitro (by using a transposase). This construct is transformed, and resistant colonies produced by homologous recombination carry the inserted transposon.

Main contributions

Pioneering studies for genetics as a whole were carried out in *E. coli*. Perhaps the greatest triumph was the elucidation of the universal 64-codon genetic code, but this achievement is far from alone on the list of accomplishments attributable to this organism. Other fundamentals of genetics that were first demonstrated in *E. coli* include the spontaneous nature of mutation (the fluctuation test, p. 461), the various types of base changes that cause mutations, and the semiconservative replication of DNA (the Meselson and Stahl experiment, p. 237). This bacterium helped open up whole new areas of genetics, such as gene regulation (the *lac* operon, pp. 305*ff.*) and DNA transposition (IS elements, p. 429). And last but not least, recombinant DNA technology was invented in *E. coli*, and the organism still plays a central role in this technology today.

Other areas of contribution:

- Cell metabolism
- Nonsense suppressors
- Colinearity of gene and polypeptide
- The operon
- Plasmid-based drug resistance
- Active transport

Saccharomyces cerevisiae

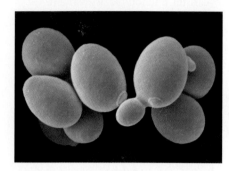

Genetic "Vital Statistics"	
Genome size:	12 Mb
Chromosomes:	$n = 16$
Number of genes:	6000
Percentage with human homologs:	25%
Average gene size:	1.5 kb, 0.03 intron/gene
Transposons:	Small proportion of DNA
Genome sequenced in:	1996

The ascomycete *S. cerevisiae*, alias "baker's yeast," "budding yeast," or simply "yeast," has been the basis of the baking and brewing industries since antiquity. In nature, it probably grows on the surface of plants, using exudates as nutrients, although its precise niche is still a mystery. Although laboratory strains are mostly haploid, cells in nature can be diploid or polyploid. In approximately 70 years of genetic research, yeast has become "the *E. coli* of the eukaryotes." Because it is haploid, unicellular, and forms compact colonies on plates, it can be treated in much the same way as a bacterium. However, it has eukaryotic meiosis, cell cycle, and mitochondria, and these features have been at the center of the yeast success story.

Yeast cells, *Saccharomyces cerevisiae*.

Special features

As a model organism, yeast combines the best of two worlds: it has much of the convenience of a bacterium, but with the key features of a eukaryote. Yeast cells are small (10 micrometers) and complete their cell cycle in just 90 minutes, allowing them to be produced in huge numbers in a short time. Like bacteria, yeast can be grown in large batches in a liquid medium that is continuously shaken. And, like bacteria, yeast produces visible colonies when plated on agar medium, can be screened for mutations, and can be replica plated. In typical eukaryotic manner, yeast has a mitotic cell-division cycle, undergoes meiosis, and contains mitochondria housing a small unique genome. Yeast cells can respire anaerobically by using the fermentation cycle and hence can do without mitochondria, allowing mitochondrial mutants to be viable.

Genetic analysis

Performing crosses in yeast is quite straightforward. Strains of opposite mating type are simply mixed on an appropriate medium. The resulting *a*/α diploids are induced to undergo meiosis by using a special sporulation medium. Investigators can isolate ascospores from a single tetrad by using a machine called a micromanipulator. They also have the option of synthesizing *a*/*a* or α/α diploids for special purposes or creating partial diploids by using specially engineered plasmids.

Because a huge array of yeast mutants and DNA constructs is available within the research community, special-purpose strains for screens and selections can be built by crossing various yeast types. Additionally, new mutant alleles can be mapped by crossing to strains containing an array of phenotypic or DNA markers of known map position.

The availability of both haploid and diploid cells provides flexibility for mutational studies. Haploid cells are convenient for large-scale selections or screens because mutant phenotypes are expressed directly. Diploid cells are convenient for obtaining dominant mutations, sheltering lethal mutations, performing complementation tests, and exploring gene interaction.

LIFE CYCLE

Yeast is a unicellular species with a very simple life cycle consisting of sexual and asexual phases. The asexual phase can be haploid or diploid. A cell divides asexually by budding: a mother cell throws off a bud into which is passed one of the nuclei that result from mitosis. For sexual reproduction, there are two mating types, determined by the alleles *MAT*α and *MAT*a. When haploid cells of different mating type unite, they form a diploid cell, which can divide mitotically or undergo meiotic division. The products of meiosis are a nonlinear tetrad of four ascospores.

Total length of life cycle: 90 minutes to complete cell cycle

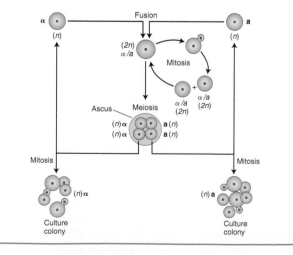

Techniques of Genetic Manipulation	
Standard mutagenesis:	
Chemicals and radiation	Random somatic mutations
Transposons	Random somatic insertions
Transgenesis:	
Integrative plasmid	Inserts by homologous recombination
Replicative plasmid	Can replicate autonomously (2μ or ARS origin of replication)
Yeast artificial chromosome	Replicates and segregates as a chromosome
Shuttle vector	Can replicate in yeast or *E. coli*
Targeted gene knockouts:	
Gene replacement	Homologous recombination replaces wild-type allele with null copy

Genetic engineering

Transgenesis. Budding yeast provides more opportunities for genetic manipulation than any other eukaryote (see p. 366). Exogenous DNA is taken up easily by cells whose cell walls have been partly removed by enzyme digestion or abrasion. Various types of vectors are available (see table above). For a plasmid to replicate free of the chromosomes, it must contain a normal yeast replication origin (ARS) or a replication origin from a 2μ plasmid found in certain yeast isolates. The most elaborate vector, the yeast artificial chromosome (YAC), consists of an ARS, a yeast centromere, and two telomeres. A YAC can carry large transgenic inserts, which are then inherited in the same way as Mendelian chromosomes. YACs have been important vectors in cloning and sequencing large genomes such as the human genome.

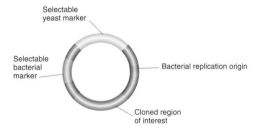

A simple yeast vector. This type of vector is called a yeast integrative plasmid (YIp).

Targeted knockouts. Transposon mutagenesis (transposon tagging) can be accomplished by introducing yeast DNA into *E. coli* on a shuttle vector; the bacterial transposons integrate into the yeast DNA, knocking out gene function. The shuttle vector is then transferred back into yeast, and the tagged mutants replace wild-type copies by homologous recombination. Gene knockouts can also be accomplished by replacing wild-type alleles with an engineered null copy through homologous recombination. By using these techniques, researchers have systematically constructed a complete set of yeast knockout strains (each carrying a different knockout) to assess null function of each gene at the phenotypic level.

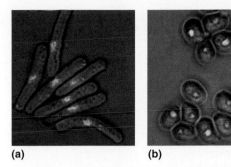

Cell-cycle mutants. (a) Mutants that elongate without dividing. (b) Mutants that arrest without budding.

Main contributions

Thanks to a combination of good genetics and good biochemistry, yeast studies have made substantial contributions to our understanding of the genetic control of cell processes.

Cell cycle. The identification of cell-division genes through their temperature-sensitive mutants (*cdc* mutants) has led to a powerful model for the genetic control of cell division. The different Cdc phenotypes reveal the components of the machinery required to execute specific steps in the progression of the cell cycle. This work has been useful for understanding the abnormal cell-division controls that can lead to human cancer (see p. 551).

Recombination. Many of the key ideas for the current molecular models of crossing-over (such as the double-strand-break model) are based on tetrad analysis of gene conversion in yeast (see p. 474). Gene conversion (aberrant allele ratios such as 3 : 1) is relatively common in yeast genes, providing an appropriately large data set for quantifying the key features of this process.

Gene interactions. Yeast has led the way in the study of gene interactions. The techniques of traditional genetics have been used to reveal patterns of epistasis and suppression, which suggest gene interactions (see p. 205). The two-hybrid plasmid system for finding protein interactions was developed in yeast and has generated complex interaction maps that represent the beginnings of systems biology (see p. 413). Synthetic lethals—lethal double mutants created by intercrossing two viable single mutants—also are used to plot networks of interaction (see p. 206).

Mitochondrial genetics. Mutants with defective mitochondria are recognizable as very small colonies called "petites." The availability of these petites and other mitochondrial mutants enabled the first ever detailed analysis of mitochondrial genome structure and function in any organism.

Genetics of mating type. Yeast *MAT* alleles were the first mating-type genes to be characterized at the molecular level. Interestingly, yeast undergoes spontaneous switching from one mating type to the other. A silent "spare" copy of the opposite *MAT* allele, residing elsewhere in the genome, enters into the mating-type locus, replacing the resident allele by homologous recombination. Yeast has provided one of the central models for signal transduction during detection and response to mating hormones from the opposite mating type.

Other areas of contribution

- Genetics of switching between yeastlike and filamentous growth
- Genetics of senescence

Neurospora crassa

Key organism for studying:

- Genetics of metabolism and uptake
- Genetics of crossing-over and meiosis
- Fungal cytogenetics
- Polar growth
- Circadian rhythms
- Nucleus–mitochondrial interactions

Genetic "Vital Statistics"	
Genome size:	43 Mb
Chromosomes:	7 autosomes ($n = 7$)
Number of genes:	10,000
Percentage with human homologs:	6%
Average gene size:	1.7 kb, 1.7 introns/gene
Transposons:	rare
Genome sequenced in:	2003

Neurospora growing on sugarcane.

Neurospora crassa, the orange bread mold, was one of the first eukaryotic microbes to be adopted by geneticists as a model organism. Like yeast, it was originally chosen because of its haploidy, its simple and rapid life cycle, and the ease with which it can be cultured. Of particular significance was the fact that it will grow on a medium with a defined set of nutrients, making it possible to study the genetic control of cellular chemistry. In nature, it is found in many parts of the world growing on dead vegetation. Because fire activates its dormant ascospores, it is most easily collected after burns—for example, under the bark of burnt trees and in fields of crops such as sugar cane that are routinely burned before harvesting.

Special features

Neurospora holds the speed record for fungi because each hypha grows more than 10 cm per day. This rapid growth, combined with its haploid life cycle and ability to grow on defined medium, has made it an organism of choice for studying biochemical genetics of nutrition and nutrient uptake.

Another unique feature of *Neurospora* (and related fungi) allows geneticists to trace the steps of single meioses. The four haploid products of one meiosis stay together in a sac called an ascus. Each of the four products of meiosis undergoes a further mitotic division, resulting in a linear octad of eight ascospores (see pp. 100–101). This feature makes *Neurospora* an ideal system in which to study crossing-over, gene conversion, chromosomal rearrangements, meiotic nondisjunction, and the genetic control of meiosis itself. Chromosomes, although small, are easily visible, and so meiotic processes can be studied at both the genetic and the chromosomal levels. Hence, in *Neurospora*, fundamental studies have been carried out on the mechanisms underlying these processes (see p. 120).

Genetic analysis

Genetic analysis is straightforward (see p. 100). Stock centers provide a wide range of mutants affecting all aspects of the biology of the fungus. *Neurospora* genes can be mapped easily by crossing them to a bank of strains with known mutant loci or known RFLP alleles. Strains of opposite mating type are crossed simply by growing them together. A geneticist with a hand-held needle can pick out a single ascospore for study. Hence analy-

LIFE CYCLE

N. crassa has a haploid eukaryotic life cycle (see p. 96). A haploid asexual spore (called a conidium) germinates to produce a germ tube that extends at its tip. Progressive tip growth and branching produce a mass of branched threads (called *hyphae*), which forms a compact colony on growth medium. Because hyphae have no cross walls, a colony is essentially one cell containing many haploid nuclei. The colony buds off millions of asexual spores, which can disperse in air and repeat the asexual cycle.

In *N. crassa*'s sexual cycle, there are two identical-looking mating types MAT-*A* and MAT-*a*, which can be viewed as simple "sexes." As in yeast, the two mating types are determined by two alleles of one gene. When colonies of different mating type come into contact, their cell walls and nuclei fuse. Many transient diploid nuclei arise, each of which undergoes meiosis, producing an octad of ascopores. The ascospores germinate and produce colonies exactly like those produced by asexual spores.

Length of life cycle: 4 weeks for sexual cycle

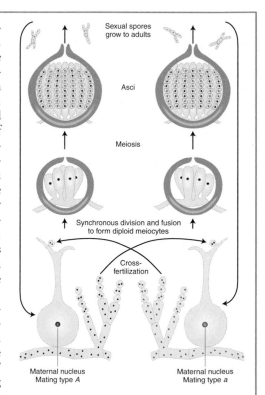

Sexual spores grow to adults

Asci

Meiosis

Synchronous division and fusion to form diploid meiocytes

Cross-fertilization

Maternal nucleus Mating type *A*

Maternal nucleus Mating type *a*

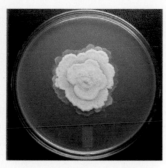

Wild-type (left) and mutant (right) *Neurospora* **grown in a petri dish.**

ses using either complete asci or random ascospores are rapid and straightforward.

Because *Neurospora* is haploid, newly obtained mutant phenotypes are easily detected with the use of various types of screens and selections. For example, filter enrichment can be used to select for nongrowing mutants (see p. 525). A favorite system for study of the mechanism of mutation is the *ad-3* gene, because *ad-3* mutants are purple and easily detected.

Although vegetative diploids of *Neurospora* are not readily obtainable, geneticists are able to create a "mimic diploid," useful for complementation tests and other analyses requiring the presence of two copies of a gene (see p. 201). Namely, the fusion of two different strains produces a heterokaryon, an individual containing two different nuclear types in a common cytoplasm. Heterokaryons also enable the use of a version of the specific locus test, a way to recover mutations in a specific recessive allele. (Cells from a +/m heterokaryon are plated and m/m colonies are sought.)

Techniques of Genetic Manipulation

Standard mutagenesis:

Chemicals and radiation	Random somatic mutations
Transposon mutagenesis	Not available

Transgenesis:

Plasmid-mediated transformation	Random insertion

Targeted gene knockouts:

RIP	GC → AT mutations in transgenic duplicate segments before a cross
Quelling	Somatic posttranscriptional inactivation of transgenes

---

Genetic engineering

Transgenesis. The first ever eukaryotic transformation was accomplished in *Neurospora*. Today, *Neurospora* is easily transformed with the use of bacterial plasmids carrying the desired transgene, plus a selectable marker such as hygromycin resistance, to show that the plasmid has entered. There are no plasmids that replicate in *Neurospora*, and so a transgene is inherited only if it integrates into a chromosome.

Targeted knockouts. In contrast with yeast, *Neurospora* does not frequently integrate transgenes by homologous recombination. Hence a transgenic strain normally has the resident gene

plus the homologous transgene, inserted at a random ectopic location. Because of this duplication of material, if the strain is crossed, it is subject to RIP, a genetic process that is unique to *Neurospora*. RIP is a premeiotic mechanism that introduces many GC-to-AT transitions into both duplicate copies, effectively disrupting the gene. RIP can therefore be harnessed as a convenient way of deliberately knocking out a specific gene (see p. 532).

Main contributions

George Beadle and Edward Tatum used *Neurospora* as the model organism in their pioneering studies on gene–enzyme relations, in which they were able to determine the enzymatic steps in the synthesis of arginine (see p. 187). Their work with *Neurospora* established the beginning of molecular genetics. There followed many comparable studies on the genetics of cell metabolism with the use of *Neurospora*.

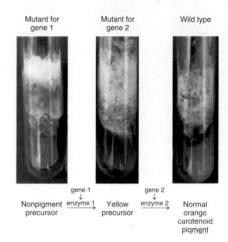

Pathway synthesizing orange caroteinoid pigment in *Neurospora*.

Pioneering work has been done on the genetics of meiotic processes, such as crossing-over and disjunction, and on conidiation rhythms. Continuously growing cultures show a daily rhythm of conidiospore formation. The results of pioneering studies using mutations that alter this rhythm have contributed to a general model for the genetics of circadian rhythms.

Neurospora serves as a model for the multitude of pathogenic filamentous fungi affecting crops and humans because these fungi are often difficult to culture and manipulate genetically. It is even used as a simple eukaryotic test system for mutagenic and carcinogenic chemicals in the human environment.

Because crosses can be made by using one parent as female, the cycle is convenient for the study of mitochondrial genetics and nucleus–mitochondrial interaction. A wide range of linear and circular mitochondrial plasmids has been discovered in natural isolates. Some of them are retroelements that are thought to be intermediates in the evolution of viruses.

Other areas of contribution

- Fungal diversity and adaptation
- Cytogenetics (chromosomal basis of genetics)
- Mating-type genes
- Heterokaryon-compatibility genes (a model for the genetics of self and nonself recognition)

Arabidopsis thaliana

Key organism for studying:

- Development
- Gene expression and regulation
- Plant genomics

Genetic "Vital Statistics"

Genome size:	125 Mb
Chromosomes:	diploid, 5 autosomes ($2n = 10$)
Number of genes:	25,000
Percentage with human homologs:	18%
Average gene size:	2 kb, 4 introns/gene
Transposons:	10% of the genome
Genome sequenced in:	2000

Arabidopsis thaliana, a member of the Brassicaceae (cabbage) family of plants, is a relatively late arrival as a genetic model organism. Most work has been done in the past 20 years. It has no economic significance—it grows prolifically as a weed in many temperate parts of the world. However, because of its small size, short life cycle, and small genome, it has overtaken the more traditional genetic plant models such as corn and wheat and has become the dominant model for plant molecular genetics.

***Arabidopsis thaliana* growing in the wild.** The versions grown in the laboratory are smaller. [Dan Tenaglia, www.missouriplants.com.]

Special features

In comparison with other plants, *Arabidopsis* is small in regard to both its physical size and its genome size—features that are advantageous for a model organism. *Arabidopsis* grows to a height of less than 10 cm under appropriate conditions, and hence it can be grown in large numbers, permitting large-scale mutant screens and progeny analyses. Its total genome size of 125 Mb made the genome relatively easy to sequence compared to other plant model organism genomes, such as the maize genome (2500 Mb) and the wheat genome (16,000 Mb).

Genetic analysis

The analysis of *Arabidopsis* mutations through crossing relies on tried and true methods—essentially those used by Mendel. Plant stocks carrying useful mutations relevant to the experiment in hand are obtained from public stock centers. Lines can be manually crossed with each other or self-fertilized. Although the flowers are small, cross pollination is easily accomplished by removing undehisced anthers (which are sometimes eaten by the experimenter as a convenient means of disposal). Each pollinated flower then produces a long pod containing a large number of seeds. This abundant production of offspring (thousands of seeds per plant) is a boon to geneticists searching for rare mutants or other rare events. If a plant carries a new recessive mutation in the germ line, selfing allows progeny homozygous for the recessive mutation to be recovered in the plant's immediate descendants.

LIFE CYCLE

Arabidopsis has the familiar plant life cycle, with a dominant diploid stage. A plant bears several flowers, each of which produces many seeds. Like many annual weeds, its life cycle is rapid: it takes only about 6 weeks for a planted seed to produce a new crop of seeds.

Total length of life cycle: 6 weeks

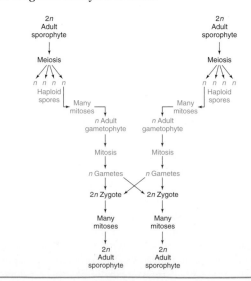

Arabidopsis mutants. (a) Wild-type flower of *Arabidopsis*. (b) The *agamous* mutation (*ag*), which results in flowers with only petals and sepals (no reproductive structures). (c) A double-mutant *ap1, cal*, which makes a flower that looks like a cauliflower. (Similar mutations in cabbage are probably the cause of real cauliflowers.) [Photos from George Haughn.]

Techniques of Genetic Modification

Standard mutagenesis:

Chemicals and radiation	Random germ-line or somatic mutations
T-DNA itself or transposons	Random tagged insertions
Transgenesis:	
T-DNA carries the transgene	Random insertion
Targeted gene knockouts:	
T-DNA or transposon-mediated transgenesis	Random insertion; knockouts selected with PCR
RNAi	Mimics targeted knockout

Genetic engineering

Transgenesis. *Agrobacterium* T-DNA is a convenient vector for introducing transgenes (see p. 368). The vector–transgene construct inserts randomly throughout the genome. Transgenesis offers an effective way to study gene regulation. The transgene is spliced to a reporter gene such as *GUS*, which produces a blue dye at whatever positions in the plant the gene is active. **Targeted knockouts.** Because homologous recombination is rare in *Arabidopsis*, specific genes cannot be easily knocked out by homologous replacement with a transgene. Hence, in *Arabidopsis*, genes are knocked out by random insertion of a T-DNA vector or transposon (maize transposons such as *Ac-Ds* are used), and then specific gene knockouts are selected by applying PCR analysis to DNA from large pools of plants. The PCR uses a sequence in the T-DNA or in the transposon as one primer and a sequence in the gene of interest as the other primer. Thus PCR amplifies only copies of the gene of interest that carry an insertion. Subdividing the pool and repeating the process leads to the specific plant carrying the knockout. Alternatively, RNAi may be used to inactivate a specific gene.

Large collections of T-DNA insertion mutants are available; they have the flanking plant sequences listed in public databases; so, if you are interested in a specific gene, you can see if the collection contains a plant that has an insertion in that gene. A convenient feature of knockout populations in plants is that they can be easily and inexpensively maintained as collections of seeds for many years, perhaps even decades. This fea-

ture is not possible for most populations of animal models. The worm *C. elegans* can be preserved as frozen animals, but fruit flies *(D. melanogaster)* cannot be frozen and revived. Thus, lines of fruit fly mutants must be maintained as living organisms.

Main contributions

As the first plant genome to be sequenced, *Arabidopsis* has provided an important model for plant genome architecture and evolution. In addition, studies of *Arabidopsis* have made key contributions to our understanding of the genetic control of plant development (see p. 603). Geneticists have isolated homeotic mutations affecting flower development, for example. In such mutants, one type of floral part is replaced by another. Integration of the action of these mutants has led to an elegant model of flower whorl determination based on overlapping patterns of regulatory gene expression in the flower meristem. *Arabidopsis* has also contributed broadly to the genetic basis of plant physiology, gene regulation, and the interaction of plants and the environment (including the genetics of disease resistance). Because *Arabidopsis* is a natural plant of worldwide distribution, it has great potential for the study of evolutionary diversification and adaptation.

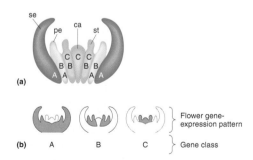

The establishment of whorl fate. (a) Patterns of gene expression corresponding to the different whorl fates. From outermost to innermost, the fates are sepal (se), petal (pe), stamen (st), and carpel (ca). (b) The shaded regions of the cross-sectional diagrams of the developing flower indicate the gene-expression patterns for the genes of the A, B, and C classes.

Other areas of contribution

- Environmental stress response
- Hormone control systems

Caenorhabditis elegans

Key organism for studying:
- Development
- Behavior
- Nerves and muscles
- Aging

Genetic "Vital Statistics"

Genome size:	97 Mb
Chromosomes:	5 autosomes ($2n = 10$), X chromosome
Number of genes:	19,000
Percentage with human homologs:	25%
Average gene size:	5 kb, 5 exons/gene
Transposons:	Several types, active in some strains
Genome sequenced in:	1998

Caenorhabditis elegans may not look like much under a microscope, and, indeed, this 1-mm-long soil-dwelling roundworm (a nematode) is relatively simple as animals go. But that simplicity is part of what makes *C. elegans* a good model organism. Its small size, rapid growth, ability to self, transparency, and low number of body cells have made it an ideal choice for the study of the genetics of eukaryotic development.

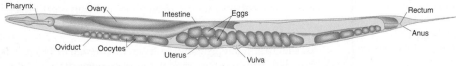

Photomicrograph and drawing of an adult *C. elegans*.

Special features

Geneticists can see right through C. *elegans*! Unlike other multicellular model organisms, such as fruit flies or *Arabidopsis*, this tiny worm is transparent, making it efficient to screen large populations for interesting mutations affecting virtually any aspect of anatomy or behavior. Transparency also lends itself well to studies of development: researchers can directly observe all stages of development simply by watching the worms under a light microscope. The results of such studies have shown that C. *elegans*'s development is tightly programmed and that each worm has a surprisingly small and consistent number of cells (959 in hermaphrodites and 1031 in males). In fact, biologists have tracked the fates of specific cells as the worm develops and have determined the exact pattern of cell division leading to each adult organ. This effort has yielded a lineage pedigree for every adult cell (see p. 554).

Genetic analysis

Because the worms are small and reproduce quickly and prolifically (selfing produces about 300 progeny and crossing about

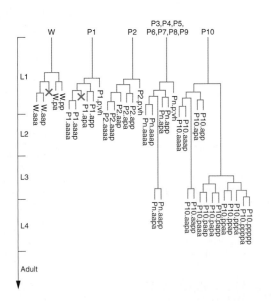

A symbolic representation of the lineages of 11 cells. A cell that undergoes programmed cell death is indicated with a blue X at the end of a branch of a lineage.

LIFE CYCLE

C. *elegans* is unique among the major model animals in that one of the two sexes is hermaphrodite (XX). The other is male (XO). The two sexes can be distinguished by the greater size of the hermaphrodites and by differences in their sex organs. Hermaphrodites produce both eggs and sperm, and so they can be selfed. The progeny of a selfed hermaphrodite also are hermaphrodites, except when a rare nondisjunction leads to an XO male. If hermaphrodites and males are mixed, the sexes copulate, and many of the resulting zygotes will have been fertilized by the males' amoeboid sperm. Fertilization and embryo production take place within the hermaphrodite, which then lays the eggs. The eggs finish their development externally.

Total length of life cycle: $3\frac{1}{2}$ days

1000), they produce large populations of progeny that can be screened for rare genetic events. Moreover, because hermaphroditism in *C. elegans* makes selfing possible, individual worms with homozygous recessive mutations can be recovered quickly by selfing progeny of treated individual worms. In contrast, other animal models, such as fruit flies or mice, require matings between siblings and take more generations to recover recessive mutations.

Techniques of Genetic Modification	
Standard mutagenesis:	
Chemical (EMS) and radiation	Random germ-line mutations
Transposons	Random germ-line insertions
Transgenesis:	
Transgene injection of gonad	Unintegrated transgene array; occasional integration
Targeted gene knockouts:	
Transposon-mediated mutagenesis	Knockouts selected with PCR
RNAi	Mimics targeted knockout
Laser ablation	Knockout of one cell

Genetic engineering

Transgenesis. The introduction of transgenes into the germ line is made possible by a special property of *C. elegans* gonads. The gonads of the worm are syncitial, meaning that there are many nuclei in a common cytoplasm. The nuclei do not become incorporated into cells until meiosis, when formation of the individual egg or sperm begins. Thus, a solution of DNA containing the transgene injected into the gonad of a hermaphrodite exposes more than 100 germ-cell precursor nuclei to the transgene. By chance a few of these nuclei will incorporate the DNA (see p. 370).

Transgenes recombine to form multicopy tandem arrays. In an egg, the arrays do not integrate into a chromosome, but transgenes from the arrays are still expressed. Hence the gene carried on a wild-type DNA clone can be identified by introducing it into a specific recessive recipient strain (functional complementation). In some but not all cases, the transgenic arrays are passed on to progeny. To increase the chance of inheritance, worms are exposed to ionizing radiation, which can induce the integration of an array into an ectopic chromosomal position, and, in this site, the array is reliably transmitted to progeny.

Targeted knockouts. In strains with active transposons, the transposons themselves become agents of mutation by inserting into random locations in the genome, knocking out the interrupted genes. If we can identify organisms with insertions into a specific gene of interest, we can isolate a targeted gene knockout. Inserts into specific genes can be detected by using PCR if one PCR primer is based on the transposon sequence and another one is based on the sequence of the gene of interest. Alternatively, RNAi can be used to nullify the function of specific genes. As an alternative to mutation, individual cells can be killed by a laser beam to observe the effect on worm function or development (laser ablation).

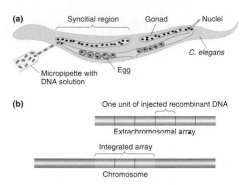

Creation of *C. elegans* transgenes. (a) Method of injection. (b) Extrachromosomal and integrated arrays.

Main contributions

C. elegans has become a favorite model organism for the study of various aspects of development because of its small and invariant number of cells. One example is programmed cell death, a crucial aspect of normal development. Some cells are genetically programmed to die in the course of development (a process called apoptosis). The results of studies of *C. elegans* have contributed a useful general model for apoptosis, which is also known to be a feature of human development (see p. 554).

Another model system is the development of the vulva, the opening to the outside of the reproductive tract. Hermaphrodites with defective vulvas still produce progeny, which in screens are easily visible clustered within the body. The results of studies of hermaphrodites with no vulva or with too many have revealed how cells that start off completely equivalent can become differentiated into different cell types (see p. 598).

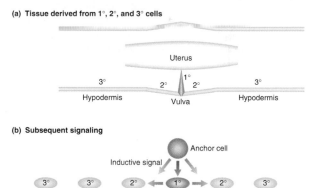

Production of the *C. elegans* vulva. (a) The final differentiated tissue. (b) Method of differentiation. The cells begin completely equivalent. An anchor cell behind the equivalent cells sends a signal to the nearest cells, which become the vulva. The primary vulva cell then sends a lateral signal to its neighbors, preventing them from becoming primary cells, even though they, too, have received the signal from the anchor cell.

Behavior has also been the subject of genetic dissection. *C. elegans* offers an advantage in that worms with defective behavior often can still live and reproduce. The worm's nerve and muscle systems have been genetically dissected, allowing behaviors to be linked to specific genes.

Other areas of contribution

- Cell-to-cell signaling

Drosophila melanogaster

Key organism for studying:

- Transmission genetics
- Cytogenetics
- Development
- Population genetics
- Evolution

Genetic "Vital Statistics"

Genome size:	180 Mb
Chromosomes:	Diploid, 3 autosomes, X and Y ($2n = 8$)
Number of genes:	13,000
Percentage with human homologs:	~50%
Average gene size:	3 kb, 4 exons/gene
Transposons:	*P* elements, among others
Genome sequenced in:	2000

Polytene chromosomes.

The fruit fly *Drosophila melanogaster*—loosely translated as "dusky syrup-lover"—was one of the first model organisms to be used in genetics. It was chosen in part because it is readily available from ripe fruit, has a short life cycle of the diploid type, and is simple to culture and cross in jars or vials containing a layer of food. Early genetic analysis showed that its inheritance mechanisms have strong similarities to those of other eukaryotes, underlining its role as a model organism. Its popularity as a model organism went into decline during the years when *E. coli*, yeast, and other microorganisms were being developed as molecular tools. However, *Drosophila* has experienced a renaissance because it lends itself so well to the study of the genetic basis of development, one of the central questions of biology. *Drosophila*'s importance as a model for human genetics is demonstrated by the discovery that approximately 60 percent of known disease-causing genes in humans, as well as 70 percent of cancer genes, have counterparts in *Drosophila*.

Special features

Drosophila came into vogue as an experimental organism in the early twentieth century because of features common to most model organisms. It is small (3 mm long), simple to raise (originally in milk bottles), quick to reproduce (only 12 days from egg to adult), and easy to obtain (just leave out some rotting fruit). It proved easy to amass a large range of interesting mutant alleles that were used to lay the ground rules of transmission genetics. Early researchers also took advantage of a feature unique to the fruit fly: polytene chromosomes (see p. 87). In salivary glands and certain other tissues, these "giant chromosomes" are produced by multiple rounds of DNA replication without chromosomal segregation. Each polytene chromosome displays a unique banding pattern, providing geneticists with landmarks that could be used to correlate recombination-based maps with actual chromosomes. The momentum provided by these early advances, along with the large amount of accumulated knowledge about the organism, made *Drosophila* an attractive genetic model.

Genetic analysis

Crosses in *Drosophila* can be performed quite easily. The parents may be wild or mutant stocks obtained from stock centers or as new mutant lines.

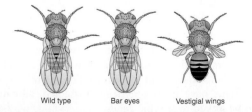

Wild type   Bar eyes   Vestigial wings

Two morphological mutants of *Drosophila*, with the wild type for comparison.

LIFE CYCLE

Drosophila has a short diploid life cycle that lends itself well to genetic analysis. After hatching from an egg, the fly develops through several larval stages and a pupal stage before emerging as an adult, which soon becomes sexually mature. Sex is determined by X and Y sex chromosomes (XX = female, XY = male), although, in contrast with humans, the number of X's in relation to the number of autosomes determines sex (see p. 48).

Total length of life cycle: 12 days from egg to adult

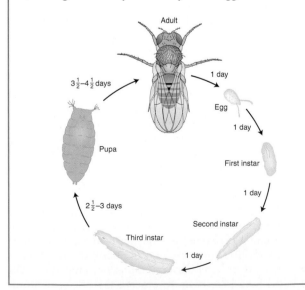

Adult

$3\frac{1}{2}$–$4\frac{1}{2}$ days

1 day

Egg

1 day

First instar

1 day

Pupa

Second instar

$2\frac{1}{2}$–3 days

Third instar

1 day

To perform a cross, males and females are placed together in a jar, and the females lay eggs in semisolid food covering the jar's bottom. After emergence from the pupae, offspring can be anesthetized to permit counting members of phenotypic classes and to distinguish males and females (by their different abdominal stripe patterns). However, because female progeny stay virgin for only a few hours after emergence from the pupae, they must immediately be isolated if they are to be used to make controlled crosses. Crosses designed to build specific gene combinations must be carefully planned, because crossing-over does not take place in *Drosophila* males. Hence, in the male, linked alleles will not recombine to help create new combinations.

For obtaining new recessive mutations, special breeding programs (of which the prototype is Muller's ClB test) provide convenient screening systems. In these tests, mutagenized flies are crossed to a stock with a balancer chromosome (see p. 500). Recessive mutations are eventually brought to homozygosity by inbreeding for one or two generations, starting with single F_1 flies.

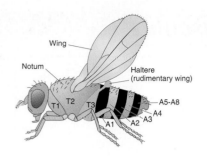

The normal thoracic and abdominal segments of *Drosophila*.

117, 124). Their discoveries opened the door to other pioneering studies:

- Early studies on the kinetics of mutation induction and measurement of mutation rates were performed with the use of *Drosophila*. Muller's ClB test and similar tests provided convenient screening methods for recessive mutations (see p. 523).
- Chromosomal rearrangements that move genes adjacent to heterochromatin were used to discover and study position-effect variegation.
- In the last part of the twentieth century, after the identification of certain key mutational classes such as homeotic and maternal effect mutations, *Drosophila* assumed a central role in the genetics of development, a role that continues today (see Chapter 18). Maternal-effect mutations that affect the development of embryos, for example, have been crucial in the elucidation of the genetic determination of the *Drosophila* body plan; these mutations are identified by screening for abnormal developmental phenotypes in the embryos from a specific female. Techniques such as enhancer trap screens have enabled the discovery of new regulatory regions in the genome that affect development. Through these methods and others, *Drosophila* biologists have made important advances in understanding the determination of segmentation and of the body axes. Some of the key genes discovered, such as the homeotic genes, have widespread relevance in animals generally.

Techniques of Genetic Modification	
Standard mutagenesis:	
Chemical (EMS) and radiation	Random germ-line and somatic mutations
Transgenesis:	
P element mediated	Random insertion
Targeted gene knockouts:	
Induced replacement	Null ectopic allele exits and recombines with wild-type allele
RNAi	Mimics targeted knockout

Genetic engineering

Transgenesis. Building transgenic flies requires the help of a *Drosophila* transposon called the *P* element. Geneticists construct a vector that carries a transgene flanked by *P*-element repeats. The transgene vector is then injected into a fertilized egg along with a helper plasmid containing a transposase. The transposase allows the transgene to jump randomly into the genome in germinal cells of the embryo (see p. 371).

Targeted knockouts. Targeted gene knockouts can be accomplished by, first, introducing a null allele transgenically into an ectopic position and, second, inducing special enzymes that cause excision of the null allele. The excised fragment (which is linear) then finds and replaces the endogenous copy by homologous crossing-over. However, functional knockouts can be produced more efficiently by RNAi.

Main contributions

Much of the early development of the chromosome theory of heredity was based on the results of *Drosophila* studies. Geneticists working with *Drosophila* made key advances in developing techniques for gene mapping, in understanding the origin and nature of gene mutation, and in documenting the nature and behavior of chromosomal rearrangements (see pp.

(a) *Bicoid* mRNA   (b) *nos* mRNA

Photomicrographs showing gradients of body axis determinants.
(a) mRNA for the gene *bcd* is shown localized to the anterior (left) tip of the embryo. (b) mRNA of the *nos* gene is localized to the posterior (right tip of the embryo). The distribution of the proteins from these genes and other genes determines the body axis.

Other areas of contribution

- Population genetics
- Evolutionary genetics
- Behavioral genetics

Mus musculus

Key organism for studying:

- Human disease
- Mutation
- Development
- Coat color
- Immunology

Genetic "Vital Statistics"

Genome size:	2600 Mb
Chromosomes:	19 autosomes, X, Y ($2n = 40$)
Number of genes:	30,000
Percentage with human homologs:	99%
Average gene size:	40 kb, 8.3 exons/gene
Transposons:	Source of 38% of genome
Genome sequenced in:	2002

Because humans and most domesticated animals are mammals, the genetics of mammals is of great interest to us. However, mammals are not ideal for genetics: they are relatively large in size compared with other model organisms, thereby taking up large and expensive facilities, their life cycles are long, and their genomes are large and complex. Compared with other mammals, however, mice *(Mus musculus)* are relatively small, have short life cycles, and are easily obtained, making them an excellent choice for a mammal model. In addition, mice had a head start in genetics because mouse "fanciers" had already developed many different interesting lines of mice that provided a source of variants for genetic analysis. Research on the Mendelian genetics of mice began early in the twentieth century.

An adult mouse and its litter.

Special features

Mice are not exactly small, furry humans, but their genetic makeup is remarkably similar to ours. Among model organisms, the mouse is the one whose genome most closely resembles the human genome. The mouse genome is about 14 percent smaller than that of humans (the human genome is 3000 Mb), but it has approximately the same number of genes (current estimates are just less than 30,000). A surprising 99 percent of mouse genes seem to have homologs in humans. Furthermore, a large proportion of the genome is syntenic with that of humans; that is, there are large blocks containing the same genes in the same relative positions (see p. 700). Such genetic similarities are the key to the mouse's success as a model

organism; these similarities allow mice to be treated as "stand-ins" for their human counterparts in many ways. Potential mutagens and carcinogens that we suspect of causing damage to humans, for example, are tested on mice, and mouse models are essential in studying a wide array of human genetic diseases.

Genetic analysis

Mutant and "wild type" (though not actually from the wild) mice are easy to come by: they can be ordered from large stock centers that provide mice suitable for crosses and various other types of experiments. Many of these lines are derived from mice bred in past centuries by mouse "fanciers." Controlled crosses can be performed simply by pairing a male with a nonpregnant female. In most cases, the parental genotypes can be provided by male or female.

Human chromosomes

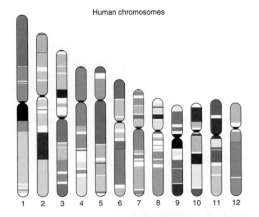

1 2 3 4 5 6 7 8 9 10 11 12

A mouse–human synteny map of 12 chromosomes from the human genome. Color coding is used to depict the regional matches of each block of the human genome to the corresponding sections of the mouse genome. Each color represents a different mouse chromosome.

LIFE CYCLE

Mice have a familiar diploid life cycle, with an XY sex-determination system similar to that of humans. Litters are from 5 to 10 pups; however, the fecundity of females declines after about 9 months, and so they rarely have more than five litters.

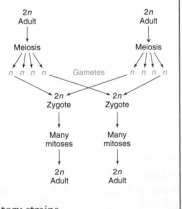

Total length of life cycle: 10 weeks from birth to giving birth, in most laboratory strains

Most of the standard estimates of mammalian mutation rates (including those of humans) are based on measurements in mice. Indeed, mice provide the final test of agents suspected of causing mutations in humans. Mutation rates in the germ line are measured using the specific locus test: mutagenize +/+ gonads, cross to *m/m* (*m* is a known recessive mutation at the locus under study), and look for *m*/m* progeny (*m** is a new mutation). The procedure is repeated over seven sample loci. The measurement of somatic mutation rates uses a similar setup, but the mutagen is injected into the fetus. Mice have been used extensively to study the type of somatic mutation that gives rise to cancer.

Techniques of Genetic Modification

Standard mutagenesis:

Chemicals and radiation	Germ-line and somatic mutations

Transgenesis:

Transgene injection into zygote	Random and homologous insertion
Transgene uptake by stem cells	Random and homologous insertion

Targeted gene knockouts:

Null transgene uptake by stem cells	Targeted knockouts selected

---

Genetic engineering

Transgenesis. Creating transgenic mice is straightforward but requires careful manipulation of a fertilized egg (see p. 372). First, mouse genomic DNA is cloned in *E. coli* by using bacterial or phage vectors. The DNA is then injected into a fertilized egg, where it integrates at ectopic (random) locations in the genome or, less commonly, at the normal locus. The activity of the transgene's protein can be monitored by fusing the transgene with a reporter gene such as *GFP* before the gene is injected. With the use of a similar method, the somatic cells of mice also can be modified by transgene insertion: specific fragments of DNA are inserted into individual somatic cells and these cells are, in turn, inserted into mouse embryos.

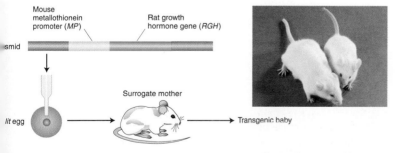

Producing a transgenic mouse. The transgene, a rat growth-hormone gene joined to a mouse promoter, is injected into a mouse egg homozygous for dwarfism (*lit/lit*).

Targeted knockouts. Knockouts of specific genes for genetic dissection can be accomplished by introducing a transgene containing a defective allele and two drug-resistance markers

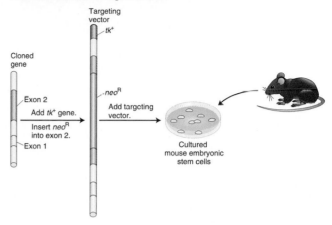

Producing a gene knockout. A drug-resistance gene (*neo*^R) is inserted into the transgene, both to serve as a marker and to disrupt the gene, producing a knockout. (The *tk* gene is a second marker.) The transgene construct is then injected into mouse embryo cells.

into a wild-type embryonic stem cell (see p. 372). The markers are used to select those specific transformant cells in which the defective allele has replaced the homologous wild-type allele. The transgenic cells are then introduced into mouse embryos. A similar method can be used to replace wild-type alleles with a functional transgene (gene therapy).

Main contributions

Early in the mouse's career as a model organism, geneticists used mice to elucidate the genes that control coat color and pattern, providing a model for all fur-bearing mammals, including cats, dogs, horses, and cattle (see p. 195). More recently, studies of mouse genetics have made an array of contributions with direct bearing on human health:

- A large proportion of human genetic diseases have a mouse counterpart—called a "mouse model"—useful for experimental study.
- Mice serve as models for the mechanisms of mammalian mutation.
- Studies on the genetic mechanisms of cancer are performed on mice.
- Many potential carcinogens are tested on mice.
- Mice have been important models for the study of mammalian developmental genetics (see p. 601). For example, they provide a model system for the study of genes affecting cleft lip and cleft palate, a common human developmental disorder.
- Cell lines that are fusion hybrids of mouse and human genomes played an important role in the assignment of human genes to specific human chromosomes. There is a tendency for human chromosomes to be lost from such hybrids, and so loss of specific chromosomes can be correlated with loss of specific human alleles.

Other areas of contribution

- Behavioral genetics
- Quantitative genetics
- The genes of the immune system

Appendix A

Genetic nomenclature

There is no universally accepted set of rules for naming genes, alleles, protein products, and associated phenotypes. At first, individual geneticists developed their own symbols for recording their work. Later, groups of people working on any given organism met and decided on a set of conventions that all would use. Because *Drosophila* was one of the first organisms to be used extensively by geneticists, most of the systems now being used are variants of the *Drosophila* system. However, there has been considerable divergence. Some scientists now advocate a standardization of this symbolism, but standardization has not been achieved. Indeed, the situation has been made more complex by the advent of DNA technology. Whereas most genes had previously been named for the phenotypes produced by mutations within them, the new technology has shown the precise nature of the products of many of these genes. Hence it seems more appropriate to refer to them by their cellular function. However, the old names are still in the literature, so many genes have two parallel sets of nomenclature.

The following examples by no means cover all the organisms used in genetics, but most of the nomenclature systems follow one of these types.

Drosophila melanogaster (insect)

ry	A gene that when mutated causes rosy eyes
ry^{502}	A specific recessive mutant allele producing rosy eyes in homozygotes
ry$^+$	The wild-type allele of *rosy*
ry	The rosy mutant phenotype
ry$^+$	The wild-type phenotype (red eyes)
RY	The protein product of the *rosy* gene
XDH	Xanthine dehydrogenase, an alternative description of the protein product of the *rosy* gene; named for the enzyme that it encodes
D	*Dichaete*; a gene that when mutated causes a loss of certain bristles and wings to be held out laterally in heterozygotes and causes lethality in homozygotes
D^3	A specific mutant allele of the *Dichaete* gene
D$^+$	The wild-type allele of *Dichaete*
D	The Dichaete mutant phenotype
D$^+$	The wild-type phenotype
D	(Depending on context) the protein product of the *Dichaete* gene (a DNA-binding protein)

Neurospora crassa (fungus)

arg	A gene that when mutated causes arginine requirement
arg-1	One specific *arg* gene
arg-1	An unspecified mutant allele of the *arg* gene
arg-1 (1)	A specific mutant allele of the *arg-1* gene
arg-1$^+$	The wild-type allele
arg-1	The protein product of the *arg-1*$^+$ gene

Arg$^+$	A strain not requiring arginine
Arg$^-$	A strain requiring arginine

Saccharomyces cerevisiae (fungus)

ARG	A gene that when mutated causes arginine requirement
ARG1	One specific *ARG* gene
arg1	An unspecified mutant allele of the *ARG* gene
arg1-1	A specific mutant allele of the *ARG* gene
ARG1$^+$	The wild-type allele
ARG1p	The protein product of the *ARG1*$^+$ gene
Arg$^+$	A strain not requiring arginine
Arg$^-$	A strain requiring arginine

Homo sapiens (mammal)

ACH	A gene that when mutated causes achondroplasia
ACH^1	A mutant allele (dominance not specified)
ACH	Protein product of *ACH* gene; nature unknown
FGFR3	Recent name for gene for achondroplasia
FGFR3^1 or *FGFR3**1 or *FGFR3*<1>	Mutant allele of *FGR3* (unspecified dominance)
FGFR3 protein	Fibroblast growth factor receptor 3

Mus musculus (mammal)

Tyrc	A gene for tyrosinase
$+^{Tyrc}$	The wild-type allele of this gene
Tyrc^{ch} or *Tyrc-ch*	A mutant allele causing chinchilla color
Tyrc	The protein product of this gene
TYRC	The wild-type phenotype
TYRCch	The chinchilla phenotype

Escherichia coli (bacterium)

lacZ	A gene for utilizing lactose
lacZ$^+$	The wild-type allele
lacZ1	A mutant allele
LacZ	The protein product of that gene
Lac$^+$	A strain able to use lactose (phenotype)
Lac$^-$	A strain unable to use lactose (phenotype)

Arabidopsis thaliana (plant)

YGR	A gene that when mutant produces yellow-green leaves
YGR1	A specific *YGR* gene
YGR1	The wild-type allele
ygr1-1	A specific recessive mutant allele of *YGR1*
ygr1-2D	A specific dominant (D) mutant allele of *YGR1*
YGR1	The protein product of *YGR1*
Ygr$^-$	Yellow-green phenotype
Ygr$^+$	Wild-type phenotype

Appendix B

Bioinformatic Resources for Genetics and Genomics

"You certainly usually find something, if you look, but it is not always quite the something you were after." — *The Hobbit*, J. R. R. Tolkien

The field of bioinformatics encompasses the use of computational tools to distill complex data sets. Genetic and genomic data are so diverse that it has become a considerable challenge to identify the authoritative site(s) for a specific type of information. Furthermore, the landscape of Web-accessible software for analyzing this information is constantly changing as new and more powerful tools are developed. This appendix is intended to provide *some* valuable starting points for exploring the rapidly expanding universe of online resources for genetics and genomics.

1. Finding Genetic and Genomic Web Sites

Here are listed several central resources that contain large lists of relevant Web sites.

- The scientific journal called *Nucleic Acids Research (NAR)* publishes a special issue every January listing a wide variety of online database resources at: http://nar.oupjournals.org/
- The Virtual Library has Model Organisms and Genetics subdivisions with rich arrays of Internet resources at: http://ceolas.org/VL/mo/ and http://www.ornl.gov/TechResources/Human_Genome/organisms.html
- The National Human Genome Research Institute (NHGRI) maintains a list of genome Web sites at: http://www.nhgri.nih.gov/10000375/
- The Department of Energy (DOE) maintains a Human Genome Project site at the Oak Ridge National Laboratory at: http://public.ornl.gov/hgmis/
- SwissProt maintains Amos' WWW links page at: http://www.expasy.ch/alinks.html

2. General Databases

Nucleic Acid and Protein Sequence Databases By international agreement, three groups collaborate to house the primary DNA and mRNA sequences of all species: the National Center for Biotechnology Information (NCBI) houses GenBank; the European Bioinformatics Institute (EBI) houses the European Molecular Biology Laboratory (EMBL) Data Library; and the National Institute of Genetics in Japan houses the DNA DataBase of Japan (DDBJ).

Primary DNA sequence records, called accessions, are submitted by individual research groups. In addition to providing access to these DNA sequence records, these sites provide many other data sets. For example, NCBI also houses RefSeq, a summary synthesis of information on the DNA sequences of fully sequenced genomes and the gene products that are encoded by these sequences.

Many other important features can be found at the NCBI, EBI, and DDBJ sites. Home pages and some other key Web sites are:

- **NCBI** http://www.ncbi.nlm.nih.gov/
- **NCBI-Genomes** http://www.ncbi.nlm.nih.gov/Genomes/index.html
- **NCBI-RefSeq** http://www.ncbi.nlm.nih.gov/LocusLink/refseq.html
- **EBI** http://www.ebi.ac.uk/
- **DDBJ** http://www.nig.ac.jp/

The harsh reality is that, with so much biological information, the goal of making these online resources "transparent" to the user is not fully achieved. Thus, exploration of these sites will entail familiarizing yourself with the contents of each site and exploring some of the ways the site helps you to focus your queries so you get the right answer(s) to your queries. For one example of the power of these sites, consider a search for a nucleotide sequence at NCBI. Databases typically store information in separate bins called "fields." By using queries that limit the search to the appropriate field, a more directed question can be asked. Using the "Limits" option, a query phrase can be used to identify or locate a specific species, type of sequence (genomic or mRNA), gene symbol, or any of several other data fields. Query engines usually support the ability to join multiple query statements together. For example: retrieve all DNA sequence records that are from the species *Caenorhabditis elegans* **AND** that were published after January 1, 2000. Using the "History" option, the results of multiple queries can be joined together, so that only those hits common to multiple queries will be retrieved. By proper use of the available query options on a site, a great many false positives can be computationally eliminated while not discarding any of the relevant hits.

Because protein sequence predictions are a natural part of the analysis of DNA and mRNA sequences, these same sites provide access to a variety of protein databases. One important protein database is SwissProt/TrEMBL. TrEMBL sequences are automatically predicted from DNA and/or mRNA sequences. SwissProt sequences are curated, meaning that an expert scientist reviews the output of computational analysis and makes expert decisions about which results to accept or reject. In addition to the primary protein sequence records, SwissProt also offers databases of protein domains and protein signatures (amino acid sequence strings that are characteristic of proteins of a particular type). The SwissProt home page is http://www.ebi.ac.uk/swissprot/.

Protein Domain Databases The functional units within proteins are thought to be local folding regions called domains. Prediction of domains within newly discovered proteins is one way to guess at their function. Numerous protein domain databases have emerged that predict domains in somewhat different ways. Some of the individual domain databases are Pfam, PROSITE, PRINTS, SMART, ProDom, TIGRFAMs, BLOCKS, and CDD. InterPro allows querying multiple protein domain databases simultaneously and presents the combined results. Web sites for some domain databases are:

- **InterPro** http://www.ebi.ac.uk/interpro/
- **Pfam** http://www.sanger.ac.uk/Software/Pfam/index.shtml
- **PROSITE** http://www.expasy.ch/prosite/
- **PRINTS** http://www.bioinf.man.ac.uk/dbbrowser/PRINTS/
- **SMART** http://smart.embl-heidelberg.de/
- **ProDom** http://prodes.toulouse.inra.fr/prodom/doc/prodom.html
- **TIGRFAMs** http://www.tigr.org/TIGRFAMs/
- **BLOCKS** http://blocks.fhcrc.org/
- **CDD** http://www.ncbi.nlm.nih.gov/Structure/cdd/cdd.shtml

Protein Structure Databases The representation of three-dimensional protein structures has become an important aspect of global molecular analysis. Three-dimensional structure databases are available from the major DNA/protein sequence database sites and from independent protein structure databases, notably the Protein DataBase (PDB). NCBI has an application called Cn3D that helps in viewing PDB data.

- **PDB** http://www.rcsb.org/pdb/
- **Cn3D** http://www.ncbi.nlm.nih.gov/Structure/CN3D/cn3d.shtml

3. Specialized Databases

Organism-Specific Genetic Databases In order to mass some classes of genetic and genomic information, especially phenotypic information, expert knowledge of a particular species is required. Thus, MODs (model organism databases) have emerged to fulfill this role for the major genetic systems. These include databases for *Saccharomyces cerevisiae* (SGD), *Caenorhabiditis elegans* (WormBase), *Drosophila melanogaster* (FlyBase), the zebra fish *Danio rerio* (ZFIN), the mouse *Mus musculus* (MGI), the rat *Rattus norvegicus* (RGD), *Zea mays* (MaizeGDB), and *Arabidopsis thaliana* (TAIR). Home pages for these MODs can be found at:

- **SGD** http://genome-www.stanford.edu/Saccharomyces/
- **WormBase** http://www.wormbase.org/
- **FlyBase** http://flybase.org/
- **ZFIN** http://zfin.org/
- **MGI** http://www.informatics.jax.org/
- **RGD** http://rgd.mcw.edu/
- **MaizeGDB** http://www.maizegdb.org
- **TAIR** http://www.arabidopsis.org/

Human Genetics and Genomics Databases Because of the importance of human genetics in clinical as well as basic research, a diverse set of human genetic databases have emerged. Among them are a human genetic disease database called Online Mendelian Inheritance in Man (OMIM), a database of brief descriptions of human genes called GeneCards, a compilation of all known mutations in human genes called Human Gene Mutation Database (HGMD), a database of the current sequence map of the human genome called the Golden Path, and some links to human genetic disease databases:

- **OMIM** http://www3.ncbi.nlm.nih.gov/Omim/
- **GeneCards** http://mach1.nci.nih.gov/cards/index.html
- **HGMD** http://www.hgmd.org
- **Golden Path** http://genome.ucsc.edu/goldenPath/hgTracks.html
- **Online Genetic Support Groups**
  http://www.mostgene.org/support/index.html
- **Genetic Disease Information**
  http://www.geneticalliance.org/diseaseinfo/search.html

Genome Project Databases The individual genome projects also have Web sites, where they display their results, often including information that doesn't appear on any other Web site in the world. The largest of the publicly funded genome centers include:

- **Whitehead Institute/MIT Center for Genome Research**
  http://www-genome.wi.mit.edu/
- **Washington University School of Medicine Genome Sequencing Center** http://genome.wustl.edu/
- **Baylor College of Medicine Human Genome Sequencing Center**
  http://www.hgsc.bcm.tmc.edu/
- **Sanger Institute** http://www.sanger.ac.uk/
- **DOE Joint Genomics Institute** http://www.jgi.doe.gov/

4. Relationships of Genes Within and Between Databases

Gene products may be related by virtue of sharing a common evolutionary origin, sharing a common function, or participating in the same pathway.

BLAST: Identification of Sequence Similarities Evidence for a common evolutionary origin comes from the identification of sequence similarities between two or more sequences. One of the most important tools for identifying such similarities is BLAST (Basic Local Alignment Search Tool), which was developed by NCBI. BLAST is really a suite of related programs and databases in which local matches between long stretches of sequence can be identified and ranked. A query for similar DNA or protein sequences through BLAST is one of the first things that a researcher does with a newly sequenced gene. Different sequence databases can be accessed and organized by type of sequence (reference genome, recent updates, nonredundant, ESTs, etc.), and a particular species or taxonomic group can be specified. One BLAST routine matches a query nucleotide sequence translated in all six frames to a protein sequence database. Another matches a protein query sequence to the six-frame translation of a nucleotide sequence database. Other BLAST routines are customized to identify short sequence pattern matches or pairwise alignments, to screen genome-sized DNA segments, and so forth, and can be accessed through the same top-level page:

- **NCBI-BLAST** http://www.ncbi.nlm.nih.gov/BLAST/

Function Ontology Databases Another approach to developing relationships among gene products is by assigning these products to functional roles based on experimental evidence or prediction. Having a common way of describing these roles, regardless of the experimental system, is then of great importance. A group of scientists from different databases are working together to develop a common set of hierarchically arranged terms—an ontology—for *function* (biochemical event), *process* (the cellular event to which a protein contributes), and *subcellular location* (where a product is located in a cell) as a way of describing the activities of a gene product. This particular ontology is called the Gene Ontology (GO), and many different databases of gene products now incorporate GO terms. A full description can be found at:

- http://www.geneontology.org/

Pathway Databases Still another way to relate products to one another is by assigning them to steps in biochemical or cellular pathways. Pathway diagrams can be used as organized ways of presenting relationships of these products to one another. Some of the more advanced attempts at producing such pathway databases include Kyoto Encyclopedia of Genes and Genomes (KEGG), Signal Transduction Database (TRANSPATH), and What Is There—Interactive Metabolic Database (WIT):

- **KEGG** http://www.genome.ad.jp/kegg/
- **TRANSPATH** http://transpath.gbf.de/
- **WIT** http://wit.mcs.anl.gov/WIT2/

Glossary

A   *See* adenine; adenosine.

A site   *See* aminoacyl-tRNA binding site.

Ac* element**   *See* ***Activator element.

accessory protein   A protein associated with DNA polymerase III of *E. coli* that is not part of the catalytic core.

acentric chromosome   A chromosome having no centromere.

acentric fragment   A chromosome fragment having no centromere.

acridine orange   A mutagen that tends to produce frameshift mutations.

acrocentric chromosome   A chromosome having the centromere located slightly nearer one end than the other.

activated transcription   In eukaryotes, a higher rate of transcription than basal transcription requiring the binding of transcription factors to cis-regulatory DNA elements.

activator   A protein that, when bound to a cis-regulatory DNA element such as an operator or an enhancer, activates transcription from an adjacent promoter.

***Activator* (*Ac*) element**   A class 2 DNA transposable element so named by its discoverer Barbara McClintock because it is required to activate chromosome breakage at the *Dissociation* (*Ds*) locus.

active site   The part of a protein that must be maintained in a specific shape if the protein is to be functional—for example, in an enzyme, the part to which the substrate binds.

adaptation   In the evolutionary sense, some heritable feature of an individual's phenotype that improves its chances of survival and reproduction in the existing environment.

adaptive landscape   A surface plotted in a three-dimensional graph with all possible combinations of allele frequencies for different loci plotted in the plane and with mean fitness for each combination plotted in the third dimension.

adaptive peak   A high point (perhaps one of several) on an adaptive landscape. Selection tends to drive the genotype composition of the population toward a genotype corresponding to an adaptive peak.

adaptive surface   *See* adaptive landscape.

addition   *See* indel mutation.

additive effect   A difference in phenotype resulting from the substitution of an *a* allele for an *A* allele when averaged over all the individual members of a population.

additive genetic variance   Genetic variance associated with the average effects of substituting one allele for another.

adenine (A)   A purine base that pairs with thymine in the DNA double helix.

adenine methylase   An enzyme that methylates adenines before replication, thereby distinguishing old strands from the newly synthesized complementary strands.

adenosine (A)   A nucleoside containing adenine as its base.

adenosine triphosphate   *See* ATP.

adjacent segregation   In a reciprocal translocation, the passage of a translocated and a normal chromosome to each of the poles.

ADP   Adenosine diphosphate.

AFB₁   *See* aflatoxin B₁.

aflatoxin B₁ (AFB₁)   A mutagen that cleaves the bond between guanine and deoxyribose, producing an apurinic site.

alkylating agent   A chemical agent that can add alkyl groups (for example, ethyl or methyl groups) to another molecule; many mutagens act through alkylation.

allele   One of the different forms of a gene that can exist at a single locus.

allele frequency   A measure of the commonness of an allele in a population; the proportion of all alleles of that gene in the population that are of this specific type.

allelic series   The set of known alleles of one gene.

allopatric   Belonging to populations that are separated in space and that do not exchange genes.

allopatric speciation   The formation of new species from populations that are geographically isolated from one another.

allopolyploid   *See* amphidiploid.

allosteric effector   A small molecule that binds to an allosteric site.

allosteric site   A site on a protein to which a small molecule binds, causing a change in the conformation of the protein that modifies the activity of its active site.

allosteric transition   A change from one conformation of a protein to another.

alternate segregation   In a reciprocal translocation, the passage of both normal chromosomes to one pole and both translocated chromosomes to the other pole.

alternative splicing   A process by which different messenger RNAs are produced from the same primary transcript, through variations in the splicing pattern of the transcript. Multiple mRNA "isoforms" can be produced in a single cell or the different isoforms can display different tissue-specific patterns of expression. If the alternative exons fall within the open reading frames of the mRNA isoforms, different proteins will be produced by the alternative mRNAs.

Alu   A short transposable element that makes up more than 10 percent of the human genome. *Alu* elements are class 2 retroelements that do not encode proteins and as such are nonautonomous elements.

amber codon   The codon UAG, a nonsense codon.

amino acid   A peptide; the basic building block of proteins (or polypeptides).

amino end   The end of a protein having a free amino group. A protein is synthesized from the amino end at the 5′ end of the mRNA to the carboxyl end near the 3′ end of the mRNA during translation.

aminoacyl-tRNA binding site   A site in the ribosome that binds incoming aminoacyl-tRNAs. The anticodon of each incoming aminoacyl-tRNA matches the codon of the mRNA. Also called the A site.

aminoacyl-tRNA synthetase   An enzyme that attaches an amino acid to a tRNA before its use in translation. There are 20 different aminoacyl-tRNAs, one for each amino acid.

amniocentesis   A technique for testing the genotype of an embryo or fetus in utero with minimal risk to the mother or the child.

AMP   Adenosine monophosphate.

amphidiploid   An allopolyploid; a polyploid formed from the union of two separate chromosome sets and their subsequent doubling.

amplification   The production of many DNA copies from one master region of DNA.

analysis of variance   A statistical methodology that assigns the proportion of variance in a population to different causes and their interactions.

anaphase   An intermediate stage of nuclear division during which chromosomes are pulled to the poles of the cell.

anaphase bridge   In a dicentric chromosome, the segment between centromeres being drawn to opposite poles at nuclear division.

anchor cell   A cell contributing to the development of the vulva in the nematode *Caenorhabditis elegans*. The anchor cell secretes an EGF-like ligand that induces neighboring cells of the vulva equivalence group to adopt different fates.

aneuploid cell   A cell having a chromosome number that differs from the normal chromosome number for the species by a small number of chromosomes.

annealing   Spontaneous alignment of two single DNA strands to form a double helix.

antibody   A protein (immunoglobulin) molecule, produced by the immune system, that recognizes a particular substance (antigen) and binds to it.

anticodon   A nucleotide triplet in a tRNA molecule that aligns with a particular codon in mRNA under the influence of the ribosome; the amino acid carried by the tRNA is inserted into a growing protein chain.

antiparallel   A term used to describe the opposite orientations of the two strands of a DNA double helix; the 5′ end of one strand aligns with the 3′ end of the other strand.

apoptosis   The cellular pathways responsible for cell death and subsequent removal of the remains of the dead cell. Cell death may be induced by intracellular damage or by signals from neighboring cells.

apurinic site   A DNA site that has lost a purine residue.

apyrimidinic site   A DNA site that has lost a pyrimidine residue.

ARS   *See* **autonomous replication site.**

artificial selection   Breeding of successive generations by the deliberate human selection of certain phenotypes or genotypes as the parents of each generation.

ascospore   A sexual spore from certain fungus species in which spores are found in a sac called an ascus.

ascus   In a fungus, a sac that encloses a tetrad or an octad of ascospores.

ATP (adenosine triphosphate)   The "energy molecule" of cells, synthesized mainly in mitochondria and chloroplasts; energy from the breakdown of ATP drives many important cell reactions.

attachment site   Region at which prophage integrates.

autonomous phenotype   A genetic trait in multicellular organisms in which only genotypically mutant cells exhibit the mutant phenotype. Conversely, a *nonautonomous* trait is one in which genotypically mutant cells cause other cells (regardless of their genotype) to exhibit a mutant phenotype.

autonomous replication sequence (ARS)   A segment of a DNA molecule needed for the initiation of its replication; generally a site recognized and bound by the proteins of the replication system.

autonomous transposable element   A transposable element that encodes the protein(s) (for example, transposase or reverse transcriptase) necessary for its transposition and for the transposition of nonautonomous elements in the same family.

autopolyploid   A polyploid formed from the doubling of a single genome.

autoradiogram   A pattern of dark spots in a developed photographic film or emulsion in the technique of autoradiography.

autoradiography   A process in which radioactive materials are incorporated into cell structures, which are then placed next to a film or photographic emulsion, thus forming a pattern on the film (an autoradiogram) corresponding to the location of the radioactive compounds within the cell.

autosome   Any chromosome that is not a sex chromosome.

auxotroph   A strain of microorganisms that will proliferate only when the medium is supplemented with a specific substance not required by wild-type organisms (*compare* **prototroph**).

BAC   Bacterial artificial chromosome; an F plasmid engineered to act as a cloning vector that can carry large inserts.

back mutation   *See* **reversion.**

bacterial artificial chromosome   *See* **BAC.**

bacteriophage (phage)   A virus that infects bacteria.

balanced rearrangements   Those in which there is no net loss or gain of genetic material.

balancer   A chromosome with multiple inversions, used to retain favorable allele combinations in the uninverted homolog.

balancing selection   Natural selection that results in a stable intermediate equilibrium of allele frequencies.

bands   Transverse stripes on chromosomes revealed by various stains.

Barr body   A densely staining mass that represents an inactivated X chromosome.

basal transcription apparatus   Contains RNA polymerase II and associated proteins.

base analog   A chemical whose molecular structure mimics that of a DNA base; because of the mimicry, the analog may act as a mutagen.

base-excision repair   One of several excision-repair pathways. In this pathway, subtle base-pair distortions are repaired by the creation of apurinic sites followed by repair synthesis.

base-pair substitution   *See* **nucleotide-pair substitution.**

bimodal distribution   A statistical distribution having two modes.

binary fate decision   A developmental decision in which the presence or absence of a specific gene product determines which of two opposing fates a cell or group of cells adopts.

binary fission   A process in which a parent cell splits into two daughter cells of approximately equal size.

bioinformatics   Computational information systems and analytical methods applied to biological problems such as genomic analysis.

bivalents   Two homologous chromosomes paired at meiosis.

blastoderm   In an insect embryo, the stage in which there is a single layer of nuclei (*syncitial blastoderm*) or cells (*cellular blastoderm*) surrounding the central yolk.

blastula   An early developmental stage of lower vertebrate embryos, in which the embryo consists of a single layer of cells surrounding the central yolk.

branch migration   A process by which a single "invading" DNA strand extends its partial pairing with its complementary strand as it displaces the resident strand.

broad heritability (H^2)   The proportion of total phenotypic variance at the population level that is contributed by genetic variance.

C   *See* **cytosine; cytidine.**

CAF-1 (chromatin activator protein)   A protein that binds to histones and delivers them to the replication fork for their incorporation into new nucleosomes.

cAMP   *See* **cyclic adenosine monophosphate.**

canalized character   A characters that is constant in development despite environmental and genetic disturbances.

cancer   A class of disease characterized by the rapid and uncontrolled proliferation of cells within a tissue of a multitissued eukaryote. Cancers are generally thought to be genetic diseases of somatic cells, arising through sequential mutations that create oncogenes and inactivate tumor-suppressor genes.

candidate gene   A sequenced gene of previously unknown function that, because of its chromosomal position or some other property, becomes a candidate for a particular function such as disease determination.

candidate-gene approach   An approach to the molecular identification of a gene (and its encoded gene products) defined by one or more mutant alleles of that gene, based on plausible molecular functions, gene expression patterns, or other properties that might account for the mutant phenotypes or disease syndrome.

CAP   *See* **catabolite activator protein.**

cap   A special structure, consisting of a 7-methylguanosine residue linked to the transcript by three phosphate groups, that is added in the nucleus to the 5′ end of eukaryotic mRNAs. The cap protects an mRNA from degradation and is required for translation of the mRNA in the cytoplasm.

carboxyl end   The end of a protein having a free carboxyl group. The carboxyl end is encoded by the 3′ end of the mRNA and is the last part of the protein to be synthesized in translation.

carboxyl tail domain (CTD)   The protein tail of the B subunit of RNA polymerase II; it coordinates the processing of eukaryotic pre-mRNAs including capping, splicing, and termination.

carcinogen   A substance that causes cancer.

cardinal genes   Those pattern-formation genes in *Drosophila* that are the zygotically acting genes directly responding to the gradients of the anterior–posterior and dorsal–ventral positional information created by the maternally expressed pattern-formation genes. *See also* **gap gene.**

carrier   An individual organism that possesses a mutant allele but does not express it in the phenotype, because of a dominant allelic partner; thus, an individual of genotype *A/a* is a carrier of *a* if there is complete dominance of *A* over *a.*

caspase   A member of a family of related proteases that take part in initiating and carrying out the cell-death response.

catabolite activator protein   A protein that unites with cAMP at low glucose concentrations and binds to the *lac* promoter to facilitate RNA polymerase action.

catabolite repression   The inactivation of an operon caused by the presence of large amounts of the metabolic end product of the operon.

CDK   *See* **cyclin-dependent protein kinase.**

CDK–cyclin complex   Any of a series of heterodimeric protein complexes that are activated during specific segments of the cell cycle and that phosphorylate specific residues on complex-specific target proteins.

cDNA   *See* **complementary DNA.**

cDNA library   A library composed of cDNAs, not necessarily representing all mRNAs.

cell cycle   A set of events that take place in the divisions of mitotic cells. The cell cycle oscillates between mitosis (M phase) and interphase. Interphase can be subdivided in order into G_1, S phase, and G_2. DNA synthesis takes place in S phase. The length of the cell cycle is regulated through a special option in G_1, in which G_1 cells can enter a resting phase called G_0.

cell division   A process by which two cells are formed from one.

cell fate   The ultimate differentiated state to which a cell has become committed.

cell–cell communication   The communication of physiological or developmental information between cells, mediated by cell-surface or secreted ligands from signaling cells that interact with specific receptors on target cells.

centimorgan (cM)   *See* **map unit.**

central tendency   The value of a continuous variable around which the distribution of individual values cluster.

centromere   A specialized region of DNA on each eukaryotic chromosome that acts as a site for the binding of kinetochore proteins.

change in phase   A term used by McClintock to describe the situation whereby a transposable element (such as Ac) is reversibly inactivated.

chaperone   A protein that assists in the correct folding of newly synthesized proteins. Chaperones are conserved in all organisms from bacteria to plants to humans.

chaperonin folding machine   A multisubunit protein complex in which a newly synthesized protein folds into its active conformation.

character   An attribute of individual members of a species for which various heritable differences can be defined.

character difference   Alternative forms of the same attribute within a species.

charged tRNA   A transfer RNA molecule with an amino acid attached to its 3′ end. Also called aminoacyl-tRNA.

chase   *See* **pulse-chase experiment.**

checkpoint   A stage of the cell cycle at which the completion of a certain event of the cell cycle, such as chromosome replication, must have been successfully completed in order for the cell cycle to progress to the next stage.

chemical genetics   The use of libraries of small chemical ligands to interfere specifically with individual protein functions, thereby phenocopying the effects of mutations in the genes encoding these functions.

chemigenomics   The use of synthetic chemical libraries to screen for specific molecules that act as protein-specific inhibitors or activators, thereby producing cellular phenocopies of mutant phenotypes.

chiasma (plural, **chiasmata**)   A cross-shaped structure commonly observed between nonsister chromatids in meiosis; the site of crossing-over.

chimera   *See* **mosaic.**

chi-square (χ^2) test   A statistical test used to determine the probability of obtaining observed proportions by chance, under a specific hypothesis.

chorionic villus sampling (CVS)   A placental sampling procedure for obtaining fetal tissue for chromosome and DNA analysis to assist in prenatal diagnosis of genetic disorders.

chromatid   One of the two side-by-side replicas produced by chromosome division.

chromatid inference   A situation in which the occurrence of a crossover between any two nonsister chromatids can be shown to affect the probability of those chromatids taking part in other crossovers in the same meiosis.

chromatin   The substance of chromosomes; now known to include DNA, chromosomal proteins, and chromosomal RNA.

chromatin remodeling   Changes in nucleosome position along DNA.

chromocenter   The point at which the polytene chromosomes appears to be attached together.

chromomere   A small beadlike structure visible on a chromosome during prophase of meiosis and mitosis.

chromosome   A linear end-to-end arrangement of genes and other DNA, sometimes with associated protein and RNA.

chromosome aberration   Any type of change in chromosome structure or number.

chromosome bands   Transverse stripes on the chromosomes of many organisms, revealed by special staining procedures.

chromosome mutation   Any type of change in chromosome structure or number.

chromosome painting   The use of a battery of fluorescently labeled in situ hybridization probes such that the probes for each chromosome contain compounds with distinct flourescence properties. After hybridization, each chromosome fluoresces in a different color.

chromosome rearrangement   A chromosome mutation in which chromosome parts are in new juxtaposition.

chromosome set   The group of different chromosomes that carries the basic set of genetic information of a particular species.

chromosome theory of inheritance   A unifying theory stating that inheritance patterns may be generally explained by assuming that genes are located in specific sites on chromosomes.

chromosome walking   A method for the dissection of large segments of DNA, in which a cloned segment of DNA, usually eukaryotic, is used to screen recombinant DNA clones from the same genome bank for other clones containing neighboring sequences.

cis conformation   In a heterozygote having two mutant sites within a gene or within a gene cluster, the arrangement A_1A_2/a_1a_2.

cis dominance   The ability of a gene to affect genes next to it on the same chromosome.

cis-acting element   A site on a DNA (or RNA) molecule that functions as a binding site for a sequence-specific DNA- (or RNA-) binding protein. The term *cis-acting* indicates that protein binding to this site affects only nearby DNA (or RNA) sequences on the same molecule.

cis-regulatory element   *See* **cis-acting element.**

cis–trans test   A test for determining whether two mutant sites of a gene are in the same functional unit or gene.

class 1 element   A transposable element that moves through an RNA intermediate. Also called an *RNA element*.

class 2 element   A transposable element that moves directly from one site in the genome to another.

clone   (1) A group of genetically identical cells or individual organisms derived by asexual division from a common ancestor. (2) *(colloquial)* An individual organism formed by some a sexual process so that it is genetically identical with its "parent." (3) *See* **DNA clone.**

clone contig   *See* **contig.**

cloning   (1) In recombinant DNA research, the process of creating and amplifying specific DNA segments. (2) The production of genetically identical organisms from the somatic cells of an individual organism.

cloning vector   In cloning, the plasmid or phage chromosome used to carry the cloned DNA segment.

cM (centimorgan)   *See* **map unit.**

coactivator   A special class of eukaryotic regulatory complex that serves as a bridge to bring together regulatory proteins and RNA polymerase II.

coding strand   The nontemplate strand of a DNA molecule having the same sequence as that in the RNA transcript.

codominance   A situation in which a heterozygote shows the phenotypic effects of both alelles equally.

codon   A section of DNA (three nucleotide pairs in length) that encodes a single amino acid.

coefficient of coincidence   The ratio of the observed number of double recombinants to the expected number.

cointegrate   The product of the fusion of two circular transposable elements to form a single, larger circle in replicative transposition.

colinearity   The correspondence between the location of a mutant site within a gene and the location of an amino acid substitution within the polypeptide translated from that gene.

colony   A visible clone of cells.

combinatorial interaction   Interactions between different combinations of transcription factors that result in distinct patterns of gene expression.

comparative genomics   Analysis of the relations of the genome sequences of two or more species.

complementary (base pairs)   "Lock and key" fit of shape and charge between adenine and thymine and between guanine and cytosine.

complementary DNA (cDNA)   Synthetic DNA transcribed from a specific RNA through the action of the enzyme reverse transcriptase.

complementation   The production of a wild-type phenotype when two different mutations are combined in a diploid or a heterokaryon.

complementation test   *See* **cis–trans test.**

composite transposon   A type of bacterial transposable element containing a variety of genes that reside between two nearly identical insertion sequence (IS) elements.

conditional mutation   A mutation that has the wild-type phenotype under certain (permissive) environmental conditions and a mutant phenotype under other (restrictive) conditions.

conjugation   The union of two bacterial cells during which chromosomal material is transferred from the donor to the recipient cell.

conjugation tube   The tube joining two bacteria that permits the transfer of DNA.

consensus sequence   The inferred most likely real sequence of a segment of DNA based on multiple sequence reads of the same segment.

conservative replication   A disproved model of DNA synthesis suggesting that one-half of the daughter DNA molecules should have both strands composed of newly polymerized nucleotides.

conservative substitution   Nucleotide-pair substitution within a protein-coding region that leads to replacement of an amino acid with one of similar chemical properties.

conservative transposition   A mechanism of transposition that moves the mobile element to a new location in the genome as it removes it from its previous location.

constitutive   Refers to a biological function that is always in one state, regardless of environmental conditions. (1) In regard to gene expression, it refers to a gene that is always expressed. (2) In regard to chromosome structure, it refers to parts of the chromosome that are always heterochromatic.

contig   A set of ordered overlapping clones that constitute a chromosomal region or a genome.

continuous variation   Variation showing an unbroken range of phenotypic values.

cooperativity   The requirement for proteins to interact to function. Cooperativity of transcription factors is a feature of eukaryotic gene regulation.

coordinately controlled genes   Genes whose products are simultaneously activated or repressed in parallel.

copia-like element   A transposable element (retrotransposon) of *Drosophila* that is flanked by long terminal repeats and typically encodes a reverse transcriptase.

core enzyme   As used in Chapter 8, all of the subunits of RNA polymerase except sigma.

correction   The production (possibly by excision and repair) of a properly paired nucleotide pair from a sequence of hybrid DNA that contains an illegitimate pair.

correlation   The tendency of one variable to vary in proportion to another variable, either positively or negatively.

correlation coefficient A statistical measure of the extent to which variations in one variable are related to variations in another.

cosegregation Parallel inheritance of two genes due to their close linkage on a chromosome.

cosmid A cloning vector that can replicate autonomously like a plasmid and be packaged into a phage.

cosuppression An epigenetic phenomenon whereby a trangene becomes reversibly inactivated along with the gene copy in the chromosome.

cotranscriptional processing The simultaneous transcription and processing of eukaryotic pre-mRNA.

cotransduction The simultaneous transduction of two bacterial marker genes.

cotransformation The simultaneous transformation of two bacterial marker genes.

covariance A statistical measure used in computing the correlation coefficient between two variables; the covariance is the mean of $(x - \bar{x})$ $(y - \bar{y})$ over all pairs of values for the variables x and y, where $\bar{x}$ is the mean of the x values and $\bar{y}$ is the mean of the y values.

cpDNA Chloroplast DNA.

cross The deliberate mating of two parental types of organisms in genetic analysis.

crossing-over The exchange of corresponding chromosome parts between homologs by breakage and reunion.

crossover products Meiotic product cells with chromosomes that have engaged in a crossover.

CTD *See* **carboxyl tail domain.**

culture Tissue or cells multiplying by asexual division and grown for experimentation.

"cut and paste" A descriptive term for a transposition mechanism in which a class 2 (DNA) transposon is excised (cut) from the donor site and inserted (pasted) into a new target site.

C-value The DNA content of a genome.

C-value paradox The observation that sometimes evolutionarily closely related (and phenotypically very similar) species can have DNA contents that are dramatically different.

CVS *See* **chorionic villus sampling.**

cyclic adenosine monophosphate (cAMP) A molecule containing a diester bond between the 3′ and 5′ carbon atoms of the ribose part of the nucleotide. This modified nucleotide cannot be incorporated into DNA or RNA. It plays a key role as an intracellular signal in the regulation of various processes.

cyclin A member of a family of labile proteins that are synthesized and degraded at specific times within each cell cycle and regulate the progression of the cell cycle through their interactions with specific cyclin-dependent protein kinases.

cyclin-dependent protein kinase (CDK) A member of a family of protein kinases that, on activation by cyclins and an elaborate set of positive and negative regulatory proteins, phosphorylate certain transcription factors whose activity is necessary for a particular stage of the cell cycle.

cytidine (C) A nucleoside containing cytosine as its base.

cytogenetics The cytological approach to genetics, mainly consisting of microscopic studies of chromosomes.

cytohet A cell containing two genetically distinct types of a specific organelle.

cytoplasm The material between the nuclear and the cell membranes; includes fluid (cytosol), organelles, and various membranes.

cytoplasmic inheritance Inheritance through genes found in cytoplasmic organelles.

cytoplasmic segregation Segregation in which genetically different daughter cells arise from a progenitor that is a cytohet.

cytosine (C) A pyrimidine base that pairs with guanine.

cytoskeleton The protein cable systems and associated proteins that together form the architecture of a eukaryotic cell.

Darwinian fitness The relative probability of survival and reproduction for a genotype.

daughter cells Two identical cells formed by the asexual division of a cell.

daughter molecule One of the two products of DNA replication composed of one template strand and one newly synthesized strand.

deamination The removal of amino groups from compounds. Deamination of cytosine produces uracil. If this deamination is not repaired, C-to-T transitions can occur.

decoding center The region in the small ribosomal subunit where the decision is made whether an aminoacyl-tRNA can bind in the A site. This decision is based on complementarity between the anticodon of the tRNA and the codon of the mRNA.

degenerate code A genetic code in which some amino acids may be encoded by more than one codon each.

deletion The removal of a chromosomal segment from a chromosome set.

deletion loop The loop formed at meiosis by the pairing of a normal chromosome and a deletion-containing chromosome.

deletion mapping The use of a set of known deletions to map new recessive mutations by pseudodominance.

deletion mutation *See* **indel mutation.**

denaturation The separation of the two strands of a DNA double helix or the severe disruption of the structure of any complex molecule without breakage of the major bonds of its chains.

deoxyribonucleic acid *See* **DNA.**

deoxyribose The pentose sugar in the DNA backbone.

determinant A spatially localized molecule that causes cells to adopt a particular fate or set of related fates.

development The process whereby a single cell becomes a differentiated organism.

developmental field A group of developmentally equivalent cells that together organize and form a particular part of the body plan.

developmental noise Variation in the outcome of development as a consequence of random events in cell division, cell movement, and small differences in the number and location of molecules within cells.

developmental pathway A chain of molecular events that take a set of equivalent cells and produce the assignment of different fates among those cells.

dicentric bridge In a dicentric chromosome, the segment between centromeres being drawn to opposite poles at nuclear division.

dicentric chromosome A chromosome with two centromeres.

dideoxy sequencing The most popular method of DNA sequencing. It uses dideoxynucleotide triphosphates mixed with standard nucleotide triphosphates to produce a ladder of DNA strands whose synthesis is blocked at different lengths. This method has been incorporated into automated DNA-synthesis machines. Also called Sanger sequencing after its inventor, Sir Fred Sanger.

differentiation The changes in cell shape and physiology associated with the production of the final cell types of a particular organ or tissue.

dihybrid A double heterozygote such as $A/a \cdot B/b$.

dihybrid cross A cross between two individuals identically heterozygous at two loci—for example, $A B/a b \times A B/a b$.

dimorphism A polymorphism with only two forms.

dioecious plant   A plant species in which male and female organs are on separate plants.

diploid   A cell having two chromosome sets or an individual organism having two chromosome sets in each of its cells.

directed mutagenesis   Altering some specific part of a cloned gene and reintroducing the modified gene back into the organism.

directional selection   Selection that changes the frequency of an allele in a constant direction, either toward or away from fixation for that allele.

discontinuous variation   Variation having distinct classes of phenotypes for a particular character.

disomic   An abnormal haploid carrying two copies of one chromosome.

dispersion   The variation of values of some variable in a population or sample.

dispersive replication   A disproved model of DNA synthesis suggesting more or less random interspersion of parental and new segments in daughter DNA molecules.

***Dissociation* (*Ds*) element**   A nonautonomous transposable element named by Barbara McClintock for its ability to break chromosome 9 of maize but only in the presence another element called *Activator* (*Ac*).

distribution   *See* **statistical distribution.**

distribution function   A graph of some precise quantitative measure of a character against its frequency of occurrence.

DNA (deoxyribonucleic acid)   A double chain of linked nucleotides (having deoxyribose as their sugars); the fundamental substance of which genes are composed.

DnaA   *E. coli* DNA replication begins when the DnaA protein binds to a 13-bp sequence (called the DnaA box) at the origin of replication.

DnaB   An *E. coli* protein that unzips the double helix at the replication fork. DnaB is known as a DNA helicase.

DNA clone   A section of DNA that has been inserted into a vector molecule, such as a plasmid or a phage chromosome, and then replicated to form many copies.

DNA element   A class 2 transposable element found in both prokaryotes and eukaryotes and so named because the DNA element participates directly in transposition.

DNA fingerprint   The autoradiographical banding pattern produced when DNA is digested with a restriction enzyme that cuts outside a family of VNTRs (variable number of tandem repeats) and a Southern blot of the electrophoretic gel is probed with a VNTR-specific probe. Unlike true fingerprints, these patterns are not unique to each individual organism.

DNA glycosylase   Enzyme that removes altered bases, leaving apurinic sites.

DNA ligase   An important enzyme in DNA replication and repair that seals the DNA backbone by catalyzing the formation of phosphodiester bonds.

DNA polymerase   An enzyme that can catalyze the synthesis of new DNA strands from a DNA template; several such enzymes exist.

DNA polymorphism   A naturally occurring variation in DNA sequence at a given location in the genome.

DNA technology   The collective techniques for obtaining, amplifying, and manipulating specific DNA fragments.

DNA transposon   *See* **DNA element.**

DNA-binding domain   The site on a DNA-binding protein that directly interacts with specific DNA sequences.

domain   A region of a protein associated with a particular function. Some proteins contain more than one domain.

dominance variance   Genetic variance at a single locus attributable to dominance of one allele over another.

dominant allele   An allele that expresses its phenotypic effect even when heterozygous with a recessive allele; thus, if *A* is a dominant over *a*, then *A/A* and *A/a* have the same phenotype.

dominant phenotype   The phenotype of a genotype containing the dominant allele; the parental phenotype that is expressed in a heterozygote.

donor DNA   Any DNA to be used in cloning or in DNA-mediated transformation.

donor organism   An organism that provides DNA for use in recombinant DNA technology or DNA-mediated transformation.

dosage compensation   The process in organisms using a chromosomal sex-determination mechanism (such as XX versus XY) that allows standard structural genes on the sex chromosome to be expressed at the same levels in females and males, regardless of the number of sex chromosomes. In mammals, dosage compensation operates by maintaining only a single active X chromosome in each cell; in *Drosophila*, it operates by hyperactivating the male X chromosome.

double crossover   Two crossovers in a chromosomal region under study.

double helix   The structure of DNA first proposed by Watson and Crick, with two interlocking helices joined by hydrogen bonds between paired bases.

double infection   Infection of a bacterium with two genetically different phages.

double mutant   Genotype with mutant alleles of two different genes.

double transformation   Simultaneous transformation by two different donor markers.

double-strand break   A DNA break cleaving the sugar–phosphate backbones of both strands of the DNA double helix.

Down syndrome   An abnormal human phenotype, including mental retardation, due to a trisomy of chromosome 21; more common in babies born to older mothers.

drift   *See* **random genetic drift.**

duplicate genes   Two identical allele pairs in one diploid individual.

duplication   More than one copy of a particular chromosomal segment in a chromosome set.

dyad   A pair of sister chromatids joined at the centromere, as in the first division of meiosis.

E site   *See* **exit site.**

ectopic expression   The occurrence of gene expression in a tissue in which it is normally not expressed. It can be caused by the juxtaposition of novel enhancer elements with a gene.

ectopic integration   In a transgenic organism, the insertion of an introduced gene at a site other than its usual locus.

EF-G   *See* **elongation factor G.**

EF-Tu   *See* **elongation factor Tu.**

electrophoresis   A technique for separating the components of a mixture of molecules (proteins, DNAs, or RNAs) in an electric field within a gel.

elongation   The stage of transcription that follows initiation and precedes termination.

elongation factor G (EF-G)   A nonribosomal protein that interacts with the A site and promotes translocation of the ribosome.

elongation factor Tu (EF-Tu)   A nonribosomal protein that binds to an aminoacylated tRNA, forming the ternary complex.

embryoid   A small dividing mass of monoploid cells, produced from a cell destined to become a pollen cell by exposing it to cold.

embryonic stem cells (ES cells)   Cultured cell lines that are established from very early embryos and that are essentially totipotent. That is, these cells can be implanted into a host embryo and populate many or all tissues of the developing animal, most importantly including the germ line. Manipulations of these ES cells are used extensively in mouse genetics to produce targeted gene knockouts.

endocrine signal   A hormone (ligand) secreted into the circulatory system from a gland (that is, endocrine organ). This hormone will bind to receptors within a target cell or on its surface.

endogamy   Mating between individuals within a group or subgroup, rather than at random in a population.

endogenote   *See* **merozygote.**

endonuclease   An enzyme that cleaves the phosphodiester bonds within a nucleotide chain.

enforced outbreeding   Deliberate avoidance of mating between relatives.

enhanceosome   The macromolecular assembly responsible for interaction between enhancer elements and the promoter regions of genes.

enhancer element   A cis-regulatory sequence that can elevate levels of transcription from an adjacent promoter. Many tissue-specific enhancers can determine spatial patterns of gene expression in higher eukaryotes. Enhancers can act on promoters over many tens of kilobases of DNA and can be 5′ or 3′ to the promoter that they regulate.

enhancer mutation   A modifier mutation that makes the phenotype of a mutation of another gene more severe.

enhancer trap   A transgenic construction inserted into a chromosome and used to identify tissue-specific enhancers in the genome. In such a construct, a promoter sensitive to enhancer regulation is fused to a reporter gene, such that expression patterns of the reporter gene identify the spatial regulation conferred by nearby enhancers.

enol form   *See* **tautomeric shift.**

environment   A combination of all the conditions external to an organism that could affect the expression of its genome.

environmental variance   The variance due to environmental variation.

enzyme   A protein that functions as a catalyst.

epigenetic inheritance   Heritable modifications in gene function not due to changes in the base sequence of the DNA of the organism. Examples of epigenetic inheritance are paramutation, X-chromosome inactivation, and parental imprinting.

epigenetically silenced   The reversible inactivation of a gene due to chromatin structure.

epimutation   A reversible mutation due to changes in epigenetic regulation rather than to changes in the DNA sequence.

epistasis   A situation in which the differential phenotypic expression of a genotype at one locus depends on the genotype at another locus; a mutation that exerts its expression while canceling the expression of the alleles of another gene.

equal segregation   Equal numbers of progeny genotypes attributable to the separation of two alleles of one gene at meiosis.

equilibrium distribution   An array of genotypic or phenotypic frequencies in a population that remain constant over time.

equivalence group   A set of immature cells that all have the same developmental potential. In many cases, cells of an equivalence group end up adopting different fates.

ES cells   *See* **embryonic stem cell.**

EST   *See* **expressed sequence tag.**

euchromatin   A chromosomal region that stains normally; thought to contain the normally functioning genes.

eugenics   State-controlled human breeding in an attempt to improve future generations.

eukaryote   An organism having eukaryotic cells.

eukaryotic cell   A cell containing a nucleus.

euploid   A cell having any number of complete chromosome sets or an individual organism composed of such cells.

excise   Describes what a transposable element does when it leaves a chromosomal location. Also called transpose.

excision repair   The repair of a DNA lesion by removal of the faulty DNA segment and its replacement with a wild-type segment.

exconjugant   A female bacterial cell that has just been in conjugation with a male and contains a fragment of male DNA.

exit (E) site   The site on the ribosome where the deacylated tRNA can be found.

exogenote   *See* **merozygote.**

exon   Any nonintron section of the coding sequence of a gene; together, the exons correspond to the mRNA that is translated into protein.

exonuclease   An enzyme that cleaves nucleotides one at a time from an end of a polynucleotide chain.

expressed sequence tag (EST)   A sequence-tagged site derived from a cDNA clone; used to position and identify genes in genomic analysis.

expression library   A library in which the vector carries transcriptional signals to allow any cloned insert to produce mRNA and ultimately a protein product.

expression vector   A vector with the appropriate bacterial regulatory regions located 5′ to the insertion site, allowing the transcription and translation of a foreign protein in bacteria.

expressivity   The degree to which a particular genotype is expressed in the phenotype.

F⁻ cell   In *E. coli*, a cell having no fertility factor; a female cell.

F⁺ cell   In *E. coli*, a cell having a free fertility factor; a male cell.

F factor   *See* **fertility factor.**

F′ factor   A fertility factor into which a part of the bacterial chromosome has been incorporated.

F′ plasmid   *See* **F′ factor.**

F₁ generation   The first filial generation, produced by crossing two parental lines.

F₂ generation   The second filial generation, produced by selfing or intercrossing the F₁ generation.

facultative heterochromatin   Heterochromatin located in positions that are composed of euchromatin in other members of the same species or even in the other homolog of a chromosome pair.

familial trait   A trait common to members of a family.

family selection   A breeding technique of selecting a pair on the basis of the average performance of their progeny.

fate refinement   The process by which decisions are fine-tuned so that all necessary cell types are fated in the proper spatial locations and cell numbers.

fertility factor (F factor)   A bacterial episome whose presence confers donor ability (maleness).

fibrous protein   A protein with a linear shape such as the components of hair and muscle.

filial generations   Successive generations of progeny in a controlled series of crosses, starting with two specific parents (the P generation) and selfing or intercrossing the progeny of each new (F₁, F₂, . . .) generation.

filter enrichment   A technique for recovering auxotrophic mutants in filamentous fungi.

fingerprint   The characteristic spot pattern produced by electrophoresis of the polypeptide fragments obtained by denaturing a particular protein with a proteolytic enzyme.

first-division segregation pattern   A linear pattern of spore phenotypes within an ascus for a particular allele pair, produced when the alleles go into separate nuclei at the first meiotic division, showing that no crossover has occurred between the allele pair and the centromere.

FISH   *See* **fluorescence in situ hybridization.**

fixed allele   An allele for which all members of the population under study are homozygous, and so no other alleles for this locus exist in the population.

fluctuation test   A test used in microbes to establish the random nature of mutation or to measure mutation rates.

fluorescence in situ hybridization (FISH)   In situ hybridization with the use of a probe coupled with a fluorescent molecule.

fMet   *See* **formylmethionine.**

foreign DNA   DNA from another organism.

formylmethionine (fMet)   A specialized amino acid that is the very first one incorporated into the polypeptide chain in the synthesis of a protein.

forward genetics   The classical approach to genetic analysis, in which genes are first identified by mutant alleles and mutant phenotypes and later cloned and subjected to molecular analysis.

forward mutation   A mutation that converts a wild-type allele into a mutant allele.

founder effect   A random difference from the parental population in the frequency of a genotype in a new colony that results from a small number of founders.

frameshift mutation   The insertion or deletion of a nucleotide pair or pairs, causing a disruption of the translational reading frame.

frequency histogram   A "step curve" in which the frequencies of various arbitrarily bounded classes are graphed.

frequency-dependent fitness   Fitness differences whose intensity changes with the relative frequency of genotypes in the population.

frequency-independent fitness   Fitness that does not depend on interactions with other members of the same species.

functional complementation   The use of a cloned fragment of wild-type DNA to transform a mutant into wild type; used in identifying a clone containing one specific gene.

functional genomics   The study of the patterns of transcript and protein expression and of molecular interactions at a genomewide level.

functional RNA   An RNA type that plays a role without being translated.

G   *See* **guanine; guanosine.**

G_0 phase   The resting phase that can occur in G_1 of interphase of the cell cycle.

G_1 phase   The part of interphase of the cell cycle that precedes S phase.

G_2 phase   The part of interphase of the cell cycle that follows S phase.

gain-of-function mutation   A mutation that results in a new functional ability for a protein, detectable at the phenotypic level. Gain-of-function mutations are often dominant, because only one copy of the novel gene is needed to produce the new function.

gamete   A specialized haploid cell that fuses with a gamete of the opposite sex or mating type to form a diploid zygote; in mammals, an egg or a sperm.

gametophyte   The haploid gamete-producing stage in the life cycle of plants; prominent and independent in some species but reduced or parasitic in others.

gap gene   In *Drosophila*, a class of cardinal genes that are activated in the zygote in response to the anterior–posterior gradients of positional information. Through regulation of the pair-rule and homeotic genes, the patterns of expression of the various gap-gene products lead to the specification of the correct number and types of body segments. Gap mutations cause the loss of several adjacent body segments.

gel electrophoresis   A method of molecular separation in which DNA, RNA, or proteins are separated in a gel matrix according to molecular size, with the use of an electric field to draw the molecules through the gel in a predetermined direction.

gene   The fundamental physical and functional unit of heredity, which carries information from one generation to the next; a segment of DNA composed of a transcribed region and a regulatory sequence that makes transcription possible.

gene balance   The idea that a normal phenotype requires a 1 : 1 relative proportion of genes in the genome.

gene conversion   A meiotic process of directed change in which one allele directs the conversion of a partner allele into its own form.

gene disruption   Inactivation of a gene by the integration of a specially engineered introduced DNA fragment.

gene dose   The number of copies of a particular gene present in the genome.

gene family   A set of genes in one genome, all descended from the same ancestral gene.

gene frequency   *See* **allele frequency.**

gene fusion   A novel gene produced by juxtaposition of DNA sequences of two separate genes. Gene fusions can be the result of chromosomal rearrangements, transposable-element insertion, or genetic engineering.

gene interaction   The collaboration of several different genes in the production of one phenotypic character (or related group of characters).

gene knockout   Inactivation of a gene by the integration of a specially engineered introduced DNA fragment. In some systems, such inactivation is random, with the use of transgenic constructs that insert at many different locations in the genome. In other systems, it can be carried out in a directed fashion. *See also* **targeted gene knockout.**

gene locus   The specific place on a chromosome where a gene is located.

gene map   A linear designation of mutant sites within a gene, based on the various frequencies of interallelic (intragenic) recombination.

gene mutation   A change in DNA that is entirely contained within a single gene, such as a *point mutation* or an *indel mutation* of several nucleotide pairs.

gene pair   The two copies of a particular type of gene present in a diploid cell (one in each chromosome set).

gene replacement   The insertion of a genetically engineered transgene in place of a resident gene; often achieved by a double crossover.

gene tagging   The use of a piece of foreign DNA or a transposon to tag a gene so that a clone of that gene can be readily identified in a library.

gene therapy   The correction of a genetic deficiency in a cell by the addition of new DNA and its insertion into the genome. Different techniques have the potential to carry out gene therapy only in somatic tissues or, alternatively, by correcting the genetic deficiency in the zygote, thereby correcting the germ line as well.

gene-dosage effect   A change in phenotype caused by an abnormal number of wild-type alleles (observed in chromosomal mutations).

general transcription factor (GTF)   Eukaryotic protein complex that does not take part in RNA synthesis but binds to the promoter region to attract and correctly position RNA polymerase II for transcription initiation.

generalized transduction   The ability of certain phages to transduce any gene in the bacterial chromosome.

genetic code   A set of correspondences between nucleotide-pair triplets in DNA and amino acids in protein.

genetic correlation   Phenotypic similarity between different groups because the groups have inherited the same genes.

genetic dissection   The use of recombination and mutation to piece together the various components of a given biological function.

genetic drift   Change in the frequency of an allele in a population resulting from chance differences in the actual numbers of offspring of different genotypes produced by different individual members.

genetic engineering   The process of producing modified DNA in a test tube and reintroducing that DNA into host organisms.

genetic load   The total set of deleterious alleles in an individual genotype.

genetic marker   An allele used as an experimental probe to keep track of an individual organism, a tissue, a cell, a nucleus, a chromosome, or a gene.

genetic screen   *See* **screen.**

genetic selection   *See* **selective system.**

genetic variance   The phenotypic variance associated with the average difference in phenotype among different genotypes.

genetically modified organism (GMO)   A popular term for a transgenic organism, especially applied to transgenic agricultural organisms.

genetics   (1) The study of genes. (2) The study of inheritance.

genome   The entire complement of genetic material in a chromosome set.

genome project   A large-scale, often multilaboratory effort required to sequence a complex genome.

genome sequencing   A large-scale project to sequence a complete genome.

genomic library   A library encompassing an entire genome.

genomics   The cloning and molecular characterization of entire genomes.

genotype   The specific allelic composition of a cell—either of the entire cell or, more commonly, of a certain gene or a set of genes.

genotype frequency   The proportion in a population of individual members of a particular genotype.

germ layer   One of the primary embryonic cell sheets that are formed as a result of gastrulation. The primary germ layers in higher animals are the endoderm, mesoderm, and ectoderm.

germ line   The cell lineage in a multitissued eukaryote from which the gametes derive.

germinal mutation   Mutations in the cells that are destined to develop into gametes.

germ-line gene therapy   *See* **gene therapy.**

globular protein   A protein with a compact structure, such as an enzyme or an antibody.

GMO   *See* **genetically modified organism.**

gradient   A gradual change in some quantitative property over a specific distance.

growth factors   A class of paracrine signaling molecules, usually secreted polypeptides, that are mitogenic—that is, that promote cell division in cells receiving these signals.

GTF   *See* **general transcription factor.**

GU-AG rule   So named because the GU and AG dinucleotides are almost always at the 5′ and 3′ ends, respectively, of introns, where they are recognized by components of the spliceosome.

guanine (G)   A purine base that pairs with cytosine.

guanosine (G)   A nucleoside containing guanine as its base.

hairpin loop   A single-stranded loop in RNA or DNA formed by complementary hydrogen bonding within a single strand of RNA or DNA.

haploid   A cell having one chromosome set or an organism composed of such cells.

haploinsufficient   Describes a gene that, in a diploid cell, cannot promote wild-type function in only one copy (dose).

haplosufficient   Describes a gene that, in a diploid cell, can promote wild-type function in only one copy (dose).

haplotype   A genetic class described by a sequence of DNA or of genes that are together on the same physical chromosome.

Hardy-Weinberg equilibrium   The stable frequency distribution of genotypes A/A, A/a, and a/a, in the proportions of p^2, $2pq$, and q^2, respectively (where p and q are the frequencies of the alleles A and a), that is a consequence of random mating in the absence of mutation, migration, natural selection, or random drift.

helicase   An enzyme that breaks hydrogen bonds in DNA and unwinds the DNA during movement of the replication fork.

helix-loop-helix protein   *See* **HLH proteins.**

hemizygous gene   A gene present in only one copy in a diploid organism—for example, an X-linked gene in a male mammal.

heredity   The biological similarity of offspring and parents.

heritability in the broad sense (H^2)   *See* **broad heritability.**

heritability in the narrow sense (h^2)   *See* **narrow heritability.**

heritable   A character is heritable if offspring resemble parents as a result of genetic similarity.

hermaphrodite   (1) A plant species in which male and female organs are in the same flower of a single plant (*compare* **monoecious plant**). (2) An animal having both male and female sex organs.

Het-A   A retrotransposon that, with another retrotransposon, *TART*, makes up the telomeres of *Drosophila*.

heterochromatin   Densely staining condensed chromosomal regions, believed to be for the most part genetically inert.

heterodimer   A protein consisting of two nonidentical polypeptide subunits.

heteroduplex   A DNA double helix formed by annealing single strands from different sources; if there is a structural difference between the strands, the heteroduplex may show such abnormalities as loops or buckles.

heteroduplex DNA model   A model that explains both crossing-over and gene conversion by assuming the production of a short stretch of heteroduplex DNA (formed from both parental DNAs) in the vicinity of a chiasma.

heterogametic sex   The sex that has heteromorphic sex chromosomes (for example, XY) and hence produces two different kinds of gametes with respect to the sex chromosomes.

heterokaryon   A culture of cells composed of two different nuclear types in a common cytoplasm.

heteroplasmon   A cell containing a mixture of genetically different cytoplasms and generally different mitochondria or different chloroplasts.

heterozygosity   A measure of the genetic variation in a population; with respect to one locus, stated as the frequency of heterozygotes for that locus.

heterozygote   An individual organism having a heterozygous gene pair.

heterozygous gene pair   A gene pair having different alleles in the two chromosome sets of the diploid individual—for example, A/a or A^1/A^2.

hexaploid   A cell having six chromosome sets or an organism composed of such cells.

high-frequency recombination (Hfr) cell   In *E. coli*, a cell having its fertility factor integrated into the bacterial chromosome; a donor (male) cell.

histone   A type of basic protein that forms the unit around which DNA is coiled in the nucleosomes of eukaryotic chromosomes.

histone code   Refers to the pattern of modification (for example, acetylation, methylation, phosphorylation) of the histone tails that may carry information required for correct chromatin assembly.

histone tail   The end of a histone protein protruding from the core nucleosome and subjected to posttranslational modification. *See also* **histone code.**

HLH (helix-loop-helix) proteins   A family of proteins in which a part of the polypeptide forms two helices separated by a loop (the HLH domain); this structure acts as a sequence-specific DNA-binding domain. HLH proteins are thought to act as transcription factors.

homeobox (homeotic box)   A family of quite similar 180-bp DNA sequences that encode a polypeptide sequence called a homeodomain, a sequence-specific DNA-binding sequence. Although the homeobox was first discovered in all homeotic genes, it is now known to encode a much more widespread DNA-binding motif.

homeodomain   A highly conserved family of sequences, 60 amino acids in length and found within a large number of transcription factors, that can form helix-turn-helix structures and bind DNA in a sequence-specific manner.

homeologous chromosomes   Partly homologous chromosomes, usually indicating some original ancestral homology.

homeosis   The replacement of one body part by another. Homeosis can be caused by environmental factors leading to developmental anomalies or by mutation.

homeotic gene   In higher animals, a gene that controls the fate of segments along the anterior–posterior axis.

homeotic gene complex   One of the tightly linked clusters of genes controlling anterior–posterior segment determination in segmented animals. All these genes encode homeodomain transcription factors.

homeotic mutation   A mutation that leads to the replacement of one part of the body by another.

homogametic sex   The sex with homologous sex chromosomes (for example, XX).

homolog   A member of a pair of homologous chromosomes.

homologous chromosomes   Chromosomes that pair with each other at meiosis or chromosomes in different species that have retained most of the same genes during their evolution from a common ancestor.

homology-dependent repair system   A mechanism of DNA repair that depends on the complementarity or homology of the template strand to the strand being repaired.

homozygosity by descent   Homozygosity that results from the inheritance of two copies of one gene that was present in an ancestor.

homozygote   An individual organism that is homozygous.

homozygous   Refers to the state of carrying a pair of identical alleles at one locus.

homozygous dominant   Refers to a genotype such as *A/A*.

homozygous recessive   Refers to a genotype such as *a/a*.

hormone   A molecule that is secreted by an endocrine organ into the circulatory system and acts as a long-range signaling molecule by activating receptors on or within target cells.

host range   The spectrum of strains of a given bacterial species that can be infected by a given strain of phages.

hot spot   A part of a gene that shows a very high tendency to become a mutant site, either spontaneously or under the action of a particular mutagen.

hybrid   (1) A heterozygote. (2) An individual progeny from any cross between parents of different genotypes.

hybrid dysgenesis   A syndrome of effects including sterility, mutation, chromosome breakage, and male recombination in the hybrid progeny of crosses between certain laboratory and natural isolates of *Drosophila*.

hybrid–inbred method   A method of plant breeding that produces large numbers of genetically identical heterozygous plants by crossing different lines that have been made homozygous by many generations of inbreeding.

hybridize   (1) To form a hybrid by performing a cross. (2) To anneal complementary nucleic acid strands from different sources.

hydrogen bond   A weak bond in which an atom shares an electron with a hydrogen atom; hydrogen bonds are important in the specificity of base pairing in nucleic acids and in the determination of protein shape.

hypermorphic mutation   A mutation that confers a mutant phenotype through the production of more gene product by mutant individuals than is produced by wild-type individuals.

hyperacetylation   An overabundance of acetyl groups attached to certain amino acids of the histone tails. Transcriptionally active chromatin is usually hyperacetylated.

hypha (plural, hyphae)   A threadlike structure (composed of cells attached end to end) that forms the main tissue in many fungus species.

hypoacetylation   An underabundance of acetyl groups on certain amino acids of the histone tails. Transcriptionally inactive chromatin is usually hypoacetylated.

hypomorph   A mutation that confers a mutant phenotype but still retains a low but detectable level of wild-type function.

ICR compound   One of a series of quinacrine–mustard compounds synthesized at the Institute for Cancer Research (Fox Chase, Pennsylvania) that act as *intercalating agents*.

imino form   *See* **tautomeric shift.**

in situ hybridization   The use of a labeled probe to bind to a partly denatured chromosome.

in vitro   In an experimental situation, outside the organism (literally, "in glass").

in vitro mutagenesis   The production of either random or specific mutations in a piece of cloned DNA. Typically, the DNA will then be repackaged and introduced into a cell or an organism to assess the results of the mutagenesis.

in vivo   In a living cell or organism.

inborn error of metabolism   Hereditary metabolic disorder, particularly in reference to human disease.

inbreeding   Mating between relatives.

inbreeding coefficient   The probability of homozygosity that results because the zygote obtains copies of the *same* ancestral gene.

incomplete dominance   A situation in which a heterozygote shows a phenotype quantitatively (but not exactly) intermediate between the corresponding homozygote phenotypes. (Exact intermediacy means no dominance).

indel mutation   A mutation in which one or more nucleotide pairs is added or deleted.

independent assortment   *See* **Mendel's second law.**

inducer   An environmental agent that triggers transcription from an operon.

induction   (1) The relief of repression of a gene or set of genes under negative control. (2) An interaction between two or more cells or tissues that is required for one of those cells or tissues to change its developmental fate.

inductive interaction   The interaction between two groups of cells in which a signal passed from one group of cells causes the cells in the other group to change their developmental state (or fate).

infectious transfer   The rapid transmission of free plasmids (plus any chromosomal genes that they may carry) from donor to recipient cells in a bacterial population.

initiation   The first stage of transcription whose main function is to correctly position RNA polymerase before the elongation stage.

insertion sequence (IS) element   A mobile piece of bacterial DNA (several hundred nucleotide pairs in length) capable of inactivating a gene into which it inserts.

insertional duplication   A duplication in which the extra copy is not adjacent to the normal one.

insertional mutation   A mutation arising by the interruption of a gene by foreign DNA, such as from a transgenic construct or a transposable element.

insertional translocation   The insertion of a segment from one chromosome into another nonhomologous one.

intercalating agent   A mutagen that can insert itself between the stacked bases at the center of the DNA double helix, causing an elevated rate of *indel mutations*.

interference   A measure of the independence of crossovers from each other, calculated by subtracting the coefficient of coincidence from 1.

interphase   The cell-cycle stage between nuclear divisions, when chromosomes are extended and functionally active.

interrupted mating   A technique used to map bacterial genes by determining the sequence in which donor genes enter recipient cells.

interstitial region   The chromosomal region between the centromere and the site of a rearrangement.

intervening sequence   An intron; a segment of largely unknown function within a gene. This segment is initially transcribed, but the transcript is not found in the functional mRNA.

intragenic deletion   A deletion within a gene.

intron   *See* **intervening sequence.**

inversion   A chromosomal mutation consisting of the removal of a chromosome segment, its rotation through 180°, and its reinsertion in the same location.

inversion heterozygote   A diploid with a normal and an inverted homolog.

inversion loop   A loop formed by meiotic pairing of homologs in an inversion heterozygote.

inverted repeat (IR) sequence   A sequence found in identical (but inverted) form—for example, at the opposite ends of a transposon.

IR sequence   *See* **inverted repeat sequence.**

IS element   *See* **insertion sequence element.**

isoaccepting tRNAs   The various types of tRNA molecules that carry a specific amino acid.

isotope   One of several forms of an atom having the same atomic number but differing atomic masses.

junk DNA   The name given to transposable elements by those who believe that they are parasitic DNA and perform no useful function for the host.

karyotype   The entire chromosome complement of an individual organism or cell, as seen during mitotic metaphase.

kb   *See* **kilobase.**

keto form   *See* **tautomeric shift.**

kilobase (kb)   1000 nucleotide pairs.

kinase   An enzyme that adds one or more phosphate groups to its substrates. If the substrates are specific proteins, the enzyme is called a protein kinase.

kinetochore   A complex of proteins to which a nuclear spindle fiber attaches.

Klinefelter syndrome   An abnormal human male phenotype due to an extra X chromosome (XXY).

knockout   Inactivation of one specific gene. Same as *gene disruption*.

lagging strand   In DNA replication, the strand that is synthesized apparently in the 3′-to-5′ direction by the ligation of short fragments synthesized individually in the 5′-to-3′ direction.

λ (lambda) phage   One kind ("species") of temperate bacteriophage.

λdgal   A λ phage carrying a *gal* bacterial gene and defective (d) for some phage function.

lateral inhibition   The signal produced by one cell that prevents adjacent cells from adopting the same developmental fate as that of the first cell.

leader sequence   The sequence at the 5′ end of an mRNA that is not translated into protein.

leading strand   In DNA replication, the strand that is made in the 5′-to-3′ direction by continuous polymerization at the 3′ growing tip.

leaky mutant   A mutant (typically, an auxotroph) that results from a partial rather than a complete inactivation of the wild-type function.

leaky mutation   A mutation that confers a mutant phenotype but still retains a low but detectable level of wild-type function.

least-squares regression line   The straight line that passes through a cluster of points relating a variable x with a variable y such that the sum of the squared deviations of all the y values from the line is as small as possible.

lesion   A damaged area in a gene (a mutant site), a chromosome, or a protein.

lethal gene   A gene whose expression results in the death of the individual organism expressing it.

lethal mutation   *See* **lethal gene.**

library   A collection of DNA clones obtained from one DNA donor.

license   In yeast, the binding of a helicase is said to *license* the replication origin and allow the assembly of the replisome and subsequent DNA replication.

ligand   *See* **ligand–receptor interaction.**

ligand–receptor interaction   The interaction between a molecule (usually of extracellular origin) and a protein on or within a target cell. One type of ligand–receptor interaction can be between steroid hormones and their cytoplasmic or nuclear receptors. Another can be between secreted polypeptide ligands and transmembrane receptors.

ligase   An enzyme that can rejoin a broken phosphodiester bond in a nucleic acid.

LINE   *See* **long interspersed nuclear element.**

linear tetrad   A tetrad that results from the meiotic and postmeiotic nuclear divisions in such a way that sister products remain adjacent to one another (with no passing of nuclei).

linkage   The association of genes on the same chromosome.

linkage disequilibrium   Deviation in the frequency of haplotypes in a population from the frequency expected if the alleles at different loci are associated at random.

linkage equilibrium   Distribution of haplotype frequencies in a population such that the frequencies are the arithmetic products of the frequencies of the alleles at the different loci in the haplotype.

linkage group   A group of genes known to be linked; a chromosome.

linkage map   A chromosome map; an abstract map of chromosomal loci that is based on recombinant frequencies.

locus (plural, **loci**)   *See* **gene locus**.

lod score   Cumulative statistic that summarizes evidence supporting a specific linkage value.

LOH   *See* **loss of heterozygosity**.

long interspersed nuclear element (LINE)   A type of class 1 transposable element that encodes a reverse transcriptase. LINEs are also called non-LTR retrotransposons.

long terminal repeat (LTR)   Sequence identical regions at the 5′ and 3′ ends of retroviruses and *copia*-like elements.

loss-of-function mutation   A mutation that partly or fully eliminates normal gene activity. Hypomorphic and null mutations are loss-of-function mutations.

loss of heterozygosity (LOH)   Mutation, mitotic crossover, or deletion that converts a heterozygous site into a homozygous one.

LTR   *See* **long terminal repeat**.

LTR-retrotransposon   A type of class 1 transposable element that terminates in long terminal repeats and encodes several proteins including reverse transcriptase.

lysis   The rupture and death of a bacterial cell on the release of phage progeny.

lysogen   *See* **lysogenic bacterium**.

lysogenic bacterium   A bacterial cell capable of spontaneous lysis due, for example, to the uncoupling of a prophage from the bacterial chromosome.

M cytotype   The name given to a line of *Drosophila* lacking *P* transposable elements.

M phase   The mitotic phase of the cell cycle.

macromolecule   A large polymer such as DNA, a protein, or a polysaccharide.

major groove   The larger of the two grooves in the DNA double helix.

map unit (m.u.)   The "distance" between two linked gene pairs where 1 percent of the products of meiosis are recombinant; a unit of distance in a linkage map.

mapping function   A formula expressing the relation between distance in a linkage map and recombinant frequency.

marker   *See* **genetic markers**.

maternal effect   The environmental influence of the mother's tissues on the phenotype of the offspring.

maternal-effect gene   A gene that produces an effect only when present in the mother.

maternal inheritance   A type of uniparental inheritance in which all progeny have the genotype and phenotype of the parent acting as the female.

mating types   The equivalent in lower organisms of the sexes in higher organisms; the mating types typically differ only physiologically, not in the physical form.

mature RNA   Eukaryotic mRNA that is fully processed and ready to be exported from the nucleus and translated in the cytoplasm.

Mb   *See* **megabase**.

mean   The arithmetic average.

medium   Any material on (or in) which experimental cultures are grown.

megabase (Mb)   One million nucleotide pairs.

meiocyte   Cell in which meiosis takes place.

meiosis   Two successive nuclear divisions (with corresponding cell divisions) that produce gametes (in animals) or sexual spores (in plants and fungi) that have one-half of the genetic material of the original cell.

meiospore   A cell that is one of the products of meiosis in plants.

meiotic recombination   Recombination from assortment or crossing-over at meiosis.

melting   Denaturation of DNA.

Mendelian ratio   A ratio of progeny phenotypes corresponding to the operation of Mendel's laws.

Mendel's first law   The two members of a gene pair segregate from each other in meiosis; each gamete has an equal probability of obtaining either member of the gene pair.

Mendel's second law   The law of independent assortment; unlinked or distantly linked segregating gene pairs assort independently at meiosis.

merozygote   A partly diploid *E. coli* cell formed from a complete chromosome (the endogenote) plus a fragment (the exogenote).

messenger RNA   *See* **mRNA**.

metabolism   The chemical reactions that take place in a living cell.

metacentric chromosome   A chromosome having its centromere in the middle.

metaphase   An intermediate stage of nuclear division at which chromosomes align along the equatorial plane of the cell.

methylation   Modification of a molecule by the addition of a methyl group.

microarray   Set of clones of all or most genes in a genome deposited on a small glass chip.

microfilament   Part of the smallest-diameter cable system of the cytoskeleton. Microfilament cables are composed of actin polymers.

microsatellite DNA   A type of repetitive DNA consisting of very short repeats such as dinucleotides.

microsatellite marker   A difference in DNA at equivalent locations in two genomes that is due to different repeat lengths of a microsatellite.

microtubule   Part of the largest-diameter cable system of the cytoskeleton. A microtubule is composed of polymerized tubulin subunits forming a hollow tube.

midparent value   The mean of the values of a quantitative phenotype for two specific parents.

minimal medium   Medium containing only inorganic salts, a carbon source, and water.

minimum tiling path   The minimal set of clones (with the least overlap) in a physical map that recapitulate the entirety of a genome.

minisatellite DNA   A type of repetitive DNA sequence consisting of short repeat sequences with a unique common core; used for DNA fingerprinting.

minisatellite marker   Heterozygous locus representing variable number of minisatellite repeats.

minor groove   The smaller of the two grooves in the DNA double helix.

mismatch-repair system   A system for repairing damage to DNA that has already been replicated.

missense mutation   Nucleotide-pair substitution within a protein-coding region that leads to replacement of one amino acid by another amino acid.

mitogen   *See* **growth factors**.

mitosis   A type of nuclear division (occurring at cell division) that produces two daughter nuclei identical with the parent nucleus.

mitotic crossover   A crossover resulting from the pairing of homologs in a mitotic diploid.

mixed infection   The infection of a bacterial culture with two different phage genotypes.

mobile genetic element   *See* **transposable element**.

mode   The single class in a statistical distribution having the greatest frequency.

model organism A species chosen for use in studies of genetics because it is well suited to the study of one or more genetic processes.

modifier gene A gene that affects the phenotypic expression of another gene.

modularity A feature of eukaryotic gene regulation in which the interaction of different combinations of a small set of transcription factors effect diverse patterns of transcription.

molecular clock The constant rate of substitution of amino acids in proteins or nucleotides in nucleic acids over long evolutionary time.

molecular genetics The study of the molecular processes underlying gene structure and function.

molecular marker A site of DNA heterozygosity, not necessarily associated with phenotypic variation, used as a tag for a particular chromosomal locus.

monohybrid A single-locus heterozygote of the type *A/a*.

monohybrid cross A cross between two individuals identically heterozygous at one gene pair—for example, *A/a* × *A/a*.

monoploid A cell having only one chromosome set (usually as an aberration) or an organism composed of such cells.

monosomic A cell or individual organism that is basically diploid but has only one copy of one particular chromosome type and thus has chromosome number $2n + 1$.

morph One form of a genetic polymorphism; the morph can be either a phenotype or a molecular sequence.

morphogen A molecule that can induce the acquisition of different cell fates according to the level of morphogen to which a cell is exposed.

morphological mutation A mutation affecting some aspect of the appearance of an individual organism.

mosaic A chimera; a tissue containing two or more genetically distinct cell types or an individual organism composed of such tissues.

motif A short DNA sequence associated with a particular functional role.

mRNA (messenger RNA) An RNA molecule transcribed from the DNA of a gene; a protein is translated from this RNA molecule by the action of ribosomes.

mtDNA Mitochondrial DNA.

m.u. *See* **map unit.**

multigenic deletion A deletion of several adjacent genes.

multimer A protein consisting of two or more subunits.

multiple allelism The existence of several known alleles of a gene.

multiple-cloning site *See* **polylinker.**

multiple-factor hypothesis A hypothesis that explains quantitative variation by assuming the interaction of a large number of genes (polygenes), each with a small additive effect on the character.

mutagen An agent capable of increasing the mutation rate.

mutagenesis An experiment in which experimental organisms are treated with a mutagen and their progeny are examined for specific mutant phenotypes.

mutant An organism or cell carrying a mutation.

mutant allele An allele differing from the allele found in the standard, or wild type.

mutant hunt The process of collecting different mutants showing abnormalities in a certain structure or in a certain function, as preparation for mutational dissection of that function.

mutant rescue *See* **functional complementation.**

mutant site The damaged or altered area within a mutated gene.

mutation (1) The process that produces a gene or a chromosome set differing from that of the wild type. (2) The gene or chromosome set that results from such a process.

mutation event The actual occurrence of a mutation in time and space.

mutation frequency The frequency of mutants in a population.

mutation rate The number of mutation events per gene copy in a population per unit of time (for example, per cell generation).

mutational dissection The study of the components of a biological function through a study of mutations affecting that function.

mutational specificity The constellation of mutational damage that characterizes a particular mutagen.

n The haploid number; the number of chromosomes in the genome.

narrow heritability The proportion of phenotypic variance that can be attributed to additive genetic variance.

nascent Newly synthesized; used to describe proteins.

native Refers to a protein that is folded correctly.

natural selection The differential rate of reproduction of different types in a population as the result of different physiological, anatomical, or behavioral characteristics of the types.

negative assortative mating Preferential mating between phenotypically unlike partners.

negative control Regulation mediated by factors that block or turn off transcription.

negative inbreeding Preferential mating between individuals who are unrelated.

neomorph A mutation with phenotypic effects due to the production of a novel gene product or novel pattern of gene expression.

Neurospora A pink mold, commonly found growing on old food.

neutral DNA sequence variation Variation in DNA sequence that is not under natural selection.

neutral evolution Nonadaptive evolutionary changes due to random genetic drift.

neutral mutation A mutation that has no effect on the survivorship and reproductive rate of the organisms that carry it.

nitrocellulose filter A type of filter used to hold DNA for hybridization.

nitrogen bases Types of molecules that form important parts of nucleic acids, composed of nitrogen-containing ring structures; hydrogen bonds between bases link the two strands of a DNA double helix.

NLS *See* **nuclear localization sequence.**

nonautonomous element A transposable element that relies on the protein products of autonomous elements for its mobility. *Ds* is an example of a nonautonomous transposable element.

nonconservative substitution Nucleotide-pair substitution within a protein-coding region that leads to the replacement of an amino acid by one having different chemical properties.

nondisjunction The failure of homologs (at meiosis) or sister chromatids (at mitosis) to separate properly to opposite poles.

nonlinear tetrad A tetrad in which the meiotic products are in no particular order.

non-Mendelian ratio An unusual ratio of progeny phenotypes that does not conform to the simple operation of Mendel's laws; for example, mutant : wild ratios of 3 : 5, 5 : 3, 6 : 2, or 2 : 6 in tetrads indicate that gene conversion has taken place.

nonnative Refers to a protein that is folded incorrectly.

nonparental ditype (NPD) A tetrad type containing two different genotypes, both of which are recombinant.

nonsense codon A codon for which no formal tRNA molecule exists; the presence of a nonsense codon causes the termination of translation (ending of the polypeptide chain). The three nonsense codons are called amber, ocher, and opal.

nonsense mutation Nucleotide-pair substitution within a protein-coding region that changes a codon for an amino acid into a termination (nonsense) codon.

nonsense suppressor A mutation that produces an altered tRNA that will insert an amino acid in translation in response to a nonsense codon.

nonsynonymous substitution Mutational replacement of an amino acid with one having different chemical properties.

norm of reaction The pattern of phenotypes produced by a given genotype under different environmental conditions.

Northern blotting The transfer of electrophoretically separated RNA molecules from a gel onto an absorbent sheet, which is then immersed in a labeled probe that will bind to the RNA of interest.

NPD *See* **nonparental ditype.**

nuclear localization sequence (NLS) Part of a protein required for its transport from the cytoplasm to the nucleus.

nuclear spindle The set of microtubules that forms between the poles of a cell during nuclear division; the function of the nuclear spindle is to segregate the chromosomes or chromatids to the poles.

nuclease An enzyme that can degrade DNA by breaking its phosphodiester bonds.

nucleoid A DNA mass within a chloroplast or mitochondrion.

nucleolar organizer A region (or regions) of the chromosome set that is physically associated with the nucleolus and contains rRNA genes.

nucleolus (plural, nucleoli) An organelle found in the nucleus, containing rRNA and multiple copies of the genes encoding rRNA.

nucleoside A nitrogen base bound to a sugar molecule.

nucleosome The basic unit of eukaryotic chromosome structure; a ball of eight histone molecules that is wrapped by two coils of DNA.

nucleotide A molecule composed of a nitrogen base, a sugar, and a phosphate group; the basic building block of nucleic acids.

nucleotide pair A pair of nucleotides (one in each strand of DNA) that are joined by hydrogen bonds.

nucleotide excision-repair system An excision-repair pathway that breaks the phosphodiester bonds on either side of the damaged base, removing that base and several on either side followed by repair replication.

nucleotide-pair substitution The replacement of a specific nucelotide pair by a different pair; often mutagenic.

null allele An allele whose effect is the absence either of normal gene product at the molecular level or of normal function at the phenotypic level.

null hypothesis A hypothesis that proposes no difference between two or more data sets.

null mutation A mutation that results in complete absence of function for the gene.

nullisomic Refers to a cell or individual organism with one chromosomal type missing, with a chromosome number such as $n - 1$ or $2n - 2$.

ocher codon The codon UAA, a nonsense codon.

octad An ascus containing eight ascospores, produced in species in which the tetrad normally undergoes a postmeiotic mitotic division.

Okazaki fragment A small segment of single-stranded DNA synthesized as part of the lagging strand in DNA replication.

oligonucleotide A short segment of synthetic DNA.

oncogene A gain-of-function mutation that contributes to the production of a cancer.

oncoprotein The protein product of an oncogene mutation.

one-gene–one-polypeptide hypothesis A mid-twentieth-century hypothesis that originally proposed that each gene (nucleotide sequence) encodes a polypeptide sequence; generally true, with the exception of untranslated functional RNA.

one-gene–one-protein hypothesis A hypothesis proposed by Beadle and Tatum, subsequently refined as the one-gene–one-polypeptide hypthesis.

opal codon The codon UGA, a nonsense codon.

open reading frame (ORF) A gene-sized section of a sequenced piece of DNA that begins with a start codon and ends with a stop codon; it is presumed to be the coding sequence of a gene.

operator A DNA region at one end of an operon that acts as the binding site for a repressor protein.

operon A set of adjacent structural genes whose mRNA is synthesized in one piece, plus the adjacent regulatory signals that affect transcription of the structural genes.

ORC *See* **origin recognition complex.**

ordered clone sequencing The sequencing of a set of clones such as a *minimum tiling path* that corresponds to a genomic segment, a chromosome, or a genome.

ORF *See* **open reading frame.**

organelle A subcellular structure having a specialized function—for example, the mitochondrion, the chloroplast, or the spindle apparatus.

origin of replication The point of a specific sequence at which DNA replication is initiated.

origin recognition complex (ORC) A protein complex that binds to yeast origins of replication.

overdominance A phenotypic relation in which the phenotypic expression of the heterozygote is greater than that of either homozygote.

oxidatively damaged base An origin of spontaneous DNA damage.

P cytotype The name given to a line of *Drosophila* with P transposable elements.

P element A transposable short-inverted-repeat element in *Drosophila* that has been used as a tool for insertional mutagenesis and for germ-line transformation.

P granule A cytoplasmic granule associated with germ-line formation in *C. elegans.*

P site *See* **peptidyl site.**

PAC (P1-based artificial chromosome) A derivative of phage P1 engineered as a cloning vector for carrying large inserts.

pair-rule gene In *Drosophila*, a member of a class of zygotically expressed genes that act at an intermediary stage in the process of establishing the correct numbers of body segments. Pair-rule mutations have half the normal number of segments, owing to the loss of every other segment.

paired-end reads In whole genome shotgun sequence assembly, the DNA sequences corresponding to both ends of a genomic DNA insert in a recombinant clone.

palindrome A DNA sequence that has 180° rotational symmetry.

paracentric inversion An inversion not involving the centromere.

paracrine signal The process by which a secreted molecule binds to a receptor on or within nearby cells, thereby inducing a signal-transduction pathway within the receiving cell.

paralogous genes Two genes in the same species that have evolved by gene duplication.

paramutation An epigenetic phenomenon in plants, in which the genetic activity of a normal allele is heritably reduced by virtue of that allele having been heterozygous with a special "paramutagenic" allele.

parental ditype (PD) A tetrad type containing two different genotypes, both of which are parental.

parental generation The two strains or individual organisms that constitute the start of a genetic breeding experiment; their progeny is the F_1 generation.

parental imprinting An epigenetic phenomenon in which the activity of a gene depends on whether it was inherited from the father or the mother. Some genes are maternally imprinted, others paternally.

parthenogenesis The production of offspring by a female with no genetic contribution from a male.

partial diploid *See* merozygote.

pattern formation The developmental processes resulting in the complex shapes and structures of higher organisms.

PCNA *See* proliferating cell nuclear antigen.

PCR *See* polymerase chain reaction.

PD *See* parental ditype.

pedigree A "family tree," drawn with standard genetic symbols, showing inheritance patterns for specific phenotypic characters.

penetrance The proportion of individuals with a specific genotype who manifest that genotype at the phenotype level.

pentaploid An individual organism with five sets of chromosomes.

peptide *See* amino acid.

peptide bond A bond joining two amino acids.

peptidyl (P) site The site in the ribosome to which a tRNA with the growing polypeptide chain is bound.

peptidyltransferase center The site in the large ribosomal subunit at which the joining of two amino acids is catalyzed.

pericentric inversion An inversion that involves the centromere.

permissive condition An environmental condition under which a conditional mutant shows the wild-type phenotype.

phage *See* bacteriophage.

phage recombinant Phage progeny from a double infection, bearing alleles from both infecting strains.

phenocopy An environmentally induced phenotype that resembles the phenotype produced by a mutation.

phenotype (1) The form taken by some character (or group of characters) in a specific individual. (2) The detectable outward manifestations of a specific genotype.

phenotypic correlation Phenotypic similarity between groups irrespective of whether that similarity is a result of genetic similarity.

Philadelphia chromosome A translocation between the long arms of chromosomes 9 and 22, often found in the white blood cells of patients with chronic myeloid leukemia.

phosphodiester bond A bond between a sugar group and a phosphate group; such bonds form the sugar–phosphate backbone of DNA.

photolyase An enzyme that cleaves thymine dimers.

phyletic evolution The formation of new, related species as a result of the splitting of previously existing species.

physical map The ordered and oriented map of cloned DNA fragments on the genome.

physical mapping Mapping the positions of cloned genomic fragments.

PIC *See* preinitiation complex.

piebald A mammalian phenotype in which patches of skin are unpigmented because of the lack of melanocytes; generally inherited as an autosomal dominant.

pilus (plural, pili) A conjugation tube; a hollow hairlike appendage of a donor *E. coli* cell that acts as a bridge for the transmission of donor DNA to the recipient cell during conjugation.

plaque A clear area on a bacterial lawn, left by lysis of the bacteria through progressive infections by a phage and its descendants.

plasmid Autonomously replicating extrachromosomal DNA molecule.

plate (1) A flat dish used to culture microbes. (2) To spread cells over the surface of solid medium in a plate.

pleiotropic mutation A mutation that affects several different characters.

ploidy The number of chromosome sets.

point mutation A mutation that can be mapped to a specific locus.

Poisson distribution A mathematical expression giving the probability of observing various numbers of a particular event in a sample when the mean probability of an event on any one trial is very small.

poky A slow-growing mitochondrial mutant in *Neurospora*.

polar granule Cytoplasmic granule localized at the posterior end of a *Drosophila* oocyte and early embryo. Polar granules are associated with the germ-line and posterior determinants.

polar mutation A mutation that affects the transcription or translation of the part of the gene or operon on only one side of the mutant site. Examples are nonsense mutations, frameshift mutations, and IS-induced mutations.

pole cell One of the cells that is located at the posterior pole of the early *Drosophila* embryo and will form the *Drosophila* germ line.

poly(A) tail A string of adenine nucleotides added to mRNA after transcription.

polyacrylamide A material used to make electrophoretic gels for the separation of mixtures of macromolecules.

polycistronic mRNA An operon mRNA that codes for more than one protein.

polydactyly More than five fingers or toes or both. Inherited as an autosomal dominant phenotype.

polygenes *See* multiple-factor hypothesis.

polylinker A vector DNA sequence containing multiple unique restriction-enzyme-cut sites that is convenient for inserting foreign DNA. Sometimes also referred to as an MCS (multiple cloning site).

polymerase chain reaction (PCR) An in vitro method for amplifying a specific DNA segment that uses two primers that hybridize to opposite ends of the segment in opposite polarity and, over successive cycles, prime exponential replication of that segment only.

polymerase III holoenzyme (DNA pol III holoenzyme) In *E. coli*, the large multisubunit complex at the replication fork consisting of two catalytic cores and many accessory proteins.

polymorphism The occurrence in a population (or among populations) of several phenotypic forms associated with alleles of one gene or homologs of one chromosome.

polypeptide A chain of linked amino acids; a protein.

polyploid A cell having three or more chromosome sets or an organism composed of such cells.

polysaccharide A biological polymer composed of sugar subunits—for example, starch or cellulose.

polytene chromosome A giant chromosome in specific tissues of some insects, produced by an endomitotic process in which the multiple DNA sets remain bound in a haploid number of chromosomes.

population A group of individuals that mate with each other to produce the next generation.

population genetics The study of the frequencies of different genotypes in populations and the changes in those frequencies that result from patterns of mating, natural selection, mutation, migration, and random chance.

position effect   Describes a situation in which the phenotypic influence of a gene is altered by changes in the position of the gene within the genome.

positional cloning   The identification of the DNA sequences encoding a gene of interest based on knowledge of its genetic or cytogenetic map location.

positional information   The process by which chemical cues that establish cell fate along a geographic axis are established in a developing embryo or tissue primordium.

position-effect variegation   Variegation caused by the inactivation of a gene in some cells through its abnormal juxtaposition with heterochromatin.

positive assortative mating   A situation in which like phenotypes mate more commonly than expected by chance.

positive control   Regulation mediated by a protein that is required for the activation of a transcription unit.

posttranscriptional processing   Modifications of amino acid side groups after a protein has been released from the ribosome.

postzygotic isolation   The failure of two species to exchange genes because the zygotes formed by the mating between them are either inviable or infertile.

preinitiation complex (PIC)   A very large eukaryotic protein complex comprising RNA polymerase II and the six general transcription factors (GTFs), each of which is a multiprotein complex.

pre-mRNA   *See* primary transcript.

prezygotic isolation   The failure of two species to exchange genes because of barriers either to mating between them or to fertilization of the gametes of one species by the gametes of the other species.

primary structure of a protein   The sequence of amino acids in the polypeptide chain.

primary transcript   Eukaryotic RNA before it has been processed.

primase   An enzyme that makes RNA primers during DNA replication.

primer   A short single-stranded RNA or DNA that can act as a start site for 3′ chain growth when bound to a single-stranded template.

primer walking   The use of a primer based on a sequenced area of a genome to sequence into a flanking unsequenced area.

primosome   A protein complex at the replication fork whose central component is *primase*.

probe   Labeled nucleic acid segment that can be used to identify specific DNA molecules bearing the complementary sequence, usually through autoradiography or fluorescence.

processive enzyme   In Chapter 7, describes the behavior of DNA polymerase III, which can perform thousands of rounds of catalysis without dissociating from its substrate (the template DNA strand).

product of meiosis   One of the (usually four) cells formed by the two meiotic divisions.

product rule   The probability of two independent events occurring simultaneously is the product of the individual probabilities.

proflavin   A mutagen that tends to produce frameshift mutations.

programmed cell death   *See* apoptosis.

prokaryote   An organism composed of a prokaryotic cell, such as a bacterium or a blue-green alga.

prokaryotic cell   A cell having no nuclear membrane and hence no separate nucleus.

proliferating cell nuclear antigen (PCNA)   Part of the replisome, PCNA is the eukaryotic version of the clamp protein.

promoter   A regulator region that is a short distance from the 5′ end of a gene and acts as the binding site for RNA polymerase.

promoter-proximal element   The series of transcription-factor binding sites located near the core promoter.

prophage   A phage "chromsome" inserted as part of the linear structure of the DNA chromosome of a bacterium.

prophase   The early stage of nuclear division during which chromosomes condense and become visible.

propositus   In a human pedigree, the person who first came to the attention of the geneticist.

protein   Macromolecule composed of one or more amino acid chains; main component of phenotypic expression.

protein kinase   An enzyme that phosphorylates specific amino acid residues on specific target proteins.

proteome   The complete set of protein-coding genes in a genome.

proteomics   The systematic study of the proteome.

proto-oncogene   The normal cellular counterpart of a gene that can be mutated to become a dominant oncogene.

protoplast   A plant cell whose wall has been removed.

prototroph   A strain of organisms that will proliferate on minimal medium (*compare* auxotroph).

provirus   The chromosomally inserted DNA genome of a retrovirus.

pseudodominance   The sudden appearance of a recessive phenotype in a pedigree, due to deletion of a masking dominant gene.

pseudogene   A mutationally inactive gene for which no functional counterpart exists in wild-type populations.

pseudolinkage   The appearance of linkage of two genes on translocated chromosomes.

pulse-chase experiment   An experiment in which cells are grown in radioactive medium for a brief period (the pulse) and then transferred to nonradioactive medium for a longer period (the chase).

pulsed field gel electrophoresis   An electrophoretic technique in which the gel is subjected to electrical fields alternating between different angles, allowing very large DNA fragments to "snake" through the gel and hence permitting efficient separation of mixtures of such large fragments.

Punnett square   A grid used as a graphical representation of the progeny zygotes resulting from different gamete fusions in a specific cross.

pure-breeding line or strain   A group of identical individual organisms that always produce offspring of the same phenotype when intercrossed.

purine   A type of nitrogen base; the purine bases in DNA are adenine and guanine.

pyrimidine   A type of nitrogen base; the pyrimidine bases in DNA are cytosine and thymine.

QTL   *See* quantitative trait locus.

quantitative genetics   The study of the genetic basis of continuous variation in phenotype.

quantitative trait locus (QTL)   A gene affecting the phenotypic variation in continuously varying traits such as height and weight among others.

quantitative variation   The existence of a range of phenotypes for a specific character, differing by degree rather than by distinct qualitative differences.

quaternary structure of a protein   The multimeric constitution of a protein.

R plasmid   A plasmid containing one or several transposons that bear resistance genes.

random genetic drift   Changes in allele frequency that result because the genes appearing in offspring are not a perfectly representative sampling of the parental genes.

random mating Mating between individuals in which the choice of a partner is not influenced by the genotypes (with respect to specific genes under study).

randomly amplified polymorphic DNA (RAPD) A set of several genomic fragments amplified by a single PCR primer; somewhat variable from individual to individual; +/− heterozygotes for individual fragments can act as markers in genome mapping.

RAPD *See* randomly amplified polymorphic DNA.

reading frame Codon sequence determined by reading nucleotides in groups of three from a specific start codon.

reannealing Spontaneous realignment of two single DNA strands to re-form a DNA double helix that had been denatured.

rearrangement The production of abnormal chromosomes by breakage and incorrect rejoining of chromosomal segments; examples are inversions, deletions, and translocations.

receptor *See* ligand–receptor interaction.

receptor tyrosine kinase (RTK) A transmembrane receptor whose cytoplasmic domain includes a tyrosine kinase enzymatic activity. In normal situations, the kinase is activated only on binding of the appropriate ligand to the receptor.

recessive allele An allele whose phenotypic effect is not expressed in a heterozygote.

recessive phenotype The phenotype of a homozygote for the recessive allele; the parental phenotype that is not expressed in a heterozygote.

reciprocal crosses A pair of crosses of the type genotype *A* (female) × genotype *B* (male) and *B* (female) × *A* (male).

reciprocal translocation A translocation in which part of one chromosome is exchanged with a part of a separate nonhomologous chromosome.

recombinant Refers to an individual organism or cell having a genotype produced by recombination.

recombinant DNA A novel DNA sequence formed by the combination of two nonhomologous DNA molecules.

recombinant frequency (RF) The proportion (or percentage) of recombinant cells or individuals.

recombination (1) In general, any process in a diploid or partly diploid cell that generates new gene or chromosomal combinations not previously found in that cell or in its progenitors. (2) At meiosis, the process that generates a haploid product of meiosis whose genotype is different from either of the two haploid genotypes that constituted the meiotic diploid.

recombinational repair The repair of a DNA lesion through a process, similar to recombination, that uses recombination enzymes.

reduction division A nuclear division that produces daughter nuclei each having one-half as many centromeres as the parental nucleus.

redundant DNA *See* repetitive DNA.

regression coefficient The slope of the straight line that most closely relates two correlated variables.

regression line of *y* on *x* The straight line passing through a cluster of points that relates a variable *y* to a variable *x* and minimizes the deviations of the points from the line in the *y* direction.

regulatory gene A gene that turns on or off the transcription of structural genes.

regulatory region Upstream (5′) end of a gene to which bind various proteins that cause transcription of the gene at the correct time and place.

relative frequency The number of occurrences of some class in a population, expressed as a proportion of the total.

release factor (RF) A protein that binds to the A site of the ribosome when a stop codon is in the mRNA.

repeat-induced point mutation (RIP) Before *Neurospora* meiosis, multiple GC-to-AT substitutions in duplication from transgenesis.

repetitive DNA Redundant DNA; DNA sequences that are present in many copies per chromosome set.

replacement substitution A nucleotide change in the encoding part of a gene that results in a change in the amino acid sequence of the encoded protein.

replica plating In microbial genetics, a way of screening colonies arrayed on a master plate to see if they are mutant under other environments; a felt pad is used to transfer the colonies to new plates.

replication DNA synthesis.

replication fork The point at which the two strands of DNA are separated to allow the replication of each strand.

replicative transposition A mechanism of transposition that generates a new insertion element integrated elsewhere in the genome while leaving the original element at its original site of insertion.

replisome The molecular machine at the replication fork that coordinates the numerous reactions necessary for the rapid and accurate replication of DNA.

reporter gene A gene whose phenotypic expression is easy to monitor; used to study tissue-specific promoter and enhancer activities in transgenes.

repressor A protein that binds to a cis-acting element such as an operator or a silencer, thereby preventing transcription from an adjacent promoter.

repressor protein A molecule that binds to the operator and prevents transcription of an operon.

resistant mutant A mutant that can grow in a normally toxic environment.

restriction enzyme An endonuclease that will recognize specific target nucleotide sequences in DNA and break the DNA chain at those points; a variety of these enzymes are known, and they are extensively used in genetic engineering.

restriction fragment A DNA fragment resulting from cutting DNA with a restriction enzyme.

restriction fragment length polymorphism (RFLP) A difference in DNA sequence between individuals or haplotypes that is recognized as different restriction fragment lengths. For example, a nucleotide-pair substitution can cause a restriction-enzyme-recognition site to be present in one allele of a gene and absent in another. Consequently, a probe for this DNA region will hybridize to different-sized fragments within restriction digests of DNAs from these two alleles.

restriction map A map of a chromosomal region showing the positions of target sites of one or more restriction enzymes.

restrictive condition Environmental condition under which a conditional mutant shows the mutant phenotype.

retinoblastoma A childhood cancer of the retina.

retro-element The general name for class 1 transposable elements that move through an RNA intermediate.

retrotransposition A mechanism of transposition characterized by the reverse flow of information from RNA to DNA.

retrotransposon A transposable element that uses reverse transcriptase to transpose through an RNA intermediate.

retrovirus An RNA virus that replicates by first being converted into double-stranded DNA.

reverse genetics An experimental procedure that begins with a cloned segment of DNA or a protein sequence and uses it (through directed mutagenesis) to introduce programmed mutations back into the genome to investigate function.

reverse mutation *See* reversion.

reverse transcriptase   An enzyme that catalyzes the synthesis of a DNA strand from an RNA template.

reversion   The production of a wild-type gene from a mutant gene.

RF   *See* **recombinant frequency; release factor.**

RFLP   *See* **restriction fragment length polymorphism.**

RFLP mapping   A technique in which DNA restriction fragment length polymorphisms are used as reference loci for mapping in relation to known genes or other RFLP loci.

rho-dependent mechanism   One of two mechanisms used to terminate bacterial transcription.

ribonucleic acid   *See* **RNA.**

ribose   The pentose sugar of RNA.

ribosomal RNA   *See* **rRNA.**

ribosome   A complex organelle that catalyzes the translation of messenger RNA into an amino acid sequence. Composed of proteins plus rRNA.

ribozyme   An RNA with enzymatic activity—for instance, the self-splicing RNA molecules in *Tetrahymena*.

RIP   *See* **repeat-induced point mutation.**

RNA (ribonucleic acid)   A single-stranded nucleic acid similar to DNA but having ribose sugar rather than deoxyribose sugar and uracil rather than thymine as one of the bases.

RNA element   An element that moves through an RNA intermediate. Also called a *class 1 element*.

RNAi   *See* **RNA interference.**

RNA interference (RNAi)   A way of assessing the function of a gene by introducing special transgenic constructs to inactivate its mRNA.

RNA polymerase   An enzyme that catalyzes the synthesis of an RNA strand from a DNA template. Eukaryotes possess several classes of RNA polymerase; structural genes encoding proteins are transcribed by RNA polymerase II.

RNA polymerase holoenzyme   The bacterial multisubunit complex composed of the four subunits of the core enzyme plus the sigma factor.

RNA processing   The collective term for the modifications to eukaryotic RNA, including capping and splicing, that are necessary before the RNA can be transported into the cytoplasm for translation.

RNA World   The name of a popular theory that RNA must have been the genetic material in the first cells because only RNA is known to both encode genetic information and catalyze biological reactions.

robotics   The application of automation technology to large-scale biological data collection efforts such as genome projects.

rolling-circle replication   A mode of replication used by some circular DNA molecules in bacteria (such as plasmids) in which the circle seems to rotate as it reels out one continuous leading strand.

rRNA (ribosomal RNA)   A class of RNA molecules, coded in the nucleolar organizer, that have an integral (but poorly understood) role in ribosome structure and function.

RTK   *See* **receptor tyrosine kinase.**

S phase   The part of interphase of the cell cycle in which DNA synthesis takes place.

safe haven   A site in the genome where the insertion of a transposable element is unlikely to cause a mutation, thereby harming the host.

sample   A small group of individual members or observations meant to be representative of a larger population from which the group has been taken.

Sanger sequencing   *See* **dideoxy sequencing.**

SAR   Scaffold attachment region; SARs are positions along DNA at which the DNA is anchored to the central scaffold of the chromosome.

satellite   A terminal section of a chromosome, separated from the main body of the chromosome by a narrow constriction.

satellite chromosome   A chromosome that seems to be an addition to the normal genome.

satellite DNA   Any type of highly repetitive DNA; formerly defined as DNA forming a satellite band after cesium chloride density-gradient centrifugation.

saturation mutagenesis   Induction and recovery of large numbers of mutations in one area of the genome or in one function, in the hope of identifying all the genes in that area or affecting that function.

scaffold   (1) The central framework of a chromosome to which the DNA solenoid is attached as loops; composed largely of topoisomerase. (2) In genome projects, an ordered set of contigs in which there may be unsequenced gaps connected by paired-end sequence reads.

screen   A mutagenesis procedure in which essentially all mutagenized progeny are recovered and are individually evaluated for mutant phenotype; often the desired phenotype is marked in some way to enable its detection.

secondary structure of a protein   A spiral or zigzag arrangement of the polypeptide chain.

second-division segregation pattern   A pattern of ascospore genotypes for a gene pair showing that the two alleles separate into different nuclei only at the second meiotic division, as a result of a crossover between that gene pair and its centromere; can be detected only in a linear ascus.

sector   An area of tissue whose phenotype is detectably different from the surrounding tissue phenotype.

segment   In higher animals, such as annelids, arthropods, and chordates, one of the repeating units along the anterior–posterior axis of the body.

segment identity   The process by which the anterior–posterior fates of the segments are established.

segmentation   The process by which the correct number and types of segments are established in a developing higher animal.

segment-polarity gene   In *Drosophila*, a member of a class of genes that contribute to the final aspects of establishing the correct number of segments. Segment-polarity mutations cause a loss of a comparable part of each of the body segments.

segregating line   A group of genetically varying individuals that are the offspring of a hybrid population that was the result of crossing two different genetically uniform lines.

segregation   (1) Cytologically, the separation of homologous structures. (2) Genetically, the production of two separate phenotypes, corresponding to two alleles of a gene, either in different individuals (meiotic segregation) or in different tissues (mitotic segregation).

selection   (1) Experimental procedure in which only a specific type of mutant can survive. (2) The production of different average numbers of offspring by different genotypes in a population as a result of the different phenotypic properties of those genotypes.

selection differential   The difference between the mean of a population and the mean of the individual members selected to be parents of the next generation.

selection response   The amount of change in the average value of some phenotypic character between the parental generation and the offspring generation as a result of the selection of parents.

selective system   A mutational selection technique that enriches the frequency of specific (usually rare) genotypes by establishing environmental conditions that prevent the growth or survival of other genotypes.

self   To fertilize eggs with sperms from the same individual.

self-assembly   The ability of certain multimeric biological structures to assemble from their component parts through random movements of the molecules and formation of weak chemical bonds between surfaces with complementary shapes.

self-splicing intron   The first example of catalytic RNA; in this case, an intron that can be removed from a transcript without the aid of a protein enzyme.

semiconservative replication   The established model of DNA replication in which each double-stranded molecule is composed of one parental strand and one newly polymerized strand.

semisterility (half-sterility)   The phenotype of an organism heterozygotic for certain types of *chromosome aberration;* expressed as a reduced number of viable gametes and hence reduced fertility.

sequence assembly   The compilation of thousands or millions of independent DNA sequence reads into a set of contigs and scaffolds.

sequence contig   A group of overlapping cloned segments.

sex chromosome   A chromosome whose presence or absence is correlated with the sex of the bearer; a chromosome that plays a role in sex determination.

sex linkage   The location of a gene on a sex chromosome.

Shine-Dalgarno sequence   A short sequence in bacterial RNA that precedes the initiation AUG codon and serves to correctly position this codon in the P site of the ribosome by pairing (through base complementarity) with the 3′ end of the 16S RNA in the 30S ribosomal subunit.

short interspersed nuclear element (SINE)   A type of class 1 transposable element that does not encode reverse transcriptase but is thought to use the reverse transcriptase encoded by LINEs. *See also* **Alu.**

short-sequence-length polymorphism (SSLP)   The presence of different numbers of short repetitive elements (mini- and microsatellite DNA) at one particular locus in different homologous chromosomes; heterozygotes serve as useful markers for genome mapping.

shotgun technique   The cloning of a large number of different DNA fragments as a prelude to selecting one particular clone type for intensive study.

shuttle vector   A vector (for example, a plasmid) constructed in such a way that it can replicate in at least two different host species, allowing a DNA segment to be tested or manipulated in several cellular settings.

sigma (σ) factor   A bacterial protein that, as part of the RNA polymerase holoenzyme, recognizes the −10 and −35 regions of bacterial promoters, thus positioning the holoenzyme to initiate transcription correctly at the start site. Sigma factor dissociates from the holoenzyme before RNA synthesis.

signal sequence   The N-terminal sequence of a secreted protein, which is required for transport through the cell membrane.

signal-transduction cascade   A series of sequential events, such as protein phosphorylations, that pass a signal received by a transmembrane receptor through a series of intermediate molecules until final regulatory molecules, such as transcription factors, are modified in response to the signal.

silenced gene   A gene that is reversibly inactivated owing to chromatin structure.

silencer element   A cis-regulatory sequence that can reduce levels of transcription from an adjacent promoter.

silent mutation   A mutation that has no effect on the function of a gene product.

simple transposon   A type of bacterial transposable element containing a variety of genes that reside between short inverted repeat sequences.

SINE   *See* **short interspersed nuclear element.**

single-copy DNA   DNA sequences present in only one copy per haploid genome.

single-nucleotide polymorphism (SNP)   A nucleotide-pair difference at a given location in the genomes of two or more naturally occurring individuals.

single-strand-binding (SSB) protein   A protein that binds to DNA single strands and prevents the duplex from re-forming before replication.

sister chromatids   The juxtaposed pair of chromatids arising from replication of a chromosome.

site-directed mutagenesis   The alteration of a specific part of a cloned DNA segment followed by the reintroduction of the modified DNA back into an organism for assay of the mutant phenotype or for production of the mutant protein.

sliding clamp   An accessory replication protein in bacteria that encircles the DNA like a donut.

small nuclear RNA (snRNA)   Any of several short RNAs found in the eukaryotic nucleus, where they assist in RNA processing events.

SNP   *See* **single-nucleotide polymorphism.**

snRNA   *See* **small nuclear RNA.**

solenoid structure   The supercoiled arrangement of DNA in eukaryotic nuclear chromosomes that is produced by coiling of the continuous string of nucleosomes.

somatic cell   A cell that is not destined to become a gamete; a "body cell," whose genes will not be passed on to future generations.

somatic gene therapy   *See* **gene therapy.**

somatic mutation   A mutation that arises in a somatic cell and consequently is not passed through the germ line to the next generation.

SOS (repair) system   An error-prone process whereby gross structural DNA damage is circumvented by allowing replication to proceed past the damage through imprecise polymerization.

Southern blot   The transfer of electrophoretically separated fragments of DNA from a gel to an absorbent sheet such as paper; this sheet is then immersed in a solution containing a labeled probe that will bind to a fragment of interest.

spacer DNA   DNA found between genes; its function is unknown.

specialized (restricted) transduction   The situation in which a particular phage will transduce only specific regions of the bacterial chromosome.

speciation   The process of forming new species by the splitting of an old species into two or more new species incapable of exchanging genes with one another.

species   A group of organisms that can exchange genes among themselves but are reproductively isolated from other such groups.

specific-locus test   A system for detecting recessive mutations in diploids. Normal diploids treated with mutagen are mated to testers that are homozygous for the recessive alleles at a number of specific loci; the progeny are then screened for recessive phenotypes.

spindle   A set of microtubular fibers that appear to move eukaryotic chromosomes during division.

spliceosome   The ribonucleoprotein processing complex that removes introns from eukaryotic mRNAs.

splicing   A reaction that removes introns and joins together exons in RNA.

spontaneous mutation   A mutation occurring in the absence of exposure to mutagens.

spore   (1) In plants and fungi, sexual spores are the haploid cells produced by meiosis. (2) In fungi, asexual spores are somatic cells that are cast off to act either as gametes or as units of dispersal.

sporophyte   The diploid sexual-spore-producing generation in the life cycle of plants—that is, the stage in which meiosis takes place.

SRY gene   The maleness gene, residing on the Y chromosome.

SSB   _See_ **single-strand-binding protein.**

SSLP   _See_ **short-sequence-length polymorphism.**

standard deviation   The square root of the variance.

statistic   A computed quantity characteristic of a population, such as the mean.

statistical distribution   The array of frequencies of different quantitative or qualitative classes in a population.

stem cell   A cell that divides, generally asymmetrically, to give rise to two different progeny cells. One is a blast cell just like the parental cell and the other is a cell that enters a differentiation pathway. In this manner, a continuously propagating cell population can maintain itself and spin off differentiating cells.

steroid hormones   A class of hormones synthesized by glands of the endocrine system that, by virtue of their nonpolar nature, are able to pass directly through the plasma membrane of a cell. Steroid hormones act by binding to transcription factors called steroid hormone receptors and activating them.

strain   A pure-breeding lineage, usually of haploid organisms, bacteria, or viruses.

structural gene   The part of a gene encoding the amino acid sequence of a protein.

structural genomics   The large-scale analysis of three-dimensional protein structures.

structural protein   A protein whose function is in cellular organism structure.

subunit   As used in Chapter 9, a single polypeptide in a protein containing multiple polypeptides.

sublethal allele   An allele that causes the death of some proportion (but not all) of the individuals that express it.

sum rule   The probability that one or the other of two mutually exclusive events will occur is the sum of their individual probabilities.

supercoil   A closed, double-stranded DNA molecule that is twisted on itself.

supercontig   _See_ **scaffold** (2).

suppression   The production of a phenotype closer to wild type by the addition of another mutation to a phenotypically abnormal genotype.

suppressor   A secondary mutation that can cancel the effect of a primary mutation, resulting in wild-type phenotype.

survivor factor   A ligand that, when bound to a receptor, blocks the activation of the apoptosis pathway.

synapsis   Close pairing of homologs at meiosis.

synaptonemal complex   A complex structure that unites homologs during prophase of meiosis.

syncytium   A single cell having many nuclei.

synergistic effect   A feature of eukaryotic regulatory proteins for which the transcriptional activation mediated by the interaction of several proteins is greater than the sum of the effects of the proteins taken individually.

synonymous mutation   A mutation that changes one codon for an amino acid into another codon for that same amino acid. Also called _silent mutation._

synonymous substitution   _See_ **synonymous mutation.**

synteny   A situation in which genes are arranged in similar blocks in different species.

synteny map   A diagram showing the location of continuous pieces of the genome of one species in the genome of another species.

synthetic lethal   Refers to a double mutant that is lethal, whereas the component single mutations are not.

systems biology   An attempt to interpret a genome as a holistic interacting system.

T   (1) _See_ **thymine; thymidine.** (2) _See_ **tetratype.**

tagging   _See_ **gene tagging.**

tandem duplication   Adjacent identical chromosome segments.

targeted gene knockout   The introduction of a null mutation in a gene by a designed alteration in a cloned DNA sequence that is then introduced into the genome through homologous recombination and replacement of the normal allele.

targeting   A feature of certain transposable elements that facilitates their insertion into regions of the genome where they are not likely to insert into a gene causing a mutation.

target-site duplication   A short direct-repeat DNA sequence (typically from 2 to 10 bp in length) adjacent to the ends of a transposable element that was generated during integration into the host chromosome.

TART   A retrotransposon that, with another retrotransposon, _Het-A_, makes up the telomeres of _Drosophila._

TATA binding protein (TBP)   A general transcription factor that binds to the TATA box and assists in attracting other general transcription factors and RNA polymerase II to eukaryotic promoters.

TATA box   A DNA sequence found in many eukaryotic genes that is located about 30 bp upstream of the transcription start site.

tautomeric shift   The spontaneous isomerization of a nitrogen base from its normal keto form to an alternative hydrogen-bonding enol (or imino) form.

tautomers   Isomers of a molecule such as the DNA bases that differ in the positions of the atoms and bonds between atoms.

TBP   _See_ **TATA binding protein.**

T-DNA   A part of the Ti plasmid that is inserted into the genome of the host plant cell.

telocentric chromosome   A chromosome having the centromere at one end.

telomerase   An enzyme that, with the use of a special small RNA as a template, adds repetitive units to the ends of linear chromosomes to prevent shortening after replication.

telomere   The tip (or end) of a chromosome.

telophase   The late stage of nuclear division when daughter nuclei re-form.

temperate phage   A phage that can become a prophage.

temperature-sensitive mutation   A conditional mutation that produces the mutant phenotype in one temperature range and the wild-type phenotype in another temperature range.

template   A molecular "mold" that shapes the structure or sequence of another molecule; for example, the nucleotide sequence of DNA acts as a template to control the nucleotide sequence of RNA during transcription.

termination   The last stage of transcription; it results in the release of the RNA and RNA polymerase from the DNA template.

terminus   The end represented by the last added monomer in the unidirectional synthesis of a polymer such as RNA or a polypeptide.

ternary complex   A complex containing an aminoacylated tRNA (tRNA plus amino acid) and elongation factor-Tu (EF-Tu). The ternary complex binds to the A site of a ribosome.

tertiary structure of a protein   The folding or coiling of the secondary structure to form a globular molecule.

testcross   A cross of an individual organism of unknown genotype or a heterozygote (or a multiple heterozygote) with a _tester._

tester An individual organism homozygous for one or more recessive alleles; used in a testcross.

tetrad (1) Four homologous chromatids in a bundle in the first meiotic prophase and metaphase. (2) The four haploid product cells from a single meiosis.

tetrad analysis The use of *tetrads* (definition 2) to study the behavior of chromosomes and genes in meiosis.

tetramer A protein consisting of four polypeptide subunits.

tetraploid A cell having four chromosome sets; an organism composed of such cells.

tetratype (T) A tetrad type containing four different genotypes, two parental and two recombinant.

theta (θ) structure An intermediate structure in the replication of a circular bacterial chromosome.

three-point testcross A testcross in which one parent has three heterozygous gene pairs.

thymidine (T) A nucleoside having thymine as its base.

thymine (T) A pyrimidine base that pairs with adenine.

thymine dimer A pair of chemically bonded adjacent thymine bases in DNA; the cellular processes that repair this lesion often make errors that create mutations.

Ti plasmid A circular plasmid of *Agrobacterium tumifaciens* that enables the bacterium to infect plant cells and produce a tumor (crown gall tumor).

Tn *See* **transposon**.

topoisomerase An enzyme that can cut and re-form polynucleotide backbones in DNA to allow it to assume a more relaxed configuration.

totipotency The ability of a cell to proceed through all the stages of development and thus produce a normal adult.

totipotent Refers to the state of a cell lineage that is able to give rise to all possible cell fates found within a given organism.

trans conformation In a heterozygote with two mutant sites within a gene or gene cluster, the arrangement $a_1 +/+ a_2$.

trans-acting factor A diffusible regulatory molecule (almost always a protein) that binds to a specific cis-acting element.

transcript The RNA molecule copied from the DNA template strand by RNA polymerase.

transcription The synthesis of RNA from a DNA template.

transcription bubble The site at which the double helix is unwound so that RNA polymerase can use one of the DNA strands as a template for RNA synthesis.

transcription factor A protein that binds to a cis-regulatory element (for example, an enhancer) and thereby, directly or indirectly, affects the initiation of transcription.

transduction The movement of genes from a bacterial donor to a bacterial recipient with a phage as the vector.

transfection The process by which exogenous DNA in solution is introduced into cultured cells.

transfer RNA *See* **tRNA**.

transformation Directed modification of a genome by the external application of DNA from a cell of different genotype.

transgene A gene that has been modified by externally applied recombinant DNA techniques and reintroduced into the genome by germ-line transformation.

transgenic organism An organism whose genome has been modified by externally applied new DNA.

transient diploid The stage of the life cycle of predominantly haploid fungi (and algae) in which meiosis takes place.

transition A type of nucleotide-pair substitution in which a purine replaces another purine or in which a pyrimidine replaces another pyrimidine—for example, G–C to A–T.

translation The ribosome- and tRNA-mediated production of a polypeptide whose amino acid sequence is derived from the codon sequence of an mRNA molecule.

translocation The relocation of a chromosomal segment to a different position in the genome.

transmembrane receptor A protein that spans the plasma membrane of a cell, with the extracellular part of the protein having the ability to bind to a ligand and the intracellular part having an activity (such as protein kinase) that can be induced on ligand binding.

transmission genetics The study of the mechanisms for the passage of a gene from one generation to the next.

transposable element A general term for any genetic unit that can insert into a chromosome, exit, and relocate; includes insertion sequences, transposons, some phages, and controlling elements.

transposase An enzyme encoded by transposable elements that undergo conservative transposition.

transposition A process by which mobile genetic elements move from one location in the genome to another.

transposon (Tn) A mobile piece of DNA that is flanked by terminal repeat sequences and typically bears genes coding for transposition functions. Bacterial transposons can be simple or composite.

transposon tagging A method used to identify and isolate a host gene through the insertion of a cloned transposable element in the gene.

transversion A type of nucleotide-pair substitution in which a pyrimidine replaces a purine or vice versa—for example, G–C to T–A.

trinucleotide repeat *See* **triplet expansion**.

triplet Three nucleotide pairs that compose a codon.

triplet expansion The expansion of a 3-bp repeat from a relatively low number of copies to a high number of copies that is responsible for a number of genetic diseases, such as fragile X syndrome and Huntington disease.

triploid A cell having three chromosome sets or an organism composed of such cells.

trisomic Basically a diploid with an extra chromosome of one type, producing a chromosome number of the form $2n + 1$.

trivalent Refers to the meiotic pairing arrangement of three homologs in a triploid or trisomic.

tRNA (transfer RNA) A class of small RNA molecules that bear specific amino acids to the ribosome in the course of translation; an amino acid is inserted into the growing polypeptide chain when the anticodon of the corresponding tRNA pairs with a codon on the mRNA being translated.

true-breeding line or strain *See* **pure-breeding line or strain**.

truncation selection A breeding technique in which individual organisms in which quantitative expression of a phenotype is above or below a certain value (the truncation point) are selected as parents for the next generation.

tumor-suppressor gene A gene encoding a protein that suppresses tumor formation. The wild-type alleles of tumor-suppressor genes are thought to function as negative regulators of cell proliferation.

Turner syndrome An abnormal human female phenotype produced by the presence of only one X chromosome (XO).

two-hybrid system A pair of *Saccharomyces cerevisiae* (yeast) vectors used for detecting protein–protein interaction. Each vector carries the gene for a different foreign protein under test; if these vectors unite physically, a reporter gene is transcribed.

2-μm (2-micrometer) plasmid   A naturally occurring extragenomic circular DNA molecule found in some yeast cells, with a circumference of 2 μm; engineered to form the basis for several types of gene vectors in yeast.

Ty element   A yeast LTR retrotransposon, the first isolated from any organism.

U   *See* **uracil; uridine.**

unbalanced rearrangement   A rearrangement in which chromosomal material is gained or lost in one chromosome set.

underdominance   A phenotypic relation in which the phenotypic expression of the heterozygote is less than that of either homozygote.

univalent   A single unpaired meiotic chromosome as is often found in trisomics and triploids.

universe   The extremely large population from which a sample is taken for statistical purposes.

unselected marker   In a bacterial recombination experiment, an allele scored in progeny for the frequency of its cosegregation with a linked selected allele.

unstable mutation   A mutation that has high frequency of reversion; a mutation caused by the insertion of a controlling element whose subsequent exit produces a reversion.

unstable phenotype   Phenotype characterized by frequent reversion either somatically or germinally or both due to the interaction of transposable elements with a host gene.

3′ untranslated region (3′ UTR)   The region of the RNA transcript at the 3′ end downstream of site of translation termination.

5′ untranslated region (5′ UTR)   The region of the RNA transcript at the 5′ end upstream of the translation start site.

UPE   *See* **upstream promoter element.**

upstream   Refers to a DNA or RNA sequence located on the 5′ side of a point of reference.

upstream promoter element (UPE)   A DNA binding site for transcription regulatory protein that is located about 100 to 200 bp upstream of the transcription start site. Also called a *promoter-proximal element.*

uracil (U)   A pyrimidine base that appears in RNA in place of the thymine found in DNA.

uridine (U)   A nucleoside having uracil as its base.

UTR   *See* **3′ untranslated region; 5′ untranslated region.**

variable number tandem repeat (VNTR)   A chromosomal locus at which a particular repetitive sequence is present in different numbers in different individuals or in the two different homologs in one diploid individual.

variance   A measure of the variation around the central class of a distribution; the average squared deviation of the observations from their mean value.

variant   An individual organism that is recognizably different from an arbitrary standard type in that species.

variation   The differences among parents and their offspring or among individual members of a population.

variegation   The occurrence within a tissue of sectors with differing phenotypes.

vector   *See* **cloning vector.**

viability   The probability that a fertilized egg will survive and develop into an adult organism.

virulent phage   A phage that cannot become a prophage; infection by such a phage always leads to lysis of the host cell.

virus   A particle consisting of nucleic acid and protein that must infect a living cell to replicate and reproduce.

VNTR   *See* **variable number tandem repeat.**

Western blot   Membrane carrying an imprint of proteins separated by electrophoresis. Can be probed with a labeled antibody to detect a specific protein.

WGS   *See* **whole genome shotgun sequencing.**

whole genome shotgun (WGS) sequencing   The sequencing of ends of clones without regard to any information about the location of the clones.

wild type   The genotype or phenotype that is found in nature or in the standard laboratory stock for a given organism.

wobble   The ability of certain bases at the third position of an anticodon in tRNA to form hydrogen bonds in various ways, causing alignment with several different possible codons.

X chromosome   One of a pair of sex chromosomes, distinguished from the Y chromosome.

X-chromosome inactivation   The process by which the genes of an X chromosome in a mammal can be completely repressed as part of the dosage-compensation mechanism. *See also* **dosage compensation; Barr body.**

X linkage   The inheritance pattern of genes found on the X chromosome but not on the Y chromosome.

X-and-Y linkage   The inheritance pattern of genes found on both the X and the Y chromosomes (rare).

X-ray crystallography   A technique for deducing molecular structure by aiming a beam of X rays at a crystal of the test compound and measuring the scatter of rays.

Y chromosome   One of a pair of sex chromosomes, distinguished from the X chromosome.

Y linkage   The inheritance pattern of genes found on the Y chromosome but not on the X chromosome (rare).

YAC   *See* **yeast artificial chromosome.**

yeast artificial chromosome (YAC)   A cloning-vector system in *Saccharomyces cerevisiae* employing yeast centromere and replication sequences.

yeast two-hybrid system   *See* **two-hybrid system.**

zygote   A cell formed by the fusion of an egg and a sperm; the unique diploid cell that will divide mitotically to create a differentiated diploid organism.

zygotic induction   The sudden release of a lysogenic phage from an Hfr chromosome when the prophage enters the F⁻ cell followed by the subsequent lysis of the recipient cell.

zymogen   The inactive precursor form of an enzyme. Zymogens are typically activated by proteolytic cleavage.

Answers to Selected Problems

Chapter 1

2. DNA determines all the specific attributes of a species (shape, size, form, behavioral characteristics, biochemical processes, etc.) and sets the limits for possible variation that is environmentally induced.

5. If the DNA is double stranded, A = T and G = C and A + T + C + G = 100%. If T = 15%, then C = [100 − 15(2)]/2 = 35%.

6. If the DNA is double stranded, G = C = 24% and A = T = 26%.

10. a. Yes. Because A = T and G = C, the equation A + C = G + T can be rewritten as T + C = C + T by substituting the equal terms.

   b. Yes, the percentage of purines will equal the percentage of pyrimidines in double-stranded DNA.

14. The simplest definition is that a *gene* is a chromosomal region capable of making a functional transcript. However, this definition does not take into account the regulatory regions near the gene necessary for the proper expression of the gene or the regions that help control transcription, which can be quite distant. In addition, many eukaryotes have large regions of noncoding sequences (introns) interspersed within the regions that encode the product (exons).

17. mRNA size = gene size − (number of introns × average size of introns)

20. a. A plant unable to synthesize red pigment would be blue.

   b. A plant unable to synthesize blue pigment would be red.

   c. A plant unable to synthesize both blue and red pigments would be white.

24. a. Recessive. The one normal allele provides enough enzyme for normal function (the definition of haplosufficiency).

   b. There are many ways to mutate a gene to destroy enzyme function. One possible mutation might be a frameshift mutation within an exon of the gene. With the assumption that a single base pair was deleted, the mutation would completely alter the translational product 3′ to the mutation.

   c. Hormone replacement could be given to the patient.

   d. If the hormone were required before birth, it would be supplied by the mother.

28. Phenotypic variation within a species can be due to genotype, environmental effects, and pure chance (random noise). Showing that variation of a particular trait has a genetic basis, especially for traits that show continuous variation, therefore requires very carefully controlled analyses, as discussed in detail in Chapter 20.

Chapter 2

1. Mendel's first law states that alleles are in pairs and segregate in the course of meiosis; his second law states that genes assort independently in the course of meiosis.

4. a. This is simply a matter of counting genotypes; there are 9 genotypes in the Punnett square. Alternatively, you know that there are three genotypes possible per gene—for example, *R/R*, *R/r*, and *r/r*—and, because both genes assort independently, there are 3 × 3 = 9 total genotypes.

   b. Again, simply count. The genotypes are:

1 *R/R* ; *Y/Y*	1 *r/r* ; *Y/Y*	1 *R/R* ; *y/y*	1 *r/r* ; *y/y*
2 *R/r* ; *Y/Y*	2 *r/r* ; *Y/y*	2 *R/r* ; *y/y*	2 *R/R* ; *Y/y*
4 *R/r* ; *Y/y*			

   c. To find a formula for the number of genotypes, first consider the following:

Number of genes	Number of genotypes	Number of phenotypes
1	$3 = 3^1$	$2 = 2^1$
2	$9 = 3^2$	$4 = 2^2$
3	$27 = 3^3$	$8 = 2^3$

6. You are told that the disease being followed in this pedigree is very rare. If the allele that results in this disease is recessive, then the father would have to be homozygous and the mother would have to be heterozygous for this allele. On the other hand, if the trait is dominant, then all that is necessary to explain the pedigree is that the father is heterozygous for the allele that causes the disease. This is the better choice because it is more likely, given the rarity of the disease.

10. $\frac{5}{8}$

15. a. *Pedigree 1:* The best answer is recessive because two unaffected individuals had affected progeny. In addition, the disorder skips generations and appears in a mating between two related individuals.

   Pedigree 2: The best answer is dominant because two affected parents have an unaffected child. Additionally, it appears in each generation, roughly half the progeny are affected, and all affected individuals have an affected parent.

   Pedigree 3: The best answer is dominant for many of the reasons stated for pedigree 2. Inbreeding, though present in the pedigree, does not allow an explanation of recessive, because it cannot account for individuals in the second generation.

   Pedigree 4: The best answer is recessive. Two unaffected individuals had affected progeny.

   b. Genotypes of pedigree 1:
   Generation I: *A/−, a/a*
   Generation II: *A/a, A/a, A/a, A/−, A/−, A/a*
   Generation III: *A/a, A/a*
   Generation IV: *a/a*
   Genotypes of pedigree 2:
   Generation I: *A/a, a/a, A/a, a/a*
   Generation II: *a/a, a/a, A/a, A/a, a/a, a/a, A/a, A/a, a/a*
   Generation III: *a/a, a/a, a/a, a/a, a/a, A/−, A/−, A/−, A/a, a/a*
   Generation IV: *a/a, a/a, a/a*
   Genotypes of pedigree 3:
   Generation I: *A/−, a/a*
   Generation II: *A/a, a/a, a/a, A/a*
   Generation III: *a/a, A/a, a/a, a/a, A/a, a/a*
   Generation IV: *a/a, A/a, A/a, A/a, a/a, a/a*
   Genotypes of pedigree 4:
   Generation I: *a/a, A/−, A/a, A/a*
   Generation II: *A/a, A/a, A/a, a/a, A/−, a/a, A/−, A/−, A/−, A/−, A/−*
   Generation III: *A/a, a/a, A/a, A/a, a/a, A/a*

18. a. Yes. It is inherited as an autosomal dominant trait.

   b. Susan is highly unlikely to have Huntington disease. Her great-grandmother (individual II-2) is 75 years old and has yet to develop it, although nearly 100% of people carrying the allele will have developed the disease by that age. If her great-grandmother does not have it, Susan cannot inherit it.

   Alan is somewhat more likely than Susan to develop Huntington disease. His grandfather (individual III-7) is only 50 years old, and approximately 20% of people with the allele have yet to develop the disease by that age. Therefore, the grandfather has an estimated 10% chance of being a carrier (50% chance that he inherited the allele from his father × 20% chance that he has not yet developed symptoms). If

Alan's grandfather eventually develops Huntington disease, then there is a probability of 50% that Alan's father inherited it from him, and a probability of 50% that Alan received that allele from his father. Therefore, Alan has a $\frac{1}{10} \times \frac{1}{2} \times \frac{1}{2} = \frac{1}{40}$ current probability of developing Huntington disease and a $\frac{1}{2} \times \frac{1}{2} = \frac{1}{4}$ probability if his grandfather eventually develops it.

20. a. A son inherits his X chromosome from his mother. The mother has earlobes, whereas the son does not. If the allele for earlobes is dominant and the allele for lack of earlobes is recessive, then the mother could be heterozygous for this trait and the gene could be X linked.

b. It is not possible from the data given to decide which allele is dominant. If lack of earlobes is dominant, then the father would be heterozygous and the son would have a 50% chance of inheriting the dominant "lack of earlobes" allele. If lack of earlobes is recessive, then the trait could be autosomal or X linked, but, in either case, the mother would be heterozygous.

22. $(\frac{1}{2})^{10}$.

26. Let H = hypophosphatemia and h = normal. The cross is $H/Y \times h/h$, yielding H/h (females) and h/Y (males). The answer is 0%.

28. a. Because none of the parents are affected, the disease must be recessive. Because the inheritance of this trait appears to be sex specific, it is most likely X linked. If it were autosomal, all three parents would have to be carriers, and, by chance, only sons and none of the daughters inherited the trait (which is quite unlikely).

b. I $A/Y, A/a, A/Y$
II $A/Y, A/-, a/Y, A/-, A/Y, a/Y, a/Y, A/-, a/Y, A/-$

32. a. From cross 6, Bent (B) is dominant to normal (b).

b. From cross 1, it is X linked.

c. In the following table, the Y chromosome is stated; the X is implied.

| | Parents | | Progeny | |
Cross	Female	Male	Female	Male
1	b/b	B/Y	B/b	b/Y
2	B/b	b/Y	$B/b, b/b$	$B/Y, b/Y$
3	B/B	b/Y	B/b	B/Y
4	b/b	b/Y	b/b	b/Y
5	B/B	B/Y	B/B	B/Y
6	B/b	B/Y	$B/B, B/b$	$B/Y, b/Y$

34. a. Because photosynthesis is affected and the plants are yellow, rather than green, the chloroplasts are likely defective. If the defect maps to the DNA of the chloroplast, the trait will be maternally inherited.

b. If the defect maps to the DNA of the chloroplast, the trait will be maternally inherited. Use pollen from sickly, yellow plants and cross to emasculated flowers of a normal dark-green-leaved plant. All progeny should have the normal dark-green phenotype.

c. The chloroplasts contain the green pigment chlorophyll and are the site of photosynthesis. A defect in the production of chlorophyll would give rise to all the stated defects.

37. The genetic determinants of R and S are showing maternal inheritance and are therefore cytoplasmic. It is possible that the gene that confers resistance maps either to the mtDNA or to the cpDNA.

40. The hypothesis is that the organism being tested is a dihybrid with independently assorting genes and that all progeny are of equal viability. The expected values would be that phenotypes occur with equal frequency. There are four phenotypes, and so there are 3 degrees of freedom.

$$\chi^2 = \Sigma \, (\text{observed} - \text{expected})^2/\text{expected}$$
$$\chi^2 = [(230 - 233)^2 + (210 - 233)^2 + (240 - 233)^2$$
$$+ (250 - 233)^2]/233$$
$$= 2.215; p > 0.50, \text{nonsignificant; hypothesis cannot be rejected}$$

43. a. Galactosemia pedigree:

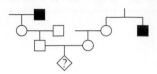

b. $\frac{1}{4} \times \frac{2}{3} \times \frac{1}{2} = \frac{1}{12}$

c. All future children have a $\frac{1}{4}$ chance of inheriting the disease.

45. a. The inheritance pattern for red hair suggested by this pedigree is recessive because most red-haired persons are from parents who do not have this trait.

b. By observing those around us, we find that the allele appears to be somewhat rare.

48. a. Because each gene assorts independently, each probability should be considered separately and then multiplied together for the answer.

For (1) = $\frac{9}{128}$ For (2) = $\frac{9}{128}$
For (3) = $\frac{9}{64}$ For (4) = $\frac{55}{64}$

b. For (1) = $\frac{1}{32}$ For (2) = $\frac{1}{32}$
For (3) = $\frac{1}{16}$ For (4) = $\frac{15}{16}$

51. If the historical record is accurate, the data suggest Y linkage. Another explanation is an autosomal gene that is dominant in males and recessive in females, which has been observed for other genes in both humans and other species.

53. Note that only males are affected and that, in all but one case, the trait can be traced through the female side. However, there is one example of an affected male having affected sons. If the trait is X linked, this male's wife must be a carrier, which suggests that the disorder is caused by an autosomal dominant with expression limited to males, depending on how rare this trait is in the general population.

56. a. The complete absence of male offspring is the unusual aspect of this pedigree. In addition, all progeny that mate carry the trait for lack of male offspring. If the male lethality factor were nuclear, the male parent would be expected to alter this pattern. Therefore, cytoplasmic inheritance is suggested.

b. If all females resulted from chance alone, then the probability of this result is $(\frac{1}{2})^n$, where n = the number of female births. In this case, n is 72. Chance is an unlikely explanation for the observations. The observations can be explained by cytoplasmic factors by assuming that the proposed mutation in mitochondria is lethal only in males. Mendelian inheritance cannot explain the observations, because all the fathers would have had to carry the male-lethal mutation in order for such a pattern to be observed. This would be highly unlikely.

Chapter 3

1. The key function of mitosis is to generate two daughter cells genetically identical with that of the original parent cell.

5. As cells divide mitotically, each chromosome consists of identical sister chromatids that are separated to form genetically identical daughter cells. Although the second division of meiosis appears to be a similar process, the "sister" chromatids are likely to be different. Recombination in earlier meiotic stages has swapped regions of DNA between sister and nonsister chromosomes such that the two daughter cells of this division are typically not genetically identical.

9. Each diploid meiocyte undergoes meiosis to form four products, and then each product undergoes a postmeiotic mitotic division to give a total of 8 ascospores per meiocyte. In total then, 100 meiocytes will give rise to 800 ascospores.

10. The genotype of the resulting cells will be identical with that of the original cell: $A/a \, ; B/b$.

15. PFGE separates DNA molecules by size. When DNA is carefully isolated from *Neurospora* (which has seven different chromosomes), seven bands should be produced with the use of this technique. Similarly, the pea has seven different chromosomes and will produce seven bands (homologous chromosomes will comigrate as a single band).

18. Yes. Half of our genetic makeup is derived from each parent, each parent's genetic makeup is derived half from each of their parents, etc.

22. All daughter cells will still be *A/a* ; *B/b* ; *C/c*. Mitosis generates daughter cells that are genetically identical with the original cell.

26. (5) chromosome pairing (synapsis)

30. When results of a cross are sex specific, sex linkage should be considered. In moths, the heterogametic sex is actually the female, whereas the male is the homogametic sex. With the assumption that dark (*D*) is dominant to light (*d*), the data can be explained by the dark male being heterozygous (*D/d*) and the dark female being hemizygous (*D*). All male progeny will inherit the *D* allele from their mother and therefore be dark, whereas half the females will inherit *D* from their father (and be dark) and half will inherit *d* (and be light).

34. Progeny plants inherited only normal cpDNA (lane 1); only mutant cpDNA (lane 2); or both (lane 3). To get homozygous cpDNA, seen in lanes 1 and 2, chloroplasts had to segregate.

39. First, examine the crosses and the resulting genotypes of the endosperm:

Female	Male	Polar nuclei	Sperm	Endosperm
f'/f'	*f"/f"*	*f'* and *f'*	*f"/f"*	*f'/f'/f"* (floury)
f"/f"	*f'/f'*	*f"* and *f"*	*f'/f'*	*f"/f"/f'* (flinty)

As can be seen, the phenotype of the endosperm correlates to the predominant allele present.

Chapter 4

2. P *A d/A d* × *a D/a D*
F$_1$ *A d/a D*
F$_2$ 1 *A d/A d*   Phenotype: A d
    2 *A d/a D*   Phenotype: A D
    1 *a D/a D*   Phenotype: a D

5. Because only parental types are recovered, the two genes must be tightly linked and recombination must be very rare. Knowing how many progeny were looked at would give an indication of how close the genes are.

8. (a) 4%; (b) 4%; (c) 46%; (d) 8%.

11. a. Because you cannot distinguish between *a⁺ b⁺ c⁺/a⁺ b⁺ c⁺* and *a⁺ b⁺ c⁺/a b c*, use the frequency of *a b c/a b c* to estimate the frequency of *a⁺ b⁺ c⁺* (parental) gametes from the female.

Parentals:	730	(2 × 365)
CO *a–b*:	91	(*a b⁺ c⁺, a⁺ b c* = 47 + 44)
CO *b–c*:	171	(*a b c⁺, a⁺ b⁺ c* = 87 + 84)
DCO:	9	(*a⁺ b c⁺, a b⁺ c* = 4 + 5)
	1001	

a–b: 100%(91 + 9)/1001 = 10 m.u.
b–c: 100%(171 + 9)/1001 = 18 m.u.

b. Coefficient of coincidence = (observed DCO)/(expected DCO) = 9/[(0.1)(0.18)(1001)] = 0.5.

14. Let *F* = fat, *L* = long tail, and *Fl* = flagella.
L–Fl: 100%(72 + 67 + 9 + 5)/1000 = 15.3 m.u.
F–L: 100%(44 + 35 + 9 + 5)/1000 = 9.3 m.u.

20. a. and **b.**

c. Interference = 0.86

24. (a) meiosis I; (b) impossible; (c) meiosis I; (d) meiosis I; (e) meiosis II; (f) meiosis II; (g) meiosis II; (h) impossible; (i) mitosis; (j) impossible.

27. (a) *f*(0) = 13.5%; (b) *f*(1) = 27%; (c) *f*(2) = 27%.

30. Cross 1:
recombinant frequency = 4% + ½ (45%) = 26.5%
uncorrected map distance = [4% + ½(45%)]/100% = 26.5 m.u.
corrected map distance = 50[45% + 6(4%)]/100% = 34.5 m.u.
Cross 2:
recombinant frequency = 2% + ½(34%) = 19%
uncorrected map distance = [2% + ½(34%)]/100% = 19 m.u.
corrected map distance = 50[34% + 6(2%)]/100% = 29 m.u.
Cross 3:
recombinant frequency = 5% + ½(50%) = 30%
uncorrected map distance = [5% + ½(50%)]/100% = 30 m.u.
corrected map distance = 50[50% + 6(5%)]/100% = 40 m.u.

31. a. All four genes are linked.
b. and **c.** The map is:

<table>
<tr><td>B</td><td>A, D</td><td></td><td>C</td></tr>
<tr><td></td><td>10 m.u.</td><td>30 m.u.</td><td></td></tr>
</table>

The parental chromosomes were actually *B* (*A,d*) *c/b* (*a,D*) *C*, where the parentheses indicate that the order of the genes within is unknown.

d. Interference = 0.5

34. For (1), the gene order is *b a c*.
For (2), the gene order is *b a c*.
For (3), the gene order *b a c*.
For (4), the gene order *a c b*.
For (5), the gene order *a c b*.

37. a. To obtain a plant that is *a b c/a b c* = 0.0049.
b. For 1000 progeny, the expected results are:

A b c	280	*A b C*	120
a B C	280	*a B c*	120
A B C	70	*A B c*	30
a b c	70	*a b C*	30

c. For 1000 progeny, the expected results are:

A b c	274	*A b C*	126
a B C	274	*a B c*	126
A B C	76	*A B c*	24
a b c	76	*a b C*	24

41. a. 3.36%; **b.** 18.2%.

Chapter 5

1. An Hfr strain has the fertility factor F integrated into the chromosome. An F⁺ strain has the fertility factor free in the cytoplasm. An F⁻ strain lacks the fertility factor.

4. Generalized transduction occurs with lytic phages that enter a bacterial cell, fragment the bacterial chromosome, and then, while new viral particles are being assembled, improperly incorporate some bacterial DNA within the viral protein coat. Because the amount of

DNA, not the information content of the DNA, is what governs viral particle formation, any bacterial gene can be included within the newly formed virus. In contrast, specialized transduction occurs with improper excision of viral DNA from the host chromosome in lysogenic phages. Because the integration site is fixed, only those bacterial genes very close to the integration site will be included in a newly formed virus.

6. —M—Z—X—W—C—N—A—L—B—R—U—

9. The most straightforward way would be to pick two Hfr strains that are near the genes in question but are oriented in opposite directions. Then, measure the time of transfer between two specific genes, in one case when they are transferred early and in the other when they are transferred late. For example,

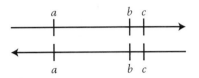

13. a. and b.

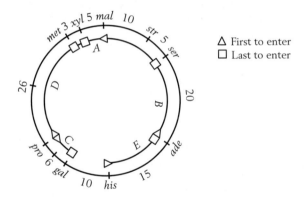

△ First to enter
☐ Last to enter

c. A: Select for *mal*⁺
B: Select for *ade*⁺
C: Select for *pro*⁺
D: Select for *pro*⁺
E: Select for *his*⁺

15. Interference = 1 − (observed DCO/expected DCO) = 1 − 5/2 = −1.5. By definition, the interference is negative.

18. In a small percentage of the cases, *gal*⁺ transductants can arise by recombination between the *gal*⁺ DNA of the λd*gal* transducing phage and the *gal*⁻ gene on the chromosome. This will generate *gal*⁺ transductants without phage integration.

21. a. Owing to the medium used, all colonies are *cys*⁺ but either + or − for the other two genes.

b. (1) *cys*⁺ *leu*⁺ *thr*⁺ and *cys*⁺ *leu*⁺ *thr*⁻ (supplemented with threonine)
(2) *cys*⁺ *leu*⁺ *thr*⁺ and *cys*⁺ *leu*⁻ *thr*⁺ (supplemented with leucine)
(3) *cys*⁺ *leu*⁺ *thr*⁺ (no supplements)

c. 39% of the colonies.

d.

24. No. Closely linked loci would be expected to be cotransduced; the greater the cotransduction frequency, the closer the loci are. Be-

cause only 1 of 858 *metE*⁺ was also *pyrD*⁺, the genes are not closely linked. The lone *metE*⁺ *pyrD*⁺ could be the result of cotransduction or it could be a spontaneous mutation of *pyrD* to *pyrD*⁺ or it could be the result of coinfection by two separate transducing phages.

26.

Agar type	Selected genes
1	c+
2	a+
3	b+

b. The gene order is *c b a*. The three genes are roughly equally spaced.

c. The gene order is *c b d a* (or *a d b c*).

d. With no A or B in the agar, the medium selects for *a*⁺ *b*⁺, and the first colonies should appear at about 17.5 minutes.

29. a. and b.

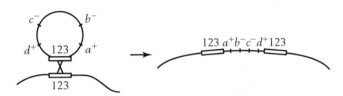

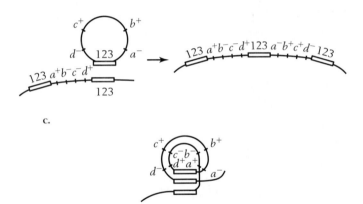

c.

d.

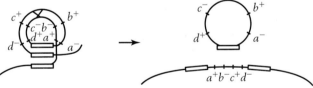

32. To isolate the specialized transducing particles of phage φ80 that carried *lac*⁺, the researchers would have had to lysogenize the strain with φ80, induce the phage with UV, and then use these lysates to transduce a Lac⁻ strain to Lac⁺. Lac⁺ colonies would then be used to make a new lysate, which should be highly enriched for the *lac*⁺ transducing phage.

Chapter 6

2. With the assumption of homozygosity for the normal gene, the mating is $A/A \cdot b/b \times a/a \cdot B/B$. The children would be normal, $A/a \cdot B/b$.

5. a. Red **b.** Purple

c.		
9	$M_1/- ; M_2/-$	purple
3	$m_1/m_1 ; M_2/-$	blue
3	$M_1/- ; m_2/m_2$	red
1	$m_1/m_1 ; m_2/m_2$	white

d. The mutant alleles do not produce functional enzyme. However, enough functional enzyme must be produced by the single wild-type allele of each gene to synthesize normal levels of pigment.

9. a. The original cross was a dihybrid cross. Both oval and purple must represent an incomplete dominant phenotype.

b. A long, purple $\times$ oval, purple cross is as follows:

P L/L ; R/R' $\times$ L/L' ; R/R''

$$F_1 \quad \tfrac{1}{2} L/L \times
\begin{cases}
\tfrac{1}{4} R/R & \tfrac{1}{8} \text{ long, red} \\
\tfrac{1}{2} R/R' & \tfrac{1}{4} \text{ long, purple} \\
\tfrac{1}{4} R'/R' & \tfrac{1}{8} \text{ long, white}
\end{cases}$$

$$\tfrac{1}{2} L/L' \times
\begin{cases}
\tfrac{1}{4} R/R & \tfrac{1}{8} \text{ oval, red} \\
\tfrac{1}{2} R/R' & \tfrac{1}{4} \text{ oval, purple} \\
\tfrac{1}{4} R'/R' & \tfrac{1}{8} \text{ oval, white}
\end{cases}$$

12.

	Parents	Child
a.	AB $\times$ O	B
b.	A $\times$ O	A
c.	A $\times$ AB	AB
d.	O $\times$ O	O

16. a. The sex ratio is expected to be $1:1$.

b. The female parent was heterozygous for an X-linked recessive lethal allele, which would result in 50% fewer males than females.

c. Half of the female progeny should be heterozygous for the lethal allele and half should be homozygous for the nonlethal allele. Individually mate the F_1 females and determine the sex ratio of their progeny.

19. a. The mutations are in two different genes because the heterokaryon is prototrophic (the two mutations complemented each other).

b. $leu1^+$; $leu2^-$ and $leu1^-$; $leu2^+$

c. With independent assortment, expect

$\tfrac{1}{4}$ $leu1^+$; $leu2^-$
$\tfrac{1}{4}$ $leu1^-$; $leu2^+$
$\tfrac{1}{4}$ $leu1^-$; $leu2^-$
$\tfrac{1}{4}$ $leu1^+$; $leu2^+$

23. a. P A/a (frizzle) $\times$ A/a (frizzle)
F_1 1 A/A (normal) : 2 A/a (frizzle) : 1 a/a (woolly)

b. If A/A (normal) is crossed to a/a (woolly), all offspring will be A/a (frizzle).

25. It is possible to produce black offspring from two pure-breeding recessive albino parents if albinism results from mutations in two different genes. If the cross is designated A/A ; b/b $\times$ a/a ; B/B, all offspring would be A/a ; B/b, and they would have a black phenotype because of complementation.

29. The purple parent can be either A/a ; b/b or a/a ; B/b for this answer. Assume that the purple parent is A/a ; b/b. The blue parent must be A/a ; B/b.

32. The cross is gray $\times$ yellow, or $A/-$; $R/-$ $\times$ $A/-$; r/r. The F_1 progeny are

$\tfrac{3}{8}$ yellow $\tfrac{1}{8}$ black $\tfrac{3}{8}$ gray $\tfrac{1}{8}$ white

For white progeny, both parents must carry an r and an a allele. Now the cross can be rewritten as A/a ; R/r $\times$ A/a ; r/r.

35. The original brown dog is w/w ; b/b and the original white dog is W/W ; B/B. The F_1 progeny are W/w ; B/b and the F_2 progeny are:

9 $W/-$; $B/-$ white
3 w/w ; $B/-$ black
3 $W/-$; b/b white
1 w/w ; b/b brown

39. Pedigrees such as this one are quite common. They indicate lack of penetrance due to epistasis or environmental effects. Individual A must have the dominant autosomal gene.

41. a. Let W^o = oval, W^s = sickle, and W^r = round. The three crosses are

Cross 1: W^s/W^s $\times$ W^r/Y $\rightarrow$ W^s/W^r and W^s/Y
Cross 2: W^r/W^r $\times$ W^s/Y $\rightarrow$ W^s/W^r and W^r/Y
Cross 3: W^s/W^s $\times$ W^o/Y $\rightarrow$ W^o/W^s and W^s/Y

b. W^o/W^s $\times$ W^r/Y

$\tfrac{1}{4}$ W^o/W^r female oval
$\tfrac{1}{4}$ W^s/W^r female sickle
$\tfrac{1}{4}$ W^o/Y male oval
$\tfrac{1}{4}$ W^s/Y male sickle

44. a. The genotypes are:

P B/B ; i/i $\times$ b/b ; I/I
F_1 B/b ; I/i hairless
F_2 9 $B/-$; $I/-$ hairless
      3 $B/-$; i/i straight
      3 b/b ; $I/-$ hairless
      1 b/b ; i/i bent

b. The genotypes are: B/b ; I/i $\times$ B/b ; i/i.

47. A total of 159 progeny should be distributed in a $9:3:3:1$ ratio if the two genes are assorting independently. You can see that

Observed	Expected
88 $P/-$; $Q/-$	90
32 $P/-$; q/q	30
25 p/p ; $Q/-$	30
14 p/p ; q/q	10

50. a. Cross-feeding is occurring, whereby a product made by one strain diffuses to another strain and allows growth of the second strain.

b. For cross-feeding to occur, the growing strain must have a block earlier in the metabolic pathway than is the block in the strain from which it is obtaining the product for growth.

c. The data suggests that the metabolic pathway is $trpE \rightarrow trpD \rightarrow trpB$.

d. Without some tryptophan, there would be no growth at all, and the cells would not have lived long enough to produce a product that could diffuse.

52. a. The best explanation is that Marfan's syndrome is inherited as a dominant autosomal trait.

b. The pedigree shows both pleiotropy (multiple affected traits) and variable expressivity (variable degree of expressed phenotype).

c. Pleiotropy indicates that the gene product is required in a number of different tissues, organs, or processes. When the gene is mutant, all tissues needing the gene product will be affected. Variable expressivity of a phenotype for a given genotype indicates modification by one or more other genes, random noise, or environmental effects.

55. a. This type of gene interaction is called *epistasis*. The phenotype of e/e is epistatic to the phenotypes of $B/-$ or b/b.

b. The following are the inferred genotypes:

I   1 (B/b E/e)   2 (B/b E/e)
II   1 (b/b E/e)   2 (B/b E/e)   3 ($-/-$ e/e)   4 (b/b $E/-$)
    5 (B/b E/e)   6 (b/b E/e)
III 1 (B/b $E/-$)   2 ($-/b$ e/e)   3 (b/b, $E/-$)   4 (B/b $E/-$)
    5 (b/b $E/-$)   6 (B/b $E/-$)   7 ($-/b$ e/e)

58. a. A multiple allelic series has been detected: superdouble > single > double.

b. Although this explanation does rationalize all the crosses, it does not take into account either the female sterility or the origin of the superdouble plant from a double-flowered variety.

60. a. A trihybrid cross would give a 63:1 ratio. Therefore, three R loci are segregating in this cross.

b. P R_1/R_1 ; R_2/R_2 ; R_3/R_3 $\times$ r_1/r_1 ; r_2/r_2 ; r_3/r_3

F$_1$ R_1/r_1 ; R_2/r_2 ; R_3/r_3

F$_2$			
27	$R_1/-$; $R_2/-$; $R_3/-$	red
9	$R_1/-$; $R_2/-$; r_3/r_3	red
9	$R_1/-$; r_2/r_2 ; $R_3/-$	red
9	r_1/r_1 ; $R_2/-$; $R_3/-$	red
3	$R_1/-$; r_2/r_2 ; r_3/r_3	red
3	r_1/r_1 ; $R_2/-$; r_3/r_3	red
3	r_1/r_1 ; r_2/r_2 ; $R_3/-$	red
1	r_1/r_1 ; r_2/r_2 ; r_3/r_3	white

c. (1) To obtain a 1:1 ratio, only one of the genes can be heterozygous. A representative cross would be R_1/r_1 ; r_2/r_2 ; r_3/r_3 $\times$ r_1/r_1 ; r_2/r_2 ; r_3/r_3.

(2) To obtain a 3 red:1 white ratio, two alleles must be segregating and they cannot be within the same gene. A representative cross would be R_1/r_1 ; R_2/r_2 ; r_3/r_3 $\times$ r_1/r_1 ; r_2/r_2 ; r_3/r_3.

(3) To obtain a 7 red:1 white ratio, three alleles must be segregating, and they cannot be within the same gene. The cross would be R_1/r_1 ; R_2/r_2 ; R_3/r_3 $\times$ r_1/r_1 ; r_2/r_2 ; r_3/r_3.

d. The formula is $1 - (\frac{1}{2})^n$, where n = the number of loci that are segregating in the representative crosses in part c.

66. a. Intercrossing mutant strains that all have a recessive phenotype in common is the basis of the complementation test. This test is designed to identify the number of different genes that can mutate to a particular phenotype. In this problem, if the progeny of a given cross still express the wiggle phenotype, the mutations fail to complement and are considered alleles of the same gene; if the progeny are wild type, the mutations complement and the two strains carry mutant alleles of separate genes.

b. Five complementation groups (genes) are identified by these data.

c. Mutant 1: $a^1/a^1 \cdot b^+/b^+ \cdot c^+/c^+ \cdot d^+/d^+ \cdot e^+/e^+$ (although only the mutant alleles are usually listed)

Mutant 2: $a^+/a^+ \cdot b^2/b^2 \cdot c^+/c^+ \cdot d^+/d^+ \cdot e^+/e^+$

Mutant 5: $a^5/a^5 \cdot b^+/b^+ \cdot c^+/c^+ \cdot d^+/d^+ \cdot e^+/e^+$

1/5 hybrid: $a^1/a^5 \cdot b^+/b^+ \cdot c^+/c^+ \cdot d^+/d^+ \cdot e^+/e^+$

phenotype: wiggles

1 and 5 are both mutant for gene A

(the relevant cross: $a^+/a^+ \cdot b^2/b^2 \times a^2/a^2 \cdot b^5/b^5$ gives)

2/5 hybrid: $a^+/a^5 \cdot b^+/b^2 \cdot c^+/c^+ \cdot d^+/d^+ \cdot e^+/e^+$

phenotype: wild type

2 and 5 are mutant for different genes

Chapter 7

1. The DNA double helix is held together by two types of bonds: covalent and hydrogen. Covalent bonds form within each linear strand and strongly bond the bases, sugars, and phosphate groups (both within each component and between components). Hydrogen bonds form between the two strands; a hydrogen bond is between a base from one strand and a base from the other strand, in complementary pairing. These hydrogen bonds are individually weak but collectively quite strong.

4. Helicases are enzymes that disrupt the hydrogen bonds that hold the two DNA strands together in a double helix. This breakage is required for both RNA and DNA synthesis. Topoisomerases are enzymes that create and relax supercoiling in the DNA double helix. The supercoiling itself is a result of the twisting of the DNA helix that occurs when the two strands separate.

6. No. The information in DNA depends on a faithful copying mechanism. The strict rules of complementarity ensure that replication and transcription are reproducible.

9. The chromosome would become hopelessly fragmented.

12. b. The RNA would be more likely to contain errors.

16. If the DNA is double stranded, A = T and G = C and A + T + C + G = 100%. If T = 15%, then C = $[100 - 15(2)]/2 = 35\%$.

17. If the DNA is double stranded, G = C = 24% and A = T = 26%.

20. Yes. DNA replication is also semiconservative in diploid eukaryotes.

22. 3′ ....GGAATTCTGATTGATGAATGACCCTAG.... 5′

26. Without functional telomerase, the telomeres would shorten at each replication cycle, leading to eventual loss of essential coding information and death. In fact, some current observations indicate that decline or loss of telomerase activity plays a role in the mechanism of aging in humans.

28. Chargaff's rules are that A = T and G = C. Because these equalities are not observed, the most likely interpretation is that the DNA is single stranded. The phage would first have to synthesize a complementary strand before it could begin to make multiple copies of itself.

Chapter 8

1. Because RNA can hybridize to both strands, the RNA must be transcribed from both strands. Transcription from both strands does not mean, however, that each strand is used as a template *within each gene*. The expectation is that only one strand is used within a gene but that different genes are transcribed in different directions along the DNA. The most direct test would be to purify a specific RNA coding for a specific protein and then hybridize it to the λ genome. Only one strand should hybridize to the purified RNA.

2. In prokaryotes, translation is beginning at the 5′ end while the 3′ end is still being synthesized. In eukaryotes, processing (capping, splicing) is taking place at the 5′ end while the 3′ end is still being synthesized.

4. Sigma factor, as part of the RNA polymerase holoenzyme, recognizes and binds to the −35 and −10 regions of bacterial promoters. It positions the holoenzyme to correctly initiate transcription at the start site. In eukaryotes, TBP (TATA binding protein) and other GTFs (general transcription factors) have an analagous function.

8. Yes. Both replication and transcription are performed by large, multisubunit molecular machines (the replisome and RNA polymerase II, respectively), and both require helicase activity at the fork of the bubble. However, transcription proceeds in only one direction and only one DNA strand is copied.

10. a. The original sequence represents the −35 and −10 consensus sequences (with the correct number of intervening spaces) of a bacterial promoter. Sigma factor, as part of the RNA polyermase holoenzyme, recognizes and binds to these sequences.

b. The mutated (transposed) sequences would not be a binding site for sigma factor. The two regions are not in the correct orientation to each other and therefore would not be recognized as a promoter.

14. a. Yes. The exons encode the protein, and so null mutations would be expected to map within exons.

b. Possibly. Sequences near the boundaries of introns and within them are necessary for correct splicing. If these sequences are altered by mutation, correct splicing will be disrupted. Although they may be transcribed, their translation is unlikely.

c. Yes. If the promoter is deleted or altered such that GTFs cannot bind, transcription will be disrupted.

d. Yes. Sequences near the boundaries of introns and within them are necessary for correct splicing.

Chapter 9

2. a. and b. 5′ UUG GGA AGC 3′

c. and d. With the assumption that the reading frame starts at the first base:

NH$_3$ – Leu – Gly – Ser – COOH

For the bottom strand, the mRNA is 5' GCU UCC CAA 3' and, with the assumption that the reading frame starts at the first base, the corresponding amino acid chain is NH_3 – Ala – Ser – Gln – COOH.

5. It suggests very little evolutionary change between *E. coli* and humans with regard to the translational apparatus. The code is universal, the ribosomes are interchangeable, the tRNAs are interchangeable, and the participating enzymes are interchangeable.

7. a. By studying the table listing the genetic code provided in the textbook, you will discover that there are eight cases in which knowing the first two nucleotides does not tell you the specific amino acid.

b. If you knew the amino acid, you would not know the first two nucleotides in the cases of Arg, Ser, and Leu.

9. a. The chance of UUU is $\frac{1}{8}$.

b. The total probability is $\frac{1}{4}$.

c. It has a probability of $\frac{1}{8}$.

d. It has a probability of $\frac{1}{8}$.

11. Quaternary structure is due to the interactions of a protein's subunits. In this example, the enzyme activity being studied may be carried out by a protein consisting of two different subunits. The polypeptides of the subunits are encoded by separate and unlinked genes.

15. No. The enzyme may require posttranslational modification to be active. Mutations in the enzymes required for these modifications would not map to the isocitrate lyase gene.

18. With the assumption that the three mutations of gene *P* are all nonsense mutations, three different possible stop codons might be the cause (amber, ochre, or opal). A suppressor mutation would be specific to one type of nonsense codon. For example, amber suppressors would suppress amber mutants but not opal or ochre.

22. Single amino acid changes can result in changes in protein folding, protein targeting, or posttranslational modifications. Any of these changes could give the results indicated.

26. If the anticodon on a tRNA molecule also were altered by mutation to be four bases long, with the fourth base on the 5' side of the anticodon, it would suppress the insertion. Alterations in the ribosome can also induce frameshifting.

27. f, d, j, c, c, i, b, h, a, g

30. a. $(GAU)_n$ codes for Asp_n $(GAU)_n$, Met_n $(AUG)_n$, and $stop_n$ $(UGA)_n$. $(GUA)_n$ codes for Val_n $(GUA)_n$, Ser_n $(AGU)_n$, and $stop_n$ $(UAG)_n$. One reading frame in each contains a stop codon.

b. Each of the three reading frames contains a stop codon.

c. If wobble is taken into consideration, Glu may also be GAA, which leaves Lys as AAG. From line 14, CUU is Leu. Therefore, UUC is Phe, and UCU is Ser. In line 13, if UCU is Ser (see above), then Ile is AUC, Tyr is UAU, and Leu is CUA.

Chapter 10

2. O^c mutants are changes in the DNA sequence of the operator that impair the binding of the *lac* repressor. Therefore, the *lac* operon associated with the O^c operator cannot be turned off. Because an operator controls only the genes on the same DNA strand, it is cis (on the same strand) and dominant (cannot be turned off).

5. A gene is turned *off* or inactivated by the "modulator" (usually called a *repressor*) in negative control, and the repressor must be removed for transcription to take place. A gene is turned *on* by the "modulator" (usually called an *activator*) in positive control, and the activator must be added or converted into an active form for transcription to take place.

6. The *lacY* gene produces a permease that transports lactose into the cell. A cell containing a *lacY*⁻ mutation cannot transport lactose into the cell, and so β-galactosidase will not be induced.

9. The term *epigenetic inheritance* is used to describe heritable alterations in which the DNA sequence itself is not changed. Parental imprinting is an example.

13. The inheritance of chromatin structure is thought to be responsible for the inheritance of epigenetic information. This inheritance is due to the inheritance of the histine code and may include the inheritance of DNA methylation patterns.

15. b. The interaction of three factors to recruit a coactivator is synergistic.

20. Although reduced levels of H4 resulted in fewer nucleosomes that were less tightly packed and the activation of certain inducible genes, it is not obvious that overproduction of H4 would have the opposite effect. A nucleosome is an octamer of two subunits each of H2A, H2B, H3, and H4. Overproduction of just one histone would not be expected to result in more nucleosomes and changes in chromatin structure.

22. Construct a set of reporter genes, with the promoter region, the introns, and the region 3' to the transcription unit of the gene in question containing different alterations that do not disrupt transcription or processing. Use these reporter genes to make transgenic animals by germ-line transformation. Assay for expression of the reporter gene in various tissues and the kidneys of both sexes.

26. The *S* mutation is an alteration in *lacI* such that the repressor protein binds to the operator regardless of whether inducer is present. In other words, it is a mutation that inactivates the allosteric site that binds to inducer, while not affecting the ability of the repressor to bind to the operator site. The dominance of the *S* mutation is due to the binding of the mutant repressor even under circumstances in which normal repressor does not bind to DNA (that is, in the presence of inducer). The constitutive reverse mutations that map to *lacI* are mutational events that inactivate the ability of this repressor to bind to the operator. The constitutive reverse mutations that map to the operator alter the operator DNA sequence such that it will not permit binding to any repressor molecules (wild-type or mutant repressor).

29. a. D through J—the primary transcript will include all exons and introns.

b. E, G, I—all introns will be removed.

c. A, C, L—the promoter and enhancer regions will bind various transcription factors that may interact with RNA polymerase

Chapter 11

2. Reverse transcriptase polymerizes DNA with the use of RNA as a template. This RNA : DNA hybrid is then treated with sodium hydroxide to degrade the RNA. (NaOH hydrolyzes RNA by catalyzing the formation of a 2',3'-cyclic phosphate.)

4. Ligase is an essential enzyme within all cells that seals breaks in the phosphate–sugar backbone of DNA. In DNA replication, it joins Okazaki fragments to create a continuous strand and, in cloning, it is used to join the various DNA fragments with the vector. If it were not added, the vector and cloned DNA would simply fall apart.

6. Each cycle takes 5 minutes and doubles the DNA. In 1 hour, there would be 12 cycles, and so the DNA would be amplified $2^{12} = 4096$-fold.

8. Plasmid (15–20 kb) < cosmid (35–45 kb) < BAC (150–300 kb) < YAC (300 kb and larger).

12. The fetus is heterozygous for sickle-cell anemia. The sickle-cell mutation affects an *MstII* restriction site such that, when digested with this enzyme, the restriction fragment is larger (1.3 kb) than normal (1.1 kb). Because both bands are present, the person is heterozygous.

15. The commercial cloning of insulin was into bacteria. Bacteria are not capable of processing introns. Genomic DNA would include the introns, whereas cDNA is a copy of processed (and thus intron-free) mRNA.

16. This problem assumes a random and equal distribution of nucleotides. *Alu*I: $(\frac{1}{4})^4 = $ on average, once in every 256 nucleotide pairs. *Eco*RI: $(\frac{1}{4})^6 = $ on average, once in every 4096 nucleotide pairs. *Acy*I: $(\frac{1}{4})^4(\frac{1}{2})^2 = $ on average, once in every 1024 nucleotide pairs.

18.

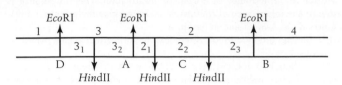

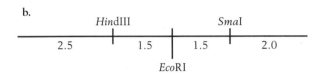

21. a.

b.

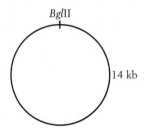

24. Size, translocations between known chromosomes, and hybridization to probes of known location can all be useful in identifying which band on a PFGE gel corresponds to a particular chromosome.

27. a. There is one *Bgl*II site, and the plasmid is 14 kb.

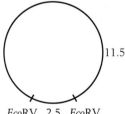

b. There are two *Eco*RV sites.

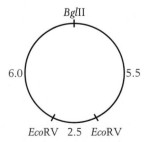

c. The 11.5-kb *Eco*RV fragment is cut by *Bgl*II. The arrangement of the sites must be as follows:

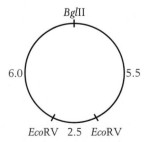

d. The *Bgl*II site must be within the *tet* gene.

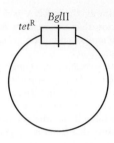

e. There was an insert of 4 kb.

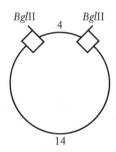

f. There was an *Eco*RV site within the insert.

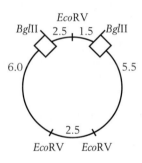

31. a. 5′ TTCGAAAGGTGACCCCTGGACCTTTAGA 3′

b. By complementarity, the template was

3′ AAGCTTTCCACTGGGGACCTGGAAATCT 5′

c. The double helix is

5′ TTCGAAAGGTGACCCCTGGACCTTTAGA 3′
3′ AAGCTTTCCACTGGGGACCTGGAAATCT 5′

d. There are a total of four open reading frames of the six possible.

35. The promoter and control regions of the plant gene of interest must be cloned and joined in the correct orientation with the gene encoding glucuronidase. This orientation places the reporter gene under the same transcriptional control as the gene of interest. Figure 8-33 in the text depicts the methodology used to create transgenic plants. Transform plant cells with the reporter-gene construct and, as described in Figure 8-33, grow into transgenic plants. The gene encoding glucuronidase will now be expressed in the same developmental pattern as the gene of interest, and its expression can be easily monitored by bathing the plant in an X-Gluc solution and assaying for the blue reaction product.

39. a. If the plasmid never integrates, the linear plasmid will be cut once by *Xba*I and two fragments will be generated that will both hybridize to the *Bgl*II probe. The autoradiogram will show two bands whose combined length will equal the full length of the plasmid.

b. If the plasmid integrates occasionally, most cells will still have free plasmids, and these plasmids will be indicated by the two bands mentioned in part a. However, when the plasmid is integrated, two bands will still be generated, but their sizes will vary, depending on where other genomic *Xba*I sites are relative to the insertion point (see the illustration for Problem 38b in the Solutions Manual for similar logic). If integration is random, many other bands will be observed, but, if it is at a specific site, only two other bands will be detected.

Chapter 12

1. Contig describes a set of adjacent DNA sequences or clones assembled by using overlapping sequences or restriction fragments. Because the pieces are assembled into a continous whole, it makes sense that contig is derived from the word contiguous.

3. a., b., and **c.** Detection of hybridization at one end of every chromosome would not be indicative that the probe used is from a unique gene (because FISH indicates more than one region of homology) or from the telomere (because there is a telomere at each end of a chromosome and not just at one end). However, it is possible that the probe is from the centromere because it hybridizes to one location on each chromosome.

5. Assume that the average BAC contains 200 kb. It would take 5000 BAC clones to hold a one-gigabase genome. For fivefold coverage, you would be handling 25,000 clones.

7. The minimal tiling path is the minimum number of clones that represents the entirety of the genome. For reasons of economy and efficiency, it is desirable to sequence clones with as little overlap as possible.

9. A scaffold is also called a supercontig. Contigs are sequences of overlapping reads assembled into units, and a scaffold is a collection of joined-together contigs.

13. Yes. The clone hybridizes to and spans a translocation breakpoint that entails the X chromosome and an autosome. Because the DMD gene normally maps to the X chromosome, it is of interest that a translocation of the X is found in a patient with DMD. If the translocation is the cause of DMD mutation, the clone identifies a least a part of the gene or its location or both.

15. You still need to verify that the gene with the amino acid change is the correct candidate for the phenotype under study. (See Problem 14.)

17. Yes. The operator is the location at which repressor functionally binds through interactions between the DNA sequence and the repressor protein.

19. A level of 35% or more amino acid identity at comparable positions in two polypeptides is indicative of a common three-dimensional structure. It also suggests that the two polypeptides are likely to have at least some aspect of their function in common. However, it does not prove that your particular sequence does in fact encode a kinase.

22. a. To determine the physical map showing the STS order, simply list the STSs that are positive, using parentheses if the order is unknown, and align them with one another to form a consistent order.

YAC A:			1		4	3	
YAC B:		5	1				
YAC C:					4	3	7
YAC D:	(6 2)	5					
YAC E:						3	7

b. When the sequence of STSs is known, the YACs can be aligned as follows, although precise details of overlapping and the locations of ends are unknown:

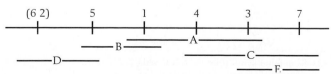

25. From low to highest resolution, the order would be: f, c, (a, d), (e, h), b, g.

28. You can determine whether the cDNA clone was a monster by alignment of the cDNA sequence against the genomic sequence. (Computer programs are available to do so.) Is it derived from two different sites? Does the cDNA map within one (gene-sized) region in the genome or to two different regions? Note that introns may complicate the matter.

32. a. The *cys-1* locus is in this region of chromosome 5. If it were not in this region, linkage to either of the RFLP loci would not be observed.

b. *cys-1* to *RFLP-1* = (2 + 3)/100 × 100% = 5 map units
cys-1 to *RFLP-2* = (7 + 5)/100 × 100% = 12 map units
RFLP-1 to *RFLP-2* = (2 + 3 + 7 + 5)/100 × 100% = 17 map units

```
 ├─5 m.u.─┼──────12 m.u.──────┤
RFLP-1    cys-1                RFLP-2
```

c. A number of strategies could be tried. Because this mutant is auxotrophic, functional complementation can be attempted. Positional cloning or chromosome walking from the RFLPs also is a very common strategy.

34. a. Of the regions of overlap for cosmids C, D, and E, region 5 is the only region in common. Thus, gene *x* is localized to region 5.

b. The common region of cosmids E and F, or the location of gene *y*, is region 8.

c. Both probes are able to hybridize with cosmid E because the cosmid is long enough to contain part of genes *x* and *y*.

37. Assessing whether a short sequence constitutes an exon is difficult. Identification of consensus donor and acceptor splice site sequences can be tried, as can the use of comparative genomics—that is, the conservation of the predicted amino acid encoded by the microexon in the same or other genomes.

Chapter 13

1. Mutations in *gal* can be generated and, from these strains, λd*gal* phage can be isolated. Through hybridization of denatured λd*gal* DNA containing the mutation with wild-type λd*gal* DNA, some of the molecules will be heteroduplexes between one mutant and one wild-type strand. If the mutation was caused by an insertion, the heteroduplexes will show a "looped out" section of single-stranded DNA, confirming that one DNA strand contains a sequence of DNA not present in the other.

If the *gal* genes are cloned, direct comparison of the restriction maps or even the DNA sequence of mutants compared with wild type will give specific information about whether any are the results of insertions.

4. Boeke, Fink, and their coworkers demonstrated that transposition of the *Ty* element in yeast entailed an RNA intermediate. They constructed a plasmid by using a *Ty* element that had a promoter that could be activated by galactose and an intron inserted into its coding region. First, the frequency of transposition was greatly increased by the addition of galactose, indicating that an increase in transcription (and production of RNA) was correlated to rates of transposition. More importantly, after transposition, they found that the newly transposed *Ty* DNA lacked the intron sequence. Because intron splicing takes place only during RNA processing, there must have been an RNA intermediate in the transposition event.

7. The staggered cut will lead to a 9-bp target site duplication that flanks the inserted transposon.

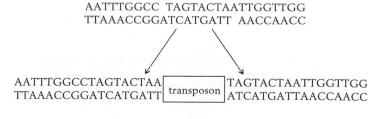

10. It would not be surprising to find a SINE element in an intron of a gene, rather than in an exon. Processing of the pre-mRNA would remove the transposable element as part of the intron and translation of the FB enzyme would not be affected.

Chapter 14

2. The mutant has a deletion of one base and this deletion will result in a frameshift (-1) mutation.

4. With the assumption of single-base-pair substitutions, CGG can be changed to CGU, CGA, CGG, or AGG and would still code for arginine.

6. The enol form of thymine forms three hydrogen bonds with guanine.

9. Apurinic sites have lost either an A or a G. If thymine is preferentially inserted opposite these sites, GC · AT transitions should be produced.

11. One of the mechanisms for repair of UV-induced photodimers is the enzyme photolyase. This enzyme requires visible light for function and would therefore be expected to be more active on bright sunny days.

14. The formation of heteroduplex DNA is part of the double-strand-break model. If the allelic mismatch is not repaired, a 5:3 ascus will result. However, if the mismatch is repaired before cell division, a 6:2 ascus results. A greater frequency of 6:2 asci than that of 5:3 asci indicates that repair of the heteroduplex mismatch is more likely to happen than not.

17. Frameshift mutations arise from the addition or deletion of one or more bases in other than multiples of three. When translated, this mutation will alter the reading frame and therefore the amino acid sequence from the site of the mutation to the end of the protein product. Additionally, frameshift mutations often result in premature stop codons in the new reading frame, leading to shortened protein products. A missense mutation changes only a single amino acid in the protein product.

19. Depurination results in the loss of the adenine or guanine base from the DNA backbone. Because the resulting apurinic site cannot specify a complementary base, replication is blocked. Under certain conditions, replication proceeds with a near random insertion of a base opposite the apurinic site. In three-fourths of these insertions, a mutation will result.

Deamination of cytosine yields uracil. If left unrepaired, the uracil will be paired with adenine in replication, ultimately resulting in a transition mutation. Deamination of 5-methylcytosine yields thymine and thus frequently leads to C-to-T transitions.

Oxidatively damaged bases, such as 8-oxodG (8-oxo-7-hydrodeoxyguanosine) can pair with adenine, resulting in a transversion.

Errors in the course of DNA replication (see Figure 10-12 in the text) can lead to spontaneous indel mutations.

23. Yes. It will cause CG-to-TA transitions.

27. a. and **b.** *Mutant 1:* Most likely a deletion. It could be caused by radiation.

Mutant 2: Because proflavin causes either additions or deletions of bases and because spontaneous mutation can result in additions or deletions, the most probable cause was a frameshift mutation by an intercalating agent.

Mutant 3: 5-BU causes transitions, which means that the original mutation was most likely a transition. Because HA causes GC-to-AT transitions and HA cannot revert it, the original mutation must have been a GC-to-AT transition. It could have been caused by base analogs.

Mutant 4: The chemical agents cause transitions or frameshift mutations. Because there is spontaneous reversion only, the original mutation must have been a transversion. X-irradiation or oxidizing agents could have caused the original mutation.

Mutant 5: HA causes transitions from GC to AT, as does 5-BU. The original mutation was most likely an AT-to-GC transition, which could be caused by base analogs.

c. The suggestion is a second-site reversion linked to the original mutant by 20 map units and therefore most likely in a second gene. Note that auxotrophs equal half the recombinants.

29. O^{-6}-methyl G leads predominantly to high levels of GC $\rightarrow$ AT transitions. 8-OxodG gives rise predominantly to high levels of G $\rightarrow$ T transversions. Finally, C–C photodimers will most often cause C $\rightarrow$ T transitions, but some transversions also are possible.

31. Yes. Because DNA is a double-stranded molecule, replication of the DNA strand with a T-to-T* (altered T that base-pairs with G) change produces an A-to-G transition in the newly replicated complementary DNA strand. If the mRNA is transcribed from the strand with the A-to-G change (the template strand), a U-to-C change is produced in the corresponding mRNA.

original: leu C<u>U</u>N; mutant: pro C<u>C</u>N (where N = any base)

34. Asci 3, 4, and 6 show gene conversion.

Chapter 15

1. MM N OO would be classified as $2n - 1$; MM NN OO would be classified as $2n$; and MMM NN PP would be classified as $2n + 1$.

4. There would be one possible quadrivalent.

7. Seven chromosomes.

9. Cells destined to become pollen grains can be induced by cold treatment to grow into embryoids. These embryoids can then be grown on agar to form monoploid plantlets.

11. Yes.

14. No.

16. An acentric fragment cannot be aligned or moved in meiosis (or mitosis) and is consequently lost.

19. Very large deletions tend to be lethal, likely owing to genomic imbalance or the unmasking of recessive lethal genes. Therefore, the observed very large pairing loop is more likely to be from a heterozygous inversion.

21. Williams syndrome is the result of a deletion of the 7q11.23 region of chromosome 7. Cri du chat syndrome is the result of a deletion of a significant part of the short arm of chromosome 5 (specifically bands 5p15.2 and 5p15.3). Both Turner syndrome (XO) and Down syndrome (trisomy 21) result from meiotic nondisjunction. The term syndrome is used to describe a set of phenotypes (often complex and varied) that are generally present together.

26. The order is $b\,a\,c\,e\,d\,f$.

Allele	Band
b	1
a	2
c	3
e	4
d	5
f	6

27. The data suggest that one or both breakpoints of the inversion are located within an essential gene, causing a recessive lethal mutation.

30. a. When they are crossed with yellow females, the results would be:

X^e/Y^{e+}	gray males
X^e/X^e	yellow females

b. If the e^+ allele was translocated to an autosome, the progeny would be as follows, where "A" indicates autosome:

P	A^{e+}/A ; X^e/Y $\times$ A/A X^e/X^e	
F$_1$	A^{e+}/A ; X^e/X^e	gray female
	A^{e+}/A ; X^e/Y	gray male
	A/A ; X^e/X^e	yellow female
	A/A ; X^e/Y	yellow male

32. Klinefelter syndrome	XXY male
Down syndrome	trisomy 21
Turner syndrome	XO female

36. a. If a 6x were crossed with a 4x, the result would be 5x.

b. Cross *A/A* with *a/a/a/a* to obtain *A/a/a*.

c. The easiest way is to expose the *A/a** plant cells to colchicine for one cell division. This exposure will result in a doubling of chromosomes to yield *A/A/a*/a**.

d. Cross 6x (*a/a/a/a/a/a*) with 2x (*A/A*) to obtain *A/a/a/a*.

e. In culture, expose haploid plants cells to the herbicide and select for resistant colonies. Treat cells that grow with colchicine to obtain diploid cells.

38. a. The ratio of normal to potato will be 5 : 1.

b. If the gene is not on chromosome 6, there should be a 1 : 1 ratio of normal to potato.

42. a. The aberrant plant is semisterile, which suggests an inversion. Because the *d*–*f* and *y*–*p* frequencies of recombination in the aberrant plant are normal, the inversion must involve *b* through *x*.

b. If recombinant progeny are obtained when an inversion is involved, either a double crossover occurred within the inverted region or single crossovers occurred between *f* and the inversion, which occurred some place between *f* and *y*.

44. The original plant had two reciprocal translocation chromosomes that brought genes *P* and *S* very close together. Because of the close linkage, a ratio suggesting a monohybrid cross, instead of a dihybrid cross, was observed, both with selfing and with a testcross. All gametes are fertile because of homozygosity.

original plant: *P S/p s*
tester: *p s/p s*

F_1 progeny: heterozygous for the translocation:

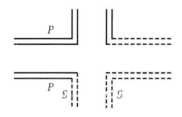

The easiest way to test this is to look at the chromosomes of heterozygotes in meiosis I.

50. The original parents must have had the following chromosome constitution:

G. hirsutum   26 large, 26 small
G. thurberi   26 small
G. herbaceum   26 large

G. hirsutum is a polyploid derivative of a cross between the two Old World species, which could easily be checked by looking at the chromosomes.

52. a. Loss of one X in the developing fetus after the two-cell stage.

b. Nondisjunction leading to Klinefelter syndrome (XXY), followed by a nondisjunctive event in one cell for the Y chromosome after the two-cell stage, resulting in XX and XXYY.

c. Nondisjunction of the X at the one-cell stage.

d. Fused XX and XY zygotes (from the separate fertilizations either of two eggs or of an egg and a polar body by one X-bearing and one Y-bearing sperm).

e. Nondisjunction of the X at the two-cell stage or later.

Chapter 16

1. It would be considered a screen for auxotrophs. If it were a selection, only the desired mutant phenotype would survive.

5. Growth on this plate is a control to check viability of the progeny.

7. The *cdc* mutants affect the cell cycle. Comparative genomics has shown that these same genes are at work in the human cell cycle, and further study has shown that many of these genes are defective in cancers.

10. These phrases are used as "place holders" so that protocols, experiments, and results may be discussed generically without concern for any specific gene or unique situations.

12. The *lacZ* construct is carried on a transposon. Crosses are made that mobilize the construct so that it is transposed to various, random sites in the genome. Flies with the inserted transgene are then screened for *lacZ* expression. Those flies with the desired expression pattern have insertions near genes that are candidates for involvement in some aspect of head development.

14. The goal of PCR mutagenesis is to amplify the DNA sequence of a gene with an altered base within its sequence. The long primer is necessary because it contains both the "mutant" base and all the sequence from that point to one end of the gene. (The other primer of this reaction only has to define the other end of the gene but does not have to stretch to the mutant base.)

16. Gene position is not a necessary requirement for using RNAi. What is needed is the sequence of the gene that you intend to inactivate, because you need to start with double-stranded RNA that is made with sequences that are homologous to part of the gene.

19. On the basis of the results, the gene is haplosufficient. The phenotype of the heterozygote is wild type and the mutant is recessive.

23. Starting with a yeast strain that is *pro-1*, plate the cells on medium lacking proline. Only those cells that are able to synthesize proline will form colonies. Most of them will be revertants; however, some will have second-site suppressors. (Treating the cells with a mutagen before plating them will significantly increase the yield.)

25. The commission was looking for induced recessive X-linked lethal mutations, which would show up as a shift in the sex ratio. A shift in the sex ratio is the first indication that a population has sustained lethal genetic damage. Although other recessive mutations might have occurred, they would not be homozygous and therefore would go undetected. All dominant mutations would be immediately visible, unless they were lethal. If they were lethal, there would be lowered fertility, an increase in detected abortions, or both, but the sex ratio would not shift as dramatically.

26. The mutants can be categorized as follows:

Mutant 1: an auxotrophic mutant
Mutant 2: a nonnutritional, temperature-sensitive mutant
Mutant 3: a leaky, auxotrophic mutant
Mutant 4: a leaky, nonnutritional, temperature-sensitive mutant
Mutant 5: a nonnutritional, temperature-sensitive, auxotrophic mutant

28. The allele for neurofibromatosis must have arisen spontaneously in one of the parents' germ lines. Their chance of having another affected child would range from 0% to 50% (the latter number if the entire germ line were mutant), depending on when this mutation happened (the size of the mutant clone of germ-line tissue).

31. A forward mutation is any change away from the wild-type allele, whereas a reverse mutation is any change back to the wild-type allele.

35. Neomorphic mutations result in novel gene activity and are dominant. Reversion of the dominant phenotype will commonly be the result of introducing another mutation in the already mutant gene that now eliminates its function completely. Most gene knockout mutations are recessive, and so most "revertants" are likely to actually be recessive loss-of-function mutations.

38. Leaky mutants are mutants with an altered protein product that retains a low level of function. Enzyme activity may, for instance, be reduced rather than abolished by a mutation.

40. 15% are essential gene functions (such as enzymes required for DNA replication or protein synthesis). 25% are auxotrophs (enzymes required for the synthesis of amino acids or the metabolism of sugars, etc.). 60% are redundant or pathways not tested (genes for histones, tubulin, ribosomal RNAs, etc., are present in multiple copies; the yeast may require many genes under only unique or special situations or in other ways that are not necessary for life in the "lab").

Chapter 17

2. The activities of proteins controlling the cell cycle or cell-death pathways are controlled by phosphorylation/dephosphorylation, protein–protein interactions, and proteolysis.

4. Oncoproteins may be the result of point mutations, gene fusions, or deletions of key regulatory domains.

7. As caspases (cysteine-containing aspartate-specific proteases) are first translated, they contain sequences that prevent their activity. These inactive versions are called zymogens. Part of the polypeptide must then be removed by enzymatic cleavage for the caspase to become active.

9. Tumor-promoting mutant alleles of tumor-suppressor genes inactivate the proteins that they encode. Because only when both copies of one of these genes are inactivated are tumors observed, these loss-of-function mutations are by definition recessive. Examples of such genes are the *p53* gene and the *RB* gene.

13. Once apoptosis is initiated, a self-destruct switch has been thrown: endonucleases and proteases are released, DNA is fragmented; and organelles are disrupted. This state is "terminal," a state from which the cell will not have a need or chance to reuse the machinery of destruction. On the other hand, the various proteins needed for the regulation and execution of the cell cycle will be needed again if the cell continues to divide. By recycling many of these proteins, the cell conserves its resources (proteins are energetically expensive to make) and recycling allows for rapid divisions, because the cell does not have to spend time remaking all the pieces.

15. a. It would be dominant. The misexpression of FasL from one allele would be dominant to the normal expression of the wild-type FasL allele. In this case, each liver cell would signal its neighboring cells to undergo apoptosis.

  b. No. It would lead to excess cell death, not proliferation.

18. a. Mutations in a tumor-suppressor gene are recessive and due to loss of function. That function can be restored by the introduction of a wild-type allele.

  b. Mutations in an oncogene are dominant and due to gain of function (overexpression or misexpression). The normal function will not inhibit these mutants, and the introduced gene would be ineffective in restoring the normal phenotype.

22. a. In the absence of functional Rb protein, E2F will be in the nucleus.

  b. No. The absence of functional E2F would be epistatic to the absence of functional Rb protein.

  c. A mutation that causes permanent sequestering of E2F in the cytoplasm will likely inhibit the cell cycle.

Chapter 18

2. Microfilaments are linear polymers of the protein actin. The arrangement of the actin proteins in the polymer gives a polarity similar to that observed in polypeptide chains of nucleic acids. Some proteins that interact with microfilaments are able to distinguish the two ends (called "+" and "−") or to distinguish the direction in which they travel along the filament.

5. The term syncitium is used to describe a multinucleate cell. In early *Drosophila* embryogenesis, the first 13 mitoses are nuclear divisions without any cell division. In essence, the embryo at this stage is a single cell with thousands of nuclei. Eventually, each of these nuclei will be surrounded by membrane and become an individual cell. So, in this sense, these future cells share one cytoplasm at the syncitial stage of development.

8. A DNA probe with sequence complementarity to *bicoid* mRNA was used. This probe needs to be labeled (fluorescence, radioactivity) so that, after it hybridizes to the *bicoid* mRNA, the probe's localization can be observed and, by inference, the localization of the mRNA can be known.

11. The DL protein is found in the cytoplasm of cells in the dorsal region of the blastoderm. The DL protein is bound to CACT protein in these cells and therefore remains sequestered in the cytoplasm.

14. The primary pair-rule gene *eve (even-skipped)* would be expressed in seven stripes along the A–P axis of the late blastoderm.

16. Geneticists say that cells "communicate" because many cell–cell interactions have been demonstrated. Cells "talk" to one another though the various molecules secreted as signals that then bind to receptors on or in other cells and, in doing so, deliver the "message."

18. By sequence comparison, the ancestor of the Hox gene B6 is the *Antp* gene in the *Drosophila* HOM set.

21. In humans, a single copy of the Y chromosome is sufficient to shift development toward normal male phenotype. The extra copy of the X chromosome is simply inactivated. Both mechanisms seem to be all-or-none rather than to be based on concentration levels.

25. a. A diffusible substance produced by the anchor cell must affect the development of the six cells. The primary fate has the strongest response to the substance, and the lack of response of the tertiary fate is due to a low concentration or absence of the diffusible substance.

  b. Remove the anchor cell and the six equivalent cells. Arrange the six cells in a circle around the anchor cell. All six cells will develop the same phenotype, which will depend on the distance from the anchor cell.

27. Proper *ftz* expression requires *Kr* in the fourth and fifth segments and *kni* in the fifth and sixth segments.

29. A number of experiments could be devised. A comparison of amino acid sequence between mammalian gene products and insect gene products would indicate which genes are most similar to each other. Using cloned cDNA sequences from mammalian genes for hybridization to insect DNA also would indicate which genes are most similar to each other.

31. If you diagram these results, you will see that the deletion of a gene that functions posteriorly allows the next most anterior segments to extend in a posterior direction. Deletion of an anterior gene does not allow extension of the next most posterior segment in an anterior direction. The gap genes activate *Ubx* in both thoracic and abdominal segments, whereas the *abd-A* and *Abd-B* genes are activated only in the middle and posterior abdominal segments. The functioning of the *abd-A* and *Abd-B* genes in those segments somehow prevents *Ubx* expression. However, if the *abd-A* and *Abd-B* genes are deleted, *Ubx* can be expressed in these regions.

33. The anterior–posterior axis would be reversed.

Chapter 19

1. The frequency of an allele in a population can be altered by natural selection, mutation, migration, nonrandom mating, and genetic drift (sampling errors).

3. The frequency of *B/B* is $p^2 = 0.64$, and the frequency of *B/b* is $2pq = 0.32$.

5. a. $p' = 0.5[(0.5)(0.9) + 0.5(1.0)]/[(0.25)(0.9) + (0.5)(1.0) + (0.25)(0.7)] = 0.53$

  b. 0.75

8. 0.65

10. a. The population is in equilibrium.

  b. The mating is random with respect to blood type.

12. a. The frequency in the population is equal to $q^2 = 0.01$.

b. There would be 10 color-blind men for every color-blind woman (q/q^2).

c. 0.018.

d. All children will be phenotypically normal only if the mother is homozygous for the non-color-blind allele ($p^2 = 0.81$). The father's genotype does not matter and therefore can be ignored.

e. The frequency of color-blind females will be 0.12 and that of color-blind males will be 0.2.

f. From analysis of the results in part e, the frequency of the color-blind allele will be 0.2 in males (the same as in the females of the preceding generation) and 0.4 in females.

15. Before migration, $q^A = 0.1$ and $q^B = 0.3$ in the two populations. Because the two populations are equal in number, immediately after migration, $q^{A + B} = \frac{1}{2}(q^A + q^B) = \frac{1}{2}(0.1 + 0.3) = 0.2$. At the new equilibrium, the frequency of affected males is $q = 0.2$, and the frequency of affected females is $q^2 = (0.2)^2 = 0.04$. (Colorblindness is an X-linked trait.)

18. Many genes affect bristle number in *Drosophila*. The artificial selection resulted in lines with mostly high-bristle-number alleles. Some mutations may have occurred in the 20 generations of selective breeding, but most of the response was due to alleles present in the original population. Assortment and recombination generated lines with more high-bristle-number alleles.

Fixation of some alleles causing high bristle number would prevent complete reversal. Some high-bristle-number alleles would have no negative effects on fitness, and so there would be no force pushing bristle number back down because of those loci.

The low fertility in the high-bristle-number line could have been due to pleiotropy or linkage. Some alleles that caused high bristle number may also have caused low fertility (pleiotropy). Chromosomes with high-bristle-number alleles may also carry alleles at different loci that caused low fertility (linkage). After artificial selection was relaxed, the low-fertility alleles would have been selected against through natural selection. A few generations of relaxed selection would have allowed low-fertility-linked alleles to recombine away, producing high-bristle-number chromosomes that did not contain low-fertility alleles. When selection was reapplied, the low-fertility alleles had been reduced in frequency or separated from the high-bristle loci, and so this time there was much less of a fertility problem.

21. a. Genetic cost $= sq^2 = 0.5(4.47 \times 10^{-3})^2 = 10^{-5}$

b. Genetic cost $= sq^2 = 0.5(6.32 \times 10^{-3})^2 = 2 \times 10^{-5}$

c. Genetic cost $= sq^2 = 0.3(5.77 \times 10^{-3})^2 = 10^{-5}$

Chapter 20

1. Many traits vary more or less continuously over a wide range. For example, height, weight, shape, color, reproductive rate, metabolic activity, etc., vary quantitatively rather than qualitatively. Continuous variation can often be represented by a bell-shaped curve, where the "average" phenotype is more common than the extremes. Discontinuous variation describes the easily classifiable, discrete phenotypes of simple Mendelian genetics: seed shape, auxotrophic mutants, sickle-cell anemia, etc. These traits show a simple relation between genotype and phenotype.

4. The following information is unknown: (1) norms of reaction for the genotypes affecting IQ, (2) the environmental distribution in which the individuals developed, and (3) the genotypic distributions in the populations. Even if this information were known, because heritability is specific to a specific population and its environment, the difference between two different populations cannot be given a value of heritability.

6. a. $p(\text{homozygous at 1 locus}) = 3(\frac{1}{2})^3 = \frac{3}{8}$

$p(\text{homozygous at 2 loci}) = 3(\frac{1}{2})^3 = \frac{3}{8}$

$p(\text{homozygous at 3 loci}) = 2(\frac{1}{2})^3 = \frac{2}{8}$

b. $p(\text{0 capital letters}) = p(\text{all homozygous recessive}) = (\frac{1}{4})^3 = \frac{1}{64}$

$p(\text{1 capital letter}) = p(\text{1 heterozygote and 2 homozygous recessive}) = 3(\frac{1}{2})(\frac{1}{4})(\frac{1}{4}) = \frac{3}{32}$

$p(\text{2 capital letters}) = p(\text{1 homozygous dominant and 2 homozygous recessive})$

or

$p(\text{2 heterozygotes and 1 homozygous recessive}) = 3(\frac{1}{4})^3 + 3(\frac{1}{4})(\frac{1}{2})^2 = \frac{15}{64}$

$p(\text{3 capital letters}) = p(\text{all heterozygous})$

or

$p(\text{1 homozygous dominant, 1 heterozygous, and 1 homozygous recessive}) = (\frac{1}{2})^3 + 6(\frac{1}{4})(\frac{1}{2})(\frac{1}{4}) = \frac{10}{32}$

$p(\text{4 capital letters}) = p(\text{2 homozygous dominant and 1 homozygous recessive})$

or

$p(\text{1 homozygous dominant and 2 heterozygous}) = 3(\frac{1}{4})^3 + 3(\frac{1}{4})(\frac{1}{2})^2 = \frac{15}{64}$

$p(\text{5 capital letters}) = p(\text{2 homozygous dominant and 1 heterozygote}) = 3(\frac{1}{2})^2(\frac{1}{4}) = \frac{3}{32}$

$p(\text{6 capital letters}) = p(\text{all homozygous dominant}) = (\frac{1}{4})^3 = \frac{1}{64}$

8. The population described would be distributed as follows:

3 bristles	$\frac{19}{64}$
2 bristles	$\frac{44}{64}$
1 bristle	$\frac{1}{64}$

The 3-bristle class would contain 7 different genotypes, the 2-bristle class would contain 19 different genotypes, and the 1-bristle class would contain only 1 genotype. Determining the underlying genetic situation by doing controlled crosses and determining progeny frequencies would be very difficult.

10. a.

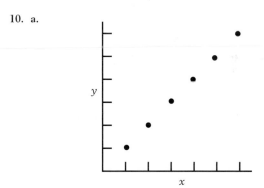

b.

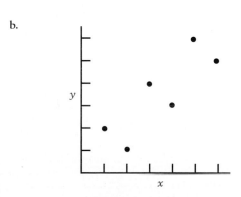

c.

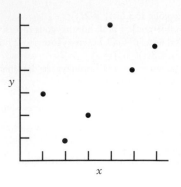

d.

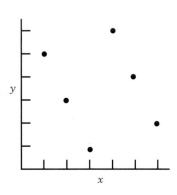

a. cov $xy = \frac{1}{6}[(1)(1) + (2)(2) + (3)(3) + (4)(4) + (5)(5) + (6)(6)]$ $- (21/6)(21/6) = 2.92$

Standard deviation $x = s_x = \sqrt{1/N\Sigma(x_i - \bar{x})^2} = 1.71$

Standard deviation $y = s_y = \sqrt{1/N\Sigma(y_i - \bar{y})^2} = 1.71$

Therefore, $r_{xy} = 2.92/(1.71)(1.71) = 1.0$. The other correlation coefficients are calculated in a like manner.

b. 0.83

c. 0.66

d. −0.20

14. a. The regression line shows the relation between the two variables. It attempts to predict one (the son's height) from the other (the father's height.) If the relation is perfectly correlated, the slope of the regression line should approximate 1. If you assume that individuals at the extreme of any spectrum are homozygous for the genes responsible for these phenotypes, then their offspring are more likely to be heterozygous than are the original individuals. That is, they will be less extreme. Additionally, there is no attempt to include the maternal contribution to this phenotype.

b. For Galton's data, regression is an estimate of heritability (h^2), *assuming* that there were few environmental differences between all fathers and all sons both individually and as a group. However, there is no evidence given to determine if the traits are familial but not heritable. This data would indicate genetic variation only if the relatives do not share common environments more than nonrelatives do.

Chapter 21

2. The three principles are: (1) organisms within a species vary from one another, (2) the variation is heritable, and (3) different types leave different numbers of offspring in future generations.

5. A population will not differentiate from other populations by local inbreeding if:

$$\mu \geq 1/N$$

and so

$$N \geq 1/\mu$$
$$N \geq 10^5$$

7. A population will not differentiate from other populations by local inbreeding if the number of migrant individuals ≥ 1 per generation.

For part a, migration is not sufficient to prevent local inbreeding, and so the results are roughly the same as seen in Problem 6. For part b, there is one migrant per generation, and so the populations will not differentiate and allelic frequencies will remain the same in all populations.

9. The mean fitness for population 1 $[p(A) = 0.5, p(B) = 0.5]$ is 0.825. The mean fitness for population 2 $[p(A) = 0.1, p(B) = 0.1]$ is 0.856. There are four adaptive peaks: at $p(A) = 0.0$ or 1.0 and at $p(B) = 0.0$ or 1.0. (The mean fitness will be 0.90 at any of these points.) With population 1, the direction of change for both $p(A)$ and $p(B)$ will be random. Both higher or lower frequencies of either allele can result in increased mean fitness (although there are some combinations that would lower fitness). Because population 2 is already near an adaptive peak of $p(A) = 0.0$ and $p(B) = 0.0$, both $p(A)$ and $p(B)$ should decrease to increase the mean fitness.

12. All human populations have high i, intermediate I^A, and low I^B frequencies. The variations that do exist among the different geographical populations are most likely due to genetic drift. There is no evidence that selection plays any role regarding these alleles.

14.

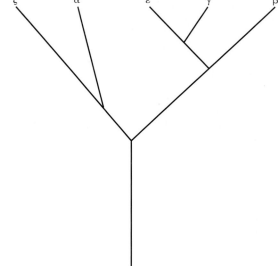

Index

Note: Page numbers followed by f indicate figures; those followed by t indicate tables. **Boldface** page numbers indicate Key Terms.

Abd-B, 321–322
Abelson tyrosine kinase, in leukemia, 569
Aberrant euploidy, 483, 484
  gene balance and, 494–496
abl proto-oncogene, 561t, 563, 563f
ABO blood groups, 194–195, 194f, 195f
Abraxas, wing color in, 76–77
Accessory proteins, in DNA replication, **242**–243
Ac elements, **425**–429, 437–438, 438f. *See also* Transposable elements
  change in phase in, 445, 445f
  as vectors, 439–440
Acentric chromosomes, **497**
Acentric fragment, **499**, 499f
Achillea millefolium (yarrow)
  genotype-phenotype interaction in, 20–21, 20f
  norm of reaction for, 20–21, 20f
Achondroplasia, 45, 45f
Acridine orange, as mutagen, **458**, 459f
Acrocentric chromosomes, 84t, **85**
Activated transcription, **317**
Activator element. *See Ac* elements
Activators, **304**, 304f
Active site, **10**, **277**, 277f
Adaptation, 685–687, 686f, 687f
Adaptive landscape, **686**, 686f
Adaptive peaks, 685–687, 686f
Adaptive surface, **686**, 686f
Addition mutations, 454, 456, 465–466, 465f, 466f
Additive effect, **660**–661
Additive genetic variation, **661**
Adelberg, Edward, 163
Adenine, 232–233, 232f, 233t. *See also* Base(s)
Adenine methylase, in mismatch repair, **471**
Adenosine 5'-monophosphate (AMP), 258, 258f
Adjacent-1 segregation, **501**, 501f
Aflatoxin B₁, as mutagen, **461**, 461f
A gene, in *lac* system, 305, 305f, 306, 307
  polar mutations in, 311–312
Agriculture. *See* Breeding
Agrobacterium tumefaciens, Ti plasmid of, 368–370, 369f
Albinism, 8–9, 9f, 10, 11f, 44–45, 44f
  complementation and, 202–203, 230f
Algae, blue-green, 152
Alkaptonuria, 362–364, 441f, 442
Alkylating agents, as mutagens, **458**–459, 459f
Alkyltransferases, in DNA repair, 468–469
Allele(s), **2**–3, **8**, **33**, 34–36. *See also* Gene(s)

dominant, 33, 34, 192–193, 193f. *See also* Dominance
  fixed, **626**
  genotype and, 9
  lethal, **195**–197, 196f, 197f
    synthetic, **206**–207, 207f
  letter code for, 8, 35
  marker, 153, 154t
  molecular properties of, 34–35
  multiple, **192**
  mutant, 35, 79. *See also* Mutation(s)
  naming of, 79
  null, **10**, **192**
  phenotype and, 8–9
  recessive, 33, 34, 192–193, 193f. *See also* Recessiveness
  replacement of. *See also* Genetic drift
    rate of, 695–696
  segregation of, 13, 33–34, 35–36, 91f, 93–95, 94f, 95f, 97–103, 98f, 99f, 101f, 102f. *See also* Segregation
  sublethal, 197
  wild-type, 35, 79
    complementation and, 199–203, 200f
Allelic frequency, **614**, 614t, 615t. *See also* Variation
  calculation of, 615
  fitness and, 631–634
  genetic drift and, 636–637
  mutations and, 626–628, 627t
  random change in, 636–637
  rate of change in, 633, 633f, 634f
  selection and, 631–633
Allelic series, **192**
Allelic variation. *See* Variation
Allopatric speciation, **690**
Allopolyploidy, **485**, 487–489, 487f–489f
  in plant breeding, 490
Allosteric binding sites, 304
Allosteric site, **304**
Allosteric transition, **306**
α helix, 276f, 277
Alternate segregation, **501**, 501f
Alternative splicing, 267–268, 267f
Alu, 424, **441**
Amber codon, 282
Amber mutant, 282, 282f
Amino acids, **275**
  codons for, 5, 278–282, 282f. *See also* Genetic code
  degeneracy for, 284–285
  in mutation, 9–10
  peptide bonds of, 275, 275f, 276f
  sequence of, 279–282. *See also* Genetic code
  side chains of, 5–7, 6f, 275. *See also* Polypeptide(s); Protein(s)
  posttranslational modification of, 275

structural formula for, 275
structure of, 275
tRNA and, 282–285, 283f, 284f
Amino acid sequence polymorphisms, 615–617, 616f, 617t
Amino acid sequences, 5
Aminoacyl-tRNA, **283**, 284, 289–290, 289f
Aminoacyl-tRNA synthetases, 283, 284
Amino end, **275**, 275f, 276f
Amniocentesis, **364**
AMP (adenosine 5'-monophosphate), 258, 258f
Amphidiploid, **487**, 487f
Analysis of variance, **657**
Anaphase, **92**
  meiotic, 94f, 95
  mitotic, 91f, 92, 93f
Anaphase bridge, **497**
Anchor cells, 599, 599f
Androgen insensitivity syndrome, 54, 602
Aneuploidy, **485**, 490–494
  in animals, 494
  deleterious effects of, 494–496
  gene balance and, **494**–496
  gene-dosage effect and, **495**
  in humans, 491–494, 492f–494f
  inviability and, 485
  monosomy, 483t, **490**, 491, 491f
  nomenclature for, 483t, 490
  nondisjunction and, **491**, 491f
  in plants, 76, 77f, 494
  trisomy, 483t, **490**, 492–494, 493f, 494f
  vs. polyploidy, 494–495
Animal(s)
  cloning of, 325–326, 325f
  polyploid, 490
Animal breeding
  continuous variation in, 10
  family selection in, **664**
  hybrid-inbred method of, **664**
  Mendelian inheritance and, 13–14, 29–45
  narrow heritability in, 662–664
Antennapedia complex, 595, 596f
Anterior-posterior axis, in pattern formation, 583–587, 583f, 586f, 588f, 589–590, 590f
Anthocyanin, 200–201
Antibiotic resistance
  causative mutations in, 524–525
  composite transposons and, **431**, 431f
Antibody, **353**
Antibody probes, 353, 353f
Anticodon, **283**, 283f
Antigens, blood group
  genetic variation in, 614, 615, 615t
  racial distribution of, 688, 689t
  Rh incompatibility and, 634–635

ANTP-C, 595, 596f
Apaf, 553, 556, 556f
AP endonuclease, in DNA repair, **469**, 469f
Apoptosis, 547–548, **552**–553
  Bcl proteins in, 553
  in cancer, 561t, 563–564, 564f, 565–566
  caspases in, **552**–553, 553f
  in cell cycle, 557–558
  events in, 552, 552f
  FAS system in, 556, 556f
  in nematodes, 554
  signaling in, 556, 556f
  survival factors and, **556**
Apoptotic bodies, 552
Apurinic site, **461**, 461f
Arabidopsis thaliana
  development in, 603–605, 604f
  as model organism, 13, 14f, 604
ara operon, 315–316, 315f
Architectural proteins, 318
ar gene/AR protein, 602, 603f
Artificial selection, **629**. *See also* Breeding; Selection
  narrow heritability and, 662–664
  variation from, 613f, 629–633
Aruncus dioicus, sex chromosomes in, 49f
Ascomycetes, meiosis in, 100
Ascus, **100**, 328, 328f
A site, **286**, 287t, 289–290
Aspergillus, genetic screens for, 528, 528f
Assortative mating, 613f, 625–626
  negative, **625**–626
  positive, **625**–626
Astrachan, Lawrence, 257
Asymmetrical distribution, 668, 688f
Asymmetry, developmental, 579–582, 579f–582f
  in pattern formation, 582–587, 582f
Attachment site, **173**
Autonomous elements, **427**
Autophosphorylation, in signaling, 556
Autopolyploidy, **485**–487, 485f, 486f
  in plant breeding, 490, 490f
Autoradiogram, **351**–352
Autosomal chromosomes, **29**
Autosomal dominant disorders, pedigree analysis of, 45–46, 46f, 47f
Autosomal inheritance, **29**–48. *See also* Mendelian inheritance
  in humans, 42–48
  meiosis in, 97–103, 98f, 99f, 101f, 102f
  pedigree analysis for, 42–48
Autosomal polymorphisms, pedigree analysis of, 46–48, 47f
Autosomal recessive disorders, pedigree analysis of, 42–44, 43f, 44f

Autotretraploid plants, breeding of, 490, 491f
Autotriploid plants, breeding of, 490
Auxotrophic mutants, **153**
Auxotrophy, reversion of to prototrophy, 525–526, 526f
Avery, Oswald T., 166, 230

Bacteria. *See also Escherichia coli*
culture of, 153, 153f
DNA repair in, 468–470
DNA replication in, 3–4, 3f, 237–246, 239, 239f
drug-resistant
causative mutations for, 524–525
composite transposons and, **431**, 431f
gene regulation in, 302f, 303–316. *See also* Gene regulation, in prokaryotes
genome of, 15, 16f
genotypic symbols for, 154t
insertion sequences in, **429–430**, 429f, 430f
laboratory methods for, 153, 153f
lysogenic, **170**
mapping of, 161–162, 163f, 171–172, 171f, 174–175, 174f, 175f
as model organisms, 13, 14f, 154
phage infection of, 167–168, 167f
prototrophic, **153**
recombination in, 152–177, 152f, 153. *See also* Bacterial conjugation; Bacterial transformation; Transduction
overview of, 152–153, 152f
transcription in, 261–263, 262f, 263f
translation in, 274f, 275
transposable elements in, 434–440
wild-type, 153, 154f
Bacterial artificial chromosome (BAC) vectors, **349**
Bacterial chromosome
circular shape of, 150f, 159–161, 161f, 239, 239f
mapping of, 161–162, 163f
Bacterial clone, **153**. *See also* Clone(s)
Bacterial colonies, **153**, 154f
Bacterial conjugation, 155–165, **156**
crossing-over in, 157, 159, 159f, 160–161, 160f, 162, 162f
definition of, 153
discovery of, 155–156, 155f
donor in, **156**
endogenote in, 162, 162f
exogenote in, 162, 162f
fertility factor in, 156–157, 156f, 157f
Hfr strain in, 157–161, 157f
insertion sites in, 159–161, 160f, 161f
merozygote in, 162, 162f
origin in, 158, 158f, 161
pili in, 157, 157f
plasmid transfer in, 156–157, 156f, 157f
recipient in, **156**
rolling circle replication in, 157
sequence of events in, 160–161, 160f, 161f
terminus in, **161**
vs. transformation, 166

Bacterial lysis, 167, 168, 168f
Bacterial mutants, 153
Bacterial transduction, 153. *See also* Transduction
Bacterial transformation, 166, 166f. *See also* Transformation
definition of, 153
Bacteriophages. *See* Phage(s)
BAC (bacterial artificial chromosome) vectors, 349
Balanced polymorphisms, 634–636
Balanced rearrangements, **497**
Balancer chromosomes, **500–501**
Balancing selection, **685**
Banding, chromosome, 86–88, 86f–88f
Barr body, **323**
Basal transcription apparatus, **317**
Base(s), **232**
alteration of, mutations due to, 458–459, 459f
damage to, mutations due to, 459–461, 460f, 461f, 464
in DNA, 3–4, 4f, 232–233
in codons, 5
complementary, 4, 4f, 5f, 237, 237f
composition of, 232–233
molar properties of, 232–233, 233t
paired. *See* Base pairs/pairing
purine, **232**, 232f
pyrimidine, **232**, 232f
enol form of, **457**, 457f
in genetic code, 278–282, 282f. *See also* Genetic code
imino form of, **457**, 457f
keto form of, **457**, 457f
replacement of, mutations due to, 10, **10**, 454, 456, 456f, 457–458, 458f, 459f, 464–465
in RNA, 258
Base analogs, **457**
Base-excision repair, 468f, **469**–470
Base pairs/pairing
in DNA, 4, 4f, 5f, 234–235, 235f–237f. *See also* Double helix
mismatched, mutations due to, 10, 454, 456, 457–461, 458f–461f, 464–465
in replication, 237–240, 237f–240f
in RNA, 237–240, 237f–240f
in transcription, 259–261, 260f, 261f, 263, 263f
variation in, 619–620
Base-substitution mutations, **10**, 454, 456, 456f, 457–458, 458f, 459f, 464–465
Bateson, William, 77, 117
bcd gene/BCD protein, 583–587, 583f, 585f–588f
B-cell lymphoma, 566–568, 567f
chromosomal rearrangements in, 563–564, 564f
B-cell oncogenes, 563–564, 564f
Bcl proteins
in apoptosis, 553
in lymphoma, 563–564, 564f
Bcr 1–Abl fusion oncoprotein, 563, 563f
Beadle, George, 187–189, 428, 428f

Beetles, transgenic, 538
Benzer, Seymour, 169–170
bicoid gene/BCD protein, 583–587, 583f, 585f–588f
Bimodal distribution, **668**, 688f
Binary fate decision, **578**
Bioinformatics, **12**, 406–410
BLAST search and, 407–408, 409f
cDNA sequences and, 407, 407f, 409f
codon bias and, 408, 409f
database searches and, 407, 409f, 410
docking site prediction and, 407, 408f
in mRNA detection, 407–410
in open reading frame detection, 407–410
in polypeptide sequence prediction, 407–410
Biological machines, 187
replisome as, 242
ribosome as, 285
Bithorax complex, 595, 596f
Bivalents, **95**, 485
Blakeslee, Alfred, 76
BLAST search, 407–408
Blood group antigens
genetic variation in, 614, 615, 615t
racial distribution of, 688, 689t
Rh incompatibility and, 634–635
Blue-green algae, 152
Boeke, Jef, *435*
Bonds
hydrogen, in double helix, 234–235, 235f
peptide, 275, 275f, 276f
Boveri, Theodor, 75
Branch diagrams, 37, 39
Brassica, allopolyploids of, 487–488, 487f, 488f
Breeding
continuous variation in, 10
family selection in, **664**
hybrid-inbred method of, **664**
Mendelian inheritance and, 13–14, 29–45. *See also* Autosomal inheritance
narrow heritability in, 662–664
vs. evolution, 681
Brenner, Sidney, 279, 282, 291
Bridges, Calvin, 79–80
Broad heritability (H^2), **657**–660
estimation of, 657–658
meaning of, 658–660
Bubble-boy disease, gene therapy for, 375–379
Bulbar muscular atrophy, 467–468
Burnham, Charles, 428f
BX-C, 595, 596f

Caenorhabditis elegans (nematode)
apoptosis in, 554
cytoskeletal asymmetry in, 581, 581f
genomic sequencing for, 397–398, 398f
germ-line determination in, 581, 581f
as model organism, 14, 14f, 554, 581
RNA interference in, 534, 534f
transgenesis in, 370–371, 370f

vulva development in, 598–599, 598f
CAF-1 (chromatin assembly factor 1), 244, 244f
Cairns, John, 239
Calico cats, X inactivation in, 323–324, 324f
cAMP, in gene regulation, 312–315, 312f–315f
Campanula (harebell), complementation in, 199–202, 199f, 200f
Campbell, Allan, 160, 173
Canalized character, **688**, 688f
Cancer, **558**–569
apoptosis inhibitors in, 561t, 563–564, 564f
complexity of, 568–569, 569f
diagnosis and treatment of, genomic approaches in, 566–569, 567f
familial, 560–561
gene fusion in, 563–564
gene inactivation in, 568–569, 569f
as genetic disease, 559
as multistep process, 559, 560f, 569
mutations in, 559–564, 560f, 561t, 562–564, 562f–564f, 562t
detection of, 566
oncogenic, 559–560, 560f, 562–564, 562f–564f
in tumor-suppressor genes, 560, 560f
oncogenes in, 559–564
classes of, 561–564, 561t
mutations in, 559–560, 560f, 562–564, 562f–564f
overview of, 547
proto-oncogenes in, **559**, 561–562
signaling in, 562–563
tumor-suppressor genes in, mutations in, 560–561, 560f, 564–566, 565f
Cancer cells
characteristics of, 558–559, 559f
growth of, 546f
loss-of-heterozygosity in, **565**, 566, 567f
mutations in, 559–561, 560f
single nucleotide polymorphisms in, 566, 567f
Candidate gene approach, **405**–406, **664**–665
CAP (catabolite activator protein), **312**–315, 312f–314f
Cap, 5′ end, **266**–267, 266f
Carboxyl end, **275**, 275f, 276f
Carboxyl tail domain (CTD), **265**, 266, 266f
Cardinal genes, in pattern formation, **591**–594, 592t
Carothers, Elinor, 76
Carriers, genetic, 52, 52f
Caspases, **552**–553, 553f
Cat(s)
tailless, 197, 197f
X inactivation in, 323–324, 324f
Catabolite activator protein (CAP), **312**–315, 312f–314f
Catabolite repression, of *lac* operon, 312–315, 312f–315f
Cavalli-Sforza, Luca, 157
cdc mutations, 551
CDK (cyclin-dependent protein kinase), **549**–550, 549f

CDK-cyclin complexes, **548**–550, 548f, 549f
cDNA (chloroplastid DNA), 8, 103–106, 103f–106f
cDNA (complementary DNA), **343**
  in open reading frame detection, 407, 407f
  in recombinant DNA production, 343, 344f
  synthesis of, 343–344, 344f
cDNA library, 350–357, **351**
  clone identification in, 351–357. *See also* Clone(s), identification of
  creation of, 350–351
  definition of, 351
  expression, 353
Cell(s)
  anchor, **599**, 599f
  cancer
    characteristics of, 558–559, 559f
    growth of, 546f
    loss-of-heterozygosity in, **565**, 566, 567f
    mutations in, 559–561, 560f
    single nucleotide polymorphisms in, 566, 567f
  daughter, 93, 93f, 98f
  in developmental fields, **577**–578
  differentiated, **547**
  diploid, 83
  haploid, 83
  pole, **582**, 582f
  stem, **547**
  totipotent, **578**
Cell–cell communication, **553**–558. *See also* Signaling
Cell clone, 153
Cell cycle, 90–95, 91f
  anaphase in, **92**
    meiotic, 94f, 95
    mitotic, 91f, 92, 93f
  apoptosis in, 552–553, 557–558
  cell proliferation in, 548–552
  checkpoints in, 547, **550**–551, 550f
  in diploid organisms, 96, 96f
  G1 phase of, 548
  G2 phase of, 548
  in haploid-diploid organisms, 96f, 97
  in haploid organisms, 96–97, 96f
  interphase in, **91**, 91f, 548
  meiosis in, 74f, 75–76, 76f, 91f, 93–95, 94f, 95f. *See also* Meiosis
  metaphase in, **91**
    meiotic, 94f, 95
    mitotic, 87
  mitosis in, 74f, 75, 90–93, 91f–93f. *See also* Mitosis
  model organism for, 551
  M phase of. *See* Mitosis
  prophase in
    meiotic, 94f, 95
    mitotic, 91, 91f, 93f
  regulation of, 548–551, 548f–550f
    apoptosis in, 556, 556f, 557–558
    CDK-cyclin complexes in, **548**–550, 548f, 549f
    growth factors in, **555**
    phosphorylation in, 549–550, 549f, 556
    signaling in, 552f–558f, 553–558. *See also* Signaling

replication in, 91–95, 91f–94f, 97–103, 98f, 99f, 101f, 102f, 246–247, 247f. *See also* DNA replication
S phase of, **91**, 91f, 93, 98f, 548
stages of, 91–95, 91f, 548
telophase in, **93**
  meiotic, 94, 95f
  mitotic, 91f, 93, 93f
  in yeast, 551
Cell-cycle regulation, TGF-β in, 555–556
Cell division. *See also* Cell cycle
  chromosome segregation in, 35–36. *See* Chromosome(s), segregation of
  in diploids, 96, 96f, 98f
  in haploids, 96–97, 96f, 98f
  in meiosis, 74f, 75–76, 76f, 93–95, 94f, 95f, 931f. *See also* Meiosis
  in mitosis, 74f, 75, 90–93, 91f–93f. *See also* Mitosis
Cell fate, **577**
  memory systems for, 597–598, 598f
  refinement of, **578**
Cell number, regulation of, 546–558
  in diploids, 96, 96f, 98f
  in haploids, 96–97, 96f
  in meiosis, 74f, 75–76, 76f, 93–95, 94f, 95f, 97, 98f, 931f
  in mitosis, 74f, 75, 90–93, 91f–93f, 97, 98f
  overview of, 546–547, 546f
Cell proliferation
  apoptosis and, 547–548. *See also* Apoptosis
  development paths in, 577–578
  mechanisms of, 548–551, 548f–550f
Centimorgan, 124
Central tendency, 648
Centrifugation, cesium chloride gradient, 238–239, 238f
Centromere, 75, 85
  kinetochore and, 92, 92f
Centromere mapping, 129–131, 131f
Centromeric satellite DNA, 85, 85f
Cesium chloride gradient centrifugation, 238–239, 238f
Change in phase, **445**, 445f
Chaperones, **292**
Chaperonin folding machines, **292**
Character, 30–34. *See also* Phenotype; Traits
  canalized, **688**, 688f
Character difference, **31**
Character form, 31
Chargaff's rules of base composition, 232–233, 233t
Charged tRNA, **283**
Chase, Martha, 230–231
Checkpoints, in cell cycle, 547, **550**–551, 550f
Chemical genetics, **534**–535, 534f
Chemigenomics, **534**–535, 534f
Chiasma, **95**, 119
Chips, DNA, **413**, 414f. *See also* DNA microarrays
Chi-square test, 40–42, 41f
  for gene interaction, 209
  in linkage analysis, 131–133
Chloroplast(s)

cytoplasmic inheritance and, 55–56, 56f, 57f
genes/chromosomes of, 55, 56, 75, 103
Chloroplast DNA (cpDNA), 8, 103
Chorionic villus sampling, 364–365
Chromatids
  crossing-over of, 95, 119–121, 119f, 120f. *See also* Crossing-over
  recombinant, in centromere mapping, 131
  sister, **91**
    crossing-over of, 121
    in meiosis, 91f, 93–95, 94f, 97–102, 98f, 99f, 101f, 102f
    in mitosis, 91–93, 91f–93f, 98f
Chromatin, **84**, 88, 89f, 322–333
  condensation of, 84, 322, 323f
  histones and, 328–329, 329f
  epigenetic inheritance and, 323–326, 331–332, 332f
  euchromatin, 84, 322, 323f
  in gene silencing, 323–327
  heterochromatin, 84, 322, 323f
  imprinting and, 324–326, 325f
  nucleosomes in, 322, 323f
  repositioning of, 327–329, 327f
  position-effect variegation and, 326, 326f, 332–333, 332f, **502**
  remodeling of, **327**–331
  silent, **327**
  structure of, 322, 323f
  X inactivation and, 323–324, 324f
Chromatin assembly factor 1 (CAF-1), **244**, 244f
Chromocenter, **87**–88, 87f
Chromosome(s), 73–107
  acentric, **497**
  acrocentric, 84t, **85**
  arm-length ratio of, 84t, 85
  inversions and, 500
  autosomal, **29**
  bacterial
    artificial, **349**
    circular shape of, 150f, 159–161, 161f, 239, 239f
    mapping of, 161–162, 163f
  balancer, **500**–501
  breakage of, 496–497, 496f. *See also* Chromosome rearrangements
  chloroplast, 75
  chloroplastid, 103
  dicentric, **497**
  differential regions of, 49, 49f
  DNA in, 80–81, 80f, 81f
    amount of, 80–81, 88
    packaging of, 80–81, 88–90, 89f, 90f
  early studies of, 75–80. *See also* Mendelian inheritance
  gene arrangement on, 81–82, 81f–83f
  gene location on, 76–80. *See also* Gene locus
    gene function and, 129
    mapping of, 124–131. *See also* Maps/mapping
  homeologous, **485**
  homologous, **3**, 74f, 75, 76, 76f
    pairing of, 485–486, 485f, 486f
  metacentric, 84t, 85
  mitochondrial, 75
  nondisjunction of, **491**, 491f

organelle, 103–106, 103f–106f
pairing of, 74f, 75, 76, 76f
  regions of, 49, 49f
peudoautosomal regions of, 49, 49f
Philadelphia, 561, 563, 563f, 569
polymorphisms and, 617–618, 618f, 618t, 619f
polytene, **87**–88, 87f
properties of, 80–90
rearrangements of. *See* Chromosome rearrangements
replication of, 74f, 75
segregation of, 13, 33–36, 49, 74f, 75–78, 76f, 77f, 91–93, 91f–93f
  abnormal, 79–80, 79f
  adjacent-1, **501**, 501f
  alternate, **501**, 501f
  in autopolyploids, 485, 485f
  balanced translocations and, **501**–502, 501f
  equal, 13, 33–36
  in meiosis, 91f, 93–95, 94f, 95f, 97–102, 98f, 99f, 101f, 102f
  in mitosis, 91–93, 91f–93f
  sex. *See* Sex chromosome(s); X chromosome; Y chromosome
size of, 83, 84t
structure of, 80–90, 80f, 81f, 85f–88f
  changes in, 496–506. *See also* Chromosome rearrangements
  three-dimensional, 88–90, 89f, 90f
telocentric, 84t, **85**
visible landmarks of, 83–90, 85f–88f
yeast artificial, 367f, 368
Chromosome 7, sequence map for, 410–411, 411f
Chromosome banding, 86–88, 86f–88f
Chromosome mapping. *See* Maps/mapping
Chromosome mutations, **482**–509. *See also* Mutation(s)
Chromosome number, 83, 83t
  aneuploid, 483, 483t, **485**, 490–496
  changes in, 483–496
    in parts of chromosome sets, 483, 490–496
    in whole chromosome sets, 483–490
  diploid, 74, 83, 483, 483t. *See also under* Diploid
  euploid, 483, 483t
    aberrant, 483, 484
  gene balance and, **494**–496
  gene-dosage effect and, **495**–496
  haploid, 83, 483, 483t. *See also under* Haploid
  hexaploid, **483**, 483t
  monoploid, **483**, 483t, 484
  monosomy, 483t, 490, 491–492, 492f
  organism size and, 484, 485f, 490, 490f, 494
  pentaploid, **483**, 483t
  phenotype and, 76, 77f, 494–496

Chromosome number *(cont)*
  polyploid, 483, 483t, 484–496. *See also* Polyploidy
  tetraploid, 483, 483t, 484
  triploid, 483, 483t, 485
  trisomy, 483t, 490, 492–494, 493f, 494f
Chromosome painting, 389f, **403**, 481f
Chromosome rearrangements, 496–506
  balanced, **497**
  breakage-induced, 496–497, 496f
  during crossing-over, 474–476, 475f, 496f, 497, 497f
  deletions, 10, 454, 456, 465–466, 465f, 466f, 496, 496f, **497**, 502–505. *See also* Deletion(s)
    mapping of, 403–404, 404f
  duplications, 10, 496f, 497, **497**–498, 505–506, 691–692. *See also* Duplication(s)
  in humans, frequency of, 506, 507f
  identification of, by comparative genomic hybridization, 506, 506f
  incidence of, in humans, 506, 507f
  inversions, 496, 496f, **497**, 498–501, 498f–500f, 617–618, 618f. *See also* Inversion(s)
  in lymphoma, 563–564, 564f
  mapping of, 403–404, 404f. *See also* Maps/mapping
  numerical, 483–496. *See also* Chromosome number; Polyploidy
  origins of, 496f
  polymorphic, 617–618, 618f, 618t
  structural, 496–506
  translocations. *See* Translocations
  transposable elements and, 425, 426f. *See also* Transposable elements
  unbalanced, **497**
Chromosome theory of heredity, **75**–80
  cytologic evidence for, 75–76
  linkage mapping and, 125
  originators of, 75–76
  sex linkage evidence for, 76–78
Chromosome walking, **356**–357, 356f
Chronic myelogenous leukemia
  Philadelphia chromosome in, 561, 563, 563f, 569
  treatment of, 569
Cis-acting sequences, in gene regulation, **308**
  in eukaryotes, 317–318, 317f, 318f
  in prokaryotes, 308, 308f
Cissa conformation, of linked genes, **118**, 118f
Clone(s), **153**
  bacterial, 153
  in cDNA libraries, 350–351
  in genomic libraries, 350–351
  identification of
    by functional complementation, 355
    by gel electrophoresis, 354, 354f
    by Northern blotting, 354
    by probes, 351–354
    by Southern blotting, 354, 355f

  sequencing of. *See* Genome sequencing
Clone contigs, 396, 397f
Clone fingerprint mapping, 396, 397f
Cloning, 343–350, 347f, 349f, 350f. *See also* DNA technology
  of alkaptonuria gene, 362–364
  of animals, 325–326, 325f
  cutting and pasting in, 344–346, 344f, 345f
  donor DNA for, 343–344
  donor DNA in, 343–344
  in medicine, 362–364
  P elements in, 439–440, 439f
  positional, 129, 346f, 356–357, **356**–357, 357f
  in recombinant insulin production, 343, 349
  transposable elements in, 439–440, 439f
  vectors for, 346–349
  vectors in, 346–349, 348f, 349f
Clover, leaf color in, 195, 195f
CMP (cytidine 5′-monophosphate), 258, 258f
Coactivators, **320**
Coat color
  gene interactions and, 195–197, 196f, 204, 205f
  X inactivation and, 323–324, 324f
Coding strand, 260f, **261**, 261f
Codominance, **194**–195, 194f, 195f
Codon(s), **5**, 278–279, 278f
  amber, 282
  stop, 5, 6, 282, 282f
    suppressor mutations and, 282, 291, 292f
  synonymous, 408
Codon-anticodon pairing
  in translation, 284–285, 285f, 286–287, 289–290
  wobble in, **284**–285, 285f
Codon bias, 408
Coefficient of coincidence (c.o.), **127**
Cognate tRNAs, 287
Coiling, of DNA, 89–90, 89f, 90f
Cointegrate, **433**, 433f
Colchicine, in autopolyploid production, 486
Colinearity, gene-protein, **278**, 278f
Colony, bacterial, **153**, 154f
Colorblindness, 52
Combinatorial interaction, in gene regulation, **317**
Comparative genomic hybridization, 506, 506f
Comparative genomics, 391, **411**–412, 412f, 698–701. *See also* Genomics
Comparative proteomics, 698–701
Complementary bases, **4**, 4f, 5f, **237**, 237f. *See also* Base(s)
Complementary DNA (cDNA). *See* cDNA (complementary DNA)
Complementation, functional, **355**
Complementation test, **199**–203, 200f–203f
  in gene counting, 535, 535f
Complete dominance, **193**
Composite transposons, **431**, 431f
Concentration gradients, in pattern formation, 583–589, 585f–589f, 591f

Congenital insensitivity to androgen (CIAS), 54, 602
Conjugation, 154–165, **156**. *See also* Bacterial conjugation
Consensus sequence, **261**, 392, 393
Conservative replication, **237**, 238f
Conservative substitutions, **456**
Conservative transposition, **432**, 433, 433f
Constitutive heterochromatin, **322**
Constitutive mutations, in gene regulation, 308–309
Contigs
  clone, **396**, 397f
  sequence, 395, 395f
Continuous variation, 8f, **10**, 644–651. *See also* Variation
  gene number and, 8f, 10, 644–651
Controlling elements. *See* Transposable elements
Cooperativity, in gene regulation, **317**
Coordinately controlled genes, **307**
Copia-like elements, **435**
Core enzymes, **261**, 261f, 317
Core promoter, **261**, 261f, 317
Corn. *See* Maize (*Zea mays*)
Corn snake, skin color in, complementation and, 202–203, 203f
Correlation, 648, **670**–671, 670f
  equality and, 671–672
  genetic, **657**
  phenotypic, **658**
Correlation coefficient, **671**
Cosmids, 349, 350f
Cosuppression, 445–**446**, 445f
Cotranscriptional processing, 266–268, 266f–268f
Cotransductants, **171**, 171f
Covariance, **671**
  correlation and, 670–671, 670f
  variance of sum and, 672
cpDNA (chloroplast DNA), 8, **103**
Creighton, Harriet, 119
Crick, Francis, 228, 233–237, 234f, 279, 280, 281, 282–283
Cri du chat syndrome, 504, 505f
Crosses, **29**–32, 31f. *See also* Crossing-over; Testcrosses
  branch diagrams for, 37, 39
  dihybrid, 36–39, 36f–38f, 117, 126, 126f, 136
    in mapping, 117. *See also* Linkage mapping
  Mendelian ratios for, 39–42, 136
  monohybrid, 36
  phage, in mapping, 168–170, 169f
  in plant breeding. *See* Plant breeding
  Punnett squares for, 38–39, 38f
  quantitative differences in, 645–646, 645f
  reciprocal, **31**–32, 31f, 32t
    in X-linked inheritance, 50–52, 51f
  results of
    prediction of, 39–40
    product rule for, 39–40
    sum rule for, 40
    variation in, 645–646, 646f. *See also* Variation
  trihybrid, 125–126, 136

Crossing-over, **95**, 117, **119**–121, 119f, 473–476. *See also* Crosses; Recombination
  chromosome rearrangements during, 496f, 497
  in conjugation, 157, 159, 159f, 160–161, 160f, 162, 162f
  definition of, 119
  dicentric bridge in, **499**, 499f
  double crossovers and, 117, 126, 126f
  double-strand breaks in, 474–476, 475f, 497, 497f. *See also* Chromosome rearrangements
  at four-chromatid stage, 120, 120f
  gene conversion in, **474**
  interference in, 127
  in inversion loop, 499–500, 499f, 500f
  mismatch repair and, 476
  molecular mechanism of, 119
  multiple, 120–121, 120f, 125–126
  of mutant alleles, 476
  nondisjunction and, 419f, 491
  in phage mapping, 168–170, 169f
  products of, 119, 119f
  recombinant frequency from, 122–124, 124f. *See also* Recombinant frequency (RF)
  between repetitive DNA segments, 496f, 497. *See also* Chromosome rearrangements
  in sister chromatids, 121
  in tetrads, 120, 120f
  triple crossovers and, 125–126
Crossover(s), 116f
  definition of, 119
  double-strand, 117, 126, 126f, 474–476, 475f
    in conjugation, 162, 162f
  single-strand, 476
  triple-strand, 125–126
Crossover products, **119**
Crossovers, multiple
  compensating for in linkage analysis, 135–136
  in linkage analysis, 134–136
Culture, bacterial, 153, 153f
Cut and paste transposition, **432**
C-value, **440**
C-value paradox, **440**, 442
Cyanobacteria, 152. *See also* Prokaryotes
Cyclic adenosine monophosphate (cAMP), in gene regulation, 312–315, 312f–315f
Cyclin, **549**, 549f
Cyclin-dependent protein kinase (CDK), **549**–550, 549f
Cystic fibrosis, 44
Cytidine 5′-monophosphate (CMP), 258, 258f
Cytochrome-Apaf complex, 553, 556, 556f
Cytogenetic map. *See also* Maps/mapping
  for chromosome 7, 410–411, 411f
  molecular markers in, 401–404, 403f, 404f
Cytohets, 56, **103**, 106
Cytoplasm, protein synthesis in, 8
Cytoplasmic inheritance, 29, 55–56, 56f, 57f

Cytoplasmic segregation, 56, 56f, 105
Cytosine, 232–233, 232f, 233t. *See also* Base(s)
Cytoskeleton, **578**
  asymmetry of, 579–582, 579f–582f
  developmental consequences of, 579–582, 579f–582f
  in pattern formation, 582–588
  structure of, 578–579, 579f

Danio rerio (zebra fish), genetic screens for, 529–530, 529f
Darwin, Charles, 680–683
Darwinian fitness, **629–633**. *See also* Fitness
Database searches, in gene prediction, 407, 409f, 410
Datura stramonium (jimsonweed), ploidy in, 76, 77f, 494
Daughter cells, 93, 93f, 98f
Daughter molecules, **246**, 247f
Davis, Bernard, 156
Deamination, **464**, 464f
Decoding center, **287**
Degeneracy, of genetic code, **281**, 284–285
Degrees of freedom, 41
Delbrück, Max, 461–463, 462f
Deletion(s), **10**, 454, 456, 465–466, 465f, 466f, 496, 496f, **497**, 502–505
  disease-causing, 504–505, 504f, 505f
  in *Drosophila*, 503–504, 503f, 504f
  in humans, 504–505
  identification of, 503–504
    by comparative genomic hybridization, 506, 506f
  intragenic, **502–503**
  lethality of, 503
  mapping of, 403–404, 404f
  multigenic, **503**
  in plants, 505
  pseudodominance and, **504**
Deletion loop, **503**, 503f
Deletion mapping, **504**
Denaturation, of DNA, 235, 245, 245f
Deoxyadenosine 5′-monophosphate (dAMP), 232, 232f
Deoxycytidine 5′-monophosphate (dCMP), 232, 232f
Deoxyguanosine 5′-monophosphate (dGMP), 232, 232f
Deoxyribonucleic acid. *See* DNA
Deoxyribose, in DNA, **232**, 232f, 233–234, 235f, 236f
Deoxythymidine 5′-monophosphate (dTMP), 232, 232f
Dephosphorylation, of carboxyl tail domain, 266
Depurination, mutations and, **463**
Development, 575–607
  algorithmic nature of, 577–578
  anchor cells in, **599**
  in *Arabidopsis thaliana*, 603–605, 604f
  asymmetry in, 578, 579–582, 579f–582f
  binary fate decision in, **578**
  cell fate in, **577**
    memory systems for, 597–598, 598f
    refinement of, 578

cell lineage-dependent mechanisms in, 578
of cytoskeleton, 578–582, 579f–582f
epidermal growth factor receptor in, 590, 590f
fate allocation in, 598–599, 598f, 599f
fate refinement in, **577, 578**, 594–595, 596f
feedback loops in, 597–598, 598f
gene-environment interactions in, 16–22, 17f–22f. *See also* Gene-environment interactions
genomic analysis of, 605–606, 605f
germ-line determination in, 581–582, 581f, 582f
homeotic, **595**, 596f
imprinting in, 324–326, 325f
inductive interaction in, **599**
information transfer in, 576f, 577
in insects *vs.* vertebrates, 600–601, 600f–602f
lateral inhibition in, **594**
levels of, 22
logic of body plan and, 577–578
mutational analysis for, 584–585
neighborhood-dependent mechanisms in, 578
overview of, 576–577, 576f
parthenogenic, **484**
pattern formation in, 582–602. *See also* Pattern formation
P granules/P cells in, 581, 581f
polarity in, 592, 593f, 595
positional information in, **577**. *See also* Positional information
position-effect variegation in, 326, 326f, **332–333**, 332f, 502
random events in, 21–22, 22f
regulatory cascade in, 595–597, 597f
segmentation in, **583**, 592–595, 593f, 594f
sex determination in, 602, 603f
signaling in, 588–589, 588f, 589f, 590f, 597–599, 598f
trafficking systems in, 579–580, 579f, 580f
Developmental field, **577**
  equivalence group as, **598**, 599f
Developmental noise, **21–22**
Diagnostics, molecular genetic, 364–366
Dicentric bridge, **499**
Dicentric chromosomes, **497**
Dideoxy sequencing, 357–359, 358f, 359f
Differentiated cells, **547**
Diffuse large B-cell lymphoma, diagnosis and treatment of, 566–568, 567f
Dihybrid(s), **36**
  linked genes in, 117–119
Dihybrid crosses, 36–39, 36f–38f
  in mapping, 117. *See also* Linkage mapping
Dihybrid selfed, ratio for, 136
Dihybrid testcross, ratio for, 136
Dioecious plants, 48–49, 49f
Diploid organisms, **3, 83**

complementation in, 199–201, 200f
life cycle of, 96, 96f
meiosis in, 96, 96f, 97, 98f, 102–103, 102f
mitosis in, 96, 96f, 97, 98f
Diploidy, **3**, 74, **83, 483**, 483t. *See also* Chromosome number; Polyploidy
  doubled, **487**, 487f, 489f
Directional selection, **685**
Discontinuous variation, 8–10, 8f, 9f, 28–29, 644, 649–651. *See also* Variation
Disomy, **490**
Dispersion, **648**
Dispersive replication, **237**, 238f
Dissociation element. *See* Ds elements
Distribution(s)
  asymmetrical, 668, 688f
  equilibrium, **621**
  genotypic, 649–650, 649f
  phenotypic, 649–652, 649f. *See also* Variance
  statistical, **647–648**, 648f, 667–673
    asymmetrical, 668, 688f
    bimodal, **668**, 688f
    standard deviation of, **669**, 670f
Distribution function, **647–648**
Distributive enzyme, **243**
Diversification, species, 681, 682f
dl gene/DL protein, 588–589, 588f, 589f
DNA, **2**, 227–250
  amount of in chromosome, 80–81, 88
  bases of, 232–233, 232f. *See also* Base(s); Base pairs/pairing
  building blocks of, 232
  cesium chloride gradient centrifugation of, 238–239, 238f
  chemically synthesized, 344
  chloroplast, 8, **103–106**, 103f–106f
  in chromosome, 80–81, 80f, 81f
  cloning of, 343
  coding strand of, 260f, 261, 261f
  complementary. *See* cDNA (complementary DNA)
  damaged
    checkpoints for, **550**, 550f
    repair of. *See* DNA repair
  daughter, 246
  denaturation of, 235, 245, 245f
  deoxyribose in, 232, 232f
  direct analysis of, 12
  donor, **156, 343**
  double-strand breaks in
    in crossing-over, 474–476, 475f, 497, 497f. *See also* Chromosome rearrangements
    repair of, 472–473, 472f, 473f
  duplication of, 691–693, 692f, 693f. *See also* Duplication(s)
  early studies of, 229–231
  flow of information from to RNA, 6–8, 7f
  from genome duplication, 691–692, 691f
  horizontal transfer of, 694. *See also* Transposable elements
  identification of as genetic material, 229–231, 229f–231f

imported, 693–694
information content of, 406
intergenic, 74
junk, **444**
key properties of, 231
loop, 318, 318f
melting of, 235, 245, 245f
mitochondrial, 8, 103–106, 103f–106f
mobile. *See* Transposable elements
for new genes, 691–695
nucleotides of, 232, 232f
overview of, 227
packaging of, 80–81, 80f, 81f, 88–90, 89f, 90f
palindromic, 344–345
phosphate in, 232, 232f
from polyploidization, 691, 691f
recombinant, 343. *See also* DNA technology
repetitive, 85, 85f, 410, 440–442, 441f
  in genomic sequencing, 394, 395–396
satellite, **85**, 85f
single-copy, 394
size of, 80, 81f
sources of, 691–695
structure of, 3–4, 4f, 231–240. *See also* Double helix
  early studies of, 231–233
supercoiling of, 89–90, 89f, 90f
  topoisomerases and, 243, 243f
synthesis of. *See* DNA replication
template strand of, 259–261, 259f, 260f, 261f, 262
in transcription, 264
  in eukaryotes, 264
  in prokaryotes, 259–261, 260f, 261f
transfer, 368–370, 370f
unwinding of
  in replication, 237, 237f, 240, 240f, 242f, 243, 245, 245f
  in transcription, 262, 263f
X-ray diffraction studies of, 233, 234f
DnaA, 245–246, 245f
DNA amplification, 342–350, 342f. *See also* DNA technology
  in vitro, 360–362, 361f
  in vivo, 342f, 343–350
DnaB, **245**
DNA-binding domains, in gene regulation, 304
  in eukaryotes, 304
  in prokaryotes, 304
DNA chips, 413, 414f. *See also* DNA microarrays
DNA elements, 424–**425**, 424f, 436–438, 437f–439f. *See also* Transposable elements
DNA fingerprinting, **400**, 401f
DNA glycosylases, **469**, 469f
DNA gyrase, 243, 243f
DNA hybridization, **345**
DNA ligase, 241, 241f, 242f, **346**
DNA microarrays, 413, 414f
  in cancer diagnosis and treatment, 567f, 568
  in comparative genomic hybridization, 506, 506f
DNA polymerases, 214f, 239–241, 239–242, 240f–242f

DNA polymerases *(cont)*
in replisome, 242, 245, 245f
DNA probes, 12, 13f, 351–354,
352f–354f. *See also* Probes
for mixtures, 354, 355f
in Southern blotting, 354, 355f
DNA repair, 468–473
base-excision, 468f, **469**–470
by direct reversal, 468–469, 468f
DNA glycosylases in, **469**, 469f
of double-strand breaks
by homologous recombination,
473, 473f
by nonhomologous end-joining,
472, 473f
error-prone polymerases in,
459–460, 460f
in eukaryotes, 470–473
homology-dependent, 468f,
469–471
mismatch, **471**–472, 471f, 472f
crossing-over and, 476
nucleotide-excision, **470**, 470f
overview of, 452–453, 452f
photorepair, 468
postreplication, 471, 471f, 472f
in prokaryotes, 468–470, 471
SOS system in, **459**–461,
472–473, 472f, 473f
transcription-coupled, 470–471
DNA replication, 3–4, 3f, 237–249
accessory proteins in, 242–243
accuracy of, 241–242
in cell cycle, 91–95, 91f–94f,
97–103, 98f, 99f, 101f, 102f,
246–247, 247f
chromatin assembly factor 1 in,
244, 244f
conservative, **237**, 238f
directionality of, 240f–242f, 241
dispersive, **237**, 238f
DNA ligase in, 241, 241f, 242f
DNA polymerases in, 239–242,
240f–242f
epigenetic inheritance and,
323–326, 331–332, 332f
errors in
mutations due to, 457–459,
464–468
repair of, 471–472, 471f, 472f
in eukaryotes, 245f, 246–249,
247f
helicases in, 242f, 243
helix unwinding in, 237, 237f, 240,
240f, 242f, 243, 245, 245f
initiation of, 244–246, 245f
lagging strand in, 240f–242f, **241**
leading strand in, **240**, 240f, 242f
in meiosis, 93–95, 94f, 97–103,
98f, 99f, 101f, 102f, 246–247,
247f
Meselson-Stahl experiment and,
237–239, 238f
in mitosis, 91–93, 91f–93f, 98f,
246–247, 247f
nucleosome assembly in, 244, 244f
Okazaki fragments in, 241, 241f,
242f
origins in, 244–247, 245f, 246f
licensing of, 247
overview of, 228, 228f, 240–242,
242f
primase in, **241**, 242

primer in, **241**, 241f, 242, 242f
process of, 4, 5f, 240–242,
240f–242f
in prokaryotes, 3–4, 3f, 237–246,
239f
proliferating cell nuclear antigen in,
244, 244f
proofreading in, 241
replication fork in, 239, 239f,
240–241, 240f–242f
replisome in, 242–247. *See also*
Replisome
rolling circle, **157**
semiconservative, **237**–240
sliding clamp in, 243, 243f
speed of, 242, 246
strand melting in, 235, 245, 245f
supercoiling in, 243, 243f
template for, 4, 5f, **237**
termination of, 247–248, 247f, 248f
3′ growing tip in, 240, 240f, 242f
topoisomerases in, 243, 243f
translesion, 459–460, 460f
of transposable elements, 432–433,
433f
DNA:RNA hybrids, in transcription,
262–263, 263f
DNA sequence variation, 617–621
DNA technology, **342**–386
amplification techniques in,
343–349, 347f
cDNA libraries in, 350–357
chromosome walking in, 356–357,
356f
cloning in, 343–349, 347f, 349f,
350f. *See also* Cloning
definition of, 342
dideoxy sequencing in, 357–359,
358f, 359f
functional complementation in, 355
gel electrophoresis in, 354, 354f
gene knockout in, 372–375, 374f
gene targeting in, 372–375, 374f
gene therapy in, 375–378
genetic engineering in, 366–379.
See also Genetic engineering
genomic libraries in, 350–357
molecular genetic diagnostics in,
364–366
mutant rescue in, 355
Northern blotting in, 12, 13f, 354
overview of, 342–343, 342f
polymerase chain reaction in,
360–362, 361f, 365–366
positional cloning in, 346f,
356–357, 357f
probes in, 351–354
restriction enzymes in, 344–346,
344f, 345f
restriction mapping in, 357, 357f
Sanger sequencing in, 357–359,
358f, 359f
Southern blotting in, 12, 13f, 354,
355f, 365, 365f
transgenic organisms in, 366–375
transposable elements in, 439–440,
439f
DNA transposons, 424–**425**, 424f,
436–438, 437f–439f
Docking sites
genes as, 406, 406f
prediction of, 407, 408f
Domains, protein, **277**, 302–303

Dominance, **31**–32, 34, 192–193,
193f
codominance, 194–195, 194f, 195f
full (complete), **193**
haploinsufficiency and, 192
homozygous, 34
incomplete, 193–**194**, 193f
law of, 13–14
Dominance variance, **661**
Dominant mutations, 192–193, 193f,
537
functional causes of, 536f
Doncaster, L., 76
Donor, in conjugation, **153**
Donor DNA, **343**
dorsal, 588–589, 588f, 589f
Dorsal-ventral axis, in pattern formation, 588–590, 588f–590f
Dosage compensation, **323**, 495
Double crossovers, 117, 126, 126f,
134, 474–476, 475f
in conjugation, 162, 162f
Doubled diploid, **487**, 487f, 489f
Double helix, 3–4, 4f, **233**–236,
234f–236f
antiparallel orientation of, **234**, 235f
base pairs in, 233–234, 234–235,
235f–237f
discovery of, 228, 233
significance of, 235–236
hydrogen bonds in, 234–235, 235f
major groove of, **234**, 236f
minor groove of, **234**, 236f
models of, 235f, 236f
polarity of, 233–234, 235f
unwinding of
in replication, 237, 237f, 240,
240f, 242f, 243, 245, 245f
in transcription, 262, 263f
Watson-Crick model of, 228,
233–236, 234f–236f
Double infection, **169**, 169f
Double mutants, **199**
epistasis and, 203–204, 204f, 205f
lethal, 206
Double-strand breaks, in DNA
in crossing-over, 474–476, 475f,
497, 497f
repair of, **472**–473, 472f, 473f
Double-stranded RNA, in RNA interference, 533–534, 533f
Double transformation, **166**
Down syndrome, 482, 482f, **493**–494,
493f, 494f, 495, 506
Robertsonian translocation in, 506,
507
dpp enhancers, 320, 321f
Drosophila melanogaster (fruit fly)
Abd-B in, 321–322
crossing-over in, 125–127
deletions in, 503–504, 503f, 504f
development in, 18, 19–20, 19f, 21,
582–600. *See also*
Development
vs. in vertebrates, 600–601,
600f–602f
DNA in, 80, 81f
embryology of, 591–592, 592f
enhancers in, 320, 321f
eye color in, 49–52, 50f, 51f,
77–80, 79f, 326, 326f,
332–333, 332f, 436, 439–440,
439f, 503f, 506

eye size in, 19–20, 19f, 21
genomic sequencing in, 395–396
genotype-phenotype interaction in,
19–20, 19f, 21
germ-line development in,
581–582, 582f
hybrid dysgenesis in, **436**–437,
437f
life cycle of, 50f
light response in, 526, 526f
as model organism, 13, 50, 428,
584–585
mutational analysis in, 584–585
norm of reaction in, 19–20, 19f, 21
pattern formation in, 583–599
pole cells in, **582**
polytene chromosomes in, 87–88,
87f
position-effect variegation in, 326,
326f, **502**
scutellar bristles of, 688, 688f
sex determination in, 48, 48t
Tab mutation in, 321–322, 322f
transgenesis in, 371, 371f
transposable elements in, 435,
436–440, 437f, 438f,
444–445, 444f
X-linked inheritance in, 49–52, 50f,
51f
Drosophila pseudoobscura, protein
polymorphisms in, 616–618,
616f, 617t, 618f, 619f
Drug resistance
causative mutations in, 524–525
composite transposons and, **431**,
431f
Ds elements, **425**–429, 425f, 438,
438f. *See also* Transposable
elements
Dt element, 427
dTMP (deoxythymidine 5′-mono-
phosphate), 232, 232f
Duchenne muscular dystrophy, 54
Duplication(s), **10**, 496, 496f,
497–498, 505–506, 691–693,
692f, 693f
as DNA sources, 691–692
genomic, 691–692
identification of, by comparative genomic hybridization, 506, 506f
insertional, **505**–506
tandem, **505**
target-site, **432**
DuPraw, Ernest, 80
Dwarfism, 45, 45f
Dyads, **95**

E2F protein, in cell cycle regulation,
549–550, 550f, 556
Ears, hairy, 55, 55f
Edwards syndrome, 494
EGF. *See* Epidermal growth factor
(EGF)
80S ribosome, 285, 286f, 289
Electrophoresis, 12
pulsed field gel, 81
Elongation factor G (EF-G), **289**
Elongation factor Tu (EF-Tu), **289**
Embryoid, **489**, 489f
Emerson, Rollins, 428, 428f
Endocrine signaling, 553f, **554**. *See
also* Signaling
Endogamy, **624**

Endogenote, **162**, 162f
Enforced outbreeding, **625**
Enhanceosomes, **319–320**, 320f
  β-Interferon, 320, 320f, 331, 331f
Enhancer(s), transcriptional, 317, **318**
  dpp, 320, 321f
  gene fusion and, 321–322, 322f
  tissue-specific, 320
Enhancer/inhibitor (En/In) element, 427
Enhancer traps, **530**, 530f
En/In element, 427
Enol form, of base, **457**, 457f
env, in viral replication, 435, 435f
Environment
  distribution of, phenotypic distribution and, 651–653, 651f
  phenotypic determination and, 17–18, 17f, 651–653, 651f
Environmental variance, 649–652, 650f, **656–657**. *See also* Variance
Environment-gene interactions, 16–22, 17f, 18f, 651–653, 651f
Enzyme(s), **5**, 8, 10
  active site of, **10**, **277**, 277f
  core, 261, 261f, 317
  distributive, **243**
  in DNA replication, 239–242, 240f–242f
  functional changes in, 694–695
  one-gene–one-enzyme hypothesis and, 42–44, 187, 190–192, 277–278
  in posttranslational processing, 292
  processive, **243**
  restriction, 12, 48, **344**
    in recombinant DNA production, 344–346, 344f, 345f
  structure of, 277, 277f
  zymogen form of, **552**
Enzyme induction, in *lac* system, 307–312. *See also under lac*
Epidermal growth factor (EGF), 555, 556
  in cancer, 562–563, 562f
Epidermal growth factor receptor (EGFR), 555, 556
  in cancer, 562–563, 562f
  in pattern formation, 590, 590f
Epigenetic inheritance, 323–326, 331–332, 332f
  imprinting and, 324–326, 325f
  X inactivation and, 323–324, 324f
Epigenetic silencing, **326**, 445
Epimutations, **445**, 445f
Epistasis, 203–204, 204f, 205f
  vs. suppression, 205
Epistatic mutation, 187
EP polymerases, 459–460, 460f
Equality, correlation and, 671–672
Equal segregation, 13, 33–36, **34**
Equilibrium distribution, **621**
Equivalence group, **598**, 599f
Error-prone polymerases, 459–460, 460f
Escherichia coli. See also Bacteria
  DNA replication in, 3–4, 3f, 237–246
  genetic screens for, 528–529, 528f

insertion sequences in, **429–430**, 429f, 430f
lac system of. *See also under lac*
  mapping of, 171–172, 171f, 174–175, 174f, 175f
  as model organism, 13, 14f, 154
  transcription in, 261–263, 262f, 263f
E site, **286**, 287f, 290
Euchromatin, **84**, **322**, 323f. *See also* Chromatin
Eukaryotes, **6**
  gene structure in, 6, 6f
  as model organisms, 13–14, 14f
Euploidy, **483**
  aberrant, 483, 484
  gene balance and, 494–496
Evolution
  adaptive peaks in, 685–687, 686f
  common ancestry in, 696–698
  Darwinian theory of, 680–682
  directional selection in, **685**
  diversification in, 681–682, 682f
  DNA sources for, 691–695
  fitness in, **629–633**. *See also* Fitness
  founder effect in, **684**
  functional *vs.* genetic change in, 694–696
  gene conservation in, 696–698
  genetic drift in, 103, **636**, 637f, 684, 684t, 695–696
  genomic, gene mapping and, 129
  isolating mechanisms in, 690
  molecular, rate of, 695–696, 695f, 696f
  natural selection in, 613f, **629–633**. *See also* Natural selection
  neutral, 695–696. *See also* Genetic drift
  new genes in, 691–695
  nucleotide substitutions in, 695–696
    nonsynonymous, **696**
    synonymous, **696**
  overview of, 680–682, 680f
  phyletic, **681**, 682f
  of plants, allopolyploidy in, 488, 488f
  polyploidization in, 691
  principles of, 680f, 681–682
  RNA world and, 268–269
  selection in, 681
  speciation in, 689–690
  variation in, 681. *See also* Variation
    heritability of, 681, 687–688
    maintenance of, 684–685, 684f
    within *vs.* between populations, 688
  vs. breeding, 681
Evolutionary genetics, overview of, 680–682
Excision, gene, **426**
Exconjugant, **158–159**, 159f
Exogenote, **162**, 162f
Exons, **6**, 6f, **267**
  identification of, 406–410, 407f
  in splicing, 267–269, 268f, 269f
Expressed sequence tags (ESTs), **407**, 407f
Expression library, 353
Expression vectors, 349
Expressivity, **208**

Extrachromosomal arrays, 370, 370f
Eye color, in *Drosophila melanogaster*, 49–52, 50f, 51f, 77–80, 79f, 326, 326f, 332–333, 332f, 436, 439–440, 439f, 503f, 506
Eye size, in *Drosophila melanogaster*, 19–20, 19f, 21

F. *See* Fertility factor (F)
Familiality, *vs.* heritability, 654–656
Familial traits, **655**
Family selection, **664**
Fas ligand, 556, 556f
Fate allocation, 598–599, 598f, 599f
Fate refinement, **578**, 594–595, 596f
Fertility (F) factor, **156**, 162–164, 164f
  carrying genomic fragments, 162–164
  discovery of, 156
  Hfr strain and, 157–161, 157f–161f
  insertion-sequence elements in, 430, 430f
  insertion sites for, 159–161, 160f, 161f
  in integrated state, 161
  in plasmid state, 161
  replication of, 157, 157f
Fibrous proteins, **277**
50S subunit, 285, 286, 286f, 288
Filamentous fungi. *See Neurospora crassa* (filamentous fungi)
Filtration enrichment, 525, 525f
Finches, of Galapagos Islands, 682–684
Fink, Gerald, 434, 435
First-division segregation (M_I) patterns, **131**
First filial generation, **31**, 31f
FISH. *See* Fluorescent in situ hybridization (FISH)
Fitness, **629–634**
  differences in, measurement of, 630–631
  frequency-dependent, **630**
  frequency-independent, **630**
  mutations and, 695
  overdominance in, **634**
  underdominance in, **634**
  viability and, **630–631**, 631f
5′ end, capping of, **266–267**, 266f
5′ untranslated region (5′ UTR), 261
Fixed alleles, **626**
Fluctuation test, 461–463
Fluorescent in situ hybridization (FISH), **83**, **403**
  in chromosome painting, **403**, 403f
  in mapping, 403, 403f
FMR-1, in fragile X syndrome, 467, 467t, 468
Forward genetics, **522**, 523–530. *See also* Mutational analysis
  enhancer traps in, 530, 530f
  genetic screens in, **524**, 525f, 527–529, 527f, 528f
  genetic selections in, 524–526, 525f, 526f
  mutagens for, 523–524, 524f
  mutant hunt and, 523, 523f
  mutational assay systems for, 524
  transposon tagging in, **524**
Founder effect, **636**, 684

Four-o'clocks. *See Mirabilis jalapa* (four-o'clocks)
40S subunit, 285, 286f
F plasmid. *See* Fertility (F) factor
F′ plasmid, **163**
Fragile X syndrome, 466–467, 467f, 468
Frameshift mutations, **10**, 280, 455t, **456**, 465
Franklin, Rosalind, 233, 234f
Frequency, relative, **668**
Frequency-dependent fitness, **630**
Frequency histogram, **647**, 648f
Frequency-independent fitness, **630**
Fruit flies. *See under Drosophila*
Full dominance, **193**
Functional genomics, 391, **413–415**. *See also* Genomics
Functional RNA, 190, 258–259, 274–275
Functional *vs.* genetic change, 694–696
Fungi. *See Neurospora crassa* (filamentous fungi)

G1 phase, of cell cycle, 548
G2 phase, of cell cycle, 548
gag, in viral replication, 435, 435f
Gain-of-function mutations, **321–322**, 322f, 453, 536f, 537
  gene fusion and, 321–322, 322f
  vs. loss-of-function mutations, 535–537, 536f
β-Galactosidase, in *lac* system, 305, 305f, 306, 307, 308
Galapagos Islands, finches of, 682–684
gal operon, insertion-sequence elements in, 429–430, 429f, 430f
Gametes, **93**
Gametic diversity, 623–624, 623t
Gametophyte, 97, 97f
Gap genes, in pattern formation, **593–594**, 593f, 594f
Garrod, Archibald, 187, 187f, 362
G bands, 86–87, 86f
Gel electrophoresis, **354**, 354f
  protein, 616–617, 616f, 617t
  pulsed field, 81
Gene(s), **2**
  alleles of. *See* Allele(s)
  allelic series of, 192
  binding sites in, 304
  candidate, **405–406**, **664–665**
  cardinal, in pattern formation, **591–594**, 592t
  chloroplast, 55, 56, 75
  cloning of, 346–349
  coordinately controlled, **307**
  cosuppression of, 445–446, 445f
  discovery of, 33
  as docking site, 406, 406f, 407
  dominant, 33, 34, 192–193, 193f. *See also* Dominance
  early studies of, 187–192
  evolutionary conservation of, 696–698
  excision of, **426**
  functions of, 4–5, 6–7
  gap, in pattern formation, 593–594, 593f, 594f
  haplo-abnormal, 495
  haploinsufficient, 192, 537

Gene(s) *(cont)*
  haplosufficient, 10, **192**
  hemizygous, 49, 49f
  homeobox, 585
    in insects *vs.* vertebrates, 600–601, 600f–602f
  housekeeping, 6
  imprinted, 324–326, 325f
  independent assortment of, 13, 37–39, 38f, 39f, 102–103, 116–117, 121
    deviations from, 117–118. *See also* Crossing-over
  in information transfer, 4–5
  jumping. *See* Transposable elements
  in linkage mapping, 127–128
  linked, 117–119, **118**. *See also* Linked genes
  location of on chromosomes
    evidence for, 75–80
    gene arrangement and, 81–82, 81f–83f
  marker, 655, **665–666**
  maternal-effect, **583**
    mutations in, detection of, 584–585
  Mendel's studies of, 33
  mitochondrial, 55–56, 75
    mutations in, 103–106, 105f, 106f
  mobile. *See* Transposable elements
  mutant, 35, 79. *See also* Mutation(s)
  naming of, 50, 79
  new, origin of, 691–695
  numbers of, quantitative traits and, 650–651
  organelle, 55–56, 103–106, 103f–106f
  paired, **33**, 74
  pair-rule, in pattern formation, 593f, **594**, 594f
  polymorphic, 615
  prediction of, 406–410, 410f. *See also* Bioinformatics
  properties of, 3
  protein-coding, identification of from genomic sequence, 406–410
  pseudolinkage of, **502**
  recessive, 33, 34, 192–193, 193f. *See also* Recessiveness
  recombination of, 121–123, 121f–124f. *See also* Recombination
  regulatory, 187
    complementation and, 202, 202f
    mutations in, down/upregulation from, 206
  reporter, **414**
  segment-polarity, 595
  segregation of, 13, 33, 34
  sex-linked, 49–52
  silencing of, 317, **318**
    epigenetic, 326, 445
    in imprinting, 324–326, 325f
    in position-effect varigation, 326, 326f
    transgenic, 445–446, 445f
    in X inactivation, 323–324, 324f
  size of, mutations and, 523
  structure of, 6, 6f
  transposition of, **426**
  triplo-abnormal, 495

  tumor-suppressor, mutations in, 560–561, 560f, 564–566, 565f
  wild-type, 35, **79, 615**
  X-linked, 49–52
Gene action, 6–8, 7f
  environmental influences on, 16–22, 17f–22f. *See also* Gene-environment interactions
  factors affecting. *See* Gene interaction
  one gene–one enzyme hypothesis and, 187–189, 188f
  via gene products, 187–192
Gene balance, 494–496
Gene clusters, paralogous, **601**
Gene conversion, **474**
Gene counting, 535, 535f
Gene-dosage effect, **495**
Gene-environment interactions, 16–22, 17f, 18f
  developmental noise and, 21–22, 22f
  environmental determination and, 17–18, 17f, 22f
  fitness and, 630–633
  genetic determination and, 16–17, 17f
  genotype-environment interaction and, 18, 18f, 22f
  genotype-phenotype interaction and, 18–19, 22f
  norm of reaction and, 19–21, 19f, 20f
Gene expression, 6–8, 7f, 258
Gene function, gene location and, 129
Gene fusion, 321–322, 322f
  in cancer, 563–564
Gene interaction
  between alleles of one gene, 192–197
  chi-square test for, 209
  complementation test for, 199–203
  dominant, 192–193, 193f
  early studies of, 187–192
  epistatic, 203–204, 204f, 205, 205f
  expressivity in, 208–209, 208f
  identification of, 198–209
  modes of, 187
  modifiers in, 206
  mutational, 187, 188f
  mutations and, 190–192
  overview of, 187
  penetrance in, 207–209, 208f
  with proteins, 197–209
  recessive, 192–193, 193f
  of regulatory genes, 202, 202f
  suppressive, 205–206, 206f
  synthetic lethals in, 206–207, 207f
Gene knockout, **372–375**, 374f
  targeted, 531, 531f
Gene locus, 76–80, **124**
  gene function and, 129
  identification of
    by molecular analysis, 664–667
    by mutational analysis, 9–10, 11, 521–539. *See also* Mutational analysis
  mapping of, 124–131. *See also* Maps/mapping
  multiple alleles at, 192
Gene mapping. *See* Maps/mapping
Gene mutations, 452–468. *See also* Mutation(s)

Gene pairs, **33**, 74
Gene prediction. *See also* Genome sequencing
Gene products, gene action via, 187–192
Gene-protein colinearity, 278, 278f
Generalized transduction, **170–172**, 171f
General transcription factors (GTFs), **264**, 265
Gene regulation, 6–8, 7f, 301–333, **301–333**
  complementation and, 202, 202f
  coordinated, 305, 307
  in eukaryotes, 302f, 316–333
    chromatin in, 322–333. *See also* Chromatin
    cis-acting elements in, 317–318, 317f, 318f
    coactivators in, 320
    combinatorial interactions in, **317**
    cooperativity in, **317**, 319–320, 320f
    core promoter in, 261, 261f, 317–318, 317f
    DNA-binding domains in, 318–319, 319, 319f
    dominant mutations in, 321–322, 322f
    enhanceosomes in, 319– 320, 320f, 331, 331f
    enhancers in, 317, 318, 320, 321–322
    epigenetic inheritance and, 323–326, 331–332, 332f
    gain-of-function mutations in, 321–322, 322f
    gene fusion in, 321–322, 322f
    gene silencing in, 317
    histone in, 329–331, 329f, 330f
    imprinting and, 324–326, 325f
    loop DNA in, 318, 318f
    loss-of-function mutations in, 321
    modularity in, 317
    nucleosomes in, 329–331, 329f, 330f
    overview of, 317
    position-effect variegation and, 326, 326f, 332–333, 332f
    promoter-proximal elements in, 317–318, 317f–319f
    promoters in, 261, 261f, 264–265, 265, 265f, 317–318, 317f, 318f
    silencers in, 317, 318
    syngergistic, 319–320, 320f
    tissue-specific, 320, 321f
    transcription factors in, 318–319
    vs. in prokaryotes, 317
    X inactivation and, 323–324, 324f
  overview of, 302–303, 302f
  in prokaryotes, 261, 262f, 302f, 303–316
    activators in, 304, 304f, 312–315, 312f–315f
    allosteric, 309–310
    cAMP in, 312–315, 312f–315f
    cis-acting elements in, 308
    constitutive mutations in, 308–309, 308t
    dual positive-negative, 315–316, 315f

    inducers in, 306–307
    metabolic pathways in, 316, 316f
    mutations in, 307–308
    negative, 304, 304f, 305–315, 306f, 314f, 315f. *See also* lac system
    operator mutations in, 307–309, 308f, 308t
    operators in, 304, 304f
    operons in, 306–307, 306f
    overview of, 303–305
    positive, 304, 304f, 312–315, 312f–315f
    promoters in, 261, 262f, 304, 304f, 310
    repressor-operator specificity in, 311
    repressors in, 304, 304f, 306–312, 306f. *See also* Lac repressor
      mutations in, 307–310, 309f, 309t, 310t
    trans-acting proteins in, 309
    transcription units in, 305
    vs. in eukaryotes, 317
Gene replacement, 366, **372**
  targeted, 531, 531f
Gene therapy, **375–379**, 444
  germ-line, 376–377, 377f
  in humans, 376–378
  in mice, 375–376, 376f
  somatic, 377–378, 377f
Genetically modified organisms (GMOs), **368**
Genetic analysis, 185–211. *See also* Gene interaction
  molecular, 664–667
  mutational, 522–539. *See also* Mutational analysis
  overview of, 11–13
Genetic code, **235**, 278–282
  codons in, 278–279, 278f
  cracking of, 281, 282f
  degeneracy of, **281**, 284–285
  overlapping *vs.* nonoverlapping, 278–279
  stop codons in, 282, 282f
  suppressor mutations and, 279–281, 280f
  triplet, 278–281, 278f
Genetic correlation, **657**
Genetic determination, 16–17, 17f, 18f
Genetic disorders
  autosomal dominant, pedigree analysis of, 45–46, 46f, 47f
  autosomal recessive, pedigree analysis of, 42–45, 43f, 44f
  carriers of, 52, 52f
  diagnosis of, 364–366
  examples of, 364t
  gene therapy for, 377–378
  mitochondrial, 103–106, 105f, 106f
  screening for, 364–365
  trinucleotide repeat, 466–468, 467f
  X-linked
    dominant, 54, 54f
    recessive, 49–54, 52f–54f
Genetic drift, 103, 613f, **636**, 637f, 684, 684t, 695–696
  natural selection and, 687
  neutral mutations and, 695–696
Genetic engineering, **342**, 366–379. *See also* DNA technology

in animals, 370–375, 372f–374f, 376f
definition of, 342
in humans, 375–379
in plants, 368–370, 369f
in yeast, 366–368, 367f
Genetic events, rare
screens for, 170
selective systems for, 170
Genetic load, **484**
Genetic map unit (m.u.), **124**
Genetic markers, **153**, 154t
Genetic mutations. *See* Mutation(s)
Genetic polymorphism, 9–10
Genetics
chemical, **534**–535, 534f
definition of, 2
evolutionary, 679–703
overview of, 680–682
forward, **522**, 523–530. *See also*
Forward genetics
population, 612–637. *See also*
Population genetics
quantitative, 643–674. *See also*
Quantitative genetics
reverse, **522**, 530–535. *See also*
Reverse genetics
Genetic screens, **170**, **524**, 525f,
527–529, 527f, 528f
for *Aspergillus*, 528, 528f
enhancer traps in, **530**, 530f
for *Escherichia coli*, 528–529,
528f
for *Neurospora*, 527, 527f
for yeast, 527–528, 527f
for zebra fish, 529, 529f
Genetic selections, **524**–526, 525f,
526f
Genetic symbols, 79
Genetic traits, 30–34
canalized, 688, 688f
familial, **655**
heritability of, 654–664, 687–688.
See also Heritability
quantitative
gene number and, 650–651
norm of reaction studies for,
652–653
vs. Mendelian, 649–650
Genetic variance. *See* Variance;
Variants
Genetic variation. *See* Variation
Genome, **3**, 4f
duplication of, 691–692
eukaryotic, 15, 16f
evolution of, gene mapping and,
129
human
contents of, 391
interspersed repeats in, 440–442,
441f
repetitive DNA in, 394,
395–396, 410, 440–442,
441f
structure of, 410–411, 411f
transposable elements in,
440–442, 441f
information content of, 406–410.
See also Bioinformatics
interpretation of, 406–410. *See also*
Bioinformatics
mapping of. *See* Maps/mapping
prokaryotic, 15, 16f

sequencing of. *See* Genome
sequencing; Maps/mapping
size of, 440
C-value paradox and, **440**, 442
transposable elements in, abundance of, 424, 440–445
viral, 15, 16f
yeast, 411–412, 412f
Genome projects, **392**
Genome sequencing, 82, 82f, 83f,
392–398
applications of, 391
automation of, 392–393, 394f
bioinformatics and, **406**–410
candidate gene approach in,
405–406
chromosome painting in, **403**, 404f
clone contigs in, **396**, 397f
completeness of, 393
consensus sequence in, **392**, 393
development of, 391
DNA fingerprinting in, **400**, 401f
of entire genome, 392
gene prediction from, 406–410, 410f
goals of, 393
in identification of specific genes,
398–406
interpretation of data from,
406–410. *See also*
Bioinformatics
map creation in, 392–398. *See also*
Maps/mapping
minimum tiling path in, **397**, 398f
ordered-clone, 396–398, 398f
overview of, 390–392, 390f
in polypeptide identification,
406–410. *See also*
Bioinformatics
proteome and, 406–410
repetitive DNA in, 394, 395–396
whole genome shotgun, 394–396,
395f
Genomic analysis, of development,
605–606, 605f
Genomic DNA, in recombinant DNA
production, 343
Genomic duplication, 691–692
Genomic library, 350–357, **351**
clone identification in, 351–357.
See also Clone(s), identification
of
creation of, 350–351
Genomic maps. *See* Maps/mapping
Genomics, **12**, **342**, 389–416
applications of, 391
in bioinformatics, 391
in cancer diagnosis and treatment,
566–568, 567f
comparative, 391, **411**–412, 412f,
698–701
functional, 391, **413**–415
overview of, 390–392, 390f
structural, **410**
systems biology and, 412–413, **415**
Genotype, 9, 18, **34**
environmental influences on, 18
fitness of, 629–633
norm of reaction of, 19–21, 20f,
644–645, 651–653, 651f
partial, 18
transmission of, in meiosis, 97–102,
98f, 99f, 101f, 102f
wild-type, 19–20, 19f, **615**

Genotype frequencies, **614**, 614t
Genotype-phenotype relationship,
8–9, 18–19, 646–647, 649–650
distributive, 649–650, 649f, 650f
norm of reaction and, 651–653
Genotypic distribution, phenotypic
distribution and, 649–650, 649f
Genotypic variation. *See also* Variation
vs. phenotypic variation, 649–652
Germ layers, in pattern formation,
583
Germ-line determination, in *C. elegans*, 581, 581f
Germ-line gene therapy, 376–377,
377f
GFTs (general transcription factors),
264, 265
Gilbert, Walter, 310–311
Globin, synthesis of, 692, 692f
Globular proteins, **277**, 277f
GMOs (genetically modified organisms), **368**
GMP (guanosine 5′-monophosphate),
258, 258f
Goldschmidt, Richard, 78
G proteins, in signaling, 556–557,
557f
Griffith, Frederick, 166, 229–230
grk gene/GRK protein, 590f
Growth factors, in cell cycle regulation, **555**
Gryphea, shell variation in, 681, 682f
GU-AG rule, **268**
Guanine, 232–233, 232f, 233t. *See also* Base(s)
Guanosine 5′-monophosphate
(GMP), 258, 258f
gurken, 590f

H^2 (broad heritability), **657**–660
estimation of, 657–658
meaning of, 658–660
h^2 (narrow heritability), 660–664
artificial selection and, 662–663
in breeding, 662–664
estimation of, 661–662, 661f, 662f
Hairpin loop, **263**, 263f
Hairy ear rims, 55, 55f
Haldane, J.B.S., 134
Halotype, **623**–624
Halotypic diversity, 623–624, 623t
Halotypic variation, 615, 615t
Haplo-abnormal genes, 495
Haploid-diploid organisms
life cycle of, 96f, 97
mitosis in, 96f, 97
Haploid organisms, 3, **83**
life cycle of, 96–97, 96f
meiosis in, 96–101, 96f, 98f, 100,
101f
mitosis in, 96–97, 96f, 98f
Haploidy, 83, **83**, 483. *See also*
Chromosome number; Polyploidy
Haploinsufficiency, 45, **192**
Haploinsufficient gene, **192**, 537
Haplosufficient gene, 10, 192
Hardy-Weinberg equilibrium,
621–623
Harebell (*Campanula*), complementation in, 199–202, 199f, 200f
Harris, H., 664
Hartwell, Leland, 551
Hayes, William, 156

HB-M protein gradient, 583–587
Helicases, 242f, **243**
Helix
α, 276f, 277
double. *See* Double helix
Helix-turn-helix, 319, 319f
Hemizygous gene, **49**, 49f
Hemoglobin, 194–195, 195f
structure of, 276f, 277
synthesis of, 692, 692f
variant, 616f
Hemoglobin-A, 17
Hemoglobin-S, 17
Hemophilia, 52–54, 53f, 442
Heredity. *See* Heritability; Inheritance
Heritability, **654**–664. *See also*
Inheritance
among relatives, 655–656
broad, **657**–660
estimation of, 657–658
meaning of, 658–660
in experimental animals, 656,
656f
of intelligence, 656, 659–660
narrow, 660–664, **661**
artificial selection and, 662–663,
663f
in breeding, 662–664
estimation of, 661–662, 661f,
662f
quantification of, 656–664
tests for, 656, 656f
of variation, **654**–664, 687–688
vs. familiality, 654–656
Heritable disorders. *See* Genetic
disorders
Heritablity studies, 645, 645f
Hershey, Alfred, 168, 230–231
Hershey-Chase experiment,
230–231, 231f
HeT-A elements, **444**–445, 444f
Heterochromatin, **84**, **322**, 323f. *See also* Chromatin
constitutive, **322**
position-effect variegation and, 326,
326f, 332–333, 332f, **502**
Heterogametic sex, **48**
Heterokaryon, **201**, 201f
Heteromorphic pair, 76
Heteroplasmons, 56
Heterozygosity, 616–617
loss of
in cancer cells, **565**, 566, 567f
from inbreeding, 625–626, 626f,
685
variation and, 623–624, 623t
Heterozygote, **34**
inversion, 499, 499f
translocation, 501–502
Hexaploidy, **483**
Hfr strains, conjugation via, **157**–161
his mutants, 434
Histogram, frequency, **647**, 648f
Histone(s), **88**, 89f
acetylation of, 330–331
in chromatin remodeling, 328–329,
329f
conservation of, 330
in gene regulation, 329–331, 329f,
330f
Histone acetyltransferase (HAT),
330–331
Histone code, 322, 329–**330**

Histone deacetyltransferase (HDAC), 330–331
Histone tails, **330**, 331f
HLA antigen system, variation in, 615, 615t
Holliday, Robin, 476
Holliday structure, 476
HOM-C, **595**, 596f, 600–601, 600f–602f
Homeobox genes, 585
  in insects *vs.* vertebrates, 600–601, 600f–602f
Homeodomain proteins, **595**
Homeologous chromosomes, **485**
Homeosis, **595**, 596f
Homeotic gene complexes, **595**, 596f, 600, 600f, 601f
  in insects *vs.* humans, 600–601, 600f–602f
Homogametic sex, 48
Homologous chromosomes, **3**, 74f, 75, 76, 76f
  pairing of, 485–487, 485f, 486f
Homologous recombination, in DNA repair, 473, 473f
Homology-dependent repair systems, 468f, **469**–471
Homozygosity
  by descent, 625–626, 625f
  viability and, 630–631, 631f
Homozygote, **34**
Homozygous dominant, **34**
Homozygous recessive, **34**
Hopkinson, D., 664
Horvitz, Robert, 554
Host range, of plaque, 168
Hot spots, **457**, 466
Housekeeping genes, 6
Hox complex, 600–601, 600f–602f
hunchback, in pattern formation, 583–587, 583f
Huntington disease, 45–46, 46f, 467–468
Hybrid, **34**
Hybrid dysgenesis, 436–437, 437f
Hybrid-inbred method, **664**
Hybridization, DNA, **345**
Hybrid plants, 13–14
Hydrogen bonds, in double helix, 234–235, 235f
Hyperacetylated histone, **330**
Hypoacetylated histone, **330**
Hypomorphic mutations, 536f, **537**
Hypophosphatemia, 54
Hypostatic mutations, 203–204, 204f, 205f
Hypothesis, validity of, chi square test for, 40–42, 41f

ICR compounds, as mutagens, **458**, 459f
I gene, 306
  constitutive mutations in, 308–309, 308t, 309t
Ikdea, K., 170
Imino form, of base, **457**, 457f
Immunodeficiency, gene therapy for, 375–379
Immunologic polymorphisms, 615, 615t
Imprinting, 324–326, 325f
Inborn errors of metabolism, 190f, 191. *See also* Phenylketonuria (PKU)

Inbreeding, **625**–626, 685
  experimental, **650**
  genetic drift and, 636–637, 636f
  loss of heterozygosity from, 626, 626f, 685
  negative, **625**
Inbreeding coefficient, **625**, 625f
Incomplete dominance, 193–**194**, 193f
Incomplete penetrance, 207–209, 208f
Indel mutations, 454, 456, **465**–466, 465f, 466f
Independent assortment, 13, 37–39, 38f, 39f, 102–103, 116–117, 121
  deviations from, 117–118. *See also* Crossing-over
  recombination by, 121–122, 122f, 123f
Indian corn. *See* Maize *(Zea mays)*
Inducer(s), **306**
  Lac
    action of, 306–307
    genetic evidence for, 307–309
    synthetic, 307, 307f
Induction, **306**
  zygotic, **172**
Inductive interaction, **599**
Infection
  double, 169
  mixed, **169**, 169f
Inheritance. *See also* Heritability
  autosomal, 29–48. *See also* Autosomal inheritance; Mendelian inheritance
  chromosomal, 75–80. *See also* Chromosome theory of heredity
  cytoplasmic, 29, 55–56, 56f, 57f
  epigenetic, 323–326, 331–332, 332f
    imprinting and, 324–326, 325f
    X inactivation and, 323–324, 324f
  maternal, 55
  Mendelian, 13–14, 29–45. *See also* Autosomal inheritance; Mendelian inheritance
  patterns of, 28–29, 28f
    branch diagrams for, 37, 39
    Mendelian ratios and, 39–42
    product rule for, 39–40
    Punnet square for, 38–39, 38f
    sum rule for, 40
  sex-linked. *See* Sex-linked inheritance
  uniparental, 55–56
Initiation complex, 288
Initiation factors, **288**, 289
Initiator, in arabinose operon, **315**, 315f
Initiator tRNA, **287**–288
Insects. *See Drosophila melanogaster; Drosophila pseudoobscura*
Insertional duplications, **505**–506
Insertion-sequence (IS) elements, 161, **429**–430, 429f, 430f. *See also* Transposable elements
In situ hybridization mapping, 403
Insulin, recombinant, 343, 349
Intelligence, heritability of, 656, 659–660
Interactome, 413–415

DNA microarrays and, 413, 414f
  yeast two-hybrid systema and, 413–415, 415f
Intercalating agents, as mutagens, **458**–459, 459f
Interference, in crossing-over, **127**
β-Interferon enhanceosome, 320, 320f, 331, 331f
Interphase, **91**, 91f, 548
Interrupted mating, **158**, 159f
Intragenic deletions, **502**–503
Intrinsic mechanism, in transcription termination, **263**
Introns, 6, 6f, **267**
  identification of, 406–410
  self-splicing, **268**–269
  splicing of, 267–269, 268f, 269f
Inversion(s), 321, 496, 496f, **497**, 498–501, 498f–500f
  in balancer chromosomes, 500–501
  in deletion synthesis, 502
  in duplication synthesis, 502
  effects of, 398f
  in mapping, 502
  paracentric, **498**, 499, 499f
  pericentric, **498**, 499–500, 500f
  position-effect variegation and, 326, 326f, **502**
  recombination and, 499–500
Inversion heterozygote, **499**, 499f
Inversion loops, 617, 618f
  paracentric, **499**, 499f
  pericentric, **499**–500, 500f
Inverted repeat (IR) sequences, **431**
IPTG, 307, 307f
IQ, heritability of, 656, 659–660
IR sequences, **431**
IS elements, 161
Isoaccepting tRNA, **285**
Isolation
  postzygotic, **690**
  prezygotic, **690**

Jackpot mutation, 462
Jacob, François, 158, 161, 163, 305, 307–309, 311
Jimsonweed *(Datura stramonium)*, ploidy in, 76, 77f, 494
Johannsen, Wilhelm, 650
Jorgenson, Richard, 445–446
Jumping genes. *See* Transposable elements
Junk DNA, **444**

Karpechenko, G., 487
Karyotype, polymorphisms for, 617
Kavenoff, Ruth, 80
Kearns-Sayre syndrome, 56
Kennedy disease, 467–468
Keto form, of base, **457**, 457f
Khorana, H. Gobind, 281
Kinases, **549**
  in posttranslational processing, 292
Kinesin, 580
Kinetochore, **92**, 92f
Klinefelter syndrome, 492–493, 493f
knirp, 593–594, 593f, 594f
Knockout, targeted, 531, 531f
Knockout mice, **372**–375, 374f
Kornberg, Arthur, 239–240, 241
Kreitman, M., 696
krüppel, 593–594, 593f, 594f

lac gene, F′ factor and, 163, 164f
Lac inducer
  action of, 306–307
  genetic evidence for, 307–309
  synthetic, 307, 307f
lac operator, 306, 306f, 313f
  cis-acting, 308, 308f
  genetic evidence for, 307–309
  molecular characterization of, 310–311
  mutations in, 307–309
lac operon, 305f, 306–307
  cAMP-CAP complex and, 312–315, 312f–315f
  catabolite repression of, 312–315, 312f–315f
  control regions of, 313, 314f
  definition of, 307–310
  DNA-binding sites of, 312–315, 312f–315f
  model of, 313–315, 314f, 315f
  mutational analysis of, 307–310
  negative control of, 304, 304f, 306–312, 306f, 314, 314f, 315f
  overview of, 313–315
  positive control of, 312–315, 312f–315f
  RNA polymerase-binding sites of, 314f, 333
  selective activation of, 312–315, 312f–315f
lac promoter, 306, 306f, 313
  genetic analysis of, 310
Lac repressor, 304, 304f, 306–312, 306f, 314f, 315f
  action of, 306–307
  gene for, 306
  genetic evidence for, 307–309
  molecular characterization of, 310–311
  mutations in, 307–310
  trans-acting, 309, 309f
lac system, 305–315
  discovery of, 307–312
  induction of, 306–307
  operon, 306–307, 306f. *See also lac* operon
  regulatory components of, 305–306
  structural genes in, 305
    constitutive mutations in, 307–309, 308t
    polar mutations in, 311–312
Lactose, metabolism of, 305f
lacZ, in mutational analysis, 528–529, 528f
Lagging strand, 240f–242f, **241**
Lambda phage. *See also* Phage(s)
  as cloning vector, 347–348, 349f, 350f
  insertion sequences and, **429**–430, 429f, 430f
  transduction in, 170–174, 170f–173f
Lateral inhibition, **594**
Lathyrus odoratus (sweet pea), as model organism, 14
Law of dominance, 13–14
Law of equal segregation, 13, 33–36, 97–102
Law of independent assortment, 13, 37–39, 38f, 39f, 102–103, 116–117, 121

deviations from, 117–118. *See also*
  Crossing-over
recombination by, 121–122, 122f,
  123f
Leading strand, **240**, 240f, 242f
Leaf color, 195, 195f
  in *Abraxas*, 56, 57f
Leaky mutations, **192**, 536f, **537**
Least-squares regression line, **672**
Lederberg, Esther, 172, 462
Lederberg, Joshua, 154, 155–156,
  157, 170, 172, 462
Leeuwenhoek, Antony van, 154
Lethal alleles, **195–197**, 196f, 197f
Lethal mutations, 195–197, 196f,
  197f
  synthetic, **206–207**, 207f
Leukemia
  Philadelphia chromosome in, 561,
    563, 563f, 569
  treatment of, 569
Libraries
  cDNA, 350–357
  expression, 353
  genomic, 350–357
  shotgun, 394
Licensing, of replication origins, 247
Life cycle
  diploid, 96, 96f, 97
  haploid, 96–97, 96f
  haploid-diploid, 96f, 97
Ligands
  receptor binding of, 556–557, 557f
  in signaling, **553–557**, 557f, 558f
Light repair, 468
Linear tetrads, in centromere mapping, 129–131, 131f
LINEs (long interspersed nuclear elements), **440–441**, 441f
Linkage analysis
  centromere mapping in, 129–131,
    131f
  chi-square test in, 131–133
  compensating for multiple
    crossovers in, 134–136
  development of, 124
  gene ordering by inspection in, 127,
    127f
  linkage mapping in, 124–129,
    125f–127f. *See also* Linkage
    mapping
  Lod scores in, **133–134**
  mapping function in, 134–135
  pedigree analysis in, 133–134
  Perkins formula in, 135–136
  Poisson distribution in, 134–135
  quantitative, 666–667, 666f, 667f
  recombinant frequency in,
    122–124, 124f. *See also*
    Recombinant frequency (RF)
  terminology of, 118
  testcrosses in, 119, 121
    three-point, 125–126
  tetrad analysis in, 120–121, 120f,
    135–136
  transduction in, 170–172, 171f
Linkage disequilibrium, **628**
Linkage equilibrium, **628**
Linkage mapping, **124–129**,
  125f–127f. *See also* Maps/
  mapping
  bacterial, 161–162, 162f
  of *Escherichia coli*, 171–172, 171f

example of, 128–129, 130f
functions of, 128–129
of human chromosome 1, 402f
molecular markers in, 127–128,
  398–401
with multiple crossovers, 134–136
recombinant frequency and,
  124–126, 129, 134–136. *See
  also* Recombinant frequency
  (RF)
Linked genes, 117–119, **118**
  in cis conformation, 118, 118f
  discovery of, 117–118
  inheritance of, 117–118, 118f
  naming of, 118
  recombination of, 121, 122, 123f,
    124f. *See also* Crossing-over
  in trans conformation, 118
Lod scores, in linkage analysis,
  133–134
Long interspersed nuclear elements
  (LINEs), **440–441**, 441f
Long terminal repeats (LTRs), **434**
Loop DNA, 318, 318f
Loss-of-function mutations, 35, **321**,
  453, 535–537, 536f
  vs. gain-of-function mutations,
    535–537, 536f
Loss of heterozygosity (LOH)
  in cancer cells, **565**, 566, 567f
  from inbreeding, 625–626, 626f,
    685
LTR-retrotransposons, **435**
  genome size and, 442, 443f
LTRs (long terminal repeats), **434**
Luria-Delbrück fluctuation test,
  461–463, 462f
Luria, Salvador, 461–462
Lymphoma
  chromosomal rearrangements in,
    563–564, 564f
  diagnosis and treatment of,
    566–568, 567f
Lysis, **167**, 168, 168f
Lysogenic bacteria, **170**

MacDonald, J., 696
MacLeod, Colin, 166, 230
Maize *(Zea mays)*
  chromosomes of, 428, 428f
  as model organism, 428
  transposable elements in, discovery
    of, 425–429. *See also*
    Transposable elements
Major groove, **234**, 236f
Manx cat, 197, 197f
Mapping function, **134–135**
Maps/mapping, 124–131, 398–406
  bacterial, 161–162, 163f, 174–175,
    174f, 175f
  centromere, 129–131, 131f
  clone fingerprint, 396, 397f
  cytogenetic
    for chromosome 7, 410–411,
      411f
    molecular markers in, 401–404,
      403f, 404f
  deletion, **504**
  experimental foundation for,
    117–121
  fluorescent in situ hybridization in,
    403–404, 403f, 404f
  in gene counting, 535, 535f

goals of, 393
inversions in, 502
linkage, 124–129, 125f–127f, 130f,
  398–401. *See also* Linkage
  mapping
microsatellite markers in, 400–401,
  401f
minimum tiling path in, **397**, 398f
molecular markers in, **128**
ordered-clone, 396–398, 398f
overview of, 115–116, 390–391
paired-end reads in, 395–396, 395f
phage, 168–170, 169f
physical, 396–398, 397f, 398f
  molecular markers for, 404–405
  in ordered-clone sequencing, 166
primers in, 394–395
primer walking in, 395
purpose of, 393
rearrangement breakpoint,
  403–404, 403f
recombination, in bacteria, 162,
  163f
repetitive DNA in, 394, 395–396
representativeness of, 393
restriction, **357**, 357f
restriction fragment length polymorphisms in, 399, 400f
sequence, 392–406, 409f. *See also*
  Genome sequencing
  for chromosome 7, 410–411,
    411f
  creation of, 394–398
  filling gaps in, 398
  identifying gene of interest on,
    398–406, 405f
  molecular markers for, 405
  production of, 392–398
sequence contigs in, 395, 395f
sequencing assembly in, 392
sequencing reads in, 392
simple-sequence length polymorphisms in, **400–401**, 401f, 402f
single nucleotide polymorphisms in,
  399–400
  for specific genes, 398–406, 399f
  steps in, 392, 393f
  synteny, 700–701, 700f
  transformation, 166
  translocations in, 502
  of transposable elements, 425
  of whole genome, 392–406. *See also*
    Genome sequencing
  whole genome shotgun, 394–396,
    395f
Map unit (m.u.)
  in centromere maps, 131
  in linkage maps, 124
Marker genes, **665–666**
Marker gene segregation, 655
Maternal-effect genes, **583**
  mutations in, detection of, 584–585
Maternal expression, **583**
Maternal inheritance, 55
Mating. *See also* Sexual reproduction
  assortive, 613f, 625–626
  negative, 625–626
  positive, 625–626
  interrupted, **158**, 159f
  random, 624, 624t
    Hardy-Weinberg equilibrium
      and, 621–623
  variation and, 621–626

Mating-type switch, 327–328
Matthaei, Heinrich, 281
Mature RNA, **264**
McCarty, Maclyn, 166, 230
McClintock, Barbara, 119, 425–429,
  428f, 445
M cytotype, **436**
Mean, **648**, **668**
Media, culture, 153, 153f
Meiocytes, **93**
Meiosis, **35**, 74f, 75–76, 76f, 91f,
  93–95, 94f, 95f
  abnormal, primary exceptional
    progeny and, 79–80, 79f
  chromosome segregation in, 35–36,
    75–76, 76f
  crossing-over in. *See* Crossing-over
  in diploid organisms, 96, 96f, 97,
    98f, 102–103, 102f
  DNA replication in, 93–95, 94f,
    97–103, 98f, 99f, 101f, 102f,
    246–247, 247f
  genotype preservation in, 97–102,
    98f, 99f, 101f, 102f
  genotype transmission in, 97–102,
    98f, 99f, 101f, 102f
  in haploid-diploid organisms, 96f,
    97
  in haploid organisms, 96–101, 96f,
    98f, 101f
  nondisjunction in, 491, 491f
  products of, **93**, 95, 98f
  random assortment in, 76, 76f
  recombination in, **121–123**
Meiospores, 93
Melandrium album, sex chromosomes
  in, 49
Melanin, 8, 10
Melting, in replication, 235, **245**, 245f
Mendel, Gregor, 2, 2f, 13
  plant breeding experiments of,
    13–14, 29–34, 29f–31f, 32t,
    33f, 34f, 36f–38f. *See also* Plant
    breeding
Mendelian inheritance, 13–14,
  29–49, 75–77. *See also*
  Autosomal inheritance
  first law of, 13, 33–**34**, 35–36,
    97–102. *See also*
    Chromosome(s), segregation of
  meiosis in, 97–103, 98f, 99f, 101f,
    102f. *See also* Meiosis
  second law of, 13, **37–39**, 38f, 39f,
    102–103, 116–117, 121
    deviations from, 117–118. *See
    also* Crossing-over
  third law of, 13
Mendelian ratios, 39–42, 136
  chi square test for, 40–42, 41f
  for gene interactions, 209
  in pedigree analysis, 43
  product rule for, 39–40
  sum rule for, 40
Mendelian traits, *vs.* quantitative traits,
  649–650, 649–651
Merozygote, **162**, 162f
MERRF, 56
Meselson, Matthew, 237–238
Meselson-Stahl experiment, 237–239,
  238f
Messenger RNA. *See* mRNA
  (messenger RNA)
Metacentric chromosomes, 84t, 85

Metaphase, **91**
  meiotic, 94f, 95
  mitotic, 87, 91, 91f, 93f
Methyltransferase, in DNA repair,
  468–469
Mice *(Mus musculus)*
  coat color in, 195–197, 196f
  comparative genomics in, 700–701,
    700f
  gene therapy for, 375–376, 376f
  knockout, 372–375, 374f
  as model organism, 14, 196
  transgenesis in, 372–375,
    372f–374f
    by ectopic insertion, 372, 372f
    by gene targeting, 372–374, 374f
    therapeutic, 375–376, 376f
Microarrays. *See* DNA microarrays
Microfilaments, **578**–580, 579f, 580f
Microsatellite markers, **400**–401,
  401f, 402f
Microscopic analysis, 12
Microtubules, 92, 92f, **578**–579, 579f,
  580f
  mRNA tethering to, 583–586, 586f
Midparent value, **661**–662, 661f
Migration, 685
  variation from, 613f, 628–629
Minimal medium, **153**
Minimum tiling path, **397**, 398f
Minisatellite markers, **400**, 401f
Minor groove, **234**, 236f
−10 region, 261, 262f
−35 region, 261, 262f
Mirabilis jalapa (four-o'clocks)
  cytoplasmic segregation in, 56, 56f,
    57f
  incomplete dominance in,
    193–194, 193f
Mismatch repair, **471**–472, 471f, 472f
  crossing-over and, 476
Missense mutations, **455**–456, 455t
Mitochondria, extracellular origin of,
  693–694, 693t
Mitochondrial diseases, 105–106,
  105f, 106f
Mitochondrial DNA (mtDNA), 8,
  103–106, 103f–106f
Mitochondrial genes, 55–56, 75
  mutations in, 103–106, 105f, 106f
Mitogens, in cell cycle regulation, **555**
Mitosis, 74f, 75, **90**–93, 91f–93f
  definition of, 90
  in diploid organisms, 96, 96f, 97, 98f
  DNA replication in, 91–93,
    91f–93f, 98f, 246–247, 247f
  in haploid-diploid organisms, 96f,
    97
  in haploid organisms, 96–97, 96f,
    98f
  nondisjunction in, 491, 491f
  postmeiotic, 100
  in stem cells, 547
Mitotic spindle, 75
Mixed infection, **169**, 169f
Mobile genes. *See* Transposable
  elements
Mode, **648**, 668
Model organisms, **13**–15, 14f
  Caenorhabditis elegans as, 14, 14f,
    554, 581
  Drosophila melanogaster as, 13, 50,
    428, 584–585

maize as, 428
mice as, 196
mutational analysis in, 527–537.
  See also Mutational analysis
Neurospora as, 13, 14f, 100
plants as, 13, 428
yeast as, 327, 328
Modifier mutation, 206
Modularity, in gene regulation, **317**
Molecular analysis, 664–667
  marker-gene segregation in,
    665–666
  quantitative linkage analysis in,
    666–667, 666f, 667f
Molecular clock, **695**–696, 695f
Molecular genetic diagnostics,
  364–366
Molecular machines, 187
  replisome as, 242
  ribosome as, 285
Molecular markers, **128**
  in cytogenetic maps, 401–404,
    403f, 404f
  in linkage maps, 127–128,
    398–401, 401f, 402f
  microsatellite, 400–401, 401f, 402f
  minisatellite, 400, 401f
  in physical maps, 404–405
  unselected, 162
Molecular mimicry, 290–291, 291f
Monod, Jacques, 305, 307–309, 311
Monohybrid, **36**
Monohybrid cross, **36**
Monohybrids, ratios for, 136
Monoploidy, **483**, 483t, 484
Monopolyploidy, in plant breeding,
  489–490, 489f
Monosomy, 483t, **490**, 491–492,
  491f, 492f
  gene-dosage effect and, **495**
  in Turner syndrome, **491**–492, 492f,
    496
Morgan, Thomas Hunt, 77–78,
  117–118, 119, 124
Morphs, 9
Mouse. *See* Mice *(Mus muscularis)*
M phase, of cell cycle. *See* Mitosis
mRNA (messenger RNA), **5**, 258.
  See also RNA
  cellular amounts of, 275
  gene docking sites for, 406, 406f
  localization of, in pattern formation,
    583–586, 586f, 591f
  nucleotide sequence of, 5
  ribosomal binding site for, 286, 287f
  synthetic, 281
  transcription of. *See* Transcription
  in translation, in initiation, 274,
    285–289, 287f
  translation of. *See* Translation
mtDNA (mitochrondrial DNA), 8,
  103–106, 103f–106f
Müller-Hill, Benno, 310–311
Muller, H.J., 523
Multigenic deletions, **503**
Multiple alleles, **192**
Multiple crossovers
  compensating for in linkage analy-
    sis, 135–136
  in linkage analysis, 134–136
Multiple-factor hypothesis, **650**
Muscular dystrophy, 54
Mutagen(s)

acridine orange as, **458**, 459f
aflatoxin₁ as, **461**, 461f
definition of, 453
for forward genetics, 523–524, 524f
ICR compounds as, **458**, 459f
induced mutations and, 453
intercalating agents as, **458**–459,
  459f
mutational specificity of, **457**
proflavin as, **458**, 459f
radiation as, 454f, 459–461, 460f,
  468
SOS-dependent, 459–461, 460f
transposable elements as, 523
Mutagenesis, **453**
  dose response in, 454, 455f
  general *vs.* targeted, 524f
  random, 530, 531
  site-directed, **531**–532, 532f
  targeted, 531–533
Mutant(s), 9–10
  amber, 282, 282f
  auxotrophic, **153**
  bacterial, 153
  crossing-over of, 476
  double, **199**
    epistasis and, 203–204, 204f,
      205f
    lethal, 206
  ochre, 282
  opal, 282
  resistant, **153**, 154f
Mutant hunt, 523, 523f
Mutant rescue, **355**
Mutation(s), **3**, 35, 452–509
  addition, 454, 456, 465–466, 465f,
    466f
  allelic frequency and, 626–627
  background rate of, 453
  base alteration and, 458–459, 459f
  base damage and, 459–461, 460f,
    461f
  base substitution and, **10**, 454, 456,
    456f, 457–458, 458f, 459f,
    464–465
  in cancer, 559–564, 560f,
    562f–564f, 562t
    detection of, 566
    oncogenic, 559–560, 560f
  cdc, 551
  chromosome, 482–509
    changes in chromosome number
      and, 483–496
    overview of, 482–483, 482f
  constitutive, in gene regulation,
    308–309
  cytoplasmic, 56, 57f
  deamination and, **464**, 464f
  deletion, 10, 454, 456, 465–466,
    465f, 466f. *See also* Deletion(s)
  depurination and, **463**
  disease-causing, identification of,
    365–366
  dominant, 192–193, 193f, 536f,
    537
    functional causes of, 536f
  duplication, 10
  dysgenic, 436–437, 437f
  effects of
    functional, 190–192, 191f, 192f
    pleiotropic, 197–198
  enzyme malfunction and, 191–192,
    191f, 192f

epistatic, 187, 203–204, 204f, 205f
expressivity of, 208–209, 208f
fitness and, 695
frameshift, **10**, 280, 455t, **456**, 465
functional consequences of,
  455–456, 457f, 694–696
gain-of-function, **321**–322, 322f,
  453, 536f, 537
  gene fusion and, 321–322, 322f
  vs. loss-of-function mutations,
    535–537, 536f
gene, 452–468
gene interactions and, 190–192
gene size and, 523
genetic drift and, 636–637, 636f
hot spots for, **457**, 466
hypomorphic, 536f, **537**
hypostatic, 203–204, 204f, 205f
identification of. *See also*
  Mutational analysis
  genetic screens in, **524**, 525f,
    527–529, 527f, 528f
  genetic selections in, **524**–526,
    525f, 526f
incidence of, in humans, 506, 507f
indel, 454, 456, **465**–466, 465f,
  466f
induced, 453–454, 453–461, 454t.
  See also Mutagen(s);
  Mutagenesis
inheritance of. *See* Inheritance
insertion, 161, 429–430, 430f, 431
jackpot, 462
leaky, **192**, 536f, **537**
lethal, 195–197, 196f, 197f
  synthetic, 206–207, 207f
loss of, 636–637, 636f
loss-of-function, 35, **321**, 453,
  535–537, 536f
  vs. gain-of-function mutations,
    535–537, 536f
mapping of, 398, 403–406, 404f,
  405f. *See also* Maps/mapping
missense, **455**–456, 455t
in mitochondrial genes, 103–106,
  105f, 106f
modifier, 206
mutagens and, 453–454, 454f
neomorphic, 536f, **537**
neutral, 695–696
new, fate of, 636–637, 636f
in noncoding regions, 456, 456f
nonsense, **455**–456, 455t
nonsynonymous, 455–456, 455t,
  696
nucleotide-pair substitution, 10
null, 10, 192, 536f, 537
  in cancer, 566
in oncogenes, 559–560, 560f,
  562–564, 562f–564f
organelle, inheritance of, 55–56
overview of, 452–453, 452f
penetrance of, 207–209, 208f
phenotypic changes and, 456, 456f
pleiotropic, 523
point, 453–461
  classification of, 455t
  mechanisms of, 457–461
  molecular characterization of,
    454, 455t
  origin of, 453–454
  repeat-induced, **532**, 533f
polar, 311–313

premutations and, 467
prevention of. *See* DNA repair
in protein machine, 206–207, 207f
radiation-induced, 454f, 459–461
randomness of, 461–463
rate of, 626–628, 627t
recessive, 192, 193f
  complementation and, 199–203, 200f–203f
recombination and, 452–453
replication errors and, 457–459, 464–468
revertant, 205
selected, 635–636, 695. *See also* Selection
silent, 455, 455t, 456
SOS-generated, **459**
spontaneous, 461–468
  fluctuation test for, 461–463, 462f
  mechanisms of, 463–468
  rate of, 453
  replica plating and, **462**–463, 463f
  vs. induced mutations, 453
spontaneous lesions and, **463**–464, 464f
sublethal, 197
substitution, 10, 454, 456
superrepressor, 309–310, 310f, 310t
suppressor, 187, 204–206, 206f
  genetic code and, 279–281, 280f
  in translation, 282, 291, 292f
synonymous, **455**, 455t, 456, 696
synthetic lethal, 206–207, 207f
temperature-sensitive (ts), **195**
transition, **454**, 455t
transversion, **454**, 455t
trinucleotide repeat diseases and, **466–468**, 467f
in tumor suppressor genes, 560–561, 560f, 564–566, 565f
variation from, 613f, 626–628
wild-type, 9–10, 9f, **615**
Mutational analysis, 9–10, 11, 521–539
in *Aspergillus*, 528, 528f
in cancer, 566
chemigenomics in, **534**–535, 534f
complementation test in, 535, 535f
in *Drosophila melanogaster*, 584–585
enhancer traps in, **530**, 530f
in *Escherichia coli*, 528–529, 528f
filtration enrichment in, 525, 525f
forward genetics in, **522**, 523–530
gene counting in, 535, 535f
gene-specific mutagenesis in, 531–533, 531f–533f
genetic screens in, 170, **524**, 525f, 527–529, 527f, 528f
genetic selections in, **524**–526, 525f, 526f
loss-of-function *vs.* gain-of-function mutations in, 535–537, 536f
mapping in, 398, 403–406, 404f, 535, 535f. *See also* Maps/mapping
in *Neurospora*, 527, 527f
overview of, 522–523, 522f
phenocopying in, 533–535, 533f, 534f, 538

random mutagenesis in, 531
repeat-induced point mutations in, **532**, 533f
reverse genetics in, **522**, 530–535
reversion of auxotrophy to protrophy and, 525–526, 526f
RNA interference in, **533**–534, 534f
saturation in, 523
site-directed mutagenesis in, **531**–532, 532f
targeted gene replacement in, 531, 531f
transgenics in, 538–539, 539f
transposon tagging in, **439**, **524**
in yeast, 527–528, 527f
in zebra fish, 529–530, 529f
Mutational specificity, **457**
Myoclonic epilepsy and ragged red fibers (MERRF), 56

n (chromosome number), 83
Napoli, Carolyn, 445–446
Narrow heritability (h^2), 660–664, **661**
  artificial selection and, 662–663, 663f
  in breeding, 662–664
  estimation of, 660–662, 661–662, 661f, 662f
Nascent protein, folding of, **291**–292
Native protein, **291**–292
Natural selection, **629**, 681–685
  adaptive peaks for, 685–687, 686f
  balancing, **685**
  directional, **685**
  genetic drift and, 687
  mutation and, 635–636
  random sampling and, 636
  rate of, 633, 633f, 634f
  variation arising from, 613f, 629–633
  viability and, **630**–631, 631f
  vs. artificial selection, 681
Negative assortive mating, **625**–626
Negative inbreeding, **625**
Negative selection, **442**
Neighborhood effect, 129
Nematode. *See Caenorhabditis elegans* (nematode)
Neomorphic mutations, 536f, **537**
Neurospora crassa (filamentous fungi)
  auxotrophy/protrophy in, 525–526, 526f
  centromere mapping in, 129–131, 131f
  filtration enrichment in, 525, 525f
  genetic screens in, 527, 527f
  heterokaryon of, 201, 201f
  life cycle of, 101f
  maternal inheritance in, 55
  meiosis in, 101f
  as model organism, 13, 14f, 100
  mutations in
    induced, 454t
    one gene–one enzyme hypothesis and, 187–189, 188f
    poky, 55–56, 103, 104f
    repeat-induced point, **532**, 533f
Neutral mutations, 695–696
Nicotiana longiflora, corolla length in, 645–646, 646f
Nirenberg, Marshall, 281

Nonautonomous elements, **427**
Nonconservative substitutions, **456**
Nondisjunction, **79**, 80, 491, 491f
Nonhomologous end-joining, in DNA repair, 472, 473f
Nonnative protein, **292**
Nonsense codons. *See* Stop codons
Nonsense mutations, 455–456, 455t
Nonsynonymous mutations, 455–456, 455t
Nonsynonymous substitutions, **696**
Norm of reaction, 19–21, 20f, 208, **651**
  phenotypic distribution and, 651–652
Norm of reaction studies, 644–645, 645f, 651f, 652–653
  in domesticated plants and animals, 652
  in natural populations, 652–653
Northern blotting, **12**, 13f, **354**
NOS protein gradient, 586, 586f
Nuclear localization sequence (NLS), **293**
Nuclear spindle, 36, **92**, 92f
Nucleoids, 103, 103f
Nucleolar organizer (NO), **85**, 85f
Nucleolus, **85**, 85f
Nucleosomes, 88, **88**, 89f, **244**, 322, 323f
  assembly of, in replication, 244, 244f, 331–332
  in gene regulation, 329–331, 329f, 330f
  histone tails of, 330, 331f
  repositioning of, 328–329, 329f
  structure of, 330, 330f
Nucleotide(s)
  in DNA, 3–4, 4f, **232**–233, 232f, 233t. *See also* Base(s)
  in RNA, 5
Nucleotide-excision repair, **470**, 470f
Nucleotide-pair substitution, **10**, 454, 456. *See also* Mutation(s)
Nucleotide polymorphisms. *See also* Single-nucleotide polymorphisms (SNPs)
  in linkage mapping, 128
Nucleotide substitutions
  neutral, 695–696
  nonsynonymous, **696**
  synonymous, **696**
Null allele, **10**, 192
Null hypothesis, **132**
Nullisomy, **490**
Null mutations, 10, 192, 536f, 537
  in cancer, 566
Nüsslein-Volhard, Christiane, 548

Ochre codon, 282
Octad, **129**
Okazaki fragments, **241**, 241f, 242f
Oncogenes, 559–564
  classes of, 561–564, 561t
  mutations in, 559–560, 560f, 562–564, 562f–564f
  v-erbB, 526f, 562
Oncoprotein(s), **562**
  Bcr 1–Abl fusion, 563, 563f
  Ras, 562, 562f
One-gene-one-enzyme hypothesis, 42–44, 187, 277–278
  mutations and, 190–192

Nonautonomous elements, **427**
One-gene-one-polypeptide hypothesis, 188f, **189**
One-gene-one-protein hypothesis, 188f, **189**, 277–278
On the Origin of Species (Darwin), 680–681
Opal codon, 282
Open reading frame (ORF), **360**, 360f, **407**
  detection of, 407–410
  in yeast, 411–412, 412f
Operators, **304**, 304f
Operon, **306**, 307
  ara, 315–316, 315f
  lac. See lac operon
  trp, 316, 316f
Ordered-clone sequencing, 396–398, 398f
Organelle chromosomes, 103–106, 103f–106f
Organelle genes, 55–56, 103–106, 103f–106f
Organelles, extracellular origin of, 693–694, 693t
oriC, **244**
Origin (O)
  in conjugation, **158**, 159f, 160
  replication, **244**–247, 245f, 246f
  licensing of, 247
Origin recognition complex (ORC), **246**–247
Osmaronia dioica, sex chromosomes in, 49f
Outbreeding, enforced, **625**
Overdominance, **634**
Oxidatively damaged bases, as mutagens, **464**

P1 artificial chromosome vectors, **349**
p21, as cell cycle checkpoint, 550, **550f**
p53
  in cancer, 565–566
  as cell cycle checkpoint, 550, 550f, 557–558
PAC (P1 artificial chromosome), **349**
PAC vectors, 349
Paired-end reads, **395**–396, 395f
Pair-rule genes, in pattern formation, 593f, **594**, 594f
Palindrome, DNA, **344**–345
Paracentric inversions, **498**, 499, 499f
Paracrine signaling, 553f, **554**. *See also* Signaling
Paralogous gene clusters, **601**
Parental generation, **31**, 31f
Parental imprinting, **324**–326, 325f
Patau syndrome, 494
Pattern formation, 582–602
  anterior-posterior axis in, 583–587, 583f, 586f, 588f, 592t
  BCD protein gradient in, 583–587, 583f, 585f–588f
  cardinal genes in, **591**–594, 592t
  cytoskeletal asymmetry and, 582–588
  DL protein gradient in, 588–589, 588f, 589f
  in *Drosophila melanogaster*, 583–599
  epidermal growth factor receptor in, 590, 590f
  feedback loops in, 597–598, 598f

Pattern formation *(cont)*
gap genes in, 593–594, 593f, 594f
germ layers in, **583**
HB-M protein gradient in, 583–586, 583f
maternal-effect genes in, **583**
maternal expression in, **583**
mRNA localization in, 583–586, 586f, 591f
mutational analysis for, 584–585
NOS protein gradient in, 586, 586f
pair-rule genes in, 593f, **594**, 594f
polarity in, 579–580, 580f, 592, 593f, 595
positional information in, **577**. *See also* Positional information
regulatory cascade in, 595–597, 597f
segmentation in, **583**, 592–595, 593f, 594f
signaling in, 588–589, 588f, 589f, 590f, 597–599, 598f
SPZ protein gradient in, 588–589, 588f, 589f
steps in, 582–583, 582f
transcription factors in, 583–589, 585f–589f, 591f, 594–597
P cells, 581, 581f
PCR. *See* Polymerase chain reaction (PCR)
P cytotype, **436**
Pea. *See under* Pisum
Pedigree analysis, **42**–48, 42f
of autosomal dominant disorders, 45–46, 46f, 47f
of autosomal polymorphisms, 46–48, 47f
of autosomal recessive disorders, 42–44, 43f, 44f
imprinting and, 324–326, 325f
Lod scores in, 133–134, 133f
penetrance and, 208, 208f
of sex-linked dominant disorders, 54, 54f
of sex-linked recessive disorders, 52–54, 52f, 53f
P elements, 371, **436**–437
in gene tagging, 439
as vectors, 439–440, 439f
Penetrance, incomplete, **207**–209, 208f
Peptide bonds, 275, 275f, 276f
Peptidyltransferase center, **287**
Pericentric inversions, **498**, 499–500, 500f
Perkins formula, 135–136
Permease, in *lac* system, 305, 307, 308
Permissive temperature, **195**
Peterson, Peter, 427
P granules, 581, 581f
Phage(s), **152**, 166–170. *See also* Viruses
definition of, 152
in genetic analysis, 167
host range of, 168
infection cycle of, 167–168, 167f
mapping of, 168–170, 169f
naming of, 167
plaque morphology of, 168, 168f
structure of, 167
temperate, **170**
in transduction, 170–174
virulent, **170**

Phage crosses, in recombination analysis, 168–169, 169f
Phage λ
as cloning vector, 347–348, 349f, 350f
insertion sequences and, **429**–430, 429f, 430f
transduction in, 170–174, 170f–173f
Phage recombination, **153**, 168–170, 169f
Phage vectors, 347–348, 349f, 350f
Phenocopying, 530, 533–535, 533f, 534f, 538
Phenome, 413
Phenotype, 8–9, 18, **31**
continuity of, 646
continuous variation and, 8f, 9f, 10, 644–647
developmental noise and, 21–22, 22f
discontinuous variation and, 8–10, 8f, 9f, 644
dominant, 33, 34. *See also* Dominance
expression of
expressivity and, 208–209, 208f
factors affecting, 207–209, 208f
penetrance and, 207–209, 208f
gene dosage and, 495–496
gene-environment interactions and, 16–22, 17f– 22f. *See also* Gene-environment interactions
gene interactions and, 187–209. *See also* Gene interaction
heritability and, **654**–664. *See also* Heritability
intensity of, 208
lethal alleles and, 195–197, 196f, 197f
partial, 18
ploidy and, 76, 77f, 494–496
polymorphic, 615
qualitative differences in, 644
recessive, 33, 34. *See also* Recessiveness
similarity of in relatives, 655–656
unstable, **425**. *See also* Transposable elements
wild-type, 19–20, 19f, **615**
Phenotype-genotype relationship, 9, 18–19, 646–647, 649–650
distributive, 649–650, 649f, 650f
norm of reaction and, 651–653
Phenotypic correlation, **658**
Phenotypic determination
developmental noise and, 21–22
genetic-environmental model of, 18, 18f
genetic model of, 16–17, 167f
norm of reaction and, 19–21, 20f
Phenotypic distribution, 648–652, 649f
environment distribution and, 651–652
genotypic distribution and, 648–650, 649f
norm of reaction and, 651–653, 651f
Phenotypic expression
expressivity and, 208–209, 208f
penetrance and, 207–209, 208f
Phenotypic variation, 612–614. *See also* Variation

vs. genotypic variation, 649–652
Phenylalanine, in phenylketonuria, 43–44, 187. *See also* Phenylketonuria (PKU)
Phenylalanine hydroxylase (PAH), 191, 193–194
Phenylketonuria (PKU), 42–44, 187, 190–192, 190f–192f, 197–198, 631
Phenylthiocarbamide, taste sensitivity for, 47, 667
Philadelphia chromosome, 561, 563, 563f, 569
Phosphatases, in posttranslational processing, 292
Phosphate, in DNA, **232**, 232f, 233–234, 235f, 236f
Phosphorylation
of carboxyl tail domain, 266
in cell cycle regulation, 549–550, 549f, 556
Photolyase, 468
Photoproducts, 461
Photorepair, 468
Phyletic evolution, **681**, 682f
Physical maps, **404**. *See also* Maps/mapping
molecular markers for, 404–405
Piebald spotting, 46, 47f
Pili, in conjugation, 157, 157f
Pisum elastium, 35
Pisum sativum
Mendel's experiments with, 13–14, 29–49
as model organism, 13
reproductive parts of, 29–30, 29f
PKU (phenylketonuria), 42–44, 187, 190–192, 190f–192f, 197–198, 631
Plant(s). *See also specific plants*
allopolyploid, **485**, 487–488, 487–489, 487f–489f, 490
amphidiploid, **487**, 487f, 489f
aneuploid, 76, 77f, 494
autopolyploid, **485**–487, 485f, 486f
chloroplast DNA in, 8, 103–106, 103f–106f
cytoplasmic segregation in, 56, 57f
development in, 603–605, 604f
dioecious, 48–49, 49f
evolution of, allopolyploidy in, 488, 488f
genetic engineering in, 368–370, 369f
genome size in, 442–443, 443f
haploid-diploid life cycle in, 96f, 97
as model organisms, 13, 428
monopolyploid, 489–490, 489f
norm of reaction studies in, 652–653
norms of reaction for, 20–21, 20f
polyploid, 484–490, 484f–490f. *See also* Polyploidy
reproductive parts of, 29–30, 29f
sex chromosomes in, 48–49, 49f
synteny in, **442**, 443
transgenic, 368–370, 369f, 370f
variegated, 56, 57f
Plant breeding
allopolyploids and, 487–488, 487f–489f, 490
autopolyploids in, 485–487, 485f, 486f

continuous variation in, 10
crosses in, 29–32, 31f
dihybrid, 36–39, 36f–38f
family selection in, **664**
first filial generation in, 31, 31f
history of, 35
hybrid-inbred method of, **664**
Mendel's experiments in, 13–14, 29–39
monopolyploids in, 489–490, 489f
narrow heritability in, 662–664
norm of reaction in, 653, 654f
parental generation in, 31, 31f
reciprocal crosses in, 31–32, 31f, 32t
selfing and, 29–30, 39
testcrosses in, 39
Plaque, phage, 168, 168f
Plasmid(s), **153**
F. *See* Fertility factor (F)
F', 162–164, **163**, 164f
genetic determinants borne by, 165t
R, 164–**165**, 164f
Ti, 368–370, 369f
transfer of, 156–157, 156f, 157f. *See also* Bacterial conjugation
yeast
episomal, 367f, 368
integrative, 366, 367f
Plasmid vectors, 346–347, 348f, 350f. *See also* Vectors
expression, 349
Ti, 368–370, 369f
yeast, 366–368, 367f
Plating, **153**, 153f
Pleated sheet, 276f, 277
Pleiotropic effects, **197**–198
Pleiotropic mutations, 523
Ploidy. *See* Chromosome number; Polyploidy
Point mutations, **453**–461. *See also* Mutation(s)
induced *vs.* spontaneous, 453–454, 454f, 454t
rate of, 626–628, 627t
repeat-induced, **532**, 533f
Poisson distribution, **134**–135
pol, in viral replication, 435, 435f
Polar granules, 581–582
Polarity, developmental, 579–580, 580f, 592, 593f, 595
Polar mutations, 311–313
Pole cells, **582**, 582f
Pol I, 240, 241, 242, 242f
Pol III, 240, 241, 242, 242f
Pol III holoenzyme, **242**–243, 242f, 245
Polyadenylation signal, 267
Polydactyly, 46, 46f
Polymerase chain reaction (PCR), **360**–362, 361f
in genetic disease diagnosis, 365–366
Polymerases, error-prone, 459–460, 460f
Polymorphisms, 9, 47–48, **614**–621. *See also* Variation
amino acid sequence, 615–617, 616f, 617t
autosomal, pedigree analysis of, 46–48, 47f
balanced, 634–636

chromosomal, 617–618, 618f, 618t, 619f
frequency of, 47–48
genetic, 615
immunologic, 615, 615t
inversion, 617, 618f
phenotypic, 615
protein, 615–617
restriction fragment length, 48, **128**, 399, 400f, 617–618
in linkage mapping, 128
silent, 620
single-nucleotide, 399–400, **619–620**
in cancer cells, 566, 567f
in linkage mapping, 128
synonymous, 620
Polypeptide(s), **5, 275**. *See also* Amino acids; Protein(s)
identification of, 406–410. *See also* Bioinformatics
Polypeptide chain, 275, 275f
amino end of, 275, 275f, 276f
carboxyl end of, 275, 275f, 276f
Polyploidization, 691
Polyploidy, 484–488
allopolyploidy, **485**, 487–489, 487f–489f, 490
in animals, 490, 494
autopolyploidy, **485**–487, 485f, 486f
gene balance and, **494**–496
gene-dosage effect and, **495**, 496
organism size and, 484, 485f, 490, 490f, 494
in plants, 484–490, 484f–490f, 494
vs. aneuploidy, 494–495
Poly(A) tail, **267**
Polytene bands, 88, 88f
Polytene chromosomes, 87–88, 87f
Population, 612
Population genetics, **612**–637
environmental genetics and, 684
experimental, 612
overview of, 612, 613f
quantitative. *See* Quantitative genetics
scope of, 612
theoretical, 612
variation in, 612–629. *See also* Variation
Populations, samples and, **672**–673
Positional cloning, 129, 346f, **356**–357, 357f
Positional information, **577**
along anterior-posterior axis, 583–587, 583f, 586f, 588f, 589–590, 590f
along DL protein gradient, 588–589, 588f, 589f
along dorsal-ventral axis, 588–590, 588f–590f
along HB-M protein gradient, 583–586, 583f
along NOS protein gradient, 586, 586f
along SPZ protein gradient, 588–589, 588f, 589f
BCD protein gradient and, 583–587, 583f, 585f–588f
cell fates and, 591–597
concentration gradients and, 588–589, 588f, 589f

cytoskeletal asymmetry and, 582–588
extracellular, 587–589, 590f
initial interpretation of, 591–594
intracellular, 583–587, 590f
signaling of, 588–589, 588f, 589f
transcription factors and, 583–589, 585f–589f, 591f
Position effect, **372**
Position-effect variegation (PEV), **326**, 326f, **502**
Positive assortative mating, **625**–626
Posttranscriptional processing, **266**
Posttranslational processing, 291–293
in histone tails, 330–331
Postzygotic isolation, **690**
Prader-Willi syndrome, 325
Preinitiation complex (PIC), **265**
pre-mRNA, **264**
cotranscriptional processing of, 266–268, 266f–268f
Premutations, 467
Prezygotic isolation, **690**
Primary exceptional progeny, 79–80, 79f
Primary transcript, **264**
Primase, **241**
Primers
in polymerase chain reaction, 343, 360–362, 361f
in replication, **241**, 241f
in sequencing, 343, 360–362, 361f, 394–395
Primer walking, **395**
Primosome, **241**
Probes, **12, 351**
antibody, 353, 353f
for disease-causing mutations, 365, 365f
DNA, 12, 13f, 351–354, 352f–354f
for mixtures, 354, 355f
in Southern blotting, 354, 355f
protein, 13, 13f, 353, 353f, 354
RNA, 12–13, 13f, 354, 355f
in Northern blotting, 354
Processive enzyme, **243**
Product rule, 39–40
Products of meiosis, 93, 95, 98f
Proflavin, as mutagen, **458**, 459f
Programmed cell death. *See* Apoptosis
Prokaryotes, **6**, 152, **152**. *See also* Bacteria
gene structure in, 6
genome of, 15, 16f
as model organisms, 13, 14f, 154
Proliferating cell nuclear antigen, **244**
Promoter(s), **261**, 302, **309**
core, 261, 261f, 317, 317f
in eukaryotes, 261, 261f, 265, 265f, 317–318, 317f, 318f
lac, 306, 306f, 313
genetic analysis of, 310
in prokaryotes, 261, 262f, 304, 304f
Promoter-proximal elements, **317**–318, 318f
Proofreading, 241
Prophage, **170**
Prophage λ, in transduction, 172–174, 173f
Prophase, **91**
meiotic, 94f, 95
mitotic, 91, 91f, 93f

Propositus, **42**
Protein(s), **2**. *See also* Amino acids
accessory, in DNA replication, 242–243
allosteric sites on, 304
α helix of, 276f, 277
amino acids in, 275, 275f. *See also* Amino acids
sequence of, 279–282. *See also* Genetic code
architectural, 318
chaperone, 292
common ancestry of, 697–700, 698f–700f
definition of, 2
DNA binding sites on, 304
fibrous, **277**
folding of, 291–292
functional distribution of, 699–700, 699f
functional domains of, 277, 302–303
functions of, 4–5
gene docking sites for, 406
in gene regulation, 187
globular, **277**, 277f
homeodomain, 595
identification of, 406–410
mutational effects on, 10–11, 11f
nascent, **291**
native, **291**
nonnative, **292**
peptide bonds in, 275, 275f, 276f
pleated sheet conformation of, 276f, 277
polypeptide chain of, 275, 275f
amino end of, 275, 275f, 276f
carboxyl end of, 275, 275f, 276f
posttranslational processing of, 291–293
in histone tails, 330–331
probes for, 13, 13f, 353, 353f, 354
secretory, 8
shape of, 277
species differences in, 699–700, 699f, 700f
structural, 5
structure of, 5, 275–277, 275f–277f
primary, **275**, 276f
quaternary, 276f, **277**
secondary, **275**–277, 276f
tertiary, 276f, **277**
subunits of, 276f, **277**
synthesis of, 5–6, 8, 273–295. *See also* Translation
trans-acting, 309, 309f
types of, 5
Protein families, 410
Protein gel electrophoresis, 616–617, 616f, 617t
Protein-gene colinearity, 278, 278f
Protein-gene interactions, 197–209
Protein machine, mutations in, 206–207, 207f
Protein polymorphisms, 615–617
Proteome, **406**, 413
Proteomics, comparative, 698–701
Proto-oncogenes, **559**, 561–562, 563
Prototrophic bacteria, **153**
Prototrophy, reversion of auxotrophy to, 525–526, 526f
Proviruses, **434**

Pseudochondroplasia, 45, 45f
Pseudodominance, **504**
Pseudolinkage, **502**
P site, **286**, 287, 287f, 289–290
PTC, taste sensitivity for, 47, 667
Pulsed-chase experiment, **257**, 257f
Pulsed field gel electrophoresis, 81
Punnett, R.C., 117
Punnett squares, 38–39, 38f
Pure line, **30**, 30f
Purine bases, **232**–233, 232f, 232t. *See also* Base(s)
Purine ribonucleotides, 258, 258f
Pyrimidine bases, **232**–233, 232f, 233t. *See also* Base(s)
Pyrimidine ribonucleotides, 258, 258f

Qualitative variation, 8–10, 8f, 9f, 28–29, 644, 649–651
Quantitative genetics, 643–674, **647**
of complex traits, 647
heritablity studies in, 645, 645f
methodology of, 644–645, 645f
multiple-factor hypothesis in, **650**
norm of reaction studies in, 644–645, 645f, 651f, 652–653
overview of, 644–645, 645f
quantitative trait locus studies in, 645, 645f
selection studies in, 645, 645f
statistical distributions in, **647**–648, 648f
statistical measures in, 648
Quantitative linkage analysis, 666–667, 666f, 667f
Quantitative trait(s)
gene number and, 650–651
norm of reaction studies for, 652–653
vs. Mendelian traits, 649–651
Quantitative trait locus, **665**
Quantitative trait locus studies, 645, 645f, 665–666
Quantitative variation, 8f, 10, 644–651
gene number and, 8f, 10, 644–651

Racial variation, 688, 689t
Radiation, as mutagen, 454f, 459–461, 460f, 468
Random genetic drift, 103, 613f, **636**, 637f, 684, 684t, 695–696
natural selection and, 687
neutral mutations and, 695–696
Random mating, 624, 624t
Hardy-Weinberg equilibrium and, **621**–623
Raphanobrassica, 487–488, 487f
Rare genetic events
screens for, 170
selective systems for, 170
Ras oncoprotein, 562, 562f
Ras protein
in cancer, 562, 562f
in signaling, 557, 557f
Raynor, G.H., 76
Rb protein
in cell cycle regulation, 549–550, 550f, 556
in retinoblastoma, 564–565
Rearrangement breakpoint mapping, 403–404, 403f. *See also* Maps/mapping

Receptors, in signaling, 555, 555f, 556–557, 557f, 558f
Receptor tyrosine kinases (RTKs), 556–557, 557f, 558f
Recessive mutations, 192, 193f
  complementation and, 199–203
Recessiveness, **31**–32, 34, 192–193, 193f
  haplosufficiency and, 192
  homozygous, 34
Recipient, in conjugation, **153**
Reciprocal crosses, 31–32, 31f, 32t
  in X-linked inheritance, 50–52, 51f
Reciprocal translocations, 496f, **497**, 501–502
Recombinant(s), **121**–123
  from crossing-over, 122, 123f
  detection of, 121
    by testcrosses, 121, 122f
  from independent assortment, 121, 122–124, 122f, 123f
  rII$^+$, 169–170, 279, 280f
Recombinant chromatids, in centromere mapping, 131
Recombinant DNA, **343**. *See also* DNA technology
Recombinant frequency (RF), 122–**124**, 124f
  in centromere mapping, 131
  chi-square test for, 131–133
  interference and, 127
  inversions and, 500
  in linkage mapping, 124–126, 129, 134–136
    for bacteria, 161–162, 162f
  Lod scores and, 133–134
  mapping functions and, 134–135
  map units and, 124
  maximum value for, 135
  with multiple crossovers, 134–137
  Perkins formula and, 135–136
  in phage mapping, 169–170
Recombination, **121**–122, 452
  bacterial, 152f, 153
    mechanisms of, 152f, 153. *See also* Bacterial conjugation; Bacterial transformation; Transduction
  crossing-over in, 119–121, 119f, 120f. *See also* Crossing-over
  endogenote in, **162**, 162f
  exogenote in, **162**, 162f
  homologous, in DNA repair, 473, 473f
  by independent assortment, 13, 37–39, 38f, 39f, 121–122, 122f, 123f
  interference in, 127
  inversions and, 499–500, 499f, 500f
  linkage disequilibrium and, **628**
  linkage equilibrium and, **628**
  of linked genes, 121, 122, 123f, 124f. *See also* Crossing-over
  meiotic, **121**–123
  merozygote in, **162**, 162f
  mutation and, 452–453
  nondisjunction in, 80, **491**, 491f
  phage, **153**
  in testcross, 121, 122f
  of unlinked genes, 121–122, 123f. *See also* Independent assortment
  variation from, 613f, 628

Recombination analysis. *See* Linkage analysis
Recombination mapping. *See also* Maps/mapping
  bacterial, 162, 163f
  in mutational dissection, 535, 535f
Red-green colorblindness, 52
Reeve, E., 685
Regression, 672
Regression line of *y* on *x*, **672**, 672f
Regulatory genes, 187
  complementation and, 202, 202f
  mutations in, down/upregulation from, 206
Relation, statistical, **648**
Relative frequency, **668**
Release factors (RFs), **290**, 290f
Repairosome, 470, 470f
Repeat-induced point mutations (RIP), **532**, 533f
Repetitive DNA, 85, 85f, 410, 440–442, 441f
  in genome sequencing, 394, 395–396
Replica plating, **462**–463, 463f
Replication. *See* DNA replication
Replication fork, 239, 239f, **240**–241, 240f–242f
Replication origins, **244**–247, 245f, 246f
  licensing of, 247
Replication slippage, 465
Replicative transposition, **432**–433, 433f
Replisome, **242**–247, 331
  assembly of, 244–247
  definition of, 242
  in eukaryotes, 244, 244f, 331
Reporter gene, **414**
Repressors, in prokaryotes, **304**–312, 304f, 306f. *See also* Lac repressor
Reproductive fitness, 629–633. *See also* Fitness; Sexual reproduction
Resistant mutants, **153**, 154f
Restriction enzymes, 12, 48, **344**
  in recombinant DNA production, 344–346, 344f, 345f
Restriction fragment length polymorphisms (RFLPs), 48, **128**, 399, 400f, 617–618
  in linkage mapping, 128
Restriction fragments, **344**
Restriction mapping, **357**, 357f. *See also* Maps/mapping
Restriction-site differences, in genetic disease diagnosis, 365, 365f
Restriction sites, 344
Restriction-site variation, 617–618
Restrictive temperature, 195, **195**
Retinoblastoma, 564–565, 565f
Retro-elements, **424**–425, 424f, 434–436, 435f, 436f. *See also* Transposable elements
Retrotransposition, **424**–425, 424f
  in telomere lengthening, 444–445, 444f
Retrotransposons, 424–425, 424f, 434–436, **435**, 435f, 436f. *See also* Transposable elements
  genome size and, 442, 443f
  HeT-A, **444**–445, 444f
  LINE, **440**–441, 441f

LTR, **434**
SINE, **440**–442, 441f
TART, **444**–445, 444f
Retroviruses, **434**–435. *See also* Viruses
  life cycle of, 434, 434f
Reverant, **205**
Reverse genetics, **522**, 530–535. *See also* Mutational analysis
  chemigenomics in, **534**–535, 534f
  gene-specific mutagenesis in, 531–533, 531f–533f
  phenocopying in, 533–535, 533f, 534f, 538
  random mutagenesis in, 531
  repeat-induced point mutations in, **532**, 533f
  RNA interference in, **533**–534, 534f
  site-directed mutagenesis in, **531**–532, 532f
  targeted gene replacement in, 531, 531f
Reverse transcriptase, 343, **434**
  in transposition, 434, 434f
R factor, 164–165, 165f, 430–**431**
R factors, **430**–431, 432f
R groups, 275, 275f
  posttranslational modification of, 275
Rh incompatibility, 634–635
Rhoades, Marcus, 427, 428, 428f
Rho-dependent mechanism, in transcription termination, **263**
Ribonucleic acid. *See* RNA
Ribonucleotides, 258, 258f
Ribose, **257**
Ribosomal RNA (rRNA). *See* rRNA (ribosomal RNA)
Ribosomes, 6, 8, **274**, 285–291
  binding sites on, 286–287, 287f
  decoding center in, 287
  peptidyl transfer center in, 287
  properties of, 285–287
  as protein factories, 289–290, 289f
  structure of, 273f, 285, 286f
  subunits of, 285–286, 286f
Ribozymes, **258**
rII$^+$ recombinant, 169–170, 279, 280f
RIP (repeat-induced point mutations), **532**, 533f
RNA, 5, **256**–270
  classes of, 258–259
  cotranscriptional processing of, 266, 266f
  double-stranded, in RNA interference, 533–534, 533f
  flow of information from DNA to, 6–8, 7f
  functional, **190**, 258–259, 274–275
  gene docking sites for, 406, 406f
  mature, **264**
  messenger. *See* mRNA (messenger RNA)
  posttranscriptional processing of, 266
  pre-mNA, 264
  properties of, 257–258
  ribosomal. *See* rRNA (ribosomal RNA)
  small nuclear, **259**
    in spliceosome, 268

  splicing of, 267–269, 268f, 269f
  structure of, 257–258
  synthesis of. *See* Transcription
  transfer. *See* tRNA (transfer RNA)
RNA:DNA hybrids, in transcription, 262–263, 263f
RNA elements, **424**–425, 424f, 434–436, 435f, 436f. *See also* Transposable elements
RNA interference (RNAi), **533**–534, 534f
RNA polymerase(s), **259**
  sigma subunit of, 261–262, 264
  in transcription
    in eukaryotes, 264, 265, 266, 317, 318, 319f
    in prokaryotes, 259–260, 260f, 261, 261f, 262, 263, 304
    lac operon and, 313, 314f
RNA polymerase holoenzyme, **261**, 262f
RNA primer, in DNA replication, 241, 241f, 242, 242f
RNA probes, 12–13, 13f, 352f, 354, 355f. *See also* Probes
  in Northern blotting, 354
RNA processing, **264**
RNA transcript, 5, 259
  primary, 264
RNA world, **268**
Robertson, F., 685
Robertsonian translocation, in Down syndrome, 506, 507f
Rolling circle replication, **157**
R plasmid (factor), 164–**165**, 165f, 430–**431**
rRNA (ribosomal RNA), 85, **259**, 274, **285**–287, 286f
  cellular amounts of, 275
  function of, 274
RTKs (receptor tyrosine kinases), 556–557, 557f, 558f
Rubin, Gerald, 439
rut site, 263
ry$^+$, 439–440, 439f

Saccharomyces cerevisiae. *See also* Yeast
  cell cycle in, 551
  genetic engineering in, 366–368, 367f
  genetic screens in, 527–528, 527f
  genomic analysis for, 411–412, 412f
  life cycle of, 328, 328f
  mating-type switch in, 327–328
  as model organism, 13, 14f, 327, 328, 551
Safe havens, **442**–444, 443f
Salmonella typhirium, transduction in, 170
Sample, 672–673
Sanger sequencing, 357–359, 358f, 359f
Satellite DNA, **85**, 85f
Scaffold, 89, 89f, 90
  sequence-contig, 396
Scaffold attachment regions (SARs)l, 89–90
Schizosaccharomyces pombe, forward genetic screens in, 527–528, 527f
Screening, for genetic diseases, 364–365
Screens, **170**

Second-division segregation (M_{II}), **131**, 131f
Secretory proteins, 8
Segmentation, developmental, **583**, 592–594, 593f, 594f
Segment-polarity genes, 595
Segregating line, **652**
Segregation
 allelic, 13, 33–34, 35–36, 91f, 93–95, 94f, 95f, 97–103, 98f, 99f, 101f, 102f
 chromosomal, 13, 33–36, 49, 74f, 75–78, 76f, 77f, 91–93, 91f–93f
  abnormal, 79–80, 79f
  adjacent-1, **501**, 501f
  alternate, **501**, 501f
  in autopolyploids, 485, 485f
  balanced translocations and, **501**–502, 501f
  equal, 13, 33–36, **34**
  in meiosis, 91f, 93–94, 93–95, 94f, 95f, 97–102, 98f, 99f, 101f, 102f
  in mitosis, 91–93, 91f–93f
 cytoplasmic, **56**, 56f, 105
 marker gene, 655
Segregation patterns
 first-division, 131
 second-division, **131**, 131f
Seiler, J., 78
Selection, **629**, 695
 adaptive peaks for, 685–687, 686f
 artificial, **629**. *See also* Breeding
  narrow heritability and, 662–664
  variation from, 613f, 629–633
 balancing, **685**
 directional, **685**
 family, **664**
 genetic, **524**–526, 525f, 526f
 genetic drift and, 687
 mutation and, 635–636
 natural, **629**, 681–685. *See also* Natural selection
 random sampling and, 636
 rate of, 633, 633f, 634f
 truncation, 663
 variation and, 613f, 629–633
Selection differential, **662**
Selection response, **662**
Selection studies, 645, 645f
Selective system, **170**
Selfing, 29–30, 39
Self-splicing introns, 268–269
Semiconservative replication, 237–240, 237f. *See also* DNA replication
 Meselson-Stahl experiment and, 237–239, 238f
Semisterility, **501**, 501f
Sequence assembly, **392**
Sequence contigs, **395**, 395f
 paired-end reads for, 395–396, 395f
Sequence map(s), 398–406, 409f. *See also* Genome sequencing; Maps/mapping
 for chromosome 7, 410–411, 411f
 creation of, 394–398
 identifying gene of interest on, 405–406, 405f
 molecular markers for, 405
Sequencing reads, 392
70S ribosome, 285, 286f, 288

Severe combined immunodeficiency, gene therapy for, 375–379
Sex
 heterogametic, **48**
 homogametic, 48
Sex chromosome(s), **48**–52. *See also* Chromosome(s)
 aneuploid, 495–496
 differential regions of, 49, 49f
 discovery of, 77–78
 in *D. melanogaster*, 48, 48t
 dosage compensation and, 323, **495**
 homologous regions of, 49, 49f
 in plants, 48–49, 49f
 in sex determination, 602, 603f
 size of, 83, 84t
Sex determination
 in *Drosophila melanogaster*, 48, 48t
 in humans, 48, 48t, 602, 603f
Sex linkage, **49**, 76–78
Sex-linked dominant disorders, pedigree analysis of, 54, 54f
Sex-linked inheritance, 29, 48–55
 chromosome theory and, 76–80
 early studies of, 76–80
 primary exceptional progeny and, 79–80, 79f
 X-linked, 49–54, 52f–54f, 77–80
 Y-linked, 54–55, 55f, 77–80
Sex-linked recessive disorders, pedigree analysis of, 52–54, 52f, 53f
Sexual reproduction
 assortive mating and, 613f, 625–626
 enforced outbreeding and, **625**
 inbreeding and, 625–626. *See also* Inbreeding
 random mating and, 624, 624t
  Hardy-Weinberg equilibrium and, **621**–623
 variation and, 621–626
Shells, variation in, 681, 682f
Shigella, R plasmid of, 164–165, 165f
Shine-Delgarno sequence, **287**–288, 287f
Short interspersed nuclear elements (SINEs), **440**–442, 441f
Shotgun library, 394
Sickle-cell anemia, 17, 194, 365, 365f, 631
Side chains, 5–7, 6f, 275, 275f. *See also* Polypeptide(s); Protein(s)
 posttranslational modification of, 275
Sigma factor (σ), **261**–262, 264
Signaling, 7, 553–558
 autophosphorylation in, 556
 in cancer, 562–563
 in cell-cycle regulation, 552f–558f, 553–558
 common ancestry in, 697, 698f
 in development, 588–589, 588f, 589f, 590f, 597–599, 598f
 endocrine, 553f, **554**. *See also* Signaling
 G proteins in, 556–557, 557f
 in immune system, 556
 ligands in, **553**–557, 556f, 557f
 paracrine, 553f, **554**. *See also* Signaling
 in protein targeting, 292–293, 293f

 receptors in, 555, 555f, 556–557, 557f, 558f
 steps in, 556–557, 557f
 survival factors in, **556**
Signal sequence, **292** 293, 293f
Signal transduction, 187
Signal-transduction cascade, **555**, 556–557, 557f
Silencing, 317, **318**, 445
 epigenetic, 326, 445
 in imprinting, 324–326, 325f
 in position-effect varigeation, 326, 326f
 transgenic, 445–446, 445f
 in X inactivation, 323–324, 324f
Silent chromatin, **327**
Silent mutations, 455, 455t, 456
Silent polymorphisms, 620
Simple-sequence length polymorphisms (SSLPs), **400**–401, 401f, 402f
Simple transposons, **431**, 431f
SINEs (short interspersed nuclear elements), **440**–442, 441f
Single-nucleotide polymorphisms (SNPs), 399–400, **619**–620
 in cancer cells, 566, 567f
 in linkage mapping, 128
Single-strand binding (SSB) proteins, **243**
Single-strand crossovers, 476
Sister chromatids, **91**
 crossing-over in, 121. *See also* Crossing-over
 in meiosis, 91f, 93–95, 94f, 97–102, 98f, 99f, 101f, 102f
 in mitosis, 91–93, 91f–93f
Site-directed mutagenesis, **531**–532, 532f
Skin color, complementation and, 202–203, 203f
Sliding clamp, **243**
Small nuclear RNA (snRNA), **259**
 in spliceosome, 268, 268f
Snake, skin color in, complementation and, 202–203, 203f
Snapdragon (*Antirrhinum*), transposable elements in, 427f
SNPs. *See* Single-nucleotide polymorphisms (SNPs)
snRNA (small nuclear RNA), 259
 in spliceosome, 268, 268f
Solenoid, **88**, 89f, 90f
Somatic gene therapy, 377–378, 377f
SOS system, **459**–461, 472–473, 472f, 473f
Southern blotting, **12**, 13f, 44f, **354**, 355f
 in sickle cell disease, 365, 365f
spaetzle, 588–589, 588f, 589f
Specialized transduction, 170, **172**–174, 172f, 173f
Speciation, 689–690. *See also* Evolution
 allopatric, **690**
 conservation in, 696–698, 697f
Species, **689**–690
 isolation of, 690
Species diversification, **681**, 682f
S phase, **91**, 91f, 93, 98f
 of cell cycle, 548
Spindle, nuclear, 92, 92f
Spindle fibers, 75

Spliceosome, 259, **268**, 268f
Splice variants, 410
Splicing, **267**–269, 268f, 269f
 RNA world and, **268**–269
 self-splicing, 268–269
 sites of, identification of, 406–410
Spm element, 427
Spontaneous lesions, **463**–464, 464f
Spontaneous mutations, 453. *See also* Mutation(s)
Sporophyte, 97, 97f
Spradling, Allan, 439
spz gene/SPZ protein, 588–589, 588f, 589f
SRY gene, **54**–55
 in sex determination, 602, 603f
SSB proteins, 243
Stahl, Franklin, 237–238
Standard deviation, **669**, 670f
Statistical distributions, **647**–648, 648f, 667–673. *See also* Distribution(s)
Statistical measures, 648, 667–673
 of central tendency, 668
 of dispersion, 668–669
 of relationship, 670–673
Statistical relation, 648
Stem cells, **547**
Stevens, Nettie, 77
Stop codons, 5, 6, 282, 282f
 suppressor mutations and, 282, 291, 292f
Strain building, linkage mapping in, 129
Streisinger, George, 280–281
Streptococcus pneumoniae, transformation in, 229–230, 229f, 230f
Structural genomics, **410**
Structural proteins, 5
Sturtevant, Alfred, 124
Sublethal mutations, 197
Substitution mutations, **10**, 454, 456
Subunits
 protein, 276f, **277**
 ribosomal, 285–286, 286f
Sum rule, 40
Supercoiling, of DNA, 89–90, 89f, 90f
 topoisomerases and, 243, 243f
Supercontigs, **396**
Superrepressor mutations, 309–310, 310f, 310t
Suppressor mutations, 187, 204–206, 206f
 genetic code and, 279–281, 280f
 in translation, 282, 291, 292f
Suppressor/mutator (Spm) element, 427
Suppressors, 204–206, 206f
Survival factors, **556**
Sutton-Boveri theory, 75
Sutton, Walter, 75
Sweet pea (*Lathyrus odoratus*), as model organism, 14
swi-snf (switch-sniff) gene, 327, 327f
Synapsis, 485
Synaptonemal complex, **95**, 95f
Synergistic effects, **319**
Synonymous codons, 408
Synonymous mutations, **455**–456, 455t, 456, 696
Synonymous polymorphisms, 620
Synonymous substitutions, 696

Synteny, **442**

Synteny map, 700–701, 700f. *See also* Maps/mapping

Synthetic lethal, **206**–207, 207f

Synthetic mRNA, 281

Systems biology, 412–413, **415**

Tab mutation, 321–322, 322f, 595

Tandem duplications, **505**

Tandem repeats
in linkage mapping, 128
variable number, **618**–619, 620t, 637

Targeting
gene, 372–375, 374f
protein, 292–293, 293f
of transposable elements, **443**–444, 443f

Target-site duplications, **432**

TART elements, **444**–445, 444f

Taste sensitivity, for phenylthiocarbamide, 47, 667

TATA binding protein (TPB), **265**, 265f

TATA box, **265**, 265f

Tatum, Edward, 154, 155–156, 157, 187–189

Tautomeric shift, **457**

Tautomers, **457**, 465

T-DNA, 368–370, 370f

Telocentric chromosomes, 84t, **85**

Telomerase, **248**

Telomeres, **86**, 86f, **247**–248, 248f
transposable elements as, 444–445, 444f

Telophase, **93**
meiotic, 94, 95f
mitotic, 91f, 93, 93f

Temperate phage, **170**

Temperature
permissive, **195**
restrictive, **195**

Temperature-sensitive mutation, **195**

Template, in replication, 4, 5f, **237**

Tenebrio, chromosome segregation in, 77, 77f

Termination codon. *See* Stop codons

Terminus, in conjugation, **161**

Ternary complex, **289**–290, 289f

Testcrosses, **39**
dihybrid, 117, 126, 126f, 136
monohybrid, 136
ratios for, 117, 126, 126f. *See also* Mendelian ratios
for recombinants, 121, 122f
trihybrid, **125**–126, 136

Tester, **39**

Testicular feminization syndrome, 54, 54f

Testis, development of, 602, *603f*

Testis-determining factor, 54–55

Testosterone, 602, 603f

Tetrad(s), **95**, **120**, 328, 328f
crossing-over in, 120, 120f, 135–136
linear, in centromere mapping, 129–131, 131f

Tetrad analysis, in linkage analysis, 120–121, 120f, 135–136

Tetrahydrobiopterin, in phenylketonuria, 198

Tetrahymena, self-splicing in, 268–269

Tetraploidy, **483**, 483t, 484. *See also* Chromosome number; Polyploidy
pathenogenic development and, **484**

TGF-β, in cell cycle regulation, 555–556

Theta (θ) structures, **239**, 239f

30S subunit, 285, 286, 286f, 287–288

3' end
poly(A) tail addition to, 267
processing of, 266–267, 266f

3' untranslated region (3' UTR), **261**

Three-point testcrosses, **125**–126

Thymine, 232–233, 232f, 233t. *See also* Base(s)

Time-of-entry mapping, 161

Ti plasmid, **368**–370, 369f

Tomato, linkage map for, 129, 130f

Tomizawa, J., 170

Topoisomerases, **243**

Totipotent cells, **578**

Traits, 30–34
canalized, 688, 688f
familial, **655**
heritability of, 654–664, 687–688. *See also* Heritability
quantitative
gene number and, 650–651
norm of reaction studies for, 652–653
vs. Mendelian, 649–651

Trans-acting proteins, in gene regulation, **309**

Trans conformation, of linked genes, **118**, 118f

Transcript, **5**, 259
primary, 264

Transcription, **5**, 6f, 259–269
activated, **317**
base pairing in, 259–261, 260f, 261f, 263, 263f
consensus sequences in, 261, 262f
definition of, 256, 259
directionality of, 259, 260f, 261, 262f
DNA as template in
in eukaryotes, 264
in prokaryotes, 259–261, 259f, 260f, 262
down/upregulation of, regulatory gene mutations and, 206
elongation in, **261**
in eukaryotes, 266, 266f
in prokaryotes, 262–263, 263f
enhancers in, 317, **318**
dpp, 320, 321f
gene fusion and, 321–322, 322f
tissue-specific, 320
general transcription factors in, in eukaryotes, 264, 265
helix unwinding in, 262, 263f
initiation of, **261**
in eukaryotes, 264–265, 265f
in prokaryotes, 261–262, 262f
overview of, 256–257, 256f, 259–261, 260f
preinitiation complex in, 265
pre-mRNA processing in, 266–268, 266f–268f
promoters in
in eukaryotes, 265, 265f
in prokaryotes, 261, 262f
regulation of. *See* Gene regulation

RNA:DNA hybrids in, 262–263, 263f

RNA polymerases in, 259–260, 260f, 261, 261f
in eukaryotes, 264, 265, 266, 317–318
in prokaryotes, 259–260, 260f, 261, 261f, 262, 263
sigma factor in, 261–262, 264
stages of
in eukaryotes, 263–269, 265f, 266f
in prokaryotes, 261–263, 262f, 263f

TATA box in, **265**, 265f

template in, **259**–260, 260f, 262

termination of, **259**
in eukaryotes, 266
in prokaryotes, 263, 263f

Transcriptional silencing, 317

Transcription bubble, **262**, 263f

Transcription-coupled DNA repair, 470–471

Transcription factors
in eukaryotes, 318–319
in pattern formation, 583–589, 585f–589f, 591f, 594–597

Transcription unit, 305

Transcriptome, 413

Transduction, **170**–174, 170f–173f
bacterial, 153
definition of, 153, **170**
discovery of, 170
generalized, **170**–172, 171f
in linkage analysis, 170–172, 171f
signal, 187
specialized, 170, **172**–174, 172f, 173f

Transfer DNA (T-DNA), 368–370, 370f

Transfer RNA. *See* tRNA (transfer RNA)

Transformation, 153, **166**, 166f
in chromosome mapping, 166
discovery of, 229–230, 229f, 230f
double, 166
homeotic, **595**, 596f

Transforming growth factor-β, in cell cycle regulation, 555–556

Transgene, **366**
silencing of, 445–**446**, 445f

Transgenic animals, 370–375

Transgenic beetles, 538

Transgenic organism, **366**

Transgenic plants, 368–370, 369f, 370f

Transgenics, in mutational analysis, 538–539, 539f

Transgenic silencing, 445–446, 445f

Transition, **454**

Translation, **5**–6, 8, 273–274, 282–293
adapter hypothesis and, 282–283
aminoacyl-tRNA synthetases in, 283, 284
A site in, 286, 287f, 289–290
codon-anticodon pairing in, 284–285, 285f, 286–287, 289–290
definition of, 256, 274
directionality of, 289, 289f
elongation in, 287f, 289–290, 289f, 290f

E site in, 286, 287f, 290
in eukaryotes, 266, 274f, 275
initiation complex in, 288
initiation factors in, 288, 289
initiation of, 287–289, 287f, 288f
molecular mimicry and, 290–291, 291f
mRNA in, 274, 285–287, 287f
nonsense suppressor mutations in, 291
overview of, 256–257, 256f, 274f
posttranslational processing and, 291–293
in prokaryotes, 274f, 275
P site in, 286, 287, 287f, 289–290
release factors in, 290, 290f
ribosomes in, 285–291
rRNA in, 85, 259, 274, 275, 285–287, 286f
Shine-Delgarno sequences in, 287–288, 287f
sites of, 8
suppressor mutations in, 282, 291, 292f
termination of, 290, 290f, 291, 292f
tRNA in, 6, 274, 282–285, 285–287, 287f

Translesion DNA synthesis, 459–460, 460f

Translocations, **496**, 496f
in cancer, 563–564, 564f
in deletion synthesis, 502
in duplication synthesis, 502
in mapping, 502
polymorphic, 617–618, 618f, 618t
position-effect variegation and, 326, 326f, 332–333, 332f, **502**
reciprocal, 496f, **497**, 501–502
Robertsonian, 506, 507f
in Down syndrome, 506, 507f

Transposable elements, **371**, 423–447, 694
abundance of, 424, 440–445
Ac, **425**–429
autonomous, **427**
change in phase in, 445, 445f
in clone-by-clone sequencing, 397
discovery of, 425–429
disease-causing, 441f, 442
in *Drosophila melanogaster*, 435, 436–440, 437f, 438f
Ds, **425**–429, 425f, 426f
in eukaryotes, 424–425, 424f, 434–440
class I (RNA elements), **424**–425, 424f, 434
class II (DNA elements), **424**–425, 424f
in exons, 442
families of, 427, 438
in forward genetics, 523–524
functions of, 425–427, 444–445, 444f
genome size and, 440–442, 441f, 442
HeT-A, 444–445, 444f
host regulation of, 445–446
in human genome, 440–442, 441f
in hybrid dysgenesis, 436–437, 437f
inactivation of, 442
reversible, 445–446, 445f
insertion-sequence, 161, **429**–430, 429f, 430f

in introns, 441–442
as junk DNA, **444**
long interspersed nuclear elements, **440–441**, 441f
mobilization of, 432–433
mosaicism and, 427f
as mutagens, 523–524
negative selection of, **442**
nonautonomous, **427**
overview of, 424–425
in prokaryotes, 429–433
replication of, 432–433, 433f
in safe havens, **442–444**, 443f
short interspersed nuclear elements, **440–442**, 441f
spicing of, 441–442
spread of, 436–437, 437f
targeting of, **443–444**, 443f
TART, **444–445**, 444f
as telomeres, 444–445, 444f
in transgenesis, 371
transposons as, **431**. *See also* Transposon(s)
ubiquity of, 429
in yeast, 434–435, 435f, 436f, 442–444, 443f
Transposase, **371**, 430
Transpose, **426**
Transposition, **431**
change in phase in, **445**, 445f
conservative, **432**, 433, 433f
of gene, **426**
hybrid dysgenesis and, **436–437**, 437f
mechanism of, 431–433
replicative, **432–433**, 433f
Transposon(s) (Tn), **431**. *See also* Transposable elements
bacterial, 431
composite, **431**, 431f
DNA, 424–**425**, 424f, 436–438, 437f, 439f
eukaryotic, class II, 435–440, 437f–439f
retrotransposons, 424–**425**, 424f, 434–436, 435f, 436f
simple, **431**, 431f
Transposon tagging, **439**, 524
Transversion, **454**
Trifolium, leaf color in, 195, 195f
Trihybrid testcross, ratios for, 136
Trinucleotide repeat diseases, 466–468, 467f
Triplet code, 278–281, 278f, **279**. *See also* Genetic code
Triplo-abnormal genes, 495
Triploidy, **483**, 483t, 484. *See also* Chromosome number; Polyploidy
Trisomy, 483t, **490**, 492–494, 493f, 494f, 495
in Down syndrome, **493**–494, 493f, 494f, 495, 506, 507f
in Edwards syndrome, 494
gene-dosage effect and, **495**
in Klinefelter syndrome, **492–493**, 493f
in Patau syndrome, 494
Triticale, 490
Triticum aestivum, as allopolyploid, 488, 489f, 490
Trivalents, **485**
tRNA (transfer RNA), **6**, **259**, **274**, 282–285

aminoacylated, 283, 284, 289–290, 289f
cellular amounts of, 274
charged, **283**
cognate, 287
functions of, 274
initiator, **287–288**
isoaccepting, **285**
ribosomal binding site for, 286–287, 287f
structure of, 283–284, 283f, 284f
in ternary complex, 289–290, 289f
in translation, in initiation, 274, 285–289, 287f
trp operon, 316, 316f
Truncation selection, **663**
ts mutation, 195
Tubulin, 92, 92f, 579–580, 579f, 580f
Tumor-suppressor genes, **560**
mutations in, 560–561, 560f, 564–566, 565f
detection of, 566, 567f
Turner syndrome, 491–492, 492f, 496
Two-hybrid test, 413–415, 415f
Ty elements, 434–435, 435f, 436f
in safe havens, **442**, 443f
Tyrosine, in albinism, 10, 11f

Ultraviolet light, as mutagen, 454f, 459–461, 460f, 468
UMP (uridine 5'-monophosphate), 258, 258f
Unbalanced rearrangements, **497**
Underdominance, **634**
Uniparental inheritance, 55–56
Univalents, **485**
Universe, **672**
Unselected markers, **162**
Unstable phenotypes, **425**
transposable elements and, 425–429. *See also* Transposable elements
Untranslated region (UTR)
5', **261**
3', **261**
Upstream promoter elements (UPEs), **261**, 317–318, 317f
Uracil (U), **258**
excision of, in DNA repair, 469
Uracil–DNA glycosylase, in DNA repair, 469
Uridine 5'-monophosphate (UMP), 258, 258f

Van Leeuwenhoek, Antony van, 154
Variable expressivity, 207–209, 208f
Variable number tandem repeats (VNTRs), **128**, **618–619**, 620t, 637
microsatellite markers and, 400–401, 401f, 402f
minsatellite markers and, 400, 401f
Variance, **648**, 649, 649f, **656–657**, **669**
additive, 660–661
analysis of, **657**
components of, estimation of, 661–662
dominance, **661**
environmental, 649–652, 650f, **656–657**
genetic, 649, 649f, 656–657
narrow heritability and, 660–661

Variants, **28**, 452
continuous, 8f, 10
discontinuous, 8–10, 8f, 9f, 28–29
Variation, 8–11, 28–29, 612–629, 614, 649, 649f, 654, 656–657. *See also* Polymorphism
additive genetic, **661**
allelic frequency and, **614**, 614t, 615t. *See also* Allelic frequency
assortive mating and, 613f, **625–626**
balanced polymorphisms and, 634–636
in blood groups, 614, 615, 615t
chromosome rearrangements and, 617–618, 618f, 618t
complete sequence, 619–620
continuous (quantitative), 8f, **10**, 644–651. *See also* Quantitative genetics
gene number and, 8f, 10, 644–651
developmental noise and, 21–22
discontinuous (qualitative), 8–10, 8f, 9f, 28–29, 644, 649–651
DNA sequence, 617–621
environmental variance and, 649–650, 650f, **656–657**
fitness and, 629–634
founder effect and, **636**
frequency of, 616–617, 617f
gene-environment interactions and, 16–22, 17f, 18f
genetic, 8–11, 649, 649f, 650f, 654, **656–657**
genetic drift and, 103, 613f, **636**, 637f, 684, 684t
genomic, 617–620
genotype frequencies and, **614**, 614t
genotypic vs. phenotypic, 649–652
halotypic, **623–624**
Hardy-Weinberg equilibrium and, **621**, 623
heritability of, **654–664**, 687–688. *See also* Heritability
heterozygosity and, 616–617, **623–624**, 623t
inbreeding and, **625–626**, 625f, 626f, 685
maintenance of, 621–623, 684–685, 684f
migration and, 613f, 628–629
molecular basis of, 10–11, 11f
from mutation, 613f, 626–628, 687
observations of, 612–615
phenotypic, 612–614
vs. genotypic, 649–652
polymorphisms and, **614–621**. *See also* Polymorphisms
positive assortive mating and, **625–626**
qualitative, 644
racial, 688, 698t
random events in, 636–637
random mating and, 624, 624t
recombination and, 613f, 628
restriction-site, 617–618
selection and, 613f, 629–633
sexual reproduction and, 621–626
single-gene, 614
sources of, **612**, 613f
statistical measures of, 667–673

variable number tandem repeats and, **618–619**
viability, **630–631**, 631f
within vs. between populations, 688
Variegated plants, 56, 57f
Vectors, **343**, 346–349, 348f, 349f
BAC, 349
bacteriophage, 347–348, 349f, 350f
cosmid, 349, 350f
delivery of, 350, 350f
expression, 349
PAC, 349
plasmid, 346–347, 348f, 350f
Ti plasmid, 368–370, 369f
YAC, 349
yeast, 349, 366–368, 367f
v-erbB oncogene, 526f, 562
Vesicular transport, in development, 579–580, 579f, 580f
Viability, **630–631**, 631f
Virulent phage, **170**
Viruses, **152**. *See also* Phage(s); Retroviruses
genome of, 15, 16f
as model organisms, 13, 14f
replication of, 434–435
VNTRs. *See* Variable number tandem repeats (VNTRs)
Volkine, Elliot, 257
Vulva, development of, in *C. elegans*, 598–599, 598f, 599f

Watson, James, 228, 233–237, 234f
Western blot, **13**
Wichman, H., 687
Wiechaus, Eric, 548
Wild-type alleles, 35, **615**
complementation and, 199–203, 200f
Wild-type bacteria, 153, 154f
Wild-type genes, 35, **79**, **615**
naming of, 79
Wild-type genotype, 19–20, 19f, **615**
Wild-type mutations, 9–10, 9f, **615**
Wild-type phenotype, 19–20, 19f, **615**
Wilkins, Maurice, 233, 234f
Williams syndrome, 504–505, 505f
Wing color, in *Abraxas*, 76–77
wingless, 597–598
Wobble, **284**–285, 284f
Wollman, Elie, 158, 161

χ^2 test. *See* Chi-square test
X chromosome, 48–52
differential regions of, 49, 49f
discovery of, 77–78
homologous regions of, 49, 49f
inactivation of, 323–324, 324f
in sex determination, 602, 603f
size of, 83, 84t
X linkage, **49**
X-linked disorders
dominant, 54, 54f
recessive, 49–54, 52f–54f
X-linked inheritance, 49–54, 52f–54f, 77–80. *See also* Sex-linked inheritance
X-linked spinal atrophy, 467–468
X ray diffraction studies
of DNA, 233, 234f
of tRNA, 284, 284f

YAC vectors, 349
Yanofsky, Charles, 277–278
Yarrow. *See Achillea millefolium*
  (yarrow)
Y chromosome, **48**–52
  differential regions of, 49, 49f
  discovery of, 77–78
  homologous regions of, 49, 49f
  in sex determination, 602, 603f
  size of, 83, 84t
Yeast
  cell cycle in, 551
  forward genetic screens in, 527, 527f
  genetic engineering in, 366–368,
    367f

genome of, 15, 16f
genomic analysis for, 411–412,
  412f
life cycle of, 328, 328f
mating-type switch in, 327–328
as model organism, 13, 14f, 327,
  328
  for cell cycle, 551
transposable elements in, 434–435,
  435f, 436f
  in safe havens, 442–444, 443f
Yeast artificial chromosome (YAC),
  367f, 368
Yeast artificial chromosome vectors,
  349

Yeast episomal plasmids, 367f, 368
Yeast integrative plasmids, 366,
  367f
Yeast two-hybrid system, 413–415,
  415f
Yeast vectors, 366–368, 367f
Y gene, in *lac* system, 305, 305f, 306,
  307
  constitutive mutations in,
    308–309
  polar mutations in, 311–312
Y linkage, **49**
Y-linked inheritance, 54–55, 55f,
  77–80. *See also* Sex-linked
  inheritance

Zea mays. See Maize *(Zea mays)*
Zebra fish *(Danio rerio)*, genetic
  screens for, 529–530, 529f
Z gene, in *lac* system, 305, 305f, 306,
  307, 308
  mutations in, 305, 305f, 306, 307,
    308, 309, 309t
  constitutive, 308–309
  polar, 311–312
Zimm, Bruno, 80
Zinder, Norton, 170
Zygote, **3**, **33**
  totipotency of, 578
Zygotic induction, **172**–173, 172f
Zymogen, **552**